PEARSON
Prentice Hall
Upper Saddle River, New Jersey
Columbus, Ohio

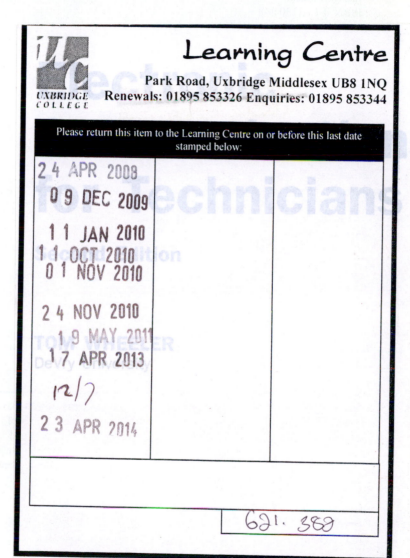

Library of Congress Cataloging-in-Publication Data

Wheeler, Tom
Electronic communications for technicians / Tom Wheeler.—2nd ed.
p. cm.
Includes index.
ISBN 0-13-113049-8
1. Telecommunication. I. Title.

TK5101.W47 2006
621.382—dc22

2005003219

Assistant Vice President and Publisher: Charles E. Stewart, Jr.
Assistant Editor: Mayda Bosco
Production Editor: Alexandrina Benedicto Wolf
Production Coordination: TechBooks/GTS, York, PA
Design Coordinator: Diane Ernsberger
Cover Designer: Linda Sorrells-Smith
Cover Art: Corbis
Production Manager: Matt Ottenweller
Marketing Manager: Ben Leonard

This book was set in Times by TechBooks/GTS, York, PA. It was printed and bound by Courier Kendallville, Inc. The cover was printed by The Lehigh Press, Inc.

Pearson Education Ltd.
Pearson Education Singapore Pte. Ltd.
Pearson Education Canada, Ltd.
Pearson Education—Japan
Pearson Education Australia Pty. Limited
Pearson Education North Asia Ltd.
Pearson Educación de Mexico, S.A. de C.V.
Pearson Education Malaysia Pte. Ltd.

PEARSON
Prentice Hall

10 9 8 7 6 5 4 3 2
ISBN: 0-13-113049-8

In memory of Dr. John Trank, NØTFW, and Les Strandt, W3EAU—accomplished practitioners and teachers. It has been a special privilege to know you.

Preface

This is a textbook written by a technician for technicians. In over 16 years of teaching electronic communications to future technicians, I have used many other fine textbooks with varying levels of success.

There's a real void in communications textbooks. Some entry-level texts, in an attempt to keep the subject matter understandable, avoid any real detailed discussion of the circuits and systems being studied. The student is left with an empty feeling; he or she can certainly answer the questions in the book correctly, but cannot apply the knowledge to real-world circuits and situations.

Other textbooks cover circuits in adequate detail, but also tend toward moderate to heavy use of mathematics in their analysis of circuits and systems. At an engineering level, this is probably an acceptable level of treatment. For technicians, the purely mathematical approach fails miserably. It's not that technicians aren't capable of grasping the mathematical concepts (they most certainly are, if the concepts are presented properly); it's just that advanced mathematical tools are not in their toolkits. In reality, people don't need heavy math to understand electronic systems. People can learn almost anything if it is presented properly.

This textbook has many features that should make it a valuable addition to any curriculum for electronic technicians:

- The level of mathematics has been moderated so that only basic algebra skills are needed to work most of the problems in the text. (The only place where basic trigonometry is required is in Chapter 18, Fiber-Optic and Laser Technology.) This has been accomplished with no sacrifice in depth.
- Each concept is first explained in plain language and is then carefully developed step by step. Every attempt is made to build on the student's already broad base of everyday experience.
- Each new concept is supported by at least one example. Examples follow a progression from simple to challenging, always emphasizing fundamental ideas.
- Real-world examples and case studies are sprinkled throughout the text.
- Real circuits, systems, and their troubleshooting are emphasized throughout the text, with discussion of necessary safety procedures where needed. A systematic method of troubleshooting is emphasized.

NEW TO THIS EDITION

Extensive feedback from users of the first edition has been solicited and necessary changes have been incorporated throughout. Much of the text has been polished to improve the readability and sharpen the definition of key concepts. The following has been added to this edition:

- Fourier series for common waveforms.
- High definition television technology and its relation to conventional analog television.

- An introduction to the use of the Smith chart.
- Link budgets.
- A new chapter covering conventional and cellular telephony, including the Qualcomm CDMA and GSM systems for PCS.
- A new chapter covering networking fundamentals with TCP/IP, including network troubleshooting.
- Expanded appendices with coverage of decibels, bipolar transistor fundamentals, amateur radio communications, and well-known TCP and UDP port numbers.

The textbook edition of Electronics Workbench Multisim 7 is included on the CD-ROM accompanying the text, along with 37 sample circuits. These circuits illustrate many concepts from the text, including signal analysis (effects of filters on complex signals), oscillators, RF amplifiers and modulators, PLL subsystems, and CDMA encoding/decoding principles. In addition, a software suite is contained in the \software folder of the accompanying CD-ROM, which includes:

- AM and FM simulators for Windows with software spectrum analyzer and scope displays.
- Arbitrary waveform generation for Windows (WaveGen) that can be used for Fourier Analysis and general waveform synthesis.
- Terminal emulation for Windows (EZ-Term).
- Smith chart emulation for the TI85/TI86 programmable calculators.
- Common emitter amplifier simulation for Windows.

In addition, an Instructor's Manual is available from Prentice Hall with answer keys for all questions, PowerPoint presentations, and a test item file.

Electronic communications is an exciting and *fun* topic. I sincerely hope that this work will help to kindle that same spirit in students, our future electronic technicians.

ACKNOWLEDGMENTS

I would like to acknowledge the assistance and advice of all those who helped during the production process. Professors Tim Morgan, Robert Diffenderfer, and Bob Pruitt of DeVry University provided valuable feedback and guidance. Thanks to Ali Ragoub, DeVry University; Jeff Rankinen, Milwaukee Area Technical College; Tim Staley, Southern Nevada Community College; and R. L. Windley, Truckee Meadows Community College for their helpful reviews. Finally, I would like to thank Mary, for being supportive throughout this project and giving up significant amounts of quality time so that it could be completed.

Tom Wheeler

Contents

1

Introduction to RF Communications

OBJECTIVES

At the conclusion of this chapter, the reader will be able to:

- explain the difference between systems and subsystems
- describe the functional blocks in a practical radio communications system
- calculate the *wavelength* of a radio wave and relate it to physical antenna length
- define *modulation* and explain why it is needed in a radio communications system
- list the three steps for *troubleshooting* systems

1–1 COMMUNICATION SYSTEMS

Today, every facet of our lives is touched by modern electronics. Being human, we take most of it for granted. It's easy to forget that for the largest part of recorded history, humankind has lived with no telephones, radios, televisions, or computers. Before the development of electronic communications, the speed of information travel was limited by the physical distance a runner or horse could cover in a day. During the colonization of America, it was accepted that a letter might take several months to reach its destination across the ocean, and several more months for the reply to make the trip back. Today, the distance across the globe is measured in fractions of a second.

Electronic Communications Is Everywhere!

Have you recently

Watched a TV broadcast or listened to the radio? These are probably the most visible applications of communications technology. Analog television uses very sophisticated electronic techniques. The latest television technology, high-definition television (HDTV), uses complex digital and software technology together with advanced analog circuit techniques. These techniques are readily understood by anyone with a firm grasp of electronic fundamentals.

Used a telephone? Your voice may be sent using many different technologies. Analog transmission carries your conversation to the central office. From there, the signal is converted to digital (digitized). The digital signal is sent (along with thousands of other calls) on a beam of light through fiber-optic cables. The process is reversed at the destination. During the process, your conversation may also travel by radio wave to and from a satellite. Cellular telephones transmit and receive *voice signal* as streams of digital data over UHF (ultra high frequency) radio-frequency carrier signals.

Used a remote control for a garage door, TV, or other appliance? Many remote controls are actually tiny radio transmitters. A small microprocessor encodes digital data onto the transmitted radio wave to represent the user's commands.

Taken a commercial flight? Aircraft use numerous types of communications to ensure flight safety. Both voice and digital (data) communications are used by aircraft. Many of the communications are computer automated. The Global Positioning System (GPS) is used to help provide accurate navigation.

Used a credit card? If so, the verification was probably done electronically. A credit card reader contains a microprocessor and a *modem* (modulator–demodulator). Your individual information record is recorded in three parallel "tracks" that are read from the card's magnetic stripe by the microprocessor. The modem allows the microprocessor to transmit the data to a host computer operated by the credit card company, typically over a telephone line.

Electronic Systems

In your previous electronic studies, you have been primarily concerned with the *theory* of circuits. For example, you might have constructed an amplifier stage with a transistor or op-amp IC (integrated circuit). The amplifier you built was studied for its own sake; it didn't fit into anything "bigger." This book will be your first study of *systems*.

> A *system* can be defined as a group of components that work together to complete a job or task.

Many technicians are a little frightened when first asked to learn a new system. Part of this might be a natural fear of the unknown. The technician might wonder if he or she is capable of learning the necessary technical details. The best way to learn a system is to break it down into functional blocks, or *subsystems*. Upon study of these parts, the

Section Checkpoint

1–1 Classify each of the following as a system or subsystem:

- **a.** A radio transmitter
- **b.** An automobile
- **c.** An automatic transmission
- **d.** A radio transmitter and receiver

1–2 What type of diagram is used to show how the systems work?

1–3 How can a technician understand a very complicated system?

technician soon recognizes familiar circuits and principles and gains an understanding of how the system actually works.

A *subsystem* is just part of a system; it helps to complete a task. A subsystem is often shown in a block diagram.

1–2 A SIMPLE RADIO SYSTEM

A simple radio system could be constructed as shown in the block diagram of Figure 1–1. This system has some severe problems, but it will serve as a good starting point. The radio system begins with a *microphone*. A microphone is a type of *transducer*. It converts the pressure variations in a sound wave (such as from a speaker or musical performer) into electrical energy. A microphone has a thin plastic or paper cone connected to a coil of wire (Figure 1–2). This coil is placed within the field of a permanent magnet. When sound strikes the cone, it vibrates, moving the coil back and forth within the magnetic field. Thus, a voltage is generated in the coil that is a copy of the sound wave that entered the microphone.

We call the electrical signal from the microphone the *intelligence* or *information* signal. The information signal is an electrical replica of the original sound wave and has the same shape.

A *transducer* is any device that converts one form of energy into another.

Figure 1–1 A simple radio system

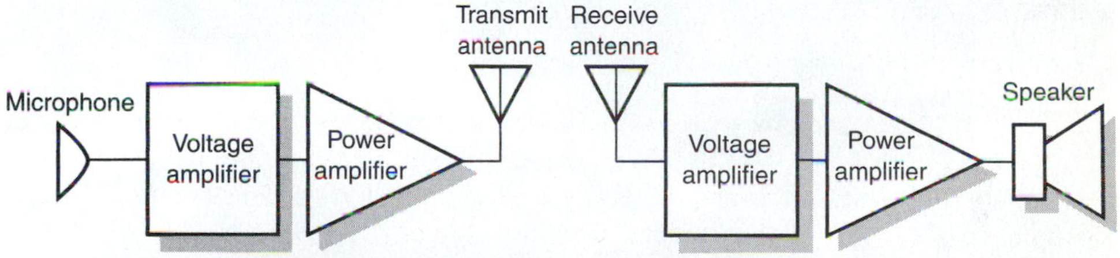

Figure 1–2 A typical dynamic microphone element

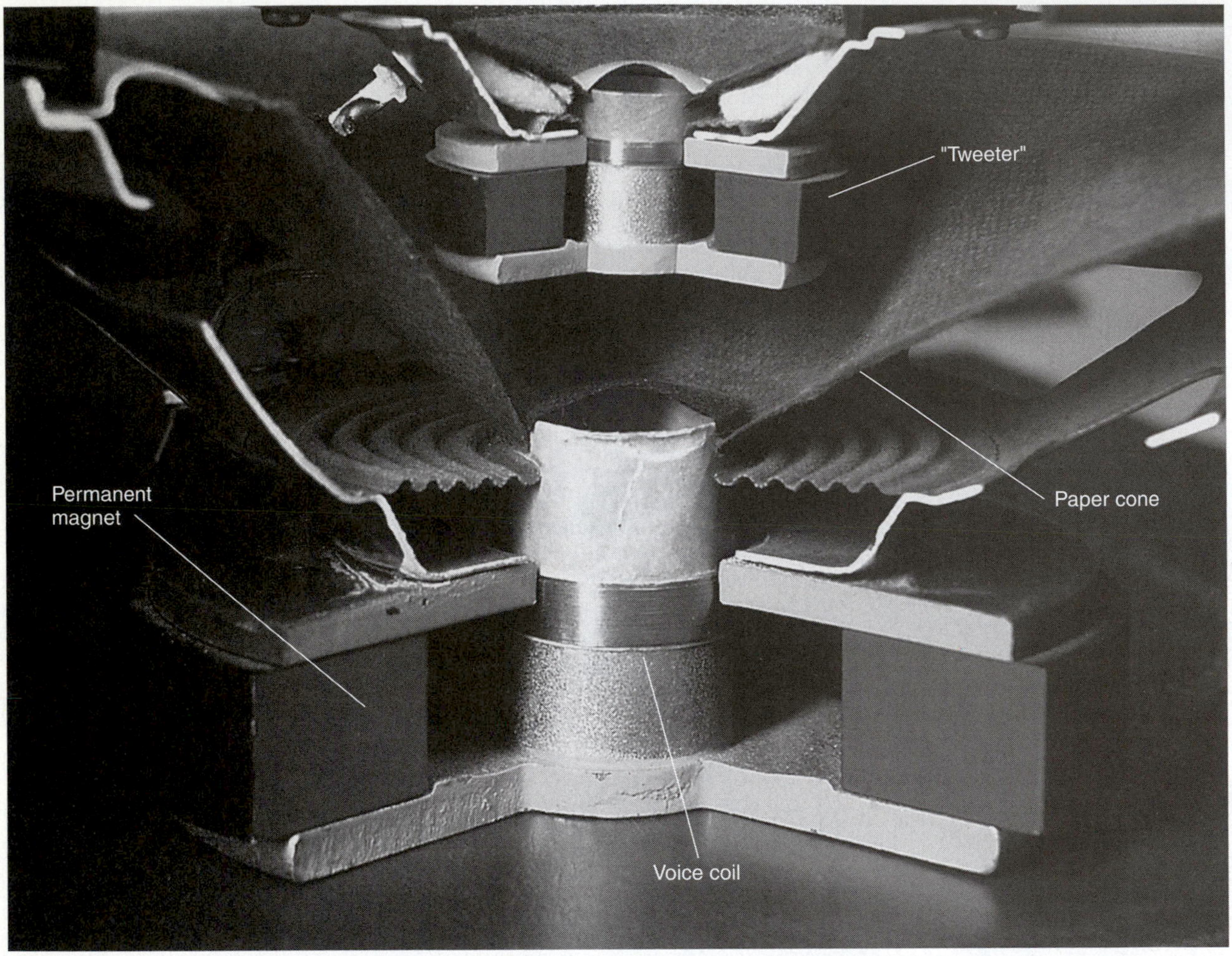

Figure 1–3 Cutaway of a coaxial loudspeaker

The signal from the microphone is quite small. Most microphones produce about 10 mV (millivolts) at a power level of about 40 μW (microwatts). This isn't enough power to cross any significant distance in space, so both voltage and power (current) amplification must take place. The final power level reached at the output of the power amplifier depends on how far we need to communicate, and under what conditions. This power level can range from a few mW (milliwatts) (personal communications devices such as walkie-talkies) to thousands of watts (military and broadcast communications).

The transmitting *antenna* next converts the amplified information signal into a new form of energy capable of traveling through space. This new energy is called *electromagnetic energy,* or a *radio wave*. Electromagnetic energy consists of two fields, a voltage or electric field, and a magnetic field. It travels through space at the speed of light; in fact, visible light is itself electromagnetic energy with a very high frequency.

The energy from the radio wave moves outward from the transmitting antenna at the speed of light, which is about 3×10^8 meters/second (m/s). It spreads out over space much like an inflating balloon. By the time it reaches the receiver's antenna, it has very little energy. Imagine the thickness of a toy balloon when it is deflated; then imagine the new thickness if the balloon were inflated to a diameter of 10 miles! This is very close to how the energy will be distributed in a radio wave. A radio receiver typically receives picowatts $[1 \times 10^{-12}$ watt (W)] or femtowatts $(1 \times 10^{-15}$ watt) of energy from its antenna!

At the receiver, the antenna receives the weak signal. It will typically be just a few microvolts, which is too small for any practical use. Therefore, voltage and current amplification will be needed to bring the signal back up to a useful level. The receiver drives a *loudspeaker,* another transducer (Figure 1–3). The loudspeaker converts the electrical signal back into sound. It works by passing electrical current through a coil (the voice coil) suspended in a strong magnetic field. The electrical current causes the voice coil to become a magnet, and it is then attracted and repelled from the permanent magnet in step with the original information signal. A paper cone attached to the voice coil pushes on the air, which recreates the original sound.

This is theoretically how a radio system *should* operate. However, there are two practical problems that prevent us from doing it this way! The next section will demonstrate these problems, and how they are avoided.

Section Checkpoint

1–4 What is a transducer?

1–5 Explain the workings behind a microphone and loudspeaker.

1–6 Why must a transmitter use a power amplifier? How is the power level decided?

1–7 What is the *information signal?*

1–8 What are the two types of energy in an electromagnetic wave?

1–9 What types of voltage and power levels are typical at a radio receiver's antenna?

1–10 Answer the following questions about power units:

- **a.** How much power is a *femtowatt?*
- **b.** How much power is a *picowatt?*
- **c.** How many femtowatts in a picowatt? Microwatt? Milliwatt?

1–3 THE NEED FOR MODULATION

A radio communication system like the one Section 1–2 would not work very well. There are two problems: First, it would be nearly impossible to build a transmitting antenna to work with the system; second, there is no way of having two (or more) transmitters on the air at the same time. There's no way of *separating* individual stations. Let's explore these problems in more detail.

Transmitting Antenna Requirements

In order to be efficient, radio transmitting antennas need to be at least one-quarter of a *wavelength* long. The term *wavelength* refers to the distance a radio wave travels in one cycle. It is measured in meters, a unit of length. Figure 1–4 represents a radio wave in

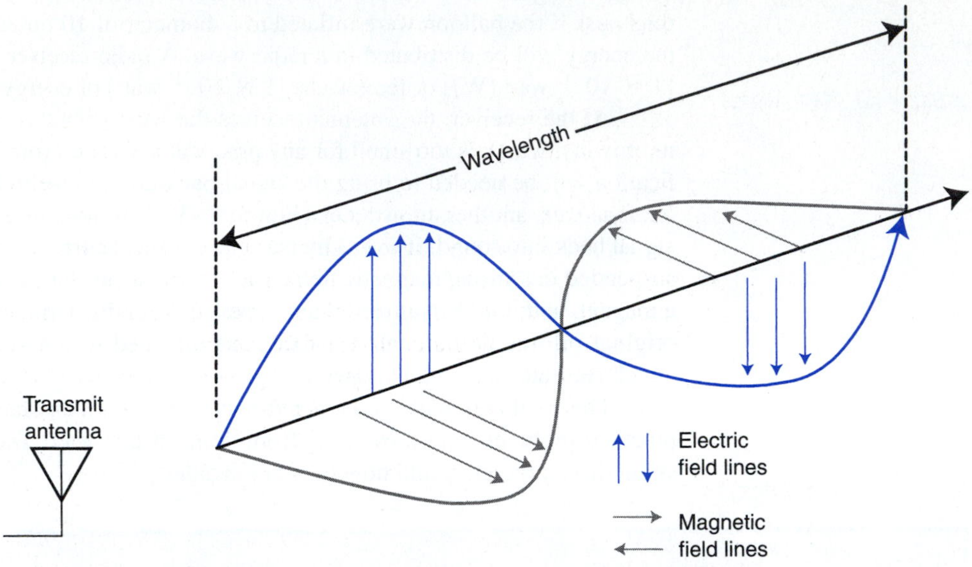

Figure 1–4 A radio wave in free space

space. It is traveling into the page. Wavelength can be easily calculated. A standard formula states that to find distance, multiply rate (speed) by time:

$$D = RT$$

The speed of a radio wave is the same as the speed of light, and we can give it the letter v (for velocity). We also know that frequency and time are *inverses:* $f = 1/T$. Therefore, we can write

$$D = v\left(\frac{1}{f}\right) = \frac{v}{f} \text{ (where } v \text{ is the speed of light, } 3 \times 10^8 \text{ m/s)}$$

It is very common to use the Greek symbol lambda (λ) to stand for the distance of one wavelength. Therefore, we get

$$\lambda = \frac{v}{f} \qquad \textbf{(1–1)}$$

Wavelength is the distance that a radio wave travels in one cycle. It is measured between two peaks or two troughs. It is always in distance units (usually meters).

Back to our problem. The radio system of Figure 1–1 needs to transmit *audio frequencies.* Audio frequencies are those that we can hear. They are from 20 Hz to 20,000 Hz (hertz). (Most people over the age of 15 have great difficulty hearing anything over 15,000 Hz.) An audio frequency that is in the "middle" of our hearing range is 1000 Hz, or 1 kHz. Let's calculate the wavelength of a 1 kHz electromagnetic wave.

According to equation 1–1,

$$\lambda = \frac{v}{f} = \frac{3 \times 10^8 \text{ m/s}}{1 \text{ kHz}} = 300{,}000 \text{ m} = \underline{\underline{300 \text{ km}}} \text{ (kilometers)}$$

300 km (kilometers) is quite a large distance—about 186 miles! Of course, the antenna doesn't have to be *this* long. It need only be *one-quarter* of this distance, or 75 km (46.6 miles). No big deal, right?

Of course, it isn't practical to build an antenna this large. Our system won't work very well. The required antenna length is too long!

Another Problem

As if the inability to build a suitable antenna weren't enough, another serious problem exists with the system of Figure 1–1. Imagine that, somehow, practical antennas were invented to work at audio frequencies. We'll want to operate several stations at the same time; having just one broadcaster "hogging" the airwaves just won't do! Do you see the problem? Yes! *All the stations will be sharing the same frequencies (the audio frequencies)*. It will be impossible to separate different stations at the radio receiver, since they'll all be on the same group of frequencies.

You can see that the system of Figure 1–1 is hardly practical. Can you think of a way to make it work? How could we get the antennas to be shorter?

Making it Shorter

Equation 1–1 gives a clue about how to proceed. It tells us that wavelength is velocity divided by frequency. We can't change the velocity—that's the speed of light in free space. But we can easily change the *frequency* of the wave. If we increase the frequency, the wavelength will get *shorter*. This looks promising!

Increasing frequency always decreases wavelength.
Decreasing frequency always increases wavelength.

Let's try changing to a 10 kHz wave. What will the wavelength be? How long will the antenna have to be? By using equation 1–1 again, we get

$$\lambda = \frac{v}{f} = \frac{3 \times 10^8 \text{ m/s}}{10 \text{ kHz}} = 30{,}000 \text{ m} = \underline{\underline{30 \text{ km}}}$$

This is much better! *The wavelength got 10 times shorter*. The minimum antenna length would be one-quarter of a wavelength:

$$L_{min} = \frac{\lambda}{4} = \frac{30 \text{ km}}{4} = \underline{\underline{7.5 \text{ km}}}$$

The antennas are still too long to build. If we continue to increase the frequency further, the wavelength will get even smaller. For example, many mobile FM radio units operate near 150 MHz. At this frequency,

$$\lambda = \frac{v}{f} = \frac{3 \times 10^8 \text{ m/s}}{150 \text{ MHz}} = 2 \text{ m}$$

and

$$L_{min} = \frac{\lambda}{4} = \frac{2 \text{ m}}{4} = \underline{\underline{0.5 \text{ m}}} = \underline{\underline{19.7''}}$$

Antennas of this type are very practical. You can often estimate the frequency of a transmitter by "eyeballing" the antenna. The longer the antenna, the longer the wavelength—and, you guessed it, the lower the operating frequency.

Frequencies above the range of hearing are called *radio frequencies*. Any frequency above 20 kHz is considered a radio frequency.

EXAMPLE 1–1

What is the wavelength of a 710 kHz AM broadcast signal? What is the minimum height of the antenna tower in *feet* if it is one-quarter of a wavelength long?

Solution

Equation 1–1 calculates wavelength:

$$\lambda = \frac{v}{f} = \frac{3 \times 10^8 \text{ m/s}}{710 \text{ kHz}} = \underline{\underline{422.5 \text{ m}}}$$

The tower need not be a full 422.5 meters high. In fact, a quarter of that length will do just fine:

$$\frac{\lambda}{4} = 422.5 \text{ m} \times \frac{1}{4} = 105.6 \text{ m}$$

So the tower height will really be 105.6 meters. However, the answer was requested in *feet,* so we need to convert:

$$L_{\text{feet}} = 105.6 \text{ m} \times \frac{3.28 \text{ ft}}{1 \text{ m}} = \underline{\underline{346.6 \text{ ft}}}$$

The physical height of the tower will be close to 346.6 feet. This is very typical of AM broadcast installations.

EXAMPLE 1–2

A quarter-wave ($\lambda/4$) whip antenna measures 108″. What is the approximate operating frequency of the transmitter?

Solution

By manipulating equation 1–1, we get

$$f = \frac{v}{\lambda}$$

We know v since it is the speed of light, but not λ (the wavelength). We have been given the dimension of *one-quarter of a wavelength* in inches. This needs to be converted into *meters* to be useful to us:

$$\frac{\lambda}{4} = 108'' \times \frac{1 \text{ m}}{39.37''} = 2.74 \text{ m}$$

This figure is one-quarter of a wavelength ($\lambda/4$) in meters. The wavelength must be equal to:

$$\lambda = 4 \times \left(\frac{\lambda}{4}\right) = 4 \times 2.74 \text{ m} = 10.97 \text{ m}$$

Now with this answer in hand, we can plug back into the manipulated equation 1–1:

$$f = \frac{v}{\lambda} = \frac{3 \times 10^8 \text{ m/s}}{10.97 \text{ m}} = \underline{\underline{27.3 \text{ MHz}}}$$

This frequency is in the middle of the class-D citizens band.

Modulation

A 150 MHz signal may transmit just fine, but there's one more problem. We can't hear a 150 MHz signal; it's too high in frequency! However, there is a solution; it's called *modulation.* When we modulate a wave, we place a low-frequency information signal onto it. The low-frequency signal is just along for the ride. It is, in effect, "carried" on top of the high-frequency signal. The high-frequency signal is therefore called the *carrier* signal.

There are three ways we can impress information onto a carrier. We can change the voltage (or power) of the wave in step with the information. This is called *amplitude modulation,* or AM. We can alter the frequency of the wave with the information; this is called *frequency modulation,* or FM. Finally, we can change the phase of the carrier wave. This is called *phase modulation,* or PM.

Section Checkpoint

1–11 What are two problems with the system of Figure 1–1?

1–12 How is the wavelength of a radio wave calculated?

1–13 Why are high frequencies used for carrier waves?

1–14 What is meant by the term *modulation*?

1–15 What are three ways a carrier can be modulated?

1–16 What is the lowest *radio frequency?*

1–4 A PRACTICAL RADIO SYSTEM

An actual radio system looks very much like the diagram of Figure 1–5.

Transmitter

Every radio transmitter is assigned to operate on a specific *carrier frequency.* The job of the carrier oscillator is to generate this frequency. You'll recall that the purpose of an oscillator is to convert the dc from the power supply into an ac signal. The output of the carrier oscillator is a nearly pure sine wave. The sine wave from the carrier oscillator carries no information at this point.

The modulator stage is a little unusual—it has *two* inputs! One of the inputs is the radio-frequency sine wave from the carrier oscillator; the other is the *information* signal. The modulator combines the carrier and information in a special way. The output of the modulator is a *modulated carrier wave.* You might wonder why a voltage amplifier is needed between the microphone and modulator. Right! The microphone produces very little voltage or power, so its signal still needs a little "boost" before it can modulate the carrier.

The output from the modulator drives the radio-frequency (RF) power amplifier. The signal at the modulator's output is too small to cover any significant distance, so additional amplification is needed. RF amplifiers use circuit techniques a little different

Figure 1–5 A practical radio system

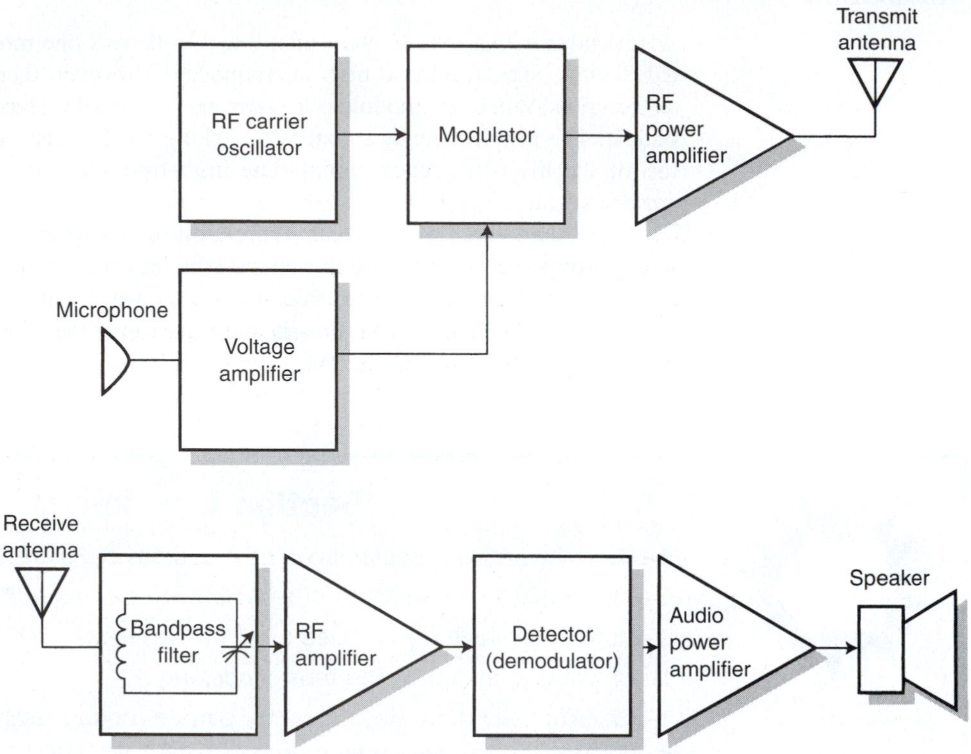

from the low-frequency amplifiers you might have studied in fundamentals courses. These techniques are necessary because of the high frequencies involved.

Figure 1–6 shows typical waveforms used to test an AM transmitter. The top waveform is the *information;* a sine wave is often used as a convenient test signal, since it's available from most benchtop signal generators. The bottom waveform shows the resultant AM signal at the modulator stage output.

Figure 1–6 Intelligence and AM carrier waves

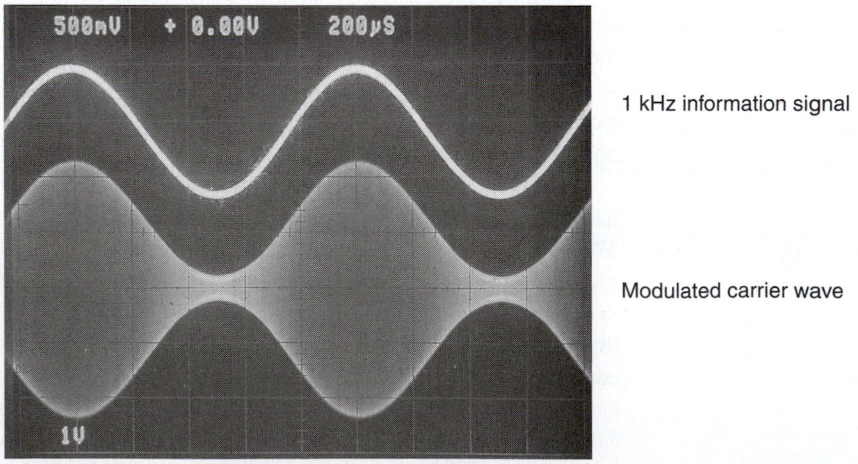

Receiver

The receiver of Figure 1–5 reverses the steps taken in the transmitter. After the incoming signal is received from space by the antenna, it is fed into a bandpass filter. Recall that a bandpass filter rejects frequencies above and below its design frequency. Most receivers use an LC resonant circuit as the bandpass filter. This filter is tuned to the *carrier frequency* of the transmitting station. There are thousands of signals reaching the receiver's antenna at any given instant, yet the receiver must reproduce only one of them. That is a tall order! Because each incoming carrier signal has a different frequency, it is possible to separate them by using a frequency-selective circuit. In other words, we use a filter to do the job of "selecting" the appropriate station.

You might have noticed the variable capacitor in the filter of Figure 1–5. Yes, that's the receiver's tuning control. The tuning knob on a receiver is coupled to a variable capacitor, which is part of the bandpass filter. When the capacitance is changed, the resonant frequency of the filter changes. When a user is tuning in a radio station, he or she is actually adjusting the resonant frequency of a tank circuit!

After selection of the appropriate carrier frequency, we will still have a very weak signal. Several stages of RF amplification are needed to bring the modulated wave up to a level that is useful. The detector or demodulator stage does exactly the *opposite* of the modulator stage. It takes the incoming modulated carrier waveform and strips away the carrier frequency, leaving a copy of the original information signal. The output of the detector will look very much like the original information signal from the microphone. This is the *recovered information* signal. The recovered information is still too weak to drive a loudspeaker (or other device), so an audio power amplifier is used to boost the signal's voltage and current. In most modern radio receivers, the audio power amplifier is likely to be on a single integrated-circuit (IC) chip. The loudspeaker completes the process, converting the amplified information signal back into sound.

Section Checkpoint

1–17 What is the purpose of an *oscillator* circuit?

1–18 How does an RF amplifier differ from an AF amplifier?

1–19 What is the purpose of the *modulator* circuit?

1–20 How many radio signals can be expected in the antenna circuit of a receiver?

1–21 How does a radio receiver "select" one signal from thousands?

1–22 What happens in a detector circuit?

1–5 THE RADIO-FREQUENCY SPECTRUM

There are many different radio frequencies available for use. Each group of frequencies has different operating characteristics. The radio spectrum is divided into "bands" to help people plan their use of the frequencies. Table 1–1 shows some common bands and what they're used for. Note how we can refer to bands by either frequency or wavelength. For example, the 10 meter band really refers to the frequencies at or around 30 MHz. A technician should memorize these; they're used often in RF work.

Table 1–1 The RF frequency spectrum

Band Name	Frequency Range	Wavelength Range	Primary Applications and Users
LF	30–300 kHz	10 km–1 km	Subterranean communications; aircraft navigation
MF	300–3000 kHz	1000–100 m	AM broadcast: amateur radio; military
HF	3 MHz–30 MHz	100 m–10 m	SW (shortwave) broadcast; amateur, military, and commercial long-distance communications; CB band at 27 MHz
VHF	30 MHz–300 MHz	10 m–1 m	FM and television broadcast; local amateur, commercial, and public safety communications
UHF	300 MHz–3 GHz	1 m–10 cm	UHF television broadcasts; amateur, commercial, public safety, and satellite communications; cellular telephones at 800 MHz
SHF	3 GHz–30 GHz	10 cm–1 cm	Microwave bands; satellite communications; radar measurement of distance and speed
EHF	30 GHz–300 GHz	1 cm–10 mm	Highest radio frequencies; high-definition radar; satellite and experimental communications modes

1–6 DIGITAL COMMUNICATIONS

Computers are an important part of modern life. Part of the power of computers arises from their ability to communicate information rapidly over a distance. A *network* is a group of computers that are connected together and share data. Computers use digital signals in their operation (recall that a digital signal has one of two possible states, 1 or 0). However, digital signals can't be directly sent over a long distance directly. Long-distance digital communication usually requires that the computer's digital signals be converted into analog form for transmission, as shown in Figure 1–7.

Figure 1–7 Computer data transmission process

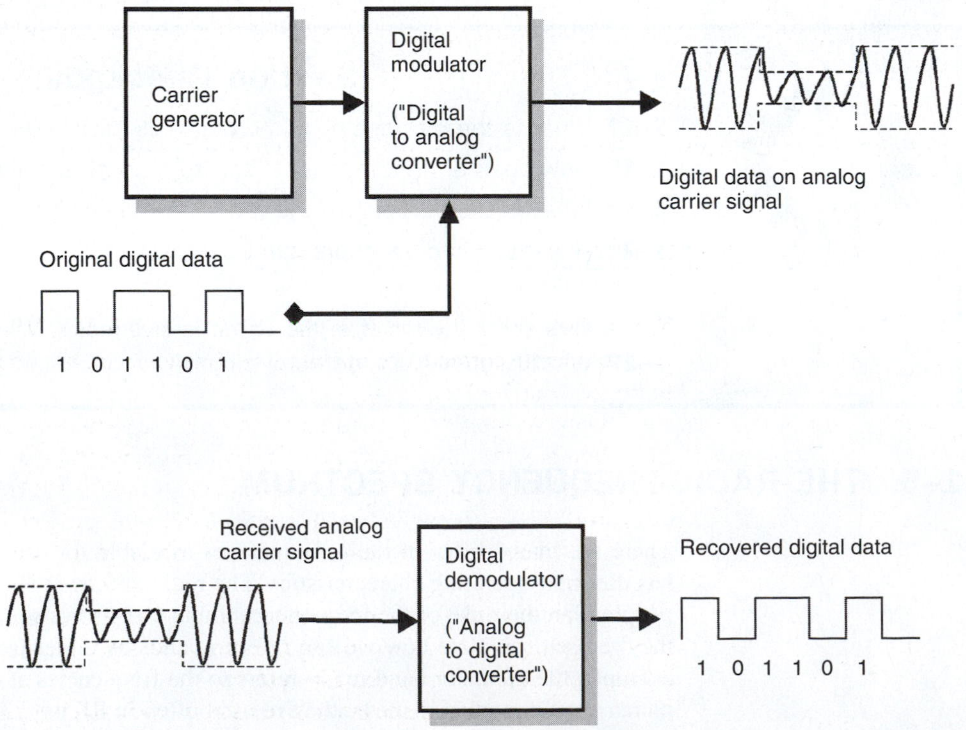

In Figure 1–7, the original computer data is transmitted in serial form (one bit at a time) to the digital modulator (which really acts as a digital-to-analog converter). Just as in voice applications, the modulator places the digital information onto a carrier signal supplied by the carrier generator circuit. The result is a digital signal that rides on top of an analog carrier wave, which can then be transmitted over a distance. The process is reversed to recover the digital data.

Figure 1–7 shows amplitude shift keying (ASK) carrying the data. ASK represents 1s and 0s by varying the amplitude of the sine wave carrier signal. It is really just amplitude modulation. FM and PM are also widely used to carry digital signals.

PCS (personal communications service) cellular telephones and HDTV both convert the sound (or picture) signals into digital representations, then transmit them on analog carriers. In the case of PCS, digital transmission enables compression of the voice signal, which allows more people to make calls from the same cellular tower. Digital transmissions can also be encrypted to prevent eavesdropping, unlike most analog mobile transmission. In HDTV, digital compression allows more stations to share the allotted frequencies, and error-correction methods allow crystal-clear picture and sound, even in the presence of interference signals.

Analog transmission technology is still important in the age of digital. Wherever there is a human interface, there will likely remain an analog device, and analog carrier transmission will likely always be needed to bridge distance.

Section Checkpoint

1–23 Explain the difference between digital and analog signals.

1–24 What type of modulation is used in Figure 1–7?

1–25 List two advantages of digital PCS phones over their analog counterparts.

1–26 Where are two places that analog components will always be needed in communication systems?

1–7 HOW TO TROUBLESHOOT SYSTEMS

Technicians are at the center of the action in electronics. When sophisticated equipment fails to operate correctly, or operating problems occur, it's often the tech who is called on to remedy the trouble.

Only a minority of technicians actually perform component-level repairs. At a job site, there is generally insufficient time for this type of work. Replacing individual components in today's high-tech electronics is best done on the bench, where proper tools are available. When trouble is found, a field technician is very likely to *substitute* working modules or assemblies for those found to be defective. Often this is referred to as "board swapping."

Some assemblies must be tested with specialized, automated equipment. This is becoming commonplace, as microprocessors and software are now an indispensable component of modern communication systems.

The successful technician follows a systematic method of troubleshooting. Many use the following three-step approach.

A Three-Step Approach to Troubleshooting

1. Visual (and other) inspection
2. Checking power supplies
3. Checking inputs and outputs

Visual inspection This means looking carefully at the equipment as well as the operating conditions. Are the controls set properly? Has someone changed a software setting? Do you smell something hot or burning? See any smoke? Is something cold that should be getting warm? Any unusual sounds? All of these things are important clues.

Be careful and use common sense. For example, a finger is not really well-suited as a "temperature probe." It's too easy to get burned or shocked. Never touch a live circuit! Also, if appropriate, use eye protection. Certain components (such as electrolytic capacitors) may fail explosively, especially in high-power circuits.

Checking power supplies This means checking the actual operating voltages *at the point in the circuit where the power is actually used.* This should *never* be directly attempted in high-voltage or high-power circuits. Appropriate (safe) test points are almost always provided to test these circuits.

More than 90% of electronic failures can be attributed to power supply troubles. It makes sense; the power supply is the most highly stressed section in electronic equipment. The power supply must pass all energy used by the equipment. It also must absorb *transients* and *spikes* from the ac power line on a regular basis. Just as a car won't run without gas, electronics won't work correctly if the power supplies aren't right.

Checking inputs and outputs When the first two steps have been followed, then it's generally safe to start on the third step, checking inputs and outputs. When you get to this step, you know that everything is getting proper power, and therefore something is stopping the signal from passing. Sometimes the signal passes but becomes distorted in some way. Waveform charts in a service manual are invaluable here when they're available. Often, you'll be on your own. You will have to devise an appropriate test procedure on the spot. Your background in fundamentals will help you to do this (plus your experience in courses like this one!).

EXAMPLE 1–3

Figure 1–8 shows a slightly more detailed picture of an AM transmitter. This unit is a 100 watt aviation transmitter. It was "swapped out" after it malfunctioned. The complaint on the repair order reads: "Unit transmits dead carrier. The listener (on the other end) cannot hear anything said into the microphone, but can tell that we are transmitting (silence)." Work out a plan for troubleshooting the unit. Which areas are suspect given the symptoms?

Solution

There are several ways of attacking the problem. First, the block diagram of the unit must be understood. It is similar to the diagram of Figure 1–5. There are a couple of new stages. The *buffer amplifier* slightly amplifies the RF carrier from the oscillator. It can be seen that the RF power amplifier has two stages, the *driver* and *final* amplifiers. The final amplifier is simply the *last* or "final" amplifier before the antenna. Note how the RF driver and final amplifiers operate on a high-voltage, high-current power supply. (You'll likely see these stages mounted on a metal *heatsink.*)

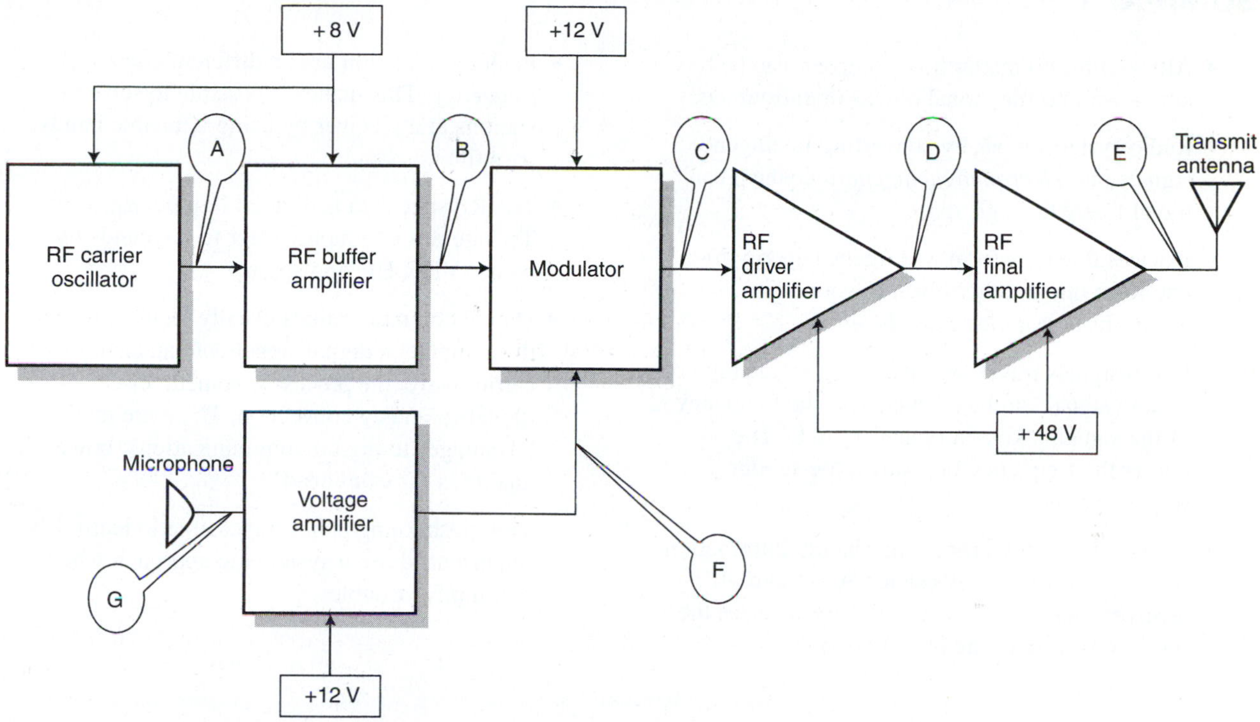

Figure 1–8 Example transmitter for troubleshooting

Since no modulation is occurring, we will be interested in the portion of the circuit that performs that task. That will include the *microphone,* the microphone *voltage amplifier (preamplifier),* and the *modulator.*

Proposed Troubleshooting Steps

Perform a visual inspection of the unit. Sometimes you'll notice a detail such as a dented cabinet corner—which could mean that the device has been *dropped* or *crushed.* The dc power supplies should be tested at the appropriate test points. There are *three* different dc levels: +48 V (volts), +12 V, and +8 V. The service manual may suggest the best place to make power supply measurements.

Once power supplies have been checked, we will check *inputs and outputs.* The original complaint is that no modulation is taking place. The *modulator* stage will be the best starting place. Test point F will be checked using an oscilloscope for an information signal. The technician may have to provide an "information" signal to do this test. Most commonly, the tech either talks or whistles into the microphone to provide a signal.

If information is seen at test point F, we know the microphone and microphone amplifier are both working, so we'll look closer into the *modulator* stage to find out why it isn't working.

If we *do not* see information at test point F, then the trouble lies either in the *voltage amplifier* or *microphone* itself. (We can check test point G with a scope to see if the microphone is producing an output).

SUMMARY

- All systems, no matter how complex, can be broken down into functional blocks or *subsystems.*
- Radio systems work by converting intelligence signals into *electromagnetic energy,* which can travel freely through space.
- Devices that transform intelligence energy from one form into another (such as sound into electricity) are called *transducers.*
- Radio signals travel as waves. The *wavelength* of a radio signal can be calculated if the frequency of the signal is known (equation 1–1). The higher the frequency of a wave, the shorter its wavelength becomes.
- *Modulation* is the process of placing information onto a radio-frequency carrier. At a radio receiver, a *demodulator* or *detector* reverses the process to recover the information.
- Each radio station uses a different *carrier frequency*. This makes it possible to separate stations at a receiver by using a tunable bandpass filter.
- The RF spectrum is divided into *bands*. Technicians commonly refer to the bands by name (VHF, HF, and so on.)
- Digital communications usually requires placing the computer's digital signal onto an analog carrier wave; the process is sometimes called digital-to-analog conversion. There are many advantages to digital communications; however, analog is far from dead!
- Troubleshooting is not impossible to learn. It's important to use a systematic approach when hunting for trouble.

PROBLEMS

1. Draw a block diagram of a practical radio transmitter. Explain the function of each block.
2. Draw a block diagram of a radio receiver. Explain what happens in each block.
3. Define *modulation*. List the three ways a carrier wave can be modulated.
4. What are two reasons why modulation is necessary in radio?
5. List all the transducers shown in Figure 1–5. For each transducer, give the type of input and output energy.
6. What is the purpose of the oscillator in a radio transmitter?
7. Define *wavelength.* How do changes in frequency affect wavelength?
8. Calculate the wavelength of the following radio signals:
   a. 1 MHz
   b. 10 MHz
   c. 2.8 MHz
   d. 54 MHz
9. How long is one-quarter wavelength at each of the frequencies of question 8?
10. Calculate the frequency that corresponds to each of the following wavelengths:
    a. 10 m
    b. 2 m
    c. 15 m
    d. 70 cm
11. What is the approximate operating frequency of each antenna listed below? The fraction of a wavelength is given for each unit.
    a. length = 56″, 1/4 λ
    b. length = 56″, 1/2 λ
    c. length = 0.5 m, 1/4 λ
    d. length = 10 m, 1/2 λ
12. Draw the block diagram of a digital communication system. Using outline form, explain the function of each major block in the system.

13. List the three steps of troubleshooting a system, in order.

14. Why are power supplies the most failure-prone portion of electronic systems?

15. A unit just like the one in Figure 1–7 has come in for servicing. The complaint on the repair order states "won't transmit." When connected to a dummy load and RF power meter, the set shows *no* power output when the microphone is keyed.

a. What will be the first step taken when this unit is troubleshot?

b. List the power supply voltage test points (by voltage) that will be measured in the second step of troubleshooting.

16. The power supply test points in the set of question 14 measured good. The following oscilloscope measurements were made *with the microphone keyed:*

- test point E, 0 V_{pp} RF
- test point C, 10 V_{pp} RF
- test point B, 5 V_{pp} RF
- test point D, 0 V_{pp} RF
- test point F, 1 V_{pp} AF (Intelligence)

What stage or stages is most likely causing the problem?

2

Signal Analysis

OBJECTIVES

At the conclusion of this chapter, the reader will be able to:

- explain the difference between the *frequency* and *time* domains
- draw a diagram of a sine wave in both the frequency and time domains
- explain how a *complex* waveform (such as a square wave) looks in the frequency domain
- define the terms *fundamental* and *harmonic*
- define *noise* and list at least two *internal* and two *external* noise sources
- calculate the *signal-to-noise ratio* if the signal and noise voltages (or powers) are known
- explain how the *noise figure* is calculated for an amplifier

Many different kinds of signals are produced and used by electronic systems. A *signal* is an electrical current or voltage that either represents information (the *information signal*) or performs some useful function (the *carrier signal* in radio). Most signals are alternating current sine waves or, as we shall see, combinations of sine waves. Signals can be viewed in either the *time* or *frequency* domains.

It's important for technicians to be able to accurately measure electrical signals. We can't see electrons flowing in circuits. Test equipment is our eyes. We will make almost all of our decisions based on what we read from test equipment—so we must read it accurately!

Many systems have signals that are not wanted. The name for any unwanted signal is *noise*. Noise is produced both inside and outside of circuits. Radio receivers are especially sensitive to noise, as they must amplify extremely tiny signals from receiving antennas.

2–1 TWO DOMAINS

Most technicians are very familiar with the instrument in Figure 2–1. It is, of course, an oscilloscope. An oscilloscope has a trace that sweeps across the screen from left to right at a selected and calibrated rate. As the voltage of a waveform varies from positive to negative and back, a waveform results.

Figure 2–1 A typical oscilloscope

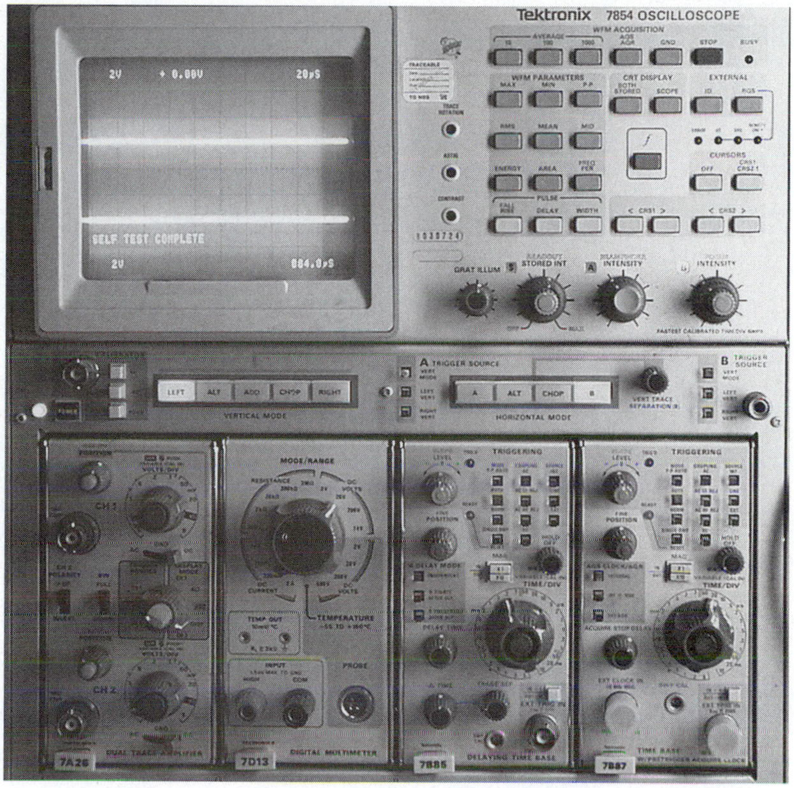

A scope works a lot like the popular "Etch-A-Sketch" toy. The *timebase* of a scope moves the dot (horizontal knob on the toy) across the face of the screen at a constant speed. The *vertical deflection* section of the scope moves the dot up and down (same as the vertical knob). Because a scope does this over and over at a rapid rate, our eyes fuse the images together into one continuous line.

We could say that a scope *draws a waveform as it happens*. In other words, an oscilloscope shows pictures of waveforms in the *time domain*. The horizontal axis on an oscilloscope is in units of *time*, in *seconds*.

EXAMPLE 2–1

What is the *frequency, peak voltage, peak-to-peak voltage,* and *RMS (root mean square or effective) voltage* of the waveform pictured in Figure 2–2 if the scope settings are as follows:

Horizontal, 100 μs (microseconds) per division

Vertical, 5 volts per division

If this voltage is being measured across a 50 Ω resistor, what power will result?

Solution

The frequency can be calculated if the *period* (T) is known:

$$f = \frac{1}{T} = \frac{1}{1\text{ ms}} = \underline{\underline{1000\text{ Hz}}} = \underline{\underline{1\text{ kHz}}}\text{ (kilohertz)}$$

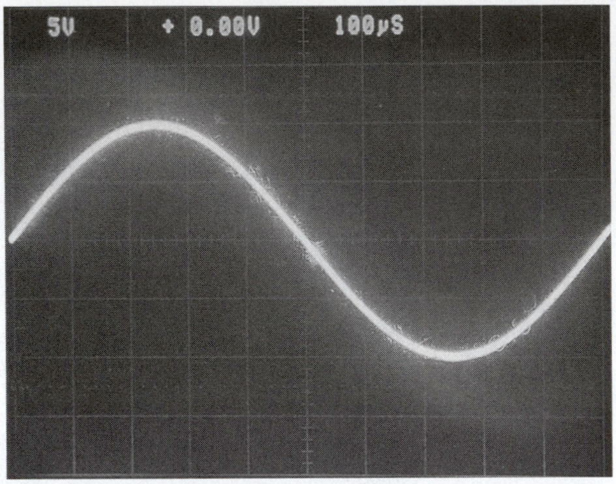

Figure 2–2 Reading the oscilloscope

The peak voltage can be calculated by observing that the trace goes two grid squares above (or below) the baseline at the peak of each cycle:

$$V_P = (2 \text{ divisions})(5 \text{ V/division}) = \underline{\underline{10 \text{ V}_P}}$$

The peak-to-peak voltage is the total height of the waveform:

$$V_{PP} = (4 \text{ divisions})(5 \text{ V/division}) = \underline{\underline{20 \text{ V}_{PP}}}$$

The RMS or *effective* value of the waveform can be calculated since the shape is a sine wave:

$$V_{RMS} = \frac{V_P}{\sqrt{2}} \approx 0.707 \text{ V}_P \approx (0.707)(10 \text{ V}) \approx \underline{\underline{7.07 \text{ V}}}$$

Note that many technicians even use 0.7 (rather than 0.707) as an "approximate" factor for calculating an RMS voltage. When reading from an oscilloscope, this is quite valid, since there may be as much as 5% measurement error just from "eyeballing" the display!

Caution: The formula above for RMS voltage is only valid for a sine wave!

The *power* can be calculated using Ohm's law:

$$P = \frac{V^2}{R} = \frac{7.07V^2}{50 \ \Omega} = \underline{\underline{1 \text{ W}}}$$

Caution: To calculate power, an RMS voltage must be used!

A Shortcut for Calculating Power

If you know the peak-to-peak reading of a *sine wave* waveform, you can also calculate power by using

$$P = \frac{V_{PP}{}^2}{8R} = \frac{20 \ V_{PP}{}^2}{(8)(50 \ \Omega)} = \underline{\underline{1 \text{ W}}}$$

Some techs like this formula, since it avoids the need to convert to RMS first. However, it can only be used for sine waves!

The Frequency Domain

There's another way of looking at electrical signals. An oscilloscope shows signals in the *time domain,* which is fine for many types of measurements. However, many times we're much more interested in what *frequency* or *frequencies* a waveform might contain. An instrument that shows the *frequency domain content* of a signal is called a *spectrum analyzer.*

Being able to measure the *frequency content* of signals is very useful to us for several reasons. You'll recall that one of a radio receiver's tasks is to separate the one desired carrier signal from all the others on the air. This is only possible because each radio transmitter uses a different frequency. *The frequency content of a signal therefore determines whether or not it will be reproduced in a receiver!*

Second, there is only a finite amount of space on the bands in which to operate radio transmitters. Transmitters use up this space in the same way that the parking lot at the local mall fills up with cars. Two radio stations can't share the same frequency, or they'll interfere with each other. *By looking at a radio transmitter's signals in the frequency domain, we can determine its bandwidth. Bandwidth* is the amount of "frequency space" taken up by a signal. It's very much like the width of a motorcycle, car, or truck to be parked in a stall.

Last, we can often tell much more about the quality of a signal by looking at it in the frequency domain. Defects or *distortions* of a waveform are often hard to spot on a scope, especially if the waveform is a sine wave. As you'll see, any distortion in a sine wave will cause new frequencies called *harmonics* to appear. *Distortion of a sine wave (such as an RF carrier) is often much easier to spot on a spectrum analyzer than on a scope.*

The Frequency Domain Is Nothing New!

Long before the dawn of electronics, humans employed the concept of the frequency domain. Take a look at the notation of Figure 2–3. The music notation of course, refers to both *pitch* and *duration* for each note to be played. *Pitch* is really the listener's mental perception of the *frequency* of the note being played. Raise the frequency, and the listener hears a higher pitch. The vertical position of each note gives a precise indication of its pitch. Pitch is related to frequency. Therefore, we're looking at a *frequency domain* picture here! (There is also time domain information, because each different type of note plays for a unique time interval.)

The Spectrogram

A *spectrogram* is a graph showing the frequency domain information of a signal. If the 1 kHz signal of Example 2–1 is perfect, it will look like Figure 2–4 on a spectrogram. This picture is troubling; it doesn't look *anything* like a sine wave! It's just like the music in Figure 2–3; the notation on paper doesn't *look* or *sound* anything like the music when it's

Figure 2–3 Old-time frequency domain notation

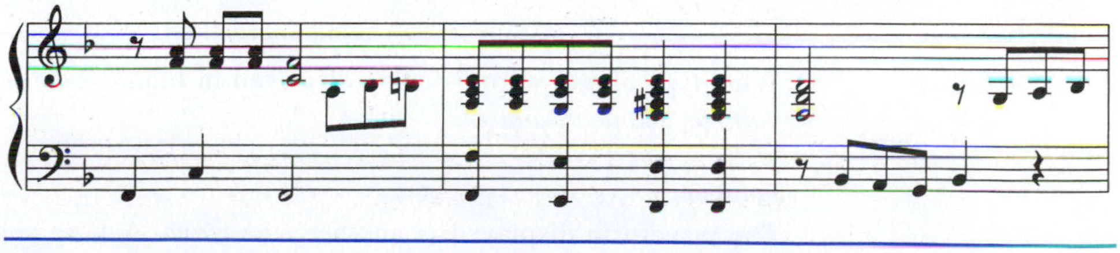

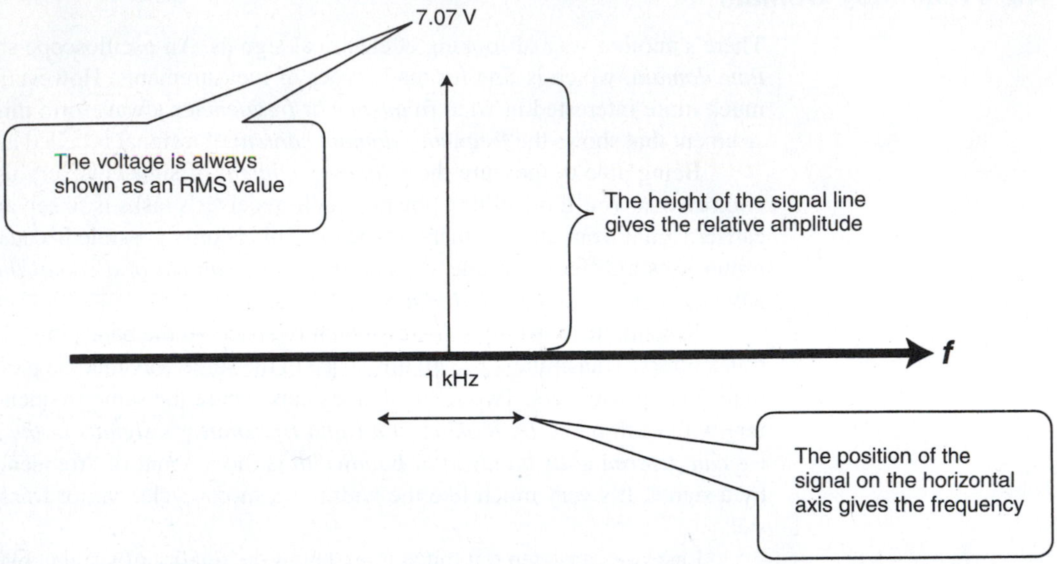
7.07 V
The voltage is always shown as an RMS value
The height of the signal line gives the relative amplitude
f
1 kHz
The position of the signal on the horizontal axis gives the frequency

Figure 2–4 The spectrogram of an ideal sine wave

Figure 2–5 Spectrum analyzer display of the pure sine wave (vertical setting, 1 V/division)

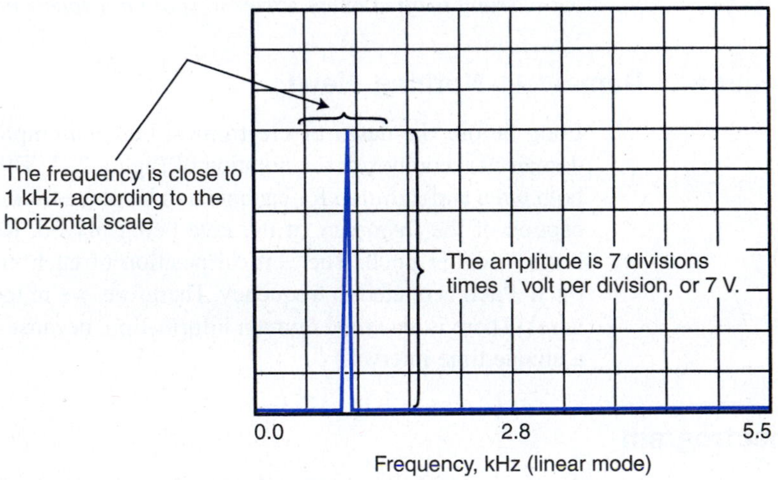
The frequency is close to 1 kHz, according to the horizontal scale
The amplitude is 7 divisions times 1 volt per division, or 7 V.
0.0
2.8
5.5
Frequency, kHz (linear mode)

being played, yet it still represents it. Figure 2–5 is the sine wave of Figure 2–1 displayed on a spectrum analyzer display.

> The sine wave is sometimes referred to as the only "pure" waveform, because a sine wave has only *one* frequency when it is viewed in the frequency domain.

EXAMPLE 2–2

What type of waveform is being displayed in Figure 2–6? What is its *frequency, RMS voltage,* and *peak voltage?*

Solution

The waveform displayed is another *sine wave.* A pure sine wave always shows up as one "line" on a spectrum analyzer display. By reading its position on the horizon-

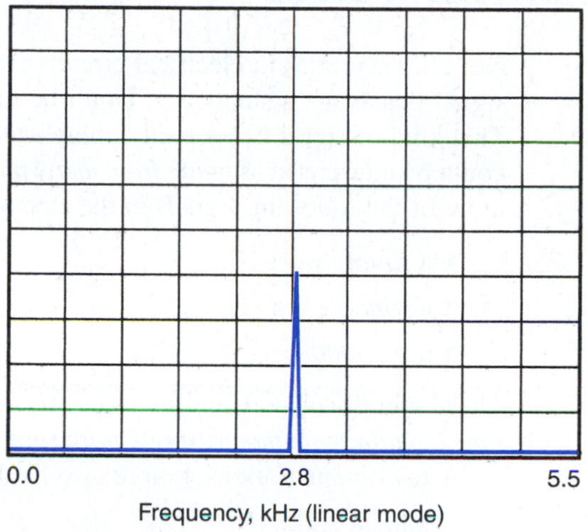

Figure 2–6 A spectrogram display (vertical setting, 1 V/division)

tal axis, we can see that the frequency is 2.8 kHz. The line is 4 units high, so its voltage is

$$V = (4 \text{ divisions})(1 \text{ V/division}) = 4 \text{ V}$$

This is an RMS voltage. Therefore,

$$V_p = V_{RMS}\sqrt{2} \approx \frac{V_{RMS}}{0.707} \approx 5.66\ V_p$$

Note that dividing by 0.707 is the same thing as multiplying by 1.41, which is approximately the square root of 2.

Section Checkpoint

2–1 What is a *signal?*

2–2 What are the two *domains* for viewing signals?

2–3 What instrument displays signals in the time domain? What are the units of its horizontal axis?

2–4 What is a spectrogram?

2–5 Give three reasons why technicians need to understand the frequency domain.

2–6 What instrument shows frequency domain information? What are the units on its horizontal axis?

2–7 What does a pure sine wave look like on a spectrogram?

2–8 Why is a sine wave referred to as "the only pure waveform"?

2–9 If the frequency of a sine wave is increased, what happens to its spectrogram picture?

2–10 If the amplitude (or voltage) of a sine wave is increased, how will its spectrogram change?

2–2 COMPLEX WAVEFORMS

Not all waveforms in electrical circuits are pure sine waves. A *complex waveform* is any signal that is not a sine wave. Don't be intimidated by the phrase "complex waveform." This kind of signal is not really complicated at all! In this application, the word *complex* could be interpreted as *made from many parts*. It's very likely that you've measured one or more of the following signals in the electronics lab:

- a *square wave*
- a *triangle wave*
- a *sawtooth wave*

These are all complex and *periodic (repeating)* signals. *We say they're complex because they contain more than a single pure sine wave.*

In communications, you're also likely to measure some of these as well:

- human voice
- music
- digital data

Yes—these signals are also complex waveforms. In fact, they're quite complicated, especially the first two. Speech and music signals are very hard to predict because they are not periodic (repeating). Fortunately, there's no need for an in-depth analysis of these signals to understand how they'll be carried by a communication system. We'll look at a few samples of these signals near the end of this section.

Fourier Analysis

A nineteenth century mathematician, Jean Baptiste Fourier (pronounced four-ee′-ay), was very interested in describing the movement of heat by using mathematics. The equations Fourier developed were *periodic* or *repeating,* but were not shaped like sine waves.

Fourier hypothesized that any periodic mathematical function (we can say "electrical signal") could be represented by the addition of an *infinite series* (sum of an infinite number of terms) of *sine* and *cosine* (sine with 90° angle) waves, plus a dc level or average. This idea is very important in communications. We can summarize it as follows. Any periodic signal that is *not a pure sine wave* can be considered to be built out of the following components:

1. a dc component or "average" (which can be zero)
2. a *fundamental* sine wave that has a frequency *exactly* the same as the frequency of the signal
3. an *infinite* number of *harmonics* (a *harmonic* is a frequency that is an exact multiple of the fundamental frequency)

That's pretty heavy stuff! Let's put it to practical application. We've seen what a sine wave looks like in both the time and frequency domains. By doing a "Fourier analysis" of any waveform, we get its *frequency domain* picture. Figure 2–7 shows a 1 kHz, 1 volt peak square wave. If we were to do a Fourier analysis of this signal, we would get the *frequency domain* version. It would look something like Figure 2–8.

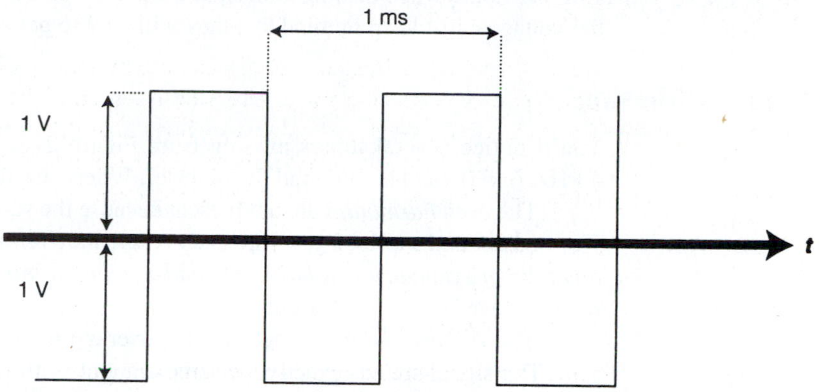

Figure 2–7 Time domain (oscilloscope) picture of a 1 kHz square wave

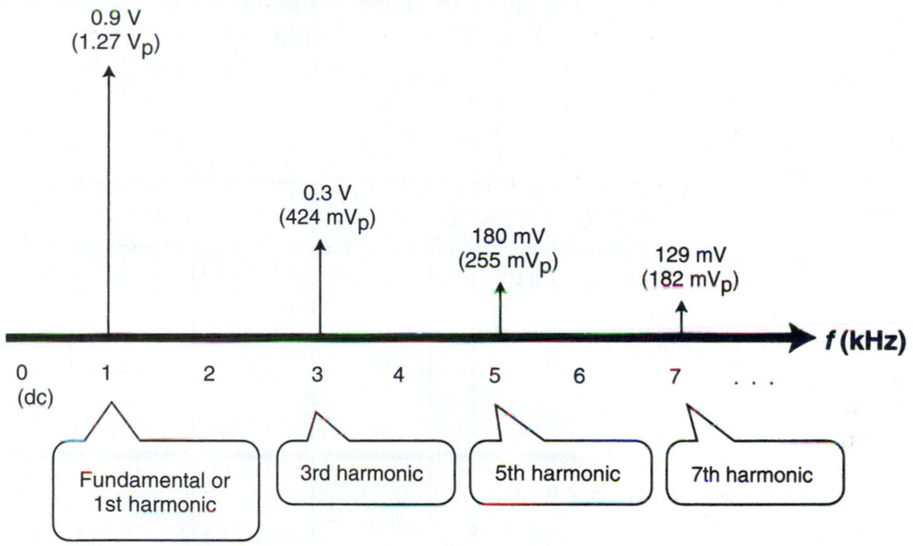

Figure 2–8 Frequency domain picture of a square wave

Again, this picture doesn't look *anything* like a square wave—but that's what it represents! By looking at this picture, we can tell that the 1 kHz square wave is really made of the following frequency components:

- a 1 kHz, 0.9 V (RMS) sine wave (the *fundamental* or *1st harmonic*)
- 3 kHz, 0.3 V sine wave (the *3rd harmonic*)
- a 5 kHz, 0.18 V sine wave (the *5th harmonic*)
- a 7 kHz, 0.129 V sine wave (the *7th harmonic*)

This sequence continues to infinity. *In a perfect square wave, there are an infinite number of frequencies present.* Fortunately, the square waves in practical circuits aren't perfect, so we don't need to analyze an infinite number of frequencies to understand them. *In fact, we only need to analyze up to the 13th harmonic in order to get a decent reproduction of a square wave.*

Notice that the voltages of the sine wave frequencies gradually get smaller as frequency is increased. Eventually, they approach zero. This is why most techs use the rule about the

13th harmonic when dealing with square waves. Don't be concerned with *how* we obtained the voltages; just keep in mind that they will tend to get smaller as frequency increases.

Something Is Missing!

You'll notice several things missing from Figure 2–8, namely, the harmonics at 2 kHz, 4 kHz, 6 kHz (and so on), and the dc level. Where are they?

The *even harmonics* are not present because the square wave is perfectly *symmetrical.* The bottom and top look just like each other; they're *mirror images. Any waveform with this type of symmetry will have only odd-numbered harmonics.* We say that the even harmonics have been canceled out.

The *dc level* is absent because the average of the signal voltages in Figure 2–7 is *zero*. The signal spends exactly the same amount of time being positive as it spends being negative, so on average, the dc voltage will be zero. If this were not true, the signal would look like Figure 2–9.

The signal of Figure 2–9 is still a 2 volt peak-to-peak square wave (1 volt peak); however, it has been pushed up or "clamped" to a level of 0.5 volt dc. The frequency domain version of Figure 2–9 looks like Figure 2–10.

Figure 2–9 A square wave riding on a dc level

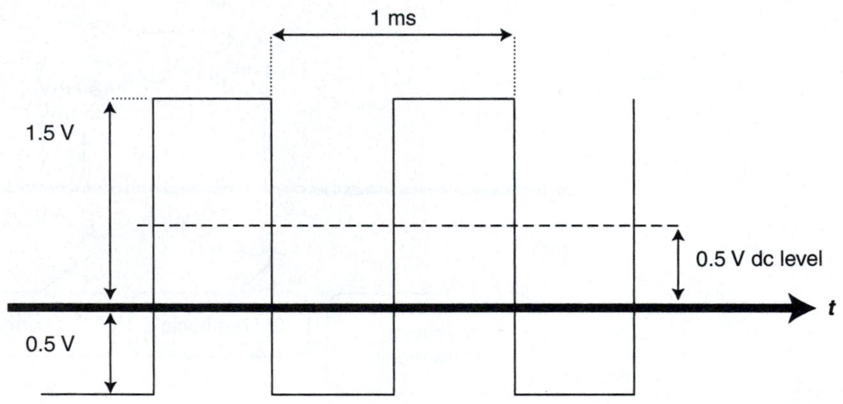

Figure 2–10 Frequency domain picture of the square wave riding on a dc level

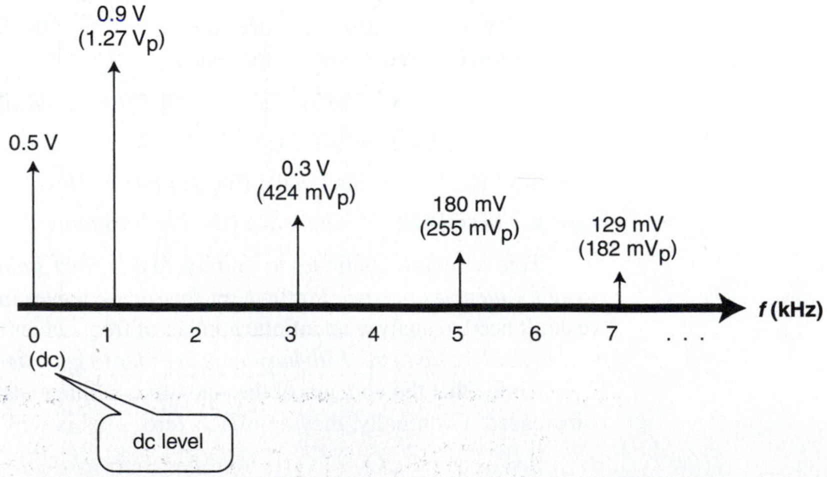

As you can see, the only difference between Figure 2–8 and Figure 2–10 is the addition of the 0.5 volt dc level. *Adding or subtracting a dc level from a complex waveform doesn't affect the sine wave frequency amplitudes.*

Square Wave Signal Measurements

Figures 2–11 and 2–12 show an oscilloscope and spectrum analyzer view of the same 1 kHz square wave that we just analyzed. How do the pictures compare to the theory?

The oscilloscope picture is exactly as we expected. There's no apparent dc level, and the period is 5 divisions, which works out to be 1000 μs (microseconds) or 1 ms (milliseconds). The frequency of this waveform is therefore 1/1 ms or 1 kHz.

The spectrum analyzer view also agrees with the theory. The fundamental frequency is 1 kHz and measures 900 mV (millivolts), the 3rd harmonic measures only 300 mV, the

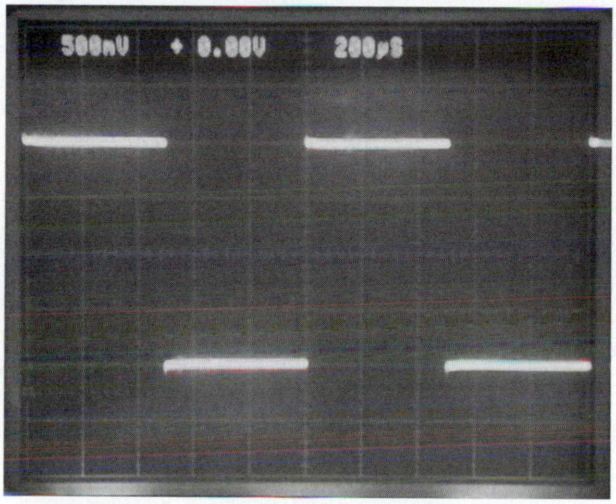

Figure 2–11
Oscillograph of 1 kHz square wave (vertical, 500 mV/division; horizontal, 200 μs/division)

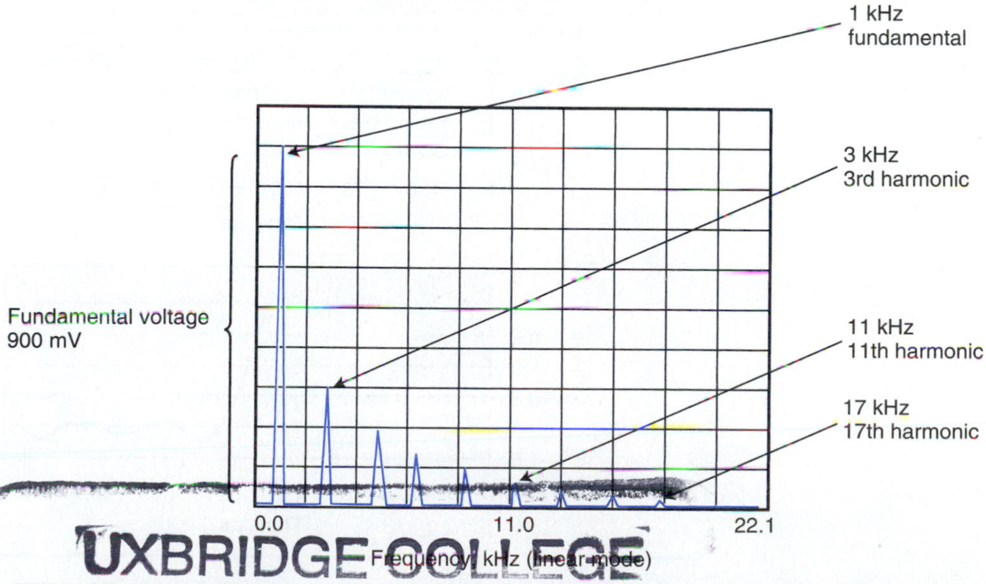

Figure 2–12
Spectrogram of a square wave (vertical setting, 100 mV/division)

5th (5 kHz) measures 200 mV, and so on. Notice how the signals start getting small around 11 kHz. *This is why most technicians stop analysis at the 13th harmonic.* The 17th harmonic is very small, and barely visible as a bump on the graph.

Separating Square Wave Signal Components

There's another way of looking at the square wave of Figures 2–7 and 2–11. A square wave is nothing more than the sum of an infinite number of sine waves, each with a frequency that is an odd multiple of the fundamental frequency. If we were to graph these sine waves and add them together point by point, we would get the square wave. Take a look at Figure 2–13.

Figure 2–13 The recipe for a square wave

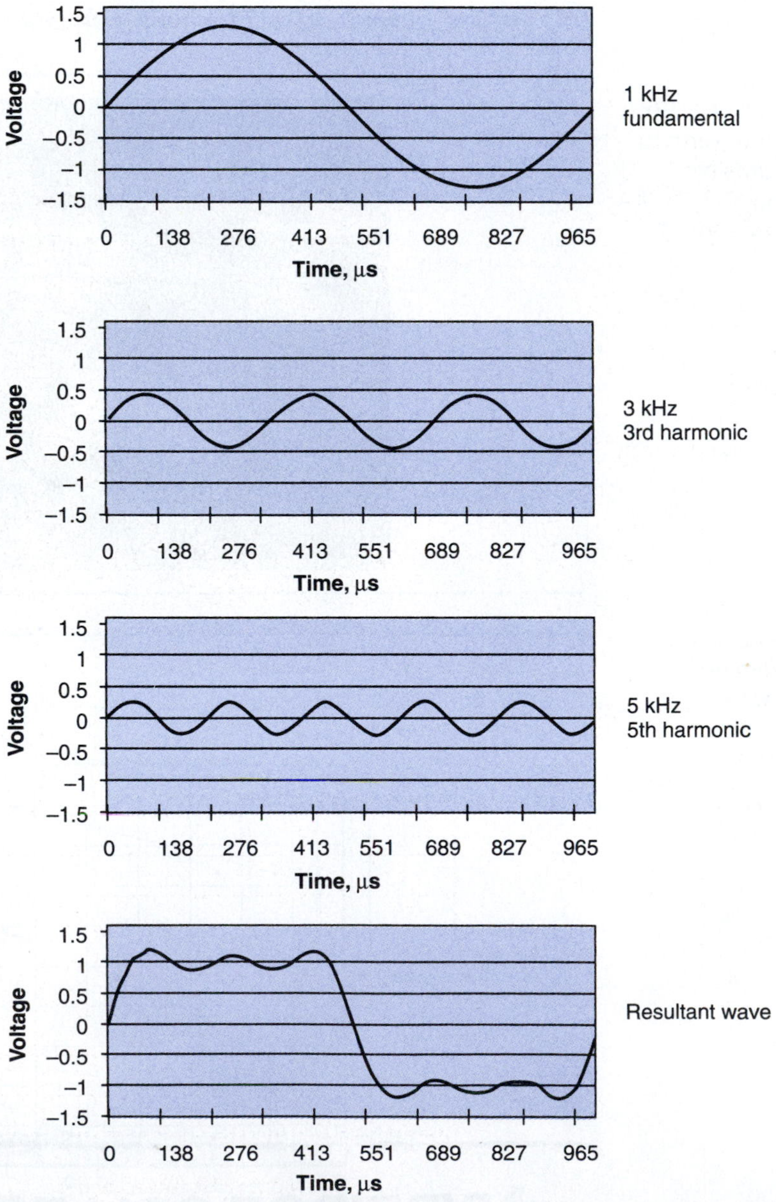

Figure 2–13 shows how a square wave can be built in the *time domain.* One way of thinking about the Fourier analysis of a waveform is to consider it a recipe for building that particular signal in the *frequency domain.* To obtain the *resultant* wave of Figure 2–13, each point in the *fundamental, 3rd harmonic, and 5th harmonic* was added together. The result was a point in the resultant waveform.

Other Complex Signals

Figure 2–14 shows a 1-kHz triangle wave on an oscilloscope. What frequencies do you expect to be in this waveform? Figure 2–15 shows the resultant *frequency spectrum.*

How does the frequency domain picture of the triangle wave compare to that of the square wave? If you compare Figures 2–15 and 2–12, you'll notice that the fundamental of the triangle wave is weaker. The harmonics are a *lot* weaker. Why would this be so?

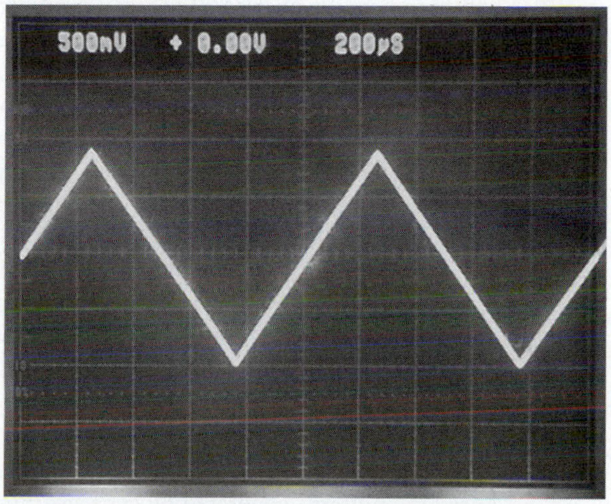

Figure 2–14 A 1 kHz triangle wave (vertical, 500 mV/division; horizontal, 200 μs/division)

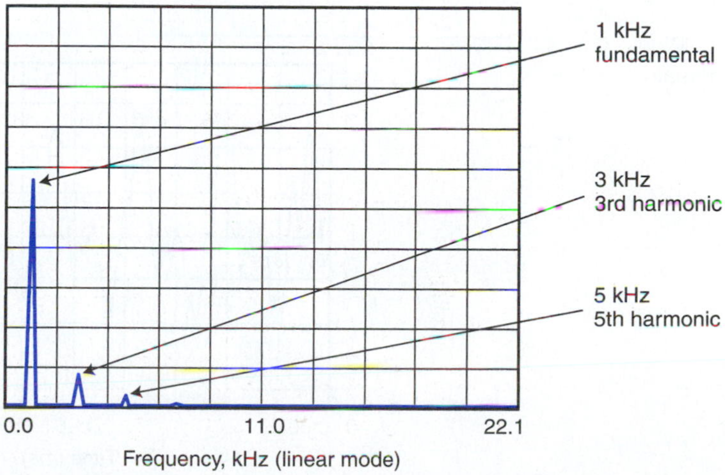

Figure 2–15 Spectrum analyzer view of a 1-kHz triangle wave (vertical setting, 100 mV/division)

You're right! The *shape* of the triangle wave is very close to the shape of a sine wave (the main difference is the addition of the "points" on top and bottom of the waveform). Therefore, it doesn't take much added energy (in the form of harmonics) to make a triangle out of a sine wave.

In contrast, a square wave is *very* different in shape than a pure sine wave. To produce the square shape requires the addition of much more high-frequency energy (sine wave harmonics) than that needed to build the triangle wave. Therefore, the square wave's harmonics remain relatively strong, way beyond the highest visible frequency of the triangle's harmonics.

Real-Life Signals

Music and voice are very complicated signals to analyze. There are two reasons for this. First, the basic waveforms are not pure sine waves and, strictly speaking, are *not* periodic. Second, these signals change in composition very rapidly and often contain many different (harmonically unrelated) *Fourier series.* A Fourier series is the collection of sine waves used to build a signal; in other words, a Fourier series contains the fundamental energy, plus the harmonics. Figure 2–16 is a snapshot of *middle C* being played on a piano keyboard. This is quite a complicated signal! You can see the *overall* period of the signal is about 3.8 ms, which corresponds to a frequency of 1/3.8 ms or 263 Hz. (The actual frequency of middle C is 261.625 Hz, so we're pretty close!) This waveform is not a perfect sine wave, so we can expect plenty of harmonics to be present. You won't be disappointed (see Figure 2–17)!

The frequency spectrum of the middle C piano note is packed with information. The harmonics all remain quite strong, right up to the 10th, after which a sudden drop in amplitude occurs. In between the harmonic "peaks" is other frequency information. *This is what makes a piano sound like a piano!* If middle C were played on a piano, a guitar, and a flute, you'd have no trouble identifying the three instruments. Why is this true, when middle C has the same pitch (frequency) on all three instruments? Yes, that's right—all three instruments will have a *fundamental* sine wave frequency of about 262 Hz, but the *harmonic patterns* generated by each instrument will be completely different. The human ear is the best spectrum analyzer on the planet; it can tell the difference between a piano and a guitar in an instant!

Figure 2–16 Middle C in the time domain

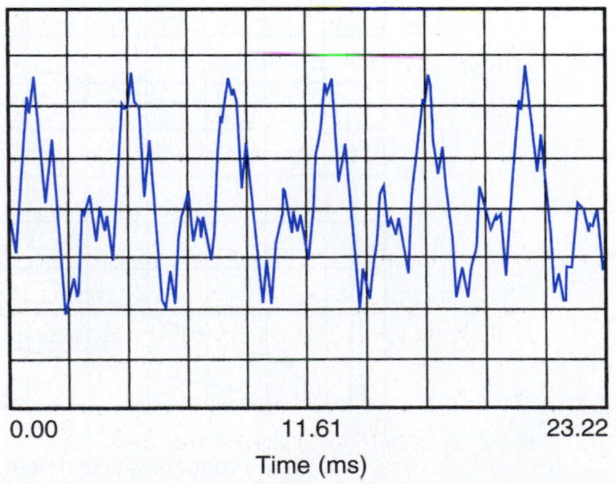

Figure 2–17 Middle C in the frequency domain

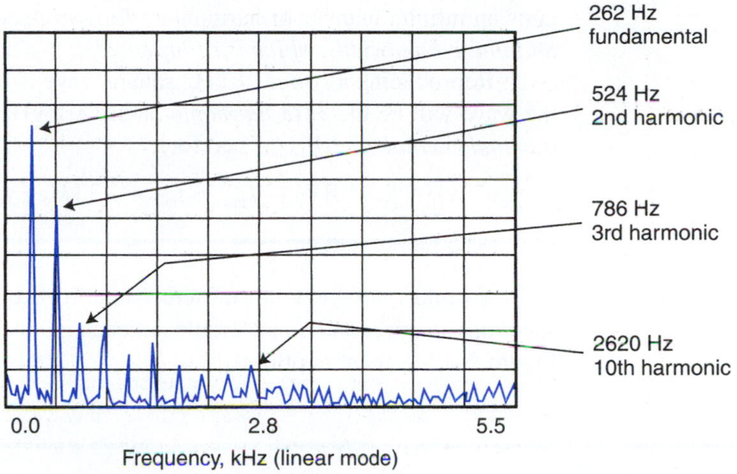

The Bandwidth of Electrical Signals

It's sometimes very important to know how much frequency or band space is required in order to reproduce a signal. In later chapters, we will refer to and calculate the *bandwidth* of many different kinds of signals. Bandwidth is easily calculated in the frequency domain:

$$BW = f_{max} - f_{min} \tag{2–1}$$

where f_{max} and f_{min} are the highest and lowest frequencies that are present in the signal, and BW is the bandwidth in Hz. Note that this is identical to the equation for the bandwidth of a filter, as you'll recall from circuit analysis. Let's do an example.

EXAMPLE 2–3

What is the bandwidth of a 1 kHz sine wave (such as the one in Figures 2–2 and 2–3)?

Solution

According to equation 2–1,

$$BW = f_{max} - f_{min} = 1 \text{ kHz} - 1 \text{ kHz} = \underline{\underline{0 \text{ Hz}}}$$

Bet you didn't see that coming! Looking at Figure 2–3, we see only *one* frequency in the sine wave. Therefore, *the maximum and minimum frequencies are the same, and NO bandwidth is required!*

How about the bandwidth of a signal that's a little more exciting . . . like a square wave?

EXAMPLE 2–4

What is the bandwidth of an *ideal* 1 kHz square wave (such as the one of Figures 2–7 and 2–8)? How does this compare to the bandwidth of a *typical* 1 kHz square wave (using the 13th harmonic as a stopping point, and ignoring the dc component?)

Solution

Again, according to equation 2–1,

$$BW = f_{max} - f_{min} = \infty \text{ Hz} - 1000 \text{ Hz} = \underline{\underline{\infty \text{ Hz}}}$$

The symbol "∞" means *infinity*. Remember that an *ideal* or *perfect* square wave contains an infinite number of harmonics. *To reproduce an ideal square wave would require an infinite bandwidth, which isn't practical!*

Reproducing a *typical* 1 kHz square wave is much easier. The highest frequency in the wave will be the *13th harmonic,* at 13 × 1 kHz, or 13 kHz. The lowest frequency is the *fundamental,* or 1 kHz. Therefore,

$$BW = f_{max} - f_{min} = 13\text{ kHz} - 1\text{ kHz} = \underline{\underline{12\text{ kHz}}}$$

To find the bandwidth of actual intelligence signals usually requires the use of a spectrum analyzer and considerable patience in measurement. Let's examine the signal of Figure 2–17 in more depth.

EXAMPLE 2–5

What is the approximate bandwidth required for the piano note of Figure 2–17?

Solution

To find bandwidth, we need to determine the highest and lowest frequencies in a signal. By inspection of Figure 2–17, the maximum frequency is approximately *2620 Hz* (the 10th harmonic) and the minimum frequency is approximately *262 Hz* (the fundamental). Again, according to equation 2–1:

$$BW = f_{max} - f_{min} = 2620\text{ Hz} - 262\text{ Hz} = \underline{\underline{2358\text{ Hz}}}$$

This is not a great deal of bandwidth; in fact, it's considerably *less* than the range of human hearing (20 Hz–20 kHz). Most of the energy in audio signals (including speech) is below 4 kHz. This fact is very useful for the designers of communication systems, for it allows them to build systems that use the *minimum bandwidth* necessary for reliable communications. Using the smallest bandwidth necessary conserves frequency spectrum and allows more stations to use a given range of frequencies. As we'll see, it also helps to reduce the effects of *noise.*

Filters and Complex Waveforms

A final note (no pun intended!) about complex waveforms: Filters can take these signals apart! Many times a frequency portion of a waveform can be used to advantage in communication systems. Take a look at Figure 2–18; can you predict what the output of the circuit will look like on an oscilloscope? Think carefully. You can do it!

Did you deduce that a 3 kHz *sine wave* would come out of the circuit? If you did, great; you've just made a great stride in understanding the frequency domain! Figure 2–19 shows the basic principle of the circuit in Figure 2–18, which we call a *frequency multiplier.* Notice how we've superimposed the bandpass curve of the *bandpass filter* on top of the harmonic pattern. This is to help us understand what the filter does to the waveform.

A bandpass filter only allows signals with frequencies within its *passband* to get through. All other frequencies are eliminated. Therefore, only the frequency of 3 kHz (and possibly other frequencies that are close to 3 kHz) can get through the filter. In terms of the frequency domain, we get the result of Figure 2–19, with an output waveform

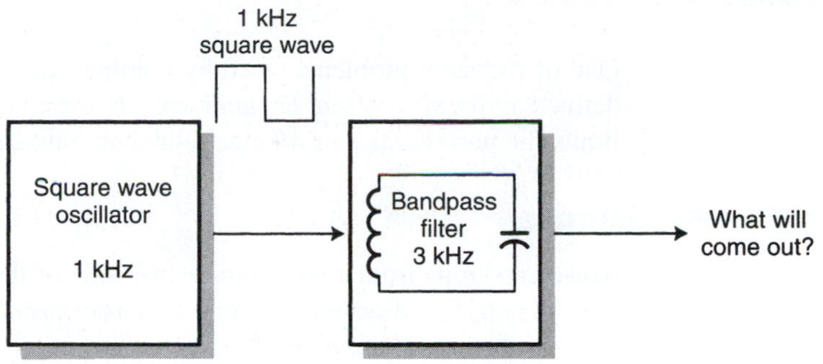

Figure 2–18 Bandlimiting a complex waveform

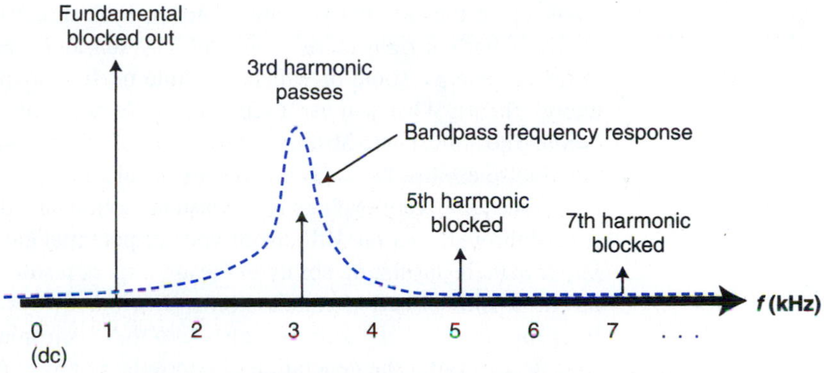

Figure 2–19 Use the frequency domain!

consisting of only the 3 kHz sine. A fancy way of saying this is that we've *bandlimited* the waveform. We've literally limited the frequency band that can get through the filter!

Note that the output is *exactly* three times the input frequency. This is why the circuit is called a *frequency multiplier.* You'll see them often in communications electronics.

Section Checkpoint

2–11 What is a *complex signal?*

2–12 What is the *fundamental* frequency in a complex signal?

2–13 What is a *harmonic?*

2–14 What waveform has a fundamental frequency and no harmonics?

2–15 According to the principles of Fourier analysis, what three things make up all periodic complex waveforms?

2–16 How many harmonics need to be reproduced (practically) to make a good square wave?

2–17 Which of these requires the most bandwidth to reproduce: a sine wave, a triangle wave, or a square wave (all at the same frequency)?

2–18 How do filters affect complex waveforms?

2–3 NOISE SOURCES

One of the main problems faced by communication system users is *noise,* which can be defined as *any signals that are undesired.* Noise comes from many sources, and it ultimately limits the performance of *all* electronic communication systems.

Classifying Noise Sources

Noise can come from either outside or inside of the electronic system. When the noise is created outside the system, we refer to it as *external noise.* Noise generated within the system is called *internal noise.*

External Noise Sources

Look up in the sky, but be careful not to look directly at this particular *external* noise source. That's it right there—the sun! The sun and other stars emit all kinds of electromagnetic energy, some of which are quite useful for life—like visible light and infrared energy (heat). What you can't see is the wide range of *radio frequencies* that the sun also emits. The atmosphere absorbs a great deal of the radio-frequency energy delivered by the sun. Noise emitted by the sun and stars is called *cosmic noise.*

The earth's atmosphere is in constant motion, and this causes electric charge to be displaced through friction. Billions of volts of potential can be built up; when the potential is larger than the insulating ability of the air, a spark results—*lightning!* At any moment, there are thousands of active thunderstorms distributed over the face of the planet, each contributing its share of lightning crashes to the electromagnetic din. A lightning discharge isn't necessary for the generation of atmospheric noise. Air moving over any object experiences friction, and electrical charge can be built up as a result. Sometimes this can cause severe radio-frequency interference! This is called *atmospheric noise.*

People also get in on the act. Many of our machines are great noise generators. Among these are spark ignition systems in motor vehicles, the power grid, fluorescent lights, and the occasional malfunctioning radio transmitter. Many mechanical devices can generate noise through mechanical friction, which can cause voltages to be created and thus tiny sparks—each releasing a bit of radio-frequency noise.

THE CASE OF THE IRATE CB SHOP OWNER

A CB shop owner in California was very angry with the local power company. He was sure that the power lines behind the store were the source of intense static on his demonstration radios, and as a result, his sales were down! A petition had even been circulated among the local CBers, since the interference was strong for several miles in any direction and seemed to be coming from the area of the shop. A radio-frequency interference (RFI) investigator from the power company arrived and verified that the static level was indeed very high. The shop owner added that the static was much worse on dry, windy days.

The investigator used a portable very high frequency (VHF) radio receiver and walked around the premises. The static became *much* stronger behind the building but not next to the power lines. Acting on a hunch, the inspector shook the CB antenna mast. The owner came out and demanded to know what the inspector had done, for the noise had just gotten much worse!

The inspector noticed that the mast was fastened to the building with iron pipe straps but not very tightly. On windy days, air friction caused static charge to build up on the mast. When the voltage got high enough to jump the fraction of an inch to the iron straps, an arc resulted (creating the static interference). Tightening down the loose straps eliminated the interference. Case closed!

What can be done about external noise? In some cases, we can *shield* circuits to prevent the noise from getting in. By *shielding,* we mean enclosing the circuit in a metal casing or cabinet to prevent electromagnetic energy from entering (or leaving, if the circuit is a potential noise producer). Unfortunately, radio receivers must be connected to an antenna. This is a direct path for noise to enter the set.

To reduce external noise in radio receivers, we generally try to locate them as far as practically possible from the noise source.

Internal Noise Sources

Circuits generate noise on their own. You may have observed how a television receiver behaves if disconnected from the antenna and tuned to a blank channel. The screen may fill up with white "snow." Where is the snow coming from? It can't be from the outside; the antenna is disconnected. The snow is the result of *internal* noise being generated in the circuits of the TV.

Internal noise comes in two flavors. *Resistors* and other passive components produce a type of noise called *thermal noise* or *Johnson noise*. Active devices such as diodes and transistors contribute *shot noise.* Field-effect transistors contribute a noise confined mostly to low frequencies known as *flicker* or *1/f* noise.

Thermal Noise

Matter is made of atoms and exists in gaseous, liquid, and solid states, depending on pressure and temperature. Even in the solid state, atoms are moving. They don't move freely as in the liquid or gaseous states, but they do vibrate back and forth while remaining in a semi-fixed position. The higher the temperature, the more rapid and energetic the vibration. This affects the path taken by charges (electrons or holes). Take a look at the current flow of Figure 2–20.

Figure 2–20 Ideal current flow

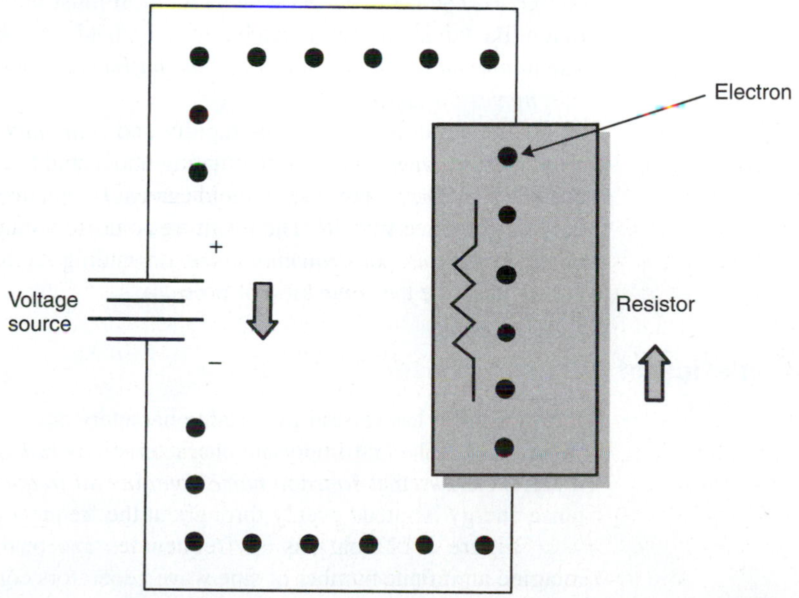

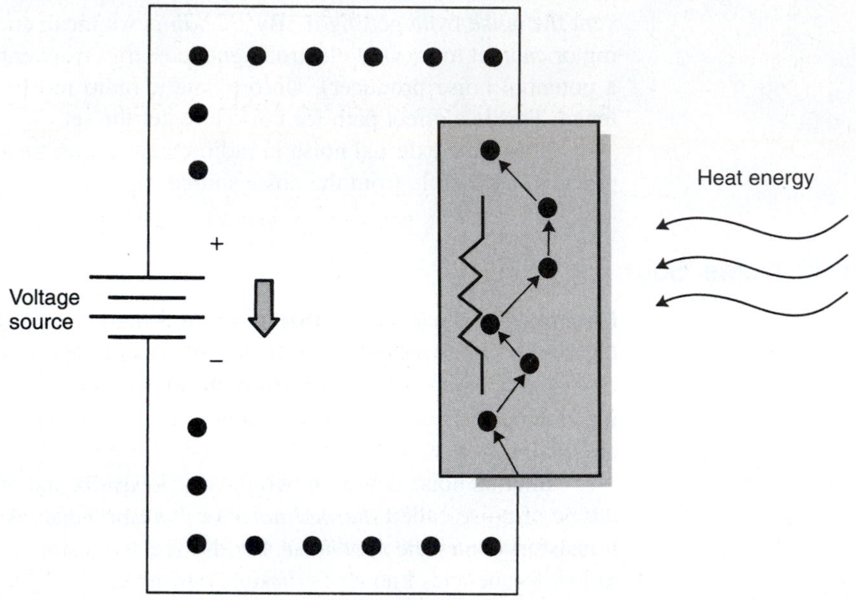

Figure 2–21 Current flow in the presence of thermal energy

In Figure 2–20, the battery pushes electrons out of its negative terminal. Ideally, these charges will flow atom-to-atom in orderly columns ("single file") down the conductors until they reach the resistor. The resistor offers opposition to the electrons, slowing them down. (Recall that current is merely the rate at which charge moves through a circuit; slowing down electron flow is the same as reducing the current.) In this circuit, there are no disruptions to the flow of electrons, and the current always flows steadily. Compare this to Figure 2–21.

Look at the current flow through the resistor of Figure 2–21. It's not very smooth! What has happened? Since heat is present, the atoms of the material are vibrating. For a charge to "catch a ride" on the next atom, it must move toward a moving target. It's very much like having to run to catch a moving bus! *The charges can't move in a straight line due to atomic vibration—and since the path is random, it will be different for each charge that moves through the device.*

The result is that charges rapidly and *randomly* speed up and slow down. In other words, the *current* is rapidly fluctuating above and below the average value that would be predicted by Ohm's law. These rapid current fluctuations induce an ac noise voltage across the resistor since $V = IR$. The resulting ac noise voltage is called *Johnson noise, thermal noise, white noise,* or *Brownian noise,* depending on the mood of the lecturer! All of these names describe the *same* kind of noise.

Characteristics of Johnson Noise

Johnson noise has several important characteristics, some of which can be deduced from Figure 2–21. The first important characteristic is *randomness.* By an advanced derivation, it can be shown that *Johnson noise contains all frequencies from near dc to infinity.* This noise energy is spread evenly throughout the frequency spectrum.

Figure 2–22 contains all frequencies except dc. One way of looking at it is to imagine an infinite number of sine wave generators connected together, each at a different

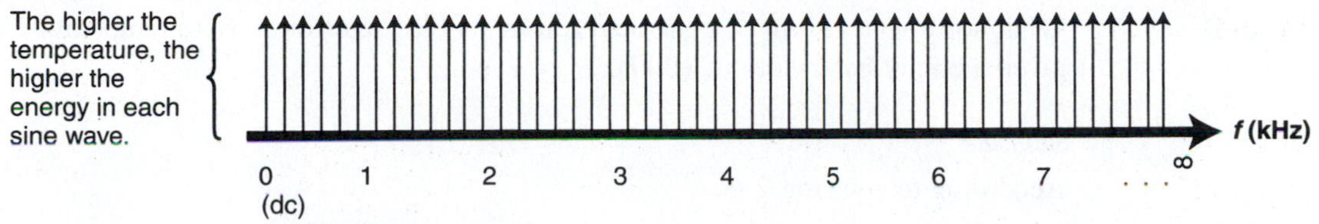

Figure 2–22 The frequency or spectral distribution of Johnson noise

frequency. The *total noise power* would then be related to the "area" of sine wave energies. *There are therefore two factors that control the power available from a Johnson noise source, the bandwidth and the temperature.* Let's modify Figure 2–22 to demonstrate why this is so.

In Figure 2–23, we see a practical modification: Real circuits do *not* have an infinite bandwidth! Can you see how bandwidth affects noise? Yes, when bandwidth is increased, noise power increases. *This is another reason why we use the minimum necessary bandwidth in electronic communications!*

The total noise power is really just an "area." We can calculate it by

$$P_n = kT\,BW \tag{2–2}$$

where k is *Boltzmann's constant* (approximately 1.38×10^{-23} J/°K), P_n is the power of the noise in watts, T is absolute temperature in degrees Kelvin, and BW is the bandwidth of the circuit. Knowing the noise power is not as useful as knowing the *noise voltage* in many cases. By the application of Ohm's law to equation 2–2, we can calculate the open-circuit noise voltage that will be generated across a resistor:

$$V_n = \sqrt{4kT\,BW\,R} \tag{2–3}$$

Equation 2–3 tells us that the voltage of a Johnson noise source depends on the following factors:

- The temperature of the circuit.
- The bandwidth of the circuit.
- The noise resistance R of the circuit. This resistance is approximately the same as the actual resistance.

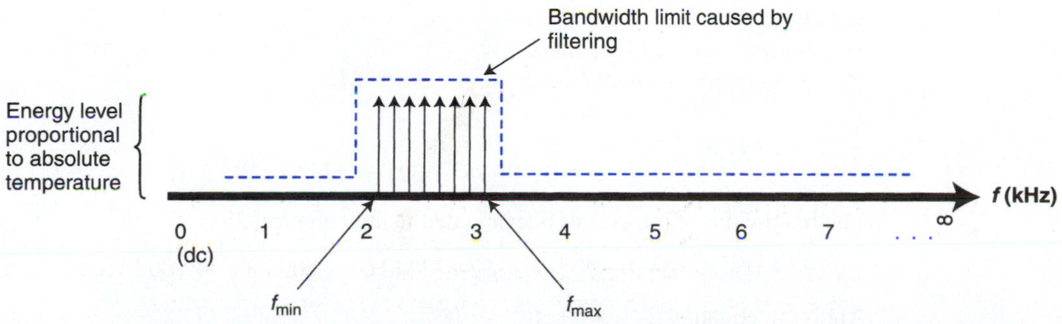

Figure 2–23 Bandlimited Johnson noise

EXAMPLE 2–6

What noise voltage will be generated across a 47 kΩ resistor at room temperature (24°C), given a circuit bandwidth of 10 kHz?

Solution

According to equation 2–3,

$$V_n = \sqrt{4\,k\,T\,\text{BW}\,R}$$

> *TIP* Many calculators will do this temperature conversion automatically.

The temperature T must be expressed in degrees Kelvin, so we must first find it:

$$°\text{K} = °\text{C} + 273$$

So,

$$T = 24°\text{C} + 273 = \underline{297°\text{K}}$$

Applying equation 2–3, we get:

$$V_n = \sqrt{4\,k\,T\,\text{BW}\,R} = \sqrt{(4)(1.38 \times 10^{-23})(297)(10\text{ kHz})(47\text{ k}\Omega)} = \underline{\underline{2.78\ \mu\text{V}}}$$

The noise voltage obtained in Example 2–6 doesn't look like much, does it? This amount of noise may or may not be a problem. Can you determine why? Right, it has something to do with the amount of *signal* in the circuit. Say that we have 1 volt of signal present. A few microvolts of noise will probably not be a severe problem. However, what if the signal is small—say, 1 μV? Ouch, big problem here! *The noise will be larger than the signal.* It would be like trying to hear someone whisper from across a crowded room.

Radio receivers must often amplify signals that are only a few microvolts. Many specialized techniques are used to keep the noise down to a minimum, as we'll soon see. Let's try another example.

EXAMPLE 2–7

What noise voltage will be generated across a 100 kΩ resistor at (a) 74°F and (b) 212°F? The circuit amplifies a frequency range of 600–700 kHz.

Solution

The temperature is given in Fahrenheit, so we must first convert it to degrees Kelvin. These two relationships are helpful:

$$°\text{C} = (5/9)(°\text{F} - 32)$$
$$°\text{K} = °\text{C} + 273$$

So,

$$T = (5/9)(74 - 32) + 273° = \underline{296.3°\text{K}}$$

The bandwidth of the circuit is calculated using equation 2–1:

$$\text{BW} = f_{max} - f_{min} = 700\text{ kHz} - 600\text{ kHz} = \underline{100\text{ kHz}}$$

Applying equation 2–3, we get:

a. $V_n = \sqrt{4\,k\,T\,\text{BW}\,R} = \sqrt{(4)(1.38 \times 10^{-23})(296.3)(100\text{ kHz})(100\text{ k}\Omega)} = \underline{\underline{12.8\ \mu\text{V}}}$

When the temperature is changed, all we must do is recalculate T:

$$T = (5/9)(212 - 32) + 273° = \underline{373°K}$$

b. $V_n = \sqrt{4\,k\,T\,BW\,R} = \sqrt{(4)(1.38 \times 10^{-23})(373)(100\text{ kHz})(100\text{ k}\Omega)} = \underline{\underline{14.3\ \mu V}}$

You might notice that the noise voltage didn't seem to change a great deal; that's because changing from 73°F to 212°F is only a 76°K difference and because of the effect of taking the square root.

Sometimes a manufacturer will specify the "equivalent input noise resistance" of an amplifier. This can be used to estimate the noise performance of such a unit.

EXAMPLE 2–8

What is the equivalent input noise resistance R of an amplifier known to produce 5 μV of equivalent input noise over a bandwidth of 500 kHz? Assume a temperature of 296°K.

Solution

Manipulation of equation 2–3 will give us what we want:

$$V_n = \sqrt{4\,k\,T\,BW\,R}$$

Solving for R, we get:

$$R = \frac{V_n^2}{4\,k\,T\,BW} = \frac{5\ \mu V^2}{(4)(1.38 \times 10^{-23})(296)(500\text{ kHz})} = \underline{\underline{3060\ \Omega}}$$

Shot Noise

Shot noise is named so because it sounds a lot like metal pellets being poured onto a hard surface when it plays through a loudspeaker. It is produced whenever there is a sudden drop or rise in potential energy within a circuit, such as occurs within a forward-biased *pn* junction. Zener diodes also generate large amounts of this type of noise, even though they normally operate in reverse bias. Let's see why this is so.

Figure 2–24 shows a forward-biased *pn* junction connected to a voltage source. Above this is an *energy level diagram*. To understand the action, it is easier to use *hole* flow. A hole is a positive charge and flows from positive to negative.

Look carefully at the energy levels in the circuit above. The energy or voltage level is highest at the positive terminal of the battery (V_{cc}, voltage source). As charges move across, the voltage (energy) is lost, and their potential energy level becomes smaller. We'd normally say that the resistor has created a voltage drop. Notice that there is no sudden change in voltage so far.

Things get exciting at the *pn* junction of the diode! The energy level (think voltage) rapidly jumps from about 0.7 volts to 0 volts (ground) in a very short physical distance. What kind of noise do you think this makes? Imagine rolling BBs over a stair step. Each BB makes the most noise when it falls over the edge and strikes the step below. The same thing is happening here. The only difference is that we are moving *charges* down an *energy hill* (the barrier potential of the diode) instead of BBs over a stair step.

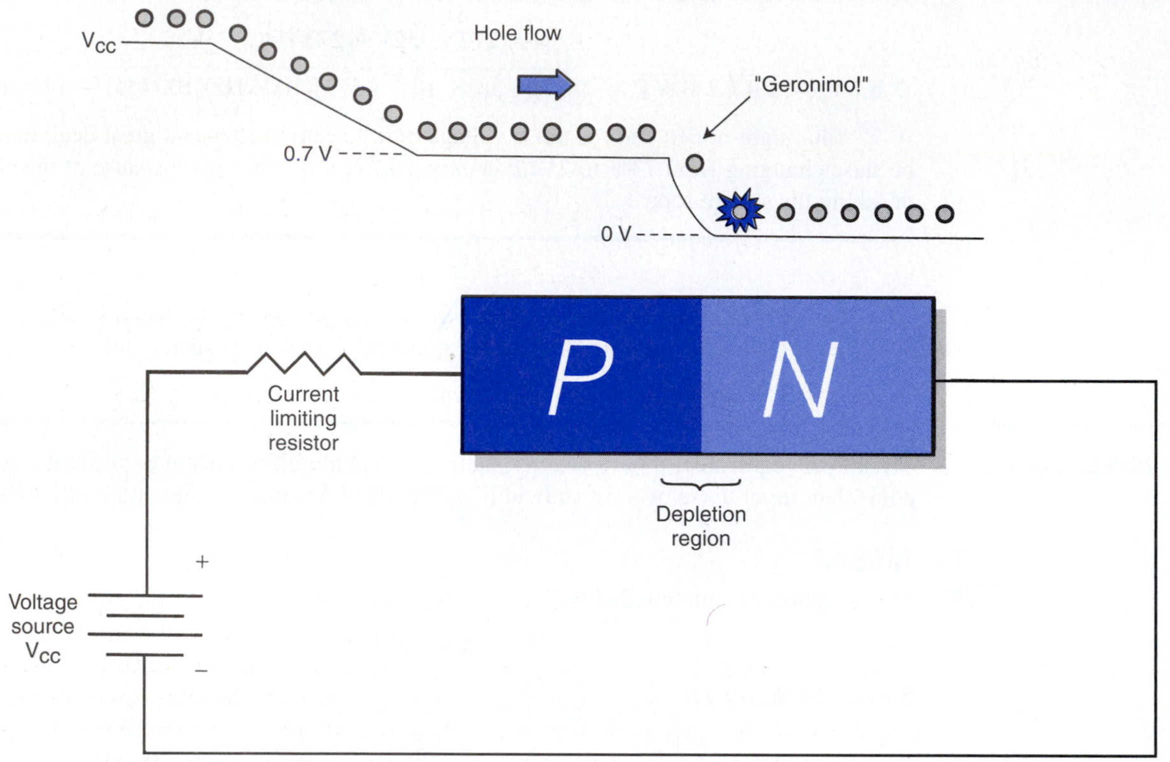

Figure 2–24 Energy level diagram for a forward-biased *pn* junction

Each charge that crosses the *pn* junction does so abruptly and thus introduces a momentary jump in the overall diode current. Because a very large number of charges are all doing this at the same time, an ac noise current results. Again, this noise current is *random,* just like the Johnson noise generated in a resistor, so all frequencies except dc are present.

In general, the following factors affect the amount of shot noise that will be produced by a transistor or diode:

- The dc current. Increasing the current increases the amount of shot noise.
- The bandwidth. Increasing bandwidth increases the amount of shot noise reproduced by the system (look again at Figure 2–23).

Devices that rely on a forward-biased *pn* junction normally act as shot-noise generators; therefore, this type of noise is greatly reduced by using field-effect transistors.

Flicker Noise

Even though field-effect transistors are largely immune to shot noise, they produce a noise of their own called *flicker noise.* Flicker noise is not equal at all frequencies, unlike Johnson and shot noise. It is strongest at low audio frequencies. For RF amplifiers, flicker noise is usually not a problem. Carbon composition resistors may also contribute this type of noise, in addition to thermal noise.

Section Checkpoint

2–19 What is *noise*?

2–20 What are two external noise sources?

2–21 What are three internal noise sources?

2–22 What three factors control the noise voltage of a Johnson noise source?

2–23 What causes shot noise?

2–24 What two factors control the amount of shot noise produced by a diode?

2–25 Can a reverse-biased diode junction generate shot noise (assuming that no current is flowing)?

2–4 SIGNAL-TO-NOISE RATIO AND NOISE FIGURE

The measurement of the amount of noise in a system is meaningless unless we also know how much *signal* is present. When the noise is much smaller than the signal, we find that the signal is much easier to reproduce accurately. As the noise approaches the signal's strength, intelligibility suffers, and eventually the signal becomes unreadable.

The *signal-to-noise ratio,* or SNR, is just the ratio of signal to noise power:

$$\text{SNR} = \frac{P_s}{P_n} \qquad (2\text{–}4)$$

TIP When expressing SNR as a power ratio, it is helpful to write it as a fraction such as "1000:1" or "200:1" to remind the reader that the number is a ratio!

where P_s is the power of the signal, and P_n is the power of the noise. Often this ratio is more conveniently expressed in decibels:

$$\text{SNR (dB)} = 10 \log\left(\frac{P_s}{P_n}\right) \qquad (2\text{–}5)$$

When the *voltages* are known instead of the powers, Ohm's law can be applied to equation 2–4 to find the signal-to-noise ratio. This formula is only valid when the resistance (or impedance) is equal for the signal and noise voltages (which is normally the case). First, we can write the power of the signal as

$$P_s = V_s^2/R$$

And of course, the power of the noise is:

$$P_n = V_s^2/R$$

By combining these two, we get

$$\frac{P_s}{P_n} = \frac{\left(\frac{V_s^2}{R}\right)}{\left(\frac{V_n^2}{R}\right)} = \frac{V_s^2}{V_n^2} = \left(\frac{V_s}{V_n}\right)^2$$

From this, we can state

$$\text{SNR} = \left(\frac{V_s}{V_n}\right)^2 \tag{2–6}$$

and

$$\text{SNR (dB)} = 20 \log\left(\frac{V_s}{V_n}\right) \tag{2–7}$$

These look like a lot of equations, but they're really straightforward to understand. There are only two quantities being calculated here:

- The SNR as a *power* ratio (watts per watt). Equations 2–4 and 2–6 will do this for you.
- The SNR in *decibels*. Equations 2–5 and 2–7 take care of this.

To decide which equation to use, look at the information you have been given and then look at what type of result you need; then choose the appropriate relationship. Let's work a practical example.

EXAMPLE 2–9

At the input of an audio amplifier, there is a noise voltage of 10 μV and a signal voltage of 500 mV.

a. What is the signal-to-noise as a power ratio?

b. What is the signal-to-noise ratio in decibels?

Solution

Since the *voltages* are given and we need the power SNR, we use equation 2–6:

$$\text{SNR} = \left(\frac{V_s}{V_n}\right)^2 = \left(\frac{500\text{ mV}}{10\ \mu\text{V}}\right)^2 = \underline{\underline{2{,}500{,}000{,}000{:}1}}$$

The signal power is 2.5 billion times larger than the noise power.

There are two ways of finding the decibel S/N ratio. Since we calculated the SNR as a power ratio in part (a), we could easily use equation 2–5:

$$\text{SNR (dB)} = 10 \log\left(\frac{P_s}{P_n}\right) = 10 \log(\text{SNR}) = 10 \log(2{,}500{,}000{,}000) = \underline{\underline{93.97\text{ dB}}}$$

We can also use equation 2–7, since the signal and noise voltages are also known:

$$\text{SNR (dB)} = 20 \log\left(\frac{V_s}{V_n}\right) = 20 \log\left(\frac{500\text{ mV}}{10\ \mu\text{V}}\right) = \underline{\underline{93.97\text{ dB}}}$$

Sometimes we are given an input signal voltage and an equivalent input noise resistance for an amplifier. The signal-to-noise ratio can also be computed for these cases.

EXAMPLE 2–10

An audio amplifier amplifies a frequency range of 20 Hz to 20 kHz and has an equivalent input noise resistance of 47 kΩ. What decibel S/N will result if an input signal of 1 mV is applied to the input of this amplifier? Assume the temperature is 24°C.

Solution

Notice that we seem to be missing some information, namely, the *noise voltage.* Since the equivalent input noise resistance is given, we can calculate the noise voltage using equation 2–3, noting that the *bandwidth* (BW) is (20 kHz–20 Hz) = 19.98 kHz and the temperature in Kelvins is 24° + 273° = 297°K:

$$V_n = \sqrt{4\,kT\,\mathrm{BW}\,R} = \sqrt{(4)(1.38 \times 10^{-23})(297)(19.98\ \mathrm{kHz})(47\ \mathrm{k\Omega})} = \underline{3.92\ \mu\mathrm{V}}$$

Now that we know the noise voltage, we can directly apply equation 2–7:

$$\mathrm{SNR\ (dB)} = 20\log\left(\frac{V_s}{V_n}\right) = 20\log\left(\frac{1\ \mathrm{mV}}{3.92\ \mu\mathrm{V}}\right) = \underline{\underline{48.1\ \mathrm{dB}}}$$

What is the actual meaning of the numbers we've just calculated? Yes, you guessed it—another example is coming!

EXAMPLE 2–11

Evaluate the performance of the systems from Examples 2–9 and 2–10. How good are they in layman's terms?

Solution

Most of us have already experienced noise in electronic systems but have no "seat of the pants" feel for it. Here are some average S/N performances for common listening equipment:

Device	Typical decibel S/N ratio
AM broadcast receiver, telephone	25 dB
Audio cassette player	35–45 dB
FM broadcast receiver	50–60 dB
Vinyl LP recordings	50–70 dB
Compact disc, digital audiotape	90–105 dB

The system of Example 2–9 has a S/N of almost 94 dB. It will sound just like a CD, meaning that the background noise will be almost inaudible, even with the volume control turned up high.

The system of Example 2–10 has a S/N of 48 dB, which is slightly better than the average audiocassette player; background noise will be evident, but not intrusive.

For the reproduction of speech, a decibel SNR of 10 dB is considered to be a minimal requirement; and even at that level, considerable background noise will be present. When the SNR falls below 3 dB, speech intelligibility becomes greatly impaired; most listeners find it difficult to make out the words under this condition.

A final point about S/N ratios: It is often difficult to separately measure the signal and noise in a real system. The reason is that when we think we're measuring *signal,* often we're really measuring *signal plus noise*. Therefore, occasionally you'll see a rating for "(S + N)/N" instead of a pure S/N rating for an amplifier. This rating takes into account the realities of measurement.

Noise Figure

There's another way of expressing the noise performance of amplifiers and systems, and that is the *noise figure,* abbreviated NF. All electronic systems introduce noise. Because of this, we can generally expect the S/N ratio at the output of a system to be *less* than the S/N at the input. This is due to internal noise. The noise figure is defined as

> *TIP* The *lower* the number for NF, the better the noise performance of the system.

$$\text{NF (dB)} = \text{Input SNR (dB)} - \text{Output SNR (dB)} \tag{2–8}$$

An ideal or "perfect" amplifier has a NF of 0 dB, meaning that *no* additional noise is introduced, and therefore the output S/N ratio would be identical to that at the input. State-of-the-art amplifiers are capable of giving noise figures below 3 dB per stage.

EXAMPLE 2–12

What decibel signal-to-noise ratio will be present in Figure 2–25 if the amplifier noise figure is

a. 3 dB,
b. 6 dB,
c. 10 dB?

Solution

By rearranging equation 2–8, we can calculate the output decibel S/N ratio:

$$\text{Output SNR (dB)} = \text{Input SNR (dB)} - \text{NF (dB)}$$

a. $\text{Output SNR (dB)} = \text{Input SNR (dB)} - \text{NF (dB)} = 50\text{ dB} - 3\text{ dB} = \underline{\underline{47\text{ dB}}}$

b. $\text{Output SNR (dB)} = \text{Input SNR (dB)} - \text{NF (dB)} = 50\text{ dB} - 6\text{ dB} = \underline{\underline{44\text{ dB}}}$

c. $\text{Output SNR (dB)} = \text{Input SNR (dB)} - \text{NF (dB)} = 50\text{ dB} - 10\text{ dB} = \underline{\underline{40\text{ dB}}}$

As the noise figure of an amplifier increases, the signal-to-noise ratio at the output decreases. Higher noise figures cause more noise to appear in amplifier outputs.

Figure 2–25 Amplifier noise figure

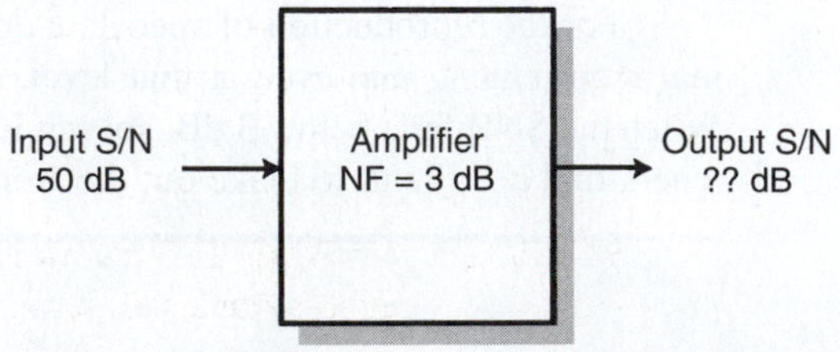

Section Checkpoint

2–26 Why is the S/N ratio an important measure of noise performance?

2–27 When an SNR is expressed as a ratio, what units are being used?

2–28 What happens to the quality of sound as the SNR becomes smaller?

2–29 What is meant by the term "noise figure"?

2–30 What is the NF in decibels for an ideal amplifier?

2–31 When the decibel NF increases, what happens to the quality of the output signal of an amplifier?

2–5 NOISE REDUCTION TECHNIQUES

As we have seen, noise is of great concern to the designers and users of communication systems. Many circuit techniques are used to reduce the effects of external and internal noise. A technician should be aware of these.

External Noise Reduction Methods

The following are commonly used design features that reduce the intake of external noise:

- metal shielding around sensitive circuit areas
- directional antennas that point away from known noise sources
- use of the minimum bandwidth necessary for communications
- use of shielded cable and connectors, to prevent accidental signal leakage
- elimination of excessive wire lengths

Internal Noise Reduction Methods

These are common low-noise circuit techniques:

- The use of low-noise resistors (typically carbon film or wire wound).
- The application of low-noise semiconductors. In particular, a type of field-effect transistor called a GaSFET (Gallium-Arsenide Field Effect Transistor) is very popular for low-noise RF amplifiers.
- Using the lowest possible resistor values in sensitive circuits (reduces thermal noise voltages).
- Special component layout techniques (such as the use of ground planes on circuit boards) to reduce unwanted circuit interactions.
- Use of the minimum bandwidth necessary for communications (this reduces both external and internal noise).
- Reducing forward-bias currents in diodes and transistors to the minimum practical values, in order to reduce shot noise.
- Cooling or refrigerating the circuit to reduce thermal noise. This is an expensive measure.

2–6 FOURIER SERIES FOR COMMON WAVEFORMS*

There are a few wave shapes that appear again and again in electronic circuits. These include *square, triangle,* and *sawtooth waves,* as well as *half* and *full-wave rectified sine waves*. In the frequency domain, these waves are calculated by using *Fourier series*. A Fourier series is a recipe for building a desired periodic signal from sine and cosine waves at each of the frequencies present in the signal. It's rare that you may need to calculate the frequency domain values for these signals. You're more likely to make a direct frequency domain measurement using a spectrum analyzer.

The Fourier Series Formula

$$f(t) = a_0 + \sum_{n=1}^{\infty}[a_n \cos(2\pi n f_0) + b_n \sin(2\pi n f_0)] \qquad (2\text{–}9)$$

This formula may look intimidating, but we can actually take it apart one piece at a time to understand it. Don't worry. There's no need to "solve" this formula to use Fourier series; you just have to know how its parts work. From left to right, here are the important parts of the formula and what they mean:

- $f(t)$: This is a time domain result. This would be read as "the function of time." What this says is that all the right-hand parts of the equation can be graphed with respect to time to reproduce the original wave shape.
- a_0: Recall from a previous discussion that a periodic waveform consists of a dc level plus an infinite series of harmonics. The quantity a_0 (pronounced "a-not") represents the dc or average level of the waveform. Solving for a_0 tells us if there is a dc level, and if so, how much there is.
- $\sum_{n=1}^{\infty}\ldots$: The "sum" of all harmonic numbers, starting at 1 and ending with infinity. This reminds us that the Fourier series is an infinite series consisting of all the harmonics of the fundamental frequency. (Remember that we don't have to solve for all harmonics up to infinity, because the bandwidth of all signals is limited practically by the bandwidth of the circuit through which they must pass.)
- $a_n \cos(2\pi n f_0)$: The portion a_n gives the *peak* voltage of the cosine (out-of-phase) energy at harmonic number n. Some waveforms can be built from cosines (or will require cosines); this tells us how much cosine voltage will be needed at each harmonic number n. Other waveforms don't need cosines, and their a_n coefficients are zero. The portion "$2\pi n f_0$" tells us that we're working with the nth harmonic whose frequency is n times the fundamental frequency, f_0.
- $b_n \sin(2\pi n f_0)$: b_n tells us how much in-phase (sine) energy is needed at each of the harmonics. If b_n is zero, then no sine voltage exists at the particular harmonic in question.

Table 2–1 contains most of the common period signals you're likely to encounter. We'll use this table in the subsequent examples. (If the waveform you're looking for isn't in the table, you can often find it in reference books on electronics.)

*This section contains useful reference information not normally needed by technicians, and may be skipped without loss of continuity.

Table 2–1: Fourier series for common waveforms

1. Half-wave rectified sine

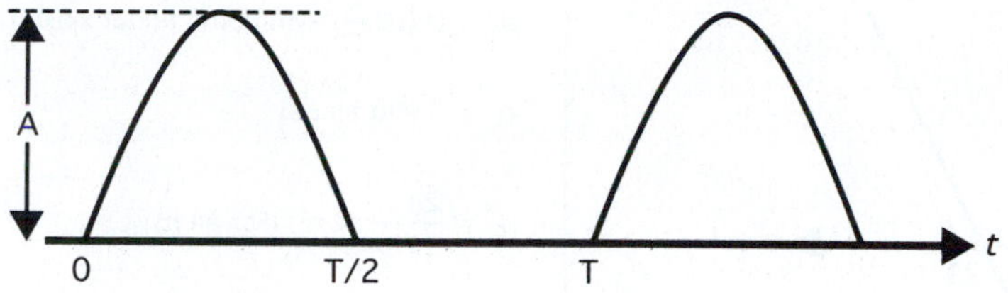

$$a_0 = A/\pi$$

$$b_1 = A/2$$

$$a_n = \frac{A[1 + (-1)^n]}{\pi(1 - n^2)} \quad \text{(For } 2 \le n \le \infty\text{)}$$

2. Full-wave rectified sine wave

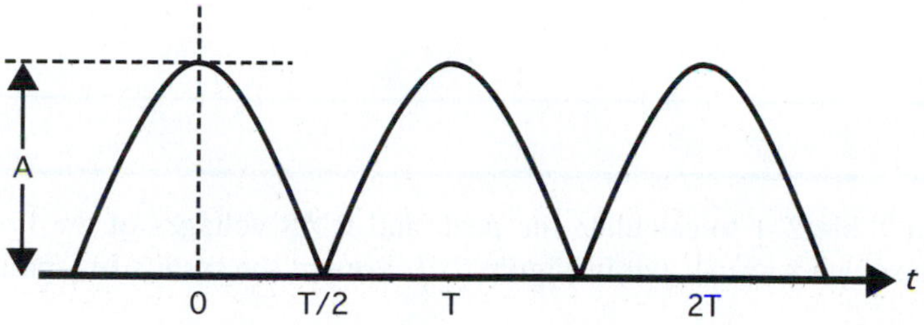

$$a_0 = 2A/\pi$$

$$b_n = 0 \quad \text{(No sines)}$$

$$a_n = \frac{4A(-1)^n}{\pi(1 - 4n^2)}$$

3. Square wave, 50% duty cycle—sine expansion

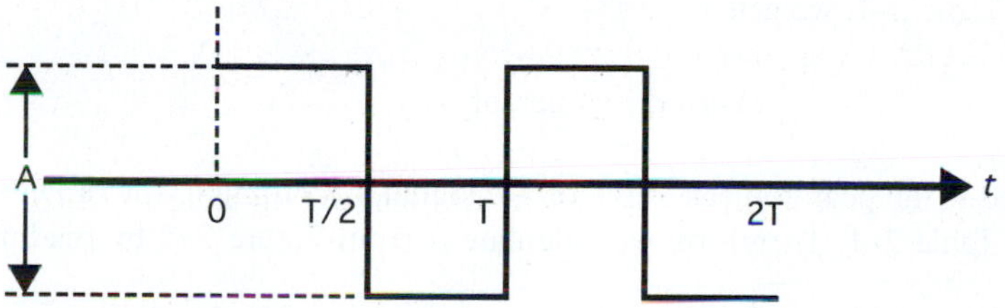

$a_0 = 0$ (When symmetric about x-axis)

$a_N = 0$ (No cosines needed)

$$b_n = \frac{2A}{n\pi} \quad \text{(For odd values of } n\text{)}$$

$b_n = 0$ (For even values of n)

4. Symmetrical triangle

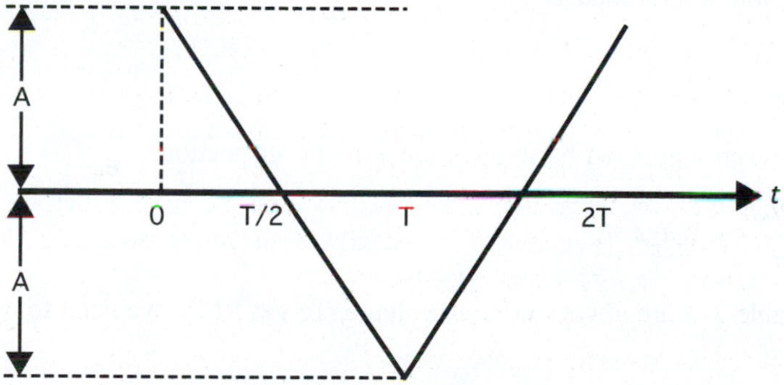

$a_0 = 0$ (When symmetric about x-axis)

$$a_n = \frac{8A}{(n\pi)^2} \quad \text{(For odd values of } n\text{)}$$

$a_n = 0$ (For even values of n)

(continued on p. 48)

Table 2–1: Fourier series for common waveforms *(continued)*

5. Sawtooth 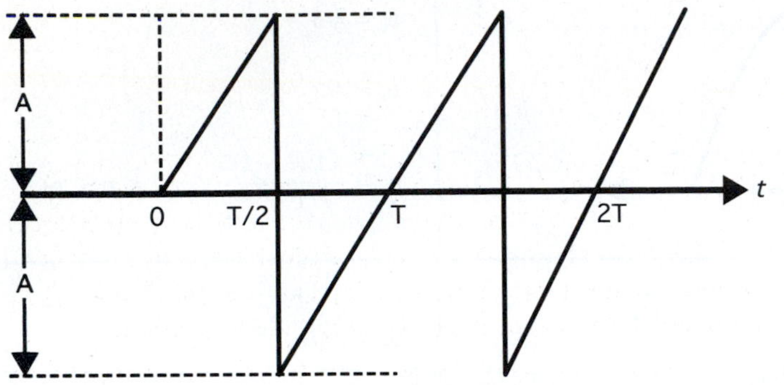	$a_0 = 0$ (When symmetric about x-axis) $b_n = 0$ (No sines) $a_n = \frac{2A}{n\pi}(-1)^{n+1}$ (For all n)

EXAMPLE 2–13 Use the information in Table 2–1 to calculate the peak and RMS voltages of the first three harmonics of the square wave shown in Figure 2–7. Present the results in tabular form.

Solution

By using Section 3 of Table 2–1, we get:

$$b_n = \frac{2A}{n\pi} \text{ (For odd values of } n\text{)}$$

The value of A required is the peak-to-peak value of the signal, according to the drawing of the square wave in Table 2–1. Therefore, we calculate A from Figure 2–7 by reading peak-to-peak voltage:

$$A = 2\ \text{V}_{\text{pp}} \text{ (by inspection)}$$

Now we substitute into the formula to get:

$$b_1 = \frac{2(2\ \text{V}_{\text{pp}})}{1\pi} = \underline{\underline{1.27\ \text{V}_{\text{p}}}}$$

$$b_2 = \underline{\underline{0\ \text{V}_{\text{P}}}} \text{ (even numbered harmonics are zero by inspection)}$$

$$b_3 = \frac{2(2\ \text{V}_{\text{pp}})}{3\pi} = \underline{\underline{0.424\ \text{V}_{\text{P}}}}$$

The coefficients from Table 2–1 are always in *peak* voltage. To get RMS, we need to divide each value by $\sqrt{2}$:

$$b_{1(RMS)} = \frac{b1}{\sqrt{2}} = \frac{1.27\ \text{V}_{\text{p}}}{\sqrt{2}} = \underline{\underline{0.900\ \text{V}}}$$

$$b_{2(RMS)} = \frac{b2}{\sqrt{2}} = \frac{0\ \text{V}_{\text{p}}}{\sqrt{2}} = \underline{\underline{0\ \text{V}}}$$

$$b_{3(RMS)} = \frac{b3}{\sqrt{2}} = \frac{0.424\ \text{V}_{\text{p}}}{\sqrt{2}} = \underline{\underline{0.300\ \text{V}}}$$

The results in tabular form look like this:

n	Frequency	b_n (Volts peak)	b_n (Volts RMS)
1	1 kHz	1.27	0.900
2	2 kHz	0	0
3	3 kHz	0.424	0.300

The result tells us that there's no energy at 2 kHz (a symmetrical, 50/50 duty cycle square wave has no even harmonic energies), and that the energy gradually becomes smaller as frequency increases. If you compare this table to Figure 2–8, you'll see that the RMS values are shown on the spectrogram, because spectrum analyzers measure power.

EXAMPLE 2–14

If we pass a 4 kHz, 10 V_{pp} sawtooth through an ideal bandpass filter tuned to a frequency of 20 kHz, what peak output signal voltage should we observe from the filter? The ideal bandpass filter has a unity gain at its center frequency.

Solution

The frequency we're looking for is the 5th harmonic of the 4 kHz sawtooth signal, which is represented by the Fourier series given in Section 5 of Table 2–1:

$$a_n = \frac{2A}{n\pi}(-1)^{n+1} \text{ (for all } n)$$

In this case A represents the *peak* value of the time domain signal; we're given a 10 V_{pp} signal, so $A = 10\ V_{pp}/2 = \underline{5\ V_p}$. To "query" the 5th harmonic, set n to 5:

$$a_n = \frac{2A}{n\pi}(-1)^{n+1} = \frac{2(5\ V_{pk})}{(5)\pi}(-1)^{(5+1)} = \underline{\underline{0.637\ V_p}}$$

The output of the filter should be about 0.637 volts peak (since it has a unity gain, it doesn't magnify the harmonic signal's energy). This type of information is sometimes very useful; for example, it helps us to determine the proper output of *frequency multiplier* circuits, which rely on one or more bandpass filters tuned to harmonics of a fundamental signal.

SUMMARY

- All electrical signals can be viewed in either the *time* or *frequency domains*.
- A *spectrum analyzer* is an instrument that is used to view signals in the frequency domain.
- The simplest signal is a pure sine wave, which has only one frequency and zero bandwidth.
- Any waveform that is not a pure sine wave will have more than one frequency present.
- Periodic or repeating signals are built out of a *fundamental* sine wave frequency, a *dc level,* and an infinite number of *harmonics*.
- Noise is any undesired signal and comes from outside a system (*external*) and inside a system (*internal*).
- Internal noise sources include *Johnson noise* (produced in conductors of electricity, especially

resistors), *shot noise* (primarily produced in bipolar transistors and forward-biased diode junctions), and *flicker noise* (produced in carbon-composition resistors and field-effect transistors).

- The *signal-to-noise ratio* is a method for determining the "goodness" of a system's noise performance.
- The *noise figure* of a system reveals how much internal noise it "adds" to incoming signals.
- Fourier series allow "exact" prediction of the frequency domain contents of periodic signals. The easiest way to employ them is by using look-up tables whenever they are available.

PROBLEMS

1. What is the *frequency domain?*
2. What are the units on the horizontal axis of a *spectrum analyzer?*
3. Draw a 10 kHz, 5 V RMS sine wave as it would be seen on the display of a spectrum analzyer.
4. What is the only "pure" waveform?
5. A square wave has a period of 2 ms. What is its fundamental frequency?
6. If a nonsinusoidal waveform has a frequency of 3 kHz, what are the frequencies of its 2nd, 3rd, and 4th harmonics?
7. What approximate (practical) bandwidth would be needed to reproduce a square wave with a frequency of 10 kHz, ignoring the dc level?
8. What type of waveform is being displayed in Figure 2–26?
9. What is the *period* and *peak voltage* of the waveform being displayed in Figure 2–26?
10. How many frequencies are present in a *perfect* square wave?
11. If the signal of Figure 2–8 is fed into an ideal *low-pass* filter with a cutoff frequency of 1 kHz, what will the filter's output signal be if viewed on an oscilloscope?
12. List and explain at least two external noise sources.
13. What factors control the voltage of a Johnson noise source?
14. What noise voltage will be generated across a 10 kΩ resistor under the following conditions?
    a. $T = 20°\text{C}, f_{min} = 1\text{ MHz}, f_{max} = 4\text{ MHz}$
    b. $T = 112°\text{F}, f_{min} = 100\text{ Hz}, f_{max} = 10\text{ kHz}$
    c. $T = 110°\text{C}$, bandwidth $= 1$ MHz

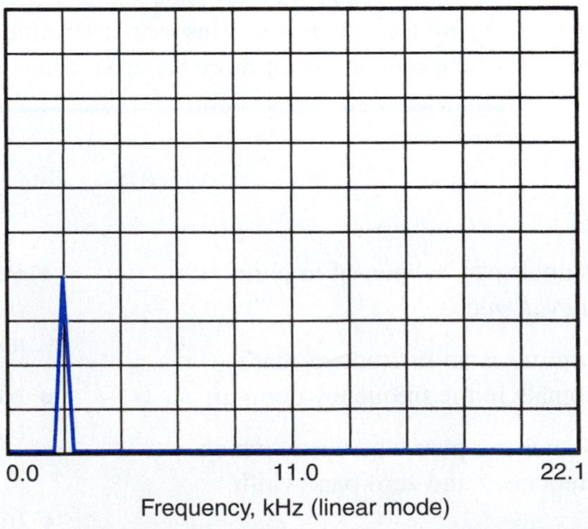

Figure 2–26 A display for analysis (vertical, 5 V/division)

15. If a certain amplifier has an input noise power of 1 μW and an input signal power of 1 mW, what is the resulting SNR power ratio?

16. What is the decibel S/N ratio for the amplifier of problem 15?

17. A certain amplifier has 1 μV of input noise and 10 μV of input signal.
    a. What is the SNR power ratio?
    b. What is the SNR ratio in decibels?

18. If the signal voltage in an amplifier is fixed at 1 V, what is the *maximum* noise voltage that can be present while maintaining a 25 dB (or better) S/N ratio?

19. What is the *noise figure* of an amplifier?

20. If an amplifier has a noise figure of 5 dB, and the input S/N is 50 dB, what will the resulting S/N at the output be?

21. A certain amplifier is rated with a noise figure of 1.5 dB. A technician measured a S/N of 30 dB at the input and 27 dB at the output. Does the amplifier meet its specification? Why or why not?

22. What are three measures used to reduce the pickup of *external noise?*

23. What are three circuit construction techniques that reduce *internal noise?*

24. Draw the spectrogram of a 10 kHz, 1.5 V peak square wave, showing the RMS values of all harmonics up to and including the 5th. Use a data table like that in Example 2–13 to corral the numbers.

25. Which has more energy at a frequency of 60 kHz, a 20 kHz, 5 V_{pp} square wave or a 12 kHz, 20 V_{pp} triangle? Calculate the peak energy of both signals at 60 kHz and compare the results. (Hint: Make sure to learn which harmonic number of each signal lines up on 60 kHz; each signal has a different fundamental frequency.)

3

Amplitude Modulation

OBJECTIVES

At the conclusion of this chapter, the reader will be able to:

- explain conceptually how an AM signal is created
- use an oscilloscope to measure the percentage of modulation of an AM signal
- predict the frequency domain characteristics of simple AM signals
- measure the various parameters of an AM signal using a spectrum analyzer

Of all the methods of impressing information onto a carrier signal, AM is the oldest. It dates back to the beginning of radio. Although it's old technology, it is still widely used in the following applications:

- local broadcast (535–1620 kHz in the United States)
- aircraft communications in the 118–138 MHz band
- shortwave broadcasts in the HF bands (3–30 MHz), which affords worldwide coverage
- analog television, in which an AM carrier is used for the picture and a separate FM carrier frequency is used to carry the sound
- data communications, in which AM and PM (phase modulation) are used together in high-speed modems (the subject of a later chapter)

With all of these applications (and more), "ancient modulation" is hardly obsolete technology. *AM is an electronic fundamental!*

3–1 GENERATING AN AM SIGNAL

As you recall, radio uses a high-frequency sine wave called a *carrier* to move information from the transmitter to the receiver. Intelligence can be impressed onto a carrier signal in three ways:

- *Amplitude Modulation (AM)*: The *amplitude* or *strength* of the carrier signal is changed in step with the information. (In place of the word *amplitude* we can substitute *voltage, power,* or *current.*)
- *Frequency Modulation (FM)*: The *frequency* of the carrier is changed with the intelligence signal. The frequency changes are normally small and hard to see on an oscilloscope (but as you have probably already guessed, they are easy to see on a spectrum analyzer!).
- *Phase Modulation (PM)*: The *phase angle* of the carrier signal is changed to convey the information. PM is very similar to FM and is very hard to observe accurately on an oscilloscope.

Figure 3–1 shows two carrier signals that have been modulated by the same information signal. Note how the *shape* of the AM signal is quite distinctive. The information

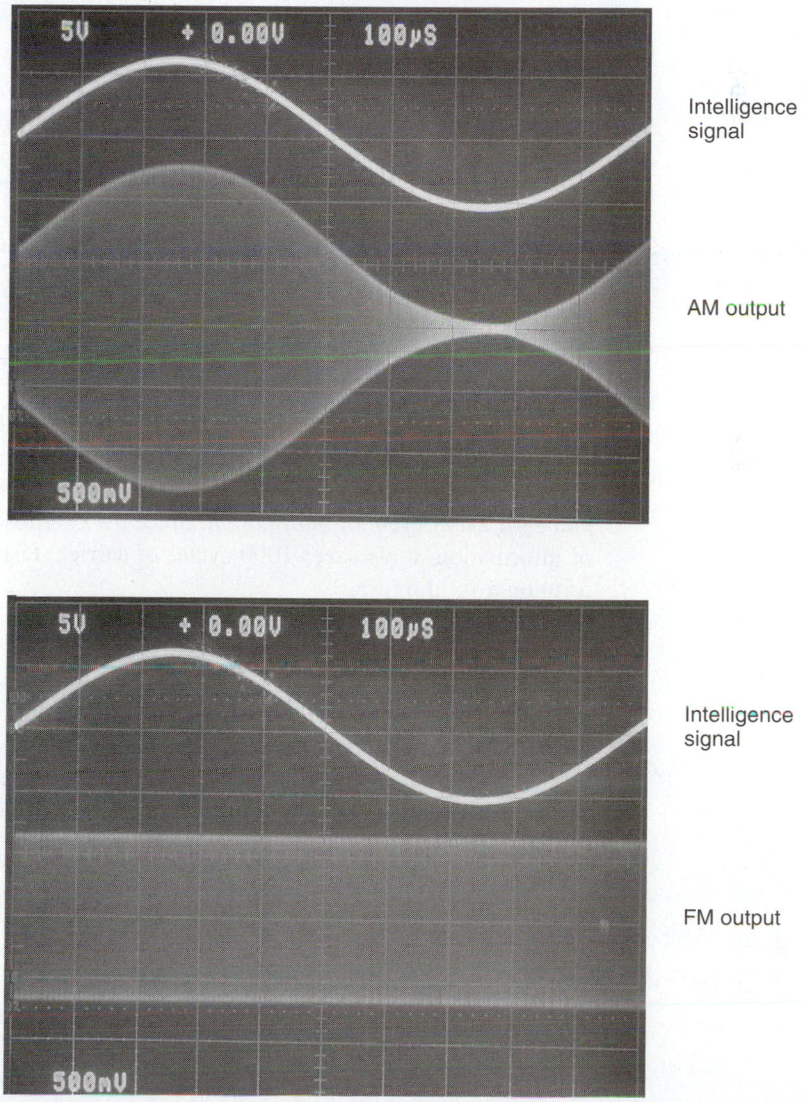

Figure 3–1 AM and FM signals on a scope

is actually contained in this shape. In contrast, the FM signal looks like a solid horizontal band. You really can't see much here at all! In fact, when the carrier frequency is much higher than the information frequency (like it is here), both FM and PM will look identical on a scope. FM and PM signals have a constant power. We need to use a spectrum analyzer to measure an FM or PM signal accurately.

EXAMPLE 3–1

What is the frequency of the *information* signals in Figure 3–1? The horizontal time base is set for 100 μs/division.

Solution

The scope is measuring in the time domain, so we must first calculate the time period of the waveform:

$$T = (10 \text{ divisions})(100\ \mu\text{s/division}) = \underline{1 \text{ ms}}$$

and the frequency is therefore

$$F = \frac{1}{T} = \frac{1}{1 \text{ ms}} = \underline{\underline{1 \text{ kHz}}}$$

EXAMPLE 3–2

Why do the modulated waveforms of Figure 3–1 appear as solid areas? Why can't we see the individual sine wave cycles of the AM and FM carriers?

Solution

The *frequency* of the RF carrier is much higher than that of the information. In fact, the carrier frequency of both the AM and FM waveforms is 1 MHz. Recall that the information frequency is 1 kHz. Since 1 MHz is the same as 1000 kHz, *1000 cycles of carrier take place for every cycle of information.* Since the oscilloscope is adjusted to show one cycle of information, it also sees 1000 cycles of carrier. The carrier sine waves blend together, forming a solid figure.

> *TIP* When observing modulated signals on a scope, it usually is best to use two scope channels. One of the scope channels is connected to the information, and the other is connected to the modulated output. *The trigger must be set to the channel providing the information, in order to obtain a stable display.* Many techs forget this and have trouble getting accurate scope readings of transmitter ouputs!

Making an AM signal

Figure 3–2 shows the conceptual process of amplitude modulation. Almost all AM transmitters work this way. When analyzing an actual circuit, it helps to keep this picture in mind.

The first stage in any transmitter is an *oscillator.* In a radio transmitter, it is usually called the *RF carrier oscillator*. The carrier oscillator converts the dc power supply energy into a radio frequency (RF) carrier wave. Oscillators will be studied in detail later.

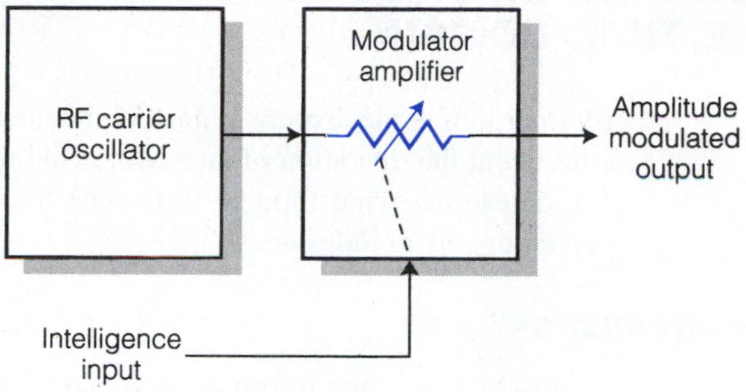

Figure 3–2 Generating an AM signal

The RF carrier wave contains no information until it is modulated. In order to modulate the carrier amplitude, its voltage (or power) must be changed. In order for the amplitude to be changed, the *voltage* or *power* gain of a subsequent stage must be changed.

The AM generator in Figure 3–2 has a special amplifier with a *variable* voltage gain called a *modulator*. This is really strange! The amplifiers you studied in fundamentals had a *constant* voltage or power gain and only one input (this one has two!). An amplifier with a constant gain is called a *linear amplifier*.

What controls the gain of this amplifier? That's right—there's a *second* input signal, the *information signal*. When the information signal goes positive, the amplifier's gain increases. This causes the output voltage (the AM signal) to swell or grow in amplitude. The opposite happens on the negative half-cycle of the information. The AM signal shrinks in amplitude because the amplifier's gain has decreased. Thus, amplitude modulation is created.

The variable-gain amplifier is a *nonlinear* amplifier because it has a gain that is not constant. One way of thinking of this amplifier is as a variable resistor that controls the amount of carrier signal that gets through. The value of the "resistor" is controlled by the instantaneous value of the information signal.

A nonlinear amplifier distorts or changes the input signal. This is normally a bad thing! However, RF engineers carefully control this nonlinearity when they design modulators, so that only a proper AM signal is produced.

A linear amplifier has a constant gain and generates no distortion of the input signal. The graph of input versus output for a linear amplifier is a straight line (hence, the word *linear*). A nonlinear amplifier has a variable gain; its input–output graph is a curve. A nonlinear amplifier is always required to generate AM.

Section Checkpoint

3–1 List three applications of AM.

3–2 Why are FM and PM hard to observe on an oscilloscope?

3–3 What instrument is preferred for measuring FM and PM signals?

3–4 Why does the variable-gain amplifier stage in Figure 3–2 generate AM?

3–5 A non ________ amplifier stage is required to generate AM.

3–2 MEASURING AM SIGNALS IN THE TIME DOMAIN

By interpreting the display of an AM signal on an oscilloscope, a technician can determine a lot about the operation of an AM transmitter. By examining the waveform, a technician can determine what type of information is modulating the transmitter as well as the *percentage of modulation*.

Where Is the Information?

In an AM signal, the information is carried on top of the RF carrier. The actual shape of the carrier is altered by the addition of the information during the process of modulation. When we look at a modulated AM carrier wave, we tend to see an overall shape. The imaginary lines that make up this shape are called the *envelope*.

Can you tell what is significant about the envelope? Take a look at Figure 3–3. Yes—the envelope is a copy or duplicate of the intelligence signal! No matter what the information is, the envelope will always imitate it. Take a look at Figure 3–4. Again, the envelope looks just like the information signal on top. Another case might look like Figure 3–5.

As you can see from Figures 3–4 and 3–5, the envelope always matches the shape of the information. Figure 3–5 is a special case. Can you tell what the source of the information might be? If you're thinking *digital,* you're on the right track. The information signal of Figure 3–5 is *digital data,* which is sent as a sequence of binary ones (highs) and zeros (lows). We'll study digital data communications in a later chapter.

Percentage of Modulation and AM Modulation Index

In an AM signal, the *percentage of modulation* is a measure of how strongly the carrier wave is being changed by the information. For radio, the higher the percentage of modulation, the louder the signal will be in the receiver's loudspeaker. Because percentage of modulation relates to sound volume, it would make sense that a broadcaster would try to attain as high a percentage of modulation as practical.

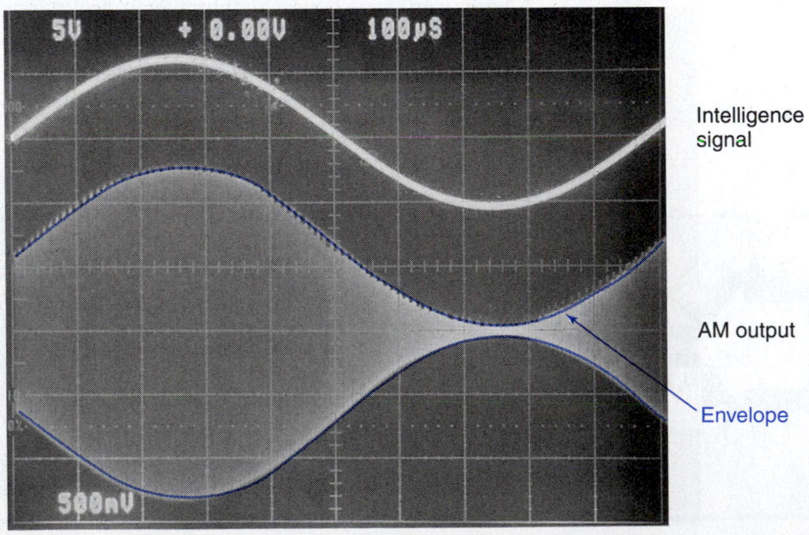

Figure 3–3 The envelope of an AM signal

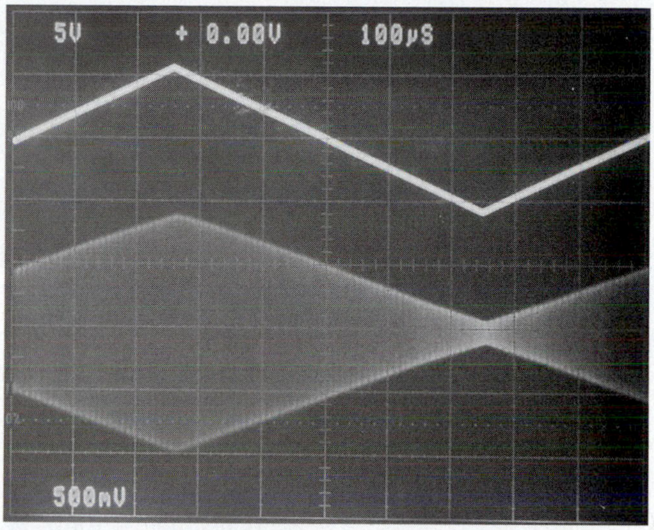

Figure 3–4 A triangular information signal

The maximum percentage of modulation is 100%, which represents maximum intelligence voltage (or volume). A 0% modulation means that no modulation is taking place; the transmitter is said to be *dead-keyed* or *unmodulated* in this case.

The *AM modulation index* is the *same* information as the percentage of modulation. It is given the symbol m and can have a value between 0 (0% modulation) and 1 (100% modulation). When we calculate percentage of modulation in an AM signal, we are really calculating the *modulation index.*

Remember that the maximum modulation index is 1 (which corresponds to 100% modulation). A signal that is over 100% modulated is said to be *overmodulated*, which is an illegal condition. Overmodulation distorts the information and causes excessive bandwidth to be used by the transmitter.

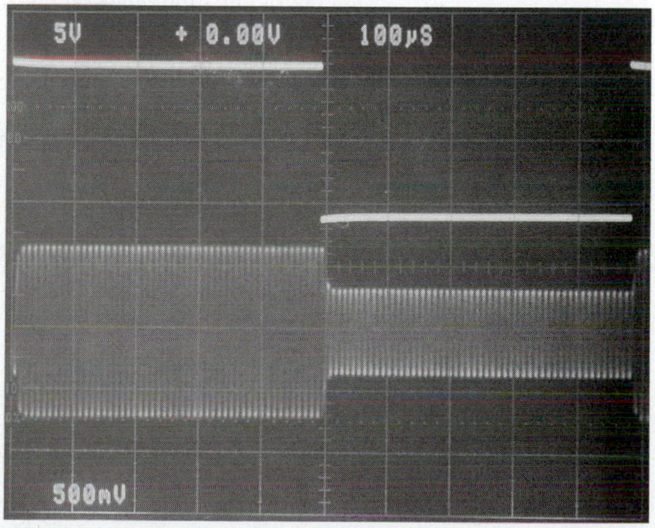

Figure 3–5 A square wave information signal

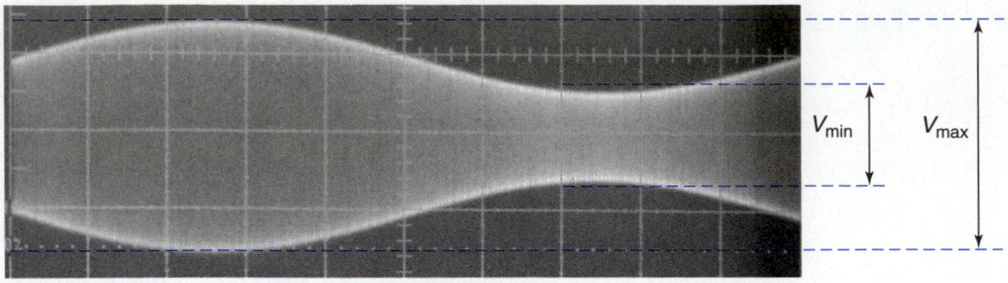

Figure 3–6 Measuring percentage modulation

EXAMPLE 3–3

What percentage of modulation corresponds to a modulation index m of 0.5 ?

Solution

Since $m = 0$ means 0% modulation and $m = 1$ means 100% modulation, we get:

$$\%\ Modulation = 100 \times m = 100 \times 0.5 = \underline{\underline{50\%}}$$

Remember that modulation index and percentage modulation are the same thing for an AM signal. (They are different quantities for FM, as we'll see later.)

Definition and Measurement of the Modulation Index

AM modulation index can be defined by the formula

$$m = \frac{V_m}{V_c} \tag{3–1}$$

where V_m is the *information voltage* and V_c is the *carrier voltage*. Equation 3–1 *defines* the modulation index but is not very helpful in making oscilloscope measurements. On a scope, it is difficult to separate the voltages V_m and V_c (but on a spectrum analyzer, it is quite easy, as we'll see).

To measure the modulation index on an oscilloscope, the following formula is used:

$$m = \frac{V_{max} - V_{min}}{V_{max} + V_{min}} \tag{3–2}$$

where V_{max} is the maximum waveform voltage (the peak), and V_{min} is the minimum waveform voltage (the trough) (see Figure 3–6).

EXAMPLE 3–4

What is the modulation index and percentage of modulation in Figure 3–6? The vertical sensitivity is 1 V/division.

Solution

Since we're measuring from an oscilloscope, we use equation 3–2:

$$m = \frac{V_{max} - V_{min}}{V_{max} + V_{min}} = \frac{3\ V_{pp} - 1\ V_{pp}}{3\ V_{pp} + 1\ V_{pp}} = \underline{\underline{0.5}}$$

The percentage modulation is the modulation index expressed as a percentage:

$$\% \text{ Mod} = 100\% \times m = 100\% \times 0.5 = \underline{\underline{50\%}}$$

TIP To adjust an AM transmitter for 50% modulation, adjust the intelligence voltage so that the ratio of V_{max} to V_{min} is 3:1. (Notice how we have $V_{max} = 3$ volts and $V_{min} = 1$ volt in this case.) Also, V_{max} and V_{min} can be measured in either peak or peak-to-peak units—as long as they are both measured the same way.

EXAMPLE 3–5

What is the percentage of modulation in Figure 3–7? What is the condition of the transmitter? The vertical sensitivity is 1 V/division.

Solution

Since we're measuring from an oscilloscope, we again use equation 3–2:

$$m = \frac{V_{max} - V_{min}}{V_{max} + V_{min}} = \frac{2.8\text{ V}_{pp} - 2.8\text{ V}_{pp}}{2.8\text{ V}_{pp} + 2.8\text{ V}_{pp}} = \underline{\underline{0}} = \underline{\underline{0\%\ \textit{Modulation}}}$$

This is an example of an *unmodulated* transmitter. No information is being conveyed. We can also say that the transmitter is "dead keyed" or is transmitting "dead air." This is normally undesirable, especially in commercial broadcasting!

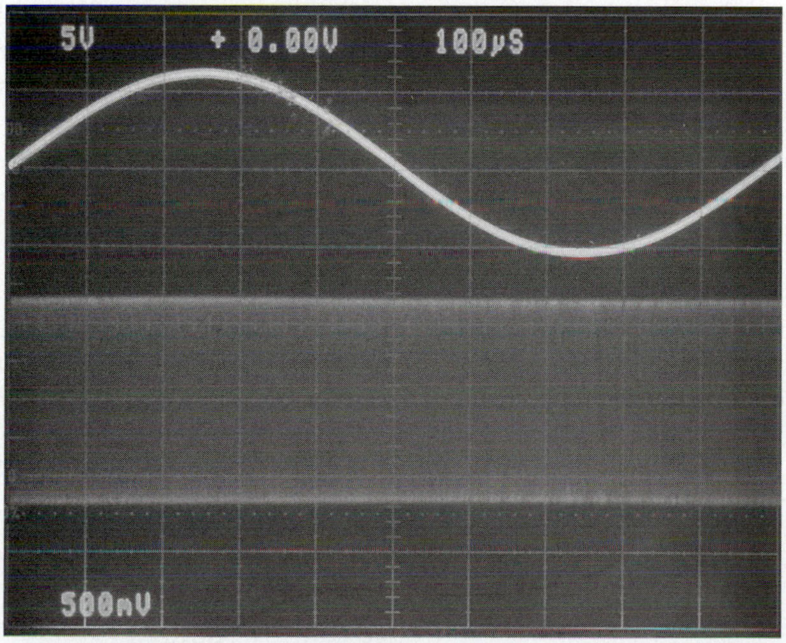

Figure 3–7 A case for examination

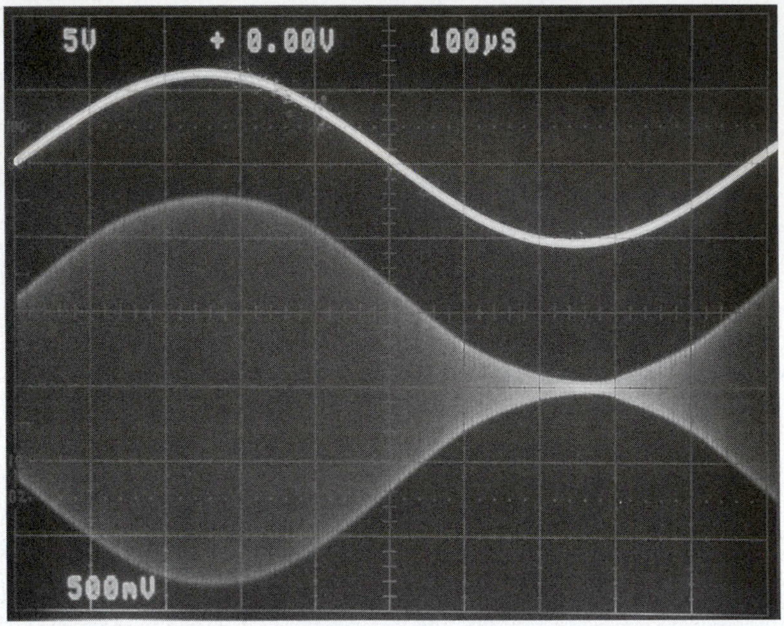

Figure 3–8 Another signal for analysis

EXAMPLE 3–6

Without doing any calculations, what is the percentage of modulation in Figure 3–8?

Solution

Look at the *trough* in Figure 3–8. It's nearly zero. *This indicates that 100% modulation (or something very close) is being achieved.* No calculations are needed!

EXAMPLE 3–7

What is wrong with the AM signal of Figure 3–9?

Solution

This is an example of *overmodulation.* The intelligence voltage is too large for the carrier, and as you can see, it has caused the trough to flatten. There are two problems here. First, the *envelope* is no longer an accurate copy of the information. The transmitted information will sound distorted. Second, *excessive bandwidth* will be used, which can cause *interference with adjacent stations on the band.* This effect is called *splatter.* It is hard to see it on a scope (but again, a spectrum analyzer gives a much clearer view of what is happening).

Overmodulation is not a good practice. It distorts the information, uses up precious bandwidth, and can be stressful on station equipment. It could have been avoided here by reducing the voltage (volume) of the information.

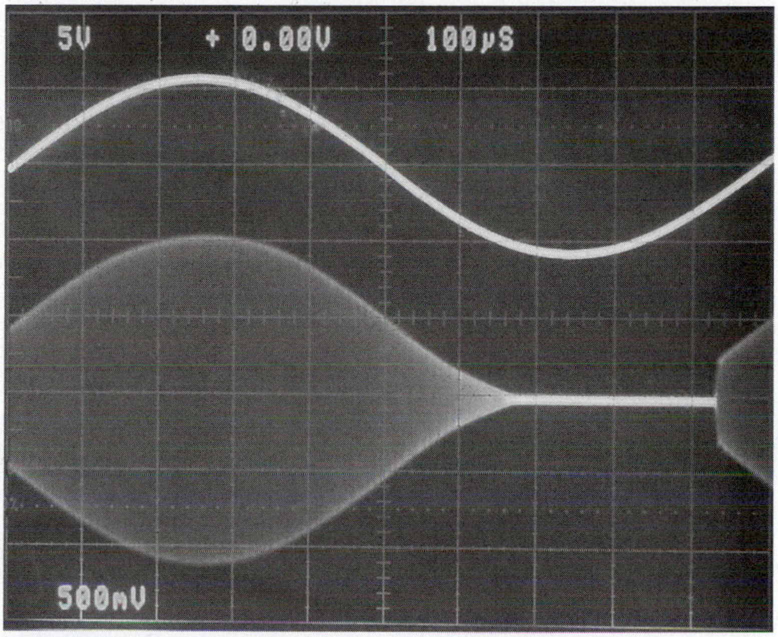

Figure 3–9 The intelligence voltage has been increased further

Section Checkpoint

3–6 What is the *envelope* of an AM signal? What is significant about the shape of the envelope?

3–7 If the information is a square-wave, what shape will the envelope of the resulting AM signal be?

3–8 To calculate the percentage of modulation of an AM signal from an oscilloscope screen, what two measurements are required?

3–9 What is the maximum allowed percentage of modulation for an AM signal?

3–10 Why do AM broadcast stations use as high a percentage of modulation as possible?

3–11 What are two consequences of *overmodulation*?

3–3 FREQUENCY DOMAIN AM ANALYSIS

As you'll recall from Chapter 2, there are two ways of examining electronic signals. We can look at signals in either the *time* or *frequency* domains. AM signals are very predictable in the frequency domain, which is important. When we know the frequency domain picture, it's easy to calculate and measure the *bandwidth* and *total power* of an AM signal.

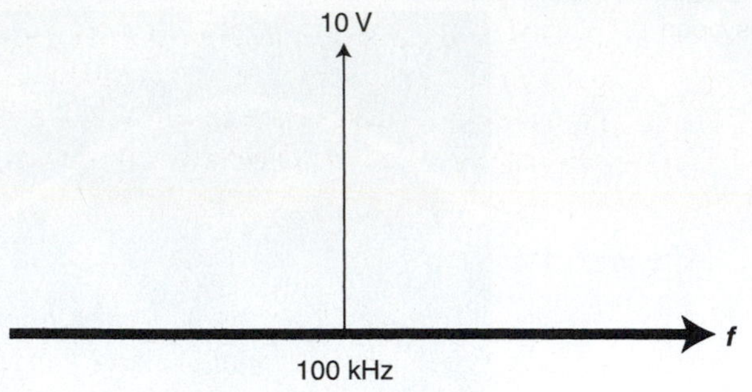

Figure 3–10 An unmodulated 100 kHz carrier

The Frequency Domain Revisited

Figure 3–10 shows a 100 kHz sine wave carrier in the frequency domain. How do we know this is a *sine* wave? Right, there's only one frequency present, 100 kHz. The sine wave is the only "pure" waveform, and it contains only one frequency. This particular sine wave is 10 volts RMS (about 14.1 volts peak).

What happens when this carrier signal is amplitude modulated? In the time domain, we know that the instantaneous voltage of the carrier will be forced to grow and shrink in step with the information. The nonlinearity of the modulator (the variable-gain stage) causes this to happen. In the frequency domain, the rapidly changing amplitude shows up as new frequencies called *sidebands.*

> *Sidebands* are new frequencies generated during the process of modulation. They are created by the nonlinear distortion introduced by the modulator amplifier. *All* modulation creates sidebands—whether it be AM, FM, or PM.

Suppose that the 100 kHz carrier of Figure 3–10 is amplitude modulated by a 10 volt, 5 kHz information signal. The resulting frequency domain picture will look like Figure 3–11.

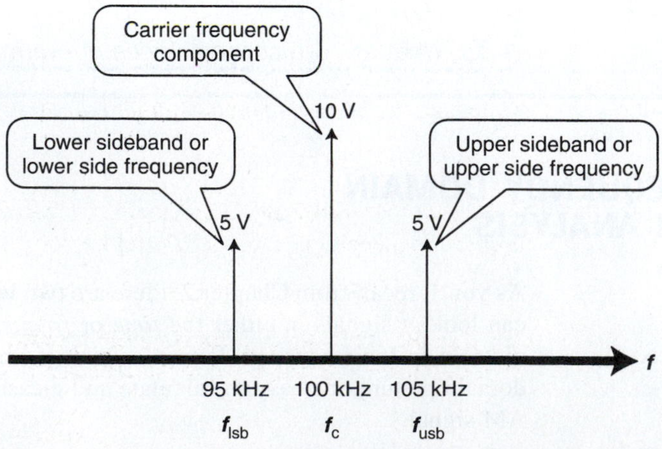

Figure 3–11 The 100 kHz carrier modulated with 5 kHz information

Note that some technicians use the term *side frequency* and others use *sideband* to describe these new frequencies. A side frequency is an *individual* frequency that can be part of a *range* of frequencies in a *sideband*. Since each of our sidebands in this example have only one frequency each (95 kHz for the lower sideband and 105 kHz for the upper sideband), we can use either term to correctly describe the signal.

What Are the Sidebands?

The result of Figure 3–11 leads to some important questions. For example, what has happened to the 5 kHz *information* signal? To get the AM modulated signal of Figure 3–11, two signals had to be applied to the modulator stage. These were the 100 kHz sine wave carrier and the 5 kHz information signals. The 5 kHz information signal seems to have disappeared! In reality, *the information energy (5 kHz) has been transformed into sideband energy (95 kHz and 105 kHz).* This transformation takes place with the help of the nonlinear transfer characteristic of the modulator stage. Therefore:

> The sidebands contain the information. They can be thought of as an "encoded" representation of the intelligence.

To calculate the frequencies of the sidebands, we use the following:

$$f_{usb} = f_c + f_m \quad (3\text{–}3)$$

$$f_{lsb} = f_c - f_m \quad (3\text{–}4)$$

In these equations, f_c is the carrier frequency and f_m is the information frequency.

Sideband Voltages

The sidebands are the information expressed in a new way. Notice that we got *two* sidebands from *one* information frequency. Our original information *voltage* was 10 volts; yet the sidebands are only 5 volts each in Figure 3–11. Why is this so? Right—because the information voltage causes *two* sidebands to form, its voltage must be *divided by two* in order to form each sideband. Therefore, we get

$$V_{usb} = V_{lsb} = \frac{V_m}{2} \quad (3\text{–}5)$$

Here, V_{usb} and V_{lsb} are the sideband voltages, and V_m is the information voltage.

EXAMPLE 3–8

Calculate the *bandwidth* and *modulation index* of the AM signal shown in Figure 3–11.

Solution

We know that bandwidth is calculated by finding the difference between the highest and lowest frequencies in a signal. Therefore, we get

$$BW = f_{max} - f_{min} = f_{usb} - f_{lsb} = 105\text{ kHz} - 95\text{ kHz} = \underline{\underline{10\text{ kHz}}}$$

Finding the modulation index is a little trickier, but not too difficult. Since we're working in the frequency domain, we'll use equation 3–1:

$$m = \frac{V_m}{V_c}$$

We know that V_m and V_c are given in the original problem definition; however, we are asked to get the information from Figure 3–11. Finding V_c is easy; we just read it off the spectrum analyzer display as 10 volts.

Remember that each sideband comes from the information. Therefore, each sideband voltage is one-half of the information voltage. By using equation 3–5,

$$V_{usb} = V_{lsb} = \frac{V_m}{2}$$

$$V_m = 2V_{usb/lsb}$$

where $V_{usb/lsb}$ is the voltage of either sideband (since they're identical).

$$V_m = 2V_{usb/lsb} = 2(5\text{ V}) = 10\text{ V}$$

Since we now know both V_m and V_c, we can calculate the modulation index:

$$m = \frac{V_m}{V_c} = \frac{10\text{ V}}{10\text{ V}} = \underline{\underline{1}}$$

The signal is 100% modulated.

TIP You'll often see a basic equation such as 3–1 or 3–2 used in many different ways. Don't try to memorize all of these variations! Instead, just learn the basic equation and concentrate on studying *how* we have applied it. *Look for basic principles.* For example, the basic idea here is that the information voltage "splits" into two sidebands during modulation. This will always happen in amplitude modulation.

EXAMPLE 3–9

For the AM signal pictured in Figure 3–12, calculate the following:

a. the information frequency, f_m

b. the voltage in the upper sideband, V_{usb}

c. the bandwidth

d. m and *percent modulation*

Solution

a. The information frequency is carried in the *difference* between either sideband and the carrier. The lower sideband frequency f_{lsb} is missing; however, we do know f_{usb}. By manipulating equation 3–3, we get

$$f_m = f_{usb} - f_c = 813\text{ kHz} - 810\text{ kHz} = \underline{\underline{3\text{ kHz}}}$$

b. The sidebands are identical twins; their voltages are always equal. Therefore,

$$V_{usb} = V_{lsb} = \underline{\underline{5\text{ V}}} \quad \text{(The 5 volt value is given in Figure 3–12.)}$$

c. As we've done before, bandwidth can be calculated by

$$\text{BW} = f_{max} - f_{min} = f_{usb} - f_{lsb}$$

Figure 3–12 An AM signal

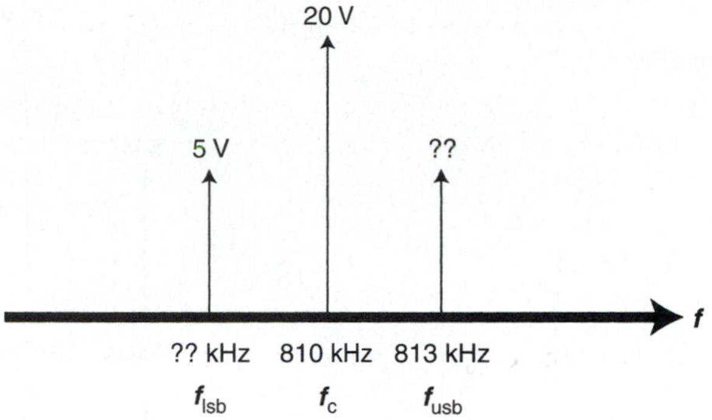

However, f_{lsb} is missing! We do know f_m, though, so we can readily complete this puzzle:

$$f_m = 3 \text{ kHz}$$

from step (a). Equation 3–4 tells us that

$$f_{lsb} = f_c - f_m$$

Therefore,

$$f_{lsb} = f_c - f_m = 810 \text{ kHz} - 3 \text{ kHz} = 807 \text{ kHz}$$

$$\text{BW} = f_{max} - f_{min} = f_{usb} - f_{lsb} = 813 \text{ kHz} - 807 \text{ kHz} = \underline{\underline{6 \text{ kHz}}}$$

TIP There are better methods for calculating bandwidth. You may already have one in mind. A shortcut will be introduced momentarily.

d. Since we're working in the frequency domain, equation 3–1 should be used:

$$m = \frac{V_m}{V_c}$$

From Figure 3–12, we can see that $V_c = 20\text{V}$ and $V_{lsb} = 5$ V. As before, we can get V_m

$$V_m = 2V_{usb/lsb} = 2(5 \text{ V}) = 10 \text{ V}$$

Finally, we can substitute:

$$m = \frac{V_m}{V_c} = \frac{10 \text{ V}}{20 \text{ V}} = \underline{\underline{0.5}} \qquad \text{(The signal is } \underline{\underline{50\% \text{ modulated}}}.)$$

A Shortcut for Bandwidth Calculation

You may have already noticed that much of the work of part (c) above is unnecessary. There's a much simpler way of calculating bandwidth. If we look at the relative locations of the parts of the AM signal, this "shortcut" becomes clear.

In Figure 3–13, *bandwidth* is the total frequency "distance" between the *maximum* (f_{usb}) and *minimum* (f_{lsb}) frequencies in the signal. *This distance is always twice the information frequency:*

$$\text{BW} = 2 f_m \tag{3–6}$$

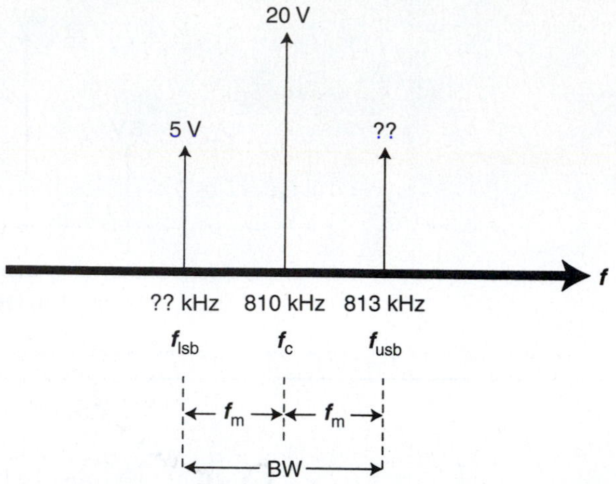

Figure 3–13 Calculating AM bandwidth the easy way

EXAMPLE 3–10 A certain AM signal has the following characteristics:

$$V_c = 100\,\text{V},\ f_c = 2182\ \text{kHz},\ V_m = 25\ \text{V},\ f_m = 2.5\ \text{kHz}$$

a. What bandwidth is required by the signal?

b. What is the modulation index m and the percentage of modulation?

c. Draw a spectrogram of the signal, showing all voltages.

Solution

a. Using equation 3–6,

$$\text{BW} = 2f_m = 2(2.5\ \text{kHz}) = \underline{\underline{5\ \text{kHz}}}$$

b. Using equation 3–1,

$$m = \frac{V_m}{V_c} = \frac{25\ \text{V}}{100\ \text{V}} = \underline{\underline{0.25}} \qquad \textit{(The same as 25\% modulation.)}$$

c. To draw the spectrogram, we must find the sideband frequencies and voltages. Using equation 3–5,

$$V_{\text{usb}} = V_{\text{lsb}} = \frac{V_m}{2} = \frac{25\ \text{V}}{2} = \underline{12.5\ \text{V}}$$

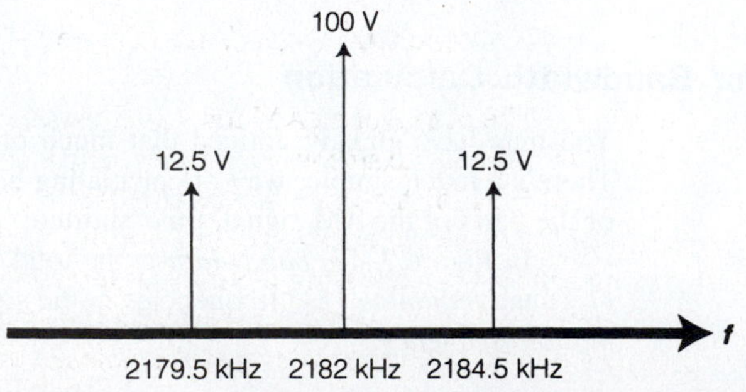

Figure 3–14 Solution to Example 3–10

Using equation 3–3,

$$f_{usb} = f_c + f_m = 2182 \text{ kHz} + 2.5 \text{ kHz} = \underline{2184.5 \text{ kHz}}$$

Using equation 3–4,

$$f_{lsb} = f_c - f_m = 2182 \text{ kHz} - 2.5 \text{ kHz} = \underline{2179.5 \text{ kHz}}$$

The spectrogram is "filled in" with the resultant values as in Figure 3–14.

Section Checkpoint

3–12 What is the name given to the new frequencies generated during the process of modulation?

3–13 What are the three frequency components of an AM signal?

3–14 What property of the modulator circuit causes sidebands to be generated?

3–15 If the intelligence voltage V_m is zero, what happens to the sideband voltages? Why?

3–16 The sidebands are a transformed version of the ________ signal.

3–17 To form the sideband voltages, the intelligence voltage is divided by ________.

3–18 What determines the bandwidth of an AM signal?

3–4 POWER AND EFFICIENCY

In many cases, we are interested in the total power developed by an AM transmitter. In the real world, a *wattmeter* is used to measure the power being produced by a transmitter.

In the frequency domain, an AM signal consists of a *carrier frequency component* (or "carrier" for short) and a pair of *sidebands*. The total power is the sum of all the individual *powers* in the signal. We can express the total power P_t as follows:

$$P_t = P_{lsb} + P_c + P_{usb} \quad (3–7)$$

By developing a *model,* as in Figure 3–15, we can gain a better understanding of how this works.

The output of an AM transmitter, when modulated by a single sine wave, consists of *three* different ac sine wave voltages. These are the carrier, upper sideband, and lower sideband. Each of the three ac voltages has a different frequency, voltage, and phase angle. Curiously, the antenna looks like a *resistor* to the transmitter! This is because the antenna accepts RF energy from the transmitter, and converts it into electromagnetic energy. The antenna is a *transducer.* A resistor is a transducer too; it converts electrical energy into heat energy. Since energy goes into the antenna and doesn't return to the circuit, it appears as a pure resistance.

Figure 3–15 The equivalent circuit of a modulated AM transmitter

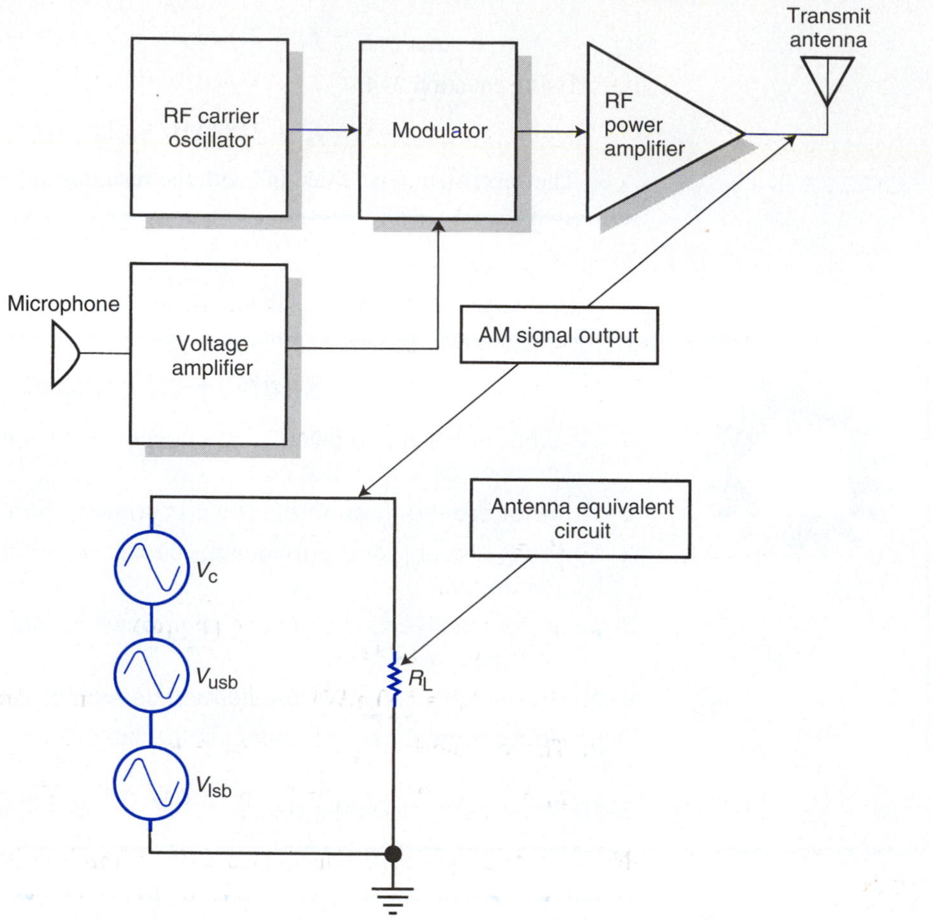

The most common antenna "resistance" (RF people often use the word *impedance*) is 50 Ω. Other common values are 25, 75, and 300 Ω.

You might be tempted to calculate the *total power* of this transmitter as follows:

$$P_t = \frac{V_{total}^{\,2}}{R_L}$$

where $V_{total} = V_{lsb} + V_c + V_{usb}$, the total of the circuit voltages.

This approach is valid. However, the individual voltages are *phasors*. To add the three signals, we must include the phase angle of each one. This is not the easiest approach!

The approach used by technicians is to rely on Ohm's law rather than phasor analysis. Equation 3–7 tells us to add the *powers* of the individual signals. To find the power of each ac signal component, we just use Ohm's law:

$$P_c = \frac{V_c^{\,2}}{R_L} \qquad \textbf{(3–8)}$$

Notice that the carrier power is *constant* in an AM signal. Modulation increases the signal power by the addition of *sideband* energy:

$$P_{usb} = P_{lsb} = \frac{V^2_{usb/lsb}}{R_L} \quad (3\text{–}9)$$

These principles are best illustrated through example.

EXAMPLE 3–11 The AM signal pictured in Figure 3–12 is being measured across a 50 Ω antenna "load." Calculate the following:

a. the carrier power, P_c
b. the power in the lower sideband, P_{lsb}
c. the power in the upper sideband, P_{usb}
d. the total power in the sidebands
e. the total power in the signal, P_t

Solution

a. $P_c = \frac{V_c^2}{R_L} = \frac{20\ V^2}{50\ \Omega} = \underline{\underline{8\ W}}$

b. $P_{lsb} = \frac{V_{lsb}^2}{R_L} = \frac{5\ V^2}{50\ \Omega} = \underline{\underline{0.5\ W\ (500\ mW)}}$

c. $P_{usb} = P_{lsb} = \underline{\underline{500\ mW}}$ (The sidebands are identical twins.)

d. The total sideband power $= P_{lsb} + P_{usb} = \underline{\underline{1\ W}}$

e. $P_t = P_{lsb} + P_c + P_{usb} = 0.5\ W + 8\ W + 0.5\ W = \underline{\underline{9\ W}}$

Notice that no "new" equations were used to get these results. Only Ohm's law was needed!

A Shortcut for Total Power

By combining equations 3–8 and 3–9, and doing some algebraic reductions, we can obtain the following relationship:

$$P_t = P_c + P_c\left(\frac{m^2}{2}\right) = P_c\left(1 + \frac{m^2}{2}\right) \quad (3\text{–}10)$$

Notice that in equation 3–10, if P_c (the carrier power) is already known, then it is not necessary to know R_L. All that is needed is the carrier power P_c and the modulation index m.

EXAMPLE 3–12 A certain AM transmitter measures 8 watts on a wattmeter when it is dead-keyed (unmodulated). If an intelligence signal is then applied that results in 100% modulation, calculate the following:

a. total power
b. sideband power

Solution

To use equation 3–10, we need to know P_c and m. These are given in the problem data. A "dead-keyed" transmitter only produces a carrier. Therefore, the carrier power P_c is

8 watts. It is then stated that the transmitter is going to be 100% modulated. This means that $m = 1$. We can now use equation 3–10 to get a result.

a. $$P_t = P_c + P_c\left(\frac{m^2}{2}\right) = P_c\left(1 + \frac{m^2}{2}\right) = 8\text{ W}\left(1 + \frac{1^2}{2}\right) = \underline{\underline{12\text{ W}}}$$

b. There are at least two ways of finding the sideband power. First, look at the difference between answer (a) and the carrier power. The power has increased by 4 watts. This power resulted from modulating the transmitter. *The sideband power is therefore 4 watts.*

We can get the same result mathematically. Equation 3–10 has been written in two forms. The first form looks like this:

$$P_t = P_c + P_c\left(\frac{m^2}{2}\right)$$

This form is "unfactored." It shows that the total power comes from two terms—the first is P_c, the *carrier* power. The second term, $P_c(\frac{m^2}{2})$, gives the *total sideband power.* Therefore, the sideband power can be calculated as

$$P_{\text{side}} = P_c\left(\frac{m^2}{2}\right) = 8\text{ W}\left(\frac{1^2}{2}\right) = \underline{\underline{4\text{ W}}}$$

The same result is obtained from both approaches. You should use the approach that makes the best sense. You may even devise a method of your own—as long as you obey Ohm's law, your solution will work!

EXAMPLE 3–13

Assuming that the transmitter output of Figure 3–12 is appearing across a 50 Ω antenna load, calculate the total power by using equation 3–10.

Solution

In order to use equation 3–10, we must know m and P_c. It is easiest to first find P_c:

$$P_c = \left(\frac{V_c^2}{R_L}\right) = \left(\frac{20\text{ V}^2}{50\ \Omega}\right) = \underline{\underline{8\text{ W}}}$$

By using voltages and equation 3–1, we can get m:

$$m = \frac{V_m}{V_c} = \frac{V_{\text{lsb}} + V_{\text{usb}}}{20\text{ V}} = \frac{5\text{ V} + 5\text{ V}}{20\text{ V}} = \underline{\underline{0.5}}$$

Now that m and P_c are known, we can apply equation 3–10:

$$P_t = P_c\left(1 + \frac{m^2}{2}\right) = 8\text{ W}\left(1 + \frac{0.5^2}{2}\right) = \underline{\underline{9\text{ W}}}$$

Efficiency of AM Transmission

In communications, the efficiency of a system is measured as follows:

$$\eta = \text{efficiency} = \frac{\text{power of information}}{\text{total power used}} \tag{3–11}$$

The value obtained from equation 3–11 is a percentage. 100% efficiency means that *all* radiated (or transferred) energy conveys information. This is the ideal goal for communications systems designers. Actual communications systems generally fall short of this goal.

Can you rewrite equation 3–11 for AM? Let's do so. Think about the three parts of an AM signal: lower sideband, carrier, and upper sideband. Which parts of the signal carry the information? Right—the *sidebands.* The total power is the sum of the sideband and carrier powers. Therefore, we get

$$\eta_{AM} = \text{efficiency} = \frac{P_{lsb} + P_{usb}}{P_{lsb} + P_c + P_{usb}} = \frac{P_{lsb} + P_{usb}}{P_t} \tag{3–12}$$

EXAMPLE 3–14 What is the efficiency of the AM transmitter of Example 3–11?

Solution

To calculate efficiency, apply equation 3–11:

$$\eta_{AM} = \text{efficiency} = \frac{P_{lsb} + P_{usb}}{P_t} = \frac{1\text{ W}}{9\text{ W}} = 0.111 \approx \underline{\underline{11\%}}$$

This is *terrible* efficiency. Only 11% of the radiated power is actually doing useful work—carrying information. Another way of putting this: Say that the monthly electric bill is \$100,000 for a radio station. Under these conditions, only \$11,000 (11%) of the electricity actually carries information—\$89,000 is being spent to emit energy components (the carrier frequency component) that essentially carry *no* information!

If AM is so inefficient, why is it used at all, when more efficient methods are certainly available? Primarily because AM receivers are simple to design and inexpensive. Also, because AM uses less bandwidth than FM, more stations can be operated in a given band (range of frequencies).

The *carrier frequency component* is needed by an AM receiver in order to demodulate or detect the signal; however, it is not absolutely necessary to *transmit* the carrier. If we eliminate the carrier and send only the sidebands, we get a *double-sideband suppressed-carrier* (DSB-SC) signal. This improves efficiency greatly, but the receiver must now include an extra circuit called a *beat frequency oscillator* to successfully detect the DSB-SC signal.

Even further improvement can be made to the DSB-SC signal by eliminating one of the sidebands (leaving one sideband intact). This works because the sidebands are identical twins and carry the same information. The resulting signal is called a *single-sideband suppressed-carrier* (SSB) signal.

Maximum Efficiency for an AM Transmitter

A final note about efficiency: For an AM transmitter, the best efficiency is obtained when the transmitter is modulated 100%. This is another reason why commercial broadcasters tend to run as close to 100% modulation as possible.

At 100% modulation, the total sideband (information) power can be expressed by the second term of equation 3–10:

$$P_{sideband} = P_c\left(\frac{m^2}{2}\right) = P_c\left(\frac{1^2}{2}\right) = \frac{P_c}{2}$$

The total power at 100% modulation will be

$$P_t = P_c\left(1 + \frac{m^2}{2}\right) = P_c\left(1 + \frac{1^2}{2}\right) = \frac{3}{2}P_c$$

Substituting these equations into the efficiency equation yields:

$$\eta_{AM-max} = \frac{P_{lsb} + P_{usb}}{P_t} = \frac{(P_c/2)}{\left(\frac{3}{2}P_c\right)} = \underline{\underline{1/3 = 33.333\%}}$$

The maximum efficiency of any AM transmitter is close to 33%. In other words, even when an AM transmitter is operating at maximum efficiency (100% modulation), 67% of the transmitted energy carries no information!

Section Checkpoint

3–19 What three frequency components contribute to the total power of an AM signal?

3–20 What is the most common impedance value for antennas?

3–21 Why does a transmitting antenna appear as a resistor to the transmitter?

3–22 What condition is present when a transmitter is dead-keyed?

3–23 What is the ratio of *total power* to *carrier power* when an AM transmitter is 100% modulated? (Use equation 3–10.)

3–24 What is the definition of *efficiency* for communications systems?

3–25 Give at least two reasons why AM is used in spite of its poor efficiency.

3–26 What is the difference between a SSB-SC signal and an AM signal?

3–5 SPECTRUM ANALYZER MEASUREMENTS

Oscilloscopes are fairly limited in the amount of frequency domain information they can present. A technician who is working with RF systems can expect to use a spectrum analyzer on the job. Typical applications for a spectrum analyzer include frequency and bandwidth verification, signal strength measurements, filter frequency response measurements, and distortion measurements.

Spectrum analyzers come in even more varieties than oscilloscopes. It is important that the working technician read and understand *all* instrument instructions. There are many more variable controls and settings on a spectrum analyzer than on any other piece of test equipment in a shop.

Analyzers come in three general varieties: analog, digital, and "digilog." *Analog* spectrum analyzers are the oldest type. An analog spectrum analyzer (Figure 3–16) functions as a swept-tuned radio receiver. The horizontal sweep circuit in an analog spectrum analyzer moves the electron beam left-to-right on the display tube and also tunes the radio receiver section in a like manner. The output from the radio receiver section is detected and sent to the vertical deflection plates of the display tube. The result is a display with *frequency* on

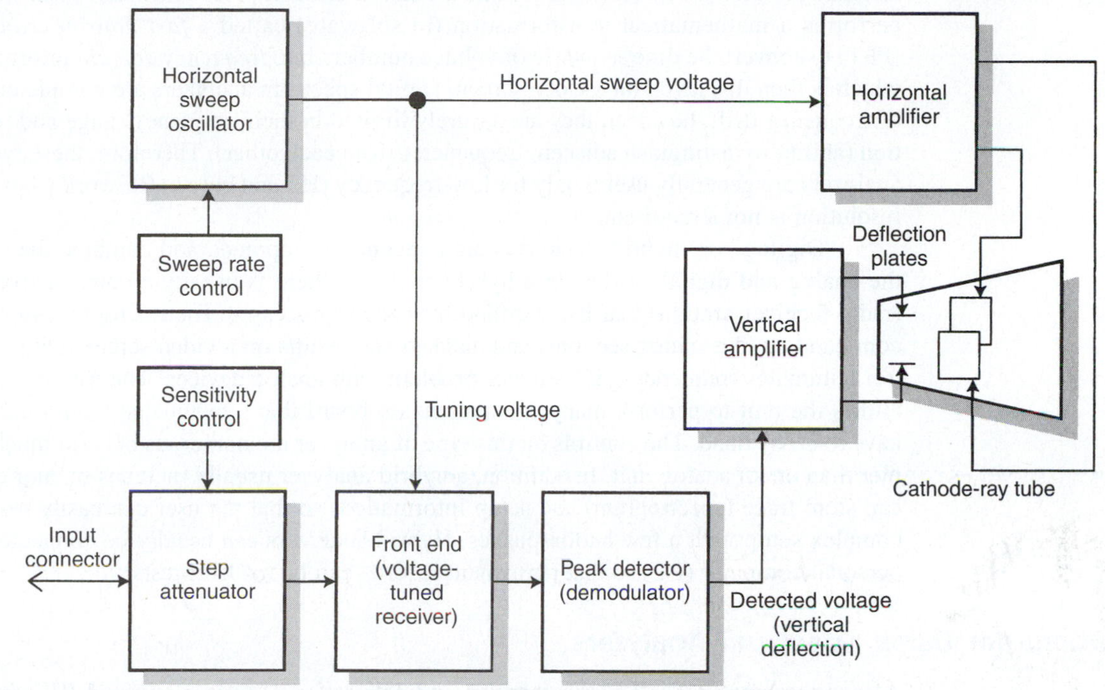

Figure 3–16 Analog spectrum analyzer block diagram

the horizontal axis and *amplitude* on the vertical axis. A well-designed analog spectrum analyzer can measure almost anything a technician needs to see. The primary limitation of the analog analyzer is *frequency drift,* a common problem with analog circuitry.

Digital spectrum analyzers (Figure 3–17) work by reading a fixed number of voltages from the input signal. These voltage readings are converted into binary numbers by

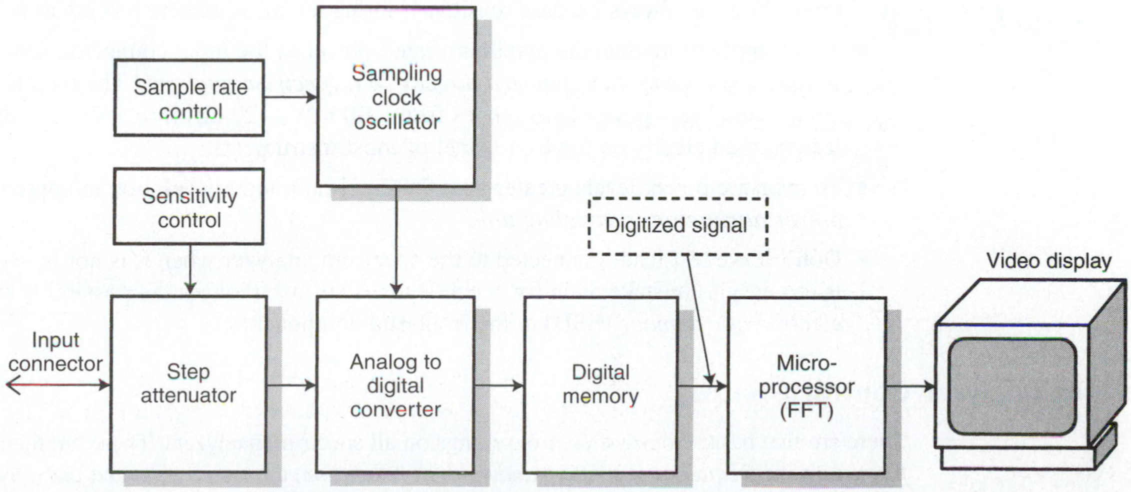

Figure 3–17 A digital spectrum analyzer

an *analog to digital converter* (ADC), and are stored in a computer's memory. The computer performs a mathematical transformation (in software) called a *fast Fourier transform* (FFT) to convert the time sequence of voltage numbers into *frequency domain* information, which is then displayed on a video screen. Digital spectrum analyzers are completely free of frequency drift; however, they are severely limited in their frequency range and resolution (ability to distinguish adjacent frequencies from each other). Therefore, these types of analyzers are generally useful only for low-frequency (HF and below) RF work where high resolution is not a requirement.

"Digilog" or "hybrid" analyzers are a recent development, and combine the best of the analog and digital worlds. In a hybrid analyzer, there is usually an analog front end (radio receiver section) that is controlled by a microprocessor. The microprocessor gives commands to the analog sections, and displays the results on a video screen. Digital control eliminates frequency drift, a major problem with analog devices. The use of software allows the unit to perform many calculations on board that a technician would normally have to do by hand. The controls on this type of analyzer are numerous but still much simpler than on an analog unit. In addition, a hybrid analyzer usually includes memories that can store trace (spectrogram) and setup information, so that the user can easily restore a complex setup with a few button pushes. Hybrid analyzers can usually be connected to a personal computer (PC) so that permanent records can be made of tests.

Precautions for Using Spectrum Analyzers

A spectrum analyzer is a very expensive and delicate tool. Even a minimal RF spectrum analyzer can cost over $10,000. The radio receiver "front end" is very susceptible to damage. Here are some things to avoid. If you're ever in doubt about how to safely perform a measurement, the best thing to do is *ask*. You will be avoiding risk to the equipment, as well as demonstrating that you understand your own limitations (we all have them).

Here are some basic precautions:

- *Never* connect a spectrum analyzer input to any point in a circuit where a dc voltage is present unless a dc block is installed on the analyzer. *Failure to observe this precaution can cause instant failure of the input section of the analyzer.*
- Always start measuring with the analyzer's *attenuator* section set for *minimum* input sensitivity. (This normally means, higher "dBm" or "decibel-milliwatt" attenuator settings.) You can always increase sensitivity during the measurement process as needed.
- *Never* apply more than the maximum rated power to the input connector. *Most transmitters cannot be connected directly to a spectrum analyzer!* The typical "maximum" safe power level ranges from 100 mW (+20 dBm) to 1 W (+30 dBm). It is marked clearly on the front panel of most instruments.
- To measure power levels greater than the maximum input level, use an approved *power attenuator* or *sampling unit.*
- Don't leave test leads connected to the spectrum analyzer when it is not in use. It is too easy to mistake them for scope leads. This also reduces the possibility of *electrostatic damage* (ESD) to the front-end components.

Spectrum Analyzer Control Groups

There are five basic groups of control settings on all spectrum analyzers. These are the *center frequency,* the *frequency span,* the *bandwidth* or *resolution,* the *sweep rate,* and the *sensitivity* or *reference level.* (Many hybrid analyzers automatically choose a correct sweep rate for you.)

Center frequency The *center frequency* is the frequency that appears in the *middle* of the horizontal axis on the display. Some analyzers place this information on the display; others use a separate LED display or calibrated analog dial scale.

Frequency span The *frequency span* is the total frequency distance from the left to right on the display. It is usually shown in units of *Hz/division, kHz/division,* or *MHz/division*. Some high-performance analyzers can also display in units of *GHz/division*. (Recall that 1 GHz = 1000 MHz.)

To locate a frequency on the display, start at the *center* and count the number of divisions left or right; add or subtract the frequency difference from the center frequency. Some analyzers have *cursors* that can be moved over a point on the display, giving an instant readout of the frequency at that point.

Bandwidth The *bandwidth* control adjusts the frequency resolution of the display. Many analyzers can automatically choose this setting based on the *frequency span*. This control is needed primarily because the analyzer may be asked to cover many different frequency ranges. For example, to cover a frequency span of 10 MHz (10 divisions at 1 MHz/division), a bandwidth of 30–300 kHz is reasonable to begin with. One way of thinking about the bandwidth control is as follows: Imagine that you're searching the length of one city block for a lost item. If the item is a car (a relatively large object), you can scan the street fairly quickly. Conversely, if the item is a lost paper clip, your search will take much longer, and you might even have to use a magnifier of some kind to help in your search.

The bandwidth control operates in a similar fashion. Setting it too narrow causes the analyzer to probe for very fine details. These details probably won't show up on the screen, since a screen can only display a certain level of detail. Setting the bandwidth control to a very wide position causes details to blend together. How can a tech know where to set this control? Sometimes, the service literature will give the settings. Otherwise, it is a matter of experience developed over years of practice.

Sweep rate The *sweep rate* control sets the frequency of the horizontal sweep oscillator. This controls the amount of time it takes for the electron beam to move from left to right on the display. This control is adjusted to the fastest rate that gives a comfortable and *calibrated* display. It interacts with the *bandwidth* control. Sweeping very rapidly while using a narrow bandwidth will cause the radio receiver section of the analyzer to misinterpret details of the signal being analyzed. Some analyzers have a warning lamp that lights when the sweep rate is set too fast. Setting this control too slow makes the display hard to read (on purely analog analyzers), and it can take a long time to get a reading. With hybrid analyzers, the sweep rate is usually set automatically by the instrument.

Sensitivity At the input of a spectrum analyzer is a precision circuit called a *step attenuator* (see Figure 3–16). This circuit controls how much of the input signal is allowed into the voltage-controlled radio receiver (front end). The radio receiver and detector in a spectrum analyzer are carefully calibrated to give a full-screen deflection when an input signal is equal to the *reference level*. The step attenuator increases the working range of the spectrum analyzer by allowing the *reference level* to be varied by the operator.

The reference level is usually given in units called "dBm." These are *decibels of power with respect to 1 mW*. (dBm and other decibel units will be the subject of a future lesson.) In short, 0 dBm is 0 decibels stronger than a milliwatt; in other words, 0 *dBm* = 1 mW. Since a 3 dB change would be a doubling of power, therefore, 3 dBm = 2 mW. A 10 dB change is a tenfold increase in power. Therefore, 10 dBm = 10 mW. A reading of 20 dBm would be *two* 10 dB increases over 1 mW—which would be 1 mW × 10 × 10 = 100 mW.

The sensitivity control of a spectrum analyzer should always be started at the highest available reference level when the signal level is unknown.

Measuring the Performance of a CB Transmitter

Harry and his neighbor Bill live several doors apart on Main Street. They are both CBers. Harry likes to talk to the truckers on channel 19 (Main Street is only a few blocks from the interstate). Bill operates a landscaping business from his home, and uses channel 14 to communicate with his trucks. Lately, Bill has been having trouble hearing; a gravelly, scratchy voice would come on when his mobile units were trying to call the base. On turning the channel selector during one episode of the interference, he found the "voice" on all 40 channels. He also heard Harry, loud and clear, on channel 19. Could Harry be the source of the interference?

Bill gave Harry a call and explained what he was hearing. Harry agreed to get his radio checked at the local radio shop.

At the shop, Susan, the service technician, listened carefully to Harry's description of the problem. "You two guys are very close. That in itself could be the problem. It won't hurt to check your transmitter, though. Would you like to help me set up the test gear?" Harry grinned back. "Sure! Let's do it!"

Harry and Susan had the equipment connected together in a few minutes. A 12 volt bench power supply provided power for the transceiver. Susan connected a coaxial cable to the antenna connector on the CB. The black coaxial cable went into a small box about the size of a tissue box with black metal fins (a heatsink). A second cable connected the metal-finned box to the input connector of a spectrum analyzer. (See Figure 3–18.)

Harry had never seen so many electronic gadgets in one place before. He was in nerd heaven! Susan explained how the spectrum analyzer was going to measure the transmitted signal from Harry's CB. "What does the little black box do that is hooked between your analyzer and my CB?" "That's an *attenuator,* Harry. An attenuator makes a signal weaker.

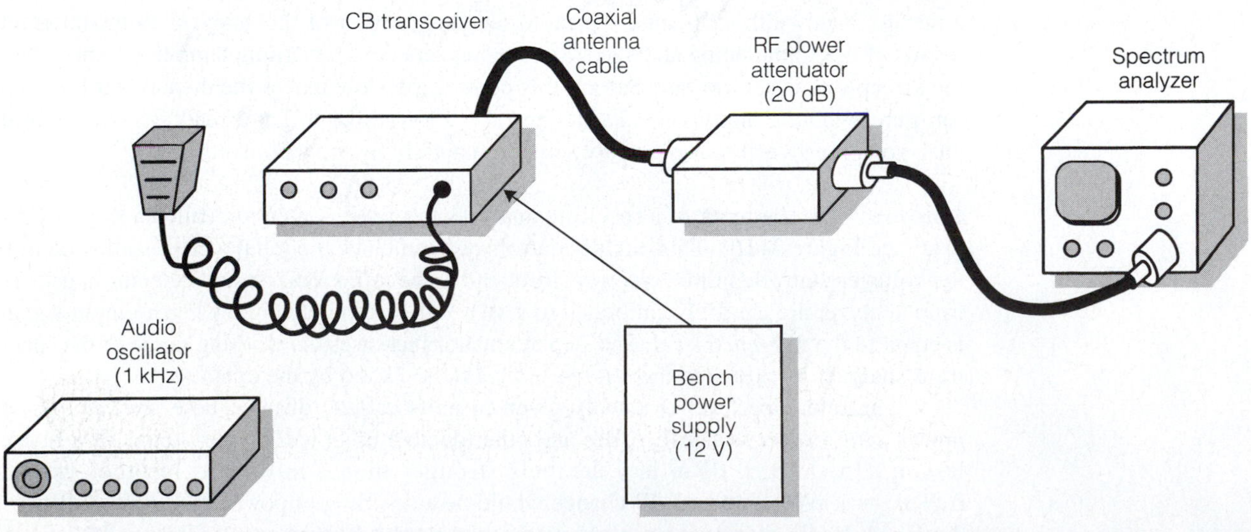

Figure 3–18 Measuring the spectral cleanliness of a CB transmitter

Your CB can provide 4 watts of carrier power, and when we modulate it 100%, the total power it can deliver is close to 6 watts. That's too much power for the spectrum analyzer, so we have to cut it down. The little black box reduces the power by 20 decibels. That is the same as saying the power is divided by 100. That way, if your CB puts out 6 watts, my spectrum analyzer only sees 6 watts/100 or 0.06 watts (60 mW). That is a safe power level."

Susan began to carefully adjust the controls on the spectrum analyzer. "Harry, you were transmitting on channel 19, right?" Harry nodded. "Then you were sending a carrier frequency of 27.185 MHz. I'll set the analyzer *center frequency* to match your transmitter." Harry suddenly looked puzzled. "How can we tell if I'm interfering on Bill's frequency when the screen is set to read *my* frequency in the middle?"

"Take a look at the screen, Harry. We're set for 27.185 MHz in the *center,* and our *frequency span* is set at 20 kHz per horizontal division. You can count frequency right and left of the center in 20 kHz jumps. Key your transmitter but don't talk into it." Harry pressed the button on the microphone. Figure 3–19 shows what Harry and Susan saw.

Susan continued: "You can see that we're putting out our signal exactly where it is supposed to be, 27.185 MHz, which is channel 19. You can also see that so far, you're not causing any problem on your friend's frequency, channel 14 (27.125 MHz). Let's modulate the transmitter and make sure it's really OK."

Susan flipped a switch on the *audio oscillator* and it sprang to life. A steady 1 kHz tone came from a nearby loudspeaker. "This is our test signal, Harry, a 1000 Hz sine wave. One thousand Hertz is right in the middle of our hearing range. It's a good signal to modulate the transmitter with. If we're careful, we can adjust the generator to get around 90 to 100% modulation. The *monitor scope* (Figure 3–20) will help us get it right."

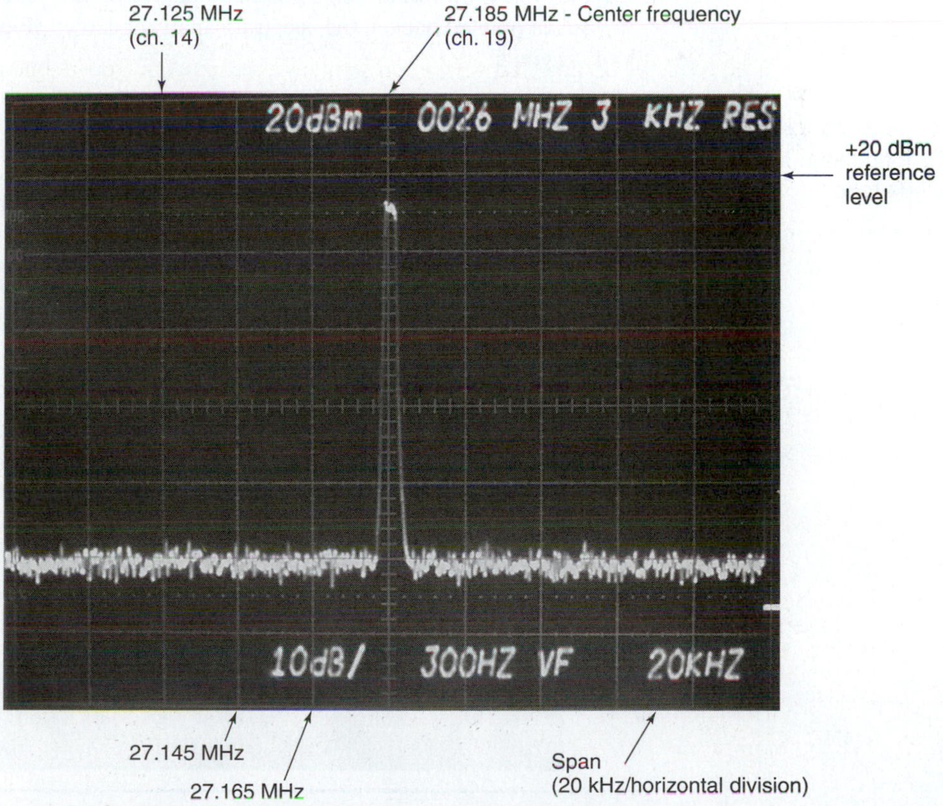

Figure 3–19 Harry's transmitter, dead-keyed on channel 19

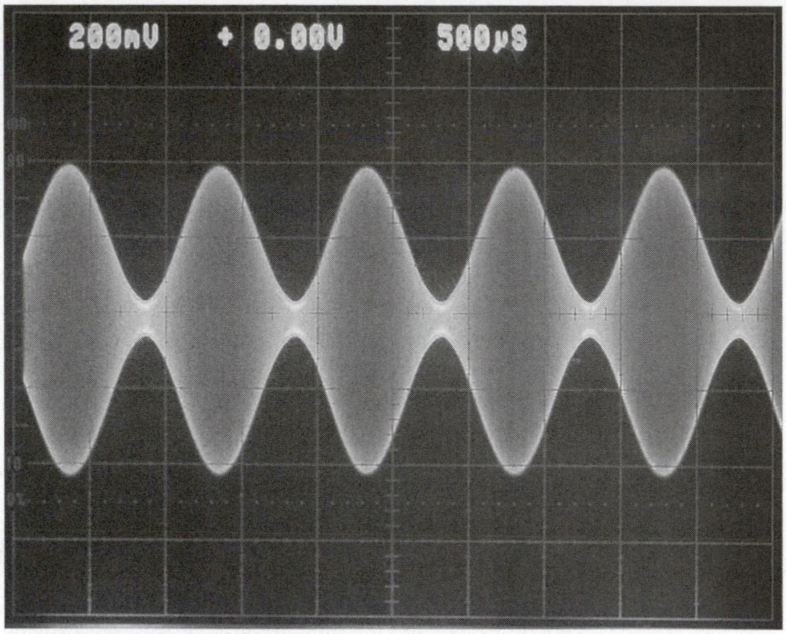

Figure 3–20 The CB signal on a monitor oscilloscope

Harry wanted to know how Susan could tell when the modulation was correct. "Look at the center of the waveform. See how it is getting small? When it becomes nearly invisible (but not flat), we're at 100% modulation. We're very close now."

Susan pointed back at the spectrum analyzer screen (Figure 3–21). "Harry, how does it look to you?"

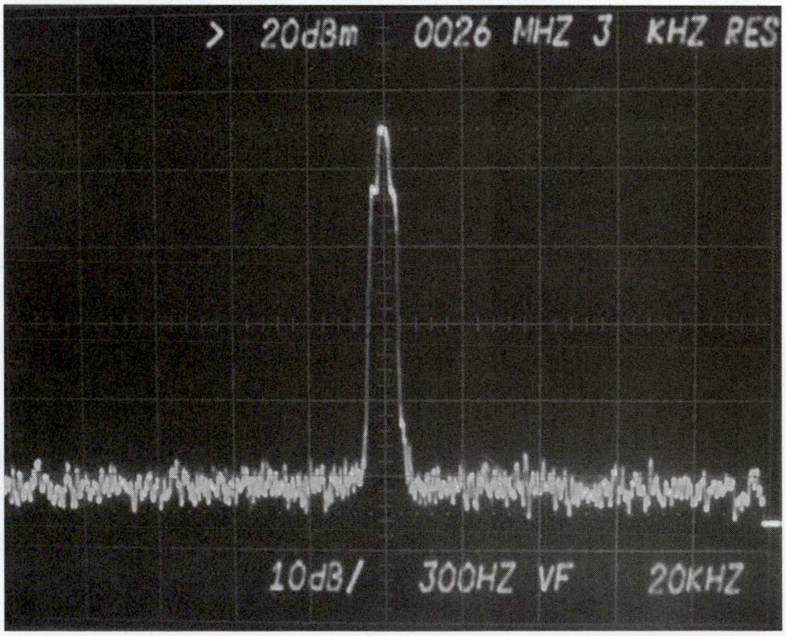

Figure 3–21 Harry's CB modulated by a 1 kHz tone

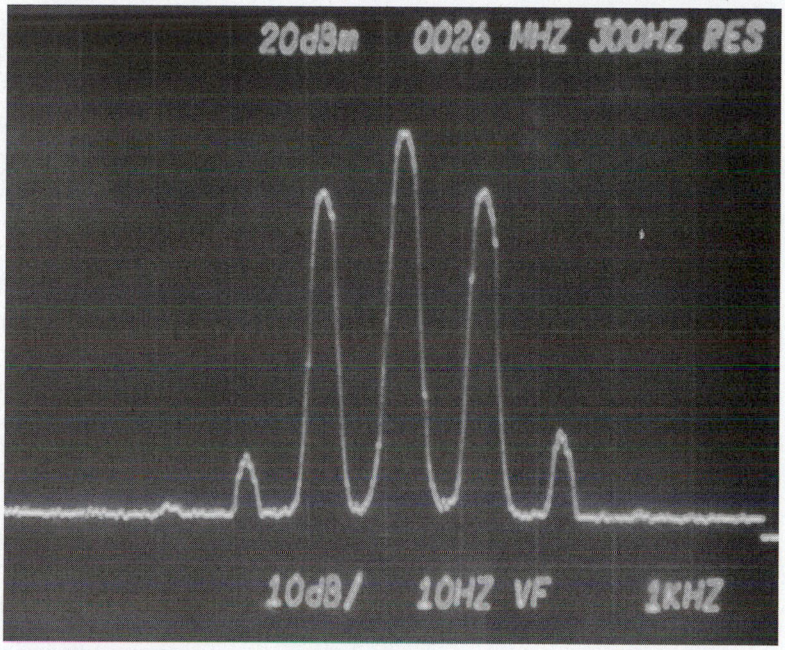

Figure 3–22 Out comes the magnifying glass

Harry hesitated. He wasn't sure if the display was OK or not! "How can I tell if it is right or not?"

"Remember where your friend's frequency was? Try looking there." Harry carefully counted backwards from the center of the analyzer display: "27.185, 27.165, 27.125, ... Yep, Bill's frequency is still clear. So I guess *I'm* not Bill's problem! But the spike in the middle of the display is shaped funny now. It looks a little wider than before. Is that bad?"

Susan smiled. "Harry, you're right. Your transmitter is perfectly clean! And you're also right about the spike getting *wider*. The *bandwidth* of the transmitter is what you're seeing. Let's zoom in on your carrier frequency and see what's really happening."

Susan quickly adjusted the *span* control on the spectrum analyzer. The display now looked like Figure 3–22.

"Susan, it looks like you've zoomed in on my frequency . . . but I'm confused. There seem to be *three* peaks now, plus two little ones on the side. I guess that's why the other peak looked fatter a minute ago. We're looking at it much more closely now. But where did these other peaks come from?"

Susan started to explain: "The two little peaks on the right and left are *distortions* caused by the transmitter. They're normal, and in this case, they're more than 30 dB (one thousand times) below the sideband power level; after all, no transmitter is perfect. The big peak in the middle is the carrier from your transmitter, and—"At that moment, Susan's pager began beeping. Harry jumped a little. She picked the pager up from the bench, looked at it, and rolled her eyes.

"Sorry to run Harry, but I'd better return this call. I hope this helped you guys out. Your transmitter looks fine!" Harry thanked her as they dismantled the equipment on the bench.

- Can you finish explaining to Harry what he is seeing in Figure 3–22?
- What should Harry and Bill try next to eliminate the interference?

Section Checkpoint

3–27 What are the three main types of spectrum analyzers?

3–28 A spectrum analyzer is an electronically tuned ______ receiver.

3–29 What is the main problem with analog spectrum analyzers?

3–30 What does FFT stand for? Where is it used?

3–31 List at least three precautions for safely using a spectrum analyzer.

3–32 What are the five basic control groups on a spectrum analyzer?

3–33 Why did Susan use an RF power attenuator when connecting Harry's CB to the spectrum analyzer (Figure 3–18)?

3–34 Why is the peak of Figure 3–21 "fat?"

3–35 Explain what each peak in Figure 3–22 is (don't worry about the tiny distortion products on the outer left and right). Which ones are the sidebands?

3–36 From looking at Figure 3–22, can you tell what type of spectrum analyzer is in Susan's shop? How can you tell? (Hint: Look at the relative horizontal position of the signals between Figures 3–21 and 3–22.)

Further Discussion: Measuring the Performance of a CB Transmitter

Interference is probably the number one complaint in RF communications installations, and Harry and Bill were wise to seek out Susan's help. Susan's measurements didn't show any outright problems with Harry's transmitter, so what could be the problem? Here are some possibilities.

- *Receiver Overload*: Since Harry and Bill live only a few doors apart, Harry's signal is extremely strong at Bill's antenna. Very strong signals can overwhelm receivers, especially consumer-grade devices such as CB radios. A receiver that is being overloaded loses its ability to reject undesired frequency energy (such as off-channel signals). To fix this problem, either Harry would need to reduce his transmit power level, or not operate when Bill needs to listen on the air. Another possibility would be for Bill to use a directional antenna (aimed away from Harry's house).
- *Unintended Mixing*: Objects such as metal gutters, fences, and even guy wires on antenna masts sometimes corrode and become a problem. The junction of two dissimilar (or corroded) metals forms a diode junction that can generate undesired frequencies from the original RF signal. The object then re-radiates the signal back into space as interference. This is a common problem around high-power transmitters (such as broadcasters). Improper grounding (for example, using a chain-link fence as an RF ground) can also lead to this problem, even at the relatively low powers used in CB.
- *Receiver Faults*: Receivers fail in strange ways, and it is entirely possible that Bill's receiver has some type of problem around strong signals. Bill may also have a control set incorrectly (for example, RF gain) adding to the problem.

- *Coincidental Interference*: Maybe there's no problem between Harry's and Bill's stations at all; perhaps another signal source is at work causing Bill's receiver interference.

In summary, interference problems can be very tough to resolve. An open and inquisitive mind is required along with a good understanding of how to interpret the test equipment's results.

RESTAURANT ON THE AIR!

A wireless Internet provider was having severe troubles in one of the sectors of a market in California. Customers in the sector were getting very slow service due to errors, and in some cases were being disconnected! The problem had been present for several weeks and the number of customer complaints was rapidly mounting. For some odd reason, the interference peaked around 12:00 p.m. and 6:00 p.m. each day. Tests at the head-end (on a nearby mountain top where the wireless signal originated) indicated that the sector's receiver was operating fine, but the interference persisted, wiping out almost all the upstream (customer to base station) channels in the 2.4 GHz band segment used by the sector. Something obviously had to be done.

Ground-based attempts to locate the jamming signal failed. In desperation, the company hired a helicopter to find the source of the phantom signal. An RF technician loaded a directional antenna and portable spectrum analyzer onto the chopper, and the search was on!

The signal was very strong in the middle of town. In fact, it seemed to be strongest over a block with a fast-food restaurant. The tech asked the pilot to fly over the restaurant several times; each time, the signal peaked as they passed overhead. At least they had an idea of where the signal came from.

The next day, two technicians visited the restaurant with the portable spectrum analyzer in hand. The RF level inside the place was incredible. How could such a high level of RF be in a restaurant? Then it clicked. The establishment must have a defective microwave oven. Microwave ovens operate at 2.45 GHz, the same frequency band as the wireless Internet service. The technicians spoke to the manager and inspected the microwave oven. The door had been removed, and the two interlock switches had been bypassed. *The microwave was transmitting a 1000 watt signal right into the restaurant (and outside through every window)!*

"Why did you do this to the microwave?" asked the technicians. "Oh, we cut 5 seconds off our order time by removing the doors!" replied the manager, apparently unaware of the hazard. The techs carefully explained the danger (and also the law, as the device was not an authorized RF radiator). The offending microwave was scrapped, taking the restaurant off the air.

SUMMARY

- In AM, the shape (envelope) of the carrier wave is a replica of the information.
- The process of amplitude modulation requires a nonlinear stage.
- The AM modulation index is a measure of how strongly the carrier is being changed by the information. The maximum value is 1.
- Percentage of modulation is the same information as modulation index.
- Modulation index of an AM signal can be easily read on an oscilloscope.
- Sidebands or side frequencies are new frequencies generated during the process of modulation.
- The sidebands of an AM signal are a modified form of the information signal.
- The bandwidth of an AM signal is controlled by the frequency of the information.
- The total power of an AM signal depends on the carrier power and modulation index. The power at 100% modulation is always 3/2 the unmodulated carrier power.

- Efficiency measures the percentage of energy actually conveying information.
- A spectrum analyzer gives a great deal more information about a transmitter's performance that is lacking in an oscilloscope presentation.
- Great care must be exercised in connecting a spectrum analyzer into a circuit.
- Spectrum analyzers are very handy for evaluating the quality of signals, especially where interference is suspected.

PROBLEMS

1. Draw a block diagram illustrating how an AM signal is generated.
2. What property of a modulator circuit causes AM to be generated?
3. What signals are applied to the two inputs of a modulator?
4. What is the *envelope* of an AM signal?
5. Define the *modulation index* of an AM signal. What is the maximum legal value?
6. What is the percentage of modulation for an AM signal with a modulation index of 0.3?
7. List two consequences of *overmodulation*.
8. For the following AM signals, calculate the modulation index and percentage of modulation:
   a. $V_{max} = 300$ V, $V_{min} = 100$ V
   b. $V_{max} = 150$ V, $V_{min} = 0$ V
   c. $V_{max} = 75$ mV, $V_{min} = 50$ mV
   d. $V_{max} = 5$ V, $V_{min} = 5$ V
9. What is the percentage of modulation of a dead-keyed transmitter?
10. What is the name given to the new frequencies generated during the process of modulation?
11. How many sidebands are generated by each information frequency?
12. An AM signal is being produced with the following characteristics: $f_c = 560$ kHz, $V_c = 1000$ V; $f_m = 3$ kHz, $V_m = 300$ V. Calculate the following:
    a. f_{usb}
    b. f_{lsb}
    c. m
    d. Percent modulation
    e. V_{usb}
    f. V_{lsb}
    g. Bandwidth
13. Draw a spectrogram of the AM signal of problem 12, showing all frequencies and voltages.
14. An AM transmitter is operating on a carrier frequency of 9260 kHz with a carrier voltage of 20 V. For each of the information voltages and frequencies below, calculate f_{usb}, f_{lsb}, m, percent modulation, V_{usb}, V_{lsb}, and bandwidth.
    a. $V_m = 10$ V, $f_m = 2$ kHz
    b. $V_m = 20$ V, $f_m = 4$ kHz
    c. $V_m = 5$ V, $f_m = 1.5$ kHz
15. Draw a spectrogram for each of the AM signals of problem 14, showing all frequencies and voltages.
16. What characteristic of an information signal determines the bandwidth of an AM transmitter?
17. In the United States, the FCC (Federal Communications Commission) regulates radio communications. In the FCC band plan for the AM broadcast band, each broadcaster is allowed to use 8 kHz of bandwidth in a 10 kHz wide "slot." What is the maximum information frequency that can modulate a broadcaster's transmitter?
18. What is the information frequency f_m and percentage of modulation for the AM signal in Figure 3–23?
19. What is the bandwidth of the signal pictured in Figure 3–23?
20. The signal of Figure 3–23 is being sent into a 50 Ω antenna system. Calculate:
    a. P_{lsb}
    b. P_c
    c. P_{usb}
    d. P_t
21. The AM transmitter of problem 14 is feeding a 75 Ω antenna. For each case, calculate P_{lsb}, P_c, P_{usb}, and P_t.

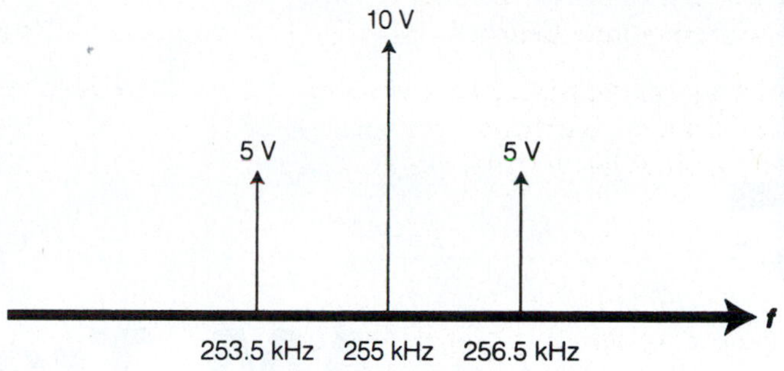

Figure 3–23 An AM signal for analysis

22. What is the ratio of P_t to P_c at 100% modulation in an AM transmitter?
23. A certain AM transmitter is operating on 760 kHz, and when unmodulated, delivers 5000 W to the antenna circuit. What will the *total power* be when it is
    a. 25% modulated
    b. 50% modulated
    c. 75% modulated
    d. 100% modulated
24. Define the *efficiency* of a communication system.
25. Calculate the efficiency of the AM transmitter of problem 23 for each of the four conditions given.
26. Why does a radio transmitting antenna appear as a resistance in a circuit?
27. What is the maximum efficiency of an AM transmitter?
28. What are the three main types of spectrum analyzers?
29. What type of spectrum analyzers are subject to frequency drift?
30. What section of a spectrum analyzer is most susceptible to damage from misconnection?
31. What is a *typical* maximum safe input power rating for a spectrum analyzer?
32. What is an *attenuator?* Why is it needed when measuring transmitters with a spectrum analyzer?

4

AM Transmitters

OBJECTIVES

At the conclusion of this chapter, the reader will be able to:

- draw a block diagram of a high- or low-level AM transmitter, giving typical signals at each point in the circuit
- discuss the relative advantages and disadvantages of high- and low-level AM transmitters
- identify an RF oscillator configuration and the components that control its frequency
- describe the physical construction of a quartz crystal
- calculate the series and parallel resonant frequencies of a quartz crystal, given manufacturer's data
- identify the resonance modes of a quartz crystal in typical RF oscillator circuits
- describe the operating characteristics of an RF amplifier circuit, given its schematic diagram
- explain the operation of modulator circuits
- identify the functional blocks (amplifiers, oscillators, etc.) in a schematic diagram
- list measurement procedures used with AM transmitters
- develop a plan for troubleshooting a transmitter

In Chapter 3, we studied the theory of amplitude modulation, but we never actually built an AM transmitter. To construct a working transmitter (or receiver), a knowledge of RF circuit principles is necessary. A complete transmitter consists of many different *stages* and hundreds of electronic components.

When beginning technicians see the schematic diagram of a "real" electronic system for the first time, they're overwhelmed. A schematic contains much valuable information, but to the novice, it's a swirling mass of resistors, capacitors, coils, transistors, and IC chips, all connected in a massive web of wires! *How can anyone understand this?*

All electronic systems, no matter how complex, are built from functional *blocks* or *stages*. A *block diagram* shows how the pieces are connected to work together. *To understand an electronic system, a technician first reads the block diagram.*

After studying a block diagram, a tech has a good idea of how an electronic device works. However, a block diagram usually doesn't have enough information for in-depth troubleshooting and analysis. For detailed work, a schematic diagram is a must.

There's no magic in electronics. Engineers design systems by using combinations of basic circuits. In RF electronics, there are only four fundamental types of circuits: *amplifiers, oscillators, mixers,* and *switches.* Once a technician learns to recognize these circuits, he or she can begin to rapidly and accurately interpret the information on schematic diagrams.

A final note: The RF circuit techniques described in this chapter are used in *receivers* as well. Understanding them will be very helpful when studying receivers.

4–1 LOW- AND HIGH-LEVEL TRANSMITTERS

There are two approaches to generating an AM signal. These are known as *low-* and *high-level* modulation. They're easy to identify: A low-level AM transmitter performs the process of modulation near the *beginning* of the transmitter; a high-level transmitter performs the modulation step *last,* at the last or "final" amplifier stage in the transmitter. Each method has advantages and disadvantages, and both are in common use.

Low-Level AM Transmitter

Figure 4–1 shows the block diagram of a low-level AM transmitter. It's very similar to the AM transmitter we studied in Chapter 1.

There are two signal paths in the transmitter, AF and RF. The RF signal is created in the *RF carrier oscillator.* At test point A the oscillator's output signal is present. The output of the carrier oscillator is a fairly small ac voltage, perhaps 200 to 400 mV RMS.

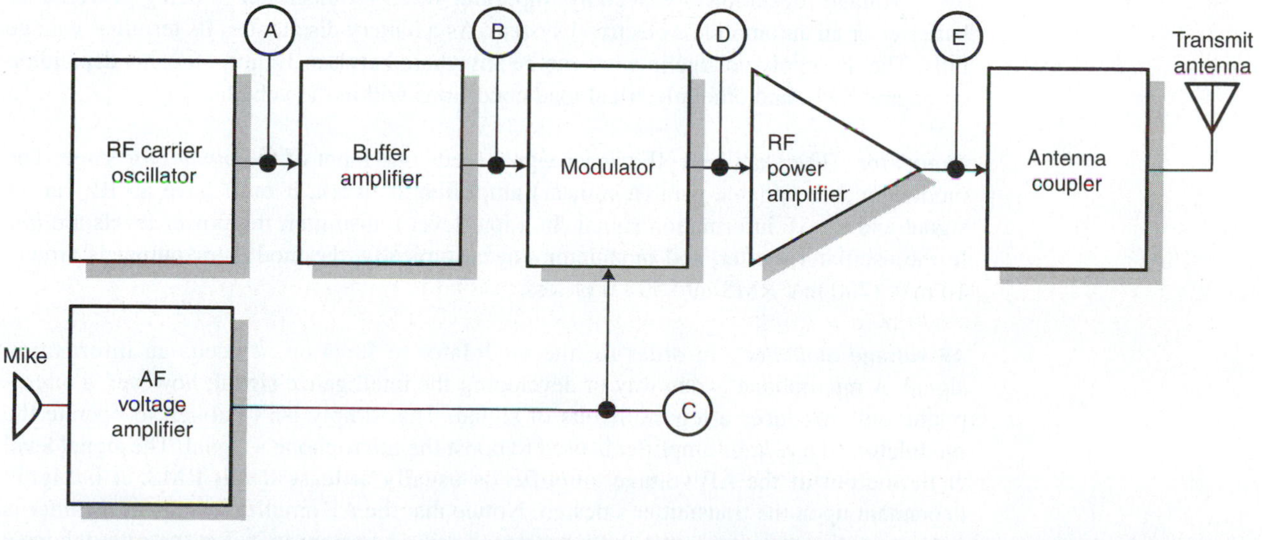

Figure 4–1 Low-level AM transmitter block diagram

The oscillator is a critical stage in any transmitter. It must produce an accurate and steady frequency. You might recall that every radio station is assigned a different carrier frequency. The dial (or display) of a receiver displays the carrier frequency. If the oscillator drifts off frequency, the receiver will be unable to receive the transmitted signal without being readjusted. Worse yet, if the oscillator drifts onto the frequency being used by *another* radio station, interference will occur. This is hardly desirable!

Two circuit techniques are commonly used to stabilize the oscillator, *buffering* and *voltage regulation.*

Buffer amplifier You might have guessed that the *buffer amplifier* has something to do with buffering or protecting the oscillator. It does! An oscillator is a little like an engine (with the speed of the engine being similar to the oscillator's frequency). If the load on the engine is increased (the engine is asked to do more work), the engine will respond by slowing down. An oscillator acts in a very similar fashion. If the *current* drawn from the oscillator's output is increased or decreased, the oscillator may speed up or slow down slightly. We would say that its frequency has been *pulled.*

The *buffer amplifier* is a relatively low-gain amplifier that follows the oscillator. The buffer amplifier has a constant *input impedance* (resistance). Therefore, the buffer amplifier always draws the same amount of current from the oscillator. This helps to prevent "pulling" of the oscillator frequency.

The buffer amplifier is needed because of what's happening "downstream" of the oscillator. Right after the buffer amplifier is the *modulator.* Because the modulator is a nonlinear amplifier, it may not have a constant input resistance—especially when information is passing into it. But since there is a buffer amplifier between the oscillator and modulator, the oscillator sees a steady load resistance, regardless of what the modulator stage is doing.

Voltage regulation An oscillator can also be pulled off frequency if its power supply voltage isn't held constant. In most transmitters, the supply voltage to the oscillator is *regulated* at a constant value. The regulated voltage value is often between 5 and 9 volts; zener diodes and three-terminal regulator ICs are commonly used voltage regulators.

Voltage regulation is especially important when a transmitter is being powered by batteries or an automobile's electrical system. As a battery discharges, its terminal voltage falls. The dc supply voltage in a car can be anywhere between 12 and 16 volts, depending on engine RPM and other electrical load conditions within the vehicle.

Modulator The stabilized RF carrier signal feeds one input of the *modulator* stage. The modulator is a variable-gain (nonlinear) amplifier. To work, it must have an RF carrier signal and an AF information signal. In a low-level transmitter, the power levels are *low* in the oscillator, buffer, and modulator stages; typically, the modulator output is around 10 mW (700 mV RMS into 50 Ω) or less.

AF voltage amplifier In order for the modulator to function, it needs an information signal. A microphone is one way of developing the intelligence signal; however, a microphone only produces a few millivolts of signal. This simply isn't enough to operate the modulator, so a *voltage* amplifier is used to boost the microphone's signal. The signal level at the output of the AF voltage amplifier is usually at least 1 volt RMS; it is highly dependent upon the transmitter's design. Notice that the AF amplifier in the transmitter is only providing a *voltage* gain, and not necessarily a *current* gain for the microphone's signal. The power levels are quite small at the output of this amplifier—a few mW at best.

RF power amplifier At test point D, the modulator has created an AM signal by impressing the information signal from test point C onto the stabilized carrier signal from test point B at the buffer amplifier output. This signal (test point D) is a complete AM signal but has only a few milliwatts of power.

The RF power amplifier is normally built with several stages. These stages increase both the *voltage* and *current* of the AM signal. We say that *power amplification* occurs when a circuit provides a current gain.

In order to accurately amplify the tiny AM signal from the modulator, the RF power amplifier stages must be *linear*. You might recall that amplifiers are divided up into "classes" according to the *conduction angle* of the active device within. Class A and class B amplifiers are considered linear amplifiers, so the RF power amplifier stages will normally be constructed using one or both of these types of amplifier. Therefore, the signal at test point E looks just like that of test point D—it's just much bigger in voltage and current.

Antenna coupler The antenna coupler is usually part of the last or *final* RF power amplifier, and as such, is not really a separate active stage. It performs no amplification and has no active devices! It performs two important jobs: impedance matching and filtering.

For an RF power amplifier to function correctly, it must be supplied with a load resistance equal to that for which it was designed. This may be nearly any value. Ohms would be an optimal value, since most antennas and transmission lines are 50 Ω. What if the RF power amplifier needs to see 25 Ω? Then we must somehow *transform* the antenna impedance from 50 Ω down to 25 Ω. Are you thinking *transformer?* If so, great—because that's one way of doing the job. A transformer can step an impedance up (higher voltage) or down (lower voltage). Special transformers are used at radio frequencies. Transformers aren't the only circuits used for impedance matching. LC resonant circuits can also be used in many different forms to do the job.

There's nothing mysterious about impedance matching. The antenna coupler does the same thing for the RF final power amplifier that the gears in a car's transmission do for the engine. To climb a steep hill, a lower gear must be chosen in order to get maximum mechanical power transfer from the engine to the wheels. Too high a gear will stall the motor—think of it as a mechanical impedance mismatch! The engine speed is *stepped down* to help the car climb the hill.

The antenna coupler also acts as a *low-pass filter*. This filtering reduces the amplitude of *harmonic energies* that may be present in the power amplifier's output. (All amplifiers generate *harmonic distortion,* even "linear" ones.) For example, the transmitter may be tuned to operate on 1000 kHz. Because of small nonlinearities in the amplifiers of the transmitter, the transmitter will also produce *harmonic energies* on 2000 kHz (2nd harmonic), 3000 kHz (3rd harmonic), and so on. Because a low-pass filter passes the fundamental frequency (1000 kHz) and rejects the harmonics, we say that *harmonic attenuation* has taken place. (The word *attenuate* means "to weaken.")

High-Level AM Transmitter

The high-level transmitter of Figure 4–2 is very similar to the low-level unit. The RF section begins just like the low-level transmitter; there is an oscillator and buffer amplifier. The difference in the high-level transmitter is *where* the modulation takes place. Instead of adding modulation immediately after buffering, this type of transmitter amplifies the *unmodulated* RF carrier signal first. Thus, the signals at points A, B, and D in Figure 4–2 all

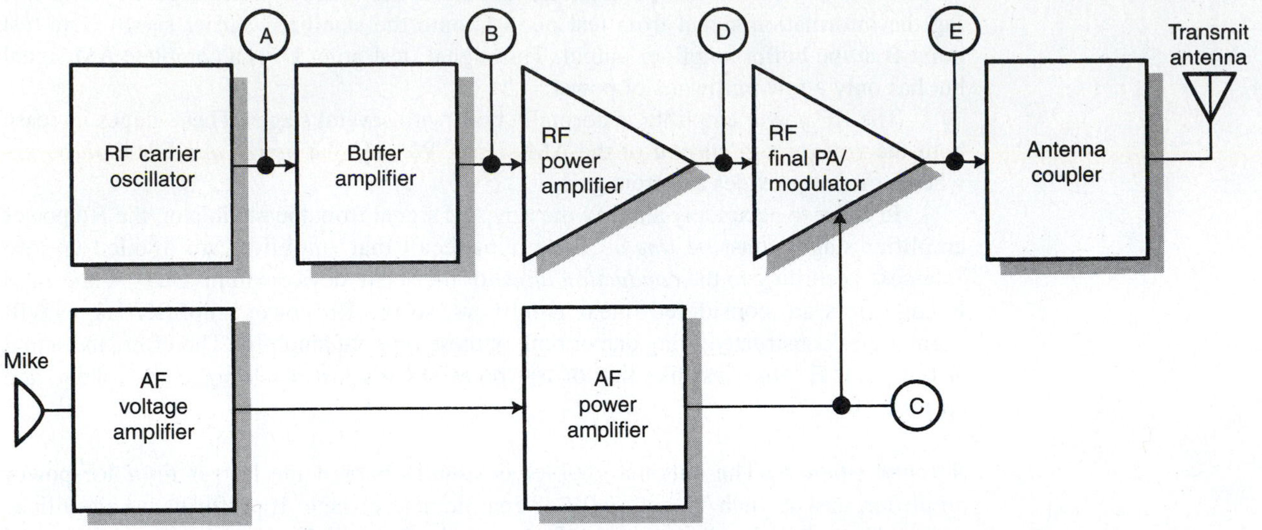

Figure 4–2 A high-level AM transmitter

look like unmodulated RF carrier waves. The only difference is that they become bigger in voltage and current as they approach test point D.

The modulation process in a high-level transmitter takes place in the last or *final* power amplifier. Because of this, an additional audio amplifier section is needed. In order to modulate an amplifier running at power levels of several watts (or more), comparable power levels of information are required. Thus, an *audio power amplifier* is required.

The final power amplifier does double duty in a high-level transmitter. First, it provides power gain for the RF carrier signal, just like the RF power amplifier did in the low-level transmitter. In addition to providing power gain, the final PA also performs the task of *modulation.* If you've guessed that the RF power amplifier operates in a *nonlinear class,* you're right! Classes A and B are considered linear amplifier classes. *The final power amplifier in a high-level transmitter usually operates in class C, which is a highly nonlinear amplifier class.*

Figure 4–3 shows the relative location of the quiescent operating point ("Q point") for several different classes of amplifier. Note that as we move away from class A operation, efficiency increases but distortion (caused by nonlinearity) also increases!

Low- and High-Level Transmitter Efficiency

You might wonder why two different approaches are used to build AM transmitters, when the results of both methods are essentially the same (a modulated AM carrier wave is sent to the antenna circuit).

The answer to this question lies in examining the relative cost, flexibility, and dc efficiency of both approaches. The *dc efficiency* of a transmitter can be defined as follows:

$$\eta = \frac{P_{\text{out-RF}}}{P_{\text{in-dc}}} \tag{4–1}$$

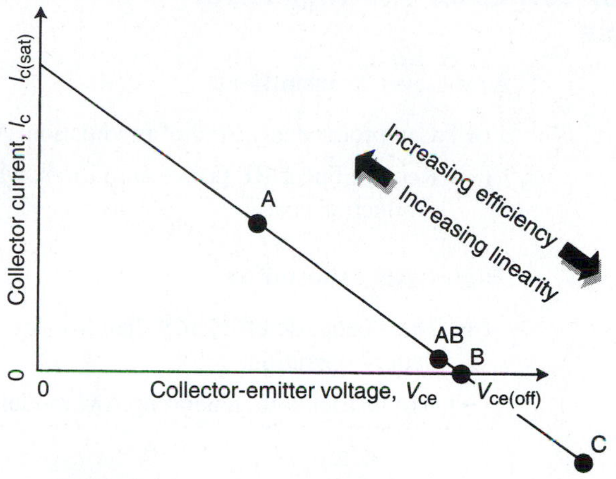

Figure 4–3 The Q point of various amplifier classes

For example, suppose that a certain transmitter requires 36 W of power from its dc power supply and produces 18W of RF at the antenna connector. The efficiency of the transmitter will be

$$\eta = \frac{P_{\text{out-RF}}}{P_{\text{in-dc}}} = \frac{18\text{ W}}{36\text{ W}} = \underline{\underline{50\%}}$$

This transmitter converts 50% of the battery power to useful RF energy at the antenna, and 50% is converted to *heat* (and lost).

Naturally, we'd like all of our electronic devices to be as efficient as possible, especially in certain cases. Suppose that a transmitter is operated from battery power—as in a walkie-talkie or aircraft ELT (emergency locator transmitter). We would want to get maximum life from the batteries, and we would use the most efficient approach possible. Broadcasting uses tremendous amounts of electricity due to the high power levels. It makes good economic sense to use the most efficient transmitter layout available.

Overall, the high-level transmitter sports better dc efficiency than the low-level approach and is normally the first choice in battery-operated AM transmitters and commercial AM broadcast. This is because the high-level transmitter is able to use class C RF power amplifiers, which are more efficient than the class A or B RF amplifiers required for a low-level transmitter. A high-level transmitter still requires a linear power amplifier, but it is an *audio frequency (AF)* type. It is much easier to build efficient linear amplifiers for audio than it is for RF, so *the high-level approach wins in efficiency contests.*

If efficiency is so important, then why use a low-level approach at all, since it uses "wasteful" linear RF power amplification techniques? This is a very good question. The high-level approach performs its modulation at the very last stage. At such high power levels, the only practical method of modulation is AM—in other words, *it's just about impossible to achieve FM or PM in a high-level transmitter. The high-level transmitter can only produce AM.*

A low-level transmitter can generate any type of modulation; all that must be done is to switch *modulator* circuits. Since the power amplifiers are of linear type in a low-level transmitter, they can amplify AM, FM, or PM signals. *The low-level method is very flexible: when a transmitter must produce several different types of modulation, this is the method that is generally used.*

A Summary of Low-Level and High-Level Characteristics

Low-Level Transmitters

(+) Can produce any kind of modulation: AM, FM, or PM.

(−) Require linear RF power amplifiers, which reduces dc efficiency and increases production costs.

High-Level Transmitters

(+) Have better dc efficiency than low-level transmitters and are very well suited for battery operation.

(−) Are restricted to generating AM modulation only.

EXAMPLE 4–1

Calculate the dc efficiency of an AM transmitter with the following ratings: $P_{out} = 4$ watts into 50 Ω load while drawing 1 amp from a 12 volt supply.

Solution

We need equation (4–1) and Ohm's law. The input power is given indirectly in the specifications, since $P = VI = (12\text{ V})(1\text{ A}) = 12$ *watts*. With this information in hand, we can calculate efficiency:

$$\eta = \frac{P_{out-RF}}{P_{in-dc}} = \frac{4\text{ W}}{12\text{ W}} = \underline{\underline{33.3\%}}$$

Section Checkpoint

4–1 What are the two main types of AM transmitters?

4–2 How can a low-level transmitter be identified?

4–3 What signal appears at test point C in Figure 4–1?

4–4 Why do transmitters use a *buffer amplifier?*

4–5 What is done with the power supply to oscillators in radio transmitters, and why?

4–6 The power amplifiers in a low-level transmitter will be in class ______ or class ______.

4–7 List two functions of an *antenna coupler*.

4–8 What are the advantages of a high-level transmitter?

4–9 The final power amplifier in a high-level transmitter operates in class ______.

4–2 OSCILLATOR THEORY

Oscillators are a key ingredient in both radio transmitters and receivers. An *oscillator* is a circuit that converts dc power supply energy into an ac output signal. Since the frequency of the oscillator in a transmitter determines the carrier frequency, it is important that the frequency produced be very stable and steady.

How Oscillators Work

Everyone has heard a public address system howl and whistle when the performer's microphone has been placed too close to a speaker. The microphone picks up a bit of the sound from the loudspeaker, which is again amplified by the PA system. The sound reenters the microphone again, resulting in *oscillation* as the sound makes repeated trips through the loop. We would say that *positive feedback* has occurred. This is an example of an undesired oscillation. It illustrates that for an oscillator to work, two things are needed, *power gain* and *positive feedback.*

Any linear oscillator can be broken into two parts, the *gain* and *feedback* blocks, as shown in Figure 4–4. An electronic oscillator has a carefully controlled gain and feedback. In Figure 4–4, a small sine wave signal is entering the amplifier. At the output of the amplifier, the sine wave has increased in size. Some of this is used as the output signal. The rest enters the feedback block, where it is reduced in size in preparation for another trip around the circuit.

A little arithmetic can reveal some very interesting things about how the oscillator of Figure 4–4 will behave. For example, suppose that the feedback signal (at the input of the amplifier) is 100 mV and that the amplifier voltage gain A is 10. What will the output voltage be?

Did you calculate about 1 volt? That's right—$V_{out} = (V_{in})(A) = (100\text{ mV})(10\text{ V/V}) = 1\text{ V}$. This 1 volt signal must then pass back into the feedback portion. Suppose now that the feedback block has a gain of 1/10 (0.1). What is the resulting output *feedback signal* voltage?

That's strange—the feedback network has reduced the signal back to 100 mV. We've recreated a signal just like the one that originally entered the amplifier! This new signal

Figure 4–4 The block diagram of an oscillator

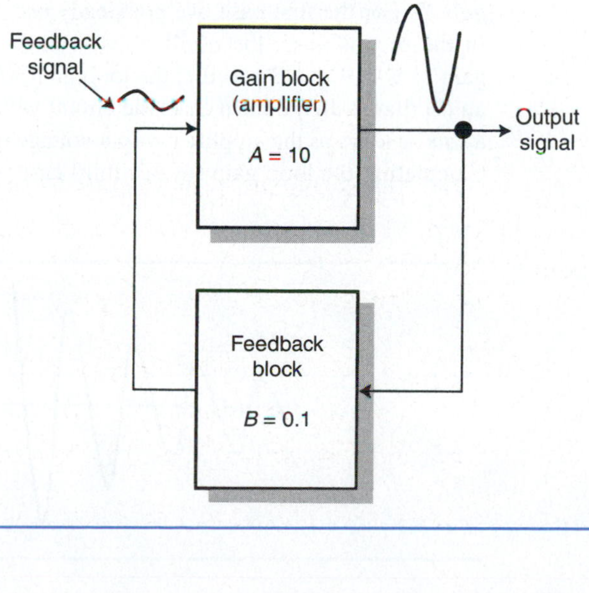

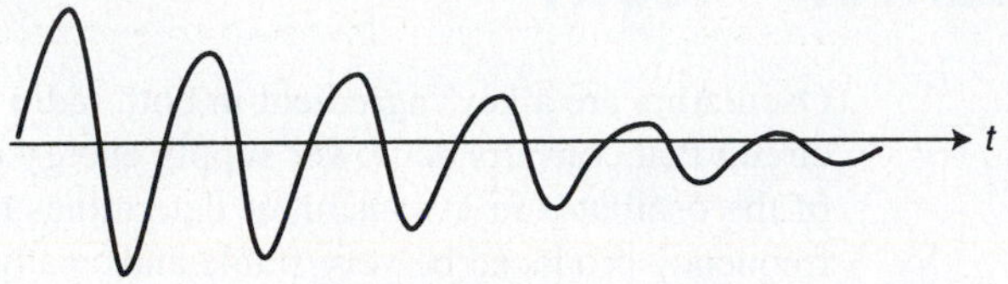

Figure 4–5 The output with gain reduced

will enter the amplifier, be amplified again to 1 volt, and be again reduced to 100 mV for the next trip around the block. The action will repeat again and again. *The circuit produces steady oscillations!*

The conditions for getting steady oscillations are tricky to maintain. Let's again put 100 mV into the amplifier but reduce its voltage gain to 9 (it was originally 10). The output voltage now becomes 900 mV—hmm, smaller than before! The 900 mV signal now flows through the feedback network, with its gain of 1/10. The resulting feedback signal is now . . . 90 mV. Something seems odd—*the feedback signal has shrunk from 100 mV to 90 mV.* Do you see what is happening? Follow the signal around the loop another time. The 90 mV "new" feedback signal is amplified times 9 again at the amplifier, giving an output signal of 810 mV (hmm, it was bigger last time around!) and after passing through the feedback loop again, becomes 81 mV. *The oscillations are dying out because there isn't enough gain around the loop to sustain them!* The output signal will look like Figure 4–5.

Let's try a third case. Let's adjust the amplifier gain to 11 and again insert a 100 mV signal into its input. The amplifier output becomes 11 times 100 mV, or 1.1 V (bigger than before). The 1.1 volt signal passes into the feedback block, where it is again multiplied by 1/10, giving a feedback signal of (1.1 V/10), or 110 mV (at least we're not dying out here!). Can you see what this case will do? The signals will grow . . . and grow . . . until something *limits* their growth. (No, you don't need to run away from this circuit!) That *something* is the power supply voltage. The amplifier can only produce an output voltage less than or equal to the power supply voltage; therefore, the output signal of the circuit will now look like Figure 4–6.

The *loop gain* of an oscillator is the product of the *amplifier gain A* and the *feedback gain B*. For the first case, we got steady oscillation. The loop gain *AB* was (10)(0.1) or 1. In the second case, the oscillations soon died away because there was insufficient loop gain. *A* was 9 and *B* was 0.1; the loop gain *AB* was (9)(0.1) or 0.9. This isn't promising as an oscillator! In the third case, the circuit went bonkers. In fact, the output rapidly became a *square wave* as the amplifier output voltage reached the limits of the power supply voltage. Calculating the loop gain for this third case, we get $AB = (11)(0.1) = 1.1$. Although this

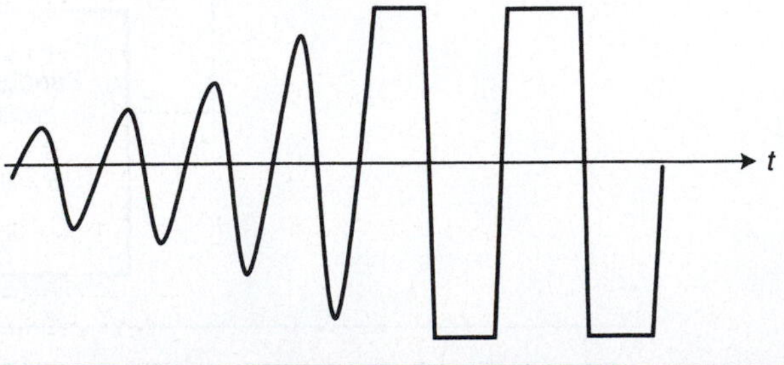

Figure 4–6 The loop gain is too large!

circuit continues oscillating, the waveform is distorted. A square wave isn't useful as a carrier signal, since it contains more than one frequency (remember harmonics?).

These observations lead to what is known as the *Barkhausen criteria for oscillators.* Don't worry too much about number crunching these. They are presented only to give a better idea about what makes oscillators work. A technician troubleshooting a "dead" oscillator needs to know what to look for. The Barkhausen criteria give the technician a way of understanding what features in a circuit cause it to oscillate. The criteria are as follows:

1. If the loop gain $AB < 1$, the circuit will *not* oscillate.
2. If the loop gain $AB = 1$, the circuit will oscillate if it is doing so already. (If it is not oscillating, it will *not* start doing so with $AB = 1$.)
3. For an oscillator to *start,* the loop gain *AB must be greater than 1*. ($AB > 1$ ensures oscillator starting.)

The designers of oscillators have quite a problem on their hands, because a successful oscillator must start reliably ($AB > 1$), but must also produce nice clean sine waves ($AB = 1$). These are seemingly contradictory conditions. Most RF oscillators rely on the nonlinearity of the active device (usually a transistor) to "roll off the loop gain" once the oscillator has started. When the oscillator is first powered up (and is producing little signal), the loop gain *AB* is designed to be more than 1, which causes the amplitude of the sine wave oscillations to rapidly increase. As the amplitude gets larger, the gain of the active device folds back, reducing the loop gain *AB* to unity (1), thus providing a stable output amplitude.

Positive and Negative Feedback

Not only must an oscillator have sufficient loop gain to run, it must also receive *positive feedback*. Positive feedback means that the signal from the feedback block is *in phase* with the original input signal to the amplifier.

For this to be true, the sum of all the phase shifts around the loop must be 0°, 360°, or some exact multiple of 360°. Figure 4–7 illustrates this point. In Figure 4–7, the amplifier

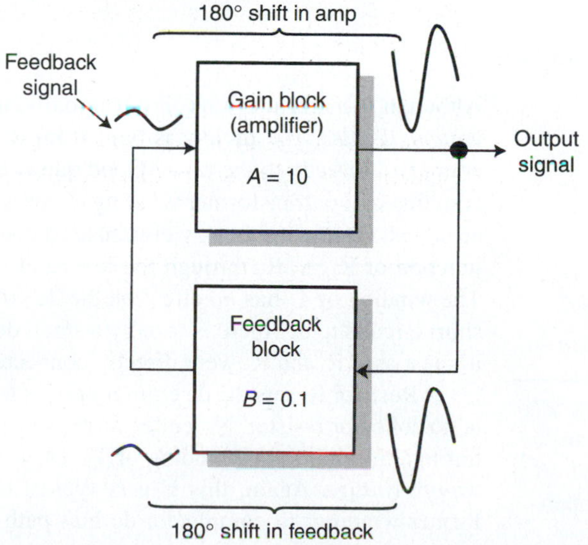

Figure 4–7 An oscillator with included phase shifts

provides a voltage gain of 10, but also inverts the signal by 180°. *If the signal were fed back in this form, it would cancel at the amplifier's input, causing the oscillation to stop!*

The feedback network in Figure 4–7 corrects the "problem" by inserting its own 180° phase shift. Since $(180 + 180) = 360$, the feedback signal at the input of the amplifier remains *in phase* and the circuit receives *positive feedback,* which allows it to oscillate.

When we look at oscillator circuits, portions of the circuit can contribute phase shift. For example, a common-emitter transistor amplifier produces a 180° phase shift. Therefore, if this type of amplifier is being used in an oscillator, the feedback network must also produce a 180° phase shift to ensure that positive feedback will take place.

Section Checkpoint

4–10 What is the purpose of an oscillator circuit?

4–11 An oscillator can be divided into two parts. What are these parts?

4–12 What is meant by the term *loop gain* when referring to oscillators?

4–13 For an oscillator to continue running, what must the loop gain *AB* be equal to?

4–14 What loop gain *AB* is required for an oscillator to start?

4–15 For an oscillator to run, what type of feedback is needed?

4–16 What loop phase shifts provide positive feedback?

4–3 THREE OSCILLATORS

There are three basic oscillator configurations used in transmitters. These are the *Armstrong, Hartley,* and *Colpitts* circuits. These circuits are named after their inventors. Many other circuits are derived from these three. Let's take a look at the *Armstrong* circuit first.

dc Circuit Analysis

Although it might not look like it initially, transistor Q_1 in Figure 4–8 is biased using a *voltage divider*. The divider is built from R_1 and R_2. Why isn't the junction of R_1 and R_2 connected directly to the base? Good question. A common trick in RF circuitry is to use a coupling coil or transformer winding to "pass" the dc bias voltage (along with the intended ac signal). Doing it this way eliminates a coupling capacitor. The voltage travels from the junction of R_1 and R_2 through the left-hand winding of T_1 and down into the base of Q_1. The winding of T_1 has *no* effect on the dc voltage for all practical purposes. Inductors are short circuits to dc (there is usually a small dc resistance due to the wire itself). Therefore, it's just as if R_1 and R_2 were directly connected to the base.

TIP When doing a dc circuit analysis, don't forget to short inductors, and open capacitors!

Resistor R_3 sets the dc emitter current for Q_1. You may be disturbed to learn that there is no collector resistor. No collector resistor is needed! The collector of Q_1 gets its dc bias through the right-hand winding of T_1. *The collector dc voltage will be equal to the power supply voltage.* Again, this is *very* typical of RF circuitry. Using an inductor (or transformer winding) to complete a dc bias path eliminates a coupling capacitor, and in this case, also eliminates a resistor.

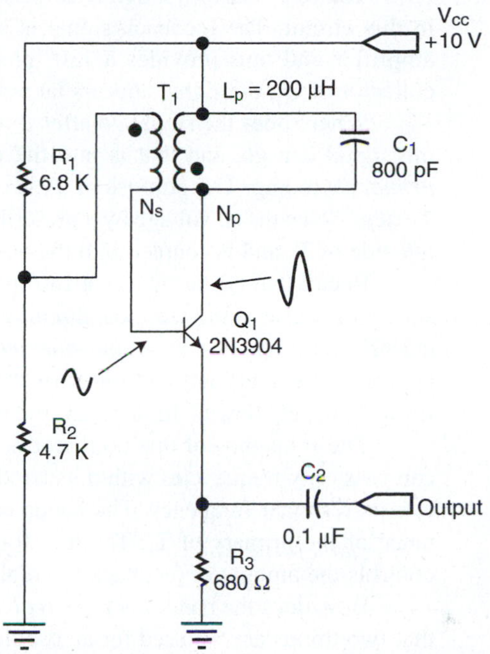

Figure 4–8 An Armstrong oscillator

EXAMPLE 4–2

What should the voltages V_B, V_E, and V_C read in the circuit of Figure 4–8?

Solution

The circuit of Figure 4–8 is voltage divider biased, so again, Ohm's law (and a little transistor theory) comes to the rescue:

$$V_B = V_{CC}\left(\frac{R_2}{R_1 + R_2}\right) = 10\ \text{V}\left(\frac{4.7\ \text{K}}{4.7\ \text{K} + 6.8\ \text{K}}\right) = \underline{\underline{4.09\ \text{V}}}$$

Since this is an *NPN* transistor, the emitter voltage is lower (more negative) than the base voltage:

$$V_E = V_B - V_{BE} = 4.09\ \text{V} - 0.7\ \text{V} = \underline{\underline{3.39\ \text{V}}}$$

Finally, the collector is "shorted" to the power supply rail by the primary of T_1, so we get

$$V_C = V_{CC} = \underline{\underline{10\ \text{V}}}$$

ac Circuit Analysis

In performing an ac circuit analysis of an oscillator, a technician normally looks for three things:

- What provides the gain?
- Where is the feedback?
- What controls the frequency of the oscillator?

You're probably not overly surprised that the transistor Q_1 provides the voltage gain in this circuit. The feedback signal is sent into the base. Q_1 acts as a *common-emitter* amplifier and thus provides a 180° phase shift on the signal as it amplifies it. At the collector of Q_1, the signal appears larger but also *out of phase*.

Where does the signal go after it leaves the collector of Q_1? There's only one place the signal can go, and that is into the right-hand side of T_1, the oscillator transformer *primary* winding. The purpose of T_1 is to control the amount of feedback. In this circuit, T_1 *steps down* the ac voltage by a prescribed amount. The feedback voltage emerges on the *left* side of T_1 and is coupled into the base of Q_1, completing the feedback path.

Recall that Q_1 introduces a 180° phase shift into the signal. Unless that is corrected somewhere else in the loop, the circuit will not oscillate! In a transformer, it's easy to get a 180° phase shift by *reversing one set of winding connections*. The *phasing dots* that appear at the upper left and lower right of T_1 specify that *one* winding's connections are flipped "upside down." In this way, the proper loop phase shift is obtained.

TIP Most oscillators have a bandpass filter that controls oscillation frequency, and it is commonly an LC circuit (or equivalent, like a quartz crystal, as we'll soon see).

The frequency of this oscillator is controlled by a bandpass filter. A bandpass filter can pass only frequencies within its bandwidth; it rejects frequencies that are above or below its resonant frequency. The bandpass filter is formed by capacitor C_1 and the inductance of the primary of T_1. The transformer does double duty in this circuit; it not only controls the amount of feedback but it also acts as part of a bandpass filter.

How does the bandpass filter *really* control the frequency of oscillation? Remember that two things are required for an oscillator to run. These are *gain* and *positive feedback*. Anything that removes the positive feedback will stop the oscillator. The bandpass filter formed by C_1 and T_1 is in the feedback path. Therefore, any frequency that cannot get through this filter *can't* be a feedback signal. The circuit can only oscillate at or near the resonant frequency of C_1 and T_1, which is a LC resonant circuit.

EXAMPLE 4–3 What is the frequency of oscillation for the Armstrong oscillator of Figure 4–8?

Solution

To calculate the frequency of oscillation, identify the components in the *bandpass filter*. In this circuit, C_1 and the primary of T_1 form a parallel-resonant circuit. This is the bandpass filter.

On the schematic diagram, the equivalent inductance of the primary winding of T_1 is given as 200 μH, and the value of C_1 is given as 800 pF. (Note that manufacturers usually don't give winding inductances on schematics!) From electronic fundamentals, we know that the frequency of resonance is given by

$$f = \frac{1}{2\pi\sqrt{LC}} = \frac{1}{2\pi\sqrt{(200\ \mu\text{H})(800\ \text{pF})}} = \underline{\underline{397.9\ \text{kHz}}}$$

The Armstrong oscillator is one of the earliest-developed electronic oscillators. *It is usually identified as using a transformer within the feedback loop*. The primary limitation of an Armstrong oscillator is its *maximum operating frequency*. Above about 2 MHz, the Armstrong oscillator tends to become unstable because at high radio frequencies, the transformer begins to operate in a nonideal manner. Other circuits are used when higher frequencies are required. These circuits eliminate the coupling transformer.

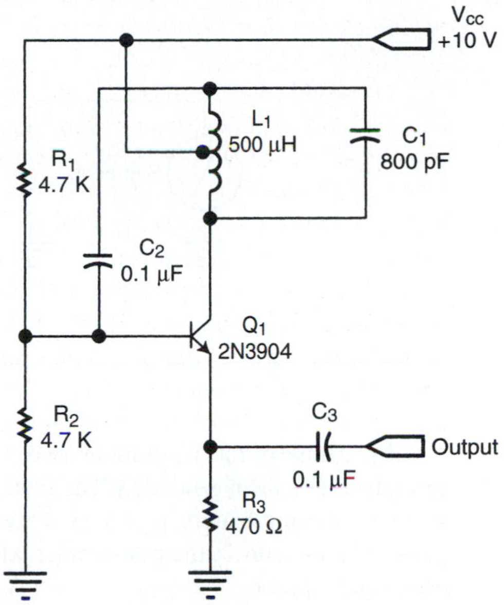

Figure 4–9 The Hartley oscillator

The Hartley Oscillator

Figure 4–9 shows a *Hartley* oscillator. It is simpler than the Armstrong oscillator because the coupling transformer has been eliminated. A Hartley oscillator can be made workable up to 30 MHz or so. A Hartley oscillator is usually identified by a *tapped inductor,* which is L_1 in this case.

dc analysis The dc analysis of the Hartley oscillator is very similar to that of the Armstrong unit. Again, voltage divider biasing is used, with R_1 and R_2 setting the base bias voltage of Q_1. R_3 sets the emitter current, and again, there's no collector resistor. The inductor L_1 "shorts" the collector of Q_1 to the positive (V_{cc}) power supply, so the full power supply voltage (10 V) appears at the collector.

ac analysis The circuit of Figure 4–9 is again using a common-emitter amplifier. ac signal flows into the base, and the transistor causes a 180° phase shift to appear (as well as providing a voltage gain). The enlarged signal exits the collector of the transistor and moves up into the bottom of inductor L_1.

Since a 180° phase shift has occurred in the amplifier, there *must* be another 180° phase shift somewhere in the loop. This one is hidden in L_1's wiring. See the center tap of L_1? It goes to V_{cc}. The power supply of an electronic circuit is normally an ac ground. In ac analysis, dc sources are replaced with a short. Therefore, the center tap of L_1 is at *ground* potential as far as ac is concerned!

This doesn't really answer the phase shift question, though; look at Figure 4–10. The circuit of Figure 4–10 is very similar to that of a full-wave dc power supply with the rectification left out. Note how the signals at top and bottom differ. *Because of the grounded center tap on the transformer, the top and bottom appear 180° out of phase*. The reason for this is straightforward: Suppose that at some instant in time, there is a *total* voltage drop of 10 volts across the secondary of T_1. This means that each half of the winding must have

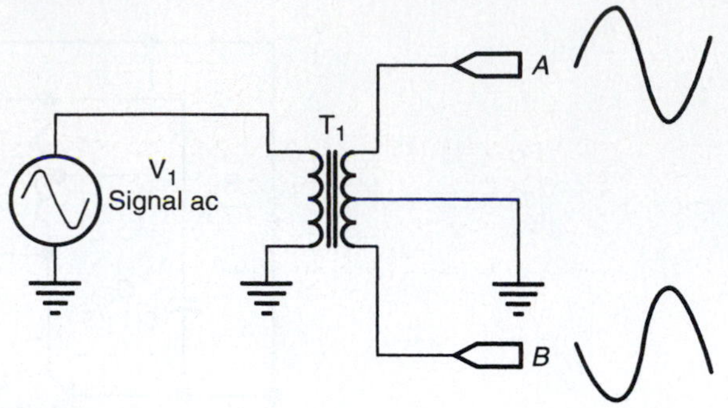

Figure 4–10 The phases of a center-tapped transformer winding

5 volts. Think of the winding as two 5 volt batteries wired in series, with the common point between them grounded. The total voltage is 10 volts, but with respect to the ground (center), the top battery produces +5 volts and the bottom one produces −5 volts. The same effect occurs in the transformer winding, with the result that the two outputs appear 180° out of phase.

Whenever a coil has a center tap at ac ground potential, it is a safe starting assumption to look for a 180° phase shift between its ends. Because of the center tap on L_1, there is a 180° phase shift across its end terminals. The bottom terminal receives the amplifier output from the collector of Q_1; the top terminal provides a 180° phase-shifted signal for feedback.

In Figure 4–10, imagine the effect of moving the transformer tap up or down. When the tap is in the middle, voltages *A* and *B* are equal. As the tap is moved upward, voltage *A* will grow smaller, since there will now be fewer turns on the top, and voltage *B* will get larger. *In a Hartley oscillator, the amount of feedback is controlled by the position of the tap on the coil.*

Capacitor C_2 completes the feedback loop. It is a *coupling* capacitor; it is there to block dc currents between the base and collector circuits of Q_1. You might wonder how a technician can tell that C_2 is merely for coupling; why isn't C_2 part of the frequency-determining part of the circuit? The answer lies in the relative *value* of C_2. At radio frequencies, a 0.1 μF capacitor has a very low reactance (opposition to ac); it therefore appears as a *short* to ac signals.

A bandpass filter again controls the frequency of the oscillator. The LC "tank" formed by L_1 and C_1 is the filter. Only frequencies close to resonance can get through the feedback filter, so the circuit can only oscillate close to the resonant frequency of the tank.

EXAMPLE 4–4 What is the frequency of oscillation for the Hartley oscillator of Figure 4–9? At this frequency, what is the capacitive reactance of capacitor C_2?

Solution

C_1 and L_1 form the bandpass filter, so we'll again use the resonant frequency formula:

$$f = \frac{1}{2\pi\sqrt{LC}} = \frac{1}{2\pi\sqrt{(500\ \mu\text{H})(800\ \text{pF})}} = \underline{\underline{251.6\ \text{kHz}}}$$

To find the reactance of C_2, we use the capacitive reactance formula:

$$X_c = \frac{1}{2\pi f_c} = \frac{1}{2\pi(251.6\text{ kHz})(0.1\,\mu\text{F})} = \underline{\underline{6.3\,\Omega}}$$

The reactance of C_2 is *very* small indeed! Compared with other values in the circuit, 6.3 Ω is practically a *short circuit*. The ac feedback signal passes through C_2 with little opposition on its way to the base of Q_1.

The Colpitts Oscillator

Figure 4–11 illustrates a *Colpitts* oscillator. This type of oscillator can be made to operate at frequencies well into the UHF region; it's one of the most common configurations in UHF and VHF communications gear. A Colpitts oscillator uses a "tapped capacitor" (actually two capacitors in series) to obtain its feedback.

dc analysis The dc operation of the oscillator of Figure 4–11 is very similar to that of the previous two oscillators. R_1 and R_2 set the base voltage, the *series* combination of R_3 and R_4 sets the emitter current, and the collector is (surprise!) tied directly to V_{cc} through inductor L_1.

Figure 4–11 A Colpitts oscillator

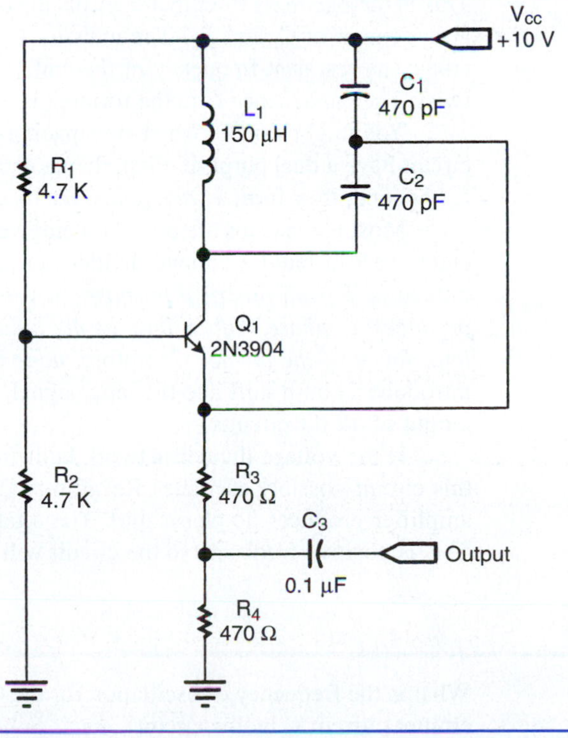

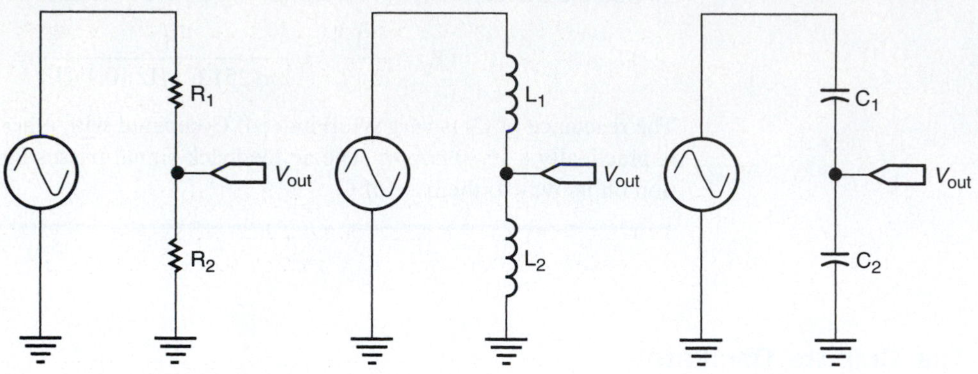

Figure 4–12 Three voltage dividers

ac analysis In the previous two oscillators, the transistor operated as a common-emitter amplifier, with signal going into the base (input) and coming out of the collector (output). This circuit is different; there are *no* ac signals being applied to the base circuit!

If you look carefully, you'll notice that the ac signals for the feedback circuit are in fact in the collector and emitter circuits. *This time, the transistor is operating as a common-base amplifier.* So which transistor lead is the amplifier "input" and "output?" There is a simple rule that applies: *The collector of a transistor is almost* never *an input, except in a few very unusual (esoteric) applications.* Therefore, the amplifier input is the emitter lead, and the output is the collector.

The collector ac output flows into the resonant "tank" formed with L_1, C_1, and C_2. This is the bandpass filter that controls the oscillator frequency. Did you notice the special arrangement of C_1 and C_2? The inductor L_1 "sees" these two capacitors in *series.* To calculate the resonant frequency of the tank, we use the resonant frequency formula, with a twist: The capacitance C in the formula is the *series* equivalent of C_1 and C_2.

You might wonder why two capacitors are used in this circuit. The capacitors in this circuit have a dual purpose. First, they are part of the resonant bandpass filter formed with L_1. Second, they form a *voltage divider* that controls the amount of feedback.

Most technicians are used to seeing voltage dividers made from resistors. For ac circuits, we can build a voltage divider out of any combination of resistors, capacitors, or inductors. *For all practical purposes, a voltage divider made out of two like components produces no phase shift as long as the load resistance is much higher than the Thévenin impedance of the divider.* Therefore, *none* of the voltage divider circuits in Figure 4–12 introduce a phase shift into the input signal. There is a 0° difference between the input and output of *all* the circuits.

If the voltage divider network built from C_1 and C_2 makes no phase shift, how can this circuit possibly oscillate? Recall that Q_1 is wired as a *common-base* amplifier. A CB amplifier produces no phase shift. The total phase shift walking around the loop is zero. This is positive feedback, so the circuit will oscillate!

EXAMPLE 4–5

What is the frequency of oscillation for the Colpitts oscillator of Figure 4–11? What is the emitter current I_E in this circuit?

Solution

L_1, C_1, and C_2 form the bandpass filter. Before we can apply the resonant frequency formula, we must first find the equivalent series capacitance of C_1 and C_2:

$$C = \frac{1}{1/C_1 + 1/C_2} = \frac{C_1C_2}{C_1 + C_2} = \frac{(470\text{ pF})(470\text{ pF})}{(470\text{ pF} + 470\text{ pF})} = \underline{235\text{ pF}}$$

(Remember, series capacitance works using the same formulas as parallel resistance. The "product over sum" formula is most convenient when there are only two capacitances.)

Now that the equivalent capacitance is known, we can directly find the resonant frequency:

$$f = \frac{1}{2\pi\sqrt{LC}} = \frac{1}{2\pi\sqrt{(150\ \mu\text{H})(235\text{ pF})}} = \underline{\underline{847.7\text{ kHz}}}$$

To find the emitter current, we must revert to dc analysis techniques:

$$I_E = \frac{V_E}{R_E}$$

R_E is the total dc emitter resistance, which is the sum of R_3 and R_4. Therefore, $R_E = 470\ \Omega + 470\ \Omega$, or $R_E = 940\ \Omega$.

The quantity V_E is unknown, but we can calculate it:

$$V_E = V_B - V_{BE} = V_{CC}\left(\frac{R_2}{R_1 + R_2}\right) - 0.7\text{ V} = 10\text{ V}\left(\frac{4.7\text{ K}}{4.7\text{ K} + 4.7\text{ K}}\right) - 0.7\text{ V} = \underline{4.3\text{ V}}$$

The dc emitter current is therefore

$$I_E = \frac{V_E}{R_E} = \frac{4.3\text{ V}}{940\ \Omega} = \underline{\underline{4.6\text{ mA}}}$$

Other Oscillator Configurations

There are additional ways of building RF oscillators. Most of these are based on one or more of the three basic circuits above and contain refinements that help to stabilize their frequency (especially at VHF and UHF frequencies). Let's look at two that are based on the Colpitts configuration.

The Clapp oscillator Figure 4–13 shows a *Clapp* oscillator. It's very similar to the Colpitts circuit, with only one component added to the resonant LC bandpass portion of the circuit.

One problem with oscillators is frequency stability. The frequency of the three basic RF oscillator circuits depends on an LC resonant circuit. A transistor, vacuum tube, or other active component adds its own *interelectrode capacitance* to the LC tank circuit. This can cause the frequency to drift because the built-in capacitance of the transistor isn't constant. It varies with applied voltage and temperature. The Clapp circuit addresses this problem by adding a *third* capacitance to the tank, C_1 in Figure 4–13. This capacitance is much smaller than the voltage divider capacitors C_1 and C_2, which causes it to dominate the tank circuit. C_1, C_2, and the interelectrode capacitance of Q_1 still effect the oscillator's frequency but much less so than in a standard Colpitts circuit. Thus, the Clapp circuit reduces frequency instability.

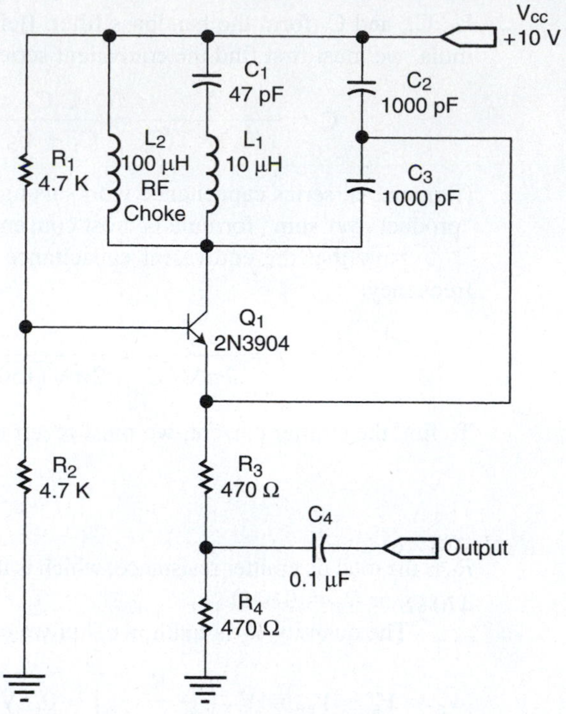

Figure 4–13 A Clapp oscillator

Note the added inductor L_2 in Figure 4–13. This is an *RF choke*. When C_1 is added to the tank to stabilize the frequency, it breaks the dc circuit for biasing the collector of Q_1! L_2 allows power supply voltage to reach Q_1's collector, while blocking the collector's ac signals from getting into the power supply. L_2 opposes or "chokes" the ac current. L_2 is not part of the resonant bandpass filter, and for practical purposes does not greatly effect the oscillator's frequency.

EXAMPLE 4–6

What is the equivalent tank circuit capacitance and operating frequency of the *Clapp oscillator* of Figure 4–13?

Solution

L_2 is not part of the LC bandpass circuit, so we can ignore it. Only L_1, C_1, C_2, and C_3 control the frequency. The inductor L_1 "sees" capacitors C_1, C_2, and C_3 in series. (Think about where the circulating tank currents must flow.) Therefore,

$$C = \frac{1}{{}^1\!/_{C_1} + {}^1\!/_{C_2} + {}^1\!/_{C_3}} = \frac{1}{{}^1\!/_{1000\text{ pF}} + {}^1\!/_{1000\text{ pF}} + {}^1\!/_{47\text{ pF}}} = \underline{\underline{42.96\text{ pF}}}$$

Notice how this is very close to the value of C_1. C_2 and C_3 have little effect because they're so large in comparison to C_1. The operating frequency is therefore

$$f = \frac{1}{2\pi\sqrt{LC}} = \frac{1}{2\pi\sqrt{(10\ \mu\text{H})(42.96\text{ pF})}} = \underline{\underline{7.6787\text{ MHz}}} = \underline{\underline{7678.7\text{ kHz}}}$$

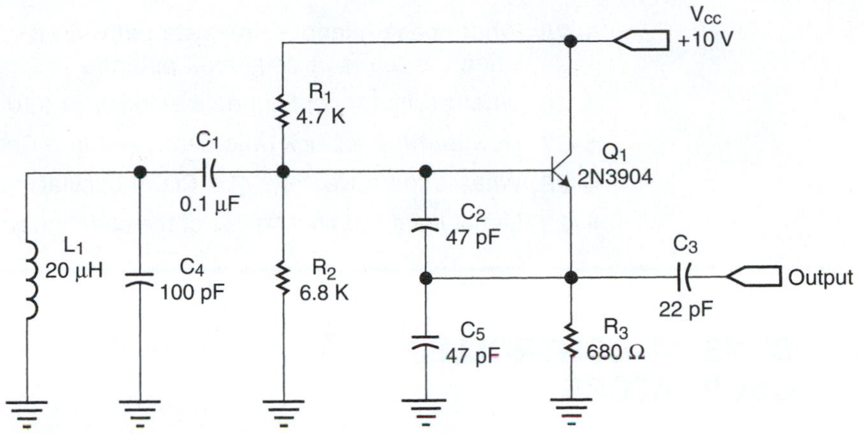

Figure 4–14 The common-collector Colpitts oscillator

The common-collector Colpitts Another variation of the Colpitts circuit is often seen in communications circuits; for lack of a better name, this can be thought of as a *common-collector* version of the Colpitts. It is shown in Figure 4–14.

The common-collector Colpitts circuit is very popular because it places one end of the resonant tank circuit (L_1 and C_4 in Figure 4–14) at ground potential. This reduces noise and also makes it much easier to electronically switch different L or C values into the circuit for *frequency synthesis.*

The circuit uses C_2 and C_5 to determine the feedback ratio. A common-collector amplifier produces no voltage gain ($A \approx 1$), so it would appear that the circuit couldn't work. However, a *current gain* is produced between the base and emitter of Q_1, while C_2 and C_5 (together with the base input impedance of Q_1) form a *current divider* that controls the percentage of *current* that gets fed back into the base of Q_1.

The common-collector Colpitts configuration is very often seen at VHF and UHF frequencies and is also popular for use with *quartz crystal control,* as we'll soon see.

Section Checkpoint

4–17 When performing dc analysis on a circuit, capacitors become ______ and inductors become ______.

4–18 Why is the collector voltage of Q_1 in Figure 4–8 equal to V_{cc}?

4–19 What three things does a technician look for when doing an ac analysis of an oscillator circuit?

4–20 What phase shift is introduced in a common-emitter (CE) amplifier?

4–21 How is a 180° phase shift introduced within the feedback loop of Figure 4–8?

4–22 What components control the frequency of the oscillator of Figure 4–8?

4–23 What is the highest practical frequency for an Armstrong oscillator?

4–24 How is a Hartley oscillator identified? What is the maximum practical frequency for a Hartley oscillator?

(*continued on p. 104*)

4–25 What phase relationship exists between the ends of a tapped inductor when the tap is at ac ground potential?

4–26 What oscillator configuration works well into the UHF frequency range?

4–27 How is the feedback ratio controlled in a Colpitts oscillator?

4–28 What is the advantage of a Clapp oscillator over a Colpitts circuit?

4–29 List at least two advantages of the common-collector Colpitts configuration.

4–4 CRYSTAL-CONTROLLED OSCILLATORS

Frequency stability is important in RF oscillators. It can be defined as *the ability to resist changes in frequency.* Unfortunately, oscillators built with discrete (individual) L and C components are often not stable enough for most transmitting and receiving requirements.

Frequency stability is measured in *parts per million (ppm) frequency error.* By definition, 1 ppm error in a 1 MHz oscillator means 1 Hz of error. A 1 ppm error in a 10 MHz oscillator produces 10 Hz of error. Oscillators built using discrete LC bandpass elements seldom have better stability than 300–500 ppm, and sometimes much worse! A frequency error of 500 ppm at 150 MHz would translate into a frequency shift of 75 kHz, which would be totally unusable if the channel bandwidths were only 15 kHz (the frequency would be 5 channels off!).

Quartz crystals are used to construct exceptionally stable oscillators; even an inexpensive crystal oscillator usually has less than 10 ppm error, which is much better performance than an LC circuit can deliver. State-of-the-art crystal oscillators can deliver less than 0.5 ppm frequency error over a wide range of temperatures.

Quality Factor and Bandwidth of a Tuned Circuit

You might recall that the *quality factor* or Q of a tuned circuit can be expressed as follows:

$$Q = \frac{f}{\text{BW}} \tag{4–2}$$

Where f is the operating frequency of the tuned circuit, and BW is its bandwidth. (This relationship is most valid when the Q is at least 10.) The higher the Q, the better the tuned circuit will be at staying near its intended frequency. With discrete components, it's difficult to get a quality factor of more than 200 or 300. This limits the frequency stability. Let's do an example.

EXAMPLE 4–7 Suppose that the Q of the tank in Figure 4–11 is 200. What is the bandwidth of the tank?

Solution

Equation 4–2 is manipulated to solve for bandwidth. We know that the frequency of resonance is 847.7 kHz from a previous calculation, so we get

$$\text{BW} = \frac{f}{Q} = \frac{847.7\text{ kHz}}{200} = \underline{\underline{4.24\text{ kHz}}}$$

Roughly speaking, this means that it is entirely possible for the circuit to drift over *approximately* a 4.2 kHz range (±2.1 kHz). This range is centered about the resonant frequency of 847.7 kHz. Compared to the oscillator frequency, this is a small amount; however, compared to the channel spacing employed at this frequency (it's in the AM broadcast band, where channels are spaced 10 kHz apart), it's terrible!

In order to get good frequency stability, a *quartz crystal* is often added to an oscillator. A quartz crystal is a very precise frequency-determining device.

Quartz Crystal Construction and Operation

The construction of a crystal is shown in Figure 4–15. It consists of a thin slab of quartz crystal sandwiched between two metal contact plates. The metal plates are used to establish contact with the terminals of the device. The schematic symbol is very much like a capacitor, with the addition of "something" in between the plates.

The Piezoelectric Effect

Quartz, among other natural materials, exhibits what is known as *piezoelectricity* or *pressure electricity*. When a force is applied to a quartz crystal, it bends very slightly. When the crystal bends, the internal atomic structure (which is very much like the frame of a building) is displaced. As a result, a small voltage is developed across the device. Increasing the force increases the voltage—at least until the crystal breaks!

This principle is used in many devices. A *crystal microphone* uses a Rochelle salt slab as the active material and converts sound pressure variations into a voltage. *Vibration and stress sensors* sometimes use a piezoelectric material to measure small movements in materials. The piezoelectric effect taken to extremes is the *push-button igniter* seen on modern gas grills: When the button on the igniter is pressed, a spring-loaded hammer strikes a piece of piezoelectric material. The intense pressure on the crystal creates a very short-lived high voltage, which is enough to create a spark for igniting the gas.

The piezoelectric effect also works in an opposite way. If a voltage is applied to a quartz crystal, the crystal will bend a small amount, just as if something pushed on it. This is used commonly in *piezoelectric speakers*. Because the piezoelectric effect works in both directions, we often call it a *reciprocal* effect. Figure 4–16 shows several popular styles of quartz crystals, and their internal construction.

Figure 4–15 Quartz crystal construction and schematic symbol

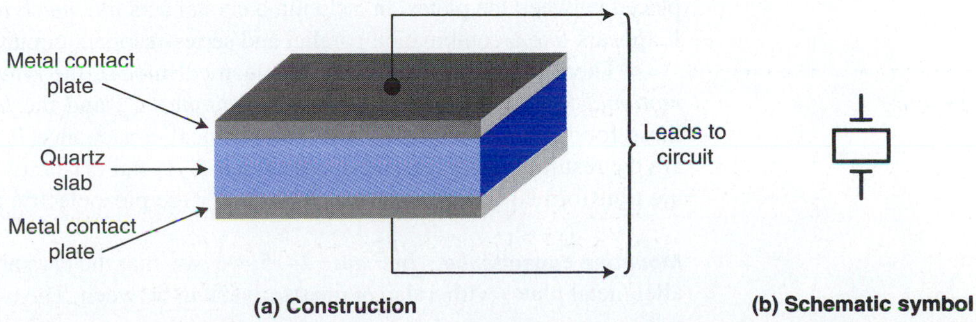

Figure 4–16 Typical crystals and internal view

The Crystal as a Tuning Fork

Everyone recognizes how a musical tuning fork operates. The device must be initially supplied with energy by striking it against a firm surface. In turn, the fork begins vibrating at a precise frequency. The vibrations decay (grow smaller) as the fork gives up the mechanical energy to the air (and friction). The fork vibrates at a given frequency because it has a precise *mass* and *elasticity* (springiness). If energy could be put back into the fork at the right time in each of its vibration cycles, it would vibrate indefinitely.

A quartz crystal can be made to vibrate very much like a tuning fork. Like a tuning fork, a crystal has both a *mass* and a certain resistance to bending or *elasticity*. By cutting the crystal to precise dimensions, it can be made to vibrate at very precise frequencies. The smaller the size of the quartz slab, the higher the frequency of the crystal. Crystals can be made to vibrate in a *fundamental* mode up to around 15 MHz.

In order to keep the crystal in oscillation (vibration), energy must be supplied periodically. This is a perfect job for an amplifier. By applying positive feedback, we can make sure the crystal will continue to vibrate.

The Electrical Model of a Crystal

The schematic symbol of a crystal is like that for a capacitor, with "something special" placed between the plates. In a circuit, a crystal acts like *much* more than a mere capacitor. It appears like a combination parallel and series-resonant circuit, as shown in Figure 4–17.

There are four elements in the electrical model, the *mounting capacitance* C_m, the *motional inductance* L_s, the *elastic capacitance* C_s, and the *frictional resistance* R_s. Of these four elements, only C_m is a real, physical capacitance. The other three components are the result of the properties of crystal mass, crystal elasticity, and crystal friction, which are transformed into electrical equivalents by the piezoelectric action of the crystal.

Mounting capacitance In Figure 4–15, we saw that the crystal is really just a pair of parallel metal plates with a slab of quartz crystal in between. The two metal plates form a real,

Figure 4–17 The equivalent circuit of a quartz crystal

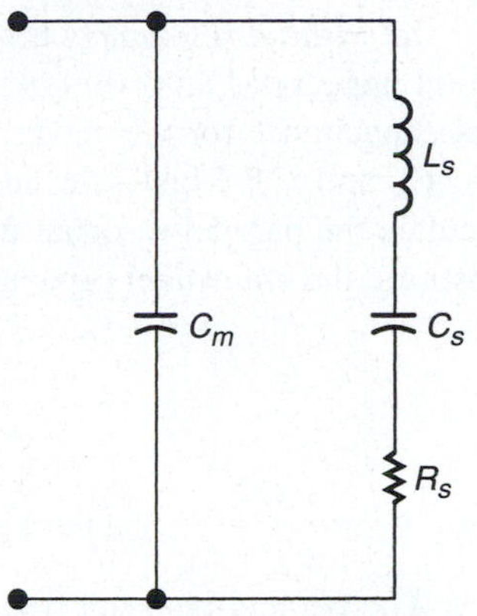

physical capacitance called the *mounting capacitance* (since the crystal is "mounted" to the metal plates). Typical values for mounting capacitance are in the 2 to 10 pF range.

Equivalent components The remaining three components are not physical parts. Rather, it *appears* (from the point of view of the circuit connected to the crystal) that these three components are present. The physical properties of the quartz crystal slab control these values. The manufacturer of a crystal can normally supply the equivalent component values if they are needed.

Resonance modes There are actually two distinct types of resonance possible in a quartz crystal. These modes are *series* and *parallel* resonance. In *series resonance,* the reactance of the motional inductance L_s and the elastic capacitance C_s are equal (but opposite). Therefore, the series combination of L_s and C_s appears as a short circuit, as in Figure 4–18.

In series resonance, the crystal ideally looks like a *short circuit* between its terminals. To a second approximation, the crystal terminal impedance at series resonance is the same as the frictional resistance, R_s. The series-resonant frequency is calculated using the standard resonant frequency formula, using L_s and C_s:

$$f_s = \frac{1}{2\pi\sqrt{L_s C_s}} \qquad (4\text{–}3)$$

Figure 4–18 Series resonance action

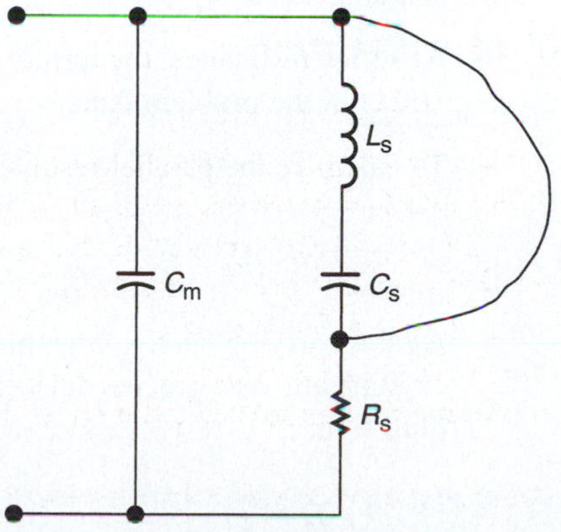

The series-resonant frequency is sometimes referred to as the *zero* of the crystal.

In *parallel resonance,* the total capacitance of the circuit resonates with L_s, and the crystal appears as an *open circuit*. Looking at Figure 4–17, can you see how to find the total capacitance for this mode? How can L_s circulate a current between both the capacitors, C_m and C_s? Right—the inductor "sees" the two capacitors in *series*. Therefore, to calculate the parallel-resonant frequency, we again use the resonance formula; we merely substitute the *equivalent* capacitance for the series combination of C_s and C_m:

$$f_p = \frac{1}{2\pi\sqrt{L_s C_t}} \qquad \textbf{(4–4)}$$

where $C_t = \dfrac{C_s C_m}{C_s + C_m}$, the total equivalent capacitance of the circuit.

The parallel-resonant frequency is often referred to as the *pole* of the crystal.

EXAMPLE 4–8

A crystal has been characterized by a manufacturer as follows: $L_s = 1$ H, $C_s = 0.02$ pF, $R_s = 100\ \Omega$, and $C_m = 5$ pF. Calculate the following:

a. the series-resonant frequency
b. the crystal terminal resistance at series resonance
c. the parallel-resonant frequency
d. the crystal Q (quality factor) at series resonance
e. the pole-zero spacing, in Hz
f. the approximate bandwidth (BW) at series resonance

Solution

a. Series resonance involves only L_s and C_s:

$$f_s = \frac{1}{2\pi\sqrt{L_s C_s}} = \frac{1}{2\pi\sqrt{(1\text{ H})(0.02\text{ pF})}} = \underline{\underline{1.125395\text{ MHz}}}$$

Note: Unlike other electronic components with typical tolerances of 5% or 10%, crystals are high-precision devices. Calculations for frequency should be carried out to at least the nearest Hz. *Be careful to use an accurate value for pi. (Use the PI key on a calculator; don't just enter 3.14!)*

b. At series resonance, the terminal resistance is very close to R_s, which is given as $\underline{\underline{100\ \Omega}}$ in the problem data.

c. To calculate the parallel-resonant frequency, we must first find the total capacitance:

$$C_t = \frac{C_s C_m}{C_s + C_m} = \frac{(0.02\text{ pF})(5\text{ pF})}{(0.02\text{ pF} + 5\text{ pF})} = 0.0199203187\text{ pF}$$

Again, carrying such a large number of decimal places looks strange; remember, crystals are very precise devices. Now that the capacitance is known, we can plug it in:

$$f_p = \frac{1}{2\pi\sqrt{L_s C_t}} = \frac{1}{2\pi\sqrt{(1\text{ H})(0.0199203187\text{ pF})}} = \underline{\underline{1.127644\text{ MHz}}}$$

d. To calculate the Q for series resonance, we dig into our electronic fundamentals warehouse and obtain

$$Q_s = \frac{X_s}{R_s} = \frac{X_L}{R_s} = \frac{2\pi f_s L_s}{R_s} = \frac{(2\pi)(1.125395\,\text{MHz})(1\,\text{H})}{100\,\Omega} = \underline{\underline{70{,}711}}$$

This is a *staggering* value for Q when compared to the measly 200 to 300 obtainable with discrete LC circuits. Most crystals have a Q between 50,000 and 200,000.

e. The *pole-zero spacing* is a fancy way of asking for the difference between series and parallel resonance frequencies:

$$\text{Pole} - \text{Zero} - \text{Spacing} = f_p - f_s = 1.127644\ \text{MHz} - 1.125395\ \text{MHz} = \underline{\underline{2249\ \text{Hz}}}$$

In other words, the series and parallel resonant frequencies are quite closely spaced! Note that the parallel resonant mode is always the higher of the two frequencies.

f. Equation 4–2 can be manipulated to estimate bandwidth:

$$Q = \frac{f}{\text{BW}}, \quad \text{so BW} = \frac{f}{Q} = \frac{1.125395\ \text{MHz}}{70{,}711} = \underline{\underline{15.9\ \text{Hz}}}$$

Compare this answer to that from Example 4–7. The quartz crystal has a much more narrow bandwidth due to its higher quality (Q) factor. *Therefore, the quartz crystal is likely to be a very precise frequency controller in a circuit!*

Controlling Oscillators with Crystals

A crystal appears like a very sharp bandpass filter and, depending on the applied frequency, can appear as either a short circuit (series resonance), an open circuit (parallel resonance), or something in between.

Suppose that we add a crystal to the feedback loop of our Hartley oscillator. We might end up with the circuit of Figure 4–19. In Figure 4–19, crystal Y_1 has been placed in series with the feedback connection (C_2). Although the crystal is open to dc, capacitor C_2 is still needed. A dc voltage across the crystal will stress it mechanically and lead to early failure of the device.

We know that an oscillator needs gain and feedback in order to operate. What will happen to this circuit if Y_1 appears as an *open* circuit? An open in the feedback path will stop the oscillator. If we place a short across Y_1, the feedback signals are free to pass no matter what. Therefore, the circuit of Figure 4–19 needs the crystal to be in series-resonant mode, where it will have a low resistance to the feedback current. Although there is an external LC tank (L_1 and C_1), the crystal will tend to dominate this circuit's frequency. This is because the crystal's Q is much higher than that of L_1 and C_1. *Only the series-resonant frequency can get through the crystal.*

Note that the tank circuit (L_1 and C_1) must be tuned close to the crystal's frequency. Adjusting the tank has little effect on the circuit frequency; the crystal has a much higher Q and keeps a tight "grip" on the oscillator's operation. However, if the tank is tuned *too* far from the crystal's frequency, the oscillator will simply stop. This is because both the crystal and tank circuit are filters. The feedback signal must be able to pass through *both* filters in order to allow oscillation. If the two filters don't agree on frequency, then *no* frequency can be fed back. This is very much like an analog AND gate!

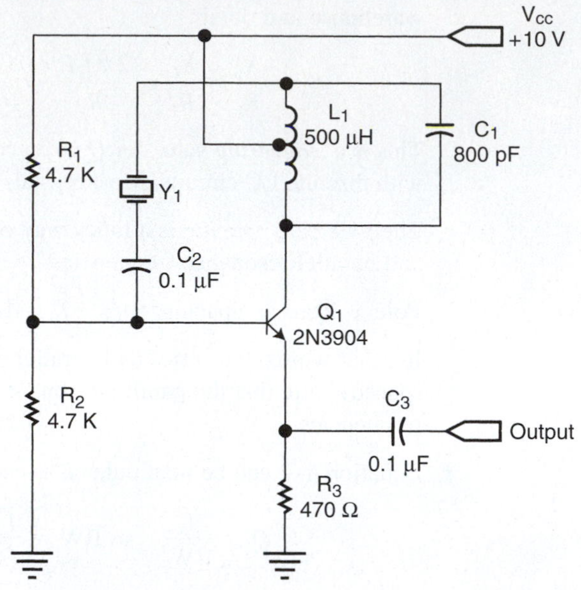

Figure 4–19 A crystal-controlled Hartley oscillator

If you can replace a crystal with a short circuit (wire across it) and a circuit continues to oscillate, it is a fairly safe bet that the circuit is using the crystal in the series-resonant mode.

By moving the crystal to a different location, we can get it to be *parallel resonant*. In the circuit of Figure 4–20, the crystal "straddles" the resonant tank (L_1 and C_1). Try mentally replacing it with an ac short—what happens?

Shorting the crystal kills this oscillator! Recall that for this circuit to oscillate, a 180° phase difference must be developed across the ends of the tank (because of the tap

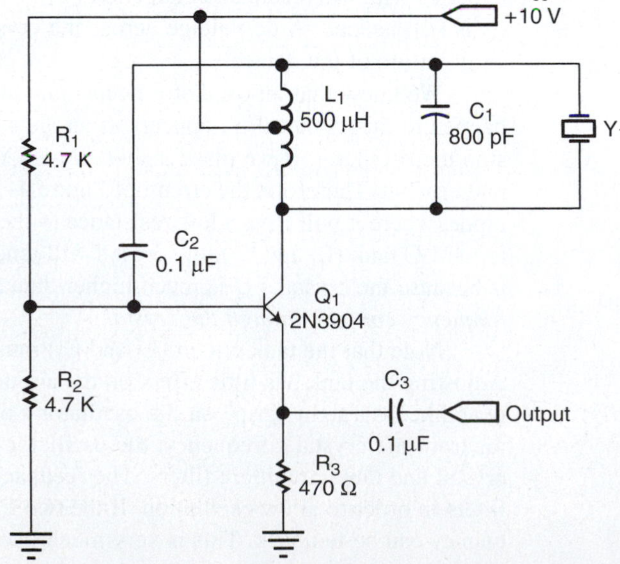

Figure 4–20 A crystal applied in parallel mode

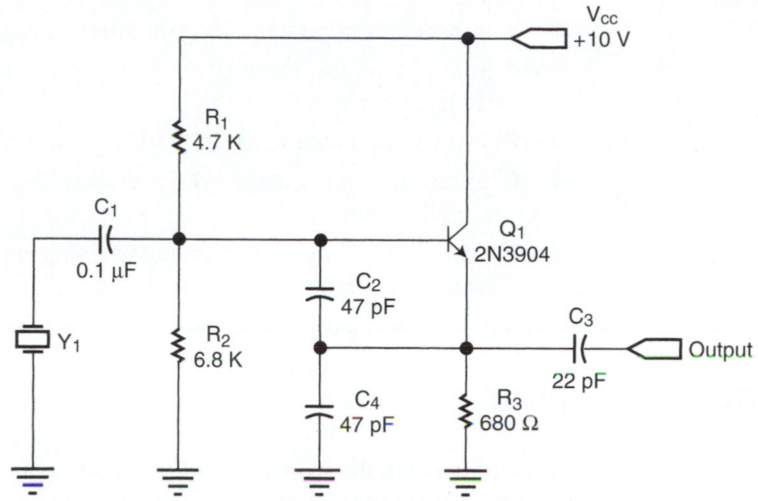

Figure 4–21
Crystal-controlled Colpitts oscillator

on L_1). A short circuit on the tank causes a 0° difference to appear; the total phase shift in the circuit becomes 180° (from Q_1). This is negative feedback; the oscillator stops.

If we replace the crystal with an open circuit, we get the original Hartley oscillator circuit of Figure 4–9. The circuit will continue to oscillate, even without the crystal! *This is a strong clue that the crystal is operating in its parallel-resonant mode.*

If you can replace a crystal with a short and the circuit continues oscillating, the circuit probably uses the crystal in series-resonant mode. If removing or opening the crystal allows oscillation, the crystal is probably operating in the parallel-resonant mode.

An Oscillator with No Visible Tank

Because a crystal is itself a resonant circuit, it is possible to build oscillators that use only the crystal to determine frequency. This is an advantage because it reduces component count by eliminating an inductor and capacitor and allows different crystals to be switched in or out for changing frequency—all without requiring the circuit to be realigned.

One of the most popular circuits used with crystals is the *common-collector Colpitts* configuration (Figure 4–21). In Figure 4–21, the crystal replaces the original parallel-resonant tank formed from L_1 and C_4 in Figure 4–14. The crystal is thus forced to operate in its parallel-resonant mode. With this circuit, nearly any crystal could be substituted; since there is no discrete LC tank to adjust, the circuit will automatically run at the crystal's parallel-resonant frequency.

Section Checkpoint

4–30 Why is frequency stability important in oscillators? What are its units of measure?

4–31 What is the *piezoelectric effect?*

4–32 List the two resonance modes of a crystal.

(continued on p. 112)

4–33 In series resonance, a crystal ideally appears as a ______ circuit.

4–34 In a crystal, the parts L_s, R_s, and C_s are equivalent components. What does this mean?

4–35 What is a typical value of Q for a quartz crystal?

4–36 To control an oscillator with a crystal, the crystal is often placed in the f______ path.

4–37 Can an oscillator be constructed using only a crystal to control its frequency (using no external LC tank)?

4–5 RF AMPLIFIERS

RF amplifiers are the workhorses of transmitting and receiving equipment. They're used any time the strength or power of a signal needs to be increased. RF amplifiers are designed in many different ways, but most of them have a few common features.

Common Features of RF Amplifiers

Because of the high frequencies involved, RF amplifiers use several common design and construction techniques. Among these are the use of *tuned circuits, high-frequency active devices,* and *special circuit construction and layout techniques.*

Most (but not all) RF amplifiers use at least one *tuned circuit.* The tuned circuit is used as a *bandpass filter* so that the amplifier can pass only a limited range of frequencies. This is helpful for two reasons. First, reducing bandwidth reduces noise, and second, it helps to eliminate interference on undesired frequencies.

RF semiconductors and tubes are used in RF amplifiers. Transistors that work well as switches or audio amplifiers often do not have sufficient gain to operate in the HF, VHF, and UHF frequency regions. Many RF transistors have special packaging (for example, short and flat terminals) to reduce lead inductance, which becomes a real problem at RF frequencies. Figure 4–22 shows several popular RF transistor packages.

Figure 4–22 Typical RF transistors

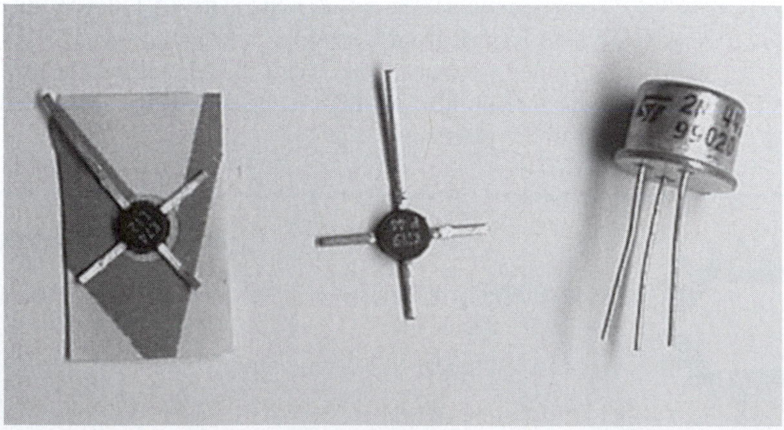

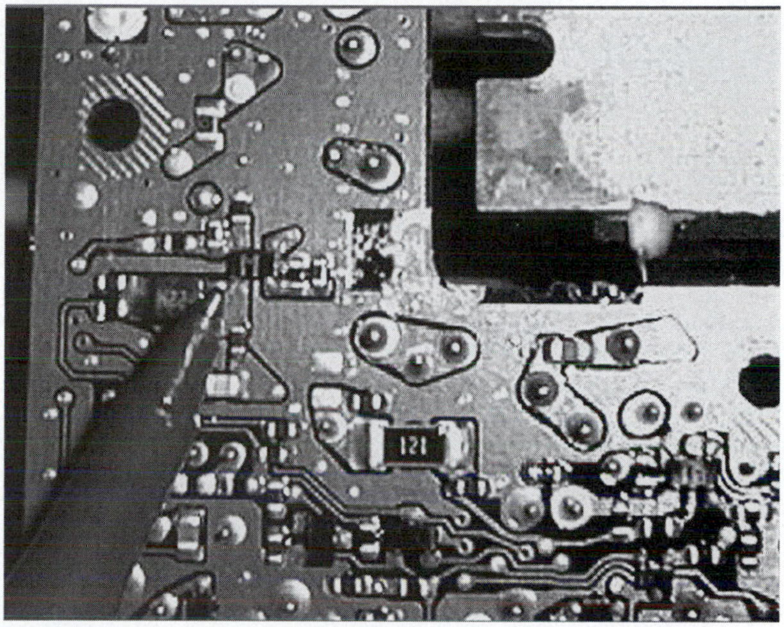

Figure 4–23 Components surface-mounted in a receiver RF amplifier

Finally, component *layout* or *dress* is important at radio frequencies. Even an extra inch of wire can cause a VHF or UHF circuit to misbehave. Component leads must be kept as short as possible. Long component leads can act as *antennas* (highly undesirable!), or can contribute extra inductance or capacitance to the RF circuit, knocking it out of tune.

Metal shields are often necessary to keep energy from leaking into or out of RF circuits. Where signals must travel more than an inch or so, *shielded coaxial cables* or other *transmission lines* are used to carry them to prevent signal loss. The use of *surface mount circuitry* as shown in Figure 4–23 is rapidly becoming a standard RF construction technique because the component leads can be made very short, which eliminates most of the undesired lead inductance.

Tuned-Input and Tuned-Output Amplifiers

RF amplifiers can be further classified as being either a *tuned-input* or *tuned-output* type. A tuned-input amplifier has a resonant LC filter (or equivalent device) in its input circuit, and a tuned-output amplifier uses an LC filter in its output circuit. The *tuned-output* amplifier is far more common. It is also possible to have tuned circuits in both the input and the output; this is called a *tuned-input–tuned-output* amplifier.

Small- and Large-Signal RF Amplifiers

There are two major groups of RF amplifiers; small and large signal. A *small-signal* RF amplifier is likely to operate at power levels below 10 mW and is almost always biased in class A. Recall that a class A amplifier produces the least distortion in a signal but is also the least efficient of all the biasing classes.

Figure 4–24
Small-signal class A RF amplifier

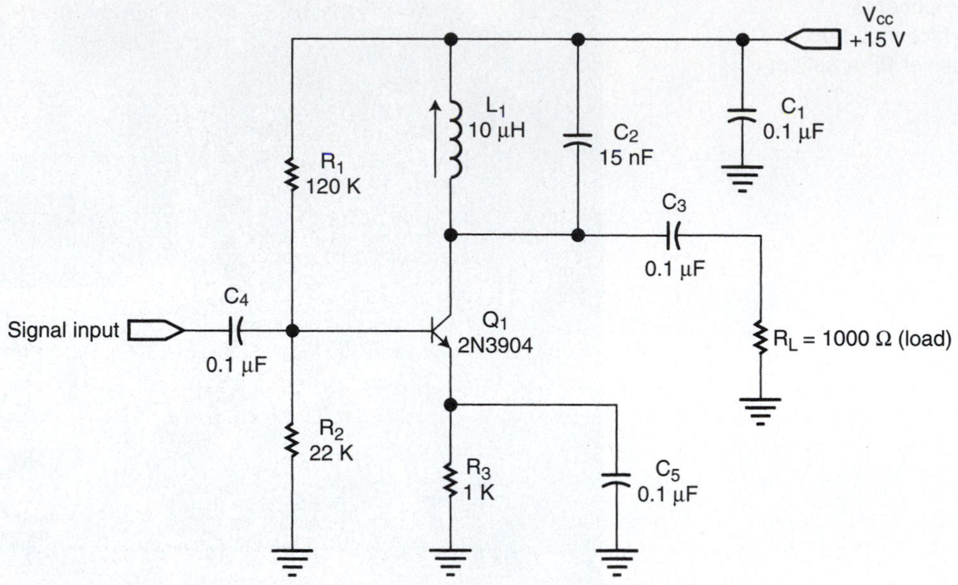

Large-signal amplifiers are those that operate above the 10 mW power level (an arbitrary choice). Up to around the 500 mW power level, most designs are still biased in class A; above 500 mW, class B or C amplification is used, depending on the circuit requirements. (Where linear amplification is required, class A or B must be used; class C amplifiers are highly nonlinear, and well suited for use as modulators.)

It's often difficult to tell the difference between small- and large-signal RF amplifiers on a schematic diagram. One giveaway is the size of *biasing resistors*. Power amplifiers, which are large-signal RF amplifiers, will often have fairly small (less than 50 Ω) resistors in their circuit. An RF *power transistor* is usually mounted to a metal heatsink, in order to help it dissipate the heat arising from normal operation.

A small-signal class A RF amplifier Figure 4–24 shows a common-emitter, tuned-output RF amplifier. In this amplifier, Q_1 is voltage divider biased by R_1 and R_2; R_3 sets the dc emitter current. The input and output are *capacitively coupled* by C_4 and C_3. The main difference between this amplifier and one you might have studied in fundamentals is the addition of the resonant tank circuit, L_1 and C_2, in the collector. (Also, notice again that there is *no* collector resistor; the collector dc bias is supplied through the coil L_1.)

The arrow by L_1 means that it is an adjustable coil. RF coils are usually adjusted using a special nonmetallic screwdriver made of plastic. Adjusting a coil with a metal tool is undesirable; the metal in the tool will throw the circuit off frequency.

Capacitor C_1 performs an important function. It is an *RF bypass capacitor*. Its job is to route any stray RF signal on the power supply bus to ground. It appears as a short to the ac RF signal and as an open to the dc power supply.

EXAMPLE 4–9

Estimate the operating frequency for the amplifier of Figure 4–24. What symptom will be produced if C_5 becomes an open circuit?

Solution

L_1 and C_2 form the bandpass filter for this unit and determine the operating frequency of the unit:

$$f = \frac{1}{2\pi\sqrt{LC}} = \frac{1}{2\pi\sqrt{(10\,\mu\text{H})(15\,\text{nF})}} = \underline{\underline{410.93\text{ kHz}}}$$

C_5 controls the *gain* of the amplifier. You might recall that the gain of a common-emitter amplifier is controlled by the collector ac resistance and the emitter ac resistance:

$$A = \frac{r_c}{r_e + r'_e}$$

In this circuit, r_c is the load (1000 ohms), r'_e is the dynamic emitter resistance of the transistor (around 16 to 32 Ω under the given conditions), and r_e is approximately zero because R_3 is *bypassed* by C_5. If C_5 *opens,* r_e will increase to 1000 Ω (the value of R_3), and the gain will *fall* to a very low value.

Whenever the gain of a common-emitter amplifier has fallen very low, the emitter bypass capacitor is a good thing to check! This is a very common failure.

Transformer coupling The amplifier of Figure 4–24 works, but because most loads in RF work tend to be small (50 Ω), it is difficult for designers to use that approach. Small-signal transistors like to work into collector resistances in the range of 500 Ω to 20 kΩ; and at most RF frequencies, best coil Q can be obtained in this reactance range. *Transformers* are often used to step impedances up and down as needed. The amplifier of Figure 4–25 can be made to operate quite well into low impedance loads. You're very likely to see amplifiers like it in real circuitry.

The primary change for the amplifier of Figure 4–25 is the *output coupling transformer* T_1. The transformer performs two functions. First, it acts as part of the bandpass

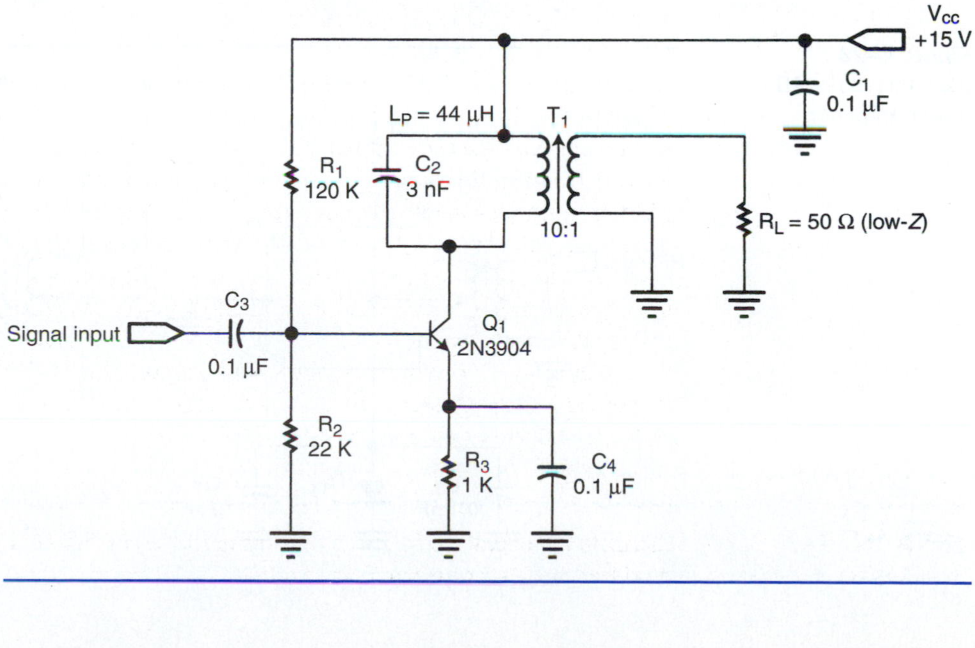

Figure 4–25
Transformer coupling the output to the load

filter, in combination with C_2. Second, it transforms the impedance (ac resistance) of the collector circuit down to the value of 50 Ω. *Most well-designed RF equipment uses 50 Ω as the interstage impedance value.*

Large-Signal RF Amplifiers

Up to about the 500 mW level, it is common to see amplifiers like those of Figures 4–24 and 4–25. Above this power level, class A operation becomes too wasteful in terms of radiated heat. A very common RF linear power amplifier is the *push-pull* amplifier of Figure 4–26. Amplifiers of this type can easily produce power levels of 100 watts or more.

dc analysis In an ideal class B amplifier, each active device conducts for exactly 180° of the input cycle; one transistor amplifies the top half-cycle, the other the bottom. A real-life class B amplifier suffers from *crossover* or *notching* distortion, since both active devices shut off as the input signal crosses through zero (and bipolar transistors need about 0.7 volts to get turned on). The input signal is lost near the zero crossings on the waveform, which distorts it.

In order to eliminate crossover distortion, the amplifier in Figure 4–26 uses *trickle bias*. Under trickle bias, both transistors remain on (with a small collector current, typically less than 1% of the collector saturation current) under no-signal conditions. This amplifier uses diode D_1 and the base-emitter junctions of Q_1 and Q_2 as a *current mirror*. The current in each collector exactly equals the current flowing into D_1, which is set by resistor R_1. This technique is necessary, since using even a small emitter-resistor would waste too much power and reduce the power gain of the unit.

With this type of biasing, it is important that the diode be thermally coupled to the cases of the output transistors so that it will thermally track them. *Bias circuit failures* in this type of amplifier are generally catastrophic; the transistors will quickly self-destruct (thermal runaway) if the diode fails open. Very impressive levels of smoke and noise are created—the power supply for a power amplifier of this type is capable of generating many amperes of current!

When a technician replaces transistors in this type of amplifier, the cause of failure must always be found—or a new (and expensive) set of RF power

Figure 4–26 A push-pull class AB RF power amplifier

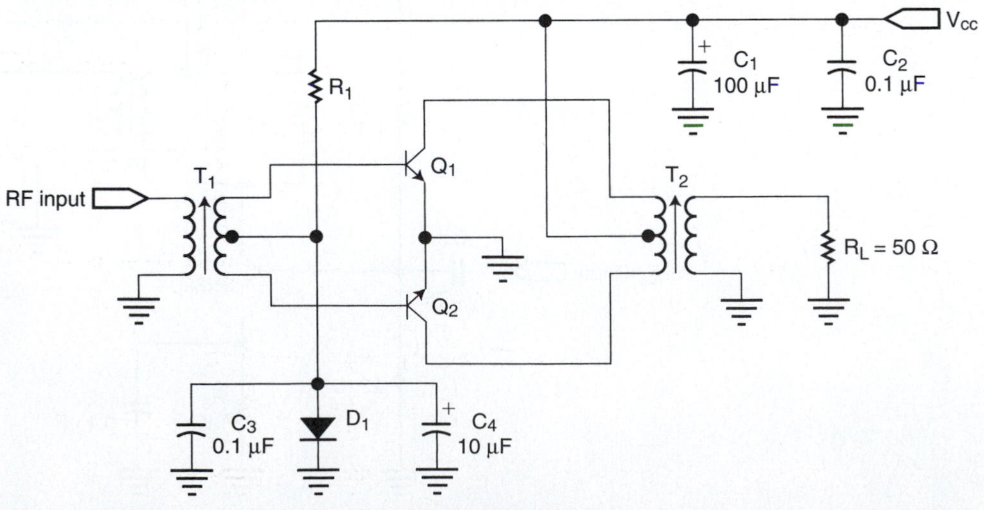

transistors will also be destroyed. Many techs always replace the bias diode, even if it checks good with an ohmmeter, since these diodes can fail "intermittent open," which can cause a failure after the unit has been returned to service.

ac analysis Transformer T_1 is center-tapped and splits the input signal into opposite phases for driving the bases of Q_1 and Q_2. T_1 also impedance-matches the input signal to the bases of Q_1 and Q_2. This is very important, because the base input impedance of RF power transistors is often less than 5 Ω, which is a very low impedance.

Q_1 and Q_2 amplify opposite halves of the signal, and these halves are recombined in transformer T_2. T_2 also impedance-matches the collectors of the transistors to the load. The impedance level at the collectors is determined by the maximum power rating of the amplifier and the power supply voltage.

Note the liberal use of filtering: C_3 and C_4 bypass D_1 to prevent the diode shot noise from being amplified; and C_1 and C_2 filter the power supply. It's very common to see an electrolytic capacitor in parallel with a nonelectrolytic (like a ceramic disc type). The electrolytic is good at filtering out low-frequency noise but is a poor RF filter; the ceramic capacitor is insufficient in size to eliminate audio noise but is an excellent RF filter.

The amplifier shown here is *untuned* or *broadbanded.* This is an advantage; an amplifier of this type can amplify a wide range of frequencies without retuning. For limited-frequency-range application, it can also be operated as a tuned RF amplifier by adding capacitance across transformer T_2.

The Class C Power Amplifier

A class A RF amplifier has a theoretical efficiency of 50%, and a class B amplifier has a theoretical efficiency of 78%. Even better efficiency can be had by using a class C amplifier (better than 90% theoretical efficiency). It must be remembered that the class C amplifier is nonlinear. It cannot accurately amplify a modulated waveform. However, amplification of an *unmodulated* carrier signal works just fine for the class C amplifier. The class C amplifier is also well suited as an amplitude modulator. Figure 4–27 is an idealized class

Figure 4–27 A class C amplifier

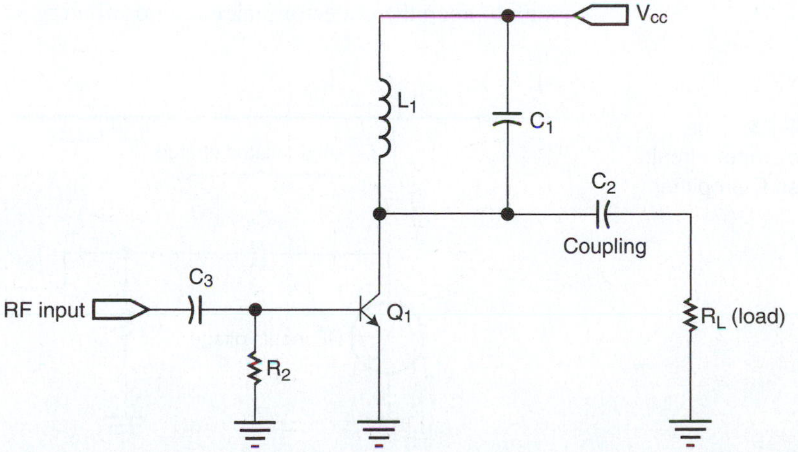

C amplifier. There isn't much in the circuit—but don't be fooled. There are some very interesting events taking place.

dc analysis The class C amplifier is unique among amplifiers in that the dc bias for the base circuit is provided by the *input signal.* If no input signal is provided to the amplifier in Figure 4–27, Q_1 will remain "off" since resistor R_2 pulls the base toward ground. (Recall that an *npn* transistor is turned "on" by a positive voltage of 0.7 volts or more between the base and emitter terminals.)

A class C amplifier is very efficient because the transistor is turned off most of the time. With no collector current flowing most of the time, the transistor doesn't dissipate very much power. This means that most of the power from the dc power supply ends up in the load as RF energy.

In order to get this action, a class C amplifier is biased slightly *beyond* cutoff, as shown in Figure 4–3. For this to happen, a slight *negative* dc voltage must be developed across the base-emitter junction. C_3, R_2, and the base-emitter diode form a *clamper* that provides this negative dc voltage.

Clamping action A *clamper* is a circuit that adds or subtracts a dc level from a signal. Figure 4–28 shows the equivalent circuit formed by C_3, R_2, and the base-emitter junction of Q_1.

In the clamper of Figure 4–28, the base-emitter diode is reverse-biased during the negative half-cycle, and only a small current flows from the input signal source through C_3 and R_2 (R_2 is a relatively large value). During the positive half-cycle, the base emitter of Q_1 is forward-biased. Capacitor C_3 rapidly charges to the source peak voltage, less the 0.7 volt base-emitter drop of Q_1. Figure 4–29 shows this graphically: The RF input voltage is 4 volts peak (8 V_{pp}). This means that C_3 can charge to a value of (4 V–0.7 V) or about 3.3 V. The voltage on C_3 appears as an extra negative bias voltage; thus the base voltage is clamped negatively.

R_2 is important. Without it, C_3 would just charge up—and hold its charge! With C_3 fully charged, there's no reason for the base-emitter diode of Q_1 to conduct. Q_1 would turn on only on the first cycle or two and then would remain off. R_2 is a *bleeder resistor.* It makes C_3 act like a tire with a slow leak by siphoning or "bleeding" a small amount of charge from C_3 on each cycle. This ensures that Q_1 will get turned on for *every* input cycle! The net effect, as you can see from Figure 4–29, is that the base emitter of Q_1 only gets to conduct during a small portion of the ac input cycle. The transistor only conducts long enough to keep the collector tank circuit oscillating.

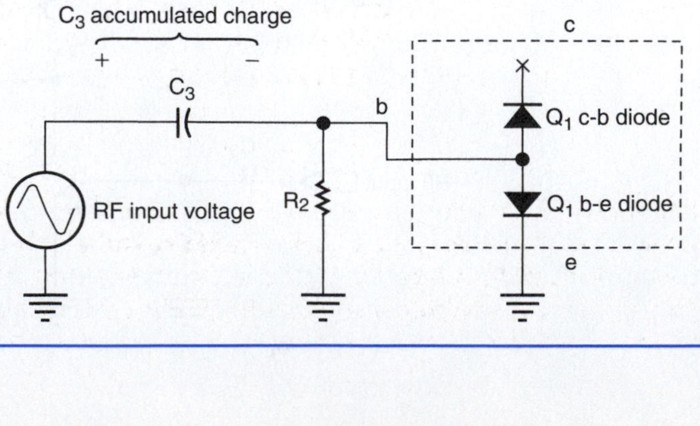

Figure 4–28 The hidden clamper circuit in a class C amplifier

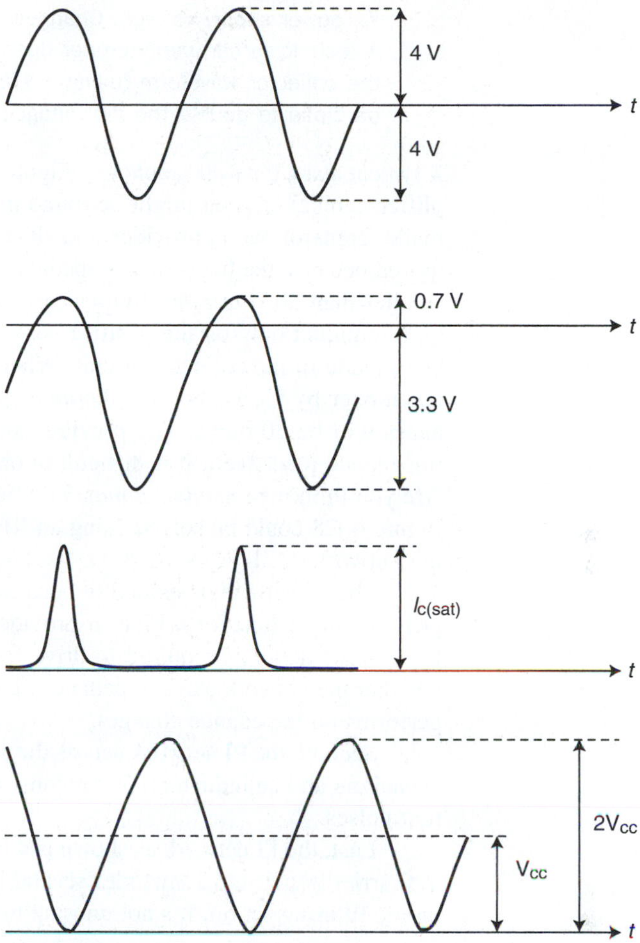

Figure 4–29 Class C circuit waveforms

ac analysis In Figure 4–29, the transistor conducts for only a small portion of the ac input cycle, and then only on the very positive tip of the input signal. The collector current appears as a series of narrow "spikes," yet the collector voltage is a nice clean sine wave. How can this be?

The answer is the *flywheel effect* provided by the LC resonant tank circuit in the collector portion of the circuit formed by L_1 and C_1. The collector current "spikes" can be thought of as individual "pushes" on the LC tank circuit. Since the LC tank is tuned to the same frequency as the input signal, the "pushes" arrive at just the right time to keep the tank oscillating. Once the tank is set in motion, it tends to produce a sine wave. The periodic pushes from the transistor keep it going.

There are a couple of important points concerning the collector voltage of Figure 4–29. First, notice that it rides on a *dc level* that is equal to the power supply voltage V_{cc}. Capacitor C_2 prevents this dc level from reaching the load. Second, look at the *amplitude* of the sine wave. It is *twice the V_{cc} power supply voltage*. As long as there is sufficient input signal to turn on the transistor, the peak-to-peak output voltage of the class C amplifier will stay approximately the same, about *twice* V_{cc}. For example, if this amplifier were operating off a 15 volt V_{cc} supply, the output voltage would be close to 30 V_{pp}.

In a class C amplifier, the ac output voltage is primarily determined by the V_{cc} power supply voltage. Changes in input signal voltage have little effect! A technician can verify proper operation of a class C amplifier by looking at the collector waveform (using a 10:1 probe); its peak-to-peak voltage should be close to double the V_{cc} voltage.

A typical class C power amplifier Figure 4–30 shows a more realistic class C power amplifier, typical of what might be found in an actual circuit. A few refinements have been made. Transformer T_1 provides impedance-matching for the base input circuit. This is required because the base-input impedance of a bipolar junction transistor is usually much smaller than 50 Ω at radio frequencies. R_1 and C_1 still function as the clamper, allowing Q_1 to conduct only on the positive peaks of the input signal. More drastic changes have been made in the collector circuit. Where did the tank circuit go? Its function has been taken over by C_3, L_2, and C_4, the *antenna coupler.* Recall that most antenna load resistances will be 50 ohms. The previous amplifier (Figure 4–28) will not drive such low-impedance loads well; it is difficult to obtain appropriate values for L and C in the tank. Are you thinking *impedance-matching time?* If so, you're right. The matching problem in Figure 4–28 could be solved using an RF transformer or by the use of a π ("PI") matching network (C_3, L_2, C_4).

The PI network is named so because it looks like the Greek letter π. The PI network provides three functions. First, it provides *impedance-matching,* stepping the impedance level up or down as required to drive the load. (The relative sizes of C_3 and C_4 control whether the network steps impedance *up* or *down;* when C_3 and C_4 are *equal,* the network performs no impedance change).

Second, the PI network acts as the resonant tank circuit (after all, it's built out of two capacitors and an inductor!) that reconstructs the carrier sine wave from the collector current pulses.

Last, the PI network as shown is a *low-pass filter.* It serves to attenuate *harmonics* of the carrier frequency. Sometimes several PI sections are cascaded in order to get better harmonic filtering action. It's not unusual to see two-, three-, or four-section PI networks!

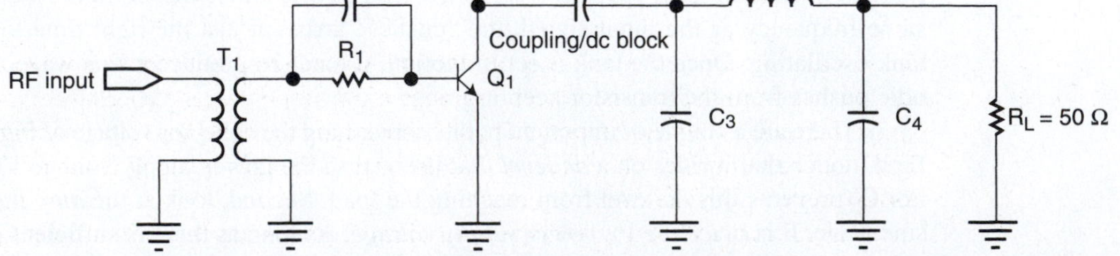

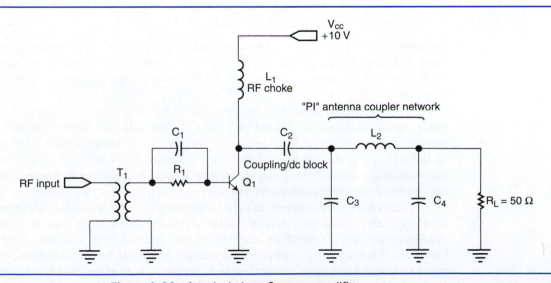

Figure 4–30 A typical class C power amplifier

The collector dc bias in most class C amplifiers is supplied through an *RF choke*. The purpose of an RF choke is to act as an open to ac RF currents yet let dc current pass. The choke is *not* part of the resonant circuit. L_1 acts as the collector RF choke in this circuit.

Section Checkpoint

4–38 List three common features that make RF amplifiers different from low-frequency amplifiers.

4–39 Why is component layout critical at radio frequencies?

4–40 What is the difference between small- and large-signal RF amplifiers?

4–41 What effect is caused by an open emitter-bypass capacitor (C_5 in Figure 4–24)?

4–42 What two functions are performed by the transformer T_1 in Figure 4–25?

4–43 Above the power level of 500 mW, what class of linear amplifiers is used?

4–44 Why is a class C amplifier more efficient that a class A or class B unit?

4–45 What controls the peak-to-peak output voltage of a class C amplifier?

4–46 What is the function of components L_2, C_3, and C_4 in Figure 4–30?

4–6 RF MODULATORS

The same circuits that function as RF amplifiers can be used to generate AM as well, with some minor additions and changes. The primary requirement for generating AM is a *variable* gain. This implies a nonlinear amplifier.

A transistor amplifier can be used to generate AM in three ways. In all three methods, the carrier signal is introduced into the transistor in a normal fashion, on either the emitter or base terminal. The *information* can be placed into any of the three terminals; thus the three methods are called *base injection, emitter injection,* and *collector injection*.

We will look at circuits using *collector injection* and *base injection,* as they are by far the most common.

Collector-Injected Class C Modulator

Figure 4–31 shows a collector-injected class C AM modulator. Since this is a class C circuit, it performs both as a modulator and as an efficient power amplifier. *This type of modulator is usually used as the final RF amplifier stage in a high-level AM transmitter.* This circuit is almost exactly like Figure 4–30; the only difference is the addition of T_2 and bypass capacitor C_5 in the collector circuit.

Recall that the output voltage of a class C amplifier depends on the V_{cc} supply voltage. For example, if V_{cc} is 12 volts, the collector output signal will be approximately 24 volts peak-to-peak (twice V_{cc}). If V_{cc} is increased, the output signal voltage will

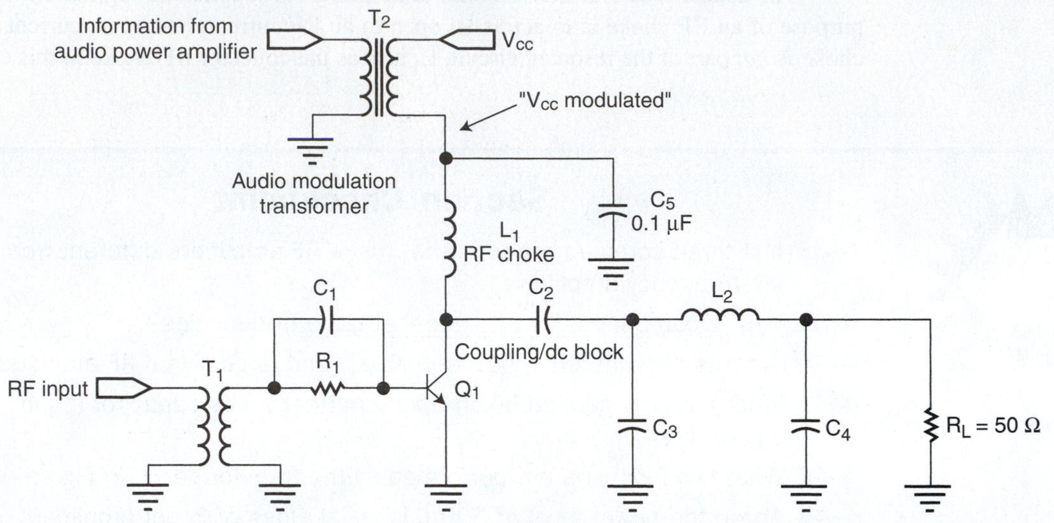

Figure 4–31 A collector-injected AM modulator

increase; if V_{cc} is lowered, the output signal shrinks in a like manner. In other words, *changing V_{cc} changes the equivalent voltage gain of the amplifier, creating amplitude modulation.*

Have you figured out what transformer T_2 is doing? We know that somehow, the information signal must end up causing a carrier amplitude change. Transformer T_2 allows the information voltage to ride on top of the V_{cc} supply voltage at the collector of Q_1. On the schematic, this voltage is marked "V_{cc} modulated." Figure 4–32 shows how it works.

In the figure, you can see the RF input being applied to the base of Q_1, which has a steady amplitude. Also, you can see the audio (information) signal that is being applied into the collector circuit through transformer T_2. Because the secondary of T_2 is effectively wired in *series* with the V_{cc} supply for Q_1, the instantaneous ac voltage developed across T_2 can either *aid* or *oppose* the V_{cc} voltage. Thus, the "modulated V_{cc}" signal is merely the information riding on top of the V_{cc} potential. Notice how the information voltage is slowly changing when compared to the carrier voltage.

The class C amplifier responds to the varying V_{cc} potential by changing its output peak-to-peak voltage. When the "modulated V_{cc}" signal gets larger, the *modulated waveform* in the collector gets correspondingly larger. Likewise, when the "modulated V_{cc}" signal falls (on the negative half-cycle of the information), the modulated waveform shrinks in direct proportion. *Thus, amplitude modulation is created.*

Capacitor C_5 is an *RF bypass*. Large ac RF currents are circulating in the collector circuit of Q_1. To prevent the RF currents from flowing upward into T_2 (and the power supply), RF choke L_1 is used. Any RF currents not stopped by L_1 are sent to ground through C_5. C_5 has little effect at audio frequencies; its reactance is much larger there.

Since this is a power amplifier stage, considerable information power is necessary in order to raise and lower the V_{cc} supply at the collector. Thus, an *audio power amplifier* is needed in order to provide signal for the primary (left-hand) side of transformer T_2.

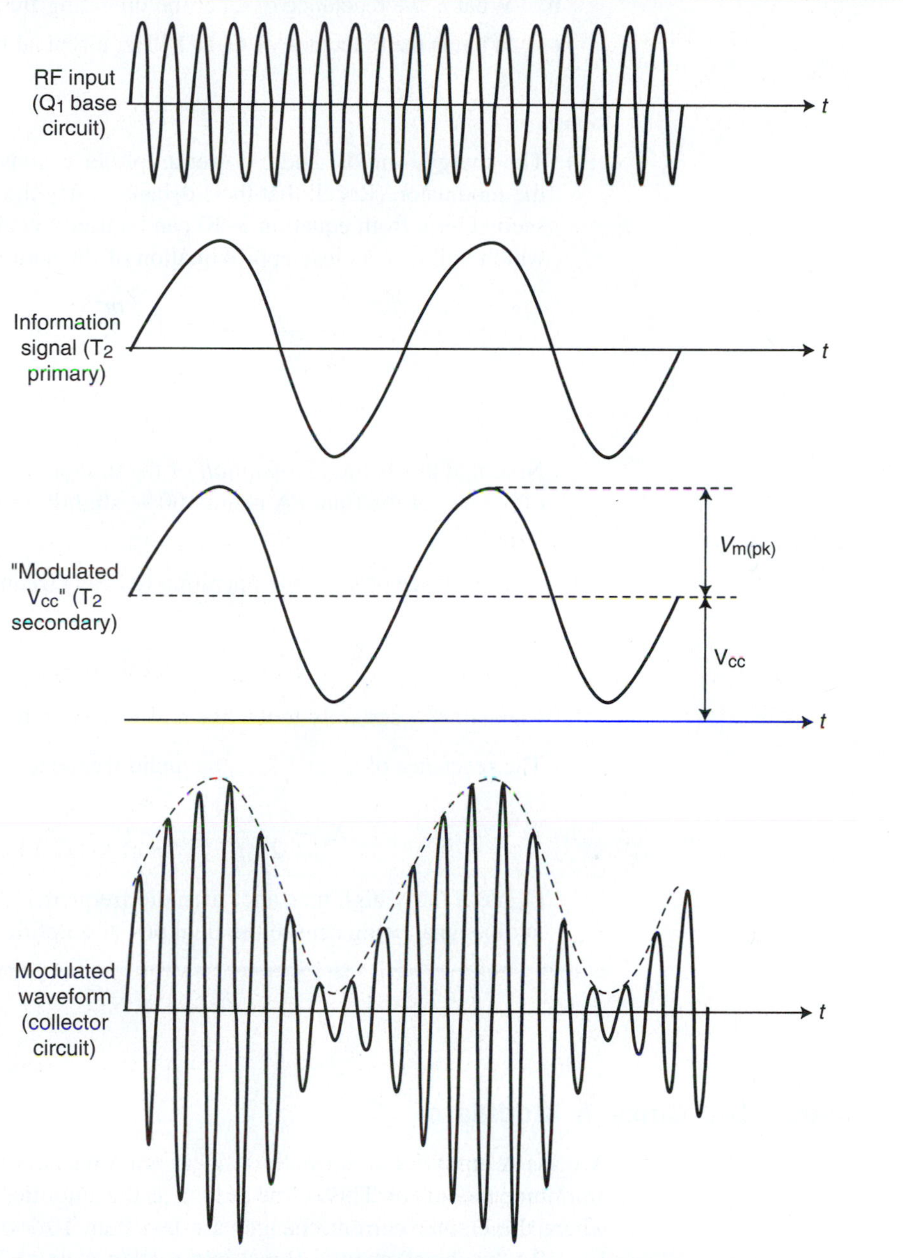

Figure 4–32 Waveforms for the collector-injected modulator

EXAMPLE 4–10 The modulator of Figure 4–32 produces 50 watts into the load when dead-keyed (unmodulated). The unit is operating at 138 MHz.

a. How much power is required from the *audio power amplifier* in order to produce 100% modulation?

b. What is the reactance of C_5 at the operating frequency?

c. What is the reactance of C_5 at 1 kHz, a typical information frequency?

Solution

a. The energy from the audio power amplifier is converted into *sideband* power by the modulator. (Recall that the sidebands carry the information.) Therefore, the second term from equation 3–10 can be used to calculate the sideband power, which will give a close approximation of the audio power required:

$$P_t = P_c + P_c\left(\frac{m^2}{2}\right) = P_c\left(1 + \frac{m^2}{2}\right) \quad (3\text{–}10)$$

$$P_{\text{sidebands}} = P_c\left(\frac{m^2}{2}\right) = 50\text{ W}\left(\frac{1^2}{2}\right) = \underline{\underline{25\text{ W}}}$$

Note that this is merely *one-half* of the unmodulated carrier power. Since the efficiency of the final PA is not 100%, slightly more than 25 watts will be required.

b. The reactance of C_5 at the operating (carrier) frequency is

$$X_c = \frac{1}{2\pi fC} = \frac{1}{2\pi(138\text{ MHz})(0.1\ \mu\text{F})} = \underline{\underline{0.012\ \Omega}}$$

Capacitor C_5 certainly looks like a *short circuit* to the RF energy!

c. The reactance of C_5 at 1 kHz, an audio frequency, is

$$X_c = \frac{1}{2\pi fC} = \frac{1}{2\pi(1\text{ kHz})(0.1\ \mu\text{F})} = \underline{\underline{1591\ \Omega}}$$

C_5 has a fairly high reactance at audio frequencies. It does not interfere with the audio signal being sent to the final power amplifier stage.

Base-Injected Class A Modulator

A class A amplifier is normally a linear amplifier and therefore is not well suited as an amplitude modulator. This is true as long as the amplifier is operated in *small-signal mode,* where the emitter current changes are less than 10% of the collector saturation current $I_{c(sat)}$. By "pushing" enough signal into a class A amplifier, it will become nonlinear and will therefore produce amplitude modulation. Figure 4–33 shows a typical base-injected modulator.

In the *base-injected modulator,* both the RF carrier signal and the AF information signal are sent into the base circuit at the same time. The AF signal is much larger than the RF signal and causes the emitter current of Q_1 to change in step. On the positive peaks of the audio, the emitter current increases. Because the emitter is ac bypassed by C_5, the transistor's gain depends on r'_e, which in turn is controlled by the emitter current I_E. The result is that the voltage gain of the circuit *increases* during the positive half-cycle of the audio and *decreases* during the negative half-cycle. Because the output voltage is equal to the

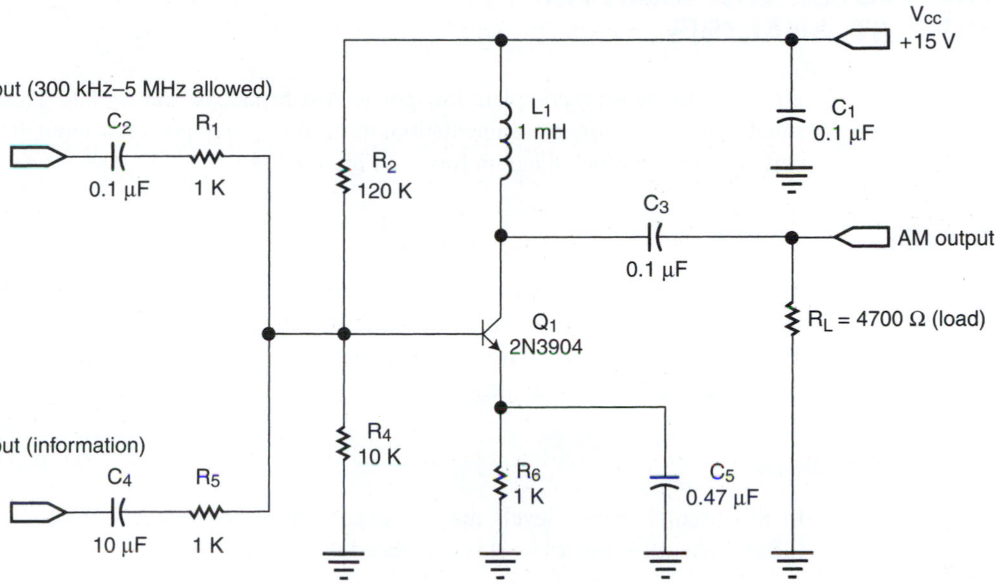

Figure 4–33 A base-injected modulator

product of voltage gain and input RF voltage, the output is an *amplitude-modulated* RF carrier wave.

To prevent AF energy from exiting the amplifier, a highpass filter is used at the output. L_1 and the load resistance combine to form this filter. Therefore, only a modulated RF waveform leaves the circuit.

Section Checkpoint

4–47 For an amplifier to produce AM, what is the primary requirement?

4–48 List three ways AM can be produced in a transistor amplifier.

4–49 What class of amplifier is used for *collector injection*?

4–50 In a collector-injected modulator, what determines the voltage on the "V_{cc} modulated" line?

4–51 Why does changing the V_{cc} voltage change the output of a class C amplifier?

4–52 What is the function of transformer T_2 in Figure 4–31?

4–53 Why is an AF power amplifier required for providing the information signal in a collector-injected modulator?

4–54 Why does the circuit of Figure 4–33 produce AM?

4–55 What causes the amplifier of Figure 4–33 to have a variable gain?

4–7 HIGH-LEVEL TRANSMITTER CIRCUIT ANALYSIS

Figure 4–35 shows a complete low-power AM broadcast transmitter. This circuit is very suitable for laboratory experimentation. Its carrier output power is about 400 mW. The circuit is shown in block diagram form in Figure 4–34.

Power Supplies

The transmitter has two power supplies. One is the main dc supply of +12 volts; the second is "hidden." Can you find it? That's right—the zener diode, D_1, is a *voltage regulator*. The regulated +6.2 volts from D_1 supplies the oscillator and buffer amplifier stages to keep their supply voltages steady.

Absolute Decibel Units

In RF circuits, power levels may be expressed either in watts or in *decibel-milliwatts* (dBm). The dBm power level is calculated as

$$\text{dBm} = 10\log\left(\frac{P}{1\text{ mW}}\right) \qquad (4\text{–}5)$$

The dBm is an *absolute* or *power* decibel unit, in contrast to standard dB units, which are *relative* or *gain* units. Relative (or regular) dB units *compare* two power levels. For example, when we say that an amplifier has a +3 dB gain, we're really saying that the output

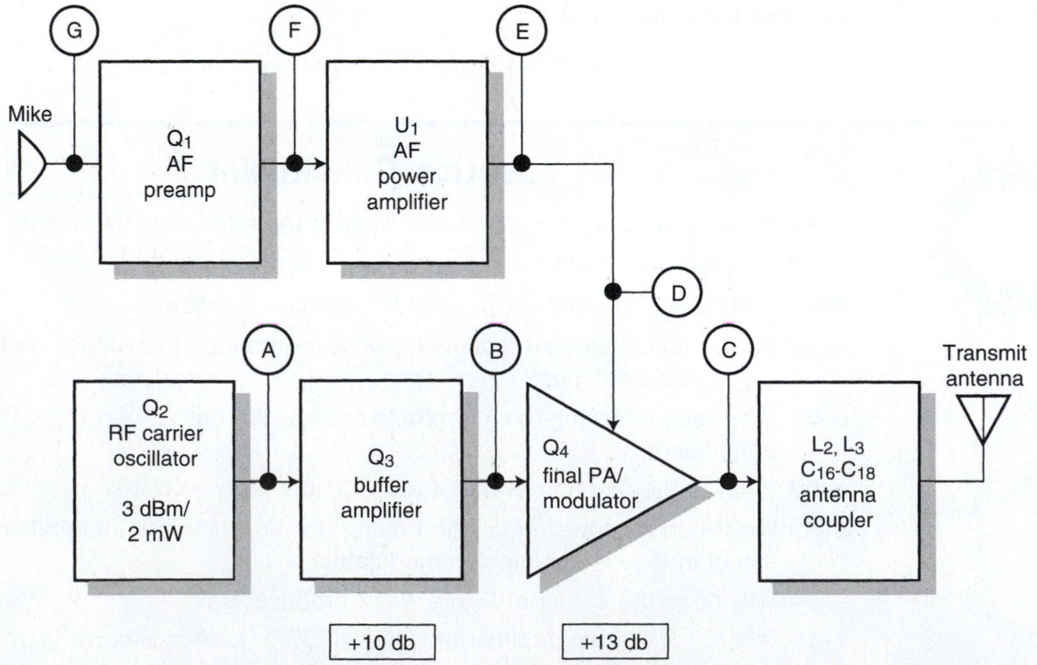

Figure 4–34 Low-power broadcast transmitter block diagram

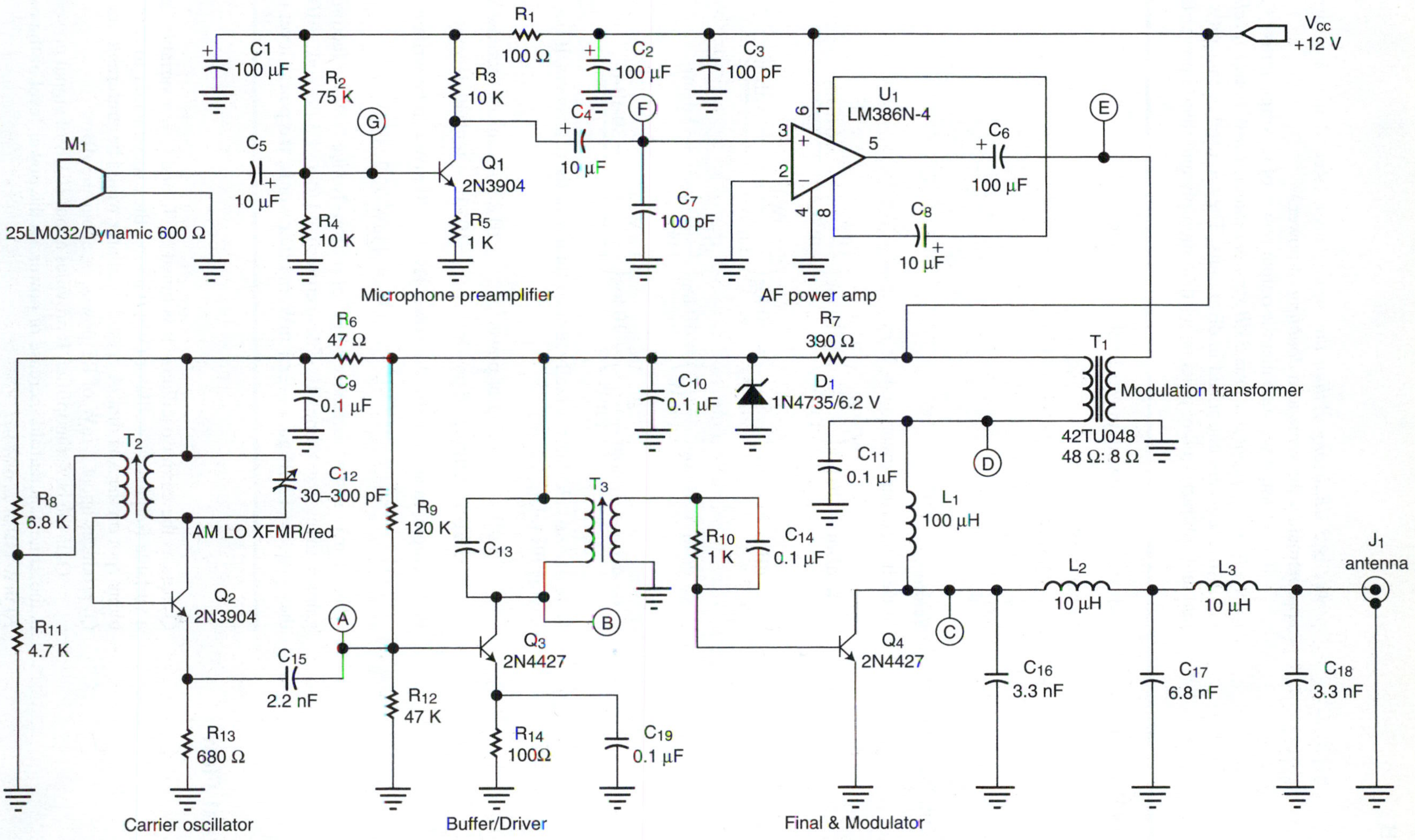

Figure 4–35 A low-power AM broadcast transmitter

power is *3 dB stronger than the input power.* (Recall that a positive 3 dB change is approximately the same as a *doubling* of power.)

In contrast, if we state that the output power of an amplifier is +3 dBm, we're really saying that the *output power is 3 dB stronger than a 1 mW power level, or 2 mW.* Most RF lab instruments are calibrated in dBm units. They're used because it's more convenient to add and subtract decibel units rather than multiply and divide power levels.

EXAMPLE 4–11

Express the following power levels in dBm units:

a. 4 mW
b. 20 mW
c. 20 W
d. 1 μW

Solution

In each case, use equation 4–5:

a. $$\text{dBm} = 10\log\left(\frac{P}{1\text{ mW}}\right) = 10\log\left(\frac{4\text{ mW}}{1\text{ mW}}\right) = \underline{\underline{+6\text{ dBm}}}$$

b. $$\text{dBm} = 10\log\left(\frac{P}{1\text{ mW}}\right) = 10\log\left(\frac{20\text{ mW}}{1\text{ mW}}\right) = \underline{\underline{+13\text{ dBm}}}$$

c. $$\text{dBm} = 10\log\left(\frac{P}{1\text{ mW}}\right) = 10\log\left(\frac{20\text{ W}}{1\text{ mW}}\right) = \underline{\underline{+43\text{ dBm}}}$$

d. $$\text{dBm} = 10\log\left(\frac{P}{1\text{ mW}}\right) = 10\log\left(\frac{1\ \mu\text{W}}{1\text{ mW}}\right) = \underline{\underline{-30\text{ dBm}}}$$

Note that *no* calculator is really needed for any of these answers, if one remembers the following dB relationships:

+3 dB = doubling of power	+10 dB = tenfold increase in power
−3 dB = halving of power	−10 dB = tenfold decrease in power

For example, in answer (b), the power level of 20 mW can be written as follows:

$$20\text{ mW} = \underline{1\text{ mW}} \times 2 \times 10$$

where 1 mW is the "reference level" of 0 dBm. Notice how we factored the gain into *two* gains, a times 2 and times 10. These correspond to a +3 dB and +10 dB gain. Together, they give a +13 dB gain over 1 mW; in other words, the power level is +13 dBm.

Signal Flow

Q_2 acts as the carrier oscillator. It is tuned by T_2 and C_{12}, a variable capacitor. The use of a variable frequency oscillator (VFO) allows the experimenter to tune the transmitter to a blank spot on the AM dial. A crystal oscillator would be preferable for frequency stability. Q_2 produces about 2 mW of RF power, or about +3 dBm.

Q_3 is the buffer amplifier. It provides about 10 dB of signal power gain. A 10 dB gain is the same as a tenfold increase of power, so the power level at the output of Q_3 is about 20 mW (10 × 2 mW).

Q_4, the final amplifier, is driven from Q_3 through impedance-matching transformer T_3. Q_4 provides an additional 13 dB of power gain, so the output to the antenna is close to (20 mW × 10 × 2) or 400 mW. C_{16}–C_{18}, L_2, and L_3 form a double-PI section for the antenna coupler. Q_4 is likely to be *heatsinked*. At the relatively low power of 400 mW, a clip-on heatsink would likely be used.

Q_4 is collector modulated by U_1, the audio power amplifier. T_1 couples the audio output of U_1 into the V_{cc} power supply of Q_4.

The microphone, M_1, provides the information signal. This signal is voltage amplified by Q_1, and both voltage- and current amplified by U_1, the audio power amplifier.

There are many power supply filters in the transmitter. C_1 filters the audio preamplifier supply; C_2 and C_3 filter the supply for U_1; C_9 and C_{10} filter the power supplies for the oscillator and buffer amplifier stages.

Section Checkpoint

4–56 What are the two power supply voltages in the transmitter of Figure 4–35?

4–57 What is the purpose of amplifier stage Q_1 in Figure 4–35?

4–58 What is the effect of the variable capacitor C_{12}?

4–59 What is the function of transformer T_1?

4–60 Explain the difference between dB and dBm units.

4–61 If a signal is +3 dBm in strength, what is its actual power level?

4–62 How much is a power of 50 mW in dBm units?

4–8 MEASURING TRANSMITTER PERFORMANCE

There are several measurements that a technician might be required to make on an AM transmitter. Most of these measurements require specialized equipment; however, they're not difficult to learn. The most common performance measurements include *power output, frequency accuracy, modulation acceptance,* and *spectral purity.*

Power Output

To measure power output, a *directional wattmeter* and *dummy load* are used, as shown in Figure 4–36.

A *directional wattmeter* (Figure 4–37) measures the flow of RF energy. Most directional wattmeters have a switch or knob to select whether *forward* (toward the load) or *reverse* (reflected from the load) RF power flow is to be measured. A correctly matched dummy load will produce little or no reflected RF power. For power measurements, the switch on the directional wattmeter is moved to the forward position, and the transmitter is keyed. The power output is read from the calibrated meter scale.

A *dummy load* is a noninductive power resistor that is meant as a temporary load for transmitters during testing. Most dummy loads have a resistance of 50 Ω. A low-power dummy load may look like Figure 4–38(a). Figure 4–38(b) shows a higher-power version

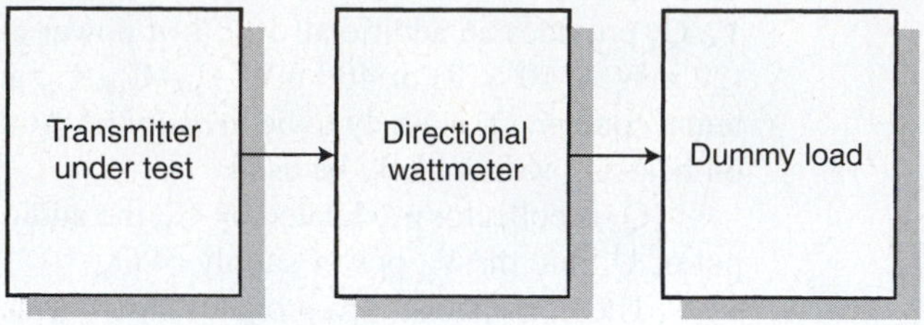

Figure 4–36 Measuring transmitter output

of the same thing. The "can" in Figure 4–38(b) contains the load resistor immersed in transformer oil. The oil helps dissipate the heat. An oil-filled dummy load can safely handle more than 1000 watts from a transmitter.

Frequency Accuracy

A frequency counter (Figure 4–39) can be used to check the carrier frequency accuracy of an AM transmitter. The counter must be coupled to the transmitter through an *RF sampling unit* (a device connected between a transmitter and load that siphons off a tiny amount of the radio signal for sampling purposes—such as checking frequency).

Specialized *station monitor receivers* can also be used to check transmitter frequency without any direct connection to the transmitter at all. These are very commonly used for servicing VHF and UHF transmitters.

Modulation Acceptance

For AM transmitters, it is usually a good idea to verify that modulation is occurring and that the unit operates as close to 100% as possible without overmodulation. Some wattmeters include a modulation measurement function. An even better method is to use an RF sampler to couple an oscilloscope to the transmitter and directly view the modulated waveform on the scope. Scope measurement has the advantage of being able to verify the *shape* of the envelope. Any envelope distortion will be immediately visible on a scope.

Figure 4–37 A typical directional wattmeter

Figure 4–38 Two types of dummy loads (not to scale)

Spectral Purity

There's no substitute for a *spectrum analyzer. An RF sampling unit or RF power attenuator must always be used between the transmitter and spectrum analyzer.* Failure to observe this precaution will cause instant failure of the RF front end of the spectrum analyzer—an expensive repair!

There are two types of problem signals that a transmitter can produce. *Harmonics* are frequencies appearing at multiples of the original transmitter frequency. For example,

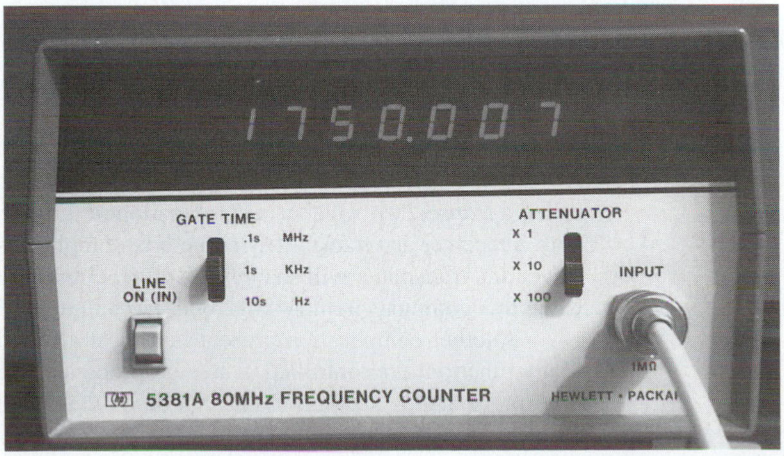

Figure 4–39 Measuring a transmitter's frequency

a 1 MHz transmitter will produce harmonics at 2 MHz, 3 MHz, 4 MHz, and so on. *All transmitters produce harmonics.* The object is to make sure that the harmonics are at no more than the specified strength, which is measured in dBc (decibels with respect to the carrier signal). A modern transmitter should produce 2nd and 3rd harmonics less than −50 dBc (negative dB means *smaller*).

The second type of problem signal is *spurious emissions,* or *spurs*. Spurs are not harmonically related to the carrier. Instead, they often appear *adjacent* to the carrier signal on the spectrum analyzer display. Spurs can cause adjacent channel interference, even when the transmitter is operating perfectly on frequency. Spurious emissions are often caused by faulty power supply filtering capacitors, missing or loose metal compartment covers, or loose RF connectors. *All transmitters produce spurious emissions.* The manufacturer will give the maximum acceptable spurious emission levels, again in units relative to the carrier strength, dBc.

Section Checkpoint

4–63 What equipment must be connected to measure *power output* of a transmitter?

4–64 What is a *dummy load?*

4–65 How much reflected power should there be from a dummy load?

4–66 Why are some dummy loads filled with oil?

4–67 Describe the purpose of an *RF sampler.*

4–68 What could happen if a spectrum analyzer input is directly connected to a transmitter output?

4–69 What are two pieces of equipment that can be used to measure *frequency accuracy?*

4–70 Why is an oscilloscope superior to a wattmeter for measurement of *modulation acceptance?*

4–71 Explain the two main "problem signals" that transmitters can produce. How are they different?

4–9 TROUBLESHOOTING TRANSMITTERS

An AM transmitter built as one free-standing unit is pretty much a thing of the past. Most AM transmitters (with the exception of broadcast units) are part of a combination unit called a *transceiver,* which is a combination of the words *transmitter* and *receiver*. Having both a receiver and transmitter in one box complicates matters a little. Sometimes the receiver and transmitter will share some of the circuitry in order to reduce component count; when this common circuitry fails, both transmitter and receiver stop working.

Another complication present in almost all modern equipment is *microprocessors*. Many functions are controlled by a microprocessor in today's communications gear. There is a mix of analog and digital that can be quite intimidating to the new tech. When something malfunctions, it can sometimes be hard to tell if the digital side is "glitched" and sending improper commands to the analog "works," or if the analog side really is sick. Some manufacturers are very good at providing service information to help the tech out of these fixes, while others provide only marginal information (at best) about what's going on inside their black boxes.

Servicing transmitters also involves two *safety* problems: RF exposure and high voltages.

RF Exposure

An *RF burn* can result when the skin comes in contact with a conductor carrying RF energy. The heat is generated just below the surface of the skin. Even a few watts can produce a painful burn. Transmitters at the kilowatt level can produce severe burns. Technicians must use caution when working near a live RF source.

Exposure to intense *RF energy fields* is another potential hazard. You cannot feel an RF field, yet it may have the potential to harm you. The most sensitive part of your body is your eyes. To avoid exposure, don't operate transmitters without shield covers in place and stay away from "live" transmitting antennas.

High Voltages

Some transmitters use high voltages. Even relatively low voltages like 24 volts can be lethal under the wrong circumstances. Many tube transmitters and amplifiers use over 1000 volts in their circuitry. One of the best methods of avoiding shock is to avoid working on live equipment. This isn't often realistic; if measurements *must* be made on live equipment, first check to see if the manufacturer has recommended a safe method. When measuring high-voltage test points, it's best to use just one hand (keep your free hand in a pocket).

Troubleshooting Method

Regardless of how a transmitter is built, the three-step troubleshooting method can be used to find problems. The steps of the method are

1. visual (and other) inspection
2. power supply checks
3. input and output checks

The details of visual inspection have been previously discussed. Power supplies should *always* be checked before doing any detailed circuit checks. *More than 90% of electronic problems are power supply related!*

EXAMPLE 4–12

Develop a plan for troubleshooting the transmitter of Figure 4–35. The symptoms given on the repair order are as follows: "While operating, the RF power output dropped suddenly from the normal 400 mW level to about 1 mW [barely readable in a nearby receiver]."

Solution

The symptoms described could be a *loss of RF power gain.* According to the schematic, Q_3 and Q_4 are responsible for providing this gain. However, before doing *any* checks on these stages, the power supply voltages should be checked. Therefore, the plan might look like this:

1. Perform a visual inspection (especially control settings); if these look OK, open the cabinet and inspect for physical damage.
2. Check the *main* +12 volt supply (upper right) input to make sure the circuit has proper voltage. (An incorrect reading here could mean a defective power supply [off page], or perhaps something in the unit itself is drawing excessive current, causing the power supply to "pull down.")
3. Check the 6.2 volt regulated supply at the anode of D_1, the zener diode (±10% is a typical tolerance for zener voltage regulators). If this is OK, proceed to the next step.
4. If the power supplies are OK, measure the RF test points at A, B, and C. These should be measured with a 10:1 oscilloscope probe. Since a severe signal output loss has occurred, one or more of the test points will have a greatly reduced signal voltage.

   Suppose that we measured the following signal voltages:

$$TP(A) = 2\ V_{pp}, \quad TP(B) = 12\ V_{pp}, \quad TP(C) = 1\ V_{pp}$$

Where is the problem most likely to be? You can see that the voltage progressively gets *larger* between (A) and (B), but at (C), something is amiss. In fact, you know that the peak-to-peak voltage on the collector of a class C stage is normally about *twice* V_{cc}, so we should be seeing about 24 V_{pp} at test point (C).

There is definitely a problem with the Q_4 stage! What do we do now? We would now normally look at the *dc bias voltages* on the three terminals of Q_4 and compare them with manufacturer's specs. (What we're *really* doing is looking at the *"power supply"* of Q_4—its bias voltages.) This sort of investigation will normally point to the component at fault.

When attempting to troubleshoot a known defective stage, it is always a good idea to measure the dc bias voltages of that stage.

Section Checkpoint

4–72 Why are stand-alone transmitters rare?

4–73 What device controls the functions in most modern equipment?

4–74 List at least two safety precautions to prevent exposure to excessive RF energy.

4–75 What are the three basic steps of troubleshooting?

4–76 Why were the Q_3 and Q_4 stages suspect in Example 4–11?

4–77 When a stage is isolated as the problem, what should be done next to localize the problem within the stage?

SUMMARY

- Systems are built out of building blocks. Understanding a block diagram is an important first step to understanding a system.
- AM transmitters can be low- or high-level designs. Each has its own set of advantages and disadvantages.

- The dc efficiency of a transmitter is the ratio of RF output power to dc input power.
- An oscillator needs sufficient power gain and positive feedback in order to work.
- The three fundamental oscillators are the Armstrong (transformer coupled), Hartley (tapped coil), and Colpitts (tapped capacitor). Most circuits use one of these or a derivative (Clapp).
- Crystals provide accurate frequency control due to their high *Q* but lack frequency agility: In order to change frequency, the crystal must be changed.
- All crystals have two resonance modes, series and parallel, sometimes known as the *zero* and *pole* frequencies.
- RF amplifiers use special circuit techniques and components. Circuit layout is critical for proper operation.
- RF amplifiers can be categorized in four ways: small or large signal, and linear and nonlinear.
- RF bypass capacitors are numerous in radio equipment; they send undesired RF currents to ground.
- A transistor can generate AM by having information injected into any of its three leads, depending on the circuit configuration.
- dBm units are *absolute* decibels with respect to a reference power of 1 mW.
- Specialized equipment is needed to accurately measure transmitter performance. Care must be exercised when connecting spectrum analyzers to transmitters.
- When troubleshooting a transmitter, the three basic steps (visual inspection, power supply checks, input/output checks) should be applied.
- Transmitters pose safety hazards. Technicians must be aware of these and work in a careful manner.

PROBLEMS

1. Draw a block diagram of both low- and high-level AM transmitters.
2. How can a high-level AM transmitter be identified when examining the schematic or block diagram?
3. What is meant by the term *frequency stability* when referring to oscillators?
4. What two design features are used in AM transmitters to enhance oscillator stability?
5. What are the two functions of an *antenna coupler?*
6. A certain AM transmitter requires 1 A from a 48 V dc supply. It produces 36 W of RF output. What is its dc efficiency?
7. If an oscillator is built like the block diagram of Figure 4–4 and has a feedback gain *B* of 0.01, what *minimum* gain *A* would be needed for it to keep running (if already running)?
8. What will the oscillation frequency of the oscillator in Figure 4–8 be if $L_p = 100\ \mu H$ and $C_1 = 1$ nF?
9. Calculate the frequency of the oscillator of Figure 4–9 if $L_1 = 1$ mH and $C_1 = 400$ pF.
10. What frequency will be produced by the oscillator of Figure 4–11 if $L_1 = 500\ \mu H$, $C_1 = 330$ pF, and $C_2 = 470$ pF?
11. What is added to a Colpitts oscillator in the Clapp configuration? How does the change increase frequency stability?
12. How are *Q* and bandwidth of a tuned circuit related? If *Q* is increased, what happens to bandwidth?
13. A certain quartz crystal has the following parameters: $L_s = 1.38$ H, $C_s = 0.019$ pF, $R_s = 68\ \Omega$, and $C_m = 4.9$ pF. Calculate the following:
    a. The series-resonant frequency
    b. The crystal's terminal resistance at series resonance
    c. The parallel-resonant frequency
    d. The *Q* at series resonance
    e. The pole-zero spacing, in Hz
14. Repeat the calculations of problem 13 for a crystal with $L_s = 1.05$ H, $C_s = 0.029$ pF, $R_s = 73\ \Omega$, and $C_m = 5.5$ pF.
15. How can it be determined whether a crystal is being used in *series* or *parallel* mode in a crystal oscillator circuit?

16. Why must component leads and wires be kept as short as possible in RF circuits?

17. What would the operating frequency of the amplifier of Figure 4–24 be if $L_1 = 100\ \mu H$ and $C_2 = 470$ pF?

18. Why is *trickle bias* used in a class AB amplifier?

19. Draw the schematic diagram of a PI low-pass section as used in an antenna coupler.

20. What three functions are accomplished by the PI network following a class C power amplifier?

21. What peak-to-peak collector voltage should be present in a properly functioning class C power amplifier?

22. If the modulator of Figure 4–31 produces 30 W when dead-keyed, what audio information power is required in order to achieve 100% modulation?

23. Convert the following power levels into dBm units:

    a. 1 mW
    b. 2 mW
    c. 10 mW
    d. 20 mW
    e. 200 mW
    f. 4 W
    g. 100 W

24. A certain stage has a power gain of 10 dB and an input power of 23 dBm. Express the output power in dBm units.

25. What power, in watts, corresponds to a level of 33 dBm?

26. What are the two hazards associated with transmitter repair work?

27. What is a *transceiver?*

28. Draw a diagram showing how a directional wattmeter can be used to measure transmitter power output.

29. What must be connected between a transmitter and a spectrum analyzer (or frequency counter) in order to prevent equipment damage?

5

AM Receivers

OBJECTIVES

At the conclusion of this chapter, the reader will be able to:

- explain the steps necessary in the reception of a radio signal
- explain how a diode AM detector works
- draw a block diagram of a TRF radio receiver
- list at least two limitations of TRF radio receivers
- draw a block diagram of a superheterodyne AM receiver, explaining the signals at each point
- given a carrier and local oscillator frequency, calculate the frequency of the various tuned circuits in a superheterodyne receiver
- calculate the image frequency of a signal
- given the schematic diagram, recognize the functional blocks and signal flow in a superheterodyne receiver
- describe the alignment procedures for AM receivers
- given a receiver with a problem, isolate the fault to a particular AF or RF stage

A radio receiver completes a communications system. Without a receiver, a transmitter is useless! Receivers come in many different forms. They can be designed to receive voice, digital data, and many other kinds of signals. Receivers of all types share many common features. The understanding of AM receivers will be an important foundation for the study of more advanced systems.

THE SECRETS INSIDE EVERYDAY OBJECTS

Technology and its products surround us, and we're accustomed to being dazzled by the magic inside the latest gadgets. There's nothing flashy about a transistor radio, a digital wristwatch, or a pocket calculator. These products are inexpensive, readily available, and have been around for a long time. But look inside them—what does it take to *build* a radio, watch, or calculator? Perhaps this question is what led you to the study of electronics.

When we look at the *details* inside such mundane devices, we're overwhelmed. How did anyone ever figure all of this out? There has to be some *magic* in there somewhere. But there isn't; all electronic devices must obey the laws of physics.

There are many years of electronics knowledge built into the humblest widget. Historically many men and women have contributed to this store of knowledge for more than a century. No one can learn it all in one night. Don't be intimidated if you don't understand a principle immediately. Most of us need to review and study technical material repeatedly in order for it to "stick." Be persistent. Be patient. You can learn any area of electronics if you apply yourself!

5–1 RECEIVER OPERATION

The process of receiving a radio signal can be broken down into a series of five steps. Not every receiver will perform every step, but most do. Figure 5–1 shows this in block diagram form.

Steps in Reception

1. *Signal Acquisition* To *acquire* a signal means to get it. Radio signals are in the form of electromagnetic energy traveling through space at the speed of light. In order for a radio signal to be useful in an electronic circuit, it must first be converted back into an electrical signal. This is the job of the *antenna*.
2. *Signal Selection* There are thousands of radio signals in the air at any instant in time. An antenna combines many of them in its electrical output to a receiver. Reception of more than one signal at a time would be annoying to the listener; it would be like listening in a crowded room. How can one signal be extracted from the pile? Right—every radio transmitter uses a different *carrier frequency*. The receiver's *bandpass filter* is tuned to the frequency of the radio station we wish to receive. Ideally, *only* the desired carrier will get through this filter. In reality, there are problems with this approach; filters are not perfect, and interfering signals can get through.
3. *RF Amplification* The distance between a radio transmitter and receiver can be very small or many miles. The transmitted power can be a fraction of a watt or millions of watts. In general, the signal received at a receiver's antenna is very small. At a receiving antenna, the amplitude of a "strong" received signal is usually 100 μV or less. Many receivers must deal with signals less than 1 μV

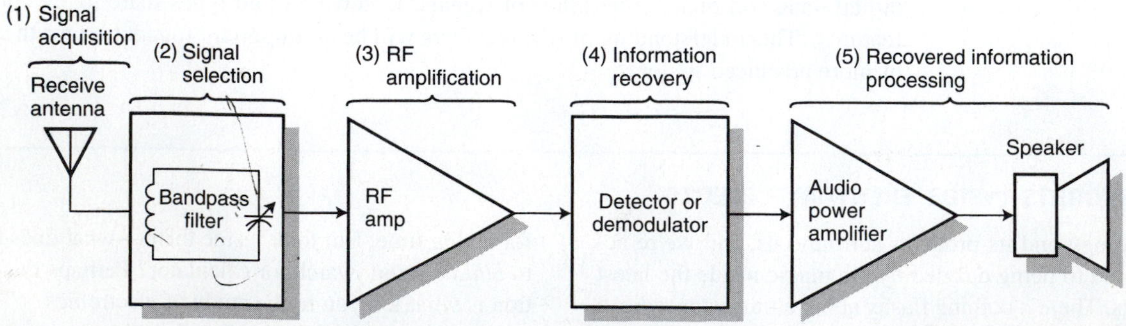

Figure 5–1 The steps in receiving a radio signal

in size. Before such small signals can be processed, they must be amplified. The RF amplifiers developed in Chapter 4 can be adapted and used in receivers as well.

4. *Information Recovery* The actions in the first three steps resulted in reproduction of the *modulated carrier wave* that was sent from the transmitter. The modulated carrier wave holds the information; in order to recover the information, we use a *detector* or *demodulator* circuit. Both words have the same meaning. When we detect a signal, we are extracting the information from the modulated carrier wave. The information is saved and used, and the carrier portion of the wave is discarded.
5. *Recovered Information Processing* This is a general way of saying that we'll be doing something useful with the information the detector extracted. The type of receiver will determine what needs to be done with the information. In a radio receiver, the detected information is an *audio* signal with insufficient voltage and current to drive a loudspeaker. Therefore, the last stage in a radio receiver is an audio power amplifier, which provides the voltage and current needed to operate the loudspeaker. For example, a *television* receiver differs from a radio receiver only in how the detected information (a *video signal* in analog TV, or a *data signal* in digital TV) is processed.

These are the basic steps a receiver needs to perform. In simple radio receivers, some of the tasks can be omitted. For example, you may have built a *crystal radio receiver.* A crystal receiver uses only the energy from the incoming radio wave to produce the sound. A long wire antenna is usually required in order to receive sufficient signal. Figure 5–2 shows the schematic of one type of crystal receiver. There are thousands of different possible crystal receiver designs. Many active enthusiast groups build and study these simple receivers.

There isn't much to the circuit of Figure 5–2! A *long wire* antenna is connected to the upper input terminal. (A length of wire 50′ or longer will work well.) The *earth ground* terminal must be connected to a good earth ground in order to provide an ac return path for the antenna signal. (A ground rod or metal cold water pipe could be used.)

The process of *selection* is accomplished with L_1 and C_1, which form a parallel resonant bandpass filter. The output of the filter is sent to the detector diode, D_1. Two types of variable capacitors for receivers are shown in Figure 5–3. The miniature type is very commonly used in portable receivers.

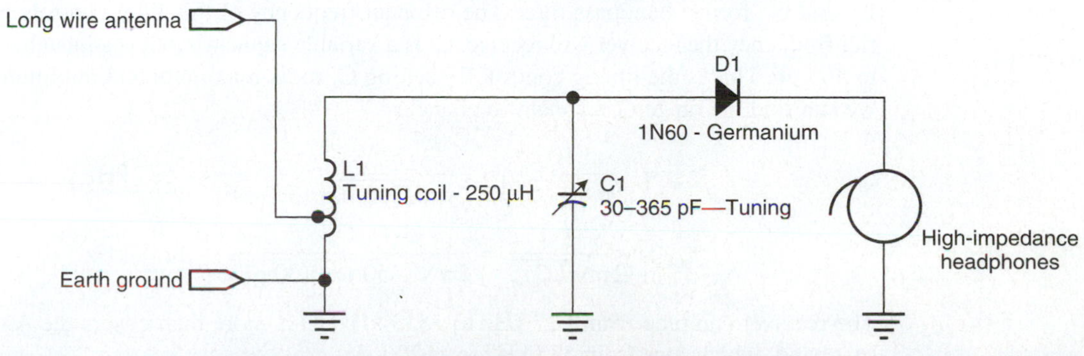

Figure 5–2 A crystal radio receiver

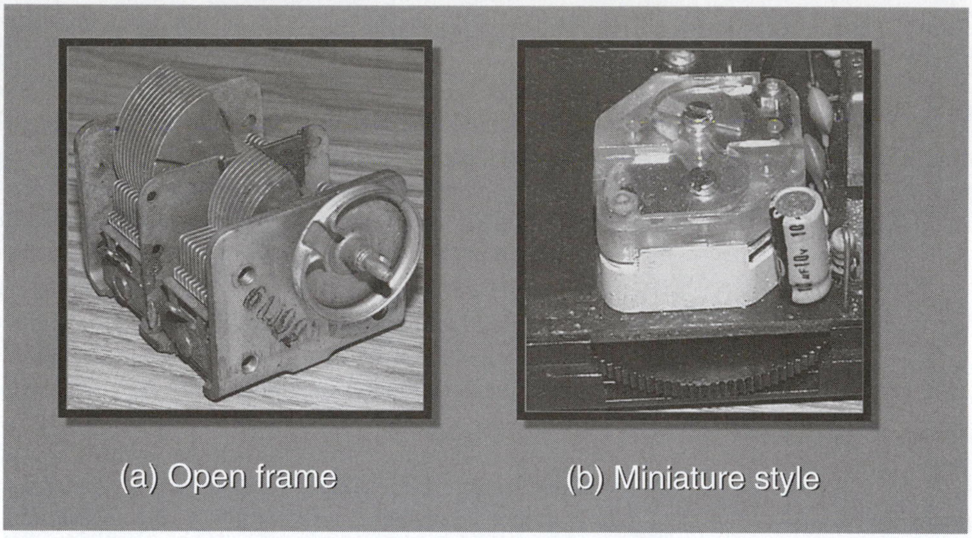

Figure 5–3 Typical receiver variable capacitors

D_1 is a *germanium* diode. The use of this type of diode is common in AM detectors. A silicon diode requires a forward bias of 0.7 volt in order to conduct. In contrast, a germanium diode will begin conduction at only 0.2 to 0.3 volts. This greatly increases the sensitivity of the receiver. *The sensitivity of a receiver is its ability to process small input signals.*

The detected signal is a copy of the original information and leaves the cathode of D_1. It consists of a dc component and two ac components, the AF information, and the RF carrier. The headphones receive this signal, but can only respond to the audio information—the RF energy changes polarity too quickly for the headphones to respond. Thus, the headphones produce sound that is a copy of the original sound from the transmitter.

EXAMPLE 5–1

What is the approximate tuning range of the receiver of Figure 5–2?

Solution

L_1 and C_1 form a bandpass filter. The resonant frequency of this filter controls what carrier frequency the receiver will receive. C_1 is a variable capacitor that is adjustable from 30 to 365 pF. This is the tuning control. By setting C_1 to its maximum and minimum values, we can find the range of resonant frequencies:

$$f_{min} = \frac{1}{2\pi\sqrt{LC_{max}}} = \frac{1}{2\pi\sqrt{(250\ \mu\text{H})(365\ \text{pF})}} = \underline{\underline{527\ \text{kHz}}}$$

$$f_{max} = \frac{1}{2\pi\sqrt{LC_{min}}} = \frac{1}{2\pi\sqrt{(250\ \mu\text{H})(30\ \text{pF})}} = \underline{\underline{1838\ \text{kHz}}}$$

The receiver can tune from 527 kHz to 1838 kHz. This more than covers the AM broadcast band, which runs from 535 kHz to 1620 kHz.

Section Checkpoint

5–1 What are the five steps in the reception of a radio signal?

5–2 What type of circuit is used for *selection* of a signal?

5–3 Why are RF amplifiers necessary in receivers?

5–4 What steps of reception are performed in the circuit of Figure 5–2?

5–5 Why are germanium diodes used in AM detectors?

5–6 How does the end user control the frequency of the bandpass filter in the receiver of Figure 5–2?

5–2 AM DETECTION

There are several types of AM detectors in use. The most common of these is the *diode detector* of Figure 5–4. The diode detector generates some distortion of the information but is the simplest and least expensive approach. A diode detector works because a diode is a nonlinear device. In general, a nonlinear device is required to both *modulate* and *demodulate* (detect) AM signals.

The waveforms for the diode detector are shown in Figure 5–5. The primary action that takes place in a diode detector is *rectification*. The AM detector is very similar to a half-wave power supply in this regard. When the AM signal is applied to the input, the diode cuts off the negative half cycles, since it can only conduct when it is forward-biased. The third waveform of Figure 5–5 illustrates this action. This waveform would be obtained at test point B if capacitor C_1 were removed from the circuit. Although it doesn't look like it, there are actually *three* primary frequency components in the half-wave rectified signal at test point B. They are the *modulated AM carrier signal and sidebands,* the *information frequency,* and a *dc level.*

Time Domain Analysis

The rectified waveform from D_1 is passed to capacitor C_1, which charges up to the voltage of each positive peak of the rectified RF wave. Resistor R_1 is a bleeder resistor. Without R_1 to continually discharge C_1, C_1 would just charge up and hold the highest positive dc level! The time constant of R_1 and C_1 is designed to be as short as possible so that the voltage on C_1 will follow the AM envelope as closely as possible.

The jagged shape of the recovered information is due to the charging and discharging of C_1. A capacitor cannot change its voltage instantaneously. When the envelope is

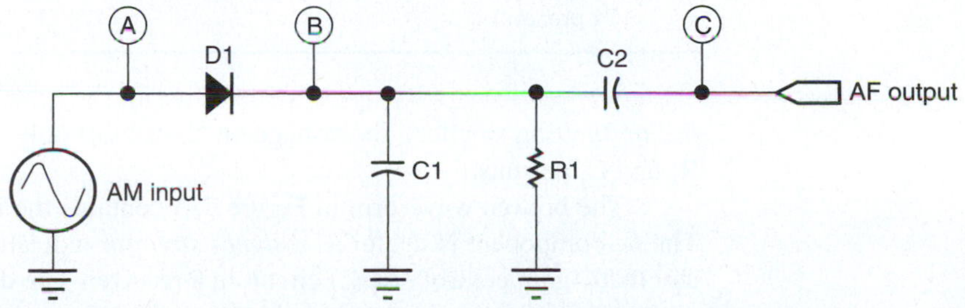

Figure 5–4 The diode detector

Figure 5–5 Waveforms for the diode detector

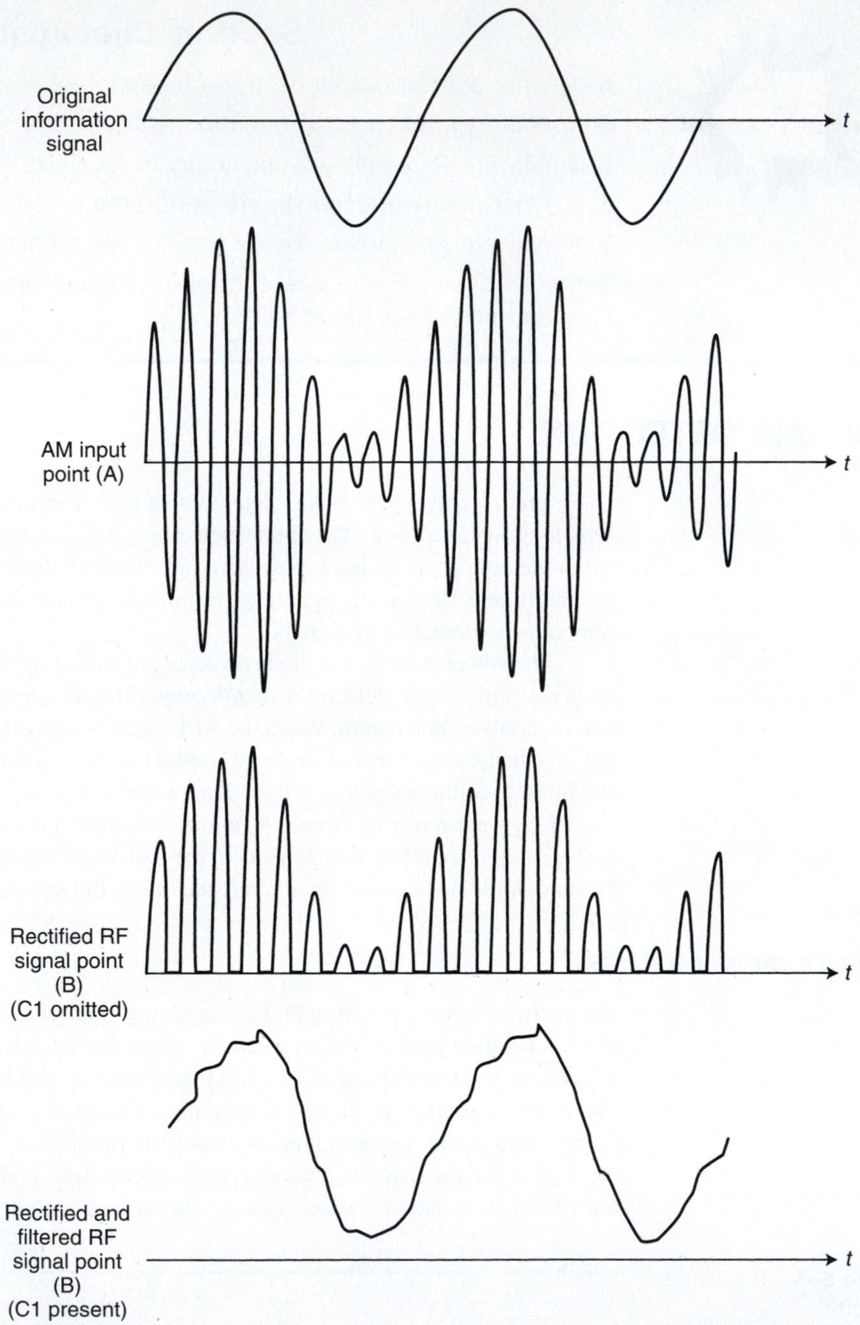

falling (getting smaller), the voltage on C_1 can fall only as fast as the RC time constant of R_1 and C_1 permits.

The bottom waveform in Figure 5–5 contains the information and a dc component. The dc component is useful as a *signal strength* indicator and can be used to operate the automatic gain control (AGC) circuit in a receiver. The dc component is *not* useful for audio amplification, so it is removed by capacitor C_2.

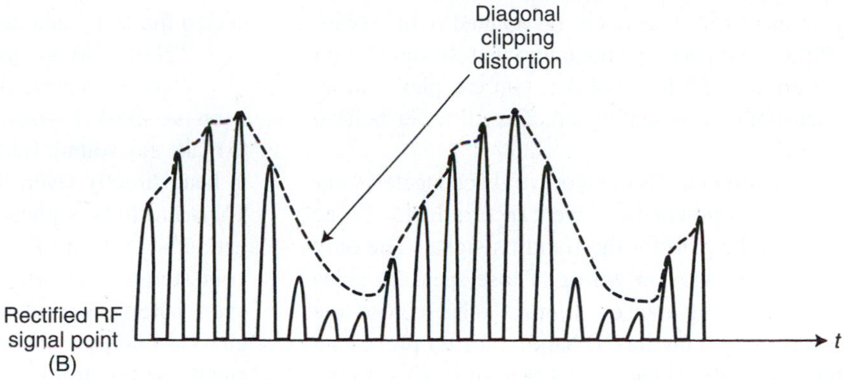

Figure 5–6 Diagonal clipping distortion

Frequency Domain Analysis

The detector's operation can also be explained in the frequency domain. The signal at test point B contains the AM carrier (an RF signal), the information frequency (an AF signal), and a dc level. We desire to recover the information frequency and discard the AM carrier signal. This job is performed by a *low-pass filter* built with C_1 and R_1. The low-pass filter works because the carrier frequency is much higher than the information frequency.

For example, the receiver might be tuned to 810 kHz, with an information frequency of 4 kHz. The low-pass filter can easily reject the 810 kHz (carrier) component and pass the 4 kHz (information) component. The bottom waveform results. The jagged shape of the recovered information is a result of imperfect filter action. A small amount of the 810 kHz carrier signal gets through the filter and shows up as a high-frequency ripple in the detected output.

Distortion in the AM Detector

A diode AM detector doesn't produce a perfect copy of the information. C_1 can't discharge (through R_1) rapidly enough to keep up when the envelope falls rapidly. The recovered information is distorted as a result.

The dotted trace in Figure 5–6 represents the voltage on C_1. C_1 rapidly charges when the peaks get larger, but it can't follow closely enough on the downward slopes. The result is called *diagonal clipping distortion,* and is shown in Figure 5–6. This type of distortion is most likely to show up when a high percentage of modulation is present. The use of a more sophisticated detector called a *synchronous detector* can nearly eliminate this distortion, but it is much more complicated and costly to construct.

THE CASE OF THE BANDIT (UNINTENDED AM SIGNAL DETECTION)

Jim was getting more than a little frustrated. For the third night in a row, he sat down to eat dinner and listen to music when an anonymous voice suddenly blared over the speakers: "Breaker 14, this is the Bandit, this is the Bandit. . . ."

Jim got up and snapped off the stereo's power switch. What happened next stunned him. The voice wouldn't go away! Jim looked at the stereo in disbelief. Even pulling the power plug from the outlet had no effect. This was really spooky. "Neil has got to see this," Jim thought to himself as he dialed the phone. "Hey, Neil, you've got to come over here. This 'Bandit' character comes over my stereo even when I unplug it. I think I might know who he is!"

In his 15 years as a ham radio operator, Neil hadn't seen any case of interference quite like this. "Jim, did you notice that your neighbor across the street has a huge *beam* antenna—and it seems to be pointed right at your place? If that's him, he'll be pounding that RF signal right into your equipment. It's cool that you can pick him up even when your set is unplugged. They'll never believe this at work!"

Neil continued: "Since you've disconnected your stereo from the power line, we can eliminate the ac power line as the path for the Bandit's signal. The only thing left is your speaker wiring. Those speaker cables are about 8 feet long apiece—hmm . . . that's close to a quarter-wavelength on the CB band. Let's try putting an RF bypass capacitor at each speaker output jack on your receiver."

Neil had brought an assortment of capacitors and in no time had connected a 0.1 μF capacitor in parallel with each speaker connector. "These 0.1 μF caps will look like an open circuit to the audio but a short to the RF from Bandit's transmitter."

The voice disappeared midsentence as Neil connected the last capacitor. Jim was amazed.

"Neil, how did you know to use capacitors?"

"Well, since we figured that the signal was coming in on the speaker wires, that in itself wouldn't be enough to make any sound. Radio frequencies are too high for us to hear directly, even if they get into the loudspeakers. There has to be some sort of *rectification* somewhere to get audio from an RF signal. The *output transistors* in your stereo were still connected to the speaker leads, even when the power was turned off. Therefore, I guessed that the speaker wires were acting as antennas, and the output transistors in the stereo were acting as an AM detector. The speakers were just responding to the rectified RF signal. The 0.1 μF capacitors prevented the RF signal from getting back into the stereo where it would be rectified by the output transistors."

The Bandit was no longer the dinner speaker at Jim's house!

Section Checkpoint

5–7 How is an AM detector similar to a half-wave dc power supply?

5–8 Why is the diode detector popular for AM?

5–9 What component in Figure 5–4 "stores" the envelope voltage between RF peaks?

5–10 Why is bleeder resistor R_1 needed?

5–11 What type of filter does R_1 and C_1 form?

5–12 What is the primary type of distortion caused by a diode AM detector?

5–13 What causes diagonal clipping distortion?

5–3 THE TRF RECEIVER

The signal processing approach of Figure 5–1 was the first electronic approach used to build radio receivers. It is known as the *tuned radio frequency* or *TRF* receiver. The TRF receiver isn't used much in modern practice, but it is still constructed by hobbyists. Figure 5–7 shows a modern version of the TRF that provides fairly good performance.

Circuit Analysis

The integrated circuit U_1 does most of the work. U_1 is designed as a TRF radio receiver on a chip; it contains the following stages:

- nine stages of RF amplification, with over 70 dB of available gain
- an active AM detector using a transistor as a "diode"
- an automatic gain control (AGC) circuit to compensate for signal fading

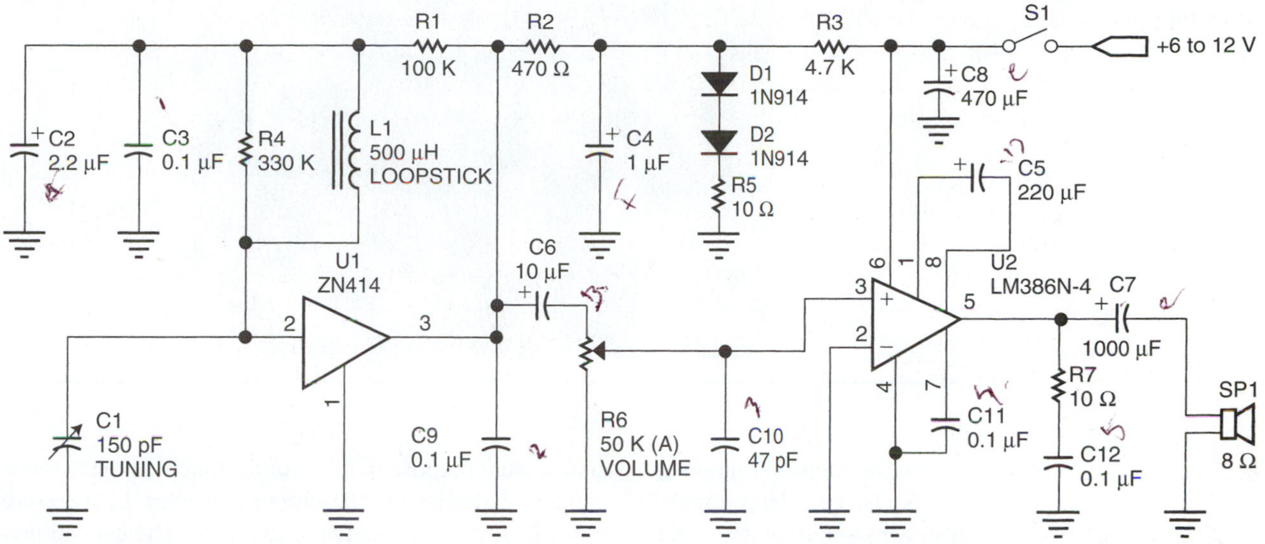

ONSTRUCTION NOTES:

ZN414 available from ALL ELECTRONICS CORP. 1-800-826-5432
All capacitors 16 WV or better.
LOOPSTICK L1 can be constructed as 100 turns #30 wire on a 5/16" dia ferrite rod. Rod length not critical.

4. C10 should be located close to U2.
5. C9 should be close to U1.
6. Keep leads on pin 2 of U1 as short as possible, and awa from the speaker leads.
7. R4 used to prevent U1 from oscillating. May not be needed in all units, depending on choice of L1 and U1 lot number.

Figure 5–7 A TRF receiver built with two IC chips

All that is needed to utilize U_1 is a 1.5 volt dc power supply, an RF input signal, and a detector filter capacitor. The output of U_1 is a detected audio signal, which needs only power amplification in order to drive a speaker. Diodes D_1 and D_2 are forward-biased by R_3 and R_5 to provide a 1.5 volt operating supply for U_1. These components act as a simple voltage regulator. C_4 filters the dc power supply for U_1.

There seems to be no antenna on this schematic! How does U_1 get an RF signal? The answer is that a special antenna called a *loopstick antenna* is used. The loopstick antenna is coil L_1 on the schematic. A loopstick antenna is a coil of wire wound on a ferrite rod. Ferrite is a magnetic material like iron but is essentially an insulator. Iron cores are very lossy at radio frequencies because they are good conductors of electricity. Eddy currents induced in iron cores cause RF energy to be lost. Ferrite is essentially *rust!* It contains iron in *oxide* form, which is essentially an insulator. The iron retains much of its magnetic properties in ferrite. Because ferrite isn't a good conductor, it works well at radio frequencies.

A radio wave contains both an electric and a magnetic field. The ends of the ferrite rod in Figure 5–8 are open, which allows the magnetic field of the radio wave to enter the rod. The ferrite rod intensifies the magnetic field, increasing the receiver's sensitivity. Since the magnetic field fluctuates back and forth at the carrier frequency, it induces an ac voltage in the coil. The ac voltage in the coil is the carrier voltage and can be sent to an RF amplifier for amplification and eventual detection.

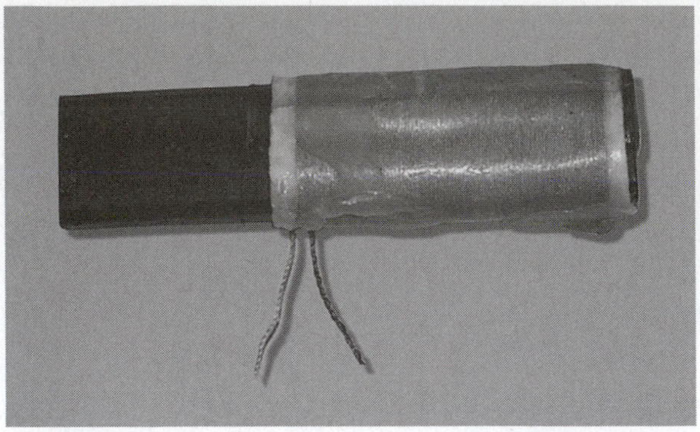

Figure 5–8 A loopstick antenna

The loopstick antenna, L_1 serves a dual purpose in this circuit. It acts as the antenna but is also parallel-resonated by variable capacitor C_1, the tuning capacitor. L_1 therefore also acts as part of the *selector* circuit; L_1 and C_1 form a *bandpass filter*. The user adjusts C_1 to select the station frequency. The filtered output from L_1 and C_1 is the carrier frequency. It is a very tiny signal—perhaps 100 μV at best!

The carrier signal from L_1 and C_1 enters pin 1 of U_1. (U_1 has only three pins and is in a plastic transistor package.) U_1 amplifies the RF signal and detects it. The recovered information leaves U_1 on pin 3. Pin 3 of U_1 is used to provide the dc power for the IC (through R_2) and extract the information (through dc block C_6). C_9 completes the detector low-pass filter (the "resistor" for the filter is inside U_1).

The recovered information signal is passed through the *volume control,* R_6, which is really just a variable voltage divider. Rotating the volume control moves the wiper of R_6 up or down on the resistor. The higher the wiper position of R_6, the more AF voltage that will be passed into pin 3 of U_2, the audio power amplifier, and the louder the sound in the speaker will become.

U_2, an LM386, is a complete audio voltage and power amplifier on one chip. C_5 sets the gain of the LM386 to maximum, and C_7 ac couples the amplified audio signal into the speaker. C_{11} and C_8 are power supply filters, which are very important in a receiver. Radio receivers have tremendous gain, which can lead to oscillation problems if the power supplies are not adequately filtered.

Problems with the TRF Receiver

The TRF receiver is not used in many applications, although it is certainly the most logical approach. There are two problems with TRF receivers. The *selectivity* of TRF receivers is not constant, and it is nearly impossible to design a TRF receiver for operation at HF or above (3 MHz or above).

Selectivity Selectivity can be defined as the ability of a receiver to *select* or choose a desired signal while rejecting all others. (Compare this with *sensitivity,* the ability to work with weak signals). As you might guess, the *bandwidth* of the receiver's circuits control its selectivity. The bandwidth of a receiver should be just wide enough to accept the desired signal—and no more! Excessive receiver bandwidth reduces selectivity and increases the noise level (recall that noise power is proportional to bandwidth). A closer examination of the tuning circuit of the TRF receiver reveals this problem, as the following example shows.

EXAMPLE 5–2 Calculate the bandwidth for the modeled tuning circuit of Figure 5–9 at the following frequencies: 540 kHz (bottom of AM dial); 980 kHz (middle of dial); 1620 kHz (top of dial).

Solution

To calculate bandwidth, we must calculate the quality factor (Q) of the resonant circuit at each frequency. From electronic fundamentals, we know that for a reactive circuit with the resistance in *shunt* with the circuit, the Q is calculated by:

$$Q_p = \frac{R_p}{X_p}$$

(where R_p is the parallel/shunt resistance, and X_p is the capacitive or inductive reactance—whichever is easiest to calculate).

We also know that the bandwidth of a circuit can be calculated if we know frequency and Q:

$$\text{BW} = \frac{f}{Q}$$

At 540 kHz, we get

$$X_p = X_C = X_L = 2\pi fL = 2\pi(540\text{ kHz})(500\ \mu\text{H}) = 1696\ \Omega$$

$$Q_p = \frac{R_p}{X_p} = \frac{100\text{ k}\Omega}{1696\ \Omega} = 59$$

Therefore, we can find the bandwidth at 540 kHz:

$$\text{BW} = \frac{f}{Q} = \frac{540\text{ kHz}}{59} = \underline{\underline{9.152\text{ kHz}}}$$

This is a reasonable bandwidth (perhaps a little wide), since an AM broadcast uses 8 kHz of bandwidth. By repeating the procedure at 980 kHz, we get

$$X_p = X_C = X_L = 2\pi fL = 2\pi(980\text{ kHz})(500\ \mu\text{H}) = 3079\ \Omega$$

$$Q_p = \frac{R_p}{X_p} = \frac{100\text{ k}\Omega}{3079\ \Omega} = 32.5$$

$$\text{BW} = \frac{f}{Q} = \frac{980\text{ kHz}}{32.5} = \underline{\underline{30.2\text{ kHz}}}$$

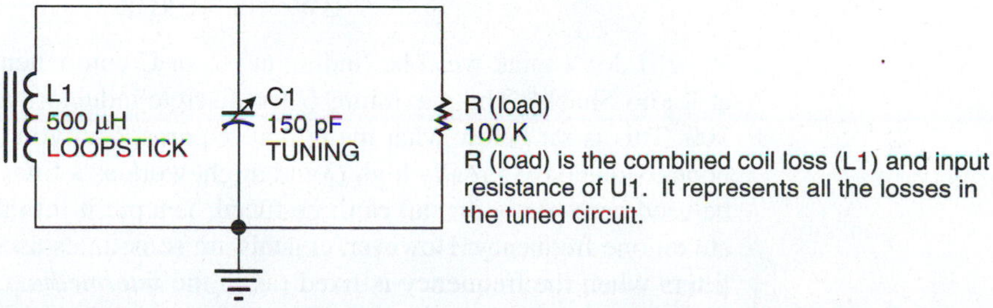
L1
500 μH
LOOPSTICK
C1
150 pF
TUNING
R (load)
100 K
R (load) is the combined coil loss (L1) and input resistance of U1. It represents all the losses in the tuned circuit.

Figure 5–9 Model of TRF receiver tuning circuit

The bandwidth seems to be getting larger. Let's try it at 1620 kHz:

$$X_p = X_C = X_L = 2\pi fL = 2\pi(1620 \text{ kHz})(500 \ \mu\text{H}) = 5089 \ \Omega$$

$$Q_p = \frac{R_p}{X_p} = \frac{100 \text{ k}\Omega}{5089 \ \Omega} = 19.7$$

$$\text{BW} = \frac{f}{Q} = \frac{1620 \text{ kHz}}{19.7} = \underline{\underline{82.2 \text{ kHz}}}$$

Wow! The receiver's bandwidth has increased from 9 kHz to more than 80 kHz over its tuning range! *This is unacceptable performance*. More than eight stations might be received at the same time! (Stations are spaced every 10 kHz on the AM broadcast band.)

An ideal receiver should have constant bandwidth and therefore constant selectivity. The TRF approach fails miserably here!

Why doesn't the TRF receiver have a constant bandwidth? The reason is within the tuning circuit. When we change the tuning dial position, we are changing the resonant frequency of the tuned circuit. The bandwidth of the tuned circuit depends upon its *frequency* and Q. The Q is furthermore determined by *circuit losses,* which depend on frequency as well. Even if the Q were held constant in Example 5–2, the bandwidth would still increase as we moved up the dial because the frequency would be changed! That Q is also decreasing makes the bandwidth degrade even faster.

The bandwidth of any tuned circuit depends on the frequency it is tuned to. In general, for a parallel-loaded capacitively tuned circuit, the higher the tuning frequency, the greater the bandwidth and the *poorer* the selectivity.

For any tuned circuit to have a constant selectivity, the tuned frequency of the circuit must not be changed.

High-frequency operation of TRF receivers Operation at frequencies above a few MHz also pose problems for *TRF* receivers. It becomes impossible to find discrete L and C components that will satisfy the tank requirements.

EXAMPLE 5–3

What Q is needed for the tuned circuit of a TRF receiver designed to operate on CB channel 19 (27.185 MHz)? The required bandwidth is 10 kHz.

Solution

To calculate Q, we simply manipulate the formula for bandwidth and Q:

$$Q = \frac{f}{\text{BW}} = \frac{27.185 \text{ MHz}}{10 \text{ kHz}} = \underline{\underline{2{,}718.5}}$$

Wow! I don't think we'll be finding any L or C components with this high a Q, not even at Radio Shack! The maximum Q for discrete inductors and capacitors is around 200 to 300. This is far below what the circuit requires. Can you think of any other circuit components that have a really high Q and might work as a filter? Crystals? Yes, a crystal *might* be used here. But a crystal can't be tuned. You put it into the circuit and essentially operate on one frequency. However, crystals are sometimes used in high-performance receiver filters when the frequency is fixed (as in the *intermediate frequency* or IF amplifier of a *superheterodyne receiver,* as we'll soon see).

A practical TRF receiver is nearly impossible to build at high frequencies; the required *Q* for the selector tank components is much higher than can be obtained with real-world components.

The problem of receiving different carrier frequencies while maintaining a constant bandwidth perplexed early radio designers. After all, moving the selector dial of a radio receiver changed the frequency of the internal tuned circuits, which upset the bandwidth. Many different and elaborate circuits were devised in attempts to cure the problem. These receiver circuits were difficult to align and operate, and weren't very reliable.

The true solution to the selectivity problem is to operate the tuned circuits at one constant frequency. This is the approach developed by Armstrong, and it is called the *superheterodyne* receiver. The superhet is the king of modern receivers; nearly all modern radio receivers are built using Armstrong's circuitry.

Section Checkpoint

5–14 How is the selection process achieved in a TRF receiver?

5–15 What is a *loopstick* antenna?

5–16 Why is a ferrite, rather than iron, core used in a loopstick antenna?

5–17 What are the two functions of L_1 in Figure 5–7?

5–18 What are the two limitations of TRF receiver designs?

5–19 What causes the bandwidth and selectivity of a tuned circuit to change as its frequency is adjusted?

5–4 THE SUPERHETERODYNE RECEIVER

The superheterodyne receiver was a breathtaking advance in the radio art. Prior to its development, engineers spent a great deal of time and effort getting reasonable performance from the tuning (selection) circuits in a receiver. The engineers were in effect "moving the receiver to the signals" by changing the operating frequency of the receiver's tuned circuits to match the incoming signals.

The superhet receiver does exactly the opposite. Instead of trying to tune all of its tuned circuits to the incoming carrier frequency, it instead moves the carrier frequency to the frequency of its own tuned circuits! This process is called *frequency conversion* and is shown in block diagram form in Figure 5–10.

In a superheterodyne receiver, the incoming signal is applied at the left side of the *frequency converter* section. The frequency of this signal is f_c, the original carrier frequency. The output of the frequency converter is always a *constant* frequency, regardless of the value of f_c. The constant frequency produced by the frequency converter is called the *intermediate frequency,* or f_{if}. For AM broadcast receivers, the value of f_{if} is fairly well standardized at 455 kHz, though other values are occasionally used.

A fixed IF frequency provides several important advantages. First, since the IF frequency is constant, the tuned circuits in the IF amplifier will have a constant bandwidth and selectivity, regardless of what carrier frequency the receiver is tuned to. The TRF receiver's variable-tuned circuits could not achieve this.

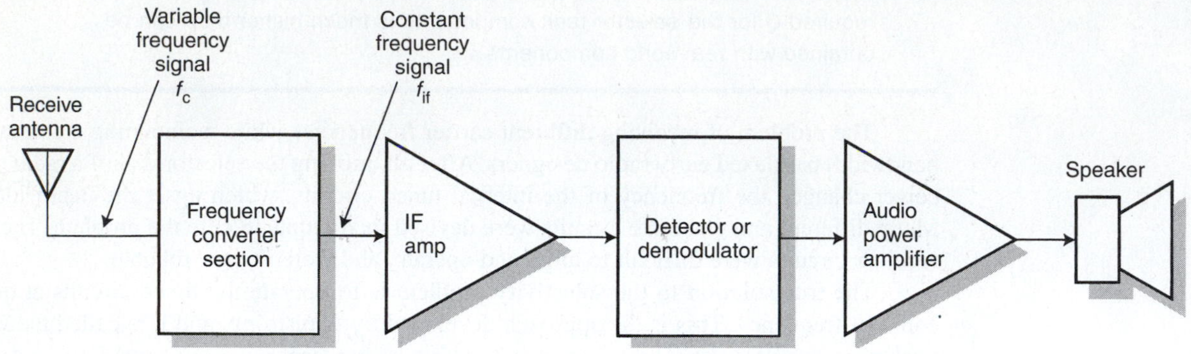

Figure 5–10 The superheterodyne concept

Second, the IF amplifier usually operates at a lower frequency than the carrier frequency. It is easier to build low-noise RF amplifiers at lower frequencies. Providing most of the RF signal gain at the IF frequency improves the receiver's noise figure.

Finally, since the IF amplifiers operate at only one frequency, receiver alignment is simplified. The IF amplifiers need be adjusted only once, at the factory.

The IF amplifier in a superhet receiver provides most of the receiver's gain and sets the receiver's bandwidth.

The Frequency Converter Section

The details of the frequency converter section are shown in Figure 5–11. The frequency converter consists of the *preselector,* the *mixer,* and the *local oscillator.* Note that there are *two* tuned circuits that are mechanically connected. These are the *preselector* filter tank and the *local oscillator* tank. The dotted line between the two variable capacitors is used to show this on a schematic. The tuning knob turns both of these capacitors at the same time; they share a common shaft. The variable capacitor of Figure 5–3(a) is actually a dual-section unit.

The purpose of the frequency converter is to produce a constant output frequency, f_{if}, regardless of what the input carrier frequency f_c is.

Before we can completely understand what is happening in the converter, we need to examine the mixer's operation in more detail.

Mixer Operation

The converter contains a nonlinear device called a *mixer*. A mixer is usually constructed with diodes or transistors. A mixer causes nonlinear distortion of the two signals being applied to it. When signals are distorted in this way, new frequencies are created (just as in the process of modulation).

Suppose that we applied two frequencies to an ideal mixer, $f_1 = 700$ kHz and $f_2 = 100$ kHz, as shown in Figure 5–12. What output frequencies could be expected?

Given two input frequencies f_1 and f_2, an ideal mixer will produce the following outputs:

- The *original* two frequencies, f_1 and f_2
- The *sum* of the two frequencies, $(f_1 + f_2)$
- The *difference* of the two frequencies, $(f_1 - f_2)$

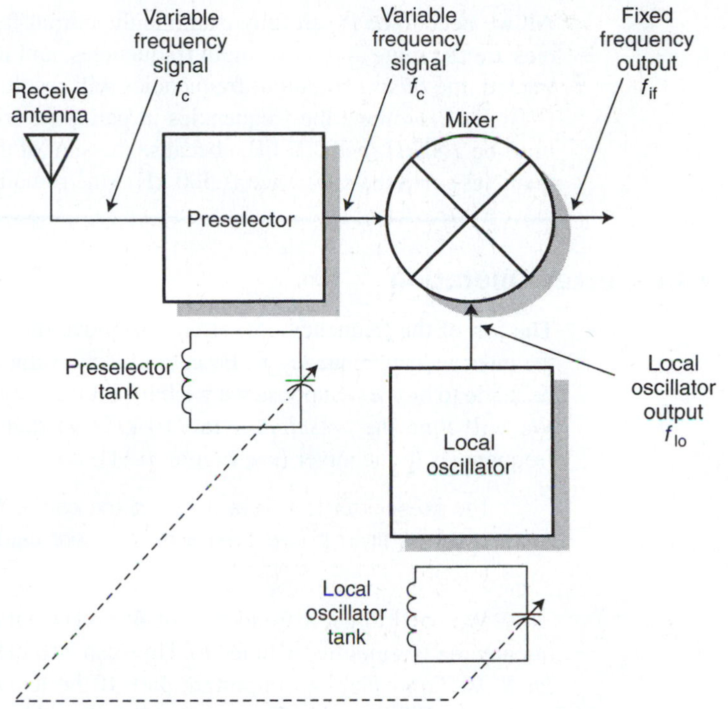

Figure 5–11 The frequency converter section

With 700 kHz and 100 kHz applied, the output frequencies of an ideal mixer would therefore be

- 700 kHz and 100 kHz (the *original* frequencies)
- 800 kHz (the *sum* of the two frequencies)
- 600 kHz (the *difference* of the two frequencies)

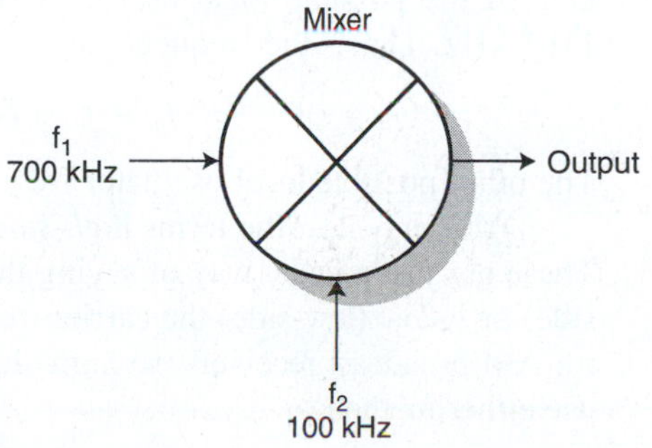

Figure 5–12 Ideal mixer operation

EXAMPLE 5–4

A certain mixer is producing the following output frequencies: 900 kHz, 700 kHz, 500 kHz, and 200 kHz. What are the two most likely input frequencies?

Solution

All we need to do is carefully examine the output frequencies. Two of the output frequencies are the same as the two input frequencies, and if these frequencies are added and subtracted, the *other* two output frequencies will result.

By examining the frequencies in pairs, we will find that the two input frequencies must be 700 kHz and 200 kHz, because the sum of these is 900 kHz (one of the output frequencies), and the difference is 500 kHz (the remaining output frequency).

Frequency Converter Operation

The job of the frequency converter is to move the incoming carrier frequency f_c down to the intermediate frequency f_{if}. By carefully choosing a local oscillator frequency, f_{lo}, this can be made to happen. Suppose we wish to receive a signal on a carrier frequency of 710 kHz. We will tune the *preselector* to 710 kHz so that 710 kHz becomes one of the input frequencies to the mixer (see Figure 5–11).

The preselector is always tuned to the carrier frequency. The preselector is not a narrow filter; it may even allow adjacent carrier frequencies to pass into the mixer.

We need to get a frequency of 455 kHz out of the mixer, because that is the frequency the IF amplifier is tuned to. How can 710 kHz be converted to 455 kHz? Right, the *local oscillator* plays an important part. If the local oscillator is tuned to the correct frequency, we'll get a 455 kHz output. All we need to do is make the local oscillator operate at a frequency that is *different* from the desired carrier (710 kHz) by the IF frequency (455 kHz). In other words, the local oscillator must operate 455 kHz *above* the carrier or 455 kHz *below* the carrier. There are *two* local oscillator frequencies that will work, but only one of them is used in an actual circuit.

The preselector and local oscillator circuits are mechanically connected so that they will "track" each other. When the two circuits properly track, they maintain a 455-kHz (f_{if}) difference.

The two possible frequencies for the local oscillator are given in equation 5–1.

$$f_{lo} = f_c + f_{if} \quad \text{(high-side injection)} \qquad (5\text{–}1)$$

One of the possible local oscillator frequencies is therefore 710 kHz + 455 kHz, or 1165 kHz. This is the frequency that will be used in a broadcast receiver.

$$f_{lo} = f_c - f_{if} \quad \text{(low-side injection)} \qquad (5\text{–}2)$$

The other possible local oscillator frequency is 710 kHz – 455 kHz, or 255 kHz.

You may hear the terms *high-side injection* and *low-side injection* used in practice. These are just a fancy way of saying that the local oscillator operates either *above* (high-side) or *below* (low-side) the carrier frequency. Low-side injection causes some problems for AM broadcast receivers, so high-side injection is generally used. Other receivers may use either method.

EXAMPLE 5–5

The frequency converter of Figure 5–11 is tuned to receive 810 kHz. Calculate:

a. the frequency of the preselector bandpass filter

b. the frequency of the local oscillator

c. the frequencies that will be produced at the mixer output

d. which mixer frequencies can pass through the IF amplifier and be detected

Solution

a. The *preselector* is always tuned to the carrier frequency. Therefore, $f_{pr} = f_c = \underline{\underline{810 \text{ kHz}}}$.

b. High-side injection is used in all broadcast receivers. Therefore, equation 5–1 is used:

$$f_{lo} = f_c + f_{if} = 810 + 455 \text{ kHz} = \underline{\underline{1265 \text{ kHz}}}$$

c. The mixer has the frequencies of 810 kHz and 1265 kHz being applied. Therefore, the two originals, the sum, and the difference frequencies will appear at the mixer output:

Mixer Outputs:

810 kHz, 1265 kHz—originals

2075 kHz—sum

455 kHz—difference

d. The IF amplifier is tuned to 455 kHz, and has a bandwidth of approximately 10 kHz in an AM broadcast receiver. Any signals in the range of 450 kHz to 460 kHz can be passed by the IF amplifier. *The only signal in part (c) that can pass is the 455 kHz difference signal.* The IF amplifier will reject all the other frequencies.

When we say that a superhet receiver is tracking properly, we mean that the preselector and local oscillator keep a constant 455 kHz (f_{if}) difference as the receiver is tuned across the dial. If a 455 kHz difference isn't produced, no signal will pass through the IF amplifier and the receiver will appear to be low in sensitivity or dead.

The Preselector and Image Frequency

If the mixer and local oscillator work together to convert the incoming carrier frequency into the IF frequency, why is the preselector needed? After all, the preselector is tuned to the carrier frequency anyway; with the preselector removed, the carrier frequency will get into the mixer with no problem.

There's a subtle problem with the process of *frequency conversion,* and we need to look at the picture in the frequency domain to understand it. Figure 5–13 shows all of the input signals being applied to the mixer from Example 5–4.

In Example 5–4, a carrier frequency of 810 kHz was converted to an IF frequency of 455 kHz by "beating" against a local oscillator frequency of 1265 kHz. The IF frequency was produced because of the 455 kHz difference between the carrier and local oscillator. Is there any *other* frequency that is 455 kHz different from 1265 kHz? You bet. It is 1720 kHz on the spectrogram in Figure 5–13. As you can see from the figure, this new frequency is *twice* the IF frequency away from the carrier frequency:

$$f_{image} = f_c + 2f_{if} \quad \text{(high-side injection)} \qquad (5\text{–}3)$$

The frequency 1720 kHz is called the *image frequency* of 810 kHz. If 1720 kHz mixes with 1265 kHz, what frequency components will be produced? We would see the originals

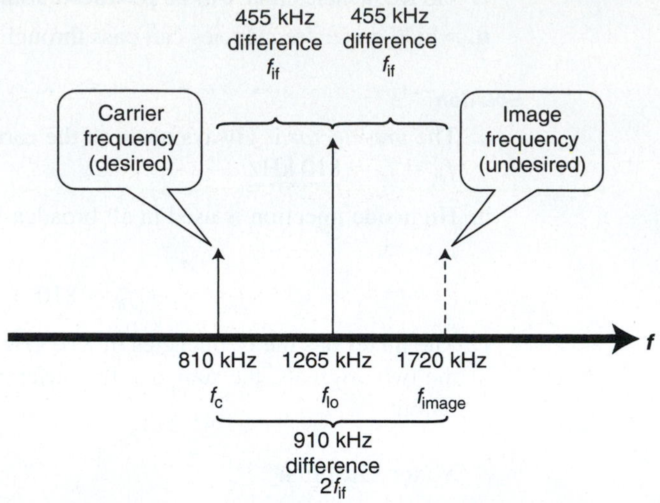

Figure 5–13 Spectrogram of mixer input signals from Example 5–4

(1720 kHz, 1265 kHz), the sum (2985 kHz), and the difference (455 kHz). This is very troubling! There are *two* frequencies that can beat with the 1265 kHz local oscillator frequency to produce the 455-kHz IF frequency. How can this be fixed?

There's no way of preventing a signal on 1720 kHz from mixing with the 1265 kHz local oscillator *once the 1720-kHz image signal gets into the mixer.* We need to prevent the 1720 kHz signal from getting into the mixer, while allowing the desired 810 kHz carrier signal to pass. This means we need a *filter*. The preselector performs this function.

Figure 5–14 demonstrates how the preselector performs this action. The preselector filter is tuned to the carrier frequency. Notice that the preselector is not a very narrow filter; it is entirely possible that *adjacent* station frequencies will pass through it. The important action occurs at the image frequency. The response of the preselector filter is very low at the image frequency. If there is a transmitter on 1720 kHz, the preselector will prevent most of the unwanted transmitter's energy from reaching the mixer.

The primary function of the preselector in a superheterodyne receiver is to reduce or eliminate the image frequency. A fringe benefit of the preselector's bandwidth limiting action is the reduction of receiver noise. The preselector does *not* control the receiver's bandwidth. That is the job of the IF amplifier.

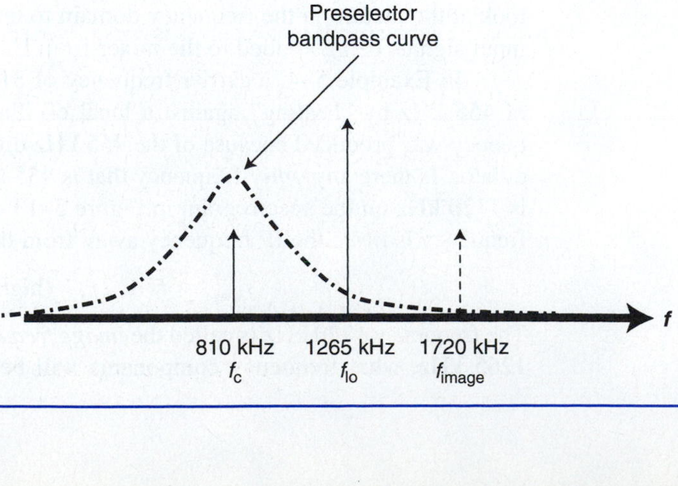

Figure 5–14 The preselector's filtering action

Adjacent channels can get through the preselector in many receivers. The following example demonstrates that in the ideal case, this isn't a problem at all!

EXAMPLE 5–6

A superhet receiver is tuned to 710 kHz. The preselector is tuned to 710 kHz, yet an adjacent station on 730 kHz is also being passed to the mixer. The local oscillator is producing 1165 kHz. Demonstrate that the 730 kHz signal is eventually rejected by the IF amplifier.

Solution

First, let's look at the results of mixing the 710 kHz (desired carrier) and 1165 kHz (local oscillator) signals. The mixer will produce 710 kHz, 1165 kHz, 1875 kHz, and 455 kHz outputs from these two signals. The 455 kHz difference passes easily through the IF amplifier.

Next, let's try the "adjacent" signal. Mixing 730 kHz (adjacent signal) with 1165 kHz (local oscillator) results in 730 kHz, 1165 kHz, 1895 kHz, and 435 kHz signals. The difference produced by the adjacent signal is 435 kHz, which cannot pass through the 455 kHz amplifier (which passes 450–460 kHz, approximately).

It works! There is *one* case where there will be trouble, though. If there is an extremely strong adjacent signal, it is possible that the mixer could be overwhelmed by the second signal. If a receiver must be operated under such conditions, the preselector bandwidth must be narrowed enough to knock down the adjacent signal as well. Commercial receivers that operated in crowded bands are often designed in this manner.

A Complete Superhet Receiver

Figure 5–15 is a block diagram of a complete superhet receiver. It shows the final detail, the *AGC subsystem.*

Automatic Gain Control (AGC)

The signal strength arriving at a receiver's antenna terminals depends on many factors, including

- the transmitter's output power and antenna configuration
- the distance between the transmitter and receiver
- the type of receiver antenna
- any physical obstructions (such as buildings or hills) between the transmitter and receiver

The signal strength at a receiver is hardly constant, even for a receiver in a fixed location such as a home. For receivers in mobile applications, received signals often vary wildly as the vehicle travels up and down hills and passes other obstructions.

To the listener, such signal strength variations would be very annoying. The receiver's volume would be very unstable; in weak signal conditions, the user would have to increase the volume control setting, only to get "blasted out" when the signal strength increased again. The solution to the problem is *automatic gain control,* or AGC.

The purpose of the AGC system in an AM receiver is to keep the volume approximately constant in the face of signal strength variations.

AGC operation A diode AM detector produces two signals that are useful. One of these is the ac information signal. In a receiver, the ac information signal is extracted from the

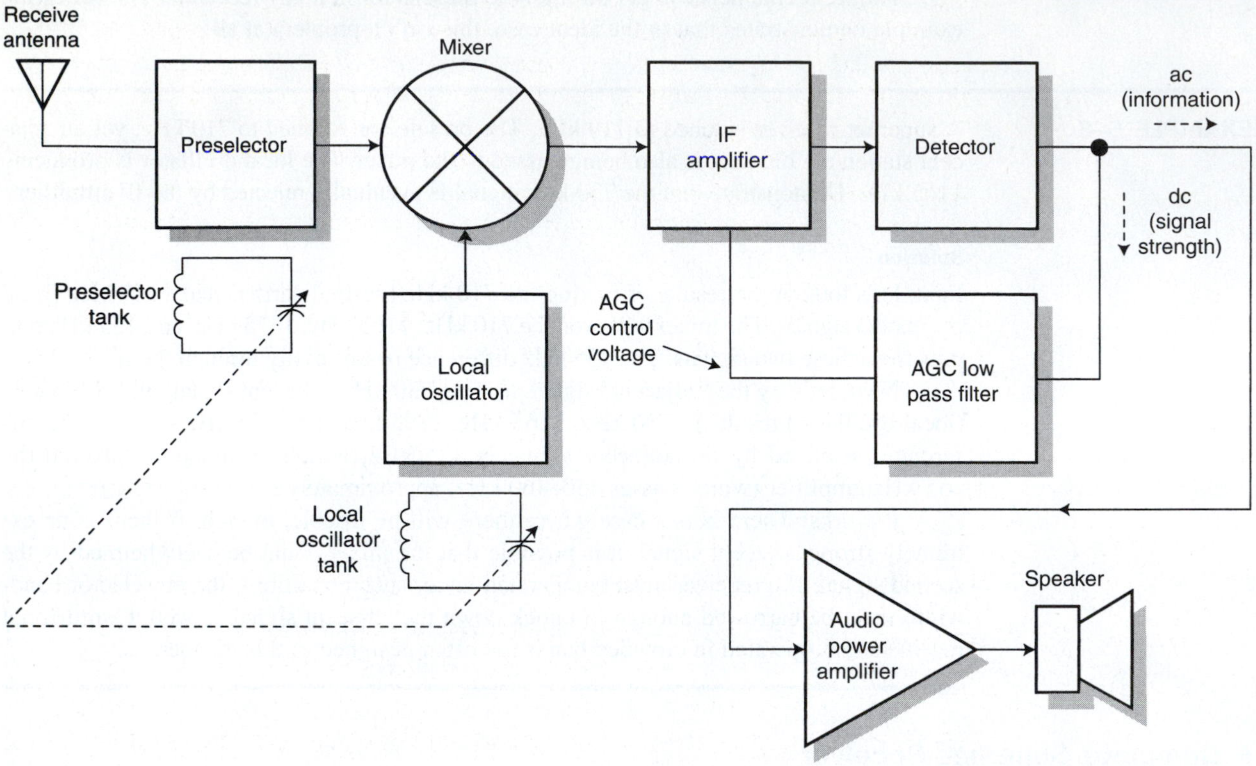

Figure 5–15 A complete superhet receiver

diode by the use of a coupling or dc block capacitor. The other signal produced by the AM detector is a dc component.

Where does this dc level come from? Remember that the diode detector is a *rectifier*. A rectifier converts ac signals to dc. The rectifier receives an ac waveform from the IF amplifier. The ac waveform is a copy of the modulated waveform from the transmitter. The stronger the received signal, the larger the ac IF waveform will be. The detector diode converts the ac IF waveform into pulsating dc. When the received signal increases in size, the average value of the pulsating dc increases. *The net effect is that the detector produces an average dc voltage proportional to signal strength.*

In an AM receiver, the detector's pulsating dc voltage is smoothed by a *low-pass filter*. This filter is often referred to as the *AGC filter*. It has a time constant of around 200 ms, which is a relatively long time period when compared to the audio frequencies coming from the detector. This means that the audio pulsations will be virtually eliminated, leaving a nearly pure dc level. *The output of the AGC filter is a steady dc voltage that is proportional to signal strength.*

The dc voltage from the AGC filter is used to *oppose* the normal dc bias of the IF amplifier stages. By *oppose* we could also say *reduce*. The stronger the rectified and filtered detector voltage, the lower the dc bias will be at the IF amplifier stages. This lowers their voltage gain.

For example, suppose that a user tunes a receiver from a very weak station (where the receiver gain would need to be high) to a strong station (where the gain must be smaller). How does the receiver circuitry compensate for the change? As the receiver is tuned onto the strong station, the detector almost instantly begins producing a much larger dc voltage component than it did before. After a few milliseconds, the filtered dc at the AGC low-pass filter increases in magnitude. Since the AGC dc component is designed to buck or oppose the dc bias on the IF amplifiers, the dc bias voltage on the IF stages is reduced. In turn, the IF amplifier power gains get smaller; this causes the detector dc output to become *smaller* in response.

This action continues, round and round the loop, until the receiver gain finally settles at some final value where the feedback loop is satisfied. The action is fast enough that the user can barely perceive it. Some receivers deliver a slightly audible "thump" as they make the adjustment.

When tuning a receiver from a strong signal back to a weak one (or none at all—a blank spot on the dial), the opposite occurs. The AGC system cranks up the gain of the IF amplifiers to compensate. Receivers with especially slow AGC can be heard "breathing" as they make the transition from low gain to high gain. The gradually increasing level of background noise in the loudspeaker causes the "breathing" sound.

Section Checkpoint

5–20 What is the *intermediate frequency* in a superhet receiver?

5–21 Explain the advantages of having a constant IF frequency.

5–22 What section of a superhet receiver provides most of the gain and controls bandwidth?

5–23 What section of a superhet converts incoming carriers into the IF frequency?

5–24 If two frequencies are applied to a mixer, what output frequencies will result?

5–25 How is the local oscillator frequency chosen in a broadcast receiver?

5–26 What is the difference between high- and low-side injection?

5–27 What must the difference between the preselector and local oscillator frequencies be in a superhet receiver?

5–28 Define *tracking* as it applies to a superhet receiver.

5–29 What is the *image* frequency?

5–30 What section of a superhet receiver is responsible for eliminating the image frequency?

5–31 If the preselector allows adjacent channels to get into the mixer, why don't the adjacent channels get through the IF amplifier?

5–32 What is the purpose of the AGC system in a receiver?

5–33 Where does the AGC system get its dc control voltage?

5–5 SUPERHETERODYNE RECEIVER CIRCUIT ANALYSIS

Figure 5–16 is a typical portable AM broadcast receiver. It can be broken down into power supply, RF, and AF stages. It can be viewed in block diagram form as Figure 5–15.

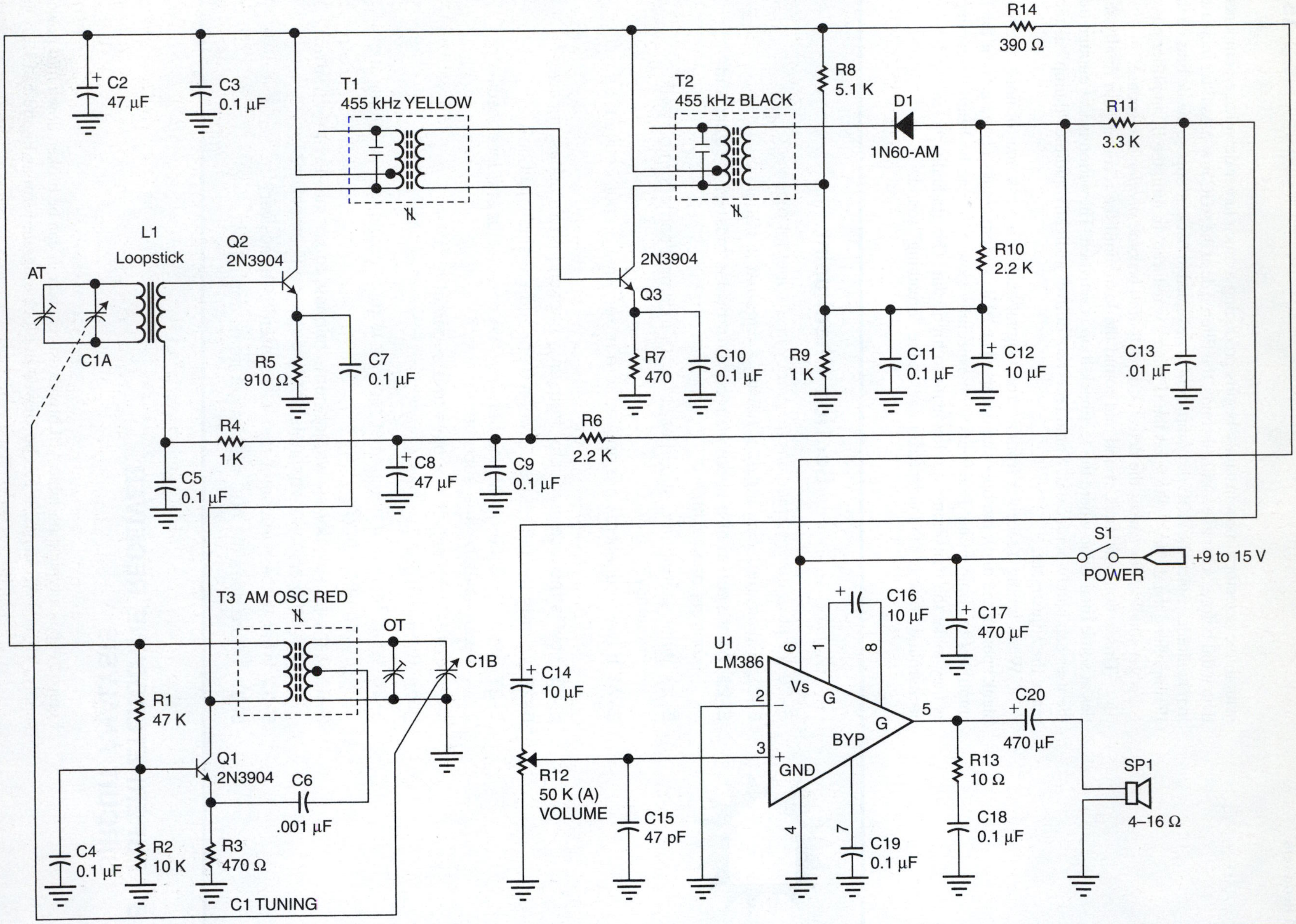

Figure 5–16 A complete superheterodyne receiver

Power Supply

There is no voltage regulation in this receiver, but there is extensive *filtering* of the power supply as it is distributed to the various portions of the receiver.

The dc supply enters through the power switch S_1 and is immediately filtered by C_{17}. C_{17} is a *low-frequency* filter for U_1, the audio power amplifier. The audio amplifier can draw relatively large ac currents from the power supply. C_{17} ensures that the ac audio currents created on the V_s pin of U_1 go to ground rather than to other sections of the receiver. The symptom of low-frequency (audio) feedback through a power supply bus is often a low-pitched oscillation called *motorboating* (it often makes a "put-put" sound). Open filter capacitors such as C_{17} should be suspected if this symptom appears!

The supply to the RF sections passes through R_{14} (top right) before being filtered by capacitors C_2 (low frequency) and C_3 (high frequency). The use of R_{14} increases the *isolation* between the AF power supply to U_1 and the RF power supply for Q_1, Q_2, and Q_3. This isolation is important; the RF sections are dealing with very small signal levels and can be easily upset by feedback through the power supply bus.

An open in capacitor C_2 may produce motorboating, since it is intended to bypass low-frequency (AF) signals to ground. C_3 is an RF bypass capacitor. An open C_3 may allow RF currents to flow on the upper power supply bus. RF feedback usually produces strange (and hard to find!) oscillations in several stages at the same time. The symptoms of RF feedback include odd whistles, screeches, and unstable tuning in general. Open RF filter capacitors (such as C_3) should be suspected when these symptoms appear.

Power supply filter capacitors commonly short circuit as well. A shorted capacitor such as C_2 or C_3 will cause the upper bus voltage to drop to zero (the entire power supply voltage will be dropped across R_{14}, which may become warm to the touch.) This will disable all the RF stages but leave the AF stage (U_1) intact. A short at C_{17} is much more exciting; C_{17} is directly across the power supply. If the power supply is capable of high current, C_{17} may heat up and explode, the connections to C_{17} may melt, or a fuse may blow. This set is meant to operate from batteries. A shorted C_{17} will quickly discharge a transistor radio battery!

RF Section Analysis

The RF section consists of the *local oscillator* (Q_1), the *mixer and first IF amplifier* (Q_2), the *second IF amplifier* (Q_3), and the *detector and AGC* (D_1).

Q_1 operates as an Armstrong oscillator. A special transformer, T_3, is used as the oscillator transformer. T_3 is parallel-resonated by C_{1B}, the oscillator half of ganged tuning capacitor C_1. The capacitor marked "OT" is also part of C_1, the *oscillator trimmer*. The oscillator trimmer allows fine adjustments to be made to the local oscillator frequency for tracking and dial calibration adjustment.

The local oscillator signal from Q_1 is coupled into the emitter of the *mixer transistor,* Q_2, by capacitor C_7. Transistor Q_2 is driven into *large-signal mode* by the local oscillator signal from Q_1. There is no secret as to why Q_2 is driven into large-signal mode; it is simply supplied with a rather large signal (2 V_{pp} typical) from the collector circuit of Q_1. The relatively "huge" input signal forces Q_2 to move back and forth across most of its load line—which is large-signal operation. *In large-signal mode, Q_2 is nonlinear and will therefore act as a mixer.*

Q_2 is supplied with carrier frequency on its base from the *secondary* winding of loopstick L_1. The secondary winding helps impedance match L_1 to the low input impedance of the base of Q_2. The primary of L_1 is parallel-resonated by C_{1A}, the other half of the ganged tuning capacitor.

L_1 performs two primary functions. First, it is the receiving antenna (a loopstick antenna is sensitive to the magnetic field of the radio wave). Second, L_1 functions as part of

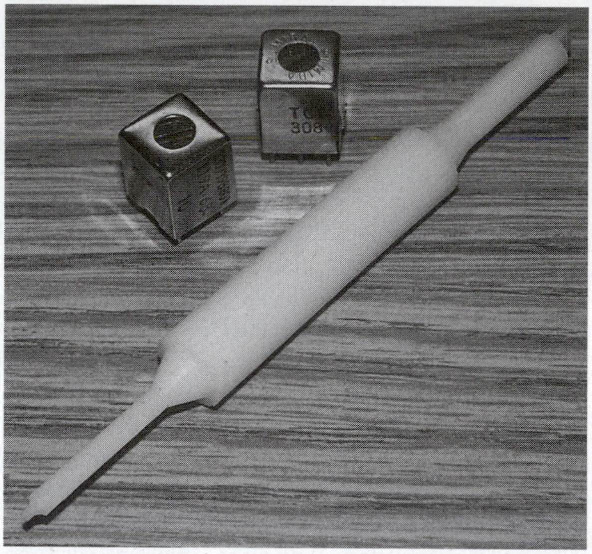

Figure 5–17 Typical RF and IF transformers, and adjustment tool

the *preselector* with the help of C_{1A}, the tuning capacitor. The output of L_1 contains the carrier frequency and adjacent frequencies but not the image frequency.

Because Q_2 is being fed both local oscillator and carrier frequencies, its collector circuit will contain the original frequencies plus the sum and difference frequencies. T_1, the first IF transformer, accepts the signal from Q_2. T_1 is designed to operate at 455 kHz; it contains an internal capacitor (inside its metal shield can, shown as a dotted line on the schematic) that resonates its primary at 455 kHz. T_1, T_2, and T_3 are *slug tuned* and are intended to be adjusted with a plastic tool (see Figure 5–17).

The IF signal passes into Q_3, where further RF amplification takes place. Q_3 is coupled to T_2, another IF transformer. T_2 also has an internal capacitor to self-resonate it at 455 kHz and is optimized for driving a diode AM detector.

The modulated IF signal passes into D_1, the AM detector. D_1 produces two signal components—a dc level and the ac recovered information, which is filtered by R_{11} and C_{13} (to remove any RF signal) and is coupled into the audio amplifier by C_{14} (bottom middle of drawing)—to remove any dc component.

At first glance, it appears that D_1 has been installed *backwards* in the circuit! This is done so that the dc signal strength voltage from the detector will be *negative*. This signal strength voltage is used to develop the dc AGC signal.

Resistors R_8 and R_9 form a voltage divider. With a 9 volt supply, the voltage at the junction of R_8 and R_9 will be about 1.5 volts. The secondary of T_2 and diode D_1 form a variable dc source. The voltage of the dc source is proportional to signal strength. This dc source *series opposes* the 1.5 volts at the junction of R_8 and R_9. The resulting dc voltage taken from the anode of D_1 is the series combination of the voltage divider voltage and the opposing rectified dc signal strength voltage from D_1.

As signal strength increases, the opposing voltage will get larger; this will cause the voltage at the anode of D_1 to become smaller. *This is exactly what is needed for AGC action.* The resulting dc voltage is sent back to bias both Q_2 and Q_3 (the IF amplifier stages) through R_6 and C_8, the AGC averaging (low-pass) filter. C_9 is an RF bypass capacitor that prevents RF feedback through the AGC signal line.

The resulting AGC system works like this: Under a no-signal condition, the AGC voltage measured at the positive end of C_8 will be about 1.5 volts, because there is no rectified voltage from D_1 to oppose the voltage divider output of R_8 and R_9. This 1.5 volts biases the bases of both Q_2 and Q_3. As the received signal strength increases, the (negative) rectified output of D_1 increases. This opposes the 1.5 volts from the divider, so the base-bias voltage of Q_2 and Q_3 drops. When the bias on Q_2 and Q_3 decreases, their gains both fall. This causes the rectified output of D_1 to decrease, completing the AGC loop.

AF Stages

The audio amplifier in this receiver is identical to the one in the TRF receiver of Figure 5–7. The LM386 contains several stages of voltage amplification and a push-pull emitter-follower power amplifier on one IC.

The ac signal from the detector is coupled through C_{14} into the top of the volume control, R_{12}. The volume control acts as a variable voltage divider. As the user rotates the control clockwise, the wiper moves up towards the top, which increases the amount of ac information signal available to U_1, which increases the volume.

Capacitor C_{15} is an RF bypass capacitor. It prevents U_1 from responding to RF signals should any "leak" past the volume control.

The output signal of U_1 is ac coupled into the speaker through C_{20}. C_{20} is needed because there is a dc potential of about 1/2 V_{cc} (about 4.5 volts in this circuit) at pin 5 of U_1. This dc voltage is a result of the centered-Q biasing of the push-pull output stage within the IC. A simple and quick "health check" for audio power ICs is to read the output pin with a dc voltmeter. For ICs that run on a single supply (like this one), the voltage at the output pin should read close to 1/2 of the V_{cc} supply voltage.

The dc test on the power IC output pin is not a perfect guarantee of correct operation for U_1, but it is *likely* that U_1 is OK if the test passes.

Section Checkpoint

5–34 Why must power supplies be filtered, even if they're not regulated?

5–35 What circuit problem could cause *motorboating*?

5–36 List at least two symptoms that could be caused by an open RF bypass capacitor in the power supply of a receiver.

5–37 Why are the two sections of C_1 mechanically connected?

5–38 How is Q_2, the mixer, made to be nonlinear in operation?

5–39 What are the two functions of the antenna coil L_1?

5–40 What is the proper tool for adjusting IF transformers and coils?

5–41 Why is diode D_1 installed "backward?"

5–42 How does the dc voltage rectified with D_1 compare with the dc voltage present at the junction of R_8 and R_9?

5–43 Explain the operation of the *volume control* circuit, R_{12}.

5–44 What test point could be checked with a dc voltmeter to quickly determine whether U_1 was operating correctly? Is this test fail proof?

5–6 INTEGRATED CIRCUIT RECEIVERS

A popular trend in electronics is miniaturization, and radio is no exception. The move to transistor technology from vacuum tubes in the 1960s was just the beginning; today, an entire receiver can be placed on an IC chip.

One of the limitations of IC technology is that large inductances and capacitances (such as might be needed in an RF circuit) cannot be practically built onto a chip. Therefore, most receiver ICs require a collection of transformers and coils similar to those that would be found in a discrete transistor receiver. These transformers and coils still use up a substantial amount of space.

In order to make circuits even smaller, *surface-mount* RF coils and transformers have been developed. With the use of surface-mount components, a complete receiver can easily be built into a tiny package (such as a wristwatch). The Philips TDA-1072A is one example of the "receiver on a chip" and is shown in Figure 5–18.

Circuit Analysis

The TDA-1072A has all necessary receiver stages, with the exception of the audio power amplifier. The signal from the external preselector enters pin 14, where it is amplified

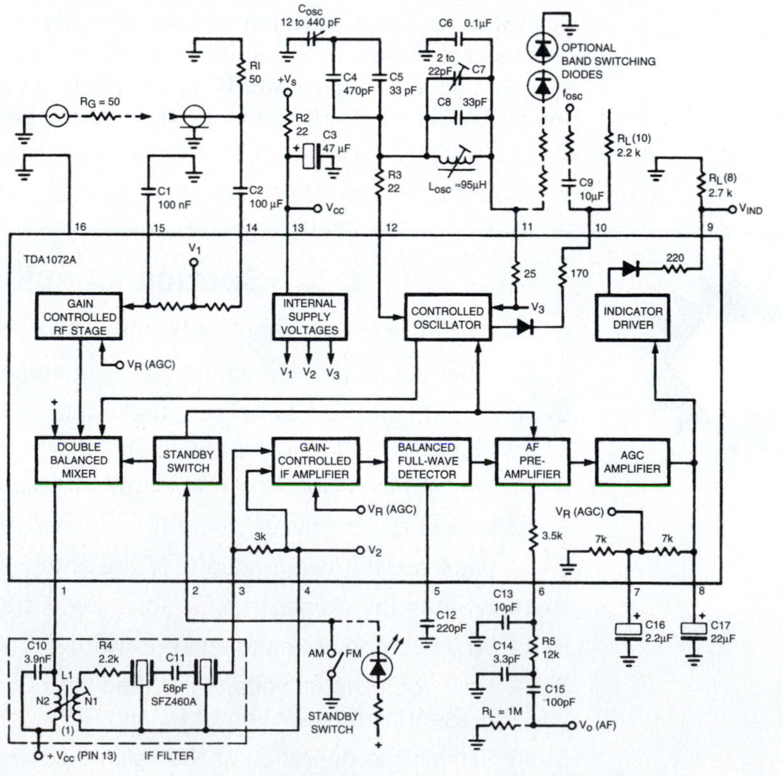

Figure 5–18 TDA-1072A circuit diagram (Source: Philips Semiconductor. Used with permission)

NOTES:
1. Coil Data: TOKO sample no. 7XNS-A7523DY; L1. N1/N2 = 12/32; $Q_O = 65$; $Q_B = 57$.
2. Filter Data: $Z_F = 700\Omega$ at $R_{3\text{-}4} = 3k\Omega$; $Z_I = 4.8k\Omega$.

before being sent to the mixer. The IC uses a doubly balanced mixer, which improves mixer performance. The oscillator is built-in too, and uses an external tank, L_{osc}, C_7, and C_8, to control its frequency. The mixer IF output is sent out pin 1 to an external IF filter (lower left) and is then amplified further after passing back into pin 3, the IF amplifier input. The IF amplifier feeds an internal detector. The detector output is taken from pin 6, and after low-pass filtering (R_5, C_{13}, C_{14}) is ready for audio amplification. The chip also features a built-in AGC. Not shown is the internal connection between pin 7 and the on-chip RF amplifier that operates the AGC. C_{16} and C_{17} control the AGC time constant.

5–7 ALIGNMENT OF RECEIVERS

When we speak of *aligning* a receiver, we are really talking about the process where each tuned circuit is adjusted for operation on the correct frequency. Because RF coils have a typical value tolerance of $\pm 10\%$, and because the stray capacitance within circuits affects their resonant frequency, almost all receivers require alignment adjustments. It would be prohibitively expensive to tighten these component tolerances in production.

Even so-called digital receivers may require alignment; the procedures for digitally controlled receivers are conceptually the same as the analog procedures described here. The main difference is that there is a microprocessor instead of a tuning mechanism; the mechanical variable capacitors are replaced by *varactor diodes,* a special type of diode that can act as a variable capacitor.

There are two principle types of alignment adjustments in AM receivers, *IF alignment* and *tracking alignment.*

IF Alignment with Instruments

The IF amplifiers determine a receiver's bandwidth and provide most of the signal gain. A misaligned IF section reduces the available gain, thus lowering the receiver's sensitivity. In AM broadcast receivers, the IFs are *synchronously tuned*. That's a high-tech way of saying that *all the IF-tuned circuits resonate at the same frequency*. That frequency is usually 455 kHz.

To align the IF amplifier, a weak 455 kHz signal source is loosely coupled to the input of the first IF amplifier while the detected output is monitored on an oscilloscope. The usual source of the 455 kHz test signal is an *RF signal generator*. The RF signal generator is usually set to produce a 455 kHz carrier, with a 1 kHz modulating tone.

To loosely couple the generator to the first IF amplifier, a small capacitor (1 pF) can be used, or the generator lead can be placed adjacent to the first IF amplifier input. Loose coupling is important; it prevents the generator's output impedance from loading the RF circuits being aligned. This loading would cause the circuit alignment to change in such a way that the alignment would seem OK until the generator was disconnected!

While keeping the generator signal as small as possible (producing a readable scope trace), the IF slugs should be carefully adjusted for maximum output. Since these adjustments interact, it may be necessary to go over them more than once. In order to prevent the AGC system from changing the readings, the generator signal must be kept low.

IF Alignment by the Seat of the Pants

Even with no test instruments in sight, it is possible to perform a crude IF alignment. The radio set is tuned to a very weak station, and each IF transformer is adjusted in succession to produce maximum signal from the speaker. If the signal gets too strong, a weaker station should be located. A strong signal is useless for alignment because the AGC becomes active, masking small alignment changes. If you find yourself turning a coil and getting no difference in output, either you're turning the wrong coil (it happens!) or the signal is too strong!

This method is OK in a pinch, but it doesn't guarantee that the final IF frequency will be 455 kHz. If the IF frequency is not close to 455 kHz, the front end of the receiver may not properly "track" as designed, and thus sensitivity may suffer at the ends of the dial.

Tracking Alignment

Once the IF amplifiers are correctly operating, the front end of the receiver can be aligned. Tracking alignment involves adjustment of the *local oscillator* and *preselector* circuits.

The first step in tracking alignment is to adjust the local oscillator. There are two local oscillator adjustments in the receiver of Figure 5–16. One is T_3, the oscillator coil; T_3 provides a *coarse* adjustment of the oscillator frequency. Fine oscillator adjustments are made with OT, the oscillator trimmer capacitor (located on the back of the tuning capacitor).

The local oscillator defines the dial calibration of the unit, so the first step is to establish a correct oscillator frequency. This is usually done by setting a signal generator to a known frequency, such as 1000 kHz, and adjusting the tuning dial of the receiver so that it also reads 1000 kHz. The signal generator is loosely coupled to the *preselector* input. If the signal generator is *not* heard at this point, the slug in T_3 (oscillator) is slowly adjusted until it is audible. Fine adjustments are finally made with OT in order to "center" the signal generator in the IF.

Alternatively, the unit can be tuned to a known radio station frequency; if the dial doesn't agree with the station frequency, the local oscillator calibration (T_3 and OT) is nudged until it does. Once the local oscillator is adjusted, it should not be touched again!

The final portion to be adjusted is the *preselector*. There are two parts to this adjustment, a low- and high-frequency tracking adjustment. Not all receivers can be adjusted at both ends.

The *high-frequency* preselector tracking adjustment is performed first. The receiver is set to a high frequency, such as 1500 kHz, and the generator is again set to match with a weak signal. The AT trimmer capacitor (part of the tuning capacitor) is carefully adjusted for maximum output. The preselector capacitor is adjusted because the tank circuit is dominated by the capacitance at high frequencies.

The *low-frequency* tracking adjustment is made next. The receiver is set for a low frequency such as 540 kHz, and with the generator set to match, the coil L_1 is adjusted. A loopstick is adjusted by sliding the coil back and forth over the ferrite rod. (Some receivers have a tuning slug in place of the loopstick.) The inductance in the preselector dominates at low frequencies, so this adjustment has maximum effect at the bottom of the dial.

Finally, the two checks for tracking may need to be alternately checked a couple more times, since there is some interaction between them.

Once an alignment has been completed, a receiver may be checked for compliance with the manufacturer's specifications. For broadcast receivers, this may mean that the unit "works!" For commercial receivers, more sophisticated test equipment and procedures (usually outlined by the manufacturer) are generally required in order to determine whether a unit is operating correctly. These procedures help to ensure that the receiver will be likely to function correctly in the environment where it will actually be located, such as in an aircraft communication system.

Section Checkpoint

5–45 Why must receiver tuned circuits be aligned?

5–46 What are the two types of alignment adjustments to be made on an AM receiver?

5–47 Why should the test signal amplitude be kept to a minimum when doing any receiver alignment procedures?

5–48 Can an IF amplifier be aligned when there are no test instruments available?

5–49 What two stages are adjusted during the *tracking* portion of an alignment?

5–50 If the local oscillator is off frequency, what aspect of receiver operation will be affected?

5–51 Once an alignment has been completed, how can a technician tell that it is working correctly?

5–8 TROUBLESHOOTING RECEIVERS

Very few of today's technicians will be working on radio receivers, especially at the component level. If that is true, then why should a technician bother to learn about troubleshooting receivers? There are two very compelling reasons.

First, RF technology is a fundamental part of most areas of electronics. Many products and systems use RF technology for transmitting data, remotely monitoring and controlling industrial processes, communicating safety and navigational information, and so on. It is *very* likely that a technician will be working with these kinds of devices. Radio is just one piece of the big picture, but it is also a fundamental one.

Second, the techniques learned in troubleshooting a humble AM radio receiver will take the aspiring technician far in his or her studies in electronics. No matter how complex a system is, we use the same methods to explore and diagnose it. The most powerful piece of equipment in a shop isn't a scope, spectrum analyzer, or digital multimeter; it is the inquisitive mind of a technician!

How to Find Little Troubles in Big Systems

Most electronic troubles are simple in nature: a bad solder joint, a shorted or open capacitor, or a failed semiconductor device. To a beginner, *finding* the fault is the hard part! The consistent use of a *troubleshooting method* helps many technicians. Our three-step method works very well:

1. perform visual (and other) inspection
2. check power supplies
3. check inputs and outputs

It is important that these steps be performed in order.

Visual inspection In a previous chapter, we talked about how and how not to look for trouble. For portable devices like radio receivers, troubles come in interesting varieties. Units can be dropped—look for cracked circuit boards when a cabinet is banged up. Much worse things can happen, such as being dropped in a lake. One customer did exactly that with a transceiver and proceeded to dry the unit out in his home oven, set to 300°. The metal exterior looked fine, but things were a little "toasty" (unrepairable) inside!

Power supplies Power supplies are the most heavily stressed portion of equipment, and there is usually a reason when one fails. A mental picture of "who gets what power" is important when troubleshooting equipment. For example, there are *two* power supply voltages in the receiver of Figure 5–16. One is the *main* power supply, which comes right off the power switch at the positive side of C_{17}. This supply runs everything. The *RF/IF* power supply is at the left-hand side of R_{14} (same as the positive side of C_2).

A totally dead unit is often the easiest to fix; this usually indicates a failure of the main power supply. Sometimes symptoms are confined to one area in a unit. For example, suppose that R_{14} becomes an open circuit. Everything that takes power from R_{14} (Q_1, Q_2, and Q_3 stages) will be disabled; all else (U_1 stage) will work just fine. This is like circuit theory in *reverse*—we ask what happens when something *breaks,* instead of asking how it works.

Inputs and outputs Once it is known that power supplies are operational, signal flow can be checked. Theory is helpful, since it tells us where and how the signals move. It is often convenient to pick a point halfway through the system and check the signal there. For example, suppose that we have a dead receiver with no sound. What are the two "halves" of the radio receiver? The first half is the RF and IF stages, and the second half is the AF stage. The output of the detector is a good halfway point to check for signal. Our findings there will tell us which way to move in our investigation.

Manufacturer's Data

In a service manual, a maker of equipment often provides typical circuit voltages and waveforms. Often the controls must be set a certain way to get the specified readings. A service manual can't substitute for troubleshooting technique; a technician must have a basic idea of where to look, and why, before attempting repairs.

Let's try a couple of scenarios, using the receiver of Figure 5–16. At each step, look at the readings and decide whether or not they're OK. If they're bad, make a mental list of the components that could be causing the trouble. *Use a card to cover up the following explanations, which give the solution for each step.*

Problem 1

Unit is totally dead. No reception of any kind.

Step 1: Visual (and other) inspection—No problems noted. This unit hasn't been dropped, or "laked and baked."

Step 2: Check power supplies: main test point, $C_{17}(+) = 8.8$ V, tuner test point, $C_2(+) =$ 0.3 V. *Is this OK?*

Answer: This is not OK. The main test point is close to 9 volts (batteries get weaker with time), but the tuner power supply test point only reads 0.3 volts. The tuner should be getting close to 8 volts on its power supply rail. Perhaps a volt or so is normally dropped across R_{14}. Possible component failures: R_{14}—open; C_2—short; C_3—open. At this point, it would be quickest to remove power from the unit, and test these three components with an ohmmeter.

Note: Also look for *solder ball shorts* between the pins of components on PC boards. These are often very hard to spot until you've focused on an area of interest.

Problem 2

Unit is totally dead. No reception of any kind.

Step 1: Visual (and other) inspection—Case has a slight crack on one side. *What should the technician be looking for?*
Answer: Look for cracked circuit boards or other damaged components inside the unit. Perhaps a wire has been knocked off the speaker (or other internal component). *No problems were found by visual inspection in this case.*

Step 2: Check power supplies: Main test point, $C_{17}(+) = 8.8$ V, tuner test point, $C_2(+) = 7.9$ V. *Is this OK?*
Answer: The power supplies are both OK. All sections of the unit apparently have power.

Step 3: Check inputs and outputs. *Where would be a good place to check initially, and how?*
Answer: Pick a point halfway through the unit, such as the detector output on either side of C_{14}, and measure it with a scope. The detector should be producing an audio signal of at least 100 mV_{pp}. *There is no audio signal present; where should we check next?*
Answer: The trouble lies in either the detector or an RF/IF stage preceding the detector. Using a signal generator, inject a 455 kHz modulated RF signal into the base of Q_3 (the halfway point between the antenna and detector) and listen for speaker output (the scope lead can be left on the detector output if desired). *There is no response; what stages likely have the problem, and where should we inject an RF signal next?*
Answer: We're putting 455 kHz into Q_3 base and getting nothing out of the detector. Either Q_3, the IF transformer T_2, or detector D_1 has a problem. We can inject a 455 kHz signal into the collector of Q_3 to see if that is the problem. *There is now a weak 1 kHz tone in the loudspeaker when injecting the RF signal into Q_3's collector. What is the most likely problem?*
Answer: The signal is being lost as it passes through Q_3. Either Q_3 is bad or something in the Q_3 stage is faulty (like a biasing resistor, coupling capacitor, and so forth). To investigate further, the technician would read the voltages on the emitter, base, and collector of Q_3 and compare them to the service manual's values—or lacking that, theoretical values from circuit analysis of the schematic data.

Note: There's no substitute for hands-on practice in the area of troubleshooting. You should take every opportunity you get to participate.

SUMMARY

- The process of receiving a radio signal can be thought of as a series of steps.
- Detection of an AM signal can be accomplished with a diode, which is a nonlinear circuit element.
- A diode detector produces an ac information output and a dc signal strength voltage that is useful for an automatic gain control system.
- AM diode detectors often cause diagonal clipping distortion when the modulation index approaches 100%.
- The *sensitivity* of a receiver is its ability to process weak signals; the *selectivity* of a receiver is its ability to separate a desired signal from a group of undesired stations. The bandwidth of a receiver controls its selectivity.

- The TRF receiver was the earliest electronic receiver type, but it did not have constant selectivity (bandwidth) and was unsuitable for high-frequency use.
- A *loopstick antenna* is a ferrite rod with many turns of fine wire wound upon it. It is sensitive to the magnetic field of a radio wave.
- The superheterodyne receiver is the king of modern receivers. Most every application uses it.
- The superhet receiver is based on the process of frequency conversion.
- The IF in a superhet receiver controls bandwidth and selectivity and provides most of the receiver's gain.
- A preselector is necessary in a superhet to attenuate the image frequency.
- The *purpose* of the AGC system in a receiver is to keep the speaker volume constant under changing signal strength conditions.
- The AGC system works by adjusting the gain of the RF and IF amplifiers. The control voltage for the AGC is developed at the detector.
- Complete radio receivers can be built on IC chips and are very common.
- The alignment of an AM receiver requires two types of adjustments, IF and tracking.
- In troubleshooting equipment, it's important to follow a logical sequence, such as the "three-step" method.

PROBLEMS

1. List the five steps in the process of receiving a radio signal.
2. In the crystal radio receiver of Figure 5–2, which of the five steps are being performed?
3. Why is RF amplification needed in a radio receiver?
4. How is a *variable capacitor* used in the tuning mechanism of a receiver?
5. A certain TRF radio receiver uses a variable capacitor with a range of 40 pF to 300 pF in combination with a fixed inductor of 200 μH. What is the tuning range of this receiver?
6. A receiver's bandpass filter can be tuned by changing the inductor instead of the capacitor. Such an arrangement is called *permeability tuning*. If the selector circuit of this receiver uses an inductor with a range of 20 μH to 200 μH in combination with a fixed capacitance of 270 pF, what will the tuning range be?
7. Compare and contrast an AM diode detector and a half-wave dc power supply.
8. What are the three frequency components present in the output of a diode detector circuit?
9. Why is a low-pass filter circuit used in a diode AM detector circuit?
10. What causes *diagonal clipping distortion* in an AM detector? What operating conditions are most likely to produce this form of distortion?
11. What is the function of inductor L_1 in Figure 5–7?
12. Explain the operation of a *loopstick* antenna. What portion of the radio wave is such an antenna sensitive to?
13. Explain the difference between the *sensitivity* and *selectivity* of a radio receiver. What circuit characteristic controls selectivity?
14. What are the two main problems with TRF receivers?
15. A certain tuned circuit has a resonant frequency of 710 kHz and a Q of 100. What is its bandwidth?
16. What is the Q of an inductor having an inductive reactance X_L of 10 Ω and equivalent parallel resistance R_P of 100 Ω?
17. A certain TRF tuning circuit consists of a 300-μH loopstick in combination with a 30 to

300 pF variable capacitor for tuning. The equivalent tank parallel resistance is 100 kΩ (see Figure 5–9).

a. What is the tuning range of the receiver?
b. What is the circuit bandwidth when tuned to 540 kHz?
c. What is the bandwidth at 1500 kHz?

18. How does a superheterodyne receiver achieve a constant bandwidth?

19. What is the function of the *converter* section in a superhet receiver?

20. An ideal mixer is being supplied with the frequencies 600 kHz and 400 kHz. What output frequencies will be produced?

21. A certain mixer is producing the following output frequencies: 760 kHz, 1760 kHz, 1000 kHz, and 240 kHz. What are the two most likely *input* frequencies?

22. What is the standard IF frequency for AM broadcast receivers?

23. In order to convert a 870 kHz carrier down to a 455 kHz IF frequency using high-side injection, what local oscillator frequency is required?

24. What local oscillator frequency would be needed in question 23 if low-side injection was used?

25. The superhet receiver of Figure 5–15 is tuned to a carrier frequency of 1080 kHz.

a. What frequency is the *preselector* tuned to?
b. What frequency is the *local oscillator* tuned to?
c. List all the frequencies produced by the *mixer*.
d. Which of the mixer output frequencies is amplified by the IF amplifier?

26. The superhet receiver of Figure 5–15 is tuned to an unknown carrier frequency. The preselector tank consists of a 300 μH inductor and a 30 to 300 pF variable capacitor, which is set for a capacitance of 88 pF.

a. What frequency is the receiver tuned to?
b. What frequency should the local oscillator be producing?
c. If the local oscillator has a 250 μH inductor, what capacitance will set the local oscillator to the correct frequency?

27. An AM superhet receiver is tuned to 1270 kHz and uses a 455 kHz IF. If high-side injection is being used, what is the image frequency?

28. What section of a superhet receiver is responsible for eliminating the image frequency?

29. Why is AGC needed in AM receivers?

30. What stage of an AM receiver develops the AGC signal?

31. What stages of an AM receiver are *controlled* by the AGC signal?

32. Why are C_{1A} and C_{1B} mechanically ganged together in the receiver of Figure 5–16?

33. What could cause a receiver to *motorboat?*

34. If a receiver produces odd hissing or whistling sounds, what could be wrong?

35. Using outline form, give the details for the alignment of a typical AM broadcast receiver.

6

Single Sideband Systems

OBJECTIVES

At the conclusion of this chapter, the reader will be able to:

- explain the advantages of SSB over conventional AM transmission
- describe the various types of SSB transmission
- explain the operation of circuits used to generate SSB signals
- given the block diagram of a SSB transmitter, describe the nature of the signal at each point in the circuit
- draw a block diagram of a SSB receiver
- explain how SSB signals are demodulated
- given the block diagram of a SSB transceiver, follow the signal flow for receive and transmit conditions
- describe typical troubleshooting procedures for SSB transceivers

AM is a widely used method of communication. AM transmitters are relatively simple, and AM receivers can be built very inexpensively with just a handful of components. Despite this, AM suffers from very poor efficiency. As was demonstrated in Chapter 3, at best only 33% of the transmitted energy in an AM signal actually carries information (the sidebands). The remainder (about 67%) of the energy is used to transmit the *carrier frequency component* (CFC), which itself carries no information.

Furthermore, an AM signal contains two sidebands, each of which carry the same information. Having two sidebands causes the required transmitter bandwidth to be twice as wide as absolutely necessary.

By modifying the AM signal in various ways, we can get various types of *sideband* transmissions. Transmitters and receivers for sideband are more complex than their AM equivalents. However, there are situations where the added complexity is justified. Where many stations must share a limited range of frequency space, sideband techniques are useful, since they can reduce bandwidth. Where voice communications must be carried out over a noisy path, sideband methods prove superior to AM and FM.

6–1 SSB VERSUS AM: TYPES OF SIDEBAND SIGNALS

Figure 6–1 shows a typical AM signal, which can be thought of as a DSB-FC (double-sideband, full-carrier) signal. A conventional AM signal consists of a *carrier frequency component* or *CFC,* a *lower sideband,* and an *upper sideband.* Notice how we have shown the sidebands as *ranges* of frequency rather than individual frequencies. This is because a real information signal has a *range* of frequencies rather than just a single frequency.

The minimum and maximum information frequencies are marked as $f_{m(\text{min})}$ and $f_{m(\text{max})}$. Therefore, the bandwidth of a conventional AM signal can be expressed as

$$BW_{AM} = 2f_{m(\text{max})} \tag{6–1}$$

The Minimum Bandwidth for Speech

In the 1960s, Bell Laboratories conducted research to determine the minimum range of frequencies needed for understandable voice communications. The research was important because the telephone companies were sending many different conversations at the same time through a single copper telephone wire. The technique of sending multiple pieces of information at the same time is called *multiplexing*. Each conversation would be assigned a different range of frequencies. If a certain amount of bandwidth were available on the wire, how many conversations could be sent at once? That would depend, of course, on the bandwidth required for each person's voice. Bell's research efforts demonstrated that most of the information power in human speech is contained in the frequency range 300 Hz to 3000 Hz. Frequencies below 300 Hz add bass "presence" to voice but little intelligibility. Most of the energy above 3000 Hz is from the unvoiced speech sounds, such as *s, f,* and so on. *Therefore, most systems that are intended to send only human voice are designed to reproduce the frequency range 300 Hz* [$f_{m(\text{min})}$] *to 3000 Hz* [$f_{m(\text{max})}$].

Figure 6–1 A conventional AM (DSB-FC) signal

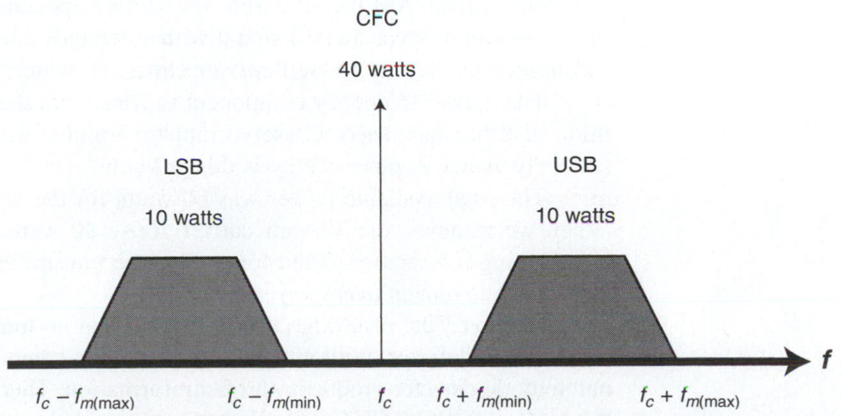

EXAMPLE 6–1 Speech is to be sent using an AM transmitter. The available carrier power is 40 watts, as in Figure 6–1. The transmitter is operating at 100% modulation.

a. What is the *total power?*

b. What is the power of the information?

c. What bandwidth will be needed?

Solution

a. The total power can be calculated as

$$P_t = P_c\left(1 + \frac{m^2}{2}\right) = 40\text{ W}\left(1 + \frac{1^2}{2}\right) = \underline{\underline{60\text{ W}}}$$

The 60 watts is the total available power from the transmitter.

b. The sideband power is the same as the information power. Since the carrier power is 40 watts, the *sideband* power will be

$$P_{\text{info}} = P_{\text{side}} = P_c\left(\frac{m^2}{2}\right) = 40\text{ W}\left(\frac{1^2}{2}\right) = \underline{\underline{20\text{ W}}}$$

Notice how the information power is only 33% of the total power being transmitted!

c. Sending speech requires a bandwidth from 300 Hz to 3000 Hz. The highest information frequency $f_{m(\text{max})}$ is 3000 Hz. Therefore,

$$\text{BW} = 2f_{m(\text{max})} = 2(3000\text{ Hz}) = \underline{\underline{6\text{ kHz}}}$$

Note that the minimum information frequency doesn't affect bandwidth at all in a conventional AM transmission.

DSB-SC Operation

The carrier frequency component (CFC) uses up most of the available transmitter power in a conventional AM transmission. By using a special circuit called a *balanced modulator,* we can produce an AM signal with sidebands but no CFC. Such a signal is called a *double-sideband suppressed-carrier* emission, which is pictured in Figure 6–2.

The carrier frequency component is *gone* from the signal in Figure 6–2, but something else has happened. Closely compare Figures 6–1 and 6–2. The *sidebands* have suddenly grown in power! Why is this possible?

The total available power was 60 watts for the original AM signal of Figure 6–2. When we removed the 40 watt carrier, those 40 watts of power became available for transmitting information. Therefore, *all* of the transmitted power is information power—a great improvement over conventional AM.

However, the resulting DSB-SC signal can no longer be properly demodulated by a diode AM detector. Without a carrier frequency component to act as a "reference" signal, a diode detector produces garbled information. Therefore, an oscillator circuit called a *beat frequency oscillator,* or *BFO,* must be added to the detector circuit to reinsert the missing carrier signal.

Figure 6–2 A DSB-SC signal

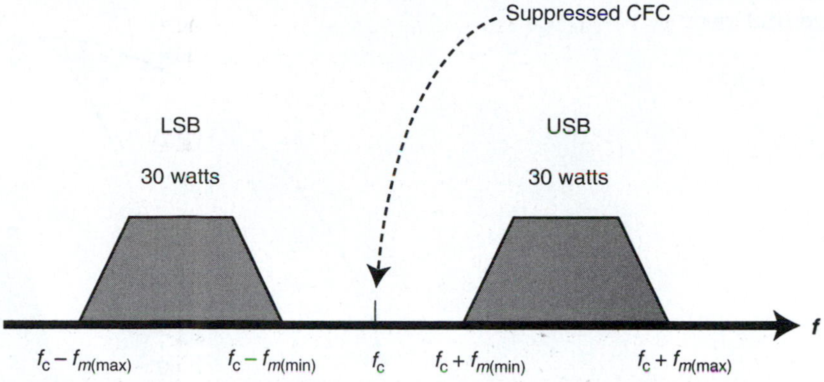

Since we are now transmitting additional information power, we say that we have gained a *decibel power advantage* over a conventional AM transmitter. The decibel power advantage can be computed as follows:

$$\text{Decibel power advantage:} \qquad \text{dB} = 10 \log\left(\left\langle\frac{P_2}{P_1}\right\rangle\left\langle\frac{\text{BW}_1}{\text{BW}_2}\right\rangle^2\right) \qquad \textbf{(6–2)}$$

where P_2 is the new information power, P_1 is the original information power, BW_2 is the new bandwidth, and BW_1 is the original bandwidth. Note that the squared bandwidth ratio takes into account the *combined* advantage for both the transmitter and receiver. The ratio of powers is *not* squared!

EXAMPLE 6–2

If the total power of the transmission of Figure 6–2 is 60 watts (same as Figure 6–1), and again human speech is to be transmitted, calculate

a. The bandwidth of the DSB-SC signal

b. The decibel power advantage of the DSB-SC signal over the AM-FC signal

Solution

a. As you can see from Figure 6–2, the bandwidth of the signal will remain unchanged, since we're still sending two sidebands. The bandwidth will therefore remain at (2)(3000 Hz) or 6 kHz.

b. Equation 6–2 can be used to calculate the decibel advantage. The information power in the AM signal is 20 watts, and the information power in the DSB signal is 60 watts. There's no change in bandwidth between the two modes:

$$\text{dB} = 10 \log\left(\left\langle\frac{P_2}{P_1}\right\rangle\left\langle\frac{\text{BW}_1}{\text{BW}_2}\right\rangle^2\right) = 10 \log\left(\left\langle\frac{60\text{ W}}{20\text{ W}}\right\rangle\left\langle\frac{6\text{ kHz}}{6\text{ kHz}}\right\rangle^2\right) = \underline{\underline{+4.77\text{ dB}}}$$

In other words, for the conventional AM transmitter to be as effective as the DSB-SC transmitter, the AM transmitter would have to be operating at a power level that is +4.77 dB (three times) stronger than the DSB unit. That is, the *total* power level (P_t) of the AM transmitter would have to be *180 watts* to equal the information-carrying capacity of the DSB-SC unit. This is a great power savings!

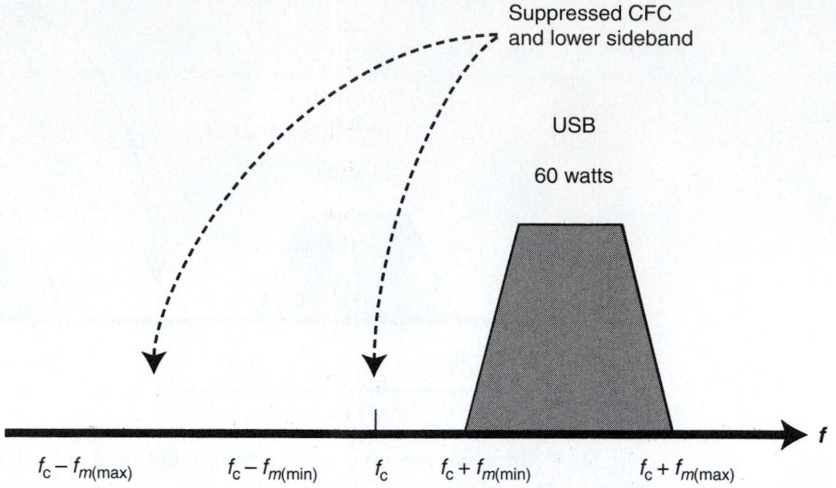

Figure 6–3 A SSB-SC signal in USB mode

SSB-SC Operation

We can improve the efficiency of transmission even further by eliminating one of the redundant sidebands. It does not matter which one is removed; when the upper sideband is kept, we say that we're operating in *USB* mode; when the lower sideband is kept, we're in *LSB* mode (see Figure 6–3). Most people simply refer to this mode as *SSB;* they assume that the carrier is suppressed.

A SSB signal has a bandwidth that is slightly less than *half* of a corresponding AM or DSB-SC signal. Look how the entire 60 watts is now concentrated in a much more narrow "slice" of spectrum. This characteristic gives a SSB transmitter much more "talk power" than an AM transmitter of the same power level!

EXAMPLE 6–3

If the SSB signal of Figure 6–3 is sending a voice signal (300 Hz–3 kHz), calculate

a. the bandwidth

b. the decibel power advantage of the SSB signal over the AM signal of Figure 6–1

c. the power level an AM transmitter would need in order to have equivalent performance

Solution

a. Since we're only sending one sideband, equation (6–1) doesn't apply. Instead, we can fall back on the basic definition of bandwidth:

$$BW = f_{max} - f_{min}$$

In Figure 6–3, f_{min} is equal to $f_c + f_{m(min)}$, and f_{max} is equal to $f_c + f_{m(max)}$, so we can state:

$$BW = f_{max} - f_{min} = [f_c + f_{m(max)}] - [f_c + f_{m(min)}] = f_{m(max)} - f_{m(min)}$$
$$BW = f_{m(max)} - f_{m(min)} = 3000 \text{ Hz} - 300 \text{ Hz} = \underline{\underline{2.7 \text{ kHz}}}$$

Notice that the bandwidth is just slightly less than one half of the bandwidth required for an AM-FC transmission.

b. Equation 6–2 calculates decibel power advantage:

$$\text{dB} = 10\log\left(\left(\frac{P_2}{P_1}\right)\left(\frac{\text{BW}_1}{\text{BW}_2}\right)^2\right) = 10\log\left(\left(\frac{60\text{ W}}{20\text{ W}}\right)\left(\frac{6\text{ kHz}}{2.7\text{ kHz}}\right)^2\right) = \underline{\underline{+11.7\text{ dB}}}$$

Note that we're not transmitting any additional information power; we're just packing the information into a smaller bandwidth "space."

c. The power ratio expressed by a 11.7 dB advantage can be found by simply expressing the terms inside the dB formula for power advantage:

$$G_p = \left(\frac{P_2}{P_1}\right)\left(\frac{\text{BW}_1}{\text{BW}_2}\right)^2 = \left(\frac{60\text{ W}}{20\text{ W}}\right)\left(\frac{6\text{ kHz}}{2.7\text{ kHz}}\right)^2 = \underline{\underline{14.8:1}}$$

The AM transmitter would need about 14.8 times the power of the SSB transmitter in order to have equivalent performance. That is a total power of (14.8)(60 watts), or 888 watts.

SSB is much more efficient than AM. In general, most people agree that there is better than a 10 dB (10:1) power advantage for SSB transmission over AM.

Vestigial Sideband (VSB) Mode

One of the disadvantages of DSB-SC and SSB transmission is that neither of these signals can be demodulated with a diode detector. A mode that combines the advantage of conventional AM with the reduced bandwidth requirements of SSB is called *vestigial sideband,* or *VSB*. A VSB signal is shown in Figure 6–4.

In VSB, the carrier is left intact (thus allowing detection with a diode detector), and most of the lower sideband is filtered out. The resulting signal has a reduced bandwidth when compared to a conventional AM signal. Analog television uses VSB to transmit the picture because the information frequencies contained within the video signal range from near dc (30 Hz) to approximately 4 MHz. Transmitting a video signal by conventional AM would therefore require an 8 MHz bandwidth just for the picture! By using

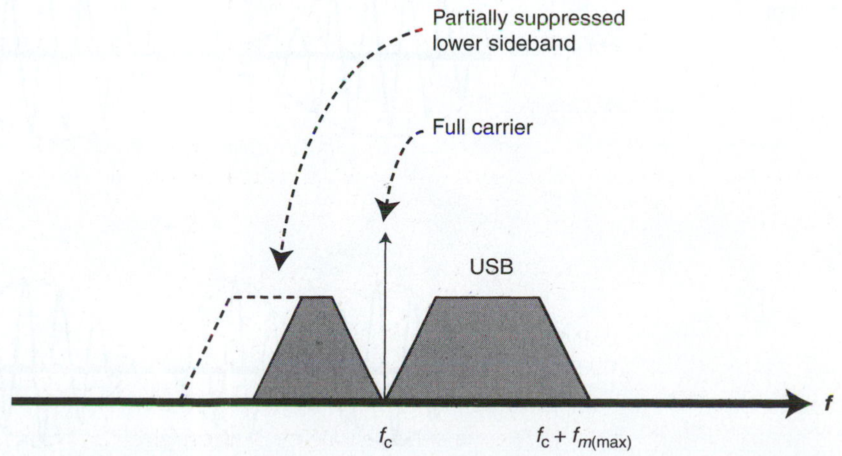

Figure 6–4 VSB operation (USB mode)

VSB, the bandwidth is reduced to slightly less than 5 MHz—which leaves room to squeeze in an FM sound carrier. The details of television transmission will be covered in a later chapter.

Percentage of Modulation and Other Myths

You might recall that for an AM signal, the modulation index (and hence percentage of modulation) could be calculated as follows:

$$m = \frac{V_m}{V_c}$$

In a DSB-SC or SSB-SC signal, there is no carrier transmitted; only sidebands are sent. Using the preceding formula for SSB or DSB signals causes a problem, for we can't divide by a *zero* V_c to get an answer. Percentage of modulation is meaningless!

SSB and DSB transmitter outputs are measured according to their *peak envelope power,* or *PEP*. PEP is calculated by finding the average (RMS) power delivered during one RF cycle at the *peak* or *maximum* of the modulation envelope. Sideband transmitters are always rated in this manner. So long as the operator of the transmitter does not exceed the rated PEP level of the unit, no "overmodulation" occurs and a clean modulated signal will be delivered to the antenna.

EXAMPLE 6–4 What is the PEP of the AM and DSB-SC signals in Figure 6–5? Each is being measured across a 50 ohm dummy load.

Solution

To calculate PEP, find the RMS power during one RF cycle at the crest of the modulation envelope. For the AM signal, the crest voltage is 40 V_p, so we can find the RMS voltage:

$$V_{rms} = \frac{V_p}{\sqrt{2}} = \frac{40\ V_p}{\sqrt{2}} = 28.28\ V$$

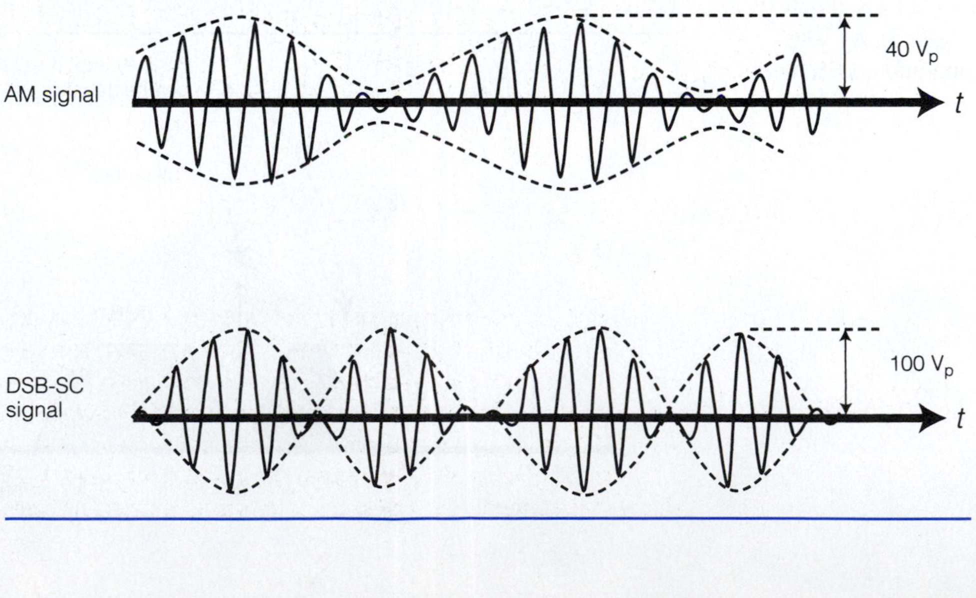

Figure 6–5 AM and DSB-SC signals for PEP measurement

Now Ohm's law is applied to find the PEP power:

$$\text{PEP} = \frac{V_{\text{rms}}^2}{R_L} = \frac{28.28\ \text{V}^2}{50\ \Omega} = \underline{\underline{16\ \text{watts}}}$$

For the DSB transmitter, the exact same procedure is used, and we get

$$V_{\text{rms}} = \frac{\text{V}_\text{p}}{\sqrt{2}} = \frac{100\ \text{V}_\text{p}}{\sqrt{2}} = 70.7\ \text{V}$$

$$\text{PEP} = \frac{V_{\text{rms}}}{R_L} = \frac{70.7\ \text{V}^2}{50\ \Omega} = \underline{\underline{100\ \text{watts}}}$$

For SSB mobile and fixed-station transmitters, 100 watts PEP is a fairly standard value for power output.

Section Checkpoint

6–1 What frequency component uses most of the power in a conventional AM signal?

6–2 What determines the bandwidth of an AM signal?

6–3 What is the minimum frequency range for reproduction of speech?

6–4 How is a DSB-SC signal different from an AM-FC signal?

6–5 What must be reinserted at the detector circuit in a receiver designed to receive DSB and SSB signals?

6–6 What is meant by the term *decibel power advantage*?

6–7 What type of emission results when the CFC and one sideband are removed?

6–8 How is the bandwidth of an SSB signal computed?

6–9 What is the advantage of VSB over SSB and DSB-SC?

6–10 Why is percentage of modulation meaningless for SSB?

6–11 How is PEP calculated?

6–2 SSB SIGNAL GENERATION: FILTER METHOD

SSB signals can be generated using several methods. These methods are known as the *filter method, phasing method,* and a relatively new approach, the *digital signal processor (DSP) method.* Most modern transceivers use the filter method; as the power of DSP chips rises (and their price tumbles), it is expected that the DSP method will continue to grow in popularity. Figure 6–6 illustrates the filter method.

In the filter method, two steps are taken to generate a SSB signal. First, a DSB-SC signal is generated in a *balanced modulator.* Second, the undesired sideband is removed

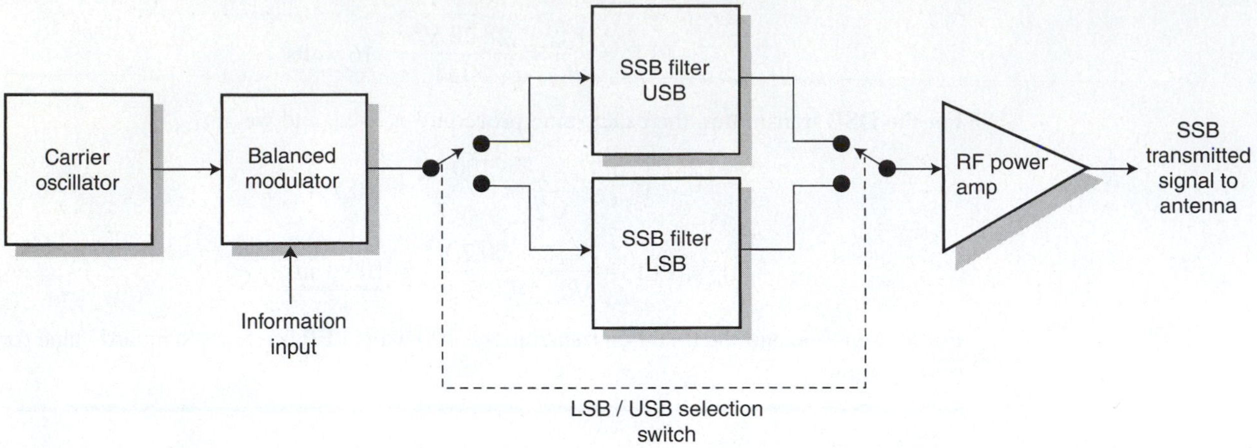

Figure 6–6 Filter method of SSB generation

by filtration. An actual SSB transceiver isn't likely to use two filters; it's simply easier to picture the process this way. We'll see soon how it can be done with a single filter.

The resulting SSB signal must be amplified before being sent to the antenna. A *linear* RF power amplifier is required. All SSB transmitters can therefore be considered to be *low-level* transmitters.

Balanced Modulator with Two JFETs

Figure 6–7 shows a balanced modulator constructed from a pair of JFETs. There are two inputs to the circuit, the *carrier* input and the *information* input.

The carrier signal enters the balanced modulator through transformer T_2. The RF signal flows left after leaving the secondary (top coil) of T_2, where it skips the secondary of T_1 (right coil) by passing through RF bypass capacitors C_2 and C_5. This seems self-defeating. Why isn't T_1 part of the RF circuit? T_1 is in the circuit for the introduction of an audio signal, as we'll soon see.

The RF carrier signal flows in-phase to the gates of transistors Q_1 and Q_2. This is a little strange; an amplifier built like this *usually* has the input signals fed 180° out of phase to the two transistors. This one has the signals being applied *in phase*. The result is that both Q_1 and Q_2 "push" the signal out their drain pins *in phase* to the ends of transformer T_3, the output transformer.

Because of the center tap on T_3, the ends of the center-tapped primary would normally be expected to be 180° out of phase. When Q_2 and Q_3 supply the in-phase RF signal to both halves of T_3, the RF signal *cancels* inside T_3 and *no output is produced!* This is known as "push-push" operation. It's very similar to a teeter-totter. If the same thing happens on both sides of a teeter-totter, it doesn't move. This push-push action *cancels* the RF carrier signal, which is exactly what we want.

That's great, you think; all this amplifier analysis just to get *nothing* out! Let's unbalance the "teeter-totter." Transformer T_1 is an *audio coupling transformer,* and an audio information signal is passed into the left side of T_1. The audio signal is slowly changing with respect to the RF signal. When the audio signal goes positive, the gate of Q_1 becomes more positive, and Q_1's RF gain increases; at the same time, the gate of Q_2 is becoming more *negative,* so Q_2's gain *decreases*. Guess what? Now Q_1 and Q_2 aren't "pushing"

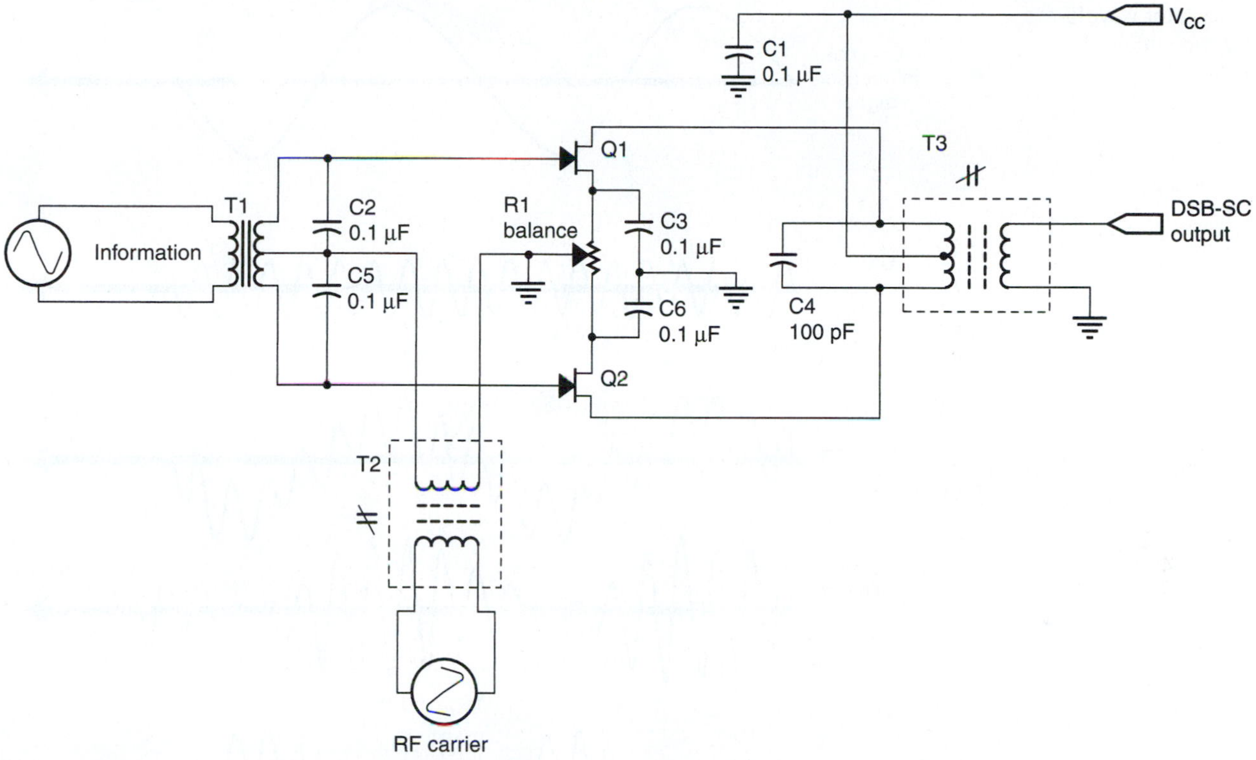

Figure 6–7 A balanced modulator employing two JFETs

exactly alike anymore; the audio signal from T_1 has *unbalanced* them. Since Q_1 and Q_2 are now producing unequal signals, they no longer cancel within T_3, and an output results. The exact opposite happens on the negative half-cycle.

Variable resistor R_1 is included as a *balance* control. Because of all the component tolerances, it's likely that some minor adjustment will be necessary to get Q_1 and Q_2 to have the same gain. R_1 is also known as a *carrier null* control.

> To properly align a balanced modulator, a technician removes any audio signal and adjusts the carrier null (balance) control for *minimum* output from the circuit. Some balanced modulators also include a trimmer capacitor; the two controls are adjusted alternately for the deepest possible null in the output.

Figure 6–8 shows typical waveforms for the balanced modulator in Figure 6–7. Note how the gain of Q_1 and Q_2 alternate—when Q_1's gain is being increased by the intelligence, Q_2's gain is being reduced. This is what causes the imbalance in T_3.

The output waveshape of a DSB-SC signal from a balanced modulator circuit is different from the original information. The envelope has a "cat's-eye" shape, with a period *one-half* of the information period. (There are two complete cat's-eyes for each cycle of information.) The point of zero crossing always has a sharp wedge shape in a DSB-SC waveform. These features are important; they show proper operation of the balanced modulator.

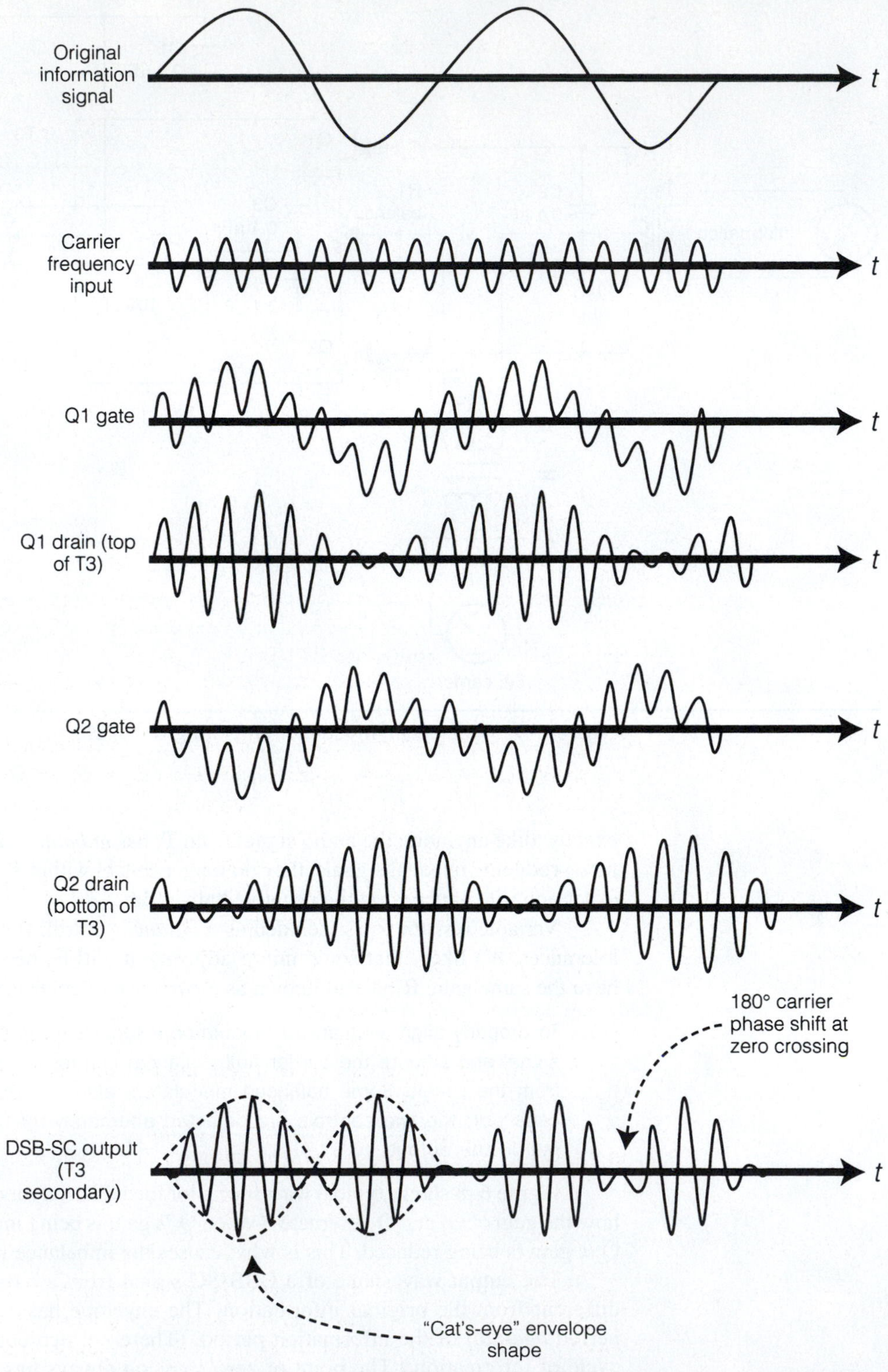

Figure 6–8 Balanced modulator waveforms for the circuit of Figure 6–7

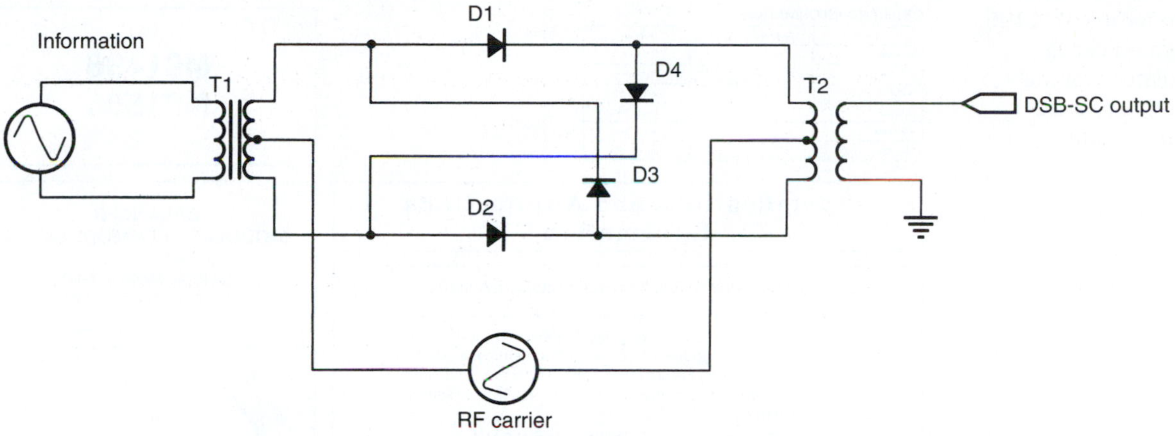

Figure 6–9 A double-balanced diode modulator

Other Balanced Modulator Circuits

Two other balanced modulator circuits are very popular; they are the *diode modulator* and *integrated circuit modulator*. The diode modulator is probably the most popular of all balanced modulator circuits; one is shown in Figure 6–9.

In the diode modulator of Figure 6–9, the RF carrier signal is applied in the middle of the circuit between T_1 and T_2. For a moment, assume that the information voltage is zero. Under this condition, diodes D_1 and D_2 will conduct equally on the positive half cycle of the RF carrier, and equal currents will flow through them to the top and bottom of T_2. You know what that means: Equal signal to the top and bottom halves of a center-tapped transformer means no output!

The same action occurs on the negative half-cycle of the RF carrier; instead, diodes D_3 and D_4 now conduct, again equally, and again with no resulting output from T_2.

In order to get output, something needs to shove diode pairs D_1-D_2 and D_3-D_4 off balance. If we apply an information voltage into T_1, that will do the trick! On the positive half-cycle of the *information,* diodes D_1 and D_4 are forward-biased, while diodes D_2 and D_3 are reverse-biased. You're not misreading this! This imbalances the diodes, and now the RF currents in the upper and lower halves of T_3 will be unequal. The exact opposite happens on the negative half-cycle of the information. The result is a double-sideband, suppressed-carrier signal at the output.

The diode modulator is very popular; commercial equipment often uses a hybrid form in a sealed metal can, which provides very high carrier rejection and excellent reliability.

Integrated Circuit Modulator

The Analog Devices AD834, Signetics NE602A, and Motorola MC1496 are examples of integrated circuit balanced mixers that can function either as balanced modulators, balanced mixers, or SSB demodulators. These devices are very common in modern equipment. Figure 6–10 is the data sheet for the MC1496.

Figure 6–10 Data sheet for the MC1496 integrated circuit modulator (Copyright of Motorola, used by permission)

ORDERING INFORMATION

Device	Temperature Range	Package
MC1496D	0°C to +70°C	SO-14
MC1496G		Metal Can
MC1496L		Ceramic DIP
MC1496P		Plastic DIP
MC1596G	−55°C to +125°C	Metal Can
MC1596L		Ceramic DIP

MC1496
MC1596

Specifications and Applications Information

BALANCED MODULATOR/DEMODULATOR

SILICON MONOLITHIC INTEGRATED CIRCUIT

BALANCED MODULATOR/ DEMODULATOR

. . . designed for use where the output voltage is a product of an input voltage (signal) and a switching function (carrier). Typical applications include suppressed carrier and amplitude modulation, synchronous detection, FM detection, phase detection, and chopper applications. See Motorola Application Note AN-531 for additional design information.

- Excellent Carrier Suppression – 65 dB typ @ 0.5 MHz
  – 50 dB typ @ 10 MHz
- Adjustable Gain and Signal Handling
- Balanced Inputs and Outputs
- High Common Mode Rejection – 85 dB typ

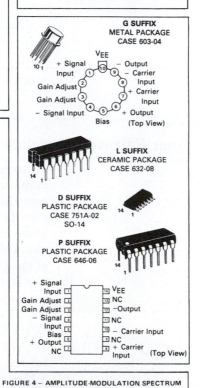

G SUFFIX
METAL PACKAGE
CASE 603-04

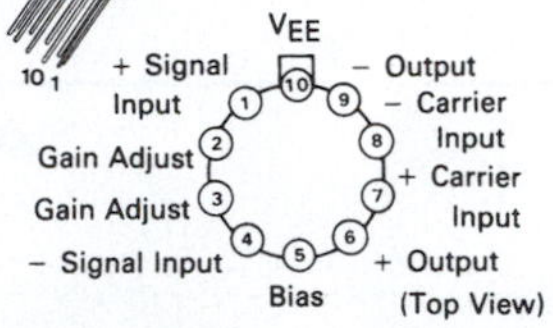

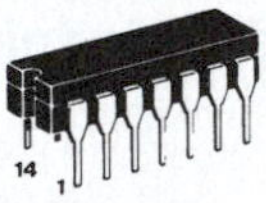

L SUFFIX
CERAMIC PACKAGE
CASE 632-08

D SUFFIX
PLASTIC PACKAGE
CASE 751A-02
SO-14

P SUFFIX
PLASTIC PACKAGE
CASE 646-06

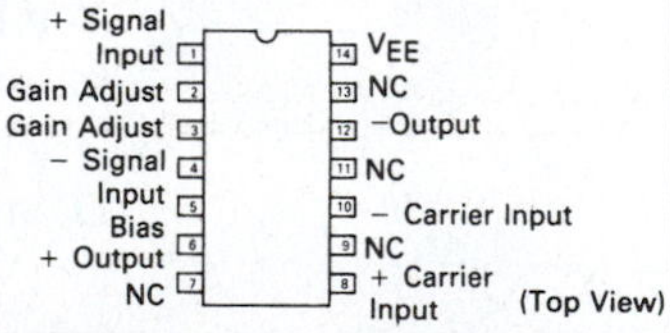

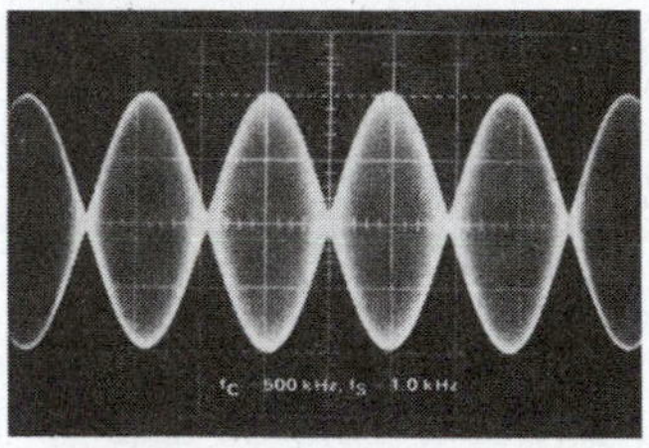

FIGURE 1 – SUPPRESSED-CARRIER OUTPUT WAVEFORM

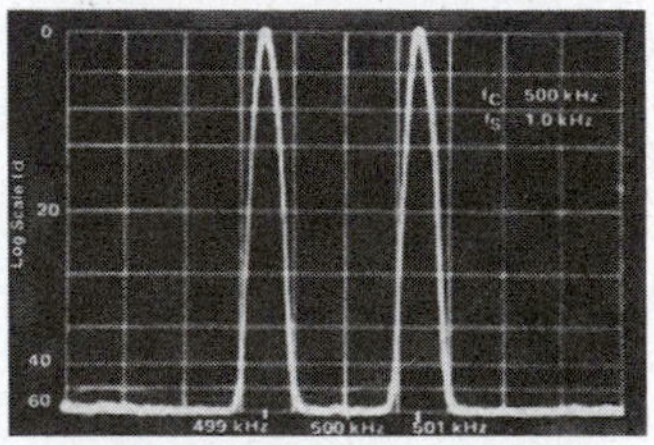

FIGURE 2 – SUPPRESSED-CARRIER SPECTRUM

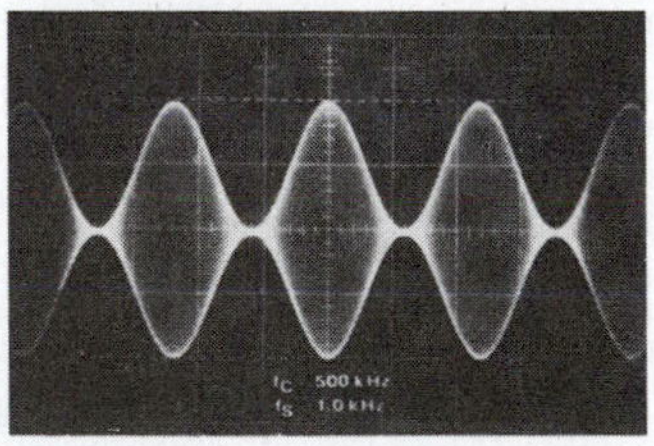

FIGURE 3 – AMPLITUDE-MODULATION OUTPUT WAVEFORM

FIGURE 4 – AMPLITUDE-MODULATION SPECTRUM

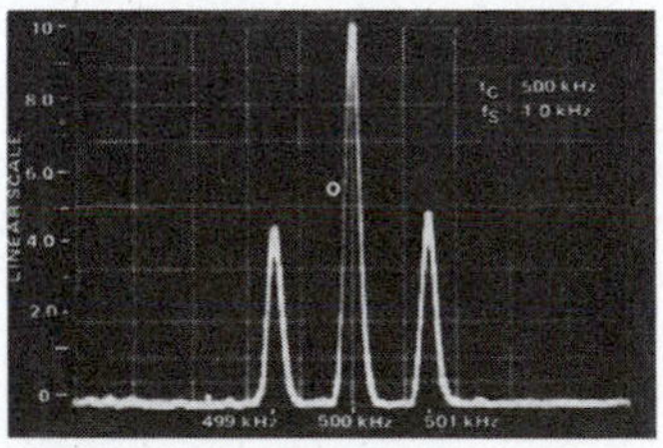

Figure 6–10
Continued

MAXIMUM RATINGS* (T_A = +25°C unless otherwise noted)

Rating	Symbol	Value	Unit
Applied Voltage ($V_6 - V_7$, $V_8 - V_1$, $V_9 - V_7$, $V_9 - V_8$, $V_7 - V_4$, $V_7 - V_1$, $V_8 - V_4$, $V_6 - V_8$, $V_2 - V_5$, $V_3 - V_5$)	ΔV	30	Vdc
Differential Input Signal	$V_7 - V_8$ $V_4 - V_1$	+5.0 $\pm(5 + I_5R_e)$	Vdc
Maximum Bias Current	I_5	10	mA
Thermal Resistance, Junction to Air	$R_{\theta JA}$		°C/W
Ceramic Dual In-Line Package		100	
Plastic Dual In-Line Package		100	
Metal Package		160	
Operating Temperature Range	T_A		°C
MC1496		0 to +70	
MC1596		−55 to +125	
Storage Temperature Range	T_{stg}	−65 to +150	°C

ELECTRICAL CHARACTERISTICS* (V_{CC} = +12 Vdc, V_{EE} = −8.0 Vdc, I_5 = 1.0 mAdc, R_L = 3.9 kΩ, R_e = 1.0 kΩ, T_A = +25°C unless otherwise noted) (All input and output characteristics are single-ended unless otherwise noted.)

Characteristic	Fig.	Note	Symbol	MC1596 Min	MC1596 Typ	MC1596 Max	MC1496 Min	MC1496 Typ	MC1496 Max	Unit
Carrier Feedthrough	5	1	V_{CFT}							μV(rms)
V_C = 60 mV(rms) sine wave and offset adjusted to zero, f_C = 1.0 kHz				—	40	—	—	40	—	
f_C = 10 MHz				—	140	—	—	140	—	
V_C = 300 mVp-p square wave:										mV(rms)
offset adjusted to zero, f_C = 1.0 kHz				—	0.04	0.2	—	0.04	0.4	
offset not adjusted, f_C = 1.0 kHz				—	20	100	—	20	200	
Carrier Suppression	5	2	V_{CS}							dB
f_S = 10 kHz, 300 mV(rms)										
f_C = 500 kHz, 60 mV(rms) sine wave				50	65	—	40	65	—	
f_C = 10 MHz, 60 mV(rms) sine wave				—	50	—	—	50	—	k
Transadmittance Bandwidth (Magnitude) (R_L = 50 ohms)	8	8	BW_{3dB}							MHz
Carrier Input Port, V_C = 60 mV(rms) sine wave, f_S = 1.0 kHz, 300 mV(rms) sine wave				—	300	—	—	300	—	
Signal Input Port, V_S = 300 mV(rms) sine wave, $\lvert V_C \rvert$ = 0.5 Vdc				—	80	—	—	80	—	
Signal Gain, V_S = 100 mV(rms), f = 1.0 kHz; $\lvert V_C \rvert$ = 0.5 Vdc	10	3	A_{VS}	2.5	3.5	—	2.5	3.5	—	V/V
Single-Ended Input Impedance, Signal Port, f = 5.0 MHz	6	—								
Parallel Input Resistance			r_{ip}	—	200	—	—	200	—	kΩ
Parallel Input Capacitance			c_{ip}	—	2.0	—	—	2.0	—	pF
Single-Ended Output Impedance, f = 10 MHz	6	—								
Parallel Output Resistance			r_{op}	—	40	—	—	40	—	kΩ
Parallel Output Capacitance			c_{oo}	—	5.0	—	—	5.0	—	pF
Input Bias Current $I_{bS} = \frac{I_1 + I_4}{2}$; $I_{bC} = \frac{I_7 + I_8}{2}$	7	—								μA
			I_{bS}	—	12	25	—	12	30	
			I_{bC}	—	12	25	—	12	30	
Input Offset Current $I_{ioS} = I_1 - I_4$; $I_{ioC} = I_7 - I_8$	7	—								μA
			$\lvert I_{ioS} \rvert$	—	0.7	5.0	—	0.7	7.0	
			$\lvert I_{ioC} \rvert$	—	0.7	5.0	—	0.7	7.0	
Average Temperature Coefficient of Input Offset Current (T_A = −55°C to +125°C)	7	—	$\lvert TC_{Iio} \rvert$	—	2.0	—	—	2.0	—	nA/°C
Output Offset Current ($I_6 - I_9$)	7	—	$\lvert I_{oo} \rvert$	—	14	50	—	14	80	μA
Average Temperature Coefficient of Output Offset Current (T_A = −55°C to +125°C)	7	—	$\lvert TC_{Ioo} \rvert$	—	90	—	—	90	—	nA/°C
Common-Mode Input Swing, Signal Port, f_S = 1.0 kHz	9	4	CMV	—	5.0	—	—	5.0	—	Vp-p
Common-Mode Gain, Signal Port, f_S = 1.0 kHz, $\lvert V_C \rvert$ = 0.5 Vdc	9	—	ACM	—	−85	—	—	−85	—	dB
Common-Mode Quiescent Output Voltage (Pin 6 or Pin 9)	10	—	V_{out}	—	8.0	—	—	8.0	—	Vp-p
Differential Output Voltage Swing Capability	10	—	V_{out}	—	8.0	—	—	8.0	—	Vp-p
Power Supply Current	7	6								mAdc
$I_6 + I_9$			I_{CC}	—	2.0	3.0	—	2.0	4.0	
I_{10}			I_{EE}	—	3.0	4.0	—	3.0	5.0	
DC Power Dissipation	7	5	P_D	—	33	—	—	33	—	mW

* Pin number references pertain to this device when packaged in a metal can. To ascertain the corresponding pin numbers for plastic or ceramic packaged devices refer to the first page of this specification sheet.

Filtering Out the Undesired Sideband

Once a DSB-SC signal has been created, the unwanted sideband must be removed. Since the upper and lower sidebands occupy different frequency ranges, a bandpass filter can be used to allow only the desired sideband through. The result will be a SSB signal.

EXAMPLE 6–5

The SSB transmitter of Figure 6–6 is to transmit a speech signal (300−3000 Hz) on a *suppressed carrier frequency* of 8.000 MHz.

a. What will the frequency of the carrier oscillator be?

b. What will the passband of the USB filter need to be?

c. What will the passband of the LSB filter need to be?

d. Comment on the bandwidth needed for the LSB and USB filters.

Solution

a. The carrier oscillator operates at the suppressed carrier frequency, so by inspection, $f_c = \underline{\underline{8.000 \text{ MHz}}}$

b. To calculate the USB filter passband, use the information in Figure 6–3. The limits of the passband can be calculated as

$$f_{\text{min}} = f_c + f_{m(\text{min})} = 8.000 \text{ MHz} + 300 \text{ Hz} = \underline{\underline{8.0003 \text{ MHz}}}$$

The upper limit is calculated in the same way:

$$f_{\text{max}} = f_c + f_{m(\text{max})} = 8.000 \text{ MHz} + 3000 \text{ Hz} = \underline{\underline{8.0030 \text{ MHz}}}$$

The USB filter needs to pass $\underline{\underline{8.0003 \text{ MHz to } 8.0030 \text{ MHz}}}$.

c. The LSB filter frequencies are calculated in the opposite manner:

$$f_{\text{max}} = f_c - f_{m(\text{min})} = 8.000 \text{ MHz} - 300 \text{ Hz} = \underline{\underline{7.9997 \text{ MHz}}}$$
$$f_{\text{min}} = f_c - f_{m(\text{max})} = 8.000 \text{ MHz} - 3000 \text{ Hz} = \underline{\underline{7.997 \text{ MHz}}}$$

The LSB filter must pass $\underline{\underline{7.997 \text{ MHz to } 7.9997 \text{ MHz}}}$.

d. The bandwidth of the USB filter is (8.003 MHz−8.0003 MHz), or $\underline{\underline{2.7 \text{ kHz}}}$. The LSB filter bandwidth is (7.9997 MHz–7.997 MHz), or $\underline{\underline{2.7 \text{ kHz}}}$.

Both USB and LSB filters use the same bandwidth, which is the same as the bandwidth of the information signal.

The job of filtering out the undesired sideband is actually much more difficult in terms of practical circuit components than it might initially seem. This is because a fairly precise filter bandpass *shape* is required. Attaining the proper shape requires tuned circuits with a very high *Q;* the *Q* required can be more than 5,000. The ideal and typical bandpass curves of the USB filter for Example 6−5 is shown in Figure 6–11.

In the filter response curves of Figure 6–11, which can be obtained by connecting the filter under test to a *network analyzer,* the vertical scale is in units of *decibel attenuation*. Recall that to attenuate a signal is to make it weaker. A zero dB attenuation means

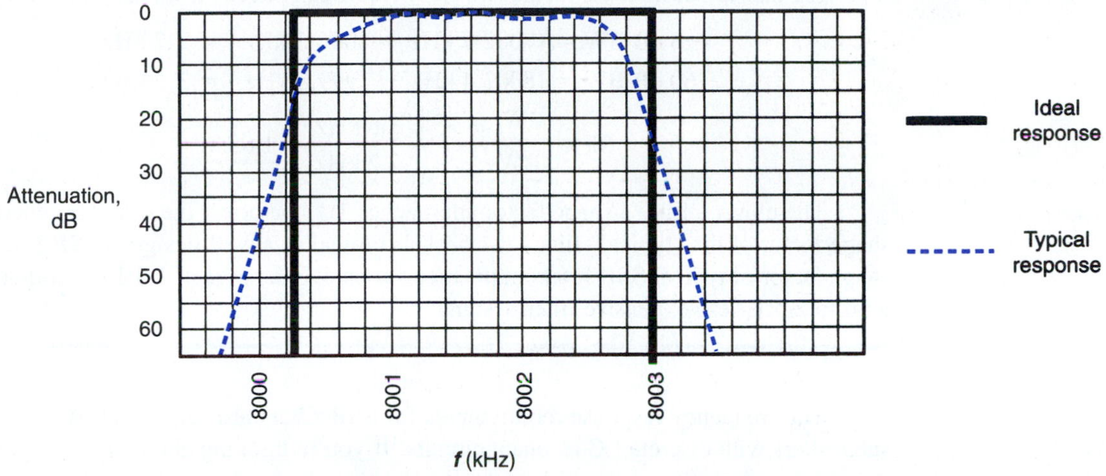

Figure 6–11 Ideal and typical responses for the USB filter of Example 6–5

that the signal is not made weaker at all; it passes with no power loss. The higher the attenuation, the greater the power loss. The solid curve is the *ideal* frequency response of the SSB filter. The ideal filter has infinite attenuation outside the desired passband (8000.3 kHz–8003 kHz) and *zero* passband attenuation. The resulting ideal response is *rectangular*.

In real life, there's no such thing as a perfect filter. The blue response curve is typical of a "real-life" filter. Note how the *skirts* (the sloped portions on the left and right) aren't perfectly vertical. The frequency response within the *passband* (the top middle portion) isn't exactly "flat" either; there is a small amount of gain variation called *passband ripple*. The passband ripple is about 2 dB in this example.

The "goodness" of a filter is often expressed as the filter's *shape factor,* or SF. The shape factor of a filter is defined as

$$SF = \frac{BW_{(-60\text{ dB})}}{BW_{(-6\text{ dB})}} \tag{6–3}$$

The shape factor is the ratio of the bandwidth at 60 dB attenuation to the bandwidth at 6 dB attenuation.

EXAMPLE 6–6 Calculate the shape factor for the ideal and typical filter curves from Figure 6–11.

Solution

For the ideal curve, the bandwidths at 6 and 60 dB attentuation are *identical,* and from the graph, these bandwidths can be read as 2.7 kHz (8003 kHz–8000.3 kHz). Therefore, the shape factor of the ideal filter is

$$SF = \frac{BW_{(-60\text{ dB})}}{BW_{(-6\text{ dB})}} = \frac{2.7\text{ kHz}}{2.7\text{ kHz}} = \underline{\underline{1:1}}$$

The shape factor of an ideal filter is 1:1.

For the *typical* filter curve, the two bandwidths can be read from the graph as follows:

$$-6\text{ dB BW} = (8002.7\text{ kHz} - 8000.5\text{ kHz}) = 2.2\text{ kHz}$$

$$-60\text{ dB BW} = (8003.4\text{ kHz} - 7999.8\text{ kHz}) = 3.6\text{ kHz}$$

$$SF = \frac{BW_{(-60\text{ dB})}}{BW_{(-6\text{ dB})}} = \frac{3.6\text{ kHz}}{2.2\text{ kHz}} = \underline{\underline{1.63:1}}$$

The closer a filter's shape factor approaches 1:1, the better the filter, in general. The shape factor of the "typical" filter, at 1.63:1, is typical of a well-designed SSB filter. [The shape factor of typical SSB filters in practice varies over a range of 1.5:1 (high-quality filters) to 2.5:1 (less-expensive filter designs).]

The frequency response requirements for SSB filters are very exact. We can't build such filters with discrete LC resonant circuits! If you're thinking about crystals, you're on the right track. A filter built with quartz crystals is called a *crystal lattice filter,* shown in Figure 6–12.

In the filter of Figure 6–12, two crystals are used. Transformer T_1 couples the input signal into both crystals, 180° out of phase. Recall that a crystal has two resonance modes, *parallel* and *series.* In parallel resonance, a crystal looks ideally like an open circuit; at series resonance, it looks like a short. In the circuit of Figure 6–12, the two crystals are manufactured with slightly different frequencies. Typically, the series resonant frequency of crystal Y_2 overlaps the parallel resonant frequency of Y_1. Crystal Y_1 passes the lower frequencies within the passband, where Y_1 is series-resonant; crystal Y_2 takes over for the upper portion of the passband. Coupling the signal into the crystals 180° out of phase cancels the effect of the *mounting capacitance* of the crystals. Crystal filters are available commercially; they're usually contained in a metal shield can and either plug into a special socket or solder directly to a printed circuit board (see Figure 6–13).

Figure 6–14 shows the response graph of the filter and how it is derived from the two crystal responses. Note that there is a definite *dip* in the passband in the center; this dip can be reduced by *cascading* more filter sections.

Mechanical SSB Filters

Crystal filters can provide a high level of performance and can be readily designed for operation in the 1–20 MHz region (and higher frequencies by using crystal overtones). However, crystal filters are somewhat susceptible to damage from mechanical shock. In applications where improved ruggedness is needed, a *mechanical filter* is often employed.

Figure 6–12 A crystal lattice filter

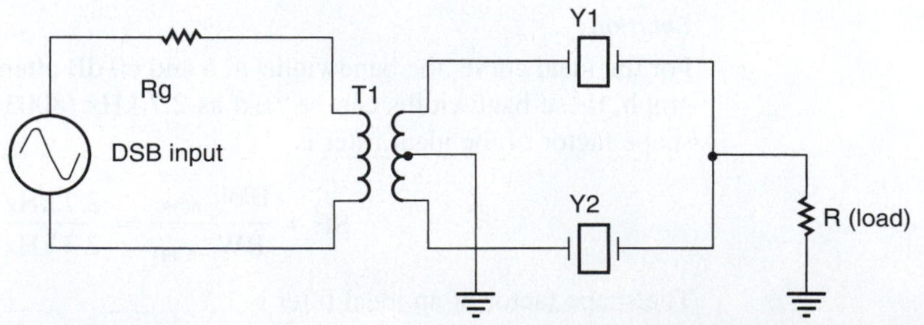

Figure 6–13 A crystal filter on a printed circuit board

A mechanical filter consist of three parts: an *input transducer,* a group of *mechanical filter elements,* and an *output transducer*. Figure 6–15 shows how it works. The electrical signal to be filtered is fed to the input pins of the filter. The input transducer converts the electrical signal into mechanical vibrations, which are transferred to the metal discs. Each metal disc, along with its associated torsion rod, acts like a single parallel-resonant LC bandpass filter. Each disc is machined to be resonant at a particular frequency within the filter's passband. The mechanical signal propagates through the discs, and reaches the output transducer, where it is converted back into an electrical signal.

Mechanical filters, though very reliable, are limited to a frequency range below 2 MHz. At higher frequencies, the dimensions of the resonant metal discs become too small for practical manufacture.

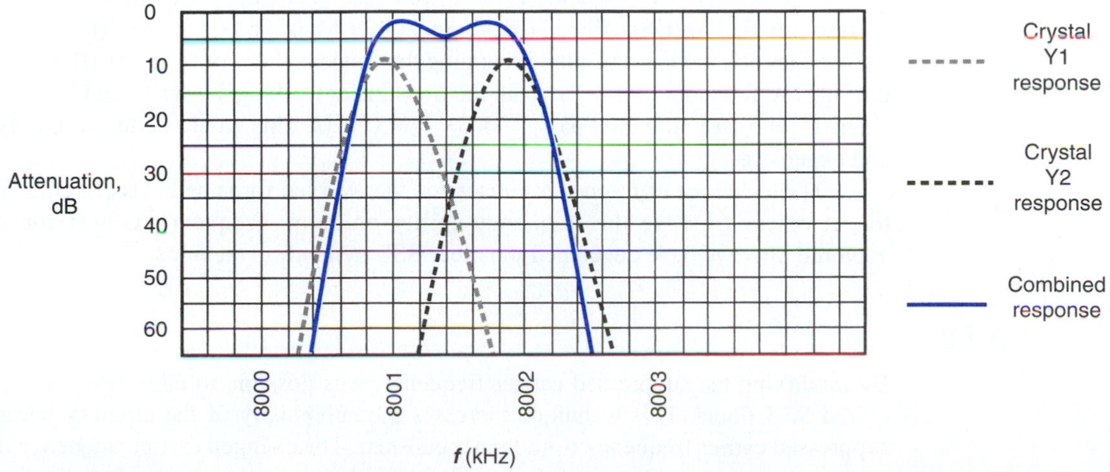

Figure 6–14 Response of the crystal lattice filter

Figure 6–15
Mechanical filter action

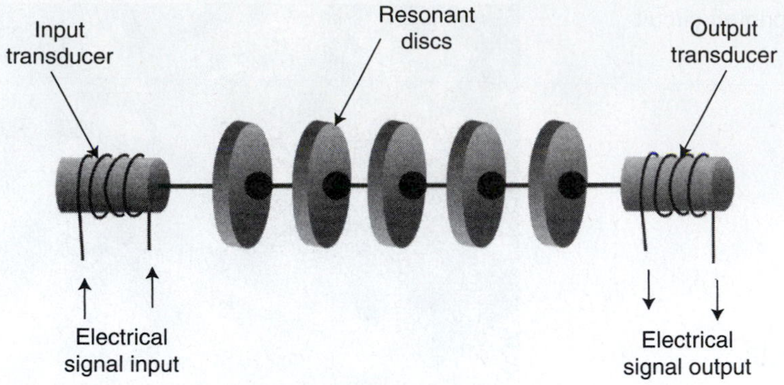

Practical Considerations for Filter-Type SSB Transmitters

In Figure 6–6, we developed a filter-type SSB exciter. The approach of Figure 6–6 works but has two big drawbacks. First, it requires two different SSB filters if operation on both USB and LSB modes will be needed; and second, the transmitter lacks *frequency agility*.

Filter requirements Dedicated SSB-type filters are expensive. The least inexpensive crystal filters can add more than $50 to the cost of a transmitter; adding *two* filters (LSB, USB) doubles that cost. Quality mechanical filters can cost over $100 each. By careful circuit design, we can eliminate one of these filters.

Lack of frequency agility *Frequency agility* is the ability to easily change operating frequency. SSB transmission is often used in long-distance HF voice communications, where the operator frequently changes the transmitter frequency. The transmitter of Figure 6–6 is *locked* onto one specific suppressed carrier frequency! Why can't we just change the carrier oscillator frequency?

Suppose that we desired to change the carrier frequency to 7.900 MHz. When this change is made, the upper sideband now occupies the frequency range 7.9003–7.903 MHz, and the lower sideband now occupies 7.897 MHz−7.8997 MHz.

Can you see the problem? Look at the filter passband frequencies we derived in Example 6–5. The USB filter is *fixed* at 8.0008 MHz−8.003 MHz. The "new" USB frequencies obtained as a result of changing the carrier oscillator to 7.900 MHz won't pass through the filter any more! To make the transmitter work again, we would have to put in a *new* USB filter that would pass 7.9003−7.903 MHz. This would be unacceptably bulky and expensive.

The technique of *frequency conversion* that worked for superhet receivers will come to our rescue to solve the frequency agility problem. Frequency conversion will be explored after we have developed the other SSB generation methods.

Eliminating a Filter

By modifying the suppressed carrier frequency, it is possible to eliminate one of the required SSB filters. This technique increases the complexity of the circuitry because the suppressed carrier frequency is no longer constant. The changed carrier frequency must be accounted for in later stages of the transmitter, either in hardware or software.

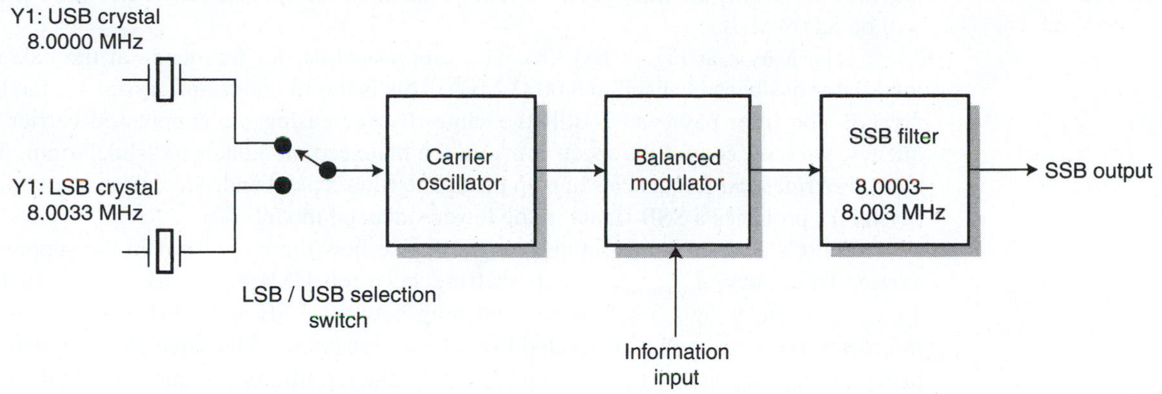

Figure 6–16 Moving the suppressed carrier frequency eliminates a filter

Figure 6–16 shows the new SSB generation scheme. We have modified the carrier oscillator to produce *two* different frequencies. This is done by simply switching one of two different crystals into the oscillator. Crystals are much less expensive than SSB filters; they cost about $1—that is much less expensive than a $100 filter!

It's hard to believe that this scheme actually works unless we take a close look at the frequency content of the balanced modulator. Figure 6–17(a) shows the balanced modulator's output with the *USB* crystal, Y_1, switched in. Assuming that there is a speech signal present (300 Hz–3000 Hz), the upper sideband frequency range will be from 8.0003 MHz $[f_c + f_{m(\text{min})}]$ to 8.003 MHz $[f_c + f_{m(\text{max})}]$. The SSB filter can pass 8.0003 MHz−8.003 MHz, so only the *upper sideband* energy passes through the filter. Therefore, when Y_1 is switched

Figure 6–17 Sliding the DSB-SC output under the SSB filter

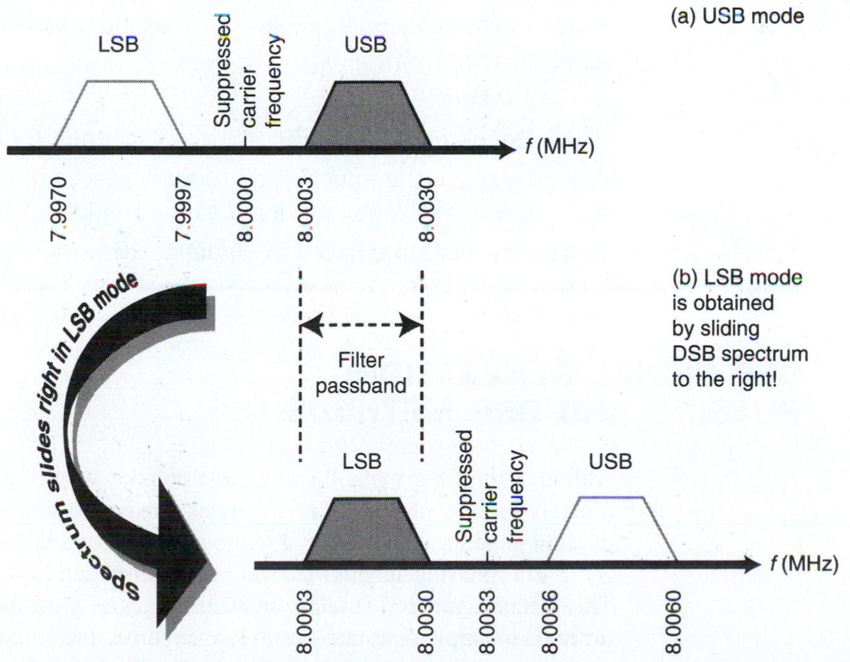

into the circuit, only the USB will pass through the filter, and the suppressed carrier frequency will be 8.000 MHz.

Take a look at Figure 6–17(b). The suppressed carrier frequency at the balanced modulator has been changed to 8.0033 MHz. This is the frequency of crystal Y_2, the LSB crystal. The filter passband is still the same. By increasing the suppressed carrier frequency, we have caused the spectrum from the balanced modulator to "slide" right. Now the *lower sideband* frequencies line up inside the filter's passband. Therefore, switching in crystal Y_2 produces a SSB signal in the lower sideband mode!

There's a catch to this setup, though. Notice how there is a shift in the suppressed carrier frequency, $f_{c(suppressed)}$, when shifting between USB and LSB modes. In fact, $f_{c(suppressed)}$ changes by 3.3 kHz when switching between LSB and USB modes. This is a *frequency error* and will be corrected later in the transmitter. The same circuitry that will provide frequency agility (ability to move freely among different frequencies) will correct the frequency error for us.

Section Checkpoint

6–12 What are the three main methods of producing a SSB signal?

6–13 What are the two steps for generating SSB in the filter method?

6–14 Give two disadvantages of the circuit of Figure 6–6.

6–15 What type of RF power amplifiers must be used with SSB transmitters?

6–16 What frequency component is *absent* in the output of a balanced modulator?

6–17 Describe the correct procedure for adjusting resistor R_1 in Figure 6–7.

6–18 What is the *shape factor* of a perfect filter?

6–19 How is the shape factor of a real-world filter calculated?

6–20 What component is used as a bandpass element in a crystal filter?

6–21 In a mechanical filter, what type of devices are used as the filtering elements?

6–22 Define *frequency agility*. Why is it needed for SSB operation?

6–23 Describe the effect on the output spectrum of a balanced modulator if the carrier frequency input to the modulator is varied up or down.

6–24 How does the circuit of Figure 6–16 work with only one filter?

6–3 SSB SIGNAL GENERATION: PHASING AND DSP METHODS

Although the filter method is the most popular way of generating a SSB signal, it unfortunately requires the use of relatively expensive crystal or mechanical filters. The *phasing* method generates a SSB signal without costly bandpass filters.

The phasing method uses a circuit approach that cancels the undesired sideband. This circuit approach requires an accurate phase-shift filter for the information, which is difficult to implement and unstable over time due to component aging. A phasing SSB transmitter has several critical adjustments that are subject to drift as well. So why study the phasing method at all?

The answer is that a special type of microprocessor called a *digital signal processor* (DSP) is able to run programs that will generate SSB signals in *software*. DSP approaches to generating SSB use the phasing method. Because digital devices aren't subject to drift (like analog systems are), a DSP phasing-type SSB generator will be rock stable and requires no adjustments for proper operation.

The Phasing Concept

Figure 6–18 shows how a phasing SSB generator works. The audio information signal is split into two parts—an original information signal, and a *90° phase-shifted information signal.* Two balanced modulators are employed; the top modulator receives the original information, and the bottom gets the phase-shifted signal. The carrier signal is likewise split into two parts, an original carrier and a 90° phase-shifted carrier. The top modulator receives the original carrier, while the bottom gets the phase-shifted version.

So far, we are just generating two DSB-SC signals at the same time. Each balanced modulator makes its own DSB-SC signal (DSB signal 1 and 2 in the figure). Figure 6–19 shows that there is something special about each of the DSB signals. The upper and lower balanced modulators generate similar DSB signals with one crucial difference: The *polarity* or *phase* of the upper sideband is opposite in the two signals. This phase difference occurs because of the phase-shifted RF and information signals that are sent to modulator 2 (bottom). When the two DSB signals are added, the result is a SSB signal with only the lower sideband present.

By switching the positions of the phase-shifted RF or information signal (but not both), the circuit can also be made to produce a SSB signal with only the USB present. In this case, the *lower sideband* outputs from the two balanced modulators will cancel, leaving only the upper sideband as an output.

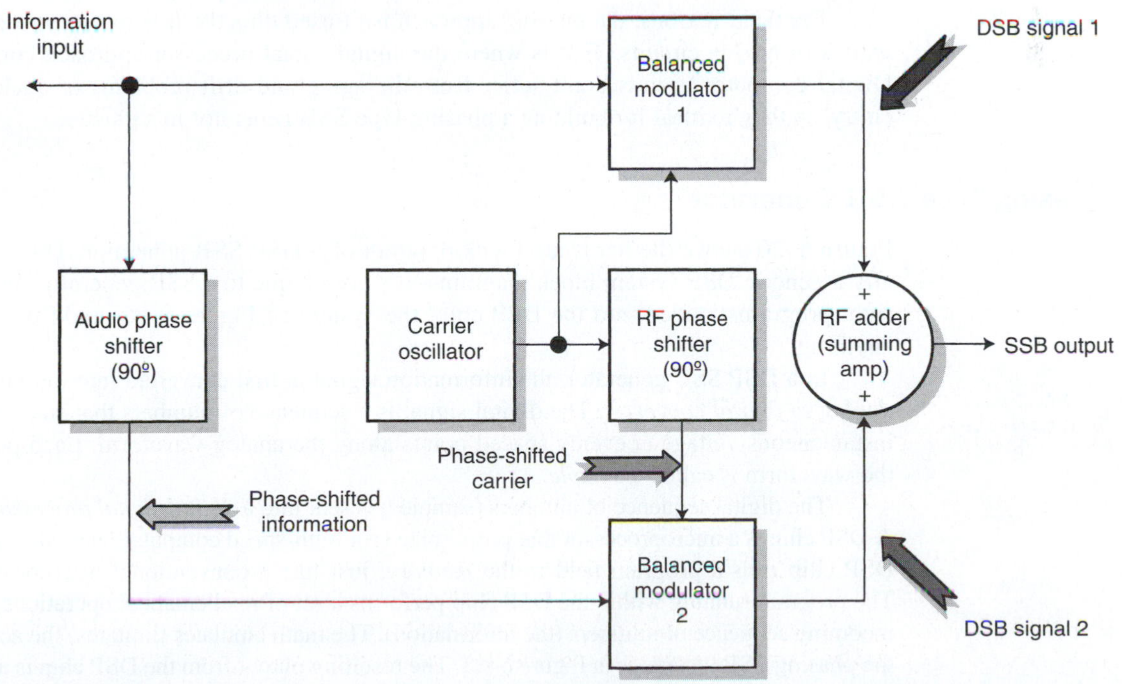

Figure 6–18 A phasing-type SSB generator

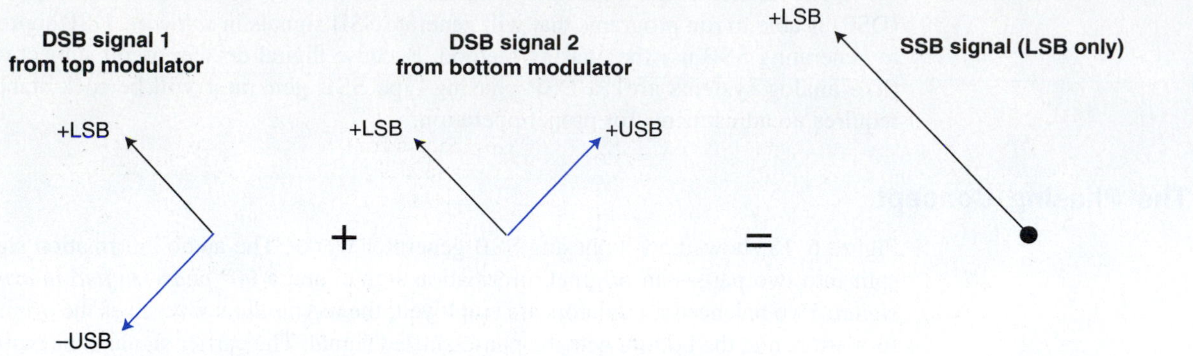

Figure 6–19 Phasor relations for DSB signals 1 and 2

Problems with the Phasing Approach

The phasing approach isn't used directly in modern SSB transmitters. It works well on paper but suffers from two practical problems:

- There are at least three analog adjustments in the circuit, all of which are prone to drift. Two of the adjustments are the nulling controls in the balanced modulators. The carrier phase shift must also be adjusted to be precisely 90°. If any of these adjustments is off, some of the energy from the wrong sideband (or carrier) will leak from the circuit output.
- The phase-shift network for the information (audio) is critical; it is difficult to get an exact 90° phase shift over the entire range of audio frequencies. If this network drifts, the circuit again leaks energy from the wrong sideband.

For these reasons, the phasing approach isn't used directly. It is too sensitive to the whims of analog circuits. This is where the digital signal processor approach comes in. Digital computer circuits don't suffer from the aging and drift problems of analog circuitry, so they're ideal for building a phasing-type SSB generator in *software*.

DSP Phasing-Type SSB Generator

Figure 6–20 shows the hardware block diagram of a DSP SSB generator. This is actually a generic DSP system block diagram—it's not unique to a SSB generator. With the right programs loaded into the DSP chip, the system of Figure 6–20 could do almost anything!

In a DSP SSB generator, the information signal is first converted into digital by an *analog to digital converter*. The digital signal is a sequence of numbers that stand for the instantaneous voltage at evenly spaced points along the analog waveform. Each point on the waveform is called a *sample*.

The digital sequence of numbers (samples) is sent into a *digital signal processor* chip. A DSP chip is a microprocessor that is optimized for high-speed computation of signals. The DSP chip runs a program held in the *memory,* just like a conventional microprocessor. The program running within the DSP chip performs a set of mathematical operations on the incoming sequence of numbers (the information). The math emulates (imitates) the action of the phasing SSB generator in Figure 6–18. The resulting output from the DSP chip is another sequence of binary numbers. These numbers represent the instantaneous voltage of the

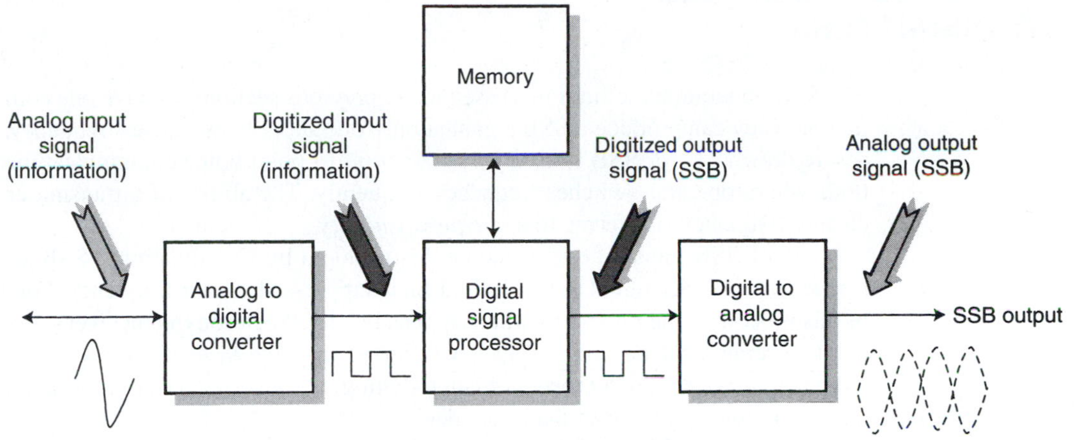

Figure 6–20 Hardware in a DSP SSB generator

resulting SSB "waveform" that was computed inside the chip. The output of the DSP chip is digital, which is not useful for transmission. The desired output is analog, which is obtained after passing the binary (digital) output of the DSP through a *digital to analog converter.*

The digital signal processor is a very busy chip. In order to "calculate" the SSB waveform, it must perform millions of operations each second. A DSP chip running at a clock speed of 50 MHz can provide a SSB signal at a suppressed carrier frequency of 10 to 20 kHz. The low-frequency SSB signal must be up-converted to the final transmit frequency, as we'll see in the next section.

The DSP approach has much going for it; not only does it eliminate the need for expensive crystal filters, it also reduces the component count needed to generate a SSB signal. In transceivers (combination transmitters and receivers), the DSP chip does double duty: It works to generate SSB during transmit and performs SSB detection during receive.

Finally, the DSP approach isn't limited to just SSB. By changing the software a little, the same chip can provide AM and FM modulation and detection as well. For "multimode" (multiple modulation-type) transceivers, this results in a further reduction of circuitry.

Section Checkpoint

6–25 What devices are eliminated by using the phasing approach?

6–26 In Figure 6–19, why does only the upper sideband appear at the output?

6–27 Why aren't phasing SSB generators popular?

6–28 What is a DSP chip?

6–29 What is the nature of the signals at the input and output of the DSP chip in Figure 6–20?

6–30 Why does the phasing approach work in a DSP chip, when it's unreliable in analog circuitry?

6–31 What types of modulation can be produced with DSP?

6–4 FREQUENCY-AGILE SSB TRANSMITTERS

The SSB generation methods discussed in the previous sections all have one common limitation: They can produce a SSB signal at only one suppressed carrier frequency. This is a severe drawback, for SSB is used for many point-to-point voice communications applications where operator switches frequency frequently. The ability of a transmitter to easily change frequency is referred to as *frequency agility.*

Most SSB transmitters overcome this problem by first creating a SSB signal, then *converting* the SSB signal to the desired final suppressed carrier frequency. You can think of this process as the reverse of what happens in a superheterodyne receiver—the SSB signal is generated at a constant frequency (referred to as the *transmit IF frequency* by many manufacturers), and by a process of heterodyning, is converted to the desired transmit frequency. Figure 6–21 shows the basic idea.

In Figure 6–21, the SSB signal is created from the original information signal. The SSB generating circuitry controls whether a USB or LSB signal is created. The circuitry of Figure 6–16 is being used to generate the SSB signal, so the low-frequency SSB signal has one of two possible suppressed carrier frequencies, either 8.0000 MHz (USB mode) or 8.0033 MHz (LSB mode).

The SSB signal enters one input of a *balanced mixer*. A balanced mixer operates in the same manner as a conventional mixer, except that the balanced circuit design prevents the two original input frequencies from leaving the mixer. A balanced mixer produces only the sum and difference of the two input frequencies. In circuits like the one in Figure 6–21, this largely eliminates the worry about filtering out the two original input frequencies.

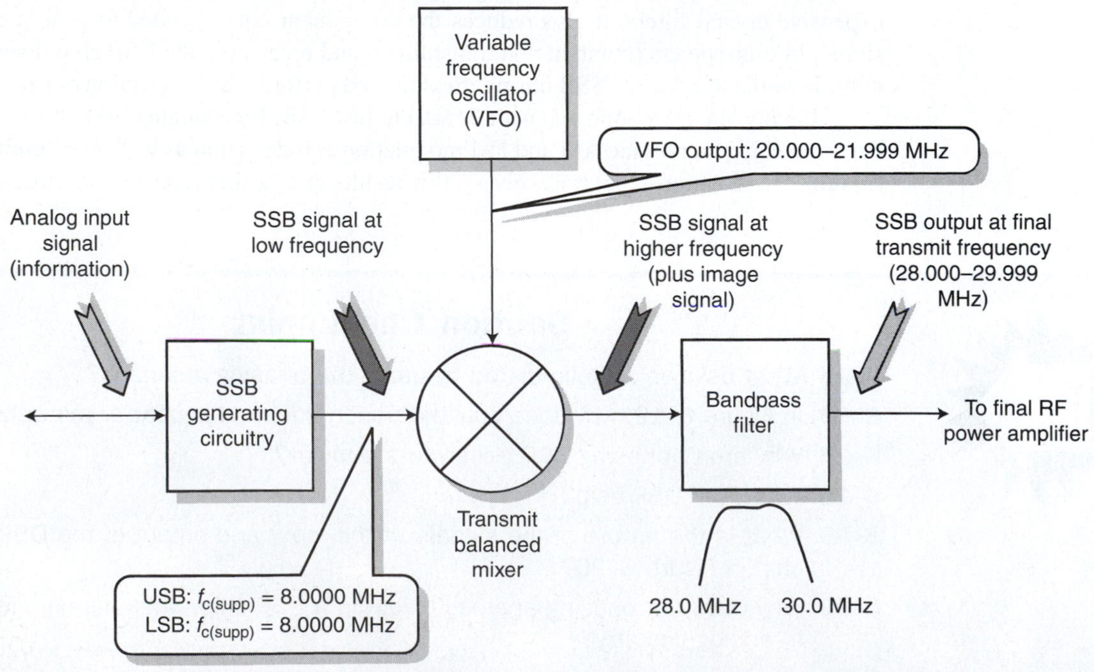

Figure 6–21 Converting the transmit SSB signal

The other input to the mixer comes from the *variable frequency oscillator,* or *VFO*. The VFO is controlled by the user. In some transmitters, the VFO has a variable capacitor for changing its frequency; the front-panel tuning knob adjusts this capacitor. Because small frequency changes are needed, the tuning knob often uses a reduction gear drive. In newer transmitters, the VFO is a *frequency synthesizer* controlled by an on-board computer. A frequency synthesizer is a digitally controlled oscillator, which is a very precise way of determining frequency. Frequency synthesizers are the subject of a later chapter. Because the output of the transmit mixer contains both sum and difference frequencies, a *bandpass filter* must be used to exclude the undesired "image" frequency. An example best shows how all of this works together.

EXAMPLE 6–7

Given that the VFO is operating on 20.350 MHz, and that USB mode is selected ($f_{c(\text{suppressed})}$ = 8.000 MHz at the SSB generator), calculate

a. The two suppressed carrier frequencies (real and image) at the output of the balanced mixer. Which suppressed carrier can make it through the output bandpass filter?

b. The range of frequencies that will be sent through the output filter. (Assume that a voice signal is modulating the transmitter, a frequency range of 300 Hz to 3000 Hz.)

Solution

a. The balanced mixer can produce only sum and difference frequencies at its output. The two input frequencies are 20.350 MHz (VFO) and 8.000 MHz (SSB $f_{c(\text{suppressed})}$). The output frequencies are therefore (20.350 MHz + 8.000 MHz) = 28.350 MHz, and (20.350 MHz − 8.000 MHz) = 12.350 MHz.

Only the 28.350 MHz signal can pass through the output filter. The 12.350 MHz signal is an "image." The output filter will not allow the 12.350 MHz signal to pass.

b. Because $f_{c(\text{suppressed})}$ = 8.000 MHz and the range of information frequencies is 300–3000 Hz, we know that the USB frequency content at the output of the SSB generator is

$$f_{\text{min}} = f_c + f_{m(\text{min})} = 8.000\text{ MHz} + 300\text{ Hz} = 8.0003\text{ MHz}$$
$$f_{\text{max}} = f_c + f_{m(\text{max})} = 8.000\text{ MHz} + 3000\text{ Hz} = 8.0030\text{ MHz}$$

Because we found that the *sum* of the VFO and SSB generator frequencies passed through the filter in part (a), we can safely say that the range of output frequencies that will be sent through the output filter will be (20.350 MHz + 8.0003 MHz) = 28.3503 MHz to (20.350 MHz + 8.003 MHz) = 28.353 MHz.

Note that the new suppressed carrier frequency, $f_{c(\text{suppressed})}$, is 28.350 MHz, and this is what the radio will display to the operator. If an analog VFO (variable capacitor or inductor) is being used, there will be a dial calibrated accordingly. If the VFO is a frequency synthesizer, the microcomputer controlling the synthesizer will show the transmit frequency on a display device (LED, LCD, and so on).

The transmitter of Figure 6–21 can generate LSB also, since the SSB generator has that capability. If the SSB generator is using a filter, then the suppressed carrier frequency at the SSB generator's output will change when it is shifted into LSB mode. This means that the VFO circuitry has to be calibrated differently for USB and LSB modes, since there is a 3.3 kHz difference in suppressed carrier frequency between the modes.

In early equipment, this was taken care of in a crude but direct fashion; *two* dial pointers would be provided, one for USB and one for LSB! In modern equipment, the microprocessor running the VFO automatically takes the 3.3 kHz frequency difference into account when displaying the operating frequency, so the operator never has to think about what is really happening.

EXAMPLE 6–8

It is desired to operate the transmitter of Figure 6–21 on a frequency of 28.475 MHz in LSB mode. Calculate

a. The correct frequency for the VFO.

b. The range of frequencies that will be sent through the SSB filter. (Assume that a voice signal is modulating the transmitter with a frequency range of 300 Hz to 3000 Hz.)

Solution

a. The *new* suppressed carrier frequency must be 28.475 MHz when we're done. The original suppressed carrier frequency at the SSB generator is 8.0033 MHz. Since we know that the *sum* of the VFO and suppressed carrier frequencies appears at the output, we can calculate the required VFO frequency:

$$f_{\text{VFO}} = f_{c(\text{final})} - f_{c(\text{original})} = 28.475\text{ MHz} - 8.0033\text{ MHz} = \underline{\underline{20.4717\text{ MHz}}}$$

b. The range of frequencies that will come from the SSB generator will be

$$f_{\max} = f_c - f_{m(\min)} = 8.0033\text{ MHz} - 300\text{ Hz} = 8.0030\text{ MHz}$$
$$f_{\min} = f_c - f_{m(\max)} = 8.0033\text{ MHz} - 3000\text{ Hz} = 8.0003\text{ MHz}$$

If you compare these results to those of Example 6–7(b), you'll see that they're identical. This is because the SSB generator uses only one filter, and these are the limits of that filter's passband! Therefore, the range of output frequencies (after mixing) will be the same as before; the *sum* of the VFO and the SSB generator output frequencies will be produced. The frequency range will be (20.4717 MHz + 8.0003 MHz) = $\underline{\underline{28.472\text{ MHz}}}$ to (20.4717 MHz + 8.003 MHz) = $\underline{\underline{28.4747\text{ MHz}}}$.

The *dial reading* of frequency on the front panel will be 28.475 MHz. You can see that the frequency range being produced is indeed *slightly below* 28.475 MHz, the suppressed carrier frequency.

Wideband Coverage

Many SSB transmitters are required to cover the entire HF band, from 3 MHz to 30 MHz, with no breaks in coverage. The heterodyning technique just shown can provide coverage only within a limited band (2 MHz or so) because a filter is required at the mixer output to suppress the image frequency. Also, the frequency range of a single VFO is usually limited.

Electronically switched sets of bandpass filters placed at the mixer output can eliminate the image problem. In modern transmitters, these filters are switched in and out by the microprocessor. In older equipment, the filters are likely to be selected by a mechanical bandswitch. In addition, the VFO range is boosted by applying *frequency conversion* or *frequency synthesis* techniques to the VFO itself so that the proper range of heterodyning frequencies can be obtained.

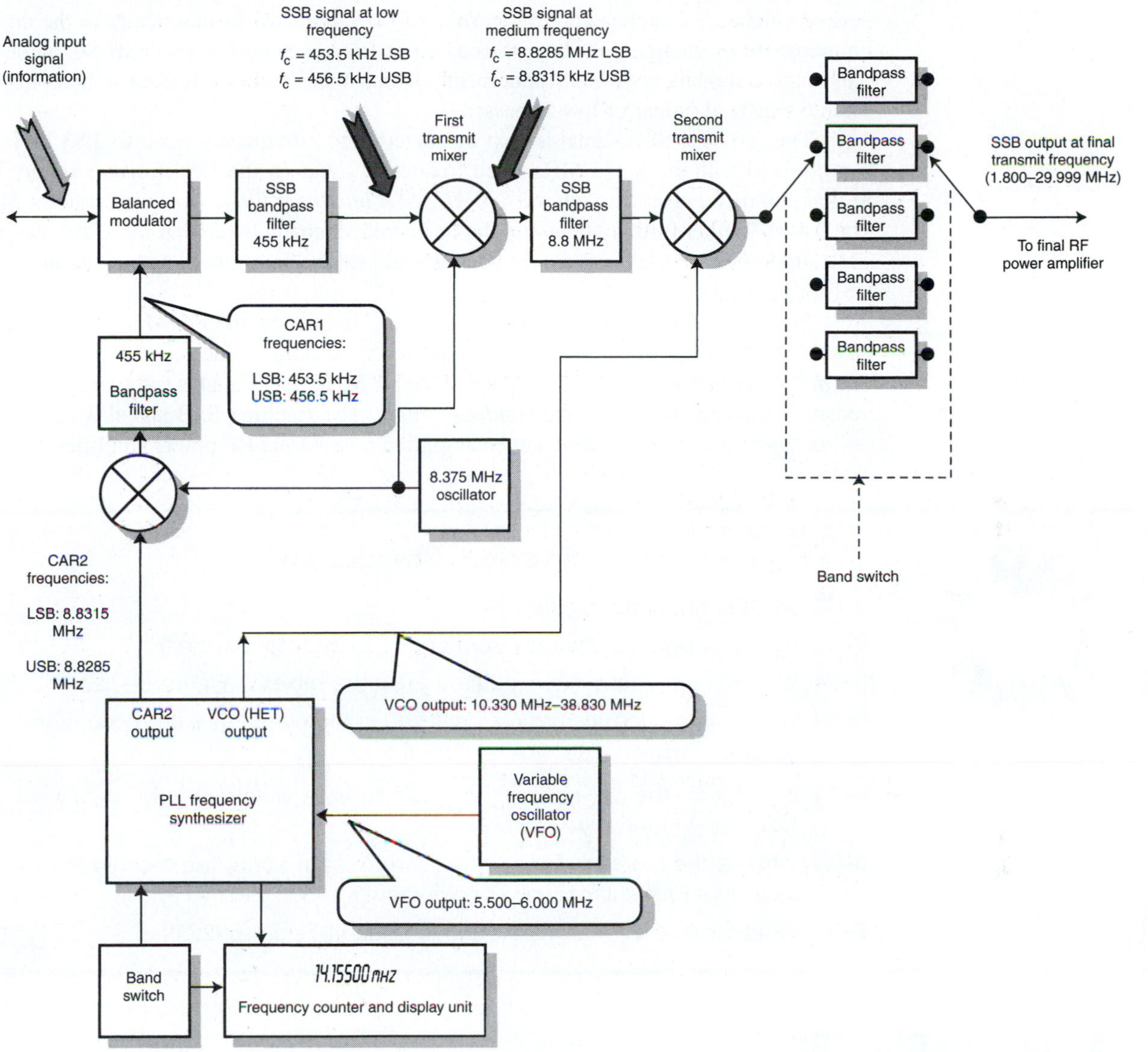

Figure 6–22 Wideband SSB generation scheme of the Kenwood TS-830

Figure 6–22 shows a simplified picture of how this is done in some Kenwood/Trio HF transceivers. Since every manufacturer uses its own method, the best way for a technician to become familiar with these techniques is to study the service literature.

There's quite a bit of activity in Figure 6–22! In the lower left, the *PLL frequency synthesizer* in combination with the *VFO* generates two different frequencies, *CAR2* and *VCO*. The *CAR2* output is one of two different frequencies, depending on whether the unit is in LSB or USB mode. The frequency at *CAR2* is mixed with a fixed (constant) 8.375 MHz oscillator; the difference frequency is in the 455 kHz range and passes through a 455 kHz bandpass filter. The difference between *CAR2* and the 8.375 MHz fixed oscillator signal is called *CAR1* and is used as the suppressed carrier frequency in the balanced modulator (upper left).

The balanced modulator output passes through a 455 kHz SSB filter, where the undesired sideband is attenuated. This portion of the circuit works identically to the one of Figure 6–16; by changing the suppressed carrier frequency *CAR1,* the DSB-SC output of the balanced modulator is "slid" underneath the passband of the SSB filter so that only the desired sideband (upper or lower) passes.

The resulting SSB signal is next upconverted to a frequency close to 8.83 MHz by being mixed with the 8.375 MHz fixed frequency signal in the *first transmit mixer*. The SSB signal then passes through the *8.8 MHz SSB bandpass filter,* which is a crystal filter. The 8.8 MHz filter further attenuates the undesired sideband. Note that there are two possible suppressed carrier frequencies here as well, depending upon whether the unit is in USB or LSB mode.

Finally, to determine the actual transmitting frequency, the 8.8 MHz SSB signal is mixed with the VCO output of the PLL frequency synthesizer in the *second transmit mixer*. The *difference* between the VCO output and the 8.8 MHz SSB signal is extracted as the final transmit frequency by one of the *bandpass filters*. The resulting SSB signal is now at the desired operating frequency and can be amplified by the final RF power amplifier.

Section Checkpoint

6–32 What is frequency agility?

6–33 How is a signal converted from one frequency to another?

6–34 Why is a bandpass filter required after the mixer in Figure 6–21?

6–35 What two frequencies are *not* produced in the output of a balanced mixer?

6–36 What is a VFO?

6–37 In order for the circuit of Figure 6–21 to have wideband coverage, what two changes must be made?

6–38 What is the best way for a technician to learn about the techniques used in a particular piece of equipment?

6–39 What method of SSB generation is used in Figure 6–22?

6–5 SSB RECEIVERS

The reception of SSB signals is very similar to the process of AM reception. An incoming suppressed carrier frequency must be acquired, selected, amplified, and detected. The superheterodyne receiver is used for SSB as well as AM.

There are two primary differences between SSB and AM reception. First, somewhere in the reception process, the appropriate sideband must be selected (filtered); and second, the detection of a SSB signal can't take place unless a carrier is "reinserted" to take the place of the one that was suppressed at the transmitter.

The same techniques that worked for transmitters for sideband selection will also work for receivers. Most receivers use the filter method. Some newer receivers using a DSP chip for the transmitter will use the same chip to select the sideband (phasing method in reverse) as well as to demodulate the signal.

A SSB detector is often called a *product detector*. A product detector can be built with a circuit very similar to a balanced modulator. Most IC balanced modulator circuits (such as the Motorola MC1496) can also perform double duty as SSB detectors. No matter

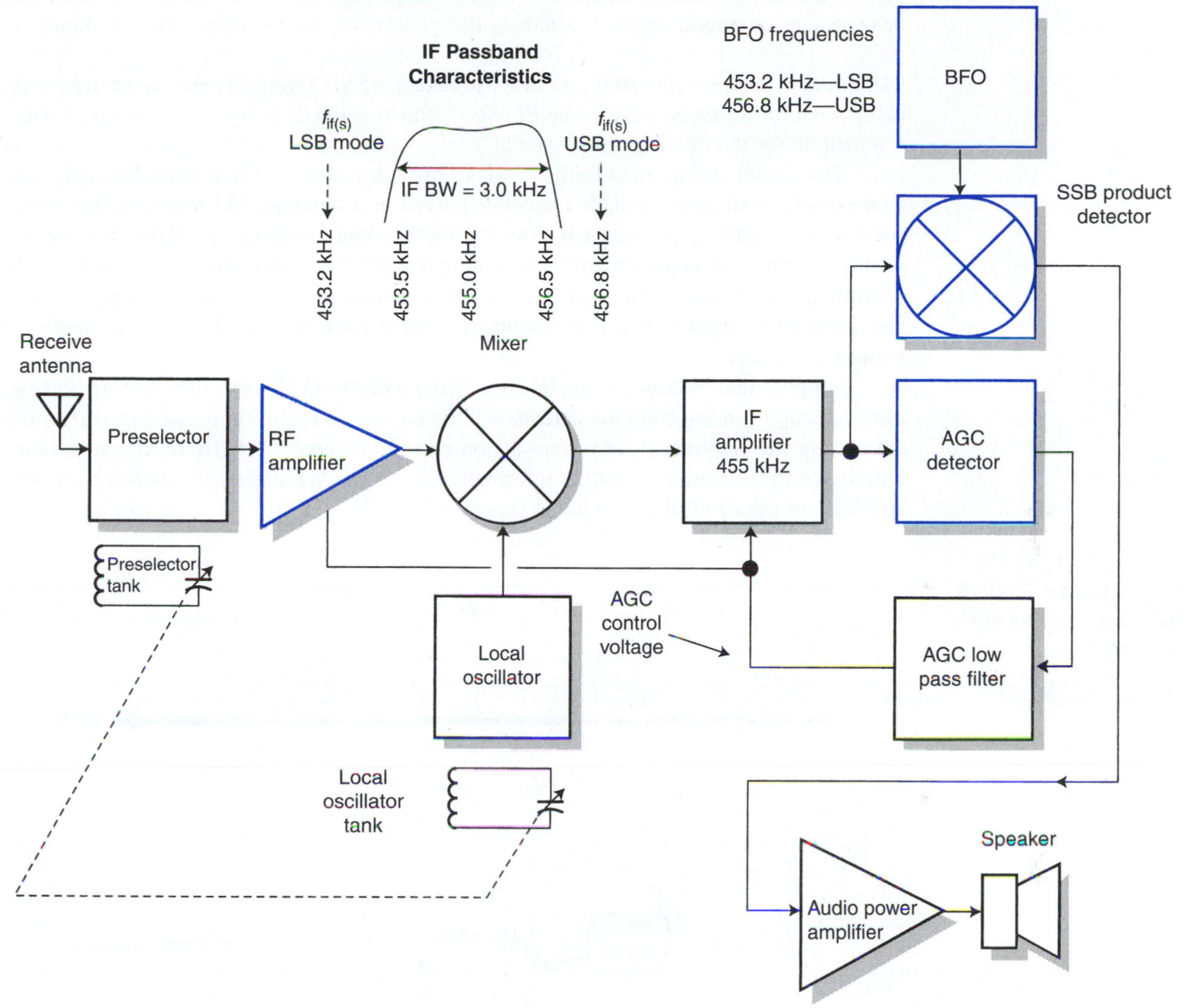

Figure 6–23 A superheterodyne SSB receiver

what circuit is used, a new carrier signal must be generated and inserted at the detector; the circuit that generates the new carrier signal is called a *beat frequency oscillator*, or *BFO*.

Figure 6–23 shows a single-conversion SSB superhet receiver. By "single conversion" we mean that there is only one intermediate frequency; in this case, that is 455 kHz.

Receiver Circuit Analysis

The sections in Figure 6–23 that are new or different from those in a conventional AM broadcast receiver are colored blue. Let's take a look at these differences first.

RF amplifier An AM broadcast receiver normally operates in the presence of strong local signals in a city. SSB is often used for long-distance communications, where signals can be very weak (just a few microvolts). The RF amplifier provides power gain for the carrier signal before the relatively noisy process of *mixing* and frequency conversion. The RF amplifier may be one stage, or several stages. It is almost always tuned as part of the preselector circuit.

SSB product detector and BFO In order to detect a SSB signal, a new carrier frequency component must first be added. The *BFO* performs this job. It is an oscillator that provides a substitute for the missing carrier signal.

The carrier is important for detection; it provides a point of reference. You may have heard what a SSB signal sounds like when played in a standard AM receiver. The speech sounds very garbled; you can tell from the rhythm that someone is talking, but the frequency information is all wrong! Most people describe the sound as *monkey chatter*. The addition of the carrier at the SSB product detector transforms the "monkey chatter" back into intelligible speech. It *seems* like magic, but it isn't. A SSB detector can really be viewed as a *mixer*.

Suppose that the *suppressed IF carrier frequency* is 453.2 kHz. (The suppressed carrier frequency coming from the antenna will be converted to the suppressed IF carrier frequency by the frequency converter section of the receiver.) A USB IF signal is being received, and the frequency of the information is 1 kHz. What does the detector see with and without the BFO? Take a look at Figure 6–24.

Figure 6–24 Product detector spectral input and output

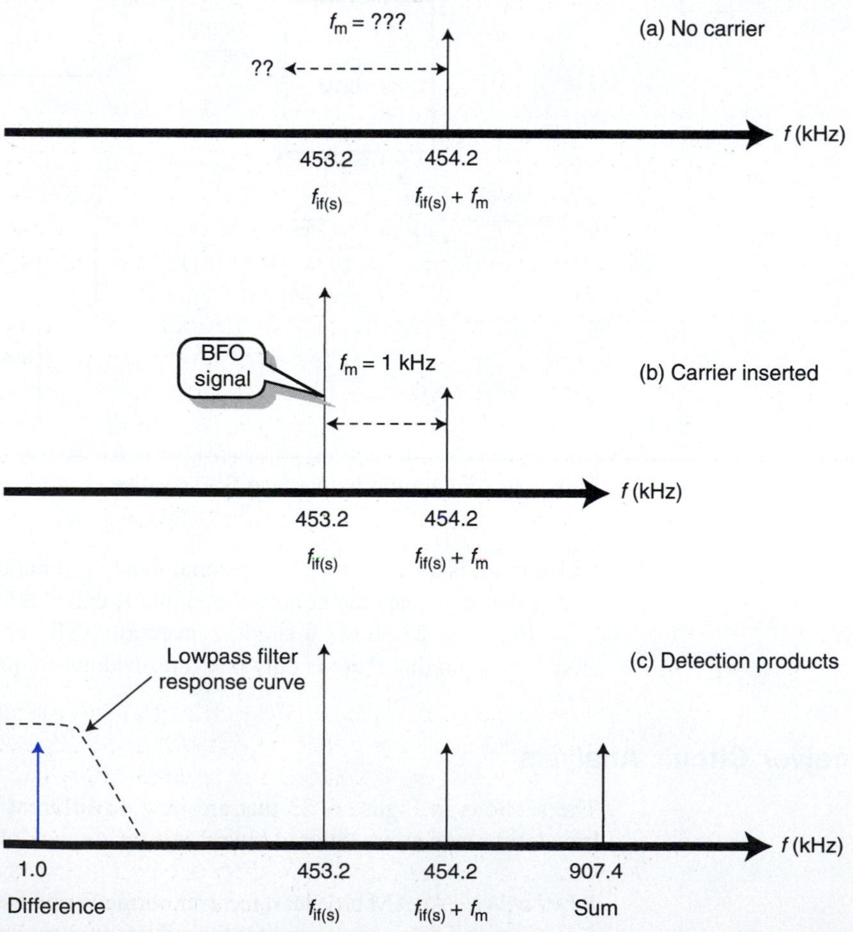

In Figure 6–24(a), no BFO signal has been inserted. The only frequency component present is the 454.2 kHz upper sideband (recall that $f_{usb} = f_c + f_m =$ 453.2 kHz + 1 kHz = 454.2 kHz). The resulting signal is a *pure sine wave* at 454.2 kHz. It would look like an *unmodulated carrier* of 454.2 kHz on an oscilloscope! *There is no way for the detector to determine the frequency of the information in Figure 6–24(a) because there is no carrier frequency component. The detector would produce only a steady dc voltage!*

In Figure 6–24(b), the BFO signal has been inserted at 453.2 kHz. The detector can now see, of all things, an AM signal! One of the sidebands is missing (the LSB), but that doesn't bother the detector in the least. The two frequencies, 453.2 kHz and 454.2 kHz, will visibly "beat" together, forming an envelope with the same frequency as that of the information!

This is much easier to see if we view the detector as a mixer. A detector is a nonlinear circuit element, and therefore it can act as a mixer. We know that given two input frequencies, a mixer produces the *originals,* the *sum,* and the *difference* frequencies. This is very evident in Figure 6–24(c). The 453.2 kHz and 454.2 kHz signals mix or *beat* together in the product detector. The two original signals, as well as the sum frequency, are much too high in frequency to pass through the low-pass filter. The *difference* frequency is an audio frequency and passes through the audio low-pass filter. This is why a detector can be thought of as a mixer.

You can see also that the frequency accuracy of the BFO is very important for SSB detection. If the BFO is off by 100 Hz, then the detected audio will also be 100 Hz off frequency. The detected pitches will all be wrong! For this reason, the BFO is usually crystal controlled.

AGC detector The output from a conventional AM detector consists of an ac information signal and a dc level that is proportional to the signal strength. In a SSB receiver, a separate detector is required for AGC, because the BFO signal provides the product detector with a steady carrier level. The steady carrier level from the BFO would cause the product detector to produce a dc voltage even when there was no input signal! This would defeat the AGC circuit.

In a SSB receiver, the AGC detector is still a diode detector. Usually, the IF signal preceeding the product detector splits into two signal paths, one for the AGC detector, and one for the product detector. This way, the interaction between the two circuits is eliminated.

SSB receivers often use very long RC time constants in the AGC circuit; 1 second is very typical. (AM receivers usually use a 200 ms AGC time constant). The longer time constant is necessary so that the AGC can properly respond to, and hold onto, the peaks of the SSB signal. If the receiver fails to do this, it may provide too much gain on signal peaks and thereby distort them.

Tuning the Receiver

In a superhet AM receiver, the frequency converter section translates the incoming RF carrier frequency down to the IF frequency. A similar but different process takes place in a SSB receiver. The key difference is that in a SSB receiver, the desired *sideband* component of the RF signal at the antenna is converted so that it *fits* within the IF amplifier's passband. The suppressed carrier frequency will *not* be the same as the IF center frequency!

Take at look at the *IF passband characteristics* graph that is part of Figure 6–23. The total bandwidth of the IF amplifier is only 3 kHz (compare this with 10 kHz for an AM broadcast receiver). In SSB voice communications, the range of information frequencies is from 300 Hz to 3 kHz, which is a 2.7 kHz bandwidth. Each sideband will therefore require 2.7 kHz of bandwidth. *The IF bandwidth is only wide enough to accept one sideband, and the frequency converter is responsible for aligning the sideband within the IF amplifier passband.*

In order to make it easier to visualize what is happening in a SSB receiver, it is useful to think of the *suppressed IF carrier frequency,* $f_{if(s)}$, as the *converted* version of the original suppressed carrier frequency, $f_{c(s)}$, of the SSB signal at the antenna. In reality, these two signals don't exist, since the carrier is not present (it is suppressed).

EXAMPLE 6–9 The receiver of Figure 6–23 needs to be tuned to a frequency of 14.155 MHz in USB mode. High-side local oscillator injection is in use. Calculate

a. The frequency of the preselector

b. The frequency of the local oscillator (use the given $f_{if(s)}$ of 456.8 kHz for USB mode).

c. If the 14.155 MHz suppressed carrier is being modulated by a 2 kHz sine wave, draw the spectrum of the original SSB signal, as well as the spectrum within the IF amplifier.

d. Demonstrate that the BFO frequency of 456.8 kHz correctly demodulates the signal.

Solution

a. By inspection, the preselector should be tuned to the suppressed carrier frequency of 14.155 MHz. (Recall that the purpose of the preselector is to reject the image frequency.)

b. Since high-side injection is being used, we can calculate the LO frequency as

$$f_{lo} = f_{c(s)} + f_{if(s)} = 14.155\text{ MHz} + 456.8\text{ kHz} = 14.6118\text{ MHz}$$

c. The original SSB signal consists of an upper sideband at (14.155 MHz + 2 kHz) = 14.157 MHz. Therefore, the original SSB signal looks like Figure 6–25.

The original SSB signal mixes with the local oscillator, producing sum and difference frequencies. However, only the mixing products that fall within the IF passband will get to the detector. The IF passband is from 453.5 kHz to 456.5 kHz, according to the data from Figure 6–23. The mixing of the LO and the SSB signal produces the following products:

Original frequencies: $f_{usb} = 14.157$ MHz, $f_{lo} = 14.6118$ MHz

Sum frequency: (14.157 MHz + 14.6118 MHz) = 28.7688 MHz

Difference frequency: (14.6118 MHz − 14.157 MHz) = 454.8 kHz

Figure 6–25 The original SSB signal

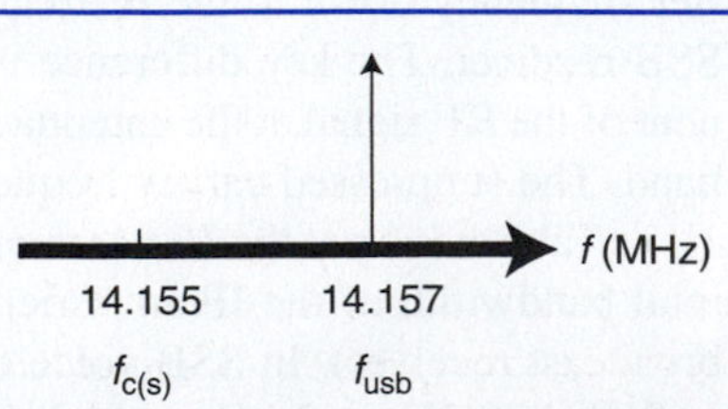

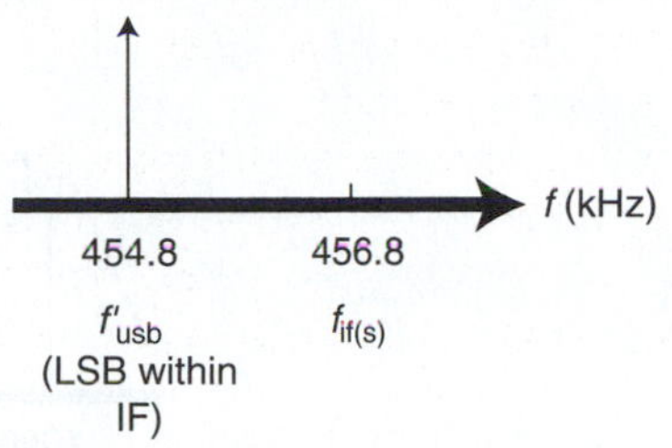

Figure 6–26 The IF signal

Only the *454.8 kHz* component fits within the IF passband. Therefore, the IF output to the detector looks like Figure 6–26.

Notice how the resulting IF output seems to be *lower sideband* energy. How can that be when we originally had a signal in USB mode? The answer is that high-side injection is being used. Because of this, the upper and lower sidebands reverse position after frequency conversion. No harm is done; the signal is merely inverted. If low-side injection had been used, the translated IF signal would remain in USB position. Also, notice that we have included the 456.8 kHz suppressed carrier frequency, even though it isn't really there. Including it helps us to better understand what we're looking at!

d. The product detector mixes the BFO frequency of 456.8 kHz with the IF spectrum of Figure 6–26. The resulting frequencies will be

$$\text{Originals:}\quad \text{BFO} = 456.8\text{ kHz}, f_{usb} = 454.8\text{ kHz}$$
$$\text{Sum:}\quad (456.8\text{ kHz} + 454.8\text{ kHz}) = 911.6\text{ kHz}$$
$$\text{Difference:}\quad (456.8\text{ kHz} - 454.8\text{ kHz}) = \underline{\underline{2\text{ kHz}}}$$

Only the 2 kHz component can pass through an audio low-pass filter. This is the original information frequency. The circuit works!

Multiple-Conversion SSB Receivers

The receiver of Figure 6–23 is a *single-conversion* receiver. There is only one frequency conversion and one IF frequency, 455 kHz. Such a receiver begins to experience image rejection difficulty at frequencies above 5 MHz or so. The *image frequency* is a second frequency that can mix with the local oscillator frequency to produce the IF frequency. In a superhet receiver, the preselector's job is to prevent the image frequency from reaching the mixer. Figure 6–27 shows two carrier frequencies and the resulting image frequencies.

In Figure 6–27(a), the preselector is doing an excellent job of eliminating the image frequency because its bandwidth is narrow compared to the distance between the carrier frequency and the image frequency ($2f_{if}$ or 910 kHz). In Figure 6–27(b), the receiver has been tuned to a higher frequency, 10 MHz. At the higher frequency, the preselector cannot maintain the same narrow bandwidth it once had. However, the frequency "distance" between the carrier (10,000 kHz or 10 MHz) and the image (10,910 kHz or 10.910 MHz) is still $2f_{if}$. *The frequency difference of 910 kHz is a lot smaller percentage of 10 MHz than 1 MHz, so a Q factor ten times higher would be needed in the 10 MHz circuit (b) to gain*

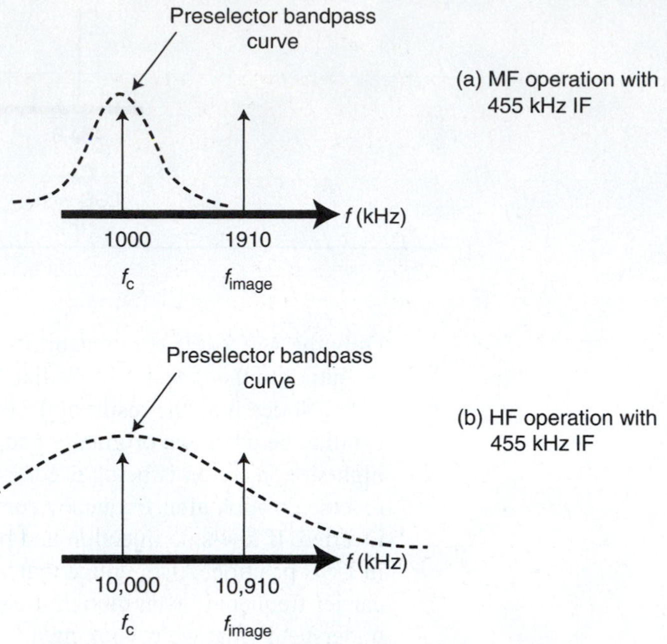

Figure 6–27 A comparison of two carriers and their images

the same image-rejection qualities. This is obviously not practical; as the frequency of most circuits increases, the Q falls and bandwidth increases. The circuit of Figure 6–27(b) is hardly rejecting the image frequency at all!

How can we improve the situation? One way would be to increase the IF frequency, as shown below in Figure 6–28. With an IF frequency of 10 MHz, the image will now be located at

$$f_{\text{image}} = f_c + 2f_{\text{if}} = 10\text{ MHz} + 2(10\text{ MHz}) = \underline{\underline{20\text{ MHz}}}$$

Increasing the IF frequency improves image rejection by moving the image frequency farther away from the carrier frequency, where it can more easily be attenuated by the preselector.

You might then ask, "Why didn't we use a 10 MHz IF to begin with?" The answer is that, in general, as we increase the frequency of an IF amplifier, its bandwidth also increases. Increasing the IF frequency to 10 MHz causes the bandwidth of the receiver to become too large. When the bandwidth is too large, selectivity (the ability to separate stations) becomes poor. We've solved one problem, but created another!

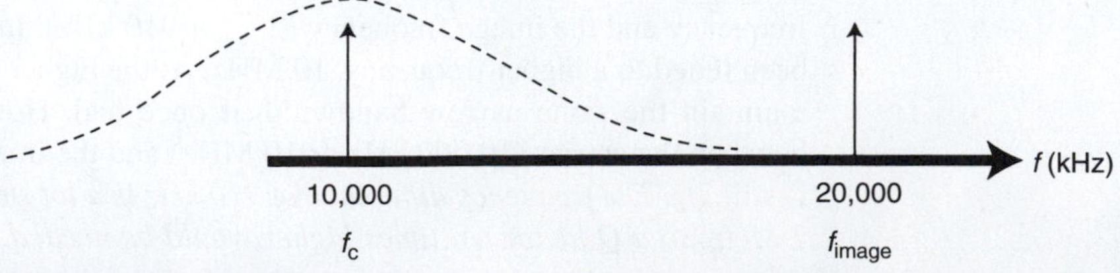

Figure 6–28 Increasing f_{if} moves the image frequency away

Multiple Conversion Saves the Day

A multiple-conversion receiver solves the bandwidth and image-rejection problems mentioned above. A *multiple-conversion* receiver is a superheterodyne receiver that performs more than one frequency conversion on the incoming RF signal. There are *two* frequency converters in the dual-conversion receiver in Figure 6–29.

Dual-conversion signal flow In the receiver in Figure 6–29, the incoming carrier is first converted down to the first IF frequency of 8.83 MHz. Since the first IF frequency is relatively high, the image frequency will be far enough away from the carrier frequency so that the preselector can easily reject it. The first IF amplifier does not have to define the receiver's bandwidth, however. In a typical commercial-grade receiver, the first IF filter is a crystal SSB filter. Some technicians refer to this as the "roofing" filter. Note how the first local oscillator tracks the preselector circuit—just as in a conventional superhet receiver.

The 8.83 MHz first IF signal is then down converted again, this time to the familiar frequency of 455 kHz. The 455 kHz IF amplifier further bandlimits the signal, and finally, the 455 kHz signal is detected at the product detector.

You might wonder why the second local oscillator (8.375 MHz) isn't ganged with the preselector and first local oscillator. The second LO frequency is *fixed*. This is because the first IF frequency is *also* fixed, at 8.83 MHz. The "second converter" is always converting 8.83 MHz down to 455 kHz. The second local oscillator is therefore locked on one frequency. In fact, the second LO is usually crystal controlled, or even better, supplied by a frequency derived from the master oscillator for the receiver, which eliminates a separate crystal (and its potential drift).

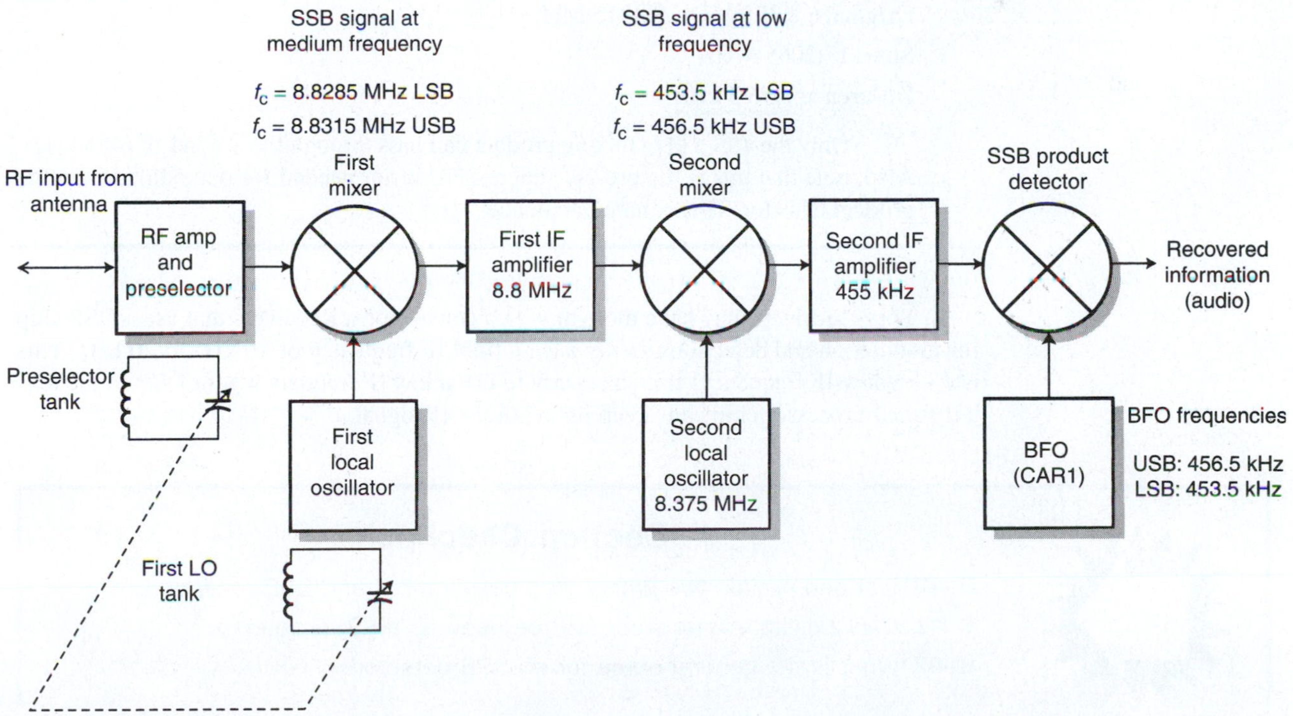

Figure 6–29 A dual-conversion receiver

Using a second IF of lower frequency has an additional advantage. It is always easier to amplify RF at lower frequencies. We can construct better low-noise amplifiers at 455 kHz than at 8.83 MHz. Reducing noise improves receiver sensitivity.

EXAMPLE 6–10 The receiver of Figure 6–29 is to be tuned to operate on 21.150 MHz in USB mode. Calculate

a. the first LO frequency, based on $f_{c(\text{supp})} = 8.8315$ MHz in the first IF (use high-side injection)

b. the image frequency

c. the resulting converted $f_{c(\text{supp})}$ that will be applied to the second IF after mixing with the 8.375 MHz second local oscillator

Solution

a. The LO frequency is calculated using the suppressed carrier frequencies:

$$f_{\text{lo}} = f_{\text{c}} + f_{\text{if}} = 21.150 \text{ MHz} + 8.8315 \text{ MHz} = \underline{\underline{29.9815 \text{ MHz}}}$$

b. The image will be located $2f_{\text{if}}$ away from the suppressed carrier frequency of 21.150 MHz:

$$f_{\text{image}} = f_{\text{c}} + 2f_{\text{if}} = 21.150 \text{ MHz} + 2(8.8315 \text{ MHz}) = \underline{\underline{38.813 \text{ MHz}}}$$

A properly designed preselector can easily reject the 38.813 MHz image.

c. The result of mixing the 8.375 MHz second LO with the 8.8315 MHz suppressed carrier in the first IF leads to the following:

Originals: 8.375 MHz, 8.8315 MHz

Sum: 17.2065 MHz

Difference: $\underline{\underline{456.5 \text{ kHz}}}$

Only the 456.5 kHz mixing product can pass through the second IF (455 kHz). Also, note that this is the precise suppressed carrier needed for demodulation in the product detector. This is no coincidence!

Some receivers may have more than two conversions. Receivers that use a DSP chip for modulation and detection usually have a final IF frequency of 10 kHz to 20 kHz. This is a very low IF frequency! It is necessary to use a low IF frequency with DSP so the digital signal processor chip can "keep up" with the IF signal.

Section Checkpoint

6–40 List and explain two differences between AM and SSB reception.

6–41 Can a digital signal processor be used as an SSB detector?

6–42 What is the general name for an SSB detector?

(continued on p. 207)

6–43 What signal must be supplied to a SSB detector in order to properly recover the information?

6–44 Why is an RF amplifier usually present in the front end of a SSB communications receiver?

6–45 What is the function of the RF carrier during detection?

6–46 Explain the operation of a SSB detector as a mixer in the frequency domain.

6–47 Which is more critical, the amplitude or the frequency of a BFO?

6–48 Why can't the AGC be derived from the product detector in a SSB receiver?

6–49 What is a typical AGC time constant for SSB reception?

6–50 List at least two reasons for the use of multiple-conversion receivers.

6–51 Why is the second local oscillator in Figure 6–29 fixed in frequency?

6–6 SSB TRANSCEIVERS

A *transceiver* is a combination transmitter and receiver. Separate transmitters and receivers are becoming increasingly rare in the communications world. Usually, any box capable of transmitting is also a receiver. The exception to this is *monitor receivers*.

A monitor receiver is a stand-alone receiver that is meant to operate in a lab environment. These receivers usually give calibrated readout of signal strength and frequency, and often can receive AM, FM, and SSB (as well as some interesting digital modes). They're commonly used in broadcast to verify station radiation patterns, and in law enforcement, to listen in on, and track, signals from various sources.

Fitting a transmitter and receiver into one package can mean fitting components into some very tight spaces. It also means that circuit reduction techniques must be utilized in order to keep the number of components to a minimum. Fortunately, there are enough similarities between transmitters and receivers that a fair amount of circuitry can be shared. Figure 6–30 is a simplified block diagram of a Kenwood TS-830 HF transceiver. There are *two* signal paths to follow: blue for receive, and gray for transmit.

By choosing IF frequencies that are identical for receive and transmit, the same set of filters can be used for sideband selection in both receive and transmit mode. This technique is used in most SSB transceivers.

Receive Signal Path

The incoming antenna signal at the original suppressed carrier frequency is amplified by RF amplifier Q_1 and sent into the first receive mixer, Q_3 and Q_4. The first mixer receives the f_{vco} signal from the *frequency synthesizer* section of the transceiver. The output of the first mixer is in the 8.83 MHz region and is sent through XF_1, the 8.83 MHz crystal IF filter.

After the 8.83 MHz receive signal leaves XF_1, it is immediately downconverted to 455 kHz in the second receive mixer, Q_2 and Q_3. The 8.375 MHz oscillator output "beats" against the 8.83 MHz IF signal to produce the 455 kHz difference signal, which is sent into the 455 kHz ceramic SSB filter CF_2.

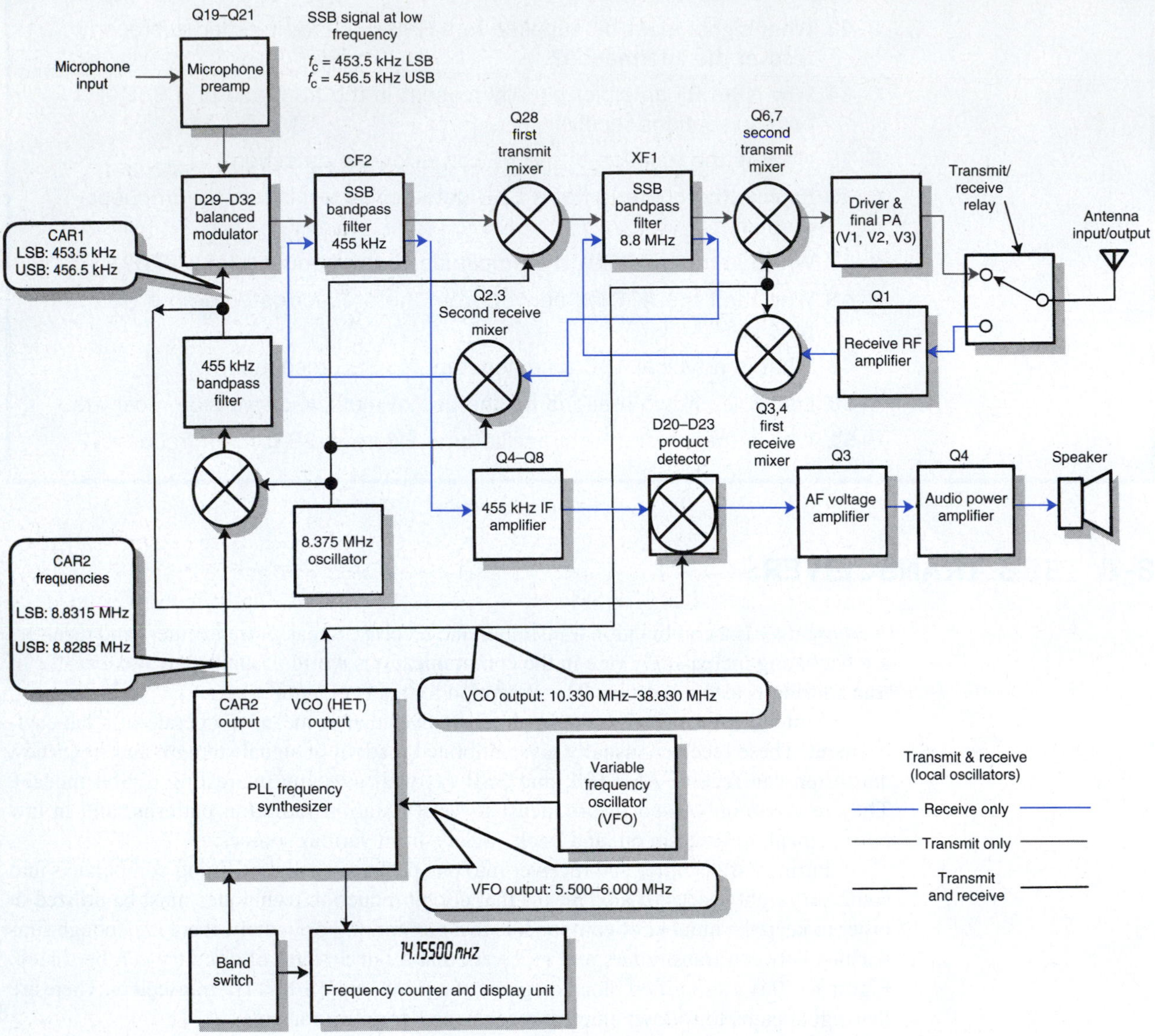

Figure 6–30 Kenwood TS-830 HF transceiver

The 455 kHz signal leaves CF_2 and is further amplified by transistors Q_4–Q_8, the 455 kHz receive amplifiers. The 455 kHz SSB signal is applied to the receive product detector, D_{20}–D_{23}. The CAR1 signal developed by mixing the CAR2 output of the frequency synthesizer with the 8.375 MHz oscillator output is used as the BFO at the product detector.

The resulting audio signal from the product detector is passed through the volume control (not shown), and through the AF voltage amplifier (Q_3), and finally through the audio power amplifier IC (Q_4) and speaker.

Transmit Signal Path

The transmit signal path is identical to that of Figure 6–22. The same set of filters is shared by the transmitter and receiver; they are electronically switched by diodes, depending on whether the set is in receive or transmit mode.

The microphone signal is voltage amplified by $Q_{19}-Q_{21}$ before being sent to the balanced modulator ($D_{29}-D_{32}$). The resulting 455 kHz DSB-SC signal is filtered into a SSB signal by CF_2, the 455 kHz ceramic filter, and then upconverted to 8.83 MHz by Q_{28}, the first transmit mixer.

The resulting 8.83 MHz SSB signal is further filtered by XF1, the 8.83 MHz crystal filter, and is then fed into Q_6 and Q_7, the second transmit mixer. The f_{vco} output of the frequency synthesizer mixes with the 8.83 MHz IF signal to produce the final transmit frequency, which is then amplified by the power amplifier section.

The transmit power amplifier consists of three vacuum tubes. The first tube, V_1, a 12BY7A, is the *driver* amplifier; the driver brings the signal level up to around 5 watts. The remaining two tubes, V_2 and V_3, a pair of 6146Bs, amplify the 5 watts signal from the driver to approximately 100 watts, which is then sent to the antenna through the transmit-receive (TR) relay. (The TR relay connects the antenna to the final PA in transmit mode and to the RF amplifier Q_1 in receive mode.)

Elements Common to the Receiver and Transmitter

In the TS-830, most of the oscillators are shared by both the receiver and transmitter. If any of these fail, the result is generally a completely dead set. The SSB filters are also shared; a failure of either filter will also affect the entire transceiver. Finally, the power supply is common to the entire radio. Even in a complicated unit such as this, the basic steps of troubleshooting still apply. A malfunctioning unit should first be inspected visually, then the major power supply points should be checked.

Troubleshooting is always carried out with the signal flow of Figure 6–30 in mind. If a unit works in transmit but not receive (or vice versa), then the tech already has a lot of information about what is (and isn't) working in the circuit.

EXAMPLE 6–11

The transceiver of Figure 6–30 transmits normally but receives nothing (speaker is totally silent). Develop a plan for troubleshooting the unit.

Solution

The three-step method will be used. These steps are (1) visual inspection, (2) power supply checks, and (3) input/output checks. The SSB filters are power supplies are likely operational, since they are shared by the transmitter and receiver. The proposed steps are as follows:

1. Remove the covers and with power off, look over the *receiver* portions of the circuit.
2. Check each of the dc power supply voltages to the receiver, using a DMM.
3. If the first two steps above reveal no problems, then proceed to find out where the signal is being lost by "cutting the circuit in half."
   a. Apply a test signal to the antenna input using an RF signal generator, making sure to tune the receiver to the same frequency as the generator.
   b. Pick a halfway point in the receiver system, such as the 8.83 MHz IF output of XF_1, and check for signal. If there is no 8.83 MHz output from XF_1, the problem is either in Q_3, Q_4 (first receive mixer), or Q_1 (RF amp), or the transmit receive relay. If the output of XF_1 is OK, proceed to step (c).

c. Check for signal at the output of the last 455 kHz IF amplifier, Q_8. If there is no 455 kHz output, the problem is in the 455 kHz IF amp (Q_4–Q_8), or the second receive mixer (Q_2, Q_3). If the 455 kHz output of Q_8 is OK, proceed to step (d).

d. Check for AF output at the product detector output (D_{20}–D_{23}). If there's no output, the problem is in the product detector (D_{20}–D_{23}). If the output is OK, the problem is in one of the AF amplifier stages (Q_3, Q_4) or in the loudspeaker and its connections. In this case, the signal can be successively tested at the output of the remaining two audio stages, which will localize the trouble.

Note: There are two active devices in the TS-830 with the designators Q_3 and Q_4 because these devices are located on different circuit boards. The block diagram does not reflect this.

SUMMARY

- AM is inherently inefficient. Sideband modes are modified AM emissions.
- The carrier frequency component in an AM signal conveys no information but is necessary for successful detection.
- Sideband signals lack a carrier component, which must be regenerated by the receiver.
- The standard frequency range for speech is usually assumed to be 300 Hz–3000 Hz.
- The decibel power advantage is used to compare the relative power advantage of one transmission mode over another, such as SSB over AM. It depends on both information power and bandwidth utilized.
- VSB is used when detection is to be performed with a conventional diode detector.
- Percent modulation is meaningless for SSB/DSB signals Instead, PEP is used to rate the transmitted output.
- PEP is calculated by finding the average (RMS) power delivered during one RF cycle at the peak of the modulation envelope.
- Practical methods of SSB generation include filtration and digital signal processing.
- Frequency conversion techniques are commonly used to reduce the number of filters needed in SSB transmitters.
- A balanced modulator is a special AM modulator that generates sidebands but no carrier frequency component.
- Filters for SSB require high-Q elements; crystal and mechanical filters are commonly used.
- The shape factor (SF) is one way of measuring the quality of a filter's frequency response.
- Practical SSB transmitters must be frequency agile; frequency conversion techniques are necessary to achieve this.
- In a SSB receiver, the BFO (beat frequency oscillator) is used to reinsert the missing carrier frequency component.
- Single-conversion receivers often suffer from poor image rejection at high frequencies; multiple conversion solves this problem.
- A complete transceiver often shares some of the IF circuitry (especially the SSB filters) between the transmit and receive sections of the unit.
- To troubleshoot a transceiver, the signal flow must be understood, and a methodical approach must be used.

PROBLEMS

1. Draw a frequency diagram of each of the following types of signals. Explain what the differences are between each and a standard AM signal: (a) DSB-SC; (b) SSB-SC; (c) VSB
2. A DSB-SC transmitter is operating on a suppressed carrier frequency of 6200 kHz, and is being fed information frequencies ranging from 50 Hz to 4000 Hz. Compute the bandwidth needed for the transmitter.
3. A SSB transmitter is using a suppressed carrier frequency of 8350 kHz, and the information frequency range of this transmitter is 150 Hz to 3500 Hz. Compute the bandwidth required by this transmitter.
4. What is considered to be the minimum bandwidth required for intelligible speech reproduction?
5. Compute the decibel power advantage of a 30 W PEP DSB-SC transmitter (using a bandwidth of 6 kHz) to a conventional AM transmitter with a carrier power P_c of 30 W operating under 50% modulation and using a 6 kHz bandwidth. (*Hint:* First compute the total sideband power in the AM transmission.)
6. If transmitter A has an information power of 10 W and a bandwidth of 10 kHz, while transmitter B has the same information power (10 W) and a bandwidth of 5 kHz, what is the dB power advantage of transmitter B? What power level could transmitter B be operated at to be equivalent to transmitter A?
7. When is it desirable to use VSB? What is the main application for VSB transmission?
8. Why is percentage of modulation not useful for SSB transmitters?
9. Describe how PEP could be measured on an oscilloscope.
10. A certain SSB transmitter produces 141 V_p at the crest of its modulation envelope while driving a 50 Ω dummy load. What is the PEP power level?
11. What peak envelope voltage would be present for a 50 W PEP level into 50 Ω?
12. What is meant by the term *push-push* when describing the operation of a balanced modulator circuit?
13. What is the purpose of potentiometer R_1 in Figure 6–7?
14. Describe a procedure for the correct adjustment of R_1 in the balanced modulator of Figure 6–7.
15. A certain SSB filter has a bandwidth of 2.3 kHz @ −6 dB down, and 4.0 kHz @ −60 dB down. What is the shape factor of the filter?
16. If the filter of problem 15 needs to have a shape factor of 1.5:1, and the −6 dB bandwidth is fixed, what would the −60 dB bandwidth be?
17. When would a mechanical SSB filter be chosen over a crystal type?
18. When the suppressed carrier frequency is changed in Figure 6–16, what is the effect on the frequency output of the balanced modulator?
19. If the SSB filter in Figure 6–16 passes 1.5530 to 1.5557 MHz and the range of information frequencies is 300 Hz to 3000 Hz, calculate the required frequency of the USB crystal.
20. Repeat problem 19 but calculate the LSB crystal frequency.
21. Why isn't the phasing method used with analog circuitry?
22. What approach to SSB generation is used in a digital signal processor?
23. Explain the operation of a DSP chip being used to generate a SSB signal.
24. Define the term *frequency agility.*
25. Draw a simple block diagram showing how frequency agility is obtained in a SSB transmitter.
26. Calculate the correct frequency for the VFO to put the transmitter of Figure 6–21 on a frequency of 24.800 MHz in USB mode. (Be sure to use the given $f_{c(\text{suppressed})}$ values from Figure 6–21.)
27. Why are the switched bandpass filters needed at the transmitter output in Figure 6–22?

28. List two differences between SSB and AM reception.

29. Why is an RF amplifier used with most SSB receivers?

30. A certain SSB product detector has an incoming frequency of 453 kHz from the IF amplifier and a frequency of 455 kHz from the BFO. List the frequencies that the product detector will produce. What AF frequency will be reproduced as the information?

31. Why is a separate AGC detector diode necessary in SSB receivers?

32. Assuming high-side injection and an IF frequency of 8.83 MHz, calculate the image frequency for 29.500 MHz.

33. What is the primary reason for using multiple-conversion receivers?

34. In order to effectively troubleshoot a SSB transceiver, what information must be understood first?

35. Some components in a transceiver are shared by the transmitter and receiver. How does knowing about these simplify the troubleshooting process?

36. Carefully examine the signal flow of Figure 6–30 and describe the symptoms that would be caused by the *failure* of the following circuit components:

   a. XF_1, the 8.83 MHz crystal IF filter
   b. Q_1, RF AMP
   c. Q_4, AF power AMP
   d. Q_{28}
   e. Q_{19}

7

Systems for Frequency Generation

OBJECTIVES

At the conclusion of this chapter, the reader will be able to:

- list the components in a phase-locked loop, explaining the purpose of each one
- describe the three operating states of a PLL
- calculate the frequencies in each part of a PLL given the parameters of the loop
- draw a block diagram of a direct PLL frequency synthesizer
- calculate the frequencies and divisors needed in a direct PLL synthesizer
- calculate the frequencies in the loop of an indirect PLL synthesizer
- describe the software events necessary for control of a PLL synthesizer
- draw a block diagram of a DDS frequency synthesizer
- calculate the various parameters for a DDS frequency synthesizer
- develop a plan for troubleshooting a frequency synthesizer, given its block and schematic diagrams

All radio transmitters and receivers use *oscillators* to provide needed frequencies. An oscillator is a stage that converts dc from the power supply into an ac output signal at a specified frequency. Up to this point, we have studied two ways of controlling the frequency of an oscillator. These two methods are *discrete LC control* and *crystal control*.

An oscillator's frequency can be controlled by a discrete LC tank. This approach is simple, and the oscillator's frequency is easily adjusted by varying either the L or C component values within the tank. However, this approach doesn't provide a very stable output frequency; the Q of the LC tank is too low to keep the frequency steady.

Crystal control of an oscillator's frequency provides rock-stable output. This would be the ultimate choice for all transmitters and receivers, except that the frequency of a crystal oscillator cannot be appreciably changed without replacing the crystal. When many different operating frequencies are required, this approach becomes very expensive.

A *frequency synthesizer* is a circuit that *synthesizes* or "builds" new frequencies. These new frequencies are based on a highly stable frequency source, which is usually a single quartz crystal oscillator. The stable frequency source is called the *reference* or *master* oscillator.

Modern frequency synthesizers are digitally controlled. They make possible all sorts of products and applications, from electronically tuned shirt pocket stereo receivers, to sophisticated commercial communication and navigation equipment. Digital control of frequency allows microprocessor control of radio features. The frequency synthesizer in a typical product is merely an input/output (I/O) device connected to the controlling central processing unit (CPU). Software calls the shots, and analog hardware does the work.

The mix of analog and digital hardware in a frequency synthesizer can be very intimidating to the technician, especially when there is a hidden piece of computer software running the show. No matter how complex, all frequency synthesizers are based on a few basic ideas. By learning these principles, you'll be well prepared for working with and troubleshooting these fundamental communications building blocks.

7–1 THE PHASE-LOCKED LOOP

The phase-locked loop, or *PLL,* is one of the most useful blocks in modern electronic circuits. It is used in many different applications, ranging from communications (FM modulation, demodulation, frequency synthesis, signal correlation), control systems (motor control, tracking controls, etc.), as well as applications such as pulse recovery and frequency multiplication. Knowing PLLs can really boost your "tech IQ"!

PLL Theory

A PLL is a *closed-loop* system, whose purpose is to lock its oscillator onto a provided input frequency (sometimes called the *reference* frequency). A closed-loop system has feedback from output to input. In a PLL, the feedback is negative, meaning that the system is *self-correcting*. When we say that the PLL's oscillator is *locked* onto the reference, we mean that there is *zero frequency difference (error) between the PLL oscillator and the reference frequency.* It might not make sense at this point as to why we would want to "lock" one oscillator onto another's frequency. Can't we just take the output from the *reference* oscillator and be done?

There are two reasons why we will want to do exactly this. First, the PLL provides *filtering* action. A PLL can lock onto a noisy reference signal, providing a filtered output that is relatively free of noise. Second, by modifying the PLL feedback loop, we can derive new frequencies from the reference signal. We can build a "tunable" frequency source based on a rock-solid crystal frequency reference. This is *frequency synthesis*. Figure 7–1 shows the basic elements in a PLL.

A PLL consists of a *voltage-controlled oscillator,* or *VCO,* a *phase detector,* and a *loop filter*. Each of these components has a special job in keeping the PLL locked onto the reference frequency.

Voltage-Controlled Oscillator

A PLL has a special oscillator, a VCO. Previous oscillators we have studied depended on either an LC resonant circuit or a crystal to determine their frequency. The output frequency of a VCO depends on an LC or RC circuit and a *control voltage*. The LC or RC

Figure 7–1 Phase-locked loop block diagram

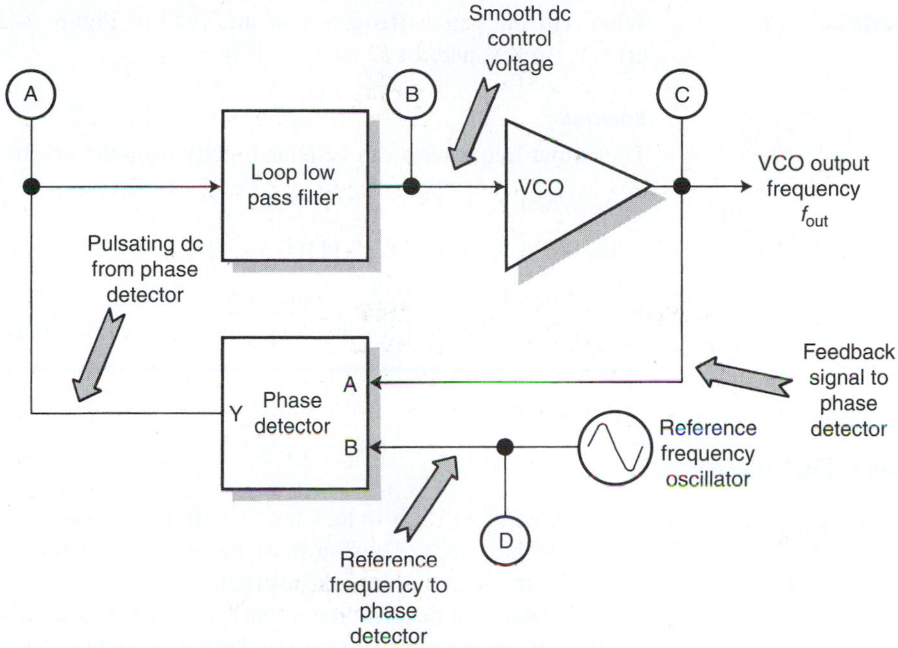

portion of the circuit determines the approximate frequency range that the VCO will operate in, and the control voltage moves the VCO frequency up or down within that range.

Most technicians think of a VCO as a *voltage to frequency converter,* since the input to a VCO is a dc control voltage, and the output of a VCO is a varying frequency. Figure 7–2 is a graph of the *transfer characteristic* of a simple VCO circuit. The transfer characteristic is the input–output relationship. Note the units on the axes in Figure 7–2. The horizontal axis has units of voltage, and the vertical shows frequency. This shows us that the output frequency of the device depends upon the input control voltage.

There are definite limits on how high and how low the frequency of the VCO can go. These limits are inherent to any VCO and are determined by the circuit designers. Most practical VCO circuits do not operate over more than about a 3:1 frequency range.

Figure 7–2 Transfer characteristic of a VCO

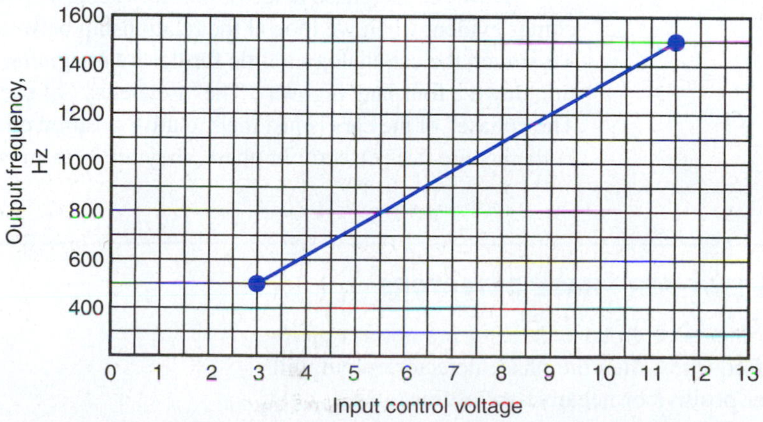

EXAMPLE 7–1

What will the output frequency of the VCO of Figure 7–2 be if the control voltage is (a) 3 V, (b) 8 V, and (c) 12 V.

Solution

The output frequencies can be read directly from the graph of Figure 7–2:

a. When $V_C = 3\text{ V}, f_{\text{out}} = \underline{\underline{500\text{ Hz}}}$.

b. When $V_C = 8\text{ V}, f_{\text{out}} = \underline{\underline{1100\text{ Hz}}}$.

c. When $V_C = 12\text{ V}, f_{\text{out}} = \underline{\underline{\text{undefined}}}$. The VCO isn't designed to accept a control voltage above 11.5 volts. We know this because the graph stops at $V_C = 11.5$ V.

Phase Detector

The purpose of a PLL is to lock the VCO frequency onto the reference frequency. In order for this to happen, a decision must be made about the VCO's frequency: Is the VCO frequency too high, too low, or just right?

The decision-making process must involve feedback. It's very much like the cruise control in a car. Suppose that you have set the cruise speed in your car to be 70 mph. Somewhere, a sensing device measures the car's speed. That speed information is fed into the cruise control unit. If the car is moving too slowly (speed < 70 mph), you know that the cruise control will respond by opening the throttle a little wider, which will bring the vehicle speed up to the desired point. The opposite will happen if the car is moving too quickly; the throttle is closed to slow down. The speed of the car is not expected to always be *exactly* the same as the set point of the cruise control. There is always a small error, usually ±2 mph. This is necessary to prevent stability problems ("hunting").

In a PLL, the VCO control voltage is like the throttle in a car, and the resulting VCO frequency is analogous to the car's speed. The *phase detector* is the decision maker that compares the VCO frequency to that of the reference. In Figure 7–1, you can see that the VCO output signal is fed back into the A input of the phase detector. The B input of the phase detector sees the reference signal. Unlike the cruise control in a car, the phase detector decision maker will not be satisfied until there is *zero frequency difference (error)* between the VCO and the reference source. In fact, this can be stated as a simple law.

How can the phase detector possibly achieve *zero* error in frequency? The answer becomes evident when we look at the relationship between frequency and phase. Suppose that we stretch the car analogy a little further by imagining two cars traveling in the same direction down a four-lane highway. The "frequency" of each car is indicated by its speedometer. The "phase" of the cars is just their relative position on the highway. In Figure 7–3, although both cars are not perfectly in phase, they are both moving at exactly the same speed. Their

FINLEY'S LAW FOR PHASE DETECTORS

If the two inputs of a phase detector are not at *exactly* the same frequency, then the phase detector output will be in either positive or negative *saturation*.

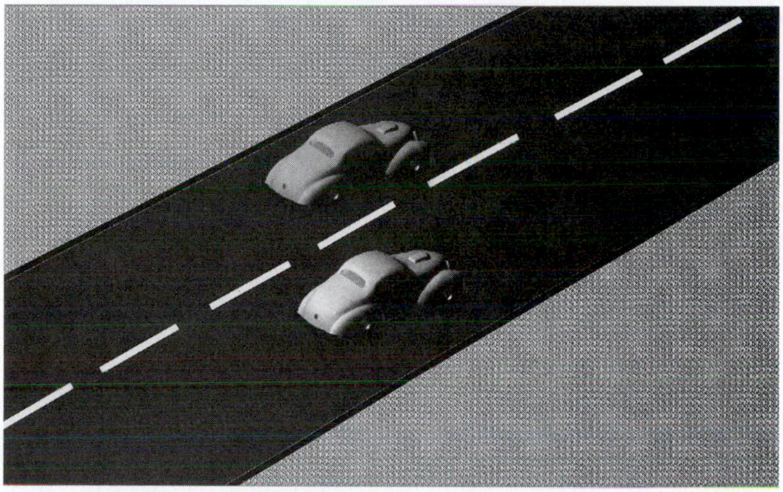

Figure 7–3 The frequencies are identical, the phases are not; the phase error is static

"frequencies" are identical. *The phase of two signals does not have to be the same for their frequencies to be identical. However, the* phase error *must be constant.*

In Figure 7–4, the left car has sped up. The two cars are no longer in a fixed relationship to each other. There is not only a phase difference between them, but also an *increasing* phase difference (error). Their speeds (frequencies) are no longer equal.

A *phase detector achieves zero frequency error by comparing phase.* When the phase difference between two signals is constant, their frequencies are identical. This is why Finley's law is true for phase detectors. This special property makes the phase detector an excellent frequency "referee" for the PLL. It also explains why there is always zero frequency error in a PLL once it is in lock.

Most phase detectors are digital. The output of a phase detector is pulsating dc with a varying duty cycle. One of the simplest possible phase detectors is an exclusive-OR logic (XOR) gate, as shown in Figure 7–5(a).

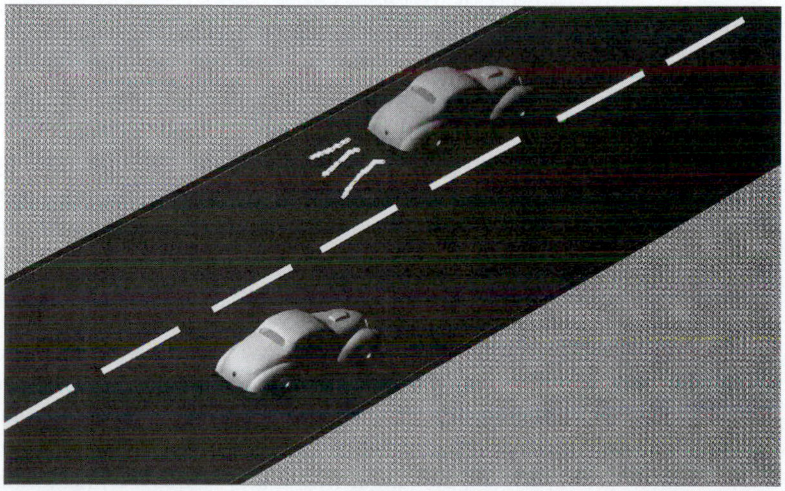

Figure 7–4 The phase error is increasing; the frequencies are unequal

Figure 7–5 XOR-type phase detector and circuit waveforms

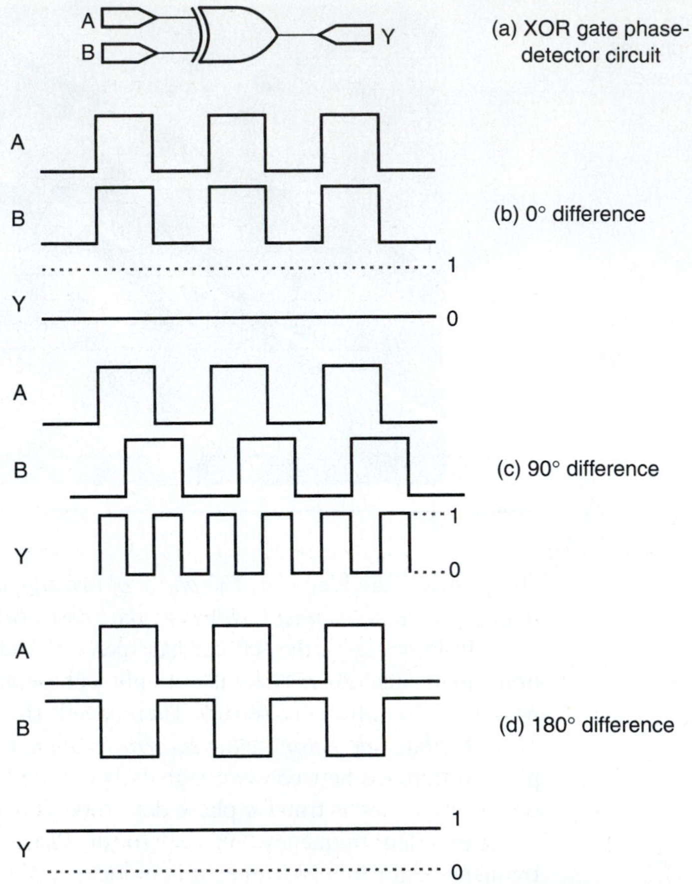

An exclusive-OR gate gives a "1" output whenever the inputs are opposites and a "0" output when the inputs are the same. In Figure 7–5(b), the two inputs are precisely in phase, which means they're the same all the time. The gate always produces a "0" output, which corresponds to 0 V.

In Figure 7–5(c), the "B" input is *leading* the "A" input by 90°. Now the gate inputs are different at parts of the cycle, and consequently, the output "Y" goes high precisely one-half of the time. We would say that its duty cycle is 50%, and that its average voltage is $V_{cc}/2$.

As we increase the phase difference to 180°, the output stays high all the time. The duty cycle is now 100%, and the average voltage is V_{cc}. The XOR gate has converted the input phase difference into an average dc voltage. This dc voltage is pulsating at twice the frequency of the input signals. Figure 7–6 shows the transfer characteristic of this phase detector.

The blue graph in Figure 7–6 reflects operation from a 0° to 180° difference. As the phase difference increases, the average output voltage increases. But something strange happens as the phase difference passes 180° (gray region of graph)—the output voltage starts *decreasing!* The XOR phase detector simply cannot operate over more than a 180° range.

In order to overcome this limitation, more complex logic circuits are used in actual PLL chips. The XOR phase detector has another problem. It is sensitive to the duty cycle of the two input signals, which are required to be square waves. If either signal varies in duty cycle, the output will falsely indicate a phase change. A practical phase detector usually includes a Schmitt trigger on each input in order to convert the signals to square waves and an edge-detector circuit to overcome the duty cycle problem.

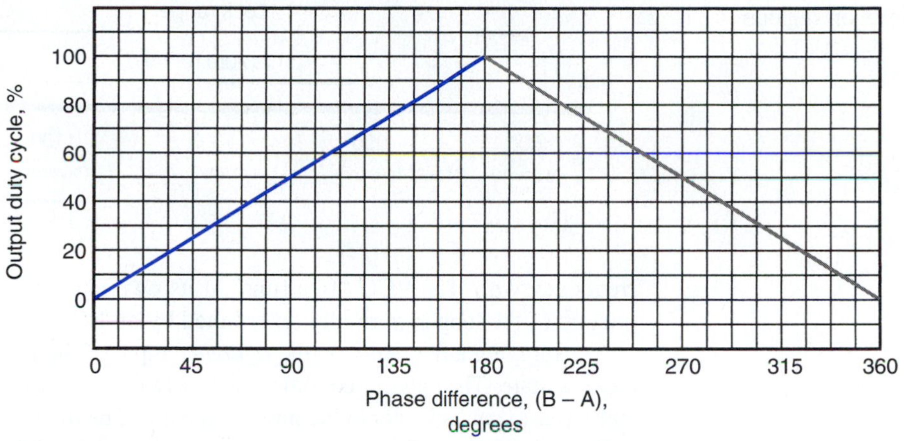

Figure 7–6 Transfer characteristic of the XOR phase detector

Loop Filter

The "Y" output of the phase detector in the PLL of Figure 7–1 is a pulsating dc voltage with a varying duty cycle. The bigger the phase difference becomes (within certain limits, of course), the larger this duty cycle becomes, and the larger the average voltage being fed back into the VCO on top. This voltage will tend to correct the frequency of the VCO, either raising or lowering its frequency. But there's a problem here.

The VCO needs a nice, steady dc voltage at its control voltage input. Can you imagine the effect of the pulsating dc on the VCO? Think of a car that only has two throttle positions, wide open and off. The desired speed is 70 mph. We're traveling 69 mph, which is too slow—so we must choose the *wide-open* throttle position. The car lurches forward with this throttle application, *overshooting* the target speed of 70 mph. Now our only choice is to totally close the throttle and jam on the brakes. The passengers are thrown forward as the car rapidly decelerates. The cycle repeats, over and over. The motion of the car isn't very smooth at all, although its average speed is very close to 70 mph!

The same thing would happen to the VCO. We'd like its output frequency to be steady, like the reference input. What we need to do is smooth out the pulsating dc from the phase detector into a steady *dc average* voltage. This is the purpose of the *loop filter*. This filter in effect smoothes the rough phase-detector output waveform into a steady dc voltage for the VCO. The VCO will then be able to smoothly track the input reference frequency.

The loop filter in a PLL is usually a *low-pass* type. It can be a simple RC time constant, or something more involved. The RC time constant within the loop filter determines several of the loop's characteristics, including how fast it can respond to changes. A long loop-filter RC time constant provides excellent filtering but very slow response. By reducing the RC time constant, we can speed up the ability of the PLL to respond to changes—but at the expense of poorer VCO control-voltage filtering, which will show up as "jitter" (time domain) or "spurious sidebands" (frequency domain) in the VCO frequency output.

PLL Operating States

A PLL has three operating states. These are the *free-run, capture,* and *locked* conditions.

In the *free-run* state, there is no reference input frequency being provided to the PLL. Under this condition, there is nothing for the phase detector to compare the VCO output

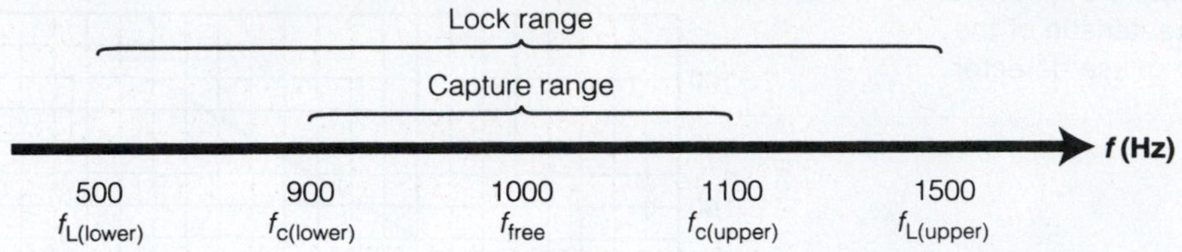

Figure 7–7 PLL operation regions

frequency with. The VCO "free runs" at its own natural frequency. The free-running frequency of the loop is normally determined by an LC or RC circuit within the VCO.

Applying a frequency to the *reference* input of the PLL causes the loop to go into the *capture* state. The capture condition normally doesn't last very long, for the PLL immediately tries to get locked onto the input frequency. The time needed to get locked depends partially on the RC time constant to the loop filter, and the difference in frequency between the VCO and the applied reference input signal. Capture is very similar to the slipping of the clutch that takes place when a manual-transmission car takes off from a stop. Initially, there is a great difference between the input and output of the clutch; as the clutch pedal is released, the car gains speed (VCO moves toward reference frequency), the input and output of the clutch eventually become *exactly* equal—the clutch is no longer slipping. When the VCO frequency exactly equals the reference frequency, the PLL has attained *lock*.

The PLL cannot lock onto all frequencies; only a certain range of frequencies within the *capture range* can be required if the PLL is initially in the free-running state. Usually, the free-running frequency is in the middle of the capture range. The width of the capture range is determined by PLL design; the loop low-pass filter is important in determining this.

The last PLL state is the desired state: *locked!* In this state, the PLL has successfully passed through the capture phase, and it has its VCO locked onto the input reference frequency. The PLL cannot remain locked for all frequencies, and if the input reference frequency moves outside the *lock range* (which is usually larger than the capture range), the PLL will drop out of lock. The VCO is the primary component in the PLL that determines lock range, for the lock range is actually just the lower and upper limits of VCO oscillation frequency.

Figure 7–7 illustrates the relationship between free-running frequency, capture range, and lock range for a typical PLL.

EXAMPLE 7–2

Suppose that the PLL of Figure 7–1 has the operation regions given in Figure 7–7. Give the PLL state and VCO output frequency for each of the cases below. Assume that initially there is *no* reference input frequency applied and that the conditions attained in each case will apply to the next case.

a. $f_{reference} = 450$ Hz
b. $f_{reference} = 800$ Hz
c. $f_{reference} = 950$ Hz
d. $f_{reference} = 1450$ Hz
e. $f_{reference} = 1550$ Hz
f. $f_{reference} = 1200$ Hz
g. $f_{reference} = 0$ Hz

Solution

a. The loop is in the *capture* state, and the VCO frequency cannot be determined. The VCO is rapidly hunting up and down in frequency trying to match the reference, but since the frequency is too low (less than $f_{c(lower)}$), the loop cannot acquire lock.

b. The loop is *still* in capture, and again, the VCO frequency is pretty much unknown.

c. The loop is in *lock*. The applied frequency has fallen inside the *capture range* (900–1100 Hz), so the VCO can "catch up" with the reference signal. *Finley's law* applies when the loop is in lock, so $f_{VCO} = f_{reference} = \underline{\underline{950 \text{ Hz}}}$.

d. Once the loop is in lock, the VCO can now follow the reference anywhere within the *lock range.* Since we attained lock already, the loop follows, and we get $f_{VCO} = f_{reference} = \underline{\underline{1450 \text{ Hz}}}$.

e. The frequency 1550 Hz is outside of the lock range—the VCO can't go that high. The loop drops out of lock, back into *capture* (since there is an applied reference signal). The VCO frequency is unknown.

f. This is weird, but true! Starting *unlocked* from condition (e), the loop will still be out of lock here. In order to gain lock, the frequency must first "dip" into the capture range. Therefore, the loop is in *capture* and the VCO frequency is still unknown.

g. The reference signal has been removed, and the loop free runs again. The VCO frequency will be *approximately* 1000 Hz. Since the VCO frequency is determined by an LC or RC circuit, this frequency is not very accurate. The loop is in the *free-run* state again.

Determining Loop State with Instruments

A technician often needs to find out whether or not a PLL is properly locked. The most common method of doing this involves the use of a dual-trace oscilloscope. Channel 1 of the scope is connected to the *reference* input of the phase detector (point D in Figure 7–1), and channel 2 is connected to the remaining phase detector input (point C). In Figure 7–1, this is exactly the same as the VCO output—but this is not true for all PLLs.

The oscilloscope is usually set to trigger off the reference, and the resulting two waveforms (reference and VCO) are compared. If they're exactly at the same frequency, the loop is in lock. On a scope, this is immediately apparent. Look at Figure 7–8.

In Figure 7–8(a), the reference signal is stable since the scope is set to trigger from it. However, the VCO looks very strange. In this photo, the VCO appears as two lines. Actually, the VCO signal looked like it was "running" left to right on the display. The action of the camera blurred this into the two lines you see. *If the VCO output cannot be easily seen, the loop is definitely not in lock!*

The next photo shows the loop in lock. The VCO display does not appear to move or "crawl" across the screen, and it is exactly the same frequency as the reference. You can verify this; both the reference and VCO have exactly the same *period* in Figure 7–8(b).

The loop state can also be verified by using a frequency counter. You've probably already guessed the two points of measurement—that's right, each phase detector input. They must read *exactly* the same frequency, in a stable manner. The scope method is more popular because many frequency counters have difficulty in properly triggering off an analog RF signal, and some frequency counters may excessively "load" an RF circuit, leading to false readings.

Figure 7–8 Assessing the PLL state with an oscilloscope

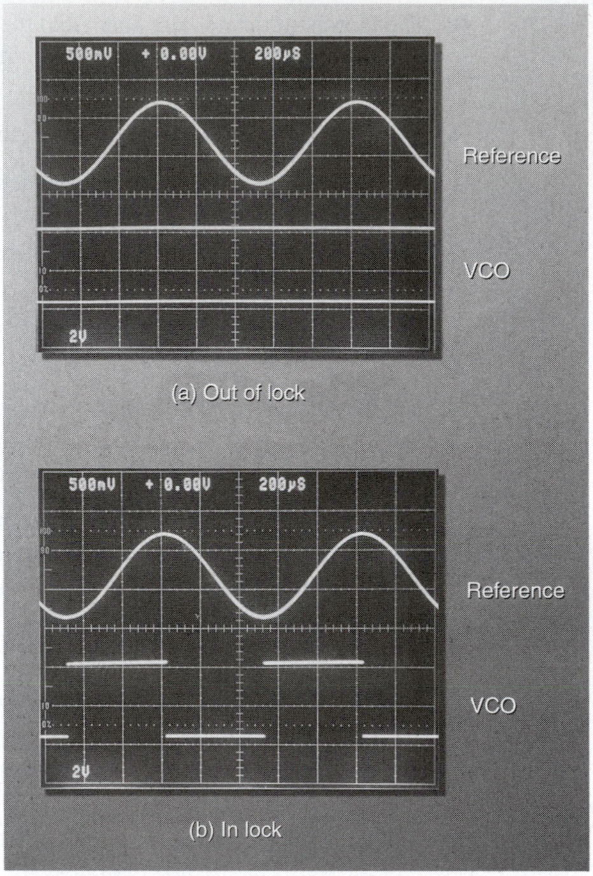

Section Checkpoint

7–1 Why are frequency synthesizers needed in modern electronics?

7–2 What is meant by the term *closed-loop system*?

7–3 List the parts of a PLL, explaining what each one does.

7–4 What is the primary action or goal of a PLL?

7–5 The VCO in a PLL converts ______ into ______.

7–6 Which part of a PLL acts as a decision maker?

7–7 Why is a low-pass filter necessary in a PLL?

7–8 State Finley's law for phase detectors.

7–9 How much frequency error is present in a *locked* PLL?

7–10 What are the three PLL operating states?

7–11 What is the difference between the *capture* and *lock* ranges of a PLL?

7–12 Explain how to determine whether or not a PLL is in lock by using benchtop instruments.

7–2 PLL SYNTHESIZERS

The basic PLL configuration (Figure 7–1) itself synthesizes nothing; the VCO frequency of the loop is always equal to the reference input (when the system is in lock). By modifying the feedback portion of the loop, we can get the loop to produce new frequencies. In other words, we can convert the PLL frequency "follower" to a frequency *synthesizer* by altering the feedback sent back to the phase detector.

Frequency Dividers

A *frequency divider* is a circuit that divides an incoming frequency by some chosen number. Frequency dividers are really nothing more than digital counters. Figure 7–9 shows a *divide-by-two* circuit with waveforms.

In Figure 7–9(a), a JK flip-flop is connected in the *toggle* mode. Recall that when both the J and K inputs are tied high, a JK flip-flop is "programmed" to toggle. The flip-flop will change state (toggle) on each active clock transition. For the circuit in Figure 7–9, we know that the JK's clock input is sensitive to the *falling edge* of the clock signal. Therefore, each time the clock input goes from high to low, the Q and /Q outputs of the flip-flop will change state.

Look at the relative frequencies of the *clock* and Q signals in Figure 7–9(b). There are *two* complete clock cycles for every Q cycle. Therefore, we can say that the frequency of Q is precisely 1/2 of the clock. *The circuit has divided the input clock frequency by two.* Another way of saying this is that the divider has a modulus of two.

Higher divisor ratios By cascading counters, we can get larger divisors. Can you determine the divisor ratio for each of the circuits of Figure 7–10?

In Figure 7–10(a), we have cascaded two JK flip-flops to form a *ripple* counter. From digital fundamentals, you'll recall that the modulus of a binary ripple counter is equal to

$$\text{mod} = 2^N \tag{7–1}$$

where N is the number of flip-flops in the circuit. Since $N = 2$, the circuit of Figure 7–10(a) divides by 2^2, or $\underline{\underline{4}}$. The top circuit therefore divides the incoming frequency by four.

Figure 7–9 A divide-by-two circuit

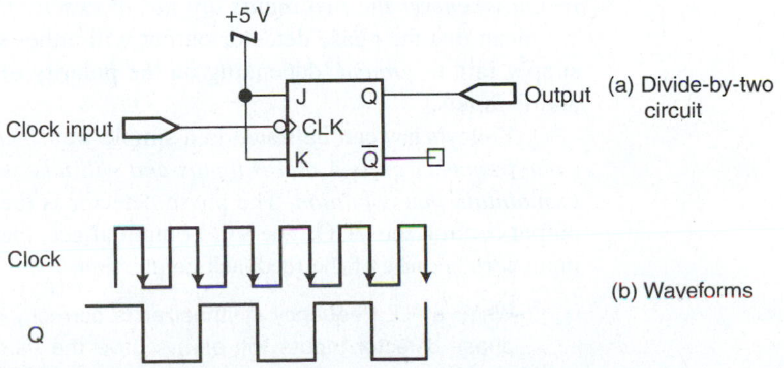

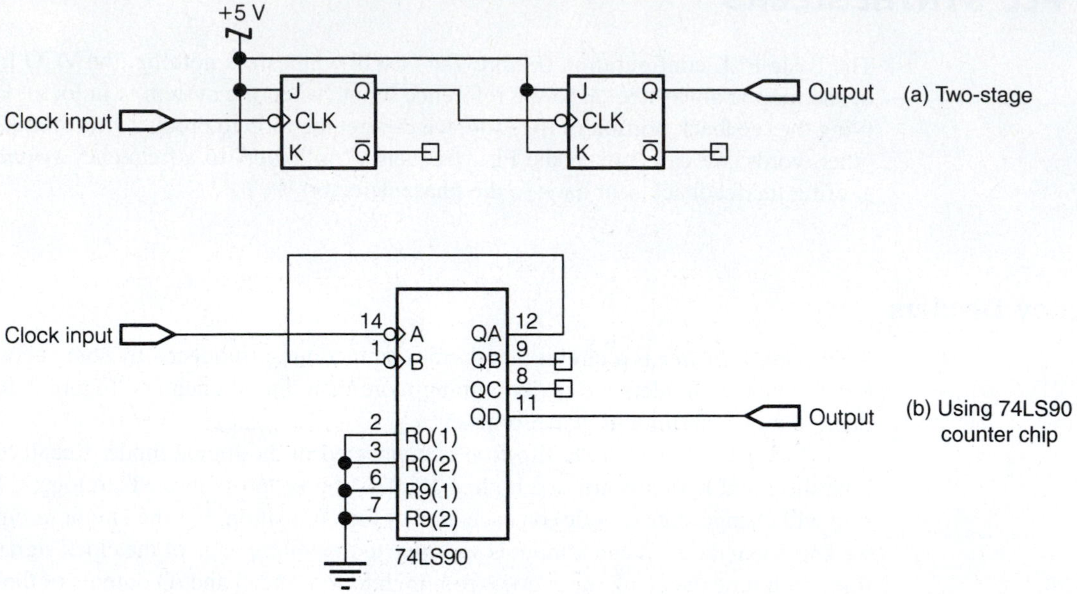

Figure 7–10 Two more divider circuits

The bottom circuit cannot be analyzed without studying the data sheet for the 74LS90. The 74LS90 is wired as a BCD (modul. 10) counter in Figure 7–10(b). This means that there are ten unique counting states, and since the Q_D output is being used as the output, there will be one pulse on Q_D for every ten input pulses on the clock input (A). The circuit divides the input frequency by 10.

Frequency Dividers within a Loop

Figure 7–11 shows a PLL with a frequency divider inserted within the feedback loop. The addition of the divider within the feedback portion of the loop changes the signal that the phase detector (the loop's decision maker) sees.

To understand what will happen in the circuit of Figure 7–11, it is helpful to keep Finley's law in mind. This law states: *The output of the phase detector will be in saturation whenever the two inputs are not at exactly the same frequency.* By "saturation" we mean that the phase detector output will either stay close to the potential of the V_{cc} supply rail, or *ground,* depending on the polarity of the frequency error between the A and B inputs.

Finley's law can be stated in a simple way: *The phase detector "likes" to have the same frequency at its A and B inputs and will take whatever action is necessary in order to maintain that condition.* The phase detector is the decision maker in the loop, and its output controls the VCO. The VCO output affects the frequency that the A phase detector input sees because of the feedback connection.

When a PLL frequency synthesizer is correctly operating (*locked*), the two phase detector inputs will *always* have the same frequency present!

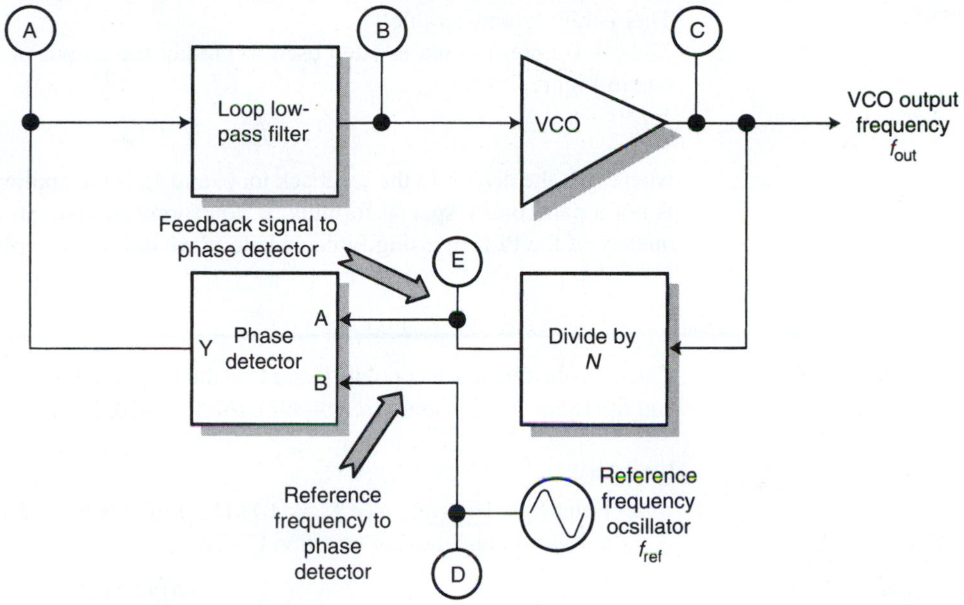

Figure 7–11 A PLL with a divider in the loop

EXAMPLE 7–3

Calculate the frequency at points D, E, and C in the loop of Figure 7–11, given the following information:

$$f_{ref} = 1 \text{ kHz, and } N = 2 \text{ (divisor)}$$

Solution

Test point D is the reference frequency input, so by inspection, this frequency will be 1 kHz.

Test point E is calculated by using Finley's law. The phase detector will not be "satisfied" until both of its inputs are at the same frequency. The frequency present at the *bottom* phase detector input is already known; it is 1 kHz. Therefore, test point E must *also* be 1 kHz, because the phase detector will give the VCO voltage "commands" to make this so.

Test point C looks a little trickier. How can we find the VCO frequency? There is a divider circuit in between point C and point E. We know the frequency at point E is 1 kHz by Finley's law. The frequency at point C can be found by thinking "backwards" about the frequency divider. *If we have a divide-by-two divider and 1 kHz is coming out, what frequency must be coming in?* That's right—the frequency at the divider input must be two times 1 kHz, or 2 kHz.

That's pretty cool! The VCO must be producing a frequency of 2 kHz in order for the divider to put out a frequency of 1 kHz (it is a divide-by-two circuit). The phase detector will not be satisfied until it sees 1 kHz at both of its inputs, and the only way that can happen is for the VCO to make 2 kHz. *The circuit has synthesized a 2 kHz signal from a 1 kHz signal!*

In Example 7–3, the VCO must be designed to be capable of producing a 2 kHz signal. This is a job for the engineer that designs the loop. If the VCO is incapable of producing the desired output frequency—you guessed it, the loop will drop out of lock. This is highly undesirable!

A simple formula is often used to predict the output of a PLL synthesizer like the one in Figure 7–11:

$$f_{out} = N \times f_{ref} \tag{7–2}$$

where N is the divisor in the feedback loop, and f_{ref} is the applied reference frequency. This is not a particularly special formula; if you forget it, you can always find the output frequency of the PLL by using Finley's law, as we did in Example 7–3.

EXAMPLE 7–4

Calculate the frequency at points E and C in the loop of Figure 7–11, given that $f_{ref} = 10$ kHz, and the following divisors: *(a)* $N = 890$, *(b)* $N = 710$, *(c)* $N = 1000$.

Solution

a. Point E = $\underline{\underline{10\text{ kHz}}}$, since $f_{ref} = 10$ kHz (Finley's law). Point C is the output node and is computed using equation (7–2):

$$f_{out} = N \times f_{ref} = (890)(10\text{ kHz}) = \underline{\underline{890\text{ kHz}}}$$

b. Again, point E is still $\underline{\underline{10\text{ kHz}}}$ due to the phase detector's self-correcting action. At point C, we'll get

$$f_{out} = N \times f_{ref} = (710)(10\text{ kHz}) = \underline{\underline{710\text{ kHz}}}$$

c. Yep, point E is *still* $\underline{\underline{10\text{ kHz}}}$. At point C, we now get

$$f_{out} = N \times f_{ref} = (1000)(10\text{ kHz}) = \underline{\underline{1000\text{ kHz}}}$$

Note that three different divisors were necessary in this example. That might suggest that three different divider circuits had to be switched in, but in reality, a special circuit is used that has a *variable* modulus. That circuit is called a *programmable divider*.

Programmable Dividers

From Example 7–4, you can see how easily the output frequency of a PLL can be changed. All that needs to be changed is the N divisor. A *programmable divider* is a special digital counter with a *programmable* or *variable* modulus. By allowing its modulus to be selected by the user, such a counter allows the feedback portion of the PLL to be changed at any time. The result is that the PLL becomes a *digitally controlled oscillator*.

The addition of a programmable divider to the PLL is a powerful enhancement. The PLL can now produce as many frequencies as the number of available divisors, and all of these frequencies will be as rock stable as the reference frequency oscillator. Figure 7–12 shows a programmable divider implemented using a 74LS192 counter chip.

The 74LS192 is a programmable up and down counter with "jam load" capabilities. In the circuit of Figure 7–12, the binary number 0111_2 (7) is present at the A, B, C, and D inputs of the chip. The input clock is connected to the DN (down) counter input. Every rising edge on the clock therefore causes the binary count on Q_A, Q_B, Q_C, and Q_D to decrease by one (decrement).

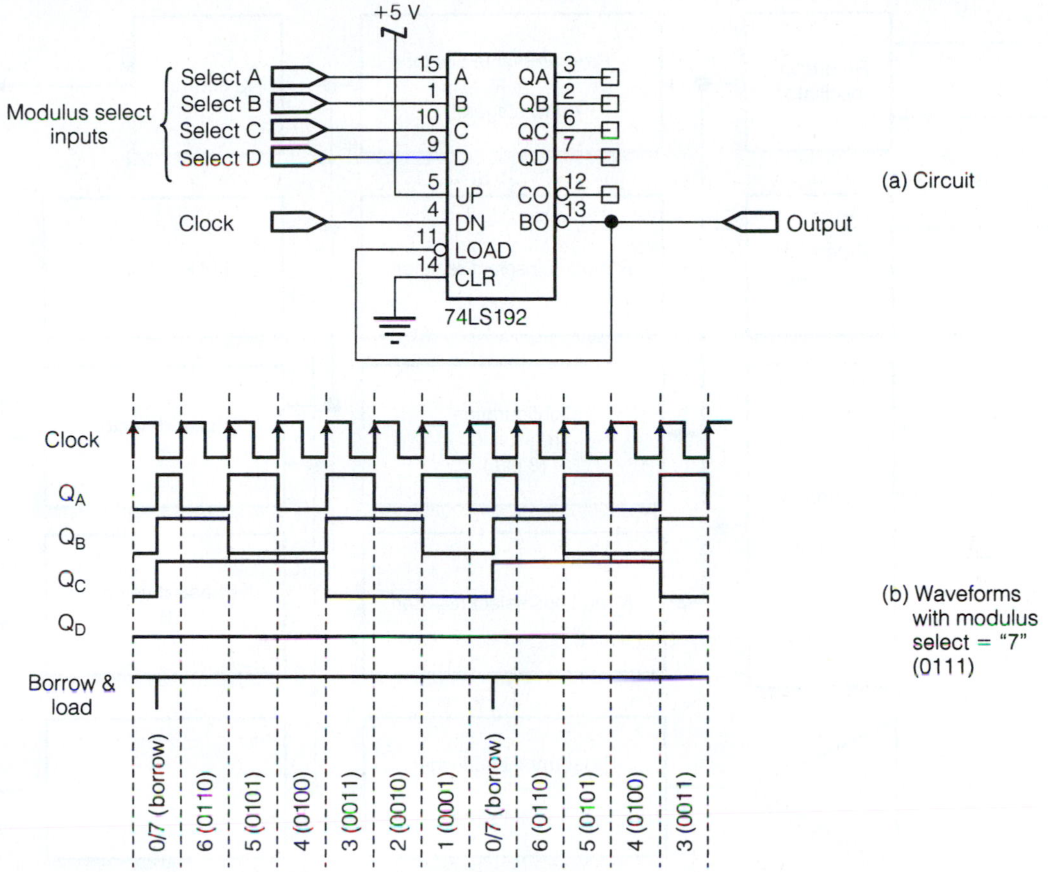

Figure 7–12 A programmable divider using the 74LS192

When the count reaches 0000_2 (zero), the counter cannot count further down without generating a *borrow*. The BO (borrow) pin goes low during the low portion of the clock signal. The BO pin is fed back to the LD (load) input. The result is that when the counter generates a "borrow," it is automatically refreshed with the count value on the ABCD inputs. In this example, the number 7 is reloaded each time the counter borrows.

By carefully studying the timing diagram of Figure 7–12(b), we can see that one *borrow* pulse is generated for every seven *clock* pulses. The circuit divides the clock signal frequency by 7. We could easily change the divisor by changing the binary number at the ABCD inputs. For example, if we load the number 0100_2 at the modulus select inputs, the counter will now divide the input clock frequency by 4.

In other words, the binary number at the ABCD input pins determines the modulus and divisor ratio of the counter. The binary number could come from a user input control (such as a BCD thumbwheel switch) or from the output pin of a microprocessor. When this counter is included in the feedback loop of the PLL, the output frequency of the PLL becomes digitally controlled. We have created a precise, digitally controlled synthesized frequency source that can easily be interfaced with a microprocessor or microcontroller.

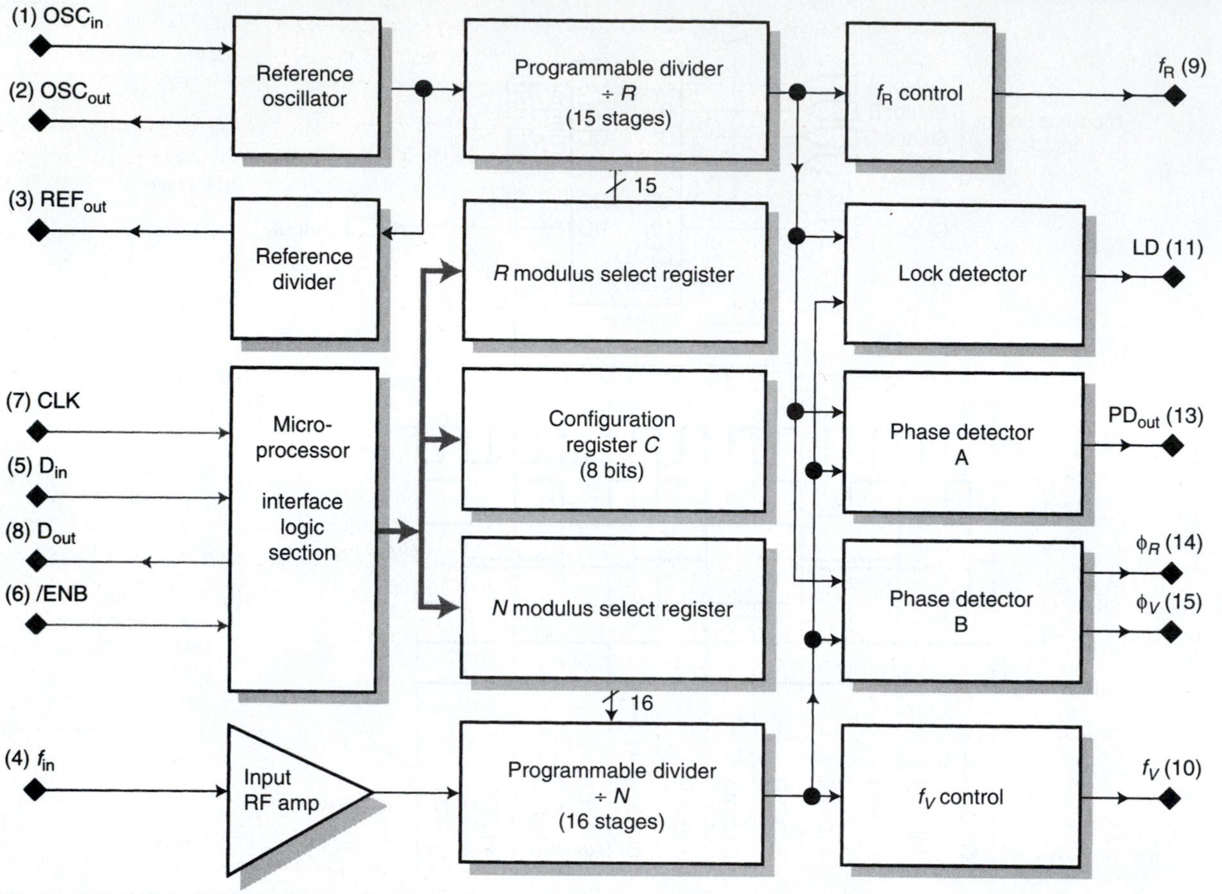

Figure 7–13 Motorola MC145170-1 internal block diagram (Copyright of Motorola, used with permission)

A PLL on a Chip

It's common for most of the elements of a PLL frequency synthesizer to be built on a single IC chip. Most new PLL ICs are designed to be interfaced with a microprocessor or microcontroller (as opposed to DIP or BCD switches). The Motorola MC145170-1 is one such device. It is shown in Figure 7–13.

The MC145170-1 is one packed chip! The following PLL blocks are contained on the IC:

- A reference oscillator. (Pins 1 and 2 are meant to be connected with a crystal to control the reference oscillator frequency.) This oscillator is usually in the 10 MHz region.
- A 15-bit (maximum modulus $= 2^{15} - 1 = 32{,}767$) programmable divider R for the reference oscillator, to divide the oscillator frequency down to the desired reference frequency.
- A 16-bit programmable N divider (maximum modulus $= 2^{16} - 1 = 65{,}535$) complete with an input RF amplifier for conditioning the VCO signal into a digital signal suitable for clocking the divider.

- Two phase detectors, A and B. The primary difference between them is that the *A* phase detector has a single-ended output, while the *B* detector has differential outputs. The B detector has better performance in high-noise environments but requires more external parts to use.
- A *lock detector,* which can be used to determine if the loop has fallen out of lock. The output of the lock detector, LD, is typically read by a microprocessor to check on the status of the PLL.

In fact, there are only two PLL blocks not provided in the IC. These are the *loop filter* and the *VCO*. These parts are added by the designer. This is not a serious shortcoming; in fact, it allows the IC to be much more versatile because including the VCO on chip might restrict the frequency range. The MC145170-1 can operate on frequencies from near dc to over 150 MHz.

A Typical PLL Synthesizer

Figure 7–14 shows a typical PLL frequency synthesizer; such a unit might be implemented with the MC145170-1. When a technician analyzes a synthesizer circuit, he or she should look for some of the basic features shown in Figure 7–14. Most modern PLL synthesizers are set up in this manner.

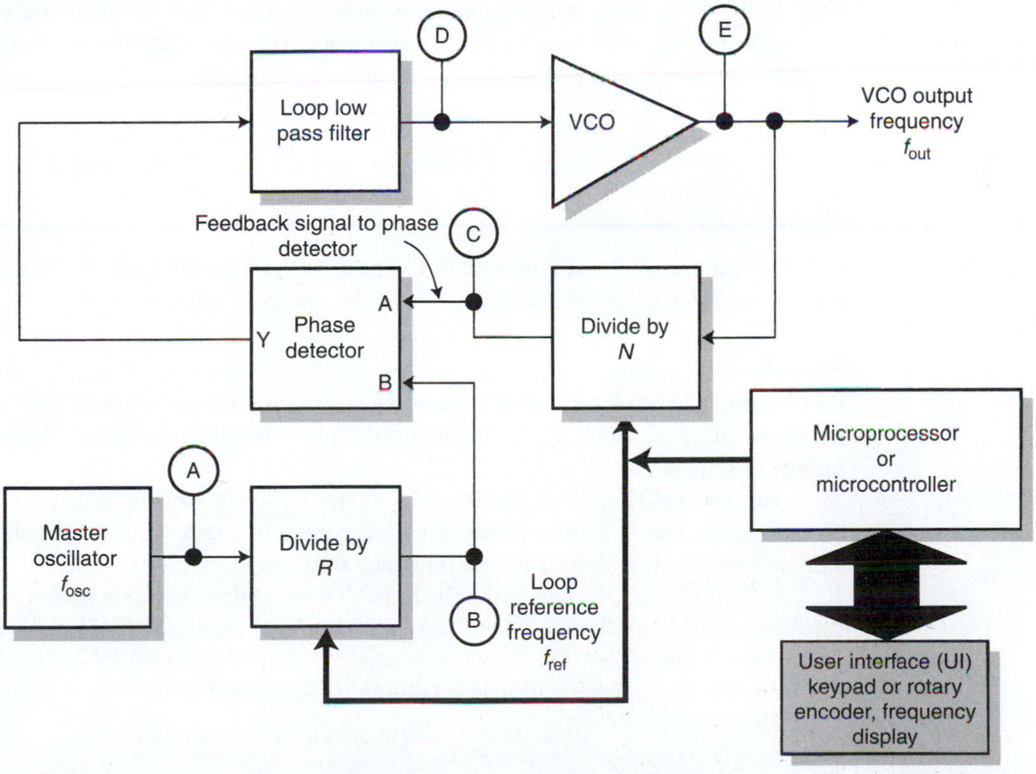

Figure 7–14 A PLL frequency synthesizer in block diagram form

The Need for the Reference Divider

Most PLL synthesizers require a relatively low yet rock-stable reference frequency. Reference frequencies from 1 kHz to 100 kHz are quite common. Unfortunately, crystals become expensive and physically large at such low frequencies.

The standard solution to this problem is to use a high-frequency crystal oscillator, called the *master oscillator,* and divide its frequency down digitally until the desired reference frequency is obtained. This oscillator is usually in the 10 MHz to 20 MHz region. There's nothing magic about those frequencies; it's just easier to build stable crystal oscillators in that range, especially if we utilize crystals in their *overtone* mode.

The *reference* divider's job, then, is to bring the master oscillator's frequency down to the proper value for the PLL reference input. The reference divider is nothing more than another programmable divider.

The Software Is King

In Figure 7–14, the end-user of the product communicates with the microprocessor through a *user interface,* which can take the form of a keypad, switches, rotary controls, and so on. The microprocessor, in turn, issues commands to the various parts of the PLL by sending digital values to each of the PLL registers (holding locations). The PLL shown has two places that must be programmed by the software, the $\div R$ and $\div N$ registers. The software (or firmware, software in ROM) controls all of these actions.

In the next section, we'll take a look at what the software is *really* doing when it talks to the PLL, from a troubleshooting point of view. The involvement of a microprocessor and its software can really make your life interesting when trying to locate problems in a PLL subsystem because it can be hard to tell if a failure is due to software or hardware!

EXAMPLE 7–5

Calculate the frequency at points A, B, C, and E in the loop of Figure 7–14, given that the master oscillator $f_{osc} = 10.240$ MHz, $R = 1024$, and $N = 98$.

Solution

By inspection, test point A should be the master oscillator frequency, or 10.240 MHz. When troubleshooting a "dead" PLL, this is a good place to check (after checking the power supply, of course!).

Test point B should be 10.240 MHz/1024, or 10 kHz. This is the loop reference frequency; again, this is another place that is very good to check when troubleshooting. The loop will not work without a proper reference frequency!

If the loop is operating correctly, Finley's law applies, and test point C should read the same as test point B, or 10 kHz. *If this point does not read exactly the same as test point B, the loop is out of lock!*

Finally, test point E will read according to equation 7–2:

$$f_{out} = N \times f_{ref} = (98)(10\text{ kHz}) = \underline{\underline{980\text{ kHz}}}$$

Therefore, the loop output frequency will be 980 kHz.

Section Checkpoint

7–13 What is added to the basic PLL block diagram of Figure 7–1 to get a frequency synthesizer?

7–14 How does a *frequency divider* work?

7–15 What is the relationship between the input and output frequencies of a divider and its modulus?

7–16 When a PLL synthesizer is correctly functioning, what two circuit test points will always have the same frequency?

7–17 What is a programmable divider? How is it used in a PLL circuit?

7–18 In Figure 7–14, why is a reference divider needed?

7–19 If the microprocessor's software fails, what happens to the loop of Figure 7–14?

7–20 List at least three test points that would be useful for troubleshooting the PLL of Figure 7–14.

7–3 HOW SOFTWARE CONTROLS A PLL SYNTHESIZER

To effectively troubleshoot modern synthesizers, a technician needs to know a little about *what* the microprocessor software does to communicate with the PLL chip. The microprocessor "talks" to the PLL chip by sending it data and commands. The PLL chip talks back to the CPU to let it know whether or not commands were successfully executed. The PLL also informs the CPU about loop status. An example of status is whether or not the loop is in lock. If the PLL reports "in-lock" then it's producing the frequency requested by the microprocessor.

There are two types of information that the CPU sends to the PLL. These can be classified as *initialization data* and *command data.*

Initialization Data

When a digital circuit is first powered up, its state (condition) is generally unknown. Most digital circuits must include a *power-on reset* circuit to put them into a known starting condition. With microprocessor circuits, an RC circuit is often used to provide the power-on reset timing delay.

When a device such as PLL is powered up and reset, it has "forgotten" everything that last happened before it was turned off. You can think of it as being "born again." It needs to learn everything anew.

When the microprocessor is reset, it normally *initializes* each device connected to it. To initialize the PLL, the CPU will send a short burst of commands to set up the proper operating conditions for the loop. If the PLL doesn't get this initialization information, it sits there and does nothing! Without the right setup data, further commands from the microprocessor might as well be gibberish, since the starting assumptions are all wrong.

A tech needs to remember that the *initialization* is normally accomplished only on power-up or reset and that failure to initialize usually means a dead PLL. *Forcing the microprocessor to reset (there is usually either a switch or pair of test pads for shorting) is how to force the CPU to send new initialization data.*

Command Data

The CPU sends *command data* whenever a change is required in the operation of the PLL. A very common example of this is when the end user of the equipment decides to change the channel or frequency of operation.

Suppose that a PLL is being used as the carrier frequency source in a transmitter with a rotary frequency knob. When the user turns the knob to change the frequency, the knob sends digital pulses to the microprocessor. Usually, there's one pulse for each frequency increment. The microprocessor's software reads the pulses, interprets them as a request for frequency change, calculates the new operating frequency, then calculates the correct PLL commands to attain that frequency. After doing all this, the software sends the appropriate *command data* to the PLL. The transmission of this data is very rapid and occurs in a short burst.

In response, the PLL makes the frequency change. When a PLL is given a command to change frequency, it always drops out of lock until its VCO has reached the new frequency. The CPU (again under software control) checks back after a short time to see that the PLL accepted the change and remained in lock. The software then looks for input from the user again.

In some equipment, failure of the PLL to relock results in a unique diagnostic output. For example, late-model Kenwood/Trio transceivers may light all the decimal points on the display to indicate that a PLL has fallen out of lock. The service literature generally describes what, if any, diagnostic output is available.

How is this useful to a technician? Remember that command data is usually sent in response to input from the controls of the radio equipment. Turning a frequency dial or pushing a frequency up/down button will force the CPU to send command data to the PLL. *Therefore, a tech can force the PLL data signals to appear by merely operating the controls of the equipment.*

Data Transfer Formats

Most modern equipment uses a special type of data transfer called *serial communication* to talk to frequency synthesizers. In serial communications, digital data is sent down a single wire, one bit at a time. The advantage of this is that only one wire is needed instead of 8 or 16 (parallel communication). This type of communication, called *data communication,* will be the subject of a later chapter. To synchronize the flow of information down a wire, a second wire carrying a *clock* signal is sometimes used.

The MC145170-1 PLL of Figure 7–13 is a good example of a serially controlled frequency synthesizer chip. The serial data flows into the chip on pin 5, and the synchronizing clock is applied on pin 7. The chip is also supplied with an *enable* signal, /ENB, on pin 6. The enable signal is a "master key" that causes the PLL chip to pay attention to the data signals (ENB active) or ignore them (ENB inactive). Figure 7–15 shows the MC145170-1 interfaced to an 8-bit microcontroller chip, the 68HC05K1.

In Figure 7–15, there are four key signals for CPU–PLL communication. These are *PLL serial data, PLL enable, PLL serial clock,* and *PLL lock status*. To check on whether communication is taking place or not, the technician would normally examine both the *serial clock* and *serial data* signals with an oscilloscope, either during power up (initialization data) or frequency change (command data).

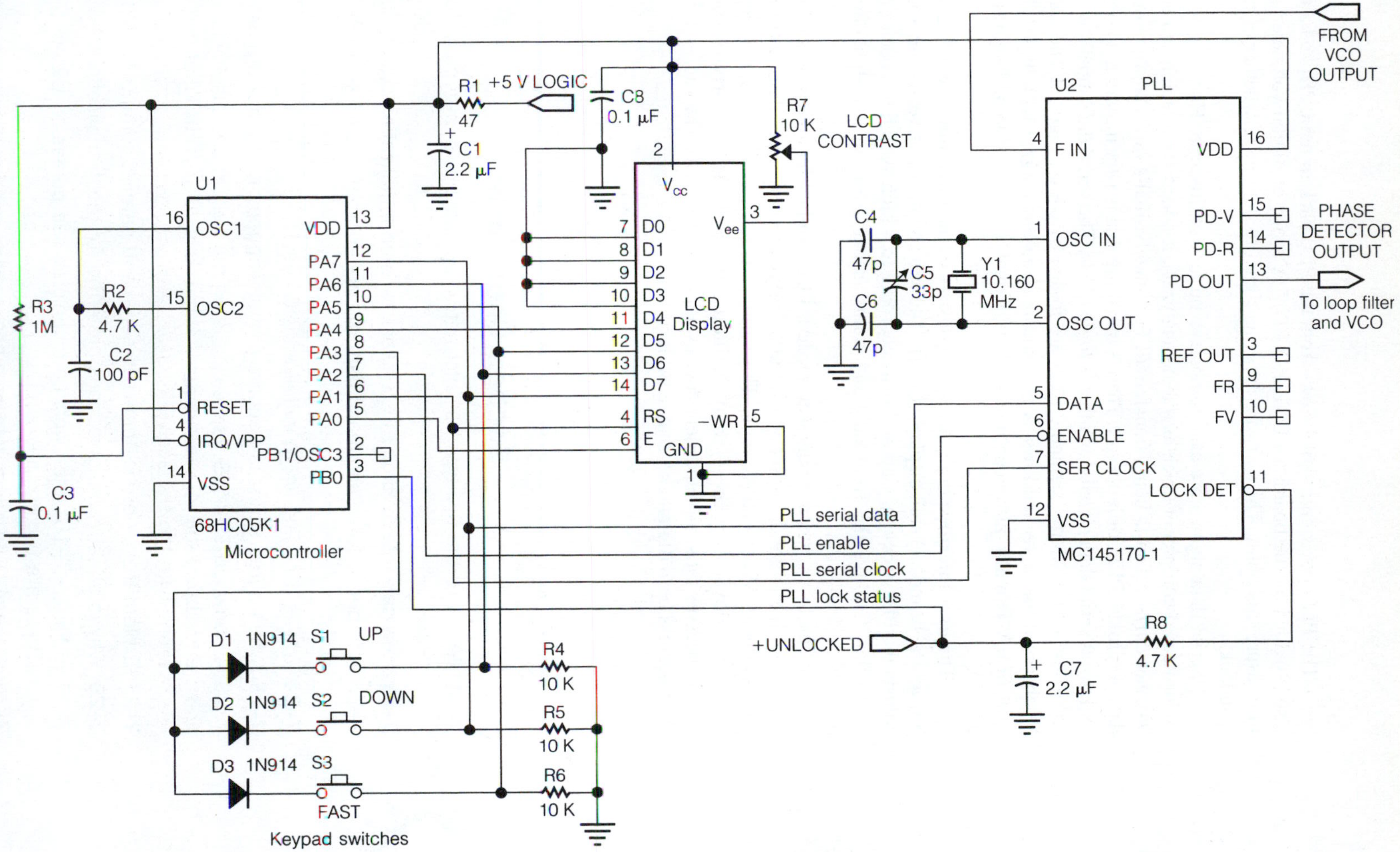

Figure 7–15 Interfacing the MC145170-1 to an 68HC05K1 microcontroller

The *PLL lock status* signal flows from the PLL back to the CPU, and a test point is also provided on the board "(+UNLOCKED)." A scope or digital voltmeter can be used to measure this point. The software uses this signal to determine whether or not the PLL is in lock.

Serial data signals are easy to observe on an oscilloscope. However, without more sophisticated equipment (such as a logic analyzer), it's hard for a technician to determine whether *correct* data is being transferred! Fortunately, that's rarely a problem. Either the data gets to the PLL, or it doesn't. There's one thing that is *very* important to observe when looking at serial data signals, and that is *logic levels*. The serial data signals will be square waves by nature. Many beginning techs are fooled by square signals! The *voltages* must be observed for highs and lows. A nice square serial data signal with highs of 0.5 V and lows of 0 V is really just a continuous "low" signal, since most logic families read 0.5 V as a "0."

The *user interface* in Figure 7–15 consists of the three input switches S1, S2, and S3, and a liquid-crystal display (LCD). The microcontroller reads the switches and adjusts the PLL frequency up or down according to which switch has been pressed. The controller displays the frequency and status information on the alphanumeric LCD display.

In summary, when it comes to modern PLL synthesizers, there is a mix of analog and digital hardware doing the work, but a fairly complex piece of software in a microprocessor's ROM calls all the shots. This software is responsible for initializing and maintaining all devices in the system, including the PLL. If the software fails, or the microprocessor fails, all dependent devices appear inoperative. The successful technician must possess both digital and analog skills to work with these circuits.

Section Checkpoint

7–21 What are the two types of information software transmits to a PLL chip?

7–22 How can a tech force the software to send initialization data?

7–23 In order to force command data to appear, what should a technician do?

7–24 Why is it necessary to initialize digital devices?

7–25 When a PLL changes frequency, what happens to its lock status?

7–26 A Kenwood TS-440S transceiver is on the bench. The display lights up and shows the following output: 1.4.1.0.0.0.0. MHz. What might be wrong with the transceiver?

7–27 Explain the difference between serial and parallel communications.

7–28 Why is serial communications preferred for most applications?

7–29 In Figure 7–15, when would a technician see pulses on U2 pin 5? (There are two conditions.)

7–30 When viewing serial data on a scope, what aspect should a technician pay attention to?

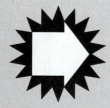

7–4 DDS SYNTHESIZERS

The PLL is an extremely popular frequency synthesizer but is a moderately complex building block. It contains analog VCO and filtering circuits. The analog circuits in a PLL can't make it drift because it is digitally locked on frequency, but they can degrade its performance in other ways as the analog components age. A PLL also cannot change frequency very rapidly; whenever a PLL changes frequency, it drops out of lock momentarily until it stabilizes on the new frequency. A *direct digital synthesis,* or *DDS* synthesizer, is an almost purely digital solution to the problem of frequency synthesis. As computer chips have increased in power, DDS has become a feasible method of frequency synthesis.

A DDS synthesizer has only one analog component, a digital-to-analog converter. Everything else is digital. There is no loop filter as in a PLL, so a DDS synthesizer can change frequency almost instantaneously. This is important in applications where frequency must be rapidly changed, such as encrypted communications. Figure 7–16 shows the basic layout inside the Harris semiconductor HSP45102, a DDS chip.

A DDS chip works by keeping track of the instantaneous *phase angle* of the sine wave being synthesized. Suppose that you were told to graph a sine wave in "real time" by plotting points 45° apart every one eighth of a second (125 ms per point). Suppose that the amplitude of your sine wave was to be 1 V (not particularly important, but nice to know). What would the frequency of the sine wave be? The resulting sine wave might look like Figure 7–17.

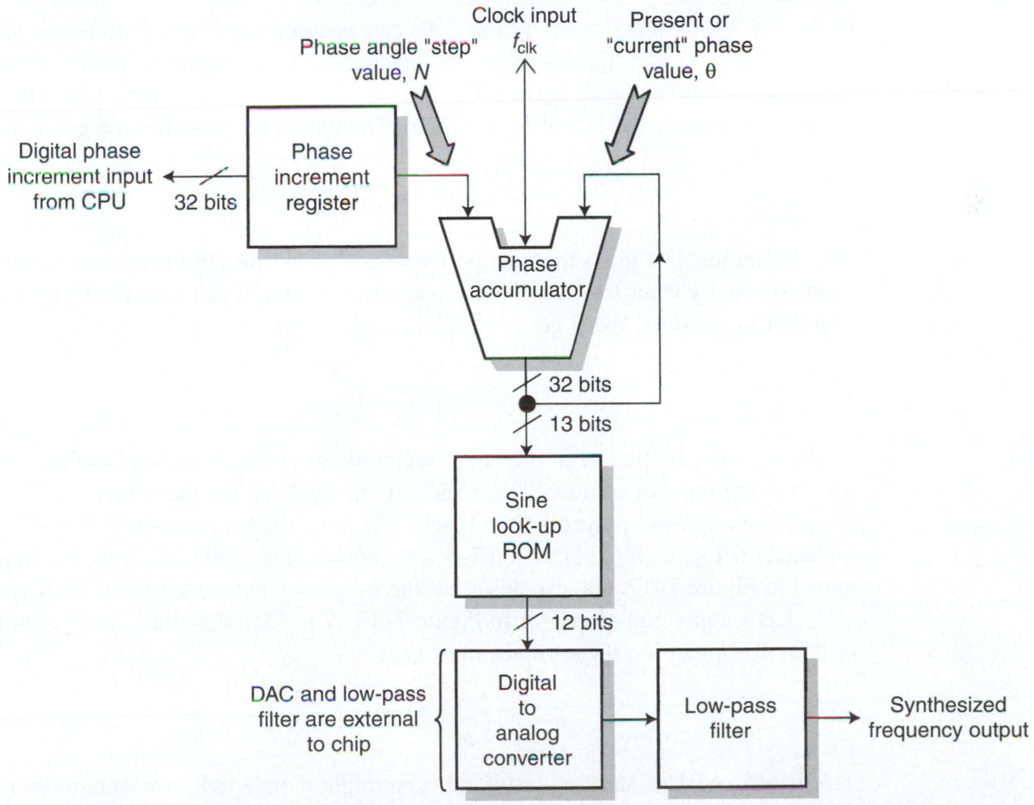

Figure 7–16 Simplified view of the Harris HSP45102 direct digital synthesizer

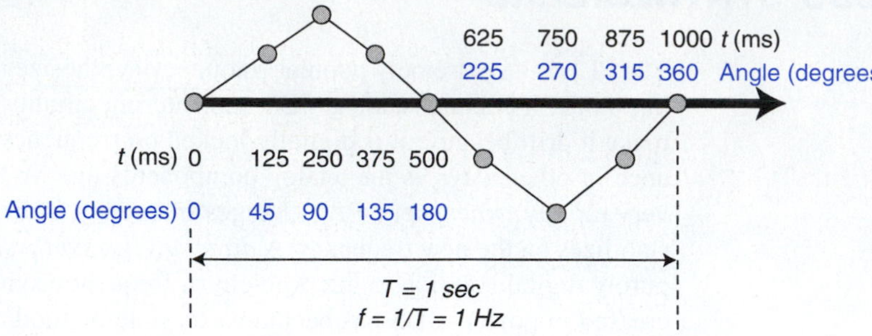

Figure 7–17 A "hand-drawn" sine wave

The points in the hand-drawn sine wave are 45° apart. This is the *phase increment.* The phase angle started at 0°, jumped to 45°, 90°, and so on. Because of the phase "jumps," the sine wave is not perfect. By using a little arithmetic, we can determine the resulting frequency. First, recall that

$$f = \frac{1}{T} \tag{7–3}$$

where f is frequency, and T is the time period of the waveform. If we can determine the time period T of the waveform of Figure 7–17, we can find its frequency.

By definition, the period T is one cycle, or 360° of the waveform. We will let the variable M be the number of degrees in a cycle; here it will be 360. We know that jumps of 45° are taking place every 125 ms. We can assign the variable N to be the *phase increment* and let it be 45°. Furthermore, the time period between points (phase increments) can be thought of as the *clock period,* T_{clk}. This is exactly the same clock signal that feeds the phase accumulator (adder) of Figure 7–16. Therefore, the time for one cycle will be

$$T = T_{clk}\left(\frac{M}{N}\right) \tag{7–4}$$

Equation 7–4 gives the time period of our waveform, but that's not exactly what we want. We really want *frequency*. That's easy to get—we'll just take the reciprocal of equation 7–4 to get *time*. We'll get

$$f = f_{clk}\left(\frac{N}{M}\right) \tag{7–5}$$

where f_{clk} is the frequency of the phase accumulator clock, N is the phase increment value, and M is the number of units ("degrees") in one cycle of the waveform.

Notice that we played a little "trick" when we turned equation 7–4 upside down; we replaced $(1/T_{clk})$ with f_{clk}. Equation 7–5 is the equation needed to describe the frequency obtained in Figure 7–17 and also describes the frequency obtained from a DDS synthesizer.

Let's apply equation 7–5: In Figure 7–17, N is 45°, M is 360°, and f_{clk} is (1/125 ms), or 8 Hz. By plugging these values in, we get

$$f = f_{clk}\left(\frac{N}{M}\right) = 8\text{ Hz}\left(\frac{45}{360}\right) = \underline{\underline{1\text{ Hz}}}$$

This is not exactly a startling result, and you might wonder why we went to such trouble to get it. In order to better understand this, let's take a look at the action of Figure 7–16.

DDS Theory of Operation

The heart of a DDS chip is a digital adder circuit called the *phase accumulator*. The phase accumulator adds two binary numbers. One of these numbers comes from the *phase increment register* (the CPU will store a number N in the phase increment register), and the other number comes from the *output* of the phase accumulator. (Positive feedback! Just what we need to build an oscillator!)

The phase accumulator is supplied with a *clock* signal. This clock signal is fixed in frequency. Every time the clock pulses, the phase accumulator *adds* the phase increment value to its own contents, then replaces its own contents with the resulting sum. If we used a phase increment of 45°, the phase accumulator contents would look like this:

Phase Accumulator Contents Versus Clock

Clock number	Phase accumulator contents
0 (no clock yet)	0°
1	45°
2	90°
3	135°
4	180°
5	215°
6	270°
7	315°
8	0° (wrap-around)
9	45°

Notice how the phase accumulator "wraps around" back to zero when the sum equals or exceeds 360°.

The phase accumulator keeps track of "where we're at" on the sine wave being produced. It is just an angle counter. The next portion of the DDS chip, the *sine look-up ROM,* converts the angles into the binary equivalents (numbers) representing the voltages on the various parts of a sine wave. The sine look-up ROM does the same function as the *"SINE"* key on a scientific calculator; it converts the given angle from the phase accumulator into a digital value representing the sine of that angle.

The digital sine wave value is finally converted into analog by a digital-to-analog converter, or DAC. This is not part of the 45102 DDS chip; it must be supplied separately by the designer. The sine wave produced by the DAC is not perfect; a low-pass filter is used to smooth the waveform. The result is a digitally synthesized sine wave.

The frequency of the synthetic sine wave from a DDS chip is controlled by changing the value of N, the phase increment. Look at the output of the *phase accumulator* in the HSP45102 DDS block diagram of Figure 7–16. The phase accumulator output is 32 bits wide. This suggests that the phase accumulator "counter" has a modulus of 2^{32} and that one complete sine wave cycle (360°) is represented by 2^{32} digital counts. In other words, the phase accumulator is really just a 32-stage (bit) binary adder circuit. *For the HSP45102, because the accumulator is 32 bits wide, the value of* M *is always* 2^{32}.

For digital circuits, it's much easier to use the system of binary numbers rather than try to represent a number like 360°. This is why equation 7–5 is useful; equation 7–5 doesn't care what the units of N and M are!

EXAMPLE 7–6

What output frequency will be produced by a Harris HSP45102 DDS chip under the following conditions:

$$f_{\text{clk}} = 30\text{ MHz},\ N = 2{,}025{,}792{,}907$$

Solution

For the 45102 DDS, we know that $M = 2^{32}$, since the accumulator is 32 bits wide. Equation 7–5 comes to the rescue:

$$f = f_{\text{clk}}\left(\frac{N}{M}\right) = 30\text{ MHz}\left(\frac{2{,}025{,}792{,}908}{2^{32}}\right) = \underline{\underline{14.150000\text{ MHz}}}$$

The output frequency is therefore $\underline{\underline{14.150000\text{ MHz}}}$.

When software talks to a DDS chip, it must calculate the number N needed to produce a particular frequency. The software uses equation 7–5, solved for N instead of f. Notice how some fairly large numbers are creeping into the calculations. These numbers are uncomfortable for calculation by hand but pose no real problem for a microcontroller.

EXAMPLE 7–7

What value of N is required to operate a Harris HSP45102 DDS at an output frequency of 2.182000 MHz, given that $f_{\text{clk}} = 50$ MHz?

Solution

Equation 7–5 is manipulated to solve for N:

$$f = f_{\text{clk}}\left(\frac{N}{M}\right)$$

$$N = f\left(\frac{M}{f_{\text{clk}}}\right) = 2.182\text{ MHz}\left(\frac{2^{32}}{50\text{ MHz}}\right) = \underline{\underline{187{,}432{,}372.797}}$$

We run into a problem here: The digital phase increment N must be an integer. To program the N counter, we just ignore the fraction "0.797." Therefore, $\underline{\underline{N = 187{,}432{,}372}}$.

We can double-check the result by simply plugging the computed value of N back into equation 7–5. We get

$$f = f_{\text{clk}}\left(\frac{N}{M}\right) = 50\text{ MHz}\left(\frac{187{,}432{,}372}{2^{32}}\right) = \underline{\underline{2.18199999072\text{ MHz}}}$$

This is very close to 2.182000 MHz—a frequency error of only 0.009 Hz resulted!

A Comparison of PLL and DDS Technologies

The development of DDS chips has hardly made the PLL obsolete. Both methods of frequency synthesis have complementary features. The PLL can be directly used at almost any radio frequency; at UHF frequencies and above, a *prescaler* is often used between the VCO and programmable divider in a PLL. A PLL changes frequency rather slowly; 10 ms to 100 ms are typical settling times for PLL synthesizers. Also, it is difficult to design a PLL with high frequency resolution.

The *frequency resolution* of a synthesizer is the smallest frequency change or "step" that it can produce. For a PLL, the resolution is always equal to the reference frequency. For example, the PLL of Example 7–5 has a 10 kHz reference, so the smallest frequency change at its output will be 10 kHz. When $N = 98$ in Example 7–5, $f_{out} = 980$ kHz. If N is moved up to the next available value, 99, f_{out} will increase to 990 kHz (a 10 kHz difference). *It is impossible for the loop to produce 981 kHz, 985 kHz, or anything in between, because the frequency step is 10 kHz.*

In a PLL, reducing the reference frequency (in order to increase resolution) creates three problems. First, the filter time constant must be larger, so the settling time between frequency changes becomes larger. Second, it becomes more difficult for the loop filter to produce a nice smooth voltage for the VCO, so the output signal from the loop isn't as clean. The level of spurious energy from the VCO increases. Last, the programmable divider counter must be built with more stages, since the divisors will have to be larger.

A DDS chip has very high frequency resolution. The frequency steps are often much less than 1 Hz for most DDS setups, which is much more than adequate for RF applications. Since DDS doesn't depend on a reference frequency and loop filter, it changes frequency almost instantly. However, since the DDS chip must produce the sine wave digitally, and in real time, the maximum frequency obtainable directly from the chip is limited to 30 MHz or less. A DDS chip can't directly produce VHF or UHF signals like a PLL can. This would seem to be a serious limitation, but it's not.

In modern frequency synthesizers that operate at VHF and require high-resolution (tuning steps less than 100 Hz), PLL and DDS techniques are often combined. The DDS section provides high-resolution ("fine") tuning steps, while the PLL section supplies the HF, VHF, or UHF coarse frequency steps. The PLL is still used in almost all other applications, especially when tuning steps of 1 kHz or larger are required.

Section Checkpoint

7–31 What components in a PLL are subject to aging?

7–32 How many analog components are in a DDS synthesizer? What are they?

7–33 Explain the two inputs of the *phase accumulator* in Figure 7–16.

7–34 How is the digital count from the phase accumulator converted into a sine wave?

7–35 Which technology changes frequency quicker, PLL or DDS?

7–36 Which technology has higher frequency resolution?

7–37 What is the approximate upper limit for direct use of DDS?

7–5 A COMPLETE FREQUENCY SYNTHESIZER

Figure 7–18 is a block diagram of a PLL synthesizer designed to cover the 10-meter amateur radio band, 28.000–29.999 MHz, in 1 kHz steps. This unit uses the Motorola MC145170-1 single-chip PLL with a few added support components to form a complete frequency synthesizer (Figure 7–19).

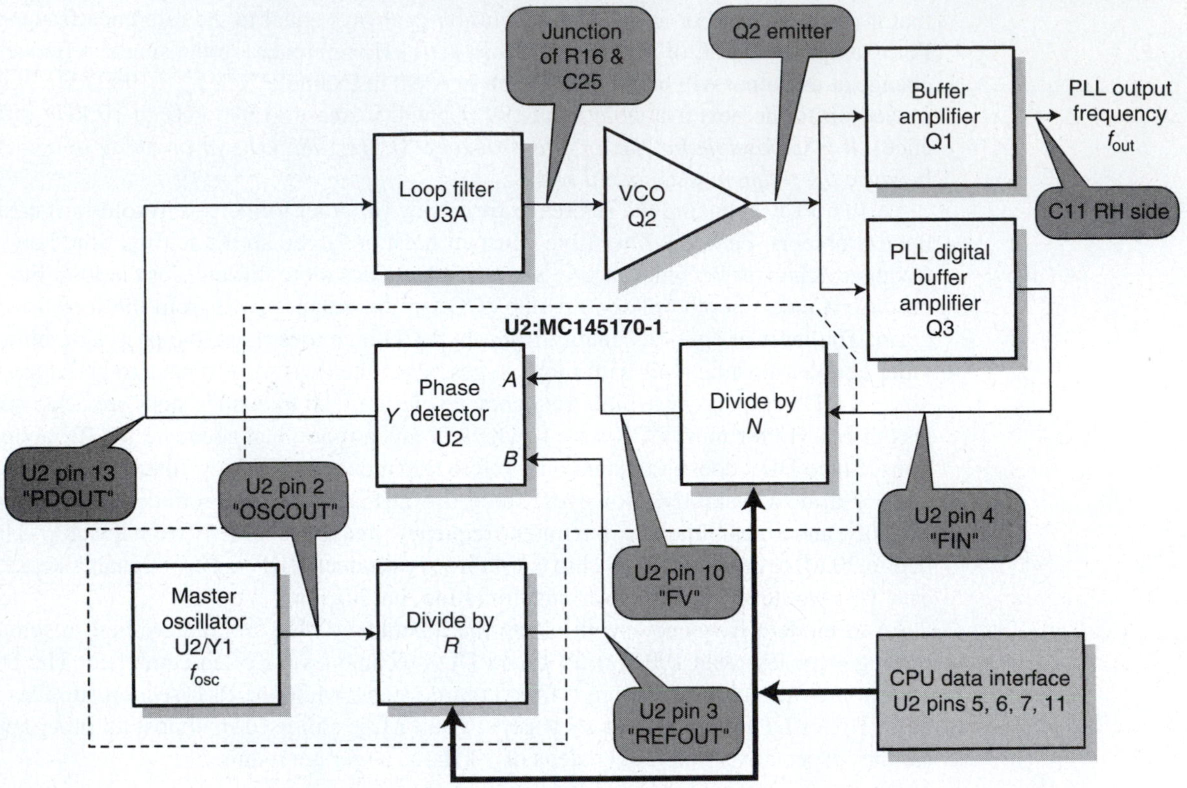

Figure 7–18 A 10 meter synthesizer

Power Supply

The synthesizer operates from a filtered 12 V dc power supply. Since the PLL, U_2, requires regulated 5 V dc instead of 12 V, a 5 V regulator, U_1, is provided. U_1, a 7805L, is a standard three-terminal voltage regulator. The "L" suffix indicates that U_1 is a *low-power* version, and in fact, it is contained in a TO-92 plastic transistor package. The PLL only draws a few mA of current from the 5 volt supply. When troubleshooting this circuit, a technician should remember to check *both* the +12 V and +5 V power supplies.

Master Oscillator, Programmable Dividers, Phase Detector

U_2, the PLL IC, contains all of these circuits. Crystal Y_1 determines the master oscillator frequency of 10.160 MHz. Variable capacitor C_{22} is included to allow fine adjustment of the master oscillator's frequency for initial calibration. Inside U_2, the master oscillator frequency is divided down by R to produce the reference frequency of 1 kHz (the frequency step value). The reference frequency is available for troubleshooting on pin 3 of U_2. U_2 also contains the N programmable divider and phase detector; the frequency input from the VCO is applied to the N divider circuit on U_2 pin 4, and the phase detector's output appears on U_2 pin 13. The out-of-lock (+UNLOCK) signal is also generated inside U_2, and appears on pin 11.

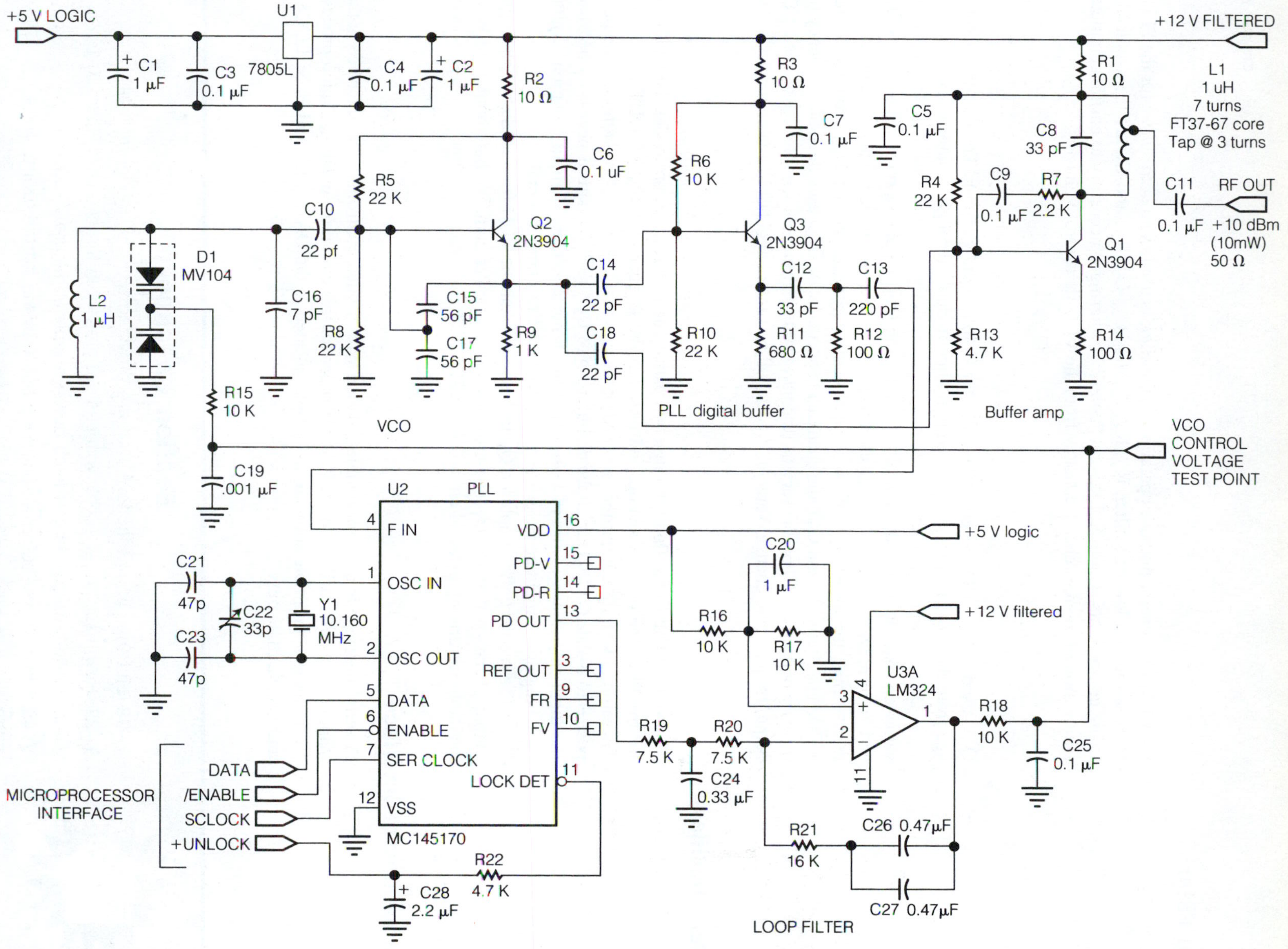

Figure 7–19 Detailed schematic of the 10 meter frequency synthesizer

Loop Filter

U_3, a LM324 general-purpose operational amplifier, is used as an *active filter* for the PLL. U_3 is set up as a low-pass filter. R_{19}, R_{20}, C_{24}, R_{21}, C_{26}, C_{27}, R_{18}, and C_{25} are the actual filtering components. The use of several "poles" of filtering is necessary to ensure complete elimination of the reference frequency from the VCO dc control voltage.

VCO

Q_2 is used as the VCO in a common-collector Colpitts configuration. The dc control voltage from the loop filter (U_3) is applied to tuning diode D_1 through isolator resistor R_{15}.

D_1 is a special diode called a *varactor* that is designed to operate as a dc-controlled variable capacitance. The dc control voltage for the loop supplies a reverse bias across D_1, and, in turn, D_1's capacitance changes in response to the varying dc voltage. The higher the reverse bias, the lower D_1's effective capacitance becomes. Because D_1 is coupled to the VCO tank circuit (L_1 and C_{16}), the changing capacitance of D_1 causes the tank frequency to increase or decrease. Thus, the varying dc control voltage from the loop filter causes the VCO output frequency to change. Varactor diodes will be studied in detail in a later chapter.

Buffer Amplifiers

The VCO output signal flows through two paths: one is digital, and provides the feedback for the PLL; the other is analog, and provides the frequency at the RF OUT connector. Therefore, two buffer amplifiers are used in the circuit. Q_3 is the *digital buffer,* and its purpose is to isolate the VCO from the digital input of U_2 (the PLL). This is necessary because the input impedance of the PLL digital input is not constant. Without isolation, digital switching noise might be superimposed onto the VCO's output signal.

The second buffer amplifier, Q_1, is there to isolate the VCO from load changes at the output of the synthesizer. Q_1 also provides a small power gain, so that about 10 mW is available for driving external circuits at the RF OUT connector.

Microprocessor Interface

A serial interface is used for communication with a microcontroller. The interface is the same as that described in Figure 7–15. In order for this circuit to function, a microprocessor running appropriate software must be connected to the DATA, /ENABLE, SCLOCK, and +UNLOCK pins.

Section Checkpoint

7–38 List the circuit functions that are provided by IC U_2 in Figure 7–18.

7–39 How many power supply voltages are present in the circuit of Figure 7–19?

7–40 Show where to check each power supply on Figure 7–19.

7–41 Why are two buffer amplifiers present in the synthesizer circuit?

7–42 What is adjusted by variable capacitor C_{22}?

7–43 Describe the function of diode D_1.

7–6 TROUBLESHOOTING FREQUENCY SYNTHESIZERS

Frequency synthesizers can be one of the most challenging troubleshooting jobs for a technician; they contain a mix of RF, audio, and digital signals that are not often found together in electronics. Fortunately, *all* synthesizers can be effectively troubleshot by using the three-step method we've used previously. The third step will be expanded upon in this section. The three steps we've used in past chapters include

1. visual (and other) inspection
2. power supply checks
3. input and output checks

When It's Not Working . . .

The chief symptom of a synthesizer that is not working correctly is an incorrect frequency output. What frequency is correct? Whatever frequency the synthesizer has been commanded to produce. In a transmitter, this may be the final transmit frequency, or the final transmit frequency minus some frequency offset. In a receiver, the synthesizer operates as the local oscillator, and thus it should be operating at whatever frequency the LO is supposed to be at. Anything that can cause the loop to fall out of lock (which includes just about everything in the loop!) can cause incorrect frequency output.

The other common synthesizer fault is *no RF output.* A dead synthesizer doesn't even produce the wrong frequency! Since the VCO circuit is the source of the synthesizer's frequency output, this is always a logical place to start when checking inputs and outputs. (It is assumed that the power supplies will always be confirmed working before proceeding to I/O checks.)

Manufacturer's service data is extremely valuable for troubleshooting. Often, they will recommend front panel control settings (such as frequency and mode) so that the factory circuit measurements can be duplicated. However, an understanding of the theory behind synthesizers is invaluable for troubleshooting. A technician who knows the "why" behind a circuit's operation is much more likely to successfully think through and understand a solution than one who merely "looks for what's in the book."

Typical Test Points in a PLL

Figure 7–18 shows the most useful measurement points in the 28-MHz PLL synthesizer of Figure 7–19. Since failure of nearly any part of the loop can cause it to unlock, the test points include all the major components in the loop. Let's try a few scenarios to see how to ferret out common problems.

EXAMPLE 7–8

The PLL synthesizer of Figures 7–18 and 7–19 has been set to 28.150 MHz by the user, yet the output frequency is "drifting" around 27.500 MHz. Develop a plan for troubleshooting the device.

Solution

This appears to be an *out-of-lock* problem. The steps of the solution might look like this: Perform a visual check of the circuit, and check the +12 and +5 V supplies with a DMM. Check for output at the *master oscillator* (U_2 pin 2); this signal should be 10.160000 MHz (crystal controlled). If there's no oscillation, the loop won't be able to develop a reference frequency, and thus there will be nothing for the phase detector to lock on to. *Failure to oscillate could be caused by a bad PLL chip (U_2), bad crystal (Y_1), or defective capacitor (C_{21}, C_{23}, or C_{22}).*

Check for proper reference frequency at U_2 pin 3, *REF OUT*. Since this synthesizer is designed to generate 1 kHz frequency steps (its resolution), *REF OUT* should be exactly 1 kHz. *Note:* Since *REF OUT* comes from a digital frequency divider, it is likely to be a very low duty-cycle signal. It will appear as a series of narrow pulses at a frequency of 1 kHz. If the *REF OUT* signal is missing, then we know the problem is likely to be in the divide by an *R* reference divider circuit.

Recall that the MC145170-1 PLL is controlled by a microprocessor. *Failure to produce a REF OUT signal could be caused by a bad PLL chip, or it could be failure in communication between the CPU and PLL.* Therefore, if *REF OUT* is malfunctioning, it would be a good idea to check the *CPU data interface signals* (*DATA,* U_2 pin 5, and SER CLOCK, U_2 pin 7). Recall that activity on these signals is *not* likely unless the CPU is being reset (initialization data is being sent) or a frequency change is being entered on the front panel (command data is being sent). If *REF OUT* is OK, then the CPU is very likely communicating with the PLL quite well, and the problem is somewhere else. The *FIN* pin of the PLL (U_2 pin 4) could be checked next for a VCO signal. *If there's no signal at FIN, then the trouble is either in the VCO (Q_2) or the PLL digital buffer amplifier (Q_3).* If the *FIN* signal is acceptable, then we know that the VCO signal is getting back to the PLL.

The *FV* signal on U_2 pin 10 can be tested next. This signal should have a frequency of 1 kHz (when the loop is operating correctly, Finley's law will be obeyed—the reference frequency is 1 kHz). With the loop "busted," the *FV* signal should be equal to the VCO frequency, divided by *N*. We know that since $f_{ref} = 1$ kHz and f_{out} is desired to be 28.150 MHz, then *N* should be (28.150 MHz/1 kHz) = 28,150. With the VCO drifting around 27.5 MHz, the *FV* signal should be drifting around (27.5 MHz/28,150) or 976.9 Hz.

This is important to observe. Often, the divide-by-*N* portion of the loop is receiving an *RF* signal from the VCO, but the signal is not strong enough to properly trigger the counter, or the counter itself is bad and is "skipping" counts. Either condition will cause the *FV* signal to be wrong, and the loop will not stay in lock. *Failure in the FV signal can be caused by a bad divide-by-*N *counter (defective PLL chip U_2), insufficient voltage level from the PLL digital buffer amplifier (Q_3), or miscommunication between the CPU and PLL (microprocessor "locked" or "crashed," and so forth).*

If the *FV* signal is OK, then the only remaining parts of the loop to investigate are the phase detector (part of U_2) and the loop filter amplifier (U_{3A}). These parts can be checked by simple replacement (they're both ICs) or by comparing voltage readings to manufacturer's specs. For example, the manufacturer may specify that the dc tuning voltage at the junction of R_{18} and C_{25} is to be 4.5 V at a frequency of 28.150 MHz. We can readily check this with a digital multimeter. *A tuning voltage that is "stuck" very high or very low (V_{cc} or GND potentials) points toward phase detector or loop filter problems.* However, any condition that drives the loop out of lock can cause the phase detector output to saturate under these conditions.

Notice how our troubleshooting procedure is really nothing more than an "inventory" of the signals in the circuit. With PLLs, it's always best to begin checking at the master oscillator and then work your way around the loop.

Troubleshooting a "dead" loop (producing no RF output) is similar, but a slightly different approach is needed.

EXAMPLE 7–9 The PLL synthesizer of Figures 7–18 and 7–19 has been set to 28.150 MHz by the user, yet there is no RF output at all on the output connector (C_{11} RH side). Develop a plan for troubleshooting it.

Solution

This is a dead loop. Either the VCO (Q_2) is not oscillating, or the buffer amplifier (Q_1) is not passing any VCO signal.

Visually inspect the PLL and associated circuitry; check the +5 V and +12 V power supplies. A failure of either power supply could certainly cause a dead loop!

Scope the VCO output at Q_2's emitter. If there is no signal, the VCO really isn't oscillating, and the trouble is known. If the signal here is OK, then the VCO signal is being "lost" in the buffer amplifier (Q_1), and the buffer amplifier stage needs examination.

Note: If the VCO is not running, this could be caused by three different problems. First, there could be some problem in the VCO itself, Q_2. Second, the VCO *control voltage* (junction of R_{18} and C_{25}) could be so far out of range that the VCO can't possibly oscillate. This can be checked with a DMM. A problem in the loop filter (U_{3A} output stuck high or low) or phase detector (U_2 pin 13 shorted high or low) could cause this to happen. *If the phase detector output seems stuck, go back and check the master oscillator and reference divider!* Third, there may be something shorting the VCO output to ground, such as a problem at Q_3 or Q_1, the buffer amplifiers. This is not a fun problem to find, because the short could be anywhere in the circuit. To find it, you might try removing C_{14} and/or C_{18} to see if the VCO resumes oscillation. This is not a common problem, but it is possible!

Many factors can kill a PLL. There isn't as much to check, since we're not getting VCO output. Once we fix the "VCO problem," the loop may still not work correctly. You guessed it—there can be multiple problems in the loop. A unit that has been directly or indirectly struck by lightning or has had improper power supply voltages connected is likely to have many problems!

Microprocessor and PLL communications faults can cause a loop to malfunction in a wide variety of ways. Fortunately, most faults involve *no* communications and are easy to find.

EXAMPLE 7–10 It is suspected that the CPU is not communicating with the PLL synthesizer of Figures 7–18 and 7–19. The radio's display shows correct information, and the control switches and knobs work to change frequency (the LCD display shows an apparent frequency change), but the PLL won't budge. It just sits there at 27.5 MHz, stuck. Demonstrate how to check for proper communications.

Solution

As we know from theory, the CPU sends the PLL two different kinds of information. *Initialization data* is sent during a CPU reset, and *command data* is sent in response to user control changes. The key to seeing the data signals is knowing how to make them appear, since they'll normally be sitting "idle" when no communication takes place. One common method is very simple: Place a scope probe on the *SER CLOCK* signal (U_2 pin 7) and issue a frequency change at the front panel (rotate the tuning knob or push one of the UP/DOWN buttons). Pushing a front panel frequency control should force the CPU to send command data to the PLL, which we should now see as pulses on the scope. If SER CLOCK is active, we proceed to check the remaining two signals in the same fashion (DATA, U_2 pin 5, and /ENABLE, U_2 pin 6). Note that the +*UNLOCK* signal is an *output* from the PLL, and is read by the CPU. We can look at this signal to determine whether or not the PLL is actually locked.

SUMMARY

- Frequency synthesizers solve the problem of choosing between an LC oscillator (unstable frequency) and a crystal oscillator (frequency unchangeable).
- A PLL is a closed-loop system utilizing negative feedback, which makes it self-correcting.
- The VCO in a PLL depends on an LC/RC circuit plus a control voltage to produce its frequency.
- Because a PLL is phase-locked, it produces zero frequency error when compared to its reference frequency. A PLL is just as accurate as its reference oscillator.
- The loop filter in a PLL is the primary determinant of capture range; the VCO characteristics affect lock range.
- A PLL frequency synthesizer uses a programmable divider (a variable modulus counter) in its feedback loop to digitally control frequency.
- Software plays a major role in modern frequency synthesizers; if the microprocessor software or hardware malfunctions, the frequency synthesizer will not be able to work.
- A technician can check for software communication by looking for initialization data (reset/power up), or command data (end-user frequency control changes). Most PLLs use serial data transmission to communicate with the CPU.
- A DDS synthesizer constructs a sine wave one point at a time. It is a truly digital oscillator, using an addition circuit called a phase accumulator to develop the sine wave phase angles in real time.
- DDS units can change frequency very rapidly but can only directly operate at around 30 MHz and below.
- The two main problems found when troubleshooting frequency synthesizers are *off frequency* and *dead* (no output). Both types of problems can be located by "walking the loop" with test instruments until the trouble is found.

PROBLEMS

1. What is a *frequency synthesizer?* When is crystal control of an oscillator's frequency inadequate?
2. Draw a block diagram of a phase-locked loop. At each major test point, describe the function and nature of the signal (digital, analog, ac, dc, etc.).
3. If a frequency of 1.2 kHz is present at point D in Figure 7–1, what frequency *should* be present at point C, and why?
4. If 1.2 kHz is present at point D in Figure 7–1 and the VCO has the transfer characteristic indicated in Figure 7–2, what dc voltage will appear at point B?
5. If the frequency at test point D in Figure 7–1 is increased from 1.22 kHz to 1.3 kHz, explain what will happen to the dc voltage at point B in the loop. Assume the VCO transfer characteristic of Figure 7–2.
6. State Finley's law for phase detectors. How does a phase detector achieve "zero frequency error"?

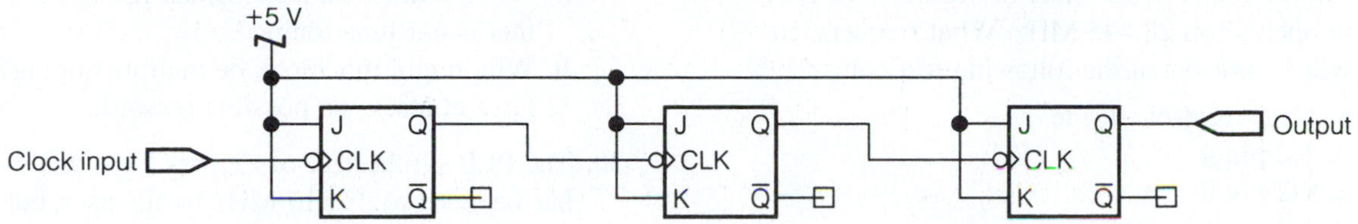

Figure 7–20 A counter for analysis

7. What is the shape of the signal from a phase detector? What loop element smoothes this signal prior to driving the VCO?

8. List the three possible operating states of a PLL, and for each one, give the input conditions necessary to attain that state.

9. Which PLL components primarily determine the (a) lock range and (b) capture range?

10. Explain where a dual-trace oscilloscope is connected in a PLL to determine its operating condition (locked or unlocked).

11. A 2 MHz signal is driving the *clock input* in Figure 7–9(a). What will the frequency at the Q output be?

12. A 32 MHz signal is driving the counter circuit in Figure 7–20. What's the divide ratio and output frequency?

13. Calculate the frequencies at points D, E, and C in the loop of Figure 7–11, given that $f_{ref} = 5$ kHz and $N = 20$.

14. If $f_{ref} = 12.5$ kHz in the loop of Figure 7–11, what is the frequency resolution of the PLL?

15. What is a programmable divider? Draw a divide-by-8 circuit using a 74LS192. Wire the divisor inputs to the correct logic levels for divide-by-8 operation.

16. What two PLL building blocks are *not* built into the MC145170-1 PLL chip of Figure 7–13?

17. Why is a reference frequency divider needed in a PLL synthesizer?

18. Calculate the frequencies at points A, B, C, and E in the loop of Figure 7–14, given that the master oscillator $f_{osc} = 10.180$ MHz, $R = 2036$, and $N = 29{,}388$.

19. What are the two types of information a CPU must communicate to a frequency synthesizer chip (such as a PLL)? When is each kind of information sent?

20. What type of communications format is most popular for PLL chips? Why is this method popular?

21. What are the major differences between PLL and DDS synthesizers?

22. Draw a block diagram of a DDS frequency synthesizer, showing the purpose and type of each signal (digital or analog) at each point.

23. A certain DDS synthesizer is operating from a 10 MHz clock and has 65,536 counts for a 360° period ($M = 65{,}536$). If the phase increment register N is set to 99, what output frequency will be produced?

24. What output frequency will be produced by a Harris HSP45102 DDS chip ($M = 2^{32}$) under the following conditions? $f_{clk} =$ 20.000 MHz.

    a. $N = 2{,}000{,}000{,}315$
    b. $N = 3{,}500{,}128{,}314$

25. What phase increment N is needed to set the DDS synthesizer of problem 24 to the following frequencies?

    a. $f = 10.000$ MHz
    b. $f = 15.000$ MHz

26. What is the frequency resolution (smallest frequency step) of the DDS synthesizer of problem 24?

27. In Figure 7–18, what value of R is needed to get a 1-kHz reference signal, given that $f_{osc} =$ 10.160 MHz?

28. The 10-meter synthesizer of Figure 7–19 is set to operate on 28.475 MHz. What frequencies will be present at the following test points?
    a. C11, right-hand side
    b. U2 pin 4
    c. U2 pin 9
    d. U2 pin 10
    e. U2 pin 2

29. The PLL synthesizer of Figures 7–18 and 7–19 has been set to 28.225 MHz by the user, but the VCO output measures 27.800 MHz (drifting slowly upward). The test points in the circuit read as follows:

    U_2 pin 2—10.160 MHz sine wave

    U_2 pin 3—no signal (steady 0 V dc)

    U_2 pin 4—27.800 MHz sine wave, drifting slowly upward

    U_2 pin 23—steady 0 V dc

    a. What is the *most likely* block in Figure 7–18 that is not functioning?
    b. Why might this block be malfunctioning? Give at least two possible reasons.

30. The PLL synthesizer of Figures 7–18 and 7–19 has been set to 28.315 MHz by the user, but the VCO output measures 27.800 MHz (drifting slowly upward). The test points in the circuit read as follows:

    U_2 pin 2–10.160 MHz sine wave

    U_2 pin 3–1 kHz narrow pulses, 5 V logic 1s, 0 V on logic 0s

    U_2 pin 4–steady dc level of 2.5 V, no ac signal

    U_2 pin 23–steady 4.9 V dc

    a. What is the *most likely* block in Figure 7–18 that is not functioning?
    b. Why might this block be malfunctioning? Give at least two possible reasons.

8

Frequency Modulation: Transmission

OBJECTIVES

At the conclusion of this chapter, the reader will be able to:

- explain how an oscillator is modulated to produce FM
- calculate the parameters of an FM signal, such as spectral content and bandwidth
- identify the topology of an FM transmitter
- predict the signals at each major point in an FM transmitter
- describe the technical characteristics of FM stereo multiplex and SCA signals
- explain how to use a station monitor to verify the operation of an FM transmitter

In AM, the voltage of the carrier signal is varied to represent the information. Because many noise sources produce amplitude disturbances, AM reception is very susceptible to them. Atmospheric noise (from lightning and other discharges) is a constant source of crackling and popping sounds in AM receivers.

Frequency modulation, or FM, was first theorized and experimented with in the 1930s. FM broadcasting on a large scale began in the late 1940s. In FM, the *frequency* of the carrier signal is changed to convey the information. The amplitude of the carrier remains constant.

There are two primary advantages to FM. First, since FM changes only the frequency and not the amplitude of the carrier wave, FM receivers can be built to ignore amplitude (voltage) changes. Therefore, FM receivers ignore most external noise sources. Second, it is much easier to design systems to reproduce *high-fidelity* sound using FM. High fidelity means accurate signal reproduction with a minimum of distortion. The reproduced information signal is a very close replica of the original in an FM system.

These advantages do come at a price. First, a typical FM broadcast station uses up to 200 kHz of bandwidth (compare this with the 10 kHz allotted for AM broadcast). Because of the high bandwidth requirements, FM broadcasting is done in the VHF band between 88 and 108 MHz and requires higher transmitter power. FM receivers and detectors are slightly more complex than those for AM, and the higher frequencies used for FM (VHF) complicate overall transmitter and receiver design.

For most serious music listeners, FM broadcast has become the medium of choice. AM broadcast has largely been relegated to talk radio, where high fidelity is not a serious concern.

Frequency and phase modulation (PM) are cousins. They are both considered *angle modulation.* PM directly changes the phase angle of the carrier wave to impart the information. FM indirectly changes the phase angle because it changes the frequency (which is the rate of phase change). PM is rarely broadcast directly for voice communications, but instead is used to gain an analytical understanding of FM. PM does find heavy application in digital (data) communications, the subject of a later chapter.

8–1 A SIMPLE FM TRANSMITTER

To create frequency modulation, the frequency of the carrier signal must be changed in step with the information signal. This suggests that the modulation process should take place at the *oscillator.* An RF oscillator's frequency can be controlled by using an LC resonant circuit, so if we can make either the L or C change in step with the information, frequency modulation will be created. Figure 8–1 shows one way of doing this. It's not an entirely practical way of building an FM transmitter, but it works!

The circuit in Figure 8–1 has a unique type of microphone connected in parallel with the LC resonant tank circuit of the oscillator. This is a *condensor* or *capacitance* microphone. The condensor mike has two metal plates separated by an insulating air space. One of the plates is very thin (like a piece of aluminum foil) and is free to vibrate back and forth when sound strikes it. When sound strikes the moveable plate of the condensor microphone, it causes the capacitance of the microphone to change in step with the sound. Recall that the capacitance of a capacitor is given by

$$C = \frac{\varepsilon A}{d} \tag{8–1}$$

where ε is the *dielectric constant* of the insulating material (air in this case), *A* is the area of one of the capacitor plates, and *d* is the distance between the plates. The vibration of the

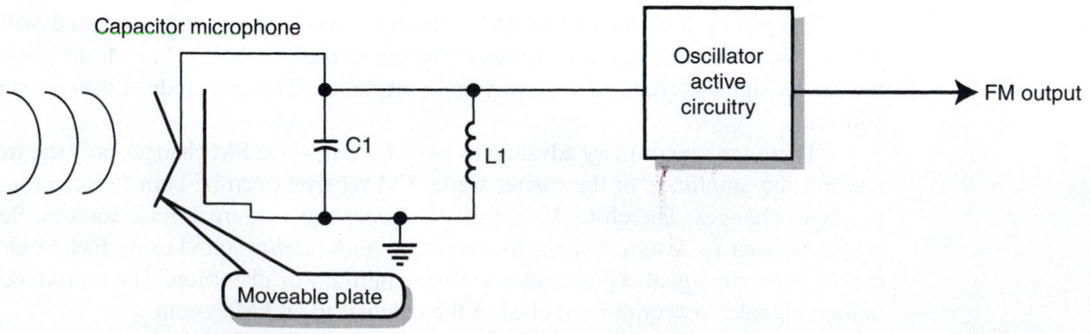

Figure 8–1 A crude FM transmitter

capacitance microphone's plate causes the distance between the plates to vary in step with the information. Thus, the total tank capacitance varies. Varying the tank capacitance changes the resonant frequency, which in turn changes the oscillator frequency.

In other words, the condensor microphone *frequency modulates* the oscillator circuit. When the microphone plates are closer together, the total tank capacitance increases, causing the resonant frequency to *decrease*. The opposite happens when the plates move farther apart: The total tank capacitance decreases, causing the oscillator to *increase* in frequency. Since it is the information (sound) that is causing the condensor microphone's moveable plate to vibrate, the frequency changes on the carrier will follow the information signal. Figure 8–2 shows this relationship.

The FM waveform in Figure 8–2 has a constant amplitude. The only thing that is being changed is the *frequency*. When the information signal is positive, the FM signal's frequency increases. The cycles of the FM waveform are squeezed closer together here. On the negative half-cycle, the opposite happens—we can see the FM waveform spreading out. Its period has become larger, indicating a lower frequency.

The capacitor microphone isn't a very practical method of generating FM. Because its wire leads are part of the tank circuit, moving the microphone would cause the capacitance of the tank circuit to shift, forcing the transmitter off frequency! In an actual FM transmitter, we use an electronic circuit called a *reactance modulator* to replace the condensor microphone. A reactance modulator converts a changing voltage into a changing capacitance (or inductance). This eliminates the need for the condensor microphone and allows an intelligence voltage to modulate the transmitter. Reactance modulators will be presented in a later section.

Center Frequency and Deviation

In Figure 8–2, the FM waveform is changing frequency in step with the information signal. Whenever the information signal is passing through zero, the frequency of the waveform is equal to the *carrier frequency* of the transmitter.

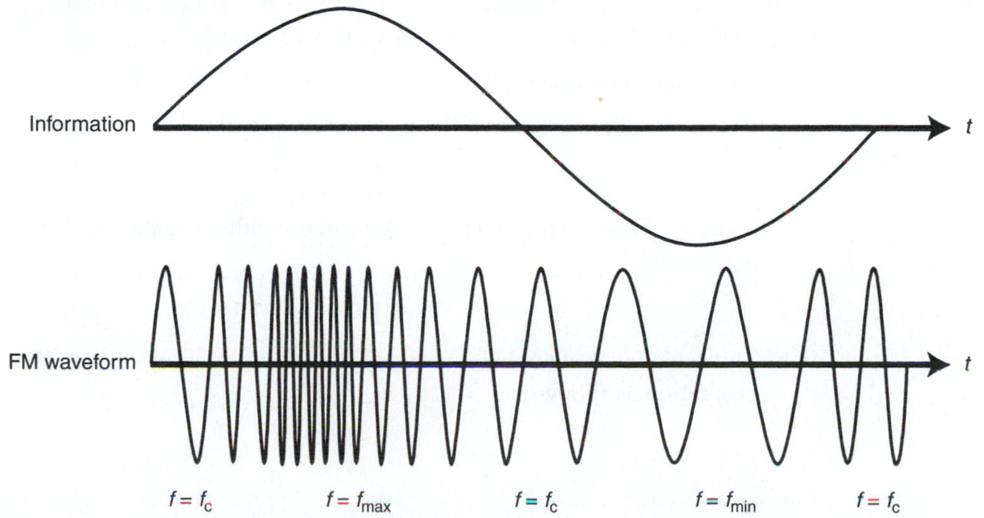

Figure 8–2 The information signal and the resulting FM waveform

The frequency produced by an FM transmitter when no information voltage is present is known by many names. Among these are the *carrier* frequency, the *center* frequency, and the *resting* frequency.

When the information signal goes to its maximum positive value, the carrier instantaneous frequency also becomes maximum. This is indicated by f_{max} in Figure 8–2. Likewise, the minimum frequency produced, f_{min} occurs at the negative peak of the information signal.

The *deviation* of an FM transmitter is equal to the *peak* frequency change produced by the information signal. The symbol δ (the lowercase Greek letter *delta*) is often used to represent the deviation. We can calculate the deviation in one of two ways:

$$\delta = f_{max} - f_c \quad (8\text{–}2)$$

$$\delta = f_c - f_{min} \quad (8\text{–}3)$$

Both equations will give the same result. The carrier frequency normally swings equally above and below the center frequency. The deviation plays a major role in determining the *bandwidth* of an FM transmitter, so it is very useful to calculate it.

What controls the amount of deviation? Since deviation is just frequency change, we can see that the *voltage* or *amplitude* of the information must be responsible. Again, look at Figure 8–2. As the information voltage increases, the frequency increases. If the information voltage were made larger, the amount of frequency change (deviation) would increase in direct proportion. The deviation will decrease if the information voltage is made smaller. Since the voltage of the information is related to its loudness, we can say that increasing the volume of the sound at the transmitter will increase the amount of deviation produced.

The quantity of deviation is *directly proportional* to the amplitude of the information. Doubling the information voltage will double the deviation; halving the information voltage cuts the deviation in half.

EXAMPLE 8–1

Suppose that the center frequency f_c in Figure 8–2 is 100 kHz and that f_{min} = 95 kHz and f_{max} = 105 kHz. The information signal V_m is 5 volts peak.

a. Calculate the deviation, δ.

b. Recalculate the deviation if the information signal is increased to 7.5 volts peak.

Solution

a. Since we know both f_{max} and f_{min}, we can use either equation 8–2 or 8–3. Substituting, we get

$$\delta = f_{max} - f_c = 105\text{ kHz} - 100\text{ kHz} = \underline{\underline{5\text{ kHz}}}$$

b. The deviation is *directly proportional* to the information voltage. We can set up a proportion as follows:

$$\frac{V'_m}{V_m} = \frac{\delta'}{\delta}$$

Don't let this scare you! We're just saying that the *new* information voltage, V'_m (the mark is pronounced "prime" and means "new value") compared to the *old* information voltage, V_m, will be the same as the *new* deviation compared to the *old*.

We know V_m, V'_m, and the original deviation. A little rearranging will give us

$$\delta' = \delta\left(\frac{V'_m}{V_m}\right) = 5\text{ kHz}\left(\frac{7.5\text{ V}}{5\text{ V}}\right) = \underline{\underline{7.5\text{kHz}}}$$

The new deviation will increase to 7.5 kHz. This will also increase the amount of frequency space, or *bandwidth,* required by the FM signal, one of the topics of the next section.

Section Checkpoint

8–1 Why are FM receivers largely unbothered by static?

8–2 List two advantages of FM when compared to AM.

8–3 Why does an FM broadcast transmitter require more power to provide the same coverage as an AM broadcast transmitter?

8–4 What frequency range is used for FM broadcast, and why?

8–5 List the two forms of *angle modulation*.

8–6 Where is the modulation performed in an FM transmitter?

8–7 How can a capacitor microphone help in generating FM?

8–8 How could the amplitude of an FM signal be described?

8–9 What does a *reactance modulator* do?

8–10 What are two other names for the *center frequency* of an FM transmitter?

8–11 Explain how to calculate *deviation* for an FM transmitter.

8–12 What controls the amount of deviation in an FM transmitter?

8–2 FM SIGNAL ANALYSIS

During the analysis of AM in Chapter 3, we looked at AM signals in both the time and frequency domains. An FM signal can be examined in the same ways. One thing we quickly discover is that an FM signal doesn't look very impressive on an oscilloscope! The frequency changes (deviation) shown in Figure 8–2 are greatly exaggerated. You will rarely be able to directly observe the frequency deviation of an FM signal on an oscilloscope. The amount of frequency change is fairly small when compared to the carrier center frequency. A "real-life" FM signal is shown in Figure 8–3.

As you might have guessed, we'll receive a lot more information by looking at an FM signal in the frequency domain with a spectrum analyzer. There are four specific quantities that a technician normally can be expected to measure or estimate in an FM signal. These are the *percentage of modulation,* the *deviation rate,* the *modulation index,* and the *bandwidth*.

Percentage of Modulation

In AM, the percentage of modulation is a practical measure of how much information voltage is being placed on top of the carrier. The more information voltage, the louder the sound

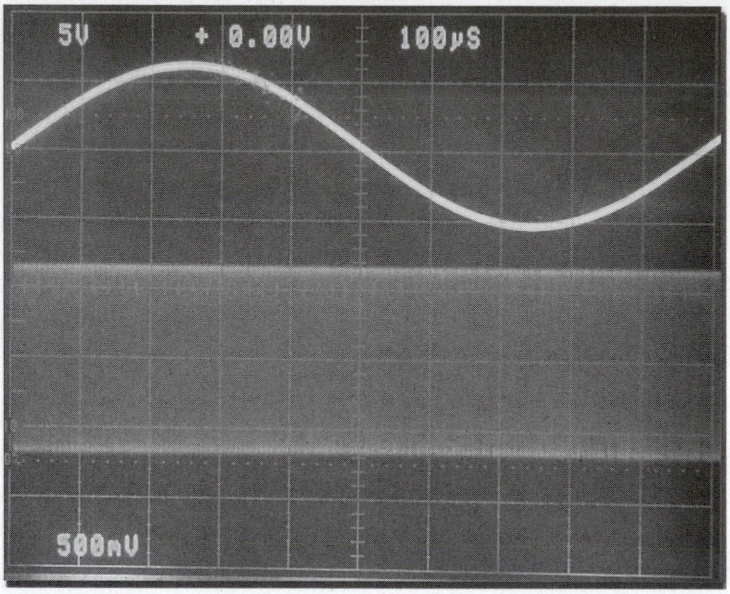

Figure 8–3 An FM signal seen on an oscilloscope (f_c = 100 kHz, δ = 5 kHz)

in the receiver. The definition holds the same practical meaning for FM, but the formula for calculating it is different. For FM, percentage of modulation is defined by

$$\%\text{mod} = \left(\frac{\delta}{\delta_{max}}\right) \times 100\% \quad (8\text{–}4)$$

In equation 8–4, δ is the deviation of the transmitter, and δ_{max} is the maximum allowed deviation for the particular type of FM transmission being used. For broadcast FM, δ_{max} is 75 kHz. For the sound portion of an analog TV signal, an FM carrier is used, with a δ_{max} of 25 kHz. Communications FM (such as emergency services and amateur radio) commonly uses a δ_{max} of 5 kHz. *To calculate FM percentage of modulation, you must know the maximum allowed deviation!*

You might wonder why "communications" applications use such a small amount of deviation when compared to FM broadcast. High fidelity isn't a strong design goal for voice communications. Instead, a large number of users must share a fixed amount of RF spectrum (frequency space). By lowering deviation, the bandwidth required for each transmitter is reduced—and therefore, more transmitters can share the airwaves.

EXAMPLE 8–2 An FM broadcaster is operating on 98.1 MHz, and the maximum frequency from the transmitter is 98.15 MHz. The information voltage is 5 V_p, and the information frequency is 1 kHz.

a. What is the percentage of modulation?

b. If the information voltage is changed to 4 V_p, what happens to the percentage of modulation?

c. What information voltage will 100% modulate the transmitter?

Solution

a. To calculate percentage of modulation, we must first know deviation:

$$\delta = f_{max} - f_c = 98.15 \text{ MHz} - 98.1 \text{ MHz} = \underline{50 \text{ kHz}}$$

Since this is an FM broadcast (88–108 MHz), we know that the maximum deviation is 75 kHz. Therefore,

$$\%\text{mod} = \left(\frac{\delta}{\delta_{max}}\right) \times 100\% = \left(\frac{50 \text{ kHz}}{75 \text{ kHz}}\right) \times 100\% = \underline{\underline{66.7\%}}$$

b. The information voltage controls the deviation. Since the voltage has *decreased* to 4 V_p, we can expect the amount of deviation to decrease in a like manner:

$$\delta' = \delta\left(\frac{V'_m}{V_m}\right) = 50 \text{ kHz}\left(\frac{4 \text{ V}_p}{5 \text{ V}_p}\right) = \underline{40 \text{ kHz}}$$

This isn't exactly the answer that was needed; we need to express it as a percentage of modulation:

$$\%\text{mod} = \left(\frac{\delta}{\delta_{max}}\right) \times 100\% = \left(\frac{40 \text{ kHz}}{75 \text{ kHz}}\right) \times 100\% = \underline{\underline{53.3\%}}$$

c. This is a rather "backward" request, but it is quite solvable. Remember that 100% modulation is 75 kHz (the deviation we desire), and substitute that into the proportion for deviation:

$$\frac{V'_m}{V_m} = \frac{\delta'}{\delta}$$

Here, V'_m is the new (unknown) information voltage, δ' is the new (desired) deviation, and V_m and δ are the original values. By manipulating the equation, we get:

$$V'_m = \left(\frac{\delta'}{\delta}\right) V_m = \left(\frac{75 \text{ kHz}}{50 \text{ kHz}}\right) 5 \text{ V}_p = \underline{\underline{7.5 \text{ V}_p}}$$

Therefore, an information voltage of $\underline{\underline{7.5 \text{ V}_p}}$ will be required to 100% modulate the transmitter.

Deviation Rate

The *deviation rate* (DR) of an FM transmitter is the number of up and down frequency changes (swings) of the RF carrier that take place per second. It is always the same number as the *information frequency*.

Suppose that we send a very low information frequency, 1 Hz, into an FM modulator. The carrier will swing up to its maximum frequency, down to its minimum frequency, then back to center again exactly *once* every second. If the information is increased to 2 Hz, then the carrier frequency changes will take place *twice* per second. It doesn't matter what the information voltage is. Only the *frequency* of the information is important.

EXAMPLE 8–3 An FM broadcaster is operating on 96.5 MHz, and the maximum frequency from the transmitter is 96.54 MHz. The information voltage is 5 V_p, and the information frequency is 1 kHz.

a. What is the deviation rate (DR)?

b. If the information voltage is changed to 2 V_p, what happens to the deviation rate?

c. If the information frequency is changed to 5 kHz, what happens to the deviation rate?

Solution

a. DR $= f_m =$ 1 kHz by inspection.

b. The DR is unchanged; it is still 1 kHz, since only the information frequency affects it.

c. The DR becomes the new f_m value or 5 kHz by inspection.

Only the frequency of the information affects the deviation rate!

FM Modulation Index

Another measure of FM transmitter performance is the *modulation index,* usually denoted by the symbol m_f. It's easy to become confused here, because the meaning of modulation index is quite different for FM than it was for AM.

Recall that for AM, modulation index and percentage of modulation are essentially the same information. The AM modulation percentage is merely the AM modulation index (a number between zero and one) expressed as a percentage. In FM, percent modulation and modulation index are *not* the same information! The FM modulation index is a measure of how much *phase shift* is being imparted to the carrier wave by the information. "Now wait," you say. "In FM, the *frequency* of the carrier is being changed. How are we getting a phase shift?"

Any time we change the frequency of a waveform, we create the same effect as an increasing phase shift. For FM, the peak or maximal amount of this phase shift is calculated by

$$m_f = \frac{\delta}{f_m} \tag{8–5}$$

where δ is the deviation (Hz), f_m is the information frequency (Hz), and m_f is the *FM modulation index,* in *radians.* Don't let this bother you. Remember that there are 2π radians in a circle, or 360°. If there is an FM modulation index of π, then the maximum carrier phase shift is simply 180°.

The relationship between phase shift and frequency can be visualized by looking at the second hand on a wall clock. In Figure 8–4(a), the body of the clock is stationary, and the second hand completes a revolution once every minute. But what if the clock is rotated *clockwise* at the same time [Figure 8–4(b)]? To the observer, the second hand appears to move *faster*—the eye sees that the "frequency" of the second hand has increased. In Figure 8–4(c), the clock now appears to be running *slower*—something is rotating the clock *counter-clockwise.*

The maximum *angle* through which the clock is rotated is analogous to the FM modulation index. Here's the catch: Suppose that all you can see is the second hand. You can't see the clock face or case. You notice the second hand speeding up (frequency increasing). Is the second hand speeding up on its own, or is the clock being turned ("phase shifted")? You have no way of knowing for sure!

Figure 8–4 Clock watching

(a) No phase shift–normal timekeeping (b) Positive phase shift–appears to run faster (c) Negative phase shift–appears to run slower

When a phase shift is taking place on a carrier wave, it causes the apparent frequency of the carrier wave to either increase (positive phase shift) or decrease (negative phase shift). For this to happen, the phase shift must be changing itself. (Stop rotating the clock but leave it at any angle and the second hand again moves at its normal speed.) Likewise, increasing the frequency of a carrier wave has the same effect as an *equivalent amount of increasing phase shift*.

That's the rub—frequency modulating the carrier wave actually creates PM, and the *modulation index* is really the *amount* of PM being generated. Likewise, you can look at it the other way—if we phase modulate a carrier, it will also appear that we've *frequency modulated* it. PM and FM are different types of modulation, but you can't have one without the other!

The FM modulation index m_f is a good measure of how strongly we're modulating the carrier. It will also be very helpful in calculating the bandwidth of the FM signal. Equation 8–5 tells us that the FM modulation index depends on two factors, the *deviation* (which is controlled by the amplitude of the information) and the *information frequency* or *deviation rate*.

EXAMPLE 8–4

An FM broadcaster is operating on a carrier frequency of 93.3 MHz, and with a 1 V_p information signal, and the transmitter is producing 10 kHz of deviation. The carrier is swinging up and down in frequency 5,000 times per second. Calculate

a. the information frequency f_m

b. the deviation rate (DR)

c. the percentage of modulation

d. the FM modulation index

Solution

a. The carrier frequency swing rate is equal to the information frequency. By inspection, $f_m = \underline{\underline{5 \text{ kHz}}}$.

b. The deviation rate is equal to the information frequency, and is also $\underline{\underline{5 \text{ kHz}}}$.

c. The percentage of modulation is

$$\%\text{mod} = \left(\frac{\delta}{\delta_{max}}\right) \times 100\% = \left(\frac{10 \text{ kHz}}{75 \text{ kHz}}\right) \times 100\% = \underline{\underline{13.3\%}}$$

d. The FM modulation index is:

$$m_f = \frac{\delta}{f_m} = \frac{10\text{ kHz}}{5\text{ kHz}} = \underline{\underline{2.0\text{ radians}}}$$

There is no "theoretical maximum" for modulation index, but most broadcasting keeps m_f at or below 15.

Changes in the information frequency or information voltage will alter the FM modulation index. Equation 8–5 tells us that as we increase the frequency of the information, the FM modulation index will fall (all else being constant). If we increase the amplitude of the intelligence (which increases deviation), the modulation index will rise in a like manner.

FM Bandwidth Analysis

AM and FM signals appear quite different in the frequency domain. You might recall that an AM signal generates one pair of sidebands for each information tone frequency being applied to the transmitter. For example, a spectrogram of a 100 kHz 10 V carrier being 100% amplitude modulated by a 5 kHz tone would look like Figure 8–5. An FM signal looks quite different. If we frequency-modulate a 100 kHz carrier with a 5 kHz information signal, we get a picture like Figure 8–6.

There's something really strange about Figure 8–6. Count the number of sidebands being generated: They go from a frequency of $-\infty$ to $+\infty$, stepping 5 kHz (f_m) as they go. In other words, *in theory, any FM signal always has an infinite number of sidebands and requires an infinite bandwidth to represent.* In practice, you might guess that we don't have to worry about *all* of the sidebands of an FM signal, and you're right. Notice that as we get far away from the carrier, the sidebands tend to get weaker. Eventually, they dip below the noise level—at which point they become insignificant.

An exact calculation of the bandwidth of an FM signal is a little involved, and for that reason, technicians often use an approximation called *Carson's rule:*

$$BW \approx 2(f_m + \delta) \qquad (8\text{–}6)$$

Carson's rule is a nice quick way of finding bandwidth. It has two "parts," which make sense when equation 8–6 is unfactored (distributed). The first part, $2f_m$, looks a lot like the equation for the bandwidth of an AM signal. This part is true, because any FM signal

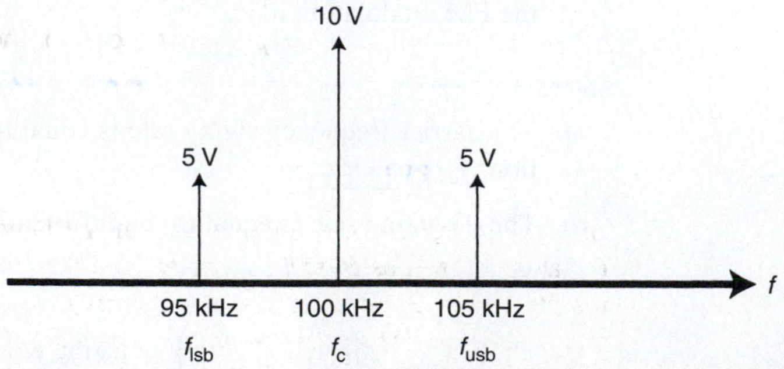

Figure 8–5 The spectrogram of an AM signal

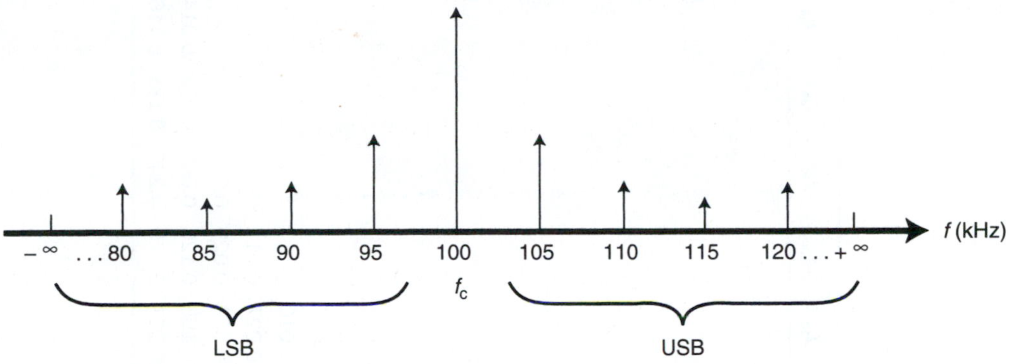

Figure 8–6 The spectrogram of an FM signal

produces *at least* two significant sidebands. The other part of Carson's rule is the intuitive part. If you cause a transmitter to deviate 5 kHz up and 5 kHz down (δ = 5 kHz), you would naturally expect the FM transmitter to use up 2 times 5 kHz, or 10 kHz of bandwidth. You might wonder why we don't just go with this part of the rule and be done; it looks simple enough. Unfortunately, FM is a *nonlinear* form of modulation. Many of the aspects of the behavior of FM defy intuition because of this. We will compare the intuitive and actual results of FM signal analysis shortly.

EXAMPLE 8–5

An FM broadcaster is operating on a carrier frequency 93.3 MHz, and with a 1 V_p information signal, the transmitter is producing 10 kHz of deviation. The information frequency is 5 kHz. Calculate the bandwidth by Carson's rule and compare it with the "intuitive" result (twice deviation).

Solution

Carson's rule states that

$$\text{BW} \approx 2(f_m + \delta) \approx 2(5\text{ kHz} + 10\text{ kHz}) \approx \underline{\underline{30\text{ kHz}}}$$

Approximately 30 kHz of bandwidth will be needed by the transmitter.

The intuitive approach (which gives a wrong answer) says that the transmitter frequency is swinging *up* 10 kHz and then *down* 10 kHz, for a total "frequency travel" of 20 kHz. Unfortunately, this approach just isn't very accurate, though. The reason is that FM generates an infinite number of sidebands, and there is nothing in this approach to account for that—which ones will be significant, and which ones can be ignored? The two terms in Carson's rule take into account an *approximate* count of the number of significant sidebands.

Exact Bandwidth of an FM Signal

In order to get an accurate estimate of the bandwidth of an FM signal, we must use an equation known as *Bessel's identity*. The formula is too complex to explore here, but the results are quite valuable. Bessel's identity gives the relative voltage of the carrier and each sideband in an FM signal *compared to the original unmodulated carrier voltage*. The numbers in the Bessel chart, then, are really just percentages.

Table 8–1 Bessel coefficients for m_f between 0 and 15.

m_f	J_0	J_1	J_2	J_3	J_4	J_5	J_6	J_7	J_8	J_9	J_{10}	J_{11}	J_{12}	J_{13}	J_{14}	J_{15}	J_{16}
0.00	1.000																
0.20	0.990	0.100															
0.25	0.984	0.124															
0.50	0.938	0.242	0.031														
1.00	0.765	0.440	0.115	0.020													
1.50	0.512	0.558	0.232	0.061	0.012												
2.00	0.224	0.577	0.353	0.129	0.034												
2.41	0.000	0.519	0.432	0.199	0.065	0.016											
2.50	−0.048	0.497	0.446	0.217	0.074	0.020											
3.00	−0.260	0.339	0.486	0.309	0.132	0.043	0.011										
4.00	−0.397	−0.066	0.364	0.430	0.281	0.132	0.049	0.015									
5.00	−0.178	−0.328	0.047	0.365	0.391	0.261	0.131	0.053	0.018								
5.52	0.000	−0.340	−0.123	0.251	0.396	0.323	0.189	0.088	0.034	0.012							
6.00	0.151	−0.277	−0.243	0.115	0.358	0.362	0.246	0.130	0.057	0.021							
7.00	0.300	−0.005	−0.301	−0.168	0.158	0.348	0.339	0.234	0.128	0.059	0.024						
8.00	0.172	0.235	−0.113	−0.291	−0.105	0.186	0.338	0.321	0.223	0.126	0.061	0.026	0.010				
9.00	−0.090	0.245	0.145	−0.181	−0.265	−0.055	0.204	0.327	0.305	0.215	0.125	0.062	0.027	0.011			
10.00	−0.246	0.043	0.255	0.058	−0.220	−0.234	−0.014	0.217	0.318	0.292	0.207	0.123	0.063	0.029	0.012		
12.00	0.048	−0.223	−0.085	0.195	0.182	−0.073	−0.244	−0.170	0.045	0.230	0.300	0.270	0.195	0.120	0.065	0.032	0.014
15.00	−0.012	0.206	0.042	−0.194	−0.119	0.130	0.206	0.034	−0.174	−0.220	−0.090	0.100	0.237	0.279	0.246	0.181	0.116

The *FM modulation index* must first be calculated in order to look up results in a Bessel table. In a Bessel table, the symbol J_0 stands for the *carrier,* J_1 the first pair of sidebands, J_2 the second pair of sidebands, and so on. Table 8–1 shows a Bessel table covering modulation indices from 0 to 15. To use the Bessel table, we first calculate the FM modulation index, m_f. Then we look down the first column of the table until we find a matching m_f value. The numbers on the same line as this value represent the *normalized voltages* in the carrier (J_0), first pair of sidebands (J_1), second pair of sidebands (J_2), and so forth.

The table does not show any values that are less than 1% (0.01) of the original carrier's voltage. *A "significant" sideband is one that will be counted when tallying the bandwidth and must have at least 1% of the original unmodulated carrier amplitude.* Technically, these sidebands are no weaker than -40 dBc (decibels with respect to the unmodulated carrier).

Take a look at the figures across the top row of Table 8–1, where m_f is 0. The value J_0 reads as 1.000 (100%), and all the other columns show blanks ($J_1 - J_{16}$). When the modulation index is zero, we have an *unmodulated carrier,* which is just a steady sine wave. With no modulation taking place, there are no sidebands, and therefore the carrier (J_0) is 100% of its original value.

EXAMPLE 8–6

Use a Bessel table to evaluate the FM transmitter from Example 8–5 and plot a spectrogram of the resulting FM signal. $V_{c(\text{unmod})} = 100$ V, and $R_L = 50\ \Omega$.

Solution

The data from Example 8–5 are: $f_c = 93.3$ MHz, $\delta = 10$ kHz, and $f_m = 5$ kHz. To use a Bessel table, we must first find the FM modulation index:

$$m_f = \frac{\delta}{f_m} = \frac{10\text{ kHz}}{5\text{ k Hz}} = \underline{\underline{2.0}}$$

From the Bessel table, we read the figures from the 2.0 row and get $J_0 = 0.224$, $J_1 = 0.577$, $J_2 = 0.353$, $J_3 = 0.129$, and $J_4 = 0.034$. The remaining figures in the row are blank, indicating that the rest of the sideband voltages ($J_5 - J_{16}$) are less than 1% of the unmodulated carrier value and are therefore insignificant.

We can now calculate the voltage of each spectral component:

$$V_c = J_0 V_{c(\text{unmod})} = (0.224)(100\text{ V}) = \underline{\underline{22.4\text{ V}}}$$
$$V_{lsb[1]} = V_{usb[1]} = J_1 V_{c(\text{unmod})} = (0.577)(100\text{ V}) = \underline{\underline{57.7\text{ V}}}$$
$$V_{lsb[2]} = V_{usb[2]} = J_2 V_{c(\text{unmod})} = (0.353)(100\text{ V}) = \underline{\underline{35.3\text{ V}}}$$
$$V_{lsb[3]} = V_{usb[3]} = J_3 V_{c(\text{unmod})} = (0.129)(100\text{ V}) = \underline{\underline{12.9\text{ V}}}$$
$$V_{lsb[4]} = V_{usb[4]} = J_4 V_{c(\text{unmod})} = (0.034)(100\text{ V}) = \underline{\underline{3.4\text{ V}}}$$

The *frequency* of each sideband is equal to the carrier plus a multiple of the original information frequency. We get a spectrogram that looks like Figure 8–7. The total bandwidth of the emission is $8f_m$, because there are four sideband *pairs,* with a frequency space of f_m between them. This is a $\underline{\underline{40\text{ kHz}}}$ bandwidth, moderately wider than the 30 kHz bandwidth estimated by Carson's rule for the same data. Note how we do *not* divide the Bessel coefficients for the sidebands by two. That has already been accounted for in the Bessel table.

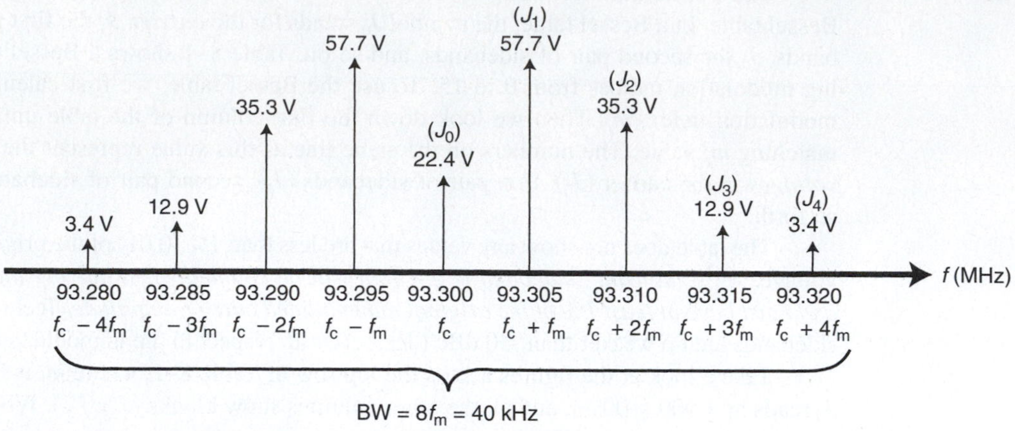

Figure 8–7 The resulting FM signal spectrogram

The Total Power Remains Constant

As we increase the modulation index, something interesting happens to the carrier energy. The power in an FM signal must remain constant. But how can the total power remain constant if we're adding *sidebands* as we modulate the carrier? The Bessel table gives the answer. The *carrier* (J_0) shrinks as the sidebands grow. In essence, the sidebands "steal" power from the carrier. This way, the total power can remain constant. Figure 8–8 shows this in graphic form.

Figure 8–8 The graph of the Bessel functions

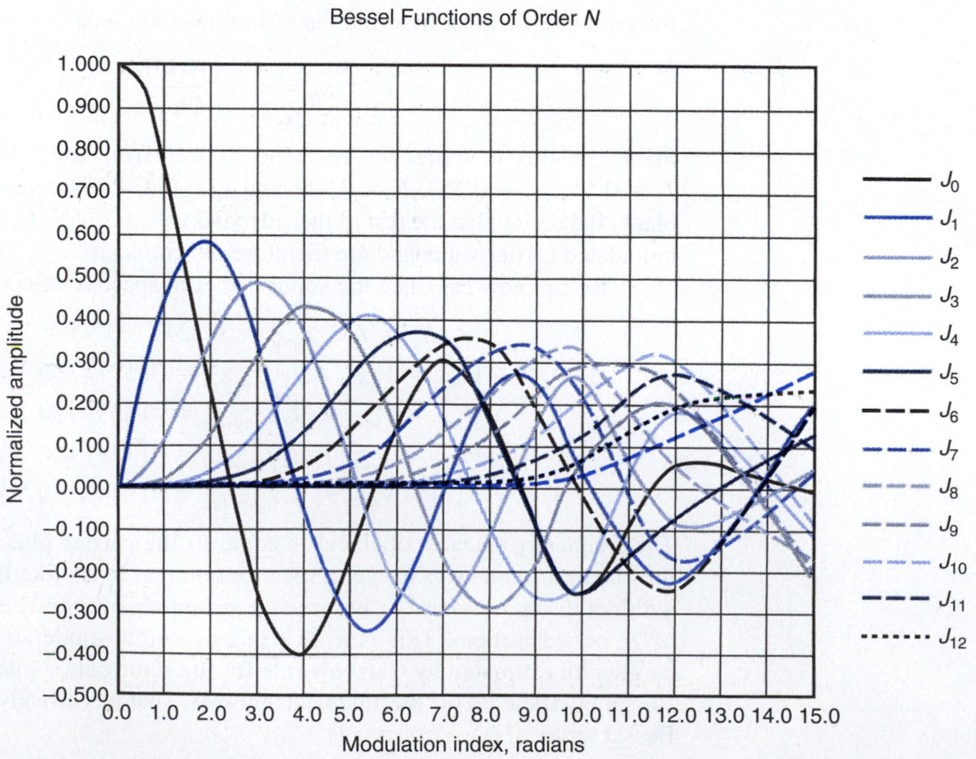

Figure 8–8 contains the same information as the Bessel table. The horizontal axis of the graph is the modulation index. The left-hand side of the graph is a modulation index of *zero,* which means an unmodulated FM carrier. Which frequency components are present when $m_f = 0$? Only the carrier, since the line representing J_0, the carrier frequency component, is at *1.00,* while all the other graph lines are at zero. As the FM modulation index is increased, you can see the carrier strength weaken as more sidebands are added. On a spectrum analyzer, the expanding pattern of sidebands looks like "grass growing" as the modulation index is increased.

EXAMPLE 8–7

Calculate the power of the transmitter in Example 8–6 when unmodulated ($m_f = 0$), and when modulated according to the conditions in the previous example. Show the power of each signal component on a spectrogram and find the total power.

Solution

When the transmitter is unmodulated, the only signal energy is the 100 volt carrier. Under this condition, the total power will be

$$P = \frac{V^2_{\text{c(unmod)}}}{R_L} = \frac{100\ \text{V}^2}{50\ \Omega} = \underline{\underline{200\ \text{W}}}$$

When the transmitter is modulated, we need to find the power in each individual frequency component by using Ohm's law, just as above. For example, when modulated, the new carrier power will be

$$P_{\text{c}} = \frac{V^2_{\text{c}}}{R_L} = \frac{[J_0 V_{\text{c(unmod)}}]^2}{R_L} = \frac{22.4\ \text{V}^2}{50\ \Omega} = \underline{\underline{10.04\ \text{W}}}$$

This process is repeated for all the frequencies in Figure 8–7, yielding the spectrogram of Figure 8–9.

The total power is therefore (2)(0.23 W) + (2)(3.33 W) + (2)(24.92 W) + (2)(66.59 W) + 10.04 W = $\underline{\underline{200.18\ \text{W}}}$. This is very close to the original value of 200 watts; the extra 0.18 watts is due to rounding errors inherent in the Bessel table.

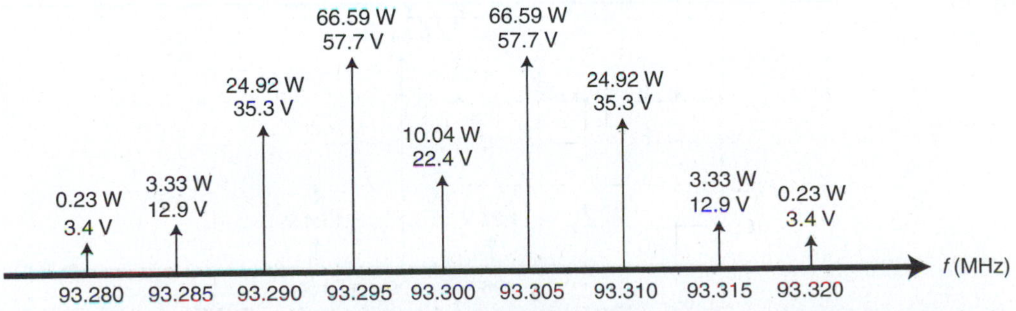

Figure 8–9 The spectrogram with power

Special Cases for FM Transmission

There are two types of FM signals that most technicians can recognize on a spectrum analyzer, *wideband* and *narrowband* FM. The preceding example is a *wideband* FM (WBFM) signal. It has more than one pair of significant sidebands. *Whenever the FM modulation index is greater than 0.25, a wideband FM signal will result.* The Bessel table verifies this; the largest index m_f where there is only one pair of sidebands (J_2 and higher are zero) is *0.25*. Any FM signal that has only *one* pair of significant sidebands is called a *narrowband* FM (NBFM) signal. *The modulation index of a NBFM signal is always less than or equal to 0.25.*

It's easy to calculate the bandwidth of a NBFM signal; just multiply the information frequency by two! The reason for this again comes from the Bessel table. In a NBFM signal, there is only one pair of significant sidebands, with a voltage given by coefficient J_1. Since the sidebands are spaced at intervals of f_m, the bandwidth becomes $2f_m$.

EXAMPLE 8–8

An FM transmitter is using a carrier frequency of 100 kHz and has a deviation of 1 kHz and an information frequency of 4 kHz. The unmodulated carrier voltage is 10 volts. Draw a spectrogram and decide whether this is a NBFM or WBFM signal.

Solution

First, calculate the modulation index:

$$m_f = \frac{\delta}{f_m} = \frac{1 \text{ kHz}}{4 \text{ kHz}} = 0.25$$

The Bessel table shows that $J_0 = 0.984$ and $J_1 = 0.124$. J_2 and higher coefficients are blank, so they are effectively zero. The spectrogram looks like Figure 8–10. This is definitely a NBFM signal, and you can see that the bandwidth is just 8 kHz ($2f_m$). It looks a *little* like an AM signal, but it is not. The phase relationship between the carrier and sidebands is quite different for AM and FM signals. Feeding this NBFM signal into an AM detector would result in little (and distorted!), if any, intelligence reproduction.

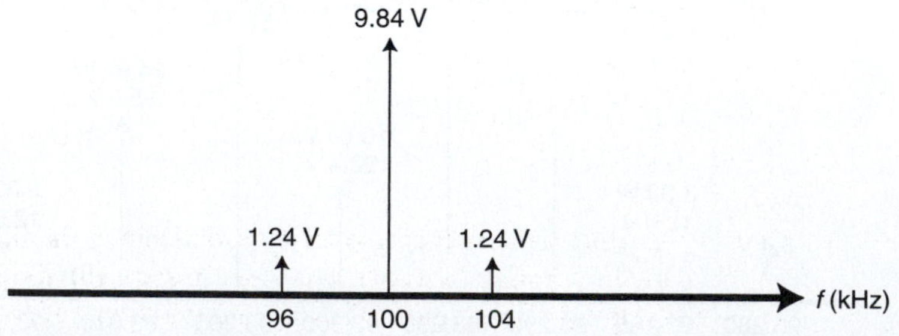

Figure 8–10 The resulting spectrogram

Carrier Voltage Nulls

There's one more special condition that sometimes occurs in FM signals. If you look carefully at the graph of the carrier (J_0) in Figure 8–8, you'll notice something peculiar. At an index of approximately 2.4, the carrier energy totally disappears. There's nothing but *sideband* energy at this particular modulation index. This is referred to as the first *null* of the carrier. A technician can observe this on a spectrum analyzer and therefore know that the index is *exactly* 2.4, which is very useful when measuring the performance of a transmitter.

As the modulation index increases above 2.4, the carrier energy comes back. The dips below zero indicate *phase inversion* in the graph of Figure 8–8. Notice that the carrier never gets stronger than 40% of its original value, and as the modulation index increases to about 5.5, another carrier null appears. As all of this is happening, the number of significant sidebands is increasing, but their individual amplitudes are tending to fall, in order to keep the total power constant.

Section Checkpoint

8–13 Why is it hard to observe much about an FM signal on a scope?

8–14 How is the percentage of modulation computed for an FM signal?

8–15 What is the maximum allowed deviation for an FM broadcast signal?

8–16 How is the deviation rate of an FM signal determined?

8–17 What does the FM modulation index measure?

8–18 What two factors control the FM modulation index?

8–19 Explain how frequency modulating a carrier creates phase modulation.

8–20 How many sidebands are generated by frequency modulating a carrier?

8–21 Explain how to use Carson's rule for estimating the bandwidth of an FM signal.

8–22 In order to use a Bessel table, what information must be known?

8–23 As m_f increases, J_0 initially decreases. Why must this be so?

8–24 What is the difference between a NBFM and WBFM signal?

8–25 What is the highest m_f a NBFM signal can have?

8–26 At what value of m_f does the first carrier null take place?

8–3 FREQUENCY MODULATED OSCILLATORS

In order to frequency modulate a transmitter, the inductance or capacitance of an oscillator circuit must be made to change in step with the information signal. A circuit that converts a voltage into a varying inductive or capacitive reactance is called a *reactance modulator*. Because the modulation takes place near the beginning in all FM transmitters, all FM transmitters can be considered low-level units.

The Reactance Modulator

The most common form of reactance modulator is the *varactor diode* of Figure 8–11. (This diode also is known as a *varicap diode* or *epicap diode.*) The varactor diode is really a normal diode that is optimized for use of its junction capacitance.

A varactor diode is normally operated in the reverse-bias region. From electronic fundamentals, we know that a diode opposes current flow when it is reverse-biased due to the removal of charges from the *depletion region*. When the varactor diode is reverse-biased, the depletion region looks like an insulator, since it has little or no free charge for conduction. It therefore acts as the dielectric of the diode "capacitor." In Figure 8–11(a), a small reverse bias is being applied; in Figure 8–11(b), the reverse bias has been increased. What happens to the total capacitance in Figure 8–11(b)?

The capacitor "plates" seem to be farther apart in Figure 8–11(b). It's as if we have spread them apart by widening the insulating region. Therefore, the total capacitance of the diode *decreases* as the amount of reverse bias is *increased.*

The relationship between reverse-bias voltage and diode capacitance is not a linear one. The graph of Figure 8–12 shows the capacitance of a typical varactor diode as a function of the applied reverse-bias voltage. The graph demonstrates that the capacitance versus voltage is a highly nonlinear function. In fact, the capacitance of a varactor diode can be approximated by the following equation:

$$C_d = \frac{C_0}{(1 + |2V_R|)^n} \tag{8–7}$$

where C_0 is the equivalent diode capacitance at zero bias, V_R is the reverse-bias voltage being applied to the device, and n is a fixed exponent value that depends on how the diode is constructed. (The usual range of n is from 0.4 to 1.0.) This is a rather unfriendly formula, and technicians normally have no need for it; however, it can be used to approximate the behavior of a varactor diode. For the MV-209 diode, the factors in the equation are

$$C_0 = 177.52 \text{ pF}, n = 0.95$$

which give the graph data of Figure 8–12. A manufacturer's data sheet rarely gives formulas for diode capacitance; usually, a characteristic curve is given.

One important note about varactor diodes is that they are normally operated with at least −2 V to −3 V of reverse bias. The reason for this is that the diode will be exposed to an ac waveform from the tank circuit it is connected to. If there is insufficient reverse

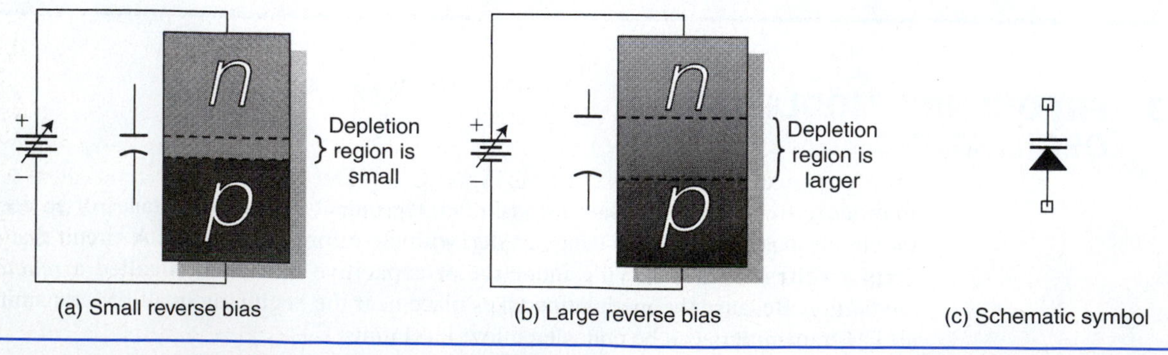

Figure 8–11 A varactor diode

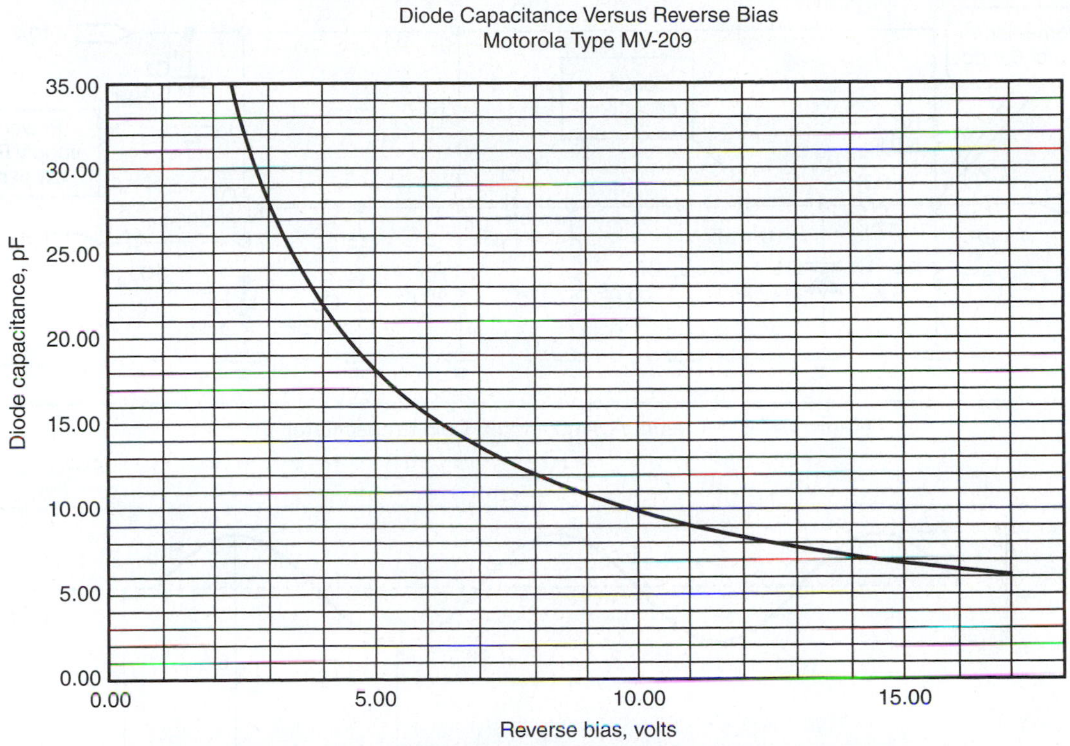

Figure 8–12 The capacitance of a varactor diode versus reverse bias

bias, the ac signal on the tank may forward bias the diode during part of its cycle, causing severe signal distortion. The circuit will produce harmonics (multiples) of the oscillator's frequency as a result of this distortion, which is normally quite undesirable! Also, the quality factor of most varactor diodes degrades rapidly as the zero-bias condition is approached, which in turn degrades the *Q* (and frequency stability) of the tank the diode may be connected to.

A Varactor Diode in an Oscillator

Figure 8–13 shows a MV-209 varactor diode modulating a common-collector Colpitts oscillator. This circuit is actually a tiny FM transmitter! Since it is based on an LC oscillator, its frequency will not be very stable, of course, but for experimentation, it is quite adequate. This same circuit can be placed within a phase-locked loop to obtain rock-solid stability; this technique will be explored later in this chapter.

The information signal is coupled into the circuit through C_2, an audio coupling capacitor. Resistors R_2 and R_4 form a voltage divider that provides a steady 6 volt dc level for the audio to ride upon, as shown in Figure 8–14. The 6 volt level is important, as it provides the reverse bias for varactor diode D_1. Figure 8–12 can be used to find the capacitance of D_1 at this voltage. The audio voltage is superimposed on the 6 volt dc level. Suppose that a 0.5 V peak information signal is present. The maximum and minimum voltage in the circuit will vary between 6.5 V and 5.5 V.

The varactor diode D_1 sees this total voltage, as it is applied through isolation resistor R_3. The isolation resistor prevents the LC tank ac currents from being shorted out by

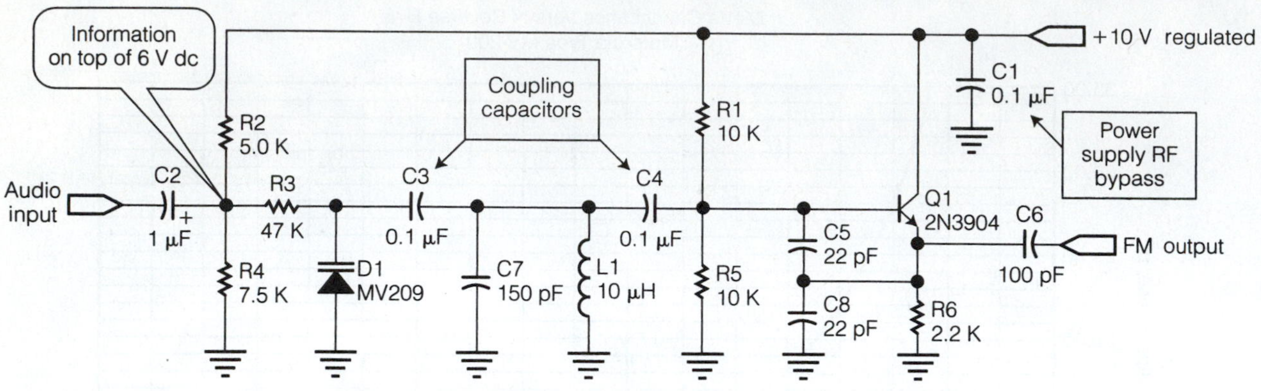

Figure 8–13 A frequency-modulated Colpitts oscillator

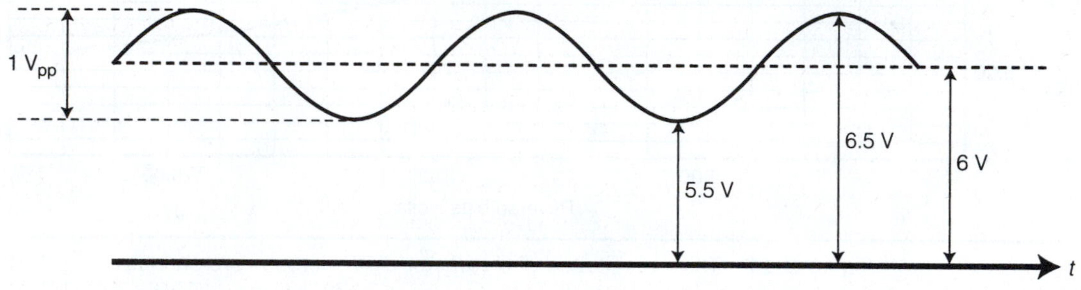

Figure 8–14 The audio riding on a dc level

the audio signal source. Because the varactor diode is reverse-biased, it draws no dc current, so there is no significant dc voltage drop across R_3. The capacitance of D_1 depends upon the voltage applied to it. The voltage applied to D_1 is the sum of the 6 V dc bias and the $1V_{pp}$ information signal. *Therefore, the capacitance of D_1 varies in step with the information signal.*

In order for D_1 to affect the frequency of the oscillator, it must be ac-coupled into the tank circuit that controls the oscillator's frequency. C_3 ac-couples D_1 to the oscillator tank circuit, which is built of L_1 and C_7. C_3 is necessary since a dc voltage is developed across D_1 for controlling its capacitance, yet there can be no dc voltage drop across inductor L_1 (an inductor is a short to dc). C_3 can therefore also be described as a dc-blocking capacitor.

The tank circuit therefore consists of L_1, C_7, D_1, and two more "hidden" capacitors. Can you see them? Yes, because the transistor base circuit is coupled to the tank by C_4, capacitors C_5 and C_8 (the feedback capacitors) are also part of the tank. This isn't really as complicated as it sounds. Figure 8–15 shows the development of the equivalent circuit for the tank portion of this oscillator.

In Figure 8–15(a), we have modeled the varactor diode as a variable capacitor. How did we get the range of capacitance? Right, we used the graph of Figure 8–12. (Our figures are rounded to the nearest 0.5 pF for simplicity.) The varactor diode sees a voltage that ranges from 6.5 V (information positive peak) to 5.5 V (information negative peak).

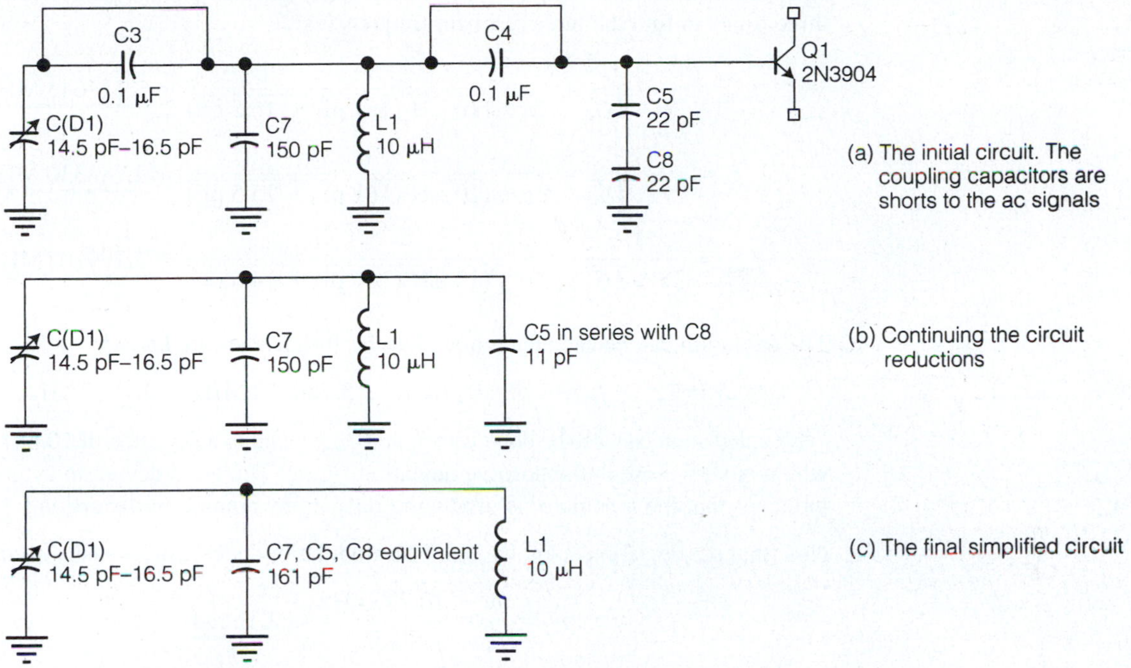

Figure 8–15 The equivalent tank circuit

When the information voltage crosses through zero, the diode sees only the divider voltage of 6 volts. Next, we have recognized that the coupling capacitors C_3 and C_4 are short circuits to the ac currents and can be replaced with wires. C_5 and C_8 appear to the tank circuit in *series* (approximately), and they can be combined into a single 11 pF capacitance. Finally, all the individual capacitances have been combined. The final circuit is a lot easier to understand than the original, and can be analyzed with good accuracy.

EXAMPLE 8–9

The oscillator of Figure 8–13 is being modulated by a 5 kHz, 1 volt peak-to-peak information signal. Determine the following quantities:

a. f_c, f_{min}, f_{max}

b. δ (deviation)

c. m_f (FM modulation index)

d. bandwidth by Carson's rule

e. bandwidth according to a Bessel table

Solution

a. To calculate the three frequencies, the capacitance of D_1 must be known. From the graph of Figure 8–12 and the circuit analysis of the oscillator, we know that the corresponding diode voltages and capacitances are 5.5 V (C_D = 16.5 pF), [center] 6.0 V (C_D = 15.5 pF), and 6.5 V (C_D = 14.5 pF). The total tank capacitance, C_t, in each case will be C_D + 161 pF. We simply apply the resonant frequency formula three times to find the three resonant frequencies:

$$f_{\min} = \frac{1}{2\pi\sqrt{LC_t}} = \frac{1}{2\pi\sqrt{(10\ \mu\text{H})(161\ \text{pF} + 16.5\ \text{pF})}} = \underline{\underline{3.777643\ \text{MHz}}}$$

$$f_c = \frac{1}{2\pi\sqrt{LC_t}} = \frac{1}{2\pi\sqrt{(10\ \mu\text{H})(161\ \text{pF} + 15.5\ \text{pF})}} = \underline{\underline{3.788330\ \text{MHz}}}$$

$$f_{\max} = \frac{1}{2\pi\sqrt{LC_t}} = \frac{1}{2\pi\sqrt{(10\ \mu\text{H})(161\ \text{pF} + 14.5\ \text{pF})}} = \underline{\underline{3.799107\ \text{MHz}}}$$

b. The deviation can be calculated now that the frequencies are known:

$$\delta = f_{\max} - f_c = 3.799107\ \text{MHz} - 3.788330\ \text{MHz} = \underline{\underline{10.777\ \text{kHz}}}$$

(This calculation can also be done with f_c and $f_{\min}$, yielding a deviation of $\underline{\underline{10.687\ \text{kHz}}}$, which is very close to the positive deviation figure. The good deviation symmetry indicates that the modulator is producing only a tiny amount of distortion.)

c. Now that the deviation is known, the FM modulation index can be easily computed:

$$m_f = \frac{\delta}{f_m} = \frac{10.777\ \text{kHz}}{5\ \text{kHz}} = \underline{\underline{2.15\ \text{rad}}}$$

d. The bandwidth by Carson's rule is

$$\text{BW} \approx 2(f_m + \delta) \approx 2(5\ \text{kHz} + 10.777\ \text{kHz}) \approx \underline{\underline{31.5\ \text{kHz}}}$$

e. The closest modulation index in the Bessel table is 2.0, and for this index, the non-zero sideband coefficients are J_1, J_2, J_3, and J_4. Therefore, the bandwidth will be close to $(2)(4)(f_m) = (2)(4)(5\ \text{kHz}) = \underline{\underline{40\ \text{kHz.}}}$

We really got a *lot* of information from the circuit! This is a very good example of the power of electronics fundamentals. We didn't break any new ground in this analysis; we just applied fundamental principles over and over.

A FET Reactance Modulator

On occasion, a transistor is used as a reactance modulator. Figure 8–16 shows a MOSFET being used in this manner.

The complexity of the circuit of Figure 8–16 is one very good reason why the varactor diode is so popular! Q_1 operates as a common-source amplifier and is midpoint-biased by resistors R_2 and R_4 (on the number 1 gate input). The audio input is applied into the number 2 (bottom) gate input through L_2, an RF choke. The drain current of Q_1 varies in step with the applied audio signal, and therefore, the *transconductance* (g_m) of the MOSFET also varies in step with the information. The gain of the "amplifier" depends entirely upon g_m because the source is totally bypassed for ac by capacitors C_6 (RF bypass) and C_7 (AF bypass).

This circuit has feedback, which is applied through R_1, C_2 (a dc block/coupling capacitor), and C_{gs}, the gate 2 to source capacitance of Q_1. The gate 2 feedback voltage phase *lags* the drain circuit current, which is the sign of a *capacitive* effect. However,

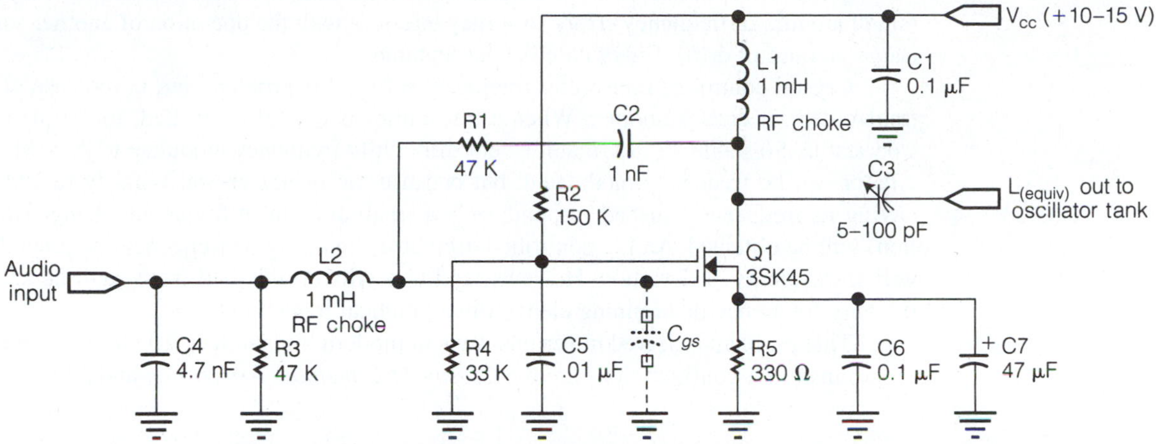

Figure 8–16 A MOSFET reactance modulator

the MOSFET produces a 180° phase shift between its gate and drain electrodes, which effectively "transforms" C_{gs} into an inductance value proportional to the transductance, which is controlled by the information signal.

Therefore, the output of the circuit appears as a variable inductance, which can frequency modulate an LC tank circuit. The circuit can also be constructed to appear as a variable capacitance by placing R_1 into the position occupied by C_{gs} and replacing C_2 with a small capacitance (high capacitive reactance) in the 10–100 pF range.

Section Checkpoint

8–27 How is an oscillator frequency modulated?

8–28 What is the function of a reactance modulator?

8–29 What causes the varactor diode to appear as a varying capacitance?

8–30 What type of bias must be used with varactor diodes?

8–31 Why are capacitors C_3 and C_4 needed in Figure 8–13?

8–32 Why is the varactor diode much more popular than the MOSFET reactance modulator of Figure 8–16?

8–4 THREE FM TRANSMITTERS

One of the basic requirements of all radio transmitters is a stable carrier frequency. By this we mean a carrier frequency that is accurate in value and free from *drift*. Frequency drift is any unwanted change in the operating frequency of the transmitter. Drift normally implies a *slow* change, occurring over a span of minutes or hours.

When a transmitter drifts off frequency, its signal may become distorted in a receiver (small amount of frequency error), or it may interfere with the operation of another station (large amount of drift). Either case is unacceptable.

Crystal control of transmitter frequency solves this problem, but in the case of FM transmitters, it creates another. When a transmitter is crystal controlled, its frequency is *very* stable. So stable, in fact, that it's very difficult to frequency modulate it! A crystal oscillator can be frequency modulated, but because the quartz crystal is fairly resistant to having its frequency "pushed" around, only a small amount of frequency change (deviation) will be obtained. An LC-controlled oscillator can easily be frequency modulated and will allow plenty of deviation. However, an LC oscillator will drift, so it can't be used as the only frequency-determining element in a practical transmitter.

This problem is solved in various ways in modern transmitters. The three most common transmitter configurations are the *Crosby AFC method, Armstrong,* and *PLL.*

Crosby Transmitter

Figure 8–17 is a block diagram of a Crosby FM exciter. It's based on a crystal *reference* oscillator and operates similar to a phase-locked loop in many respects. The heart of a Crosby transmitter is the *frequency discriminator*. The frequency discriminator is an analog circuit that converts frequency changes into voltages. If you're thinking that the discriminator is really an FM detector, you're right! The discriminator "discriminates" or *distinguishes* between different frequencies. It is tuned by an LC circuit to operate on a particular frequency, and when the input frequency varies above or below that operating frequency, the output voltage of the discriminator either goes positive (frequency above the

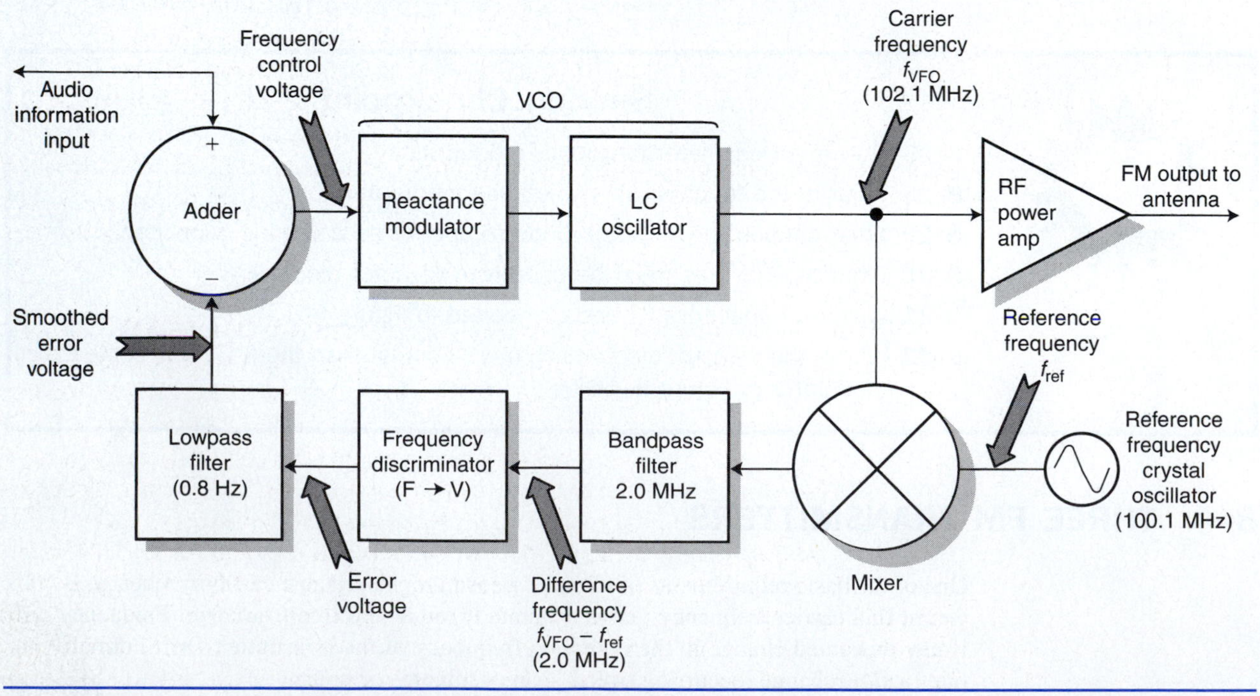

Figure 8–17 A Crosby FM transmitter

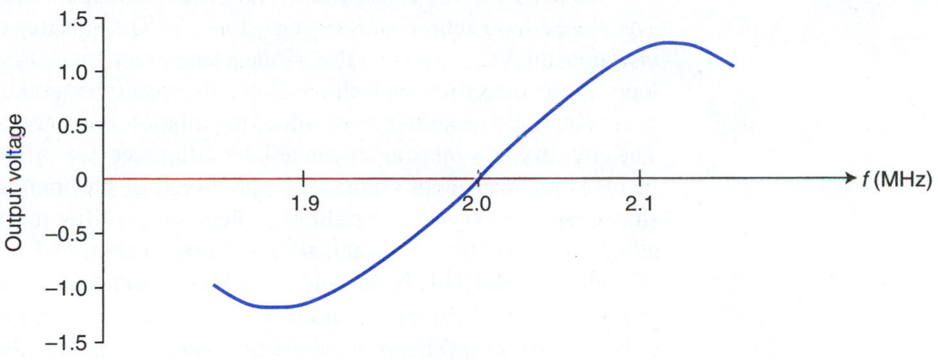

Figure 8–18 The transfer characteristic of a discriminator

tuned frequency) or negative (frequency below the tuned frequency). Figure 8–18 is an example graph of a discriminator input–output relationship.

When the input frequency of this discriminator is 2.0 MHz, the output voltage is *balanced* at 0 volts. If the input frequency increases, the output voltage rises; if the input frequency decreases, the output voltage goes negative. There's only a limited frequency range over which an analog discriminator can operate; this one can operate about 100 kHz above or below the center frequency. This is important, because the maximum deviation in the broadcast FM signal is 75 kHz, and the discriminator must have a wide enough bandwidth to accommodate this.

The carrier frequency in a Crosby transmitter is created by a voltage-controlled oscillator, or VCO, which is built by combining a reactance modulator with an LC-controlled oscillator. Together, these two circuits convert the *frequency control voltage* into an output frequency. Since the VCO is an LC oscillator, it can be readily frequency modulated, but it will tend to drift off frequency. To correct this, the Crosby exciter uses *negative feedback.* The output frequency of the VCO is downconverted to the operating frequency of the discriminator by the use of a crystal-controlled *reference oscillator* and a loop mixer.

The crystal oscillator operates 2 MHz below the desired carrier frequency of the transmitter. In the Figure 8–18, the crystal reference oscillator operates on 100.1 MHz, and the intended VCO frequency is 102.1 MHz. The loop mixer produces the sum and difference frequencies, which are 202.2 MHz and 2.0 MHz. The mixer's output passes through a bandpass filter tuned to the difference of 2.0 MHz, so only the 2.0 MHz signal passes on to the discriminator.

The discriminator sees the 2.0 MHz signal and converts it to a 0 volt level. This 0 volt level passes through the low-pass filter, which removes everything but the dc level. This dc level, the *smoothed error voltage,* passes back to the VCO through the inverting (−) portion of the *adder*. The adder algebraically adds two voltages. In this case, there is a 0 volt feedback signal, so the VCO doesn't need to make any corrections; it's right on frequency.

Suppose that the VCO frequency were to drift upward by 50 kHz (the maximum permissible frequency error in FM broadcast is 2 kHz). How will the loop respond?

The VCO frequency will be 102.15 MHz, and this will mix with the 100.1 MHz reference oscillator signal to produce a difference frequency of 2.05 MHz. Referring to the graph of Figure 8–18, we can see that the discriminator will produce about +0.5 volt under this condition. This +0.5 volt level will pass back to the VCO through the low-pass filter and inverting-input of the adder, which produces a *negative* correction voltage at the VCO. In turn, the VCO output frequency goes *down,* which corrects the error.

Like a PLL, the Crosby transmitter is a closed loop. There's a problem here, though. The closed-loop action tends to correct *any* VCO frequency errors. But wait—what if we *modulate* the VCO? Doesn't that create a temporary frequency error (deviation)? Won't the loop try to correct that as well—which will *cancel* the modulation we tried to create?

An FM transmitter must allow modulation but must also prevent frequency drift. The circuitry of a transmitter can tell the difference between the two, because *modulation* involves rapid changes (short-term variations) of the transmitter, while *drift* consists of slow changes (long-term variations). The low-pass filter in a Crosby transmitter has a relatively long RC time constant, about 200 ms. This translates to a filter cutoff frequency of about 0.8 Hz! This is far below the lowest audio frequency the transmitter might reproduce, so practically none of the ac audio energy can get through this filter. Therefore, only a steady dc level appears at the filter output, which really represents the *average* frequency of the transmitter. This is exactly what is needed to correct the VCO—a nice, smooth dc voltage!

Problems with the Crosby exciter The Crosby transmitter has two main problems. First, it is subject to drift from analog components; the frequency discriminator circuit is a purely analog circuit and is crucial to determining the transmitter's frequency. Crosby transmitters often need discriminator "touch-up" alignments after a few years of operation.

Second, the Crosby transmitter is not frequency agile. Its frequency is determined by a fixed-frequency quartz oscillator. In FM broadcast, this is not a serious problem; stations are expected to stay on the same frequency. Thus, the primary application of this type of transmitter is broadcast.

The Armstrong Wideband FM System

The Armstrong transmitter takes a unique approach to generation of wideband FM: Directly modulate a crystal oscillator (which produces a stable center frequency, but only 100 to 200 Hz of deviation), then use extraordinary amounts of frequency multiplication to increase the deviation to the required value of 75 kHz needed for broadcast FM. Both *frequency multiplication* and *frequency conversion* are used in an Armstrong transmitter.

Frequency multiplication A frequency multiplier is a special amplifier that is designed to produce a harmonic of the input frequency, as shown in Figure 8–19. The figure shows a class C amplifier. There is nothing particularly special about this unit, except the tuning of

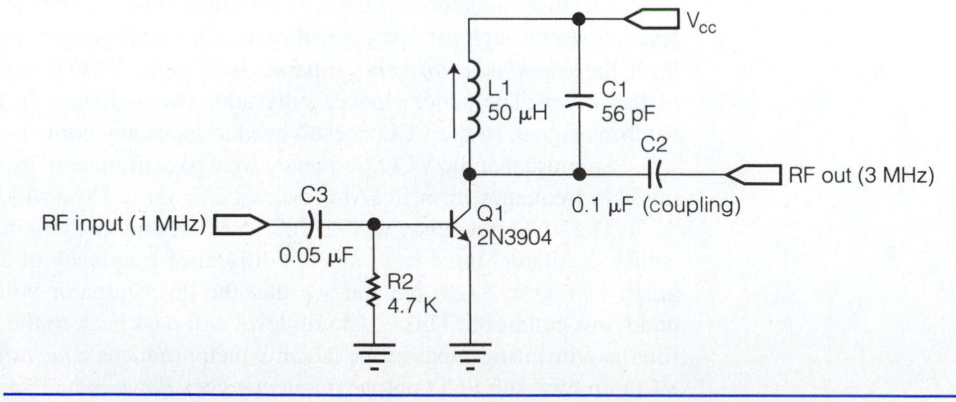

Figure 8–19 An ×3 frequency multiplier

the output tank circuit, L_1 and C_1. This circuit is designed to operate with an input frequency of 1 MHz, yet the tank is tuned to 3 MHz. How can this work?

The *flywheel* effect of the tank allows this to happen. It's very much like pushing someone in a swing. When the tank is tuned to the input frequency (1 MHz), it is analogous to giving one push per swing. However, to keep the person on the swing in motion, it really isn't necessary to push every time. You could choose to push once every three times, for example. The ratio of the swinging (the tank frequency) to your pushing (the input frequency) would be precisely 3:1. This is exactly what this circuit does, when viewed in the time domain.

The multiplier can also be explained using frequency domain concepts. The class C amplifier heavily distorts the 1 MHz input signal, resulting in a pulsating collector current. Because the collector signal is no longer a pure sine wave, it contains *harmonics,* which are multiples of the input frequency. The tank circuit (L_1 and C_1) acts as a bandpass filter, and if it is tuned on to one of the harmonics (which are sine waves), the output will be a sine wave at the frequency of the harmonic. When we look at the circuit in the frequency domain, we see that the amplifier *must* distort the input signal in order for the circuit to work. That is why a class C stage is generally used in a frequency multiplier.

Harmonics normally become weaker as their frequency increases. For this reason, the maximum practical multiplication factor for one stage is between $\times 5$ and $\times 7$. To obtain higher multiplication ratios, several multiplier stages are cascaded.

A frequency multiplier increases not only the carrier frequency, but the deviation of an FM signal. If the 1 MHz input signal had a deviation of 5 kHz, then the resulting output deviation would be *three times* that, or 15 kHz.

A frequency multiplier increases deviation by its multiplication ratio.

Figure 8–20 shows this principle as it is applied in an Armstrong WBFM transmitter. In the Armstrong transmitter, the incoming audio signal directly modulates a crystal

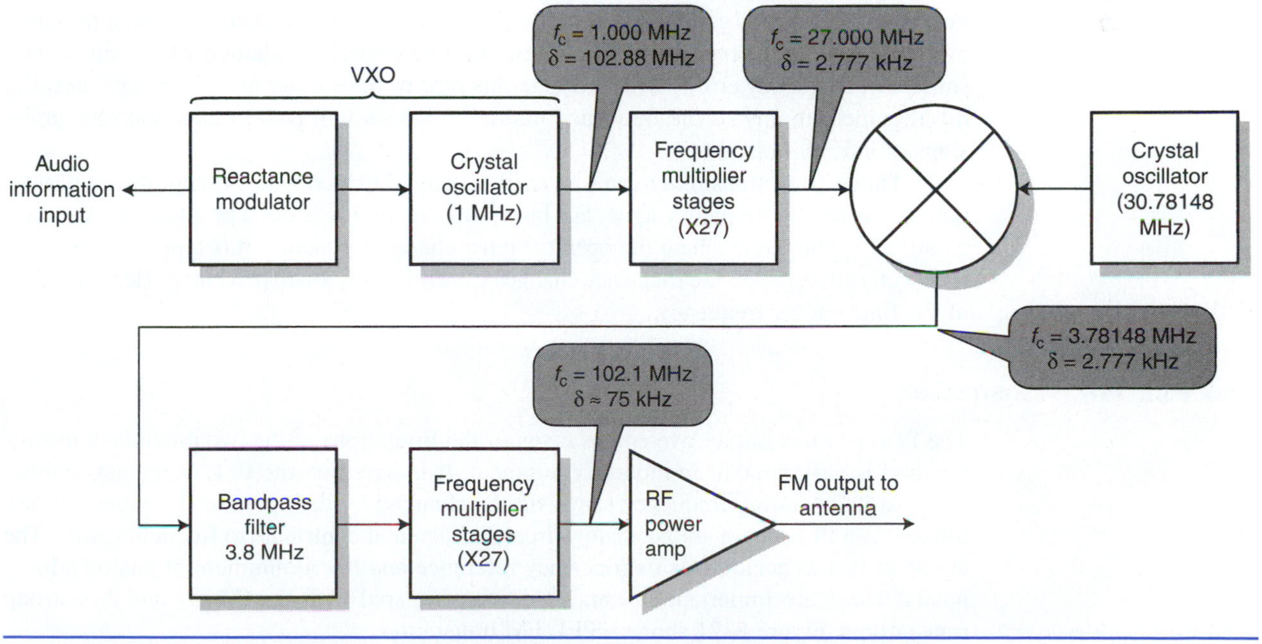

Figure 8–20 An Armstrong WBFM transmitter

oscillator. The combination of the reactance modulator and crystal oscillator is sometimes called a *VXO,* or voltage-controlled crystal oscillator. Because the crystal is very stable, its frequency deviates very little when modulated. The circuit in Figure 8–20 is producing a whopping 102.88 Hz of deviation. This is hardly enough for broadcast FM, but we know that by using frequency multiplier stages, the deviation can be increased.

The 1 MHz carrier signal is passed through a ×27 multiplication, which requires three ×3 stages (3 MHz, 9 MHz, 27 MHz tanks) to implement. The resulting deviation is now 27 times 102.88 Hz, or 2.77 kHz. This is a definite improvement—we could almost use the resulting signal for voice communication applications ($\delta = 5$ kHz) with a little more multiplication. However, we do run into a problem at this point. If we continue to multiply the frequency of the oscillator until we reach the FM broadcast band, we won't get enough deviation. For example, suppose that we sent the signal into one more ×3 multiplier stage. The resulting carrier would be 81 MHz (just below the 88 MHz limit of the band), and the resulting deviation would be (2.77 kHz)(3) = 8.31 kHz. The amount of 8.31 kHz is still very far from the 75 kHz deviation needed, but if we multiply the signal further, the carrier frequency will be far above the FM broadcast frequencies. In order to let us continue to multiply the signal, we must *downconvert* it to a lower carrier frequency first. This is the secret of the Armstrong system!

The process of frequency conversion does not affect the deviation of an FM signal. It only affects the carrier frequency. In Figure 8–20, a second oscillator at 30.78148 MHz mixes with the 27 MHz FM signal to produce a difference frequency of (30.78148 MHz − 27 MHz) or 3.78148 MHz. The deviation at the output of the mixer is still 2.77 kHz, but the carrier frequency is now much lower. Aha! Now we can continue the frequency multiplications!

Right—the 3.78148 MHz carrier, when passed through the bottom ×27 multiplier stages, becomes a 102.1 MHz carrier with 75 kHz (2.77 kHz × 27) deviation. A wideband FM signal has been generated from the humble deviation of a crystal oscillator!

Problems with the Armstrong wideband FM system The Armstrong system has two main problems. First, there are *many* analog circuits that must be aligned and are subject to drift. For example, an ×27 frequency multiplier has at least three tuned circuits, each of which must be set to the correct frequency. There are two crystal oscillators, which allows two sources of frequency error. Adjustments in this type of transmitter must be made carefully; misalignment in any of the frequency multiplier stages will cause either spurious emissions or a dead transmitter!

This type of transmitter can't be easily changed in frequency, except over a narrow range. It is hardly frequency agile. For broadcast, again this is not a problem; for communications applications where the operator must change frequency, this type of transmitter isn't well suited, unless the frequency changes are a relatively small percentage (less than 2%) of the final carrier frequency.

The PLL FM Transmitter

The PLL FM transmitter overcomes many of the limitations of the two previous transmitters and is very popular in modern equipment. It is based on the PLL frequency synthesizer, so its operating frequency can easily be changed by the operator (frequency agile). Since it is a PLL, there are no analog circuits to age and contribute to frequency drift. The circuit is just as accurate as its frequency reference and has a minimum of analog adjustments. These are important advantages when compared with the Crosby and Armstrong transmitters. Figure 8–21 shows a PLL FM transmitter.

The PLL FM transmitter in the figure is very similar to the PLL frequency synthesizer presented in Chapter 7. The primary addition to the loop is the *adder* circuit that has

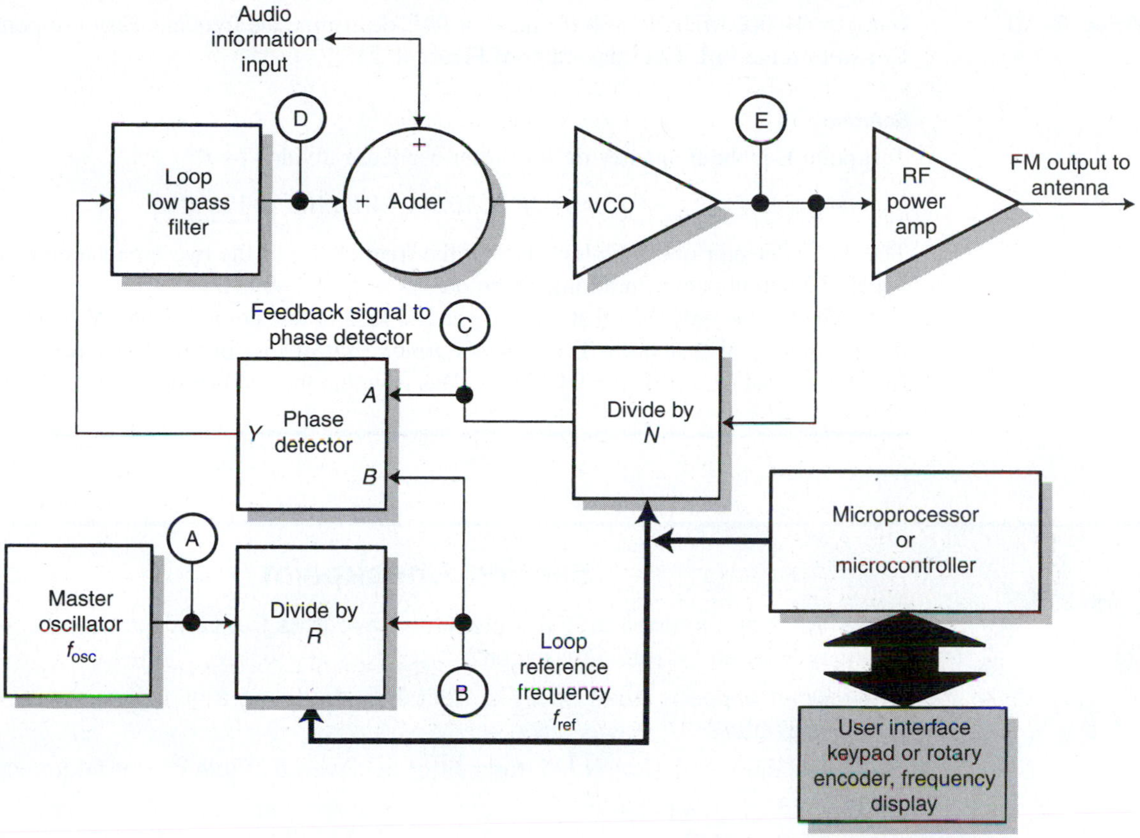

Figure 8–21 A PLL FM transmitter

been placed between the loop low-pass filter and VCO. The adder circuit allows the information signal to directly frequency modulate the VCO. The loop otherwise operates exactly as a normal PLL would. The phase detector likes to see the frequencies of the signals at test points C and B to be equal and outputs appropriate correction command voltages to the low-pass filter when this is not so, in order to lock the VCO onto a multiple (*N*) of the reference frequency. The programmable divider sets the operating frequency of the loop, and is controlled by the microprocessor (which interprets the front-panel controls in order to decide what value of *N* is needed).

The low-pass filter is very important in a PLL FM transmitter. When the information signal modulates the VCO, the phase detector immediately recognizes that the loop is out of lock (Finley's law) and makes corrections to the VCO control voltage to push the loop back on frequency. In other words, the same self-canceling effect appears here that was present in the Crosby transmitter. The low-pass filter is selected to have a long RC time constant, in order that the loop can ignore the short-term errors caused by modulation yet correct any long-term errors (drift).

A technician should be aware that proper design of the phase-locked loop involves careful balancing of many different circuit parameters. Making alterations to a PLL circuit (especially the filter sections) can have drastic effects on its operation. The loop may suddenly refuse to lock, or it may *self-oscillate*. When trouble is suspected in a PLL, don't attempt to reengineer the circuit. Replace components only with the *exact* same value and type!

EXAMPLE 8–10 If $f_{osc} = 10.000$ MHz, $R = 100$, and $N = 941$, determine the frequencies at test points B, C, and E in the PLL FM transmitter of Figure 8–21.

Solution

Test point B is the frequency of the master oscillator divided by R:

$$f_B = f_{osc}/R = 10.000 \text{ MHz}/100 = \underline{\underline{100 \text{ kHz}}}$$

The phase detector obeys Finley's law (if the frequencies of the two inputs aren't exactly equal, the output is in saturation). Therefore, $f_C = f_B = \underline{\underline{100 \text{ kHz}}}$.

Finally, the only way that we can get 100 kHz at test point C when $N = 941$ is for the frequency at test point E to be 941 times that at test point C. Therefore, $f_E = (N)(f_C) = (941)(100 \text{ kHz}) = \underline{\underline{94.1 \text{ MHz}}}$. This is within the FM broadcast band.

Section Checkpoint

8–33 What problem will an FM transmitter have if its frequency is determined only by an LC resonant circuit?

8–34 What happens when we try to frequency modulate a quartz crystal oscillator?

8–35 Explain how a Crosby FM transmitter achieves a stable carrier frequency.

8–36 Why is a low-pass filter necessary in the feedback loop of a Crosby transmitter?

8–37 What happens to the frequency and deviation of a carrier signal when it passes through a frequency multiplier?

8–38 What effect does frequency conversion have on the deviation of an FM signal?

8–39 Why is a class C amplifier usually used in a frequency multiplier stage?

8–40 What is a VXO?

8–41 In what ways is the PLL FM transmitter superior to the Crosby and Armstrong circuits?

8–42 What is added to the PLL synthesizer circuit to allow frequency modulation of the VCO?

8–5 SPECIAL TOPICS

Power Amplification in FM Transmitters

All FM transmitters are essentially *low-level* transmitters. You'll recall that this means that the modulation in the transmitter occurs before the final or last amplifier stage. A low-level AM transmitter requires linear amplification stages (class A or B) in order to preserve the shape of the modulation envelope. In contrast, an FM transmitter can use class C amplifiers for all power amplification. Class C is one of the more efficient ways of amplifying a signal, with more than 90% theoretical efficiency possible.

It is possible to use a class C amplifier to amplify an FM signal because of its constant amplitude. There is no envelope shape to distort because the envelope is flat. This gives FM transmitters an edge over AM when comparing efficiency.

Preemphasis and Deemphasis

FM is well known for its immunity to noise, but in one respect, it is inferior to PM (phase modulation). As the frequency of the information signal increases, the ability of the receiver to reject noise at that frequency *decreases*. Most listeners find high-frequency noise to be more bothersome than that at lower frequencies. The hissing sound that is often audible during quiet music passages is mostly high-frequency noise.

In FM broadcast, the techniques of *preemphasis* and *deemphasis* are used to improve the system's noise immunity at high information frequencies. In *preemphasis,* the deviation of the transmitter is boosted for high information frequencies; in *deemphasis,* the response of the detector in the receiver is attenuated at high information frequencies. Figure 8–22 shows the frequency response curves for preemphasis and deemphasis used in the United States.

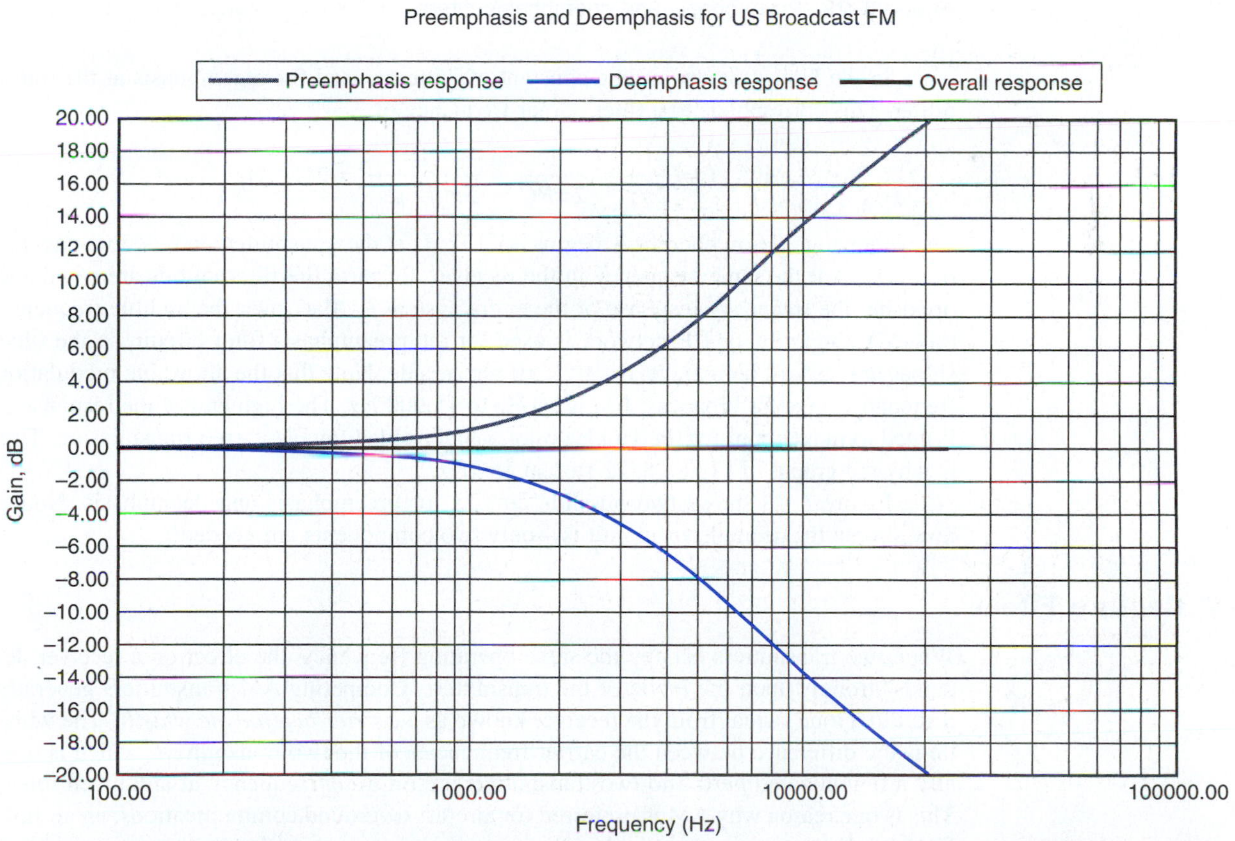

Figure 8–22 Preemphasis and deemphasis frequency response curves

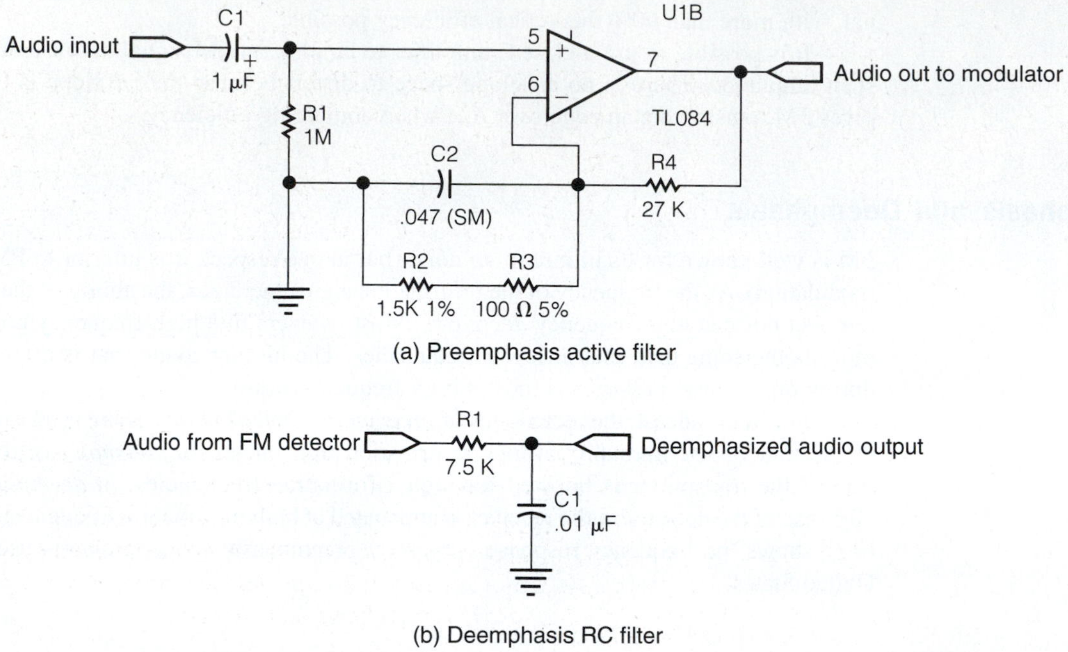

Figure 8–23 Preemphasis and deemphasis filters

In the United States, a time constant of 75 µs is used for preemphasis at the transmitter. This corresponds to a filter corner frequency of

$$f = \frac{1}{2\pi\tau} = \frac{1}{2\pi RC} = \frac{1}{2\pi(75\ \mu\text{s})} = \underline{2122\ \text{Hz}}$$

The "boosting" action, therefore, begins at 2122 Hz at the transmitter, and the low-pass action occurs at the same frequency in the receiver. Because the filter curves are equal and opposite, the frequency response of the entire system is "flat" over the audible frequency range. A single RC or RL network is used for the preemphasis filter circuit, so the filter slopes are ± 6 dB per octave, or ± 20 dB per decade. Note that the allowable modulation frequency range for broadcast FM is 50 Hz to 15,000 Hz. The high end of the response is limited to make room for the 19 kHz pilot signal needed for FM stereo transmission. This is why the graphs of Figure 8–22 stop at 15 kHz.

Figure 8–23 shows two possible circuits for preemphasis and deemphasis. Notice how simple the deemphasis circuit is—only two components are needed!

FM Capture Effect

When two transmitters occupy the same operating frequency, the effect on a receiver depends strongly upon the mode of the transmitters. Competing AM transmitters generally result in a loud squeal from the receiver known as a *carrier heterodyne whistle.* The whistle is the difference between the carrier frequencies of the two transmitters, which is usually a few hundred hertz (no two transmitters are on *exact* frequency at any given time). This is one reason why AM is preferred for aircraft-to-ground communications; an air traffic controller can instantly tell when two aircraft have transmitted at the same time, and no communication is likely to be lost.

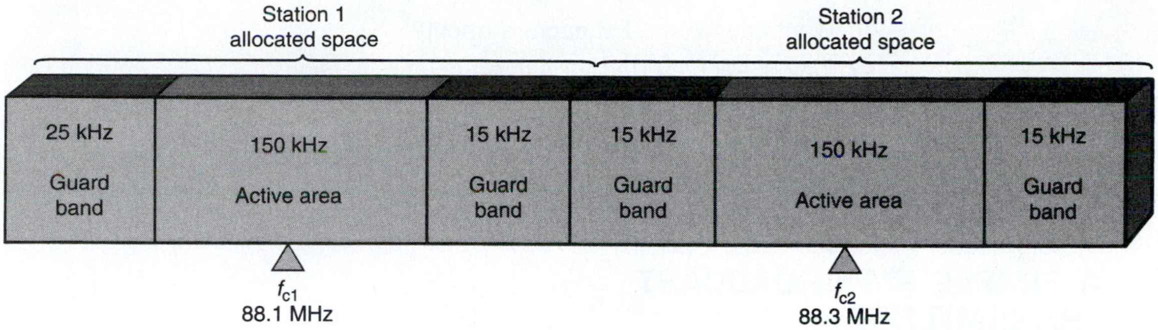

Figure 8–24 U.S. FM broadcast bandplan

FM receivers exhibit what is known as the *capture effect.* When several signals are on the same frequency, the strongest signal "captures" the receiver's detector circuit. The result is that only the strong signal is heard; there is little or no trace of the weaker signal! You can readily observe the capture effect when driving a moderate distance. FM broadcast frequencies are shared by stations in different towns. As your car reaches a point in between the two stations, you will alternately hear one, then the other, as the signals fade in and out. Whichever station is strongest will be the one you hear!

U.S. Broadcast Bandplan

The FCC assigns broadcast stations an operating frequency in a process known as *frequency coordination.* When a station is given a frequency, the geographic location of the station in relation to other stations (in other towns) that may be sharing the frequency is taken into account, so that the possibility of interference is minimized. In addition, an orderly *bandplan* is used to assign station frequencies. Graphically, the plan looks like Figure 8–24.

In Figure 8–24, you can see that the slot for a broadcast station is 200 kHz wide. Of the assigned space, 50 kHz is used by *guard bands.* The guard bands are extra space between stations to help prevent interference. You might also notice that the assigned carrier frequencies make 200 kHz (0.2 MHz) jumps. The available channels are 88.1 MHz, 88.3 MHz, 88.5 MHz, and so on. In addition, the FCC never assigns adjacent channels in the same geographic area. This means that if your town has a station on 102.1 MHz, the next usable frequency will be 102.5 MHz— *not* 102.3 MHz. This further helps to minimize interchannel interference.

Most modern receivers are much more selective and stable than the receivers that existed when the Federal Communications Commission (FCC) first assigned this bandplan, but because there are a lot of older receivers in use, it still makes good sense.

Section Checkpoint

8–43 What type of noise is most objectionable to listeners?

8–44 Explain how preemphasis and deemphasis work together to reduce the noise susceptibility of the FM system.

8–45 What time constant is used for deemphasis in the United States?

(continued on p. 282)

> **8–46** What causes the FM capture effect?
> **8–47** Why is AM preferred for aircraft communications?
> **8–48** What is a bandplan?
> **8–49** Describe the bandplan for FM broadcast in the United States.

8–6 A SIMPLE FM BROADCAST TRANSMITTER

Figures 8–25 and 8–26 are the schematic diagram of a low-powered synthesized FM broadcast exciter. The circuit is PLL frequency synthesized, and covers the entire FM broadcast band in 100 kHz steps. The power output is approximately 4 mW, which will give an operating range of several hundred feet with an appropriate antenna.

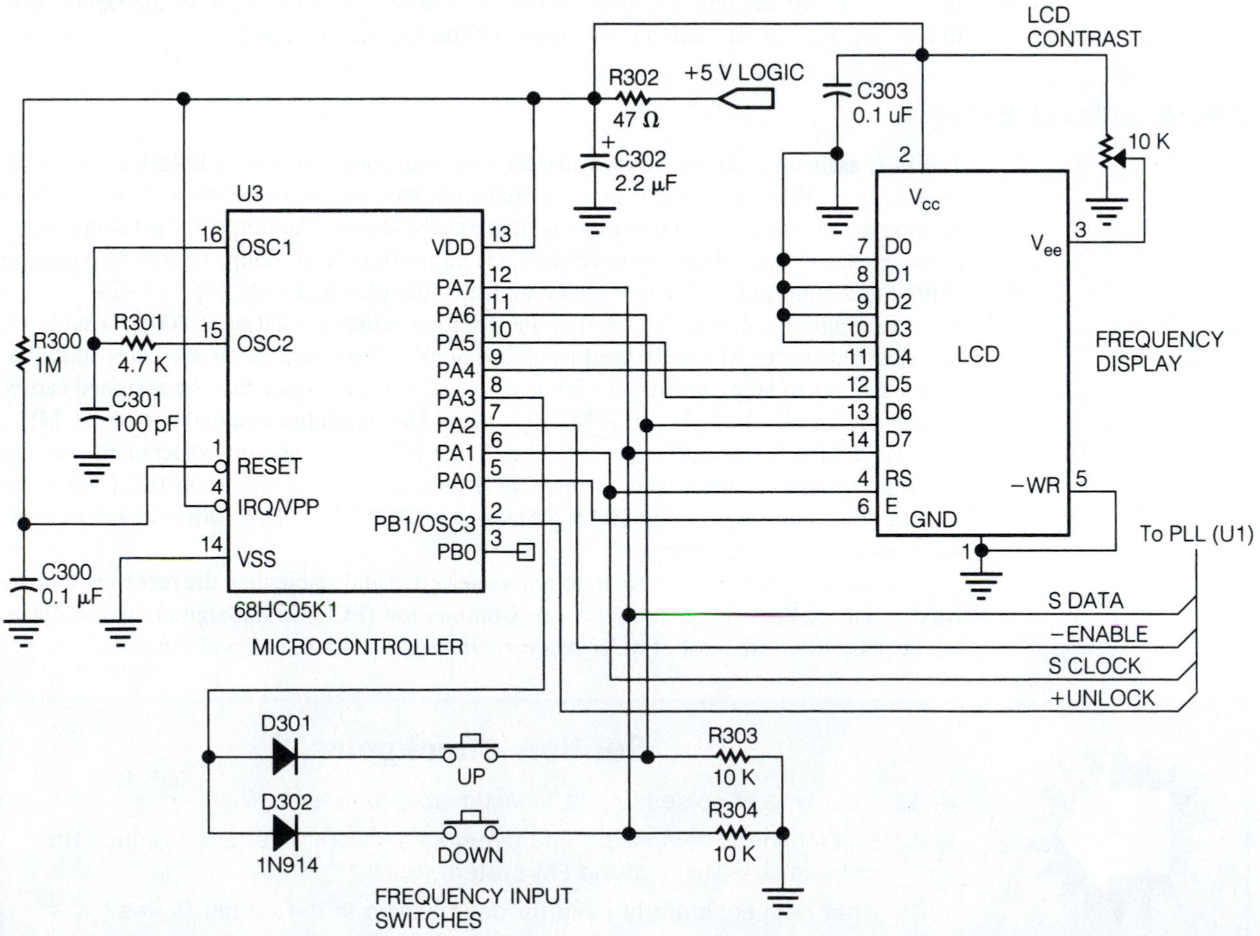

Figure 8–25 The microcontroller circuitry

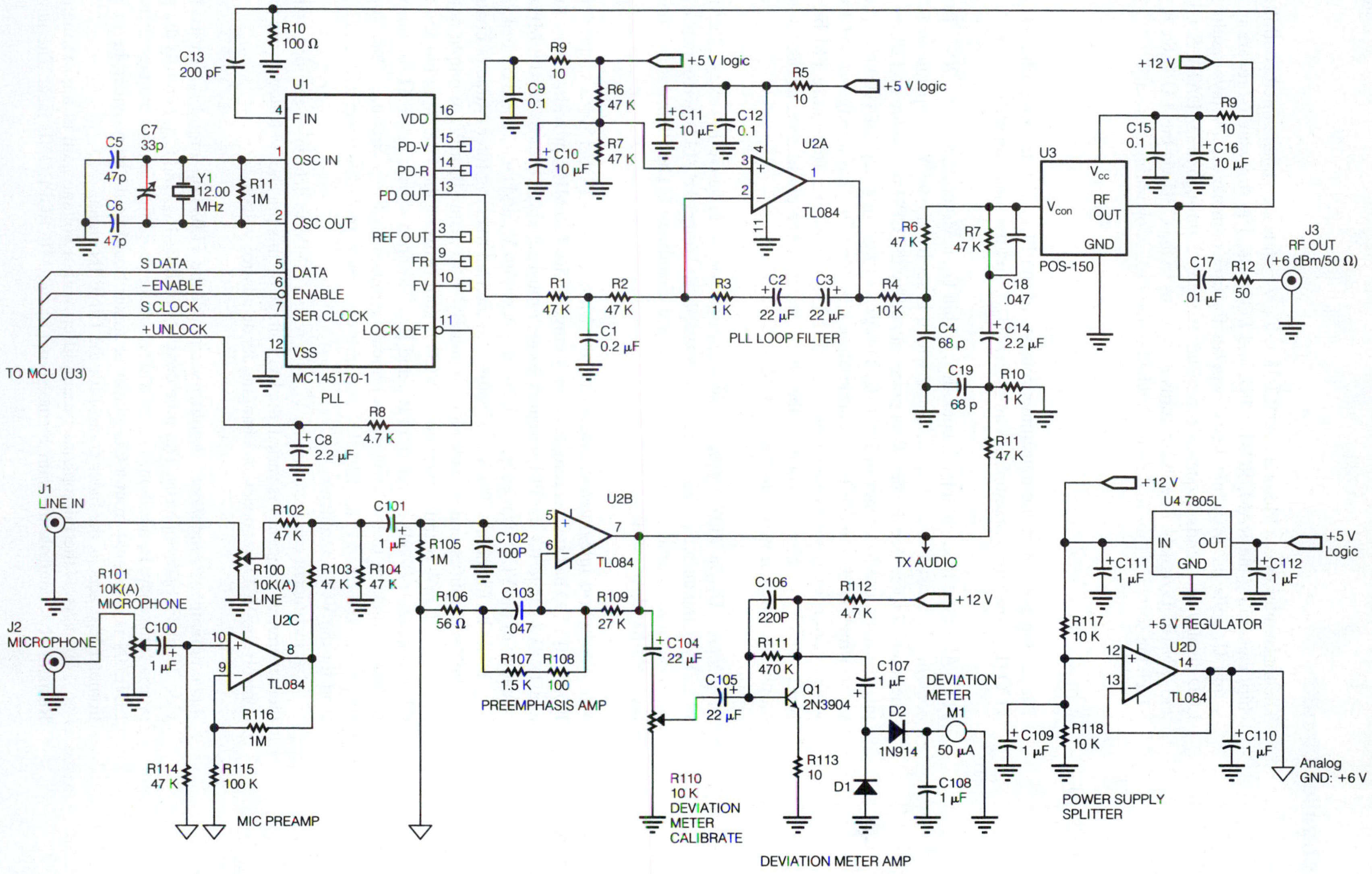

Figure 8–26 The analog transmitter circuitry

Microcontroller Circuitry

The transmitter uses a Motorola 68HC05K1, U_3, which is a complete microcontroller on a chip, containing the RAM, ROM, CPU, and I/O ports. The firmware (software in ROM) within U_3 polls the *frequency input switches* for user commands, displays results on the LCD display, and sends frequency commands to the PLL through the S_DATA, S_CLOCK, and /ENABLE signals. The PLL returns lock status to the CPU via the +UNLOCK signal, which the software uses to determine whether the PLL is operating correctly.

Analog Circuitry

The analog portion of the transmitter consists of the PLL (U_1), the loop filter (U_{2A}), the VCO (U_3), the audio preemphasis amplifier (U_{2B}), and the meter amplifier (Q_1).

PLL All of the PLL circuitry is contained within U_1, a Motorola MC145170-1. This chip contains the master reference oscillator, which is set by crystal Y_1 to operate at 12 MHz. The internal reference divider, *R*, is programmed by the CPU to a value of 120, giving an internal reference frequency of 100 kHz. This gives frequency steps of 100 kHz. The *N* divider, which is also inside U_1 and programmed by the CPU, has a range of 881 to 1081, which corresponds to a frequency range of 88.1–108.1 MHz (the entire FM broadcast band). The PLL chip also contains the phase detector. The phase detector output appears on pin 13, where it is sent to the loop filter.

Loop filter U_{2A}, a TL084 operational amplifier, forms an active low-pass filter for the PLL. By using an active filter rather than a simple RC filter, the PLL's performance can be maximized in several areas (especially spectral cleanliness). The filter output appears at the right-hand side of R_4.

VCO A hybrid integrated-circuit VCO, the Mini Circuits POS-150, is employed. The POS-150 VCO has an operating frequency range of 50–150 MHz, providing more than adequate lock range for the PLL circuitry (which must reliably cover 88 to 108 MHz). Application of the VCO IC is easy; the IC is supplied with V_{cc} and GND and a control voltage, V_{con}, to set its frequency. The VCO features an internal tracking filter (varactor controlled) and buffer amplifier and provides approximately +10 dBm (10 mW) of output power.

The 10 mW output power of the VCO is split between the PLL input pin and the RF OUT connector (load) by resistors R_{10} and R_{12}. The load should be 50 Ω for proper impedance matching to the VCO. (In a typical application of this circuit, further amplifier stages are connected to the RF OUT connector J_3, which provides power gain and a proper load for the VCO circuitry.)

A technician can measure the V_{con} pin with a DMM to determine the loop status. As the transmitter frequency is varied, the DC voltage on this pin should follow, since the PLL controls the VCO frequency by changing its dc control voltage.

Audio preemphasis amplifier Another section of the TL084 op-amp, U_{2B}, is used to provide 75 μs FM preemphasis. The time constant of R_{107}–R_{108} and C_{103} controls the preemphasis frequency. In addition, U_{2B} provides voltage amplification for the signal from the LINE IN jack (signal from a CD player or mixer board could be provided here); potentiometer R_{100} acts as the fader control for the LINE IN signal.

U_{2C} acts as the microphone preamplifier, with a voltage gain of approximately 11. Resistor R_{101} controls the gain of the signal at the MICROPHONE jack. The amplified microphone signal from U_{2C} is summed with the LINE signal by summing network R_{102} and R_{103}, then sent on to the preemphasis amplifier.

The preemphasized audio is sent directly to the VCO through the VCO summing network, consisting of R_6, R_7, R_{11}, and R_{10}. C_{14} is a dc block, and C_{18} corrects preemphasis errors due to the input capacitance at the V_{con} pin of U_3. The summing network allows the ac information signal to be superimposed on the dc control voltage from the loop filter (U_{2A}).

Meter amplifier A deviation meter is provided so that the operator can determine the correct audio level settings while the transmitter is in operation. Since the deviation of an FM signal is proportional to the information amplitude, by measuring the peak audio amplitude, deviation can be displayed. Transistor Q_1 amplifies the ac information signal, while diodes D_1 and D_2 form a full-wave rectifier for converting the ac signal level into a dc level, which operates the DEVIATION METER, M_1. Variable resistor R_{110} varies the audio level being provided to the meter amplifier, thereby allowing calibration of the deviation meter circuit.

Note that the deviation meter measures *preemphasized* audio, so it provides a true reading of deviation at all audio frequencies.

Power Supply

The transmitter is designed to operate on a single 12 volt supply. Operational amplifiers normally require either a split power supply (positive and negative voltage) or a bias power supply to center their Q points. The last section of U_{2D} the TL084 op-amp, is used as an electronic power supply splitter. U_{2D} is wired as a voltage follower (inverting input is tied directly to the op-amp output), and the positive input is connected to a voltage divider (R_{117} and R_{118}). The output of U_{2D} is therefore exactly one half of the power supply voltage. Capacitors C_{109} and C_{110} provide filtering of the bias voltage (notice the different symbol used for *analog ground;* this is an ac ground but has 6 volt dc potential).

The microprocessor and PLL chips require regulated 5 volt; a three-terminal regulator, U_4, provides the necessary 5 volts. Capacitors C_{111} and C_{112} filter the regulated 5 volt supply.

Section Checkpoint

8–50 What controls the frequency of the transmitter of Figure 8–26?

8–51 What is the function of U_3, the 68HC05K1?

8–52 Where should a scope be connected to test communication between the CPU and PLL? (Refer back to "Troubleshooting" in Chapter 7 if necessary.)

8–53 Which chip contains the *R* and *N* dividers, master oscillator, and phase detector?

8–54 What frequency should be present on U_1 pin 2?

8–55 If a dc multimeter is connected to the V_{con} pin of U3 and the transmitter frequency is stepped UP or DOWN by the keypad switches, what should be observed?

8–56 Which components near U_{2B} control the preemphasis time constant?

8–57 List the test points to be measured when testing the power supply (there are at least two). What voltages should be present?

8–7 FM STEREO AND SCA SYSTEMS

In the late 1950s, the technology for making stereo recordings was maturing rapidly. Until that point, all AM and FM broadcasting had been monaural (one channel). The migration to stereophonic (two-channel) reproduction allowed more accurate reproduction of music. One problem presented itself: How could a stereo broadcasting system be developed that wouldn't render the millions of radio receivers already in use obsolete?

The designers of FM stereo had to devise a *compatible* broadcasting format so that an FM stereo signal could be received on a monaural set just as well as the older mono stations. To achieve this, signal components were added to the FM signal to represent the additional information in a stereocast. Mono receivers ignore these extra signal components.

Multiplexing

Stereo reproduction relies on the ability to send two information signals at the same time. One represents the left channel, the other represents the right. *Multiplexing* is the transmission of multiple pieces of information at the same time over the same communications channel.

FM stereo uses *frequency division multiplexing* (FDM) to accomplish this purpose. In FDM, each information component is assigned a different *band* or *range* of frequencies. You might imagine that in FM stereo, we would use one range of frequencies for the left channel, and another for the right. This is conceptually correct but practically unworkable. Such a signal wouldn't be compatible with older receivers, which violates the principle of compatibility.

FM stereo uses frequency division multiplexing, with a trick. That trick is illustrated in Figure 8–27. In FM stereo, *all* of the signal components of Figure 8–27 modulate the carrier of the FM broadcast station. This may seem incredible, but it's true! The same group of signals also appears at the output of the detector in an FM receiver. A mono FM receiver can receive the FM stereo signal with ease by simply ignoring all the frequencies above 15 kHz. Since FM receivers use a low-pass filter for deemphasis, this is automatically taken care of. The deemphasis network in a monaural FM receiver prevents most energy above 15 kHz from reaching the audio amplifier.

Stereo receivers must process the remaining signal components. To understand how this works, let's examine how the FM stereo signal is assembled at the transmitter.

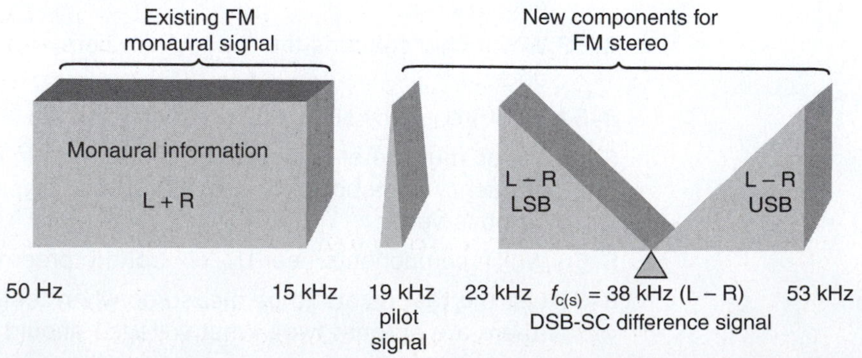

Figure 8–27 The spectrum of an FM stereo signal

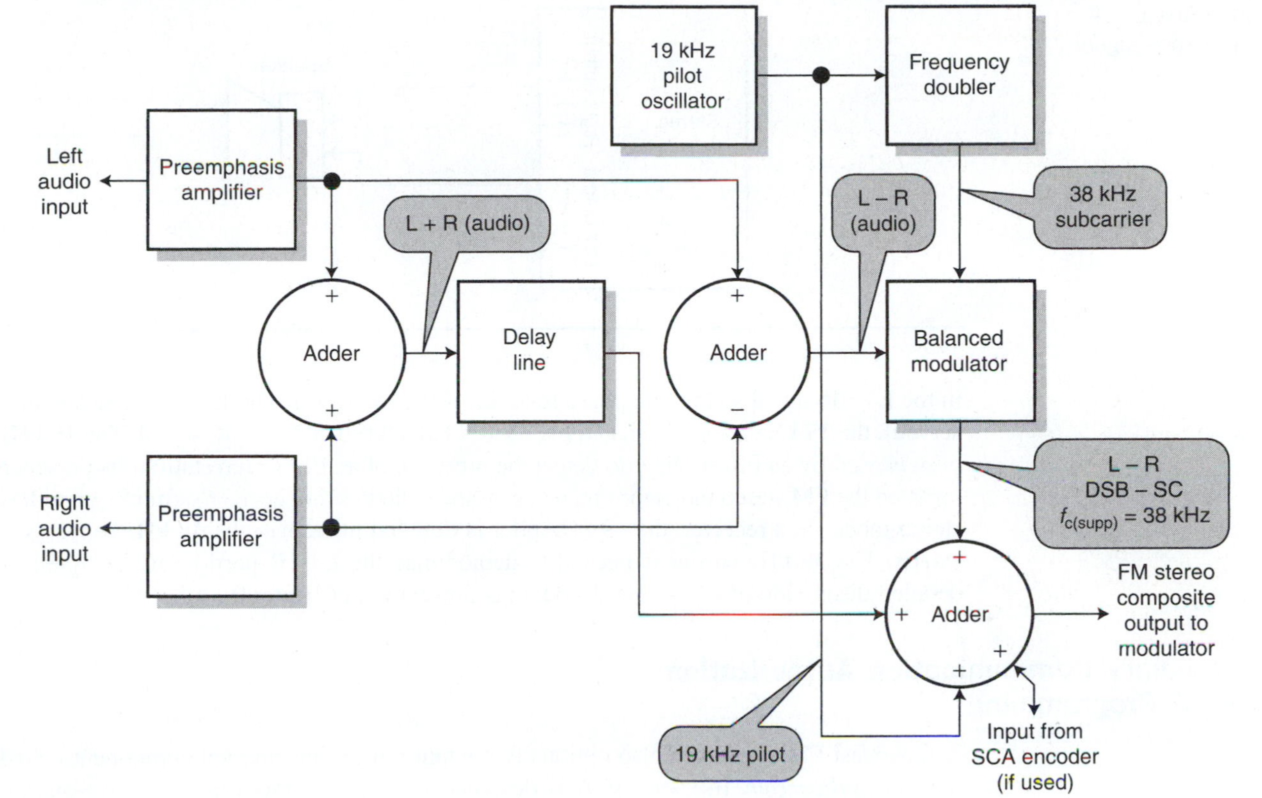

Figure 8–28 Generating an FM stereo signal

Figure 8–28 shows how it works. The two audio signals first undergo preemphasis, exactly as would be done in a monaural FM transmitter. Two *adder* circuits are used to develop the algebraic sum (L + R) and difference (L − R) signals.

The L − R signal is mysterious to most people. In fact, it contains the *difference* between the two audio channels. You can easily listen to the L – R signal by deliberately "miswiring" the speakers in a stereo receiver (Figure 8–29). (Use caution if you try this; some receivers can be damaged by misconnection of speakers to the output stages!) The L − R signal is very interesting to listen to; for example, most music is mixed with the bass tracks equally recorded on the left and right channels. Thus, when listening to the L − R signal, you'll probably notice a lack of bass! Often you'll be able to clearly hear parts of the recording that are fairly obscure in standard playback.

The L − R signal is placed on a 38 kHz suppressed subcarrier by a balanced modulator. The 38 kHz carrier is derived by doubling the frequency of an accurate 19 kHz oscillator called the *pilot oscillator*. Thus, the L − R information is contained on a 38 kHz, DSB-SC signal. Using a DSB-SC signal allows the L − R information to be represented with the best fidelity possible in the available bandwidth and improves the signal-to-noise ratio at the receiver by avoiding the wasted power that would be present if a 38 kHz AM carrier were sent.

The "final assembly" of the signal takes place in another adder. This time, three signal components are algebraically added: The delayed L + R signal (a few microseconds of delay are used to compensate for the time delay generated by the balanced modulator

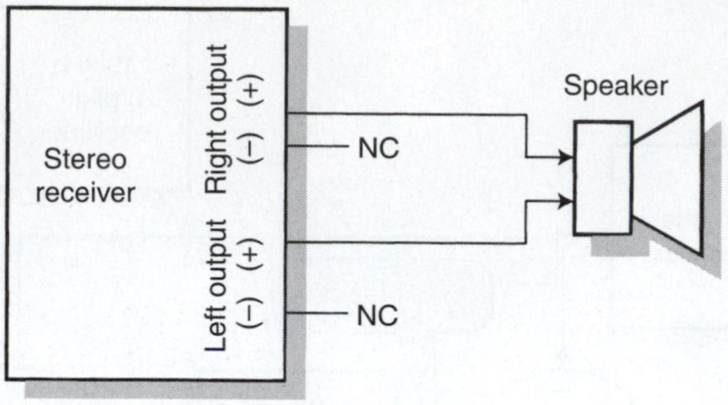

Figure 8–29 Listening in on the L − R difference signal

in the L − R signal so that the phase relationship of the L + R and L − R signals is preserved), the 19 kHz *pilot* signal, and the 23 to 53 kHz DSB-SC L − R signal. The 19 kHz pilot is used by an FM receiver to detect the presence of an FM stereo station. Its presence turns on the FM stereo indicator in a receiver and switches the receiver's circuitry into FM stereo mode. In a receiver, the 19 kHz pilot is doubled to recover the 38 kHz suppressed carrier. The 38 kHz carrier is needed to demodulate the L − R portion of the signal. A detailed discussion of FM stereo decoding is presented in Chapter 9.

Subsidiary Communication Authorization (SCA) Programming

A broadcast FM signal can also optionally contain a third information component called the *SCA subcarrier*. In short, SCA is designed for the transmission of modest-quality speech and music using the same FM carrier as a normal FM signal. The most common application of SCA is the transmission of background music for businesses. The forgettable melody you heard in the elevator in the department store may have been transmitted to the store using SCA. A broadcaster typically charges a business a monthly or yearly fee for the use of the SCA service. The business installs a special FM tuner with an SCA decoder built in; the output from the tuner drives the PA system (or telephone system, for music on hold). Some broadcasters may also transmit digital data on the SCA subcarrier. The digital data can be almost anything, from *telemetry* (remote measurement data) to stock reports. Again, the users of the service pay a fee to the broadcaster for the service. Figure 8–30 shows the components in an FM stereo signal with SCA. It's quite crowded!

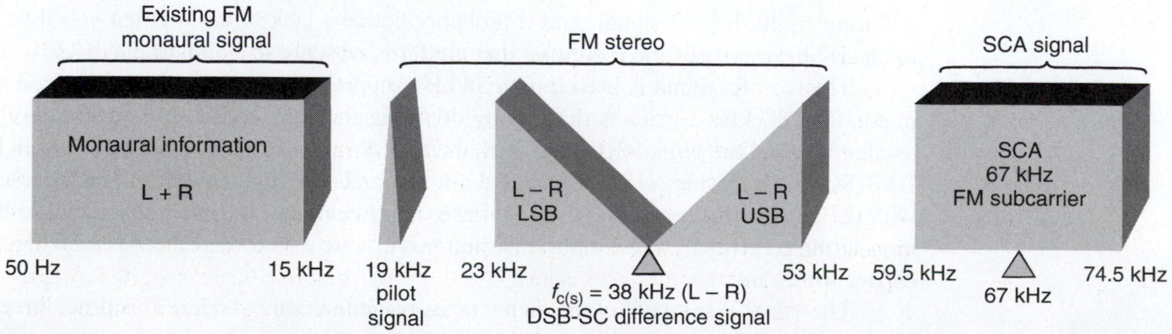

Figure 8–30 An FM stereo signal with SCA

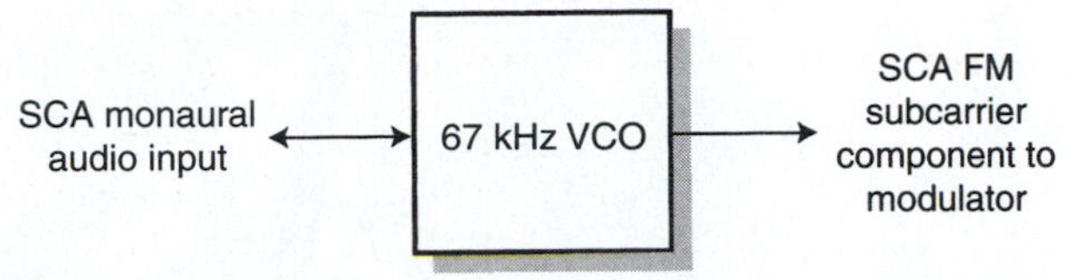

Figure 8–31 Generating the SCA signal

The SCA signal is transmitted on a 67 kHz FM carrier. Consider this carefully: All of the signals of Figure 8–30 modulate the FM broadcast transmitter. One of these signals is itself an FM carrier! When such a signal is detected in an FM receiver, a second stage of detection must follow to demodulate the 67 kHz SCA subcarrier that is buried inside the composite FM signal. The SCA carrier is not intended for true hi-fi reproduction. In fact, only 7.5 kHz of deviation is allowed, and the relative strength of the SCA signal (compared to other components in the overall FM signal) is fairly low (the SCA signal is allowed to modulate the FM carrier only 10%). For background music (or nonmusic, depending on your taste!), the fidelity is more than sufficient.

The SCA signal is generated by a 67 kHz VCO and is added to the final composite output on its way to the FM modulator. Figure 8–31 shows how it's done. The output of the circuit of Figure 8–31 is sent to the FM modulator in the transmitter. If the transmitter is an FM stereo unit, the output goes to the final adder, where it is combined with the rest of the components of the FM signal.

Demodulating the SCA signal at the receiver requires two FM detectors. The first detector reproduces the composite FM signal (Figure 8–30) with the SCA carrier "buried" inside. After detection, the composite FM signal is sent to the SCA decoder, which is an FM detector tuned to 67 kHz. The output of the SCA decoder is the "hidden" audio program.

8–8 MEASURING THE PERFORMANCE OF FM TRANSMITTERS

An RF technician is often called upon to verify the operation of transmitters and receivers. You've already seen that the spectrum analyzer is invaluable for measuring FM transmitter outputs. There are several types of measurements that are often made to verify performance. These include *transmit frequency accuracy, deviation,* and *power output.*

Frequency Accuracy and Deviation

A specialized piece of equipment called a *station monitor* is often used to measure these two parameters. A station monitor is really a calibrated FM receiver with either an oscilloscope or meter for displaying the results. Figure 8–32 shows a Motorola station monitor designed for operation in the upper VHF (130–170 MHz) and mid- UHF (440–512 MHz) bands.

The station monitor of Figure 8–32 directly measures the frequency and deviation of an FM signal. Because it is a sensitive receiver, it need not be connected directly to the transmitter. In many shops, the station monitor is simply connected to its own antenna, perhaps a collapsible rod or "rubber duck." To use the monitor, the transmitter frequency is dialed in on the *Channel 0* thumbwheel switches in the center of the instrument. The transmitter is then keyed with *no modulation* (preferably into a dummy antenna), and the *frequency error,* in kHz, is read on the right-hand display. In the photo, the VHF weather

Figure 8–32 An FM station monitor

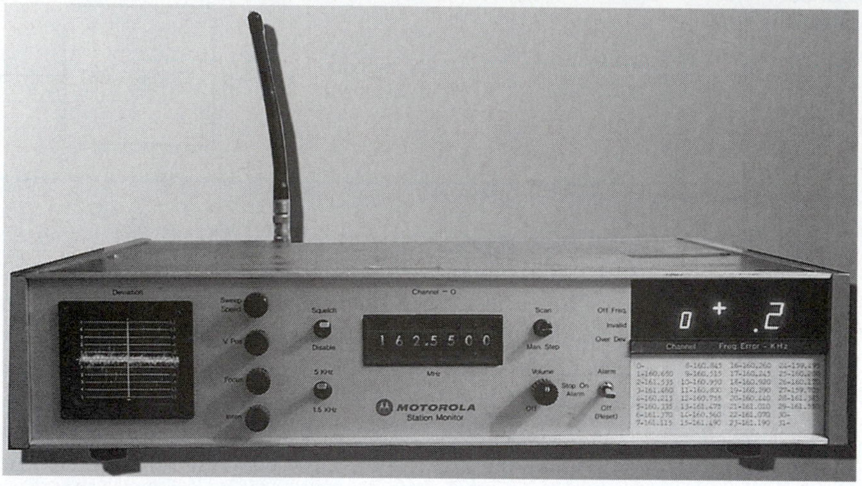

transmitter under test has a +0.2 kHz frequency error, which is well within the 1 kHz limit for VHF transmitters of its type.

The monitor receiver also has a built-in oscilloscope display that is calibrated with its vertical axis in kilohertz, to show deviation. The display in Figure 8–33 shows an FM signal with a deviation of approximately 3 kHz (reading from the left-hand vertical scale). The oscilloscope display is very useful in determining whether or not a transmitter is modulating "cleanly." By this, we mean that the transmitter is not causing excessive distortion of the information. The sine wave on the display above is slightly wide, which could indicate that the transmitter is superimposing additional noise onto the signal.

Power Output Measurement

The power output of an FM transmitter is measured in exactly the same manner as that of an AM transmitter (see Chapter 4). A *directional wattmeter* is connected between the transmitter and a dummy antenna, and the transmitter is keyed.

Figure 8–33 Reading deviation on the station monitor

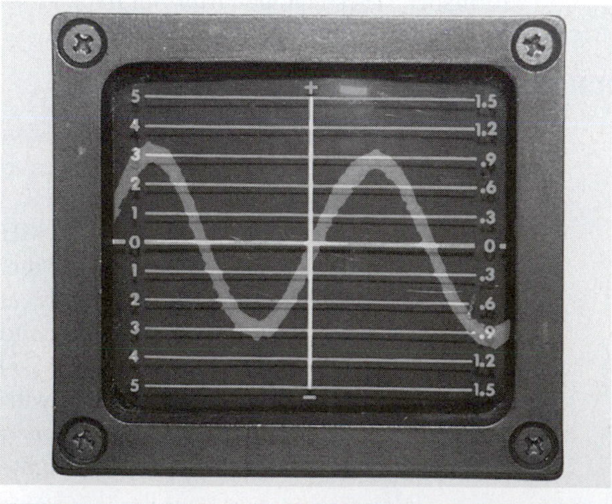

A *service monitor* is an instrument that combines the calibrated FM receiver of a station monitor with a dummy antenna, and sometimes a spectrum analyzer, in one enclosure. Some service monitors even include a signal generator for receiver testing. They're very popular for field work, since almost all major transmitter and receiver testing tools are contained in one portable box.

Measuring Deviation with a Spectrum Analyzer

A spectrum analyzer can be used to indirectly measure the deviation in an FM signal. In order to do this, the technician needs to be able to control the relative amplitude of the information signal. This method relies on the appearance of *nulls* in the carrier frequency component at certain FM modulation indices (notably 2.4 and 5.5). The following steps are used to perform the measurement:

1. Record the original information voltage or amplitude, V_m'.
2. Key the transmitter, and with the information voltage set to zero, slowly increase the information amplitude until the first carrier null occurs (the carrier frequency component will be zero on the analyzer display). *The modulation index is now 2.4.* Record the information voltage that gave the index of 2.4, V_m.
3. Use equation 8–5 to determine the amount of deviation:

$$m_f = \frac{\delta}{f_m}, \text{ so } \delta = m_f f_m = (2.4)f_m$$

   Note that many technicians use an f_m of 1 kHz from a signal generator.

4. The actual deviation is simply the deviation obtained in step 3 multiplied by the ratio of information voltages from steps 1 and 3, since the deviation is proportional to the amplitude of the information.

EXAMPLE 8–11

An FM transmitter is being modulated by a 1 kHz tone whose amplitude is 5 V_p. When the signal is viewed on a spectrum analyzer, adjusting the information amplitude to 2 V_p (starting at zero amplitude) causes the carrier frequency component to null out. Calculate the deviation and modulation index of the transmitter.

Solution

The carrier frequency reaches its first null at an index m_f of 2.4, so when $V_m = 2\text{ V}_p$, we know that

$$\delta = m_f f_m = (2.4)f_m = (2.4)(1\text{ kHz}) = 2.4\text{ kHz}$$

Since deviation is proportional to the voltage of the information, we can now calculate the original deviation:

$$\delta' = \left(\frac{V_{m'}}{V_m}\right)\delta = \left(\frac{5\text{ V}_p}{2\text{ V}_p}\right)2.4\text{ kHz} = \underline{\underline{6\text{kHz}}}$$

Now that we know the original deviation and modulating frequency, we can also find the actual modulation index:

$$m_f = \frac{\delta}{f_m} = \frac{6\text{ kHz}}{1\text{ kHz}} = \underline{\underline{6.0\text{ rad}}}$$

SUMMARY

- In FM, the frequency of the carrier is changed to convey the information. The amplitude of an FM signal is constant, which helps FM receivers reject noise.
- The most direct way of generating FM is to modulate the carrier oscillator. A reactance modulator is often used to do this.
- The deviation in an FM signal is controlled by the amplitude of the intelligence and contributes to bandwidth.
- The FM modulation index is a measure of *peak phase shift* imparted to the carrier during the process of modulation.
- Bandwidth can be estimated for an FM signal by using Carson's rule; if the FM modulation index is known, a more accurate estimation can be done with a Bessel table.
- A reactance modulator converts a voltage into a varying capacitance or inductance. The varying reactance can directly frequency-modulate an oscillator.
- All practical FM transmitters must solve the opposing goals of obtaining a stable carrier frequency yet producing adequate deviation.
- A frequency multipler increases both the carrier frequency and deviation of an FM signal. A frequency converter changes only the carrier frequency.
- Preemphasis and deemphasis are used to improve the S/N for high information frequencies in broadcast FM.
- Multiplexing is the transmission of more than one piece of information at a time on the same communications channel. Frequency division multiplexing is used for FM stereo and SCA.
- The *station monitor* and *service monitor* are two pieces of specialized equipment that are needed to verify the performance of FM transmitters.

PROBLEMS

1. What are two primary advantages of FM over AM?
2. Define *angle modulation.*
3. Where does modulation usually take place in an FM transmitter?
4. Explain the operation of a condensor microphone. How can it be used to generate FM?
5. A certain FM transmitter has a resting frequency of 200 kHz, a maximum frequency of 202.5 kHz, and a minimum frequency of 197.5 kHz. An information signal of 3 V_p, 2.5 kHz is driving the modulator. What is the *deviation* of the transmitter?
6. What is the *deviation rate* for the transmitter of question 5?
7. Recalculate the deviation for the transmitter of question 5 if the information voltage is changed to (a) 1 V_p; (b) 6 V_p
8. For each of the following FM signals, calculate the following: *deviation, percentage of modulation,* and *bandwidth by Carson's rule:*
   a. $f_c = 100$ kHz; $f_{max} = 102$ kHz; $f_{min} = 98$ kHz, $f_m = 1$ kHz, $\delta_{max} = 5$ kHz
   b. $f_c = 300$ kHz; $f_{max} = 310$ kHz; $f_{min} = 290$ kHz, $f_m = 4$ kHz, $\delta_{max} = 10$ kHz
   c. $f_c = 155.450$ MHz, $f_{max} = 155.458$ MHz, $f_{min} = 155.442$ MHz, $f_m = 8$ kHz, $\delta_{max} = 15$ kHz
   d. $f_c = 95.7$ MHz, $f_{max} = 95.760$ MHz, $f_{min} = 95.64$ MHz, $f_m = 10$ kHz, $\delta_{max} = 75$ kHz
9. A certain FM broadcast transmitter produces 25 kHz of deviation when 1 V_p of information is applied. What information voltage will produce 100% modulation?
10. Calculate the FM modulation index for each of the FM signals in question 8.
11. An FM signal is present on a carrier frequency of 100 kHz. The information frequency is 5 kHz,

and the deviation is 5 kHz. The unmodulated carrier voltage is 30 V. Using a Bessel table,

a. calculate the bandwidth of the emission
b. draw a spectrogram of the signal, showing all frequencies and voltages

12. For each signal in problem 8,

a. calculate the bandwidth using a Bessel table
b. assuming that $V_{c(unmod)} = 100$ V, draw a spectrogram, showing all frequencies and voltages

13. Calculate the total power in problem 11 under two conditions:

a. unmodulated (use the original 30 V carrier amplitude)
b. under the conditions of problem 11, calculate the total power of all frequency components. Compare this with the first result. *Use a load resistance of 50 Ω for the calculations.*

14. How many pairs of sidebands are present in a NBFM signal? How does a NBFM signal appear on a spectrum analyzer?

15. Explain what happens to the amplitude of the carrier frequency component (J_0) as the modulation index of an FM signal increases from 0 to 15. At what values of m_f does the carrier null?

16. Why does the capacitance of a varactor diode vary as the reverse bias is varied?

17. Using the graph data of Figure 8–12, what is the capacitance of an MV-209 varactor diode at a reverse bias of:

a. 10 V?
b. 11 V?
c. 12 V?

18. An MV-209 varactor diode is coupled in parallel with a 100 μH inductor. What will the frequency of resonance be at (a) $V = 8$ V; (b) $V = 12$ V; (c) $V = 15$ V. Round the diode capacitance values to the nearest pF.

19. What is meant by the term *drift* when referring to radio transmitters?

20. What problem arises when a quartz crystal oscillator determines an FM transmitter's frequency?

21. Draw a block diagram of a Crosby FM transmitter. Using outline form, explain its operation. How does it maintain a stable center frequency while allowing plenty of deviation?

22. A 5 MHz carrier with a deviation of 5 kHz is sent through the following frequency multipliers: ×3, ×5. What is the final carrier frequency and deviation?

23. Draw a block diagram of an Armstrong wideband FM exciter, showing typical carrier frequency and deviation at each point.

24. Draw a block diagram of a PLL FM transmitter. What is added to the PLL to allow it to be frequency modulated?

25. If $f_{osc} = 10.000$ MHz, $R = 200$, and $N = 1978$, determine the frequencies at test points B, C, and E in the PLL FM transmitter of Figure 8–21.

26. What class of power amplifiers are used in FM transmitters? Why is this an advantage?

27. What are *preemphasis* and *deemphasis?* Why are they necessary?

28. Explain the *FM capture effect.*

29. Draw a diagram of the U.S. FM broadcast bandplan. How much space is allocated for each station, including guard bands?

30. Define the term *multiplexing*.

31. Draw a diagram showing the frequency components in a composite FM stereo signal with an SCA subcarrier.

32. List the three primary measurements a technician might be required to make on an FM transmitter. Explain the significance of each measurement.

9

FM Receivers

OBJECTIVES

At the conclusion of this chapter, the reader will be able to:

- draw a block diagram of an FM receiver, showing the frequency and type of signal at each major test point
- explain the operation and alignment of Foster–Seeley, ratio, PLL, and quadrature FM detector circuits
- describe the features of noise-suppressing circuits in an FM receiver
- draw a block diagram of a frequency-synthesized FM receiver
- trace the signal flow through FM stereo and SCA decoder circuits
- describe the alignment procedures unique to FM receivers
- apply basic troubleshooting methods to FM receivers

FM is popular as a communications mode because of its superior noise performance and fidelity when compared to AM. The operation of AM and FM receivers is very similar; the same familiar circuit techniques are used in both. The primary differences in an FM receiver stem from the relatively high frequencies used for FM transmission (the VHF and UHF bands) and the differences in detector circuitry. FM receivers tend to be more "feature laden." The addition of circuitry to support FM stereo, SCA, and other features adds complexity to the set.

FM is a fundamental technology, like AM. Its techniques are used in satellite and data communications, telemetry (remote measurement), and a score of other nonbroadcast applications. A technician with a strong knowledge of FM can go far in communications!

9–1 FM SUPERHETERODYNE RECEIVERS

FM receivers use the superheterodyne principle, as shown in Figure 9–1. Recall that a superhet receiver operates by converting the desired incoming RF carrier frequency down to the *IF* or *intermediate frequency,* where most of the amplification is provided and receiver bandwidth is defined. The sections of the receiver that are new or different compared to an AM receiver are in blue.

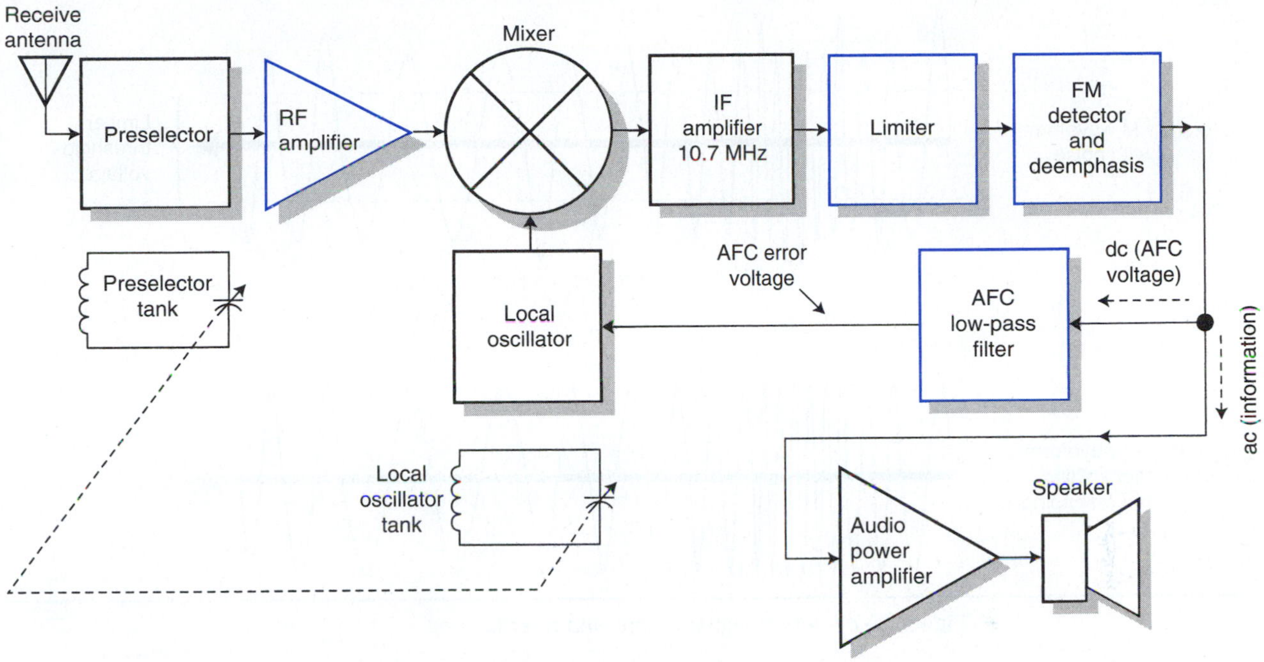

Figure 9–1 An FM superheterodyne receiver

An FM receiver contains several stages that are new or different from those in an AM set. These include the *detector, limiter, RF amplifier,* and *AFC stages.*

FM Detector

Naturally, detecting an FM signal requires circuitry different from that for demodulating AM. There are several popular types of FM detectors. All of them can be thought of as *frequency-to-voltage converters.* That is, they take a varying input frequency (a frequency-modulated carrier wave) and convert that into a varying output voltage. This is exactly the opposite of the action of the modulator in an FM transmitter, so the output of an FM detector is a replica of the original information signal.

Limiter

In FM, the information is encoded by changing the frequency of the carrier wave. Ideally, the carrier wave amplitude remains constant; in other words, the transmitter does not amplitude-modulate the carrier, and the envelope carries no information.

However, between the transmitter and receiver are various sources of external noise, such as atmospheric noise and humanmade noise sources. These noise sources add at random to the voltage of the FM signal envelope, as shown in Figure 9–2(a). An AM receiver is affected quite strongly by noise because an AM receiver recovers the envelope of the modulated wave. Not so with FM!

Because an FM signal contains information only in the wave's frequency, an FM receiver can safely ignore all amplitude changes without losing any information. The *limiter* in an FM receiver is a stage that essentially flattens the top and bottom of the modulated

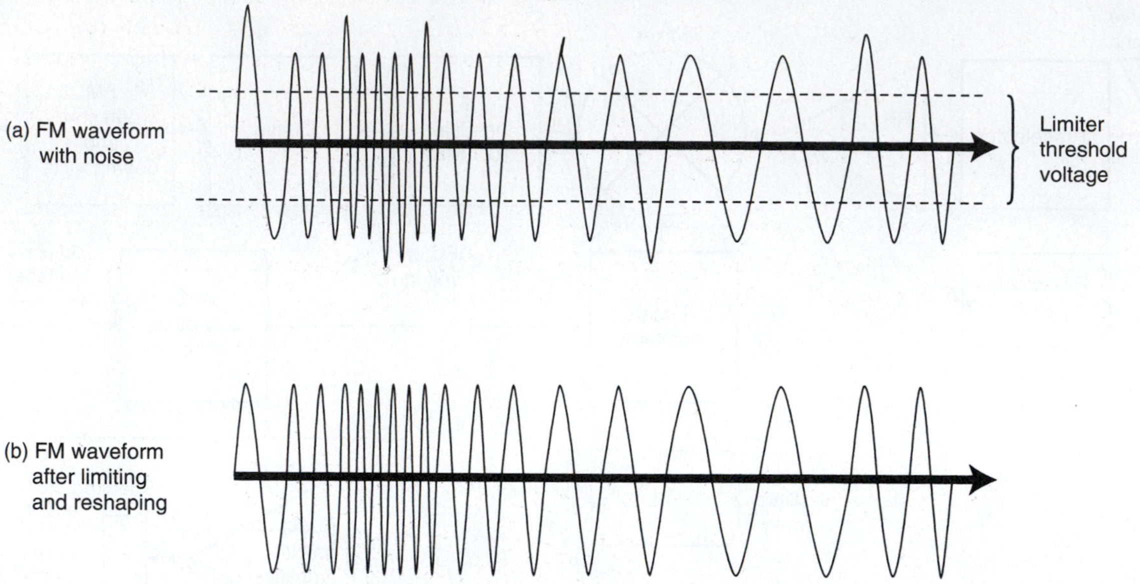

Figure 9–2 An FM signal before and after limiting

waveform prior to detection, as shown in Figure 9–2(b). Flattening or clipping the waveform eliminates most of the noise but preserves the information. This is why FM reception is virtually free of all sorts of static interference, even in the immediate presence of very strong noise signals (thunderstorms, nearby electric motors, etc). Because the limiter removes most of the amplitude changes, the detector sees only frequency changes in the modulated waveform, and therefore the output of the detector is only the original information signal.

RF Amplifier

In an AM broadcast receiver, there is seldom an RF amplifier in the front end. In an FM broadcast receiver, an RF amplifier provides two important actions, *amplification* and *local oscillator energy suppression.* The signal from the antenna in a VHF receiver (such as an FM broadcast receiver) can be very tiny. A "strong" signal may be only 50 μV, and often signals are only a few microvolts in strength. This is due to the high frequencies (VHF and UHF) that are being used, combined with the small antennas employed for reception at these frequencies (recall that wavelength and antenna size decrease as frequency increases). When such a small signal is mixed with the local oscillator (for conversion down to the IF frequency), it can be easily lost in the noise from the mixer. The mixer adds a considerable level of *internal noise* to signals that pass through it. In an AM broadcast receiver, this is not a problem; signal levels at the antenna are in the hundreds of microvolts. An RF amplifier is not needed. *The RF amplifier provides sufficient gain for the incoming RF signal to overcome the noise floor of the mixer.* The noise floor of the mixer is the noise power level, in dBm (decibels with respect to 1 mW), that the mixer produces by itself, with no RF input from the antenna.

The RF amplifier serves a second purpose: *local oscillator energy suppression.* The local oscillator in a receiver operates at a typical power level of 0 dBm (1 mW) to 10 dBm (10 mW). This doesn't sound like much energy, but think about what could happen if the local oscillator's output were coupled to an antenna. The receiver would become a *transmitter!* The wavelength at VHF is short compared to the MF frequencies used for AM broadcast. This means that even the telescoping rod antenna of a portable receiver can be a fairly effective transmitting antenna. If the local oscillator energy is allowed to couple to the antenna, the receiver can become quite a potent interference source. *The RF amplifier prevents local oscillator reradiation by allowing signals to flow in only one direction, from the antenna to the mixer circuit.* This prevents most of the energy from leaving the receiver.

EXAMPLE 9–1

What is the wavelength of a 100 MHz FM broadcast signal? Compare this to the length of a typical rod antenna (20″).

Solution

From Chapter 1, we know that

$$\lambda = \frac{v}{f} = \frac{3 \times 10^8 \text{ m/sec}}{100 \text{ MHz}} = \underline{\underline{3 \text{ m}}}$$

To compare these lengths, let's convert the wavelength into inches:

$$\lambda_{\text{inches}} = \lambda_{\text{meters}} \times \frac{39.37''}{1 \text{ m}} = 3 \text{ m} \times \frac{39.37''}{1 \text{ m}} = \underline{\underline{118.11''}}$$

Only about one quarter of a wavelength is needed for an antenna to be an efficient radiator. One quarter of 118.11″ is $\underline{\underline{29.5''}}$. You can see that the rod antenna isn't nearly this long, but it is in the ballpark (20″) and therefore could radiate significantly! (We will discuss the theory of antennas in much more detail in a later chapter.)

AFC Stages

The local oscillator in an FM receiver operates at very high frequencies. Before the advent of frequency synthesizers, the frequency of the local oscillators in FM receivers was controlled by an LC tank circuit, just as in an AM receiver. Using an LC tank allows the oscillator to drift off frequency, and as luck would have it, drift becomes much more difficult to control in a VHF oscillator. In addition, receiver frequency drift rapidly degrades the quality of FM reception. The signal becomes distorted quickly as tuning degrades. *The automatic frequency control or AFC system is built into analog FM receivers to correct local oscillator drift.* The receiver local oscillator is essentially frequency locked onto the carrier frequency from the transmitter, which is crystal controlled.

The AFC control voltage is developed at the FM detector, which is a frequency-to-voltage converter. As the receiver drifts off frequency, a positive or negative dc voltage is produced at the FM detector. This dc voltage is fed back to the local oscillator, which pushes the local oscillator back in the correct direction. A low-pass filter is included so that only steady dc is sent back to the local oscillator circuit. The local oscillator contains a *reactance modulator,* not shown in Figure 9–1, which converts the dc AFC voltage into a varying capacitance or inductance, which corrects the oscillator frequency. The AFC system operates almost exactly like the control method in a Crosby FM transmitter.

Many modern receivers are *digitally synthesized.* They lack the mechanical tuning capacitors of older sets; instead, they sport keypads, buttons, and digital frequency displays. In synthesized receivers, there is no need for AFC, since the local oscillator is actually a PLL or DDS frequency synthesizer and is locked to a stable quartz frequency reference.

Choice of IF Frequency

There's one other difference between AM and FM receivers that you may have already noticed. The standard IF frequency for AM is 455 kHz and for FM it is 10.7 MHz. Why is a higher IF frequency used?

Recall from Chapter 8 that the bandwidth allocated for each FM broadcast station is 200 kHz, including guard bands. An AM broadcast uses only 10 kHz. FM uses a *lot* more bandwidth than AM! By raising the IF frequency, the IF bandwidth increases accordingly. Accommodating a 200 kHz wide signal in a 455 kHz IF would be tough!

In general, the wider the bandwidth of the receiver, the higher the chosen IF frequency will be.

EXAMPLE 9–2

Calculate the range of local oscillator frequencies required for an FM broadcast receiver (88.1–107.9 MHz) with an IF of 10.7 MHz, assuming high-side injection.

Solution

From Chapter 5, we know that when high-side injection is being used,

$$f_{LO} = f_c + f_{if}$$

So this same relationship will be applied at the bottom and top of the FM broadcast band. At the bottom of the band, we get,

$$f_{LO(min)} = f_c + f_{if} = 88.1\,\text{MHz} + 10.7\,\text{MHz} = \underline{\underline{98.8\,\text{MHz}}}$$

And at the top of the band,

$$f_{LO} = f_c + f_{if} = 107.9\,\text{MHz} + 10.7\,\text{MHz} = \underline{\underline{118.6\,\text{MHz}}}$$

This 118.6 MHz falls within the *aviation* band. This is one reason why you can't play a portable FM radio onboard a commercial aircraft. The local oscillator energy may leak out and interfere with the sensitive communication receivers on board the plane!

ANNOY YOUR FRIENDS!

Even though an FM receiver uses an RF amplifier to prevent local oscillator energy leakage, a tiny amount of RF *does* leak out of an FM receiver. You can easily demonstrate this by "listening" to the local oscillator of a receiver with a second receiver. Analog sets in plastic cases work best.

1. Place the two receivers close together (a few inches is best).
2. Tune one receiver to a blank spot *high* on the FM dial (such as 107.3 MHz).
3. Tune the second receiver *10.7 MHz below* the frequency of the first (96.6 MHz, for example). This is the "transmitter."

If your second receiver is a good "leaker," you'll hear the sound of *dead air* in the first receiver. The local oscillator is an unmodulated carrier, after all!

Feeling adventuresome? If you've picked up the carrier, *carefully* turn up the volume on the first receiver (107.3 MHz) and at the same time, tap on the case of the second unit. What do you hear, and why?

No Need for AGC

FM receivers don't respond to amplitude variations, thanks to the operation of the limiter circuitry. Because the relative amplitude of the signal going into the limiter and detector is unimportant, most FM receivers don't have any automatic gain control (AGC) circuitry. The exceptions to this rule are certain communications-grade FM receivers that must operate over a very wide range of antenna signal voltages. In these receivers, a modified AGC called *delayed AGC* or *DAVC* is often used.

Section Checkpoint

9–1 What types of receiver are used for both AM and FM?

9–2 An FM detector converts ______ into ______.

9–3 What is the purpose of the *limiter* in an FM receiver?

9–4 Give two reasons for the use of an RF amplifier.

9–5 Why is AFC needed in analog FM receivers?

9–6 What type of receivers don't require AFC?

9–7 What governs the choice of IF frequency in a receiver?

9–8 Why isn't AGC used in FM receivers?

9–2 DETECTION OF FM SIGNALS

Over the years, many different circuits have been developed for the detection of FM signals. FM demodulation is a little more complicated than that for AM, and as technology has evolved, new FM detector circuits have been developed. The latest circuits can be almost entirely contained on an IC chip.

Slope Detection

The oldest and least practical FM detector is the *slope detector* of Figure 9–3. This detector relies on the frequency response of a tuned circuit. The slope detector works by utilizing the *slope* of a tuned circuit's frequency response. It can be best understood as a two-step detector. First, the FM signal is converted into an AM signal and then the resulting AM signal is envelope detected.

In Figure 9–3(a), the tuned circuit consists of transformer T_1 and its internal capacitor, and the AM portion of the detector is built from D_1, R_1, and C_1. This is in fact a standard AM detector circuit. In normal operation (AM detection), the input signal would be at the resonant frequency of the tank. In FM operation, the tank is deliberately mistuned, as shown in Figure 9–3(b).

By moving the tank frequency up, the carrier frequency now falls on the left-hand slope of the tuned circuit's frequency response. As the carrier frequency increases, the amplitude of the tuned circuit output increases; as the carrier frequency decreases, the amplitude falls. The amplitude of the circuit's output depends on the frequency of the FM signal! By passing the resulting AM signal into the diode detector, the information is recovered.

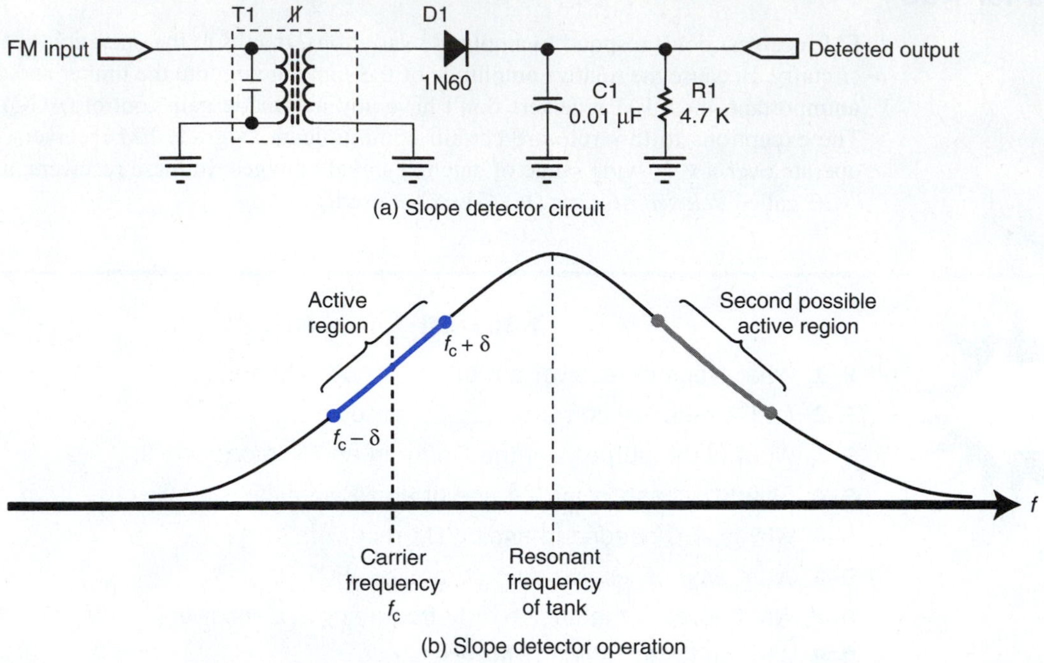

Figure 9–3 The slope detector

The slope detector isn't an entirely practical FM detector for two reasons. First, this detector has a *dual response*. There are two points on the tuning curve of the tuned circuit that will provide a detected output. This will result in ambiguous tuning characteristics (a station will appear at two adjacent points on the tuning dial). Second, the shape of the tuned circuit's frequency response is not very linear, unless only a very small portion of the curve is used. For wideband FM detection, this would not work very well; the resulting carrier frequency swings are quite large (±75 kHz), and considerable distortion of the sound would result.

The Foster–Seeley Detector

The *Foster–Seeley* detector circuit is a vast improvement over the slope detector. It provides a single response and reasonably linear characteristics for wideband FM detection. This circuit can be best understood if developed in steps, as shown in Figure 9–4.

The operation of the Foster–Seeley detector is a two-step process, just like the slope detector. Instead of converting input frequency changes to AM, the Foster–Seeley circuit converts the incoming FM signal into a series of *phase shifts*. The resulting signal, though not a true PM signal, is then phase detected to produce a varying output voltage. The circuit can be developed in three steps.

Figure 9–4(a) shows the first step. The FM signal is coupled into transformer T_1, whose secondary is parallel resonated by C_1 at the IF center frequency (usually 10.7 MHz). Capacitor C_1 and the secondary inductance of T_1 form a phase-shifter circuit. A steady sine wave at the IF frequency of 10.7 MHz, voltage V_c, is applied to the center tap of T_1.

The phase shift produced at the secondary of T_1 depends on the instantaneous frequency of the FM signal being applied. At rest ($f = f_c$), the phase shift is adjusted to be 90°

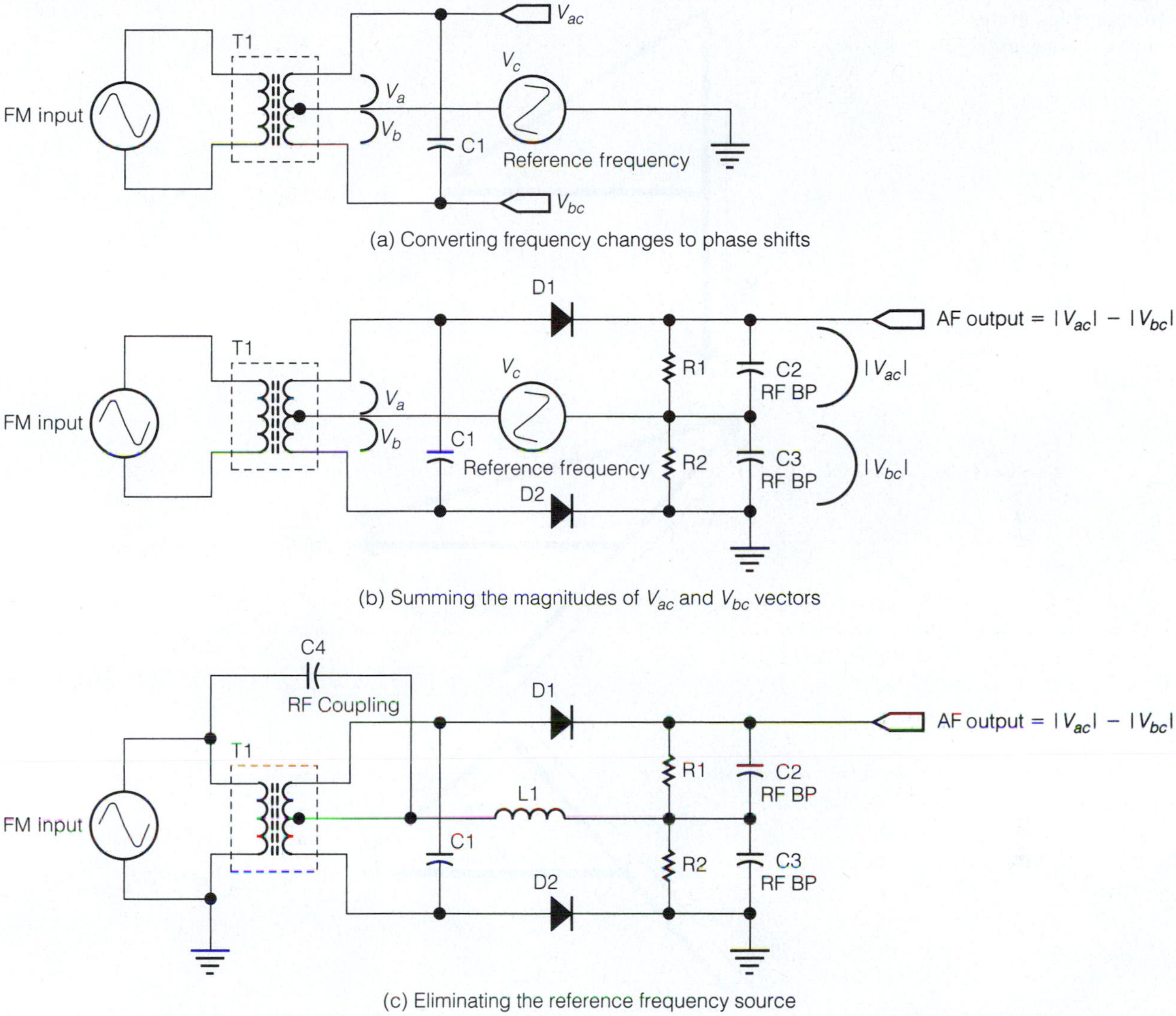

Figure 9–4 Development of the Foster–Seeley detector

by tuning T_1. The phasors appear as in Figure 9–5(a). Notice that the *top* output of the circuit is the phasor sum of the voltages V_c (the reference) and V_a (the voltage from the top of T_1). This voltage is V_{ac}. The same happens at the bottom side of T_1, and the voltage V_{bc} is developed here. Note the phase relationship between V_a and V_b: They are 180° out of phase. This is true due to the center tap on the secondary of T_1. *When $f = f_c$ the circuit is balanced.* The outputs V_{ac} and V_{bc} are equal in magnitude, and if their absolute values are subtracted, the result is zero.

What happens when the FM carrier frequency deviates upward? The phasor result is shown in Figure 9–5(b). Again, V_a and V_b maintain their 180° relationship; they both rotate in a counter clockwise direction. Since V_c is the reference, it does not move. As you can see, the phasor results V_{ac} and V_{bc} are now out of balance; V_{ac} is much larger. To detect the out-of-balance condition, the circuit of 9–4(b) is used. Diodes D_1 and D_2 serve to rectify the ac voltages V_{ac} and V_{bc}. The algebraic output of 9–4(b) is $(|V_{ac}| - |V_{bc}|)$.

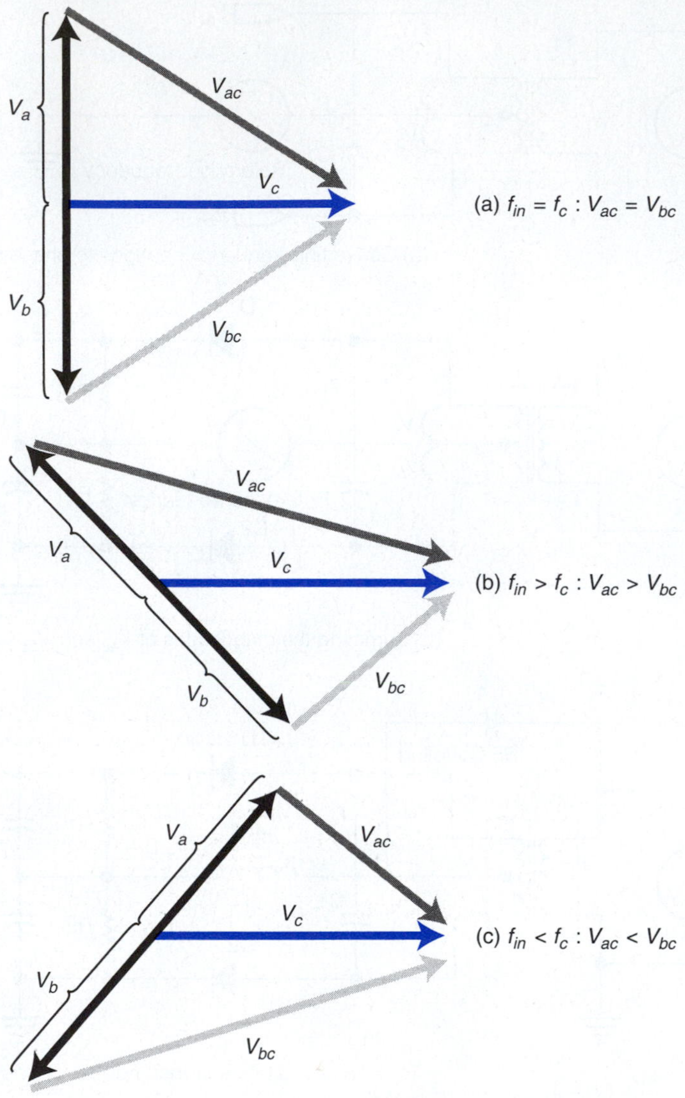

Figure 9–5 Phasor relationships in the Foster–Seeley detector

Therefore, in Figure 9–4(b), when the input frequency *rises,* the rectified output $|V_{ac}|$ is larger than the $|V_{bc}|$ output, and a positive output voltage results. The opposite effect occurs when the carrier frequency is *lowered;* now $|V_{bc}|$ becomes larger, and the output voltage becomes *negative*. The circuit has effectively converted voltage into frequency.

The final refinement of the circuit is shown in Figure 9–4(c). The first two circuits require an oscillator to provide a reference frequency, and this oscillator must be phase locked to the FM carrier signal. This is not very convenient, so the oscillator is replaced with a sample of the RF carrier voltage in Figure 9–4(c). The carrier voltage is coupled into the circuit through C_4, and inductor L_1 (which is usually adjustable) serves to allow an RF voltage drop to be developed to serve as the reference voltage.

Why does this work? The answer is that *two* rates of phase change occur in the circuit. The portion of the circuit containing L_1 changes phase relatively slowly when the input frequency changes because it contains only L_1. The resonant circuit consisting of T_1's secondary and C_1 changes phase more quickly because it has two reactive parts (T_1 and C_1). Therefore, for practical purposes, the phase change across L_1 is small enough to consider that portion of the circuit to have an approximately constant phase angle. The voltage across L_1 is approximately a "reference" voltage.

The Foster–Seeley detector can reproduce a high-quality information signal but has one limitation. In Figure 9–5, the size of the vectors V_a and V_b determines the magnitude of voltage output from the detector. If the input voltage of the FM signal changes, these vectors will change in size, and the detector's output voltage will change accordingly. In other words, a Foster–Seeley detector doesn't reject AM. Since most noise is in the form of amplitude disturbances, it will pass through a Foster–Seeley detector and be reproduced in the audio. To prevent this, a limiter stage must precede a Foster–Seeley detector to prevent AM noise from reaching the circuit.

The Ratio Detector

A modification to the Foster–Seeley circuit results in the *ratio detector* circuit of Figure 9–6. One thing most technicians notice immediately about this circuit is that "one of the diodes is in backwards!"

The operation of the ratio detector is very similar to that of the Foster–Seeley circuit. The *difference* involves how the audio output voltage is developed. Notice that the voltages V_{ac} and $-V_{bc}$ are developed on the top and bottom legs of the circuit and that both of these voltages are *ground referenced*. The bottom voltage $-V_{bc}$ appears as a negative value because D_2 is reversed. Resistors R_1 and R_2 form a voltage divider. When the two voltages V_{ac} and V_{bc} are equal, the output from the divider will be *zero,* since R_1 and R_2 are equal values. When V_{ac} and V_{bc} become unequal, the voltage divider unbalances, and a positive or negative voltage output is produced, depending upon which voltage (V_{ac} or V_{bc}) is larger, which in turn depends upon the input frequency.

In this manner, the operation is very similar to the Foster–Seeley circuit. The primary *addition* is capacitor C_5, which sees the total voltage ($|V_{ac}| + |V_{bc}|$). Even when the circuit unbalances, the sum of these two voltages is constant (when one gets smaller, the other becomes larger to compensate). Capacitor C_5 is a large electrolytic, usually around 10 μF,

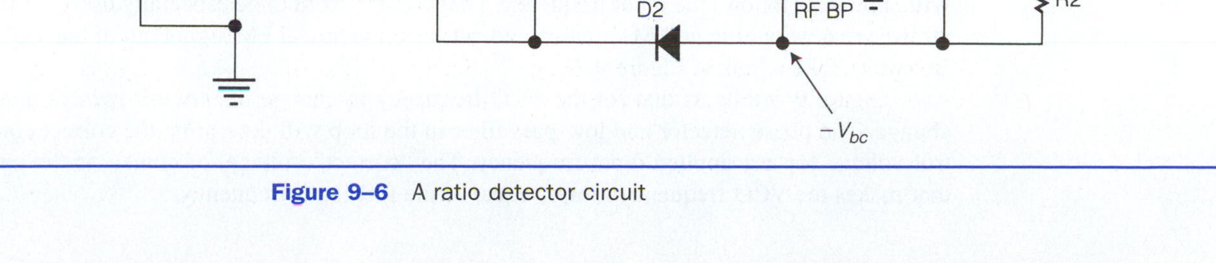

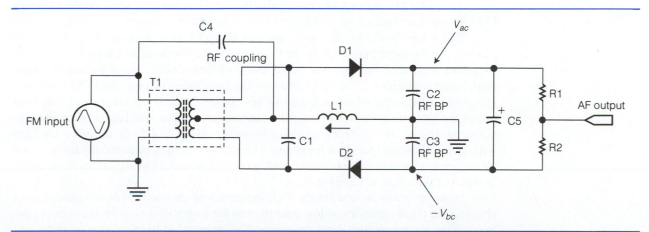

Figure 9–6 A ratio detector circuit

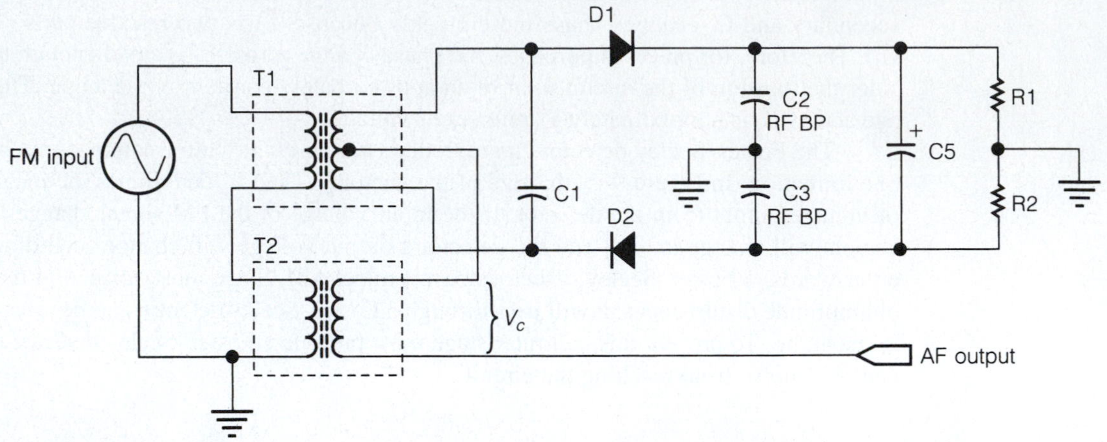

Figure 9–7 A very common form of the ratio detector

and tends to stabilize the total voltage across the circuit. The benefit of capacitor C_5 is that the circuit has a *built-in limiting action,* but only for short-term (less than a second or so) amplitude changes. *The ratio detector does a better job of rejecting AM noise than the Foster–Seeley circuit* but must still be preceded by a limiter amplifier circuit.

Figure 9–7 shows a very common form of the ratio detector that uses two adjustable transformers. The operation is identical to Figure 9–6, except that the secondary of T_2 is used to provide the reference voltage V_c. Usually T_1 and T_2 are a matched set of transformers designed especially for ratio detector duty and may even be contained in one metal can (with the two screw adjustments visible).

PLL FM Detection

The Foster–Seeley and ratio detectors are excellent FM detectors but require the use of bulky RF transformers. A phase-locked loop, or PLL, can be used as an FM detector by extracting the audio signal at the low-pass filter output, as shown in Figure 9–8.

The PLL FM detector looks too good to be true. How can such a simple circuit detect FM? The answer comes from looking at the basic action of a phase-locked loop.

You might recall that the basic function of a PLL is to lock the VCO frequency *exactly* onto the input frequency. In the PLL circuits we discussed in Chapter 7, the input frequency was called the *reference,* and because of Finley's law, the phase detector gave appropriate commands to the VCO (through the loop filter) to keep the VCO frequency *exactly* the same as the reference frequency (test points D and C). In other words, the loop causes the VCO frequency to follow the input frequency. What will happen if an FM signal is substituted for the steady *reference* frequency of the PLL? You're right—the loop will "imitate" (follow) the input frequency. That doesn't seem to be especially useful; after all, we are now putting an FM signal in and getting an identical FM signal out of the VCO. But wait! Take a look at Figure 9–9.

Figure 9–9 tells us that for the VCO frequency to change, its *control voltage* must change. The phase detector and low-pass filter in the loop will determine the correct control voltage for any applied input frequency. The "correct" voltage, of course, is the one that makes the VCO frequency *exactly* equal to the FM input frequency.

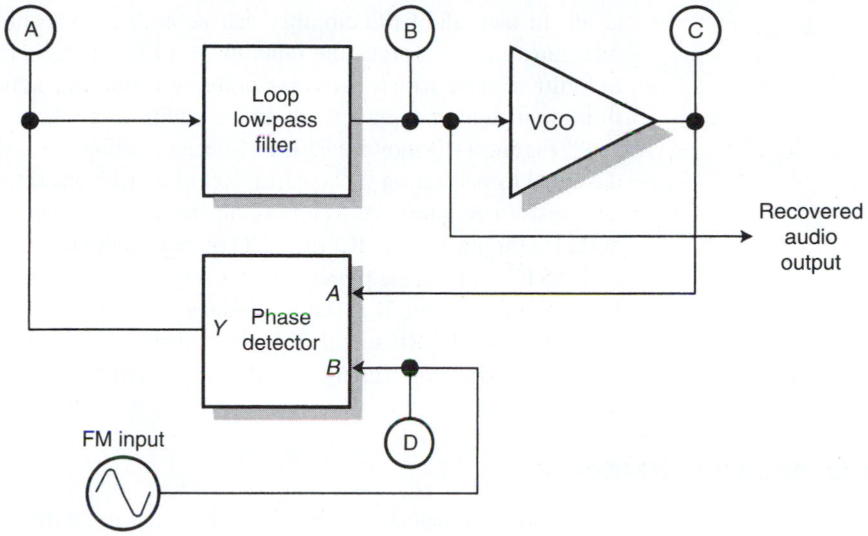

Figure 9–8 The PLL FM detector

Suppose the input frequency is 10.7 MHz (the IF center frequency). What VCO control voltage will result? According to the graph in Figure 9–9, the VCO requires approximately *3 volts dc* to produce 10.7 MHz out, so the dc control voltage will be 3 volts. *An input of 10.7 MHz is converted to a dc voltage of 3 volts.*

When the carrier frequency increases in a positive direction, the loop must follow. Suppose that an input frequency of 10.775 MHz (+75 kHz deviation) is applied. What VCO control voltage will result? Again, Figure 9–9 suggests that the VCO "needs" 4.5 volts for this to happen. *An input of 10.775 MHz is converted to 4.5 volts.*

Can you see what is happening? The loop *must* follow the input frequency changes. We're taking advantage of that behavior by "peeking" inside the loop to extract the dc control voltage for the VCO. The information signal will be riding on top of the dc control voltage; we simply block the dc with a capacitor to extract the information!

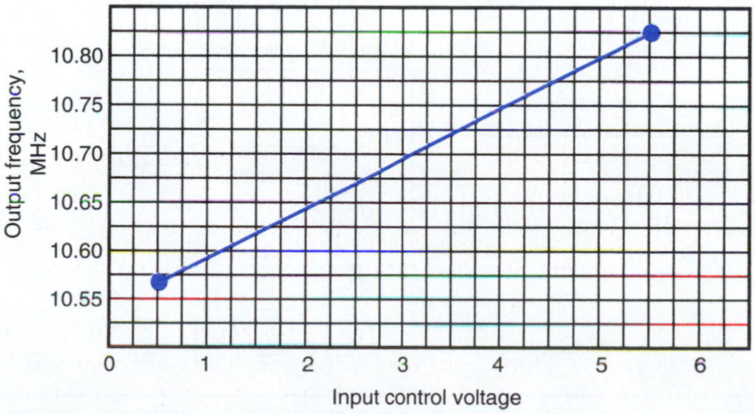

Figure 9–9 The VCO transfer characteristic

The PLL is an excellent FM detector and can be built without any RF transformers at all. In fact, all of the circuitry can be built onto a chip. This makes it well suited for miniature circuitry. Since the input to the PLL is a digital input, it can be equipped with a Schmitt trigger, which provides automatic *limiting* action. The PLL can provide true limiting action on its own, without the addition of a limiter amplifier.

Figure 9–10 shows a PLL FM detector using the LM/NE565 PLL IC. This circuit is designed to operate on a carrier of 100 kHz, with maximum deviation of ±10 kHz. Variable resistor R_{106} sets the free-running frequency of the loop, which should be adjusted to 100 kHz (at pin 4 of the IC, the VCO output) with no signal applied to the *FM INPUT*. The LM565 IC contains the phase detector, VCO, and filter resistor. C_{105} and C_{106} are part of the loop RC filter circuit; R_{108} and C_{107} also act as part of the loop filter. R_{109}, R_{110}, C_{108}, and C_{109} filter any 100 kHz RF signal from the detected FM output. C_{110} is a dc block, which leaves only the information signal at the *AF OUT* terminal.

The Quadrature Detector

The quadrature detector of Figure 9–11 is a simple digital FM detector. It is perhaps the simplest of the FM detectors. like the PLL, it provides built-in limiting action through a Schmitt-trigger input and can be implemented on an IC chip.

The quadrature detector relies on conversion of frequency to phase shift, just like a Foster–Seeley detector. In the quadrature detector in Figure 9–11, the exclusive-OR gate functions as a phase detector. The FM input signal is applied directly to the top gate input and indirectly to the bottom gate input through isolator resistor R_1 and the *quadrature coil,* L_1 (which usually contains a built-in capacitor to resonate at the IF center frequency). R_1 and L_1 form a *phase-shifter network.* As the FM input changes in frequency, the phase shift developed across R_1, L_1, and L_1's internal capacitor varies. This causes the phase of the two signals at the phase-detector XOR gate to vary, which in turn causes the duty cycle of the

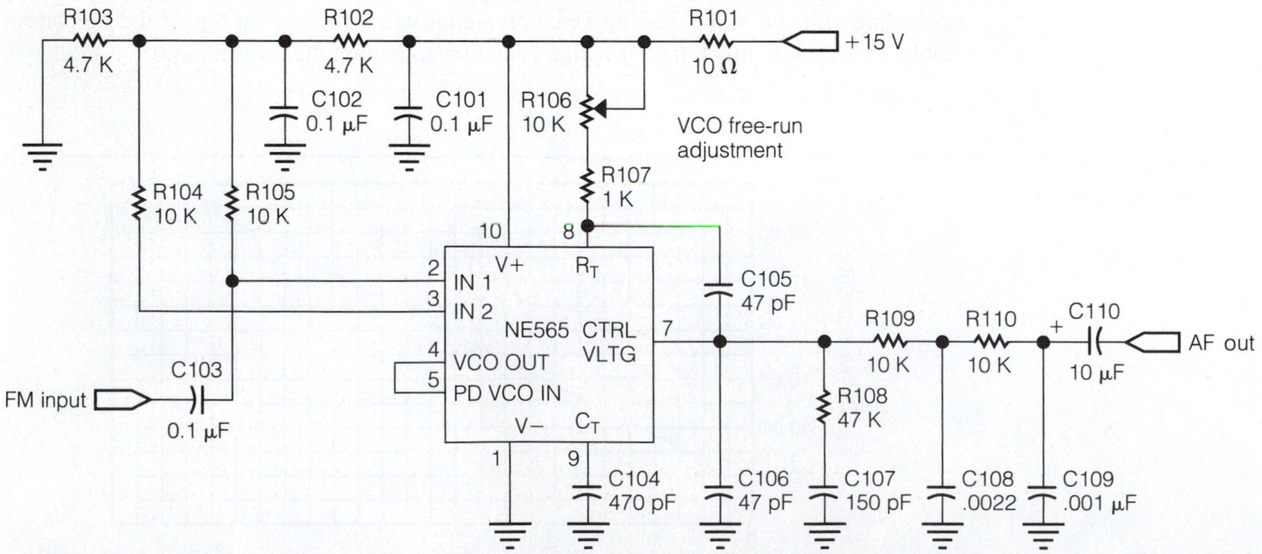

Figure 9–10 A PLL FM detector utilizing the LM/NE565

Figure 9–11 The quadrature detector

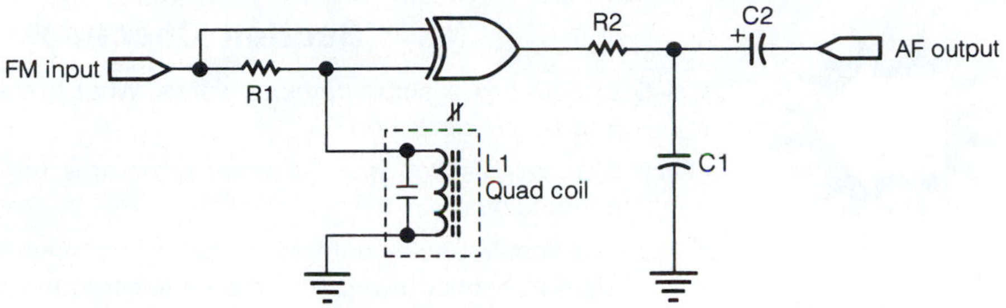

pulses at the gate's output to vary. Figure 9–12 shows the phase-versus-output frequency relationship for the XOR phase detector.

The varying duty-cycle output of the XOR gate is converted to a dc level by the filter comprised of R_2 and C_1. This dc level has the information riding on top of it. Capacitor C_2 removes the dc level, leaving only recovered audio.

A quadrature detector is very easy to align. The technician simply applies the center frequency to the detector's input (usually 10.7 MHz) using an RF generator and then monitors the dc voltage at the right-hand portion of R_2. The quadrature coil is adjusted until this voltage is one-half (50%) of V_{cc}; this provides 90° phase shift between the two gate inputs (see Figure 9–12). This adjustment is necessary to provide a centered Q point to allow both positive and negative frequency deviation. Alternatively, the technician can insert the 10.7 MHz signal from the signal generator and view the waveform at the XOR gate output pin with a scope. The quadrature coil is adjusted until a 50% duty cycle is obtained, which again signifies a 90° phase shift.

Some quadrature detectors require no alignment at all. It is possible to replace the parallel resonant circuit (L_1 and its internal capacitor) with either a quartz crystal (in parallel resonant mode) or a ceramic resonator. This technique is commonly employed in narrowband communications-grade FM receivers.

Figure 9–12 Output voltage of the XOR gate as a function of phase

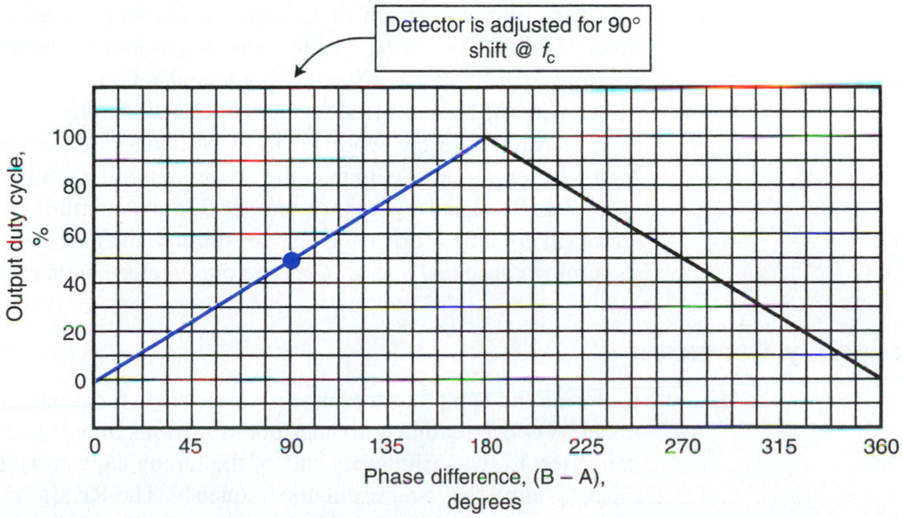

Section Checkpoint

9–9 Explain how a slope detector works. What is the FM signal converted into before detection?

9–10 Give two reasons why the slope detector is not a very practical FM demodulator.

9–11 In a Foster–Seeley detector, what two voltages are equal when the input frequency is equal to the center frequency?

9–12 What is the purpose of the diodes in a Foster–Seeley detector?

9–13 What two components are adjusted to align a Foster–Seeley detector?

9–14 What is the advantage of a ratio detector when compared to the Foster–Seeley circuit?

9–15 What is the purpose of capacitor C_5 in Figure 9–6?

9–16 Explain how a PLL can be used to detect an FM signal.

9–17 How should R_{106} be adjusted in Figure 9–10? What does it set?

9–18 Explain how to align the quadrature detector of Figure 9–11.

9–3 A COMPLETE FM RECEIVER

The circuit of a complete monaural FM receiver is shown in Figure 9–13. The receiver can also be viewed in block diagram form as Figure 9–1.

The FM receiver of Figure 9–13 is easiest to understand if analyzed one stage at a time. Let's follow each stage the incoming signal must take as it makes its way from the antenna to the loudspeaker.

RF Amplifier

The carrier signal from the rod antenna first enters the RF amplifier, Q_4, after entering the bandpass filter comprised of L_3 and C_{15}. (L_3 and C_{15} are tuned to the center of the FM broadcast band. In a better-quality unit, C_{15} would be tunable as a *third* ganged section of the tuning capacitor, C_{17}.) You might notice that Q_4 is a *common-base* amplifier. Q_4 has been drawn with the base electrode pointing down to emphasize this to the reader; this is a common convention in the world of RF. The common-base configuration gives much better high-frequency power gain than the common-emitter configuration.

The RF signal enters Q_4's emitter through coupling capacitor C_8 and leaves Q_4's collector (in amplified form), then passes through the *preselector* bandpass filter. C_{17A} (half of the tuning capacitor) and L_2 form the preselector bandpass filter.

Frequency Converter

Q_3 forms the complete frequency converter. It functions as both the local oscillator and mixer. A converter that combines both functions in one active device is called an *autodyne* converter. L_1, C_{17B} (the other half of the tuning capacitor), C_{18}, and D_3 (the varactor AFC diode) control the local oscillator frequency. The RF signal from the RF amplifier is coupled into the converter through coupling capacitor C_9.

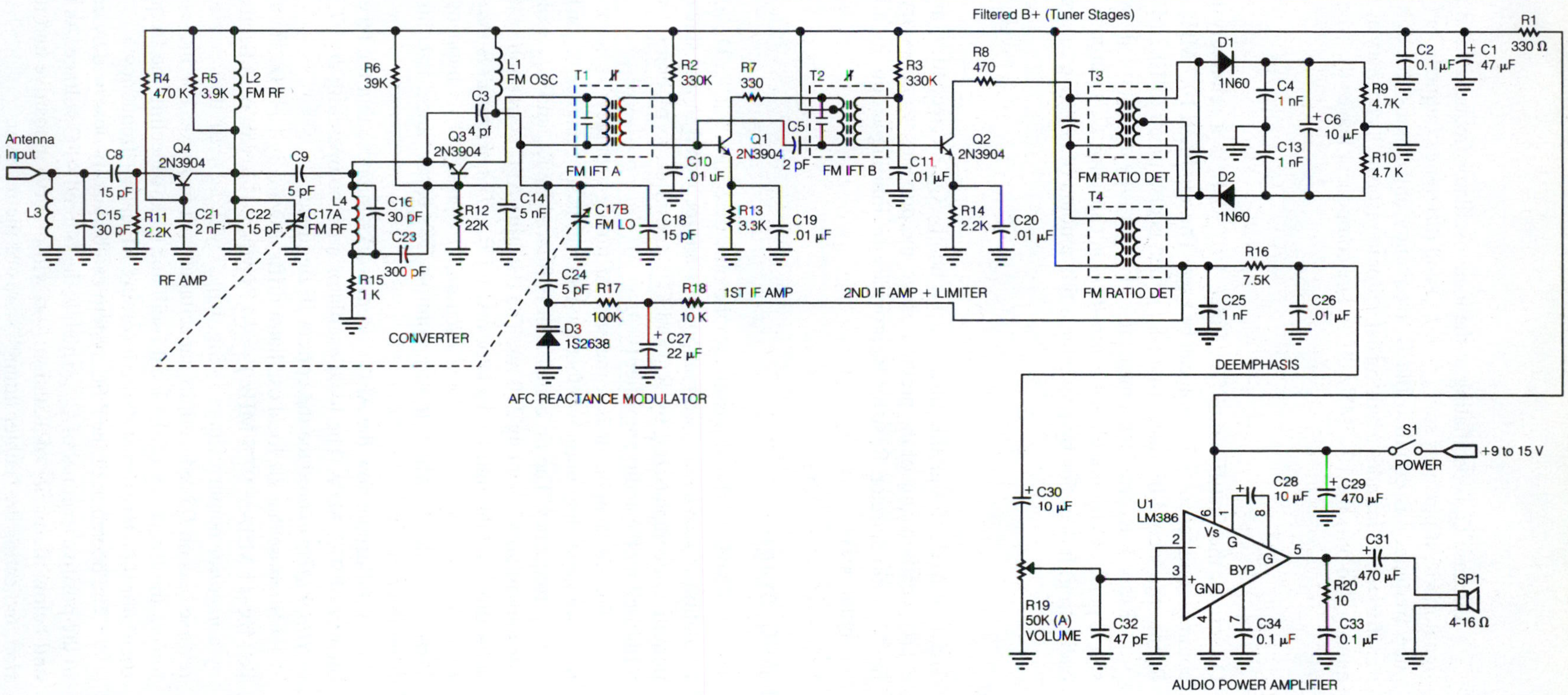

Figure 9–13 A complete FM receiver

Because Q_3 is oscillating at the local oscillator frequency and is operating in large-signal mode, it produces sum $(f_c + f_{LO})$ and difference frequencies $(f_c - f_{LO})$. The difference frequency is equal to the IF frequency and passes through T_1, the first FM IF transformer. T_1 includes a built-in capacitor to resonate its primary at 10.7 MHz (the IF frequency), so only the 10.7 MHz difference component leaves the frequency converter.

First IF Amplifier

Q_1 acts as the first IF amplifier. Since 10.7 MHz is a relatively low RF frequency, a common-emitter configuration can be used. The amplified 10.7 MHz signal passes through R_7 into the second IF transformer, T_2, which is also tuned to 10.7 MHz, just like T_1.

Resistor R_7 provides a measure of FM limiting action, by allowing Q_1 to saturate on signal peaks. Capacitor C_5 is a *neutralization* capacitor. C_5's purpose is to prevent Q_1 from oscillating. It does this by providing a tiny amount of negative feedback.

Second IF Amplifier

The 10.7 MHz IF signal is further amplified by Q_2, the second IF amplifier. This stage is nearly a carbon copy of the previous stage and provides the necessary power gain for driving the ratio detector. Resistor R_8 provides the limiting action, by allowing Q_2 to saturate on signal peaks.

Ratio Detector and AFC Stage

The 10.7 MHz IF signal passes into the ratio detector—T_3, T_4, D_1, and D_2—where the information is recovered. Two components are recovered in this stage. The ac signal component is deemphasized by R_{16} and C_{26}, and after dc blocking by C_{30}, is ready to be amplified by the audio amplifier (U_1).

The *dc* component is a measure of tuning accuracy; it becomes zero, positive, or negative depending upon whether the receiver is spot-on, above, or below the correct operating frequency. The dc component is filtered by R_{18} and C_{27}, which give an AFC time constant of approximately 200 ms. The resulting AFC correction voltage is applied to varactor diode D_3 through RF isolator resistor R_{17}. *The net result is that the dc level from the ratio detector controls the local oscillator frequency.* This happens because the bias on varactor diode D_3, which in turn controls D_3's capacitance, is controlled by the dc level from the detector.

To illustrate how the AFC system works, suppose that the receiver is tuned to a frequency of 97.7 MHz. The local oscillator must be operating at (97.7 MHz + 10.7 MHz) or *108.4 MHz* to produce the correct IF frequency.

Suppose that the local oscillator drifts upward by 10 kHz. The IF frequency will now be (108.41 MHz − 97.7 MHz) = *10.71 MHz*. This will be reflected at the ratio detector by a *negative* output voltage swing, which will *decrease* the reverse bias on D_3. With the reverse bias on *D3* reduced, its capacitance will *increase*. The increased capacitance of D_3 *lowers* the frequency of the local oscillator tank, canceling the frequency drift. The local oscillator therefore "tracks" the incoming RF carrier frequency.

The operation of the AFC system can be readily observed by connecting a voltmeter to the positive terminal of C_{27}, which is the filtered AFC voltage. As the set is tuned above and below the carrier of a station, you will observe a positive and negative meter deflection, indicating the "center tuning" status of the receiver.

Power Supply and Audio Power Amplifier

The power supply of this receiver is split into two separate supply voltages. A filtered B+ voltage operates the low-power stages in the tuner. R_1, C_1, and C_2 provide the filtering for this supply.

The audio power amplifier is identical to that in the receiver of Chapter 5; an LM386 single-chip audio amplifier is employed, which contains all the necessary voltage and current gain stages for driving the loudspeaker.

For the purpose of troubleshooting, it should be considered that there are two different power supplies in this unit. One, the main power supply, is measurable at the positive terminal of C_{29} (at the power switch); the other, the filtered B+ tuner supply, is measurable at the positive terminal of C_1. If a problem appears to be a "tuner" problem, the tuner power supply should be measured first.

Section Checkpoint

9–19 Why is a common-base configuration used for the RF amplifier, Q_4?

9–20 Which circuit elements form the preselector bandpass filter?

9–21 A converter that uses one active device as both mixer and local oscillator is called an ______ converter.

9–22 Explain how a 10.7 MHz IF signal is formed by Q_3.

9–23 Why does only the 10.7 MHz frequency component pass through T_1?

9–24 List the components that control the local oscillator frequency.

9–25 Why is *neutralization* needed in some amplifiers?

9–26 What is the purpose of resistors R_8 and R_7?

9–27 List the two signal components' output at the ratio detector.

9–28 What is the purpose of R_{18} and C_{27}?

9–29 What time constant is formed by R_{16} and C_{26}?

9–30 Explain how the AFC system corrects local oscillator drift.

9–31 What is a good test point for observing AFC action?

9–32 What is the purpose of C_1, C_2, and R_1?

9–4 SYNTHESIZED FM RECEIVERS

Analog FM receivers such as the one of Figure 9–13 are becoming increasingly rare as the price of digital hardware (microprocessors and PLL synthesizers) tumbles. A *synthesized* FM receiver uses a frequency-synthesized local oscillator instead of an LC-controlled oscillator. Sometimes such receivers are called *electronically tuned receivers.*

The advantages of this approach are many. First, the AFC feedback loop can be eliminated from the receiver, because the local oscillator frequency will be locked to a crystal oscillator and will be extremely stable. Second, mechanical tuning capacitors can be eliminated, which can allow further miniaturization as well as improvement in receiver reliability. Finally, since software will be controlling the receiver, all sorts of "features" become feasible, such as push-button station memory, seek-and-scan operation, and so on.

Figure 9–14 shows a synthesized FM receiver. Today, all of the circuitry can be built on one or two IC chips. The synthesized receiver in the figure is almost exactly like the FM receiver of Figure 9–1. A PLL synthesizer has replaced the *local oscillator* circuit, and a microcomputer and user interface has been added to control the local oscillator.

Tuning a Synthesized Receiver

To tune a superheterodyne receiver, both the preselector and local oscillator circuits must be adjusted at the same time. They must maintain a frequency difference of f_{if}, or no passable IF signal will be developed.

In Figure 9–14, you can see that the local oscillator is being precisely controlled. What about the preselector? It must be made to electronically "track" the local oscillator so that a frequency difference of f_{if} will be maintained.

In an electronically tuned receiver, the mechanical variable capacitor in the preselector is replaced with a varactor-diode reactance modulator. *The preselector is tuned by a*

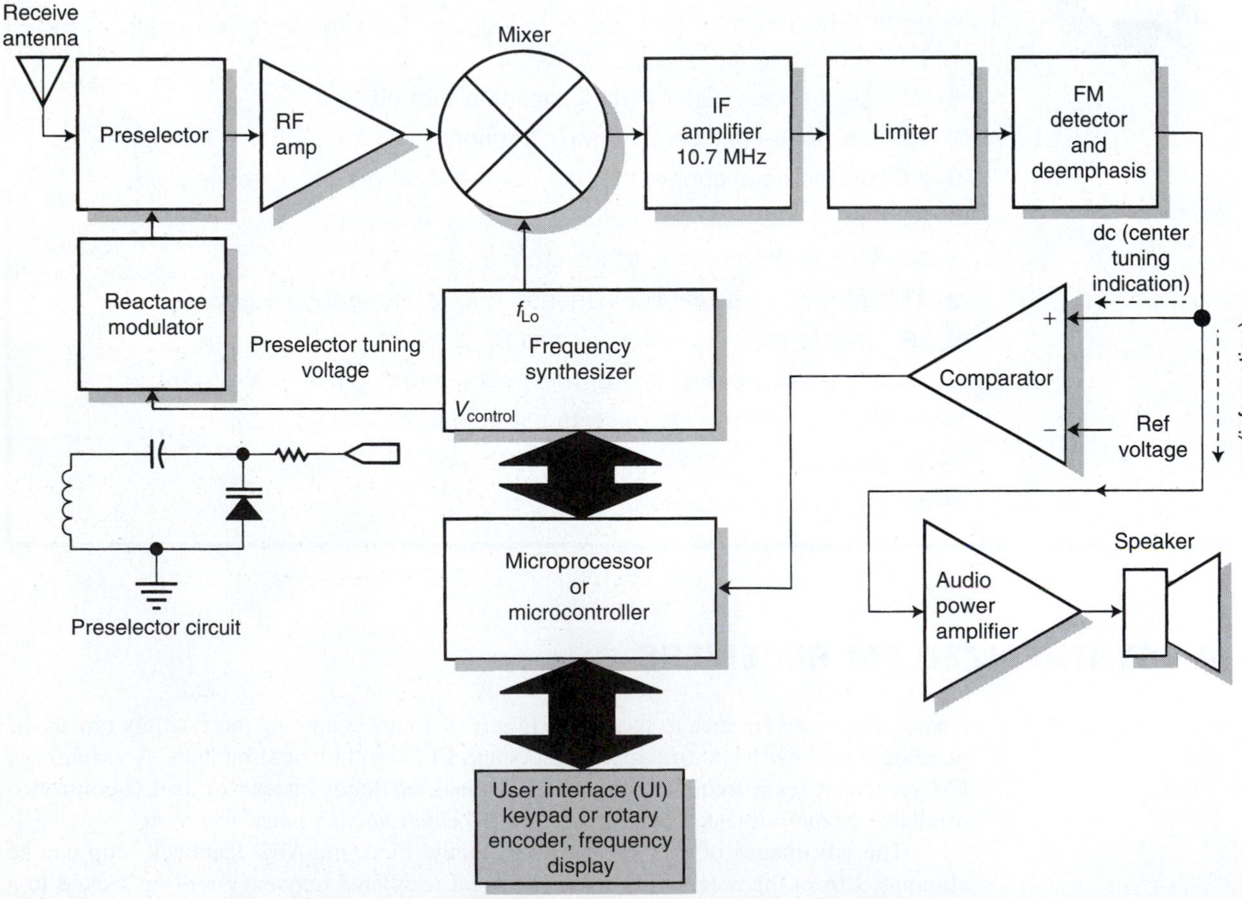

Figure 9–14 A frequency-synthesized FM receiver

dc control voltage, just like a VCO. You can think of the preselector as an electronically tuned bandpass filter; its center frequency depends on the *control voltage*.

But where can we get the control voltage? There are two ways of doing it. We can use *software* to drive a digital-to-analog converter, and use the resulting voltage to tune the preselector. This method is used in high-performance receivers. A simpler and much more common method is to "borrow" and condition the VCO *control voltage* from within the PLL synthesizer.

The VCO control voltage tracks local oscillator frequency changes. As the local oscillator frequency is increased (the receiver tuning position is changed), the VCO control voltage is increased. If this voltage is also sent to the varactor diode in the preselector, the preselector's resonant frequency will also rise. You can think of the VCO control voltage as "tuning" both the local oscillator (VCO) and preselector in unison. Thus, the synthesized receiver is made to "track" in the same manner as a conventional analog receiver. The old analog tracking adjustments remain; there are usually trimmer capacitor adjustments on both the LO and preselector in a synthesized receiver. These adjustments must be made in strict accordance with manufacturer's data to ensure proper tracking.

EXAMPLE 9–3

A synthesized FM receiver uses a 214 nH inductance and a Motorola MV-209 tuning diode in its preselector. Its PLL synthesizer has a VCO transfer characteristic as shown in Figure 9–15. The VCO tuning voltage is fed directly to the MV-209 in the preselector to attain tracking. Calculate the receiver's operating frequency when the tuning voltage is

a. 8 V
b. 9 V

Solution

In this problem, we are essentially being given the *frequency* of the local oscillator, and are being requested to work backward to deduce what the preselector is doing.

a. From the graph data of Figure 9–15, when $V_{\text{control}} = 8$ V, the VCO operates at approximately *110 MHz*. Therefore, the preselector frequency *should* be

$$f_{\text{pre}} = f_c = f_{\text{LO}} - f_{\text{if}} = 110\text{ MHz} - 10.7\text{ MHz} = \underline{\underline{99.3\text{ MHz}}}$$

The receiver *should* be tuned to 99.3 MHz when the tuning voltage is 8 volts. Let's verify that the preselector is tracking. From the MV-209 graph, we see that the diode capacitance $C_d = 12$ pf (approximately) when 8 volts is applied. This can be plugged into the resonance formula to calculate the preselector's actual operating frequency:

$$f = \frac{1}{2\pi\sqrt{LC}} = \frac{1}{2\pi\sqrt{(214\text{ nH})(12\text{ pF})}} = \underline{\underline{99.3\text{ MHz}}}$$

Yes! The preselector is "tracking" the local oscillator electronically!

b. When $V_{\text{control}} = 9$ V, the local oscillator now operates at *115 MHz,* and again, the receiver will be tuned 10.7 MHz below that:

$$f_{\text{pre}} = f_c = f_{\text{LO}} - f_{\text{if}} = 115\text{ MHz} - 10.7\text{ MHz} = \underline{\underline{104.3\text{ MHz}}}$$

We can again examine the preselector action. When 9 volts is on the diode, its capacitance becomes *11 pf* approximately, leading to

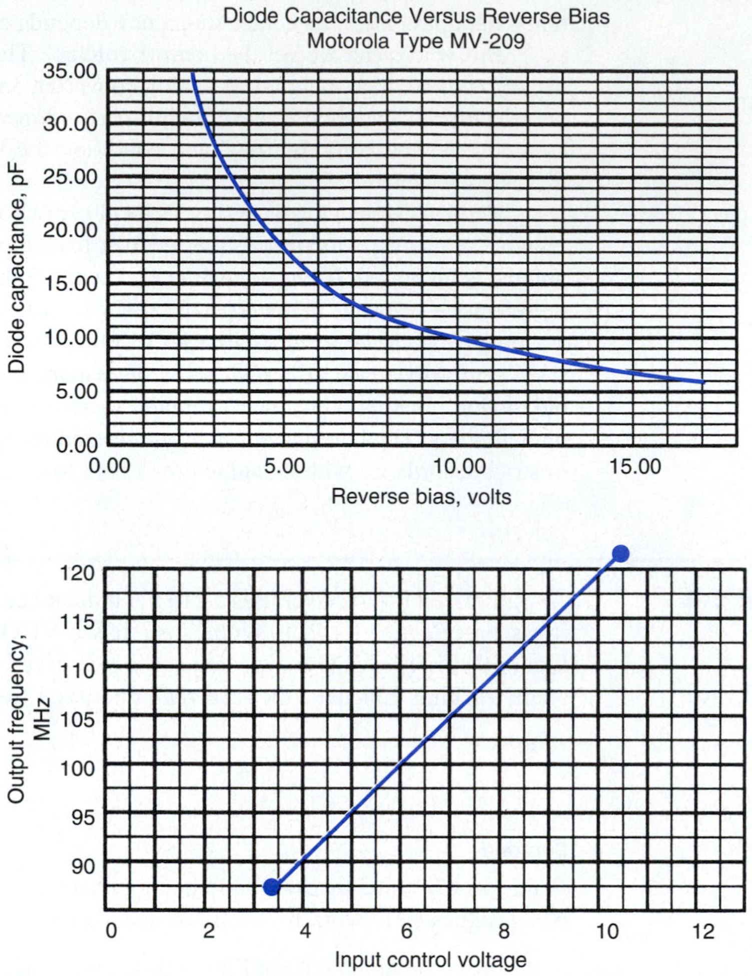

Figure 9–15
Characteristic graphs for Example 9–3

$$f = \frac{1}{2\pi\sqrt{LC}} = \frac{1}{2\pi\sqrt{(214 \text{ nH})(11 \text{ pF})}} = \underline{\underline{103.7 \text{ MHz}}}$$

Wait—this didn't work out quite right. Our receiver doesn't work! *Relax*. We have severely rounded some of our figures, especially the diode capacitances. Even taking that into account, the tracking is quite reasonable. Recall that the preselector is *not* a narrow filter, and in this application, it probably has a bandwidth of 4 MHz. A tracking error of 600 kHz is not too severe, then. Everything really is OK—the preselector correctly (and electronically) tracks the local oscillator!

Bells and Whistles

Since the software in the microprocessor runs the show, the receiver can be made to do almost anything the software designers can imagine. Almost all receivers include some sort of *scanning* capability, where the unit automatically locates the next usable channel at the touch of a button.

The *comparator* in Figure 9–14 is used for this type of operation. There are several ways of implementing this feature; here, we are simply comparing the dc voltage from the FM detector to a predetermined dc reference voltage (perhaps $V_{cc}/2$). Recall that as an FM receiver is tuned above or below a station, the FM detector outputs a voltage with a changing dc polarity. The comparator detects this condition and registers a "1" or "0" to the CPU to indicate "center of channel" (or not). Many makes use dual comparators to give a true center channel indication.

When scanning, the software sends a command to the synthesizer to go to the next frequency, then reads the "center-of-channel" bit. If the bit is 0, there is no signal, or the receiver isn't centered on the channel, and the software calculates the next frequency, then goes back to the top of the loop. When the center channel bit goes high, the FM detector has landed on a usable signal, and the software stops the search loop. All this time, the software is updating the front-panel display with the current frequency and scanning the keypad for more user commands (for example, depression of a *stop* button).

Section Checkpoint

9–33 List three advantages of electronically tuned receivers.

9–34 What provides the local oscillator signal in a synthesized receiver?

9–35 List two methods used to derive the tuning voltage for the preselector in a synthesized receiver. Which method is more popular?

9–36 The preselector in an electronically tuned receiver is really a voltage-controlled ________.

9–37 What features use the comparator circuitry in Figure 9–14?

9–5 FM STEREO AND SCA DECODING

In Chapter 8, the methods for encoding both the FM stereo and SCA (subcarrier authorization) signals were developed. Both of these are *multiplex* techniques; they allow more than one piece of information to share the same carrier signal. To decode these signals, we reverse the encoding techniques.

Decoding FM Stereo

Figure 9–16 shows the components in an FM stereo signal with an SCA carrier. The composite FM signal appears at the output of the FM detector. It contains four unique signals: The L + R or monaural signal, the 19 kHz pilot, the L − R signal, and when SCA is being broadcast, the 67 kHz SCA subcarrier and sidebands.

In order to decode FM stereo, the first three signals must be processed, and the fourth (the SCA subcarrier) must be *ignored*. Most FM stereo decoders have a *67 kHz trap* in the signal path, which is a band-stop filter tuned to the frequency of the SCA energy. The 67 kHz trap prevents any interaction between the SCA signal and the rest of the composite FM signal during processing, which could result in whistles or squeals from undesired signal mixing.

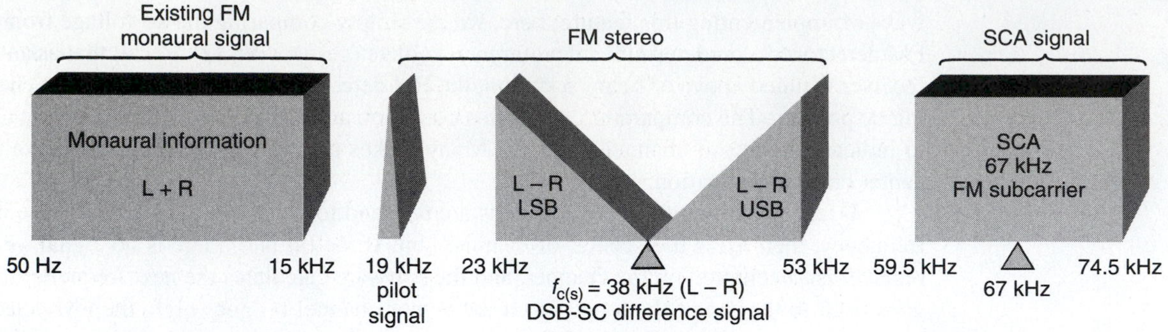

Figure 9–16 A composite FM stereo/SCA signal

Figure 9–17 shows the basic configuration of an FM stereo decoder. Notice that the steps are almost in exact reverse order when compared to the encoding process. The composite FM signal from the detector enters the decoder circuit and immediately splits in two directions. First, to isolate the monaural signal, the signal is passed through a low-pass filter; the output of the low-pass filter (bottom) is only the L + R component. The second signal path (toward the top) decodes the L − R component.

The L − R or *difference* information is carried on a 38 kHz DSB-SC signal. In order to demodulate a DSB-SC signal, the *carrier* signal must be recovered. In a SSB receiver,

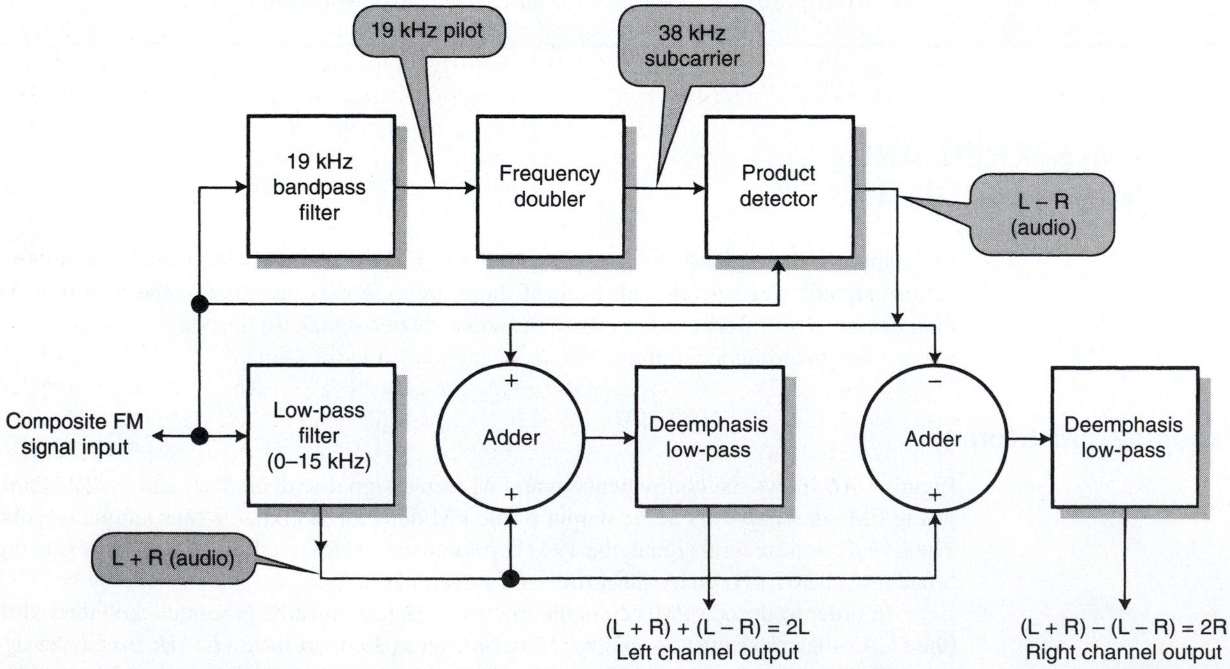

Figure 9–17 An FM stereo decoder

a BFO (beat frequency oscillator) is used to regenerate the carrier. For this application, a BFO will not be accurate enough; there must be *zero* frequency error in the reinserted 19 kHz carrier, or severe distortion will result. The *19 kHz pilot* signal is used to regenerate the 38 kHz carrier. At the transmitter, the 19 kHz signal is frequency-doubled to generate the 38 kHz subcarrier; exactly the same circuitry is used at the receiver. Thus, the 38 kHz subcarrier at the receiver will be at the same precise frequency as the 38 kHz carrier at the transmitter.

The 19 kHz pilot is first extracted by the use of a narrow bandpass filter and is then doubled in a frequency multiplier stage whose tank is tuned to 38 kHz, the 2nd harmonic of 19 kHz. At this point in the decoder, the presence or absence of the 38 kHz signal is often used to drive a *stereo indicator lamp* to let the user know that a stereo signal is being received. The 38 kHz recovered carrier is applied to a product detector, along with the L − R DSB-SC portion of the composite signal. The output of the product detector is the *audio* version of the L − R or difference signal.

The steps taken so far seem fairly involved, and we still haven't recovered the left and right channel information. However, the end is very near. The last step in decoding is to simply *add* and *subtract* the (L + R) and (L − R) signals in two *adder* circuits. The addition and subtraction in this step is purely algebraic; if we *add* the mono and difference signals, we get:

$$(L + R) + (L - R) = 2L$$

which is simply the left channel, with a doubled amplitude. The resulting left-channel output is then ready for deemphasis. By subtracting the two signals, we get

$$(L + R) - (L - R) = 2R$$

This is the *right* channel, which is again sent through a deemphasis network before passing on to the audio amplifier.

The circuitry in an FM stereo decoder looks complicated, but in reality, it can be contained entirely on a single IC chip. In modern equipment, this is the approach that is almost always used. IC versions of the FM stereo decoder generally use a PLL to recover the 38 kHz carrier, which eliminates the bulky 19 kHz and 38 kHz RF tuned circuits.

Decoding the SCA Signal

The SCA signal is a narrowband FM carrier signal that modulates the FM station carrier. This is a little strange—it's a little like the concept of "a universe within an atom." In order to demodulate the SCA signal, a *second* FM detector is needed.

Most SCA decoders use a phase-locked loop tuned to the 67 kHz PLL center frequency; the output from the FM detector contains the SCA subcarrier and is applied through a high-pass filter to the PLL. A high-pass filter is necessary, since the other components of the FM stereo signal are *below* 67 kHz (see Figure 9–16) and the 67 kHz SCA signal is small in amplitude when compared to the rest of the composite FM signal (about 10% of its voltage). Figure 9–18 shows an LM565 PLL FM SCA decoder that could be connected to the detector output of any FM receiver.

In the SCA decoder in Figure 9–18, which is similar to the PLL FM detector of Figure 9–10, the composite FM signal enters the PLL through a two-stage high-pass filter built from C_5, C_6, R_{10}, and R_6. The free-running frequency of the PLL is set to 67 kHz by C_7, R_7, and R_4. (You might compare these values with the circuit of Figure 9–10 to see how they've changed.) The PLL filter output on pin 7 contains the ac SCA information signal

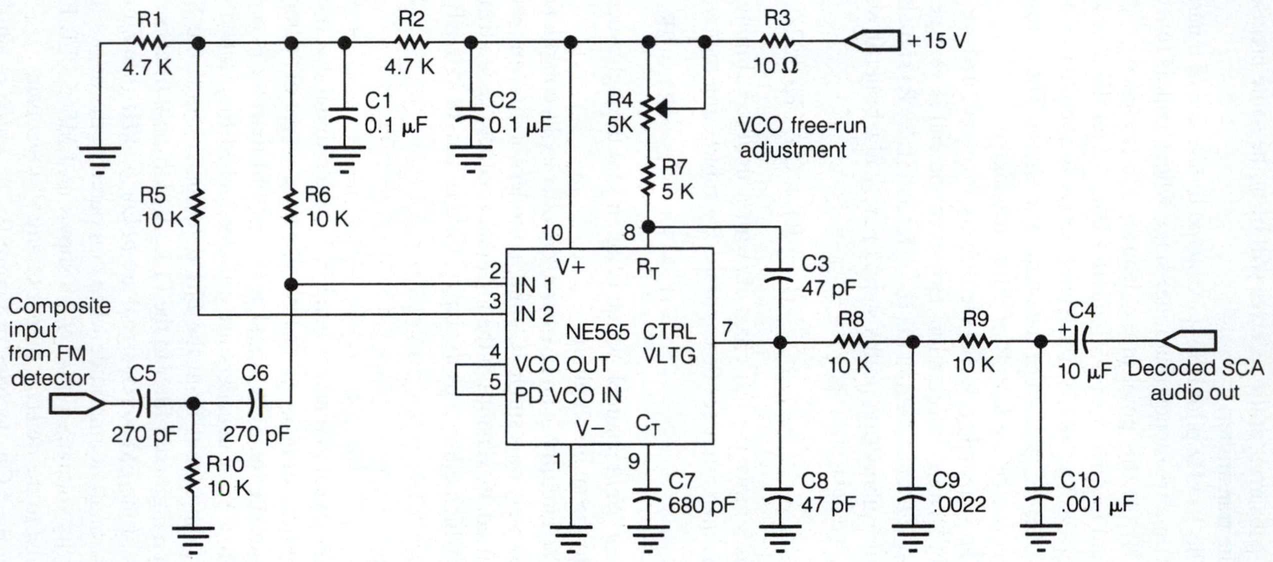

Figure 9–18 An SCA decoder employing the LM565 PLL

riding on top of the dc loop correction voltage; R_8, R_9, C_9, and C_{10} remove any residual 67 kHz RF signal, then C_4 blocks the dc, leaving only the decoded ac information signal.

Should you decide to build Figure 9–18 and connect it to an FM receiver, be sure to connect the SCA decoder input *before* the deemphasis network, or you'll get no SCA subcarrier. An inexpensive battery-operated receiver is best for experimentation of this type. You'll be very surprised to hear the "secret" signals being transmitted by broadcasters!

Section Checkpoint

9–38 What is meant by the term *multiplex*?

9–39 Draw a frequency diagram showing the parts of an FM stereo composite signal.

9–40 Why is a 67 kHz trap used in an FM stereo decoder?

9–41 What is the purpose of the *frequency doubler* in Figure 9–17?

9–42 What does the presence of a 38 kHz signal at the output of the frequency doubler in Figure 9–17 mean?

9–43 Why is a high-pass filter needed at the input of an SCA decoder?

9–44 What does R_4 adjust in Figure 9–19?

9–6 ALIGNMENT OF FM RECEIVERS

When a technician *aligns* an FM receiver, he or she is really making adjustments to the various tuned circuits in the unit so that each one operates on the correct frequency. There are several types of alignment adjustments that can be made in receivers. Non technicians often assume that all circuits are adjusted for the loudest output at the speaker, and some may even attempt adjustments. This usually results in a severely misaligned receiver! To make matters worse, amateurs often try "turning screws" when a unit isn't working right. This can make for interesting troubleshooting, as there can now be *several* problems to find (or more, if the user breaks something in the process).

There are three basic ways that tuned circuits can be aligned. Adjustable elements can be aligned for *maximum output,* correct *tracking,* or a correct *shape*. The location in a receiver dictates what type of adjustment must be made.

In order to make accurate adjustments, a calibrated FM signal generator is required. It is generally foolhardy to attempt alignment while tuned to a broadcast signal. The content of the signal is too complex to see accurately on a scope, and the signal strength is hard to control. Figure 9–19 shows one type of AM/FM signal generator, the Leader model 3215.

The generator of Figure 9–19 is PLL frequency synthesized and computer controlled, with frequency entry by keypad, rotary knob, or computer remote control through a general purpose interface bus (GPIB) on the rear of the unit. Because it is synthesized, it offers a very stable carrier frequency by which receivers can be accurately calibrated. The generator can produce accurate FM or AM modulation with either an internal 400 Hz/1 kHz audio source, or an external audio signal source. The generator also features a calibrated *attenuator* section, which is on the right. The attenuator allows the signal generator to provide an exact, calibrated power level to the receiver (indicated by the right-hand

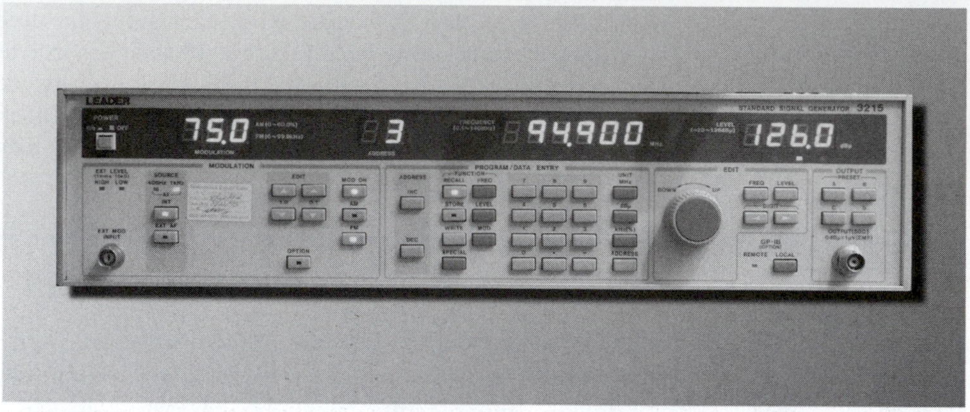

Figure 9–19 The Leader 3215 Synthesized AM/FM signal generator

LED display), which is very useful for performance evaluation. Some frequency generators can also generate a composite FM stereo signal, which is very useful for testing FM stereo tuners.

Maximum Output Adjustments

In an FM receiver, the *preselector* or *IF amplifier* tuned circuits may be tuned for maximum output. This procedure is only true when the receiver employs *synchronously tuned IF transformers.* Most narrowband FM receivers are designed in this fashion.

Because maximum output is hard to determine in an FM signal (the receiver ideally ignores amplitude changes), the signal generator must be set to the *lowest possible output level that can be readily seen at the detector output or heard at the loudspeaker.* If the FM signal generator level is too high, the limiter will begin to operate, which will obscure any amplitude changes caused by circuit tuning.

Tracking Adjustments

You might recall that in a superhet receiver, *tracking* refers to the ability of the local oscillator and preselector to maintain a difference of f_{if} across the entire tuning range of the receiver. The *preselector* and *local oscillator* circuits are adjusted to obtain the correct tuning relationship. Often the manufacturer provides a recommended procedure as part of the receiver's service literature.

The basic procedure for tracking adjustments in an analog receiver is as follows. First, the FM signal generator is connected to the antenna terminals and is set to a point in the middle of the receiver's tuning range (such as 99.7 MHz). The receiver dial is adjusted to match. If the generator is not audible, the *local oscillator* is adjusted until the receiver passes the generator signal correctly (center tuned). This sets the dial calibration. Next, the generator is tuned high (such as 107.9 MHz), and the receiver is tuned to match. The generator's output is then reduced to the minimum level that is detectable; then the *preselector trimmer capacitor* is adjusted for maximum receiver output. Finally, the generator and receiver are set at the low end of the dial (such as 88.1 MHz), and the *preselector inductor* is adjusted for maximum output, again with the minimum signal necessary. Since

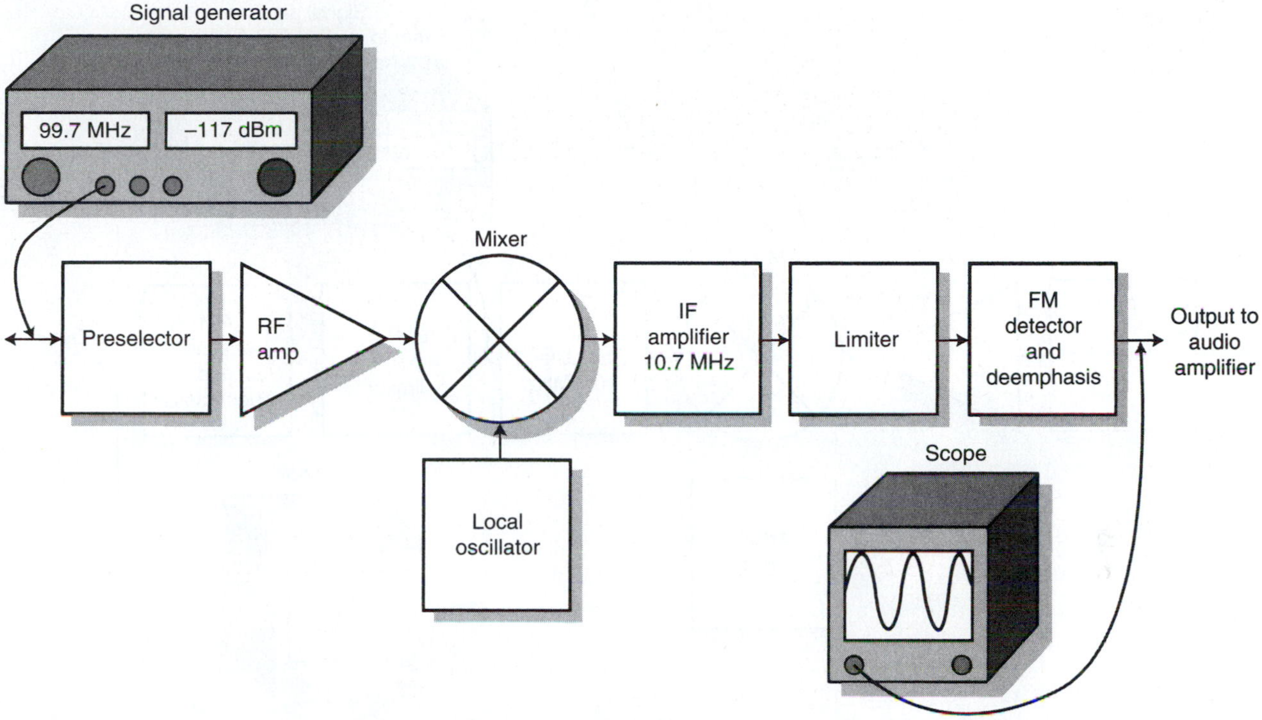

Figure 9–20 Tracking adjustments

these adjustments interact, they must be rechecked after completion and touched up as needed. Figure 9–20 shows the basic setup for tracking adjustments.

With synthesized receivers, the tracking adjustment is usually quite simple. A typical adjustment is made by tuning the FM receiver to a specified display frequency (such as 99.7 MHz), and while monitoring the dc voltage at a test point in the PLL synthesizer (usually the VCO control voltage), the local oscillator tank is adjusted until the correct dc voltage is obtained. Once this has been done, the local oscillator adjustment is complete, and a small touch-up adjustment may be made in the preselector. The procedure is very easy and can usually be done without a generator.

Shape Adjustments

In an FM receiver, the *detector* circuit must be adjusted to obtain a correct frequency-to-voltage transfer characteristic. The way this will be accomplished depends on the type of detector in use. In general, the FM signal generator is connected so that it provides signal to the last IF amplifier stage at the IF frequency (10.7 MHz). For aligning a PLL or quadrature FM detector, the generator remains unmodulated, and the PLL VCO trimmer or quadrature detector coil is adjusted until a specified dc voltage appears at the detector test point. (The manufacturer usually gives this information in the service manual.) Figure 9–21 shows how the instruments are connected for detector alignment.

For a Foster–Seeley or ratio detector to make adjustments, the generator must be *swept* up and down in frequency while the detector output is monitored with an oscilloscope.

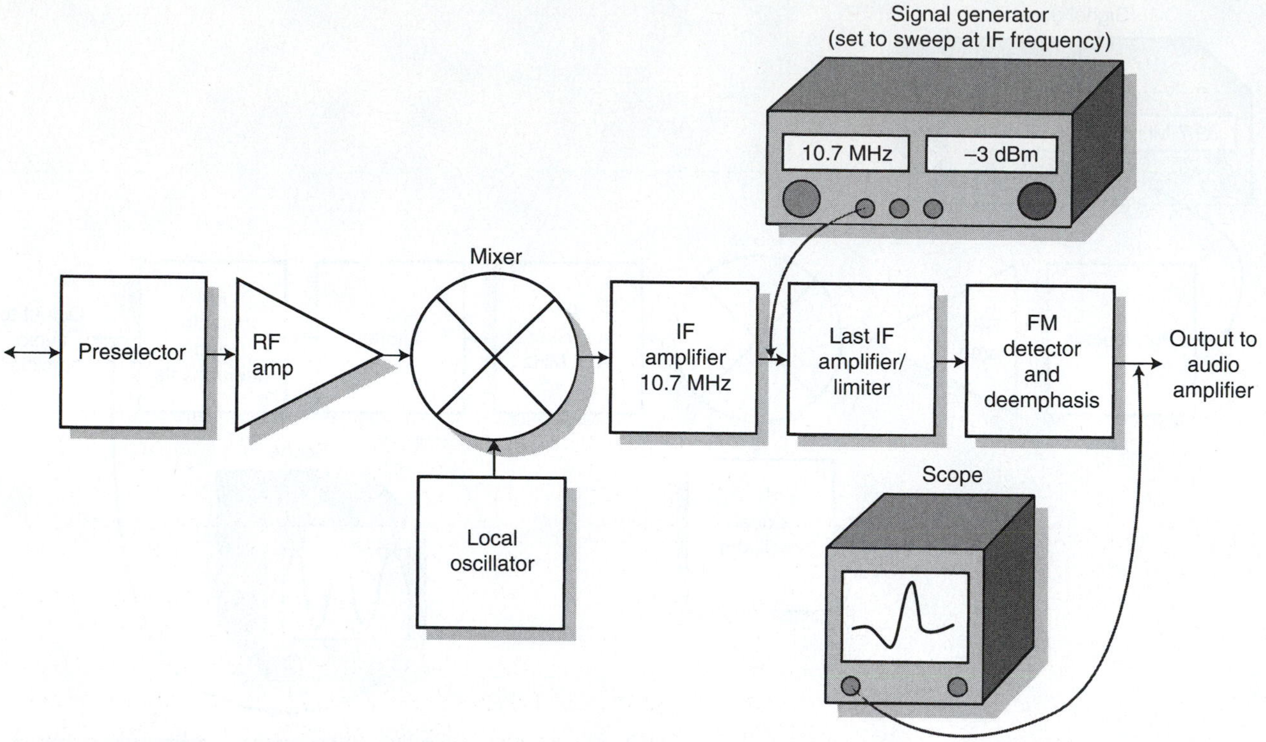

Figure 9–21 Detector alignment

Usually, the generator is set to sweep at a slow rate (such as 60 Hz) with a linear (ramp) waveshape, which produces the characteristic curve known as an *s-curve,* as shown in Figure 9–22.

In a ratio or Foster–Seeley detector, there are two interacting adjustments. There's no good way of describing how these work, except to point out that they should be made *slowly* and *carefully*. It's very easy to get lost when making these adjustments because turning either coil too far away from its optimal set point makes the detector stop working! The important portion of the s-curve is in the middle; it should be as *linear* (straight) as possible. The middle portion of the curve is what detects the FM signal.

Detector Alignment Using an Audio Spectrum Analyzer

Because minimum signal distortion occurs when an FM detector is correctly aligned, it is also possible to perform alignment by using a tone-modulated RF source (such as the Leader 3215 of Figure 9–19) and an audio spectrum analyzer (such as the FFT analyzer provided on the Agilent 54622D mixed signal oscilloscope). The instrument setup is just like that of Figure 9–21, except that the oscilloscope is set to display the spectrogram (FFT mode) of the detected audio, and the signal generator is modulated with a 400 Hz or 1 kHz tone at maximum deviation (75 kHz for broadcast receivers).

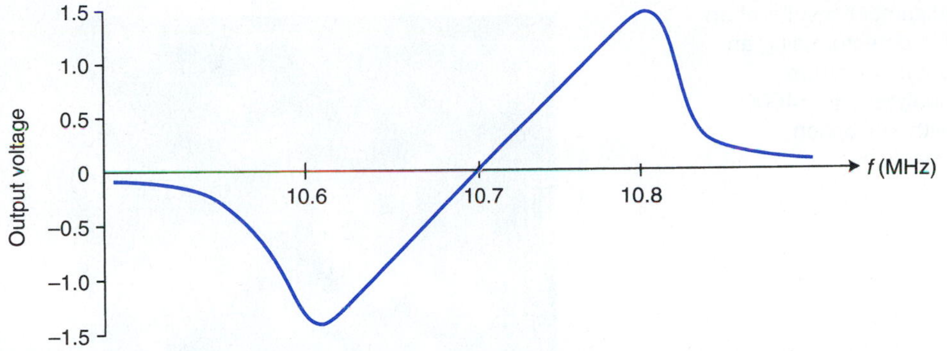

Figure 9–22 An s-curve

To correctly adjust the detector using this method, proceed as follows:

- Provide a 10.7 MHz (f_{if}) signal to the IF circuit at maximum deviation (75 kHz in a broadcast receiver). Make sure enough signal is present to put the receiver into full limiting.
- Use a 1 kHz sine for modulation. This must be a low-distortion signal for accurate measurement. (Use a spectrum analyzer to check the signal source if in doubt.)
- While observing the spectrogram at the detector's output, *slowly* adjust the detector circuits to minimize the harmonic energies (which represent distortion due to misalignment). Only a single "spike" should appear on the spectrum analyzer display (representing the demodulated 1 kHz information signal).
- If possible, repeat the measurements at several frequencies (400 Hz, 5 kHz) to double-check the results. You may also want to increase the deviation slightly beyond 100% (85–90 kHz should be accepted) and again check the distortion level.

Figure 9–23(a) shows the output of a detector that is poorly aligned, while Figure 9–23(b) shows the same detector after correct alignment. Remember to make adjustments *very slowly* when employing this method!

A Typical Alignment Procedure

A complete alignment of an FM receiver usually is carried out in the following order: detector (shape or minimum distortion adjustment), IF amplifier (maximum output and/or correct shape), and front-end (preselector and local oscillator) tracking (maximum output). This sounds backward, but it makes sense because the detector must first be properly aligned before any signals can be expected to accurately pass through it. Completion of the other alignment steps usually requires the detector to be operating correctly.

After an alignment is completed, the technician may be asked to check receiver sensitivity. Sensitivity can be checked by applying a specified input signal (from a calibrated generator) at the antenna terminals, then checking the detector or audio output at the speaker with an oscilloscope. The manufacturer typically gives the signal level at the input and the amount of *quieting* at the receiver output.

Quieting is a measure of the noise reduction capability of the FM receiver. For example, if the manufacturer's data indicates that a 5 μV input signal provides 30 dB of quieting (at 100% modulation) what this says is that the resulting demodulated signal should be at least 30 dB stronger than the received noise level. (The manufacturer may also express

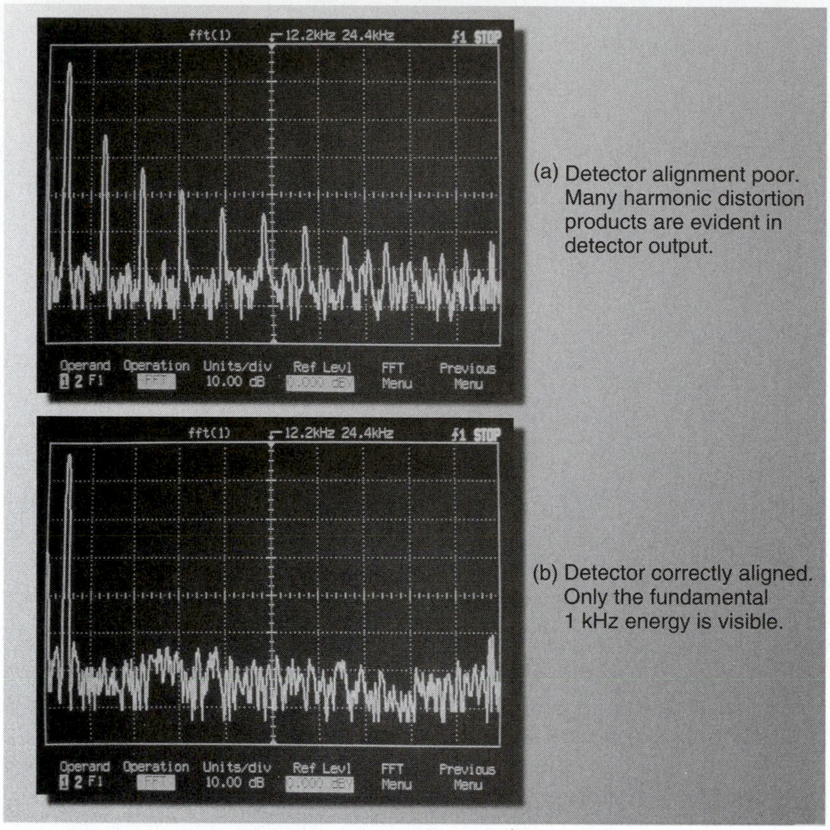

Figure 9–23 Alignment results of an FM detector using an audio spectrum analyzer (HP 54600 with FFT option installed)

(a) Detector alignment poor. Many harmonic distortion products are evident in detector output.

(b) Detector correctly aligned. Only the fundamental 1 kHz energy is visible.

the input signal in dBm.) In FM receivers, the limiter begins to operate at a certain minimum input signal level called the *limiting threshold.* Above this signal level, quieting increases rapidly because the limiter is able to clip the signal voltage, removing most of the amplitude noise. Below this level, the receiver's noise reduction capabilities fall off rapidly, until only noise remains.

Section Checkpoint

9–45 Define the term *alignment* as it applies to RF systems.

9–46 List the three types of alignment adjustments in FM receivers.

9–47 What instrument is needed to make accurate receiver alignment adjustments?

9–48 What is adjusted by the *attenuator* section of a signal generator?

9–49 When adjusting a component for maximum output, how must the signal generator be set?

9–50 When adjusting tracking in an *analog* receiver, what is adjusted first?

9–51 What is the curve of an FM detector shaped like?

9–52 Explain how to connect the instruments for detector alignment.

9–53 What is *quieting*? How is this related to the *limiting threshold* of the receiver?

9–7 TROUBLESHOOTING FM RECEIVERS

To find problems in FM receivers, a technician uses the same approach that we've developed in previous chapters, with a few enhancements that are needed for high-frequency work. The steps in this approach are

1. visual (and other) inspection
2. power supply checks
3. input and output checks

Since FM transmitters and receivers tend to be more complex than their AM counterparts, there is a lot more that can go wrong. Don't let this intimidate you; continue to follow the three-step approach. As you successfully resolve troubles, your confidence will grow.

High-Frequency Considerations

Because of the high frequencies involved in FM transmitters and receivers, new troubleshooting techniques are required. There are two problems a new technician immediately finds when he or she attempts to work on VHF and UHF FM equipment. First, circuits that operate at VHF and above are very temperamental. Simply touching a test point with a scope probe can *load* the circuit heavily enough to make it stop working. If you disable the circuit by trying to check it, you get nowhere fast! Second, many scopes can't directly view VHF and UHF signals in a meaningful manner (showing correct voltage, time, etc.). Even if you don't kill the circuit with the scope probe, all you may see is a fuzzy trace. That's not enough to tell if things are OK or not! This problem gets much worse as we move toward the front end of a receiver, where the signals are measured in *microvolts.*

The technique most technicians use to get around these two problems is *signal injection*. With signal injection, a calibrated signal generator is coupled to the circuit and adjusted to the frequency that *should* be present. If the receiver responds, everything after that point is probably OK. If not, then the injection point is moved toward the detector. Figure 9–24 shows how this works.

Suppose that we have a dead receiver. What steps should be taken to find the problem? Did you think *visual inspection* and *check power supplies* first? If so, great—you're well on your way to becoming a first-rate tech! Assuming that we've checked those things out, next we need to determine where we're losing the signal.

In Figure 9–24, you can see that there are *two* different frequencies we might have to apply from the generator, depending upon where we're at in the circuit. Where is a good place to start? The *middle* is often a good place to start. The solid gray line shows that we have injected a 10.7 MHz signal modulated with a 1 kHz audio tone into the first IF

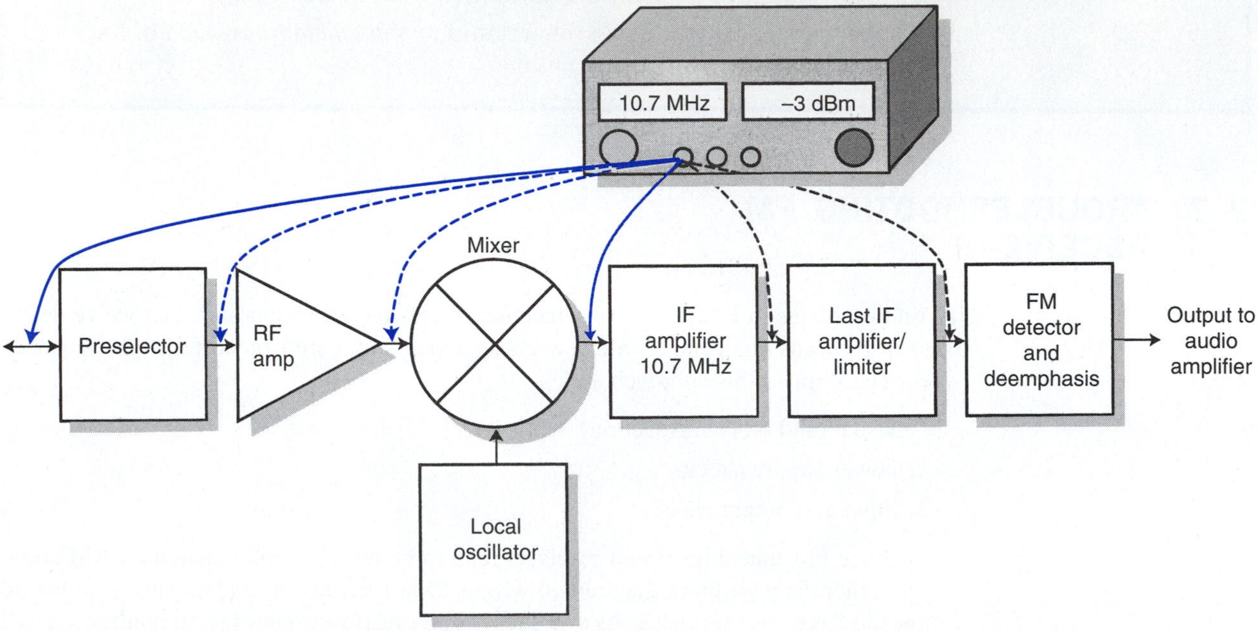

Figure 9–24 Using signal injection to hunt for problems

amplifier. Why a modulated signal? Right—we want to be able to hear something in the loudspeaker or see a waveform at the detector with a scope.

If the receiver responds to this, we know that everything to the right is functional. *We've just eliminated 50% of the circuitry from the suspect list, and the problem is to the* **left** *of the injection point.* What if nothing happens? Then we proceed to inject at the next stage to the right (dashed gray lines). We know the problem lies to the right.

A technician must keep in mind two important points when injecting signals. First, always remember what *frequency* is supposed to be present and second, keep the *amplitude* of the signal generator at a minimum appropriate value. For example, injecting 10.7 MHz into the preselector circuit of an FM receiver will have little effect, even if the receiver is working correctly; it's simply not designed to receive this frequency!

In Figure 9–24, we will have to readjust the generator frequency to the RF carrier frequency once we start injecting before the mixer, since that is the frequency expected in that portion of the circuit.

When injecting signals, always be aware of the frequency and amplitude that the circuit is designed to work with. Injecting the wrong frequency leads nowhere—fast!

A Technician's Worst Nightmare

When a well-meaning customer has tampered with the innards of equipment, it usually spells trouble. Often the technician will have no clue that this has happened. People being people, they usually find it hard to admit to their mistakes! Most people won't mess with something unless they perceive a problem. This means that a severely misaligned receiver is likely to have some other "trouble" that you will have to find to make things right again.

A technician should proceed with caution when he or she finds any of the following conditions in a unit:

- broken paint seals on adjustable coils
- cracked or broken tuning slugs on coils
- marred screws on the case of the unit
- obvious places where someone has soldered on a board

Any of these findings are a sign that someone has been there before you have, and you should *assume nothing* about the condition of the unit. Stick with the three-step approach; the original problem may have only been a power supply problem (but may be worse now!).

A stage that is not responding to signal injection may have a defective component, or it may have had its tuned circuit so far misadjusted that no signal can pass. When you see any of the conditions above, keep this in mind!

SUMMARY

- FM receivers use the superheterodyne principle, just like AM receivers.
- An FM detector can be viewed as a frequency-to-voltage converter.
- The limiter stage in an FM receiver removes most of the noise from an incoming signal by eliminating amplitude changes.
- Analog FM receivers require AFC to keep the local oscillator from drifting off frequency. Digitally synthesized receivers have no need for AFC.
- Most FM detectors operate by converting frequency changes to phase shift, which is then converted into a voltage.
- Analog FM detectors (Foster–Seeley, ratio) must be preceded by a limiter stage; digital detectors (PLL, quadrature) have built-in limiting action.
- In a synthesized FM receiver, a microprocessor controls most of the action. Alignment is simplified in synthesized receivers.
- FM stereo and SCA decoding reverses the multiplexing process of the transmitter in order to recover the added channels of information.
- Alignment of a receiver involves the adjustment of tuned circuits such that each operates at its intended frequency.
- An FM detector is aligned for proper *shape;* the response of an FM detector is an s-curve.
- In troubleshooting VHF equipment such as FM receivers, signal injection is a very useful tool. At many points in a receiver, the signals are too small to view on a scope.

PROBLEMS

1. Draw a block diagram of a superheterodyne FM receiver, showing which sections are *new* or *different* from an AM receiver.
2. What is the function of the *limiter* in an FM receiver?
3. Give two reasons for the use of an RF amplifier in FM receivers.
4. What is the purpose of the AFC system in an FM receiver? What type of FM receivers do not require AFC?

5. Why do FM receivers use a 10.7 MHz IF instead of 455 kHz as AM receivers use?

6. Calculate the range of local oscillator frequencies required for an FM weather receiver (161 MHz to 163 MHz) with an IF of 10.7 MHz, assuming high-side injection.

7. Why isn't AGC needed in FM receivers?

8. How does a slope detector work? Why isn't this type of detector used in FM receivers?

9. Draw the schematic diagram of a Foster–Seeley FM detector. Explain the purpose of each component in the circuit.

10. Draw the schematic diagram of a ratio detector. What advantage does it have over the Foster–Seeley circuit?

11. Draw a block diagram of a PLL FM detector. Explain its operation using outline format.

12. An FM signal with a center frequency of 10.7 MHz is driving the PLL FM detector of Figure 9–8. The carrier is being modulated by a 1 kHz audio signal, producing a deviation of 50 kHz. If the VCO in the PLL has the transfer characteristic of Figure 9–9, calculate the resulting peak-to-peak audio output voltage.

13. Explain the procedure for aligning a quadrature FM detector.

14. The receiver of Figure 9–13 is inoperative. *None of the RF sections seem to work,* but the audio amplifier (U_1) works quite well. A 1 kHz audio signal injected at the positive end of C_{30} is readily (and loudly!) passed at the speaker. What is probably wrong? (What should have been checked before injecting a signal at C_{30}?)

15. Draw a block diagram of a synthesized FM receiver.

16. Explain how *tracking* is achieved electronically in a synthesized FM receiver.

17. A synthesized FM receiver uses a 214 nH inductance and a Motorola MV-209 tuning diode in its preselector. Its PLL synthesizer has a VCO transfer characteristic as shown in Figure 9–15. The VCO tuning voltage is fed directly to the MV-209 in the preselector to attain tracking. The tuning voltage is 10 V. Calculate (a) The resulting LO frequency; (b) The carrier frequency, f_c, that the receiver is tuned to; (c) The preselector resonant frequency.

18. Repeat problem 17 but change the tuning voltage to 7 V.

19. Draw a block diagram of an FM stereo decoder, showing the type of signal present at each point.

20. What are the two purposes of the 19 kHz pilot signal?

21. Demonstrate, using algebra, how the (L + R) and (L − R) signals are combined to recover the L and R signals.

22. In what frequency range is the SCA signal contained? What type of filter is used to separate the SCA signal from the rest of the composite FM signal?

23. What two signal characteristics are accurately controlled in the output of an FM signal generator?

24. When performing adjustments for maximum output, why must the signal generator output level be set to the lowest possible value?

25. Using outline form, explain how to adjust tracking in an analog FM receiver.

26. Draw a block diagram showing how to connect instruments for the alignment of an FM detector circuit. What is the shape of the resulting curve?

27. Define *quieting*. How is this related to the *limiting threshold* of an FM receiver?

28. A certain FM receiver is rated as having a limiting threshold of 2 μV. A calibrated FM signal generator is connected to the antenna terminals and is set for a 4 μV output. In terms of noise, how should the resulting receiver output sound?

29. What are the two considerations when injecting signals into RF circuitry?

30. The FM receiver of Figure 9–13 is inoperative. The following voltages have been measured: Positive side of C_{29}, +12 V; Positive side of C_1, 11 V. A 10.7 MHz modulated FM signal was injected into the base of Q_2, and there was no response. What stage or stages could be at fault?

31. The FM receiver of Figure 9–13 is inoperative. The following voltages have been measured:

positive side of C_{29}, +12 V; positive side of C_1, 0 V. What is the most likely problem? (There are several components that could cause this problem.)

32. The FM receiver of Figure 9–13 is inoperative. Both AF and RF dc power supplies check OK (12 V and 11 V); injecting a 1 kHz modulated 10.7 MHz FM signal into the base of Q_1 produced a loud 1 kHz tone in the speaker. The dial of the receiver is set to 99.7 MHz. In order to check the frequency converter, a signal can be injected at the collector of Q_4, the RF amp. What frequency should the generator be set to, and why?

10

Television

OBJECTIVES

At the conclusion of this chapter, the reader will be able to:

- describe the spectral (frequency domain) composition of an analog television signal
- describe the nature of a digital television signal
- draw a block diagram of a television transmitter
- identify the parts of an NTSC (RS170A) video waveform
- draw a block diagram of an analog television receiver
- describe the characteristics of high-definition television (HDTV)
- describe the steps in HDTV transmission
- draw a block diagram of a HDTV receiver
- identify the section of a malfunctioning TV receiver that is the likely source of trouble

Among the technologies that have shaped our modern society, television ranks high in influence. Rare is the household that doesn't have a TV set; many homes have several. Broadcast TV is a daily source of news, entertainment, and other information for most Americans.

Television technology is used for more than just entertainment. The video display on a computer is closely related to the TV set in the family room—it has been redesigned for operation with the digital signals from a computer system. The same technology is used in radar screens, video surveillance systems, and dozens of other applications.

The modern TV receiver (either analog or digital) is one of the most complex electronic systems that a technician can be called upon to service. It has everything: VHF and UHF RF circuitry; PM, FM, and AM demodulation; microprocessor circuitry; high-voltage analog circuitry; and in many cases, sophisticated switching power supply technology.

Digital TV (DTV) receivers will soon be commonplace worldwide. A DTV receiver is really just a combination of an RF receiver or "tuner," digital data demodulator, computer signal processor, and a display monitor. The display monitor of a digital receiver often uses the same technology as an analog TV monitor (or computer monitor), though other display technologies (such as plasma, LCD, DLP, organic LED, and field emission) are likely to be used. The digital TV signal is carried as a high-speed computer data stream on an analog RF carrier. Because DTV uses digital error correction techniques, minor defects in the signal quality do not affect the picture or sound reproduction at the receiver.

One chapter in a textbook can't make you an expert in TV. Most successful technicians in this field have honed their skills over a period of years, often under the watchful eye of a senior technician. There's a *lot* to learn, but don't be discouraged—you can do it! Let's get started.

10–1 ANALOG TELEVISION PRINCIPLES

The signal from a television station contains information from two primary sources, the *video* information signal (picture), and the *audio* information signal (sound). When early experiments were being conducted with wireless TV transmission, it was decided that this was probably too much information to squeeze on to one carrier signal, so the system shown in Figure 10–1 was developed.

The picture is sent using AM and the sound is sent with FM. The reason for this is simple: Using two different types of modulation reduces the chance that picture and sound will interfere with each other. In early television receivers this was a real problem. Circuits had not yet been developed to separate closely spaced signals. Using AM and FM was the best technical solution available at the time.

TV broadcasts are *channelized,* just like AM and FM broadcasting. In Figure 10–1, the frequencies for channel 2 are shown; channel 2 uses the space between 54 and 60 MHz. That's a lot of bandwidth! The picture requires most of that bandwidth, because displaying moving images requires a lot of information to be sent. That translates into a large bandwidth.

The TV picture is sent using a special form of sideband emission called *vestigial sideband,* or *VSB*. You might recall that both the upper and lower sidebands of an AM signal contain identical information. The upper sideband of the picture carrier in Figure 10–1 uses a 4 MHz bandwidth (55.25–59.25 MHz) by itself. If we transmitted the entire lower sideband, the picture alone would require 8 MHz of bandwidth! The TV transmitter uses a special filter to remove part of the lower sideband, which reduces the *video* bandwidth requirement.

Figure 10–1 The spectrum of a television signal

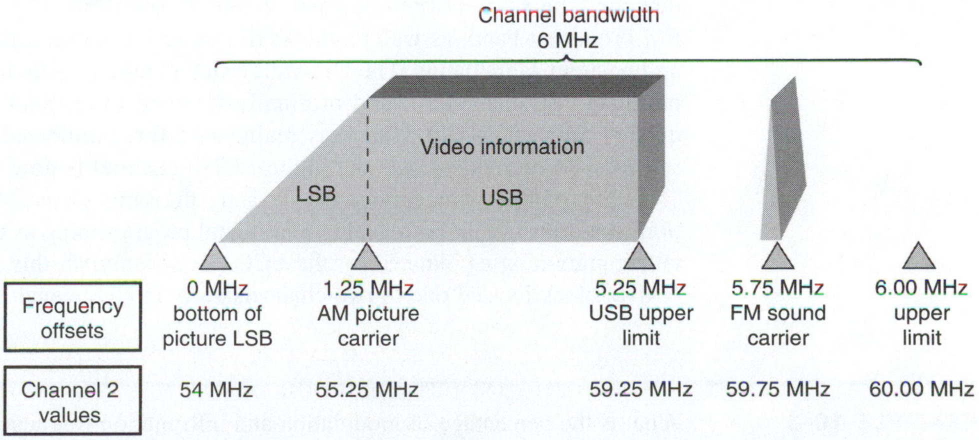

Table 10–1 VHF and UHF television bandplan for the United States

Channel number	Base channel frequency, MHz
2	54
3	60
4	66
(Skipped)	(72–76 MHz)
5	76
6	82
7	174
8–13	180–216 MHz in 6 MHz steps
14	470 MHz
15	476 MHz
16–69	482–800 MHz in 6 MHz steps
[Old channels] 70–83	806–884 MHz in 6 MHz steps

The *sound* is transmitted on a second carrier, which is always 5.75 MHz above the bottom of the channel, or 4.5 MHz above the picture carrier. (This latter fact comes in handy when we analyze a TV receiver's operation.) FM is used, and the maximum deviation is limited to 25 kHz (compare with 75 kHz maximum deviation for broadcast FM in the 88–108 MHz band). The reduced deviation means that TV sound doesn't have quite the fidelity of an FM broadcast, but with proper receiver design, good reproduction can be obtained. Furthermore, many stations now transmit a stereo sound signal; the multiplexing technique is almost exactly the same as that used for broadcast FM stereo.

Bandplans

Table 10–1 shows the plan for the VHF and UHF television channels. Notice the wide range of frequencies that must be covered; this requires specialized antennas for receiving and good circuit design throughout the TV tuner sections.

Notice that there are several skips in the channel plan. The first skip from 72 to 76 MHz is a frequency range allocated to licensed remote control and telemetry devices. The skip between channels 6 and 7 covers 88–174 MHz, which contains the FM broadcast band, as well as the VHF public service, military, aviation, commercial, and amateur radio bands. The UHF television channels, which are a later addition to the plan, start at channel 14, and originally went up to channel 83. In the 1980s the frequency range 806–890 MHz was reallocated for mobile radio and cellular telephone services, so on modern sets the highest UHF channel is now 69.

Digital television service uses exactly the same channel plan. Currently some stations are transmitting both analog and digital programming as they transition to full digital programming as planned by the FCC. To accomplish this, the station uses a second 6 MHz block located one or two channels away from the analog channel.

EXAMPLE 10–1 What is the percentage of modulation and information frequency for an FM sound carrier whose frequency swings between a minimum of 691.74 MHz (f_{min}) and a maximum of 691.76 MHz (f_{max}) 5000 times a second?

Solution

The deviation must be calculated in order to find percentage of modulation. The carrier frequency is not given, but the min and max frequencies are. The carrier frequency lies exactly in the middle between f_{min} and f_{max}. Therefore, the deviation is:

$$\delta = \frac{f_{max} - f_{min}}{2} = \frac{691.76\text{ MHz} - 691.74\text{ MHz}}{2} = \underline{10\text{ kHz}}$$

Now that the deviation is known, percentage of modulation can be calculated:

$$\%\text{ Mod} = 100\%\left(\frac{\delta}{\delta_{max}}\right) = 100\%\left(\frac{10\text{ kHz}}{25\text{ kHz}}\right) = \underline{\underline{40\%}}$$

The *information frequency* is equal to the *deviation rate* (DR), which is given as 5000 cycles per second. Therefore:

$$f_m = \text{DR} = \underline{\underline{5\text{ kHz}}}\text{ By inspection.}$$

Transmitted Power Levels

Most VHF television transmitters operate in the 15–25 kW power range, where the FCC has authorized a maximum effective isotropic radiated power (EIRP) of 100 kW. (EIRP is a measure of the effective transmitted power, and is controlled by the transmitter output power and the design of the antenna. EIRP is covered in detail in the chapter on antennas.) UHF TV uses power levels about 4 times larger (60–100 kW), with a maximum allowed EIRP of 500 kW. The FM sound carrier power must be maintained at a level between 10 and 20% of the peak video (AM) transmitter power.

The power levels developed by the AM portion of a television transmitter are tremendous. You might wonder why such power is necessary (and whether or not TV station employees glow in the dark when they go home at night!). The answer to this question comes from examining the *bandwidth* required by a TV signal. Because a TV signal uses such a wide bandwidth, more power must be applied to overcome the various external and internal noise sources in the receiver's signal path. This also explains why the FCC limits the sound carrier power to only 10–20% of the picture carrier peak power. The sound carrier uses much less bandwidth (only 25 kHz of deviation is allowed), so a much lower power level will suffice to overcome the channel noise.

EXAMPLE 10–2

What are the picture and sound carrier frequencies for UHF television channel 41?

Solution

According to Table 10–1, channels 14–69 are spaced in 6 MHz steps beginning at 470 MHz. Therefore, the beginning of channel 41 is $(41 - 14) \times 6$ MHz away from 470 MHz, or $\underline{632\text{ MHz}}$. To get the picture carrier frequency, add 1.25 MHz:

$$f_{c(pict)} = f_{ch} + 1.25\text{ MHz} = \underline{\underline{633.25\text{ MHz}}}$$

To get the sound carrier frequency, add 5.75 MHz:

$$f_{c(pict)} = f_{ch} + 5.75\text{ MHz} = \underline{\underline{637.75\text{ MHz}}}$$

This can be demonstrated very clearly if you have access to a scanning receiver with wide-band FM receive capability. If you tune the receiver to 637.75 MHz and set it for WBFM reception, you should clearly hear the channel 41 sound.

Figure 10–2 An analog television transmitter

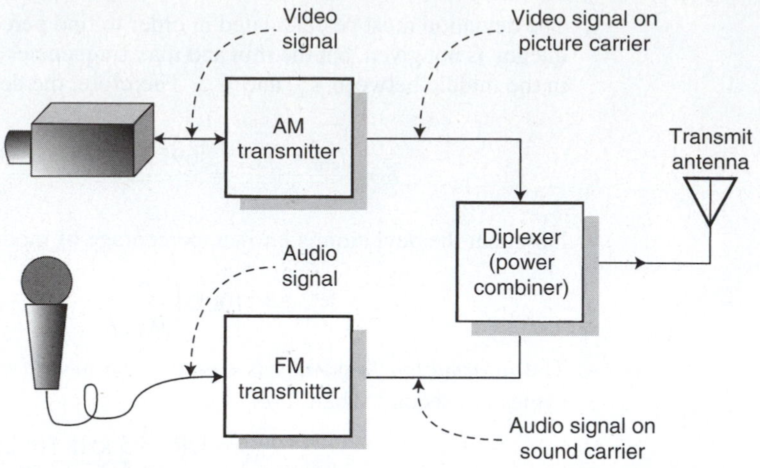

A Television Transmitter

The block diagram of a typical analog TV transmitter is shown in Figure 10–2. Notice how it is really two transmitters in one.

The AM transmitter in the figure generates the *VSB* AM modulated picture carrier; the FM transmitter produces and modulates the sound carrier. The picture and sound carrier signals are combined in a special circuit called a *diplexer*. The diplexer allows the AM and FM signal components to flow to the antenna, but prevents any signal from flowing between the two transmitters. The two transmitters are essentially isolated from each other. If the diplexer were not used, energy would attempt to flow from one transmitter's output into the other, which not only would waste considerable energy, but would likely result in the release of smoke (think transmitter tug-of-war!). The diplexer is a special filter built from *transmission line* sections. Transmission lines are the subject of a later chapter.

Section Checkpoint

10–1 What circuit technologies are found in TV receivers?

10–2 Why are two different carriers used in TV?

10–3 What type of sideband emission carries the picture information?

10–4 Why is VSB used for the television picture?

10–5 How much deviation constitutes 100% modulation of the sound carrier?

10–6 In the U.S. bandplan, how much frequency space is used for each TV channel?

10–7 Why does the bandplan "skip" between channels 6 and 7?

10–8 Why don't new TV sets receive channels 70–83?

10–9 What are typical peak transmit power levels for VHF and UHF TV transmitters?

10–10 What is the purpose of a *diplexer* circuit?

10–2 THE ANALOG VIDEO SIGNAL

The video signal is perhaps the most complex and misunderstood portion of the television system. This complexity is due, in part, to the fact that color was introduced after millions of sets were in use, and therefore a color TV signal must be compatible with black and white receivers. This is similar to the problem that arose with FM stereo broadcasting.

The color television standard in the United States is called *NTSC,* which was developed in 1953 by the national television system committee. Other countries use different systems; for example, you may run across systems designed for *PAL* (Phase Alternation Line), which is used mainly in the United Kingdom and Germany, or *SECAM* (Sequential Couleur Avec Memoire), which is used primarily in France. Once you've learned the details of one system, the others can be rapidly learned.

Motion Picture Principles

Television uses the same method as motion pictures for transmitting animated images. A movie consists of a rapidly displayed sequence of still images, each taken at a consecutive point in time. If the images are presented rapidly enough, persistence of vision causes the individual frames to fuse together into a moving picture.

The trick in generating convincing motion is to present images rapidly enough to prevent the viewer from perceiving *flicker*. For example, in commercial motion pictures, there are 24 frames for each second of film. The rate of 24 fps (frames per second) is not quite fast enough to prevent the sensation of flickering (especially near the periphery of human vision), so the movie projector rapidly projects each frame of the film *twice* in succession (producing for the eye 48 images per second). This eliminates the sensation of flickering, while conserving the amount of film needed to produce a show.

Television uses a process called *scanning* to collect the information from a scene. In the NTSC system, there are 29.97 complete *frames* (analogous to the frames of a movie) transmitted every second. This rate is too slow to prevent the perception of flicker, so each frame is transmitted twice as a set of *fields*. There are two interlaced fields for each transmitted frame, so the eye sees a total of (2)(29.97) fields per second, or 59.94 fields per second, which is high enough to prevent the sensation of flicker.

Scanning the TV Picture

To gather information from a scene, a system of *scanning* is employed, as shown in Figure 10–3.

The camera tube of Figure 10–3 is really a modified vacuum tube. It has a *heater* that heats the *cathode,* which is coated with a substance that emits electrons when heated. The cathode produces a stream of electrons, which are accelerated toward the *target plate*. The electron beam is made to scan the target by the use of two oscillators (horizontal and vertical) and two deflection coils. The scanning follows the same path your eyes are following as they read this page. It starts in the upper left, scans horizontally to the right, then rapidly returns to the left (the *horizontal retrace*) and scans the next line down. The process is repeated until the entire scene has been scanned; then the electron beam rapidly moves from the bottom up to the top (the *vertical retrace*) to scan the scene again.

The target plate is coated with a mosaic of photoresistive material. *Photoresistive* materials change their electrical resistance when light strikes them. When the electron beam strikes the target, it charges it. If light is striking the material at that point, the resistance to current flow from the electron beam decreases, and the resulting voltage drop across the

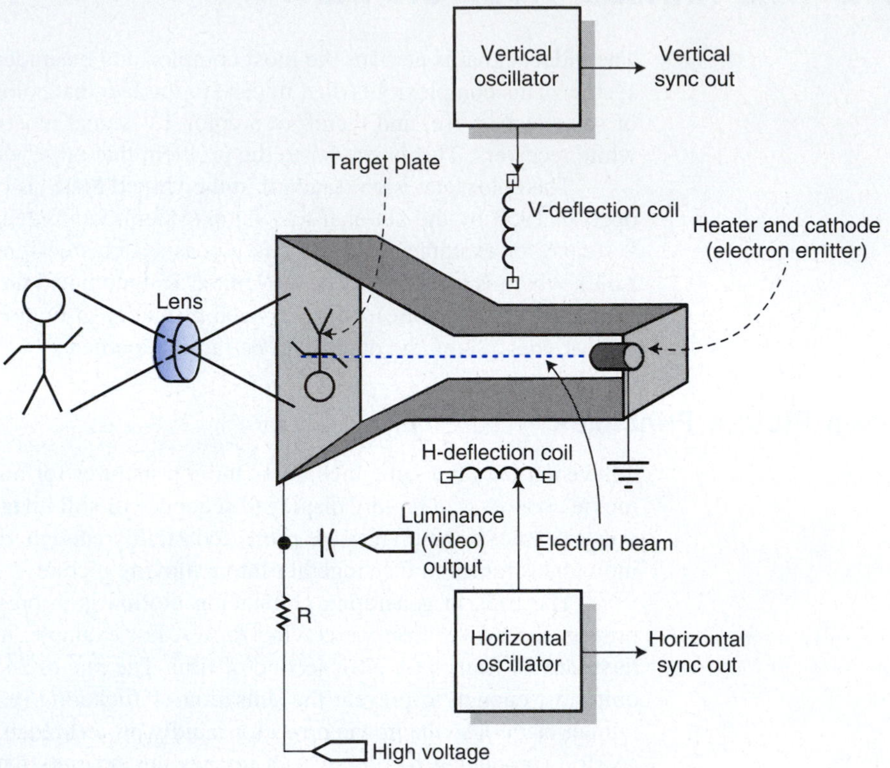

Figure 10–3 Scanning a scene with a camera

load resistor *R* increases. If light is not striking the material, its resistance increases, and as a result, the current through the load resistor *R* decreases, resulting in a decrease in voltage. The resulting output voltage variations mirror the dark and light areas of the scene, as shown in Figure 10–4.

In the United States a complete TV picture *frame* has 525 lines, and is completely scanned 29.97 times every second. There are therefore (525 lines/frame) (29.97 frames/sec) or 15,734 lines scanned per second. This also means that each frame is scanned in 33.37 ms, and that each horizontal scan line lasts for 63.5 μs.

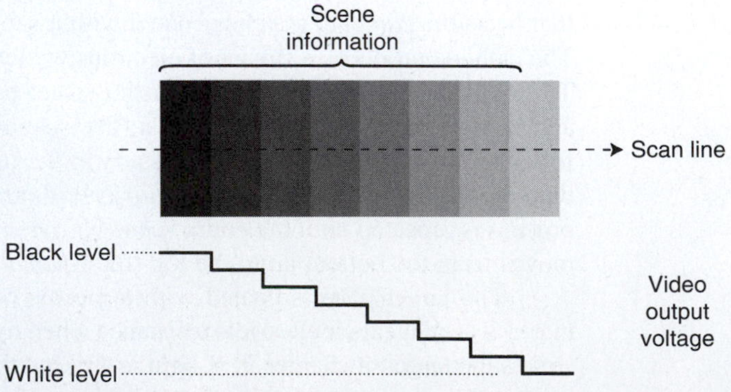

Figure 10–4 Camera output voltage

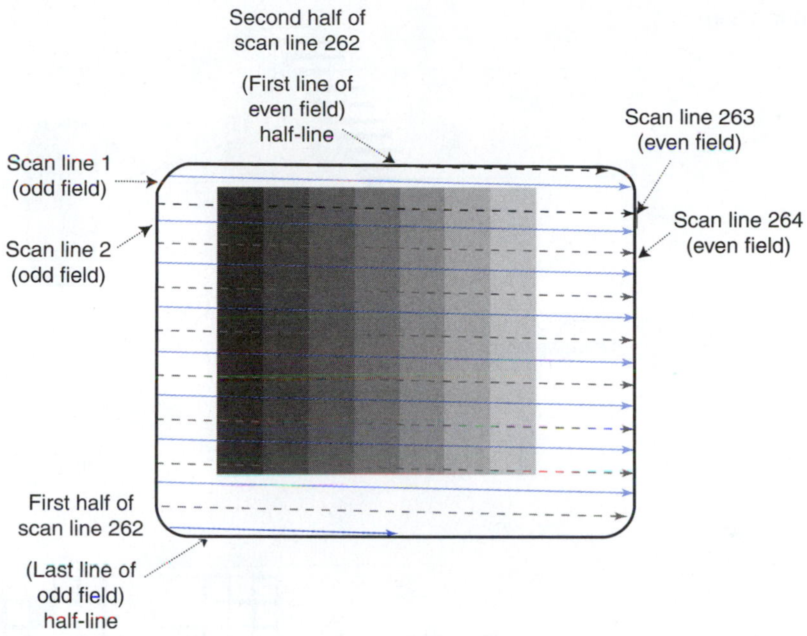

Figure 10–5
Interlaced scanning

The NTSC system uses an *interlaced* scanning method, as shown in Figure 10–5.

Interlaced scanning is used to eliminate the perception of flicker. Recall that each frame is scanned and displayed twice as two *fields* for the viewer. Since there are 525 horizontal scan lines in a frame, there are $262\frac{1}{2}$ horizontal lines in each field. Each frame of video consists of an *odd* and an *even* field. The *odd* field scans the first line, the third line, and so on; the *even* field immediately follows, and scans the remaining lines. This can be a little confusing, so it's important to remember that each *frame* is equivalent to one frame of a motion picture, and each frame contains two *fields*. These two fields are very similar in content, and share about 30% of their information.

Notice that the even and odd video fields have one peculiar characteristic that differentiates them. The last (bottom) scan line of the odd field is only a half-line; the first scan line of the even field is only a half-line also, and starts in the middle. A receiver can therefore detect which field is which by looking at this timing information. Typically, a short horizontal sync interval ("half-line") followed by a vertical sync pulse signals the *beginning* of the even field.

Solid State Imaging Devices

A modern television camera is very likely to use a CCD (charge coupled device) imager as shown in Figure 10–6(a). A CCD imager consists of a rectangular matrix or array of photosensitive sites fabricated on a silicon "die" (a section of a silicon wafer). Each of these sites has a diode junction (which is reverse-biased) and a small capacitance. The operation is quite straightforward. The CCD is placed behind a lens so that it receives incoming light from the scene to be photographed. A control circuit located on the CCD chip is commanded to charge up all the capacitors in the first (top) row of the array. All the cells in the row therefore start at the same voltage. Light striking the array causes a current to be drawn through the reverse-biased diodes at each photosite. The brighter the light, the

Figure 10–6 CCD imager operation

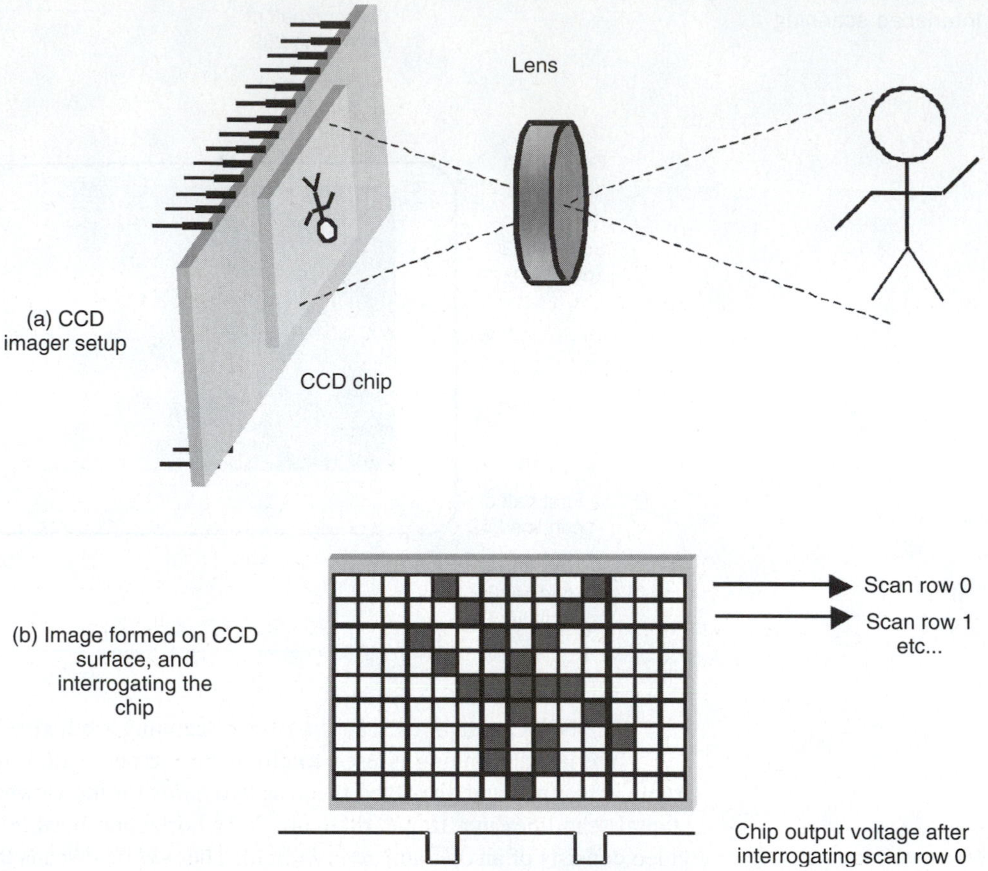

higher the discharge current, and the lower the voltage falls at that particular pixel location. After a small amount of time is allowed to pass, the voltages of the "capacitors" at each of the sites are read out (again by giving the CCD chip a command) as shown in Figure 10–6(b). These voltages represent the dark and light values for one horizontal scan line of the scene. The process is repeated for each row in the CCD imager's array until the entire scene has been scanned. In order to keep the signals compatible with NTSC standards, CCD imagers for television may be scanned at the same frequency (15.75 kHz, or 63.5 μs/line) as vacuum tube imagers, or higher (with computer signal processing).

CCD imagers have many advantages over vacuum tubes. They're much more rugged, use only a fraction of the power, and take up a lot less space. They also don't need high voltage supplies. All of these are tremendous advantages. Color CCD imagers are also common; color is detected by fabricating a color filter array (consisting of microlenses) on top of the CCD chip. The color filter array usually employs a Bayer pattern as shown in Table 10–2.

To recover the color information from a Bayer-filtered CCD requires an additional circuit called a *color encoder*. The color encoder chip mathematically converts the color values formed by the adjacent cells surrounding each photosite to approximate the hue and saturation of the pixel. Note that most of the pixels in the Bayer array are green; this is because the human eye sees the best detail under green light, and is also most sensitive to green.

Table 10–2 Bayer pattern for CCDs

R	G	R	G	R	G	R	G
G	B	G	B	G	B	G	B
R	G	R	G	R	G	R	G
G	B	G	B	G	B	G	B

Synchronizing the Receiver

In order for a legible picture to be displayed on a receiver, the horizontal and vertical scanning of the receiver's picture tube must exactly match the scanning pattern from the transmitter. In other words, the scanning in the receiver must be *synchronized* to that of the transmitter. Vertical and horizontal *synchronization pulses* are added to the video signal as it leaves the camera for this purpose. Figure 10–7 shows a portion of video signal with the horizontal synchronization pulses added.

In a video signal, different voltage levels are used to represent the lightness or darkness of parts of a scene. In addition, other voltage levels represent the *synchronization* information, as shown in Figure 10–7. The percentage levels in the figure are voltage percentages, with respect to the peak video voltage, which is nominally 1 volt. (A standard video signal has a maximum positive value of 1 volt, and a minimum close to 0 V; it is therefore close to 1 V_{pp}. It rides on a dc level.) The *black level,* which is 70%, represents the darkest possible area in a scene, while the *white level,* 12%, is the brightest possible level. Notice how the white level is a *lower* voltage, which corresponds to less transmitted power. This is used to conserve transmitter energy, since most scenes consist of light picture areas. The carrier is never allowed to reach zero amplitude, for that would result in poor noise rejection at the receiver.

The *blanking level,* 75%, represents the "blacker than black" condition. In a receiver, it is necessary to turn off or *blank* the electron beam during retrace (when the beam is sweeping back to the left side, or from the bottom back to the top). At the end of each horizontal scan line, the voltage rises up to this level, where the receiver's electron gun is guaranteed to be shut off. The electron gun must be turned off to prevent a *retrace* line from being drawn as the beam is rapidly swept right-to-left in preparation for the next scan line.

The *sync level,* also known as the *peak pulse level,* is the highest positive voltage possible. Two types of synchronization are transmitted: *horizontal* and *vertical.* The horizontal sync pulse occurs at the end of each horizontal scan line, and is used to synchronize the horizontal sweep oscillator in the receiver to the horizontal sweep of the transmitter. Since there are 15,734 horizontal lines scanned per second, the frequency of the horizontal sync pulses is 15.734 kHz.

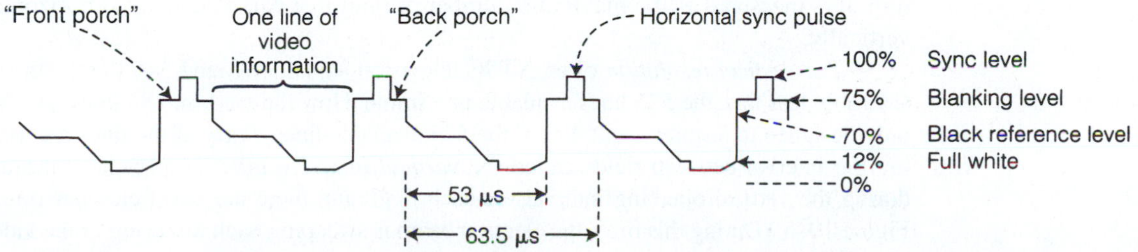

Figure 10–7 Video with horizontal synchronization pulses

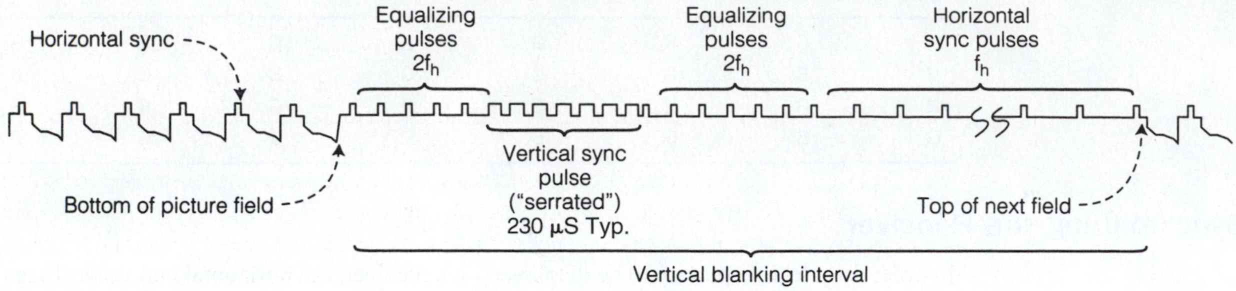

Figure 10–8 Slowing the timebase shows the vertical sync pulse

The *vertical* synchronization pulse is sent at the end of every *field;* therefore, the frequency of the vertical sync is 59.94 Hz. The vertical synchronization is also sent at the 100% (peak) level. To see the vertical sync pulses, more waveform must be examined, as in Figure 10–8.

At the end of a frame, the electron beam must be made to move from the bottom of the picture tube to the top. During this time, the electron beam in the receiver must be turned off to prevent a visible upward streak in the picture. This time period is called the *vertical blanking interval.* A vertical synchronization pulse is sent during this time. The vertical synchronization pulse occurs about three horizontal periods after the end of the frame. In order to keep the receiver's horizontal oscillator from dropping out of synchronization during the vertical sync interval, *equalizing pulses* are sent. The equalizing pulses are at *twice* the horizontal frequency. About six equalizing pulses are sent, then the vertical sync pulse is sent (which is "serrated" by six or seven additional horizontal pulses), then finally six more normal horizontal equalizing pulses are sent. For the remainder of the vertical blanking period, normal horizontal sync pulses are sent, to ensure that the horizontal oscillator in the receiver is still locked to the transmitter.

You might wonder how the two sync pulses (vertical and horizontal) can share the same voltage—how can the receiver separate them? The reason this works is that the sync pulses are at two different and widely spaced frequencies. A receiver extracts the vertical sync with a low-pass filter (sometimes called an *integrator*), and the horizontal sync is extracted with (you guessed it), a high-pass filter, or *differentiator*. Another way of saying this is that *frequency division multiplexing* is used for the horizontal and vertical synchronization signals.

Resolution and Bandwidth

Resolution is the ability of a television system to display or *resolve* small details in a picture. It is measured as the maximum number of *lines* that can be displayed horizontally or vertically.

The *vertical resolution* of an NTSC television signal is around 340 lines. This is considerably less than the 525 lines available in a frame. How did we lose 185 lines? A TV set is only designed to display about 485 of the 525 available lines. Forty of the lines are used during the interval between fields called the *vertical blanking interval.* (Twenty lines are lost during the vertical blanking interval for each *field,* and there are two fields per frame; see Figure 10–5.) During this time, the electron beam is sweeping back to the top of the video display in preparation for painting the next field on the screen. It takes time for the vertical oscillator to move the electron beam back up to the top of the screen. Furthermore, about 30%

of the information in the remaining 485 scan lines is redundant or *duplicate* information. This means that the vertical resolution is closer to 70% of 485, or about 340 vertical lines.

The *horizontal resolution* of a NTSC signal can be calculated by examining the details of one horizontal scan line from Figure 10–7. One horizontal scan line takes 63.5 μs to complete, but only 53 μs is actually available for transmitting picture information (the rest is required for transmitting the horizontal sync and blanking pulse).

The "lines" of resolution in television work are counted as consecutive vertical white and black lines, which alternate. One sine wave cycle could therefore represent a white line at its negative tip, and a black line at the positive peak. *Each sine wave cycle therefore represents two lines of resolution.* If we know the available bandwidth, this will give us the maximum sine wave video frequency; knowing this, we can figure out how many of the sine wave cycles will "fit" into the 53 μs time slot. Multiplying this number by two (there are two lines per sine wave cycle) will give the video resolution. This can be expressed as:

$$HR = BW \times T_{\text{horiz}} \times 2 \tag{10–1}$$

where HR is the horizontal resolution, in lines, BW is the available bandwidth, and T_{horiz} is the amount of horizontal time available.

EXAMPLE 10–3

Calculate:

a. The theoretical horizontal resolution for an NTSC video signal, given that the FCC limit for bandwidth is 4.2 MHz

b. The resolution that would result if 6 MHz of bandwidth were available

Solution

a. By using equation 10–1, we get:

$$HR = BW \times T_{\text{horiz}} \times 2 = (4.2\text{ MHz})(53\ \mu\text{s})(2) = \underline{\underline{445\text{ Lines}}}$$

b. Increasing the bandwidth increases the number of lines in a proportional manner:

$$HR = BW \times T_{\text{horiz}} \times 2 = (6\text{ MHz})(53\ \mu\text{s})(2) = \underline{\underline{636\text{ Lines}}}$$

Equation 10–1 yields results that are fairly generous; in real life, the horizontal resolution of an NTSC receiver is around 320 lines, instead of 445 lines. There are two reasons for this. First, generating two crisp adjacent lines on video display requires a *nonsinusoidal* signal shape, which has more bandwidth than a pure sine wave; second, the addition of the *color* or *chrominance* signal reduces the available bandwidth.

Section Checkpoint

10–11 What color television standard is used in the United States?

10–12 How is flicker prevented in motion pictures?

10–13 Explain how a television picture is scanned.

10–14 What is the difference between a *field* and a *frame*?

(continued on p. 342)

10–15 How many frames are sent per second in NTSC video?

10–16 How does a video camera tube convert light in a scene into a varying voltage?

10–17 How does interlaced scanning work?

10–18 What is unique about the odd and even fields in a video signal?

10–19 What is added to the video signal to keep the receiver scanning in step with that of the transmitter?

10–20 Why is *blanking* needed during vertical and horizontal retrace?

10–21 What types of circuits can be used to isolate vertical and horizontal sync in a receiver?

10–22 What factors control the *vertical resolution* of a video signal such as NTSC?

10–23 What factors control the *horizontal resolution* of a video signal (see equation 10–1).

10–3 TV RECEIVER OPERATION

A television receiver is a collection of subsystems that, in essence, have the task of converting the RF signal from the antenna back into picture and sound. In order to do this, three tasks must be accomplished. First, the appropriate carrier signal must be selected and amplified; second, the picture and sound carriers must be demodulated; and last, the recovered picture and sound information signals must be appropriately processed. The sound is fed to an audio amplifier and loudspeaker, just as in radio; the recovered picture information is a *video* signal, and must be disassembled into its component parts to be useful.

TV receivers use the same superheterodyne approach as AM and FM broadcast receivers, with two major differences. First, the IF frequency of a TV receiver is usually about 45 MHz, due to the large bandwidth (6 MHz) of a broadcast TV signal. Second, because the picture information is on an AM carrier, automatic gain control (AGC) is an absolute necessity—but because a video signal has a wide range of values, an averaging AGC can't be used. TV uses *keyed AGC,* which will be discussed in detail. Figure 10–9 is a simplified block diagram of a monochrome (black and white) receiver.

RF Stages

The RF section of a television receiver is very similar to that of any other superheterodyne receiver. The incoming carrier signal from the antenna passes into an RF amplifier, then into a frequency converter for translation down to the IF frequency. The RF amplifier and frequency converter stages are usually contained in a complete "tuner module" that is sealed within in a metal box. The tuner module usually accepts a *control voltage* (not shown) to select the channel, and may also provide a buffered *local oscillator output* for use with a frequency synthesizer loop (and microprocessor control). The tuner module also has an *AGC* input, which varies the gain of the RF amplifier. Figure 10–10 shows a typical tuner module.

The output of the tuner block is a 45.75 MHz intermediate frequency signal. The IF signal is a translated version of the video signal of Figure 10–1, and contains *both* the sound

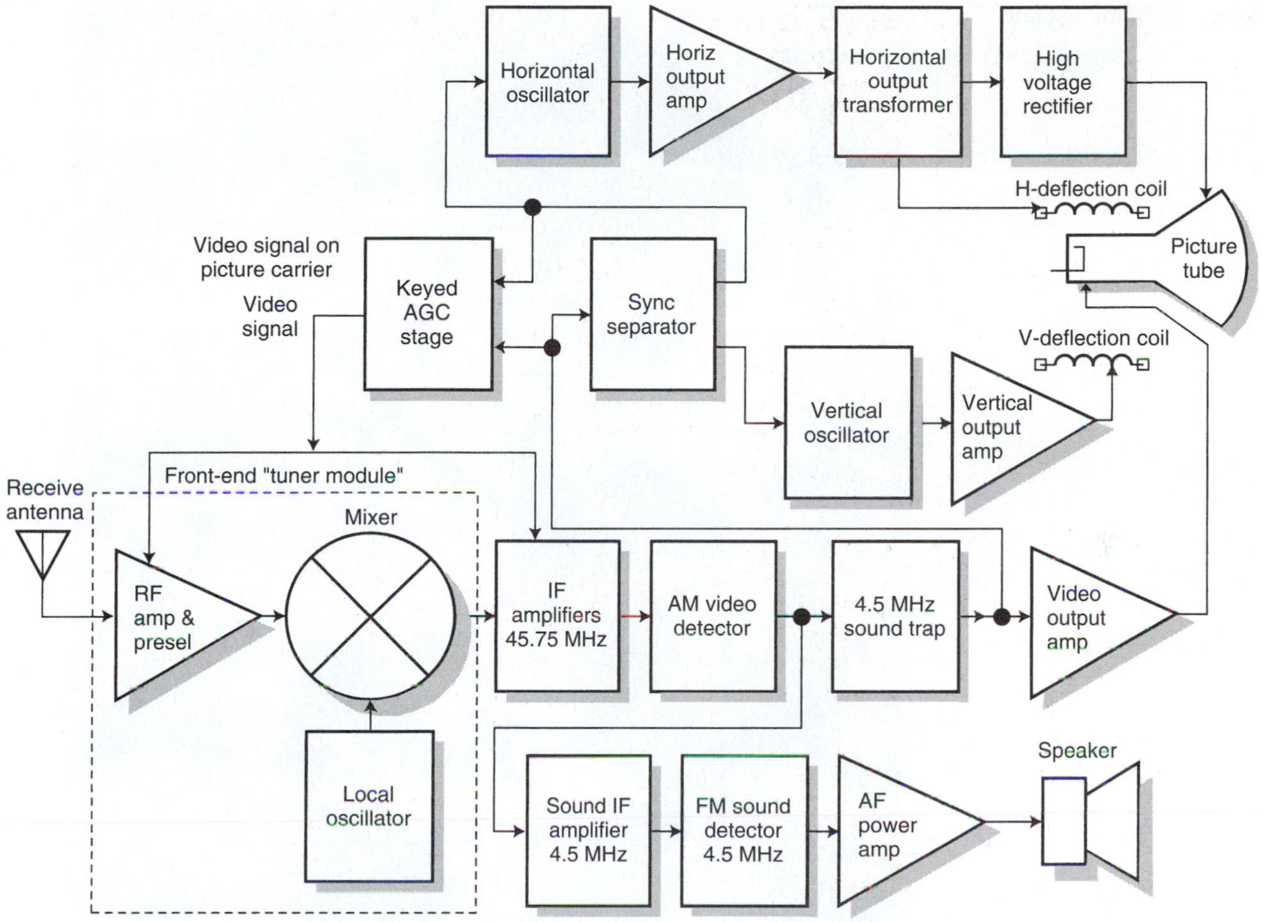

Figure 10–9 A black and white TV receiver

and picture carriers. Because high-side injection is being used ($f_{LO} > f_{IF}$), the upper and lower sideband of the picture carrier are swapped, and the sound carrier now appears 4.5 MHz *below* the picture carrier. Figure 10–11 shows the translated IF signal.

IF Amplifier

The IF amplifier in a television set must amplify signals over a relatively large bandwidth, 6 MHz. This is quite a wide bandwidth—much larger than that needed for any signals we've studied so far. The filtering requirements for the IF amplifier are made even more complex because the picture is carried on a vestigial sideband (VSB) signal. The lower sideband is complete, but only a vestige (trace) of the upper sideband is present. Because part of the upper sideband is present, detection of the video signal will result in an uneven *video frequency response* unless some correction is made. Notice that the upper sideband contains information frequencies from dc to 1.25 MHz, and the lower sideband contains dc to 4 MHz. Because the range of dc to 1.25 MHz is contained in *both* sidebands, that range of video frequencies will tend to be emphasized at the detector. The result will be a

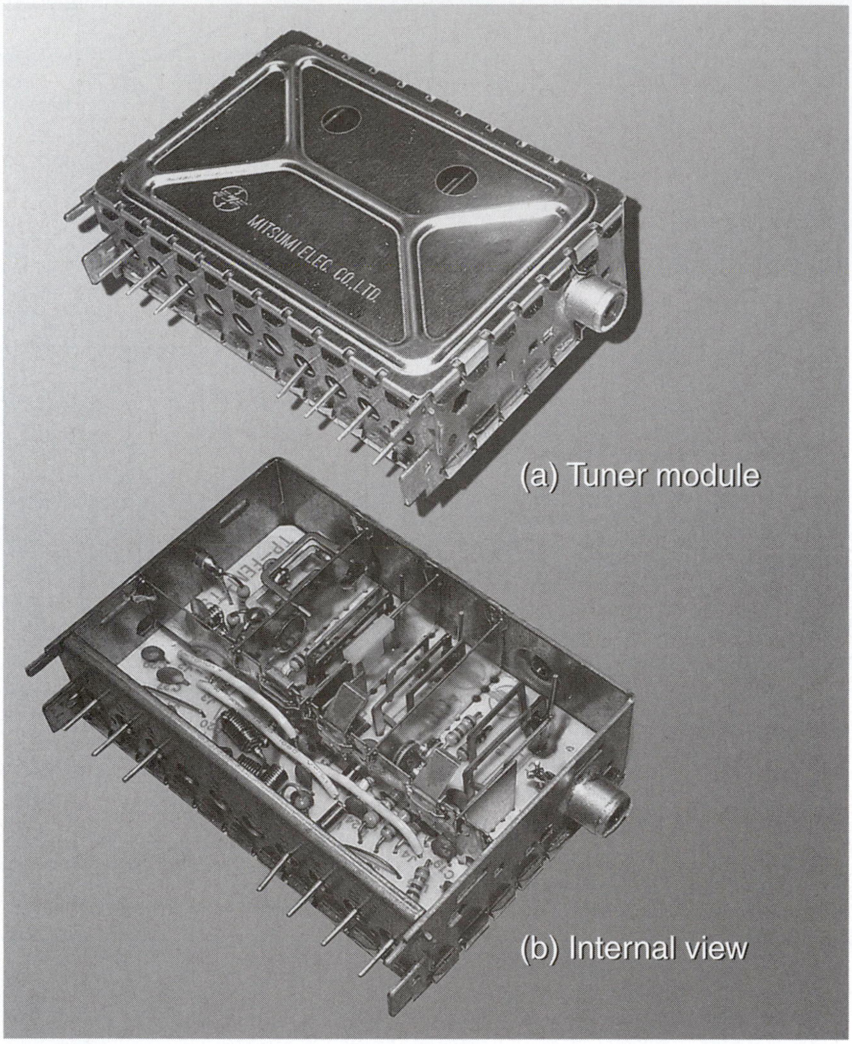

Figure 10–10 A television tuner module

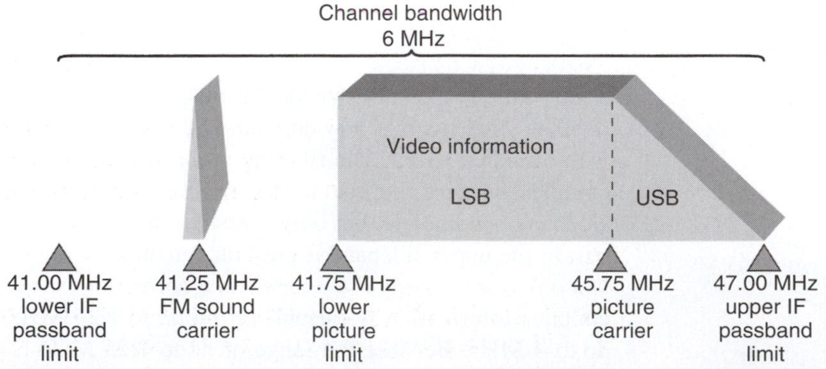

Figure 10–11 Television IF amplifier passband

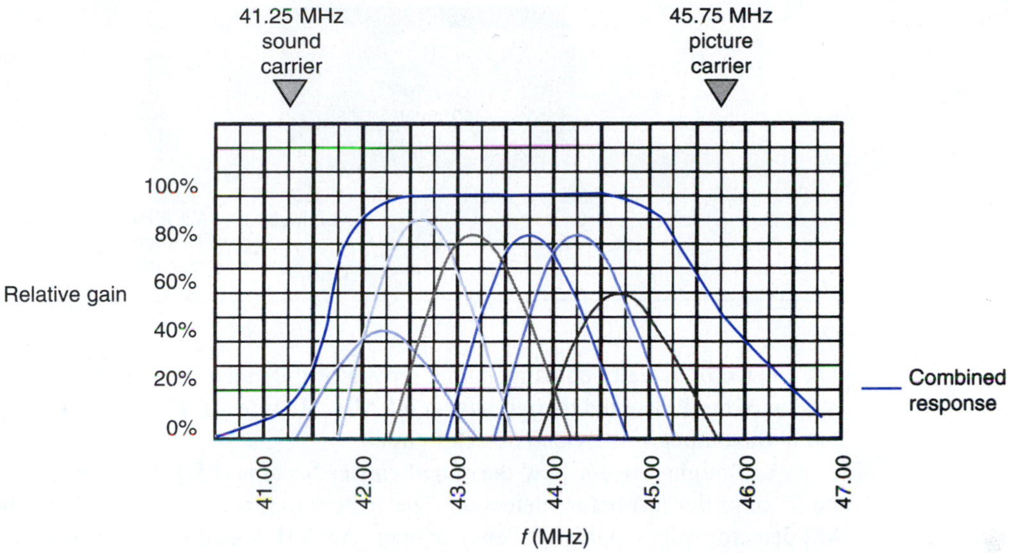

Figure 10–12 Stagger tuning in a television IF amplifier

picture with a lack of detail, because the high information frequencies contain the fine picture information.

In order to solve these two problems, a system of *stagger tuning* is used in the IF amplifier of a television set. The concept of stagger tuning is fairly straightforward: Instead of tuning each IF resonant bandpass circuit to the same frequency, each IF circuit is designated to handle a different frequency in the 6 MHz passband range. By carefully choosing the resonant frequency and Q factor of each tuned circuit, the designer can control the IF *frequency response* in any manner desired. Figure 10–12 shows a typical TV IF frequency response curve, and how it is obtained.

In Figure 10–12, six different tuned circuit responses contribute to the overall IF amplifier frequency response curve. The shape of the curve in the 45–47 MHz region attenuates the 0–0.75 MHz portion of *both* upper and lower sidebands, which corrects the frequency response error caused by the presence of the vestigial sideband.

The IF frequency response is *very* low for the 41.25 MHz sound carrier; it is only 10% of the passband gain value. This is necessary to prevent the sound from interfering with the picture. The lack of gain for the sound carrier is corrected in the sound IF amplifier and detector stages.

Modern receivers often use a prefabricated filter to produce the desired IF frequency response, instead of individual tuned circuits. Two popular filter types are the *crystal lattice,* which was discussed in Chapter 6, and the *surface acoustic wave,* or *SAW* filter. The SAW filter is built from a piezoelectric material such as quartz, and relies on both the piezoelectric effect and the movement of acoustic RF wave energy across its surface. SAW filters are inexpensive to manufacture, and thus have become very popular in television receivers.

AM Video Detector

The 45.75 MHz IF signal is next passed to a conventional diode AM detector. You might recall that one of the advantages of a VSB signal is the ability of a receiver to detect it with a simple detector circuit, while reducing the bandwidth of the signal. The output of the

Figure 10–13 Video detector output spectrum

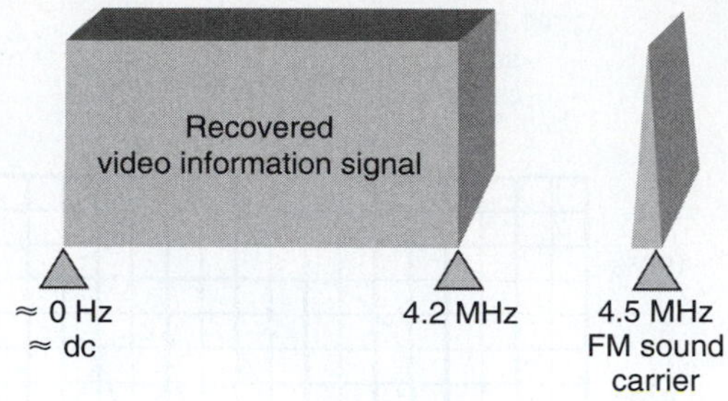

video detector actually contains *both* a replica of the video information signal and a *new* version of the FM sound carrier, now at 4.5 MHz. Figure 10–13 shows the frequency content at the output of the video detector circuit.

You might wonder how the sound carrier became 4.5 MHz; it was 41.25 MHz in the IF amplifier just before detection. The answer to this comes if we think about how an AM detector works in the frequency domain. An AM detector built from a diode is a nonlinear device. Therefore, if two frequencies are passed into it, we will get the *originals,* the *sum,* and the *difference* frequencies. The detector really acts as a *mixer;* in fact, sometimes the mixer in the RF section of a receiver is sometimes called the *first detector* for this very reason!

The 45.75 MHz *picture* carrier mixes with the 41.25 MHz sound carrier at the AM video detector stage. The difference between these frequencies is 4.5 MHz. *The net effect is that the 41.25 MHz sound carrier is downconverted to 4.5 MHz at the video detector!*

Failure of any stage in the 45.75 MHz IF kills both picture and sound, because both are carried in this circuit.

Sound Processing

The signal from the AM detector contains a replica of the original video signal and the sound information on a 4.5 MHz FM carrier. In order to recover the sound, the 4.5 MHz signal must first be amplified; the 4.5 MHz sound IF amplifier boosts the 4.5 MHz FM sound carrier, and eliminates any picture frequency components (DC to 4 MHz in Figure 10–13).

After amplification, the FM sound carrier is ready for detection; a conventional FM demodulator circuit is used to recover the sound information signal, which is further amplified by an audio power amplifier before being fed to the loudspeaker.

Video Processing

Reproducing a picture from the video signal is considerably more involved than detection of the audio, because of the need for *synchronization.* When the receiver is synchronized, the sweep of the electron beam in its picture tube exactly matches the scanning pattern from the camera in the transmitter. In order for this to happen, the *sync* must be extracted from the *luminance* (brightness) information.

The detected signal from the AM detector contains both sound and video information. The first order of business is to eliminate the 4.5 MHz FM sound carrier. A 4.5 MHz *notch (bandstop)* filter called a *trap* eliminates the FM sound carrier, leaving only the video information.

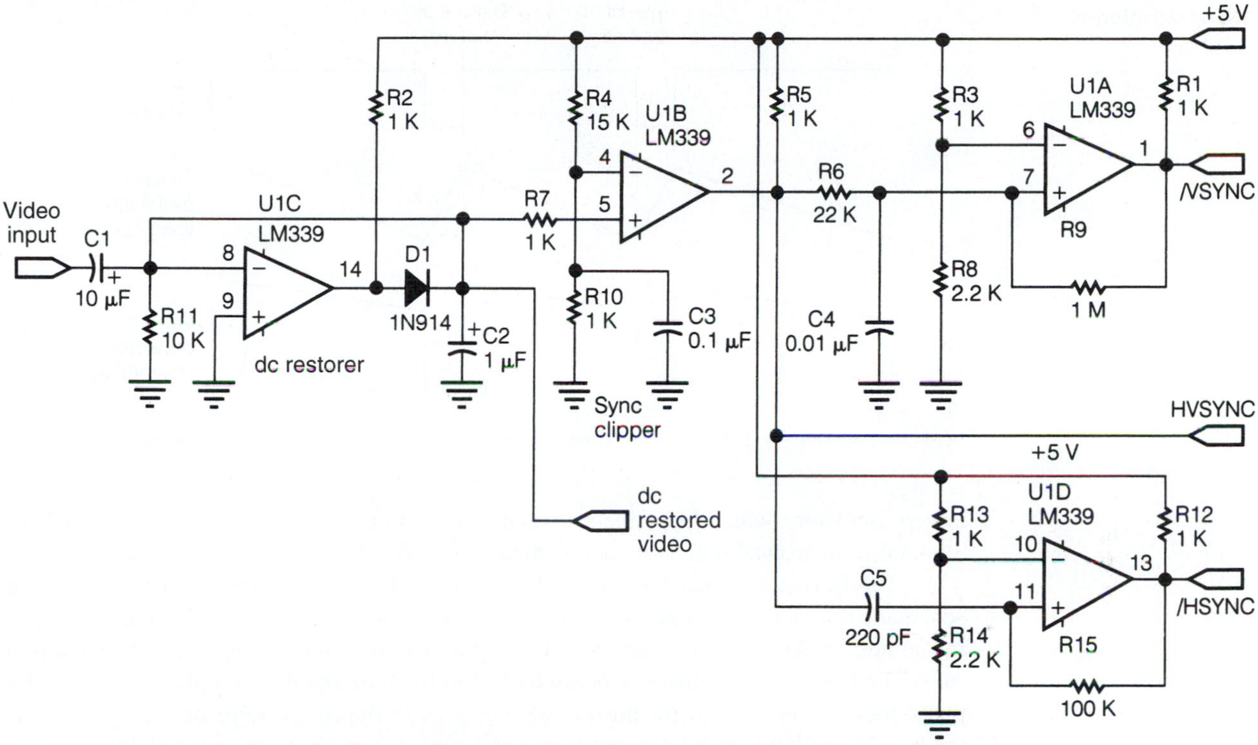

Figure 10–14 A sync separator using the LM339

The video information must then be separated into sync and luminance components. This is a fairly simple operation; if you examine Figure 10–7, you'll see that any voltages above 75% in the video waveform are *sync* pulses. The most common approach uses a voltage comparator set to trip at 75% of the peak video voltage. The output of the comparator (sometimes called a *sync clipper* in this application) is the combined horizontal and vertical sync. Figure 10–14 shows a sync separator circuit employing the LM339 quad comparator IC.

In Figure 10–14, U_{1C} and D_1 form a *clamper* circuit that restores the DC component to the video signal, and U_{1B} is the sync clipper comparator. U_{1B}'s trip point is set by R_4 and R_{10} at about 0.31 V, which corresponds to the black reference level of 70% on the video waveform. Therefore, whenever the video waveform crosses the black level (meaning that sync is being produced), U_{1B} outputs a pulse. R_5 and C_4 form the low-pass filter (integrator) for extracting the vertical sync, and U_{1A}, a Schmitt trigger, cleans the resulting vertical pulse up into a square wave. C_5 and R_{15} form the high-pass filter for differentiating the horizontal sync pulses, and U_{1D} performs the final cleanup for the horizontal sync output.

The vertical and horizontal sync pulses from the sync separator circuit drive the horizontal and vertical oscillators. You can think of the horizontal and vertical oscillators as phase-lock loop circuits; they will "lock on" to the synchronization pulses being applied to them.

When a sync pulse arrives at the vertical oscillator, it "resets" the oscillator back to the beginning of the sweep waveform. On some sets, there are *vertical hold* and *horizontal hold* control potentiometers; these pots provide a fine adjustment for the vertical and horizontal

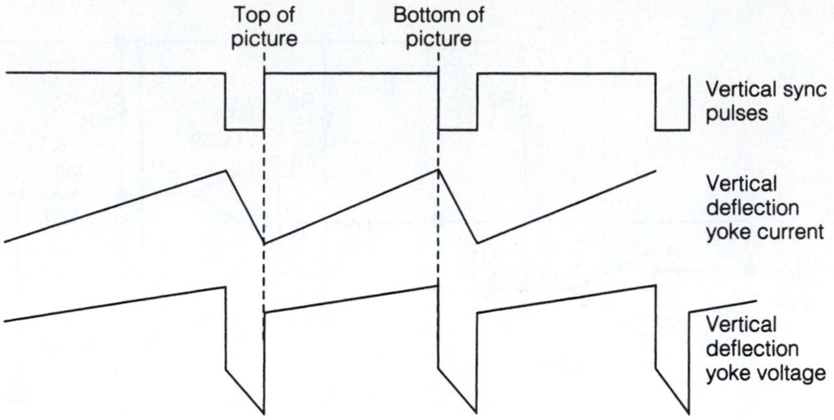

Figure 10–15 Vertical sweep waveforms

sweep oscillators, which helps them to attain lock. On newer sets, digital techniques have eliminated the manual horizontal and vertical hold controls.

The horizontal oscillator is locked using a full phase-lock loop circuit. On a schematic diagram, you might see a section marked "horizontal AFC." The term *AFC* is a misnomer; an AFC system (such as a Crosby transmitter) has a frequency error while it runs. The horizontal oscillator is phase locked to the horizontal sync pulses. The use of a phase-locked loop system for horizontal sync greatly improves noise immunity, because the high-pass filter that extracts the horizontal sync is very prone to noise pickup.

Figure 10–15 shows the vertical oscillator sweep waveforms. Notice that the arrival of a sync pulse keeps the sweep in step with the transmitter's scanning.

The outputs of the vertical and horizontal oscillators drive the vertical and horizontal deflection coils on the *yoke* of the picture tube, which causes the electron beam to be deflected in step with the camera at the transmitter.

It is important that the deflection yoke current rise in a linear manner, to avoid stretching or shrinking of any part of the picture. If an ideal (perfect) inductor is connected directly across a voltage source, its current will rise in a linear fashion.

However, the deflection coils in a TV receiver are not ideal inductors, and have a series resistance. In order to compensate for this, the vertical deflection voltage is shaped before being sent to the deflection coil. Notice that the vertical deflection coil voltage has a *trapezoidal* shape. The added slope in this shape is what compensates for the deflection coil resistance, and provides a linear vertical sweep.

Total failure of the vertical deflection circuit in a TV set causes a single, bright ***horizontal line*** to appear on the screen. A set displaying this symptom should ***not*** be left on for more than a minute or so, unless the brightness control is turned down, because the phosphor can be burned. Other failures in the vertical deflection circuit will cause the picture to be distorted ***vertically***—the picture can shrink, or be misshapen (vertical nonlinearity).

High-Voltage Generation

A television picture tube is a vacuum tube optimized for displaying images. The inside front of the tube is coated with a phosphorescent substance that glows when struck by an electron beam. At the rear of the tube is the *electron gun,* which contains a heater, a cathode

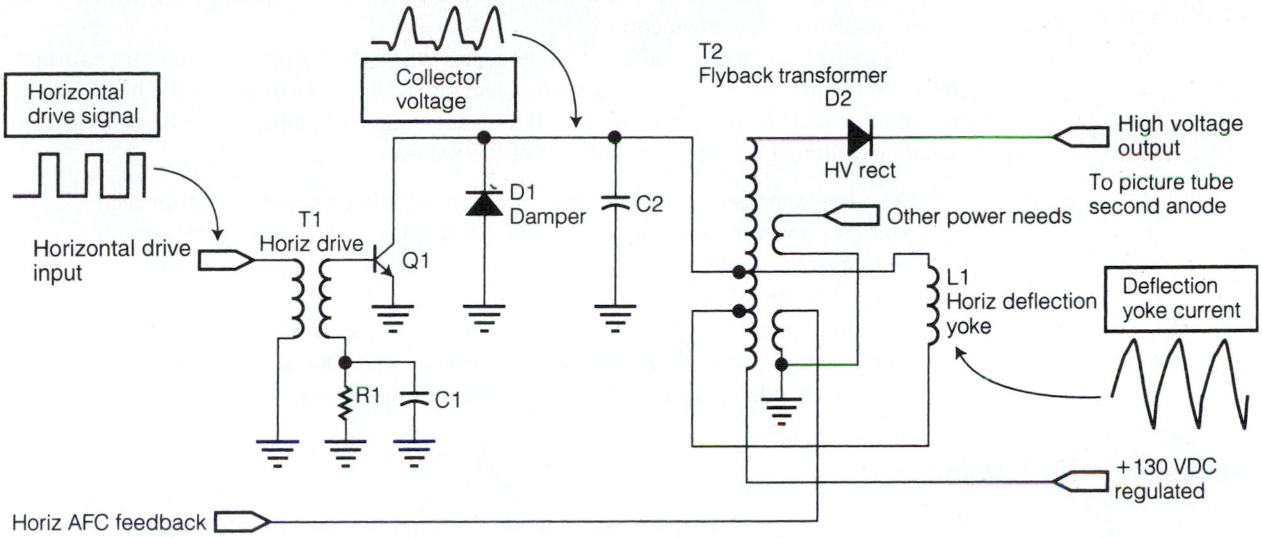

Figure 10–16 A horizontal output circuit

(coated with a thermoemissive compound), and control and focusing electrodes. A low voltage (usually 3 to 12 volts) at high current operates the heater, which causes the cathode to emit electrons.

To get the electrons to accelerate toward the face of the picture tube, a high voltage is required; usually over 10 kV is used, and in large color sets, over 25 kV is typical. The output from the horizontal oscillator is stepped up through a *flyback transformer,* and then rectified. The high-voltage positive dc is then applied to the *second anode* of the picture tube, which contacts a conductive coating inside the tube. Thus, the electrons from the cathode are attracted toward the screen, where they return to the high-voltage supply. Figure 10–16 shows a typical horizontal output circuit.

The horizontal output circuit is very busy. The *drive* signals from the horizontal oscillator are coupled into the base of Q_1 by T_1. R_1 and C_1 cause Q_1 to operate class C, so that Q_1 operates as a switch. Q_1 is a power transistor and is usually mounted on a heatsink. When the drive signal is positive, Q_1 turns on and the horizontal sweep begins, from left to right. With Q_1 on, the +130 V supply is connected across flyback transformer T_2, and part of the winding of T_2 shares its current with the horizontal deflection yoke coil. The current in the yoke steadily rises, and the electron beam sweeps right.

At the end of the sweep, Q_1 turns off. With Q_1 off, transformer T_2 must find a place for its current to flow. T_2 is parallel resonated by C_2 at around 50–70 kHz, so when Q_1 turns off, T_2 and C_2 resonate together and cause the yoke current to rapidly become *negative,* which jerks the electron beam quickly back to the left-hand side. At this point, the remaining energy in T_2 is conducted to ground through D_1, the *damper* diode. Without the damper diode, T_2 and C_2 would continue to resonate, which would cause visible picture interference from the ringing.

The horizontal output transformer also provides power for several other circuits in the TV. The top portion of the winding provides the high voltage necessary for accelerating the electron beam toward the screen; diode D_2 rectifies this into high voltage dc, and is usually contained inside the case of the horizontal output transformer. Thus, there is usually a high-tension wire (usually with red insulation) connected between the top of the horizontal output transformer and the nipple on the side of the picture tube, which is the second anode (screen) connection.

Usually, there are several low-voltage windings on the horizontal output transformer. One of these almost always is 3–12 V ac at one or two amperes to power the heater of the picture tube; and another is usually 12–30 V ac at a couple of amps, which is rectified, regulated, and filtered for powering the rest of the receiver.

In a television set, failure of the horizontal amplifier causes a loss of high voltage, which means that ***no picture*** will display, since there isn't any potential to accelerate electrons toward the screen. In many sets, the low-voltage power supplies also are derived from the horizontal output transformer; so horizontal failure also means ***no power supply*** to key portions of the set. ***There may be no sound as well. The horizontal output can be considered part of the power supply in most receivers.***

Processing the Luminance

With the electron beam properly scanning the picture tube face (in step with the camera at the transmitter), the only remaining task is to control the beam intensity so that the brightness of the resulting phosphor "dot" the beam is writing matches the light intensity of the original scene. The *video output amplifier* is responsible for amplifying the voltage and current from the *video detector* circuit in order to drive the control grid of the picture tube.

The control grid of the picture tube can either assist or buck the flow of electrons between the cathode and screen, depending on the voltage applied to it. The more electrons that flow, the higher the current, and the brighter the point on the display will be. The variations in current intensity occur very rapidly; up to 4.2 MHz in an NTSC signal.

The video amplifier is also responsible for *blanking*. Blanking is the act of turning off the electron beam. It is necessary during both vertical and horizontal retrace, which are the periods where the electron beam is being moved back to a starting position. The voltage level of the video signal waveform controls blanking; whenever it is higher than 75%, the video amplifier shuts off, preventing any retrace lines from appearing on the screen.

Video amplifier failure can cause a picture that is totally black (amplifier stuck off), totally white with visible retrace lines (amplifier stuck on), a washed-out picture (insufficient video gain, causing lack of contrast), and other similar symptoms. The other sections of the receiver continue to operate normally.

Automatic Gain Control

Because the average value of a video signal is unpredictable (scenes can vary from total darkness to full white under normal conditions), an averaging AGC such as that used in an AM broadcast receiver isn't very practical. The video signal must have precisely the right amplitude for accurate processing (such as sync extraction, white and black level determination, and so on.) If the video signal varies in amplitude even a small amount, the picture will appear to grow light and dim as the video signal voltages drift up and down. Signal levels at the antenna are almost constantly fluctuating due to changing propagation conditions. Modern receivers use what is called *keyed* or *gated* AGC to solve this problem. Although the average value of a video waveform is quite variable (due to scene content), there is one part of the waveform that's always of constant amplitude, and that is the *synchronization.*

The *sync* level corresponds to 100% or maximum transmitter power output, and this is a constant. The AGC circuit is designed to measure or "sample" the signal level *only during the horizontal synchronizing pulse*. In Figure 10–9, you'll notice that the AGC stage has *two* inputs. One of the inputs is the horizontal sync from the sync separator stage (or horizontal output transformer), and the other is the video output from the AM video detector. Because of the gating action, the AGC stage "sees" only the horizontal sync pulse of the video waveform leaving the AM detector. Usually this value is stored in a capacitor, and then sent back to the RF and IF amplifier stages in the set.

The use of keyed AGC eliminates the dependence on signal content, and provides a relatively noise-immune and rapid AGC response, since sampling only occurs on horizontal sync pulses, where maximum transmit power is occurring.

A picture whose intensity fades (with possible shriveling at extremes) in step with signal strength points to a problem in the AGC circuit of the set. Normally the sound isn't affected, since it is on an FM carrier.

Section Checkpoint

10–24 What are two differences between TV receivers and conventional superhets?

10–25 What stages are typically contained within the metal box of a tuner module?

10–26 Why do the upper and lower sidebands of the TV signal switch positions after frequency conversion?

10–27 What are the FM sound carrier and AM picture carrier frequencies in a TV IF?

10–28 Describe the system of *stagger tuning,* and give two reasons for its use.

10–29 Why is the IF response designed to be down to 10% at the frequency of 41.25 MHz in Figure 10–12?

10–30 After passing through the AM detector, why is the FM sound carrier now 4.5 MHz?

10–31 What is the difference between *luminance* and *sync* information?

10–32 Why is a 4.5 MHz notch filter present in the video section?

10–33 What type of filter is used to extract (a) vertical and (b) horizontal sync?

10–34 List at least three outputs provided by the horizontal output circuit.

10–35 During which part of the video signal does the AGC in a TV receiver become active?

10–4 COLOR TELEVISION

The development of color television created the same potential compatibility problem that FM stereo did. The NTSC color system is a *compatible* system, meaning that a color TV broadcast can be received on a black and white receiver. Adding color in a compatible manner requires that some very sneaky methods be used; these methods slightly compromise the overall quality of the video signal. A NTSC color signal can't resolve the same level of detail as a monochrome transmission; however, in terms of viewing pleasure, the addition of color more than makes up for the loss in resolution.

Parts of a Color Signal

A black and white video signal consists of two parts, *luminance* (brightness) and *sync*. In order to display color, a third signal must be sent, called the *chrominance* or *color* information. One way of visualizing how these three elements work together to create a color TV picture is to think of "painting" someone's portrait using a grid of dots:

The *sync* information tells the painter where to place each dot on the canvas.

The *luminance* information tells the painter how *bright* to make the dot.

The *chrominance* information tells the painter the *color* of the dot.

The color of the dot (chrominance) must be described in two ways. First, the *hue* must be given, which is the same thing as the color of the paint. Second, the *saturation* must be expressed. The *saturation* describes how thin or thick the "paint" is to be spread on the "canvas." For example, pink and red are the same color *hue,* but to make pink, the artist spreads the paint thinner. In other words, the *saturation* of pink is less than red. All colors will eventually become *white* if they're spread thinly enough.

The NTSC Color Model

A color model explains how the two attributes of color (hue and saturation) will be represented. The NTSC method is sometimes called an "HSL" (hue, saturation, luminance) model. This is not the only way of representing color. Computer displays (and HDTV) usually use a model called "RGB" (red, green, blue), where each dot is "painted" by the additive combination of red, green, and blue dots. In fact, at the camera, the colors are represented as RGB; in the NTSC video signal, they are converted to HSL for transmission; and at the last stage of a color receiver, the original RGB signals must be recreated, for the picture tube contains three electron guns, one for each primary color. Figure 10–17 shows the NTSC color model:

Figure 10–17 NTSC color model

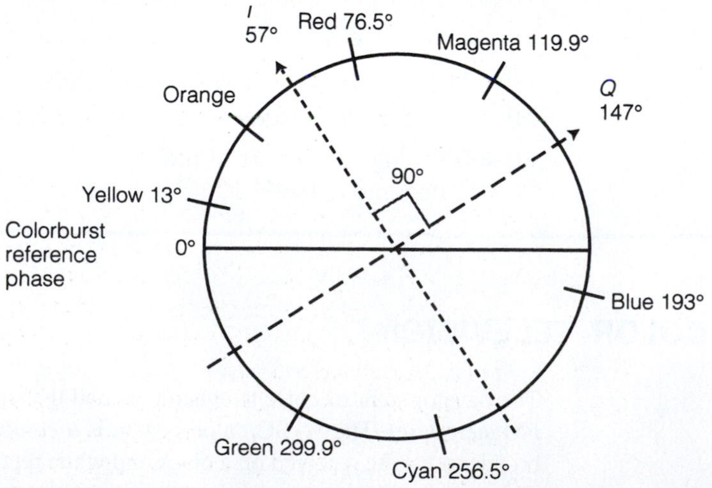

The color information is sent on a phase-modulated subcarrier in the NTSC system. The *instantaneous phase angle* of the signal represents the *hue* (green, cyan, etc), and the *amplitude* of the subcarrier represents the *saturation*. The color subcarrier therefore uses both AM and PM simultaneously!

An instrument called a *vectorscope* can be used to examine a video waveform. It produces a display much like that of Figure 10–17, which is called a signal *constellation*. A vectorscope can be used to compare the color setup of two or more cameras (or other video-generating devices); the cameras are pointed at a standard target (such as a color bar chart). The vectorscope will show the resulting video signal as a set of dots on the circle. The distance from the center of the circle represents the *saturation* of the color, and the angle of the dot represents the *hue*. The cameras are then adjusted to produce identical constellations. In TV production, this is important; often several cameras are used during filming, each showing a different aspect of the scene. Unequal camera setups will cause annoying color tint changes each time a different view is selected.

The *I* and *Q* Signals

The color subcarrier actually contains two individual carrier signals placed at 90° from each other called the *I* and *Q* signals. Each of these two signals is assigned to carry a different portion of the color information. The *I* and *Q* signals are themselves double-sideband, suppressed carrier (DSB-SC) signals. Together, their vector sum forms the *C* (chrominance) signal. We will explore exactly how that happens shortly.

In order to reduce the required bandwidth for these two signals to a bare minimum, they have been carefully chosen with regard to the characteristics of human vision.

Properties of Human Vision

The eye is an amazing instrument. However, there are some subtle limitations built into the human vision system. It has been shown experimentally that *visual acuity,* the ability to resolve closely spaced details in a scene, depends on the wavelength (color) of light. Humans can see detail better when the light color is red to orange, and worse as the light color becomes more bluish. Therefore, to present the best possible color picture for human viewing (while minimizing bandwidth), the *Q* signal (which is "pointed" at the less-resolvable colors in the blue range) is allotted a bandwidth of 0.5 MHz, and the *I* signal (which "points" toward red-orange and needs to resolve more detail), is given a bandwidth of 1.5 MHz.

The *I* and *Q* signals are each formed from the RGB (red–green–blue) output of the color camera, as shown in Figure 10–18. Notice a third signal, the *Y* or *luminance* signal. This signal records the *brightness* in terms of the sum of R, G, and B components.

The action in Figure 10–18 looks complicated, but it's really not. It's really much like the encoding of FM stereo, where a $L - R$ signal is separately sent for reconstructing the individual left and right channels in a receiver. *The red, green, and blue signals from the camera are the input signals into the "encoder" circuitry in Figure 10–18, and eventually, the receiver will "disassemble" the resulting signals back into R, G, and B components.* The *luminance* output is merely the sum of the R, G, and B components. Each color component is given a *weight* so that each color will properly show up in gray scale when shown on a black and white receiver.

The *I* signal is built by taking a weighted sum of the R, G, and B signals from the camera. *The* I *signal is actually the difference between the red signal and the* Y *(luminance) signal, and is also called the* "R − Y" *signal.* It is then placed on the 3.579545 MHz subcarrier by a balanced modulator. The output of the top balanced modulator is a DSB-SC

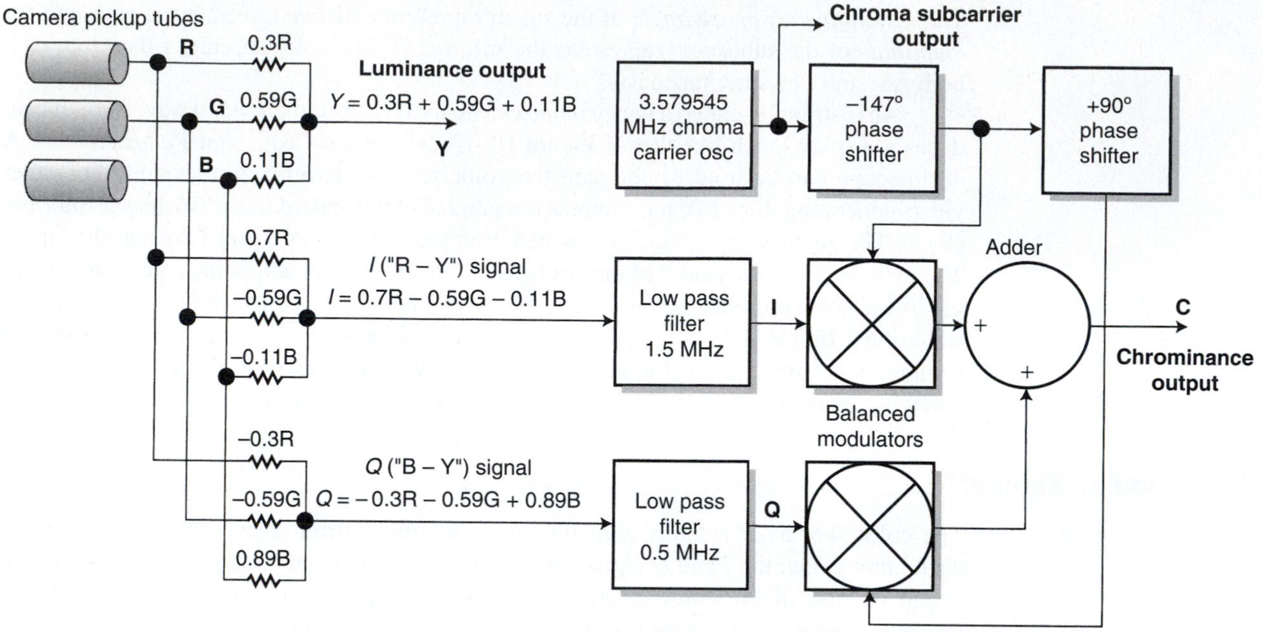

Figure 10–18 Building the *I*, *Q*, and *Y* signals

signal containing the *I* information. Note how the angle of the carrier lags by 147° from the chroma reference; this time lag compensates for the time delay through the balanced modulator, so that the *I* and *Y* signals will have the proper phase relationship in the receiver.

The *Q* signal is built in the same way, except that it represents the *difference between blue and the composite luminance* ("B − Y"). Before the 3.579545 MHz subcarrier enters the *Q* balanced modulator, a 90° phase lead is added. The 90° phase lead makes it possible to separately demodulate the *Q* and *I* signals in the receiver, even though they share the same subcarrier frequency! Also, because *I* and *Q* are 90° out of phase, *any* phase angle can be represented by adding appropriate amounts of *I* and *Q*. It works just like the system of imaginary numbers; you can think of *Q* as the *real* part and *I* as the *imaginary* part. For example, if $I = Q = 1$, then we have a number that is equivalent to $(1 + 1j)$, which is really $\sqrt{(1^2 + 1^2)} = 1.41 \angle 45°$. When we say that the chrominance signal is the "vector sum" of *I* and *Q*, that is what we mean.

If we are developing "*R* − *Y*" and "*B* − *Y*" (or *I* and *Q*) signals (which are analogous to the *L* − *R* signal used in FM stereo), why don't we also develop a "*R* − *G*" signal as well? Mainly, because there's no need to do so. By recombining the "*R* − *Y*" and "*Y*" signals at the receiver, we'll get back *red* (*R* − Y + *Y* = R = red), and by adding "*B* − *Y*" and "*Y*," we'll likewise get blue again. To get the *green* signal back in the receiver, we can combine all *three* signals, *I*, *Q*, and *Y* (in the proper proportions, of course) to recover *green*.

The *chrominance* or *C* signal is just the sum of the modulated *I* and *Q* signals. The final assembly of the color signal components works as in Figure 10–19.

In order to demodulate the *I* and *Q* subcarrier signals, the receiver needs the 3.579545 MHz subcarrier. The transmitter does not send the subcarrier continuously. Instead, it sends a short sample of the subcarrier called the *color burst* at the beginning of every horizontal scan line. The color burst lives on the "back porch" of the horizontal

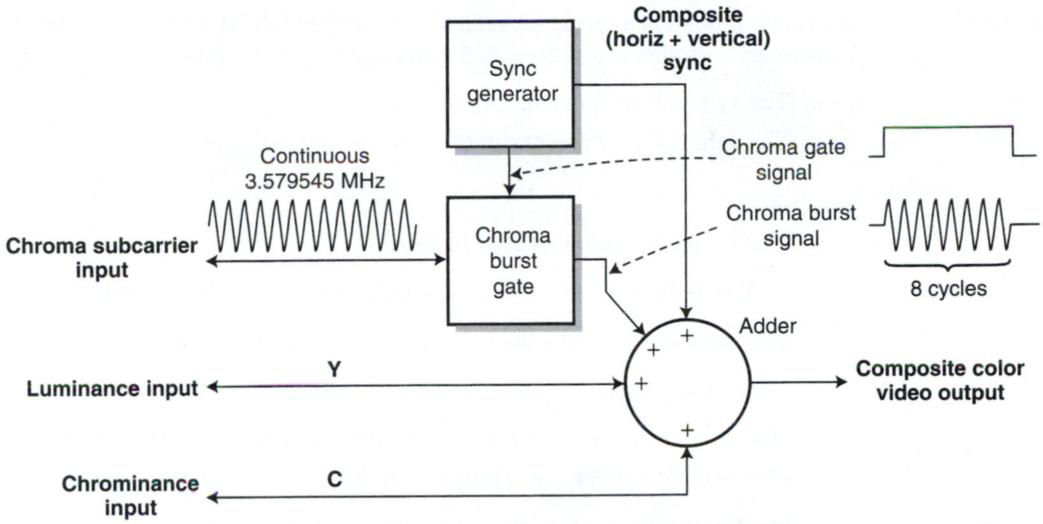

Figure 10–19 Final assembly of the color signal

synchronization pulse, and ends before the video is sent. The *chroma burst gate* in the figure is just a switch; it turns the subcarrier on for just a short time (about 8 cycles) after every horizontal sync pulse. Figure 10–20 shows how the composite color video signal looks.

How can the receiver demodulate *I* and *Q* if it only has a short burst of subcarrier? It works the same way as musicians tuning up before a performance. When a band is going to play, someone first sounds a note, and everyone quickly tunes their instrument to match that pitch. After that, all the musicians will be playing at the same pitch (frequency), at least for that session.

In a color receiver there's a circuit that looks for the color burst. A color receiver also has a special, tunable crystal oscillator that runs at 3.579545 MHz. If the burst is present (indicating that a color signal is being transmitted), the receiver quickly tunes its crystal oscillator to exactly match the frequency and phase of the burst. The receiver's crystal oscillator is then stable enough to "hold that pitch" for the remainder of the horizontal scan line, and the crystal oscillator's output is used to demodulate the *I* and *Q* signals.

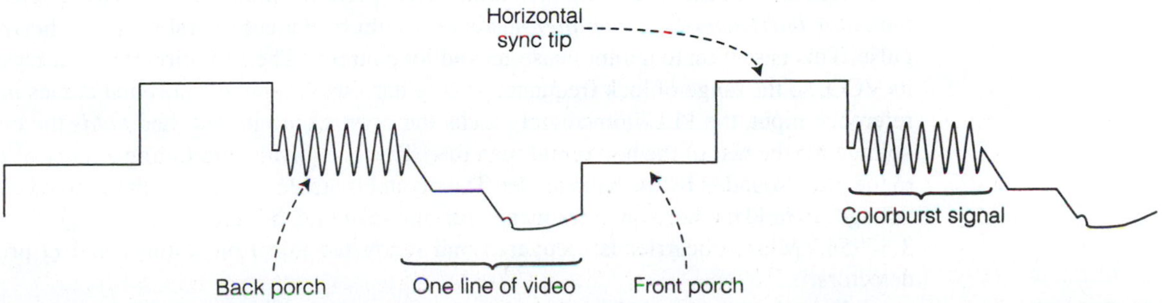

Figure 10–20 Detail of the color signal

EXAMPLE 10–4

At a certain point in a scene, the relative intensities of the R, G, and B signals from a color camera are as follows: red, 50% (0.5); green, 25% (0.25); blue, 35% (0.35). Calculate:

a. The value of the luminance or Y signal.

b. The value of the I and Q signals.

Solution

a. The Y signal is formulated as follows:

$$Y = 0.3R + 0.59G + 0.11B = 0.3(0.5) + 0.59(0.25) + 0.11(0.35) = \underline{\underline{0.336}}$$

The units of Y can be voltage, current, or just relative intensity.

b. The I and Q signals are also formulated according to predetermined equations:

$$I = 0.7R - 0.59G - 0.11B = 0.7(0.5) - 0.59(0.25) - 0.11(0.35) = \underline{\underline{0.164}}$$

$$Q = -0.3R - 0.59G + 0.89B = -0.3(0.5) - 0.59(0.25) + 0.89(0.35) = \underline{\underline{0.014}}$$

Incidentally, the angle of the color being produced (and hence its position on the color wheel) can be found by using a little trigonometry. The angle can be expressed as:

$$\phi = \tan^{-1}\left(\frac{I}{Q}\right) + 33° = \tan^{-1}\left(\frac{0.164}{0.014}\right) + 33° = 118.1°$$

From Figure 10–17, we can see that this is very close to the color *magenta*.

Demodulating a Color Signal

To demodulate a color signal, the Y (luminance) and C (chrominance, or I and Q signals) must be converted back into an RGB representation for driving the three electron guns in the color picture tube. The process is the opposite of that in the transmitter; the three signal components (Y, I, and Q) are recovered, then algebraically added in the correct order to recover the R, G, and B information for the picture. Figure 10–21 shows how the composite color signal is "magically" transformed back into R, G, and B signals. Actually, there is *no* magic, just a little algebra. Sorry to disappoint you.

The first step in color demodulation is to recover the I and Q signals. To do this, the 3.579545 MHz *subcarrier* must be recovered. Recall that a burst or "sample" of this frequency is sent by the transmitter at the beginning of every horizontal scan line as the *color burst* signal. With the help of the horizontal sync pulse from the receiver's sync separator, the *color burst separator amplifier* snatches a gulp of the colorburst on every horizontal pulse. This is sent on to a mini phase-locked loop circuit. The PLL circuit uses a crystal in its VCO, so the range of lock frequency is *very* narrow. When the colorburst comes into its reference input, the PLL immediately locks the crystal onto it, and then *holds* the control voltage for the rest of the horizontal scan line. This is equivalent to tuning a musical string to the pitch sounded by the band leader. The crystal is stable enough (with the fixed control voltage) to hold the horizontal frequency for one scan line, because of its high Q. Thus, the 3.579545 MHz subcarrier is recovered and ready for injection at the I and Q product detectors.

Note that the I and Q product detectors get a different version of the carrier. The I detector gets an in-phase version; the Q detector gets a 90° phased version. This again will

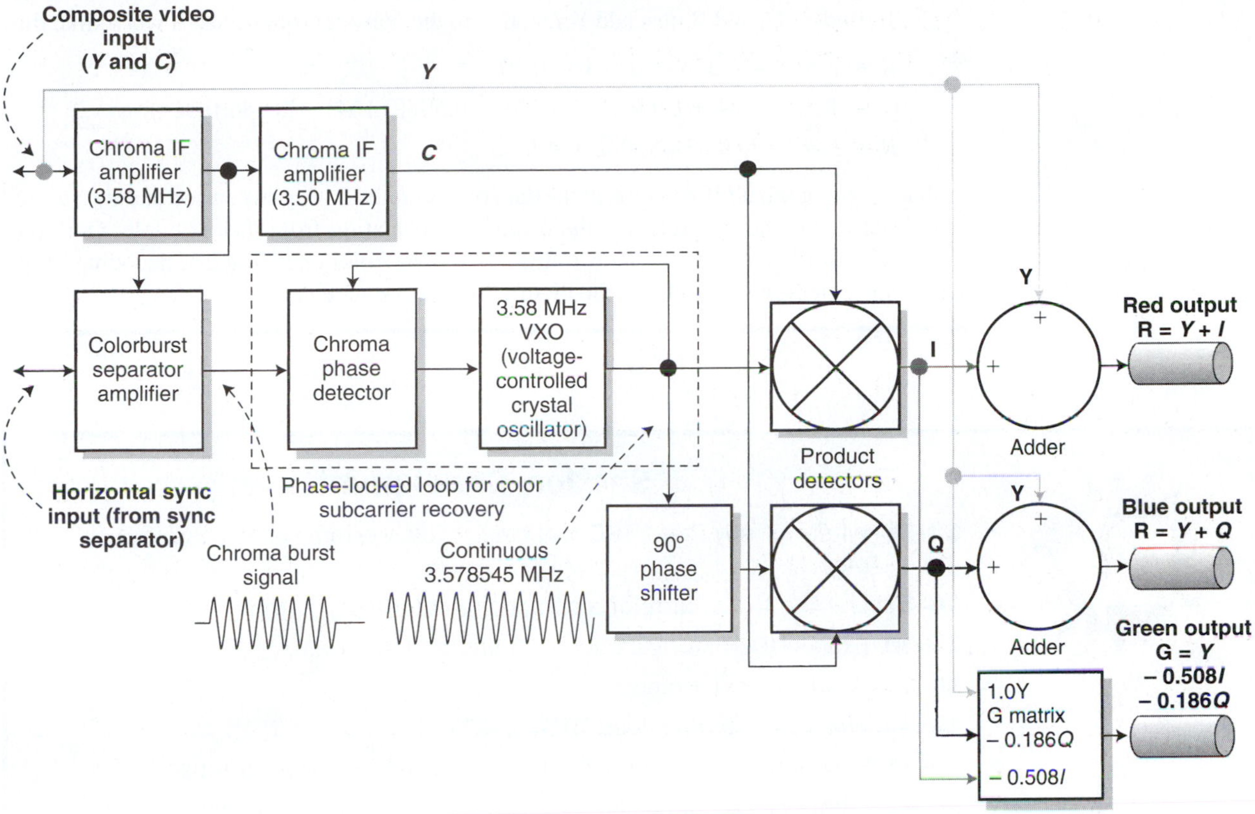

Figure 10–21 Color signal demodulation

allow separate extraction of the *I* and *Q* signals, even though they share the same subcarrier frequency.

The *sidebands* that contain the *I* and *Q* information signals are contained in a frequency range of 2.0–4.2 MHz, approximately. The *chroma IF amplifier* passes these frequencies to the *I* and *Q product detectors*. Thus, the output at the product detectors is the *I* and *Q* information.

To get the R, G, and B information signals back, we merely do a little addition. To get the *red* signal, we just add the luminance *Y* and the *I* signal; to get the *blue* signal back, we add *Y* and *Q*. Getting *green* is a little trickier. A little more algebra is required to get green, and a complete explanation of it is well beyond the scope of this text. It is a little like me telling you that one half of the sum of three numbers is 7, and two of the numbers are 3 and 5. You could easily find the third number.

To get *G*, we must add exactly the correct proportion of all three signals. In short, $G = Y - 0.508\,I - 0.186\,Q$. Any other proportion won't work, and the amount of green in the picture will be off!

EXAMPLE 10–5

At a certain point in a scene, the values of the *Y*, *I*, and *Q* signals in a color demodulator are as follows: $Y = 0.336$; $I = 0.164$; $Q = 0.014$. Find the original values of R, G, and B for driving the three guns of the picture tube.

Solution

To find R, G, and B, just add Y, I, and Q in the correct proportions. It works like this:

$$R = Y + I = 0.336 + 0.164 = \underline{\underline{0.50}}$$

$$G = Y - 0.508I - 0.186Q = 0.336 - 0.508(0.164) - 0.186(0.014) = \underline{\underline{0.25}}$$

$$B = Y + Q = 0.336 + 0.014 = \underline{\underline{0.35}}$$

Is that magic or what? If you look at the data of example 10–4, you can see that we cheated here. The R, G, and B results are the *original* information from that example. Our color TV system works! We have demonstrated a complete color encoding and decoding "trip," from the camera in the studio to the picture tube in the receiver.

Section Checkpoint

10–36 When we say that NTSC is a *compatible* system, what does that mean?

10–37 List the three parts of a composite color video signal.

10–38 Explain the difference between *hue* and *saturation*.

10–39 What is a color model?

10–40 What color model does NTSC use?

10–41 Why is a vectorscope useful in TV production? How is it used?

10–42 What two signals make up the chrominance information?

10–43 Which signal (I, Q, or Y) does a black and white receiver respond to?

10–44 What is the phase difference between the I and Q subcarriers?

10–45 Explain how a receiver recovers the color subcarrier from the *color burst* signal.

10–46 Why isn't the "$G - Y$" signal transmitted? How is green recovered at a receiver?

10–5 HIGH DEFINITION TELEVISION (HDTV)

In the late 1980s, many countries began to experiment with improved television transmission systems. Analog TV transmission in its various forms suffers from many problems. First, the resolution of all analog systems is fairly low (around 300 horizontal lines, equivalent to about 600 horizontal pixels on a computer monitor). Second, there's little tolerance for noise (the picture in analog TV is sent using AM, which has no noise reducing properties). Finally, the sound capabilities of analog television leave much to be desired (even with the addition of stereo TV, which uses the same analog multiplexing techniques as stereo broadcast FM.) Research and development of a HDTV system involved dozens of organizations worldwide; more than 23 different (and incompatible!) systems were proposed to the FCC by the end of 1988. In 1993, the FCC decided to go with all-digital technology and formed a "Grand Alliance" composed of AT&T, General Instrument Corporation, MIT, Philips, Sarnoff, Thomson, and Zenith. The Grand Alliance would take the best features from the existing HDTV proposals and combine them into a working HDTV standard.

As of September 2003, the FCC has finally decided on the standard for terrestrial HDTV broadcasts in the United States. By 2006, analog television will be phased out—broadcasters will no longer be required to transmit it.

HDTV Characteristics

HDTV has several important improvements over analog transmission. HDTV boasts improved resolution, sound, feature sets, and security. First, an HDTV video signal has at least *twice* the *horizontal resolution* of a standard NTSC analog signal. The signal is carried by completely digital means, using MPEG-2 as the transport mechanism. (We'll discuss the specifics of MPEG-2 shortly.) HDTV signal resolutions are given in computer terms (*pixels*—picture elements, or dots—of resolution, compared to the "lines" for conventional analog video). Two pixels are roughly equivalent to one line of resolution. HDTV is transmitted in either 1922 × 1080 ("1080 i" interlaced mode)—equivalent to 961 horizontal lines of resolution, or 1280 × 720 ("720 p" progressive scan mode) with equivalent resolution of 640 horizontal lines. Even in the lowest resolution mode, HDTV signals have twice the horizontal resolution of analog signals, which produce only about 320 lines of horizontal resolution under the best possible conditions!

The sound portion of HDTV is a completely digital system known as Dolby AC-3 (sometimes known as "Dolby 5.1"). AC-3 provides "five point one" channels of compressed digital audio: Front left and right (conventional stereo), front center, rear left and right ("surround"), and a low-frequency channel ("point one") for subwoofer information. Dolby AC-3 reduces the data rate needed for the audio information by using a lossy compression scheme that relies upon a model of human hearing; it analyzes the audio data in real time and reduces or eliminates sound information that is inaudible to listeners. The overall quality of the sound can easily exceed that of CDs, since AC-3 supports sampling rates up to 96 kHz (versus 44.1 kHz for CD audio), and sample quality up to 24 bits/sample (compared with 16 bits/sample in CD recordings).

Since the HDTV signal is a digital data stream, many features can be directly supported. For example, a broadcaster can directly transmit program information to users; this information may include episode information, caption information for hearing-impaired viewers, and content advisory information (similar to the "V-chip" rating identification system currently in use in the United States). HDTV is very computer compatible, so a future including interactive television (ITV) is quite possible. The line between computers and televisions may very well become blurred; televisions of the not-so-distant future may sport keyboards, mice, and other input devices as a result.

Finally, HDTV has been designed with security in mind. In particular, HDTV supports DRM (digital rights management) information. DRM is used to control what the end user is able to do with the content of the HDTV data stream. Since HDTV is a digital medium, copies of HDTV programs could theoretically be copied repeatedly with no loss in quality (just like computer programs on disks). DRM restricts copying of the content. For example, a consumer may be allowed to record a show for later viewing (using an HDTV-compatible VCR, DVD recorder, or other device), but not to make further copies of the program. Some programming may not be recordable at all; for example, a pay-per-view performance may be restricted in this manner.

Finally, note that HDTV has *not* been designed for compatibility with the existing NTSC analog system. The only characteristic that has been retained is the standard 6 MHz channel bandwidth. This is perhaps the most amazing thing about HDTV: The new system transmits *many* times more equivalent data than analog TV, yet uses the same channel bandwidth. This is possible because of the application of *compression* techniques to the HDTV data stream. Because of this incompatibility, current HDTV receivers have two separate tuner sections—one for analog NTSC reception, and the other for HDTV digital signals.

HDTV Transmission Process

The HDTV transmission process involves a very complex sequence of both analog and digital signal processing; the digital processing occurs first, and is shown in simplified form in Figure 10–22.

All of the steps in Figure 10–22 involve digital signals. Most of the actions take place in dedicated chipsets, which are contained within the TV studio equipment. The process begins with digital video and audio signals, which could originate directly from a studio camera switching system, or a digital program source (such as a DVD). Both the video and audio sources are *digital* signals to begin with. These digital signals consist of sequences of bytes that represent the information to be encoded. The data rates are impressive; uncompressed digital video (1280 × 728 pixels, 24 bits/pixel, 30 frames/sec) results in a 671 Mbps data stream. Five uncompressed audio streams (96 k samples/sec, 24 bits/sample, 5 channels) requires a 11.52 Mbps data rate. Fortunately, most of the bits in these two streams are redundant. The Dolby AC-3 encoder reduces the audio stream to a much more manageable 384 kbps. The MPEG-2 encoder compresses the video information using *DSP* (digital signal processing) algorithms. These algorithms are well beyond the scope of this text, but essentially they convert the time-domain picture information into a frequency-domain version by a procedure called a discrete cosine transform (DCT). The DCT is a computerized form of the Fourier series transformation presented in Chapter 2. Using a DCT allows the transmitter to selectively remove picture elements that use up memory space, yet won't be noticed as "missing" by the viewer. The algorithm then compares successive frames of the picture for changes (motion). Where no motion is detected (or predicted), that portion of the picture frame is not retransmitted, saving a great deal of space. The receiver is only commanded to "repaint" the portions of the picture that change from frame to frame. Since most scenes contain largely static (unchanging) backgrounds, the technique is highly effective. Similar techniques are used by the AC-3 encoder to reduce the data rate required for the audio. The final data rate for the combined video and

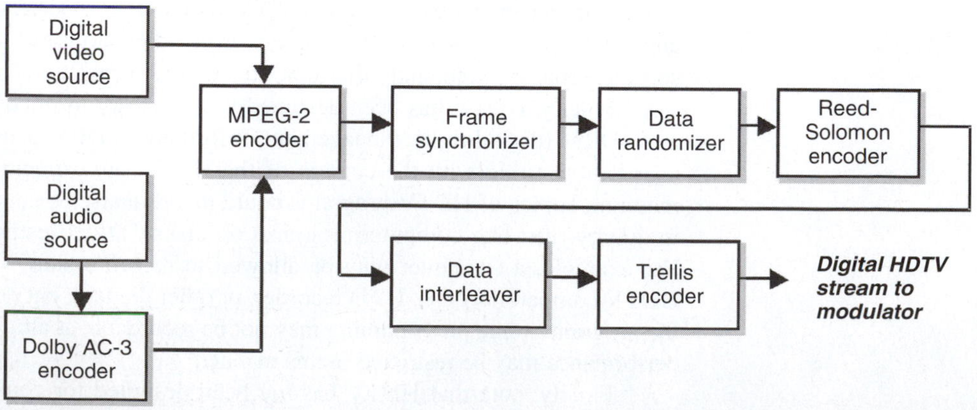

Figure 10–22 HDTV digital transmission processing steps

audio signal leaving the MPEG-2 encoder is 19.4 Mbps, a great reduction from the nearly 700 Mbps of the uncompressed data streams! The MPEG-2 data emerges in the form of *packets*. Each packet is 188 bytes long and is a self-contained unit of information.

The remaining steps in the process prepare the HDTV data stream for transmission over the air by implementing forward error correction. They're needed because interference with the signal between the transmitter and receiver is highly likely. Any interruption of the signal will cause a loss of data bytes in the "stream." These lost bytes would cause entire MPEG-2 packets (groups of 188 bytes) to be lost, which would cause *very* visible picture and sound defects! Sophisticated *forward error correction* (FEC) methods are used to minimize the effects of interference. FEC is necessary since the HDTV transmission is one-way only; the receiver can't "ask" the transmitter to resend data with errors (and at 30 complete picture frames per second, there wouldn't be time for resending anyway!)

The first step is detection of the MPEG-2 packet start and end points, and elimination of the MPEG-2 *sync byte*. The *frame synchronizer* performs this function. The reason for this is that the MPEG-2 data will be soon enclosed within a new packet structure. Next, the data set is randomized. A mathematical formula is used to do this. The receiver has the same formula and synchronizes to the transmitter, so it will be able to recover the original information. Randomizing the data isn't an error correcting step; it merely "flattens" the frequency spectra from the transmitter, which ensures that the 6 MHz frequency slot of the TV station will be evenly used.

Next, 20 bytes of parity (error-checking) information are added at the Reed–Solomon encoder. These parity bytes will allow up to 10 byte errors to be corrected within the 187-byte data block. The Reed–Solomon encoded data blocks are sent to the *interleaver*. The interleaver literally "splits" the data blocks into pieces, and mixes them with other data blocks. This further reduces the sensitivity of the system to burst errors (where groups of data bytes are lost). A burst of noise may damage *several* data frames, but the damage to each is (hopefully) limited to 10 bytes or less, which can be corrected by the Reed–Solomon decoder in the receiver.

The final digital step is *Trellis encoding*. This is yet another forward error correction layer. Trellis encoding follows groups of *bits,* instead of the groups of *bytes* followed by Reed–Solomon coding. Trellis coding divides each 8-bit byte into four 2-bit (di-bit) words. Each new 2-bit word is compared to the previous 2-bit word and then transformed into a 3-bit binary code according to the difference between the previous 2-bit words. The final output of the Trellis encoder is groups of 3-bit code words. These 3-bit words contain the forward-error-corrected HDTV signal, and they are ready for conversion to analog for transmission.

The analog steps in transmission are much more simple than the digital steps! Figure 10–23 shows the analog processing. First, the three-bit groups are converted to analog using a digital-to-analog converter. A different voltage represents each possible combination of three bits. This is *multilevel* encoding, and is discussed in more detail in

Figure 10–23 HDTV analog transmission processing steps

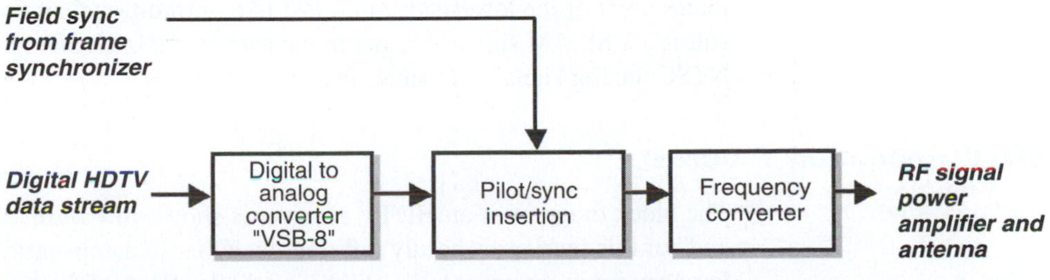

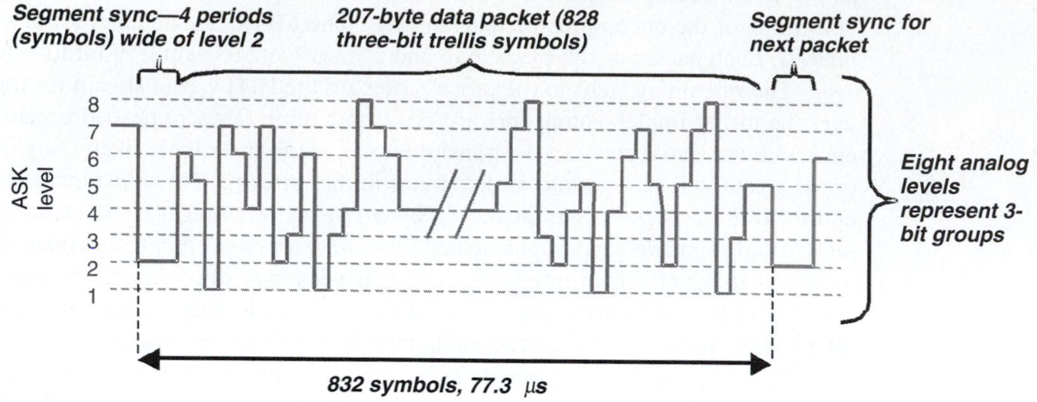

Figure 10–24 The 8-VSB signal after pilot/sync insertion

Chapter 15. Using this type of encoding reduces the bandwidth needed to transmit the HDTV signal by reducing the *symbol rate* from 19.4 MHz (one symbol per bit) to 6.46 MHz (one symbol per three bits). Figure 10–24 shows what the analog signal looks like.

The receiver needs to have a way of synchronizing itself to the incoming data. Two mechanisms are provided, a *pilot tone* and *segment sync* pulses. A pilot tone is added to the RF output by adding a DC offset to the data stream (this pilot tone is at a low level and adds only +0.3 dB to the overall transmitted power output). The frequency of the pilot is equal to the transmitted carrier frequency. The pilot tone helps the receiver accurately lock onto the frequency of the HDTV signal, which is critical for accurate demodulation. Once a receiver has locked onto the HDTV signal, it needs to be able to determine the beginning and end of each incoming data packet. Therefore, a sync pulse called the *segment sync* is added to every segment (packet). The sync pulse is four periods wide (no other portion of the data is allowed to have the same value for four periods), and is extracted by the demodulator at the receiver. The receiver can therefore get synchronized to the data flow from the transmitter. In addition, a *frame sync* data segment is sent for every 313 normal data segments. The frame sync segment also begins with a segment sync pulse. The frame sync segment contains data about the picture frame being constructed.

Note that these sync pulses do not necessarily correspond with the horizontal or vertical sync of an analog TV transmission. There is no fixed relationship between the digital segment sync pulses and the picture. The data contained within the MPEG-2 packet tells the receiver where to "paint" each part of the picture, so no other synchronization is needed.

The final result of the pilot and sync insertion is a low-frequency RF signal. This signal is upconverted to the final RF channel frequency by the frequency converter section, which contains an accurate oscillator, mixer, and output bandpass filter. The bandpass filter eliminates most of the lower sideband, just like in traditional analog TV transmission. The result is a VSB AM signal that fits in the same 6 MHz channel bandwidth as a conventional NTSC analog signal—no small feat!

HDTV Reception Process

The block diagram of an HDTV receiver is shown in Figure 10–25. Notice that the "front end" of this receiver is hardly different than that of a conventional receiver. The major differences are in the processing of the signal after IF amplification and detection. The HDTV

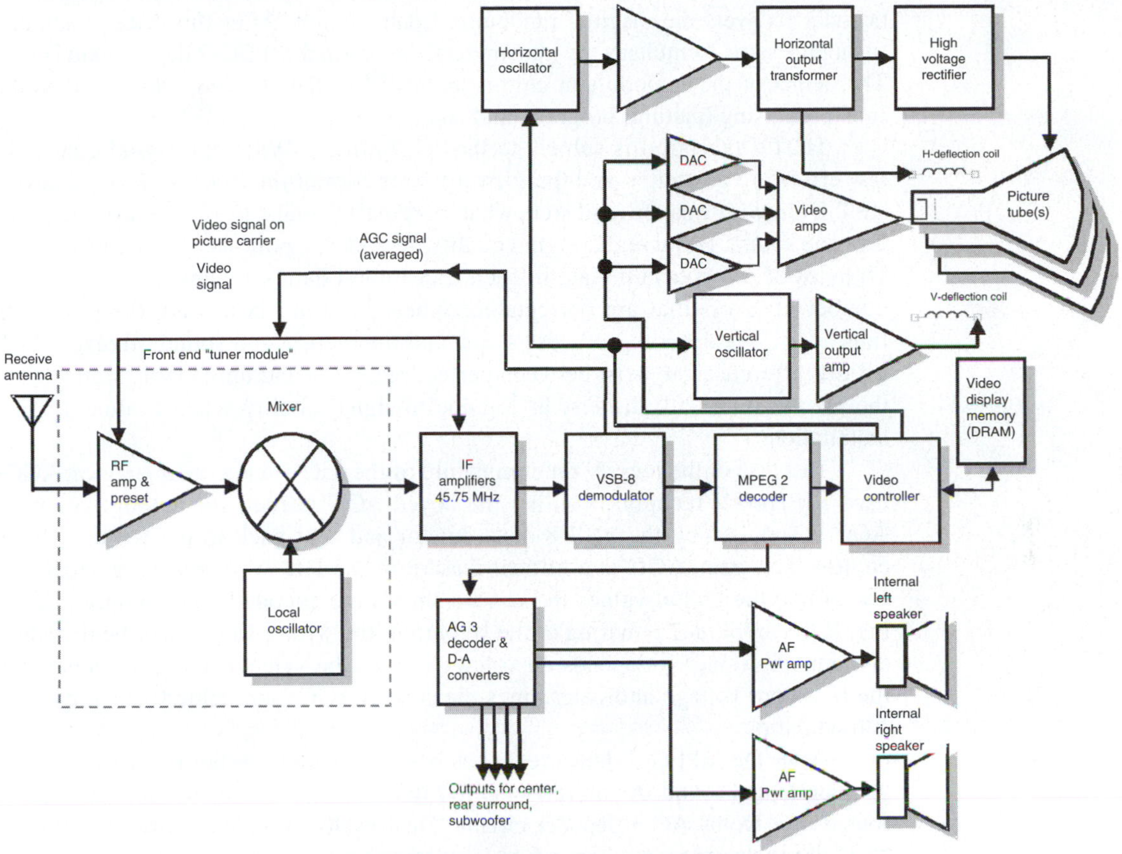

Figure 10–25 HDTV receiver block diagram

receiver processes the signal in three steps, which are analog (the tuner), digital (post detection processing), and *analog* (display).

The tuner or front-end section of an HDTV receiver is practically identical to that of a conventional analog receiver, with one exception: The local oscillator must be very accurately controlled, which means that a synthesized local oscillator using a crystal time base is required. The IF output of the tuner section passes directly to a VSB-8 demodulator circuit instead of an AM diode detector. The VSB-8 demodulator is typically contained on a single chip; one example is the Oren Semiconductor OR5122, which contains all the circuitry needed to convert the raw IF signal back into an MPEG-2 data stream. These steps are the reverse of the steps used in the digital portion of the HDTV transmitter.

The VSB-8 demodulator first locks itself onto the pilot signal present in the IF output using a phase-lock loop circuit (or DSP equivalent). This is necessary because of the possibility of minor frequency error in the IF output (even though the receiver uses a synthesized local oscillator). The demodulator's internal clock signals will then be locked to the pilot tone, which will assure accurate timing during demodulation.

Once lock has been attained, the demodulator converts the analog IF signal back to a stream of 3-bit digital values. These values are then passed into a Trellis decoder, which results in a di-bit data stream (hopefully) identical to that emitted by the interleaver and

Reed–Solomon encoder of the transmitter. The interleaving is reversed, and a Reed–Solomon decoder recovers the original randomized data stream. After this data is derandomized (using the same formula as the transmitter), the original MPEG-2 data stream is recovered. The output of the demodulator chip is the MPEG-2 data stream, which is then ready for final processing (picture, program, and sound.)

HDTV reception is subject to the "cliff effect." When the signal quality is good, few errors are generated and the forward error correction process of the receiver works well. Even with weaker (and somewhat marginal) signals, FEC will provide perfect picture and sound. However, as signal quality degrades, a point is reached where more than 10 bytes of errors accumulate in each transmitted data segment. This is the maximum number of errors that are correctable. When this limit is passed, the system literally falls apart. The decoding process stops, and the picture and sound "freeze." As little as a 1 dB difference can exist between perfect reception and unreadable signal. Therefore, the picture is literally useless as a guide to signal quality when evaluating an HDTV installation.

Because of the largely random nature of the data stream, an averaging AGC can be used for HDTV reception (unlike the keyed AGC needed for analog reception). The AGC is supplied by the VSB-8 demodulator and sent back to the RF and IF stages to control their gain. AGC is a critical feature of the HDTV receiver; recall from Figure 10–24 that the digital values in the data stream are encoded using amplitude shift keying. If the amplitude is wrong at the IF output, the wrong values will be decoded by the demodulator (since it depends on voltage to read the values). The demodulator samples the IF output voltage and determines the correct AGC gain value to ensure correct data demodulation.

After the MPEG-2 data stream has been recovered, further processing is needed to get picture and sound. Again referring to Figure 10–25, notice that the sound packets are routed to a Dolby AC-3 decoder circuit. This results in analog outputs representing the "5.1" channels of information, which can then be sent to audio amplifiers and speakers to reproduce the soundtrack. The RGB picture information is obtained by processing the picture data packets in an MPEG-2 decoder, which passes its information to a *video controller* chip. The video controller chip is very much like the display chipset in a computer video card; it accepts graphic drawing commands from the MPEG-2 data stream and renders the resulting pixel data to the *video display memory* (a dynamic RAM chip array). The video controller chip also scans the DRAM video memory periodically, producing analog outputs to drive the display circuitry.

The display circuitry in an HDTV receiver is closely related to the video monitor of a computer. In fact, many HDTV receivers have a 15-pin computer input on the rear panel since it costs little to add this item! The display accepts the horizontal and vertical synchronization pulses and analog RGB outputs of the video controller chip and renders a visible picture. Its technology is almost identical to that of an analog television display. Projection TVs often use an array of three CRTs (red, green, and blue) for final display; however, this isn't the only possible technology. Plasma, LCD, organic LED, and field emission are just a few of the possibilities.

Finally, many HDTV receivers provide direct digital output from the VSB-8 demodulator circuit. This can greatly simplify home hookups for external devices such as video recorders or audio amplifiers. For example, passing the "5.1" audio signal to a surround-sound amplifier would require *six* different shielded audio cables. Since the digital signal contains all the information, it is much more convenient to pass it directly to the external amplifier, which will contain its own AC-3 decoder for rendering the analog soundtrack.

Section Checkpoint

10–47 What two NTSC problems does HDTV address?

10–48 Is HDTV a “compatible” system with NTSC? Explain.

10–49 What standards are used for transport of audio and video in HDTV?

10–50 Why is compression necessary for the audio and video streams of HDTV?

10–51 Describe the HDTV modulation process.

10–52 What two synchronization signals are used by an HDTV receiver?

10–53 What is the “cliff effect?”

10–54 How is an HDTV receiver similar to a computer video display monitor?

10–6 TROUBLESHOOTING TV RECEIVERS

A television receiver is one of the most complex pieces of equipment that a technician may be called upon to service. Even though a TV certainly is complicated, many problems can be isolated to a stage by observing the symptoms. The picture is a particularly revealing source of information.

Servicing Precautions

There are several hazards in TV work that a tech should be aware of. The number one risk is *electrical shock* from the many high voltages in a set. Many sets use a *live chassis* configuration, where one side of the metal chassis is directly connected to the AC power line.

Connecting a test instrument ground lead directly to a live chassis can result in a shower of sparks from the resulting short circuit!

To minimize the risks, the following precautions must be taken:

1. Always use a line isolation transformer when working with sets on a bench. A line isolation transformer is a 1:1 transformer that isolates the 120 volt house current from the set under test.
2. Always work with *one* hand when testing live circuitry; keep the other hand in your pocket. Wear good-quality shoes with insulating soles. This precaution helps to prevent current from passing through your chest, should you receive a shock.
3. Do not wear watches, rings, or other jewlery when working on the bench. These things can short a circuit under test, or can accidentally put you in contact with high voltage.
4. Wear safety goggles when handling or working near a picture tube. A picture tube contains a partial vacuum, and may shatter violently (implode) if scratched, cracked, or dropped. Do not touch a picture tube with any metal tools; you may scratch it.
5. Before picking up a picture tube, short the second anode nipple (on the side of the tube) to the ground frame of the tube. The liner of a picture tube is one plate of a

capacitor. The amount of charge stored there can't hurt you, but it might startle you into dropping the tube when you pick it up.

6. Do not operate a TV set with the X-ray shielding removed. A TV set with excessive high voltage can emit X-rays. If the high voltage readings are not within the manufacturer's specifications, turn the unit off until the cause is discovered.

Common TV Problems

Remember the three-step approach to troubleshooting? Those steps were:

1. Visual (and other) inspection.
2. Check power supplies.
3. Check inputs and outputs.

In TV servicing, those three steps are more important than ever. In fact, there is a well-kept secret among TV repair people: *Many faults in television receivers are due to self-heating and accelerated aging of solder joints.* In the rush to produce consumer electronics devices of the smallest possible size, manufacturers have crammed a lot of heat-producing electronics into a tiny space. TV receivers make a lot of heat. Place your hand over the ventilation holes of a receiver and you'll get the idea. Consumers often make this problem worse by placing TVs where they don't get proper ventilation, such as in a cramped entertainment cabinet.

When doing a visual inspection of a TV chassis (with the power off and the set disconnected from the power line), look very carefully at the *horizontal output* and *power supply* sections. Very often, you will note a brown discoloration of the circuit board underneath these sections; the solder joints on the major components (horizontal output transformer, etc.) will appear dull and oxidized. The television has literally cooked itself! If you see this condition, retouch the affected solder joints (adding fresh solder). You may fix the problem!

Power Supply Checks

There are as many power supply configurations in use as there are models of TV sets; every manufacturer has its own favorite set of designs. One of the most common power supply designs looks like Figure 10–26.

In this setup, the 120 volt ac power is immediately full-wave rectified and filtered, producing a +170 volt (approximate) potential. This is different than what is normally expected; the traditional (analog) power supply method uses a power transformer to first step up or step down the voltage. This circuit skips that step entirely!

The +170 volt dc potential is unregulated; it can vary according to the house voltage, current draw, and so on. The next step is *voltage regulation.* An IC is almost always used to regulated the 170 volt potential down to approximately +130 volts. This IC is bolted to a heatsink.

> The first power supply check in the receiver would therefore be the +130 volt regulated source. The next checks should be the lower-voltage supplies (+18, +5, etc.). Pay attention to "where ground is" when measuring supplies; in the circuit shown in Figure 10–26, there are *two* grounds, the "hot" ac ground on the left, and the "cold" dc ground on the right. They're not the same!

The +130 volt supply runs *everything* in the receiver by supplying energy to the horizontal output stage. Extra secondary windings are used on the horizontal output transformer to extract energy for operating the rest of the set. The horizontal output

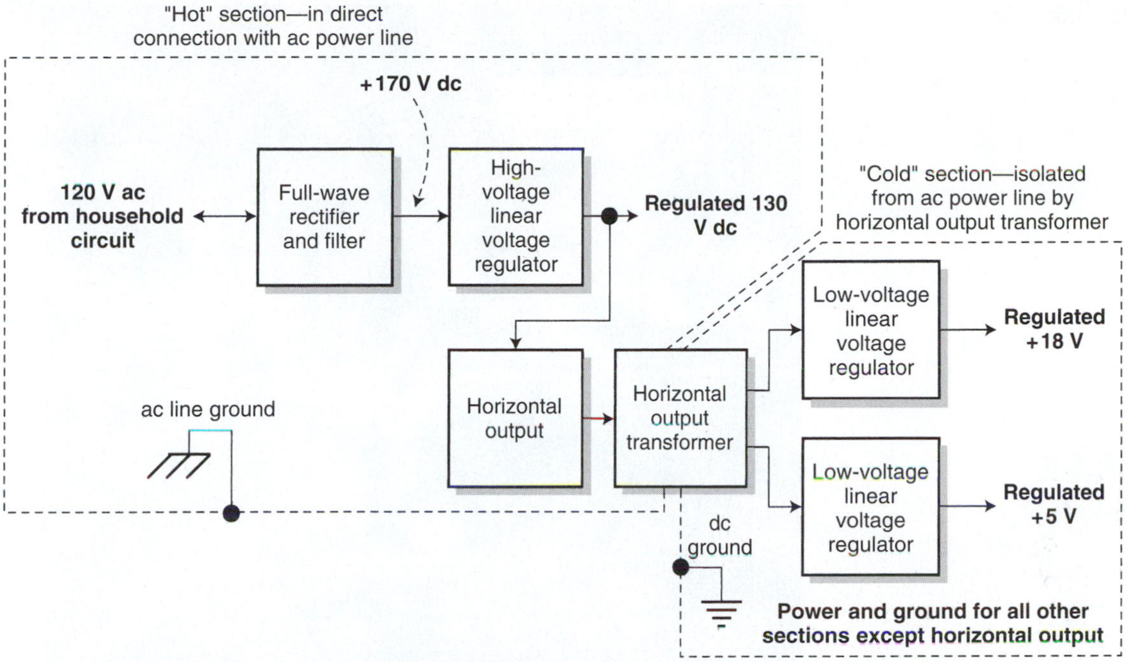

Figure 10–26 A common power supply configuration

transformer serves a dual purpose; it not only steps the +130 volt down to the lower voltages (+18, +5, etc.) needed for the rest of the circuit, it also isolates the rest of the set from the ac power line. (Of course, it also functions to create horizontal sweep voltage for the picture tube!)

> If the 130 volt supply is present, but no other supplies are active, the horizontal output stage should be strongly suspected, because it supplies the energy for every other section in the set.
>
> If the 130 volt supply is **not** working, check the 170 volt input. Perhaps there's a blown fuse. Some makes use a fusible resistor in the ac power supply input. If a fuse has blown, there's usually a reason; there could have been a power surge, or perhaps something in the power supply is shorted (rectifier diode, filter capacitor, or even the horizontal output transistor). Examine the blown fuse carefully; if the fuse wire has sagged and broken, the overload was "soft." If the fuse wire is totally vaporized, the fuse blew "hard" and you'll likely find something dead shorted in the power supply.

Troubleshooting by Symptoms

Most receiver problems produce distinct effects in the sound, picture, or both. In fact, the symptoms should *always* be carefully evaluated before attempting any circuit-level troubleshooting. Figure 10–27 depicts some common problems.

In Figure 10–27(b), the *vertical sweep* of the picture tube isn't occurring. A set in this condition should not be left on without backing down the brightness control, to prevent burning the picture tube. The most common *cause* of this symptom is failure of the vertical output amplifier.

Figure 10–27
Common problems

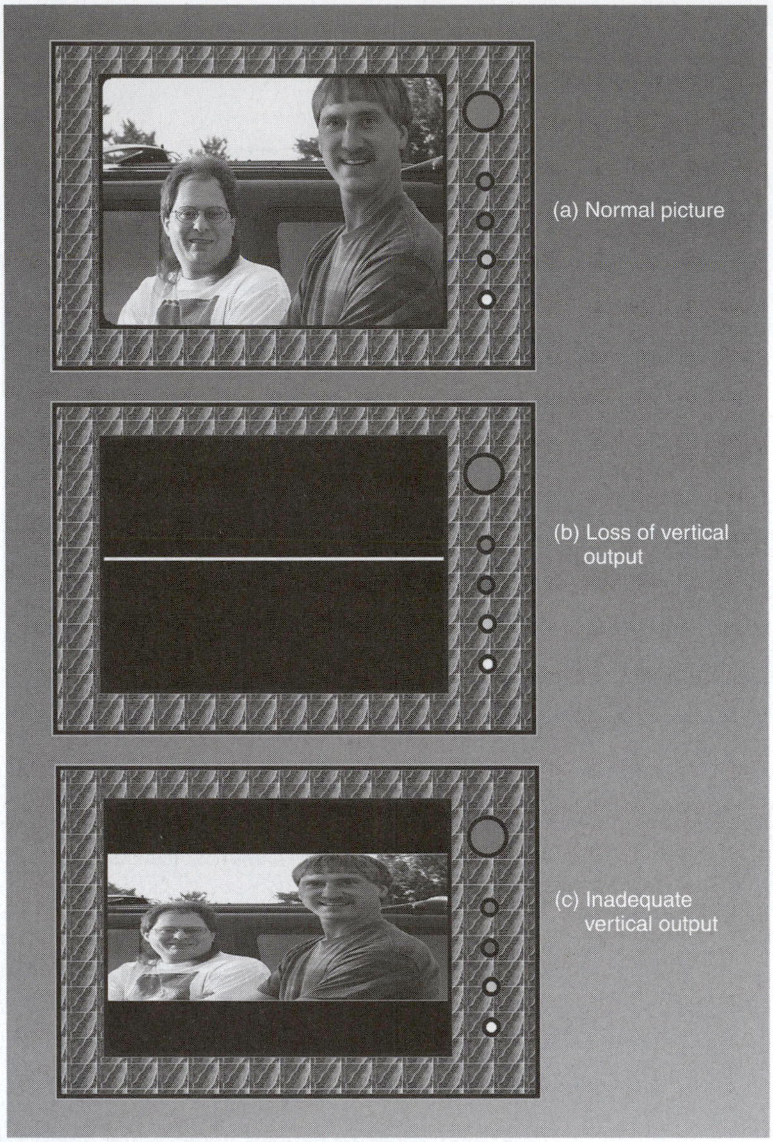

Figure 10–27(c) shows another problem that is usually caused by *inadequate power supply voltage*. The vertical output stage can also be at fault. There just isn't enough energy being supplied to the vertical sweep coils on the picture tube to get a full picture. The picture information is squeezed. If you see a picture like this, *check the power supplies* before proceeding. Many other problems can be isolated by remembering the action in the block diagram of Figure 10–9. Here are some examples:

- No picture, sound OK.

If the set is producing sound, then the RF and IF sections are probably operating fine. If the set derives the lower dc voltages from the horizontal output, then that stage is probably OK as well. Look for a problem in the *video amplifier* or *video output* stages.

- Picture, but no sound.

Most of the receiver is operating correctly. The sound section consists of the 4.5 MHz IF, which is taken from the AM picture detector; the problem is either there, in the audio amplifier, or loudspeaker.

- No color when receiving a color program. Otherwise, picture and sound are OK.

The *color demodulator* section of the receiver is at fault. One of the most common component failures in this area is the 3.579545 MHz colorburst crystal; without a local oscillator, the color can't be demodulated, and the set displays only a black-and-white image.

- Sound bars in picture.

If the picture appears to have vertical bars that move in step with the sound, there may be a misalignment in the IF amplifier section, or a problem with the 4.5 MHz trap in the video section. Don't attempt alignment without the proper equipment!

- Snowy picture, even when applying strong signal from test pattern generator.

A lack of RF gain may be causing the problem. This could be an *AGC* problem (keyed AGC circuit), an *IF* problem (lack of gain), or an *RF* problem (defective stage in tuner module, causing lack of RF gain).

Section Checkpoint

10–55 How can many TV problems be isolated to the stage level without the use of test equipment?

10–56 List at least five precautions for television servicing.

10–57 What causes many faults in TV receivers?

10–58 What two areas are likely to have "cooked" solder joints in a receiver?

10–59 What component isolates the chassis and ac line grounds in Figure 10–26?

10–60 When checking the power supplies in Figure 10–26, which one is tested first?

10–61 If the +130 volt supply is OK in Figure 10–26, but everything else appears dead, what stage is likely at fault?

10–62 What stage should be checked if the picture of Figure 10–27(b) is produced?

10–63 If the video IF amplifier in a TV receiver fails, will the sound be affected? Explain.

SUMMARY

- TV receivers are complex systems; they use both AM and FM technologies, as well as a little of everything else (digital, microprocessors, and advanced analog methods).
- A television signal contains two distinct carriers: AM for picture, and FM for sound. The use of VSB reduces the needed picture bandwidth.

- Broadcast allocations for television are channelized; three ranges of frequences are used (low VHF, high VHF, and UHF).
- A video signal is made up of *luminance* (brightness information) and *sync*. The time and voltage dimensions of video signals are very precise.
- The horizontal resolution of a TV signal is controlled by the available bandwidth and horizontal scanning frequency.
- At the block diagram level, almost all TV receivers are alike. Most use ICs and modules for the major functions.
- The IF amplifier filtering requirements for TV receivers are very strict. To align an IF amplifier requires proper service documentation and instrumentation. It can't be aligned for "maximum output."
- The sync separator and horizontal output amplifier are two examples of specialized circuits in a receiver. They operate on fundamental principles, and depend on subtle characteristics of the incoming signals in order to operate correctly.
- Analog TV receivers require the use of keyed AGC, because the average value of a video signal is widely variable, depending on the scene content. Keyed AGC relies on the fact that the sync tips are always of constant amplitude.
- Color TV works much like stereo FM; frequency division multiplexing methods are used to encode the color information while maintaining a signal compatible with black and white receivers.
- In color TV transmission, the RGB information from the camera is converted to HSL for transmission; the receiver recovers RGB from the signal by disassembling it algebraically.
- HDTV utilizes high-speed digital data transmission and compression methods to transmit picture and sound. Specialized chipsets handle most of the work.
- Like analog TV, HDTV uses VSB (vestigial sideband) to reduce the required transmit bandwidth and uses the same 6 MHz channel plan.
- An HDTV receiver is a close cousin of the computer display monitor; it contains the same analog RF components as a conventional TV signal, but specialized circuits for demodulating the HDTV signal and displaying it.
- In troubleshooting TV receivers, a technician must be aware of the dangers inside a set and work in a safe manner.
- Power supply problems plague many TV receivers; heat is a receiver's worst enemy.
- By observing picture and sound symptoms, an experienced technician can often identify a problem stage before reaching for a screwdriver.

PROBLEMS

1. Draw a spectrogram for a channel 9 television signal, showing all frequencies.
2. Why are "opposite" modulation methods (AM and FM) used for the picture and sound in a TV transmission?
3. What type of sideband transmission is used for the picture in television? How does this help?
4. What FM deviation constitutes 100% modulation for the sound portion of a TV signal?
5. In order to listen to the sound for TV channel 62, what frequency should a scanning receiver be tuned to?
6. What is the picture carrier frequency of channel 9?
7. With an analog FM broadcast receiver, it is sometimes possible to tune slightly below the bottom of the dial and hear the sound from TV channel 6. Why is this so?
8. A certain TV transmitter is operating with a peak transmitter output of 10 kW. What is the acceptable power level range for the sound portion of the transmitter?
9. Jose works two benches down from you, and has a brand new scanner. One day he is scanning the frequency range between 630 and 700 MHz, and his scanner stops on 649.75 MHz, where clear audio can be heard. What channel is Jose listening to?

10. What is the function of a *diplexer* in a TV transmitter?

11. What is interlaced scanning? Why is it used in TV transmission?

12. What is the difference between a *frame* and a *field* in a TV signal?

13. Why is a photoresistive material used in a television camera pickup tube?

14. Draw a picture of one line of video (going white to black as the beam goes left to right), showing the *sync level, blanking level, black reference level,* and *white level* percentages.

15. During what time periods is *blanking* necessary? How does the receiver know when to turn off its electron beam?

16. A certain video recorder has a bandwidth limit of 3.5 MHz; if it records an NTSC video signal with horizontal time of 53 μs, what will its horizontal resolution be?

17. Repeat problem 16, but allow a bandwidth of 5.5 MHz.

18. What receiver design is used for constructing TV sets?

19. What is the standard IF frequency for TV receivers? Why is this frequency used (instead of 455 kHz or 10.7 MHz)?

20. A TV receiver with a standard IF frequency is tuned to VHF channel 4. Calculate the frequency of the local oscillator, assuming high-side injection.

21. From the data of problem 20, demonstrate that the FM sound carrier frequency will indeed be 41.25 MHz in the IF amplifier. ("Mix" the channel 4 FM sound carrier frequency and the local oscillator frequency from problem 20.)

22. What is stagger tuning? Why is it needed in the IF section of a TV receiver?

23. What symptoms are caused by failure of any stage in the 45.75 MHz IF?

24. What type of filter is used to extract (a) horizontal sync and (b) vertical sync from a received TV signal?

25. What is the function of diodes D_1 and D_2 in Figure 10–16?

26. What are three possible symptoms of video amplifier failure?

27. Explain how the keyed AGC operates in a TV receiver. When does it take a sample of the incoming signal?

28. What is the function of the *colorburst* signal? When is it sent?

29. At a certain point in a scene, the relative intensities of the R, G, and B signals from a color camera are as follows: red, 75% (0.75); green, 15% (0.15); blue, 30% (0.30). Calculate the value of the Y, I, and Q signals for color transmission.

30. A TV receiver color decoder signal has the following voltages: $Y = 1$ V, $I = 0.5$ V, and $Q = 0.2$ V. What values will be decoded for R, G, and B at the picture tube?

31. A TV receiver color decoder signal has the following voltages: $Y = 1$ V, $I = 0$ V, and $Q = 0$ V. What color is being displayed? (Work backward to find R, G, and B first.)

32. List the digital steps in the HDTV transmission process and explain each one.

33. List the analog processing steps needed to transmit HDTV over the air.

34. Why does the "cliff effect" occur in HDTV reception?

35. A certain TV receiver uses a power supply scheme like Figure 10–22. It has no sound and no picture; the +130 V supply and +18 V supplies measure OK, but the +5 V supply measures only 0.2 V. What section is likely at fault?

36. A TV receiver has no picture and no sound; there is no voltage at the +170 V test point in the power supply (Figure 10–22). Upon looking at the fuse, there's no trace of the fuse wire left inside it. What type of fault is probably present?

37. A TV receiver has good picture, but no sound. Describe where the problem might be, and how to find it in a step-by-step fashion.

11

Transmission Lines

OBJECTIVES

At the conclusion of this chapter, the reader will be able to:

- describe the physical construction of several common types of transmission line
- identify several connector styles used with coaxial line
- define important line parameters, such as characteristic impedance, velocity factor, and loss coefficient
- describe how waves travel on a transmission line
- calculate the effects of impedance mismatches
- describe the physical construction of several impedance-matching devices
- utilize a Smith chart to perform basic transmission line calculations
- set up test equipment to verify transmission line performance

Transmission lines are the short-haul truckers of the communications field. Whenever RF energy needs to be transported between two nearby points, a transmission line is used. *The purpose of a transmission line is to get RF energy from one point to another with a minimum of signal loss.*

Signal energy can be lost in two ways. First, it can be lost to the resistance of the wires that carry it, which converts it to heat. This is undesirable and can be minimized by choosing conductors of appropriate size. Second, signal can be lost by *radiation.* When power is lost in this manner, it moves out into space as a radio wave. When a transmission line radiates, it acts as an antenna, which is not very useful; the signal leaks out where it shouldn't!

Many people find the subject of transmission lines to be mysterious, and a lot of misinformation circulates. Actually, you'll see that transmission lines operate according to the principles of physics, many of which you already know intuitively.

11–1 BASIC CONSTRUCTION OF TRANSMISSION LINES

There are only a few basic ways of building a transmission line. Figure 11–1 shows several types of common line. Let's talk about each type.

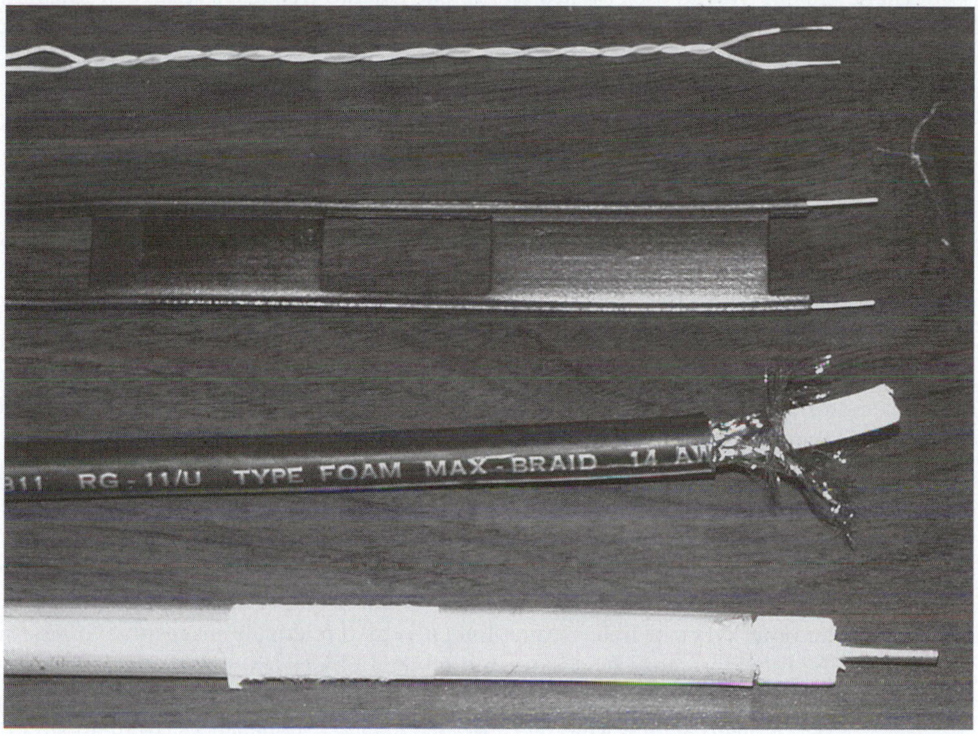

Figure 11–1 Basic transmission line types

Ladder Line

One of the simplest and lowest-cost transmission lines is the *ladder line* of Figure 11–1(b). This line consists of two parallel conductors, separated by an insulating material. You may have seen it on the back of your TV set in the form of *flat line*. This type of line has a *characteristic impedance* (which we will define in the next section) between 200 and 600 Ω, depending on how far apart we've spaced the conductors.

Ladder line is a *balanced* transmission line. Both of the wires are of equal size, which means that the paths for outgoing and incoming current are equal. This type of line has a very low loss but above 100 MHz, begins to experience an increasing loss by *radiation*. We would say that this line is not well shielded. In fact, if you install this type of line, make sure it is kept at least 6 inches from any metal objects. Because it is poorly shielded, a nearby metal object can interfere with the signal on the line.

Twisted Pair

Twisted pair, as shown in Figure 11–1(a), is also an inexpensive transmission line. Like ladder line, it too is a balanced medium. Twisting the wires closely together improves the shielding, when compared to ladder line. However, twisting the wires greatly increases the high-frequency losses of the line, mainly because the capacitance between the wires increases. Above a few MHz, twisted pair isn't very useful as a transmission line; it is too lossy there.

This type of line is primarily used for low-frequency work. It is used extensively in the telephone system to connect homes and businesses with the telephone company central office.

The telephone system uses twisted pair to carry audio frequencies; at such low frequencies, the loss in this type of line is relatively low. Twisted pair has a characteristic impedance between 25 and 200 Ω, depending on the type of insulation on the wires and how tightly they are twisted together (number of twists per unit length).

Coaxial Line

In order to improve the shielding properties without incurring the heavy high-frequency losses caused by twisting the wires together, we can employ *coaxial* transmission lines, as in Figure 11–1(c). The word *coaxial* means "shared axis." Coaxial cable consists of a center conductor inside a hollow outer metal tube. In between the two conductors is the dielectric, which can be air, plastic, or some other insulating material. The dielectric is there to insulate the two conductors from each other. It also partially controls the electrical characteristics of the line. Coaxial line is an *unbalanced* line. The two conductors are unequal, and the paths for current going in and out are different.

The silver coaxial line in Figure 11–1(d) is called *hardline*. The metal outer jacket is rigid and resists bending. Flexible cable is much more common; the black coaxial cable shown uses a braided outer conductor instead of a solid metal tube. The outer conductor of coaxial cable is approximately at ground potential (unless the cable is being used for certain special purposes). Therefore, coax cable is not only superior in shielding to the other types of transmission line, it is also relatively insensitive to *placement*. Normally, running a coax lead-in next to a metal object will not cause a problem with the line.

Coax is made with a wide range of characteristic impedances, from about 25 Ω to approximately 100 Ω. It is useful up to around 1000 MHz (1 GHz), where it begins to become fairly lossy. By "lossy" we mean that a significant portion of the transmitted energy is converted to heat within the line rather than being passed on to the load. Coaxial cable is used at even higher frequencies than 1 GHz, but only for short runs (such as for an RF signal patch cord or "jumper" between two pieces of gear).

Waveguide Above 1000 MHz, coaxial cable is not a good solution when signals need to be sent more than a meter or so. At microwave frequencies, *waveguide* is commonly used. *Waveguide* is a hollow metal tube (or box) that carries UHF and SHF (microwave) radio energy. Energy is introduced into one end of the waveguide by a small *coupling stub* (a miniature antenna) or *loop.* The RF then travels down the waveguide to the other end. Because the walls of the waveguide are metal, the RF energy can't escape. There is no center conductor or dielectric. At the receiving end, another coupling stub extracts the microwave RF energy. Figure 11–2 shows how this works.

Waveguide can also be used to feed microwave antennas, which are usually *horns* or *dishes*. In this case, the receiving end of the waveguide just opens into the dish of the antenna.

Waveguide is a very low-loss transmission line, and it is by far the most expensive of all the types discussed so far. However, it is not useful below about 1000 MHz, for its physical size is related to the *wavelength* of the signals that pass through it. At frequencies below 1 GHz, the required size for waveguide would be very large. The formula for finding the wavelength of a signal is

$$\lambda = \frac{v}{f} \tag{11–1}$$

where v is the velocity or speed of the radio wave (3×10^8 m/sec in free space), and f is the frequency of the wave.

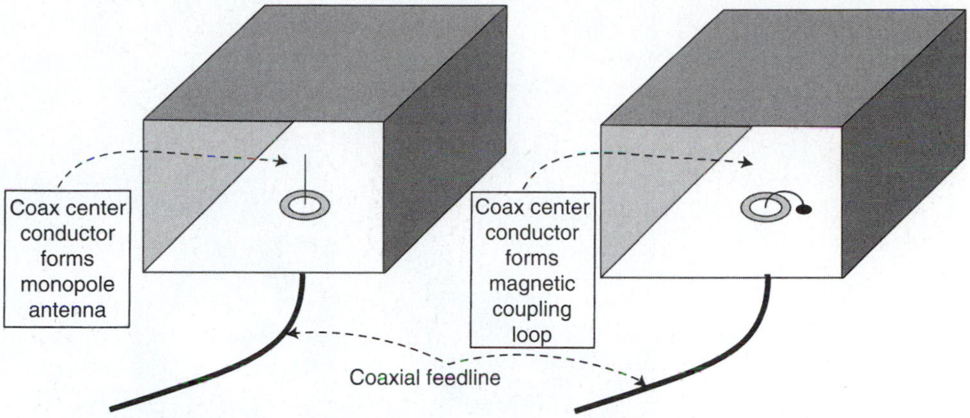

Figure 11–2 Getting energy in and out of a waveguide

EXAMPLE 11–1

What is the wavelength of the following signals?

a. 1 MHz

b. 30 MHz

c. 3 GHz

Solution

Equation 11–1 solves all these problems. We get

a. $\lambda = \frac{v}{f} = \frac{3 \times 10^8 \text{ m/sec}}{1 \text{ MHz}} = \underline{\underline{300 \text{ m}}}$

b. $\lambda = \frac{v}{f} = \frac{3 \times 10^8 \text{ m/sec}}{30 \text{ MHz}} = \underline{\underline{10 \text{ m}}}$

c. $\lambda = \frac{v}{f} = \frac{3 \times 10^8 \text{ m/sec}}{3 \text{ GHz}} = \underline{\underline{10 \text{ cm}}}\ (0.1 \text{ m})$

Notice how the wavelength becomes smaller as frequency increases. At 3 GHz, the wavelength is only 10 cm (about 3.94″). At such a high frequency, a quarter-wavelength stub "antenna" need only be about 1″!

Connectors for coaxial cable With ladder line and twisted pair, the end user can connect the cable to a device easily enough by using binding posts. Because of its concentric construction, connecting coaxial cable isn't quite so simple. In fact, simply stripping the end off a coax cable and connecting it to a circuit degrades the connection by causing an *impedance mismatch.* Special connectors are used for connecting coaxial cables, as shown in Figure 11–3.

You're likely to see all of the types of connectors shown in Figure 11–3 (plus others) at one time or another in your studies in electronics. You're probably most familiar with the *BNC* connector; it's used on almost all oscilloscopes. The BNC connector comes in both 75 and 50 Ω versions (they're nearly impossible to tell apart by sight) and is good to at least 200 MHz. The BNC connector is not weathertight. It also can't be mated with the larger coaxial cable sizes.

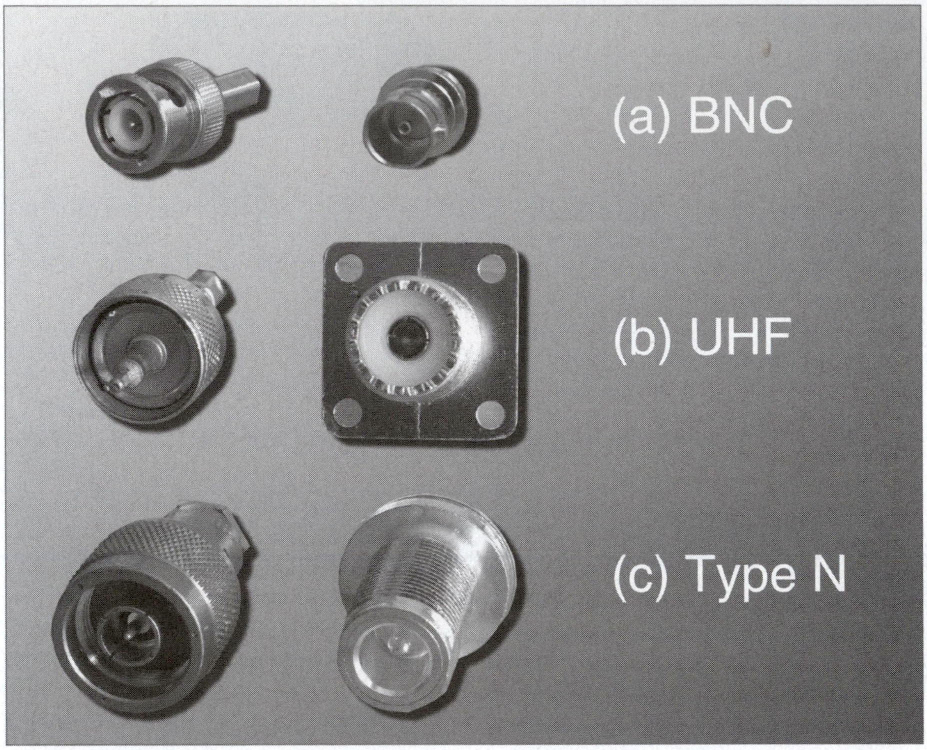

Figure 11–3 Coaxial cable connectors

The *PL-259,* or *UHF* connector, is another very common coaxial cable plug. These connectors are actually not very effective at UHF frequencies; they are a poor choice for anything operating over 100 MHz or so. Above 100 MHz, a UHF connector causes a slight impedance mismatch, which can degrade the performance of the equipment it is connected to. These connectors are also not weathertight, so if they're used outdoors, they must be sealed.

The *N* connector is the most expensive of all the connectors in Figure 11–3, but for UHF work, it is worth it. It is a true constant-impedance connector, which means that it does not introduce an impedance mismatch into the circuit it is connected to. N connectors are useful well into the microwave region (GHz frequencies); there's no real frequency limit (except for that of the attached coaxial cable). In addition, the N connector has an internal rubber gasket to seal out moisture, so it's very suitable for outdoor work. Most technicians generally use a sealant over the finished connection even though the connector has a gasket. It's very inconvenient and expensive to fix a "problem" connector on the top of a 500-foot tower that failed due to moisture entry!

Section Checkpoint

11–1 What is the purpose of a transmission line?

11–2 When signal is lost on a transmission line, what two forms of energy might it be converted to?

11–3 Define *balanced line*.

11–4 Which transmission line type has the highest useful frequency?

11–5 Which type of line would be best for operation at 100 MHz near metal objects?

11–6 Why does twisted pair have poor high-frequency response?

11–7 What connector type is usually found on oscilloscopes and other bench equipment?

11–8 What type of coax connector would be best for operation in a 450 MHz repeater system?

11–9 Why can't waveguide be used at low frequencies (30 MHz, for example)?

11–10 As the frequency of a signal is increased, what happens to its wavelength?

11–2 ELECTRICAL CHARACTERISTICS OF TRANSMISSION LINES

A transmission line may look like plain wire and insulation, but there's a lot more hidden inside when we take a closer look. Even a single wire has three basic properties: *resistance* to current flow (controlled by its cross-sectional area and length); *inductance,* controlled by its length (and any nearby magnetic objects); and *capacitance,* which the wire shares with ground.

The values of resistance, inductance, and capacitance are normally small enough to ignore for dc and low-frequency work. At radio frequencies, these effects can no longer be ignored. A transmission line can be considered to be made up of an *infinite* number of *infinitely small* sections, like Figure 11–4.

The inductances L_1 and L_2 are the inductances per *unit length* of line; the capacitance C_1 is the capacitance per unit length; and the resistances R_1 and R_2 are the resistances per unit length for each wire. There is also a leakage conductance (recall that $G = 1/R$) between the wires that represents the fact that the dielectric is not a perfect insulator. A small amount of energy is lost in the dielectric and is dissipated as heat.

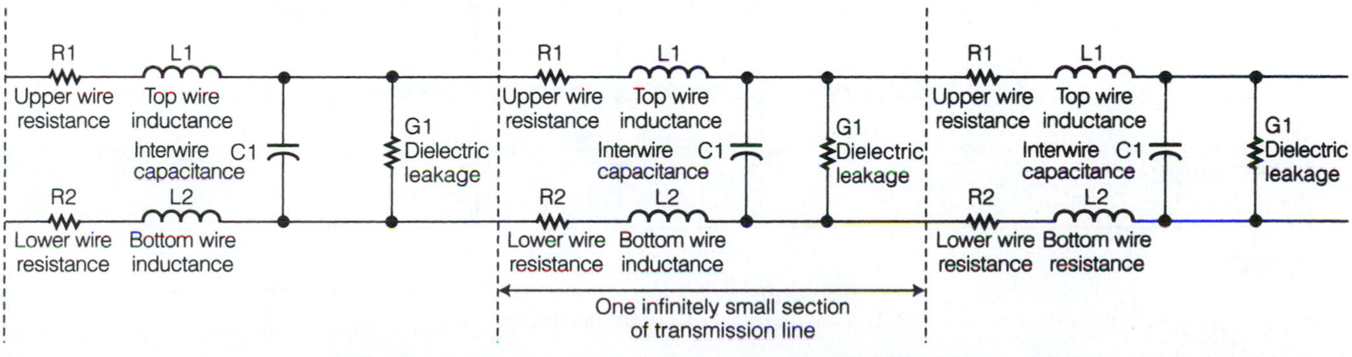

Figure 11–4 Model for a real-world transmission line

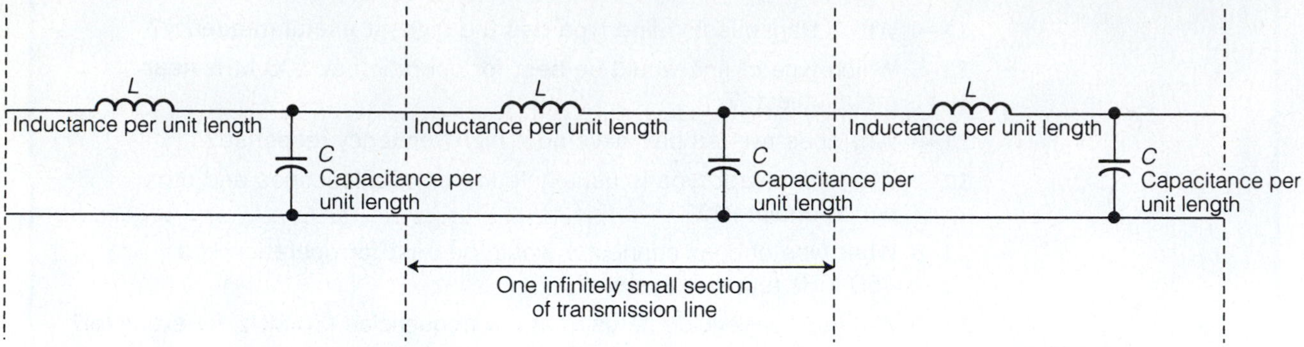

Figure 11–5 Model for a lossless transmission line

The model of Figure 11–4 is too complex for most purposes, and most technicians prefer to simplify it by lumping the inductances of the two wires together, as in Figure 11–5. Also notice that the resistors are gone from Figure 11–5; this means that the line is *lossless,* which never happens in real life. However, for short runs of transmission line, it is quite helpful to ignore loss when trying to understand how things work.

A lossless section of transmission line contains an inductance, *L* and a capacitance, *C*. If you're having trouble with the "infinitely small" concept, consider slicing a pie. You can divide a pie into 6 slices—or 12—or 24. You still have the same amount of pie. Each slice gets thinner and thinner as you divide the pie further. Eventually, you might end up with an *infinite* number of slices. Each will be "infinitely thin," which means *as thin as possible without ceasing to exist*. A transmission line is thought of in this way; each of the *L-C* sections in the figure represents an infinitely small section of the line.

To the technician, the *L* and *C* are important because they control the characteristics of the transmission line. Also, the *L* and *C* components help to explain how energy travels down the line. Let's try an experiment. Let's hook the line of Figure 11–5 to a battery, switch, and some meters and see what happens. Figure 11–6 shows the setup. In the figure, we've connected a dc ammeter in series between the dc voltage source and the transmission line and load. The test points TP_1 through TP_4 are located at consecutive points on the line. We will monitor the voltage in these places.

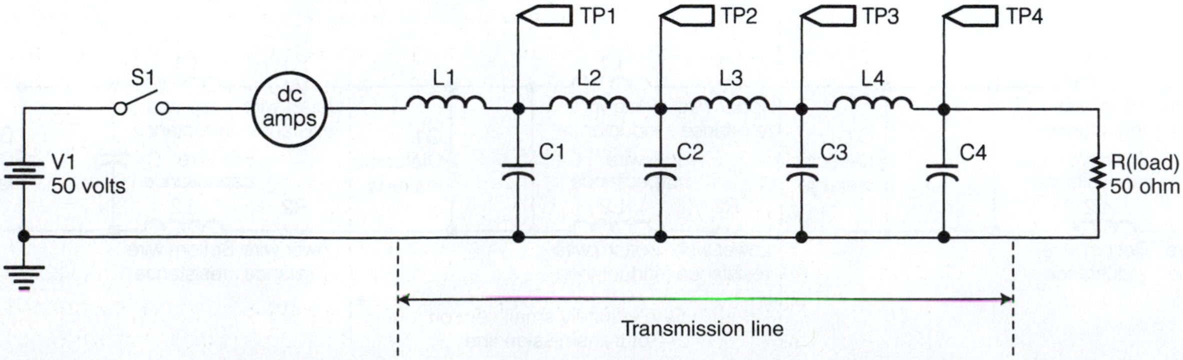

Figure 11–6 An experimental setup

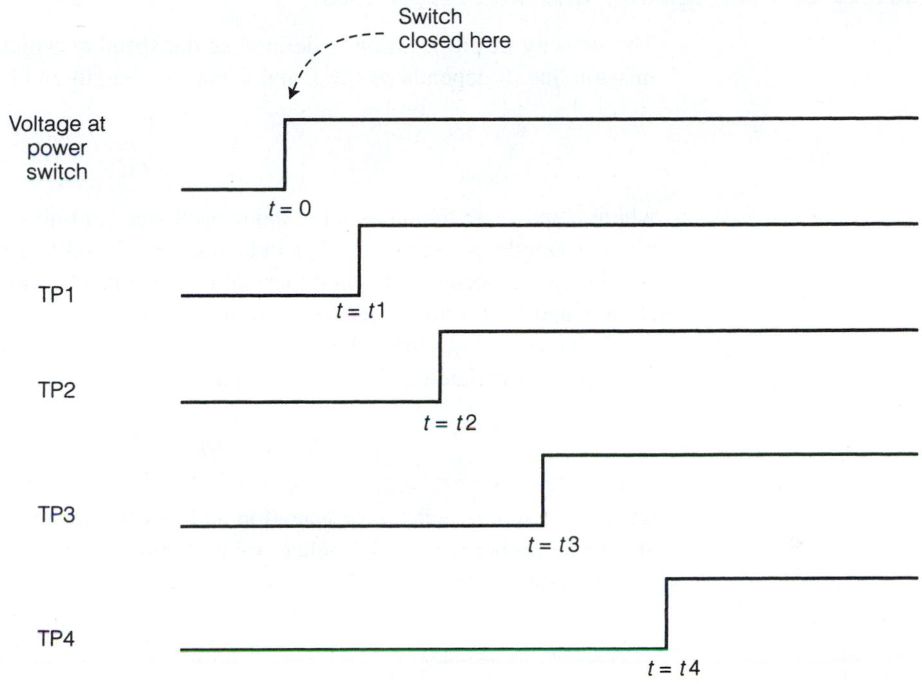

Figure 11–7 The voltages at the test points

Imagine that the switch S_1 is open and has been open for a very long time. There would be no voltage in any of the capacitors, and no voltage at the load TP_4. Now close S_1. What happens? Are you thinking of inductor and capacitor charging? Good, because that's what begins to take place. With the switch closed, capacitor C_1 can begin to charge through inductor L_1. As capacitor C_1 begins to develop some voltage (it's being charged), capacitor C_2 begins to charge through L_2. As C_2 begins to charge, capacitor C_3 now begins charging through L_3.

This looks complicated, but it isn't. Think of a line of dominoes. When the first one is tipped, it starts a chain reaction; the line of motion passes rapidly among all the dominoes in the row. This is very similar to what our little section of transmission line is doing. Each *L-C* section charges the next *L-C* section when it gets the voltage. The voltage *eventually* reaches the load and appears across it. Figure 11–7 shows the voltages at the test points along the way.

As you can see, the voltage simply takes time to reach each of the test points. It moves toward the load as a *traveling wave* with a square edge. Since they are infinitely small, the *L* and *C* components in the line do not change the shape of the edge; it remains perfectly sharp all the way down the line. Finally, the edge reaches the load, and voltage appears on the load. *The main idea here is that the voltage does not reach the load instantly*. It takes time for it to get there. *All* of the battery's 50 volts reaches the load. Because there is no resistance, there is no loss of voltage.

Curiously enough, the *ammeter* we placed in the circuit shows a *steady current* the entire time the experiment is running. As soon as the switch is closed, the ammeter needle deflects and holds steady. Even before the load gets the voltage ($t = t4$), the ammeter is showing current draw. How can this be? The answer is that the ammeter shows that current is required for charging the *L-C* sections along the line. Once all the *L-C* sections have charged, then the ammeter shows the current in the load.

Velocity of Propagation and Velocity Factor

The velocity of propagation is defined as the speed at which waves move down a transmission line. It depends on the L and C per unit length and is calculated as

$$v_p = \frac{1}{\sqrt{LC}} \tag{11–2}$$

where L and C are the inductance and capacitance per unit length. The units of the answer are unit-lengths per second. If L is in henries/meter and C is in farads/meter then the units will be meters/second. This is important to remember, because some manufacturers' data show L and C in terms of feet or some other units.

The *velocity factor* is the ratio of conduction speed to the speed of light in free space. It is often abbreviated as *VF*, and is calculated by:

$$VF = \frac{v_p}{c} \tag{11–3}$$

where v_p is the velocity of propagation and c is the speed of light. The velocity factor is always a number less than 1, since waves can't exceed the speed of light within a transmission line.

EXAMPLE 11–2

What is the velocity of propagation and velocity factor of RG-58U coaxial cable having $L = 250$ nH/m and $c = 100$ pF/m?

Solution

The velocity of propagation can be calculated as

$$v_p = \frac{1}{\sqrt{LC}} = \frac{1}{\sqrt{(250 \text{ nH/m})(100 \text{ pF/m})}} = \underline{\underline{2 \times 10^8 \text{ m/s}}}$$

The velocity factor can be computed:

$$VF = \frac{v_p}{c} = \frac{2 \times 10^8 \text{ m/sec}}{3 \times 10^8 \text{ m/sec}} = \underline{\underline{0.66666}} = 66.67\%$$

Therefore, waves travel in this cable at 66.67% of the speed of light.

Transmission lines are sometimes used to delay or slow a signal down on purpose. The time delay produced is dependent upon the length of the line and the L-C product.

EXAMPLE 11–3

How long will it take a pulse to travel through a 50′ length of RG58U coaxial cable ($VF = 0.66667$, $v_p = 2 \times 10^8$ m/sec)?

Solution

To solve this problem, we use the old standby formula:

$$D = RT$$

where R is *rate,* which means the same thing as *velocity*. Here, we must solve for *time,* so the formula is manipulated as

$$T = \frac{D}{R}$$

There's one minor adjustment we have to make. The length is given in *feet* rather than meters. Since we know the velocity in meters per second, it's probably easier to convert the 50′ value into meters first, then substitute in the formula.

$$L_{\text{meters}} = L_{\text{feet}} \times \frac{1\text{ m}}{3.28\text{ ft}} = 50\text{ ft} \times \frac{1\text{ m}}{3.28\text{ ft}} = 15.24\text{ m}$$

We can now find the time:

$$T = \frac{D}{R} = \frac{15.24\text{ m}}{2 \times 10^8\text{ m/s}} = \underline{\underline{76.2\text{ ns}}}$$

It will take 76.2 ns for the pulse to move down the line. (We hope you don't mind the wait.)

Characteristic Impedance

The *characteristic impedance* of a *transmission line* is the ratio of voltage to current at any point. Don't let this confuse you; this is *not* a resistance. The characteristic impedance of a transmission line is controlled mainly by the tiny *L-C* sections that make it. Ohm's law is obeyed, so we get

$$Z_0 = \frac{V(x)}{I(x)} \qquad \textbf{(11–4)}$$

where $V(x)$ is the voltage at some point x, and $I(x)$ is the current at the same point. The symbol Z_0 is used for the characteristic impedance, which is in ohms. This equation *defines* what characteristic impedance is, and it tells us something about what characteristic impedance *means* (a ratio of voltage to current at any point); however, it doesn't tell us what *controls* characteristic impedance.

Since the lossless transmission line is just made up of *L-C* sections, you might guess that the equation for calculating Z_0 must have an L and a C in it. You're right! It looks like this:

$$Z_0 = \sqrt{\frac{L}{C}} \qquad \textbf{(11–5)}$$

where L is the inductance per unit length, and C is the capacitance per unit length. Since we're dividing these two, it doesn't matter what the units of length are—they will cancel, leaving us with units of *ohms*. Let's try a couple of examples.

EXAMPLE 11–4

What is the characteristic impedance of RG58U coaxial line if it has a capacitance of 30.5 pF/ft and an inductance of 76.25 nH/ft?

Solution

The units don't matter for the L and C, as long as they're the same. From equation 11–5, we get

$$Z_0 = \sqrt{\frac{L}{C}} = \sqrt{\frac{76.25\text{ nH/ft}}{30.5\text{ pF/ft}}} = \underline{\underline{50\ \Omega}}$$

We would say that RG58U is "50 Ω coaxial cable." It is meant to be connected to generators and loads that have a resistance of 50 Ω.

The characteristic impedance also relates voltage and current at *any* point on the transmission line.

EXAMPLE 11–5

Suppose that a certain transmission line has a characteristic impedance of 50 Ω. If there's an ac voltage on that line of 50 volts at a distance x of 72″ from the generator, what is the current flowing at that same point?

Solution

Ohm's law can be applied, and we get

$$Z_0 = \frac{V(x)}{I(x)}$$

so

$$I(x) = \frac{V(x)}{Z_0} = \frac{50\ \text{V}}{50\ \Omega} = \underline{\underline{1\ \text{A}}}$$

A current of 1 ampere must be flowing at that point on the line.

Surge Impedance

The characteristic impedance of a transmission line is sometimes called the *surge impedance.* Here's why. Suppose that the transmission line of Figure 11–6 is infinitely long. If the line is infinitely long, the process of *charging* the line also takes an infinite amount of time—the energy never gets to the load!

How much current will flow to charge the line? Ohm's law and the characteristic impedance of the line give the answer to that. When the switch is closed in Figure 11–6, the transmission line is connected to the battery. It immediately draws a *constant* current from the battery. The current remains constant because there's always another *L-C* section to charge (remember, the line is infinitely long). Since the battery is 50 volts, and the line's characteristic impedance is 50 Ω, a current of 50 volts/50 Ω, or *1 ampere* will flow from the battery.

When a line is infinitely long, it looks like a pure "resistance" that is equal to Z_0.

Further Defining Characteristic Impedance

Let's summarize: *The characteristic impedance,* or Z_0, of a transmission line can be defined in several ways. Don't let this bother you; these ways are just based on different methods of observing the line's action.

- Z_0 is the ratio of voltage to current at any point on the line.
- Z_0 is the input impedance (resistance) that appears at the end of an infinitely long section of line.

- Z_0 is the input impedance of a line of *any* length that has a load resistance *equal* to Z_0 at the end. (We call the load a *termination* for the line.) A transmission line should always be terminated in a load equal to Z_0.
- Some data books will refer to Z_0 as the "surge impedance" of the line. That's just another way of saying characteristic impedance.

Physical Factors Controlling Velocity of Propagation and Characteristic Impedance

The physical construction of a transmission line determines both the speed of wave propagation and the characteristic impedance of the line. The following equations are rarely, if ever, used directly by technicians. However, they do show how the physical details of a transmission line's construction define its electrical properties. In the equations, ϵ_r is the *relative permittivity* (dielectric constant) of the insulating material, or dielectric (see Table 11–1), and *s*, *D*, and *d* are the dimensions from Figure 11–8.

For either type of transmission line, the *velocity factor* (*VF*) can be closely approximated by

$$VF \approx \frac{1}{\sqrt{\epsilon_r}} \tag{11–6}$$

Notice that as ϵ_r gets larger, the velocity factor gets *smaller*. For transmission lines with an air dielectric, ϵ_r is 1 and the velocity factor for that line will be close to 1. Placing a material with a high dielectric constant in between the conductors (such as polystyrene, a plastic) slows propagation even further.

In certain RF hybrid integrated circuits, the substrate (the flat insulating surface on which the components and conductors are placed) is formed from beryllium oxide, a white ceramic material with a very high dielectric constant. This makes the transmission lines formed on the material have a very slow velocity of propagation and allows them to be formed in a very small space.

For parallel-wire line (such as ladder line), the characteristic impedance is found by

$$Z_0 \approx \frac{120}{\sqrt{\epsilon_r}} \ln\left(\frac{2s}{d}\right) \tag{11–7}$$

Table 11–1 The dielectric constant for common materials

Material	Relative permittivity, ϵ_r
Air	1.0
Styrofoam	1.03
Teflon	2.25
Vinyl	2.3
Polyethylene	2.3
Wood	2–4
Polystyrene	2.7
Water	80–81
Beryllium oxide, BeO	6.7

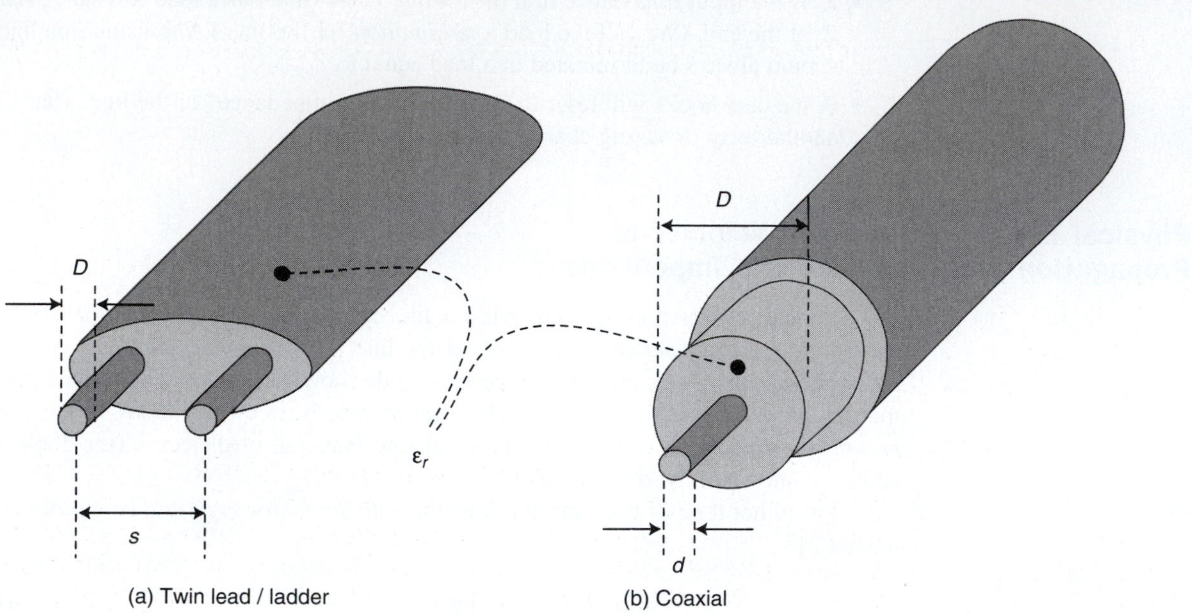

Figure 11–8 Physical measurements of transmission line dimensions

For coaxial line, the characteristic impedance is calculated using

$$Z_0 \approx \frac{60}{\sqrt{\epsilon_r}} \ln\left(\frac{D}{d}\right) \tag{11–8}$$

Equations 11–7 and 11–8 tell us that the *closer* the wires are spaced in a transmission line, the *lower* the characteristic impedance becomes. This explains why coaxial cable tends to have low values of characteristic impedance (120 Ω or less), while flat (twin lead) cable has higher impedance—the flat cable has a larger space between the conductors. The dielectric also affects characteristic impedance: the higher the dielectric constant, the *lower* Z_0 becomes. The dielectric constant, ϵ_r, of various materials is given in Table 11–1.

EXAMPLE 11–6

What is the velocity factor (VF) and characteristic impedance (Z_0) of a coaxial transmission line with a vinyl dielectric and electrode dimensions $D = 0.27''$ and $d = 0.075''$?

Solution

The dielectric constant ϵ_r for vinyl is 2.3, and can be substituted into equation 11–6:

$$VF \approx \frac{1}{\sqrt{\epsilon_r}} \approx \frac{1}{\sqrt{2.3}} \approx \underline{\underline{0.659}}$$

RF energy travels through this cable at about 66% the speed of light.

The characteristic impedance is found using equation 11–8:

$$Z_0 \approx \frac{60}{\sqrt{\epsilon_r}} \ln\left(\frac{D}{d}\right) \approx \frac{60}{\sqrt{2.3}} \ln\left(\frac{0.27''}{0.075''}\right) \approx \underline{\underline{50.6\ \Omega}}$$

This is pretty close to 50 Ω, a standard value.

Table 11–2 Specific loss in dB/100 ft for various transmission lines

Cable type	dB Loss @ 50 MHz	dB Loss @ 100 MHz	dB Loss @ 400 MHz
RG58U	3	4.5	9.5
RG59U	2.4	3.4	7.1
RG62A	1.9	2.7	5.4
9913	1.0	1.4	2.8
300 Ω twin lead	0.8	1.1	2.4
3/4″ 50 Ω hardline	0.4	0.64	1.45
RG174 (mini 50 Ω)	8.0	10.9	20.4

Physical Factors Controlling Transmission Line Losses

In any transmission line, there are two main sources of energy loss, *ohmic resistance* and *dielectric heating*.

Ohmic resistance (sometimes called I^2R loss) refers to the resistance of the wires that make up the line. In general, the larger the conductors, the lower I^2R will be. At radio frequencies, most of the current tends to flow on the outside surface of conductors (rather than being uniformly distributed throughout the cross-sectional area). This is known as the *skin effect,* and it becomes more severe as frequency is increased. At very high frequencies, the *circumference* of the conducting wire becomes much more important than its cross-sectional area, due to the skin effect. This means that the effective resistance of a wire increases as frequency increases.

When energy is lost in the dielectric, it is converted to heat. The best dielectric, as far as loss is concerned, is air. Other materials, such as plastics, heat more readily in the presence of the RF energy field inside a transmission line and lose more energy. Dielectric losses increase with frequency and are also somewhat dependent on the *voltage* being applied.

Both of these effects (ohmic loss and dielectric loss) *increase* with frequency; therefore, we can say that transmission line losses, in general, increase with frequency. Manufacturers rate each type of transmission line in *dB loss per unit length* at specific frequencies. The loss per unit length is called *specific loss*. Technicians and engineers use this data when choosing transmission lines for different applications. Table 11–2 shows the *specific loss* for several types of coaxial line at three frequencies.

EXAMPLE 11–7

What power will be delivered to an antenna from a 100 MHz, 100 watt transmitter if it supplies the antenna through a 200 feet length of Belden type 9913 transmission line?

Solution

From Table 11–2, the *specific loss* of '9913 is 1.4 dB for every 100 feet at 100 MHz. Therefore, the total power loss is

$$dB_{loss} = 200\text{ ft} \times \frac{1.4\text{ dB}}{100\text{ ft}} = \underline{2.8\text{ dB}}$$

This doesn't give us the power at the load; it just says that the load power is *2.8 dB less* than the transmitter power. A 2.8 dB *loss* is the same as a − 2.8 dB *gain,* and therefore the dB power gain formula can be used:

$$dB = 10\log\left(\frac{P_O}{P_i}\right) = 10\log G_p$$

where P_O is the power at the antenna, P_i is the power at the transmitter, and G_p is the power gain. By rearranging this equation, and substituting a *gain* of −2.8 dB (remember that a negative dB gain number is the same thing as a positive dB loss), we get

$$P_O = P_i \times 10^{(\text{dB}/10)} = 100\text{ W} \times 10^{(-2.8/10)} = \underline{\underline{52.5\text{ W}}}$$

But wait, we're technicians! Can't we get an answer without going through all this math? Sure, because we know a thing or two about dB. Every 3 dB loss *halves* the power. A 2.8 dB loss is slightly less than a 3 dB loss, so we should expect approximately *half* the power (50 W) to make it to the antenna.

Section Checkpoint

11–11 What four properties does a lossy transmission have in each unit length?

11–12 What are the two components that model a section of lossless transmission line?

11–13 List three ways that energy can be lost from a transmission line.

11–14 Explain how the *L-C* sections get charged in Figure 11–6.

11–15 What controls the velocity of propagation of a transmission line?

11–16 What is the difference between velocity of propagation and velocity factor?

11–17 Define *characteristic impedance*. Why is it important?

11–18 What is the *surge impedance* of a transmission line?

11–19 What value of resistance *should* terminate a transmission line?

11–20 Physically, what construction material controls the velocity factor of a transmission line?

11–21 Define *specific loss* as it refers to transmission lines. What are its units?

11–3 IMPEDANCE MISMATCHES AND STANDING WAVES

A transmission line should normally be connected to a load resistance equal to its characteristic impedance, Z_0. When this is done, we say that the line has been *terminated* properly and that the load is matched to the line. Under this condition, energy flows from the generator, down the line, and is totally absorbed by the load. This is normally how we want it to work!

A proper termination is always purely resistive; it has no capacitive or inductive reactance component.

An *impedance mismatch* is any other condition. *Any time the load resistance is not equal to the characteristic impedance of the transmission line, a mismatched condition exists.*

If the mismatch is small, the system may continue to operate normally with no problem; if it is a bad mismatch, the transmitter final power amplifier stage can be damaged, or other operational problems can appear.

> If you ever repair a transmitter with a "blown" final power amplifier stage, you should make sure that the transmission line and antenna systems are operating correctly. They may have caused the transmitter's death!

Traveling Waves

When energy moves down a transmission line, it does so as *traveling waves*. The action is similar to what happens when you drop a pebble in the still water of a pond. The waves move uniformly outward from the center, and if there's nothing in the way, they'll continue outward forever. The primary difference between the pond water and our transmission line is that the movement of the waves is limited to one dimension (forward or backward) on a transmission line.

Reflections

Returning to the pond, suppose again that the water is perfectly still, and you again drop in a stone. This time, there is a rock jutting out of the water a few feet away from the center of origin. What happens when the waves hit the rock? Right, they are *reflected* back in the direction from which they came. The patterns formed in the water are the result of *two* traveling waves: One is moving away from the point of origin (where you dropped the rock), and the other is moving *toward* the point of origin, having been reflected from the rock.

The original wave moving from the center outward is called the *incident wave,* and the returning wave is called the *reflected wave*. Why does a rock in the water cause a reflected wave? One way of explaining it is to think of the mechanical "impedance" of the water (not the electrical kind). When a wave is moving through water, energy is being transferred between water molecules in a sort of a bucket-brigade fashion. When mechanical energy moves from water to water, it sees the same "impedance" and therefore no change occurs in the direction of the energy flow.

When the wave of water meets rock, there is a great difference in impedance. Water moves easily, rock does not; we say that there has been an *impedance mismatch*. The motion of the water is largely reflected backward in the direction from which it came. The exact same thing happens in transmission lines, as shown in Figure 11–9.

In the figure, the line is properly terminated; the load resistance exactly equals the line's characteristic impedance. Under this condition, the wave moves from the generator (on the left) toward the load and is *totally absorbed* by the load. There is no reflection. *The average voltage is equal at all points on the line.*

When the load is mismatched, things get more interesting. Figure 11–10 shows what happens when 50% of the energy is reflected at the load. (Just how 50% of the energy can be reflected we will discuss shortly.) As you can see, there is a wave traveling in the *opposite* direction on the line of Figure 11–10. In fact, there are two waves being carried by the transmission line: the *incident* and the *reflected*.

The total voltage at any point on the line is the sum of the incident and reflected voltages. If we examine the total voltage of the line at regular time intervals, something interesting appears. An interference pattern forms between the two waves (outgoing and

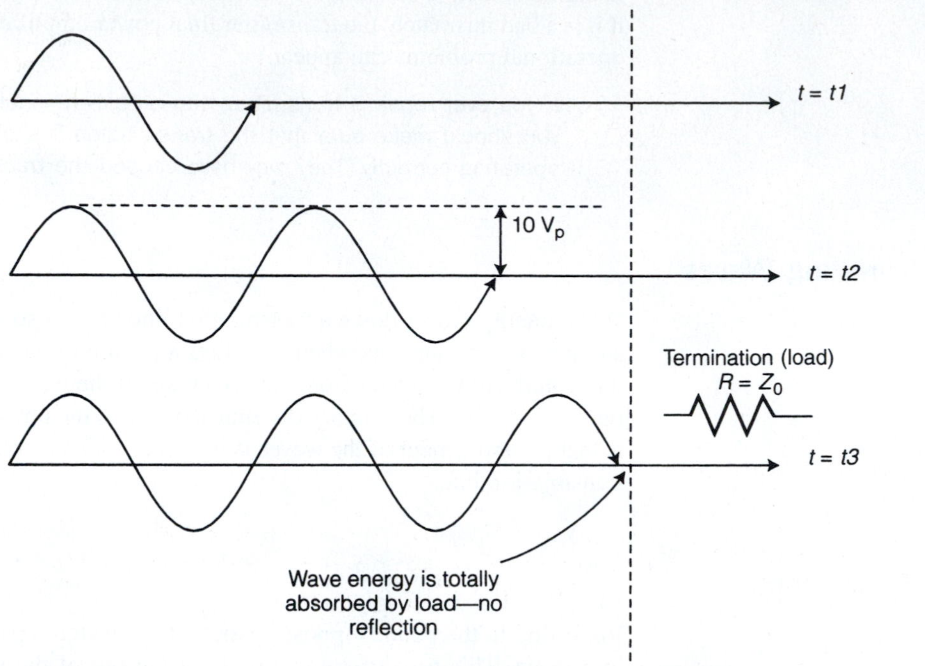

Figure 11–9 Traveling waves on a transmission line with a matched load

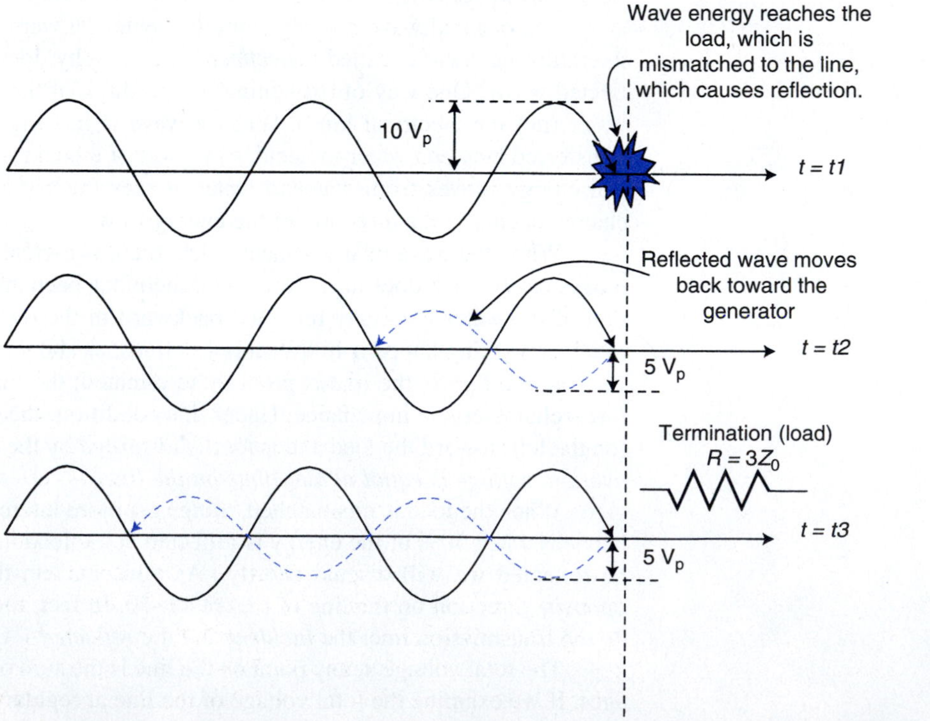

Figure 11–10 Wave conditions when 50% of the voltage is reflected

Figure 11–11
Development of standing waves

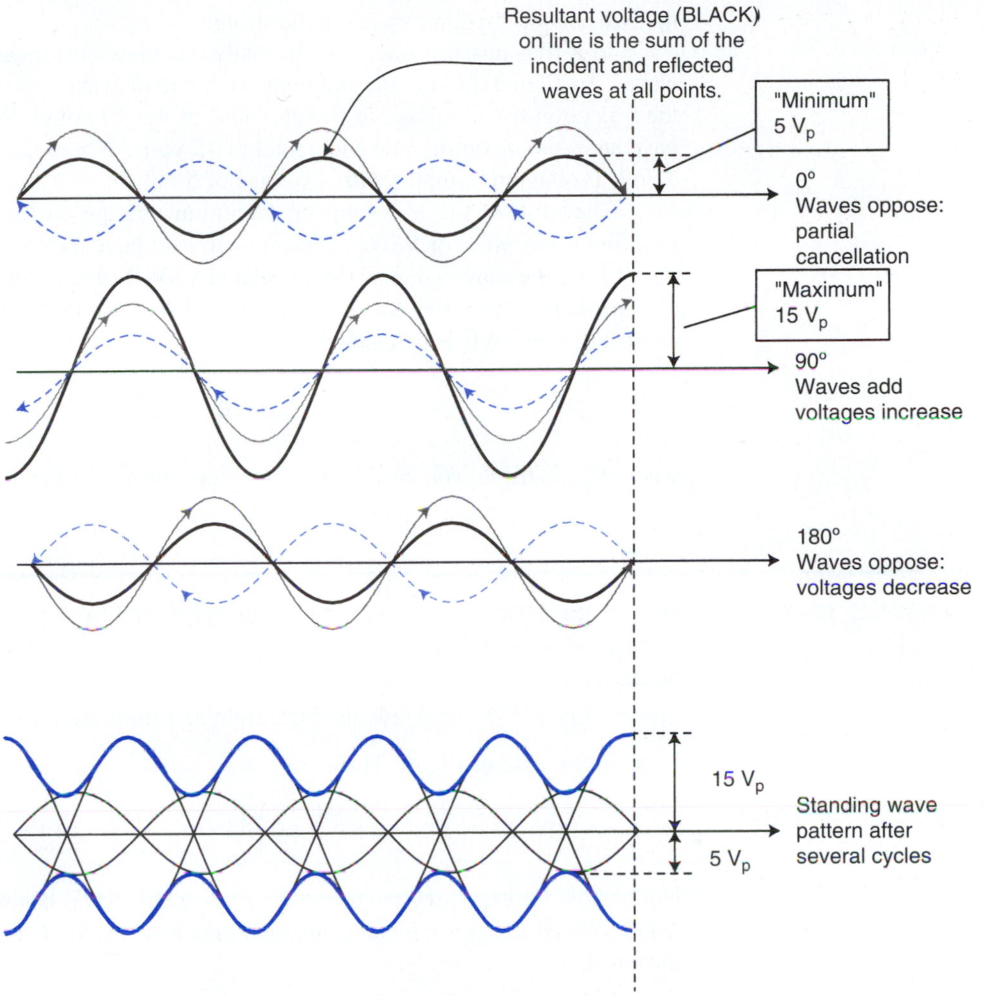

incoming). The pattern appears to stand still on the line! This is called a *standing wave* pattern for this reason. Look at Figure 11–11 to see how it develops.

Let's compare Figures 11–9 and 11–11. In Figure 11–9, the load is matched to the transmission line, and no reflections occur. More importantly, the total average voltage at all points on the line of Figure 11–9 is *constant*. For this reason, we sometimes call a transmission line with a matched load a *flat* line; all of its voltages are equal. This is hardly the case in Figure 11–11. The outgoing (incident) and incoming (reflected) waves are meeting each other, causing their voltages to *add* at some points and *subtract* at others. The result is *standing waves,* which is normally not what we want!

You can see standing waves in other ways, too. Take a close look at a musical string while it's vibrating and you'll see that there appear to be points where there's hardly any motion at all (called *nodes*) and others where the string seems to move back and forth quite a bit (called *antinodes*). You can explain the vibration of a string by traveling waves: When the string is plucked (or struck), a wave of energy moves down it until it reaches the end, where the string is tightly fastened. The wave reflects backward and moves down the string in the *opposite* direction until it meets the *other* end, where the cycle repeats. It's very much like a ping-pong ball bouncing back and forth. The standing waves are formed by the two sets of traveling waves in the string.

In a transmission line, we can easily calculate and measure the effect of standing waves. In Figure 11–11, the outgoing or incident wave has a voltage of 10 volts, and the reflection has a voltage of 5 volts (50% of the 10 volts). Where the voltages add, we have an *antinode* on the line and a total of 15 volts. Where the voltages subtract, we have a *node* (voltage minimum) with a voltage of 5 volts.

The ratio of the maximum to minimum voltage on the line is called the *voltage standing wave ratio,* or *VSWR*. (Yes, we can also have a *current* standing wave ratio, and it will have the same value as the calculated VSWR, due to Ohm's law.) Technicians usually just call it the SWR, knowing that the VSWR and ISWR will give the same numerical result. The SWR is calculated by

$$\text{SWR} = \frac{V_{\text{max}}}{V_{\text{min}}} \qquad \textbf{(11–9)}$$

where V_{max} is the *maximum* voltage on the line, and V_{min} is the *minimum* voltage on the line.

EXAMPLE 11–8

What is the *SWR* on the lines of (a) Figure 11–9 and (b) Figure 11–11?

Solution

To calculate SWR, we divide the maximum and minimum line voltages:

a. In the case of Figure 11–9,

$$\text{SWR} = \frac{V_{\text{max}}}{V_{\text{min}}} = \frac{10\text{ V}}{10\text{ V}} = \underline{\underline{1.0{:}1}}$$

Notice how we expressed the result as a ratio: 1.0:1. SWR is always expressed in this way. *A 1:1 SWR results when a line is properly matched to a load.* The voltage is constant along the entire line.

b. In the case of Figure 11–11,

$$\text{SWR} = \frac{V_{\text{max}}}{V_{\text{min}}} = \frac{15\text{ V}}{5\text{ V}} = \underline{\underline{3.0{:}1}}$$

A 3:1 SWR represents a fairly severe impedance mismatch and could cause damage to a transmitter. A reading this high indicates trouble with the transmission line, or the load!

Calculating the SWR

There are several ways of calculating or estimating SWR. Use the method that works best for you. Exact calculations of SWR are not very useful to the technician, as he or she is likely to directly measure it, as we'll soon see. However, it's very important to understand the "why" of what we're measuring!

In order to calculate the SWR, the minimum and maximum line voltages must be known. In order to do this, we must know the percentage of voltage reflection that occurs at the load.

The percentage of voltage reflection is called the *reflection coefficient* and is symbolized by an uppercase Greek letter *gamma* (Γ):

$$\Gamma = \frac{Z_R - Z_0}{Z_R + Z_0} \tag{11–10}$$

where Z_R is the load impedance (or resistance) and Z_0 is the characteristic impedance of the transmission line. Note that Γ can be a *complex* number (containing a real and imaginary part) if the load has any inductive or capacitive reactance. In real life, some loads do; we'll avoid that type of calculation here.

The magnitude or strength of the reflected voltage is simply the product of the incident voltage and the reflection coefficient:

$$V_r = \Gamma V_i \tag{11–11}$$

where V_r is the reflected voltage, and V_i is the incident voltage. The *maximum* line voltage is the sum of the incident and reflected voltages:

$$V_{max} = V_i + V_r = V_i + |\Gamma| V_i = V_i(1 + |\Gamma|) \tag{11–12}$$

The vertical bars (| |) symbol tells us to take the *absolute value* of Γ in equation 11–11. Likewise, the minimum voltage can be calculated as

$$V_{min} = V_i - V_r = V_i - |\Gamma| V_i = V_i(1 - |\Gamma|) \tag{11–13}$$

By combining equations 11–12 and 11–13, we get a formula for SWR:

$$\text{SWR} = \frac{V_{max}}{V_{min}} = \frac{V_i(1 + |\Gamma|)}{V_i(1 - |\Gamma|)} = \frac{(1 + |\Gamma|)}{(1 - |\Gamma|)} \tag{11–14}$$

Notice how V_i has cancelled in equation 11–14. This tells us that we only need to know the reflection coefficient (Γ) in order to find SWR.

EXAMPLE 11–9

A certain transmitter is providing an incident voltage of 50 volts to a 50 Ω transmission line. The line is terminated in a 75 Ω load. Calculate:

a. the reflection coefficient Γ

b. the reflected voltage

c. the minimum and maximum voltages on the line

d. the SWR

Solution

a. The reflection coefficient is calculated according to equation 11–10:

$$\Gamma = \frac{Z_R - Z_0}{Z_R + Z_0} = \frac{75\ \Omega - 50\ \Omega}{75\ \Omega + 50\ \Omega} = \underline{\underline{0.2}}$$

This means that 20% of the voltage will be reflected from the load.

b. The reflected voltage will be 20% of the original 50 volt incident value:

$$V_r = \Gamma V_i = (0.2)(50\ V) = \underline{\underline{10\ \text{V}}}$$

c. The maximum and minimum line voltages are the sum and difference of the incident and reflected voltages:

$$V_{max} = V_i + V_r = 50\ \text{V} + 10\ \text{V} = \underline{\underline{60\ \text{V}}}$$
$$V_{min} = V_i - V_r = 50\ \text{V} - 10\ \text{V} = \underline{\underline{40\ \text{V}}}$$

d. The SWR can be calculated from the data of part (c) or from equation 11–14. We'll do it both ways:

$$\text{SWR} = \frac{V_{\text{max}}}{V_{\text{min}}} = \frac{60\text{ V}}{40\text{ V}} = \underline{\underline{1.5:1}}$$

$$\text{SWR} = \frac{V_{\text{max}}}{V_{\text{min}}} = \frac{(1 + |\Gamma|)}{(1 - |\Gamma|)} = \frac{(1 + |0.2|)}{(1 - |0.2|)} = \underline{\underline{1.5:1}}$$

An SWR of 1.5:1 indicates a *moderate* load mismatch. Recall that a 1:1 SWR is a perfect match. The value of 1.5:1 is about the *highest* mismatch that a technician would allow without looking for trouble in a system. In other words, for most applications, an SWR of 1.5:1 or less is acceptable.

EXAMPLE 11–10

A 50 Ω transmission line is connected to a 40 Ω load. Calculate the resulting SWR and determine the quality of the impedance match (should the system be troubleshot?).

Solution

Here, we don't know the transmitter voltage—and we don't care! First, let's find the reflection coefficient:

$$\Gamma = \frac{Z_R - Z_0}{Z_R + Z_0} = \frac{40\ \Omega - 50\ \Omega}{40\ \Omega + 50\ \Omega} = \underline{\underline{-0.11111}}$$

That's strange, we got a *negative number*. That means that 11% of the voltage is reflected *180° out-of-phase* at the load. To find the SWR, we use the *absolute value* of the coefficient, as follows:

$$\text{SWR} = \frac{V_{\text{max}}}{V_{\text{min}}} = \frac{(1 + |\Gamma|)}{(1 - |\Gamma|)} = \frac{(1 + |-0.11111|)}{(1 - |-0.11111|)} = \frac{(1 + 0.11111)}{(1 - 0.11111)} = \underline{\underline{1.25:1}}$$

Since the SWR is less than 1.5:1, there is no indication of a severe mismatch. Unless there is something else that is not working right, this system does not need to be troubleshot.

Note: A low SWR does *not* mean that an antenna system is radiating effectively. It only means that the transmission line sees an appropriate load. Remember that a *dummy antenna* also has the proper resistance to terminate a line; it shows a 1:1 SWR, yet radiates no RF energy!

A Shortcut for Finding SWR

When the load is a pure resistance (or close to it), there's a nice shortcut method for finding SWR. The method can be derived by substituting equation 11–10 every place where Γ appears in equation 11–14 (and letting Γ be purely real). The shortcut looks like this:

$$\text{SWR} = \frac{Z_R}{Z_0} \quad (\text{when } Z_R > Z_0) \qquad \textbf{(11–15a)}$$

$$\text{SWR} = \frac{Z_0}{Z_R} \quad (\text{when } Z_0 > Z_R) \qquad \textbf{(11–15b)}$$

Use whichever equation gives a result greater than or equal to 1. This is a lot easier than all those other steps, but it works only when the load is purely resistive (or close to it).

EXAMPLE 11–11 Calculate the SWR on a 50 Ω line when it is terminated in (a) 75 Ω, (b) 40 Ω.

Solution

a. Since 75 is greater than 50, use equation 11–15a:

$$\text{SWR} = \frac{Z_R}{Z_0} = \frac{75\ \Omega}{50\ \Omega} = \underline{\underline{1.5{:}1}}$$

b. Since 40 is less than 50, use equation 11–15b:

$$\text{SWR} = \frac{Z_0}{Z_R} = \frac{50\ \Omega}{40\ \Omega} = \underline{\underline{1.25{:}1}}$$

These results agree with those that we obtained in the previous two examples for the same transmission line and load impedances. This sure is easier!

Measuring the SWR on a Transmission Line

A *directional wattmeter* is an instrument designed to measure standing waves on transmission lines. Sometimes they're referred to as *SWR meters*. A directional wattmeter is connected between a transmitter and a transmission line or between a transmission line and a load, as shown in Figure 11–12. It measures RF power flow between its input and output connectors.

A directional wattmeter can be connected at either the transmitter or antenna end of a feedline. Where the antenna connection is inaccessible or inconvenient (such as when the antenna is located high on a tower), the wattmeter is placed at the transmitter. Placement of the wattmeter at the antenna yields the most accurate reading, because the reflected wave does not have to undergo the transmission line loss to be measured. The SWR always measures *lower* when measured at the transmitter; if the loss in the transmission line is very high, the SWR may show *much* lower at the transmitter end of the line!

To read the SWR using the Welz wattmeter of Figure 11–13, the following procedure is used:

1. The SENSOR SELECT button corresponding to the system under test is selected. The wattmeter in Figure 11–13 has three independent directional RF sensors, each covering a different frequency range (HF, HF-VHF, VHF).
2. The FUNCTION knob is moved to the CAL position, and the transmitter is keyed. The CALIBRATION knob is adjusted for a full-scale reading. The instrument is reading the *incident voltage* and this is being calibrated as 100% scale.
3. With the transmitter still keyed, the FUNCTION knob is moved to the SWR position. The meter now reads the SWR on the bottom meter scale.

Reading SWR from a wattmeter Some directional wattmeters (such as the Bird model 43 of Figure 11–13) do not have a built-in SWR measurement function. Meters of this type have a switch that selects *forward* or *reflected* power; on the Bird, the selection is made by rotating the "slug" until the arrow lines up the direction of RF power flow to be measured.

Figure 11–12
Connecting a directional wattmeter

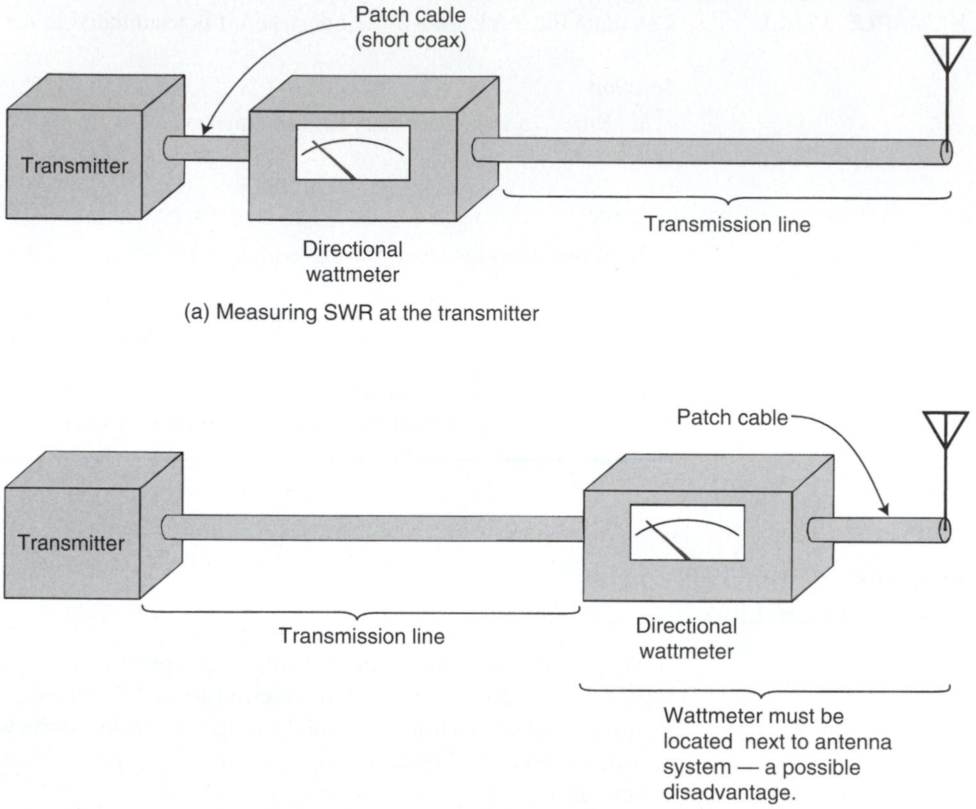

(a) Measuring SWR at the transmitter

(b) Measuring SWR at the antenna

Figure 11–13 Typical directional wattmeters: left, Bird model 43; right, Welz SP-600. The Bird uses removable cartridges ("slugs") to cover a wide frequency range; the Welz uses multiple internal RF couplers for the same purpose

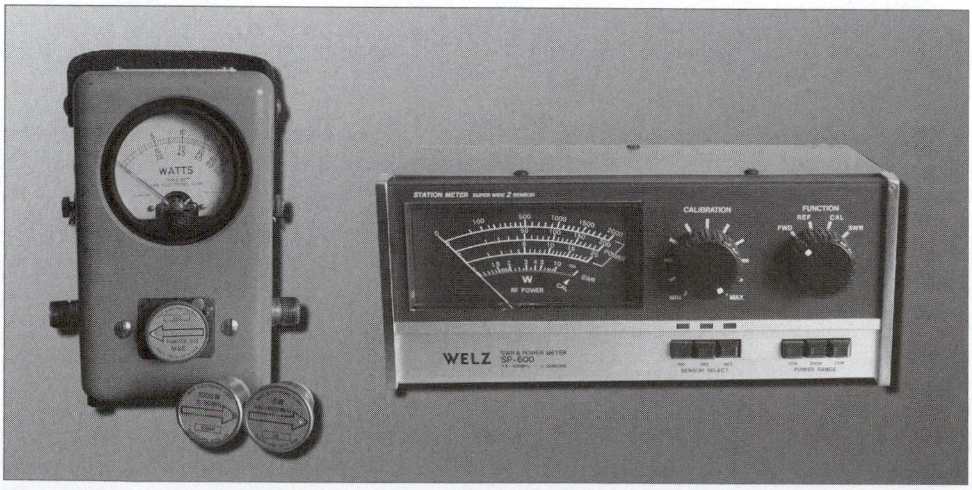

The meter in the figure is set to measure RF flow from right to left. By using Ohm's law, we can calculate SWR from the forward and reflected power readings:

$$\text{SWR} = \frac{\sqrt{P_f} + \sqrt{P_r}}{\sqrt{P_f} - \sqrt{P_r}} \tag{11–16}$$

where P_f is the forward power, and P_r is the reflected power.

EXAMPLE 11–12 The forward power on an HF communications transmitter measured 100 watts, and the reflected power measured 2 watts. What is the SWR on the feedline? Is there any indication of serious trouble?

Solution

Equation 11–16 finds SWR when the forward and reflected powers are known:

$$\text{SWR} = \frac{\sqrt{P_f} + \sqrt{P_r}}{\sqrt{P_f} - \sqrt{P_r}} = \frac{\sqrt{100\ \text{W}} + \sqrt{2\ \text{W}}}{\sqrt{100\ \text{W}} - \sqrt{2\ \text{W}}} = \underline{\underline{1.33:1}}$$

The SWR is less than 1.5:1, which again is a reasonable expectation for most antenna and feedline combinations.

Section Checkpoint

11–22 What impedance properly terminates a transmission line?

11–23 What condition constitutes an impedance mismatch?

11–24 When will reflected waves appear on a transmission line?

11–25 How is the total voltage at any point on a transmission line determined?

11–26 What is a *flat* line?

11–27 What is the SWR of a line with a matched load?

11–28 What is the maximum SWR a technician should consider acceptable?

11–29 What are the two options for connecting a directional wattmeter? Which gives a better indication of SWR at the load?

11–4 IMPEDANCE MATCHING DEVICES

Transmission line sections are often used in various combinations to *match* the impedance of a load to a transmitter or transmission line. There are three basic kinds of impedance mismatch. Suppose that we have a transmitter that wants to see a 50 Ω load, and the following loads are applied:

- 35 Ω
- 50 + *j*20 Ω
- 75 − *j*50 Ω

None of these are proper terminations! In the first case, 35 Ω is simply the wrong *magnitude* of resistance. The 35 Ω needs to be *transformed* or "stepped up" to 50 Ω for proper system performance. In the second case, the *resistance* of the load is correct, but there is an undesired *inductive reactance* of 20 Ω. If we could somehow *cancel* the 20 Ω X_L, we would be left with a proper 50 Ω termination. In the third case, the resistance is wrong, plus there is an undesired *capacitive reactance* of 50 Ω. (We know it is capacitive reactance because of the negative sign on the *j* factor.) In this case, both transformation and reactance cancellation are needed.

An impedance mismatch can consist of either a wrong resistance *magnitude,* an *unwanted reactance,* or *both* conditions. When these conditions exist, a system designer uses an *impedance matching device* to correct the problem. Sometimes an off-the-shelf component can be used to fix an impedance mismatch. Can you think of anything that could fix each of the three cases above . . . ?

(*Electronic Communications* theme music played here.)

Here are some common solutions: RF transformers can be used to step up or down an impedance. A capacitor can be added to cancel an unwanted inductive reactance. An inductor can be used to cancel an undesired capacitive reactance. *A transmission line section* can be made to act like any of these devices (at least at one frequency, and close to that frequency). The application of transmission lines to solve impedance mismatches looks like nothing short of black magic. An impedance matching device built with transmission line looks unassuming. In fact, it just looks like a few pieces of transmission line hooked together in ways that might appear to kill any signal trying to pass through them!

It is *not* the job of a technician to design impedance-matching networks and devices. However, a tech needs to be aware of the basic principles involved in them. Many of these techniques are used in hybrid RF integrated circuits, others appear in antenna designs, and a few are used in specialized applications such as broadcasting. In order to understand how impedance-matching sections work, we need to develop an understanding of three basic concepts: *electrical length, stub characteristics,* and *quarter-wave section characteristics.*

Electrical Length

The lengths of the transmission line sections that will be used to build impedance-matching devices are not usually expressed in inches, meters, or any other physical length units (although they certainly measure that way in the field.) Instead, the *electrical length* of these is given. The *electrical length* is the number of physical wavelengths contained within the length of transmission line. It can be expressed as a fraction of wavelength (λ) or as a number of *degrees*. One wavelength is 360°.

The wavelength of radio waves in a transmission line is always shorter than in free space because waves *slow down* in a transmission line (the velocity factor is less than or equal to 1). We can still use our old equation from Chapter 1 to find wavelength; we just substitute the velocity of propagation within the transmission line for V_p:

$$\lambda = \frac{v_p}{f} \qquad \textbf{(11–17)}$$

The *electrical length* of a line in fractions of a wavelength is its *physical length* divided by *wavelength:*

$$\text{EL} = \frac{L}{\lambda} \qquad \textbf{(11–18a)}$$

where EL is the electrical length, L is the physical length, and λ is the wavelength. Remember that we can also express electrical length in degrees—360° is one wavelength:

$$\text{EL} = \frac{L}{\lambda} \times 360° \qquad \textbf{(11–18b)}$$

Be careful when working with electrical length. Because it depends on wavelength, electrical length depends on frequency. If you change the frequency, electrical length changes. *Calculations with electrical length are valid at only one frequency.*

EXAMPLE 11–13

What is the electrical length (in both wavelength fractions and degrees) of a 1 meter section of RG58U coaxial cable ($VF = 0.66666$) at (a) 30 MHz and (b) 200 MHz.

Solution

a. First, we need to find the velocity of propagation:

$$v_p = c \times VF = 3 \times 10^8 \text{ m/sec} \times 0.66666 = 2 \times 10^8 \text{ m/sec}$$

Knowing this, we can now find the wavelength of the signal within the coax:

$$\lambda = \frac{v_p}{f} = \frac{2 \times 10^8 \text{ m/sec}}{30 \text{ MHz}} = 6.66667 \text{ m}$$

Does this number for wavelength look a little "off?" It should—the wavelength of the same signal in *free space* is 10 meters. The wavelength is *shorter* inside the transmission line because the velocity of propagation has decreased.

The electrical length (wavelength fraction) is

$$EL = \frac{L}{\lambda} = \frac{1 \text{ m}}{6.66667 \text{ m}} = \underline{\underline{0.15\lambda}}$$

Note how this figure is expressed as "0.15λ." This immediately tells us that we are dealing with a *fraction* of a wavelength.

The same figure can be expressed in degrees:

$$EL = \frac{L}{\lambda} = \frac{1 \text{ m}}{6.66667 \text{ m}} \times 360° = \underline{\underline{54°}}$$

b. At 200 MHz, the wavelength is now

$$\lambda = \frac{v_p}{f} = \frac{2 \times 10^8 \text{ m/sec}}{200 \text{ MHz}} = 1 \text{ m}$$

The electrical wavelength fraction is

$$EL = \frac{L}{\lambda} = \frac{1 \text{ m}}{1 \text{ m}} = \underline{\underline{1\lambda}}$$

This is of course the same thing as $\underline{\underline{360°}}$.

Notice again that the electrical length of a section of transmission line depends on two factors: the *frequency* of the signal and the *velocity factor* for that line.

Sometimes we need to know how long a section of line needs to be in order to have a certain electrical length. By reversing the process of Example 11–13 we find this information.

EXAMPLE 11–14

What is the *physical length,* in meters, of a 36° section of transmission line at 100 MHz if its velocity factor is 0.7?

Solution

First, find the wavelength of the 100 MHz signal in the transmission line:

$$\lambda = \frac{v_p}{f} = \frac{3 \times 10^8 \text{ m/sec} \times 0.7}{100 \text{ MHz}} = 2.1 \text{ m}$$

With this information in hand, we can find the fraction that is 36°:

$$L = \frac{36^\circ}{360^\circ} \times 2.1 \text{ m} = \underline{\underline{0.21 \text{ m}}}$$

Where did the value of 360° come from? Remember that one wavelength is 360°; to get back to a *fraction,* we must divide by 360.

Stub Characteristics

A *stub* is a short section of transmission line that has been terminated with an open circuit or a short circuit, as shown in Figure 11–14. A *shorted* stub is usually preferred over the open type because the open connections of an open-circuited stub can tend to radiate RF energy (they act as miniature antennas). Stubs have some very interesting behavior. They can appear as a *short circuit,* an *open circuit,* an *inductive reactance,* or a *capacitive reactance,* depending on their *electrical length.* (Remember that the electrical length of a transmission line depends on both frequency and velocity factor.)

Engineers use a tool called a *Smith chart* to perform calculations with stubs and other transmission line sections. Since we will not be doing design work, we don't need

Figure 11–14
Transmission line stubs

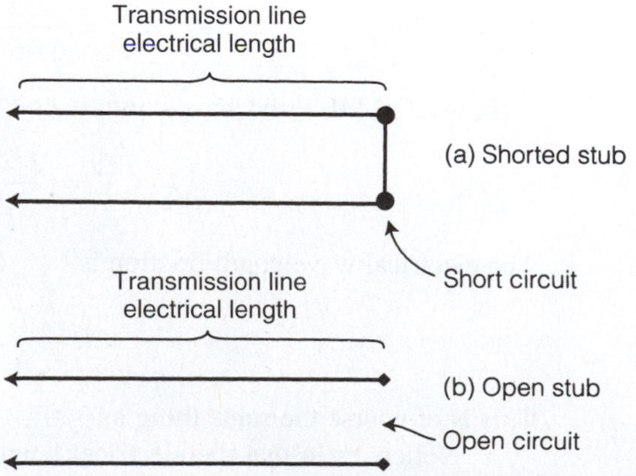

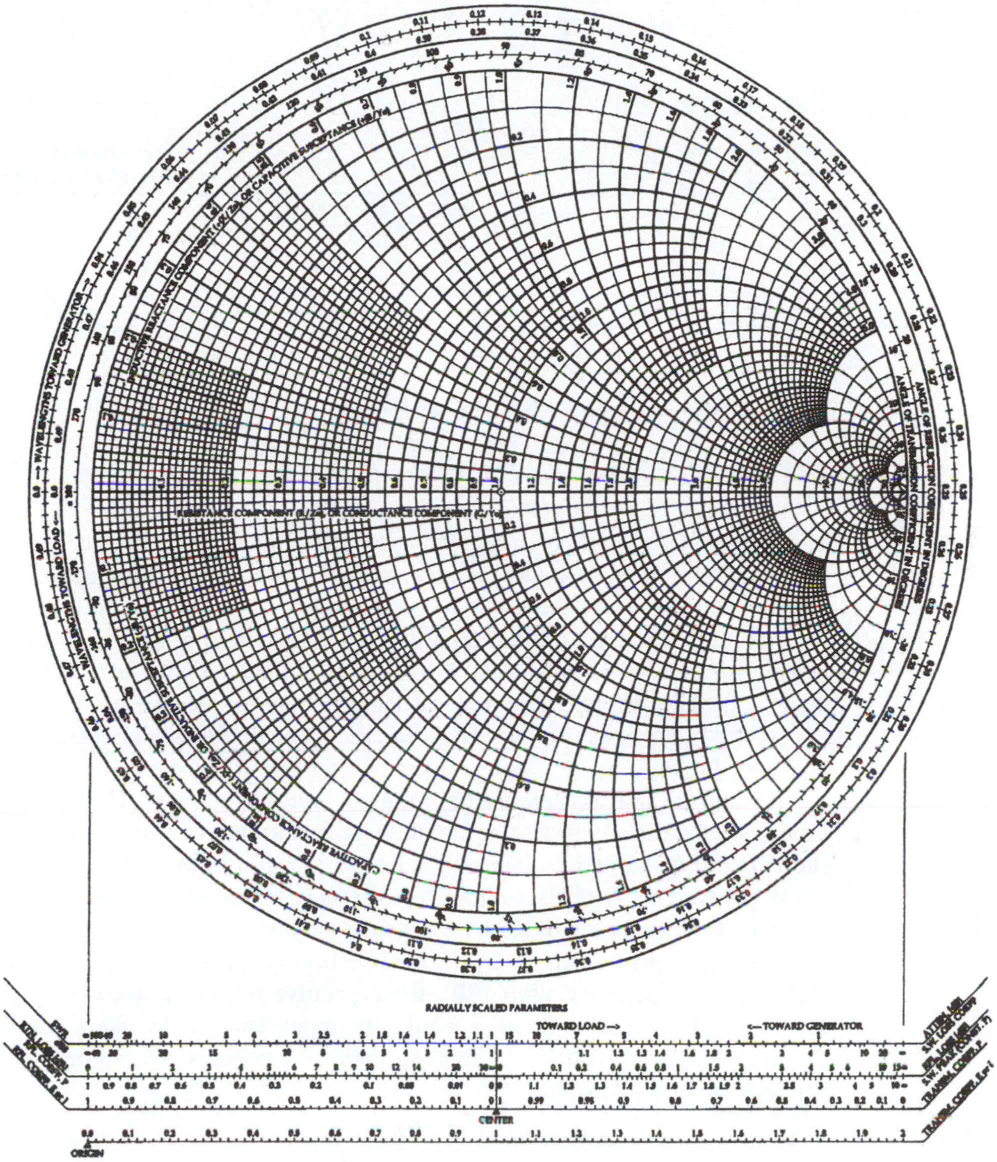

Figure 11–15 A Smith chart

this tool. A Smith chart is shown in Figure 11–15. We will do some simple calculations with it later in this chapter. The Smith chart can be used to determine the input impedance of a section of transmission line, among other things. The chart tells us that the input impedance of a transmission line is dependent upon the *termination impedance*, Z_R, and the *electrical length* of the line. If you look carefully at the chart in Figure 11–15 you'll notice that the outer rim of the circuit has two scales marked in degrees. Unlike a normal circle, there are 180 electrical degrees to a complete circle on the

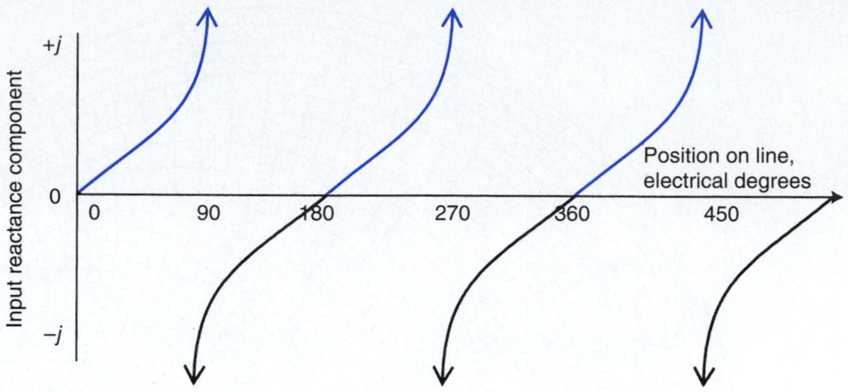

Figure 11–16 The input impedance of a shorted stub versus electrical length

Smith chart. One hundred eighty degrees is the same thing as one-half of a wavelength. *This tells us that the impedance patterns on a transmission line repeat every 180°, or half wavelength.*

Note that when a transmission line is terminated in a matched load, the input impedance of that line is *always* Z_0. If the line is terminated with anything but Z_0, then the input impedance varies with electrical length. Suppose that we were to graph the input impedance of a shorted stub versus its electrical length. We would obtain a graph like that in Figure 11–16.

When the stub has *zero* length, the input impedance is 0 Ω. That makes sense; we've just connected directly to a short circuit, with no transmission line at all. We *should* see a short circuit! As the stub is increased in length, it begins to look like a *pure inductive reactance,* $+jX_L$. This is really strange; the line now looks like an inductor! When the line reaches 90° length, the inductive reactance becomes *infinite*. In other words, the line appears as an *open circuit*. Past 90°, the line begins appearing as a *capacitive reactance,* $-jX_c$. As the line approaches 180°, the capacitive reactance decreases, becoming *zero* again at 180° in length. In other words, the input again looks like a short circuit! This action repeats every 180°. Therefore, the following conclusions can be reached about the behavior of a shorted stub:

1. It appears as an *inductance* between 0° and 90° in electrical length.
2. It appears as a *capacitance* between 90° and 180°.
3. At 0° and 180° (in fact, *any* multiple of 180°), it appears as a short circuit.
4. At 90°, 270°, and so on, it appears as an *open* circuit.

This type of behavior is exactly what we need for impedance matching. The stub can be made to appear as an X_L or an X_c. What we *can't* directly see from this data is that the stub can transform one impedance into another. Look at what happens at 90°: The input resistance is an *open* circuit. The 90° section of line has *transformed* a short (0 Ω) into an *open* (∞ Ω).

What do you think will happen if we take a stub with an *open* circuit on one end and measure it 90° away? If you're thinking that you will see a *short* circuit, you're right!

Quarter-Wave Section Characteristics

A 90° section of transmission line always transforms impedance in the following way:

$$Z_{\text{in}} = \frac{Z_0^2}{Z_R} \qquad \textbf{(11–19)}$$

where Z_{in} is the input impedance of the line and Z_R is the load impedance. Notice that when Z_R is 0 Ω, an answer of *infinity* (open circuit) is returned by equation 11–19; when an open circuit ($Z_R = \infty$) is substituted, the answer returned is *zero* (a short circuit). *Because of this property, sometimes a quarter-wave (90°) section of transmission line is used to transform an impedance up or down.* This is normally done only when a pure resistance needs to be stepped up or down, just like our first case at the beginning of this section. By manipulating equation 11–19, we can find the appropriate value of characteristic impedance for the matching section:

$$Z_0 = \sqrt{Z_{\text{in}} Z_R} \qquad \textbf{(11–20)}$$

Equation 11–20 tells us that we can *transform* an impedance of Z_R to a desired value of Z_{in} by choosing an appropriate value of characteristic impedance (Z_0) for the 90° matching section. Let's try an example.

EXAMPLE 11–15

A certain *loop antenna* built for the 6 meter amateur radio band (50–54 MHz) has an input resistance of 112.5 Ω. It is to be driven by a transmitter needing a 50 Ω load. In other words, we need to transform 112.5 Ω down to 50 Ω.

a. What value of Z_0 is needed for the quarter-wave matching transformer?
b. If the velocity factor (VF) of the quarter-wave section is 0.7, what will its length in *inches* be? Assume operation in the center of the band at 52 MHz.
c. Draw a picture showing how to build the circuit.

Solution

a. Equation 11–20 handles quarter-wave transformers:

$$Z_0 = \sqrt{Z_{\text{in}} Z_R} = \sqrt{(50\ \Omega)(112.5\ \Omega)} = \underline{\underline{75\ \Omega}}$$

A 90° section of $\underline{\underline{75\ \Omega}}$ line must be used for the matching section.

b. First, find the wavelength of the 52 MHz signal within the cable:

$$\lambda = \frac{v_p}{f} = \frac{(VF)(3 \times 10^8\ \text{m/sec})}{52\ \text{MHz}} = \frac{(0.7)(3 \times 10^8\ \text{m/sec})}{52\ \text{MHz}} = 4.04\ \text{m}$$

Now, 90° is just 1/4 of a wavelength, so the physical length is

$$L = \frac{\lambda}{4} \times \frac{4.04\ \text{m}}{4} = \underline{1.01\ \text{m}}$$

But the answer was requested in *inches,* so we need to convert:

$$L_{\text{inches}} = L_{\text{meters}} \times \frac{39.37''}{1\ \text{m}} = \underline{\underline{39.75''}}$$

The length of the matching section must be about 39.75″.

c. Figure 11–17 shows how to build this circuit.

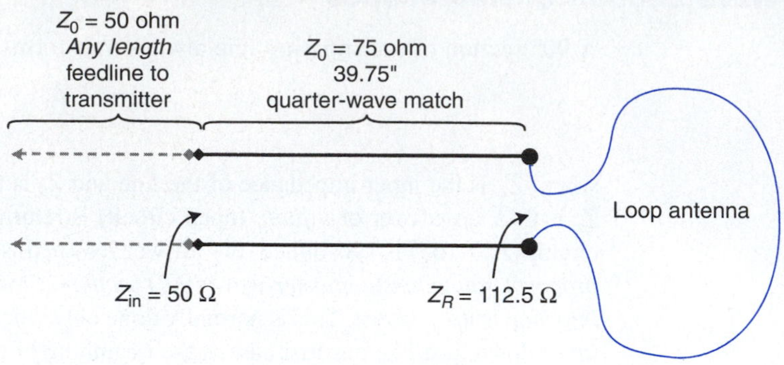

Figure 11–17 A completed quarter-wave matching transformer circuit

Other Matching Sections

The quarter-wave transformer depends on the availability of a 90° section of line with "just the right Z_0." It also will not work well if there is undesired X_L or X_C at the load. The *single-* and *double-stub* matching sections of Figure 11–18 are often used in these situations. The matching sections of Figure 11–18 can be constructed with transmission line of any characteristic impedance.

Single-stub matching section There are two actions in the single-stub section of Figure 11–18(a). Consider our example of an impedance mismatch where the resistive part is

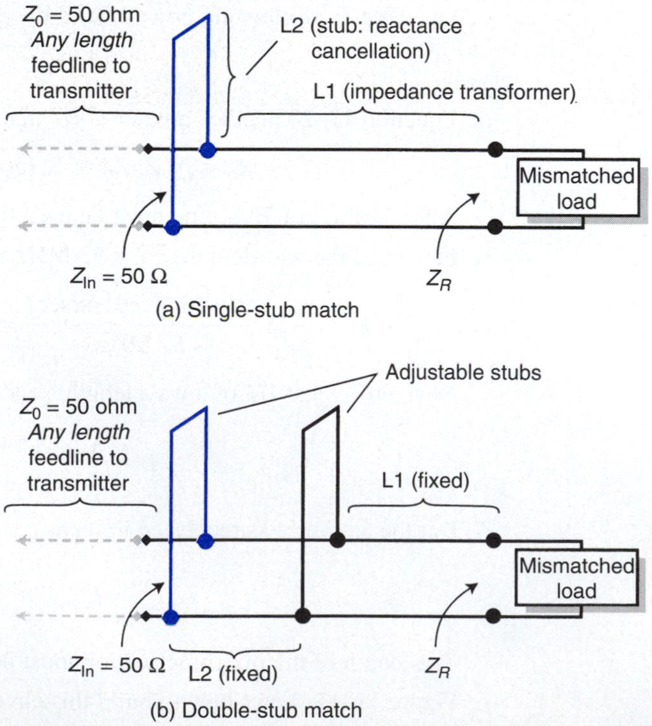

Figure 11–18 Single- and double-stub impedance matching sections

wrong and an undesired capacitive component exists as well. This type of mismatch can't be corrected with a quarter-wave transformer because there is unwanted reactance. Therefore, a two-step approach is used.

The length of line L_1 in Figure 11–18(a) transforms the load impedance Z_R to a value where the *real* (resistive) portion of the impedance is correct. The length of L_1 determines the transformation ratio. Once the impedance has been transformed, the resistive portion is correct, but there is still an undesired reactance (X_L or X_C). The *stub* L_2 has the job of providing a reactance exactly equal but opposite that of the undesired reactance left over on the line. The length of L_2 determines the amount of reactance canceled, and of course, whether it is to be capacitive or inductive.

Note that since the stub L_2 is in parallel, it actually provides a shunt *admittance* to cancel the undesired capacitive or inductive *susceptance* at the line's input. It's just easier to think about the overall action in terms of impedance units.

For the *single-stub* match, the designer must calculate *where* to place the shorting stub and *how long* the shorting stub is to be. Generally, a Smith chart is used; in modern practice, a computerized version of the chart (such as *MicroSmith,* from the American Radio Relay League, or *Smith85,* part of the software package that comes with this text) is often used.

Double-stub matching section The action in the double-stub matching section is similar. The double-stub method is used where it might be necessary to modify the settings of the impedance-matching unit. With a single-stub match, this means that the *location* L_1 would have to be changed, which is not very practical. With a double-stub match, to change the impedance transformation, only the lengths of the two stubs need be changed; this is much more convenient. The double-stub match is frequently used in broadcast and microwave applications.

Section Checkpoint

11–30 List and give an example of three kinds of impedance mismatch.

11–31 What are the two units used for expressing electrical length?

11–32 Why does the wavelength of a signal shorten within a transmission line?

11–33 Can the electrical length of a section of line be the same at two different frequencies?

11–34 What is a *stub?*

11–35 What are the four possible input impedances of a shorted or open stub?

11–36 What is a Smith chart used for?

11–37 How often do impedance patterns repeat on a transmission line?

11–38 When will the input impedance of a transmission line *always* be Z_0, regardless of electrical length?

11–39 What type of mismatches can be corrected with a quarter-wave transformer?

11–40 List two disadvantages of a quarter-wave transformer.

11–41 What is the purpose of sections L_1 and L_2 in Figure 11–18(a)?

11–5 INTRODUCTION TO THE SMITH CHART

As you saw in Section 11–4, the input impedance of a section of transmission line will be equal to the characteristic impedance, Z_0, as long as the load is properly matched to the line. A 50 Ω line has an input impedance of 50 Ω when it is terminated in the proper impedance, 50 Ω. However, this is not always the case. When a transmission line is terminated in anything other than its characteristic impedance, an impedance mismatch results and standing waves appear on the line. The input impedance of a lossless mismatched line can be derived by advanced mathematics and is described by equation 11–21.

$$Z_{in} = Z_0\left\{\frac{jZ_0 \sin\beta + Z_R \cos\beta}{jZ_R \sin\beta + Z_0 \cos\beta}\right\} \tag{11–21}$$

where j is the square root of negative 1, Z_0 is the characteristic impedance of the line, Z_R is the termination impedance, and β is the electrical length of the line in degrees. Equation 11–21 is hardly convenient to use directly; however, it is the basis of the Smith chart. The Smith chart is a graphical solution of equation 11–21. It allows us to design antenna-matching networks, calculate the input impedance of transmission line *sections,* estimate the degree of impedance mismatch by calculating the SWR, and express complicated impedance–frequency relationships graphically.

A technician does not need to know how to design with the Smith chart. However, the chart is often used by manufacturers to characterize the impedance of RF systems on data sheets. A few pieces of RF test equipment present their results using an on-screen Smith chart. Knowledge of this tool can help you to understand and correctly apply various RF system components.

Parts of the Smith Chart

The Smith chart is designed to take into account all the factors of equation 11–21. The features that assist in this are highlighted in Figure 11–19.

The starting point on the Smith chart is the *origin,* which represents an impedance equal to the characteristic impedance of the system. Note that the point is marked "1.0." That's strange. Why doesn't it read 50 Ω? The reason is that the Smith chart uses a *normalized* coordinate system. Don't let that scare you! By using a normalized coordinated system, the chart can work with any characteristic impedance. When we plot points on the chart, we divide their values by Z_0, then enter them on the chart according to equation 11–22.

$$Z' = \frac{Z}{Z_0} \tag{11–22}$$

Equation 11–22 says that to get the normalized impedance Z', simply divide the actual impedance value Z by the characteristic impedance Z_0 of the system. We reverse the process when we interpret points on the chart; the point values are *multiplied* by Z_0 to get back into actual ohm values.

The *real axis* of the chart is where all pure resistances will be plotted. The center of this axis is the origin, which again represents a pure resistance of Z_0. The *distance scale*

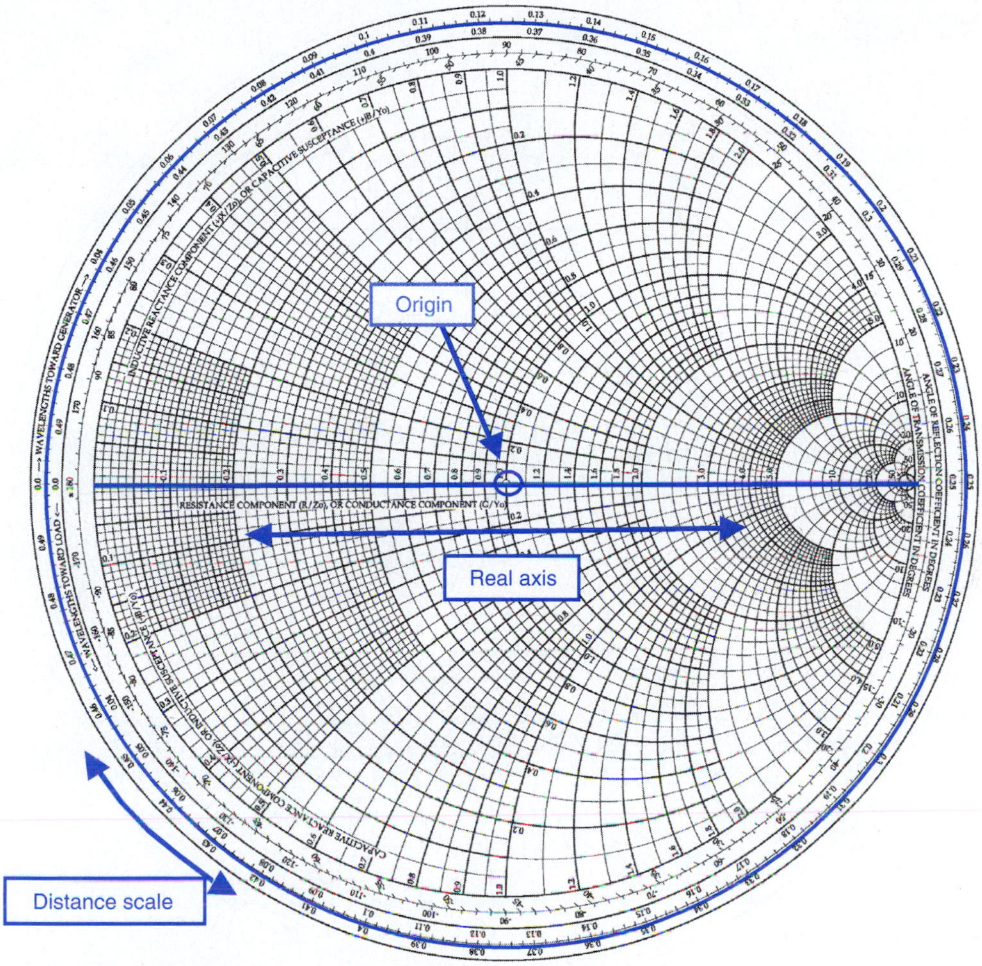

Figure 11–19 Parts of the Smith chart

runs around the rim of the chart. It's in *wavelength fractions*. It starts at 0 (zero) on the left-hand side of the chart, and increases to a maximum of 0.25 (one-quarter wavelength) on the right. *One complete circle on the chart is the same as moving one half wavelength on the transmission line.*

EXAMPLE 11–16 Plot the following pure resistances on the Smith chart for a 50 Ω system:

a. 50 Ω
b. 25 Ω
c. 75 Ω

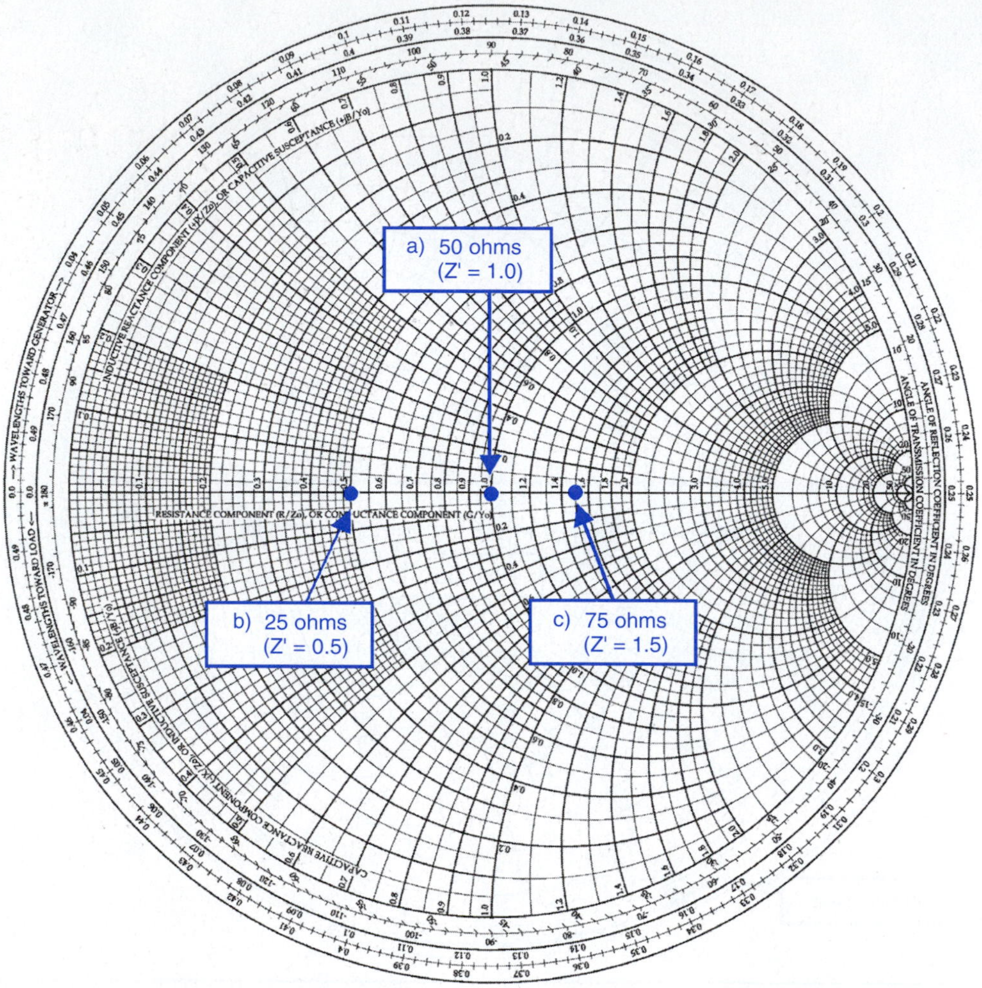

Figure 11–20 Solution for example 11–16

Solution

To plot the points, normalize them by dividing them by the Z_0 value of 50 Ω according to equation 11–22:

a. $Z' = \dfrac{Z}{Z_0} = \dfrac{50\ \Omega}{50\ \Omega} = \underline{\underline{1.0}}$

b. $Z' = \dfrac{Z}{Z_0} = \dfrac{25\ \Omega}{50\ \Omega} = \underline{\underline{0.5}}$

c. $Z' = \dfrac{Z}{Z_0} = \dfrac{75\ \Omega}{50\ \Omega} = \underline{\underline{1.5}}$

These points are plotted on the chart of Figure 11–20.

EXAMPLE 11–17 Figure 11–21 shows two points plotted on a Smith chart for a 75 Ω system. Read out the actual impedance for each point.

Solution

By rearranging equation 11–22, we get:

$$Z = Z' \times Z_0$$

This relationship can be used to find the original impedances. *Note that Z_0 must be known!*

a. $Z = Z' \times Z_0 = 0.6 \times 75\ \Omega = \underline{\underline{45\ \Omega}}$

b. $Z = Z' \times Z_0 = 2.0 \times 75\ \Omega = \underline{\underline{150\ \Omega}}$

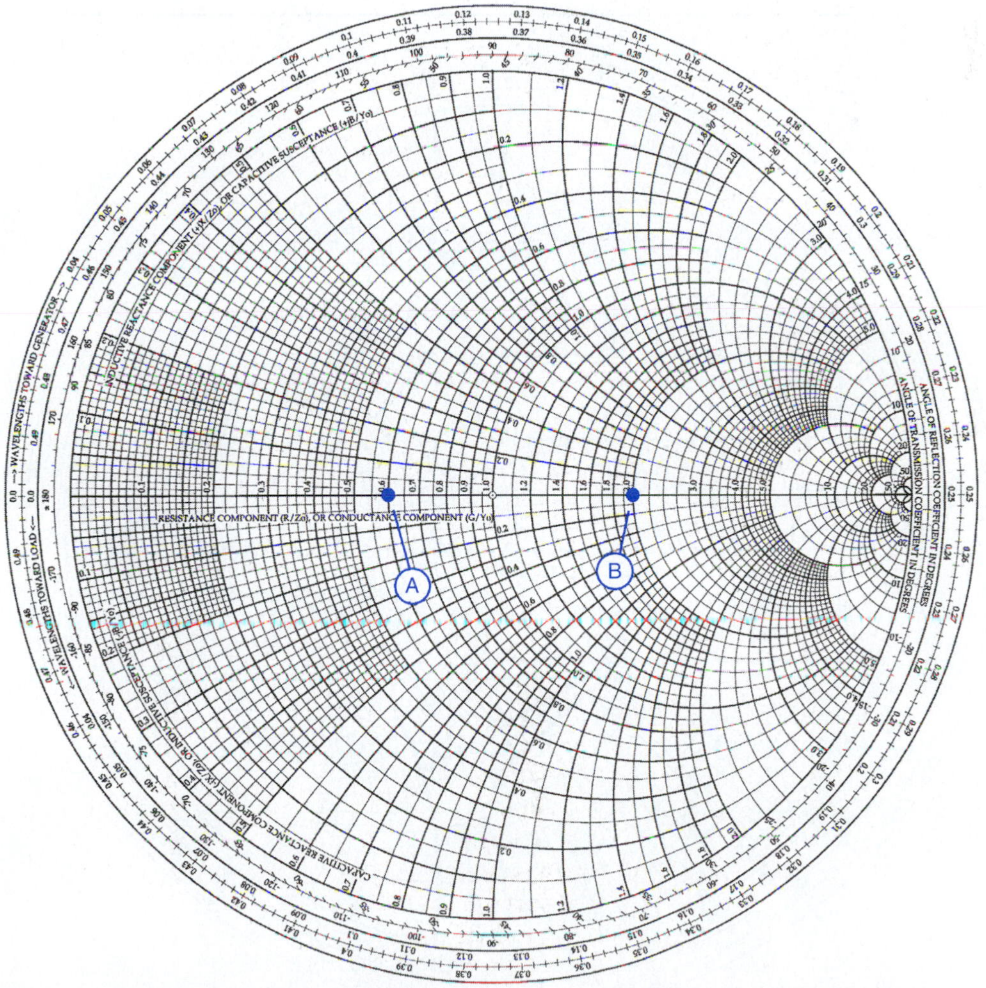

Figure 11–21 Data for example 11–17

Plotting Complex Impedances on the Smith Chart

The previous examples have shown only pure resistances on the chart. Of course, a circuit can appear *inductive* (having a positive *j* factor) or *capacitive* (having a negative *j* factor). Figure 11–22 shows where capacitive and inductive reactance values are plotted on the chart.

In the figure, the *constant reactance curves* are highlighted in blue. There are two families of these curves on the chart. Those above the real (center horizontal) axis represent inductive reactance; those below represent capacitive reactance. The impedance values on these curves are normalized using equation 11–22, just as we did with real resistance values. The normalized (relative) impedance values are on the inner rim of the chart. Be careful when reading the chart. It's easy to accidentally read the wrong value for each axis, especially since the axes curve! Patience and practice are required. Let's try an example.

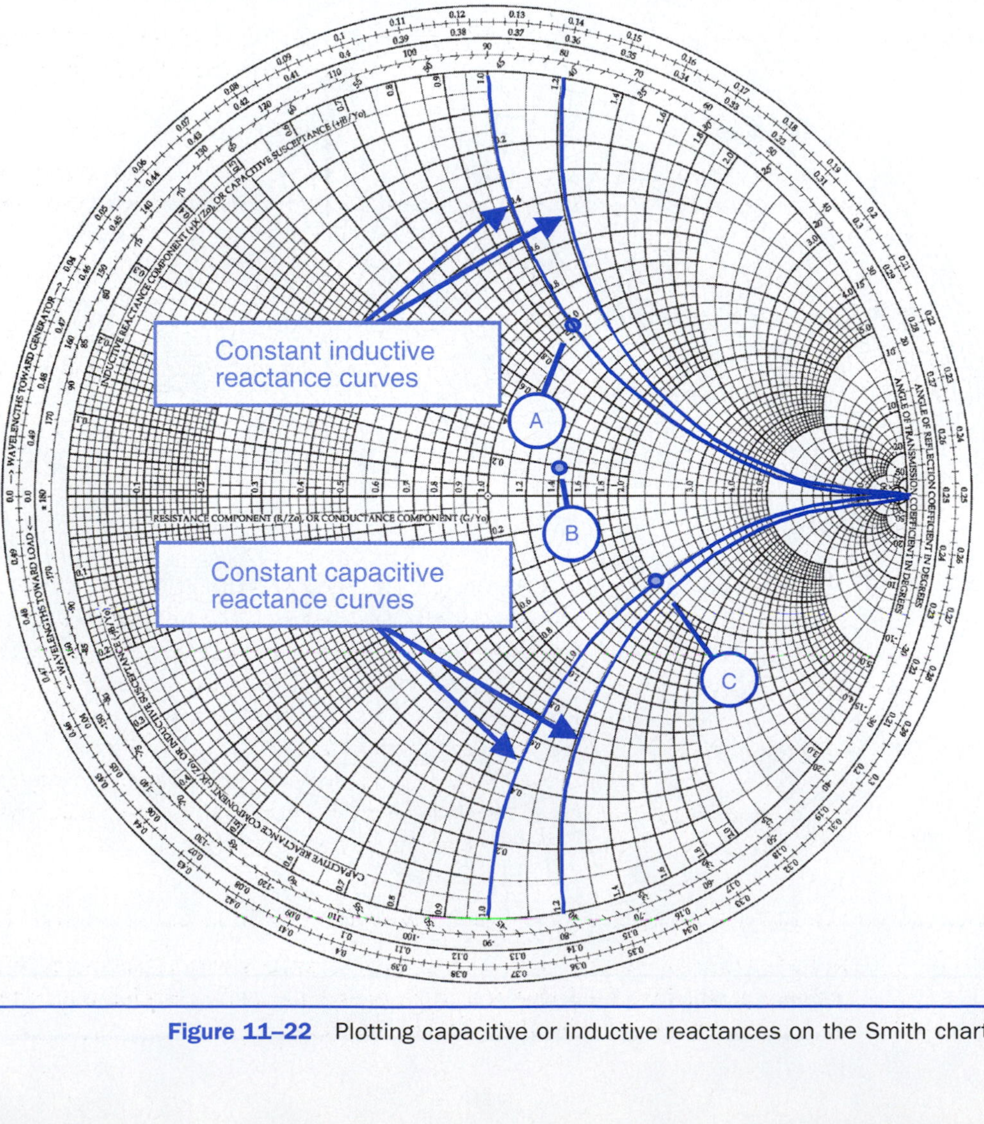

Figure 11–22 Plotting capacitive or inductive reactances on the Smith chart

EXAMPLE 11–18 Using the Smith chart of Figure 11–22, plot the impedance $(50 + 50j)\ \Omega$ for a $50\ \Omega$ transmission line.

Solution

First, we need to normalize the impedance according to equation 11–22:

$$Z' = \frac{Z}{Z_0} = \frac{50 + j50\ \Omega}{50\ \Omega} = \underline{1.0 + 1.0j}$$

Now that we've got normalized coordinates, we can plot them. We must do this in two steps. First, we locate the *real* part of the coordinate—1.0—on the *real axis*. This is right in the middle of the chart of Figure 11–22. Once this point is located, we can then move *up* to plot the *inductive reactance* component of the impedance. This is shown as point "A" in the figure.

EXAMPLE 11–19 What impedances are represented by points "B" and "C" in Figure 11–22? State whether each one is capacitive, inductive, or purely resistive. Assume a $50\ \Omega$ system ($Z_0 = 50\ \Omega$).

Solution

a. The normalized impedance at point "B" is $(1.4 + 0.2j)$. The positive j factor tells us that the impedance is *inductive* in nature. The actual impedance is obtained by applying equation 11–22 in reverse, as we did previously:

$$Z_B = Z' \times Z_0 = (1.4 + 0.2j) \times 50\ \Omega = \underline{\underline{(70 + j10)\ \Omega}}$$

b. At point "C" the normalized impedance is $(2.0 - 1.0j)$. The negative j value indicates a *capacitive* circuit. Again, the actual impedance can be calculated:

$$Z_C = Z' \times Z_0 = (2.0 - 1.0j) \times 50\ \Omega = \underline{\underline{(100 - j50)\ \Omega}}$$

Calculating SWR Using the Smith Chart

The Smith chart can be used to quickly calculate the resulting SWR for various loads on a transmission line. It's true that equations 11–15a and 11–15b can easily do this when the load is purely resistive; however, when the termination has inductive or capacitive reactance, the forms in equations 11–10 and 11–12 must be used. These are not convenient to use because they involve many calculations with complex numbers. The Smith chart makes it very easy to get an answer with minimal effort.

To find SWR on a Smith chart, plot the point corresponding to the impedance of interest. Then using the origin as the center, draw a circle on the chart that passes through the point. Finally, read the SWR where the circle crosses the real axis.

EXAMPLE 11–20 A $50\ \Omega$ transmission line is driving an antenna system with a feed-point impedance of $(75 + 25j)$. Find the SWR that will result on the line, and determine whether this will be a problem requiring additional impedance matching.

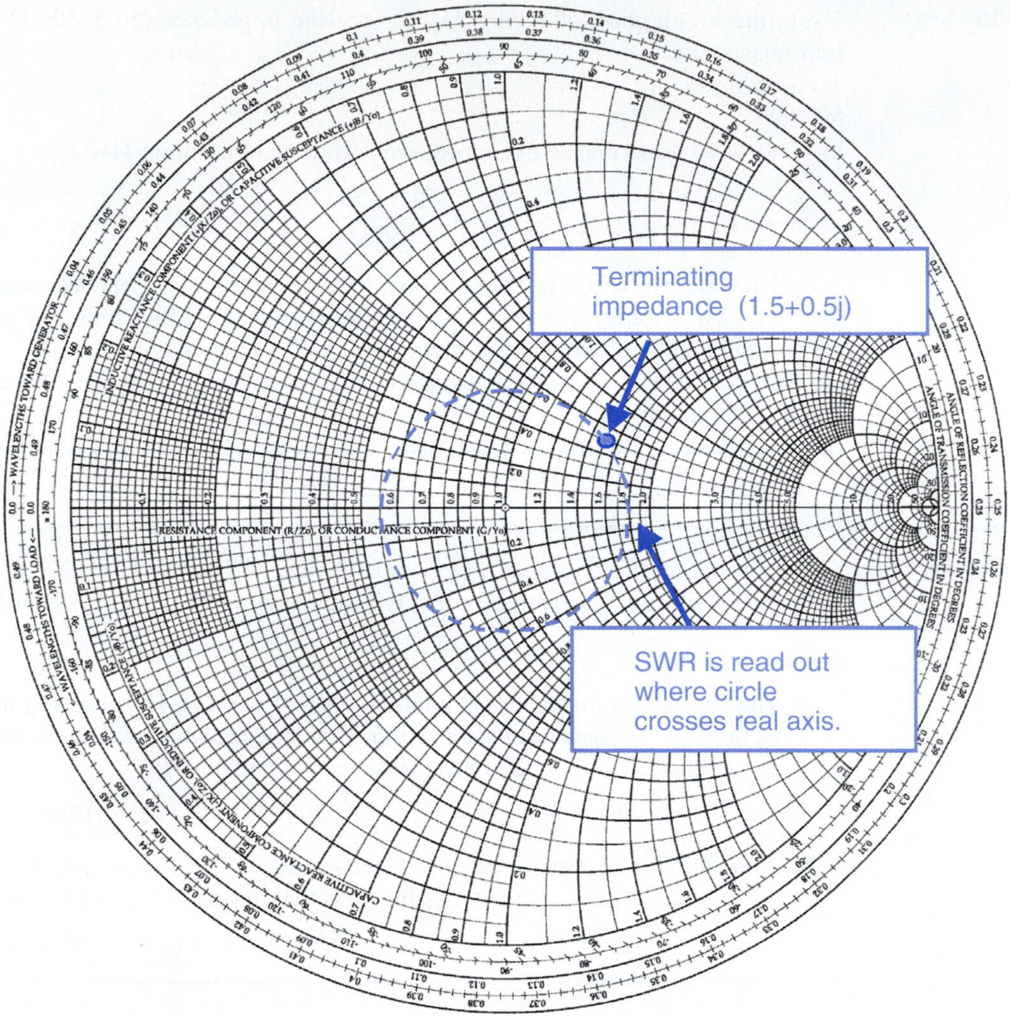

Figure 11–23 SWR circle of Example 11–20

Solution

The normalized impedance is $(1.5 + 0.5j)$, which is plotted on the chart of Figure 11–23. The circle crosses the real axis at approximately 1.8 (on the right side). The SWR is therefore 1.8:1. Since this is greater than 1.5:1, it may require additional impedance matching to make the transmitter accept the load. Since it is an antenna system, it is entirely possible that the SWR may be lowered by careful tuning of the antenna elements.

Calculating Input Impedance of Lines with Mismatched Loads

When a transmission line is terminated in a load equal to its characteristic impedance, its input impedance will always be Z_0, regardless of its length. But what about lines with mismatched loads? Equation 11–21 tells the whole story. The input impedance will depend upon the load impedance, the line characteristic impedance, and the electrical length of the line in wavelengths (or degrees). The impedance patterns repeat every half wavelength on the line.

To find the input impedance, first plot the load impedance on the chart and draw the SWR circle, as shown in Example 11–20. Then move *clockwise* on the circle the number of wavelengths corresponding to the line's length. The resulting point can be read out as the input impedance of the line.

EXAMPLE 11–21 Use the Smith chart to calculate the input impedance of the circuit of Figure 11–17. The circuit consists of a 112.5 Ω load (loop antenna) and a 1/4 wave section of 75 Ω transmission line.

Solution

$$Z' = \frac{Z}{Z_0} = \frac{112.5\ \Omega}{75\ \Omega} = \underline{1.5 + 0j}$$

(Be careful to use the correct Z_0! The matching section is 75 Ω.)

This impedance is purely real, so it is plotted directly on the real axis as point "A" in Figure 11–24. We then draw the SWR circle and move *clockwise* 1/4 (0.25) wavelength

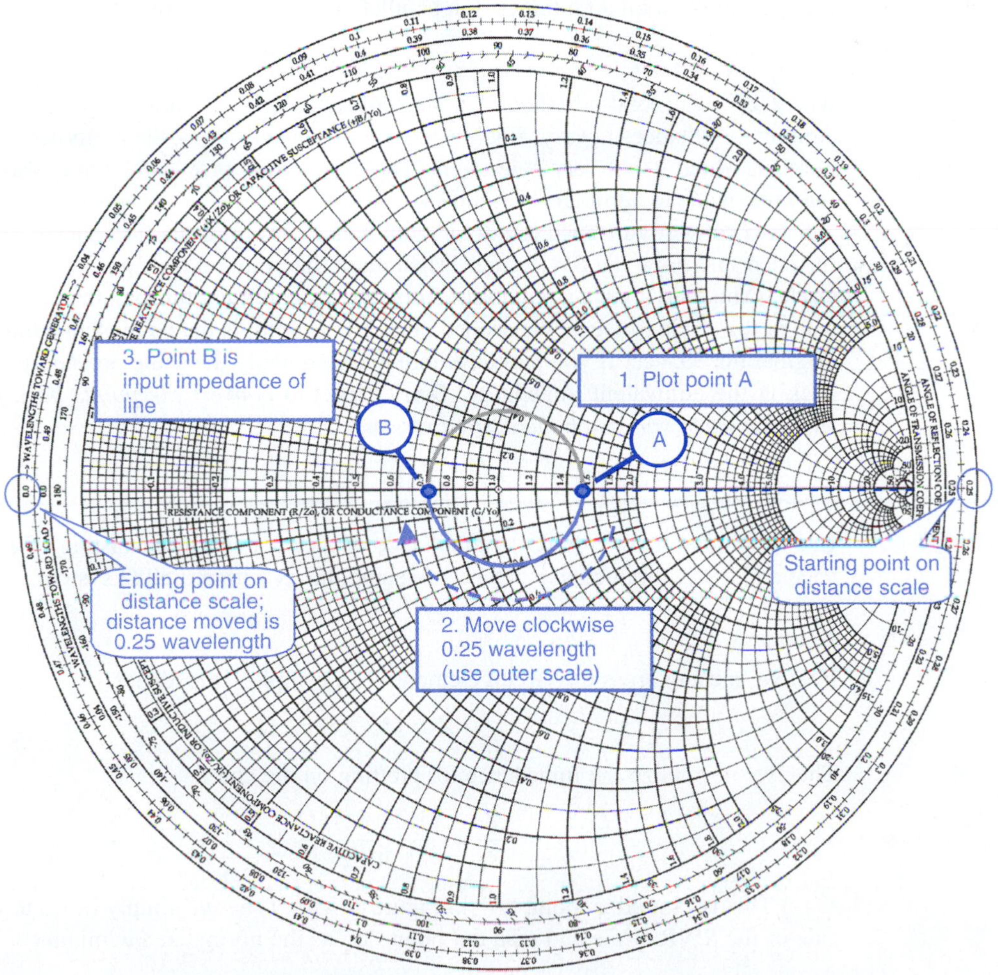

Figure 11–24 Solution for example 11–21

to point "B." *Point "B" represents the input impedance of the line.* This point reads out as $\underline{0.7 + 0j}$. Finally, we convert the value at point B back into ohms:

$$Z_B = Z' \times Z_0 = (0.7 + 0j) \times 75\ \Omega = \underline{\underline{52.5\ \Omega}}$$

This is very close to the desired 50 Ω result—the matching circuit of Figure 11–17 will work just fine!

Admittance and Impedance Conversions Using the Smith Chart

You might recall from fundamentals that the admittance Y of a circuit is simply the reciprocal of its impedance:

$$Y = \frac{1}{Z} \tag{11–23}$$

Admittance is the ability of a circuit to pass an AC current (as opposed to impedance, which is its *opposition* to the current). It's often used for calculating parallel AC circuits, since the total admittance of a parallel circuit simply adds:

$$Y_T = Y_1 + Y_2 + Y_3 + \cdots \tag{11–24}$$

where Y_T is the total admittance in siemens, and Y_1, Y_2, Y_3 are the parallel admittances. Adding admittances is easy; finding admittance means taking the reciprocal of a complex number, which is not very convenient. The Smith chart provides a direct way of converting impedance to admittance, and back.

To convert an impedance to an admittance, first plot the normalized impedance on the chart and draw the SWR circle. Then mark the point on the opposite side of the SWR circle. (You may draw a straight line from the original point through the origin to the other side of the circle to locate the second point.) This point is the equivalent admittance of the original impedance. If you plot an admittance and apply the same procedure, you'll arrive back at the equivalent impedance. *Don't forget to convert the points back to ohms (or siemens) when you reach the desired results.*

EXAMPLE 11–22

Use the Smith chart to find the admittance of a series circuit consisting of a 55 Ω resistance and a 35 Ω inductive reactance. Assume this is within a 50 Ω system.

Solution

First we must express the circuit elements as a complex impedance:

$$Z = (55 + 35j)\ \Omega$$

Then we normalize the impedance for plotting on the chart:

$$Z' = \frac{Z}{Z_0} = \frac{(55 + 35j)\ \Omega}{50\ \Omega} = \underline{1.1 + 0.7j}$$

This is plotted as point "A" in Figure 11–25. Then we simply move to the opposite side of the SWR circle and read out point "B" as the normalized admittance.

$$Y' = (0.65 - 0.4j)$$

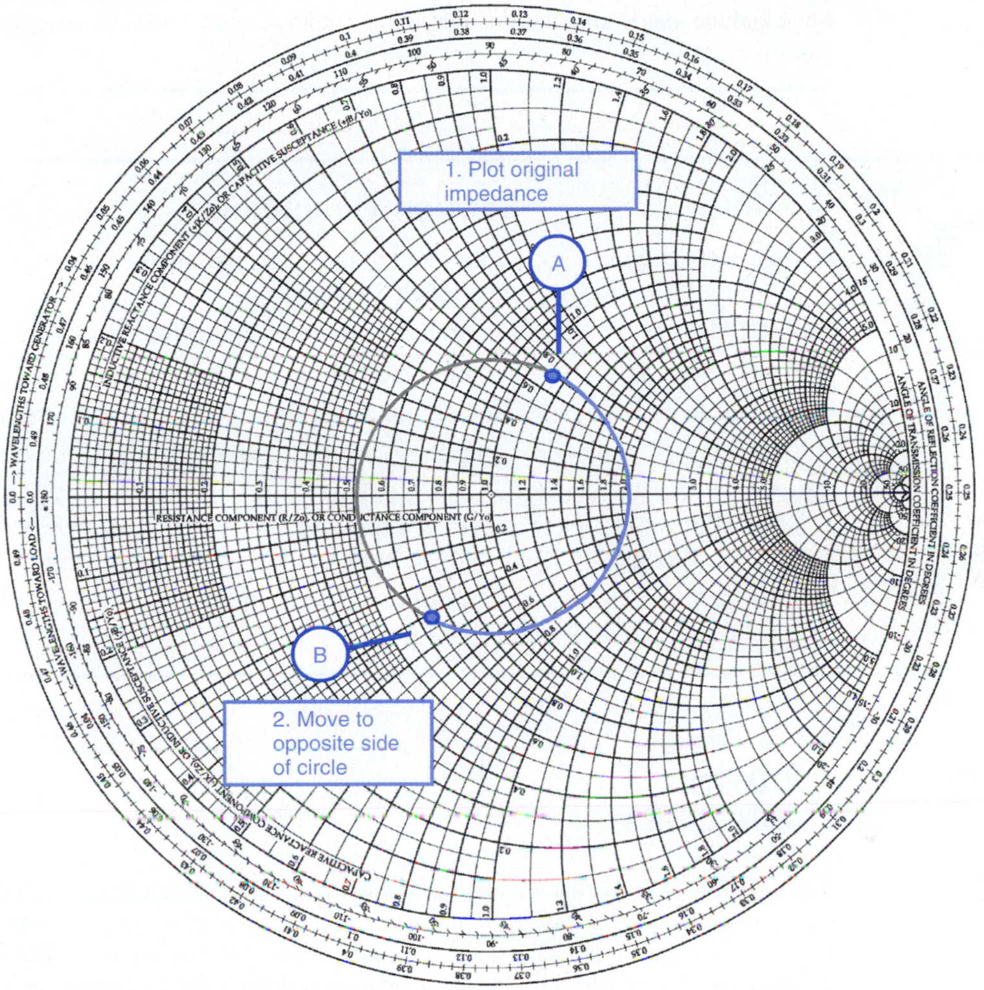

Figure 11–25 Finding admittance on the Smith chart

Finally, we need to convert normalized admittance back into siemens. This is a little tricky; recall that equation 11–22 states:

$$Z' = \frac{Z}{Z_0} \qquad \textbf{(11–22)}$$

Since we're working in admittance units, we must restate equation 11–22 in terms of admittance units (recall that $Y = 1/Z$):

$$Y' = \frac{Y}{Y_0} = \frac{Y}{1/Z_0} = Y \times Z_0 \qquad \textbf{(11–25)}$$

Therefore, we can rearrange equation 11–25 to get:

$$Y = \frac{Y'}{Z_0} = \frac{(0.65 - 0.4j)}{50\ \Omega} = \underline{\underline{(0.013 - 0.0082j)\ \mathrm{S}}}$$

In many instances it is necessary to work using admittances. Common applications include single and double-stub tuning sections and parallel to series conversion in circuits.

Section Checkpoint

11–42 What three factors determine the input impedance of a transmission line? (*Hint:* Examine equation 11–21.)

11–43 Why are normalized coordinates used on the Smith chart?

11–44 Where are real coordinates plotted on the Smith chart?

11–45 If an impedance contains X_L or X_C, where will it lie on the Smith chart?

11–46 Explain how to calculate the SWR on a line using the Smith chart.

11–47 If a certain transmission line has an input impedance of $(25 + 25j)\ \Omega$, what will its input impedance become if it is *lengthened* by one-half wavelength?

11–48 Define *admittance*. How do we convert impedance to admittance on the Smith chart?

11–6 TROUBLESHOOTING TRANSMISSION LINES

When trouble is suspected with a transmission line, there's often little a technician can do except to replace the section with the suspected fault. When the line runs only a short distance, and the ends are readily accessible, this is a reasonable approach. But what do you do when the line is 7/8″ hardline (costing several dollars per foot) running 500 feet up a tower (and the customer wants the system fixed *yesterday*)? You can't justify replacing such a line without sure knowledge of the problem!

There are three common faults that plague transmission lines: *moisture entry, short and open circuits,* and *damage due to lightning discharge.*

Moisture entry is a common problem with any device exposed to weather. Even indoor installations can take on water from condensation. Systems employing *hardline* are often pressurized with an inert gas, such as nitrogen. *Waveguides* are prone to collecting water and as a rule are almost never run perfectly horizontal. A *drain hole* is often provided at the low spot in a waveguide installation to allow water to escape. Most coaxial transmission line is permanently ruined by moisture entry and must be replaced. If green or white corrosion is present on any of the connectors, it is a safe bet that moisture has entered the line. A small amount of moisture entry may cause the feedline loss to skyrocket, even though the SWR readings at the transmitter appear normal.

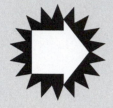

Open and short circuits can occur as a result of mechanical damage to a cable from bending, crushing, or continuous flexing. Occasionally, persons have been known to sabotage exposed coaxial cables by piercing them with a small pin, nail, or other sharp object, and then cutting off the exposed ends of the inserted item to hide the damage (in the hope that it will take someone a long time to find the problem). Occasionally, opens and shorts appear at connectors due to poor connector installation, or mechanical abuse (tugging on cables, etc.). The resistance of RF connectors increases with each insertion and removal; connectors that are frequently plugged and unplugged degrade rapidly and become unreliable. Opens and shorts can be quickly located using a *cable analyzer,* which sends a pulse down the cable and measures the return time. By knowing the velocity of propagation of the cable, the analyzer gives a direct readout of distance to the fault.

Melted components can occur due to excessive RF input power, or more commonly, due to *lightning strikes*. Often the damage is invisible until one or more connectors are dismantled. If any portion of a line has been struck by lightning, the entire line must usually be replaced.

Diagnosing Line Problems

There are several good methods for determining the extent of transmission line problems. You may devise others on your own, or your company may own some specialized gear for testing.

Ohmage testing Figure 11–26 shows an *ohmage* test on a transmission line. Shorting the far end should result in a low resistance reading; opening the far end should result in an open circuit. This is a good test to make on a fresh installation. Be careful: A feedline may look perfectly normal to an ohmmeter but may have high losses at radio frequencies.

Feedline loss evaluation There are several ways to check feedline loss. The best way, if the instrumentation is available and both ends are accessible, is shown in Figure 11–27(a). A known power at the frequency of normal operation is introduced into the line and measured there; a second power meter is placed at the other end of the line. By comparing the power readings of the two meters, the feedline loss can be calculated.

Figure 11–26 Ohmage testing

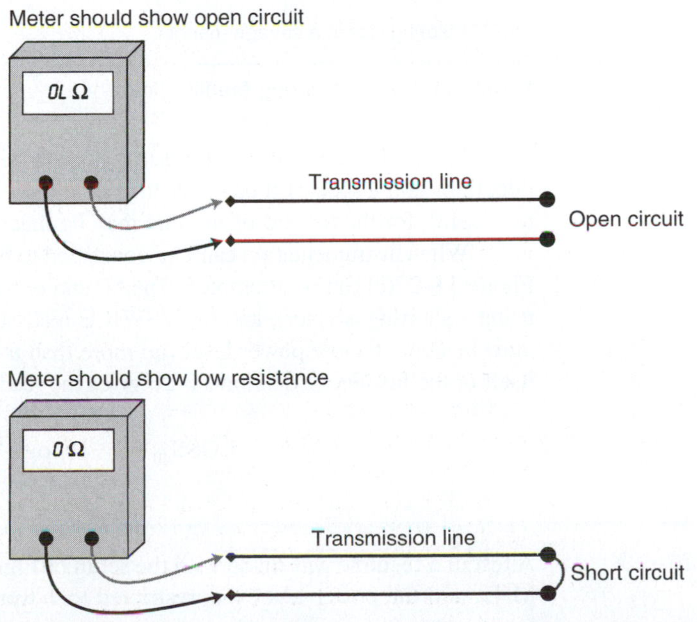

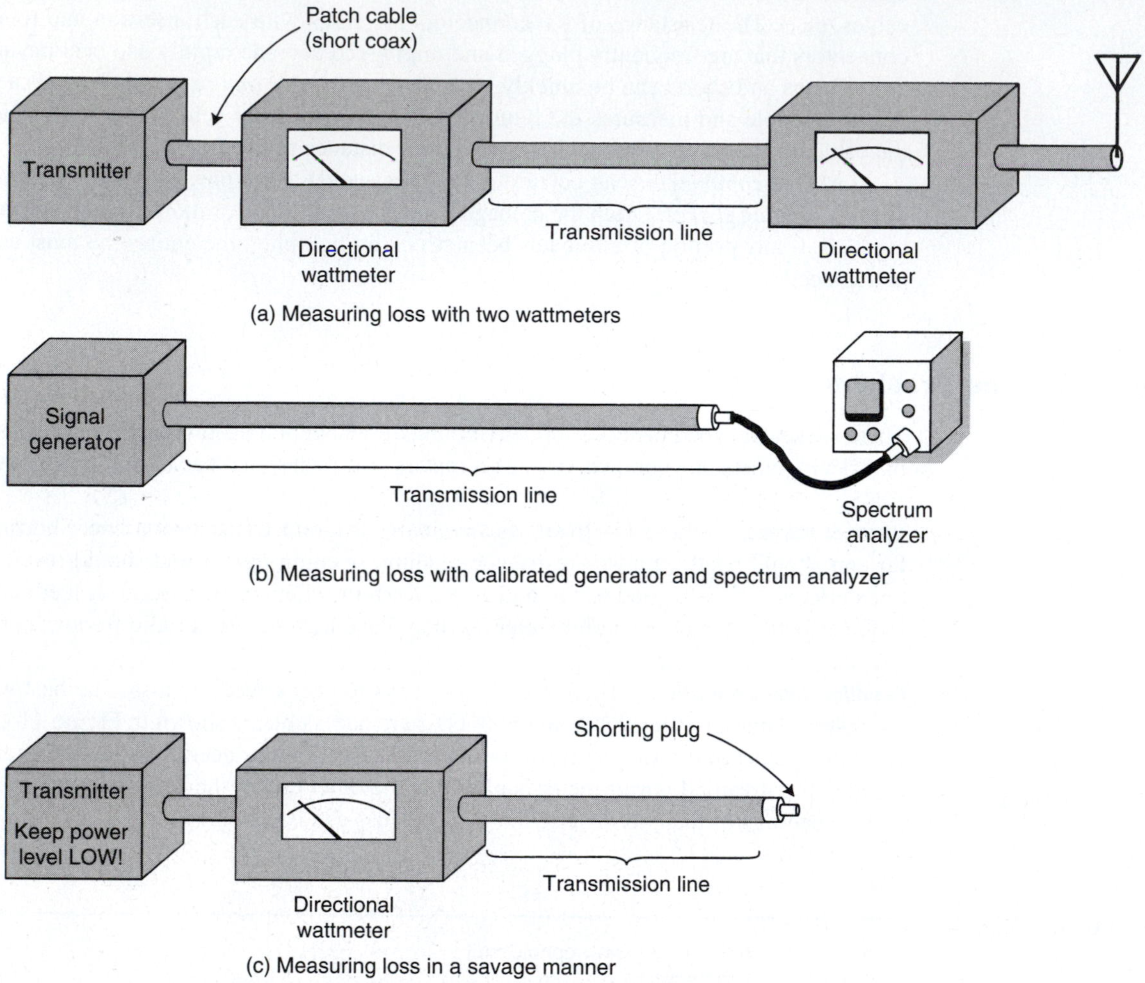

Figure 11–27 Evaluating feedline loss

Figure 11–27(b) shows the same approach, but instead uses a calibrated RF signal generator and a spectrum analyzer instead of two RF wattmeters. This approach is not often useful, for the far end of the line may be inaccessible for testing.

When instrumentation can't be connected to both ends at once, the simple test shown in Figure 11–27(c) can be attempted. The far end of the transmission line is shorted (preferably using a shorting adapter), and the VSWR is read at the near end of the line. The transmitter *must* be kept at a low power level (no more than a few watts) in order to prevent damage to itself or the feedline! Equation 11–26 gives the feedline loss in dB from the VSWR reading:

$$\text{LOSS}_{\text{dB}} = -10 \log\left(\frac{\text{SWR} - 1}{\text{SWR} + 1}\right) \tag{11–26}$$

EXAMPLE 11–23

A test of a feedline was made with the setup of Figure 11–27(c). The frequency was 444.25 MHz, and the power level was restricted to 1 watt. The SWR measured 3:1. What is the feedline loss?

Solution

Equation 11–26 gives the loss:

$$\text{LOSS}_{\text{dB}} = -10\log\left(\frac{\text{SWR}-1}{\text{SWR}+1}\right) = -10\log\left(\frac{3-1}{3+1}\right) = \underline{\underline{3\text{ dB}}}$$

It is just a coincidence that a 3:1 SWR indicates a 3 dB loss. However, it is nice to keep in mind. If the SWR measures *more* than 3:1, then the loss is *less* than 3 dB on the feedline. If the SWR measures *less* than 3:1, then the loss is *more* than 3 dB.

Several manufacturers make a device like the *MFJ-249 SWR Analyzer* as shown in Figure 11–28. The MFJ-249 combines a low-power signal generator and an accurate SWR bridge in one package and includes a digital frequency counter for confidence in frequency measurement. The MFJ-249 can be substituted for the transmitter and directional wattmeter of Figure 11–27(c) for quick and safe loss measurements. Instruments like the MFJ-249 are a snap to use; they are simply connected to the near end of the transmission line, and the frequency control is set to the operating frequency of the system. The SWR reads directly on a meter or LCD display. Because of its compact size, the MFJ-249 can be easily used for direct testing in hard-to-access locations, such as the top of radio towers.

Figure 11–28 The MFJ-249 SWR analyzer

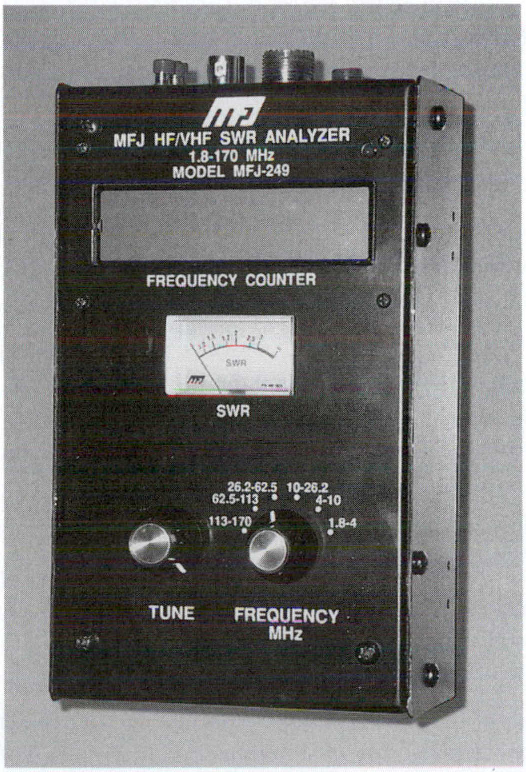

Section Checkpoint

11–49 List three types of failures that occur in transmission lines.

11–50 Why is hardline often pressurized with nitrogen gas?

11–51 What happens to RF connectors after repeated insertion and removal?

11–52 How does a cable analyzer locate short and open circuits?

11–53 Explain how to perform an ohmage test on a transmission line.

11–54 When testing a feedline for loss as shown in Figure 11–27(c), what precaution must be taken?

11–55 What is an *SWR analyzer?*

SUMMARY

- Transmission lines are used whenever RF energy is to be transported a short distance.
- The choice of transmission line type depends on the frequency of intended operation, loss tolerance, required connector type, and economics.
- A lossless section of transmission line can be thought of as an infinite number of infinitely small *L-C* sections.
- The characteristic impedance of a transmission line is the ratio of voltage to current at any point on the line.
- The *L-C* values in a transmission line determine the velocity of propagation and characteristic impedance.
- Energy moves on a transmission line in the form of traveling waves, which move at the velocity of propagation for that line.
- Energy loss in transmission lines occurs primarily as ohmic loss in the conductors and dielectric loss (heating).
- Each type of transmission is rated by the manufacturer for *specific loss,* which is the loss per unit length of line. Specific loss increases with frequency.
- Any impedance mismatch causes a reflected wave to appear on a transmission line.
- The pattern produced by the interaction of the forward and reflected waves on a transmission line is called a *standing wave pattern.*
- The maximum voltage at any point on a line, divided by the minimum voltage, is called the *voltage standing wave ratio* (VSWR), which is usually just called the SWR by technicians.
- A perfectly matched load produces a 1:1 SWR and a flat line. The maximum "normal" SWR permitted in most circumstances is 1.5:1.
- A directional wattmeter can be connected at any point on a transmission line; however, the accuracy of the SWR reading is better when connected near the antenna.
- There are many ways of matching impedances with transmission line sections. All methods rely on the inherent impedance transformation a section of line produces for a mismatched load.
- Stubs are used to simulate opens, shorts, capacitors, and inductors.
- The electrical length of a section of line is the number of wavelengths in a physical length of line. It can be measured in degrees or wavelength fractions.
- The Smith chart is a graphical solution to the transmission line equation. It can be used for many purposes, including SWR calculation,

input impedance calculation, and impedance-to-admittance conversion.

- The input impedance of a section of transmission line depends on the load impedance, the line characteristic impedance, and the distance to the load in wavelengths.
- Impedance patterns repeat every half wavelength on a transmission line. This corresponds to one revolution around the Smith chart.
- Transmission lines can fail due to moisture entry, shorts and opens, and lightning discharges.
- Testing methods need not be exotic; an ohmmeter can be used to find opens and shorts on lines, for example.
- Many specialized test instruments are available for transmission line testing. These are a good investment for organizations that do a lot of transmission line work.

PROBLEMS

1. List the ways that a transmission line can lose signal. Which of these ways could cause interference with other systems?
2. Why isn't twisted pair line useful at radio frequencies?
3. Explain the difference between *balanced* and *unbalanced* transmission lines, and give an example for each one.
4. What type of transmission line is popular at microwave frequencies?
5. Calculate the wavelength of the following signals in free space: (a) 2 MHz, (b) 10 MHz, (c) 100 MHz.
6. Which of the connectors in Figure 11–3 is most useful at UHF? Why?
7. Which of the connectors in Figure 11–3 would be cost effective for a 27 MHz CB radio system?
8. Draw the schematic diagram showing the equivalent circuit for (a) a lossy transmission line and (b) a lossless line. What types of losses do the two resistors in the lossy model represent?
9. Explain the charging action that takes place in Figure 11–6 after closure of the switch S_1. Why doesn't the voltage appear instantly at the 50 Ω load?
10. What is the velocity of propagation for a length of RG59U cable having $L = 298$ nH/m and $C = 52.5$ pF/m?
11. What is the velocity factor (VF) for a cable whose velocity of propagation v_p is 2×10^8 m/sec?
12. RG62 as manufactured by General Cable is specified as having a velocity factor of 84% and a capacitance of 44.3 pF/m. Find the equivalent inductance per meter for this cable.
13. What is the characteristic impedance of the RG62 cable of problem 12?
14. How long will it take for a signal to travel through a 10 m length of cable with a velocity factor (VF) of 0.7?
15. A *delay line* is a section of transmission line used to slow down or delay a signal by a specified amount. How long must a section of RG6U be (in inches) to develop a time delay of 10 ns?
16. A certain transmission line has a characteristic impedance of 93 Ω, and at some point a current of 2 A is flowing on the line. What is the voltage at the same point?
17. Calculate the characteristic impedance of a coaxial transmission line having the following physical characteristics: inner conductor diameter $d = 0.0071''$, dielectric diameter $D = 0.116''$, dielectric material = polyethylene.
18. What is the velocity factor for a transmission line using *teflon* as a dielectric?
19. What is the *skin effect?*
20. What is the loss of a 50 ft section of RG62A at 100 MHz?
21. An RG174 cable is being driven by a 10 W source and is terminated 25 ft away by a 50 Ω load. What power will be delivered to the load at (a) 50 MHz, (b) 400 MHz?

22. What is the proper impedance for terminating a 50 Ω transmission line?

23. If the final power amplifier stage in a transmitter fails, what is a good thing to check before placing the transmitter back in service?

24. What causes reflections on transmission lines? How can reflections be eliminated?

25. What causes standing waves to develop on transmission lines?

26. A certain transmission line has standing waves; the maximum voltage on the line is 100 V and the minimum voltage is 50 V. What is the SWR?

27. Find the reflection coefficient, Γ, that will result when a 75 Ω line is terminated in the following resistances: (a) 100 Ω, (b) 50 Ω, (c) 25 Ω, (d) 75 Ω.

28. Calculate the SWR of the line for each of the four load resistances of problem 27. Which of the SWR readings would be high enough to require further investigation?

29. A certain transmitter provides an output voltage of 100 V to a 75 Ω transmission line. The line is driving a mismatched load of 50 Ω. Determine the following: (a) the SWR, (b) the maximum and minimum voltages on the line.

30. You have been given a 100 ft length of transmission line to test. The characteristic impedance of the line is *unknown,* but according to your measurements, an SWR of 2:1 results when the line is terminated in 50 Ω, and an SWR of 3:1 results when a 75 Ω termination is substituted. What is the characteristic impedance of the mystery line? (*Hint:* Use equations 11–15a and 11–15b.)

31. A certain directional wattmeter reads 100 W in the forward position and 10 W in the reverse position while testing a transmission line and antenna for an HF transmitter. What is the SWR on the line, and should it be troubleshot?

32. To match a 173 Ω load to a 50 Ω transmitter using a quarter-wave transformer, what characteristic impedance is required?

33. What is the electrical length (in both wavelength fractions and degrees) of a 2 m section of RG59 coax ($VF = 0.82$) at a frequency of 54 MHz?

34. Repeat problem 33 for frequencies of 27 MHz and 138 MHz.

35. What is the physical length, in inches, of a 90° section of $R_{G8}(VF = 0.67)$ for a frequency of 145.29 MHz?

36. Operating at 30 MHz, a 90° section of R_{G11} is terminated in a short circuit. (It is a shorted stub.) What will the input impedance of the stub be at 60 MHz? (*Hint:* Calculate the new electrical length at 60 MHz first.)

37. Draw a schematic diagram of a single-stub match. Explain what each part of the circuit does.

38. A feedline is being tested for loss according to Figure 11–20(a). The wattmeters read 100 W and 75 W, respectively. What is the decibel loss in the line?

39. An FM broadcaster (operating near 100 MHz) is using a 100 ft length of 3/4″ hardline to feed an antenna. Upon replacing the load with a shorting plug, the SWR at the transmitter measures 1.5:1. (a) What is the loss in the transmission line? (b) Is this within acceptable limits for this type of line? What should be done next? (See Table 11–2.)

40. A 50 Ω transmission line is terminated in an impedance of $(50 + 25j)$ Ω. Calculate the SWR using the Smith chart. (If you have a Texas Instruments TI-85 or TI-86 calculator with the TI GraphLink, you can use the *Smith85* program to verify your results. The program and documentation are located in the Software folder of the Student CD supplied with this text.)

41. What is the input impedance of a 0.25 λ line with a characteristic impedance Z_0 of 100 Ω when terminated in an impedance Z_R of $(50 - j100)$ Ω?

42. What lengths of 50 Ω transmission line will give a purely real (no X_L or X_C) input impedance for a load impedance of $(25 + 25j)$ Ω? There are at least *two* answers. Give the two shortest-distance answers. *Hint:* Plot the point on the chart and move around the SWR circle until the impedance is purely real. (Using *Smith85,* you can use the Walk command to move around the chart until the result is purely real. Or you can issue a MkReal command to force the program to move to the nearest purely real answer.)

12

Antennas and Wave Propagation

OBJECTIVES

At the conclusion of this chapter, the reader will be able to:

- describe the characteristics of electromagnetic energy, such as field strength and polarization
- describe the characteristics of an isotropic point source
- explain the operation of the dipole and Marconi antennas, giving the 3-dimensional field pattern for each one
- given manufacturer's data for antenna gain and pattern, describe the performance characteristics of an antenna system
- explain the operation of the Yagi-Uda, log-periodic, and other directional antennas and give a typical application for each
- describe safety concerns and procedures for working with antenna installations
- describe three modes of electromagnetic propagation
- explain the interaction between frequency and electromagnetic wave propagation modes
- describe the basic principles of satellite communications
- calculate the values needed to provide a link budget

Antennas are a critical part of every RF communication system. Without a proper antenna to launch its energy into space, the most powerful transmitter is useless. A receiver's performance and sensitivity are very dependent on the quality of the attached antenna.

Transmitting and receiving antennas are *transducers*. A transducer converts one form of energy into another. A transmitter's antenna converts RF electrical energy into RF electromagnetic energy. The receiver's antenna reverses the process, allowing the recovery of the information on the RF carrier wave sent by the transmitter.

The design of antennas involves a large amount of scientific theory, combined with practical experience and even a little bit of art. There is no "best antenna" for any

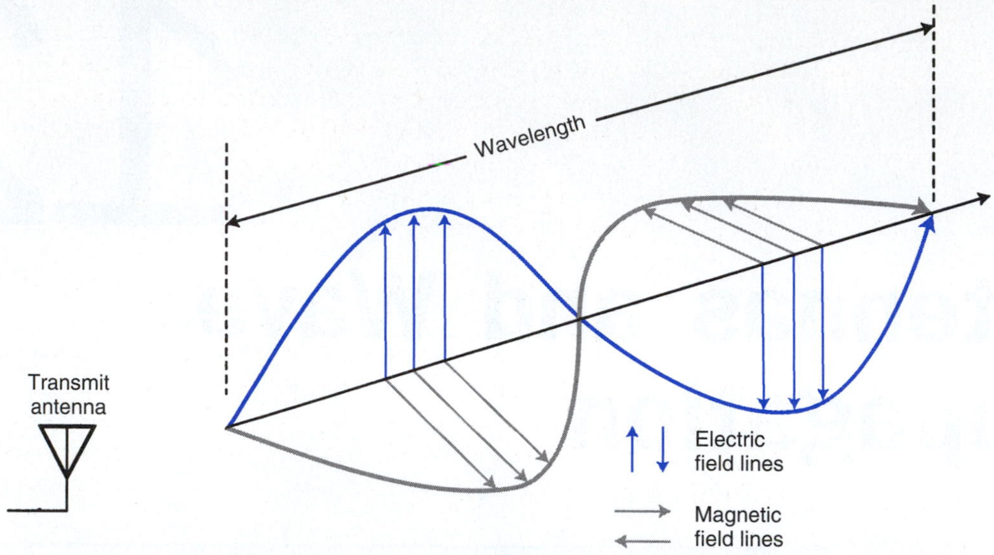

Figure 12–1 An electromagnetic wave in space

application. Often, there are many competing designs from various manufacturers, each having been optimized in one or more ways.

12–1 ELECTRICAL AND ELECTROMAGNETIC ENERGY

When RF electrical energy enters a transmitting antenna, an interesting change takes place. The energy is converted into an *electromagnetic wave* and moves away from the antenna at the speed of light (3×10^8 m/sec). An electromagnetic wave is made up of two parts, an *electric* field, and a *magnetic* field, as shown in Figure 12–1. The *wavelength* of an electromagnetic wave is the distance the wave travels in one cycle. We can calculate it if we know the velocity (speed) of the wave and frequency of the wave.

$$\lambda = \frac{v}{f} \qquad (12\text{–}1)$$

where v is the velocity (3×10^8 m/sec in free space) and f is the frequency.

EXAMPLE 12–1 What is the wavelength of a 100 MHz FM broadcast signal in free space?

Solution

Using equation 12–1, we get

$$\lambda = \frac{v}{f} = \frac{3 \times 10^8 \text{ m/sec}}{100 \text{ MHz}} = \underline{\underline{3 \text{ m}}}$$

Polarization

The radio wave of Figure 12–1 has two energy fields, the electric and magnetic. These two fields are always at right angles to each other, and they are in phase. The *polarization* of a radio wave simply refers to the orientation of the electric field component.

The electric field lines of Figure 12–1 are straight up and down with respect to the earth's surface, so we refer to this as *vertical polarization*. The transmitting antenna design and orientation controls the polarization of the wave. We can also build antennas that produce *horizontal polarization*. A horizontally polarized wave has a magnetic field that is vertical and an electric field that is horizontal.

But that's not all! The radio wave of Figure 12–1 goes straight out into space, never changing its polarization. What if the wave rotated or "corkscrewed" as it traveled outward? This would be an example of *circular* or *elliptical* polarization. And yes, the corkscrew can turn in either a right-hand or left-hand direction.

Polarization of antennas is usually important. If a receiving antenna is designed to pick up vertically polarized energy, then it will not be very sensitive to horizontally polarized signals. In fact, it might not hear them at all!

In general, the polarization of the receiving antenna should match that of the transmitter for best performance.

Circular polarization is often used in satellite communications because the relative orientation of the receiver and transmitter antennas is hard to control. With this type of polarization, both transmitter and receiver must agree on the direction of the wave rotation (right or left hand) or little signal will be received.

The polarization of a radio wave is not set in stone. As a wave makes its way from transmitter to receiver, its polarization can be changed in many ways. *Reflection* from terrain, buildings, and other obstructions can alter polarization drastically. Even waves traveling above ground are affected by the earth's magnetic field, which *rotates* the wave's polarization. This effect is called *Faraday rotation* and strongly affects radio waves in the HF region (3–30 MHz) that travel over long distances (thousands of miles).

An Ideal Radio Wave Source

In order to calculate and compare the performance of various antennas, technicians often compare them with a theoretically perfect antenna called an *isotropic point source*. No such antenna really exists. This may sound intimidating, but the idea is really quite simple. *An isotropic point source radiates equally in all directions*. This includes all three dimensions: height, width, and depth.

If you could see the radiation pattern from an isotropic source, it would look like a perfectly round balloon that expands outward at the speed of traveling radio waves (speed of light). If you could freeze this pattern for a moment of time, it might look like Figure 12–2.

Field Intensity of an Electromagnetic Wave

With the appropriate equipment, we can measure the electric and magnetic field strengths or intensities of a radio wave. Because we know about the characteristics of free space, we can also calculate them. An isotropic source is often used as the basis for calculation of field strength. There are two components in a radio wave, and we can calculate both of them:

$$\mathcal{E} = \frac{\sqrt{30P_t}}{d} \tag{12–2}$$

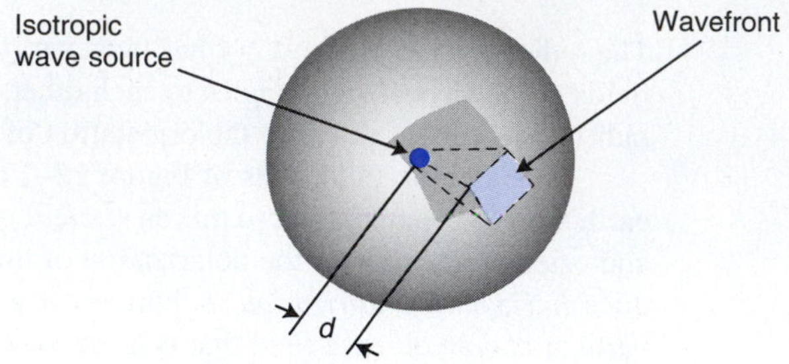

Figure 12–2 The wavefront from an isotropic point source

where $\mathcal{E}$ is the free-space *electric field intensity,* in *volts per meter, d* is the distance from the point source, and P_t is the *isotropic radiated power* (the total power being radiated from the point RF source).

An *electric field* exists any time there is a *difference in potential* between two points in space. The *electric field intensity* indicates how strong this voltage difference is. Although space is essentially an *insulator,* it can still have a voltage drop. (Think about the space between two conducting wires in a circuit; there is an electric field there too, any time there is a difference in potential between the conductors.) The voltage drops on the transmitting antenna are what creates the electric field of the outgoing radio wave.

As the distance from the point source is increased, the electric field decreases in an inversely proportional manner. This makes sense; radio waves get weaker as we move away from the transmitting source.

We can also calculate the strength of the *magnetic* field component:

$$\mathcal{H} = \frac{\sqrt{P_t}}{68.8d} \tag{12–3}$$

where $\mathcal{H}$ is the *magnetic field strength* in amperes/meter, P_t is the isotropic radiated power, and *d* is the distance from the point source. Recall that a magnetic field is created whenever a current flows in a conductor. The currents that flow in a transmitting antenna contribute to the creation of the magnetic field of the radio wave that leaves the antenna. Like the electric field, the magnetic field strength is inversely proportional to the distance from the RF energy source.

EXAMPLE 12–2

A certain transmitter has an equivalent isotropic radiated power (EIRP) of 100 watts. What are the magnetic and electric field strengths at a distance of 10 meters from the transmitting antenna?

Solution

Equations 12–2 and 12–3 calculate electric and magnetic field strengths:

$$\mathcal{E} = \frac{\sqrt{30P_t}}{d} = \frac{\sqrt{30(100\text{ W})}}{10\text{ m}} = \underline{\underline{5.47\text{ V/m}}}$$

$$\mathcal{H} = \frac{\sqrt{P_t}}{68.8d} = \frac{\sqrt{100\text{ W}}}{68.8(10\text{ m})} = \underline{\underline{0.0145\text{ A/m}}}$$

Having trouble with $\mathcal{E}$ and $\mathcal{H}$? Just think of V and I in a conventional ac circuit. $\mathcal{E}$ and $\mathcal{H}$ are the electromagnetic equivalent of V and I. In fact, Ohm's law can even be used to work problems with them, as we'll soon see.

Power Density and Ohm's Law

Suppose that a certain resistor had a 10 volt drop, and a current of 1 amp. What would the power dissipation of the resistor be? You could use Ohm's law to find the answer, since $P = VI = (10\text{ V})(1\text{ A}) = 10$ watts.

The electric and magnetic fields of a radio wave combine to form a *power field.* Look at the rectangular area of Figure 12–2. This area is a *wavefront*. The wavefront moves away from the center of the point source at the speed of light and expands just like the surface of an inflating balloon. The wavefront of the radio wave starts with a fixed amount of power. As the wavefront moves out from the source, that power is spread over a rapidly increasing area. The *power density* of a radio wave is the amount of power that is distributed over an area and is calculated by the equivalent of Ohm's law:

$$\mathcal{P} = \mathcal{E}\mathcal{H} \qquad (12\text{–}4)$$

where $\mathcal{P}$ is the power density in watts per square meter (W/m^2), $\mathcal{E}$ is the electric field intensity in volts/meter, and $\mathcal{H}$ is the magnetic field strength in amperes/meter. Notice how the units make sense; volts times amps gives watts, and of course, we now have an *area* in square meters.

By combining the terms and constants in equations 12–2, 12–3, and 12–4, we can get a formula for power density in terms of just transmitted power and distance:

$$\mathcal{P} = \mathcal{E}\mathcal{H} = \left(\frac{\sqrt{30P_t}}{d}\right)\left(\frac{\sqrt{P_t}}{68.8d}\right) = \frac{P_t}{12.56d^2} = \frac{P_t}{4\pi d^2} \qquad (12\text{–}5)$$

Equation 12–5 shows us that the power density falls off as the *square* of distance. This is one reason why so much amplification has to be performed in radio transmitters and receivers! This is sometimes referred to as the "inverse square law" of power. Equation 12–5 is valid when we are at least several wavelengths away from a transmitting antenna.

EXAMPLE 12–3

A certain transmitter has an equivalent isotropic radiated power (EIRP) of 5 kW. What are the magnetic and electric field strengths and power density at distances of (a) 10 meters and (b) 100 meters from the transmitting antenna?

Solution

a. Equations 12–2 and 12–3 calculate electric and magnetic field strengths:

$$\mathcal{E} = \frac{\sqrt{30P_t}}{d} = \frac{\sqrt{30(5000\text{ W})}}{10\text{ m}} = \underline{\underline{38.73\text{ V/m}}}$$

$$\mathcal{H} = \frac{\sqrt{P_t}}{68.8d} = \frac{\sqrt{5000\text{ W}}}{68.8(10\text{ m})} = \underline{\underline{0.103\text{ A/m}}}$$

$$\mathcal{P} = \frac{P_t}{4\pi d^2} = \frac{5000\text{ W}}{4\pi(10^2)} = \underline{\underline{3.98\text{ W/m}^2}}$$

b. The same procedures are used:

$$\mathcal{E} = \frac{\sqrt{30P_t}}{d} = \frac{\sqrt{30(5000\text{ W})}}{100\text{ m}} = \underline{\underline{3.873\text{ V/m}}}$$

$$\mathcal{H} = \frac{\sqrt{P_t}}{68.8d} = \frac{\sqrt{5000\text{ W}}}{68.8(100\text{ m})} = \underline{\underline{0.0103\text{ A/m}}}$$

$$\mathcal{P} = \frac{P_t}{4\pi d^2} = \frac{5000\text{ W}}{4\pi(100^2)} = \underline{\underline{.0398\text{ W/m}^2}}$$

Notice that the magnetic and electric field strengths became smaller by a factor of ten, but the power density shrunk one-hundred-fold!

The Characteristic Impedance of Free Space

The concept of Ohm's law is also quite handy for expressing a quality or property of space known as its *characteristic impedance*. If 10 volts appears across a resistor and 1 amp flows, then the resistor must be 10 Ω, according to Ohm's law. We can define the impedance of space in the same manner:

$$\mathcal{Z} = \frac{\mathcal{E}}{\mathcal{H}} \qquad (12\text{–}6)$$

where $\mathcal{Z}$ is the impedance of a medium (like space) in ohms, and $\mathcal{E}$ and $\mathcal{H}$ are the electric and magnetic field intensities. For any medium, $\mathcal{Z}$ is a constant, very much like the characteristic impedance of a transmission line. Different media have varying impedances.

EXAMPLE 12–4

What is the impedance of free space, given the data from Example 12–3(a)?

Solution

Since $\mathcal{Z}$ is just the ratio of $\mathcal{E}$ and $\mathcal{H}$, we get

$$\mathcal{Z} = \frac{\mathcal{E}}{\mathcal{H}} = \frac{38.73\text{ V/m}}{0.103\text{ A/m}} = \underline{\underline{376\ \Omega}}$$

This is *very* close to the theoretical value of 377 Ω that is used by most technicians. This is also very close to 120π, which is sometimes used for convenience.

There are other ways of obtaining the impedance of free space—they are probably not of much use to a technician.

The characteristic impedance of free space is the ratio of electric to magnetic field strengths in radio waves traveling through it. Here's why it's important: You'll recall that a transmission line is used to carry energy from a transmitter and has its own characteristic impedance, Z_0. A transmission line needs to see a *matched load* at its end in order to transfer 100% of its energy and avoid reflection.

One way of viewing the functioning of an antenna is as an *impedance matching device*. The antenna accepts the traveling RF wave from the feedline (operating at Z_0) and performs the necessary impedance transformation so that the wave can continue to flow out into space, a medium with a $\mathcal{Z}$ of 377 Ω. For this to happen, an *impedance transformation* must take place!

If you've ever used a megaphone, you've seen *acoustic* impedance transformation in action. A megaphone helps you to be "louder" at a distance not only because it tends to focus energy in a specific direction but also because it improves the acoustic impedance match between your voice apparatus (vocal cords, throat, mouth, and nasal passages) and the outside air. All antennas operate in this fashion. They are really just impedance transformers!

EXAMPLE 12–5

At a distance of 75 miles from a transmitter, the power density is 10 pW/m². What are the electric and magnetic field strengths? What is the EIRP of the transmitter?

Solution

Since $\mathcal{Z}$ is just the ratio of $\mathcal{E}$ and $\mathcal{H}$ and is known to be 377 Ω, we can find the unknowns by using what we already know about Ohm's law:

$$P = I^2 R \quad \text{and} \quad I = \sqrt{\frac{P}{R}}$$

So, remembering that $\mathcal{H}$ is analogous to I, and $\mathcal{Z}$ is really R, we get

$$\mathcal{P} = \mathcal{H}^2 \mathcal{Z} \quad \text{and} \quad \mathcal{H} = \sqrt{\frac{\mathcal{P}}{\mathcal{Z}}} = \sqrt{\frac{10 \text{ pW/m}^2}{377\ \Omega}} = \underline{\underline{163 \text{ nA/m}}}$$

Likewise, we can do the same thing with $\mathcal{E}$.

$$P = \frac{V^2}{R} \quad \text{and} \quad V = \sqrt{PR}$$

So,

$$\mathcal{P} = \frac{\mathcal{E}^2}{\mathcal{Z}} \quad \text{and} \quad \mathcal{E} = \sqrt{\mathcal{P}\mathcal{Z}} = \sqrt{(10 \text{ pW/m}^2)(377\ \Omega)} = \underline{\underline{61.4\ \mu\text{V/m}}}$$

The most direct way of finding the EIRP is to use equation 12–5 and solve for P_t. The distance units of *miles* must first be converted to *meters*:

$$d = 75 \text{ miles} \times \frac{5{,}280 \text{ ft}}{1 \text{ mile}} \times \frac{1 \text{ m}}{3.28 \text{ ft}} = 120{,}732 \text{ m}$$

$$\mathcal{P} = \frac{P_t}{4\pi d^2} \quad \text{so} \quad P_t = \text{EIRP} = \mathcal{P}(4\pi d^2) = (.01 \text{ pW/m}^2)4\pi(120{,}732 \text{ m})^2 = \underline{\underline{1.83 \text{ W}}}$$

This example demonstrates that everything you know about Ohm's law is useful in electromagnetics. It also shows that a lot can be learned about a transmitter's power output and radiation pattern by remote measurements. The FCC can use this information to verify that broadcasters (and other radio spectrum users) are operating under the correct power levels by accurately measuring the field strength from stations at a distance.

In real life, a small signal power like this isn't likely to go 75 miles without being blocked by terrain or some other obstacle. However, in *satellite* applications, a watt or two of power can cover an amazing area and distance, due to the lack of obstructions.

The received field intensities in this example are fairly small, but current receiver technologies are fully capable of recovering signal levels in this range. Remember that noise (both internal and external to the receiver) will be the primary limiting factor in the receiving portion of the process.

Section Checkpoint

12–1 What are the two field components in a radio wave? How are they oriented with respect to each other?

12–2 Explain what is meant by the *polarization* of a radio wave. Give several examples.

12–3 For best signal pickup, how should a receiver's antenna be oriented?

12–4 What is an *isotropic point source?*

12–5 Describe the shape of the radiation pattern from an isotropic source.

12–6 What are the units of electric and magnetic field intensity?

12–7 Give the analogous (Ohm's law) equivalents for $\mathcal{E}$, $\mathcal{H}$, $\mathcal{P}$, and $\mathcal{Z}$.

12–8 If the distance between a transmitter and receiver is *doubled,* what will happen to the power density $\mathcal{P}$?

12–9 What does *EIRP* stand for?

12–10 What is the value for the characteristic impedance of space? What two quantities can be divided to find it?

12–2 THE DIPOLE AND MARCONI ANTENNAS

The *Hertz antenna* or *dipole* is perhaps the simplest and most popular antenna found in communication systems. It consists of nothing more than two conductors fed 180° out of phase. One way of viewing a dipole antenna is as a modified section of transmission line, where the two conductors have been spread apart, resulting in the antenna of Figure 12–3.

Dipole Operating Principle

In a transmission line, an impedance mismatch causes the traveling wave from the generator to *reflect* backward from the point of mismatch. In Figure 12–4, each half of the dipole antenna is treated as a transmission line. The *voltage standing wave pattern* (heavy blue line) is developed after many wavefronts pass down the antenna. Note that there is *zero* current at the ends of the antenna (after all, they are open-circuited) but a relatively high voltage. Maximum current flows at the feedpoint. Notice the high voltage at the ends of the antenna; depending on the design of the antenna, this voltage can be many times that at the feedpoint.

The ends of a dipole are relatively high impedance points. This is why the voltages are high. The antenna can be easily detuned if the dipole ends are too close to other objects.

Figure 12–3 A half-wavelength dipole antenna

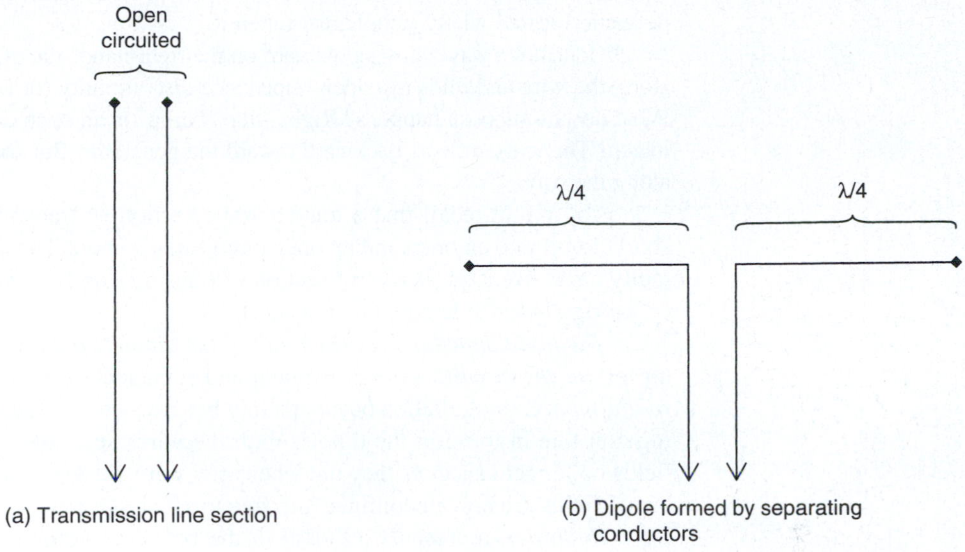

Figure 12–4 Wave movement on a dipole and resulting current and voltage profile

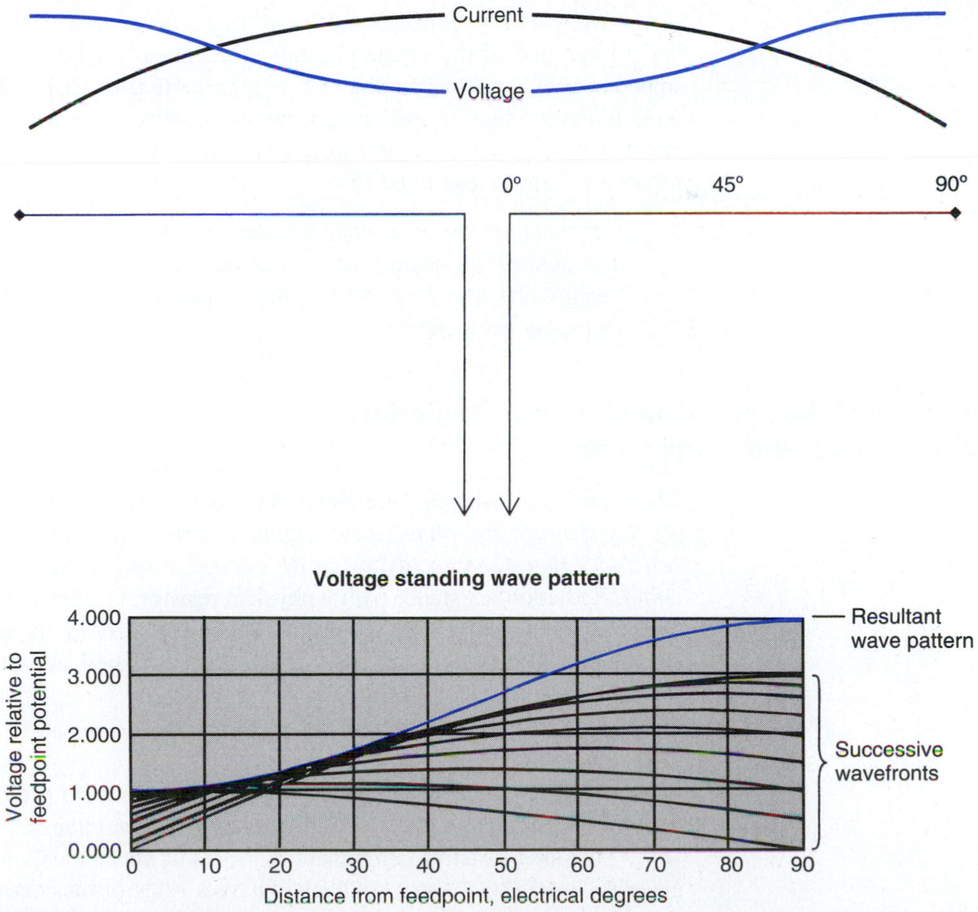

Also, the high voltages are a potential safety hazard. A transmitting dipole antenna should never be located where people can touch it.

Consider a wave moving outward on the right-hand side of the dipole. The wave travels along the wire and finds a sudden impedance discontinuity (or bump) at the end of the wire. What do you suppose happens? Right—the "bump" is an open circuit, which is a severe mismatch! The wave reflects backward toward the generator. But something interesting happens along the way.

You might recall that a quarter-wave section of transmission line can transform a short circuit into an open and an open circuit into a short. The dipole-half in the figure certainly *looks* like a quarter-wave section of transmission line. So does the input impedance of the right-hand side equal 0 Ω, a short?

The input impedance of each half of the dipole would be a short circuit if there were no loss in the circuit. As both incident and reflected waves move along the dipole wire, *radiation* occurs. Radiation occurs mainly because we separated the two wires of the transmission line in forming the dipole. With the wires separated, their magnetic and electric fields no longer cancel as they did when they were closely spaced in the transmission line. These fields are now unconfined and free to move out into space as radio waves.

Radiation represents a "loss" in the reflected voltage and current on the dipole. Therefore, the input impedance of the half-dipole is *not* a short circuit; in fact, it comes close to $(36.5 + j21.5)\ \Omega$. This is why the voltage at the feedpoint is not canceled, as shown on the graph of Figure 12–4. Of course, this is exactly what we want the antenna to do. We want it to convert the RF electrical energy into electromagnetic wave energy.

Each half of the dipole presents an identical impedance. Therefore, the total input impedance of a half-wave dipole is just twice that of each side, or about $(73 + j43)\ \Omega$. Notice that the input impedance is mostly resistive, with a small inductive reactance. By *shortening* the antenna about 5%, most of the inductive reactance is eliminated, leaving an input impedance close to 67 Ω.

For most simple antennas, the element lengths will work out to be 95% of the desired wavelength fraction at resonance. For a dipole antenna, most technicians use 73 Ω as the input impedance value and forget about the inductive component.

Resistive Antenna Components: Radiation Resistance and Ohmic Loss

There are two resistance components that appear at the input terminals of an antenna, *radiation resistance* and *ohmic loss*. *Radiation resistance* is *the resistive portion of the input resistance that when excited by an RF current, results in radiated RF energy from the antenna*. Radiation resistance isn't a physical resistor; it represents the ability of the antenna to emit radio waves. Given a constant RF current driving the antenna, a higher radiation resistance will emit a larger radio signal. It is calculated as follows:

$$R_R = \frac{P_R}{I^2} \quad \textbf{(12–7)}$$

where P_R is the total radiated power from the antenna and I is the feedpoint current. In general, the longer an antenna, the higher its radiation resistance.

The other resistive component of the input impedance, *ohmic loss,* does nothing but generate *heat*. The antenna conductors have a value of resistance determined by the material they're made of and their physical dimensions. Antenna manufacturers would like to be able to sell units with zero ohmic loss, but that's impossible! Nearby objects and even the earth's surface itself can contribute to ohmic loss since they can absorb RF energy and convert it to heat.

EXAMPLE 12–6

A certain antenna radiates 100 watts when its feedpoint current is 1.4 amps. Find the radiation resistance of the antenna.

Solution

From equation 12–7, we get

$$R_R = \frac{P_R}{I^2} = \frac{100\text{ W}}{(1.4\text{ A})^2} = \underline{\underline{51\ \Omega}}$$

Therefore, the portion of the input impedance for this antenna that contributes to radiation (the production of radio wave energy) is $\underline{\underline{51\ \Omega}}$. Be careful; we're *not* saying that the total input impedance is 51 Ω!

Antenna Efficiency

The *efficiency* of an antenna system is defined as the ratio of power radiated to total power put into the antenna (which is equal to the power loss plus the radiated power):

$$\eta = \text{efficiency} = \frac{P_R}{P_R + P_{\text{LOSS}}} \tag{12–8}$$

No antenna is 100% efficient. However, there is a great difference between different antenna designs. This is partly where the art and science of antenna design meet.

EXAMPLE 12–7

A certain antenna has a radiation resistance of 51 Ω, and an ohmic (loss) resistance of 10 Ω. If a current of 2 amps is driving the antenna, calculate the following:

a. the power radiated

b. the power lost due to heat

c. the total input resistance of the antenna

d. the efficiency of the antenna

Solution

a. By rearranging equation 12–7 to solve for radiated power, we get

$$P_R = I^2 R_R = (2\text{ A})^2(51\ \Omega) = \underline{\underline{204\text{ W}}}$$

b. The power lost due to heat can be computed by Ohm's law:

$$P_{\text{LOST}} = I^2 R_{\text{LOSS}} = (2\text{ A})^2(10\ \Omega) = \underline{\underline{40\text{ W}}}$$

c. The total input resistance of the antenna is just the sum of the ohmic and radiation components:

$$R_{\text{in}} = R_R + R_{\text{LOSS}} = 51\ \Omega + 10\ \Omega = \underline{\underline{61\ \Omega}}$$

d. The efficiency is calculated using equation 12–8:

$$\eta = \text{efficiency} = \frac{P_R}{P_R + P_{\text{LOSS}}} = \frac{204\text{ W}}{204\text{ W} + 40\text{ W}} = 0.836 = \underline{\underline{83.6\%}}$$

This is a good antenna system; 83.6% of the RF energy from the transmitter is converted into electromagnetic energy.

A Shortcut for Efficiency

The efficiency of an antenna can be found without knowing the antenna feedpoint current I by using Ohm's law to simplify equation 12–8:

$$\eta = \text{efficiency} = \frac{P_R}{P_R + P_{\text{LOSS}}} = \frac{I^2 R_R}{I^2 R_R + I^2 R_{\text{LOSS}}}$$

The common factor I^2 can be canceled between the numerator and denominator, leaving us with

$$\eta = \text{efficiency} = \frac{R_R}{R_R + R_{\text{LOSS}}} \tag{12–9}$$

Equation 12–9 can be used to find efficiency directly from the resistance values, so it is handy when these are given (or can be measured).

Dipole Polarization

The *polarization* of an antenna refers to the orientation of the electric field of the radio waves being produced (or received) by it. For a dipole antenna, the polarization is the same as the axis of the conductor.

A dipole strung horizontally produces horizontally polarized waves. A dipole hung straight up and down produces vertically polarized output.

Remember that two stations must normally be using the same type of polarization, or severe signal loss will likely result. Antennas used for special purposes often have markings or other instructions as to how to mount them so that proper polarization is achieved.

Dipole Radiation Pattern and Gain

The *radiation pattern* of an antenna is the 3-D distribution of energy leaving the device. You'll recall that an *isotropic point source* is a nonexistent, theoretical antenna that radiates equally well in all directions. The 3-D radiation pattern of an isotropic source can be thought of as a perfect sphere. The RF energy is spread evenly over the entire area of the sphere, so that from any direction, the signal is equal.

Suppose that we take a round balloon, and fill it with water so that it has plenty of room for expansion. The shape of the balloon is like the radiation pattern of the isotropic point source—it is perfectly round. The total amount of water in the balloon is analogous to the *transmitted power* from the isotropic source. How does squeezing the balloon affect

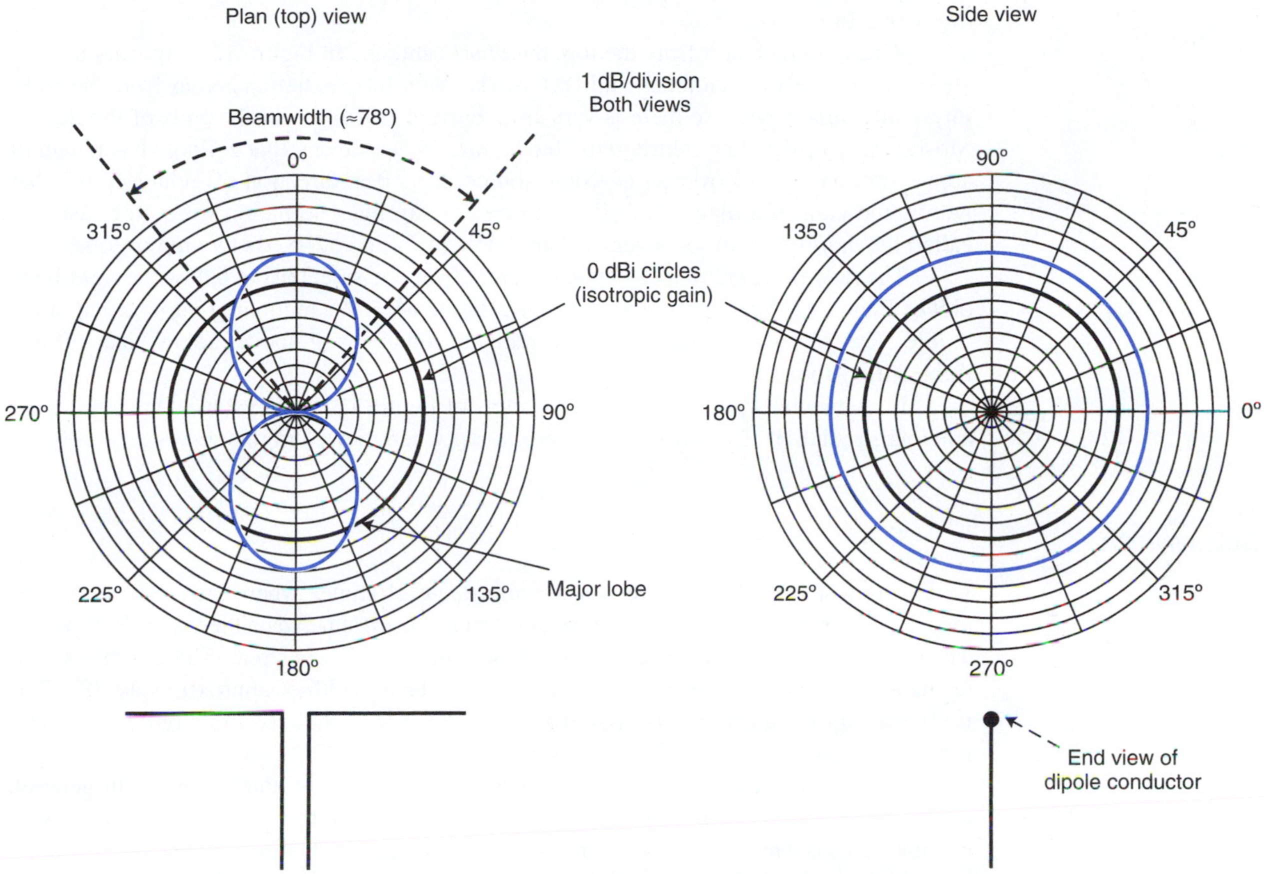

Figure 12–5 The pattern of a dipole antenna

the amount of water inside? Since water is relatively incompressible, its volume remains constant as the balloon is compressed. If we squeeze in the center, the sides bulge out; if we try pushing in one end, the other end pops out. *The total amount of water remains the same. Only its distribution is changed.*

The same thing can be said for antennas. Most antennas radiate power better in some directions and worse in others. The three-dimensional picture of this is called the *radiation pattern* of the antenna. Since it is hard to show 3-D patterns on the flat pages of books, manufacturers normally show a top and side graph of the antenna's pattern. The pattern for a half-wave dipole is shown in Figure 12–5.

There are two ways of looking at the radiation pattern, either from the top (the plan view) or from the side. Notice that the graph of each is built out of circles, one inside another. Each circle represents a 1 dB change in field strength (power). The radial lines indicate the *angle* at which the power changes were measured. On both graphs, the bold circle in the middle represents the radiation of an *isotropic* antenna. This *must* be a circle because an isotropic antenna radiates equally well in all directions. Therefore, we're just looking at the top and side views of a *sphere*. The units on the graphs are *dBi,* which means "decibels of gain with respect to an isotropic source." Recall that the water balloon was a sphere before we squeezed it. When we build an antenna, it usually has *directionality*. It radiates better in some directions at the expense of radiating *worse* in others.

When we look at it from the top, the *dipole* antenna in Figure 12–5 appears to radiate best from the *sides,* at the 0° and 180° marks. Very little radiation comes from the ends. Physically, this is because there is very little *current* flowing near the ends of the dipole. Most of the current flows close to the feedpoint. We would say that a dipole has a gain of approximately +2 dB over an isotropic source in its best direction of radiation. In other words, the gain of a dipole is 2 *dBi*. The area of strongest radiation from an antenna is called the *major lobe* of the antenna. The dipole in Figure 12–5 has two major lobes.

The radiation pattern looks quite different from the side; the dot below the right-hand plot of Figure 12–5 represents the *end* of the dipole wire. We're looking into the end of the dipole. This plot shows that the dipole radiates equally well above, below, and left and right as far as we can see from the end.

If we mentally put together the two plots of Figure 12–5, we get a 3-D picture of the antenna's radiation. The overall shape is a *donut* when viewed in this way.

Beamwidth

Returning to Figure 12–5, there is a definite range of angles where the antenna radiates most of its energy. The *beamwidth* of an antenna is the angle between the two −3 dB points. The −3dB is measured with respect to the maximum radiation strength in the major lobe of the antenna's pattern. A half-wave dipole has a beamwidth of approximately 78°. This is *somewhat* directional, but it's not effective enough to accurately aim a radio signal at a remote location.

There is a general relationship between *beamwidth* and antenna gain: In general, the more *narrow* the beamwidth of an antenna, the *larger* the gain of the antenna will become. This is why directional antennas are desirable; a highly directional antenna usually has a high gain and also focuses its energy over a narrow angle, which helps to eliminate interference.

What do you suppose the dipole's radiation pattern will look like if it is placed *vertically?* Just switch the left- and right-hand graphs of Figure 12–5, and you'll get the answer! (Note that a vertical dipole must be at least one-half wavelength off the ground because the earth will tend to capacitively "load" the lower dipole conductor if it is too close to ground, which will increase the antenna's loss as well as distort its radiation pattern.)

EXAMPLE 12–8

Calculate the length of a half-wavelength dipole for operating on a frequency of 14.150 MHz in the following ways:

a. as a raw half-wavelength figure

b. taking into account the 95% "fudge factor" for practical antenna construction

Solution

a. A half-wavelength at this frequency is

$$\lambda = \frac{v}{f} = \frac{3 \times 10^8 \text{ m/sec}}{14.150 \text{ MHz}} = 21.2 \text{ m}$$

So one-half of this wavelength is (21.2 m/2) = $\underline{\underline{10.6 \text{ m}}}$

b. Because of fringing at the end of the antenna wires (a capacitive effect), a dipole antenna resonates when it is about 95% of the actual half-wavelength length. The *practical* length is therefore

$$L = (0.95)(10.16 \text{ m}) = \underline{\underline{10.07 \text{ m}}}$$

Note: A popular formula used for a half-wavelength dipole is as follows:

$$L_{\text{feet}} = \frac{468}{f_{\text{MHz}}}$$

This formula returns the length of the antenna in feet and includes the 95% "fudge factor." The antenna above calculated in this way gives

$$L_{\text{feet}} = \frac{468}{f_{\text{MHz}}} = \frac{468}{14.150} = \underline{\underline{33.07 \text{ ft}}}$$

This is the same as 10.07 meters, which was obtained before.

In order to get an *omnidirectional* pattern, a dipole can be strung vertically. An *omnidirectional* antenna radiates equally well in all directions (in the horizontal plane, looking from above). Such antennas are popular in broadcast, where a uniform area is to be covered. Using a vertical dipole at the low frequencies of AM broadcast creates a problem, as the next example will demonstrate.

EXAMPLE 12–9

Calculate the length of a half-wavelength dipole for operating on a frequency of 810 kHz, giving the answer in feet. Give construction details for the device.

Solution

The shorthand formula works best here:

$$L_{\text{feet}} = \frac{468}{f_{\text{MHz}}} = \frac{468}{0.810} = \underline{\underline{577.7 \text{ ft}}}$$

This is quite a large antenna! And there's another problem: We can't let the bottom conductor of a vertical dipole get close to the earth, or the earth will capacitively "short out" the dipole's signal. Generally, the bottom of the dipole will be either $1/2\,\lambda$ or $1\,\lambda$ above ground. This is going to be one very tall tower!

This is a good example of why a different antenna, the *Marconi* antenna, is used for AM broadcasting.

The Marconi Antenna

A Marconi antenna is pictured in Figure 12–6. It's really nothing more than a one-quarter length radiator placed vertically over the earth. The Marconi antenna is fed from the ground and has a theoretical input impedance of half of a dipole (over perfectly conducting ground), which is about $(36.5 + j21.5)$. When it is shortened 5% (like

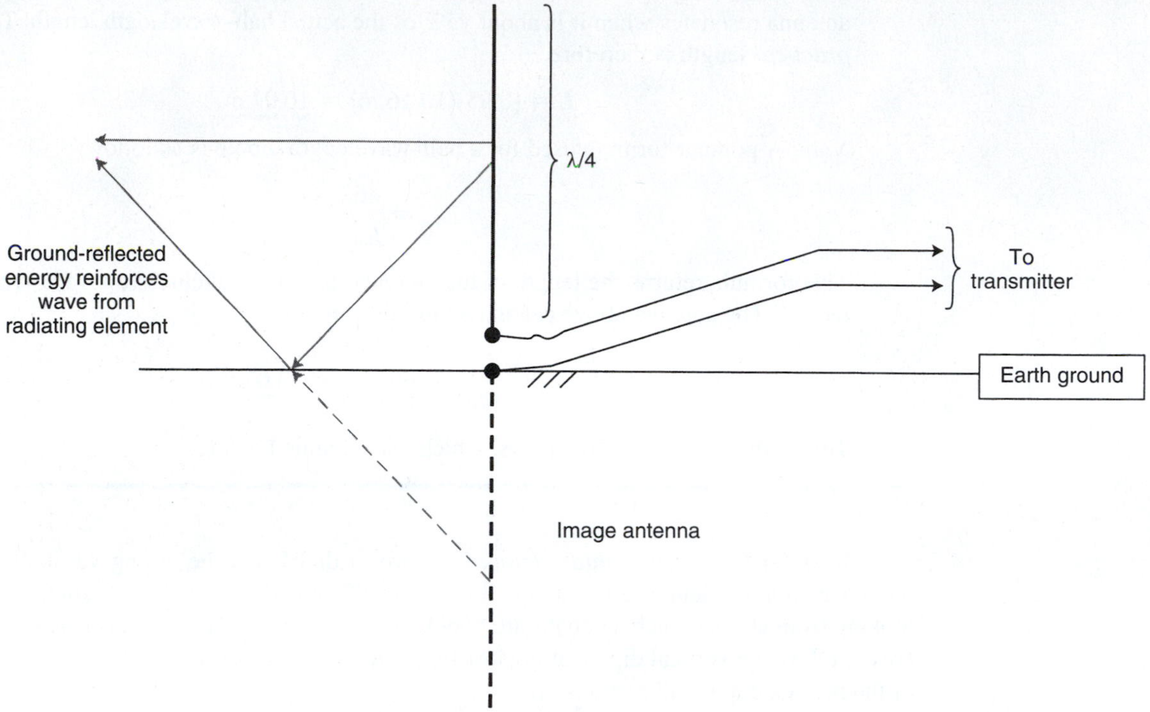

Figure 12–6 A Marconi antenna

the dipole), the input impedance becomes approximately 34 Ω. The Marconi antenna works because of the low resistance of the earth. The earth ground provides a return path for the RF currents that flow in the antenna. Part of the antenna radiation pattern is provided by the earth in the form of reflected radiation. It is as if an "image antenna" buried beneath the surface of the earth were providing the radiation, rather than reflection.

For low-frequency applications (such as AM broadcast), the Marconi has several important advantages over a vertical dipole. First, it need only be one-half of the dipole's length, which greatly reduces the cost of erecting towers. Second, since the feedpoint has a low impedance, it can be located at earth level without degrading the antenna's performance. Last, when the radiation pattern of the antenna is viewed from the top, it is *omnidirectional*—it radiates equally in all directions. Figure 12–7 shows the radiation pattern for a Marconi antenna.

There are disadvantages to the Marconi antenna. First, the gain of the antenna is about 1 dB less than a half-wave dipole. Second, the Marconi antenna is extremely dependent upon the conductivity of the earth, because the earth acts as a mirror, effectively acting as the missing half of the antenna. The earth also supplies the path for the returning currents from the antenna. In AM broadcast, the earth conductivity is improved through the use of a *radial* system. The standard practice is to use 120 wires, equally spaced and arranged like the spokes of a wheel. The radial wires improve the ground conductivity, which greatly increases the efficiency of the antenna.

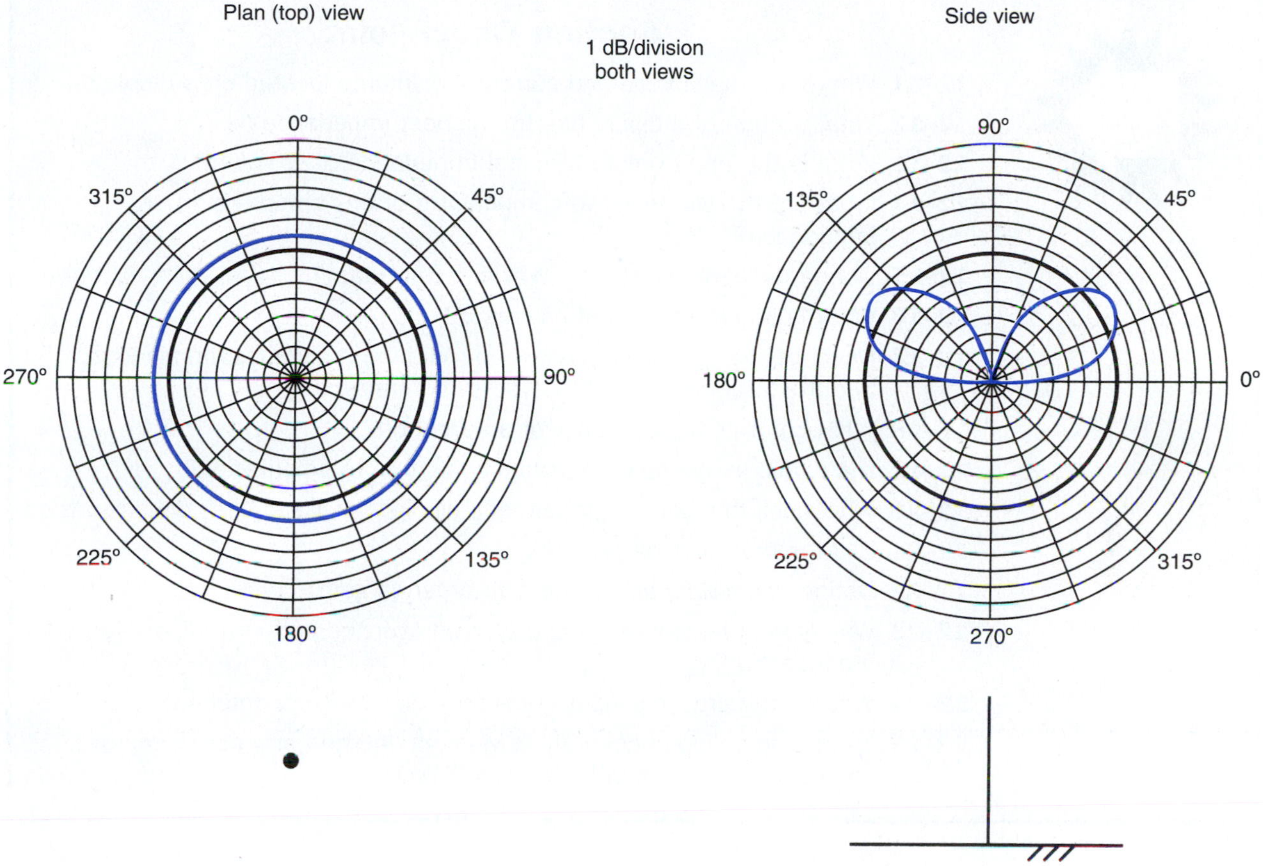

Figure 12–7 Radiation pattern of a Marconi antenna

EXAMPLE 12–10 Calculate the length of a Marconi antenna for operating on a frequency of 810 kHz, giving the answer in both meters and feet.

Solution

The practical length of the antenna will be 95% of one-quarter wavelength:

$$L = 0.95(\lambda/4) = (0.95)\left(\frac{1}{4}\right)\left(\frac{3 \times 10^8 \text{ m/sec}}{810 \text{ kHz}}\right) = \underline{\underline{87.96 \text{ m}}}$$

The length when expressed in feet becomes

$$L_{\text{feet}} = L_{\text{meters}} \times \frac{3.28 \text{ ft}}{1 \text{ m}} = 87.96 \text{ m} \times \frac{3.28 \text{ ft}}{1 \text{ m}} = \underline{\underline{288.6 \text{ ft}}}$$

This is still a physically large structure, but it is *much* easier to construct than the half-wavelength version!

Section Checkpoint

12–11 Where are the voltage and current maximums located on a dipole?

12–12 What portion of a dipole has the highest impedance?

12–13 What is the approximate terminal impedance of a dipole?

12–14 What are the two resistive components of any antenna's input impedance?

12–15 Define *radiation resistance*. Why is it important?

12–16 What is *ohmic loss?* What causes it?

12–17 Describe the relationship between radiation resistance, ohmic loss, and the total input resistance of an antenna.

12–18 What controls the *efficiency* of an antenna?

12–19 If a dipole is mounted horizontally, what is its polarization?

12–20 What does the radiation pattern of a dipole look like in three dimensions?

12–21 What is the meaning of *dBi?*

12–22 Define *beamwidth* and relate it to antenna gain.

12–23 Why is the Marconi antenna preferred over the half-wave dipole for AM broadcasting?

12–24 What is the practical input resistance of a Marconi antenna?

12–25 What is installed underneath a Marconi antenna in order to improve the earth's conductivity?

12–3 DIRECTIONAL ANTENNAS

The dipole and Marconi antennas are useful in many applications, but often an antenna that focuses its energy over a more limited area is required. Antennas that have that capability are called *directional* antennas.

There are several important advantages of directional antennas. First, because the energy is sent only in the desired direction, the possibility of interference between stations is greatly reduced. Second, the reduced beamwidth of a directional antenna translates into improved *gain*. Less transmitter power is required to cover the same distance than would be needed if nondirectional antennas were being used. Last, because the direction of the RF signal can be controlled, security can be enhanced (the signal can be aimed away from potential eavesdroppers), and frequencies can be reused by several stations in different locations.

The primary disadvantage of a directional antenna is its directionality! If one station is equipped with a directional antenna, and it (or the other station) is moved, the antenna must be physically reoriented to point in the proper direction. For situations where this does occur, a *rotator* is used on the antenna system to achieve this automatically.

Figure 12–8 shows an automated U.S. Geological Survey (USGS) stream gauging station equipped with a UHF transmitter and a directional crossed-Yagi antenna. (This antenna transmits *circular* polarization.) The station continuously collects data and periodically transmits the data to an orbiting satellite. There are many parasitic elements on the antenna, which provides a high power gain. Notice the solar panel on top of the

Figure 12–8 A remote USGS station with a UHF Yagi antenna

station. This station is designed to operate independently of commercial power, so it can be placed anywhere measurements need to be made.

The Yagi-Uda Antenna

The *Yagi-Uda antenna,* named after its Japanese inventors, is often called the *Yagi* for short. It consists of a *driven element* (which is usually a dipole antenna) and extra antenna elements called *parasitic elements,* as shown in Figure 12–9.

The parasitic elements of the Yagi antenna are not directly connected to the feedline, yet they help to focus the antenna's RF energy in the forward direction. There are two types of parasitic elements, *reflectors* and *directors*. When RF energy is applied to the driven element, the energy flows outward and excites the parasitic elements. The *reflector* elements are made electrically "long" (the first reflector is about 5% longer than the driven element) and serve to reverse the direction of the RF energy emitted from the back of the driven element. The *director* elements are in the front of the antenna, and they are made electrically "short" (the first director is about 5% shorter than the driven element). The director elements further reinforce and focus the energy leaving the front of the antenna. In general, increasing the number of parasitic elements increases the gain of the antenna. Adding additional director elements is more effective (in terms of dB gain per element added) than adding more reflectors.

A Yagi antenna is normally built for operation on a single band, although with special techniques, it can be made to resonate on several different frequencies. The Yagi antenna is therefore not considered a wideband antenna.

Yagi radiation pattern and gain Figure 12–10 shows a typical radiation pattern for a three-element Yagi antenna. Such an antenna has one large *major lobe* at the front, a *minor*

Figure 12–9 The Yagi-Uda antenna

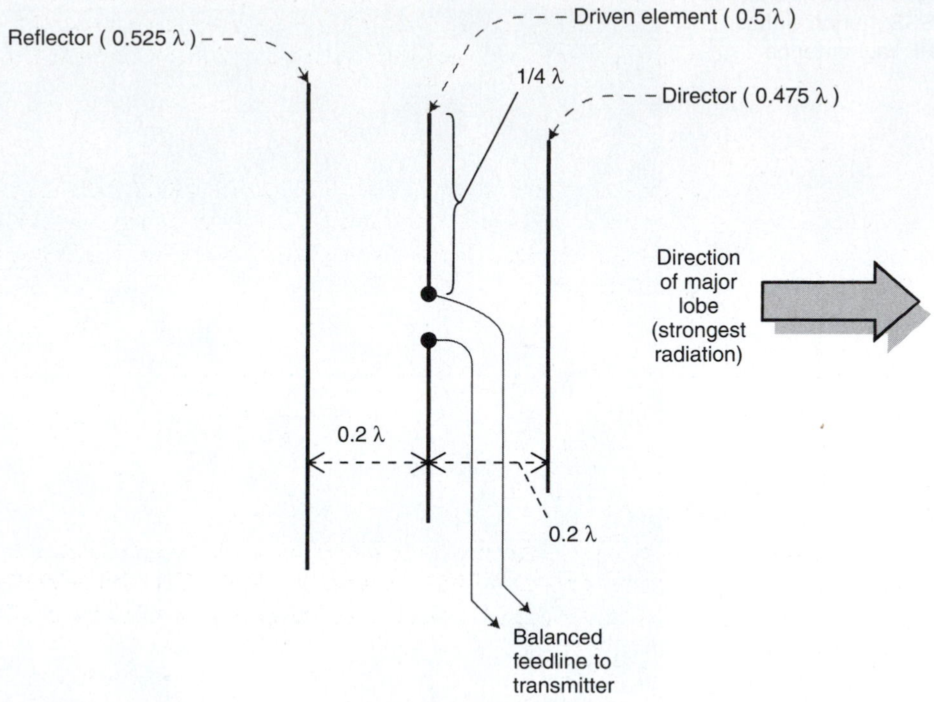

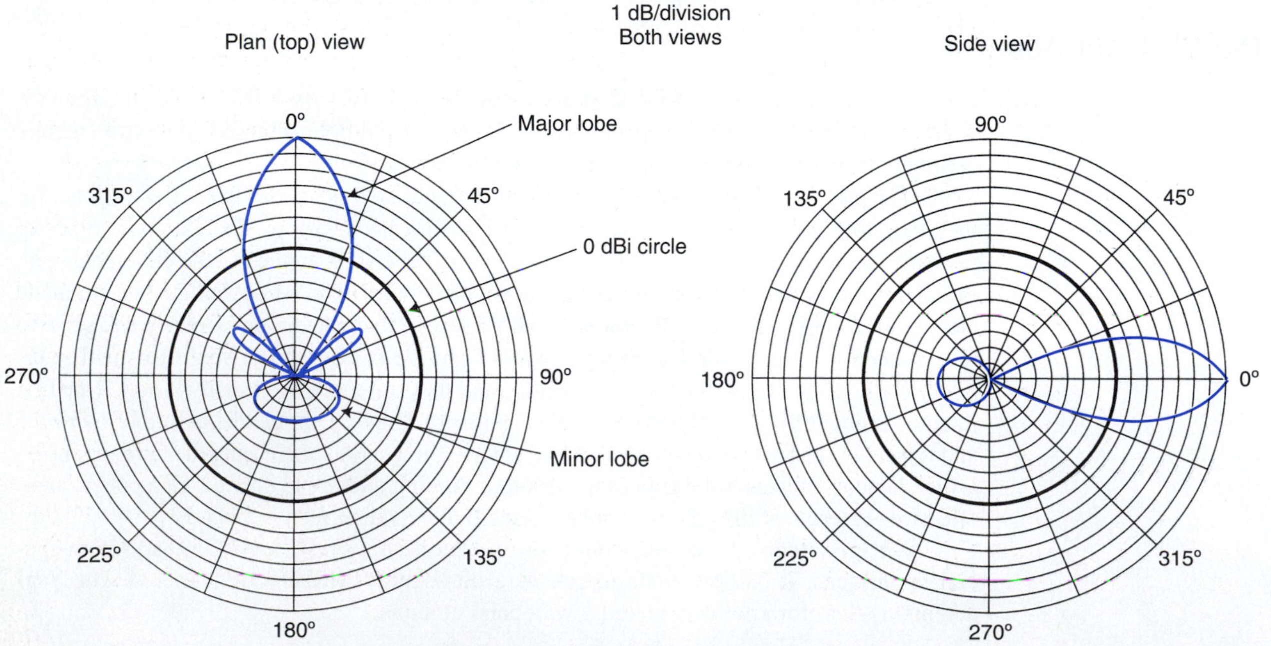

Figure 12–10 Radiation pattern of a three-element Yagi antenna

lobe at the rear, and two small *parasitic lobes*. A three-element Yagi can produce a forward gain of about 7.1 dBi, which is a great improvement over a half-wave dipole (2 dBi). The *parasitic lobes* are undesired, and by careful design and manufacture, they can be minimized.

Front to Back Ratio

One measure of the effectiveness of a directional antenna is its *front to back (FB) ratio,* which is the difference in decibel power gain between the preferred (front) and backward directions. The higher this value, the better the antenna is at controlling the energy distribution.

EXAMPLE 12–11 What is the front to back ratio of the three-element Yagi whose radiation pattern data is shown in Figure 12–10?

Solution

The front to back ratio is simply the difference between the forward and reverse gains. By inspection, the forward gain is +7 dBi (7 divisions past the 0 dBi circle). The back gain is −5 dBi (5 divisions inside the 0 dBi circle). The FB ratio is therefore (+7 dBi − (−5 dBi)) = +12 dB.

The meaning of this is as follows: The antenna has 12 dB more gain in the forward direction than in the reverse direction. That is almost a 16:1 power ratio!

Log-Periodic Antenna

A Yagi antenna produces excellent forward gain, and has good front–back characteristics, but can only operate over a limited frequency range (±5% of the center frequency is typical). In applications such as television broadcast reception, where a wide frequency range must be covered, the bandwidth limitations make the Yagi unusable. The *log-periodic antenna* retains most of the directionality of a Yagi yet permits operation over a wide frequency range. Log-periodic antennas can be constructed to cover more than an octave (2 : 1, 200%) frequency range. Figure 12–11 shows how a log-periodic antenna is built.

In contrast to a Yagi, all the elements of a log-periodic antenna are driven by the feedline. Every other element is connected 180° out of phase. The longest element, L_1 in Figure 12–11, is cut to be resonant at the lowest operating frequency, and the shortest element resonates at the highest frequency. Because there is an element near resonance for any frequency in the operating range, the antenna covers a large bandwidth without any frequency breaks.

The directionality of the antenna is obtained the same way as the Yagi. For example, if the antenna is operating near the frequency where element L_2 is resonant, then element L_1 acts as a reflector, and all the shorter elements act as directors.

The antenna is given the name "log-periodic" because of the design formulas used to construct it. The spacing between the elements progressively becomes smaller as the elements shrink; however, the *ratio* (τ) of any two adjacent element lengths is a constant:

$$\tau = \frac{L_2}{L_1} = \frac{L_3}{L_2} = \frac{L_4}{L_3} \cdots \qquad \textbf{(12–10)}$$

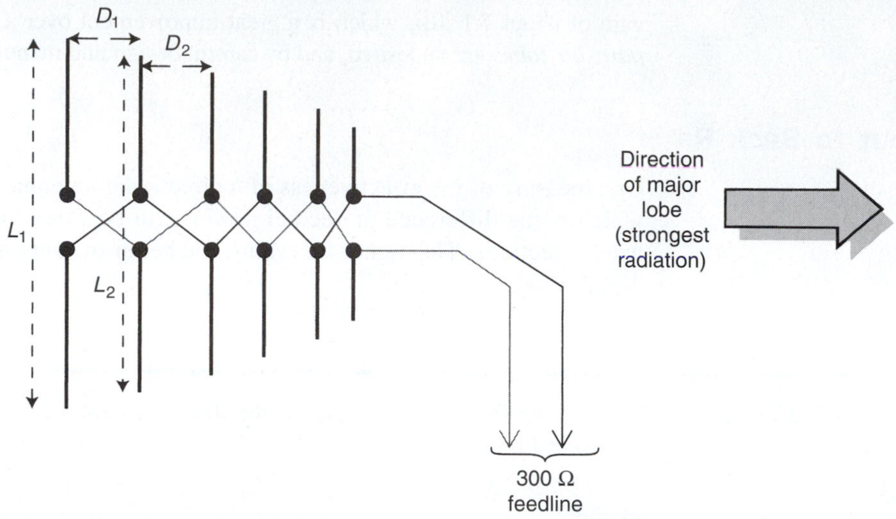

Figure 12–11 A log-periodic antenna

Usually, τ is between 0.7 and 0.9. This same ratio also holds true for the *spacing* between elements:

$$\tau = \frac{D_2}{D_1} = \frac{D_3}{D_2} = \frac{D_4}{D_3} \cdots \tag{12–11}$$

For a given physical size, a log-periodic antenna does not deliver as much forward gain as a Yagi; however, in applications where wide bandwidth is required (such as TV reception), this antenna is a standard solution.

Dish Antennas

At frequencies above 1 GHz, sometimes referred to as *microwave frequencies,* wavelength becomes progressively shorter, and for receiving antennas, this causes a sensitivity problem. A transmitting antenna, regardless of physical dimensions, radiates all of the power put into it ($P_R = I^2 R_R$).

A *receiving* antenna can collect only the electromagnetic energy that passes by it, and as wavelength (and therefore antenna physical size) shrinks, the amount of energy that a receiving antenna collects dwindles. This is one characteristic of antennas that is *not* reciprocal. It is the one important difference between transmitting and receiving antennas.

In order to compensate for the falling energy recovery of receive antennas at higher frequencies, it is common to use *gain antennas*. At the lower UHF frequencies (300 MHz to 1000 MHz), directional antennas (such as the Yagi, log-periodic, and others) can be readily employed. Because of the short wavelengths, antennas with a high gain can be constructed to make up for the receive signal loss.

At *microwave* frequencies (1 GHz and above), it is often more convenient to use a *dish* or *horn* antenna to collect the RF energy. Dish and horn antennas are also quite effective for transmitting; a tremendous amount of gain can be made available. Because of the high frequencies, *waveguide* is often used to feed a dish or horn antenna. This has the advantage of physical simplicity (technicians often refer to waveguide construction as "plumbers delight," since waveguide sections fasten together very much like household plumbing).

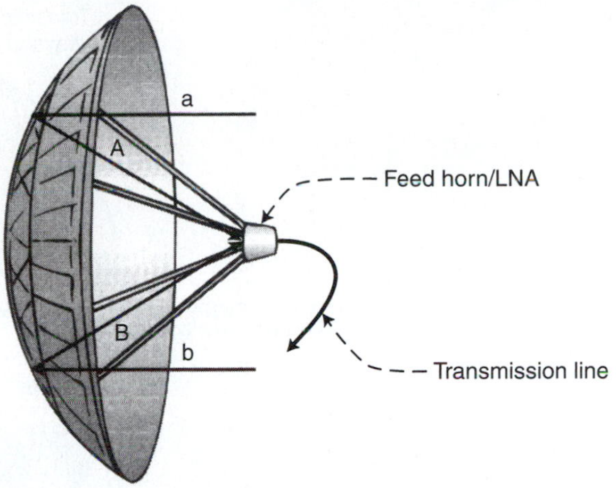

Figure 12–12 A horn-fed parabolic dish antenna

A dish antenna can be made using a variety of shapes; the *parabolic* dish is probably the most common shape. It is shown in Figure 12–12. The horn-fed antenna of Figure 12–12 is very common in television receive-only (TVRO) applications. The *feed horn* is located at the *focus* of the parabola formed by the dish. By definition, the *focus* is the point that is equidistant for all radio wave "rays" entering the antenna. Thus, with the feed horn placed at the focus, rays *aA* and *bB* travel exactly the same distance to reach it and arrive in phase. Therefore, they *reinforce* each other and increase the gain of the antenna. In receive-only applications, the feed horn often contains a low-noise amplifier (LNA) and *downconverter* unit. The downconverter contains a microwave oscillator and mixer, which converts the incoming RF signals down to a lower range of frequencies (UHF and below). This allows inexpensive coaxial cable to feed the antenna instead of waveguide.

One problem with the dish-feed arrangement above is that the feed horn (downconverter in a receive antenna) must be mounted in front of the dish. This creates a dead spot in the antenna pattern, since the feed horn is in the way of some of the RF signal. The horn-fed antenna can be fed with waveguide, but this is a physically clumsy arrangement. The waveguide occupies more of the space in front of the dish and further blocks part of the antenna's pattern. Because of this limitation, the *Cassegrain* feed system of Figure 12–13 is sometimes used.

The Cassegrain feed system is useful for both transmitting and receiving applications. In a receive application, the feed horn is likely to contain an LNA and downconverter; for transmitting, *waveguide* can directly provide signal for the feed horn. Incoming signals first reflect off the dish onto the *subreflector,* which is correctly curved and placed so that signals are all focused into the feed horn. The antenna can easily be fed with waveguide because the feedline is connected at the *back* of the dish. This is the basic advantage of the Cassegrain configuration. The subreflector still causes a dead zone in the antenna's pattern, however. This can be eliminated through the use of the *hog-horn* antenna of Figure 12–14.

The hog-horn antenna is built from two distinct shapes. The upper portion of the antenna is parabolic and is shaped to deflect rays down into the focus at the open end of the transmission line. The lower portion of the antenna is horn-shaped, and helps to focus the RF energy exiting or leaving the transmission line. The hog-horn antenna is primarily designed to be fed with waveguide, and can be used as either a transmitting or receiving antenna.

Figure 12–13
Cassegrain feed system

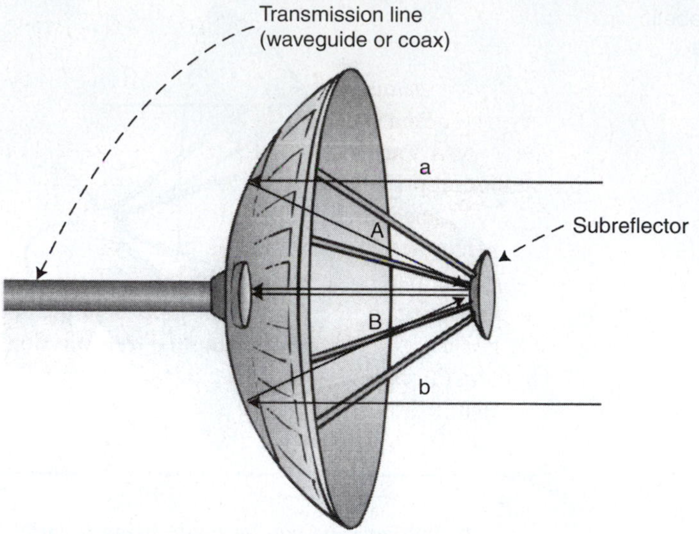

Figure 12–14
Hog-horn antenna

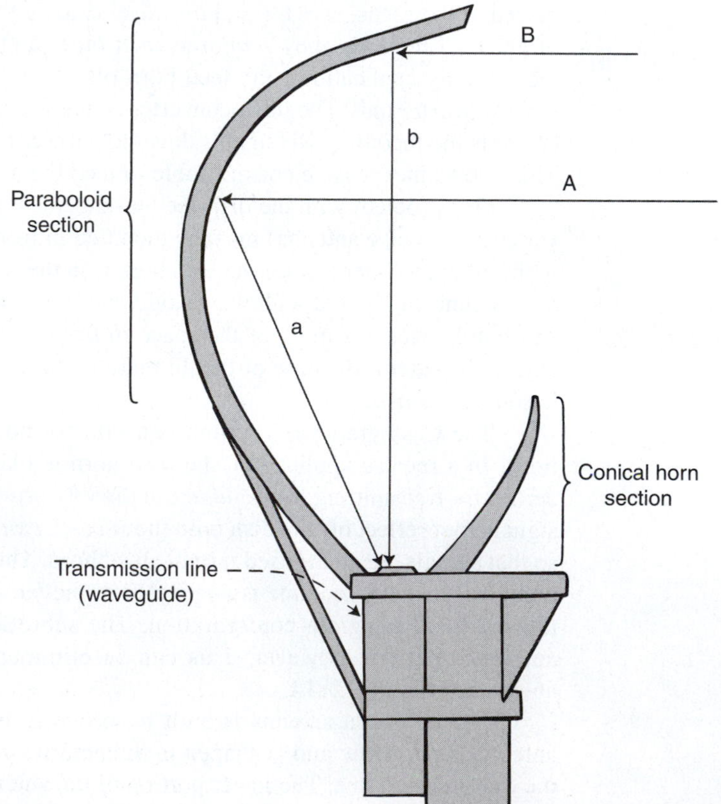

Section Checkpoint

12–26 List at least two advantages of directional antennas. What is a *disadvantage* of these types?

12–27 Which elements in a Yagi antenna are connected to the feedline?

12–28 Explain how the parasitic elements in a Yagi antenna contribute to its directionality.

12–29 What is the decibel gain of a three-element Yagi over an isotropic point source?

12–30 What are parasitic lobes?

12–31 Define *front to back ratio.*

12–32 Which antenna covers a wider bandwidth, a Yagi or log-periodic? Why?

12–33 Which elements in a log-periodic antenna are connected to the feedline?

12–34 What happens to the received energy from an antenna as its physical size gets smaller (for example, in response to shortening wavelength)?

12–35 List three antennas used at microwave frequencies.

12–36 What is the *focus* of an antenna (such as the one in Figure 12–12)?

12–37 What type of transmission line must be used to feed a hog-horn antenna?

12–4 SPECIAL ANTENNAS

There are several special antennas widely used in communications applications that aren't easily categorized. Among these are *collinear arrays, phased arrays,* and *loop antennas.*

Collinear Arrays

The word *collinear* means "sharing a common line." Collinear antennas usually have a radiation pattern very similar to a dipole (donut-shaped), but the radiation is concentrated in the horizontal plane (the donut is flattened). Figure 12–15 is the schematic diagram of a four-element collinear antenna, and Figure 12–16 is a Hy-Gain V2R, which is a two-element 5/8 λ collinear antenna for operation at 2 meters (150 MHz).

The collinear antenna provides additional signal gain when compared to a dipole. This gain comes from the interaction between the radiated fields of each of the antenna's active elements. In Figure 12–15, the fields of the individual antenna elements are forced to *add* by the addition of the 180° *phasing lines* in between elements. This is very much like series-aiding voltage sources; however, remember that because we're "adding" radiation fields in 3-D space, a detailed analysis would be very complex. The phasing lines can

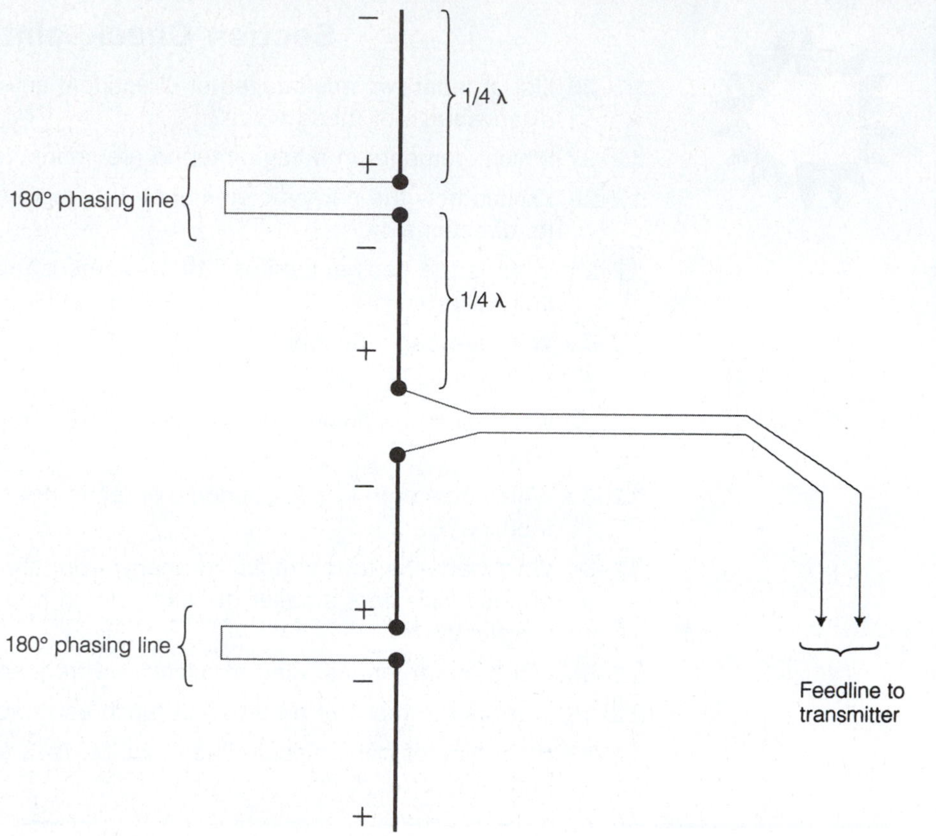

Figure 12–15
Collinear antenna schematic and typical unit

be RF transformers, or 180° sections of transmission line. Thus, the radiating conductors are spaced and phased to obtain signal addition in the horizontal plane. Figure 12–17 shows the radiation pattern of the Hy-Gain V2R collinear antenna, with the pattern of a quarter-wave Marconi superimposed on the graph for comparison.

Notice how the collinear V2R concentrates more of its RF energy toward the horizon. For most terrestrial communications, this "low angle of radiation" is helpful, since it is aiming the radio wave energy in a direction where the receivers will most likely be located (along the ground). In contrast, much of the RF energy from the quarter-wave Marconi antenna moves upward toward space.

Phased Arrays

A single Marconi antenna produces an omnidirectional radiation pattern when viewed from the top (Figures 12–7 and 12–17). Antennas can be driven in *sets* to produce directional radiation patterns. A set of such antennas is called a *phased array*. By controlling the amplitude and phase of the RF current driving each of the antennas, an almost infinite variety of patterns can be obtained. Figure 12–18(a) shows a phased array constructed with two quarter-wave Marconi antennas. The antennas are spaced one-quarter wavelength apart, and the RF current to one of the antennas is 90° lagging. The resulting radiation pattern is a *cardioid* (heart) shape.

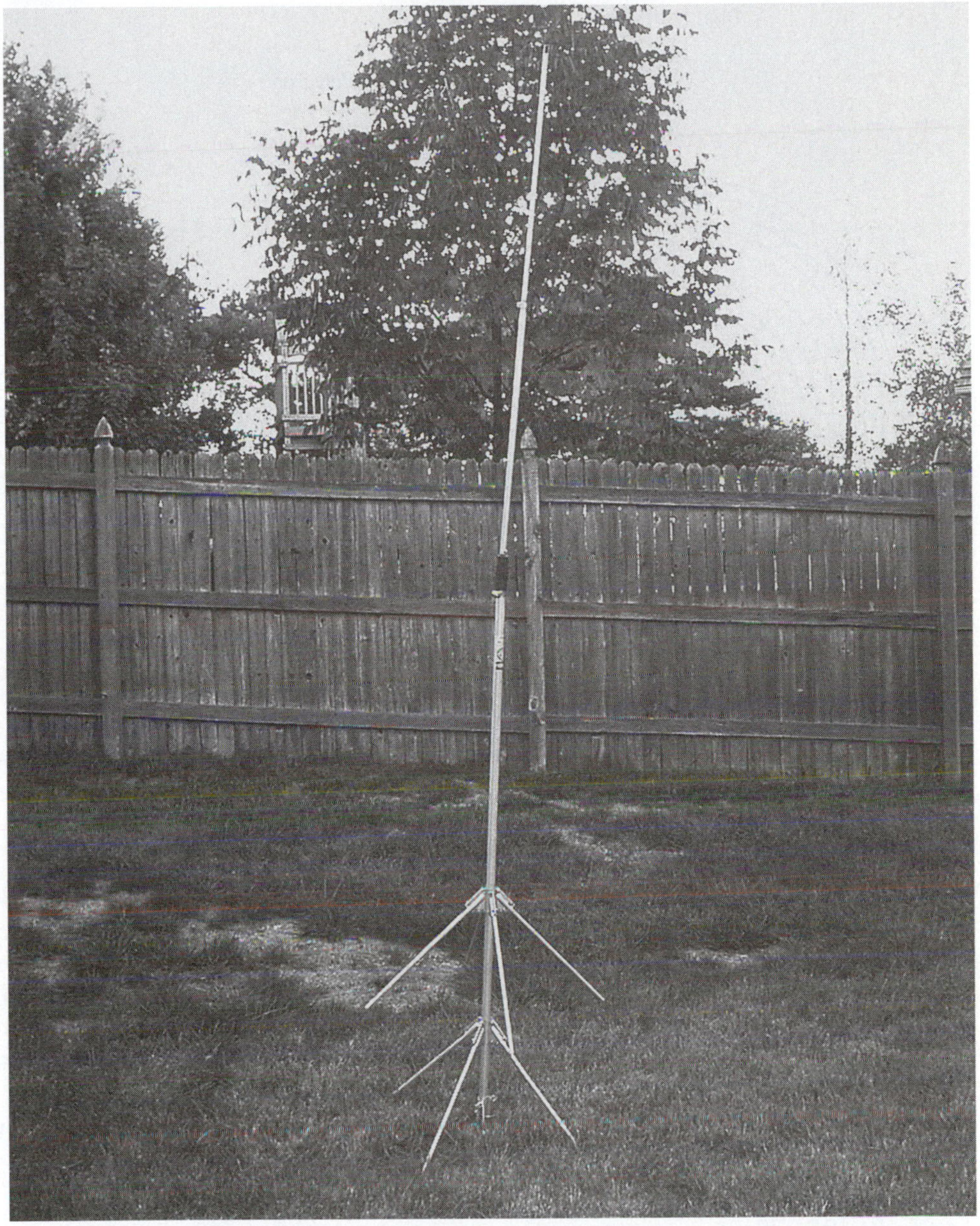

Figure 12–16 The Hy-Gain V2R collinear antenna

You may have noticed that AM broadcasters usually have two or three tall towers at the transmitting site. These towers are a phased array. When such a station is installed, the FCC typically coordinates and approves the radiation pattern for the station. Usually, a different pattern and lower power are mandated for night operation. A consulting engineer determines the phasing requirements for the antennas, and the station equipment is built to match these specifications. Most of the phasing equipment is located in a small shed known as a "dog house" located at the base of each tower. The dog house contains the impedance matching and phase shifting networks to drive the antenna.

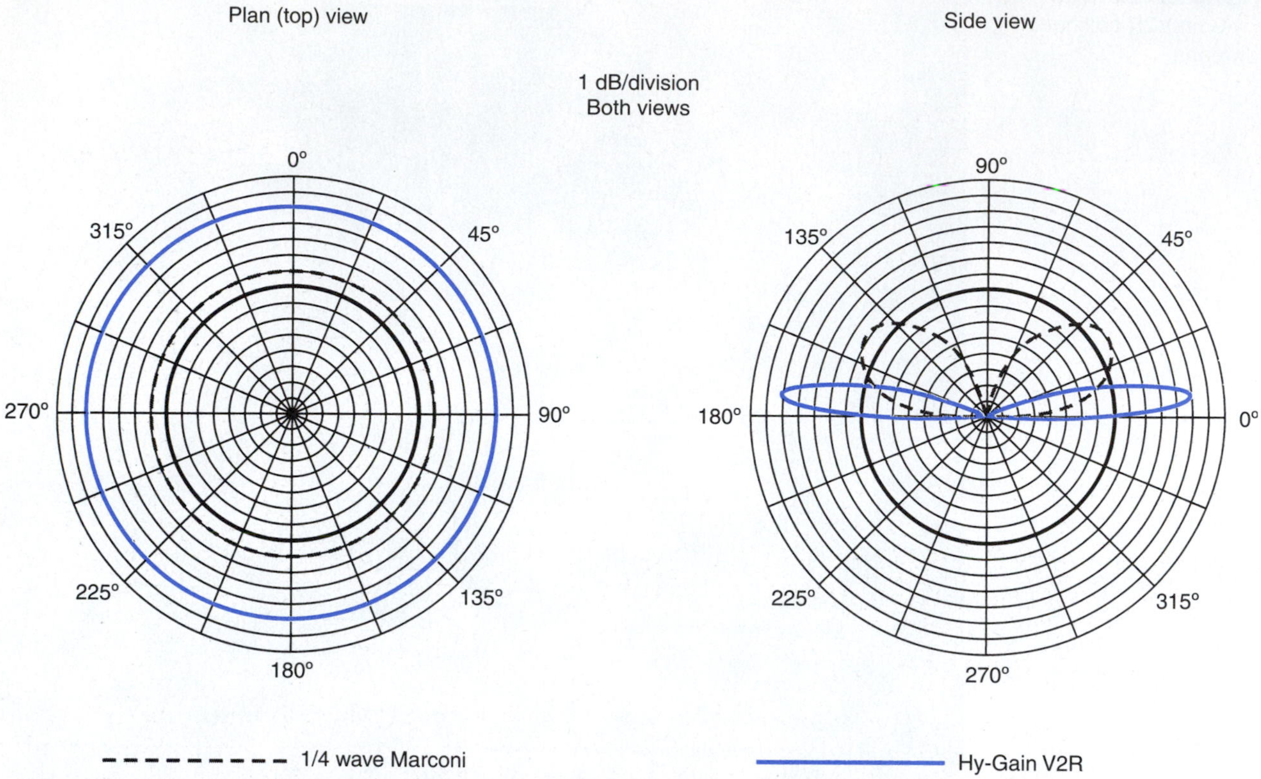

Figure 12–17 The radiation pattern of the Hy-Gain V2R

The cellular bay antenna of Figure 12–19 is really a phased collinear array consisting of three half-wave dipoles. The dipoles of this antenna are vertically oriented in normal operation, producing vertical polarization. The energy from the feedline is distributed to the three antenna elements by a tunable phasing network (the canisters). The tubing connecting the elements of the antenna is coaxial hardline. The backplane of the antenna is a grid that acts as a reflector, which causes the antenna to radiate primarily from the front (the side we're looking at). Note that the "works" of this antenna are normally contained within a plastic weathertight box, which has been removed.

Loop Antennas

When a compact antenna is required (especially at HF and VHF frequencies), a *loop* antenna can be utilized. Loop antennas are somewhat directional; the strongest radiation occurs along the axis of the loop, as shown in Figure 12–20.

A loop constructed like the one in Figure 12–20 has a total length of one wavelength and a radiation resistance of about 100 Ω. Because of their high radiation resistance, loops are efficient antennas. With the feedpoint on the bottom (as shown), the antenna will produce horizontal polarization; with the antenna fed on the side, the polarization will become vertical. A full-wavelength loop has a gain of approximately 2 dB more than a half-wave dipole. A full-wave loop constructed for operation in the 50–54 MHz (6 m) amateur radio band is shown in Figure 12–21.

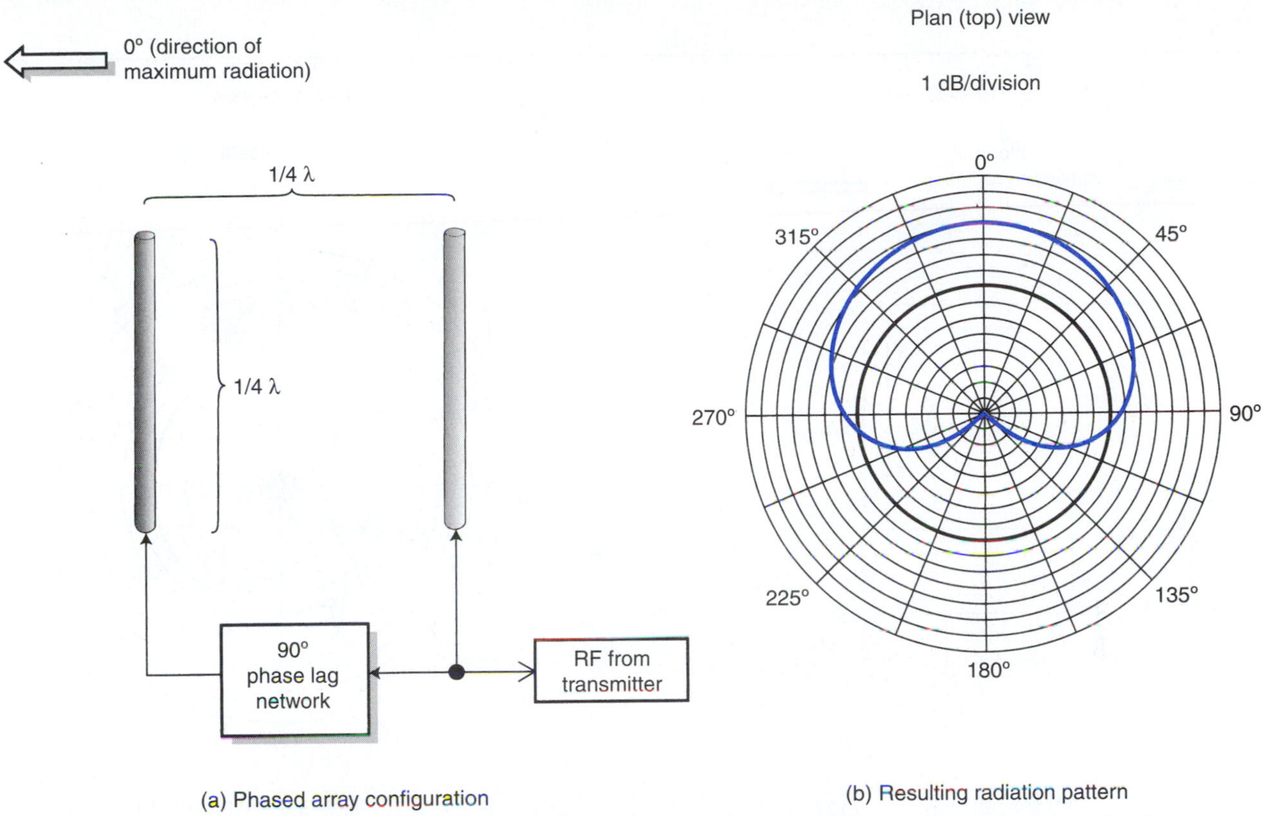

Figure 12–18 A two-element phased Marconi array, and horizontal pattern

Figure 12–19 A cellular bay antenna

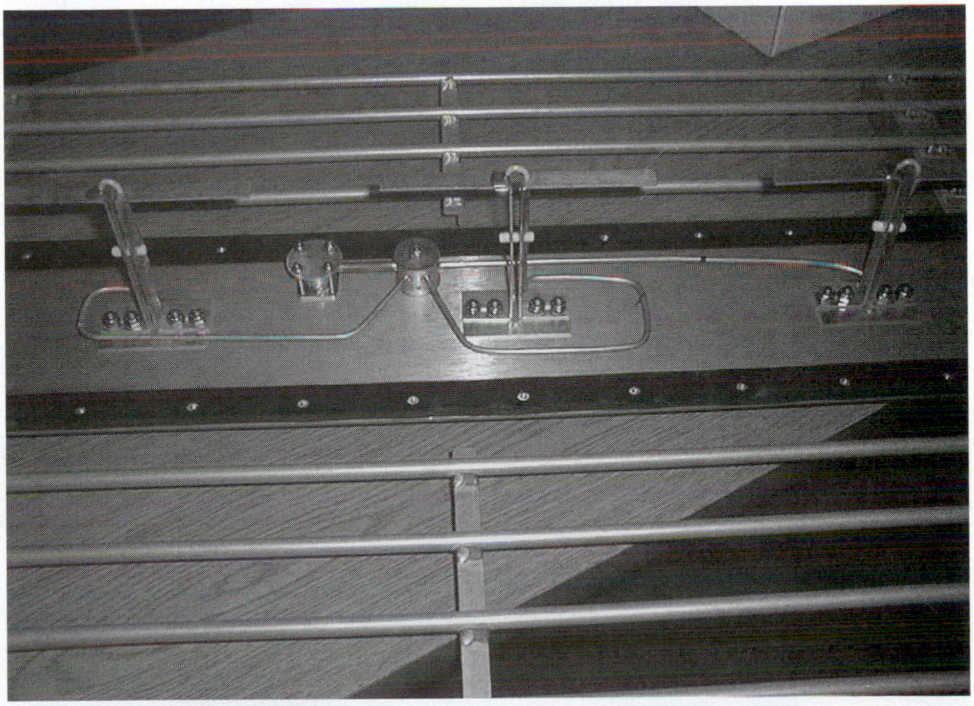

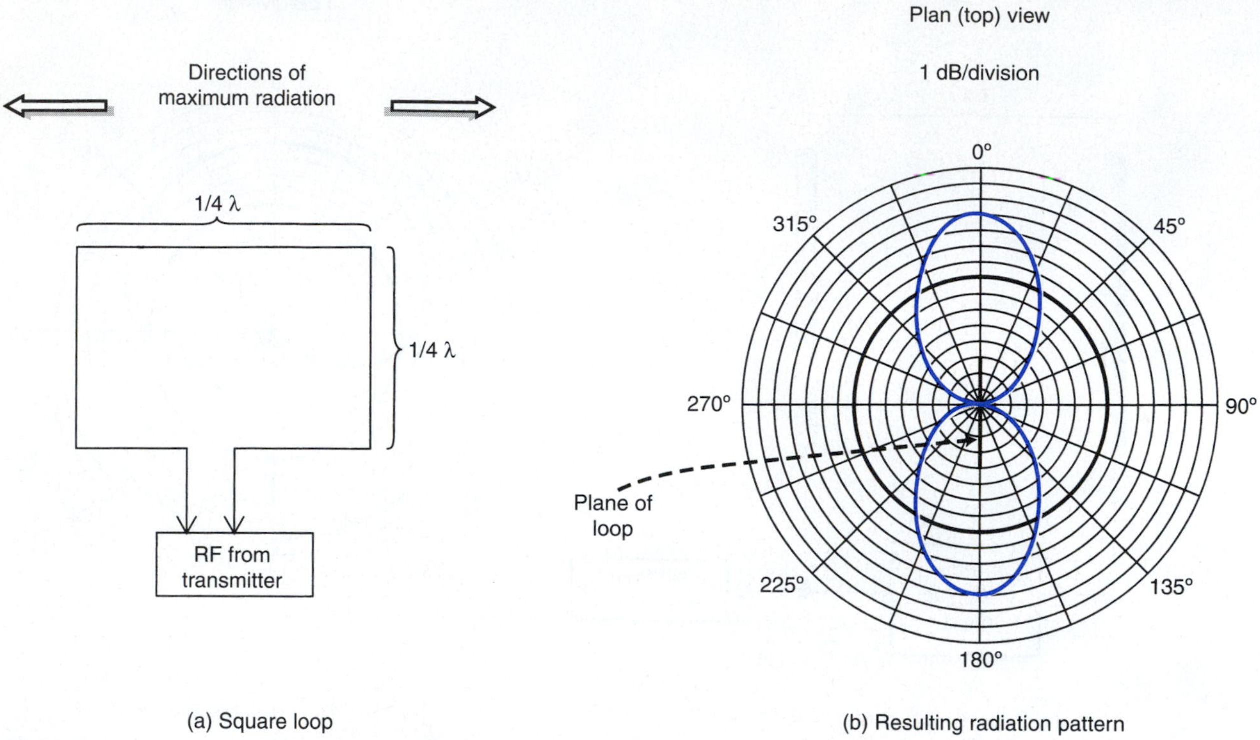

Figure 12–20 Loop antenna and radiation pattern

If we shrink the sides of a loop antenna until they are very short, we end up with a *folded dipole* antenna, as shown in Figure 12–22. The folded dipole antenna has a very high radiation resistance, on the order of 277 Ω. It is usually fed with 300 Ω twinlead or some other balanced transmission line. It is often used as an antenna for broadcast FM reception, but it can also be found as a driven element in some Yagi antennas because of its high efficiency. The folded dipole does not have as much gain as a full-wave loop, but it is much more compact.

Section Checkpoint

12–38 How are the conductors arranged in a *collinear* antenna?

12–39 What is the function of the *phasing lines* in Figure 12–15?

12–40 Why is a low angle of radiation usually desirable?

12–41 What is a *phased array?* Why is it used?

12–42 Where does the strongest radiation leave a *loop* antenna?

12–43 What is the primary advantage of a folded-dipole over a full-wave loop antenna?

Figure 12–21 A full-wave loop for 50–54 MHz

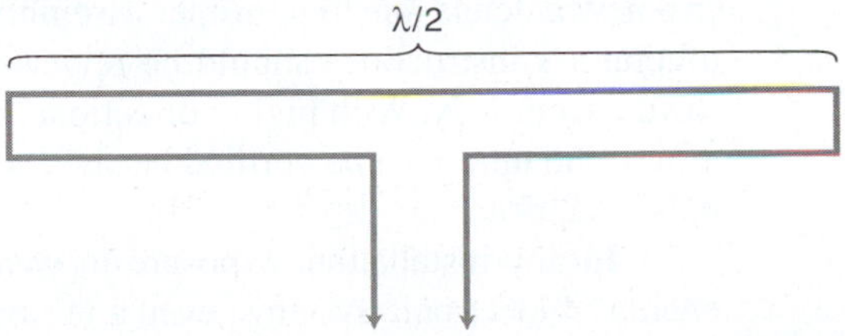

Figure 12–22 The folded dipole antenna

12–5 ANTENNA TROUBLESHOOTING AND RF SAFETY

When a technician is called upon to correct a problem in a communications installation, there are many unknowns. A failure can occur in any of the components of the station, and it is possible that multiple problems could exist. It is also possible that the problem exists off-premises (for example, failure at another station, or outside interference with the station). Operational faults at a station can include any or all of these:

- transmitter and receiver problems
- transmission line (feedline) failures
- damaged antennas
- improperly installed feedlines and antennas
- operator trouble (controls set improperly, or operator not properly instructed in the use of the equipment)

Because antennas are usually inconvenient to repair and replace, they are usually the last item to be investigated by a repair technician, and with good reason. It makes good sense to check the things that are easily corrected *first!* Ground equipment checks include:

- *Transmitter* and *receiver* evaluation. (Substitution is a powerful tool; a suspected unit can be swapped out with a known good unit.) Most commercial equipment has built-in meters and other indicators so that the operating conditions can be determined at a glance; some equipment can even be queried remotely through a telephone, radio, or Internet link, for off-site troubleshooting. Common transmitter measurements include final PA plate/collector voltages, final PA plate/collector currents, forward and reflected power on the transmission line, and SWR. If this information is in the station log, it can be very useful for later comparisons.
- Equipment settings. Someone may have tampered with the controls (often in an attempt to fix a perceived "problem").

Once the ground equipment is determined to be working correctly, the antenna and transmission line should be evaluated together. A sudden change in the SWR at the transmitter usually points to antenna and/or feedline problems. Before troubleshooting an antenna, make sure the transmission line is operating correctly (troubleshooting procedures for transmission lines are given in Chapter 11).

How to Isolate Antenna Problems

Most problems that occur with antenna systems are mechanical in nature and can be isolated with relatively simple test procedures. First and foremost, *visual inspection* often detects many antenna problems. For example, it is very common for problems to develop in a new antenna due to improper assembly by an installer. In a new installation, the manufacturer's instructions should be reviewed step-by-step to make sure they've been followed accurately. With highly directional antennas such as microwave dishes, the correct aim of the unit must be verified because a few degrees of positioning error can make signals disappear.

In any installation, exposure to weather can cause failures. *Moisture* is the prime enemy of electronic systems; even a teaspoon of water within a connector or feedline can render an installation useless. Connectors exposed to the weather should be examined for signs of corrosion and replaced as needed. *Waveguide* is particularly sensitive to moisture, and water can collect inside it if the drain holes become blocked. *Lightning discharge* can cause obvious visible damage to antennas, but it is also possible for a nearby strike to cause hidden problems. Examples of "invisible" problems include open-circuited loading coils (which are usually contained within a plastic housing), scorched insulators in areas not directly visible (such as inside a connector socket), which cease being insulators once they're "cooked," or damaged dielectric materials (which may look fine but present the wrong impedance to the circuit).

If the visual inspection reveals no trouble, then *electrical testing* can be performed. The types of electrical tests commonly performed by technicians include *ohmage, resonance,* and *field strength* tests.

Ohmage Testing

Many antenna problems can be located by simple continuity checks with an ohmmeter. Be careful not to be misled by "short circuit" readings from an ohmmeter. Many antennas should read *low* resistance at their input terminals, due to the presence of matching coils, transformers, and other networks that appear as a short to dc. Some manufacturers give the proper dc resistance values for the various components in their antenna systems in order to assist troubleshooting.

Resonance Testing

In resonance testing, the correct *operating frequency* of the antenna is verified. The resonant frequency of antennas is usually adjusted by varying the length of one or more antenna elements. These elements are fixed in place with clamps or set screws. To perform this type of testing, the transmitter frequency must be varied slightly above and below the operating frequency while the SWR of the antenna system is monitored.

The lowest SWR is *usually* obtained when the antenna is operating at or near its resonant frequency, because at this frequency, the antenna provides the appropriate input impedance to match the transmission line. Remember that *longer* antenna elements resonate at *lower* frequencies, and *shorter* elements resonate at *higher* frequencies.

A very high SWR reading (above 3:1) may indicate a problem *other* than a detuned system (such as a short or open circuit in the antenna or transmission line).

EXAMPLE 12–12 A technician is performing a resonance test on a business-band VHF whip antenna on a car. The transmitter is designed to operate on 151.625 MHz. With a directional wattmeter placed between the transmitter and transmission line, the SWR was measured at the following frequencies:

151.000 MHz, SWR = 1.5 : 1

151.625 MHz, SWR = 1.7 : 1

152.000 MHz, SWR = 1.9 : 1

Recommend a change to bring the antenna closer to resonance at 151.625 MHz.

Solution

As the frequency was lowered, the SWR decreased. This means that the antenna was closer to resonance *below* the intended operating frequency. Because lengthening an antenna lowers its operating frequency, we know that the antenna is slightly too *long* and needs to be shortened a small amount (perhaps about 1/2″ or so). Note that the SWR is not objectionable at 151.000 MHz (1.5:1 or less is normally considered acceptable), but it is too high at the operating frequency of the transmitter (1.7:1).

Field Strength Testing

This is perhaps the most expensive and difficult antenna test to perform. To accurately measure field strength, a calibrated radio receiver and reference antenna is required (a portable spectrum analyzer can take the place of a calibrated receiver in many cases). All of this equipment is quite expensive and hard to justify unless a company performs a lot of testing of this type.

Accurately measuring the field strength from an antenna is a very difficult exercise, especially for novices. A field strength indicator cannot tell if the signal is actually coming directly from an antenna, or if it is being reflected from some nearby object. For broadcast stations, periodic field strength measurement at various points in the coverage area is a requirement under FCC rules, and calibrated receivers are generally used in this application. The data are recorded in the station log.

At VHF and above, field strength can be measured with much better accuracy, but for troubleshooting, this type of information is not very useful, unless it can be compared with previous records from a station log.

If field strength measurements are used, they should be taken as far away from natural and human-made obstructions as possible in a high location that is in the direct line of sight of the radiating antenna.

RF Safety

We are continually exposed to radio frequency energy in everyday living. The effects of this exposure are not fully known, but it *is* known that direct exposure to high-intensity RF energy is not healthy! Contact with an antenna element carrying high-power RF signals can cause a *RF burn*. For antennas with a high-power input and high gain, such as microwave dishes, even being near the focal point of the antenna can result in a nasty burn injury.

Even at intensities below the "burn" level, excessive RF energy is hazardous. The eyes are particularly sensitive, and unfortunately, will not register "pain" when they are being injured by RF. Keep your head clear of strong RF fields!

The technician performing the work on any transmitting antenna is responsible for ensuring that the power to the RF energy source (the transmitter) is disabled prior to beginning the work. *Power supply circuit breakers must be tagged in the "off" position, and only the person performing the actual antenna or tower work is to remove the tags.*

The U.S. Office of Engineering and Technology (OET) *Bulletin 65* describes the technical requirements for station installations that will expose the public to RF energy. The lowest limits for radiated power levels are in the 30 to 50 MHz range, where the human body tends to become resonant (and therefore accepts the most energy from a radio wave).

Safe Work Habits

Since most antennas are mounted on towers and other high locations, technicians need to learn and follow basic rules for safely performing their work. The following list is *not* all-inclusive:

- Never work alone. Always have at least one other person as your ground crew, preferably with quick access to a telephone in case of emergency.
- Never look into an energized waveguide. Because the energy is highly concentrated within a waveguide, even a few watts of RF can cause an eye or brain injury.
- *The person who climbs the tower tags the electrical service cutoff.* The climber is the only one who restores electrical power.
- Wear appropriate clothing when working on towers. (Long denim pants and a long-sleeve shirt are recommended. Work shoes with heavy soles will give you a good footing on narrow tower rungs. Good-quality leather work gloves are a necessity.)
- Never climb a tower, even for only 5 or 10 feet, without an approved safety belt. The safety belt should be thoroughly inspected prior to each climb, paying particular attention to seams, ropes, and stitching. Replace any damaged belt. Do not attempt to repair safety belts—destroy any damaged belts immediately.
- Safety belts should be tested at ground level before each climb. Attach yourself at ground level and lean backward to double-check the belt's fit and integrity. Carefully inspect the belt and harness components during this step.
- All workers near or on a tower must wear hard hats, in case tools or other objects are dropped.
- Do not work on a tower during inclement weather. Keep an eye on the weather while on the tower.
- Plan the job before ascending the tower so that you will have the materials you need to finish. The less time you spend on a tower, the better off you will be.
- Before climbing, plan your route and look for obstructions (such as hornet's nests).
- Do not climb a tower if you are not in good health or do not feel right. Someone else can finish the job safely.

Section Checkpoint

12–44 List at least three different operational problems that are possible at a transmitting or receiving facility.

12–45 When troubleshooting a station, which equipment should be checked first? Why?

12–46 Why would a technician record final PA plate or collector voltages and currents in a station log?

12–47 What does a sudden change in SWR at a transmitter usually mean?

12–48 What troubleshooting method detects many antenna problems without the use of any additional test equipment?

(continued on p. 456)

12–49 How can an antenna system be tested with an ohmmeter?

12–50 Explain how *resonance testing* is performed. When is the lowest SWR usually obtained from an antenna installation?

12–51 If the SWR of an installation measures more than 3:1, what are some likely problems?

12–52 Who is responsible for shutting off and tagging the power switch on a transmitter when antenna work is being performed?

12–6 WAVE PROPAGATION

Once a radio wave has been generated at a transmitting antenna, it begins its journey to the receiver. *Propagation* is the process of wave movement. Radio waves propagate in a similar fashion to light waves, which are simply higher-frequency electromagnetic energy. The frequency of visible light is around 4×10^{14} Hz . The highest radio frequencies are close to 100 GHz (1×10^{11} Hz). Because of the frequency difference between radio and light waves, there are some important differences in how the two types of electromagnetic energy travel.

How Radio Waves Change Directions

As radio waves travel from point to point, they often move in paths that are *not* straight lines. There are three basic mechanisms that can change the course of a radio wave: *reflection, refraction,* and *diffraction.* In real life, *all* of these actions take place for any given radio signal. Figure 12–23 illustrates these three actions.

Reflection is the most intuitive of the propagation modes. When a radio wave strikes a conductive surface, it bounces off the surface in the same manner as light reflects from a mirror. When a wave reflects from a conductive surface, it experiences a 180° phase shift. Its polarization is normally not changed. Good conductors of electricity make good reflectors; this is why aluminum and silver are used to manufacture optical mirrors.

When a wave travels from one material into another (such as from air into water), it is *bent* or *refracted.* Be careful not to confuse *refraction* with *reflection*—it's easy to do! The index of refraction for each material and the original angle of the incoming wave (ray) determine the amount of bending and the new angle of travel in the second material.

Not all rays can be bent at the interface of two different materials. When rays are approaching at more than the *critical angle,* they don't enter the second material at all; instead, they are *reflected.* This principle is used in fiber optics to conduct light efficiently down a fiber.

A good example of the refraction of light is a pencil half-immersed in a glass of water. Where the pencil enters the water, it appears to be bent; this is due to refraction of the light waves as they enter and leave the water. You can also observe the effect of the critical angle: If you kneel down close to the water in a swimming pool while looking across the surface of the water, you will eventually reach an angle where the water acts as a mirror. What you are doing is positioning your eyes to view the light rays entering the water at more than the *critical angle;* these light rays are *reflected* off the surface of the water for you to see.

Layers of the atmosphere can refract both light and radio waves. Atmospheric refraction is responsible for the water mirage (false image) that often appears atop a hot

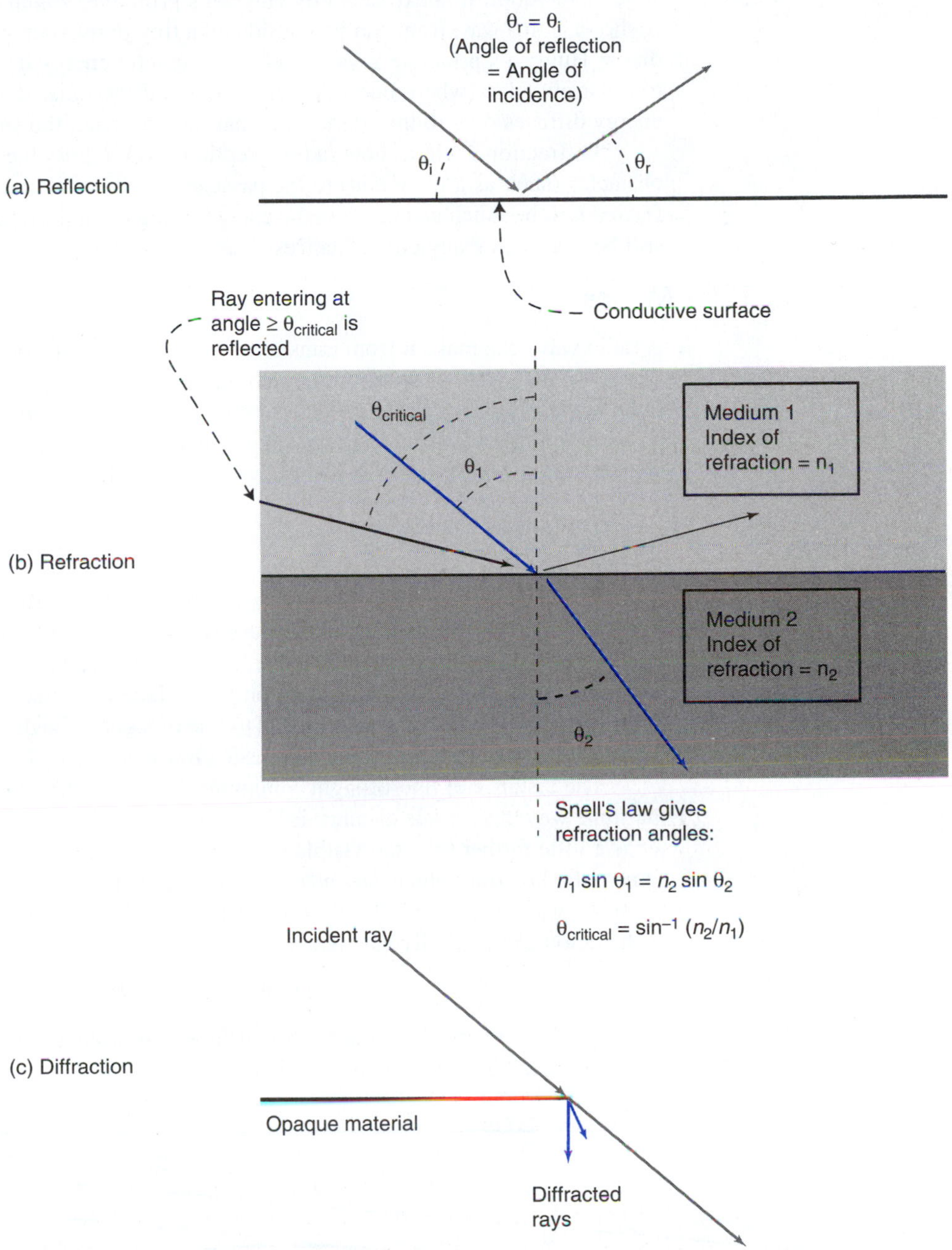

Figure 12–23
Reflection, refraction, and diffraction

$\theta_r = \theta_i$
(Angle of reflection
= Angle of
incidence)

θ_i θ_r

(a) Reflection

Ray entering at
angle $\geq \theta_{critical}$ is
reflected

$-$ Conductive surface

$\theta_{critical}$

Medium 1
Index of
refraction = n_1

θ_1

(b) Refraction

Medium 2
Index of
refraction = n_2

θ_2

Snell's law gives
refraction angles:

$$n_1 \sin \theta_1 = n_2 \sin \theta_2$$

$$\theta_{critical} = \sin^{-1}(n_2/n_1)$$

Incident ray

(c) Diffraction

Opaque material

Diffracted
rays

roadway on a sunny day. The image you see is light that has been bent toward you by two different layers of air. The layer of air just over the road's surface is warmer and less dense than the layer above it, and its index of refraction is lower than the air above it. Therefore, light waves can be bent between the two air layers.

You've observed *diffraction* many times. Suppose there is one room in your house with the light on and the door to that room is partially open. The door causes a shadow, but the edges of the shadow are not perfectly sharp. This is because of light diffraction at the edge of the door. It is explained by Huygen's principle, which states that each point on a (radio or light) wavefront can be considered a tiny point source of energy, or isotropic radiator. Huygen's principle says that electromagnetic energy doesn't "like" to be confined to a narrow edge (where one side has energy and the other does not). Some of the wave energy diffuses over to the "dark" side, partially blurring the shadow.

Diffraction explains how radio reception is often possible even when there are direct obstacles (such as a mountain) to the passage of radio waves. A signal that has been diffracted will be much weaker than the energy in the main portion of the radio wave but can still be usable in many circumstances.

Three Propagation Modes

A radio wave can make it from transmitter to receiver by one or more of three signal paths. These are the *direct* or *space wave* path, the *ground wave* path, and the *sky wave* path. A radio wave may have to be reflected, refracted, or diffracted as it follows each of these paths. If a radio wave follows *two* or more paths to get to a destination, the result is usually *multipath interference,* as we'll investigate shortly.

Direct Path

A *direct wave* or *space wave* is a radio wave that travels in a straight line between the transmitter and receiver. No bending of the wave is necessary for this mode. This is also known as *line-of-sight* communication. This is the mode of propagation for satellite communications and for short-distance communication on the surface of the earth (for example, local VHF radio such as FM broadcasting is limited to line of sight). The distance that a space wave can travel is limited by natural and human-made obstructions and the curvature of the earth.

The distance of line-of-sight communications is restricted by the earth's curvature in the same way that visible distance is limited to the horizon. The *radio horizon* actually extends a little farther than the visible horizon for VHF waves. If two stations can "see" the same radio horizon (which depends on their antenna heights), they can communicate using the space wave. Figure 12–24 illustrates this concept. The distance to the radio horizon for each station can be easily calculated:

$$D(\text{mi}) = 1.415\sqrt{H\ \text{ft}} \qquad \textbf{(12–12)}$$

This equation assumes that the earth's surface is smooth and that no objects are blocking the radio signal between the two stations.

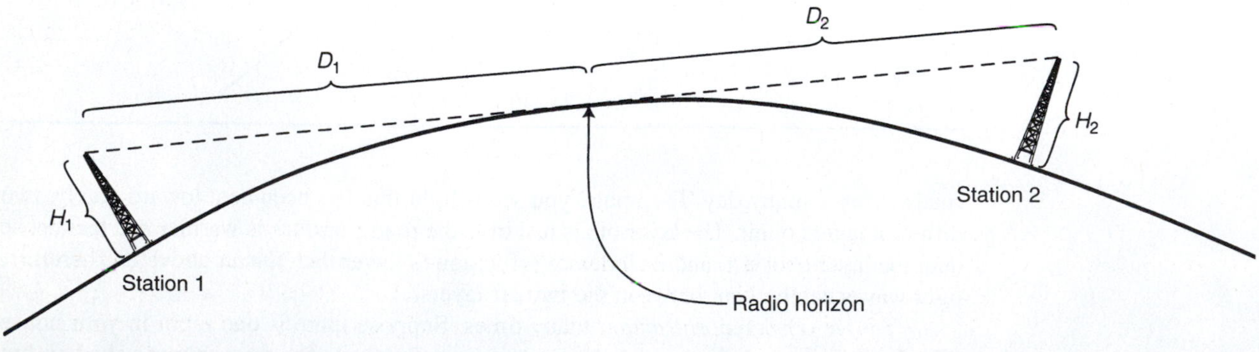

Figure 12–24 The radio horizon

EXAMPLE 12–13 What is the maximum distance two earth stations can communicate using space wave propagation if one station has an antenna height of 20 feet, and the other has an antenna height of 100 feet?

Solution

To solve this problem, calculate the distance to the radio horizon for *each* station and add the results.

The distance to the radio horizon for the first station is

$$D_1(\text{mi}) = 1.415\sqrt{H(\text{ft})} = 1.415\sqrt{20\text{ ft}} = \underline{6.32\text{ miles}}$$

And the distance for the second station is

$$D_2(\text{mi}) = 1.415\sqrt{H(\text{ft})} = 1.415\sqrt{100\text{ ft}} = \underline{14.15\text{ miles}}$$

The total distance between the two stations is the sum

$$D_{\text{total}} = D_1 + D_2 = 6.32\text{ mi} + 14.15\text{ mi} = \underline{\underline{20.5\text{ miles}}}$$

Therefore, the maximum distance the two stations can be apart while communicating line-of-sight is 20.5 miles.

Ground Waves

At frequencies below 2 MHz, signal energy can travel far beyond the radio horizon limit of space waves. At such low frequencies, a portion of the radio wave energy will tend to follow the curvature of the earth for a distance of approximately 100 miles, providing very reliable communications. Because ground wave propagation is not very strong above 2 MHz, this mode is primarily useful for services that operate at 2 MHz and below, which includes amateur and military communications at 1.8 MHz, AM broadcast, and military and commercial communications below 500 kHz.

Ground wave propagation is not dependent on weather conditions, but it *is* very dependent upon the earth's conductivity, which varies widely from region to region. Salt water is a very good conductor, and ground wave propagation is very good over the ocean for this reason. Where the ground is dry or rocky, ground wave propagation suffers.

The ground wave component of a radio signal must be *vertically polarized* in order to be propagated efficiently. The reason for this is simple: When a radio wave is horizontally polarized, its electric field component is parallel to the earth's surface, which is a conductor of electricity. The earth tends to "short out" the electric field of horizontally polarized ground waves because of its conductivity. Ground wave is the only propagation mode sensitive to wave polarization; the other modes are not affected in this way.

Sky Waves

At frequencies below about 50 MHz (depending on conditions), it is possible for radio signals to travel far beyond the limits of either space wave or ground wave propagation. Radio waves at these frequencies can travel by *skip,* where the wave is alternately *refracted* by one or more *ionospheric* layers then *reflected* back upward by the earth's surface.

The *ionosphere* is a region of the atmosphere that is electrically charged (ionized). When ultraviolet light from the sun strikes the outer layers of the atmosphere, it

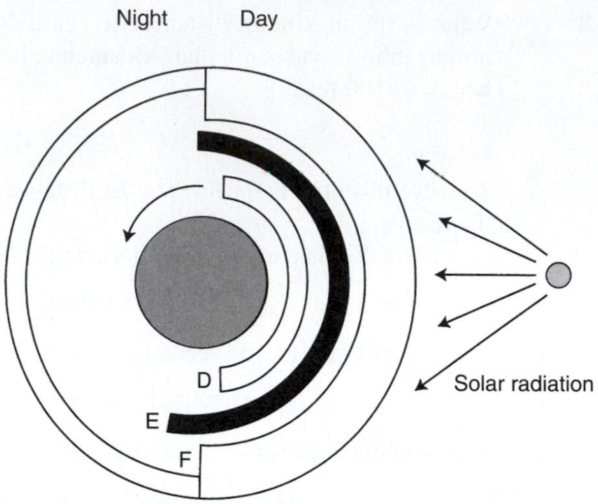

Figure 12–25 The ionospheric layers

causes electrons to be stripped from the air molecules. Because this region of the atmosphere is not very dense, the resulting free electrons and positive ions are able to move about for some time before they recombine into neutral atoms. The ionosphere has a different index of refraction than atmospheric layers below it and can therefore bend radio waves back toward the earth. The ionosphere is in a constant state of change; the exact levels of ionization are hard to predict. It can be considered to be built of a number of *layers,* though this isn't strictly true. The level of ionization depends on the time of day, the season, and the 11-year solar cycle. Figure 12–25 illustrates the ionospheric layers.

The lowest layer, the *D layer,* is about 37–57 miles above the earth's surface. It occupies denser atmosphere than the other layers, and because of this, its ionization rapidly disappears after sunset. It appears at sunrise, becomes maximum at local noon, and fades rapidly at sunset. The D layer tends to *inhibit* skip propagation for two reasons. First, because the layer is dense, it tends to absorb radio signals that pass into it. Second, the D layer is close to the earth's surface. It is difficult to launch radio waves into it at less than the required *critical angle*. Waves that enter an ionospheric layer at an angle greater than the critical angle are either absorbed or pass through to outer space. *For frequencies below approximately 7 MHz, the D layer prevents most long-distance skip propagation while it is active*. This explains why it is easier to receive distant AM broadcasts in the 535–1620 kHz band at night; the D layer absorbs most of the sky wave from these stations in the daytime.

The *E layer,* which is about 62 to 71 miles above the earth, retains its ionization longer than the D layer, but it is not involved in most skip propagation. It reaches its maximum around noon and its minimum *after* midnight local time. A single skip "hop" to and from the E layer is usually between 250 and 1,200 miles.

The *F layer* can be thought of as two layers, F1 and F2, but for our purposes, we can think of it as a single layer. It is at a height of 100 to over 310 miles and is responsible for most long-distance skip communications. Because of its increased height, skip distances from this layer range from 1,000 to more than 2,500 miles. The F layer is responsible for most long-distance skip propagation. At night, the two layers merge into a single layer.

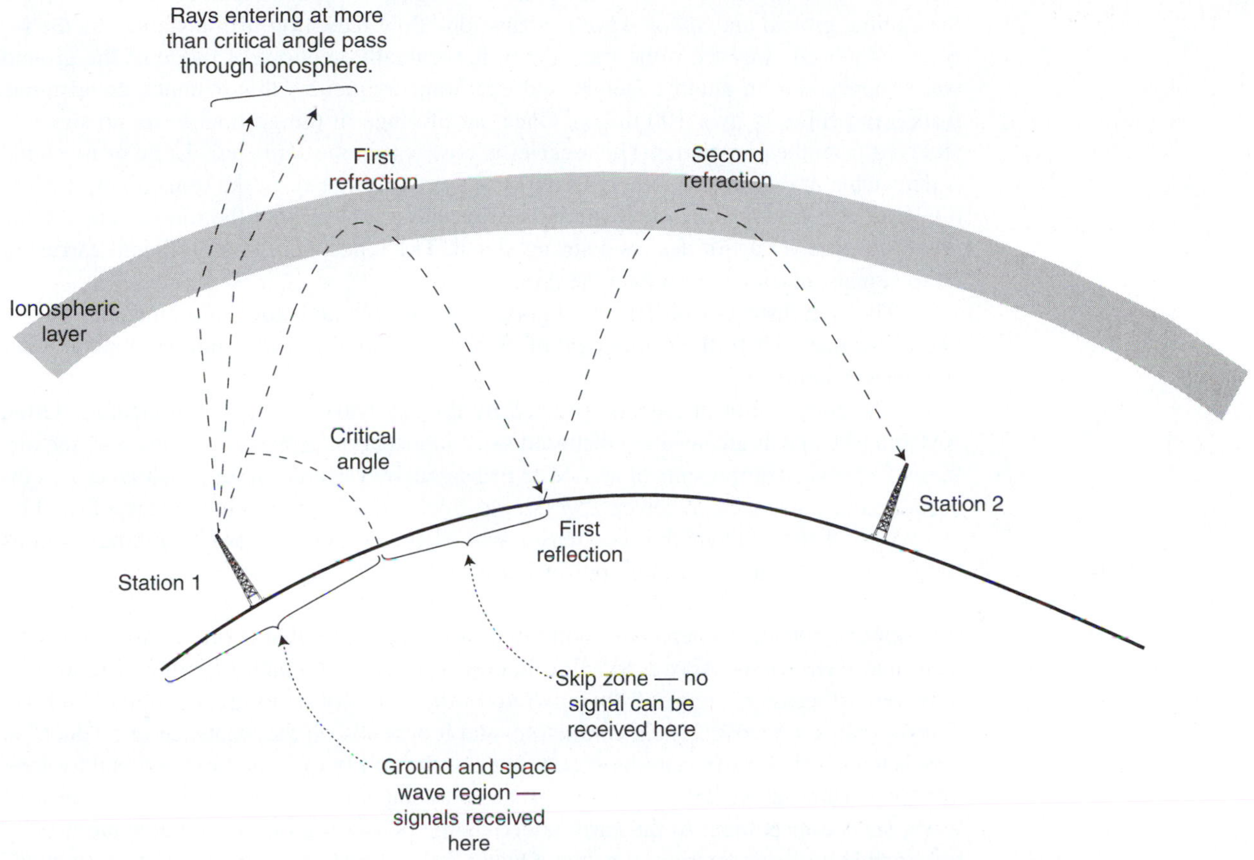

Figure 12–26 Skip (sky wave) propagation

Conditions for sky wave propagation Figure 12–26 shows a highly simplified picture of skip propagation. There are *many* variables involved, and unfortunately, this subject can't be reduced to neat little equations. Whether or not communication can be accomplished by skip depends greatly on ionospheric conditions, transmitter and receiver antennas, power levels, and the frequency in use.

For a radio signal to be refracted successfully back to earth, it must approach the ionosphere at or less than the *critical angle. The critical angle is the largest angle at which a radio wave can enter the ionosphere and still be reflected back to earth*. The critical angle is highly dependent on the wave frequency; higher frequency waves must be launched into the ionosphere at a more shallow angle in order to be refracted back to earth. For efficient skip propagation, *low* angles of radiation from radiating antennas have proven to be the most effective in most situations. If the angle is held constant and the radio frequency is increased further, *no* wave is returned to earth. The highest frequency that can be refracted for a given path is called the *maximum usable frequency,* or *MUF*. The MUF varies constantly for any given path and varies from path to path. When communicating by sky wave, the highest usable frequency provides the best received signal strength. The ionospheric layers tend to absorb longer wavelength (lower frequency) signals and let shorter wavelengths pass (refract). *Therefore, the most efficient frequency for any particular path is one just below the MUF.*

In Figure 12–26, the transmitting station's signal is not audible in all locations. Near the station, *ground* and *space* wave propagation allow reception of the signal. As the receiver is moved away from the transmitter, it eventually gets out of range of the ground wave (depending on antenna height and operating frequency, this distance could range from a few miles to over 100 miles). Once out of range of the ground wave, no signal is received from the transmitter. The receiver is now within the *skip zone.* Little or no signal is detectable in this area. As the receiver moves even further from the transmitter, it eventually moves close to a location where the wave has returned from the ionosphere. At this point, the receiver again detects a strong signal. The ionosphere is an efficient carrier of radio signals, much more so than the ground wave.

The skip distances on HF are highly unpredictable, and for radio amateurs, this is one factor that adds to the excitement of "DXing" (searching for distant stations in other countries to contact).

The polarization of signals received by the sky wave is very unpredictable. Often AM and FM signals are severely distorted since ionospheric propagation can delay the different frequency components of an AM or FM signal by various amounts, which causes inaccurate detection at the receiver. *SSB* (single-sideband) signals are much less affected by this type of distortion, which is one reason why SSB is so popular for HF communications (in addition to the increased talk-power it affords).

Tropospheric ducting There's one additional sky wave mode that is often observed on the VHF and UHF bands. Above 50 MHz, line-of-sight communications is the general rule; however, on occasion, signals from hundreds or thousands of miles are sometimes audible. The mechanism involved is called *ducting,* and it depends on the existence of a "duct." A *duct* is a portion of the lower atmosphere that has the ability to carry a radio signal for long distances. Unusual weather conditions, such as a temperature inversion (where a mass of warm air is trapped next to the earth underneath a denser region of cold air), are usually responsible for the formation of a duct. Once a radio signal enters the duct, it is "trapped" between the air layers by refraction, and travels very efficiently to its destination.

In order to communicate through a duct, both stations must be able to launch a significant portion of their transmit energy into it; this is highly dependent upon the radiation pattern of the stations. The FM broadcast band is an excellent place to monitor under ducting conditions. Stations from hundreds of miles away may suddenly pop into existence on the dial as the duct forms, with very strong signals.

Multipath Interference

Because a radio signal can reach a receiver in many ways, it is often the case that it does! When this happens, the receiver experiences *multipath* interference if the second signal is strong enough to interfere with the original signal. In urban areas, multipath can cause great problems with FM reception in automobiles. Figure 12–27 shows why this is so.

In the figure, there are *two* versions of the radio signal reaching the receiver. If the signals reach the receiver's antenna in-phase, then *constructive* interference will take place and the receiver will experience a stronger signal than normal. Moving the receiver even a few inches can change this relationship drastically. Depending on how good the "reflector" of radio waves is (and how many times it is reflected), there can be places where the reflected and direct waves totally cancel each other, leaving a *dead spot* in the coverage area of the transmitter.

When a vehicle is moving through a multipath area, its receiver experiences a varying signal strength that goes through a minimum and maximum (approximately) every half-wavelength of travel. At highway speeds, this results in the familiar sound of *picket*

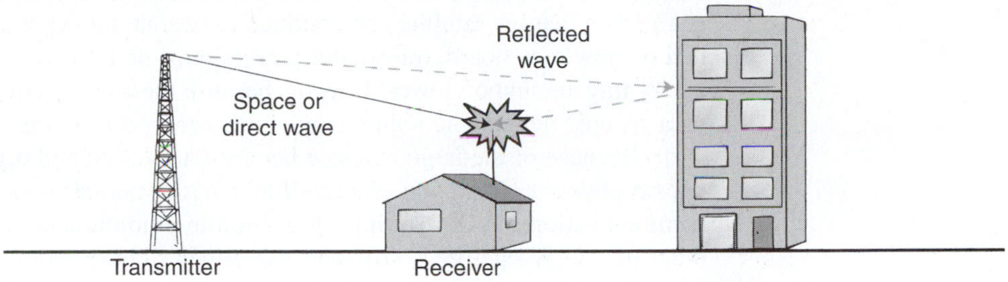

Figure 12–27
Multipath interference

fencing, where the sound is "chopped out" several times a second as the receiver rapidly passes through the zones of cancellation.

Television receivers are also affected by multipath interference. The appearance of *ghosts* on a TV screen is usually a good indication that this type of interference is taking place. (It can also signal trouble with an antenna or transmission line.) Since TV receive antennas tend to be directional, they can usually be aimed in a direction to minimize reception of the reflected signal, while still obtaining reasonable reception of the original.

Satellite Communications

Satellite communications is a special case of direct or space wave propagation, where it is intended that the signals *penetrate* the ionosphere. Satellites are specialized radio stations placed in various types of orbits around the earth and will be studied in more detail in a later chapter. A satellite usually operates as a signal "repeater," retransmitting signals back down to earth. Reliable communications over an entire continent can be achieved in this manner.

You might recall that low radio frequencies below 2 MHz tend to be *absorbed* by the D layer of the ionosphere during the daytime and that up to around 50 MHz, it is possible for signals to propagate by "skipping" back to earth from the ionosphere. What happens above 50 MHz? Above 50 MHz, the ionosphere does not easily refract radio signals. *Therefore, the VHF (and higher) frequencies are the ones used for most satellite communications, because these frequencies easily pass through the ionospheric layers to the satellite stations.*

Satellites can be classified as *orbiting* or *geostationary* types. Most commercial and government satellites are geostationary, which means that they appear fixed in space over a specific region of the earth. A geostationary satellite must be placed at an altitude of approximately 22,500 miles in order for its *orbit* to coincide with the earth's rotational period. At this altitude, the moving satellite appears "fixed" in space over a continent. An *orbiting* satellite is much less expensive to launch, for it occupies a lower orbit. Many organizations have orbiting satellites in space, including the military, which uses them for reconnaissance.

There's another reason besides ionospheric penetration that VHF and UHF frequencies are used in satellite communication, and that is *antenna gain.* The distance to a satellite in geosynchronous orbit is very large, and appreciable signal loss occurs over this path. (For comparison, the circumference of the earth at the equator is approximately 24,800 miles.) At high frequencies, it becomes very practical to use *gain antennas* at both earth stations and the satellite station. The use of gain antennas more than overcomes the signal loss between satellite and earth.

The use of gain antennas is also useful from the viewpoint of the *power budget* of the satellite. A satellite gets its dc power from solar panels, and it gets slightly more than

12 hours of sun per day in a geosynchronous orbit. During this time, on-board batteries are charged so that the satellite can continue to operate during the night. Because of the limited dc power on board, transmitter powers must be limited to a few watts or less. This is a very tiny amount of power! It works because the gain antenna on the satellite is focused on a specific geographic region (sometimes referred to as the "footprint" of the satellite).

Because of the large distance between a satellite and a ground station, and the unknown physical orientation of a satellite, *circular polarization* is always used for satellite communications. When setting up a satellite communications system, the polarization must be set. A satellite "bird" can use either right-hand or left-hand polarization; the ground station must be set to match, or no signal will be received.

Section Checkpoint

12–53 List the three mechanisms that can change the direction of a radio wave.

12–54 What is the difference between *reflection* and *refraction?*

12–55 What property must a surface have in order to be a good reflector of radio waves?

12–56 What properties of two mediums determines the amount of bending a wave will undergo when it passes between them?

12–57 Which effect explains the blurriness of shadows?

12–58 What angle is being demonstrated by viewing a reflection in a body of water?

12–59 What are the three main propagation modes between transmitters and receivers?

12–60 What limits the distance of line-of-sight communications?

12–61 Over what frequency range is *ground wave* propagation best?

12–62 Explain how the ionosphere plays a part in *sky wave* propagation.

12–63 Which layer of the ionosphere is responsible for most long-distance "skip"?

12–64 What is the *skip zone?*

12–65 Why are low angles of radiation preferable for HF long-distance communication?

12–66 What is the MUF? When the MUF is known, what frequency gives the best signal propagation?

12–67 Under what conditions might you hear an FM broadcast station from 300 miles away? Explain.

12–68 What is *multipath interference?*

12–69 Why are VHF and UHF frequencies used for satellite communications?

12–70 Give two reasons for the use of gain antennas (such as dishes) for satellite communications.

12–71 What is the difference between an *orbiting* and *geostationary* satellite?

12–72 What type of polarization is required for satellite communications?

12–7 LINK BUDGETS

Many variables contribute to the performance of a wireless system. The ultimate goal is communications over a specified distance with a guaranteed level of reliability. The goals include *distance, frequency,* and *link availability.* Engineers can specify *antennas, transmitters, receivers,* and *transmission lines* as needed to meet these goals. The level of *power output* from transmitters in the system will often be limited by law (or practicality). A *link budget* is a worksheet that demonstrates the correct functioning of a wireless system in terms of these variables. The block diagram of Figure 12–28 illustrates how these factors work together.

In order for a communications link to work, the receiver must obtain a minimum level of power from its antenna system (P_r in Figure 12–28). This power level must be high enough to overcome both internal and external noise experienced by the receiver. The

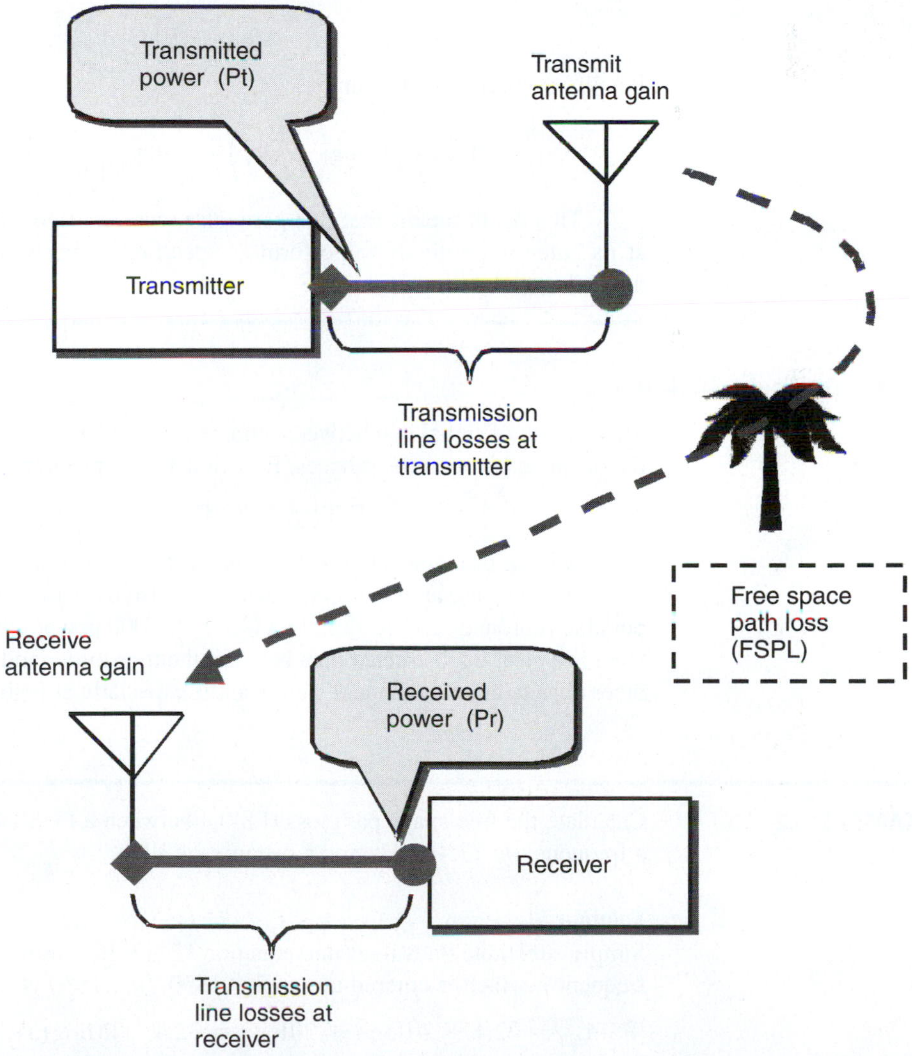

Figure 12–28 Basic factors involved in a link budget

receiver manufacturer normally specifies the sensitivity of the receiver either as an antenna circuit voltage or power needed to obtain a stated BER (bit error rate) for a digital data receiver, or S/N (signal to noise) for an analog receiver. The factors that influence the received power level include *transmitted power, transmission line losses, transmit antenna gain, free space path loss, path obstructions,* and *receiver antenna gain*. These are a lot of factors—some of which aren't necessarily predictable (such as path obstructions). These calculations are made easier by the use of decibel units; let's try an example.

EXAMPLE 12–14

A certain data receiver has a stated sensitivity of 20 μV into 50 Ω for a BER of 1×10^{-2}. Express the sensitivity in dBm units.

Solution

First convert the power level into watts by using Ohm's law:

$$P = \frac{V^2}{R} = \frac{(20\ \mu\text{V})^2}{50\ \Omega} = \underline{8\ \text{pW}}$$

Finally, convert into dBm units:

$$\text{dBm} = 10\log\left(\frac{P}{1\ \text{mW}}\right) = 10\log\left(\frac{8\ \text{pW}}{1\ \text{mW}}\right) = \underline{\underline{-80.97\ \text{dBm}}}$$

This result means that this particular receiver needs at least $\underline{\underline{-80.97\ \text{dBm}}}$ of signal at its antenna terminals to perform as specified. Current state-of-the-art receivers have noise floors close to −136 dBm.

Free Space Path Loss

The amount of signal lost between a transmitter and receiver depends on the frequency and the distance between the devices. Equation 12–13 gives the relationship.

$$\text{FSPL}_{\text{dB}} = 32.4 + 20\log d + 20\log f \qquad (12\text{–}13)$$

where d is the distance in kilometers, and f is the frequency in MHz. Equation 12–13 tells us that losses increase at higher frequencies, and also increase with distance. The minimum loss possible is around 32.4 dB (almost a factor of 2,000 power loss)! This equation assumes that there is a clear unobstructed path between the transmitter and receiver. Buildings, trees, and other objects increase the loss significantly, especially at higher frequencies.

EXAMPLE 12–15

Calculate the free space path loss (FSPL) between a PCS base station and a subscriber at a frequency of 1850 MHz and a distance of 3 km.

Solution

Simply substitute the values into equation 12–13. Be careful about the units (especially the frequency, which is entered as simply "1850," *not* 1850×10^6!).

$$\text{FSPL}_{\text{dB}} = 32.4 + 20\log d + 20\log f = 32.4 + 20\log(3) + 20\log(1850) = \underline{\underline{107.29\ \text{dB}}}$$

A Complete Link Budget

A link budget gives the minimum transmit power needed to achieve the connectivity goal of the system. The easiest way to accomplish this is to list all the parts of the system in a table and then tabulate the results. Working in dB units is preferable; it reduces the arithmetic to addition and subtraction.

EXAMPLE 12–16

What is the ideal required transmitter power for the following system? The parameters are as follows: Frequency = 450 MHz; distance = 50 km; receiver power level needs to be at least −100 dBm; transmit antenna gain is +6 dBi, receiver antenna gain is +3 dBi. Assume that transmission line losses are negligible (not a realistic assumption, but this simplifies the problem). Express the transmit power in dBm and watts.

Solution

Use a table to work out the link budget. It's much easier this way. First, calculate the free space path loss using equation 12–13:

$$\text{FSPL}_{\text{dB}} = 32.4 + 20\log d + 20\log f = 32.4 + 20\log(50) + 20\log(450) = \underline{\underline{119.44\text{ dB}}}$$

With this value in hand, Table 12–1 can be built.

Table 12–1 Total of system gains for example 12–16

Component	Gain (Loss), dB
TX antenna	6.00
Path loss	(119.44)
Rx antenna	3.00
Total gain (loss)	(110.44)

Note how the total gain (actually a loss) is calculated: The losses are *subtracted* (shown in parentheses to denote that they are in fact negative numbers), and the gains are *added*. The total of the gains in Table 12–1 is calculated as:

$$\text{Total Gain} = 6\text{ dB} - 119.44\text{ dB} + 3\text{ dB} = \underline{-110.44\text{ dB}}$$

The transmit and receive antennas overcome some of the path loss, which will help reduce the required transmitter power. Finally, the transmit power can be calculated by referring back to the block diagram of Figure 12–28. The receive power must be at least −100 dBm (according to the system specification). Therefore:

$$P_{\text{RX}} = P_{\text{TX}} + \text{Total Gain}$$

(Remember that everything is in dB units, so we add and subtract to get the results.)

And by rearranging,

$$P_{\text{TX}} = P_{\text{RX}} - \text{Total Gain} = -100\text{ dBm} - (-110.44\text{ dB}) = \underline{\underline{+10.44\text{ dBm}}}$$

Since dBm = 10log(P/1 mW), we can get the power in watts by:

$$P = 1\text{ mW} \times 10^{(\text{dBm}/10)} = 10^{(10.44/10)} = \underline{\underline{11\text{ mW}}}$$

A transmit power of at least 11 mW is needed to get the desired power level at the receiver—at least under ideal conditions.

Link Margin

In Example 12–16 we calculated a minimum transmit power needed to get a desired power level to the receiver. The calculations assumed ideal conditions. Unfortunately, if this was a terrestrial (earth-based) link, it might not be very reliable. The reason for this is *fading* due to obstructions between the transmitter and receiver. In terrestrial links, fading is a random occurrence and is hard to predict. In order to compensate for possible fading, the transmitter power is usually increased according to a chosen set of design rules. One set of rules assumes that terrestrial fading follows a Rayleigh distribution, and that to get a certain percentage of link availability, the transmit power must be increased accordingly. The *link margin* is the additional power transmitted to overcome fading. Table 12–2 shows the appropriate link margins for Rayleigh fading.

Table 12–2 Rayleigh fade margins

Link reliability (%)	Fade margin (dB)
90	8
99	18
99.9	28
99.99	38
99.999	48

Table 12–2 shows us that in order to increase the fade immunity of a link, we must increase the transmit power. There's a pattern to this data, and of course, no amount of power will guarantee 100% reliability. But we can get close!

EXAMPLE 12–17

What is the required transmitter power in Example 12–14 to get 99.9% link availability? Express in dBm and watts.

Solution

Use Table 12–2 to calculate the link margin. For 99.9% availability, we need a 28 dB link margin. The transmit power now becomes:

$$P_{TX} = P + dB_{MARGIN} = 10.44 \text{ dBm} + 28 \text{ dB} = \underline{\underline{38.44 \text{ dBm}}}$$

In watts, this power is:

$$P_{watts} = 1 \text{ mW} \times 10^{(dBm/10)} = 1 \text{ mW} \times 10^{(38.44/10)} = \underline{\underline{6.98 \text{ W}}} \approx \underline{\underline{7 \text{ W}}}$$

About 7 watts of transmit power is needed. This is vastly greater than the original 11 mW we calculated. The conclusion is that fading greatly impacts earth-based communication systems, and that failure to include appropriate link margins in calculations will lead to unreliable wireless links!

Section Checkpoint

12–73 What is a link budget?

12–74 What factors need to be tabulated in a link budget?

12–75 What factors control free space path loss (FSPL)?

12–76 Explain how gains and (losses) are accounted for in Table 12–1.

12–77 Why is *link margin* needed for most wireless systems?

12–78 How is the link margin calculated?

12–79 Is it possible to get 100% reliability in a radio link? Why or why not?

SUMMARY

- Antennas are critical for the operation of all wireless communications systems. Antennas are *transducers* that convert one form of energy into another.
- A radio wave travels at the speed of light and consists of both electric and magnetic fields. The *wavelength* is the distance the wave travels in one ac cycle.
- Polarization refers to the orientation of the electric field of a radio wave. Waves can be vertically, horizontally, or circularly polarized. Receiver and transmitter antenna polarizations must agree for best signal transfer.
- An *isotropic* source is a theoretical, ideal RF source. It doesn't exist, but it is useful as a "measuring stick" when comparing the performance of different antennas.
- The electric and magnetic fields of a radio wave are inversely proportional to the distance from the RF source.
- The *power density* of a radio wave is the product of electric and magnetic field strength and is inversely proportional to the *square* of distance from the source.
- The *characteristic impedance* of a medium (such as free space) is the ratio of electric to magnetic field strength of radio waves within it and is measured in ohms. Free space has a Z_0 of approximately 377 Ω. One way of looking at antennas is to consider them *impedance matching devices* that convert the impedance of free space to the proper electrical impedance at the antenna terminals of a transmitter or receiver.
- The half-wave dipole antenna is the basic building block for many other antennas. It has a characteristic voltage and current distribution along its conductors, and its radiation pattern looks like a donut in three dimensions.
- Radiation patterns of most antennas are measured in two planes, the horizontal and vertical. The standard of comparison is the spherical radiation pattern of the isotropic point source. The approximate input impedance of a dipole is 73 Ω.
- The *radiation resistance* of an antenna is the resistive portion of input impedance that causes signals to be radiated when excited by an RF current.
- Ohmic loss is part of any antenna installation and includes not only conductor resistance but other effects, such as ground loss.
- The *efficiency* of an antenna is the percentage of input power that gets converted to electromagnetic power.
- The *gain* and *beamwidth* of an antenna are closely related. As the antenna becomes more directional, which is measured as a reduced beamwidth, its power gain increases.
- The *Marconi* antenna is used where *omnidirectional* coverage is needed, especially at MF, where wavelength is long. It only needs to be one quarter-wavelength long, rather than the full

half-wavelength needed for a dipole. The earth acts as the "missing" half of the antenna.

- Directional antennas are used where improved gain and control of radiation pattern are needed. The Yagi and log-periodic antennas are common at HF and VHF.
- The Yagi antenna uses *parasitic elements* to act as reflectors and directors, which intensify the radiated signal in the desired direction.
- Dish antennas are used in various forms at UHF and microwave frequencies, where the wavelength is very short. They're highly directional and provide much more gain than conventional directional antenna designs.
- The most popular dish designs include the *horn-fed parabolic, Cassegrain-fed,* and *hog-horn.*
- Many other specialized antennas are used, such as *collinear arrays* (which are sets of dipoles phased to produce a particular radiation pattern) and *phased arrays* (which are sets of antennas, usually Marconi types, spaced and phased to obtain a directional pattern).
- When troubleshooting antennas, safety is an important consideration. Proper safety apparel and practices are a must when working near or on towers.
- Many antenna problems can be found by visual inspection; SWR and ohmage readings are also quite useful.
- Intense RF fields can be harmful to people. The best practice is prudent avoidance.
- *Propagation* is the movement of radio waves from point to point. Radio waves can change direction in their course of travel by *reflection, refraction,* and *diffraction.*
- When electromagnetic energy passes between two materials with different indices of refraction, it bends. The angle change depends on the relative values of the index of refraction for each of the materials.
- When a radio wave passes between two materials and enters at more than the *critical angle,* it doesn't enter the second material; instead, it is reflected back out into the first.
- Radio waves can travel from source to destination by three main paths: the *space (direct) wave,* the *ground wave,* and the *sky wave* (ionospheric propagation).
- Space wave propagation is limited by the *radio horizon.* The distance to the radio horizon is determined by the height of the antenna.
- Ground wave propagation is not very effective above 2 MHz.
- The *ionosphere* is the portion of the earth's atmosphere that is electrically charged. It refracts radio signals back to earth for *sky wave propagation* (sometimes referred to as *skip* propagation). Skip propagation occurs primarily below 50 MHz.
- Skip propagation is hard to predict, as the ionosphere is in a constant state of change. Many ionospheric characteristics change from night to day and with the 11-year solar cycle.
- The most efficient sky wave propagation occurs just below the maximum usable frequency, or MUF. Above the MUF, little or no ionospheric refraction takes place. The MUF is constantly changing for any given path.
- Above 50 MHz, tropospheric *ducts* can conduct signals for hundreds to thousands of miles. Ducts are often formed during unusual weather conditions, such as a temperature inversion.
- VHF and UHF frequencies are used for satellite communications so that signals will pass through the ionospheric layers without refraction or absorption.
- Satellites generally use circular polarization since the relative orientation of satellite and ground station are not known.
- Link budgets are used to determine the required transmitter power to provide wireless communications. They include all factors in the link, including distance, antennas, transmission lines, and receiver sensitivity.
- Free space path loss is controlled by distance and frequency. Increasing either parameter increases the loss.
- Link margins are used to assure a certain level of link reliability in the presence of fading, a common feature of terrestrial links.

PROBLEMS

1. What is a *transducer?* Why are transmitting and receiving antennas considered transducers? What types of energy are involved?
2. What are the two fields that make up a radio wave? What angle is maintained between them?
3. Calculate the wavelength of the following signals in free space: (a) 50 MHz, (b) 300 MHz, (c) 28.325 MHz.
4. List and explain the three types of *polarization* for electromagnetic waves.
5. What theoretical (nonexistent) antenna radiates equally well in all directions? Why is it needed?
6. A certain transmitter radiates an EIRP of 500 W. Calculate the electric and magnetic field strengths in volts/meter and amps/meter at a distance of 20 km away.
7. Calculate the radiated power density in watts/meter for the transmitter of question 6.
8. Repeat the calculation of electric field, magnetic field, and power density for the transmitter of question 6 at the following distances: (a) 1 mile, (b) 100 ft, (c) 1 km.
9. What is meant by the *characteristic impedance* of free space?
10. The electric field from a transmitter measures 5 V/m at a certain point in free space. What will the magnetic field strength be at the same point?
11. At a distance of 10 miles from a transmitter, the measured power density is 1 $\mu W/m^2$. What is the EIRP of the transmitter?
12. Your company is investigating an interfering signal from a CB operator on a frequency of 27.125 MHz (CB channel 14). At a distance of 4 miles from the CBer's location, the equipment measures a field strength of 10 mV/m. (a) What is the EIRP of the CB operator's station? (b) The CBer claims to be using a half-wave vertical antenna with a gain of +2 dBi. The maximum power allowed from a CB transmitter is 4 watts (carrier). What is the transmitter power level, and is the CBer operating legally?
13. Draw the schematic diagram of a dipole antenna, showing the voltage and current distribution. What is the approximate input impedance of this antenna?
14. What two resistive components make up the resistive (real) portion of an antenna's input impedance?
15. A certain antenna radiates 250 W when its feedpoint current is 2 A. What is the radiation resistance of this unit?
16. An antenna has a radiation resistance of 100 Ω, a loss resistance of 10 Ω, and a reactance of $+10j\ \Omega$. (a) What is the total input impedance of the antenna in rectangular form? (b) If the feedpoint current is 2 A, what are the *loss* and *radiated* powers?
17. If an antenna has a feedpoint current of 2 A, and an input impedance of $(73-10j)\ \Omega$, what are the *loss* and *radiated* powers if the radiation resistance is 65 Ω?
18. What is the efficiency of the antenna of problem 17?
19. What is the efficiency of an antenna with a loss resistance of 20 Ω and a radiation resistance of 10 Ω?
20. Explain the relationship between the orientation and polarization of a dipole antenna.
21. What is the *radiation pattern* of an antenna?
22. How is the radiation pattern of an isotropic point source used in comparing antenna radiation patterns?
23. Sketch a picture of a dipole and on top of the sketch, draw its radiation pattern in the horizontal plane. Where does maximum radiation occur?
24. Antenna A is marketed as having beamwidth of 78°, and antenna B claims a beamwidth of 60°. Which antenna is most likely to have the greatest forward gain, and why?
25. Calculate the length of a half-wavelength dipole for operation on a frequency of 91.9 MHz, taking into account the 95% factor for practical antenna construction. Report the total length in *feet*.

26. How should a dipole antenna be oriented if an omnidirectional radiation pattern is desired?

27. Calculate the length of a Marconi antenna for operation on 52.565 MHz. Report the answer in *inches*.

28. What is the forward gain of a three-element Yagi antenna? What EIRP will result if this antenna is fed with 10 watts of power?

29. What is one characteristic that is *not* reciprocal between transmitting and receiving antennas? What is done to compensate for the falling energy recovery of simple antennas at UHF frequencies?

30. Draw a diagram of a horn-fed, Cassegrain-fed, and hog-horn dish antenna. Which of these antennas has no dead spot in its pattern?

31. What is a *phased array* antenna? Draw a diagram of a two-element phased array.

32. List at least four safety precautions for working with antennas, RF sources, and towers.

33. List four methods that can be used to isolate problems in antenna systems.

34. Why would it be useful to record the SWR of an antenna in a station log immediately after it has been installed or serviced?

35. What is the difference between *reflection* and *refraction* of a radio wave? What is required to produce each effect?

36. List the three primary modes that a radio signal can use to get from transmitter to receiver.

37. Why does sky wave propagation depend upon the time of day? What is the maximum frequency for this type of propagation?

38. Why is sky wave propagation hard to predict?

39. What is the maximum distance two stations can communicate by direct or space wave if the antenna heights are 25 and 200 ft?

40. Two stations are located a distance of 20 miles apart and have antenna heights of 30 and 50 ft. Can they communicate by space wave? Why or why not?

41. What is the maximum frequency for useful ground wave propagation?

42. What is the ionosphere? What causes it to be electrically charged?

43. Which ionospheric layer is responsible for most skip propagation?

44. Why is it desirable to operate slightly below the MUF when operating skip?

45. What type of polarization is used in satellite work? Why?

46. Calculate the free space path loss (FSPL) for the following signal paths:

    a. Distance = 10 km, frequency = 100 MHz
    b. Distance = 1000 km, frequency = 7253 kHz
    c. Distance = 1 km, frequency = 24 GHz

47. A data receiver requires –90 dBm of power input from the antenna. If the operating frequency is 5620 MHz, the distance is 10 km, and both transmit and receive antennas have a 6 dBi gain, calculate the ideal power (no link margin) needed at the transmitter. Express in both dBm and watts. Assume that there are no significant transmission line losses.

48. Recalculate the transmit power in problem 47 if there is a 3 dB transmission line loss at the transmitter, and a 4 dB loss at the receiver. Express the ideal (no link margin) transmit power in dBm and W.

49. A receiver requires a signal level of −110 dBm to work correctly. The operating frequency is 920.1125 MHz, and the link distance is 20 km. If both transmit and receive antennas have a gain of 6 dBi, the transmitter's line losses are 3 dB, and the receiver's line losses are 1 dB, calculate the necessary transmit power for a link reliability of 99%. Express in dBm and W.

50. Recalculate the transmit power for problem 49 in order to increase the link reliability to 99.9%. Again express in dBm and W.

13

Microwave Communication Systems

OBJECTIVES

At the conclusion of this chapter, the reader will be able to:

- describe the differences between microwave and lower-frequency communications techniques
- describe the different modes of propagation within a waveguide
- explain the operation of a magnetron and other microwave oscillators
- explain the operation of traveling wave tube (TWT) and klystron amplifiers
- explain the operation of a parametric amplifier
- describe the operation of radar systems using time-of-arrival and Doppler shift methods

Microwave communications systems are those that operate above approximately 2 GHz (2000 MHz). Microwave communications applications include satellites, terrestrial (earth-based) relay links, radar, plus some consumer applications such as cordless telephones, IEEE 802.11 wireless *LANs,* and a number of emerging technologies, such as Bluetooth. At these high radio frequencies, conventional circuit construction techniques using *lumped* (discrete) L and C components are not very effective for reasons we'll soon discuss. The circuit elements that you have used in your studies of electronic fundamentals (resistors, capacitors, inductors, and so on) are referred to as *lumped* components because the "works" of each part are contained or "lumped" within a single package. You can easily identify the individual parts in a circuit containing these elements. Each part performs one function.

Microwave circuits often appear to be little more than a collection of tubing and other "plumbing parts" to the novice technician. Each component in a microwave system has a function similar to the *groups* of lumped components used in lower-frequency approaches. There's nothing mysterious or magic about microwaves, once you learn a few new principles.

Figure 13–1
Equivalent circuit of a 100 Ω resistor

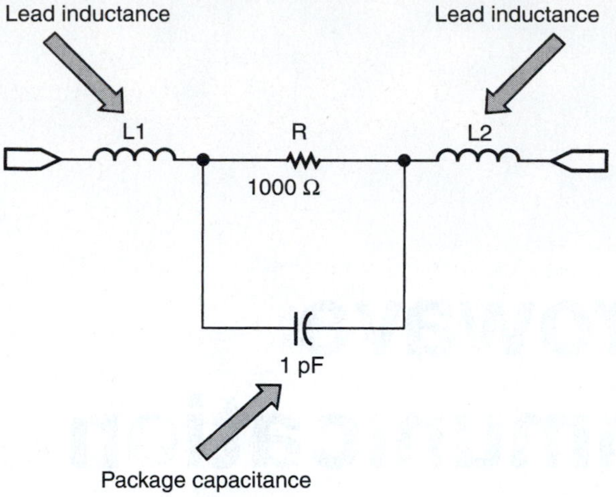

13–1 MICROWAVE CONSTRUCTION TECHNIQUES

Limitations of Conventional Components

There are undesired or *parasitic* elements in all conventional electronic components. At low frequencies, the effects of these additional elements can largely be ignored; this is not true at UHF and SHF! For example, the 100 Ω resistor shown in Figure 13–1 actually consists of the equivalent components shown. The inductors L_1 and L_2 represent the *lead inductance* of the component. These values are dependent upon the length of the resistor's leads and are generally a few nH or less. In addition, the resistor appears shunted by a *package capacitance* of approximately 1 pF (this value often ranges up to 5 pF for certain types of resistors).

You might be inclined to think that these additional values are too small to have any real effect upon the operation of the resistor; of course, this is correct at *low* frequencies. At even the "low" microwave frequency of 2 GHz, this is no longer true!

EXAMPLE 13–1 If L_1 and L_2 are 1 nH, and the operating frequency is 2 GHz, calculate the reactance of each component in Figure 13–1 and calculate the total impedance of the resistor.

Solution

The inductive reactance X_L for L_1 and L_2 can be calculated as

$$X_{L1} = X_{L2} = 2\pi f_L = 2\pi(2\text{ GHz})(1\text{ nH}) = \underline{\underline{12.57\ \Omega}}$$

Therefore, the lead inductances add about 25 Ω to the circuit impedance!

The capacitive reactance X_C shunting the resistance can be calculated as

$$X_C = \frac{1}{2\pi fC} = \frac{1}{2\pi(2\text{ GHz})1\text{ pF}} = \underline{\underline{79.5\ \Omega}}$$

The capacitance therefore reduces the impedance of the resistor, because it is in parallel with it.

The *total impedance* is the phasor total of all the reactances and resistances in the circuit. To calculate this, we have to use appropriate rectangular and polar notation for the individual reactances in the circuit. We get

$$Z_{\text{total}} = jX_{L1} + jX_{L2} + 100\ \Omega \| (-jX_C) = j25.14 + (100\ \Omega) \| (-j79.5\ \Omega)$$
$$= \underline{\underline{38.7 - j23.57\ \Omega}} = \underline{\underline{45.34\ \Omega\angle -31.3^\circ}}$$

This is very strange—the resistor hardly looks like a resistor at all! In fact, because there are both L and C components, there will also be at least one frequency where the component looks like a resonant circuit—it may look very close to a *short* at the resonant frequency.

Example 13–1 demonstrates that there are real problems with using conventional components at microwave frequencies. Conventional capacitors and inductors also have hidden resistances, capacitances, and inductances that prevent their use at microwave frequencies.

So how can we build microwave communications circuits at all? There are three important elements we must consider. First, we must abandon many of the components used at lower frequencies and adopt devices that are built for UHF and SHF application. In general, the use of *surface-mount* components greatly reduces the amount of parasitic lead inductance by eliminating the wire leads. Figure 13–2 shows several surface-mount RF semiconductors on a demonstration board. Second, we must begin thinking in terms of *distributed* circuit elements rather than *lumped* elements. *Distributed* circuit elements are contained throughout the physical circuit rather than being contained within individual locations within it (the *lumped* approach). Finally, because of the high frequencies, wavelengths are very short. Even an inch of wire can be an effective antenna at microwave

Figure 13–2
Surface-mount RF semiconductors

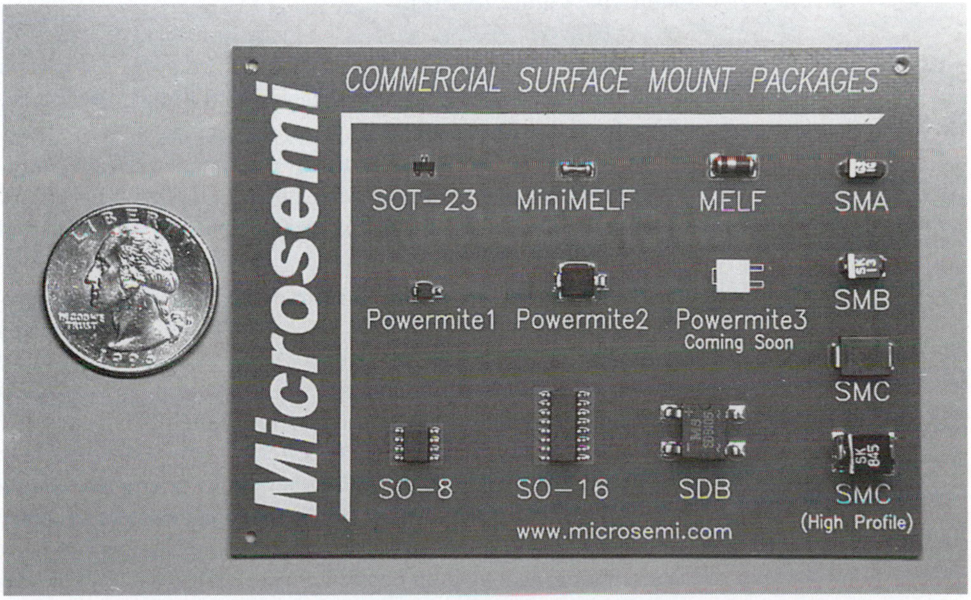

frequencies. Unless the entire circuit is treated as a *transmission line,* it is likely that unintended radiation and feedback will occur.

EXAMPLE 13–2 What is the length, in inches, of a quarter-wave antenna at a frequency of 10.24 GHz?

Solution

The wavelength equation from Chapter 1 can be used:

$$\lambda = \frac{v}{f} = \frac{3 \times 10^8 \text{ m/s}}{10.24 \text{ GHz}} = \underline{0.029 \text{ m}}$$

Only one-quarter of this wavelength is required, so we get

$$\lambda/4 = 0.029 \text{ m}/4 = 0.00732 \text{ m} = \underline{\underline{0.288''}}$$

This is a *very* short wavelength. In conventional circuits, it is not hard to imagine a lead length of a quarter-inch. In a 10.24 GHz circuit, a quarter-inch lead becomes a very effective antenna, disrupting the circuit's operation!

Distributed Circuit Techniques

By looking at electrical circuits in a new way, we can develop microwave *equivalents* to circuits that we used at lower frequencies. The microwave versions of the circuits depend on the overall layout of the conductors and can be thought of as *distributed* circuits.

Suppose that we wish to build an LC resonant circuit to act as a bandpass filter at a frequency of 10.24 GHz. We *could* try building the circuit from a discrete capacitor–inductor combination (and we can still think in those terms, of course). The circuit might look a little like Figure 13–3(a).

Now we need to reduce the leads to near zero length to get this circuit to work at microwave frequencies. Imagine that we can somehow put the inductor right inside the capacitor, in between its plates. In order to reduce the effect of the inductor's own lead lengths, we will *distribute* it evenly in four parts around the capacitor's plates. This is Figure 13–3(b). Figure 13–3(c) is the completed filter, which is implemented as a *cavity resonator*. Where did the inductor go? It is *distributed* within the walls of the cavity. The longer the cavity is built, the more inductance it will have, and the lower its resonant frequency will become.

The *capacitance* is also distributed in the cavity, between the top and bottom lids. With this arrangement, how can we get energy in and out of the filter, since the top and bottom are closed? There are two common methods. First, we can "poke" coupling wires or *stubs* into the cavity through one end. The coupling stubs act as transmitting and receiving antennas, and the cavity then filters the signal that passes in between them. Second, the cavity can be attached to a printed circuit board assembly, and the conductors on the board can be etched in a manner so that they pass underneath the filter cavity.

Finally, we might need to fine-tune the cavity resonator filter. Tuning can be accomplished by the simple addition of a screw and locknut on top of the unit. As the screw is turned, it will move in or out of the open cavity, which will slightly change the capacitance of the filter, which in turn changes the filter's resonant frequency. The cavity can also be

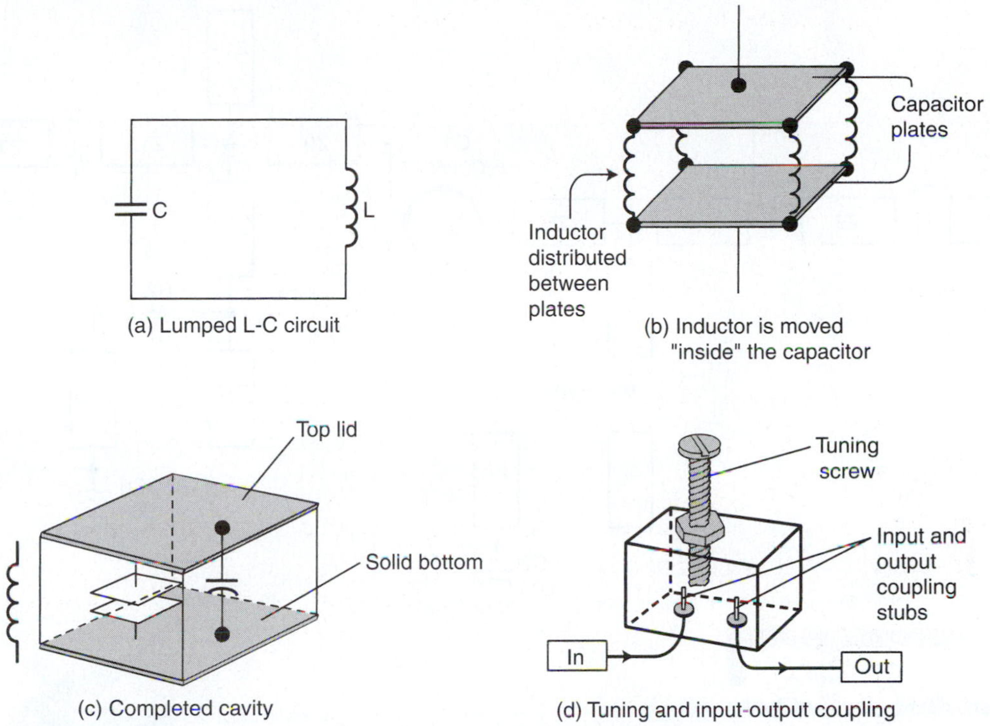

Figure 13–3 Development of a cavity resonator

tuned by the addition of a sliding piston, which acts to vary the length of the cavity and thus changes its resonant frequency.

Microwave Circuits and Transmission Lines

Since the wavelength of microwave frequencies is very short, even a conductor measuring a mere fraction of an inch must be accounted for as a *transmission line* in order for the circuit to work as intended. From Chapter 11, we know that transmission lines can perform two primary circuit functions. First, they can carry RF energy between two points in a circuit with minimum radiation and loss. Second, they can *match* or *transform* one impedance into another in order to obtain maximum power transfer between two stages. Let's take a look at how this can be done.

Figure 13–4 shows a low-noise RF preamplifier for the 10 GHz band employing a special microwave JFET called a GaSFET (gallium-arsenide field-effect transistor). The GaSFET offers far superior performance at microwave frequencies than a conventional bipolar or field-effect transistor. The schematic diagram of Figure 13–4(a) looks very strange. You can recognize the transistor and a few passive components (such as R_1, C_2, and so on), but the rest of the circuit consists of rectangles labeled "Z_1," "Z_2," and so on. What are these?

The rectangles in this circuit are *transmission line sections*. These transmission line sections are used to impedance match the input and output of the amplifier to 50 Ω and also provide bandpass filtering. They are fabricated right on the printed circuit board. Transmission lines formed on a PC board are often called *microstrip* lines. Their physical dimensions must be very accurate, and the PC board material must have a known (and

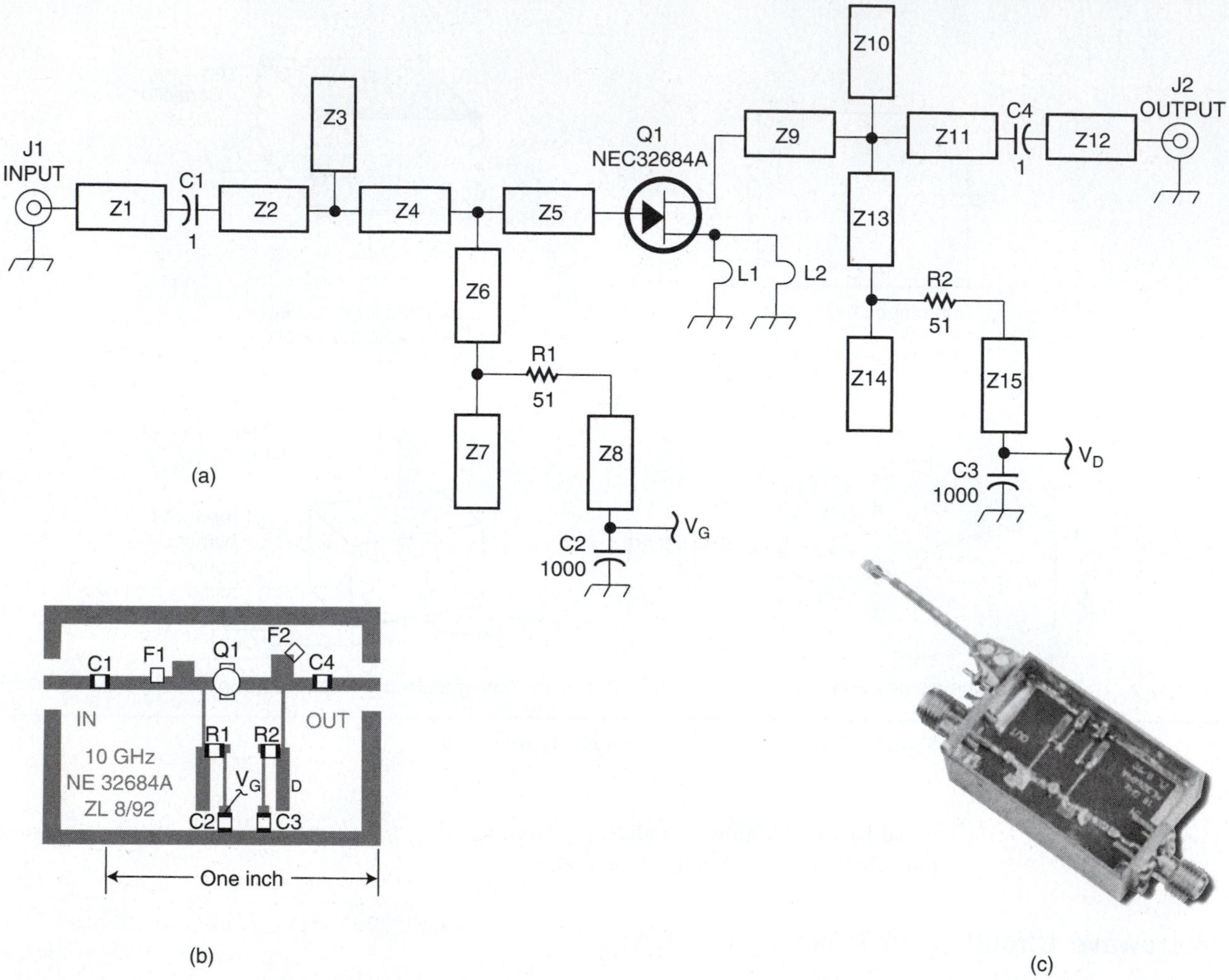

Figure 13–4 A GaSFET preamplifier for the 10 GHz band. (a) Schematic. (b) Printed circuit layout. (c) Complete amplifier. (Source: American Radio Relay League. Reprinted with Permission.)

stable) dielectric constant. Figure 13–4(b) shows the PC board layout; look carefully at the board, and you can identify each of the "Z" microstrip line elements.

Figure 13–4(c) shows a completed amplifier. Note the extensive use of surface mount components! Should you ever have to repair a unit such as this, remember that the circuit board traces are transmission lines. If they are damaged in any way (such as from overheating during soldering), their characteristic impedances will be altered, and the circuit will no longer function as designed!

Remember that these transmission line sections also depend upon the dielectric constant of the PC board material. Overheating a PC board trace can burn or melt the PC board insulating material, which changes its dielectric constant. Even a small amount of flux left behind after soldering can throw the circuit off-frequency. Proper construction and repair techniques therefore include using the minimum amount of heat for soldering and thorough cleaning (defluxing) of circuits after repairs have been completed.

Section Checkpoint

13–1 What is the lowest approximate frequency where microwave techniques are used?

13–2 Explain the concept of *lumped* components and give an example.

13–3 How can a resistor also behave as an inductor? What part of the package provides the inductance?

13–4 What three principles must be considered when constructing microwave circuitry?

13–5 What are *distributed* circuit elements?

13–6 Explain how the resonator of Figure 13–2 acts as a bandpass filter.

13–7 What are two functions carried out by transmission lines?

13–8 Describe how the transmission line sections in the amplifier of Figure 13–3 are fabricated.

13–9 What can happen if a microstrip section is overheated during soldering?

13–10 After servicing a microwave PC board assembly, solder flux must be removed. Why?

13–2 MICROWAVE TRANSMISSION LINES

At microwave frequencies, the conventional transmission lines discussed in Chapter 11 become very inefficient. There are two reasons for this. First, the *wavelength* of microwave signals is very short; and second, many of the dielectric (insulator) materials in transmission lines become lossy at SHF.

The *open* transmission lines (such as ladder line or ribbon cable) will become *antennas* at microwave frequencies because the distance between the two conductors in the line approaches a sizable fraction (more than 1/8th) of a wavelength. The fields set up by the transmission line conductors can no longer cancel, resulting in *radiation loss*. This is very undesirable!

Even coaxial cable performs miserably at microwave frequencies. In a coaxial cable, the dielectric is often made of a plastic such as polyethylene or polystyrene, in solid or foam form. The waves moving down the coax must pass through the dielectric, since they are confined within the shield. As the frequency is increased, the dielectric begins to absorb more of the RF energy and convert it to heat. In addition, the *skin effect* (where most of the RF energy flows on the outer surface of conductors) also increases at SHF, increasing the effective RF resistance of the transmission line conductors.

There is only one transmission line that can efficiently conduct microwave energy over any distance at all, and that is *waveguide*. For short runs (a few inches or less) on printed circuit boards, *microstrip* and *stripline* lines can be used. Figure 13–5 shows many of the types of waveguide that are available.

Figure 13–5 Several types of waveguide

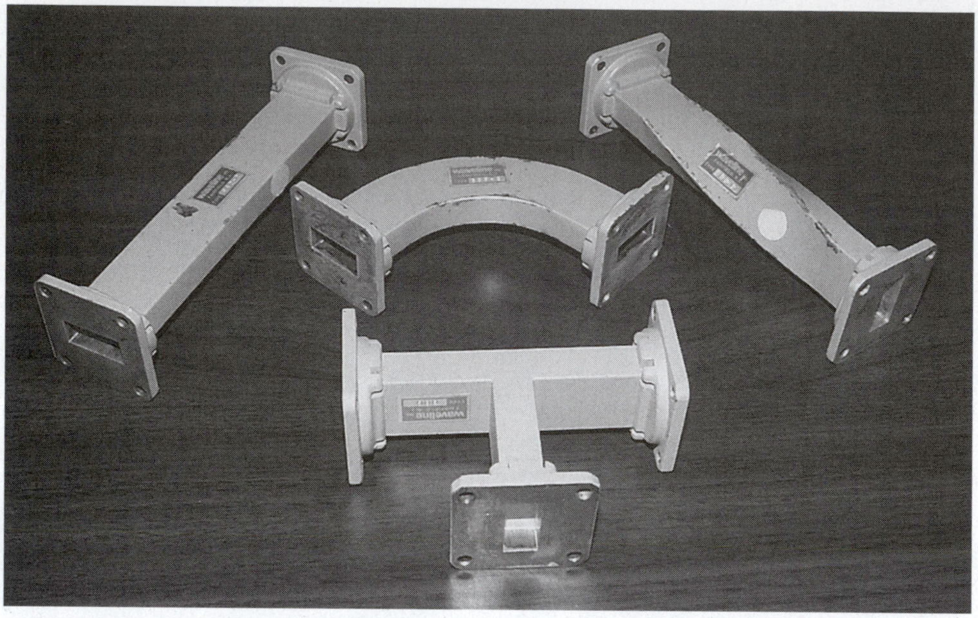

Waveguide Operation

Waveguide is a hollow metal tube that conducts RF energy between two points in a circuit. It can be built in both rectangular and round configurations. Figure 13–6 shows a section of rectangular waveguide. In Figure 13–6, the RF energy is *coupled* (passed) into the waveguide by means of a quarter-wave *stub* antenna. This is often referred to as *capacitive* coupling. Once the RF energy is inside the waveguide, it travels along inside, *reflecting* off the inside walls as it proceeds. Efficient reflection of RF energy requires that the inside surface of the waveguide have a low electrical resistance. Thus, the inside of waveguide is often plated with excellent (and expensive) conductors, such as silver or gold. The plating need not be very thick because RF currents will only flow along the "skin" of the wall.

Figure 13–6 A rectangular waveguide section

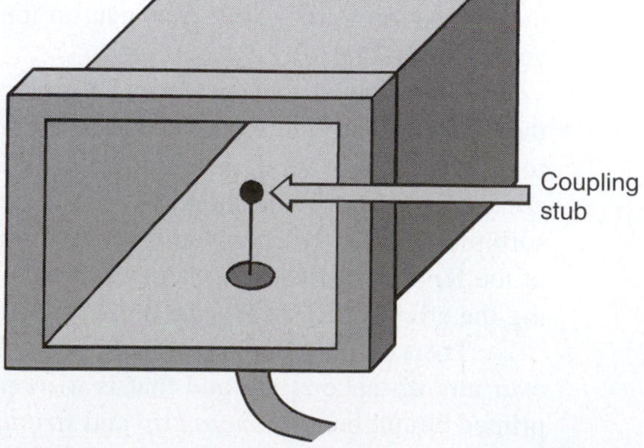

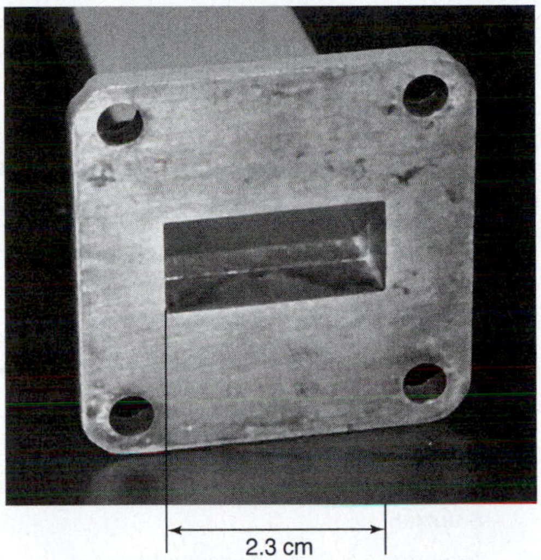

Figure 13–7
Measurement of the cutoff frequency in rectangular waveguide

Waveguide Cutoff Frequency

You might wonder why waveguide isn't used for *all* RF transmission line needs, since it's much more efficient than conventional types of line. There are two reasons for this. First, waveguide is very expensive when compared to other transmission lines. Second, there is a *minimum* frequency that can be passed by any particular type of waveguide. This frequency is referred to as the *cutoff frequency*.

The cutoff frequency is determined by the longest cross dimension of the waveguide. When the wavelength of the energy being introduced into the waveguide becomes too long, the waveguide "shorts" out the electric and magnetic fields of the wave. This occurs when one-half (or more) of the signal wavelength occupies the waveguide's cross section. Therefore, to determine the cutoff frequency of a section of waveguide, simply measure the longest cross dimension and then find the frequency where that length is one-half of a wavelength. The following equation summarizes this relationship:

$$f_{co} = \frac{v}{2d} \tag{13–1}$$

where v is the velocity of radio energy (3×10^8 m/sec), and d is the longest cross dimension, in meters. Figure 13–7 shows the measurement of the dimension d in a section of rectangular waveguide.

EXAMPLE 13–3

The long dimension of the waveguide measures 2.3 cm in Figure 13–7. Calculate the cutoff frequency, f_{co}, for the unit.

Solution

Equation 13–1 directly solves this problem. We must not forget to express the value of 2.3 cm as 0.023 m (100 cm = 1 m):

$$f_{co} = \frac{3 \times 10^8 \text{ m/s}}{2(0.023 \text{ m})} = \underline{\underline{6.52 \text{ GHz}}}$$

Therefore, frequencies below 6.52 GHz cannot pass through this section of waveguide. In effect, the waveguide acts as a *high-pass filter.* The manufacturer will typically rate waveguide in terms of recommended frequency range; these ranges are usually 10–20% higher than the cutoff frequency for the simplest possible propagation mode. We will discuss propagation modes shortly.

Waveguide is not practical for frequencies below the microwave region. The next example demonstrates this point.

EXAMPLE 13–4

How long must the long dimension of a rectangular waveguide be in order to support a lower cutoff frequency of 30 MHz?

Solution

Equation 13–1 can be manipulated to solve this problem. We must solve for the dimension *d*:

$$f_{co} = \frac{v}{2d}$$

$$d = \frac{v}{2f_{co}} = \frac{3 \times 10^8 \text{ m/sec}}{2(30 \text{ MHz})} = \underline{\underline{5 \text{ m}}}$$

This is an *enormous* physical size! The waveguide would have to be at least 5 meters (16.4 feet) wide inside to support the propagation of 30 MHz signals. This would obviously not be very practical!

Propagation Modes

There are many ways in which RF energy can move through a waveguide. Each of these is called a *mode* of propagation. There are two primary modes, called *transverse electric* and *transverse magnetic,* abbreviated as TE and TM, respectively. Recall that a radio wave is composed of electric and magnetic fields. These fields are at 90° angles with respect to each other. In TE propagation, there is no component of the *electric* (E) field aligned with the direction of propagation; the electric field is perpendicular to the line of movement. In TM propagation, the magnetic (M) field is perpendicular to the line of propagation.

A shorthand notation is used to describe TM and TE modes within a waveguide. Two numbers follow the TE or TM designator. The first indicates the number of half-wave patterns in the *long* cross dimension of the waveguide, and the second indicates the number of half-wave patterns in the *short* cross dimension of the waveguide.

Dominant propagation mode Figure 13–8 shows TE_{10} propagation, which is often called the *dominant* propagation mode for waveguide. This is the most common, and simplest, mode of energy travel. In the TE_{10} mode, there is one half-wave pattern of the electric ($\mathcal{E}$) field across the long cross dimension of the waveguide, and *no* half-wave E pattern across the short dimension. The lines of the E field are perpendicular to the direction of wave movement. The magnetic ($\mathcal{H}$) field lines are perpendicular to the E field lines. In practical work with waveguides, TE_{10} is the most commonly used mode. The TE_{10} mode also provides the lowest possible *cutoff frequency,* as predicted by equation 13–1.

Figure 13–8
TE_{10} propagation

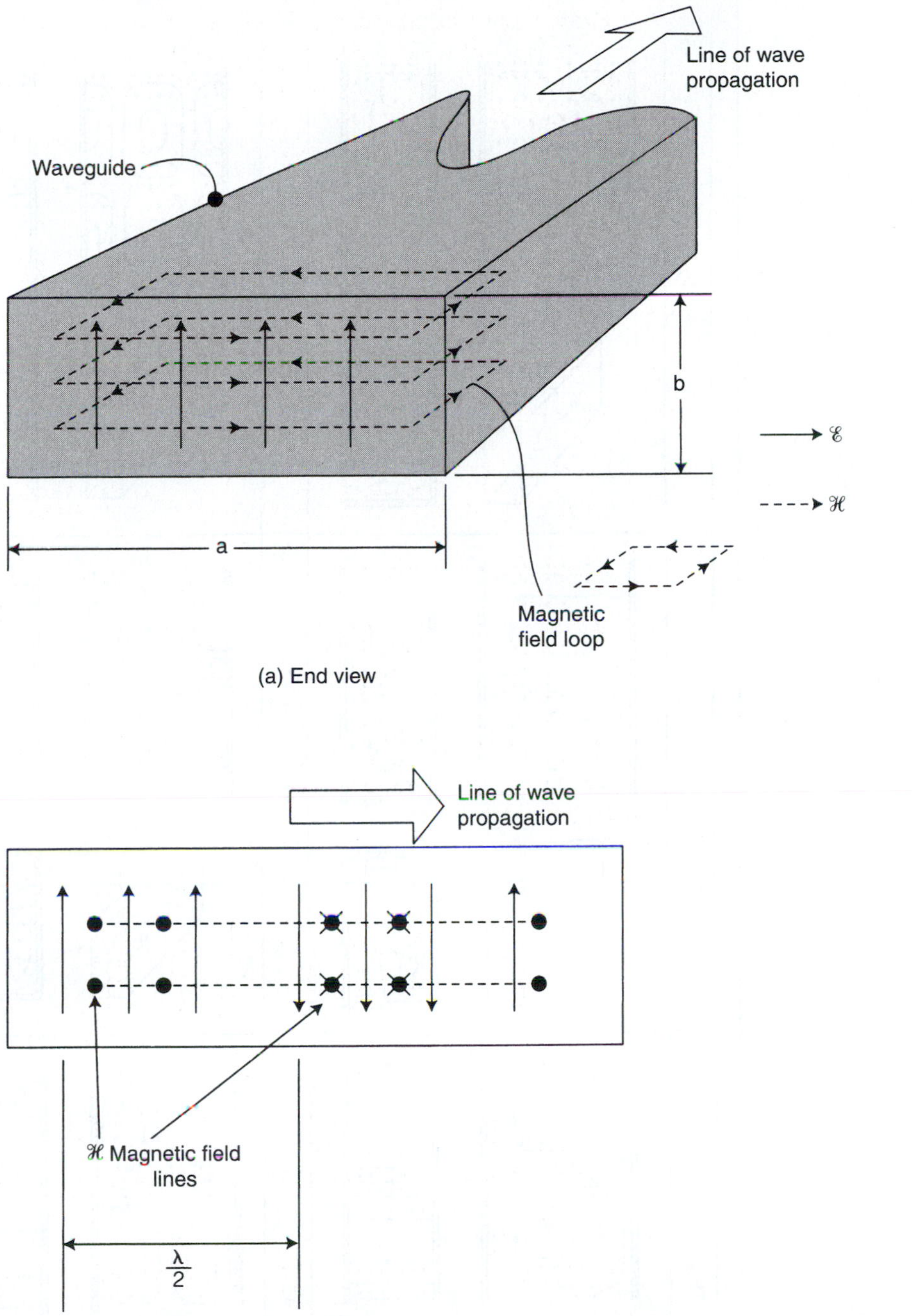

(a) End view

(b) Side view

Higher modes Higher modes of propagation are obtained when the operating frequency is increased sufficiently to allow one or more wavelengths to fit within the long cross dimension of the waveguide. The use of higher modes is generally restricted to special applications; it is much simpler to work in the dominant TE_{10} mode. Figure 13–9 is a comparison of the various propagation modes.

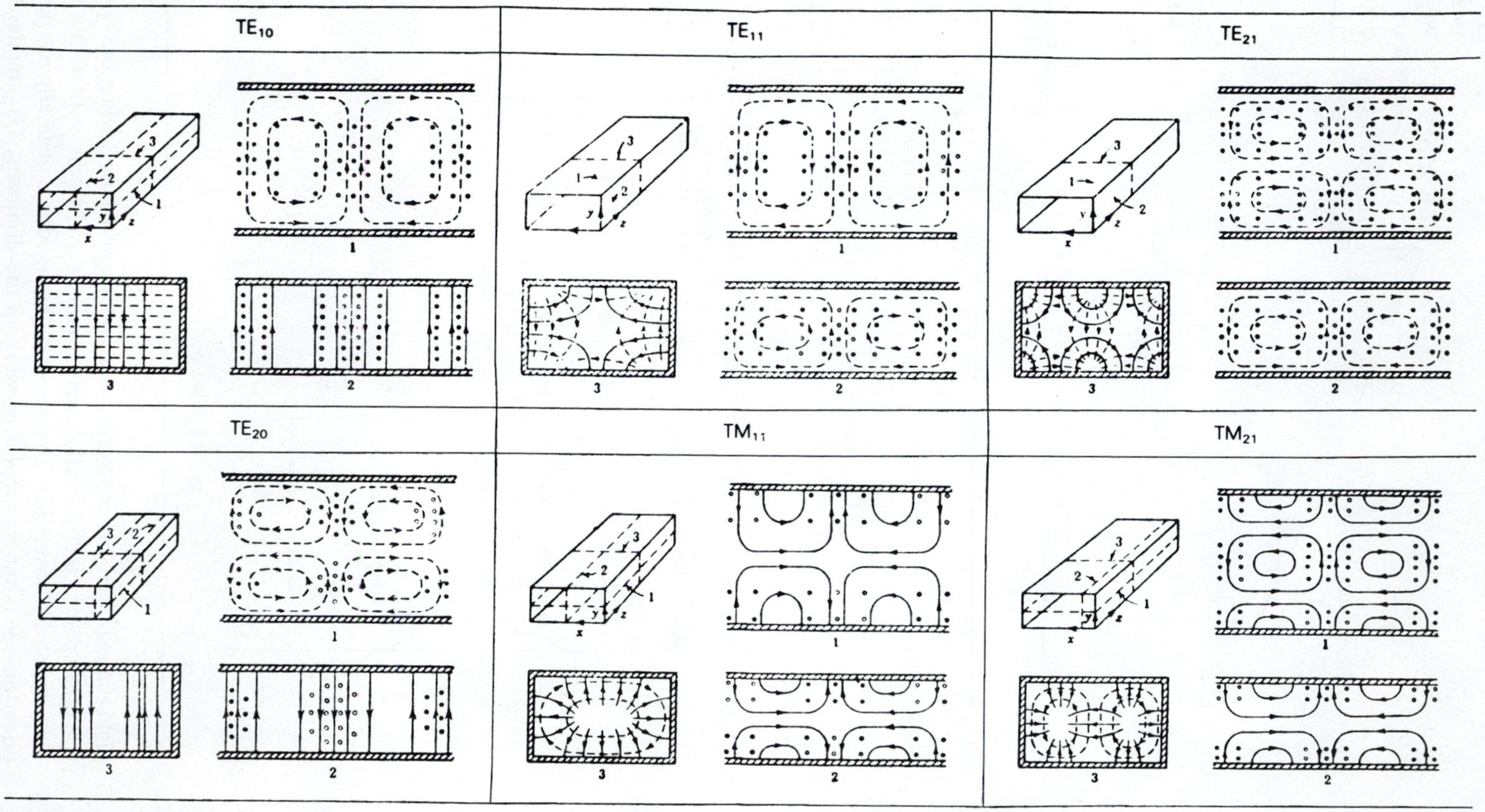

Figure 13–9 Field configurations in various propagation modes (Source: S. Ramo, *Fields and Waves in Modern Radio.* Reprinted with permission of Simon Ramo)

In the TE_{20} mode, there are now two half-wave patterns across the long dimension of the waveguide but still no patterns across the short dimension. By increasing the size of the short dimension, it is possible to get TE_{11}, where one half-wave field pattern now appears across *both* dimensions of the waveguide. Compare the pictures of TE_{10} and TE_{11} propagation. The latter is *much* more complicated in terms of where the electric and magnetic field lines will lie. This fact makes coupling (getting energy in and out) much more critical for higher modes of propagation and is one reason why the dominant or fundamental mode, TE_{10}, is so popular.

To completely describe higher propagation modes requires the use of Maxwell's equations, which describe the conditions under which electromagnetic energy can exist. These equations are well beyond what a technician needs to know.

Special Waveguide Sections

There are several special-purpose waveguide sections that are often used. Among these sections are *bends, tees, attentuators, terminators,* and *directional couplers.*

Bends The path between a microwave antenna and radio is seldom a straight line. *Bends* are sections of waveguide that can accommodate the necessary turns in an installation while preserving the desired mode of propagation. Figure 13–10 shows several of these.

The *H* bend of Figure 13–10(a) is used to turn a 90° corner. The propagation is undisturbed, except that the magnetic field is bent 90° as it passed through the section. The polarization (orientation of the electric [E] field) is unaffected. The *E* bend also completes a 90° turn in either an upward or downward direction. The magnetic field is undisturbed, but the *E* field is bent by 90° within. The *twist* of Figure 13–10(c) is used to effect a shift in the polarization of the wave. The electric and magnetic fields maintain the same orientation within the waveguide, but because of the gradual twist, their orientation is different at the opposite ends with respect to the outside world.

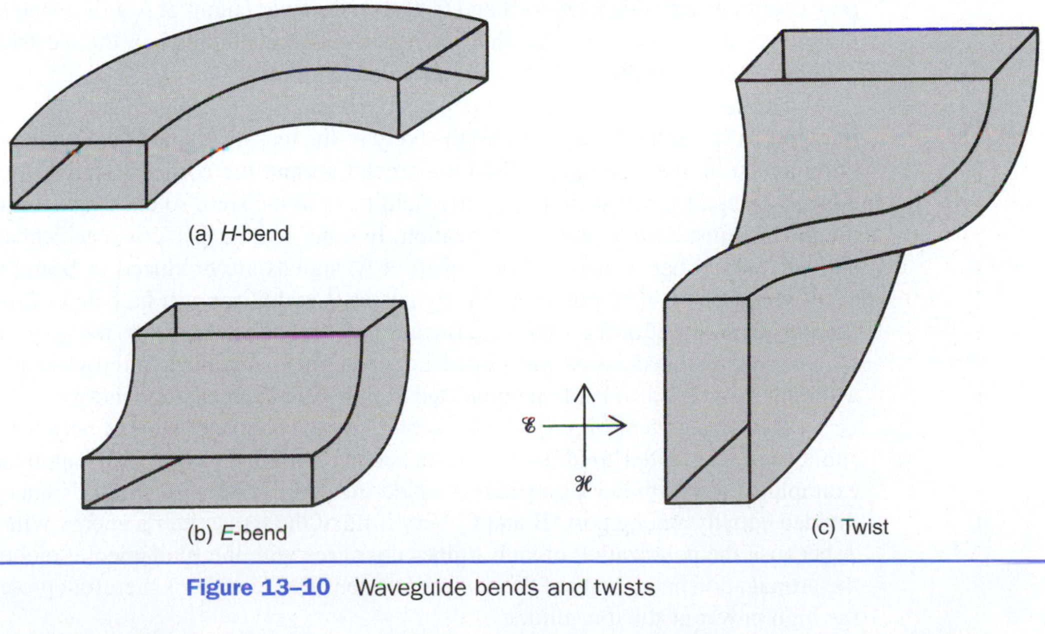

Figure 13–10 Waveguide bends and twists

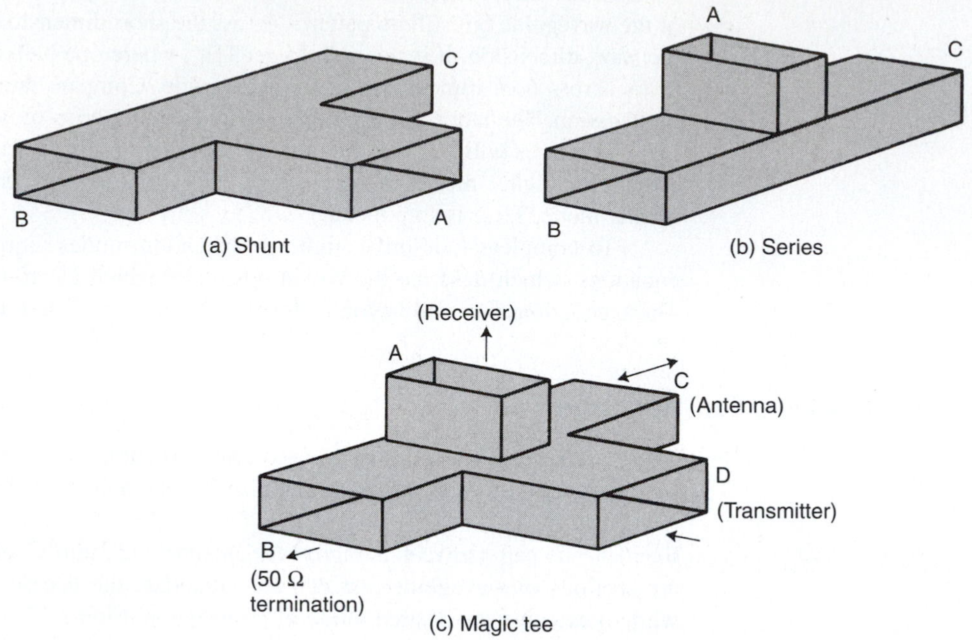

Figure 13–11
Three tees

The choice of bends and twists is largely determined by mechanical considerations; in other words, an engineer will simply specify a part that fits in a given space! By combining these three basic shapes, nearly any electrical and mechanical requirement can be met.

Tees Tees are often used to act as power combiners, power splitters, and isolators. A *power combiner* adds the RF power from two sources (which must supply in-phase energy). The *shunt tee* of Figure 13–11(a) can be used as a power combiner. If RF energy is applied to port A in Figure 13–11(a), ports B and C will each have an in phase of the input signal. The power is equally divided among ports B and C. Likewise, if two RF power sources are connected to B and C, the total power output at A will be the sum of the individual input powers. Notice that since there is no disturbance in the electric field orientation, there is no phase shift produced in the shunt tee.

The *series tee* of Figure 13–11(b) operates in a similar manner. If a signal is inserted into port A, it travels downward into the body of the tee. As it arrives at the junction of the body and stub, the electric (E) field must bend around the corner. Electrically, the body appears to be in *series* with the electric field lines in the stub, so each half of the body of the device must have *opposite* polarization. In other words, the series tee acts as a phase-splitter transformer: If a signal is applied at A, signals are produced at B and C that are 180° out-of-phase. The power of the signals at B and C are still half that of the inserted power, just like that of the shunt tee. Because of the 180° phase shift, the series tee is useful for creating transmission line impedance-matching networks at microwave frequencies; a sliding short (piston) is often connected at port C in such an application.

The *magic tee* of Figure 13–11(c) combines the characteristics of both the series and shunt tees. It is often used as a transmit-receive switch in noncritical applications. For example, if a transmitter's output is coupled into port D, the transmitter's energy will be divided equally among ports B and C. Very little of the transmitter's energy will leave port A because the polarization of stub A does not agree with the fundamental mode (TE_{10}) of the propagation in the body of the device. The sensitive receiver is therefore protected from the high power of the transmitter.

When the transmitter power enters port D, it's equally split among ports B and C. Port B must be terminated in a matched load (to be discussed shortly) to prevent reflection, and thus 3 dB (one-half) of the transmitter's power is wasted as heat. Without the matched load at port B, standing waves will form within the body of the tee, some of which will couple excessively into the receiver port (A), possibly damaging the receiver front end. In receive mode, the energy from the antenna enters port C, and it encounters a 90° shift (which is harmless) as it bends into port A to make its way to the receiver. The receiver can "see" the antenna's energy because the E-field orientations of ports C and A are compatible. Some of the antenna energy also flows into ports B and D, where it is dissipated. Thus, slightly more than 3 dB of attenuation is inserted in the receive signal. Because of the 3 dB transmit and receive losses, this approach to T/R switching isn't the most practical one for high-power or high-sensitivity applications.

Attenuators and terminators An *attenuator* is a device that weakens a signal in a controlled fashion. Often, an attenuator is used to precisely control the output power level of an oscillator or other RF power source before combining it in a mixer, bridge, or amplifier. In order to attenuate microwave signals, a resistive material (such as plastic impregnated with carbon) is placed within the waveguide. Passing electric and magnetic field components set up currents within the resistive material, which cause power to be converted to heat (I^2R loss). A *variable attenuator* can be built by using a flap of resistive material that can be lowered into the waveguide, as shown in Figure 13–12(a). The amount of attenuation is controlled by how far the flap is inserted into the waveguide; the farther it is inserted, the more power will be diverted from the output. All commercially made attenuators are

Figure 13–12 Variable attenuator and fixed termination

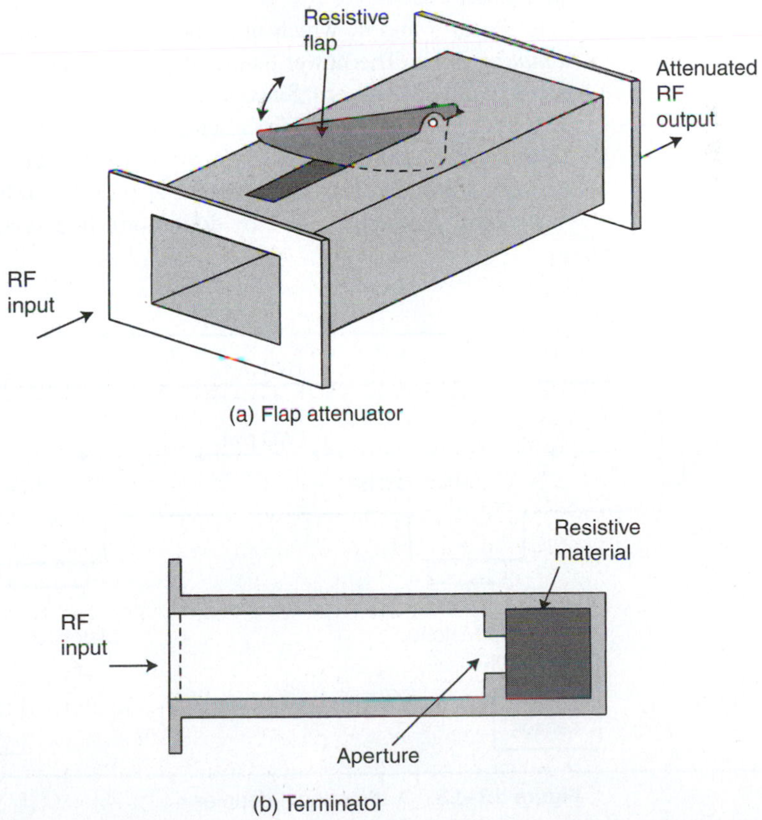

rated for a *maximum input power,* and this power level must not be exceeded, or the device will literally go up in smoke (the resistive material in the flap can only handle a limited amount of heat, depending upon its design).

A *terminator* is a device connected to the end of a transmission line in order to provide a matched load. Signals enter a terminator and do not leave! (It is also possible to provide *mismatched* loads as desired.) A terminator is therefore a dummy antenna for microwave frequencies. There is no place to hook a discrete resistor for terminating a waveguide (as might be done to terminate coaxial line), so a different approach is necessary. The *fixed termination* of Figure 13–12(b) is in fact very similar to an attenuator. Signal flows into the device and moves past the *aperture* (provided to obtain an impedance match) and into the resistive material, where it is converted to heat.

Like an attenuator, a terminator can only handle a specified level of RF power; excessive power will cause overheating, degradation of impedance characteristics (wave reflections will result due to the impedance mismatch), and possibly even fire in severe cases. Unlike conventional electronic components, a technician cannot normally eyeball an attenuator or terminator and see this type of damage (unless the device has been hot enough to discolor the metal or cause the paint to bubble).

Directional couplers A *directional coupler* is a device that can be used to measure both the *amount* and *direction* (forward or reverse) of RF energy flow. One type is shown in Figure 13–13. The coupler is designed to allow RF energy to move very easily between the *input and output ports,* labeled 1 and 2. For example, if 1 watt of RF is placed into port 1, then very nearly 1 watt will be output at port 2. The *insertion loss* of a directional coupler is the measure of how much RF power is lost between ports 1 and 2, and it's measured in dB.

Ports 3 and 4, which must both be terminated in a matched load, are called the *sampling ports.* The power output at port 3 is related to the *forward* power (flowing from port 1 to 2 by definition). Since most of the RF energy flows from 1 to 2, only a small fraction is available at port 3; the fraction is related by the *forward coupling factor,* which is given in dB. Coupling factors from 10 dB to 20 dB (10 : 1 power ratio, 100 : 1 power ratio) are very common. The power output at port 4 is related to the *reflected power* (flowing "backward" from port 2 to 1 by definition). In a system with a perfectly matched load at

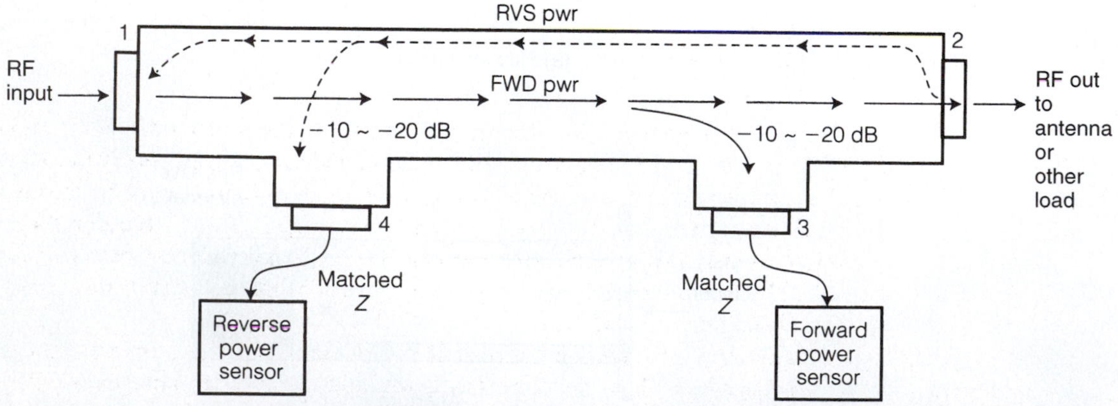

Figure 13–13 A directional coupler

port 2, there is no reflection, and the sampled output at port 4 is zero. In reality, the *directivity* of the sampling ports isn't perfect, so port 4 will read a small amount of RF energy, even with a matched load.

Directional couplers are commonly used for automatic power control (keeping a transmitter's output power constant by using feedback, which can compensate for gain variations due to component aging), protection from high SWR (the transmitter can automatically be turned off if a load fault occurs), or manual measurement of the SWR along a waveguide section. The forward sampling port can also be used for direct measurement of transmitted signal quality using a spectrum analyzer (and in-line attenuator, if needed). This sampling method has the advantage of not causing a service interruption; the connection of the measuring instruments is virtually transparent to the system.

Section Checkpoint

13–11 What is *waveguide*?

13–12 In Figure 13–6, what is the purpose of the quarter-wave stub?

13–13 What materials are often used to plate the inside walls of waveguide?

13–14 What is the *cutoff frequency* for waveguide? What type of filter does waveguide act as within a circuit?

13–15 What factors control cutoff frequency?

13–16 What is the difference between the TE and TM propagation modes?

13–17 Why is TE_{10} called the fundamental or dominant propagation mode?

13–18 How can a waveguide section be excited into higher propagation modes?

13–19 Describe the difference in internal field structure in the *E* and *H* bends of Figure 13–10.

13–20 Explain how a shunt tee can be used as an RF power combiner.

13–21 What might have happened to a terminator if the paint has bubbles?

13–22 List three uses for a directional coupler.

13–3 MICROWAVE OSCILLATORS

Just as in lower-frequency communications systems, a microwave transmitter begins with an oscillator, and from that point on, amplification, modulation, and finally radiation of the signal occur. The system concepts are exactly the same as before; only the means of producing and processing the RF signals are different. An oscillator is the first stage in any microwave transmitter and is also important in a superheterodyne receiver as the *local oscillator*. Microwave oscillators can be roughly broken into two categories, *small-signal oscillators* and *power oscillators*.

Small-signal oscillators produce an RF output power of 1 watt or less. They are used where precise AM or FM modulation is going to be introduced onto the signal, and of course, they are also commonly applied as the local oscillator in receivers. The Gunn and YIG (yittrium–iron–garnet) are two common small-signal oscillators.

Power oscillators produce more than 1 watt of RF output. They're used when sheer power is more important than transmitting information, such as in radar. The magnetron and klystron are two vacuum-tube power oscillators that can produce thousands of watts of RF output from a single stage!

Oscillator Theory Revisited

Figure 13–14 is the block diagram of an oscillator. It is the same block diagram we introduced in Chapter 4. In order for an oscillator to run, two conditions must be present. First, there must be *positive feedback*. By positive feedback, we mean that the total phase shift around the feedback loop must be 0°, 360°, and so on. When positive feedback is applied, the feedback signal reinforces what was already applied at the amplifier input, and oscillation is allowed to continue. Any other condition will cause the oscillations to gradually weaken and die out. The second condition for oscillation is sufficient *loop gain*. By this we mean that the signal must not grow smaller as it travels around the loop. For example, suppose that a 1 volt signal exists at the input of the amplifier in Figure 13–14. The amplifier has a voltage gain of 10, so the output signal is 10 volts. The signal must then travel down through the feedback block, which has a gain of 0.1; the output is *exactly* 1 volt again. The signal can flow indefinitely around the loop under these conditions, and we will see stable oscillation at the output.

At frequencies below 2 GHz, it is still possible to build oscillators using the lumped component techniques (Colpitts, Hartley, etc.) of Chapter 4. At microwave frequencies, we can no longer use discrete L and C components to control oscillators because the component values become too small to be practical. In fact, even quartz crystals can't be directly used at these frequencies; the maximum practical frequency for quartz-controlled oscillators is somewhere around 150 MHz, using overtone modes.

Figure 13–14 A linear oscillator

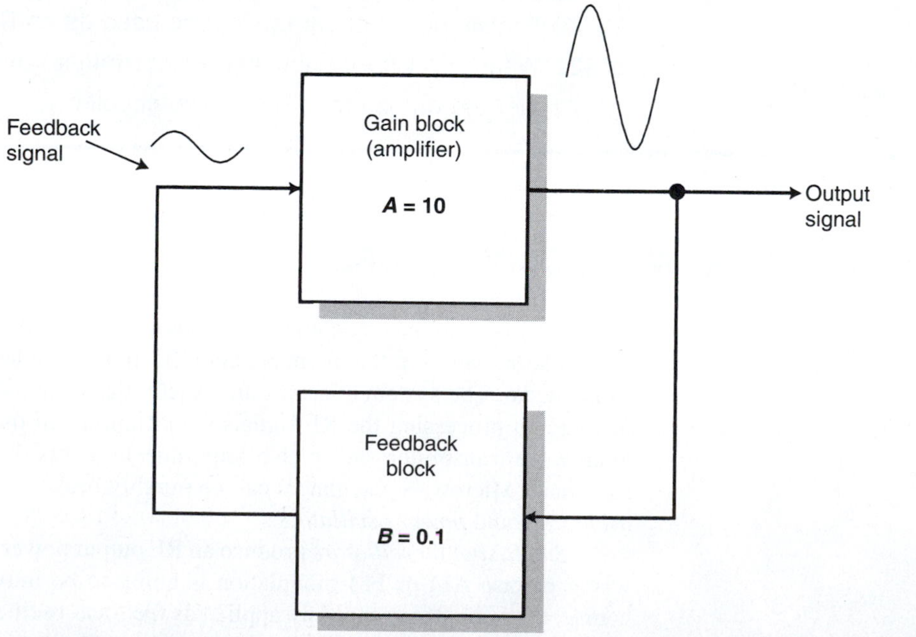

To produce a microwave oscillator, we can simply substitute the microwave equivalent of an LC circuit and construct the unit with high-frequency transistors. We can also use a few techniques unique to microwaves as well.

Most microwave oscillators are inherently unstable in frequency. However, we can still use the PLL synthesis techniques discussed in Chapter 7 to lock a microwave oscillator onto a low-frequency crystal reference, thus producing a stable SHF oscillator.

Transmission Line Oscillator

Figure 13–15 shows the schematic diagram of a *transmission line oscillator* using a high-frequency GaSFET. Such transistors are now commonly available for use at frequencies in excess of 20 GHz. A transmission line section, Z_2, replaces the LC circuit of a conventional oscillator. The circuit works this way: Q_1 (the GaSFET) is a common-source amplifier. Input signals to the gate experience a power gain and a 180° phase shift on their way to the drain. You know from oscillator theory that an additional 180° phase shift is needed somewhere in order for the oscillator to work. That phase shift is provided by the *time delay* in transmission line section Z_2. Z_2 would likely be fabricated directly on a PC board using the microstrip technique described earlier. The physical length of Z_2 is precisely one-half of a wavelength at the desired operating frequency, which corresponds to the needed 180°

Figure 13–15 A transmission line oscillator

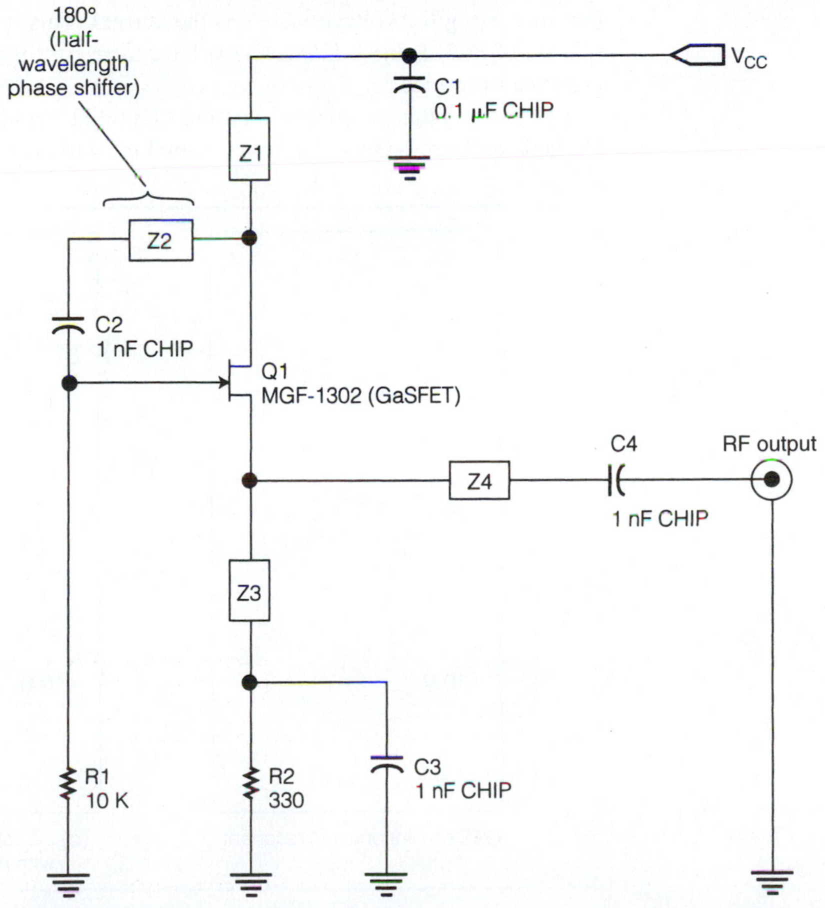

phase shift. Thus, when feedback passes through Z_2, it is inverted by 180°, completing the loop phase shift. The circuit works! The transmission line section Z_1 acts as an RF choke, to prevent the RF current from flowing into the power supply; and lines Z_3/Z_4 impedance match the oscillator output to 50 Ω, so that it can supply energy to the next stage.

This oscillator can be directly coupled into a microwave antenna, forming a complete transmitter. In fact, a device called the "Zapper" used a circuit very much like this. The Zapper was built in a case about the size of a pack of cigarettes and was tuned to transmit at 10.525 GHz, which is the frequency used by police X-band traffic radar. The device was marketed initially as a "radar detector tester," of course, with the hidden intent that users test *other's* detectors! Because operation of the Zapper in this way constituted operation of an unlicensed radio transmitter, the FCC forced the manufacturer to retune the device to bring it down into the amateur radio portion of the 10 GHz band. A jack was added for connection of a telegraph key—the Zapper was now a legal amateur 10 GHz CW transmitter!

Gunn Oscillator

The Gunn oscillator relies on the generation of microwave-frequency pulses within a semiconductor material. In 1963, J. B. Gunn of IBM discovered that microwaves were generated when an N-type block of GaS (gallium arsenide) was exposed to a constant voltage. The operation of the Gunn oscillator can be explained in terms of negative resistance and time of travel.

When the N-type semiconductor block is appropriately doped, it exhibits *negative resistance*. This is a nonlinear and counterintuitive effect. In most conductors, we expect that increasing the voltage increases the current. This is not true with a Gunn diode; for a portion of its *V-I* curve, increasing voltage *decreases* the current. In other words, it shows negative resistance.

How can this produce oscillations? Examine Figure 13–16(a). This is a conventional LC tank with a resistance *R* in series. How long will the oscillations last in Figure 13–16(a)?

Figure 13–16
Negative resistance oscillator model

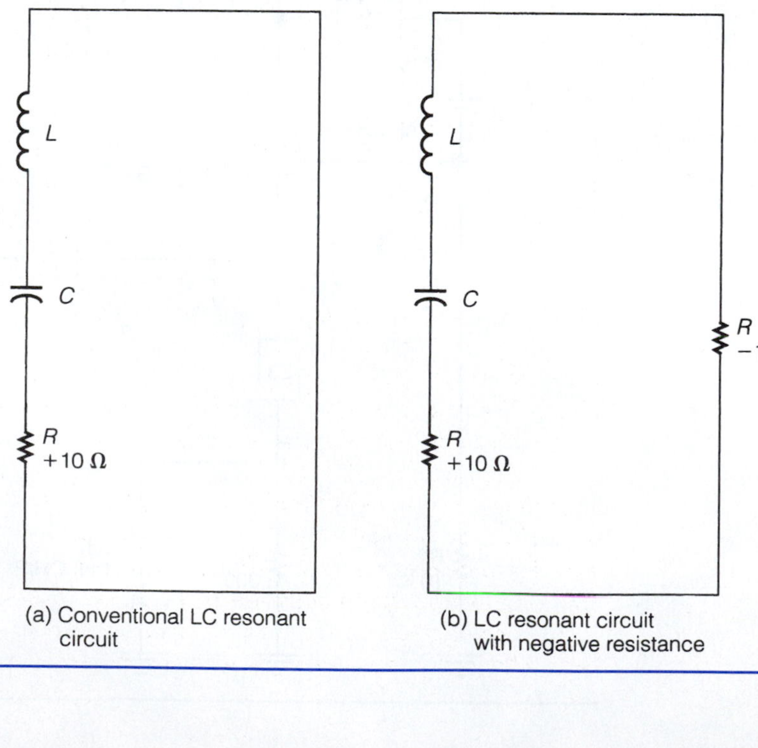

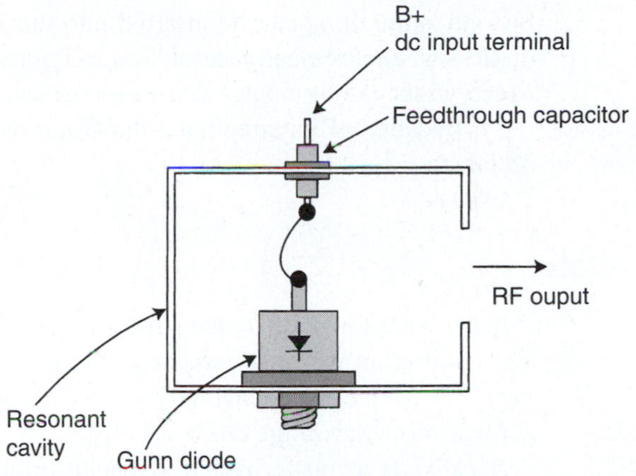

Figure 13–17 A Gunn oscillator

Well, since there is no source of energy (but certainly a *consumer* of energy, the resistor), the answer is "not long." The resistor eventually converts most of the energy in the tank to heat, and the oscillations die out. If we could eliminate *all* resistance, the oscillations would continue forever.

Try inserting a "magic" resistor of $-10\ \Omega$ into the circuit (Figure 13–16(b)) and see what happens. You didn't misread this; we can get a $-10\ \Omega$ *effect* by the use of a negative-resistance device such as the Gunn diode. What happens? The *total* resistance in the circuit is now zero. The oscillations do not die out, for there is no effective resistance to use up the energy. We now have a steady oscillation. The effect of a negative resistance device is exactly opposite of a conventional resistance. A negative resistance contributes energy back to a circuit, which, in the case of an oscillator, is enough to keep it running.

The LC circuit can be replaced by a resonant cavity, as shown in Figure 13–17. Thus, we end up with a very simple microwave oscillator that requires only one active component, the Gunn diode!

The Gunn diode's effect can also be explained by the concept of *time of travel.* When a voltage (electric field) is applied across the diode material, the electrons move in a strange way. Instead of flowing uniformly through the device from negative to positive, they instead bunch up into little clusters. The bunching up is caused by differences in electron mobility between valence (outer) and lower bands in the semiconductor material. When the voltage is applied, the electrons are excited into moving to the valence band. There it is hard for them to move (much like a car moving from a side street out onto a main street may slow down considerably if there is heavy traffic and plenty of traffic lights). The electrons collect and bunch up until they are attracted strongly enough by the positive charge ahead of them, and they then move together as a group.

Because the electrons travel in bunches, the current in the diode material is not constant. It arrives in *pulses*. The frequency of the pulses is in the microwave region and is determined by the doping profile of the crystal and its physical dimensions. If the diode is placed within a resonant cavity tuned to its operating frequency, a fairly clean sine-wave RF output will result. In other words, we get oscillation.

The Gunn oscillator can be tuned in two primary ways. First, it can be tuned over a narrow range (less than 1% of its operating frequency) by varying the applied voltage. This method is limited, because the diode will only oscillate over a limited voltage range. Second, something can be inserted into the resonant cavity (a tuning screw or a varactor diode—an electronically variable capacitance) to alter the resonant frequency of the cavity. Much wider tuning ranges (3% or more) can be obtained by these methods.

Because of its simplicity, the Gunn oscillator is very popular at frequencies up to 35 GHz.

YIG Tuned Oscillator

When a wide range of frequencies is required, the YIG tuned oscillator is often employed. The oscillators we've discussed so far are basically fixed-frequency units, allowing perhaps 5% frequency change at best. A YIG oscillator can be tuned over an octave (2 : 1, 200%) frequency range electronically.

YIG is a ferrite having a special property. Its resonant frequency depends on the magnitude of an applied magnetic field. The higher the current, the higher the resonant frequency. A sphere of YIG can be coupled into the feedback path of an oscillator to control the oscillator's frequency; you can think of the YIG sphere as a *current-controlled bandpass filter*. Figure 13–18(a) shows a YIG sphere in an oscillator with a magnetic control coil surrounding it.

The YIG oscillator is generally present as a preassembled *module* within a piece of equipment. It is replaced, not repaired, in the field. There are generally at least six connections on a YIG oscillator unit. These include power, sweep or coarse coil, FM or fine coil, and RF output (a coaxial connector). Generally, two magnetic coils are provided on a YIG oscillator. One, the *sweep* coil, has many turns and provides a widely changing magnetic field to rough- or coarse-tune the oscillator. The *fine* or *FM* (frequency modulation) coil usually has only one turn and provides much smaller variances in frequency. Because a YIG oscillator is very temperature sensitive, it usually has an internal thermostatically controlled heater.

A YIG oscillator changes frequency approximately linearly (much like a VCO). It can be thought of as a *current-controlled* oscillator. Special circuitry must be added to provide the precise levels of drive current. The YIG oscillator is an expensive unit to replace. Because it is a complete module, a known-good unit can easily be substituted in its place if it is suspected to be defective.

Power Oscillators

The *magnetron* and *klystron* are vacuum-tube oscillators. Of the two, the magnetron is still very popular and finds use in a wide variety of systems, from radar to microwave ovens, which operate at 2.45 GHz (a frequency where water readily absorbs RF energy, converting it to heat, which cooks food). The klystron is primarily an amplifier but can also be configured as an oscillator. It is discussed in the next section.

The magnetron was originally developed for use in the first radar systems, where its higher frequency meant improved military target-ranging accuracy (due to the shorter wavelength). In order to understand how it works, we must first understand how electrons flow in a vacuum tube. Figure 13–19(a) shows the operation of a vacuum-tube *diode* (a two-electrode device).

The vacuum-tube diode of Figure 13–19(a) consists of two parts, a *cathode* and an *anode*. The cathode is a source of electrons, and the anode is a collector of electrons. The two electrodes are separated within a glass or metal envelope containing a partial vacuum.

Figure 13–18 YIG sphere in an oscillator circuit

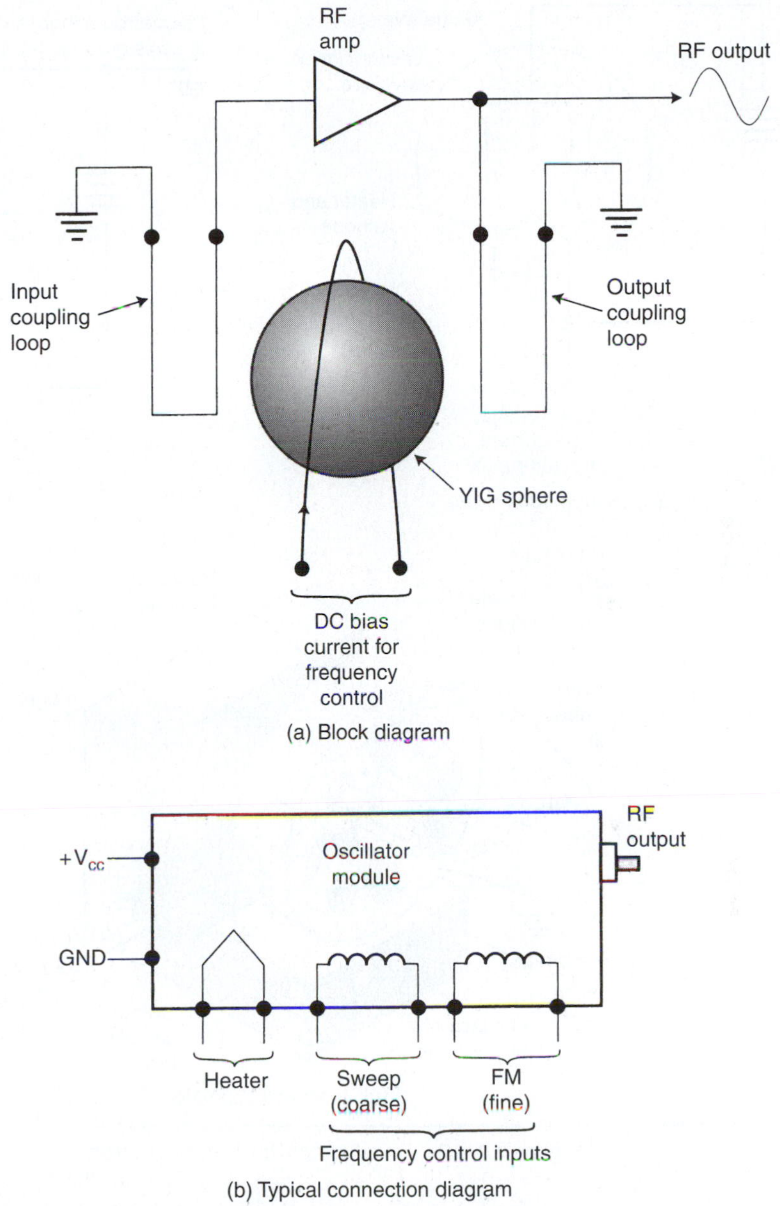

Because of the space between them, you would imagine that it would be impossible for a current to flow between them, but it can!

The cathode can be stimulated to give off electrons by *heating*. A *heater* coil can be placed underneath the cathode, which will heat the cathode red-hot and literally boil electrons from its surface. This is called an *indirectly heated cathode*. A magnetron uses the heater wire itself as the cathode. This is called a *directly heated cathode*. It simplifies the design of the tube somewhat.

Because the heater in a magnetron is also the cathode, it connects to two power sources, the A and B supplies. The A supply is a low-voltage, high-current (several amperes

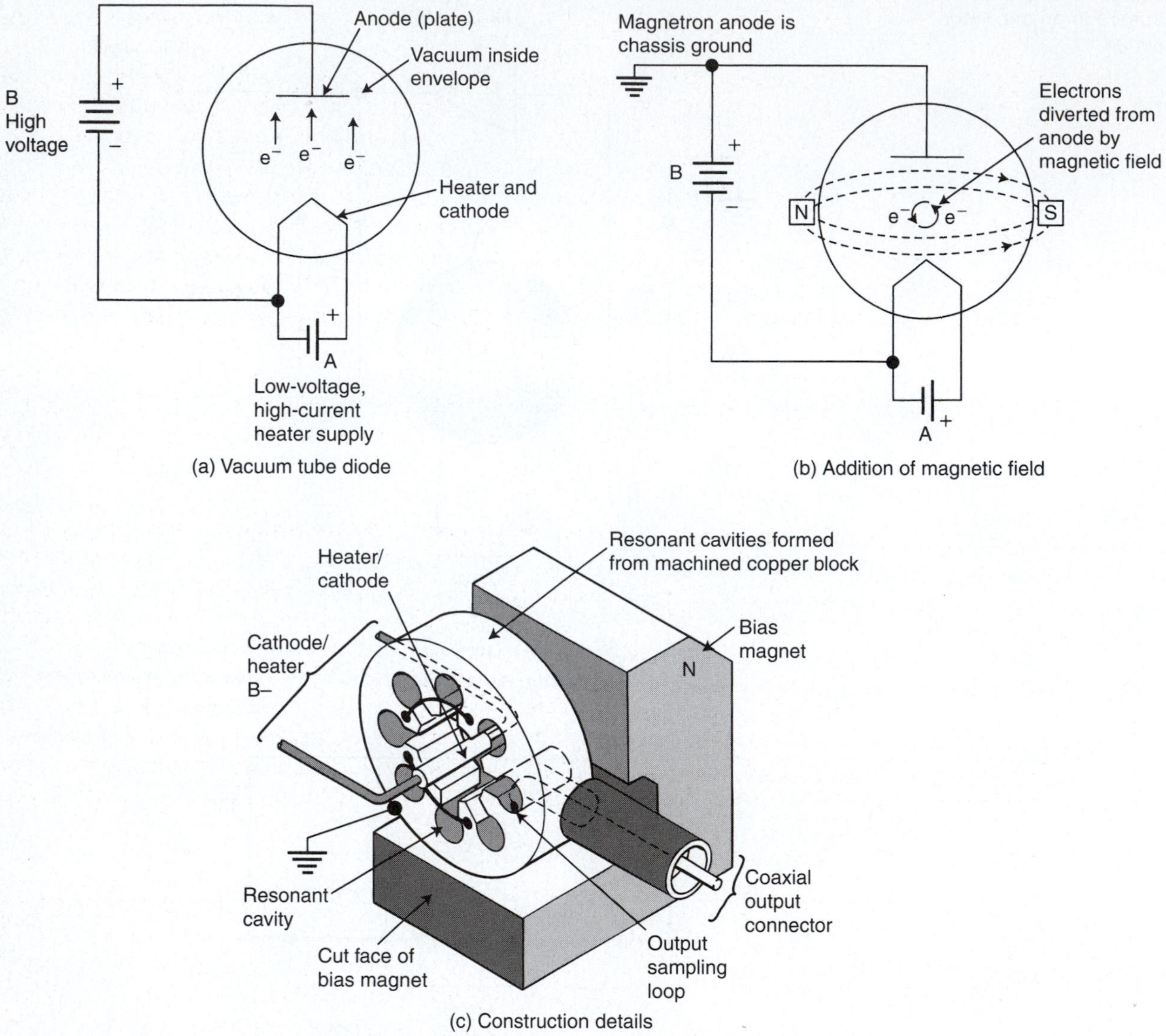

Figure 13–19 Development of the magnetron

typical) source that heats the cathode (so that electron emission can take place). The B supply is important, too; it supplies high voltage (usually at least −1000 volts) to the cathode, so that the electrons emitted by the cathode will be attracted to the anode, which rests at dc ground potential. In operation, the A supply heats the cathode of the tube. The electrons emitted from the red-hot cathode surface migrate out toward the anode because they are attracted to it by the anode's *relative* high voltage (ground potential). Ideally, the flow of electrons is continuous and in a straight line. A steady current flows in the diode tube under these conditions.

Suppose that we introduce a magnetic field as shown in Figure 13–19(b). We know that a magnetic field can alter the path of moving electrons (this is because moving electrons themselves constitute a current, which itself causes a magnetic field). The force exerted on a moving electron is at a right angle to both the magnetic field and the path of the electron. The net result is that the electrons take a *curved* path towards the anode of the tube rather than traveling in a straight line.

If the magnetic field strength is further increased, we can deflect the electrons away from the anode totally. The electrons will simply return to the cathode, where they will again be boiled off. The path of the electrons will form a complete circle under these conditions, but no current will flow in the tube. No useful work can be done under this condition, for the tube draws no current from the B supply! If we reduce the magnetic field slightly (to the point where *some* of the electrons can actually make it to the anode), we get *sustained RF oscillation* from the device. How is that? Remember that many of the electrons are still making circular paths between the cathode and anode. The periodic motion of the electrons represents an ac current (with corresponding magnetic and electric fields). These ac fields are coupled and reinforced in the *cavities* that are formed into the solid-copper anode of the magnetron.

The microwave energy is extracted from the magnetron by a pickup loop in one of the cavities. The pickup loop can terminate in a coaxial connector, a quarter-wave stub (for insertion into a section of waveguide), or a complete waveguide assembly (most common).

The magnetron is not a very frequency-stable oscillator, but it can easily produce thousands of watts of output power. Its frequency depends highly upon the size of the resonant cavities machined into its copper anode and slightly on the B power supply voltage. It is often used in pulse radar applications, where the B voltage is switched on and off quickly in order to emit a narrow burst of high-power microwave RF energy.

When working with magnetrons, a technician should be very aware that the magnetic components of the tube must not be disturbed or interfered with in any way. Magnetic items should be kept clear of the external permanent magnet. The strength of the magnetic field is critical; if there is insufficient magnetic field strength, the tube stops oscillating and draws *heavy* current from the high-voltage power supply. Under lucky circumstances, the B fuse blows.

Section Checkpoint

13–23 What are the two categories of microwave oscillators?

13–24 List the two conditions necessary for oscillation in Figure 13–14.

13–25 What element controls the oscillator frequency in Figure 13–15?

13–26 What is the effect of a negative-resistance device in a circuit?

13–27 What forms the LC tank for the Gunn oscillator of Figure 13–17?

13–28 How is the frequency of a YIG oscillator adjusted?

13–29 Why are there generally two magnetic current coils in a YIG oscillator module?

13–30 Explain the operation of a vacuum-tube diode; what direction does the current flow?

13–31 What causes the electrons in a magnetron to travel in a circular path?

13–32 Why is the magnetic field strength important in a magnetron?

13–4 MICROWAVE AMPLIFIERS

Microwave amplifiers perform the same function as those designed for low frequencies. Amplifiers can be classified as either *small-signal* or *large-signal (power amplifiers).* In general, most technicians consider amplifiers with power outputs under 100 mW as small-signal and anything else as large-signal.

Until recently, it was impossible to design a microwave amplifier with transistors and ICs because these components were either not fast enough for SHF operation or very noisy. This is no longer the case, and new designs commonly employ them for small-signal amplification. In large-signal (power amplifier applications), the vacuum tube remains king.

Small-Signal Techniques

Small-signal amplifiers can be built with discrete transistors, using transmission line elements to replace discrete L and C components, as shown in Figure 13–20. The common-emitter amplifier in Figure 13–20 uses microstrip section Z_1 to match the 50 Ω input impedance down to the low input impedance at the base of Q_1. Section Z_3 provides output impedance matching, and choke L_1 improves the ac isolation between the power supply and Q_1's collector circuit. All the capacitors in this circuit are chip capacitors; they are designed to be directly surface-mounted on the PC board. The use of chip capacitors is mandatory at microwave frequencies, since even a short component lead has considerable inductive reactance at these frequencies.

A new class of integrated circuits is also popular for small-signal amplifiers. These chips are called *MMICs,* which stands for *monolithic microwave integrated circuits.* (The

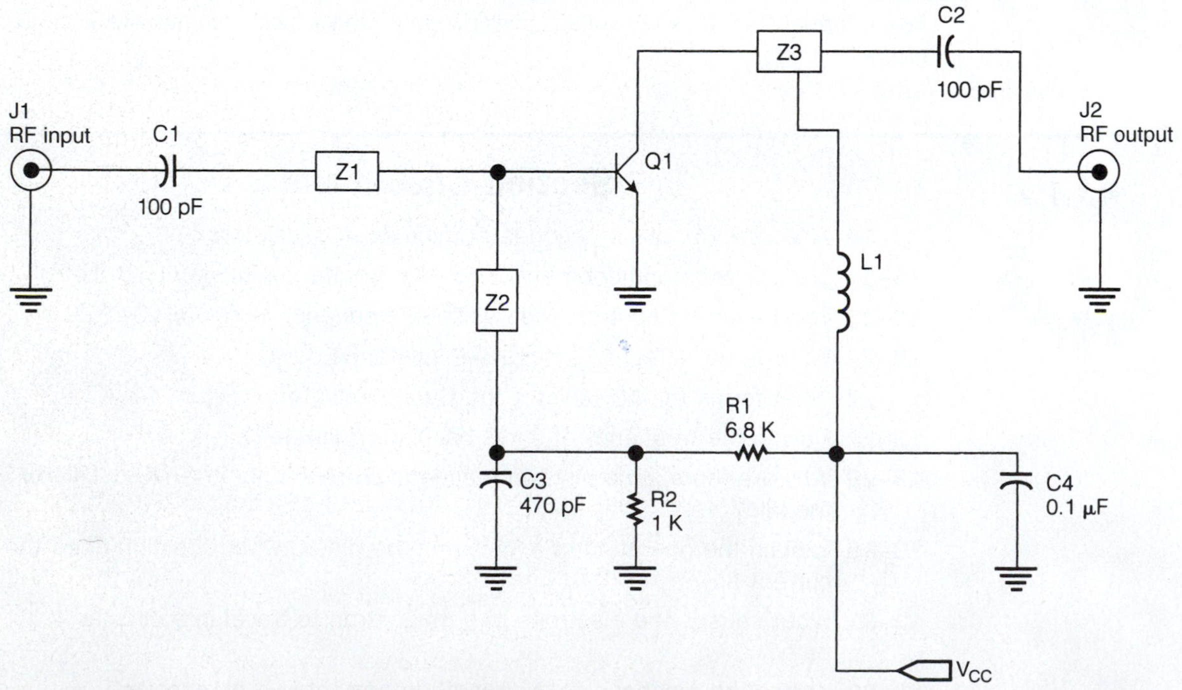

Figure 13–20 Microstrip amplifier

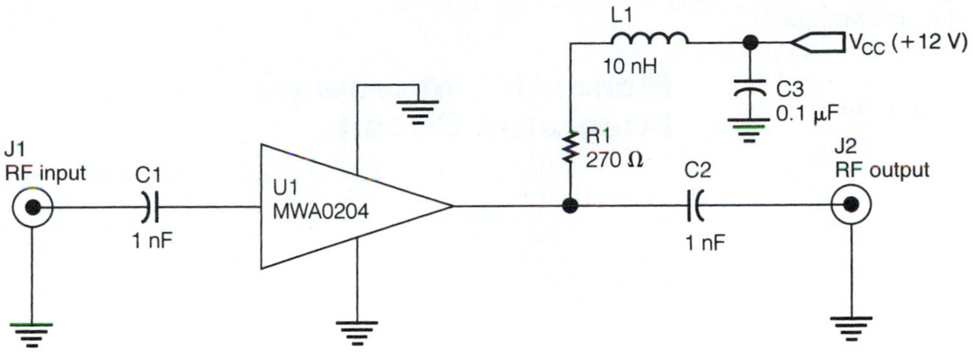

Figure 13–21 A MMIC amplifier using the MWA0270

name is often pronounced as "mimics.") A MMIC contains all the active circuitry for an amplifier stage on one chip. Typically, the only components that must be added are dc bias, power supply decoupling (an RF choke), and dc blocking capacitors at the input and output terminals of the device. Figure 13–21 shows the schematic diagram of a MMIC amplifier block.

The amplifier in Figure 13–21 looks very simple but has excellent performance. It can amplify signals from near dc to well over 1500 MHz (1.5 GHz) with little change in gain. It provides better than 12 dB of power gain over this entire frequency range. In addition, when properly constructed, this MMIC amplifier is unconditionally stable, meaning that it is highly immune to undesired oscillation.

In order to get additional power gain, MMIC stages can be *cascaded* (the output of one stage can be connected to the input of the next). Figure 13–22 is the data sheet of the Motorola MWA-0270, a popular MMIC.

Parametric Amplifiers

The small-signal amplifiers discussed so far introduce a considerable amount of noise onto the amplified signal. Noise figures of 3 dB are fairly common for MMIC devices, for example. When very weak microwave signals must be boosted, a parametric amplifier is often used, as shown in Figure 13–23.

The operation of a parametric amplifier is rather strange when compared to that of conventional amplifiers. A standard amplifier doesn't really amplify a signal at all; instead, it creates a magnified replica of the input signal using energy from the dc power supply. All of the analysis of traditional amplifiers centers on how the ac voltages and currents originate in order to produce a voltage and/or power gain. In other words, a regular amplifier takes dc energy and modifies it to look like the original ac energy that was presented at its input.

A *parametric amplifier* operates by using an external ac source as the source of energy. The ac signal source is at precisely *twice* the frequency of the signal to be amplified and is called the *pump* frequency. In Figure 13–23, the pump frequency is 20 GHz.

A varactor diode is a voltage-to-capacitance converter and is very commonly used in parametric amplifiers. It works like this: Imagine that at 90° in the ac cycle of the *input signal* of Figure 13–23, the capacitor plates are physically spread apart. Since the total charge between the plates must remain the same, the voltage between the plates will *increase*. This is because the capacitance decreased, and we know that $V = Q/C$. When the capacitance decreases, the voltage increases. The signal grows larger! Also, *work* must be performed to move the capacitor plates at this instant because there is a force acting on

Figure 13–22
Motorola MWA0270 data sheet (Copyright of Motorola, Used by Permission)

Monolithic Microwave Integrated Circuit

MWA0204
MWA0211L
MWA0270

MONOLITHIC MICROWAVE INTEGRATED CIRCUIT

. . . designed for narrow or wideband IF and RF applications in industrial and commercial systems up to 3 GHz.

- 12 dB Gain at 1000 MHz (Typ)
- Fully Cascadable
- 50 Ω Input and Output Impedance
- Choice of Package Types
    - Low Cost
    - Surface Mount
    - Hermetic
- Tape and Reel Package Options
- 4.0 dBm P_o 1 dB, at 500 MHz (Typ)
- Unconditionally stable

CASE 317-01, STYLE 3
MWA0204

CASE 318A-05, STYLE 4
MWA0211L

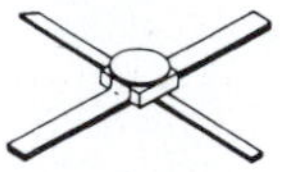

CASE 303A-01, STYLE 3
MWA0270

ABSOLUTE MAXIMUM RATINGS (T_A = 25°C)

Parameters		Symbol	Ratings	Unit
Circuit Current		I_{CC}	40	mAdc
Input Power, RF		P_{in}	+16	dBm
Bias Voltage		V_{CC}	6	Vdc
Storage Temperature	0204/0211L 0270	T_{stg}	−65 to +150 −65 to +200	°C

RECOMMENDED OPERATING CONDITIONS

Parameters	Symbol	Ratings	Unit
Operating Current	I_{CC}	25	mA
Source Impedance	Z_S	50 to 75	Ω
Load Impedance	Z_L	50 to 75	Ω

THERMAL CHARACTERISTICS

Thermal Resistance, Die to Case	MWA0204 MWA0211L MWA0270	$R_{\theta JC}$	150 200 130	°C/W

DEVICE MARKING

MWA0211,L = 06

ELECTRICAL CHARACTERISTICS (T_A = 25°C, I_{CC} = 25 mA, Z_S = Z_L = 50 Ω, unless specified otherwise)

Characteristic		Symbol	Min	Typ	Max	Unit
Gain (f = 1000 MHz) (f = 100 MHz)	MWA0204/0211L MWA0270	G_T	10 11.5	12 12.5	— 13.5	dB dB
Gain Flatness (f = DC to 800 MHz — MWA0204/0211L) (f = DC to 1500 MHz — MWA0270)			— —	1 1	— —	dB
Noise Figure (f = 100–1600 MHz)		NF	—	5.5	—	dB
Third Order Intercept Output Power (f_1 = 480 MHz) (f_2 = 500 MHz) (f_1 = 980 MHz) (f_2 = 1000 MHz)		ITO_1 ITO_2 ITO_3 ITO_4	— — — —	16 16 16 16	— — — —	dBm
Second Order Intercept Output Power (f_1 = 480 MHz) (f_2 = 500 MHz) (f_1 = 980 MHz) (f_2 = 1000 MHz)		ISO_1 ISO_2 ISO_3 ISO_4	— — — —	20 20 19 19	— — — —	dBm

them (from the electrostatic charge). Now imagine that the plates are quickly pushed back together at 180° in the cycle. This takes *no* work to do, because there is no voltage present at this instant, and therefore there is no electrostatic force acting on the plates.

The process is repeated for the negative half cycle; at 270°, the capacitor plates are again spread, causing signal amplification; and at 0°, they are pushed back together again, in preparation for the next cycle. We have amplified the signal by varying a *parameter* in the circuit, namely the *capacitance!*

The purpose of a varactor diode in a parametric amplifier is to act as the varying capacitance. It is supplied with the *pump* signal, which is at twice the input signal

MWA0204, MWA0211L, MWA0270

TYPICAL CHARACTERISTICS

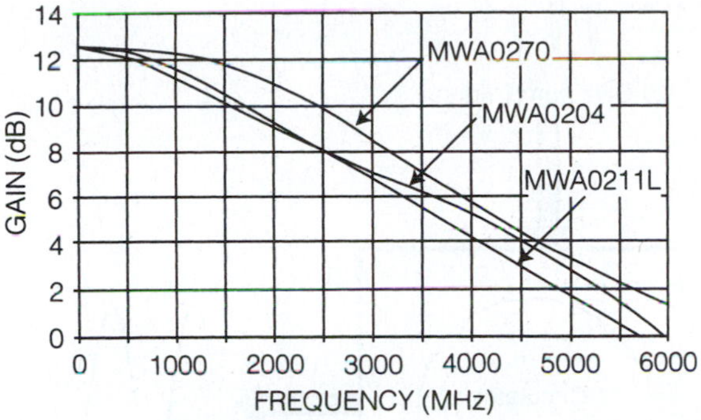

Figure 1. Gain versus Frequency

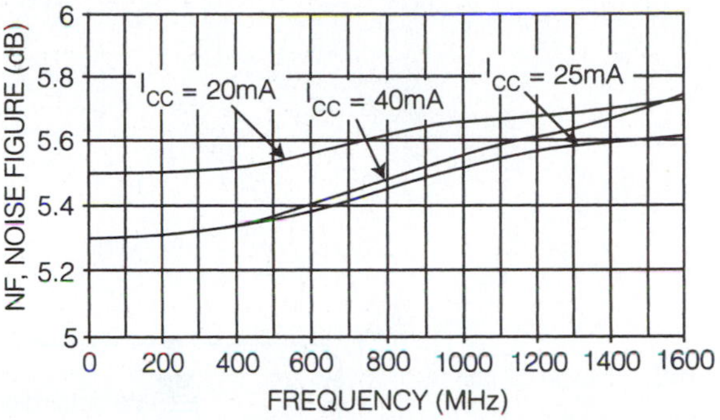

Figure 2. Noise Figure versus Frequency

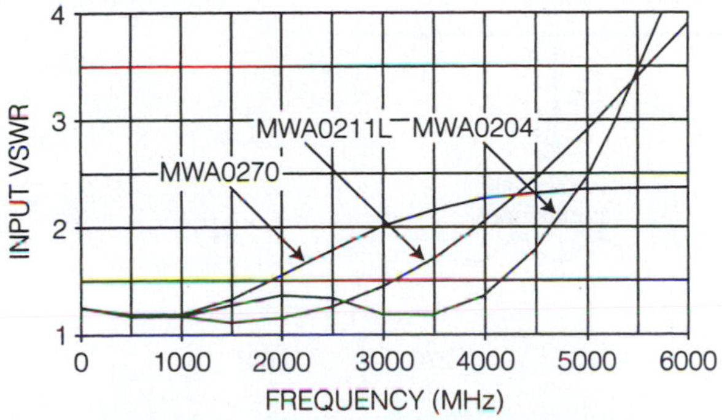

Figure 3. Input VSWR versus Frequency

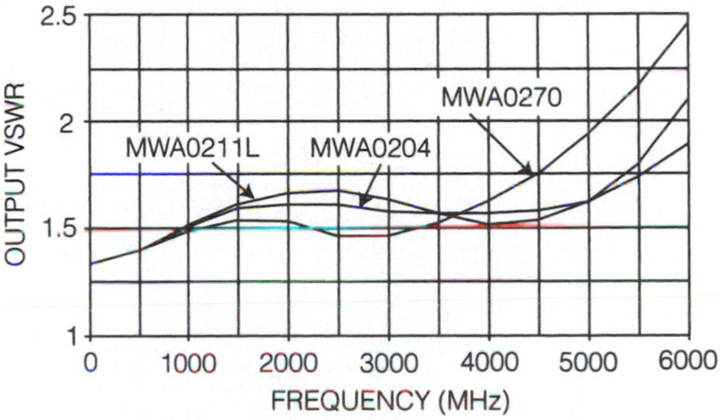

Figure 4. Output VSWR versus Frequency

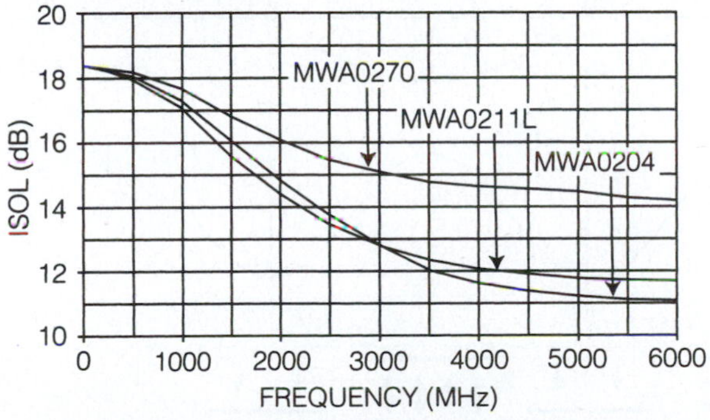

Figure 5. Reverse Isolation versus Frequency

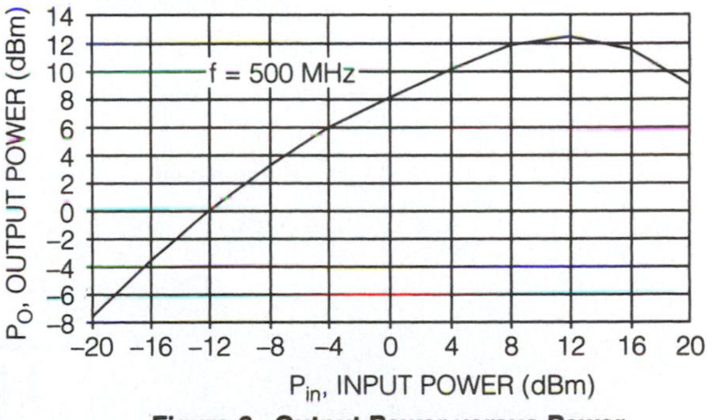

Figure 6. Output Power versus Power

Figure 13–22 *(continued)*

frequency. You can see why the pump must be at twice the input frequency—one complete "pump cycle" is needed for each half of the input signal.

A parametric amplifier can be designed with a very low noise figure; noise figures of 0.5 dB and better are readily attainable. The primary disadvantage of these amplifiers is that they aren't broadbanded; the input signal must be close to one-half of the pump

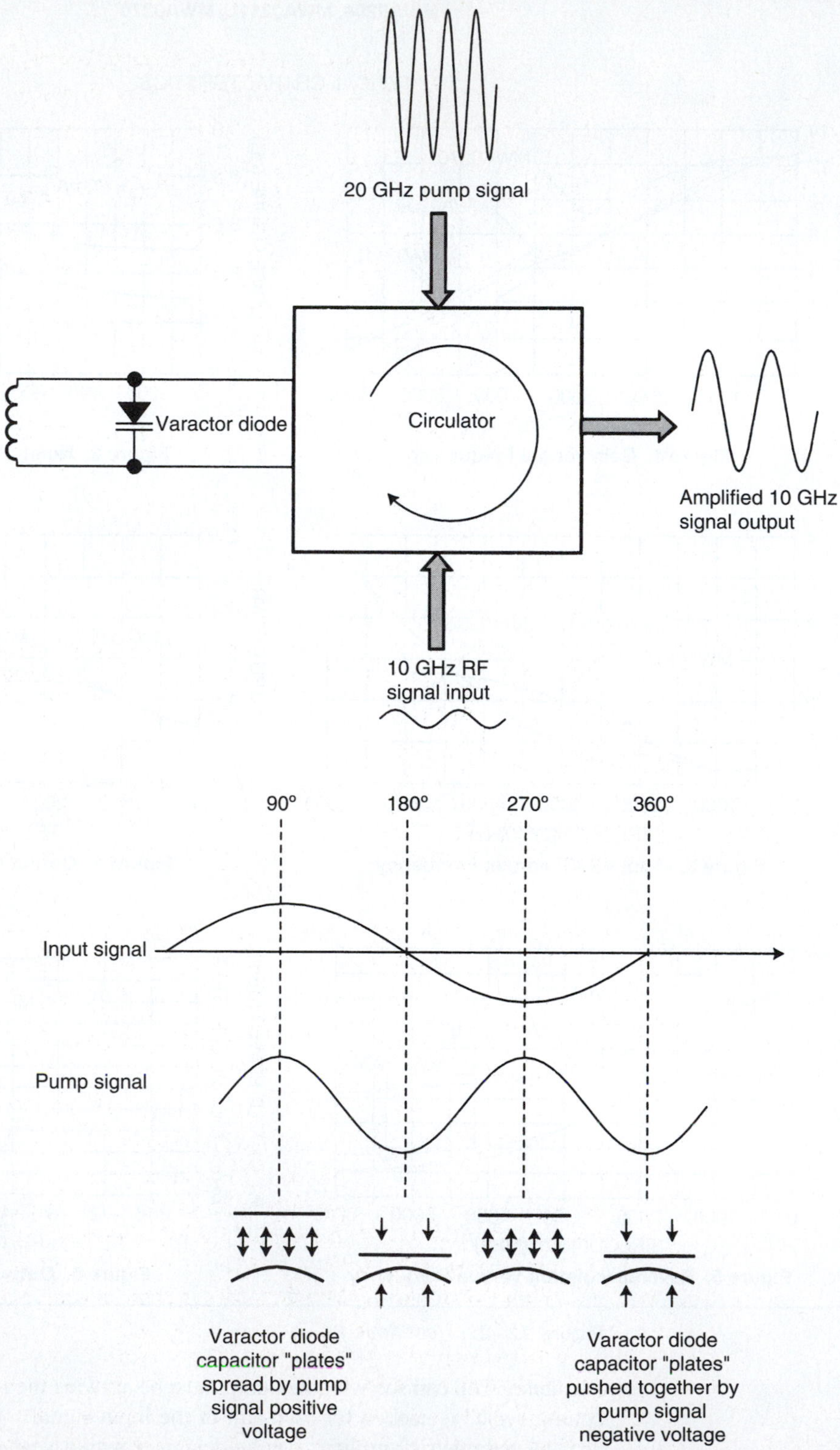

Figure 13–23 A parametric amplifier employing a varactor diode

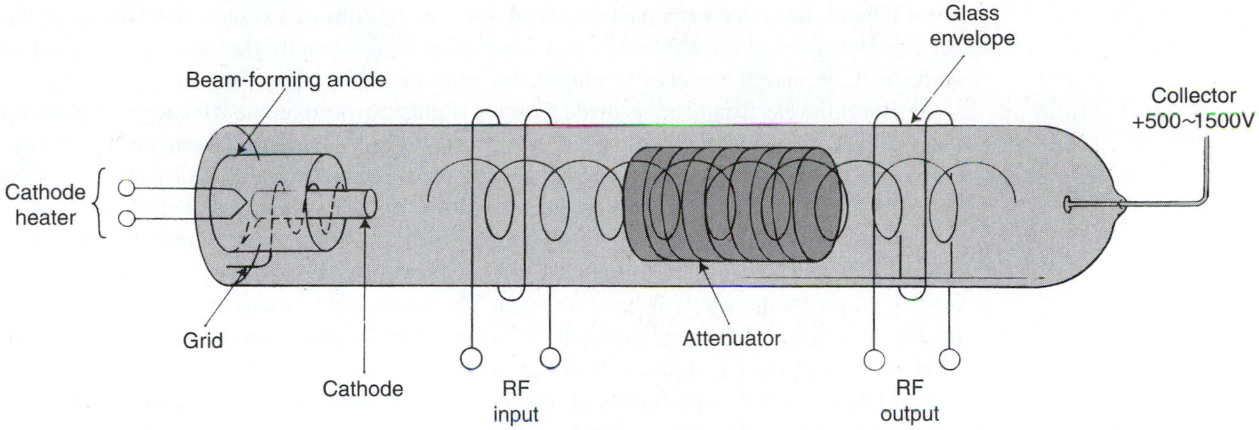

Figure 13–24 Traveling wave tube construction

frequency. A parametric amplifier can't operate over a wide frequency range. Incidentally, when the input signal isn't exactly half the pump frequency, the parametric amplifier also acts as a *mixer,* meaning that sum and difference frequencies are present in the amplifier's output. This capability is handy, for although the amplifier is producing less gain under these circumstances, it's also operating as a frequency converter, just like in a superheterodyne receiver.

Power Amplifiers

The traveling wave tube *(TWT)* and klystron are two popular microwave power amplifiers. The TWT finds application where wide bandwidth and moderate power levels are required. The klystron is used commonly in both low- and high-power applications for operation over a narrow frequency range.

Traveling wave tube The TWT is an efficient high-power microwave amplifier. It is the workhorse of the microwave industry and finds application in both earth stations and satellites. TWTs have been constructed to produce over 5 kW of continuous power output. The TWT can easily amplify signals over an octave (2 : 1) frequency range and is a very linear amplifier. Figure 13–24 shows the construction of a typical TWT.

The TWT is primarily a long vacuum tube, with a *heater and cathode* as the source of electrons and a *collector* or *anode* plate for the return of the electrons to the B supply, which is typically 500–1500 volts. The electrons are made to travel down the tube in a narrow line under the influence of a *focusing electrode* and an external magnetic field (not shown). Inside the tube, a spiral wire called the *helix* acts as the carrier of the signal being amplified.

The operation of the TWT depends upon the interaction between the electric field of the RF signal being amplified and the motion of the electron beam. The velocity of the electron beam in a TWT depends largely upon the voltage applied at the collector (anode). The forward velocity of the electrons in the beam is about 1/5 that of the speed of light (the RF wave).

The function of the helix is to conduct the RF wave slowly along the tube (it is really a modified transmission line). The pitch of the helix's windings determines the velocity at which RF energy will travel. The RF energy is coupled into the left-hand side of the tube and begins to propagate toward the right (anode). When the electron beam is accelerated to travel exactly the same speed as the RF signal, an interesting interaction occurs within the tube. The electric and magnetic fields of the propagating RF signal cause the electrons in the beam to travel toward the collector in *bunches*. Work must be performed to create this bunching, but because the speed of the RF energy and electron beam are exactly the same, no net work is performed, because the energy is returned on opposite half-cycles of the RF signal.

But if the electron beam is made to move slightly *faster* than the RF energy, it gives up some of its energy to the RF field because the total transfer of energy for each whole RF cycle is now *positive,* in favor of the RF field. The RF field intensifies as it propagates toward the right. The tube *amplifies* the RF signal by adding energy to it from the electron beam!

Note that nothing prevents the RF energy from traveling in any particular direction; in fact, it can easily travel both forward and backward along the helix. This would lead to feedback and oscillation if it were allowed. The *attenuator section* is present to prevent feedback. When an RF signal passes by the attenuator, the signal on the helix is made very weak. However, the RF signal still passes by *underneath* the attenuator, "coasting" on the electron beam (which only travels in one direction). It then continues to propagate on the helix once it passes the attenuator. Thus, signals are prevented from flowing backwards on the tube, preventing undesired oscillation.

The TWT has one additional feature that makes it very good as both a modulator and a mixer, and that is the *control grid.* By the application of a negative bias voltage to this grid, the current flowing in the electron beam can be modulated. This in turn modifies the gain of the tube. Thus, a TWT can function as an *amplifier,* a *modulator,* and a *mixer* if so desired.

Backward wave oscillator There is one special form of the TWT that can be used as a microwave power oscillator, and that is the BWO, or backward-wave oscillator. To form a BWO, a TWT is built without an input coupler and without any attenuator sections. In this form, signals can flow in both directions along the helix, and positive feedback results. The frequency of the BWO is controlled grossly by the pitch of the helix (in that regard, it is really a transmission line oscillator). The helix is, of course, fixed at manufacture and can't be changed in the field. However, the frequency can also be fine-tuned by varying the positive bias on the collector electrode.

Klystron tubes The klystron is probably the most popular power amplifier for UHF television transmitters. Klystron tubes range in size from small plug-in units that work in the milliwatt power range (these have been used as local oscillators for receivers) to mammoth 100,000 watt versions that require a room full of power supplies and cooling equipment (as well as trained personnel). Figure 13–25 shows a basic klystron tube. The tube gets its name from the German word *klyster,* which means "cluster" or "bunch." In a klystron tube, the electrons are made to move through the tube in bunches.

In the klystron tube, a narrow electron beam is formed at the cathode, and through the use of a beam-forming (focusing) electrode and electromagnets placed along the length of the tube, the beam is carefully guided through the center of the device, which is called the *drift tube*. The RF signal to be amplified is coupled into the *input cavity* by a coupling loop. This cavity is tuned to be resonant at or near the carrier frequency. As the RF signal goes though positive and negative half cycles, it alternately attracts and repels the electrons in the beam. In particular, when the signal is negative, the electrons in the beam are repelled, and they "stack up" into *bunches* within the drift tube (in the center of the klystron). On a positive half cycle, the electrons are no longer repelled and travel up the drift tube towards the anode (positive electrode). This process is very similar to that of the TWT and is called *velocity modulation*.

The bunches or clusters of electrons travel toward the anode. As a cluster of electrons passes the *output cavity,* an alternating RF field is set up in that cavity. The new RF field is

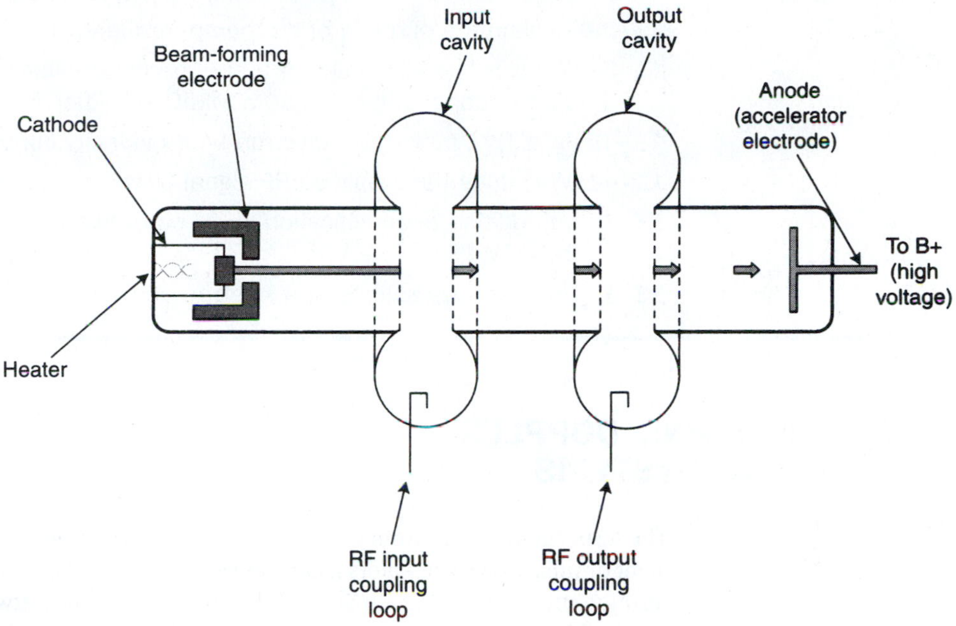

Figure 13–25 A klystron tube

much stronger than the original one in the input cavity because the electron beam's energy adds to it. Thus, *amplification* occurs; a second coupling loop extracts the amplified RF signal. The electron beam passes to the anode or *collector,* which completes the dc circuit. The output cavity of a klystron is usually tuned slightly higher in frequency than the input cavity in order to prevent undesired oscillations, which could destroy the tube in seconds.

Klystrons with three, four, and five resonant cavities are quite common in the television broadcast industry. *Extreme caution must be used when dealing with high-power klystron amplifiers.* The anode of a high-power klystron can have 25 kV potential, while supporting 5 amps of current. That is a *beam power* of (25 kV)(5 A) or 125,000 watts! At these voltage levels, even coming *close* to the power supply can be fatal. Because of the high current, arc-overs tend to be violent and explosive. *Exposure to this level of power is generally not survivable. Merely getting close to the high voltage can kill you, since the high voltage will arc to your body, completing the circuit.* For this reason, power klystrons are usually contained in metal cabinets with power supply interlock systems on the doors.

Section Checkpoint

13–33 What are the two categories of microwave amplifiers?

13–34 How are the sections Z_1, Z_2, and Z_3 constructed in the amplifier of Figure 13–21?

13–35 What is a MMIC?

13–36 Define the term *parametric amplifier.*

13–37 What circuit value changes to provide amplification in the amplifier of Figure 13–23?

(continued on p. 506)

13–38 Explain the purpose of the *pump* frequency in a parametric amplifier.

13–39 What is the advantage of a parametric amplifier over a more conventional unit (such as a MMIC amplifier)?

13–40 In a TWT, how is the electron beam velocity controlled?

13–41 Why does the amplified RF signal travel in a helix in a TWT?

13–42 In addition to amplification, what other two functions can be performed by a TWT?

13–43 How are the actions of a klystron and TWT similar?

13–5 PULSE AND DOPPLER RADAR SYSTEMS

The term *radar* comes from the phrase "radio detection and ranging." Before the development of microwave oscillators and receivers, the *resolution* of radar was quite limited. The *resolution* of a radar system is its ability to discriminate between closely spaced objects within the target area.

The use of microwave frequencies is beneficial to radar systems for two reasons. First, the wavelength of microwave signals is very small. Short wavelengths translate to improved feature discrimination. It becomes much easier to determine if the target is an enemy aircraft or a weather ballon if the radar *signature* is better defined. Second, the high frequencies used permit the use of much more compact *gain antennas* than would be possible at low frequencies (UHF and below).

There are two primary types of radar systems. These are the *TOA* (time-of-arrival) and *Doppler* systems. TOA radar is used to determine the distance and direction of a target, and Doppler radar is used to calculate the speed or *velocity* of a target relative to the observer. TOA radar is used in military installations and in air traffic control; Doppler radar is also used in these applications, as well as in weather analysis and traffic enforcement.

TOA Radar Operation

The operation of a TOA radar is shown in Figure 13–26. The system depends on the time required to receive an *echo* of a transmitted pulse of RF. Typically, the transmitter sends a "burst" of RF less than 1 ms long at intervals ranging from 1 ms to 50 ms. The latter

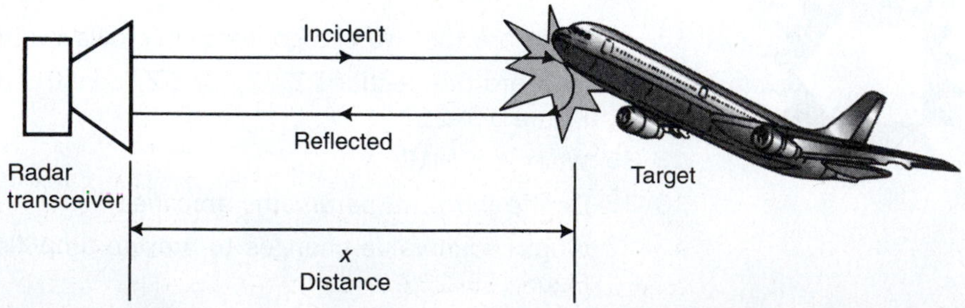

Figure 13–26 TOA radar operation

information is the time between pulses and determines the pulse repetition rate, or PRR of the radar system.

The event sequence in Figure 13–26 can be summarized as follows:

1. The radar station transmits a short burst of microwave RF energy at the target, using a directional antenna. Typical bursts are 1 ms or less in length.
2. The radio energy travels toward the target at the speed of light, and a small percentage of it is reflected back to the radar station. (The larger the target, the stronger the reflected signal. Metallic targets reflect better than nonmetallic.)
3. Upon reflection from the target, the signal travels back to the radar station, which is in receive mode and awaiting the return pulse. The time elapsed between the transmission and reception of the pulse gives the distance.

The distance to a target can be readily calculated when the total time between transmission and reception of the radar pulse is known. The formula for speed, distance, and time can be used:

$$D = RT = vT \tag{13–2}$$

where v is the velocity (speed) of light, D is the distance traveled, and T is the total time. In Figure 13–26, we can see that the total distance the wave must travel is *twice* the distance, between the radar transceiver and the target X. Therefore, we get

$$D = vT = 2X \tag{13–3}$$

Solving equation 13–3 for X, the actual distance to the target, we get

$$X = \frac{vT}{2} \tag{13–4}$$

where X is the actual distance to the target, v is the speed of light, 3×10^8 m/sec, and T is the time between transmit and echo.

EXAMPLE 13–5 An echo was received in 100 μs from a target in a TOA radar system. Determine the range to the target in meters and miles.

Solution

Equation 13–4 solves for the distance X directly:

$$X = \frac{vT}{2} = \frac{(3 \times 10^8\ \text{m/sec})(100\ \mu\text{s})}{2} = \underline{\underline{1500\ \text{m}}} = \underline{\underline{1.5\ \text{km}}}$$

In miles, the answer can be expressed as

$$X = 15\ \text{km}(0.621\ \text{mile}/1\ \text{km}) = \underline{\underline{9.320\ \text{miles}}}$$

Pulse repetition rate and duty cycle Three important factors in the operation of a TOA radar are the pulse repetition rate (PRR), pulse width T_W, and duty cycle DC. The higher the PRR, the more information that can be displayed on the radar screen; however, a high pulse repetition rate limits the maximum measurable distance to a target, since a received echo from a distant target won't arrive at the receiver soon enough before the next transmitted pulse.

The PRR is simply the reciprocal of the total time between pulses:

$$PRR = \frac{1}{T_R} \tag{13–5}$$

And the duty cycle is defined as percentage of time that the radar is actually transmitting a signal:

$$DC = \frac{T_W}{T_R} \tag{13–6}$$

The duty cycle is important—it determines the average power emitted by the radar, which is important because it determines in part how much power it takes to run the transmitter and also determines the life of major transmitter components (higher duty cycles wear out power tubes faster). The average power from the transmitter can be determined by

$$P_{av} = P_{PEAK} \times DC \tag{13–7}$$

EXAMPLE 13–6

A certain TOA radar has a total pulse period of 10 ms, a pulse width of 0.5 ms, and a peak transmit power of 50 kW. Calculate the PRR, duty cycle, and average radiated power.

Solution

Equations 13–5, 13–6, and 13–7 can be used to find these quantities:

$$\text{PRR} = \frac{1}{T_R} = \frac{1}{10\text{ ms}} = \underline{\underline{100\text{ Hz}}}$$

$$DC = \frac{T_W}{T_R} \times 100\% = \frac{0.5\text{ ms}}{10\text{ ms}} \times 100\% = \underline{\underline{5\%}}$$

$$P_{av} = P_{PEAK} \times DC = 50\text{ kW} \times 0.05 = \underline{\underline{2.5\text{ kW}}}$$

Note the very high value for peak power; this is typical of TOA radar installations. The RF power emitted from the antenna is very intense but only of short duration. Remember that *gain antennas* are typically used in a radar installation, so the peak radiated power may be much higher than 50 kW in the direction the antenna is facing!

Doppler Radar

Doppler radar is used whenever the relative speed (velocity) of a target is in question. In weather forecasting, it can readily determine the direction of motion of the parts of a storm system. For example, the rotating pattern of a tornado in formation can often be detected long before the actual funnel emerges from the cloud. And everyone is familiar with the radar speed-measurement "guns" in use by law enforcement.

The Doppler effect is the apparent shift in frequency of a signal source, depending upon the relative velocity of the observer and signal source. You've experienced this effect before: The sound of the whistle on a passing train is a common example. As the train approaches, the whistle seems higher in pitch; as it departs, the pitch falls. This action is illustrated in Figure 13–27.

Figure 13–27 The Doppler effect

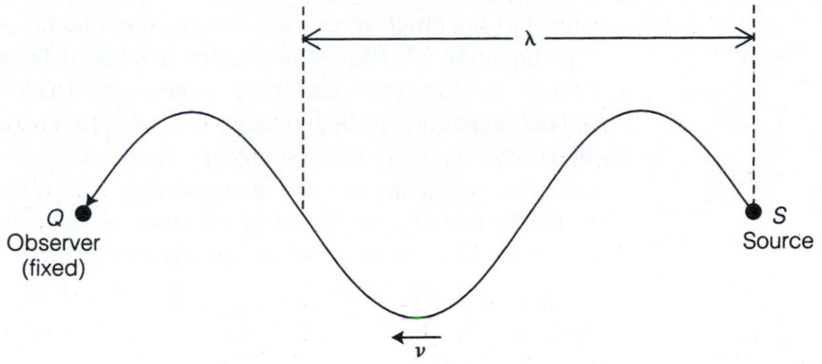

(a) Observer and source fixed in position

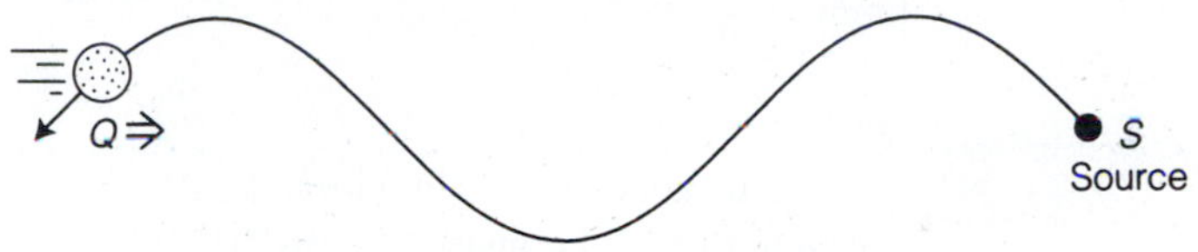

(b) *Q* is moving toward *S*, apparent frequency increases

In Figure 13–27(a), both the observer *Q* and the signal source *S* are stationary. The wave moves away from *S* toward *Q* at the speed of propagation, *v*. *Observer Q experiences a frequency as successive wavefronts pass by at velocity v.* Since we know the wavelength of the signal and the velocity of propagation, we can express the frequency "seen" by observer *Q* as follows:

$$f = \frac{v}{\lambda} \tag{13–8}$$

Does this equation look familiar? It is just the *wavelength* equation solved for *frequency* instead of wavelength. This may look a bit backward, but expressing it this way will let us more conveniently find the Doppler frequency shift.

In Figure 13–27(b), observer *Q* is now moving toward the signal source. What will happen to the frequency that *Q* experiences? The velocity of propagation is unchanged (the waves still move the same as before in the air). However, *Q* experiences *more* waves per second since it is moving *toward* the source. *By moving toward the signal source,* Q *increases the effective (but not the actual) velocity of propagation.* We can express the new frequency experienced by *Q* as follows:

$$f' = \frac{(v + v_Q)}{\lambda} \tag{13–9}$$

where v_Q is the velocity of the observer relative to the signal source. The *Doppler shift* is, then, the difference between the original frequency and the frequency perceived by the moving observer:

$$\Delta f = f' - f = \frac{(v + v_Q)}{\lambda} - \frac{v}{\lambda} = \frac{v_Q}{\lambda} \tag{13–10}$$

where v_Q is the velocity of the observer with respect to the source S, with positive motion defined as the observer moving *towards* the source.

Equation 13–10 tells us that the amount of Doppler shift is controlled by only two factors. These are the *wavelength of the source* (which is determined by the frequency of the source and the normal velocity of wave propagation), and the *relative velocity of the observer* with respect to the source, v_Q. Note that equation 13–10 assumes that the observer and signal source are moving along a single line; if this is not true, the Doppler shift will be *smaller* than that predicted by equation 13–10. (This is why a law enforcement officer must be as close to the roadway as possible in order to accurately measure the speed of vehicles; as he or she moves away from the highway, the apparent speed of the vehicles decreases. This is known as *cosine error* and is always in the motorist's favor.)

EXAMPLE 13–7

Calculate the Doppler shift expected on a 150 MHz radio signal on a LEO (low-earth-orbit) satellite downlink if the relative velocity of the satellite with respect to the observer is 6000 km/h.

Solution

Be careful about units. Equation 13–10 gives the Doppler shift when the velocity and wavelength are in consistent units. The value of 6000 km/h must be converted to km/sec in order to be useful, and the wavelength of the 150 MHz signal in meters must be found.

First, the conversion of the velocity can be done:

$$v_Q = 6000 \text{ km/h} \times \frac{1 \text{ h}}{3600 \text{ s}} = 1.6666 \text{ km/sec}$$

The wavelength of the 150 MHz signal needs to be computed next:

$$\lambda = \frac{v}{f} = \frac{3 \times 10^8 \text{ m/s}}{150 \text{ MHz}} = 2 \text{ m}$$

Now we can plug into equation 13–10:

$$\Delta f = \frac{v_Q}{\lambda} = \frac{1666.6 \text{ m/s}}{2 \text{ m}} = \underline{\underline{833.3 \text{ Hz}}}$$

The apparent frequency of the satellite's signal will be 833 Hz higher than what it actually is. For FM communication, this is not a problem. If the signal was SSB, this would represent a severe frequency error, and the receiving station would have to retune its receiver in order to demodulate the information.

There are two types of satellites, *orbiting* and *geosynchronous*. An orbiting satellite (such as the LEO) moves faster than the earth's rotation, so it completes a worldwide pass in less than 24 hours. Where it appears on the globe depends upon its "track" and the time. A geosynchronous satellite has an orbital period corresponding exactly with the rotational speed of the earth, so that it appears fixed at a point in space. Therefore, Doppler shifts can only affect *orbiting* satellites.

Doppler radar principles Figure 13–28 illustrates a Doppler radar unit. Unlike a TOA radar, a Doppler radar must transmit and receive at the same time. This might seem impossible, but it isn't. A device known as a *circulator* helps. The circulator is a three-port device that only allows RF signals to flow in one direction. By definition, signals can flow

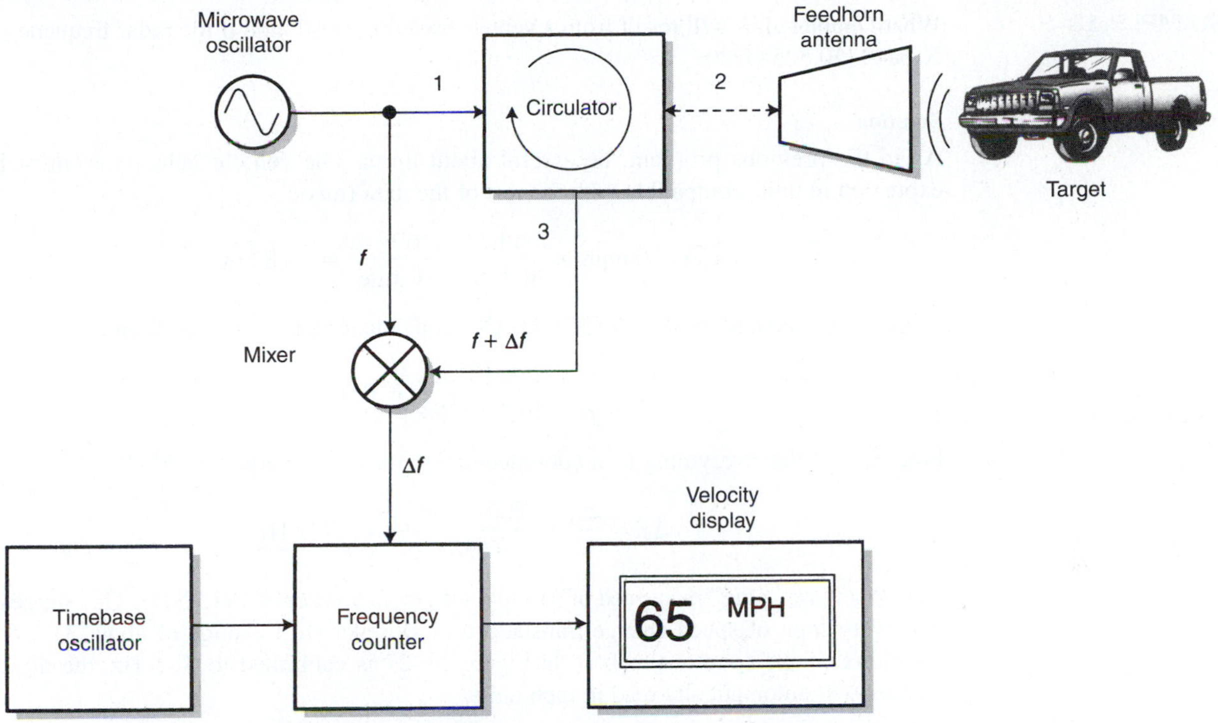

Figure 13–28 A Doppler radar

from port 1 to 2, and from port 2 to 3, and from 3 back to 1 (if there is a signal source on 3). Signals are not allowed to move in the opposite direction. This is a great help, because the transmitting oscillator in the radar (usually a Gunn oscillator) emits a very strong signal (several watts or more). Because the oscillator is connected to port 1, the oscillator's signal cannot get into the front end of the sensitive receiver. Instead, the oscillator signal leaves port 2, where it is coupled to the transmitting and receiving antenna (usually a horn antenna for traffic radar applications). The transmitted signal leaves the radar gun and travels to the target (which may be moving) and then back into the horn antenna. The circulator sees to it that the received signal flows from port 2 (antenna) to port 3 (receiver). A *mixer* (usually consisting of diodes) mixes the received signal with a sample of the transmitted signal, giving the *difference* frequency. This difference frequency is the Doppler shift and represents the relative speed of the target with respect to the radar gun.

Equation 13–10 must be modified for Doppler radar velocity calculations. The reason for this is that *two* distinct Doppler shifts occur in this application. The first occurs as the wave from the radar transmitter strikes the moving target. The target in effect experiences one Doppler shift by being in motion. The *second* shift is also caused by the target. This time, the target can be viewed as a *moving RF source*. The original frequency of this moving RF source is (you guessed it) the frequency from the first Doppler shift. Therefore, the total Doppler shift will be *twice* due to the reflection and reradiation, and we get

$$\Delta f = \frac{2v_T}{\lambda} \tag{13–11}$$

where v_T is the velocity of the target relative to the Doppler radar transmitter.

EXAMPLE 13–8

What Doppler shift will result from a vehicle traveling at 60 mph if the radar frequency is X-band (10.525 GHz)?

Solution

As in the previous problem, be careful about units. The vehicle velocity v_T must be expressed in units compatible with the rest of the data (m/s):

$$v_T = 60 \text{ mph} \times \frac{1 \text{ h}}{3600 \text{ s}} \times \frac{1609 \text{ m}}{1 \text{ mile}} = 26.82 \text{ m/s}$$

Next, the wavelength of the 10.525 GHz (X-band) radar carrier must be found:

$$\lambda = \frac{v}{f} = \frac{3 \times 10^8 \text{ m/s}}{10.525 \text{ GHz}} = 0.0285 \text{ m}$$

Finally, now that everything is in consistent units, we can use equation 13–11:

$$\Delta f = \frac{2v_T}{\lambda} = \frac{2(26.82 \text{ m/s})}{0.0285 \text{ m}} = \underline{\underline{1882 \text{ Hz}}}$$

Note that if we "plug" in a speed of 61 mph, we get an answer of 1913.5 Hz. This suggests that every mph of speed change translates to a Doppler shift change of about 31.6 Hz. Therefore, if the counter timebase in Figure 13–27 is calibrated to 31.6 Hz, the digital counter will automatically read in mph units.

EXAMPLE 13–9

A Doppler shift of 4108 Hz is being measured from a target under traffic radar surveillance. The original carrier frequency is 24 GHz, and the speed limit is 70 mph. What is the vehicle speed in mph? Will the driver get a citation?

Solution

Solve equation 13–11 for target velocity, and you get

$$v_T = \frac{\lambda \Delta f}{2}$$

The wavelength of the 24 GHz "K" band signal is

$$\lambda = \frac{3 \times 10^8 \text{ m/s}}{24 \text{ GHz}} = 0.0125 \text{ m}$$

The target velocity, in m/sec is therefore

$$v_T = \frac{\lambda \Delta f}{2} = \frac{(0.0125 \text{ m})(4108 \text{ Hz})}{2} = 25.675 \text{ m/s}$$

This is the same thing as 92,430 meters/hour, or $\underline{\underline{57.4 \text{ mph}}}$. The driver does not get a ticket today!

Section Checkpoint

13–44 What is the meaning of the term *radar?*

13–45 Explain the term *resolution* as it refers to a radar system. What controls it?

13–46 Give two reasons why microwave frequencies are used in radar.

13–47 What is the basic principle behind a TOA radar system?

13–48 What is the PRR of a radar system?

13–49 What two factors determine the average power of a radar transmitter?

13–50 Give an example of the *Doppler effect.*

13–51 What two factors control the amount of Doppler frequency shift?

13–52 Can Doppler shift affect satellite communications?

13–53 Why is a *circulator* needed in a Doppler radar?

SUMMARY

- Microwave communications systems operate above 2 GHz. At such high frequencies, the distributed circuit approach works best.
- At microwave frequencies, most discrete components are not usable unless they have been designed for UHF and SHF work.
- The cavity resonator is a popular replacement for LC tanks at microwave frequencies.
- All signal-carrying conductors must be considered as transmission lines at microwave frequencies.
- Waveguide is the most popular microwave transmission line. All waveguide has a lower cutoff frequency dependent upon physical dimensions.
- RF energy can travel in either TM or TE modes in a waveguide. By far the most popular mode is TE_{10}, often called the *fundamental* or *dominant* mode.
- Tees can be used as power combiners, splitters, and isolators. An isolator prevents energy from flowing between two circuits.
- Microwave oscillators usually use a cavity resonator to replace an LC tank. Both small- and large-signal oscillators are possible.
- Two popular small-signal oscillators include the Gunn and YIG. The Gunn oscillator is inexpensive but tunes only a narrow range. The YIG oscillator can be electronically tuned over more than an octave of range but is expensive.
- The magnetron is a power oscillator that can provide thousands of watts of RF microwave energy with just one stage, but it is not a very stable frequency source.
- Microwave amplifiers can be classified as small- or large-signal (power). Vacuum tubes such as the TWT and klystron dominate power applications, while solid-state devices such as the GaSFET are very popular in small-signal work.
- For very low noise amplification, parametric amplifiers are used. These use the power of an external ac source called the *pump* in place of a dc power supply.
- The TWT and klystron both depend upon velocity modulation of an electron beam for providing amplifier action. The TWT operates over a wide bandwidth, while a klystron must be tuned near the frequency of operation.

- The two main types of radars include pulse or time-of-arrival (TOA) and Doppler.
- A pulse or TOA radar measures the distance to a target by measuring the time between transmission of the RF pulse and the return of an echo of the pulse after it strikes a target.
- Doppler radar measures the speed of the target by measuring the frequency change or *Doppler shift* of the RF signal caused by target motion.

PROBLEMS

1. What are *microwave* frequencies?
2. Explain the difference between *lumped* and *distributed* components. Why aren't most lumped components useful at microwave frequencies?
3. Calculate the *total impedance* of the nonideal 100 Ω resistor of Figure 13–1 at the following frequencies: (a) 10 MHz (b) 10 GHz. Express the results in rectangular form.
4. Which parasitic circuit element is largely eliminated by surface-mount techniques?
5. Calculate the length of a quarter-wave stub antenna for a frequency of 5.6 GHz. Express the result in both meters and inches.
6. Draw a picture of a cavity resonator. What two lumped circuit elements does it replace?
7. Describe at least one method that is used for tuning a cavity resonator.
8. What are *microstrip* transmission lines? How are they built?
9. The long cross dimension of a waveguide measures 4.6 cm. Calculate its lower cutoff frequency f_{co}.
10. Repeat problem 9, with a dimension of 1.0 cm.
11. How large must the long cross dimension of a rectangular waveguide be in order to support a lower cutoff frequency of 1 GHz?
12. What are *propagation modes?* What is meant by the designations "TM" and "TE"?
13. For the following modes, give the number of half-wave electric field patterns across the long and short cross dimensions of the waveguide:
    a. TE_{10}
    b. TE_{11}
    c. TE_{21}
14. Which of the modes from question 13 is the most commonly applied in practice?
15. What is the difference between an *H* and *E* bend?
16. Which tee can be used to split one RF signal into two or combine two into one?
17. Which tee can be used as a transmit-receive switch? Explain its operation.
18. What is the purpose of an *attenuator?* What is placed within the attenuator to dissipate the RF energy?
19. Explain the purpose of a *terminator*. How does it differ from an attenuator?
20. What type of waveguide section is useful for measuring SWR?
21. In a Gunn oscillator, what forms the resonant bandpass filter? Explain its operation.
22. How is the frequency of a YIG oscillator determined? If such an oscillator is suspected of being defective, what is an appropriate troubleshooting method?
23. In a vacuum tube, which element operates from a low voltage and high current? What is the purpose of this element?
24. Draw a simple diagram of a magnetron tube. What causes the electrons to take circular paths in this device?
25. What circuit elements are used as impedance-matching devices in Figure 13–20?
26. What is a MMIC? Draw a schematic diagram of a typical RF amplifier employing a MMIC.
27. Explain the operation of a parametric amplifier. Which circuit parameter is often varied in order to add energy to the circuit?

28. Draw a diagram of a TWT amplifier. Explain the signal flow and principle of amplification.

29. Draw a diagram of a klystron amplifier. How does it amplify an RF signal?

30. What are the two types of radar systems?

31. A TOA (pulse) radar receives an echo from a target in 0.35 ms. What is the range to the target in meters?

32. A pulse radar receives an echo from a target in 0.25 ms. What is the distance to the target in miles?

33. A certain pulse (TOA) radar has a total pulse period of 25 ms, a pulse width of 0.7 ms, and a peak transmit power of 100 kW. Calculate the PRR, duty cycle, and average radiated power.

34. What causes the Doppler effect? Give an everyday example.

35. What Doppler frequency shift will be experienced at a ground station receiving a signal from a transmitter in a fighter jet if the jet is traveling in the direction of the earth station at a speed of 1000 mph and is using a radio frequency of 900 MHz?

36. Repeat problem 35 but let the jet move directly *away* from the earth station at the same velocity. What happens to the Doppler shift?

37. What Doppler shift will result if a fixed transmitter's signal strikes a target moving at 120 mph, assuming that the radar frequency is 10.525 GHz?

38. A Doppler shift of 1200 Hz was measured upon reflecting a 16 GHz signal from a moving target. What is the velocity of the target in meters per second?

14

Telephony and Cellular Networks

OBJECTIVES

At the conclusion of this chapter, the reader will be able to:

- describe the overall topology of the U.S. telephone network
- list the electrical characteristics of an analog POTS subscriber local loop
- describe the operation and characteristics of a typical analog CPE
- describe the DTMF signaling standard
- describe the digital signaling standards used in the telephone system
- describe the operation of digital traffic switches
- compare and contrast AMPS and PCS service characteristics
- describe how subscriber signals are carried through a PCS network
- given a frequency reuse map for a PCS market, identify the groups of frequencies utilized by each cell in the market
- describe the operation of DSL/ADSL subscriber services
- describe a typical installation for VoIP equipment

"Mr. Watson, come here, I want you." These historic words were the first speech to be successfully transmitted through wires, uttered by Alexander Graham Bell to his assistant in March of 1876. It was the beginning of the second wave of the telecommunications revolution, started by the invention of the telegraph in 1833. Now a basic part of modern life, the telephone system is one of the most creative and complex achievements of humankind. Over its lifetime, it has directly influenced and benefited from many advancements in electronic technology. The modern telephone system uses analog and digital circuitry, computers, advanced signal encoding, and of course a healthy dose of wireless technology. Let's see how it all works together.

14–1 THE SYSTEM VIEW

In the earliest days of the telephone, no more than a few hundred telephone sets existed. Customers of Bell's new telephone company usually purchased telephones in pairs, and a single dedicated wire was strung between the phones (the earth provided the ground return path). Such an arrangement was suitable for local point-to-point communications, but hardly practical for anything else. Suppose that a community has 1000 phones, each of which must be capable of talking to any of the other phones. That might mean that each residence would require 999 lines for connecting to each of the other phones in the network. This would not do at all! Furthermore, what if someone wanted to call a relative in another city? The problem just became a lot worse. At the turn of the twentieth century, cities were becoming "blackened with wires" because phone lines were being strung from every available rooftop, pole, and building. Many inventors and engineers would work over a period of decades to provide working solutions to this fundamental problem.

The modern telephone system has evolved over a century from this crude beginning, and it continues to change. It has several important features. First, the national telephone system is a hierarchical arrangement of calling centers, as shown in Figure 14–1. The word *hierarchy* means a system ranking specific kinds of things, with the most important (and usually least numerous) things at the top, and others below. Most hierarchies look like upside-down trees. The higher levels of this system serve large geographic areas, such as the class 1 regional centers in the figure. The lower levels serve a small area, such as a section of a city.

Figure 14–1 The national telephone system plan

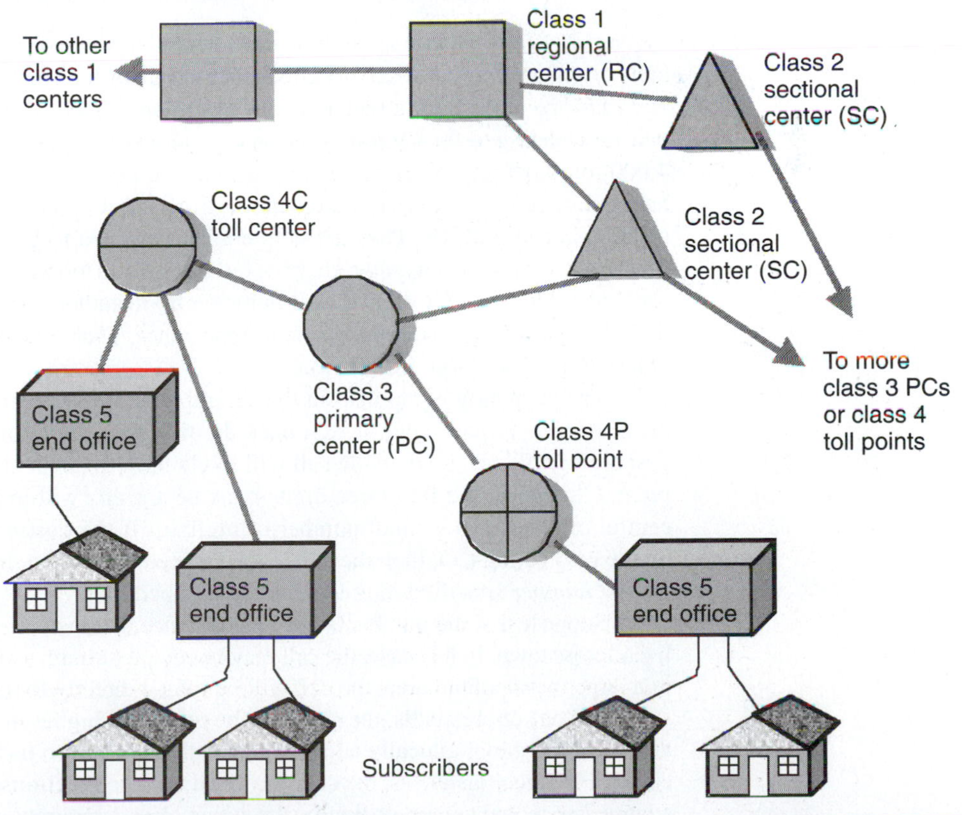

The second important characteristic of the phone system is that it is divided into regions. This has become a little more difficult to understand since the government-mandated breakup of the Bell system in 1984. Each of the Bell operating companies (sometimes referred to as the "*RBOCs*," for *regional Bell operating companies*) has a specific geographic area that they are permitted to provide service in. This is called a *LATA*, or *local access and transport area*. LATA boundaries are defined in the *Modified Final Judgement* (*MFJ*) handed down by the FCC in 1984. A region may cover a state or parts of several states.

Third, the telephone system is by and large completely automated. No human intervention is required for more than 99% of any local or long-distance calls anywhere in the country. Finally, the telephone network uses digital signaling to carry almost all conversations. The only remaining analog part in the system is the customer loop (the copper connection between a residence and the central office) and *CPE* (customer premises equipment). Prior to the use of digital speech transmission, analog facilities were needed to transmit conversations from point to point on the network. Analog signals weaken and accumulate noise over distance, and must pass through repeater-amplifiers every 10–50 miles, or they'll become too small to be useable. Every stage of processing degrades an analog signal, so that after many miles and "repeater hops" it no longer sounds very good! Because of the use of digital coding, speech signals are no longer degraded as they are repeated through the telephone network. Also, replacement of copper by *fiber optics* has also increased the distance between repeaters. We will study how speech is represented by digital codes in this chapter.

Order of the Telephone System: How a Call Is Routed

Figure 14–1 tells us a lot about how calls get from point to point in the system. Normally, calls originate from a subscriber connected to one of the central offices (sometimes called *end offices*). This office directly serves customers—each residence has a copper twisted-pair line leading to the central office. A central office is designed to handle approximately 4,000 to 10,000 customer lines. Calls within the service area of the central office can be handled directly without any intervention from the levels above. Figure 14–2 is a typical suburban central office. They are deliberately designed to blend in with the surroundings (the phone companies would rather not have everyone too aware of them) and typically are fortified against natural disaster and intrusion by unauthorized persons. Most have no windows but are otherwise unremarkable in appearance. There may be *no* staff at a central office unless it is a very large installation.

The switching equipment in the central office (CO) examines the number dialed by the customer. In particular, it examines the first four digits of the number closely. If the first digit is "0" or "1," then the call will likely have to be handed off to a *toll center* or *toll point*. Otherwise, the first three digits must be a *prefix* within the local calling area. Each central office serves a small number of prefixes. If the destination prefix is being served by the originating CO, then the call can be routed directly to the customer at the four-digit *station number* specified in the telephone number.

Suppose that the number is not long distance, but instead destined for a customer who lives across town. In this case, the call may be routed through a class 4 center, but more likely in a large metropolitan area, the call will be routed directly to the correct central office.

For out-of-area calls, the call must be routed up higher in the hierarchy. Between two areas that are geographically adjacent, the call may need to pass through a class 3 primary center. For larger distances, the call may need to "hop" sections, which may pass it through a class 2 sectional center, or finally, through a class 1 regional center. Notice that all traffic can't pass through the class 1 centers; doing so would create a bottleneck. Instead, the

Figure 14–2 A central office in south Kansas City

system is linked at several levels, so that a call bubbles only high enough through the system to make it to the destination.

NOTE: The copper wire connecting a subscriber to an end office is sometimes called the *last mile* of the telephone system. It is highly valuable real estate, for it's usually the only way of connecting to a customer. ILECs usually own the last mile, and only recently have been forced by government regulations to share it with other carriers. An ILEC is an incumbent local exchange carrier, meaning the telephone company that was there to begin with and owns the copper wire leading to the customers in the region. This copper is sometimes called the *last mile* of the telephone system. Accessing the customer premises is often called the *last mile problem* in telecom circles. In some areas cable companies have begun offering telephone service, in many cases sharing the coaxial cable carrying TV signal with the analog telephone signal.

Section Checkpoint

14–1 What was the primary difficulty with Bell's original telephone system?

14–2 List several characteristics of the modern telephone system, and explain each one.

14–3 Why is the hierarchical organization of Figure 14–1 necessary?

14–4 Explain how a call travels from house to house within a neighborhood, and contrast this with a call across the state.

14–5 What is the last mile problem?

14–2 THE LOCAL LOOP: OPERATION, SIGNALING, AND TELEPHONE CIRCUITRY

The local loop is the connection between a telephone subscriber and the central office. It includes the CPE, or telephone set, and the equipment at the end office. With the exception of wireless calls (such as *PCS*), everything begins and ends here! The electrical characteristics of the local loop are standardized; many of these standard values date back to the beginning of the Bell system. Figure 14–3 is a simplified schematic of the local loop.

Local Loop Components

The components in the figure provide minimal analog telephone service, called *POTS* (plain old telephone service) in the industry. There are two primary parts in the local loop, the subscriber line interface card *(SLIC)* at the central office, and the CPE. These are connected together by the wire pair leading to the customer's residence. Although there are usually four wires at a telephone jack in a home, only two of the wires are usually active. The red wire, also called the *ring,* carries the voice and ringing signals, while the green wire, called the *tip,* provides the ground return. Occasionally the black and yellow wires are used for a second phone line connection.

The subscriber line interface card controls the electrical activity on the loop. It is responsible for receiving and sending the voice signals on the customer's line, ringing the customer telephone when a call is incoming, and detecting when the CPE is taken off-hook when the customer is beginning to place an outgoing call. The SLIC routes the status tones (dial tone, busy signal, etc.) to the customer and also accepts the dialing information from the CPE (passing it onto the telephone network). Finally, since most telephone signal switching is now digital, the SLIC is also responsible for converting the customer's outgoing voice signal into digital for transmission through the network, and converting the incoming digital voice signal from the telephone switching network back into analog so that it can be reproduced by the customer's telephone set. Most SLICs service eight customer

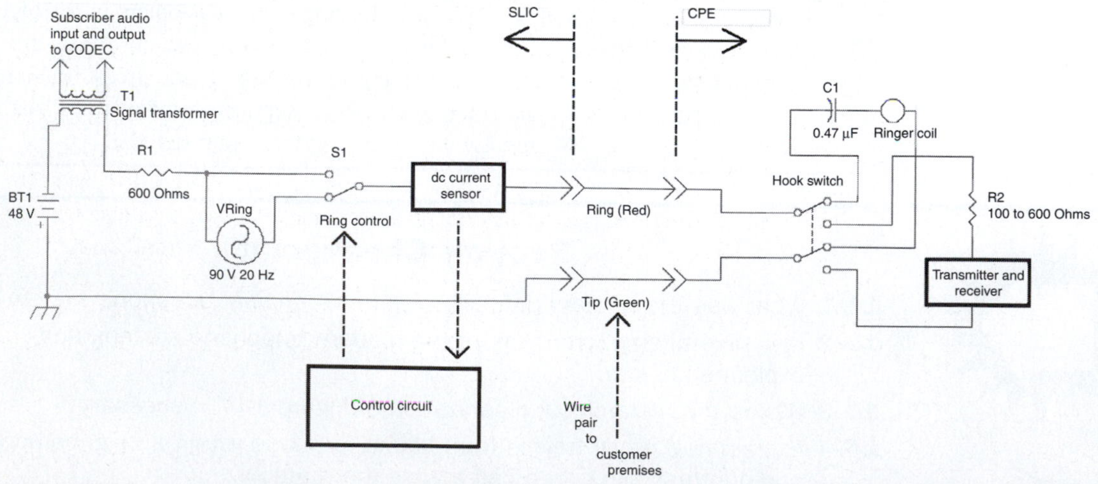

Figure 14–3 The local loop connects the customer premises to the end office

Figure 14–4
Equipment rack with SLICs at an end office

loops simultaneously. Figure 14–4 is a peek inside an end office, showing the array of subscriber cards (plus other equipment) in the equipment rack. This end office is in a small town and probably wins the "messiest installation" award!

The telephone set or CPE is a fairly simple circuit (we'll look at it in more detail shortly). It consists of a hookswitch, ringer, transmitter, and receiver. The hookswitch disconnects the transmitter and receiver when the unit is on-hook, and connects the ringer. The ringer requires 90 volts ac at 20 Hz to operate, and is coupled to the line through capacitor C_1. This is important; the capacitor blocks dc current. *The end office detects an off-hook phone because it draws a dc current.* An on-hook phone can't draw a dc current because of C_1. When the handset is picked up the hookswitch is activated and connects the transmitter-receiver circuit onto the line. The transmitter and receiver draw a small dc current (usually 10–20 mA). The subscriber line interface card at the end office sees this current and determines that the unit is off-hook. Both incoming and outgoing voice signals share the same pair of wires; a circuit called a *line hybrid* in the telephone set helps to keep them separate.

Loop Operation

The circuit of Figure 14–3 operates on −48 volts dc. This is a fairly standard value in telephony. This voltage is usually supplied by banks of batteries in a central office. These batteries are kept under continual charge, of course. Because batteries supply the operating voltages, momentary power interruptions don't disrupt the telephone service. (For longer disruptions, diesel generators are usually on-site to keep the batteries charged.) The loop current is limited by resistor R_1, which is usually between 400 and 1200 Ω. The dc loop current is needed to operate the customer telephone circuits. An important part of the SLIC is the off-hook detection circuit, which is provided by the dc current sensor in the figure. When the customer's phone is on the hook, it can draw no dc current from the line. The hookswitch in Figure 14–3 is shown in the on-hook condition, which connects the ringer and dc blocking capacitor into the circuit. Switch S_1 is in the "up" position in the figure so that no ringing voltage is present. Suppose that the customer decides to make an outgoing call. He or she picks up the handset, which moves the hookswitch to the opposite position. A dc current now flows through the current sensor on the subscriber interface card. The control circuitry thus senses that a new call is starting, and a dial tone (a low-level ac signal) is sent onto the customer's line through signal transformer T_1. The customer then hears the dial tone and can begin the call.

A similar process occurs during an incoming call. The control circuitry detects the presence of the incoming call and moves switch S_1 into the position shown, which connects the 90 volt ac ringing signal onto the line. The dc level is almost *always* present on the line so that the end office equipment can continually detect the state of the customer's phone. Note that the 90 volt ac signal appears in *series* with the −48 volt dc bias so that the dc current sensor can detect when the customer's phone is answered, even while it is ringing. The ringing signal is turned off immediately when the off-hook condition is detected; then the circuit is ready for carrying the conversation. If at any time the customer hangs up the phone, the dc current will be interrupted, and the circuit will go back to the idle condition.

EXAMPLE 14–1

Suppose that the customer short-circuits the *ring* and *tip* terminals at a residence. What is the maximum current that could flow through the telephone network, given the values in Figure 14–3?

Solution

With a short at the customer premises, R_1 will limit the total current. Ignoring the resistance of the copper wires (which may add 100 Ω or more to the total!), we get:

$$I = \frac{V}{R} = \frac{48\text{ V}}{600\ \Omega} = \underline{\underline{80\text{ mA}}}$$

Up to 80 mA could flow from the end office.

EXAMPLE 14–2

What is the maximum dc power that a telephone set could get from the telephone network given the values in Figure 14–3?

Solution

To draw maximum power from a source, connect a load resistance equal to the Thévenin resistance of the source (maximum power transfer theorem). Then use Ohm's law to calculate the power in the load. By inspection, $R_L = R_1 = 600\ \Omega$. The voltage that would appear across R_L would therefore be one-half of the 48 V dc provided at the end office, since

R_1 and R_L form a voltage divider with equal devices. Therefore, the power delivered to the load would be:

$$P = \frac{V^2}{R_L} = \frac{24\ \mathrm{V}^2}{600\ \Omega} = \underline{\underline{960\ \mathrm{mW}}}$$

Less than a watt of dc power is available from the end office. This doesn't sound like much. However, consider that a central office serves thousands of users; the total power required is controlled by the number of off-hook telephones. In general, telephone facilities are not designed for 100% utilization, so if everyone takes their phone off-hook at the same time, only about 10–20% of the users will get a dial tone, and great stress will be placed on the end office power supplies!

Signaling on the Loop: Dialing a Number

Two primary methods are supported for customer dialing, *pulse* and *dual tone multifrequency* or *DTMF*. Pulse dialing dates back to the first automatic switching system, the Strowger switch, patented in 1891. DTMF, also known as *touch tone,* is supported by most subscriber line cards, and costs the telephone companies little (if anything) to support, but many phone companies still charge extra for DTMF service.

Pulse dialing relies on the operation of a mechanical telephone dial, as shown in Figure 14–5. The dial includes a return spring, a mechanical governor, and two sets of contacts. When the customer operates the dial, the *mute* contacts close to bypass the receiver so that he or she can't hear the dialing pulses. When the dial is released, it returns under power from the return spring, while the governor keeps the speed of the dial constant. The *dialing* contacts open and close the circuit according to the digit dialed; to dial "1," the circuit is opened and closed one time; to dial "2," the circuit is opened and closed twice; and

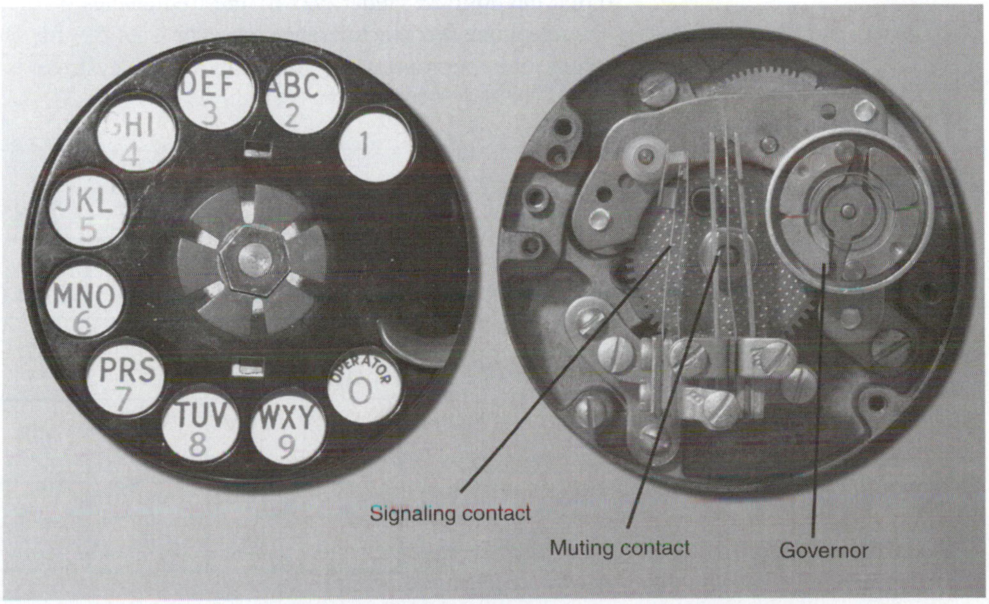

Figure 14–5 Mechanical dialing mechanism details

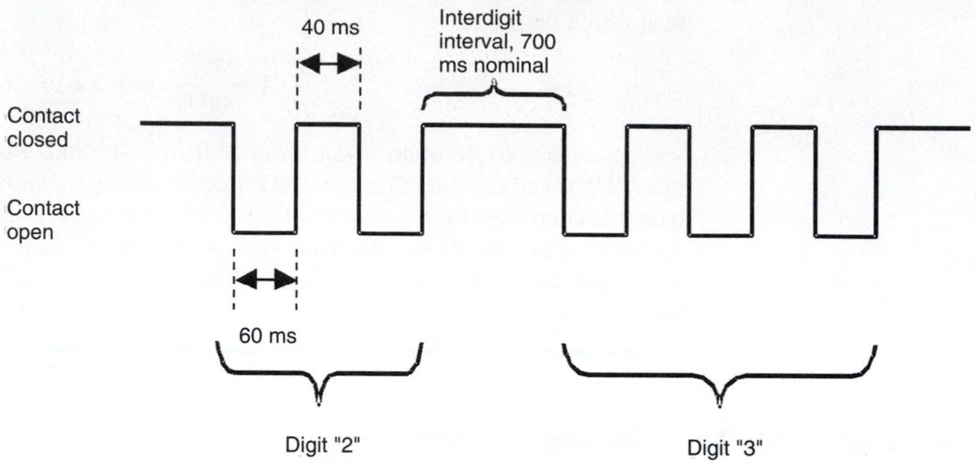

Figure 14–6 Pulse dial timing (United States)

so on. (Ten open–close cycles represent "0.") When the dial returns to the rest position, the mute contact opens to allow normal receiver operation again. In a modern telephone the mechanical dial is replaced with a microcontroller and solid-state switch; the timing remains the same. Figure 14–6 shows the timing relationships for pulse dialing. There are a nominal 10 pulses per second (pps) in the United States, with a 40% duty cycle (the circuit is broken 60% of the time).

DTMF or touch-tone signaling is much more popular since it allows quicker and easier dialing than pulse, and also allows access to more features (such as voice mail, banking by phone, and so on). In DTMF each digit is represented by two sine wave tones as shown in Figure 14–7. These frequencies were chosen carefully; none of them is a harmonic of the power line frequency (60 Hz) in the United States, and none of the tones is a harmonic of any of the other tones.

To dial the digit "2" under DTMF, the frequencies 697 Hz and 1336 Hz are generated by the telephone set. The tolerance for error is ±1.5% for acceptance of these tones at the end office; if either tone is out of tolerance, the keystroke is ignored by the telephone

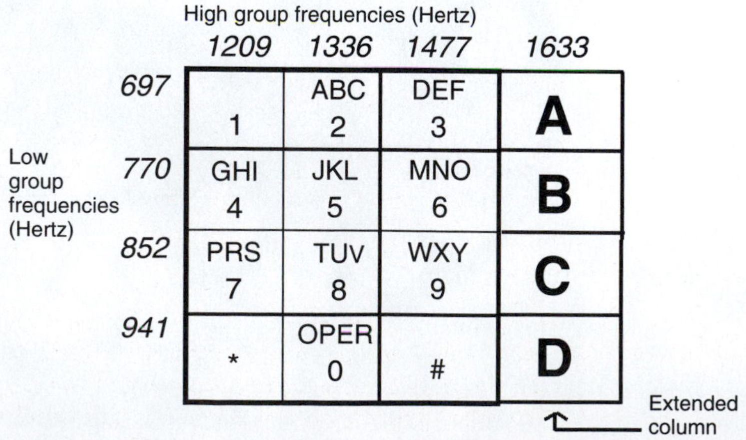

Figure 14–7 DTMF tone system

network. Modern telephone sets typically use a dedicated IC such as the National Semiconductor TP-5089 (usually under crystal control) to provide DTMF tones. Home telephones usually don't include the *extended column* of tones shown in Figure 14–7. These are used for special purposes within the telephone system, and occasionally for signaling for nontelephone applications (such as amateur radio).

DTMF tones must not only be accurate in frequency, but also in amplitude. Nominally, the transmitted level of the lower frequency group is −8 dBm into the 600 Ω impedance of the subscriber line, and the upper group's level is −6 dBm (boosted by 2 dB since higher frequencies are attenuated more in the signal path to the end office). The difference in amplitude between the two DTMF tones transmitted by a CPE is referred to as *twist*. Excessive twist can cause one of the two DTMF tones to be rejected by the SLIC decoder, resulting in a dropped digit. Twist develops from uneven frequency response in the telephone system. Most DTMF decoders can tolerate up to about 6 dB of twist (2:1 voltage ratio of the two DTMF tones).

EXAMPLE 14–3 What is the RMS and peak-to-peak voltage of a 0 dBm signal on a 600 Ω subscriber line?

Solution

Use Ohm's law to find the voltage. First, we must convert the dBm power level back into watts. By inspection, we know that 0 dBm is 1 mW of power, but we can also calculate the power by:

$$P = 1\text{ mW} \times 10^{(\text{dBm}/10)} = 1\text{ mW} \times 10^{(0/10)} = \underline{\underline{1\text{ mW}}}$$

Since we know that $P = V^2/R$ by Ohm's law, we can solve for voltage:

$$V = \sqrt{PR} = \sqrt{(1\text{ mW})(600\ \Omega)} = \underline{\underline{775\text{ mV}}}$$

In peak-to-peak units, this is:

$$\text{V}_{\text{pp}} = 2\sqrt{2} \times V_{\text{RMS}} = 2\sqrt{2} \times 775\text{ mV} = \underline{\underline{2.19\text{ V}_{\text{pp}}}}$$

The value of 0 dBm, or 2.19 V_{pp}, is also called "*0TLP*" in the telephone industry. It is the maximum signal level that a customer device may place on the telephone line.

TIP To accurately measure ac signal levels on a telephone line, you must either use an audio isolation transformer, or make sure that the test equipment is not connected to earth or power line ground (for example, use battery powered equipment). Otherwise, you'll be introducing a *ground loop* onto the line, with a great deal of power line noise! Commercial equipment is readily available to measure signal levels on POTS lines.

Telephone Line Frequency Response

During the development of frequency-division multiplexing techniques (which would allow a wire to carry more than one conversations), experiments by Bell Laboratories were performed to determine what frequencies were needed to carry a voice conversation. By restricting the bandwidth, more voice channels could be carried on a wire. The frequency response pattern of Figure 14–8 was adopted as a result. Dial-up telephone lines have this characteristic.

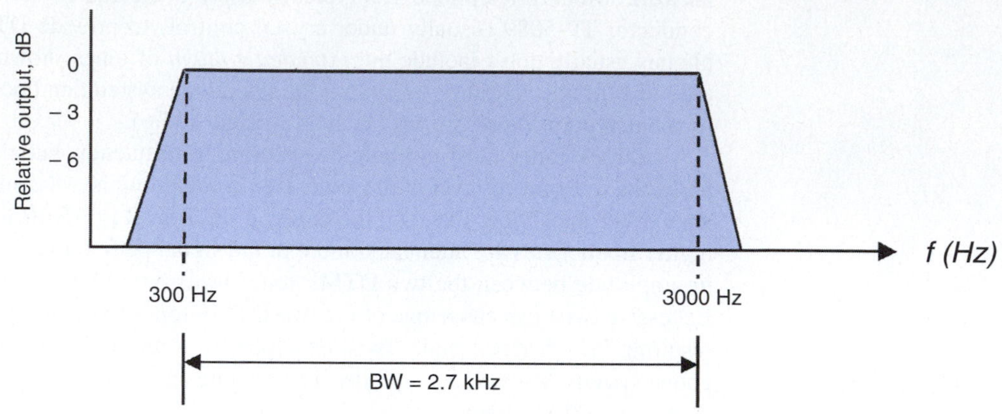

Figure 14–8 Ideal frequency response of a standard dial-up telephone line

You'll notice that the frequency response of a telephone line is quite narrow compared to the full range of human hearing (20 Hz to 20 kHz). The rolloff at 300 Hz limits the bass presence of the voice, which adds very little to intelligibility. The high frequency limit is close to 3.3 kHz (with the upper rolloff starting at 3 kHz). The limited treble response reduces the crispness of the speech signal; most of the higher-frequency energy in speech comes from unvoiced consonants such as *f, s,* and so on. The frequency response of the telephone system is controlled by end-office equipment, primarily, though the transmission lines between a residence and end-office also have a high-frequency effect as well. In the modern digital phone network it is necessary to limit the high-frequency response prior to *sampling* the analog signal during analog-to-digital conversion, as we'll see in the next section.

In addition to the limits of frequency response, there are two types of delays a signal is subjected to as it passes through the telephone system. These are *propagation delay* and *group* delays. Propagation delay is caused by the distance a signal must travel to reach the destination. On a long-distance connection this delay may amount to 10 ms or more. The telephone system employs echo cancellation equipment to prevent delayed long-distance signals from echoing back and forth across a connection.

Group delays (also called envelope delays) are a *variation* of the delay with frequency. In other words, some passband frequencies may be passed faster or slower than others through the telephone system. This causes distortion of the received signal since the phase relationship of the various energies in the signal becomes incorrect, which means that the signal will have an incorrect shape. For voice signals, the distortion may hardly be noticeable, but for data communications, excessive group delays can corrupt transmitted data. Figure 14–9 shows the graph of group delay for a typical telephone connection. Note that the vertical axis shows the *differential* or *change* in delay with respect to the minimum delay, which is found at 1500 Hz in this example.

You can simulate the effect of group delay on a square wave by using the *Wavegen* program provided in the \software folder of the CD-ROM provided with this text. To do so, proceed as follows:

- Start the *Wavegen* program. If you have speakers on your PC, turn them on. You should hear a 1 kHz sine wave playing through them.
- Click the Show/Hide Scope Window button. A scope window will pop up to show you the 1 kHz sine wave being produced.

Figure 14–9 Group delay through a telephone connection

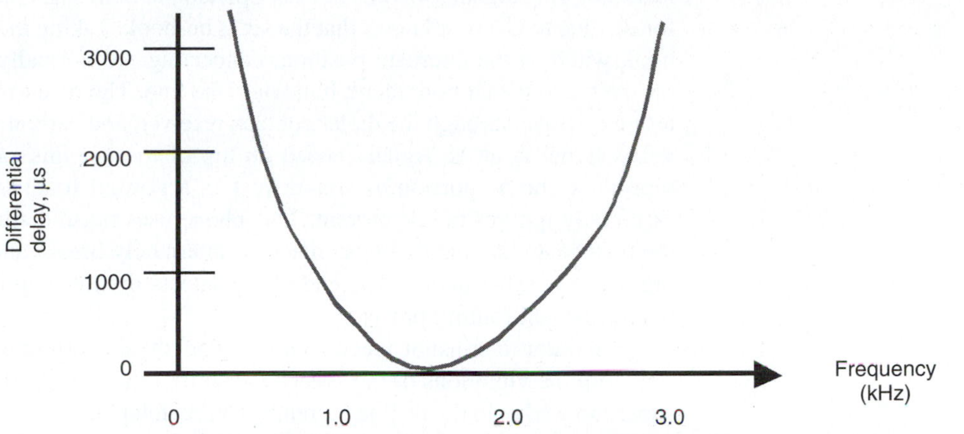

- Click the Square button. You'll see the scope window waveform change to an approximated square wave. (This could represent a digital data signal.) The sliders representing the various harmonic energies (and their phase angles) will automatically move to show you the formula for the square wave.
- You can now adjust the phase angle of any frequency in the waveform (which for a square wave would be the 1st, 3rd, or 5th harmonic *phase* sliders). This will introduce a phase shift into each signal, which is equivalent to an envelope delay. You'll see the resulting waveform change as you adjust the sliders. See if you can hear the difference between the original and distorted waveforms!

Inside the Telephone Set

The electronics within a telephone set are relatively basic. A telephone set in its simplest form consists of a hookswitch, dialing mechanism, a receiver, and a transmitter. Figure 14–10 shows a simplified telephone set employing a mechanical pulse dialing mechanism.

The operation of the set in the figure is straightforward. When the phone is on-hook, the hookswitch connects the ringer coil and capacitor C_1 to the telephone line so that the

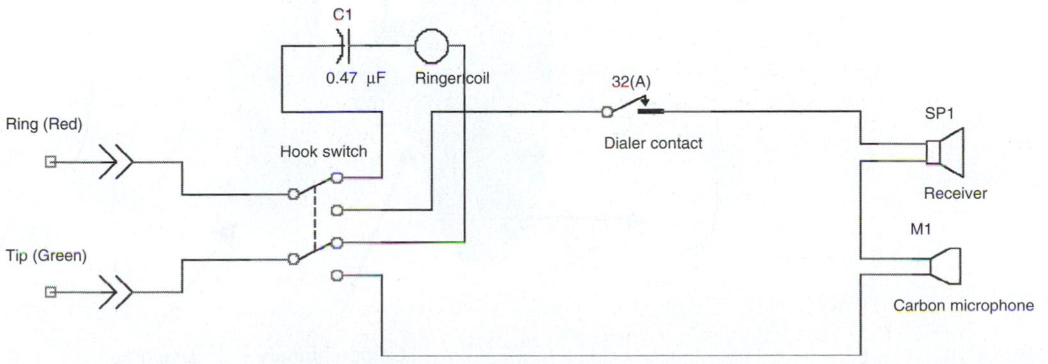

Figure 14–10 Simplified telephone set

incoming ring signal (90 volts ac) can operate the bell. The end office sees an open circuit for dc (due to C_1) so it knows that the set is on-hook. Taking the phone off-hook moves the hookswitch to the alternate position, connecting the normally closed dialer contact, the receiver, and the carbon microphone onto the line. The dc current from the end-office battery now flows through the dialer contact, receiver, and carbon microphone. The incoming voice signal is an ac signal carried on top of the dc; this causes the receiver (SP_1) to reproduce the ac portion as sound, just as a normal loudspeaker would. The receiver essentially ignores the dc current. This phone uses pulse dialing; when the user operates the telephone dial, the dialing contact S_{2A} alternately breaks and makes the circuit according to the number dialed. The end office detects the interruptions in dc loop current and begins the call routing process.

In order to transmit speech onto the line, the dc current on the line must be varied in step with the vibrations of the speech signal (in other words, the ac speech signal must be superimposed onto the dc line current). In early telephones, a *carbon microphone* was used for this purpose. It is still used in a few nonelectronic telephones. A carbon microphone converts sound into a varying electrical resistance. (This is quite different than microphones normally used in communications, which convert speech into a small ac signal.) Figure 14–11 shows a cross section of a carbon microphone disc.

In the microphone of Figure 14–11, the diaphragm collects the sound vibrations and concentrates them upon a pellet of carbon granules. This alternately compresses and releases the particles of carbon, which causes the pellet's resistance to vary in step with the information. Carbon microphones provide an immense amount of signal gain—typically 30–60 dB in excess of what a conventional microphone can provide, but they require a dc bias source to operate. The early telephone system wouldn't have been possible without the carbon microphone, since electronic amplification hadn't yet been invented! Modern

Figure 14–11 Carbon microphone action

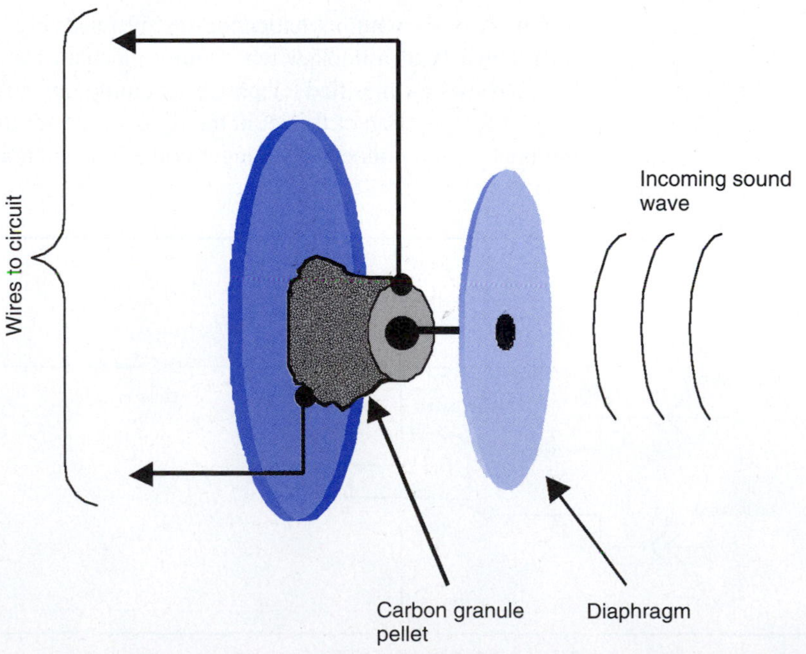

telephones use either dynamic or electret condensor microphones with solid-state amplification to achieve the same results without the inherent distortion caused by the carbon microphone.

Sidetone Control

The phone of Figure 14–10 will work to transmit and receive speech from a telephone line, but has one annoying problem. When the user speaks into the microphone, their voice will be echoed back quite loudly in the earpiece (SP_1), since the earpiece shares the same dc circuit as the carbon microphone. In response, most people will either speak more softly, or move the microphone away from their mouth, which will greatly reduce the speech level going to the line. Transmitted audio that is heard in the receiver is called *sidetone*. A correct level of sidetone is beneficial; it helps people to speak with the proper voice volume on the phone, and also lets them know that the line is "live." (You've probably blown into the mouthpiece of a telephone to learn if it was working.)

The circuit of Figure 14–10 produces too much sidetone level. You might think that we could just put a resistor across the earpiece to reduce sidetone, and you'd be right; however, this also has the very undesirable side effect of reducing the sensitivity of the telephone receiver to incoming speech! Instead, the circuit of Figure 14–12 is used to partially cancel the transmitted speech signal.

The circuit of Figure 14–12 has been modified by the addition of the *hybrid transformer* T_1 and the *balancing resistor* R_1. T_1 has three coils with equal numbers of turns; the bottom two coils are connected in series to form the primary. Each of the coils in T_1 always has the same ac voltage drop. The operation is straightforward. Incoming voice signals form an unbalanced voltage drop across the entire primary (bottom two windings) of T_1. This causes the voice signal to be reproduced on the secondary (upper) winding of T_1 and therefore it is heard in the receiver. When the user speaks into the microphone, the dc line current is modulated as before, but in the primary of T_1 two *opposing* signals are created, since the microphone signal is driving the center tap. Therefore, only a small amount of microphone signal shows up on the secondary of T_1, canceling most of sidetone. For this cancellation to take place, T_1 must see a balanced impedance on each side of its primary winding. To the left, the phone line provides an impedance of about 600 Ω, so resistor R_1 is set to 600 Ω to provide this balance. There's another enhancement in this circuit

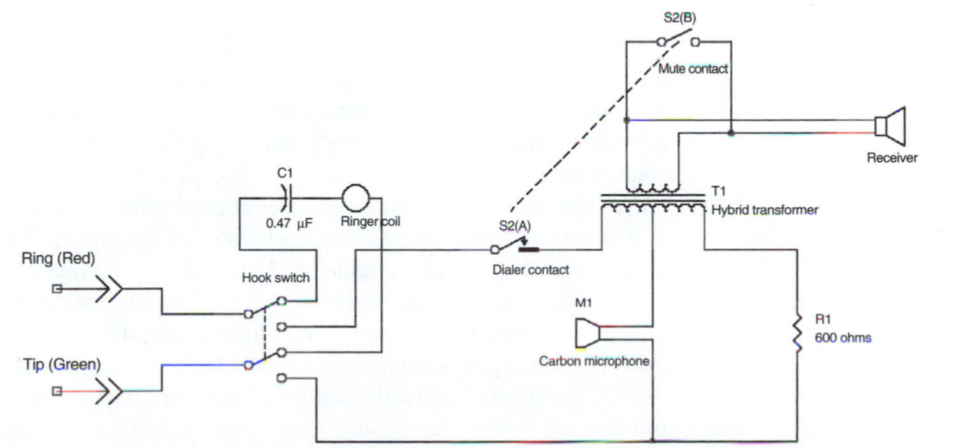

Figure 14–12 A telephone set with sidetone suppression

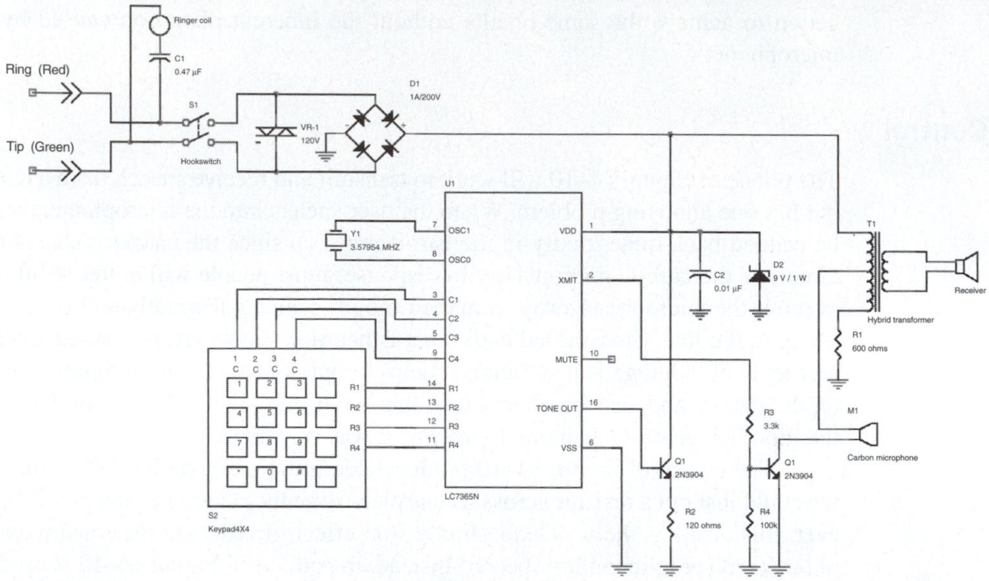

Figure 14–13 A telephone set with DTMF capability

unrelated to sidetone. In the circuit of Figure 14–10, the dialing process interrupts the dc loop current, producing a very loud clicking noise in the receiver. Mechanical dialers provide a *muting* contact that closes whenever the dialer is moved away from its mechanical stop. The muting contact bypasses the earpiece, preventing the annoying clicks.

Electronic Telephone Circuits

The circuit of Figure 14–13 is the first step toward an all-electronic telephone unit. The mechanical pulse dialing mechanism has been replaced with a Sanyo LC7365N DTMF generator IC. The carbon microphone circuit persists; it could be replaced with a condensor microphone and amplifier circuit. This is typically the case in modern units.

In the figure, you can see that we simplified the hookswitch circuit. The ringer is now permanently wired across the line, with C_1 to block the dc component. (If this connection were made with the phone of Figure 14–12, the bell might "tinkle" during the dialing process due to inductive transients caused by making and breaking the circuit.) The hookswitch enables the electronics to be coupled to the line; VR_1 is a transient suppressor, which absorbs any high voltage spikes that might appear on the line before they can damage the sensitive solid-state circuitry. Because telephone lines in a home are often wired with the wrong polarity (which would be bad for the electronics!), bridge D_1 is used to assure that the proper line voltage polarity feeds the internal works of the phone. When the phone goes off-hook, zener diode D_2 limits the supplied voltage to 9 V to protect the DTMF generator chip, U_1.

U_1 contains all the logic needed to generate the DTMF signal. It has an internal crystal oscillator (an external TV colorburst crystal is used), plus a keyboard scanning circuit. The keypad is directly connected to the row and column inputs (C_1–C_4, R_1–R_4) of the IC. The IC automatically scans the keyboard; when a key is pressed, it drops the *XMIT* pin low (to disable the local microphone, so that local sound isn't transmitted with the DTMF tones, which could cause an error in reading them at the end office) and provides

DTMF audio at the *TONE OUT* pin, which is buffered by Q_1. The DTMF audio is directly impressed across the line (power supply) by Q_1. When the key is released, the *TONE OUT* signal shuts off, and the *XMIT* pin again goes high, turning on Q_2, which again enables the local microphone. The hybrid network is identical to that in Figure 14–12, and performs exactly the same function.

Finally, note that the circuit of Figure 14–13 has been "stripped" of RF bypass capacitors (except for C_2), which would add slightly to the complexity of the circuit. A well-designed telephone should be immune to strong radio frequency fields, but unfortunately, manufacturers often omit a few pennies worth of bypass capacitors for the sake of cost cutting. Thus, many modern electronic telephones are quite sensitive to RF interference, especially from HF and VHF transmitters. In particular, the addition of a 0.01 μF/600 V capacitor across the ring and tip terminals of the unit would be particularly welcome to block RF from getting into the unit.

Section Checkpoint

14–6 What is the open-circuit voltage found on a subscriber loop?

14–7 How does the end-office detect when a customer's phone is off-hook?

14–8 What is the function of capacitor C_1 in Figure 14–10?

14–9 What functions are provided by the SLIC in the end office?

14–10 Explain the difference between pulse and DTMF dialing mechanisms. Why is DTMF preferable in most cases?

14–11 What is the frequency response of a dial-up telephone line?

14–12 What are the two types of delay introduced by telephone lines? Which one can interfere with data transmission, and why?

14–13 Why were carbon microphones popular in the early telephone sets?

14–14 What is sidetone, and why is it important to a telephone user?

14–15 What is the function of VR_1 in Figure 14–13?

14–3 SWITCHING AND ENCODING: PCM

When the first telephones were put into use, they were connected with a dedicated wire between two points, typically between a businessman's office and his home. This prevented the universal use of the telephone, and led to the development of switching technologies. The first such technology was the *telephone switchboard*. The telephone lines of subscribers were routed to a central location with the switchboard; human operators manually made telephone connections one at a time at each caller's request. The first automated switching mechanism was invented by Almon B. Strowger, an undertaker, in 1892. The story goes that he was losing business because his competitor's wife ran the local telephone exchange. The Strowger switch is a complex electromechanical device that "steps" vertically and horizontally under the control of the dialing pulses from a subscriber's telephone. A telephone exchange required thousands of the devices; the switching scheme is shown in Figure 14–14.

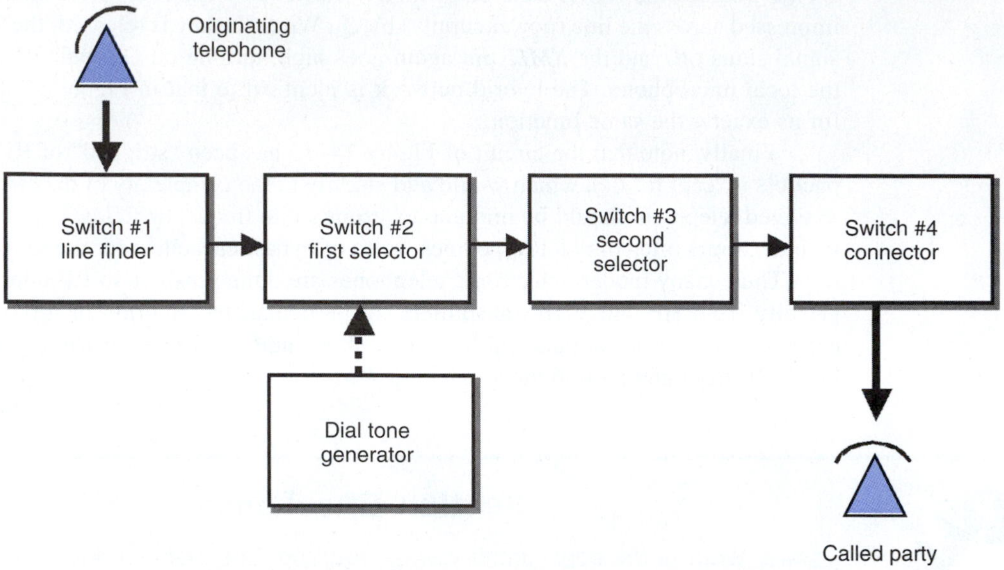

Figure 14–14 Strowger switching scheme for a four-digit phone number

In the Strowger scheme, subscriber lines are connected to a line finder switch (#1 in the figure). Each line finder can service 100 subscriber phones. When the subscriber goes off-hook, the line finder begins stepping until it finds an available line, which is represented as switch #2. The *dial tone* then appears. (If there's no available line, there's no dial tone.) The first selector then responds to the first digit (pulses), and connects to the appropriate *selector* (switch #3). The second selector switch counts the number of pulses in the second digit of the number, and connects to an appropriate *connector* switch. The connector switch responds to the last two dialed digits to connect to the correct subscriber. The Strowger switch worked well, but it required constant maintenance; the constant wiping action of the mechanical contacts produced rapid wear, and often noisy connections. Another approach to the problem is the *crossbar switch,* shown in Figure 14–15.

A crossbar switch consists of an array or matrix of switches as shown in Figure 14–15(a). Because of the arrangement of the switches, any row can be connected to any column in the switch. Telephone crossbar switches usually consist of a 10 by 20 array of switches, so that a unit contains a total of 200 mechanical switches. Connecting telephones directly using just a single matrix is inherently inefficient. To interconnect ten telephones would require a 10×10 matrix, or 100 switches—think n^2 switches, where n is the number of telephones—this would obviously be unworkable in a central office serving 10,000 telephones, which would theoretically require $10{,}000^2$ or 100 million switches! Therefore, crossbar switching is accomplished in stages in a manner similar to the Strowger system as shown in Figure 14–15(b).

Suppose that the top subscriber in group A (left side of the figure) wishes to contact the top subscriber in group B. The first stage of the switching network would connect the calling subscriber to one of the *intraoffice trunks*. Suppose that this happens by closing switch A_1 within the *first stage*. This would connect to the intraoffice trunk leading to row C in the second-stage matrix. The second stage matrix might close switch C_3, connecting the line to column 1 of the *third stage*. At the third stage, switch A1 would close to connect to the target party.

Figure 14–15
Crossbar switch action

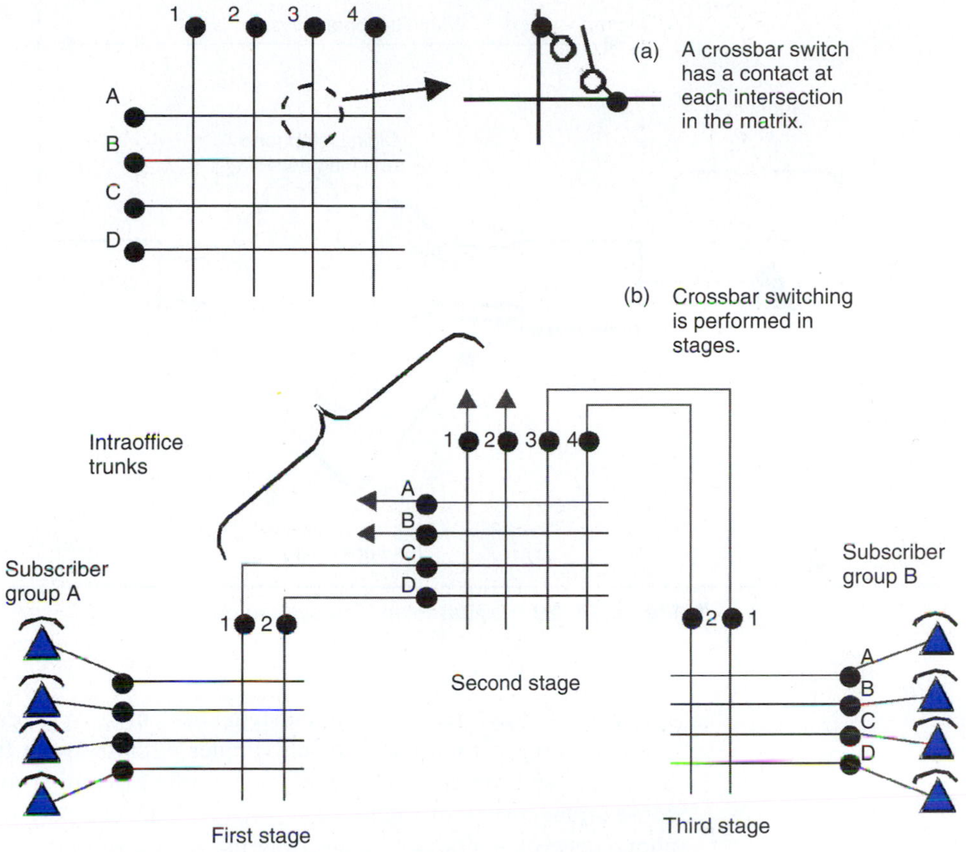

Notice that there are several inherent limitations built into this switching system. One of them is that no more than two subscribers on either end may use the phone at the same time, since there are only two trunks serving their groups. (If a third subscriber in a group tries to make a call, he or she will get no dial tone.) This is the same as 50% utilization of the group—in reality, a typical central office can only handle 10–20% utilization of its subscriber lines. It's also possible that upon arrival to the second stage, none of the outgoing intraoffice trunks to group B may be available—the call will *block* with a rapid busy signal. Finally, this switching system is still an analog one. Each time the signal is relayed through a switching matrix, the quality is degraded. Crossbar switches are much more reliable (and electrically quiet) than the Strowger devices, but they still wear and add electrical noise. In areas served by a crossbar switch, a subscriber occasionally gets a bad connection and remedies the problem by hanging up and calling back. Someone else will eventually get the "bad" intraoffice circuit!

Digital Telephony

Each time a signal is repeated within an analog system, it's degraded slightly. If the signal is sent over a distance, it weakens and picks up external noise from the environment, which means that it must be amplified back to original strength by a *repeater*. The signal-to-noise ratio suffers greatly on long-distance connections. Digital systems don't

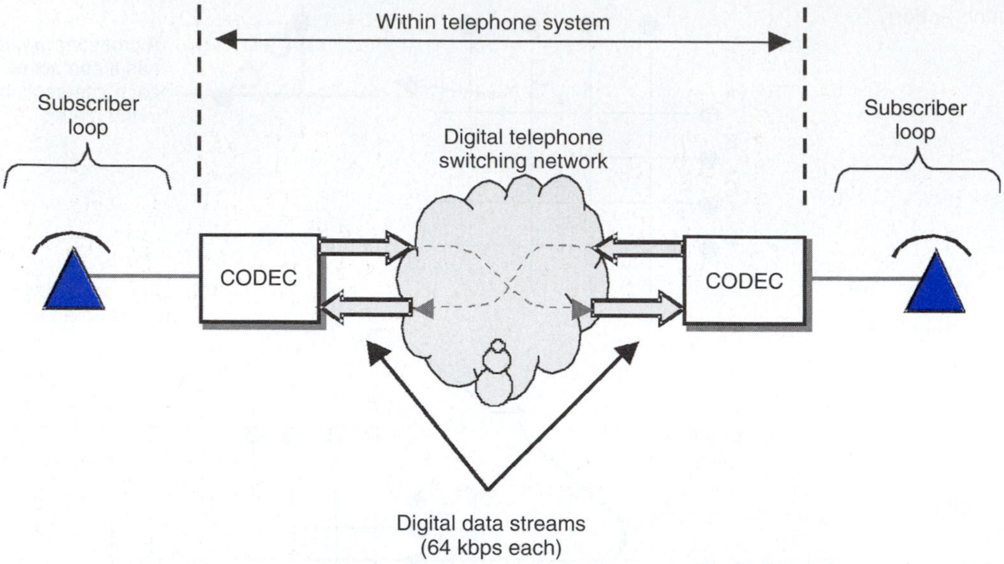

Figure 14–16 Digital telephone system operation

suffer in this way; because a digital signal has only two states, high or low, it can be repeated (or regenerated) with much simpler circuitry with little or no loss in quality. Switching and routing can be greatly simplified by going digital as well; the cumbersome arrangements of Figures 14–14 and 14–15 can be eliminated. The modern telephone network operates by converting the analog subscriber signal to digital, passing it through the telephone network as digital, and then converting it back to analog on the last part of the trip, as shown in Figure 14–16.

In the figure, the subscriber loop provides an analog signal to the *coder–decoder,* or *CODEC,* which is a component on the SLIC at the end office. The job of the CODEC is twofold: First, it converts analog signal coming from the customer on the local loop (which is still configured as shown in Figure 14–3) into a stream of digital data at 64 kbps. Second, the CODEC receives an incoming digital data stream (again at 64 kbps) from the telephone network, converts it back into an analog representation, and places it back onto the local loop. The CODEC therefore contains an *analog to digital converter* and a *digital to analog converter,* plus some necessary filtering circuitry.

EXAMPLE 14–4

What total (aggregate) data rate is required to represent both ends of a telephone conversation, assuming that each side requires 64 kbps?

Solution

The total data rate is the sum of the two bit streams. Therefore, the resulting data rate is two times 64 kbps or 128 kbps. This may not sound like much by today's standards, but consider the burden of carrying (and switching) 2,000 such conversations simultaneously in a typical end office—this is not a trivial problem!

Figure 14–17
Sampling a sine wave information signal

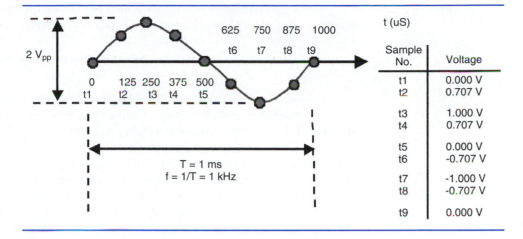

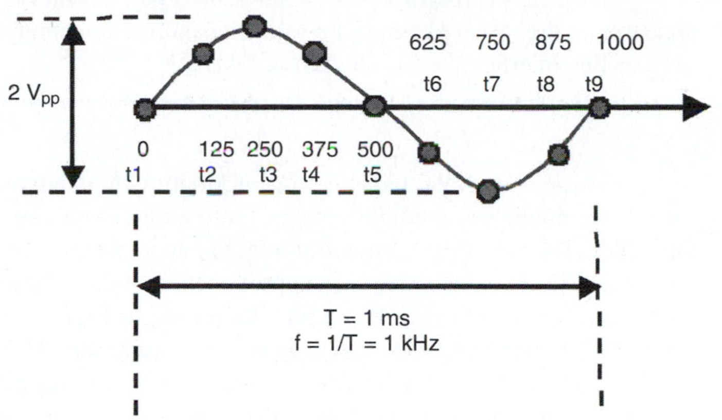

Sample No.	Voltage
t1	0.000 V
t2	0.707 V
t3	1.000 V
t4	0.707 V
t5	0.000 V
t6	-0.707 V
t7	-1.000 V
t8	-0.707 V
t9	0.000 V

Analog to Digital Conversion: Pulse Code Modulation

Pulse code modulation, or PCM, is the standard choice for representing analog signals in the modern telephone system. It is also the basis for almost all computer sound reproduction. Compact discs also use PCM to represent sound. To understand how it works, you need to understand two concepts, *sampling* and *quantization*. *Sampling* is the periodic measurement of a continuous or analog signal. The idea behind sampling is to represent an information signal in abbreviated form by taking periodic snapshots of the waveform's voltage. A motion picture works the same way as a sampled waveform—it consists of a set of picture frames (analogous to the waveform samples). Because the film's frames are shown in rapid sequence, they appear to recreate the continuous motion of the subject. Figure 14–17 illustrates this concept.

In the figure, a measurement is being made every 125 μs on the analog waveform. These samples are snapshots of the instantaneous voltage on selected parts of the waveform. The table data in Figure 14–17 represents the original waveform in abbreviated form. It doesn't completely represent the signal, but instead it presents a close approximation of it—good enough that if the waveform were a speech signal instead of a sine wave, most people wouldn't be able to hear the difference between the sampled and original signals. The *sampling rate* of a PCM signal is the reciprocal of the time between samples:

$$f_s = \frac{1}{T_s} \qquad \textbf{(14–1)}$$

where f_s is the sampling frequency, in Hz, and T_s is the time between samples. This is just another variation of the "time is one over frequency" idea.

EXAMPLE 14–5 What is the sample rate in Figure 14–17?

Solution

Equation 14–1 calculates sampling rate:

$$f_s = \frac{1}{T_s} = \frac{1}{125\ \mu\text{s}} = \underline{\underline{8\ \text{kHz}}}$$

There are 8000 samples being measured per second on the waveform. It turns out that this is the standard sampling rate for landline telephony. The CODEC on the subscriber line interface card is clocked at 8 kHz.

Quantization is the reduction of information in an analog signal sample (which has an infinite number of possible voltages) into a digital number value with a finite number of values. The waveform samples in a PCM-encoded signal are sequences of digital number values. Each digital value is an approximation of the voltage at that instant of the waveform, and has a fixed number of bits. Returning to Figure 14–17, you can see that four decimal digits are being used to represent each voltage sample. That's a lot of information—in fact, much more than we need to reproduce speech in a telephone network. Figure 14–18 shows the effect of quantization on a continuously varying voltage.

The bottom waveform of Figure 14–18 has been quantized to three bits per sample. Because there are three bits per sample, we know that there are only eight states (voltages) possible on the waveform, as described by equation 14–2:

$$\text{MOD} = 2^N \qquad (14\text{–}2)$$

where MOD is the modulus (number of states) in the system, and N is the number of bits per sample. The important idea is this: Quantization reduces the information in a sample by limiting its representation to a fixed, limited number of possible values. The quantized signal is a digital approximation of the original. Three bits per sample isn't very much information; in fact, speech reproduced with only three bits per sample would sound distorted and harsh. The telephone system uses eight-bit samples, which is more than adequate for speech (most people can't tell the difference between an eight-bit digital PCM signal and the original speech signal).

Figure 14–18
Quantization of a signal to three bits per sample

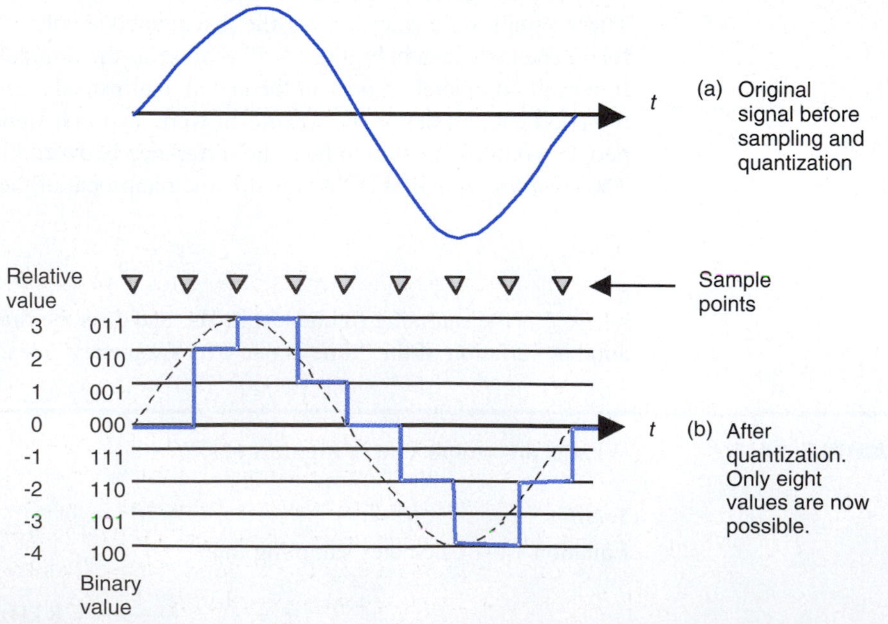

EXAMPLE 14–6 If the sampling rate is 8 kHz in Figure 14–18, what will the resulting data rate be if there are (a) three bits per sample and (b) eight bits per sample?

Solution

To find the data rate *DR,* simply multiply the number of bits per sample and samples per second so that the units work out correctly:

a. $$\text{DR} = \frac{3 \text{ bits}}{1 \text{ sample}} \times \frac{8000 \text{ samples}}{1 \text{ sec}} = \underline{\underline{24 \text{ kbps}}}$$

b. $$\text{DR} = \frac{8 \text{ bits}}{1 \text{ sample}} \times \frac{8000 \text{ samples}}{1 \text{ sec}} = \underline{\underline{64 \text{ kbps}}}$$

Increasing the number of bits per sample increases the required data rate for the signal. Telephone grade signals use a sampling rate of 8 kHz and 8 bits per sample, resulting in a data rate of 64 kbps for each side of a conversation.

Determining the Sample Rate: Nyquist Theorem

To correctly represent a waveform, we must sample it often enough, or an undesirable effect called *aliasing* will occur. Aliasing is the misrepresentation of the frequency of the reproduced signal. The minimum frequency of sampling is predicted by the *Nyquist theorem:*

$$f_s \geq 2f_{m(\text{max})} \tag{14–3}$$

where f_s is the sampling frequency, and $f_{m(\text{max})}$ is the maximum frequency in the composite information signal. You can see aliasing in movies; just watch the wheels on a car as it accelerates from a stop. They'll appear normal until a certain speed, at which point the wheels will appear to suddenly reverse direction, and as further speed is attained, they may even appear stopped momentarily. The Nyquist theorem tells us how fast we must sample analog signals.

EXAMPLE 14–7 What is the minimum sampling rate to carry a frequency range of 300 Hz to 3 kHz (normal speech range for telephony)?

Solution

Use the Nyquist theorem (equation 14–3). The *highest* information frequency is the important value; if we reproduce the highest information frequency correctly, then all lower frequencies will also be represented correctly.

$$f_s \geq 2f_{m(\text{max})} \geq 2(3 \text{ kHz}) \geq \underline{\underline{6 \text{ kHz}}}$$

Notice that we didn't get the value of 8 kHz, which is the sampling rate used in the telephone system. Why not? The reason is that we must use a low-pass *anti-aliasing filter* before the analog-to-digital converter within the central office CODEC. This filter doesn't sharply cut off at 3 kHz; instead, it gradually rolls off. Therefore, most PCM systems sample at a frequency more than twice the maximum frequency to be reproduced, so that the low-pass filter design becomes manageable.

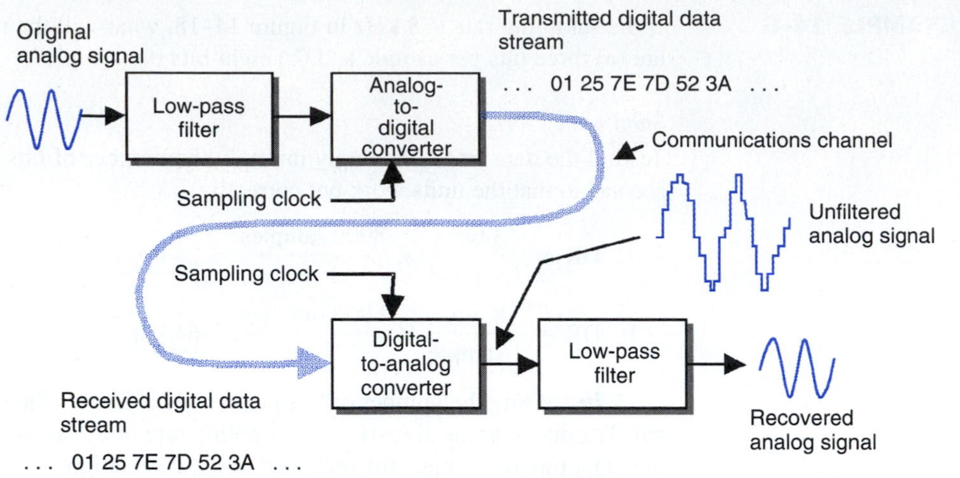

Figure 14–19 PCM encoding, transmission, and decoding

PCM System Block Diagram

Figure 14–19 shows how PCM is applied to encode and recover a voice signal. The process begins with the voice signal being passed through a low-pass filter called an *anti-aliasing filter*. This filter removes any frequency components higher than $f_s/2$, to prevent aliasing distortion.

The analog signal is then sent into the *analog-to-digital converter,* which takes a snapshot of the signal for every pulse of the *sampling clock*. The output of the ADC is a series of binary numbers (eight-bit values in this example, shown in hexadecimal). This stream of digital information can be sent along a communication channel using any of the methods described in Chapter 15, or it can be stored on a computer (hard drive, CD, or other media). The PCM demodulator reverses the process. It uses a *digital-to-analog* converter to change each digital value from the data stream back into an analog voltage. These voltages strongly resemble the original information signal, but they have square edges (sometimes called "jaggies") from the sampling and quantization process. These edges represent unwanted high frequency information, so we remove them by using a low-pass filter, sometimes called a reconstruction filter. The final result is a copy of the original information.

This process looks a lot more complex than analog transmission, so there must be some very good reasons why telephony is now done this way. First, transmitting digital signals produces little or no degradation of the original signal, no matter how far they're sent. Therefore long distance calls can have the same sound quality as local calls. Second, it's easy to construct memory circuits that can store and forward these signals, leading to advanced services such as paging and voice mail (which store the PCM signal in a computer file). Finally, it's much more efficient to route digital information signals to a destination using software (instead of the analog switching methods such as the crossbar switch). Let's take a look at how we can route digital voice signals.

Digital Voice Signal Routing

Once a voice signal has been reduced to mere digital data, it can be routed in a variety of ways, and it can also share the same communications channel with other conversations. Figure 14–20 shows how two conversations can be carried using time division multiplexing.

Figure 14–20 Time division multiplexing for two conversations

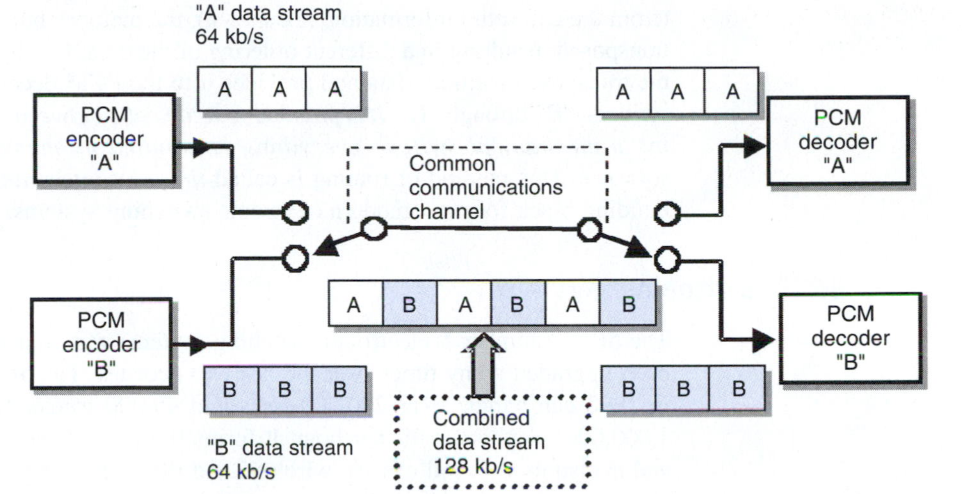

In the figure there are two PCM encoders, each supplying a 64 kbps data stream to the multiplexer switch at the left. The switch alternately transmits one packet (group of data bytes) from device A, then from B. The resulting data stream contains both "A" and "B" information. The resulting data rate is the combined total of sources A and B, resulting in a requirement of 128 kbps. At the right, a demultiplexer removes packets of data from the line in step with the multiplexer (the dashed line indicates that they are synchronized). Therefore, packets are correctly routed to each destination device.

Each half of a telephone conversation is represented by a stream of data bytes. Therefore, it is possible to route calls to correct destinations by merely rearranging the order of the data in a memory buffer, as shown in Figure 14–21.

In this figure, there are four ongoing connections between the subscribers on both sides of the picture. Multiplexer "A" reads the data from PCM encoders "A" through "D"

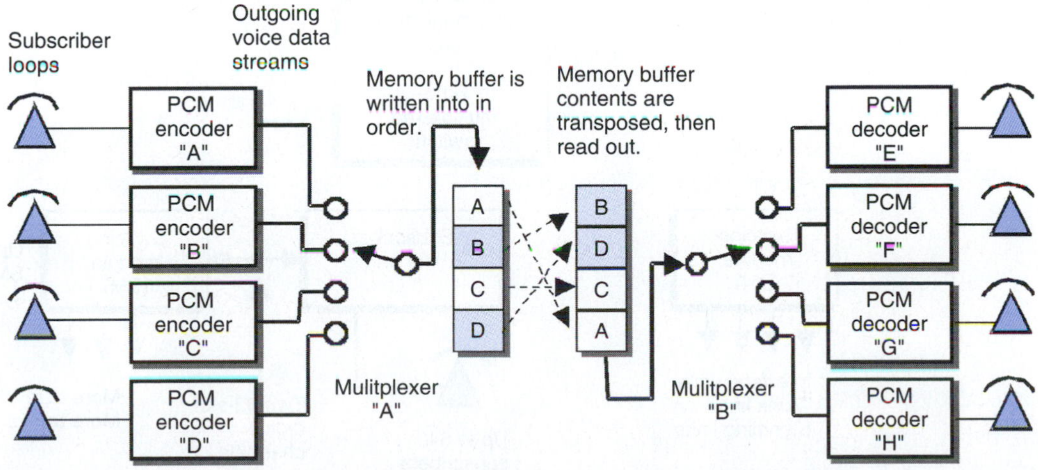

Figure 14–21 Digital call routing using time slot switching

one at a time and places it into a memory buffer. According to the connections desired (from the call setup information, not shown), the memory buffer locations are swapped or transposed, resulting in a different ordering of the data. Finally, multiplexer "B" reads out the transposed memory data and provides it to the PCM decoders for the right-hand subscribers, "E" through "H." *It is possible to route calls between subscribers without ever using a mechanical switch—everything is handled by memory buffers and computer software.* This method of routing is called *time slot interchange,* and it is a fundamental building block for most modern electronic switching systems.

Lucent 5ESS Switching System

The 5ESS (number 5 electronic switching system) was first introduced in 1982 and has been upgraded many times over the last two decades. The 5ESS-2000 switch introduced in 1998 can handle over 200,000 lines and 45,000 trunks simultaneously and support 1,000,000 telephone calls per hour. It forms the core of most Bell company end offices, and is also used in toll offices, wireless, and PBX (private branch exchange) systems for large companies. Its block diagram is shown in Figure 14–22.

The heart of the 5ESS is the *communication unit,* which contains a time-multiplexed switch that operates on the same principle as Figure 14–21 (just on a much larger scale). The communication unit is divided into two *planes,* each serving 190 ports of 512 time slots each. The communication unit supplies a *master clock* signal that synchronizes the entire switching system; it may have up to 190 *switching modules* connected. Each switching module communicates with the communication module by two pairs of fiber-optic cable (one pair for each direction of communications). Each of these cables carries a maximum of 512 *slots* every 125 μs (4.096 Mslots/sec). A slot is a single sample from a voice channel, and it consists of a 16-bit word: 8 of the bits are the PCM voice data, and

Figure 14–22 Lucent 5ESS digital switching system

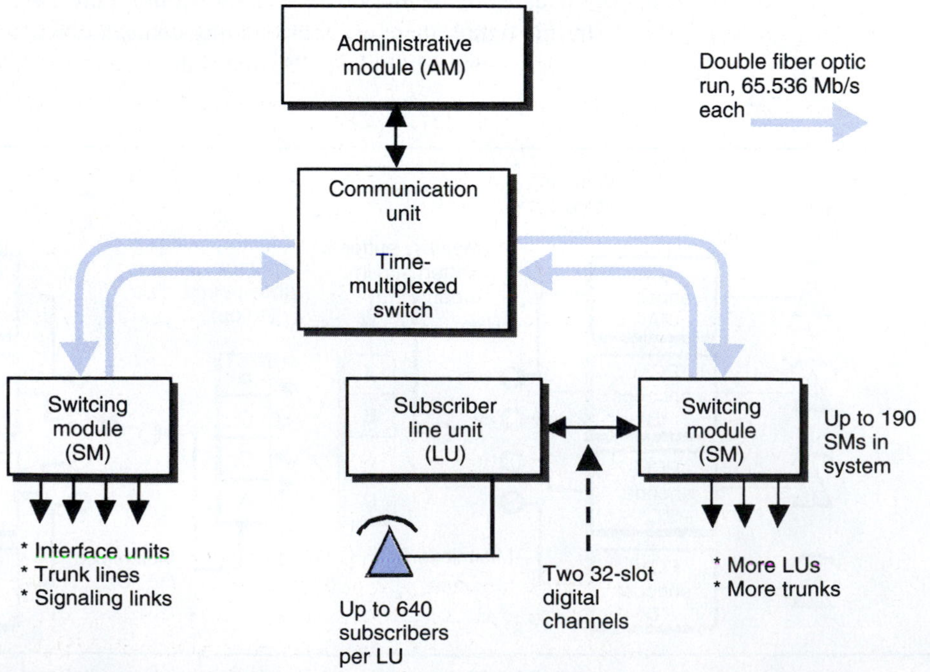

the other 8 are used for channel management (signaling, idle/busy, on/off hook status, and so on). Each of the fibers therefore carries a continuous traffic of 65.536 Mbps (mega bits per second) and a total of 1024 active conversations can be carried by each switching unit (there are 1024 incoming slots, and 1024 outgoing slots available through the fiber-optic interface to the communications unit).

A variety of options can be connected to each switching unit. For example, one or more *subscriber line units* (LUs) can be connected to the switching unit. Each LU uses 64 of the available 512 slots, and furthermore, each LU can be programmed to concentrate from 256 (4:1 concentration) to 640 (10:1 concentration) subscriber loop lines onto its pair of 32-slot communication channels. *Concentration* is used to conserve telephone company facilities. For example, with 10:1 concentration, no more than 10% of the lines leaving the LU may be active at any given time. When more than 10% of the subscribers attempt to communicate under this condition, the excess users are refused service (no dial tone). Switching units may also lead to digital trunk lines for communications outside the call center.

Finally, the *administrative unit* monitors the activity of the switching network. It's responsible for billing, call translation, trunk routing, and other functions.

Section Checkpoint

14–16 Describe the operation of a Strowger and crossbar switch. What is the primary limitation of these devices?

14–17 Before a telephone circuit can be digitally switched, what must be done with the analog voice signal coming from the subscriber?

14–18 Define *pulse code modulation*.

14–19 Explain the difference between *sampling* and *quantization*.

14–20 What is the Nyquist theorem?

14–21 What happens when a sampled system has a signal introduced at a frequency in excess of $f_s/2$?

14–22 Why is a low-pass filter needed before analog-to-digital conversion in a sampled system (such as PCM)?

14–23 Explain how time-division multiplexing can be used to route signals digitally in the telephone system.

14–24 Explain the function of each portion of the Lucent 5ESS switching system shown in Figure 14–22.

14–4 WIRELESS TELEPHONY: AMPS AND PCS, CDMA, GSM, TDMA

When the first wireless telephones were placed into service in the 1960s, service was very limited and expensive. A *supercell* model was used as shown in Figure 14–23(a). The base station equipment would be installed on a tower near the center of the city to be served, and perhaps 20 different frequencies between 150 MHz and 160 MHz were supported. This meant that only 20 different mobile phones could be in use at any time in the area,

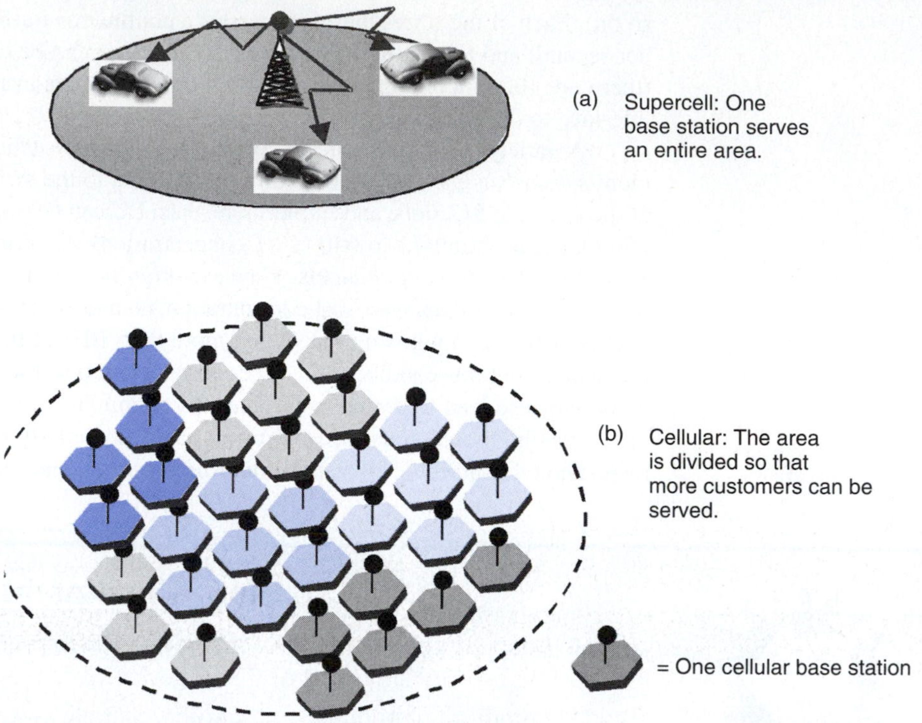
(a) Supercell: One base station serves an entire area.
(b) Cellular: The area is divided so that more customers can be served.
= One cellular base station

Figure 14–23 Supercell and cellular approaches to communication

which could hardly satisfy the wireless communication needs of a large population. There had to be a better solution!

In the 1940s, Bell Labs had formulated the basic ideas behind *cellular* telephony, but it would not be until the 1980s that the appropriate technology became available to implement the idea. A cellular telephone system works by dividing a coverage area into small working units called *cells,* as shown in Figure 14–23(b). There are several benefits to the cellular approach. First, *frequency reuse* becomes possible. More than one caller can use the same frequency as long as they are in separate, nonadjacent cells. This increases the number of users the system can support. Second, since each cell serves just a small area (in comparison with the supercell approach), the power requirements for the base station and mobile units are greatly reduced. For mobile units this translates into improved battery life. Finally, the failure of a single base station is much less critical in a cellular system; the adjacent cells will pick up most of the traffic from the dead cell until repairs can be made. In a supercell system, base station failure disables the entire system.

There are two primary cellular systems in use in the United States. These are advanced mobile phone service (AMPS), an analog system dating back to 1983, and PCS, a digital system introduced in 1996. AMPS is largely falling from use since it can't support as many users as PCS and can't provide many digital services to users.

Multiplexing Methods

All cellular telephone systems rely on *multiplexing* to service multiple customers at the same time. Multiplexing is the transmission of multiple pieces of data over a common channel. Three primary multiplexing methods are used in cell phones—*frequency division, code division,* and *time division.* Frequency division multiplexing uses a different RF carrier frequency for each conversation, just like radio uses a different carrier frequency for each station. Code division multiplexing makes use of the special properties of certain digital codes; these codes are called *orthogonal* codes because a desired code can be distinguished from a large group of undesired codes because it appears mathematically perpendicular to the other codes and thus "stands out." Finally, time division multiplexing uses the same frequency range, but assigns time slots to different streams of information. A TV station uses time division multiplexing to send its programs. Each program has a specific time slot for broadcast; thus, a viewer can find the correct program by tuning in during the correct time slot.

AMPS Telephony

In 1975 the FCC reallocated the frequency space used by UHF TV channels 70–83 for AMPS. The FCC designed this service to foster competition between various phone companies—two companies could serve each area. Thus, the AMPS frequency spectrum (Figure 14–24) is divided into "A" and "B" bands. The "B" bands are for use by the local phone company or *ILEC* (think "Bell") and the "A" bands are for use by a second carrier (think "alternative carrier"). The bands marked A′ and B′ are merely continuations of the

Figure 14–24 AMPS frequency allocations

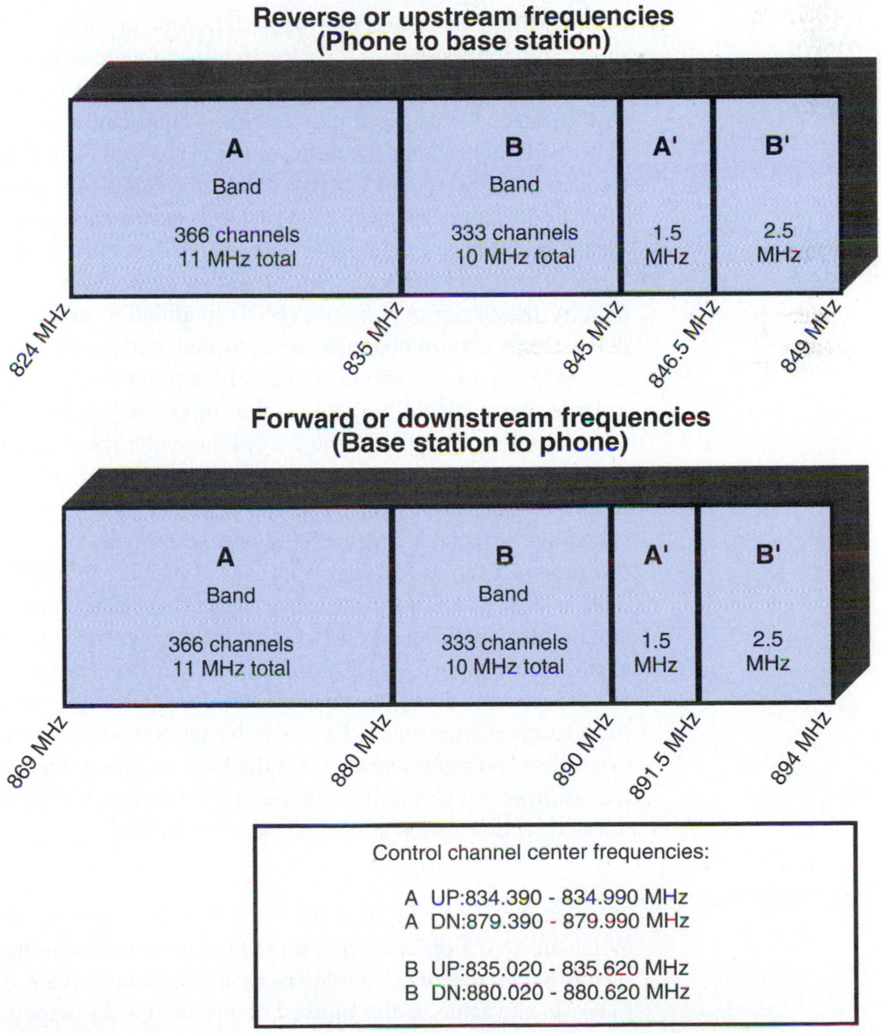

Figure 14–25 AMPS typical frequency reuse pattern

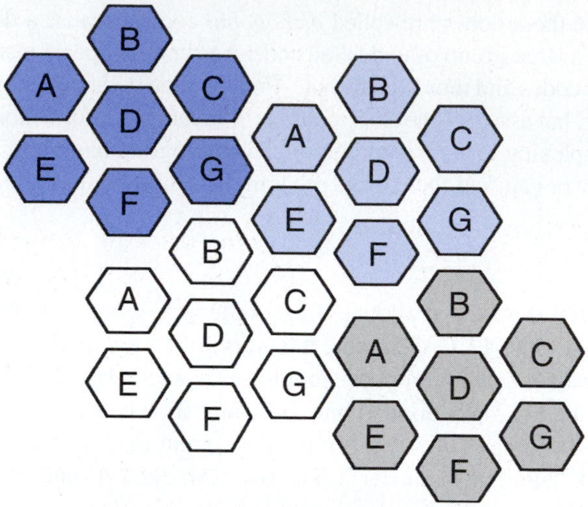
A
B
C
D
E
F
G
The seven cell pattern is repeated throughout the area of coverage, depending on the geography of the area. All cells are not necessarily hexagonal.

A and B allocations, and may be 5 MHz wide in some markets. There is a 20 MHz gap between the downstream and upstream allocations that is unused.

AMPS uses *frequency division multiplexing* to separate the conversations within a cell. In other words, each mobile unit is commanded to listen and transmit on a different pair of frequencies by the base station in the cell. The frequency "split" between the two frequencies is always 45 MHz. Within the initial A and B forward (downstream) allocations in the figure are also 21 dedicated *control channels*. The control channels provide a means for controlling the mobile phones in the cell. These channels are located at the juncture of the A and B allocations, as shown in the figure. The control channels carry digital data by *frequency shift keying* (FSK) modulation of the FM carrier signals. Both up and downstream control channels are provided, and again, they are split by 45 MHz.

Figure 14–25 shows a typical frequency reuse scheme used for AMPS. In this scheme, the available frequencies are split among groups of seven cells. These frequencies are labeled A through G. The 21 available control channels are also split seven ways, so that there are three control channels allocated for each cell.

A block diagram of an AMPS phone is shown in Figure 14–26. It contains a microcontroller, a modem for receiving and sending telecommands on the control channel, an FM receiver, and an FM transmitter. The channels in AMPS are 30 kHz wide and FM modulation is used to carry the voice signals. AMPS phones are full-duplex devices—they simultaneously listen on one forward frequency, and transmit on one reverse frequency. To allow this to happen, a duplexer is used. The *duplexer* is a filter that allows the transmitter and receiver to share the same antenna. It contains low- and high-pass filters. The low-pass filter receives transmitted RF from the set and passes it to the antenna, and the high-pass filter sends downstream RF from the base station to the receiver while blocking transmitted RF from the phone from getting into its receiver. The 45 MHz separation of the transmit and receive bands allows this scheme to work.

AMPS Operation

When an AMPS phone is first turned on, it begins listening for the strongest control signal it can find on one of the 21 downstream control channels, either in the "A DN" or "B DN" range in the table at the bottom of Figure 14–24, depending on which carrier (Bell or

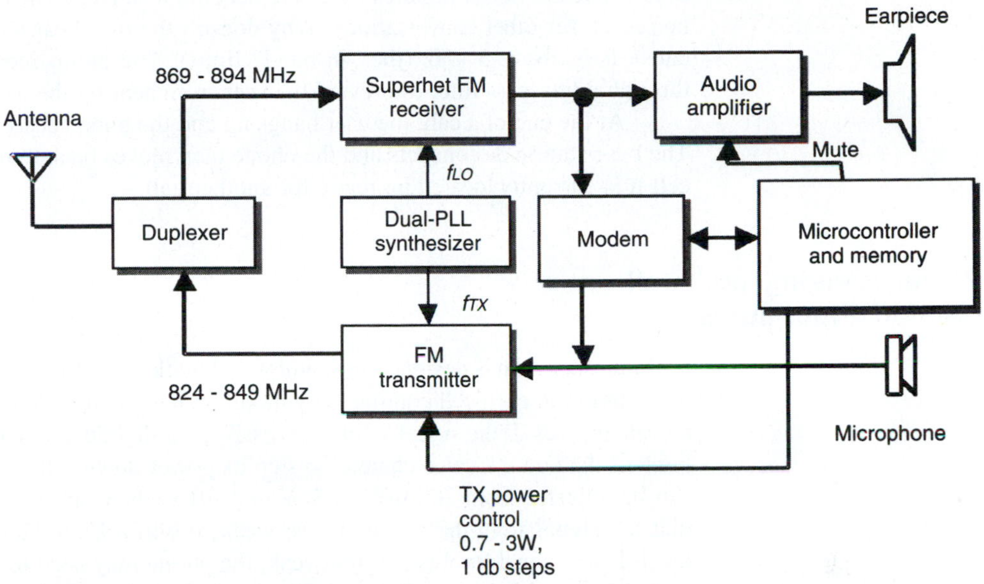

Figure 14–26 AMPS telephone set block diagram

alternative) the phone is programmed for. Presumably the strongest control signal will be the nearest cell, though this is not always so. The phone's receiver is tuned by half of a dual phase-locked loop (PLL) synthesizer chip (PLLs are discussed in detail in Chapter 7). The PLL tunes the receiver by providing the correct local oscillator frequency. The other half of the dual PLL is programmed 45 MHz "up" to provide the transmit frequency of the phone. The difference between receive and transmit frequencies is always 45 MHz.

Once the phone has located the correct downstream control channel, it collects information transmitted by the base station including the SID (system ID code, a five-digit number identifying the carrier providing service) from the base station. It also *replies* to the base station with a *registration request* containing two pieces of information, the mobile identification number or MIN (a 10-digit number derived from the phone number that identifies the mobile phone), and the electronic serial number or ESN (a unique 32-bit serial number that identifies the phone). By comparing the SID received from the base station with the SID programmed into the phone, it can be determined if the phone is "home" or "roaming." The registration request is submitted into a national database so that the telephone system as a whole learns where to route calls for the mobile.

If a call is incoming for the mobile phone, the base station transmits the appropriate commands on the control channel to the mobile, which in turn outputs a ringing signal. If the call is answered, the mobile sends this information back to the base—which then allocates a channel pair for the conversation and commands the phone to move to the new pair of frequencies. The phone remains on the control channels until the conversation starts.

Out-of-band analog signaling is used to control and monitor the progress of a call. The base station transmits a supervisory audio tone (SAT) every 250 ms on the downstream channel. This tone is either 5,970, 6,000, or 6,030 Hz, according to the SAT color code, SCC, transmitted to the mobile at the beginning of the connection. The mobile must reply to the SAT with the correct frequency or face disconnection from the network. The

SAT is used to detect mobiles that have become inactive, so that valuable frequencies can be reused for other conversations. Why doesn't the user hear the SAT? The reason is because it is above 3 kHz (the "in-band" limit). The audio feeding the earpiece passes through a low-pass filter to prevent the user from hearing the SAT data.

At the end of a call, the user hangs up and the phone ceases repeating the SAT tone. The base station disconnects and the phone then moves back to the control channel for the cell it is currently located in, ready for another call.

Power Ranging and Cell Handoff Procedures

Cellular operation is designed to seamlessly handle movement of users between cells. The base station in each cell continuously monitors the received signal from each of the active mobile phones. If the signal level is excessive, the mobile phone is instructed with a muted burst of data on the voice channel to step its power down. The power of the mobile phone can be adjusted from 700 mW to 3 W in 1 dB steps. Conversely, if the base station sees that the signal from the mobile is too weak, it will ask the mobile to increase its power level. If the signal level is still too weak, the phone may need to be handed off to an adjacent cell. The adjacent cell's base stations are queried to find out which one is getting the best signal strength from the mobile unit; the call is handed over to the adjacent cell's base station by first allowing the adjacent cell to choose the new frequency of operation (adjacent cells don't use the same sets of frequencies). Once the new frequency is known, this data is passed to the mobile unit as a rapid (< 250 ms) data burst on the voice channel. The mobile unit then shifts to the new frequency where the new base station will be waiting. The handoff process is nearly invisible to the user. Note that the AMPS telephone set includes a mute circuit. The FM detector in the set detects *both* the analog voice signal as well as the digital FSK data signal. To prevent the user from hearing the data signal, the mute circuit is enabled whenever data is to be sent.

EXAMPLE 14–8

How many active customers can be supported by *each carrier* (alternative or Bell) in each cell of Figure 14–25, assuming that the 416 available channels are divided equally among all seven cells in the pattern?

Solution

The 416 voice channels are divided in this way:

- 21 control channels (three per cell)
- 395 voice channels—(395/7) or 56 per cell (the fraction is discarded.)

Therefore, 56 active conversations can be supported in each cell by each carrier assuming that the channels are equally divided between the seven cells in the pattern. If the customers are active (talking) 10% of the time, the system can support 500 customers in each of the cells. Even though this is certainly a lot more users than a supercell system could support, it still isn't enough to meet the demand for service is most metro areas. One way of getting around this problem is to subdivide the busier cells into *microcells,* as shown in Figure 14–27. Also, *ad hoc* cells can be set up and turned on for special conditions, such as in a sports stadium parking lot to handle the extra traffic after the big game.

Figure 14–27 Cell splitting

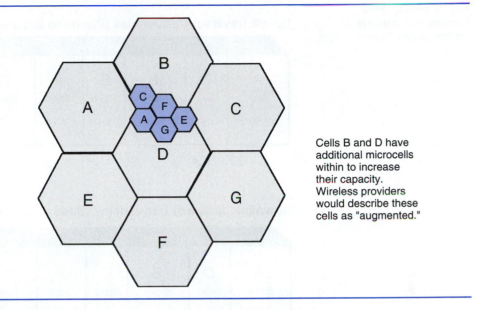

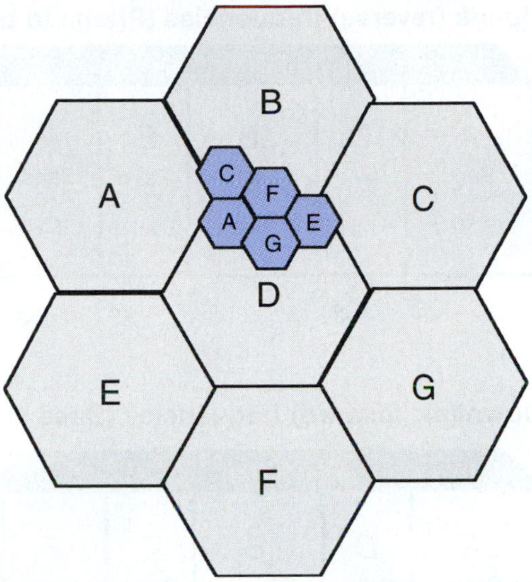
B
A
C
D
E
G
F
C
F
A
E
G
Cells B and D have additional microcells within to increase their capacity. Wireless providers would describe these cells as "augmented."

The microcells in the figure operate at lower power than the main cells. Their frequency usage is carefully chosen to prevent interference with existing cells.

Sectorization

Another way of boosting cell capacity is to *sectorize* each cell. Most AMPS cells are divided into three sectors, each with its own antenna and frequency block. (This is also why most cell towers have three groups of antennas.) The use of directional antennas (each with about 120° of beamwidth) also helps to reduce interference from other cells.

Personal Communications Service (PCS)

By 1990, the demand for cellular service in the United States exceeded the capacity of AMPS. The FCC allocated additional spectrum for a new service called the *personal communications service* or PCS. PCS differs from AMPS in several important ways. First, PCS is a digital service from the ground up. The voice signal is converted to and from digital in the handset before transmission and reception. Second, PCS uses *both* frequency division multiplexing plus either *code division multiplexing (CDMA)* or *time division multiplexing (TDMA)* to separate conversations within a cell. Finally, PCS is designed to support much more than voice service. Because it is a purely digital service, the telephone sets can serve as portable computer terminals and run networked applications, such as text messaging, games, and business applications. Software applications for mobile phones are already being created in a compact form of the Java language called *J2ME,* Java 2 micro edition. The frequency allocations for PCS are shown in Figure 14–28.

One of the basic differences between PCS and conventional AMPS cellular is that the FCC has divided the spectrum into six distinct slices per market area instead of the two used for AMPS. The idea is to further stimulate competition by allowing more providers (up to six) into a market area. The A, B, and C blocks are only slightly larger than the A

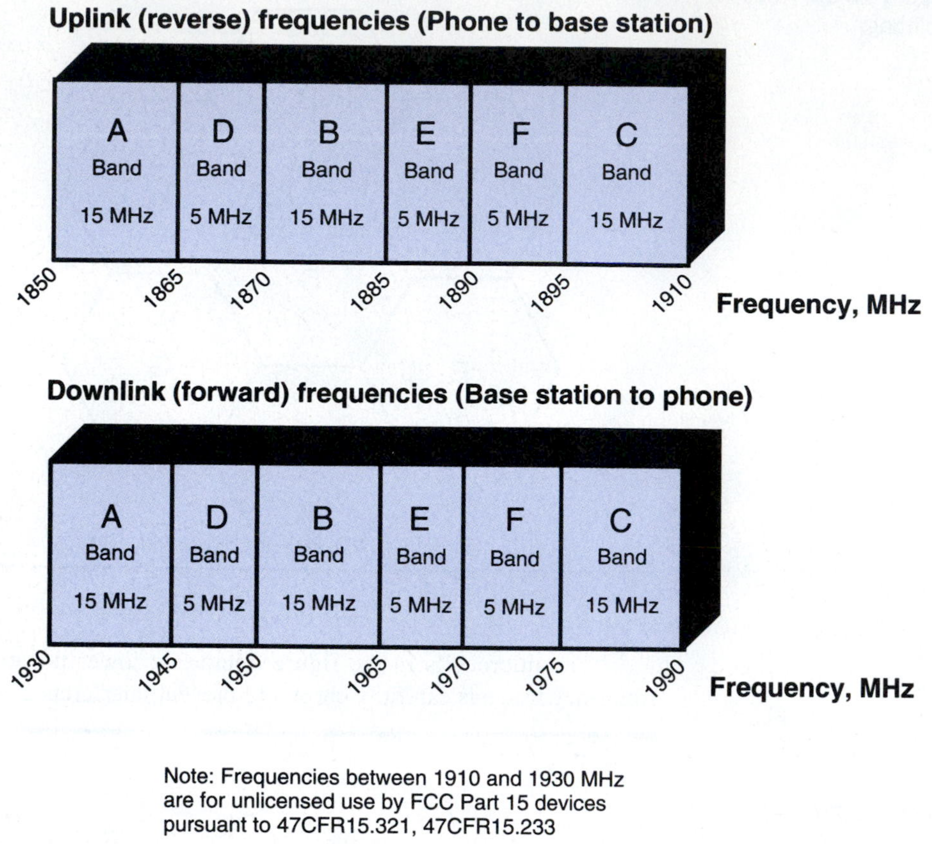

Figure 14–28 PCS frequency allocations for the United States

and B blocks of AMPS (15 MHz versus 12.5 MHz). Yet how can PCS serve so many more customers? There are two reasons for this. First, PCS cells occupy a smaller area than AMPS cells and therefore can use lower power. (This also helps extend the battery life of telephones, since they are closer to the base stations and can use lower transmit power.) A typical AMPS cell is about 5 miles in diameter; a PCS cell is usually under two miles in diameter. This is also why PCS cells are sometimes referred to as *microcells*. Second, PCS uses digital speech compression techniques prior to modulating the carrier so that more conversations can be carried using the same frequencies as compared to analog FM. Digital speech transmission also improves security by making it nearly impossible for third parties (such as scanner enthusiasts) to intercept conversations. Two incompatible PCS systems are currently in use within the United States, CDMA (code division multiple access) and GSM (global system for mobile communication), which uses TDMA.

CDMA

CDMA is a counterintuitive approach to communications. It works by *spreading* the information signal across a relatively wide frequency range with the use of a pseudo-random *signature code*. The signature code appears random, but it isn't; it's quite predictable if you know the sequence. But importantly, the code looks a lot like *wideband random noise* in the *frequency domain*. We would say that the transmitted information has been *spread* in

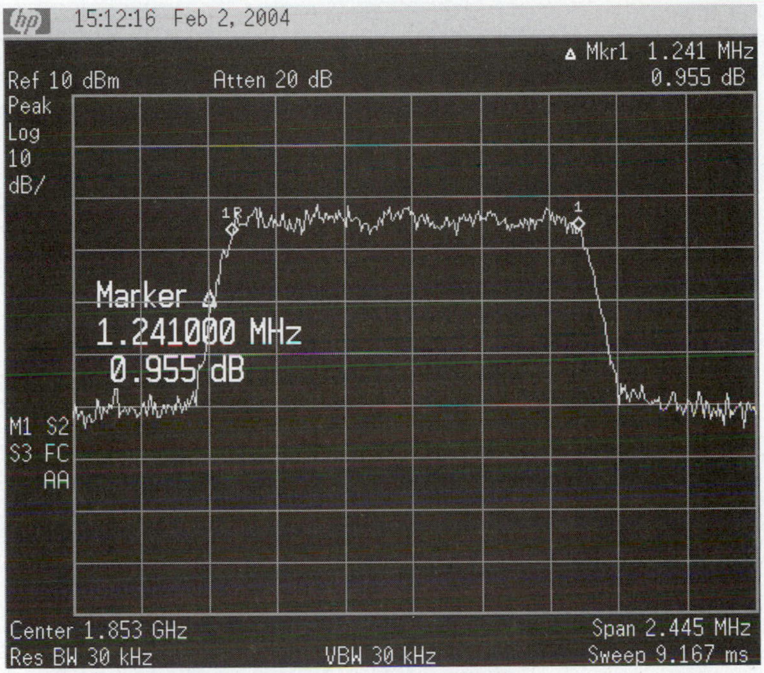

Figure 14–29 Typical spectrum from a CDMA handset. Spectrum analyzer set to peak hold mode

the frequency domain. The resulting signal is called *direct sequence spread spectrum* or DSSS. A CDMA transmitter mixes (actually, exclusive-ORs) its transmitted data stream with the signature code prior to transmission in a process called *chipping,* which results in a wideband transmission that is hard to distinguish from background noise. The spectrum from a typical CDMA handset is shown in Figure 14–29; it is about 1.25 MHz wide.

To decode a spread-spectrum signal, a CDMA receiver synchronizes itself to the transmitted signature code sequence and again exclusive-ORs the regenerated signature code with the "random" received signal. This causes the signal to *collapse* back to its original frequency domain form where it can then be processed back into the original data stream (which can then be converted back into sound, a web page, or other desired information). *It's important to understand the receiver gets nothing if it uses the wrong code, or gets the code out of sequence.* In fact, many different conversations can share the same range of frequencies by mixing each data stream with a different code. Figure 14–30 shows the downstream processing at a CDMA base station.

CDMA Signal Processing

A CDMA cellular base station can transmit up to 64 different channels of information simultaneously on each 1.25 MHz downstream channel. A special set of digital codes called *Walsh codes* are used to multiplex all of these channels onto a single carrier. Walsh codes allow the information to be placed together on the carrier and then later separated by the receiver. Sixty-four different Walsh codes are used, called W0 through W63. W0 is the *pilot* code. It contains no data, just a stream of binary 0 values mixed with the Walsh code. The mobile uses the W0 code channel to synchronize its internal Walsh code generator with the base station (remember that in order to recover any data,

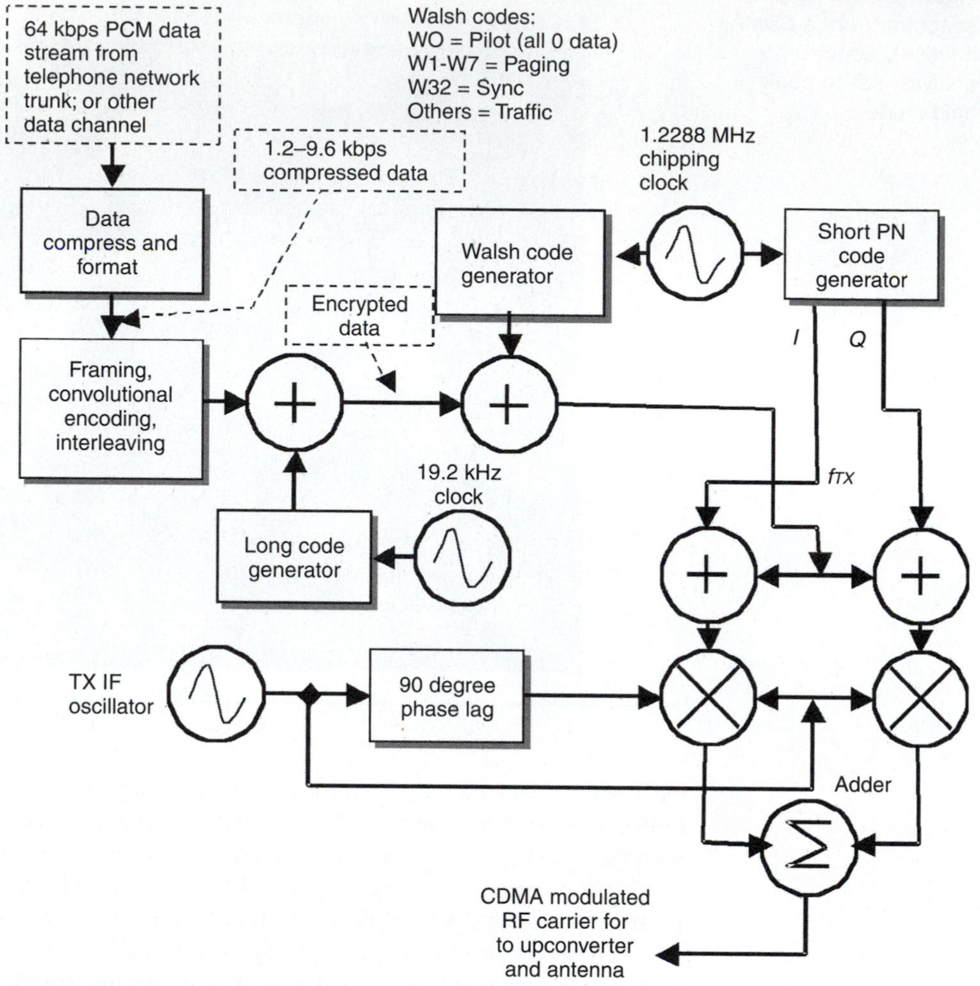

Figure 14–30 CDMA base station downstream (forward) channel processing: IS-95

the code sequences of the transmitter and receiver *must* be synchronized). Once synchronization has been obtained, code W32, the *synchronization channel,* can be read. Channel W32 continuously transmits the system time, system identification, and the state of the Long Code (which is used for encrypting both the forward and reverse channels). Channels W1–W7 carry system management messages, pages, call setup information, and other related data.

The processing at the base station is very complex—much more so that shown in Figure 14–30. This is because the base station can be sending up to 64 different Walsh-encoded data streams at the same time! The figure shows the processing for *one* of these data streams. Recall that the standard sampling rate for PCM encoding within the telephone system is 64 kbps. Sending data directly at this rate would use a great deal of channel bandwidth. Fortunately, voice signals have a great deal of redundancy. For example, pauses between words need not be encoded. Many sounds (such as a prolonged vowel sound, such as "A") are repetitive and can also be expressed in reduced (compressed) form. The data compression and formatting stage removes most of the redundancy from the speech signal, reducing it to 8 kbps or less. (That amounts to an 8:1 compression of the voice data!)

Next, the compressed speech signal is encoded for forward-error-correction (discussed in Chapter 15) and interleaving (which helps improve the chance of error recovery in case a dropout occurs during transmission). The framed signal is then mixed with the *Long Code,* which scrambles its contents to prevent unauthorized reception. The scrambled data is then mixed with one of the 64 Walsh codes, which allows it to be mixed with the other remaining 63 code channels. Finally, the Walsh-coded and scrambled signal is mixed with the *Short PN Code*. The Short Code performs the final spreading of the transmitted code to conform to the 1.25 MHz spectral mask. The *spectral mask* is an FCC definition for the shape of the final transmitted signal (Figure 14–29 is essentially the spectral mask of the reverse or upstream channel). The final result is a quadrature phase shift keyed (QPSK) RF signal (QPSK is discussed in Chapter 15) at a low IF frequency, which is then upconverted to the final RF channel frequency before transmission.

Reception of the CDMA signal is much more complex than transmission. *Three* synchronizations are required for success. You probably didn't want to hear this! The receiver (a phone) must first find the incoming CDMA signal and then synchronize its Short Code generator to the one in the base station. The Short Code repeats every 26.67 ms, so the receiver can normally synchronize to it within a second or two. (Remember that the CDMA signals pretty much look like random channel noise until this step is achieved, so this is pretty impressive in itself!)

The second synchronization in reception is to the pilot code W0. The receiver analyzes the incoming data stream using Walsh code W0, which has no data (all zeroes), so that the Walsh code generator in the receiver now becomes synchronized to that in the base station. At this stage the receiver now is synchronized to two different codes, but still sees scrambled data.

The final synchronization uses channel W32, which fortunately is not encrypted, so the receiver can start receiving W32 (synchronization) channel data. Part of the W32 data is the state of the *Long Code* sequence from the base station. The receiver quickly applies this data to synchronize its own Long Code generator to the base station so that the remaining channels can be decoded. Once the Long Code is synchronized, the receiver can now choose any of the 64 Walsh code channels to demodulate. It will examine the data in the paging channels (W1–W7) to determine call status, and can then move to a remaining traffic channel to begin demodulating voice data.

Remember that CDMA is designed to transmit digital data, so the traffic channels can carry much more information than voice. For example, in a Web-enabled phone, the traffic channel can carry TCP/IP packets to and from the mobile phone, providing mobile Internet connectivity.

Reverse or upstream CDMA operates in a similar fashion with a few changes, as shown in Figure 14–31. First, the mobile phone speech is digitized into a 64 kbps data stream by an analog to digital converter, and is then compressed down to 8 kbps and formatted into data frames, just like at the base station. Forward error correction is added (convolutional coding and interleaving) to help the base station recover damaged data bits. The data is then encoded on to *all 64 Walsh codes* by the Walsh code modulator. This is done so that the base station can synchronize to the mobile by locking on to (correlating with) all 64 of the codes. The upstream signal is then scrambled by being mixed with a modified form of the Long Code, which is chipping at 1.2288 MHz (compare to the Long Code chip rate of 19.2 kHz at the base station). The Long Code is modified by exclusive-ORing it with a 42-bit channel identification number (CID), which allows the base station to uniquely identify each transmitting mobile phone. The CID is assigned to the mobile by the base. Because each phone transmits a different CID, it can be separated by the base station. Finally, the signal is frequency spread and shaped almost identically to that of the base station by mixing with the Short Pseudo-Noise (Short PN) Code before upconversion

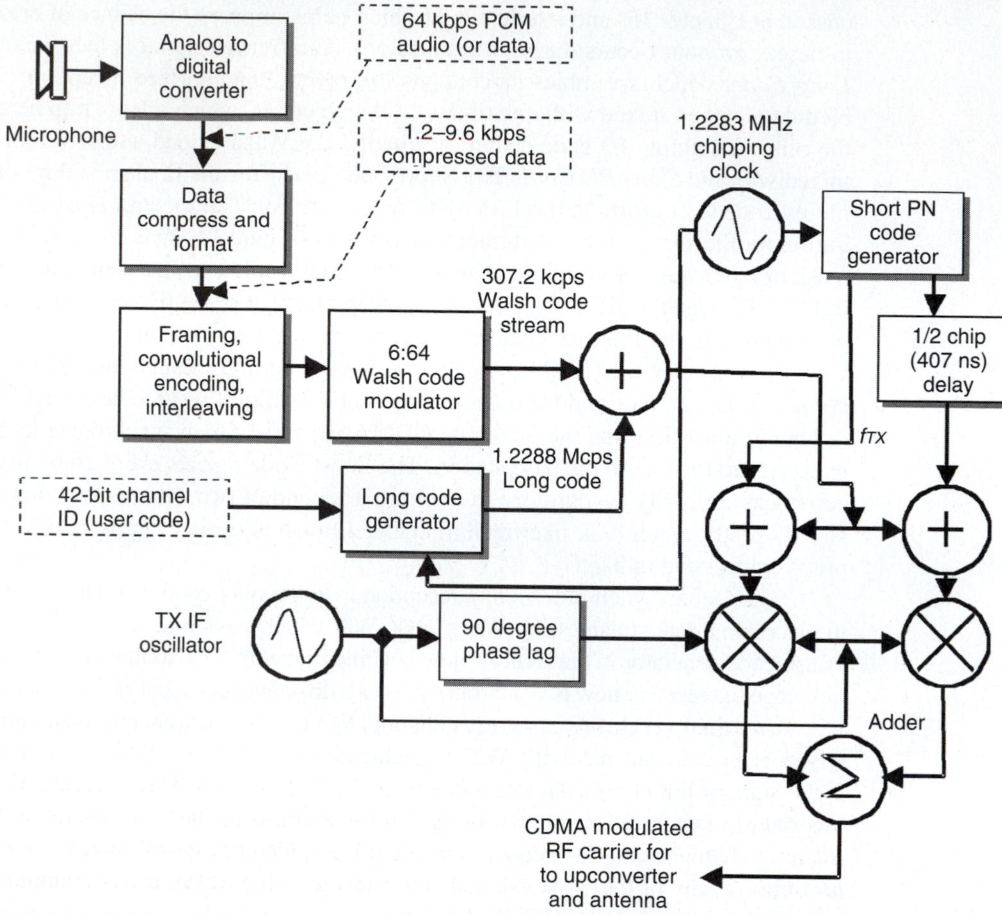

Figure 14–31 Reverse channel CDMA signal flow

to the final RF frequency. Note the addition of a half-chip delay at the Q output of the PN generator. This delay prevents the mobile signal QPSK constellation from passing through zero, which helps the base station to more accurately demodulate the mobile signal (and estimate the mobile's power level). Signal constellations are discussed in detail in Chapter 15.

CDMA has one primary weakness, domination of a receiver by the strongest signal. If a mobile unit is close to the base station, its upstream signal will tend to override the reverse channel for all the other mobiles. Therefore, the base station continuously sends *power ranging* data to all mobiles to keep all the units at the same approximate signal strength in the base station receiver.

GSM: A TDMA System

CDMA is not the only way PCS can be delivered. An alternative system, GSM, is also used in the United States. GSM uses time division multiple access to allow multiple users to share the same 200 kHz wide channel space, as shown in Figure 14–32.

Each GSM channel is divided into eight time slices, so that eight users can share a channel at any given time. This is equivalent to giving each user a (200 kHz/8) or 25 kHz

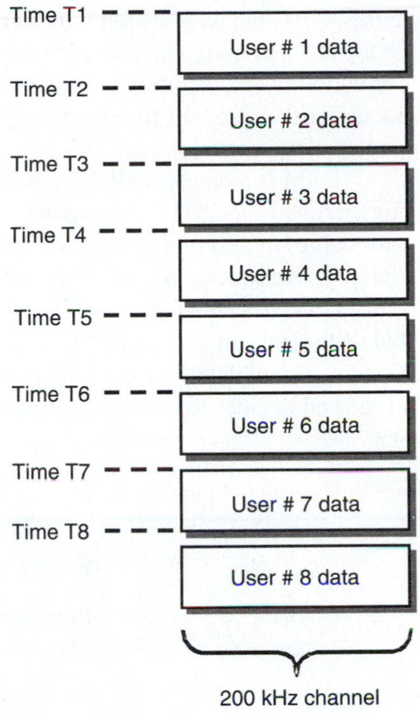
Time T1
User # 1 data
Time T2
User # 2 data
Time T3
User # 3 data
Time T4
User # 4 data
Time T5
User # 5 data
Time T6
User # 6 data
Time T7
User # 7 data
Time T8
User # 8 data
200 kHz channel

Figure 14–32 Using TDMA to share channel space

slice of frequency spectrum. Synchronization is critical in GSM; if a mobile transmits at the wrong time, it will likely "step" on the transmission of another mobile station. Therefore, the base station continuously sends timing information on the control channels. Mobiles are responsible for strictly adhering to the timing and frequency commands from the base. Figure 14–33 shows how it works.

When a GSM phone is first powered up, it listens on the designated control channels. The base station broadcasts which time slots are available on which channels within

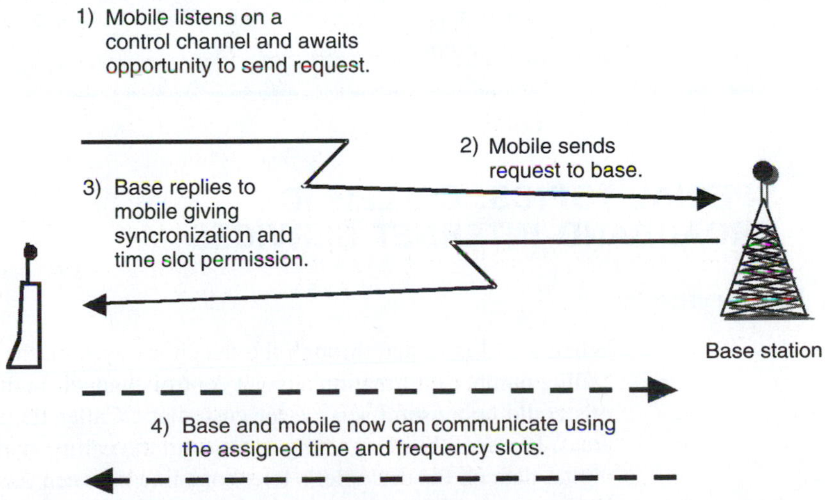
1) Mobile listens on a control channel and awaits opportunity to send request.
2) Mobile sends request to base.
3) Base replies to mobile giving syncronization and time slot permission.
Base station
4) Base and mobile now can communicate using the assigned time and frequency slots.

Figure 14–33 TDMA communications process

the cell; if a phone is new, it sends a request on one of the upstream contention channels. In response, the base station will look at the frequency and time schedule for all the slots in the cell and assign up and downstream slots for the mobile unit, and it will send this information to the mobile. The mobile can now transmit data directly to the base in one of its assigned upstream time and frequency slots, and it listens in the appropriate downstream slot.

GSM is a purely digital standard, just like CDMA. The speech encoding and decoding circuits in a GSM phone operate in a similar manner to those in a CDMA-based unit, producing streams of digital data. Therefore GSM can be used to provide the same services (paging, text messaging, and Web access) as CDMA.

GSM has one thing in its favor, and that is simplicity. It's *much* simpler than CDMA, but at the current time, CDMA is much more popular in the United States. CDMA and GSM are completely incompatible standards, so if a user signs up with a carrier using GSM and decides to switch carriers, he or she will have to buy a new phone.

Section Checkpoint

14–25 What is a supercell? What is the problem with this approach to wireless telephony?

14–26 List and describe the three methods of multiplexing used in cellular telephony.

14–27 Why are AMPS frequency allocations divided into A and B portions?

14–28 What type of modulation carries speech signals in AMPS?

14–29 How does the cellular approach increase user capacity?

14–30 What is meant by the term *frequency reuse*? How can a cellular system use the same frequency more than once?

14–31 Can a scanner listener demodulate PCS signals? Why or why not?

14–32 How does CDMA separate users' signals within a cell if they are on the same frequency?

14–33 Explain the process that a GSM mobile phone uses to open a channel to a base station.

14–5 SPECIAL TOPICS: CALLER ID, BROADBAND INTERNET SERVICES

Caller ID Service

When a call is routed through the telephone system, the switching equipment passes the calling number information along a control channel. In the early 1990s it was decided that this could be a useful service for customers. Caller ID service is offered in most service areas. It uses in-band signaling to transmit the calling party information as a short burst of digital data on the customer's telephone line between the first and second rings, as shown in Figure 14–34.

Figure 14–34 Caller ID timing for the United States

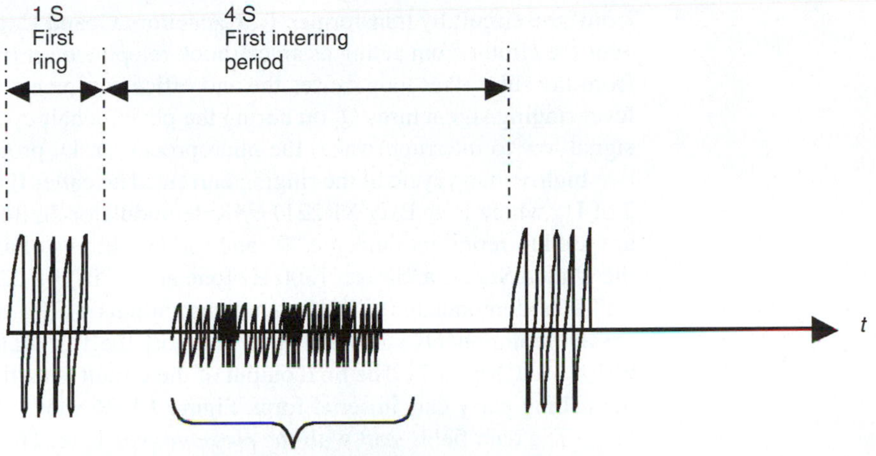

The calling number data is sent as an *FSK* data stream at 1200 bps. (Digital data modulation techniques are discussed in detail in Chapter 15.) The circuit of Figure 14–35 can demodulate this data stream and provide a "wake-up" signal for a microcontroller to read the data and show it on a display. The wake-up signal is derived from the first ringing pulses and tells the microcontroller that the FSK data is about to be sent on the line. The microcontroller then reads the data, displays it, then goes back to sleep to conserve power (most caller ID equipment is battery powered).

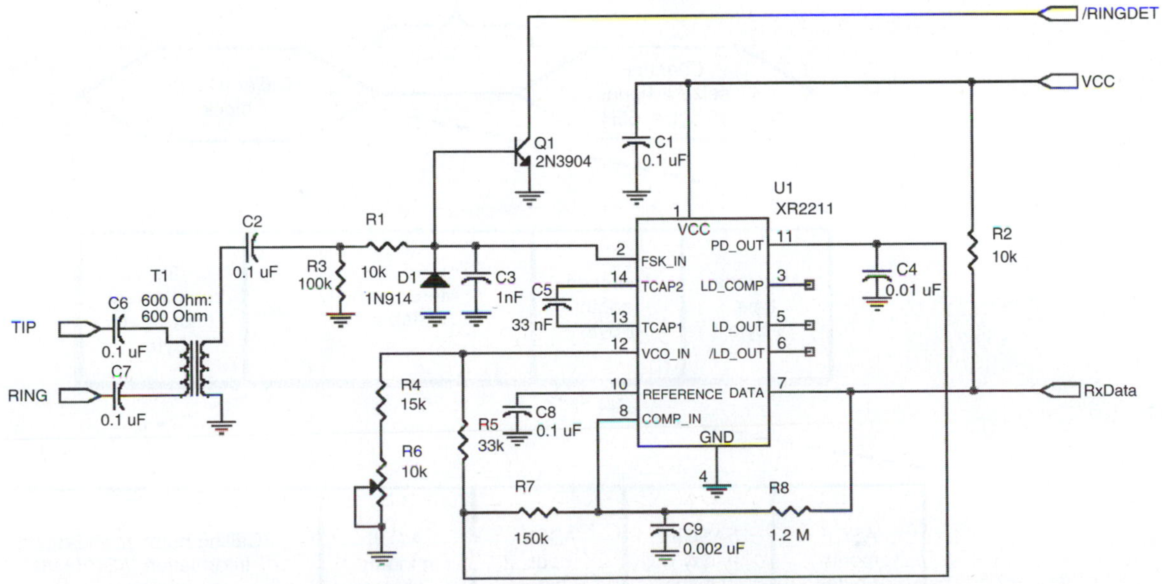

Figure 14–35 Caller ID demodulator circuit for U.S. applications

In the circuit of Figure 14–35, the telephone line is ac coupled to (and dc isolated from) the circuit by transformer T_1. Capacitors C_7 and C_6 are necessary dc blocks to prevent the circuit from acting as an off-hook telephone set (recall that drawing a dc current from the subscriber loop causes the end office equipment to activate the line). The high-level ringing signal turns Q_1 on during the positive half cycle, which pulls the/RINGDET signal low to interrupt (wake) the microprocessor. D_1 protects Q_1 and U_1 from the negative high-voltage cycle of the ringing current. The caller ID data signal is coupled into pin 2 of U_1, which is an Exar XR2211 FSK demodulator. In frequency shift keying, two tones are used to represent the logic "0" and "1" conditions of a digital signal. For caller ID in the United States, a "1" is a 1300 Hz tone and a "0" is a 2200 Hz tone. The XR2211 uses a PLL to demodulate the FSK signal and outputs a TTL serial data signal on pin 7. The passive components surrounding U_1 support the PLL action (for more information on PLLs, see Chapter 7). The final output of the circuit is on the RxData line, which contains the calling party data in serial form. Figure 14–36 shows the format of this serial data.

The data fields start with the *message type* byte. This field always holds 04H to indicate CND (calling number delivery). Next, a byte is sent to indicate the message length. This is the count of data in the message body. The message body is followed by a *checksum* for detection of reception errors. (Checksums are discussed in Chapter 15.) The message body holds the information in ASCII form.

Note that it is possible for caller ID to be blocked for various reasons. For example, if the caller dialed "*67" or "1167" to enable privacy, the calling number field in the figure will contain only a single ASCII "P" to indicate a private call. If for some other reason the number isn't available, the field contains only a single ASCII letter "O."

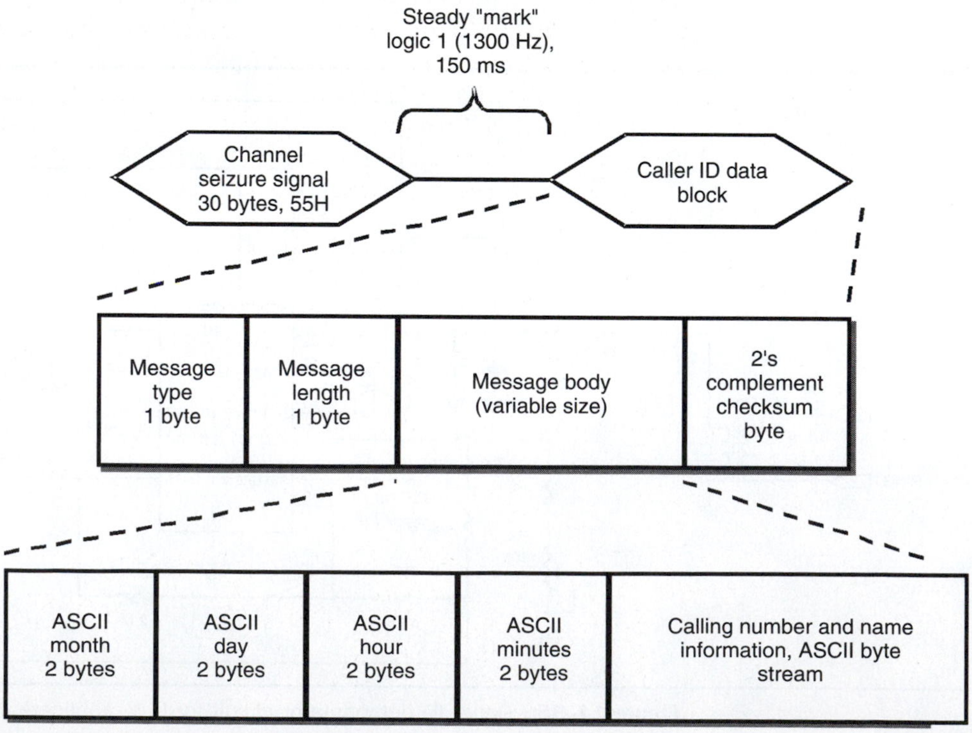

Figure 14–36 Caller ID data formatting for United States

Broadband Internet: Digital Subscriber Line Service

Another recent service offering from most telephone companies is digital subscriber line service, or *DSL*. The term "broadband" is actually misused in this context; in communications it refers to signals that have been frequency-translated (such as being modulated onto an RF carrier), but in the networking world, broadband refers to so-called high-speed Internet connections. DSL allows a subscriber's telephone line to be used for both normal analog telephone service and Internet service at the same time. There is no interaction between the services. It works by frequency division multiplexing as shown in Figure 14–37(a).

In the figure you can see that the normal POTS uses but a fraction of the available bandwidth on the copper subscriber loop. DSL takes advantage of this by using the frequency ranges from 4 kHz up to about 2.5 MHz (maximum). In practice, a DSL user generally must install a *low-pass* filter onto each telephone in the building so that the higher-frequency DSL frequency components won't be attenuated by the telephone sets (which often have a high internal capacitance).

Because the subscriber loop was never designed to carry RF signals, there are some restrictions on DSL service. First, there is a maximum allowed distance of 12,500 feet to the telephone company end office. DSL can't work reliably over a greater distance. Second, the maximum allowed data rates decrease as the subscriber gets farther from the end office. DSL uses adaptive modems (modulator–demodulators) that measure the quality of the line and make adjustments to their frequency usage and data rate accordingly.

A customer must connect a DSL modem to the telephone line to interface with the network. Often the local telephone company supplies this device as part of the service agreement. At the end office, the subscriber loop connects to a DSLAM, or digital subscriber line access multiplexer. The DSLAM terminates the DSL circuits and collects them so that they can be passed onto the Internet. It also separates out the POTS signals for normal telephone system operation.

Because most people download more than they transmit to the Internet, many phone companies offer a variant of DSL called asymmetric DSL or ADSL. ADSL divides the 4 kHz to 2.5 MHz segment of the phone line spectrum unevenly, allocating most of it to the downstream signal (end office to customer) and a small portion to the upstream signal (customer to end office). This is shown in Figure 14–37(b). Typical ADSL performance might be a maximum downstream data rate of 1 Mbps, and an upstream rate of 128 kbps.

Figure 14–37
Frequency assignments for DSL service

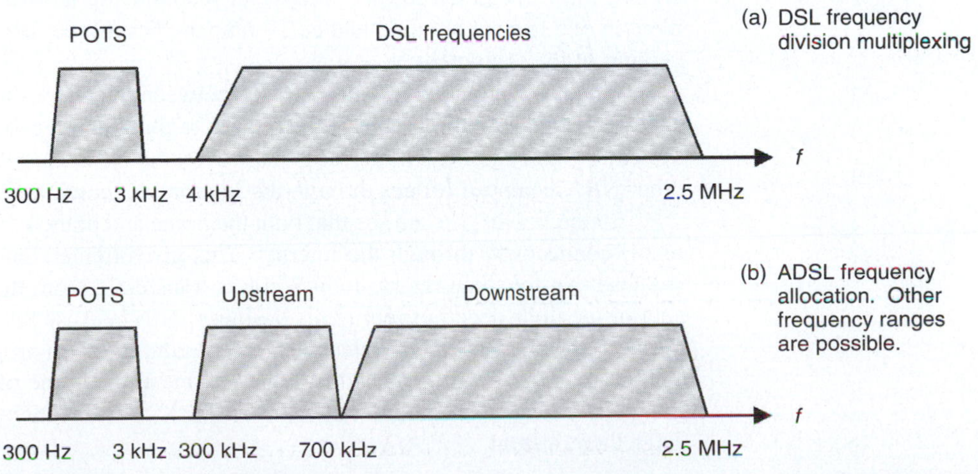

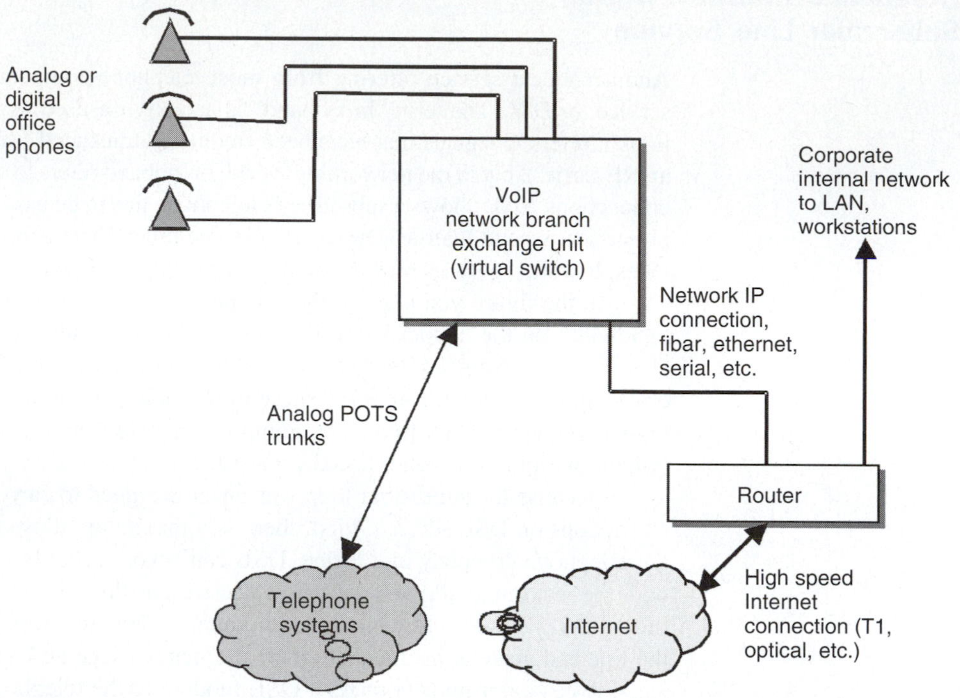

Figure 14–38 VoIP components and connections

Voice Over IP

Finally, the telephone system is likely to become entirely digital. The last remaining analog portion of the telephone system is the copper subscriber loop and CPEs connected to it. *VoIP* (pronounced "voyp") provides telephone service primarily through the Internet (with full interconnectivity to the normal telephone network). VoIP simply transports the voice and control channels (which may be compressed prior to transmission) of a telephone conversation through an internetwork (IP-based) connection instead of the telephone system. Figure 14–38 shows a block diagram of a typical VoIP installation in a business.

The central component in a VoIP network is the network branch exchange (NBX) unit. This device replaces the ordinary private branch exchange (PBX) found in a business. Nortel, 3COM, and Cisco (and others) all manufacture network branch exchange equipment in one form or another, and each company has a specific and unusual name for their version of the equipment.

The NBX provides in-house service between telephones using its internal switching capability, and also allows any of the phones to place and receive calls from the telephone system through the POTS interfaces. However, the NBX can also provide connections to other NBX-equipped offices through the Internet as shown in Figure 14–39.

In the figure, you can see that both the home and branch office have IP (Internet Protocol) connectivity through the Internet. This gives tremendous flexibility. For example, suppose that for some reason John Smith is transferred from the home office to the Sarasota office. John need not reprint his business cards because his phone number can follow him to Sarasota. The NBX system is simply reconfigured to program John's new location. Therefore, when a call is placed to John's phone at the home office, the NBX there automatically forwards his phone traffic to the NBX in the branch office; John's phone now rings there instead.

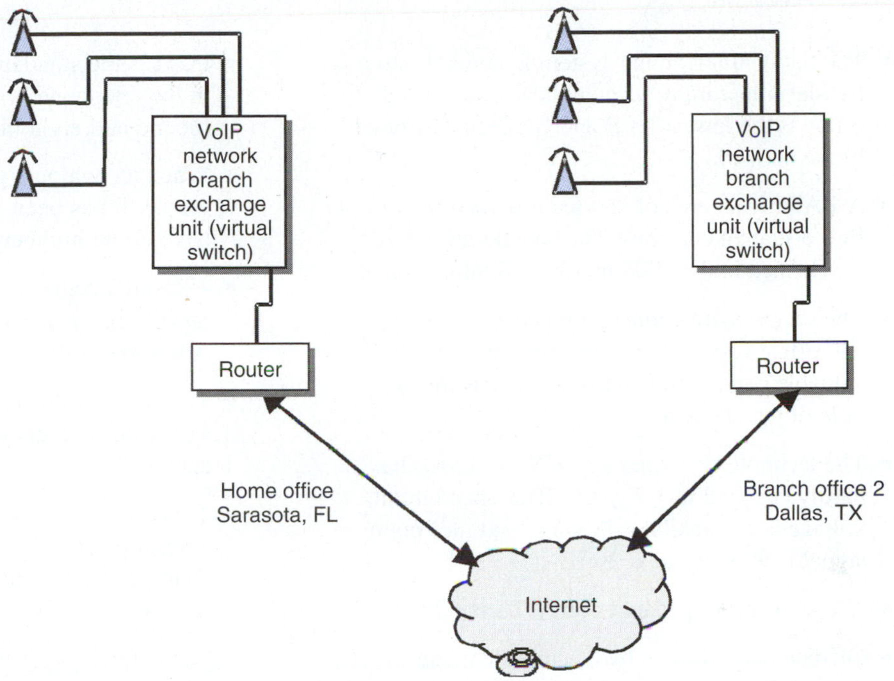

Figure 14–39 Calls can be transported through the Internet by IP

Other services are possible besides simple telephony; conference calling and video conferencing are also two strong applications of this technology. Because of the increased complexity of VoIP services, some network exchanges provide a built-in Web service that can be accessed by patrons to assist in the application of features. The Web service would be made available on a corporate intranet by a system administrator.

Finally, because VoIP uses a medium that is packet-based (the Internet) rather than connection-based (like the phone company), new protocols will have to be developed to handle the control traffic. For example, how is an Internet call established and terminated? One such protocol is *session initiation protocol,* or SIP. This protocol has been developed by the IETF (Internet Engineering Task Force) as a peer-to-peer, multimedia signaling (call routing) method. Currently, VoIP equipment and standards vary between manufacturers.

Section Checkpoint

14–34 What is the CND service?

14–35 Explain how data is delivered to the CPE by the CND service.

14–36 What is the difference between DSL and ADSL?

14–37 Explain the parts of a typical VoIP installation in an office. If a company decides to go with this technology, can it keep its old phones?

14–38 What is the purpose of SIP?

SUMMARY

- The national telephone system is divided into a five-level heirarchy. The end office or central office is the lowest level and connects customers to the network.
- A LATA is an area of service provided by a local Bell operating company. The boundaries of LATAs are defined in the 1984 modified final judgement.
- The copper wire connecting a customer to an end office is called the subscriber group. This valuable commodity is also known as the *last mile* of the network.
- The technology behind a POTS local loop has hardly changed in 100 years. The open circuit voltage is nominally −48 volts, and the ringing signal is 90 volts ac at 20 Hz.
- The standard impedance for POTS is 600 Ω.
- Off-hook detection is by dc current. When a CPE draws sufficient dc current from the loop, the end office knows that the telephone is off-hook.
- Two primary loop signal methods are used, pulse and DTMF dialing. All end offices still support pulse dialers.
- Twist is the difference in amplitude between the two tones of a DTMF signal. The maximum permissible value at a central office is 6 dB.
- The designation "0TLP" means 0 dBm (1 mW into 600 Ω) on a telephone line, which is the maximum signal level that a CPE may place on the line.
- The nominal frequency response of a dial-up POTS line is 300 Hz to 3 kHz.
- Group delays can distort data signals by varying the phase of the various frequency components in the signal. They normally don't both voice signals.
- Electronics in telephone sets are relatively simple. A hybrid transformer is usually required to reduce sidetone levels.
- The first telephone switches were complex mechanical affairs. Most switching is now completely electronic, being facilitated by the use of digital speech representation (such as PCM).
- PCM is the standard way of representing signals in the telephone system, at 8,000 samples per second and eight bits per sample.
- When an analog signal is converted to digital, we say it has been quantized. A quantized signal has a finite number of states.
- Aliasing occurs when a frequency component greater than $f_s/2$ is allowed into a sampled system. A low-pass filter is used to prevent this problem.
- The Nyquist theorem gives the minimum sampling frequency needed to represent an analog signal.
- Time division multiplexing is the transmission of several pieces of information along the same channel at different times. It is often used in digital switching.
- Early wireless telephony used a supercell model, which was totally inadequate for a large population. Modern wireless systems use a cellular model that divides an area into cells and allows frequency reuse.
- Wireless telephony uses time, frequency, and code division multiplexing, depending on the system.
- AMPS cellular usually assigns frequencies in groups of seven cells (the frequency reuse pattern is seven cells in size).
- PCS is an all-digital service. PCS uses either CDMA or GSM, depending on the service provider. CDMA and GSM are incompatible standards.
- CDMA transmits multiple conversations on the same frequency and uses code division multiplexing to separate them.
- GSM transmits multiple conversations on the same frequency but uses time division multiplexing to keep them separate.
- Caller ID service transmits an in-band data burst at the beginning of a call to identify the caller. The data is sent in normal ASCII and can be easily demodulated and displayed using simple equipment.

- DSL uses the unfarmed frequency ranges (4 kHz to 2.5 MHz) on a subscriber loop to deliver medium speed Internet services.
- VoIP allows telephone calls to be routed through either a corporate network or the Internet. It can seamlessly interface with the telephone system and allows many advanced features such as videoconferencing. Standards are still in development for this technology.

PROBLEMS

1. List and explain three important characteristics of the U.S. telephone system.
2. How many levels are in the hierarchy of the telephone system? Explain how these work together to route calls (a) within a city, and (b) between widely separated areas.
3. Define the following terms:

   a. CO b. ILEC
   c. RBOC d. MFJ
   e. LATA f. POTS
   g. AMPS h. PCS
   i. CPE j. SLIC

4. What is the open-circuit voltage on a typical subscriber loop?
5. The subscriber loop of Figure 14–3 has a telephone connected with an internal resistance of 200 Ω. What will the off-hook voltage be on the loop at the subscriber end, assuming that the copper wire in the loop has zero resistance?
6. Repeat problem 5, but this time assume that each of the two wires in the copper loop has a resistance of 50 Ω.
7. What are the names and colors of the two wires in a subscriber loop? Which one carries the −48 V line potential?
8. How is ringing signaled on a local loop? Give the voltage and frequency of the signal.
9. What is the maximum ac power a ringer could extract from the telephone network given the values in Figure 14–3? Compare your results with those of Example 14–2.
10. Explain how the following dialing systems work: (a) Pulse dialing; (b) DTMF dialing.
11. The following pairs of tones were transmitted by a CPE at the dial tone. Determine the number being dialed.

    ① 1209, 697
    ② 1209, 770
    ③ 1336, 852
    ④ 1209, 852

12. What is *twist?* How much is allowed at a central office?
13. What is the RMS voltage of the following signals on a 600 Ω line:

    a. −6 dBm
    b. −20 dBm

14. What is the frequency response of a standard dial-up line?
15. Define the term *group delay*. Explain how group delay affects (a) voice and (b) data signals.
16. If capacitor C_1 in Figure 14–10 became short, how would the telephone set behave? What impact would this have on the local loop?
17. Define *sidetone*.
18. What is the purpose of hybrid transformer T_1 in Figure 14–12?
19. What is the purpose of integrated circuit U_1 in Figure 14–13?
20. If transistor Q_1 in Figure 14–13 developed an open circuit on its collector terminal, how would the operation of the telephone set be affected?
21. Explain the operation of a Strowger switch.
22. Why is crossbar switching accomplished in stages? (What problem does this eliminate?)

23. What is the purpose of a CODEC? What is the standard data rate for telephony?

24. What data capacity is required to carry both sides of 12 phone conversations?

25. How many conversations could theoretically be carried through the CAT5 cable of a local area network (LAN) assuming it is capable of a throughput of 100 Mbps?

26. What is the minimum sampling frequency (according to the Nyquist Theorem) for digitizing the following information sources:
   a. An EEG tracing with a frequency range of 0.1 Hz to 500 Hz.
   b. An AM broadcast signal with a frequency range of 50 Hz to 4000 Hz.
   c. An FM broadcast signal with a range of 50 Hz to 15 kHz.
   d. Compact disc audio with a range of 20 Hz to 20 kHz.
   e. NTSC video with a range of 0 Hz to 4 MHz.

27. What data rate will be required to represent two channels of CD-quality audio if the sampling rate is 44.1 kHz and each sample is 16 bits? Express in kbps.

28. Explain how time division multiplexing can be used to send multiple pieces of information along the same communications channel.

29. What are three benefits of using a cellular approach to wireless telephony as opposed to a centralized supercell approach?

30. List and explain the three multiplexing methods used in cellular telephony.

31. Why are the AMPS frequency allocations divided into "A" and "B" sections?

32. In your work with a local AMPS provider you are investigating an interference complaint. You find the reverse portion of a conversation clearly received at the base station on a frequency of 824.040 MHz. What frequency should you tune the monitor receiver to if you wish to monitor the forward or downstream portion of the same conversation?

33. What is a handoff? Explain how it is handled in an AMPS system.

34. What are three differences between PCS and conventional analog cellular?

35. If you tune a conventional FM or AM monitor receiver across the PCS bands, what will you hear and why?

36. What is the channel bandwidth of PCS using CDMA in the United States?

37. Explain how CDMA supports multiple conversations on the same frequency. What is its primary weakness?

38. How many time slots are used on each channel in GSM? Explain the process that is used to allocate a frequency-time slot for a GSM mobile phone.

39. Explain how the CND service works. How does the CPE know when the CND data burst is about to arrive?

40. Draw a diagram showing the format of the data that arrives in CND.

41. How do DSL and POTS signals share the same telephone line (what type of multiplexing is used)? Give the frequency ranges used by each service.

42. What is the maximum distance between a subscriber and end office for standard DSL? Why is this restriction necessary?

43. What is ADSL? How does it differ from standard DSL?

44. Explain the operation of a VoIP system. What basic piece of equipment is needed to convert an office telephone system into a VoIP system?

15

Introduction to Data Communications

OBJECTIVES

At the conclusion of this chapter, the reader will be able to:

- draw a block diagram of a simple data communication system
- explain the difference between parallel and serial communications
- describe at least three different network topologies
- draw a block diagram of a modem, explaining the function of each part
- draw the waveform for an asynchronous character, given the data and parameters
- explain the operation of a UART
- list and describe two methods used for error detection
- list the basic communications pins of an RS232 interface and give the purpose of each
- develop plans for troubleshooting data communications networks
- follow preventive procedures to maximize network reliability

Computers are a basic ingredient of modern society. As communication tools, computers will probably replace the telephone and already are being exploited as a new broadcasting medium through the Internet. The rapid acceptance of computers in business has even led to a formation of a new discipline called *information technology,* or *IT* for short. Many companies have pooled their available talent into IT departments, which are collectively responsible for the proper operation of computers, networks, and communications equipment. The demand for IT professionals far exceeds the available supply!

A technician who wants to specialize in IT needs to understand the basics of data communications, networking fundamentals (Chapter 16), *plus* the essentials of any software (computer programs) that he or she will need to work with. A typical IT technician must have a working knowledge of many software applications and must be able to communicate clearly with nontechnical users.

Network equipment uses a staggering array of hardware and software. Hardware and software interact in an intimate fashion to make systems work. A detailed discussion of these technologies could certainly fill several volumes. Therefore, this chapter is intended as an *overview* of data communications techniques. Data communications techniques form the physical basis of all modern networks.

15–1 NATURE OF DIGITAL DATA

Digital data refers to the form of information used by computers and other digital equipment. Digital systems use *binary numbers* to represent information. A binary number is a value expressed in base-2. Binary numbers are made up of *bits,* each of which can be either *1* or *0*.

The term *bit* is an abbreviation of the term *binary unit,* which is the simplest and smallest unit of information. A *byte* is a group of eight bits.

In an analog circuit, information is represented by a continuously variable voltage or current. A good example of an analog quantity is the voltage in a circuit. There is an infinite number of possible voltages. For example, we can have potentials of 1 V, 2 V, 1.5 V, 1.505 V, and so on. There is no limit to how far this can go. The representation of digital information is quite different from that in analog circuits. A digital system can only represent a *finite* (countable) number of *states* or *conditions*. Consider the digital voltmeter of Figure 15–1. The digital voltmeter can read voltages between 0 V and 7 V. The incoming voltage is converted into a 3-bit binary code by an *analog-to-digital converter* (ADC). Therefore, the voltage is represented by a 3-bit *binary code* at the output of the ADC. Each 3-bit binary reading from the ADC is called a *sample*. It is a snapshot of the input (analog) signal at some instant in time. A collection of samples can be put together to reproduce an analog signal.

Figure 15–1 A digital voltmeter

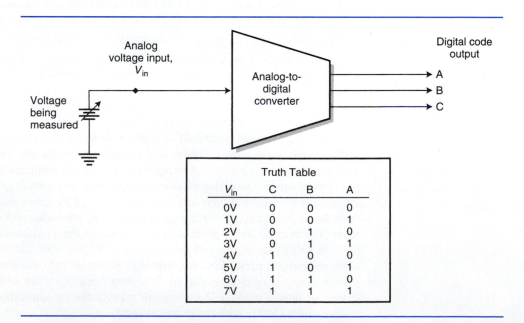

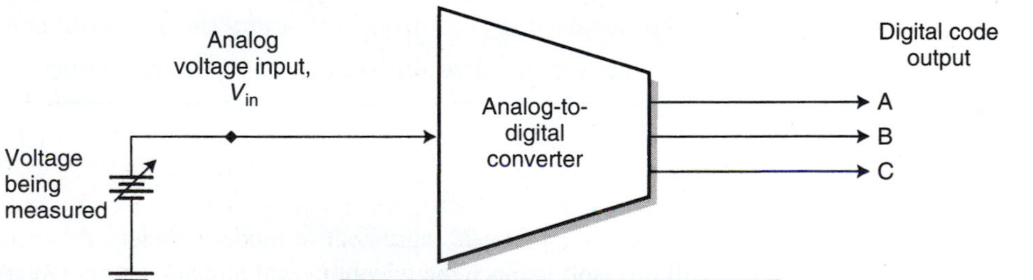

Truth Table

V_{in}	C	B	A
0V	0	0	0
1V	0	0	1
2V	0	1	0
3V	0	1	1
4V	1	0	0
5V	1	0	1
6V	1	1	0
7V	1	1	1

What will this system register if the input is 2.2 volts? You can see that it is impossible to represent this exact value. Therefore, the digital system *rounds* its reading to the nearest reading, which would be *0 1 0*. We would say that the voltage information has been *quantized*. *Quantization* is the reduction of information with an infinite number of possible values (like a voltage) into something with a *finite* (countable) number of values (like the digital output of the ADC in Figure 15–1).

A digital system can never exactly represent analog information, but it can get close. The more bits that are used, the better the digital representation becomes. This is why 16-bit sound recordings (such as those on CDs or high-quality PC sound cards) sound much better than 8-bit recordings, and 24-bit recordings (such as those made on studio digital audio mastering equipment) sound even better.

The number of possible states in a digital sample can be easily calculated:

$$\text{mod} = 2^N \qquad (15\text{–}1)$$

where mod is the *modulus* or number of possible counting states, and N is the number of bits in the digital code.

EXAMPLE 15–1

How many different analog voltages can be represented in a computer WAV (digital audio) file if the samples are (a) 8 bits and (b) 16 bits?

Solution

Since these are binary numbers, we can treat them just like digital counters. Equation 15–1 can be directly applied in each case:

a. $\text{mod} = 2^N = 2^8 = \underline{\underline{256}}$ different voltages

b. $\text{mod} = 2^N = 2^{16} = \underline{\underline{65536}}$ different voltages

The 16-bit system represents a much better picture of the original sound, at the cost of doubling the number of bits required for each sample.

A Complete Digital Communication System

Figure 15–2 shows a complete digital communication system. The system in the figure begins with a *digital data source*. This data source could be any kind of information. It could be a temperature reading at a remote location in a manufacturing plant, a count of the number of cars passing by a point in a roadway, a stream of digital audio, or even the typed characters representing an electronic mail (e-mail) message.

Digital data tends to be produced in *parallel* form. *Parallel data* is data where all the bits are sent at the same time. Computers work and communicate using 8 bits or a multiple of that number. It is common for computer data to be 8, 16, 32, 64, or even 128 bits wide. For a computer that communicates in 8-bit chunks, this poses a problem: At least eight wires are needed to carry the information, one for each data bit. For 16-bit and wider machines, the problem gets even worse. How can this problem be solved? Figure 15–3 shows how it can be done.

Having one wire for each data bit is like having one lane for each car in Figure 15–3. This is called *parallel* transmission. For short distances (such as carrying a signal a few feet to a printer), this type of transmission is useful. It is simple and easy to manage. It is possible to get all the cars to merge into a single lane. Although this will be slower, it

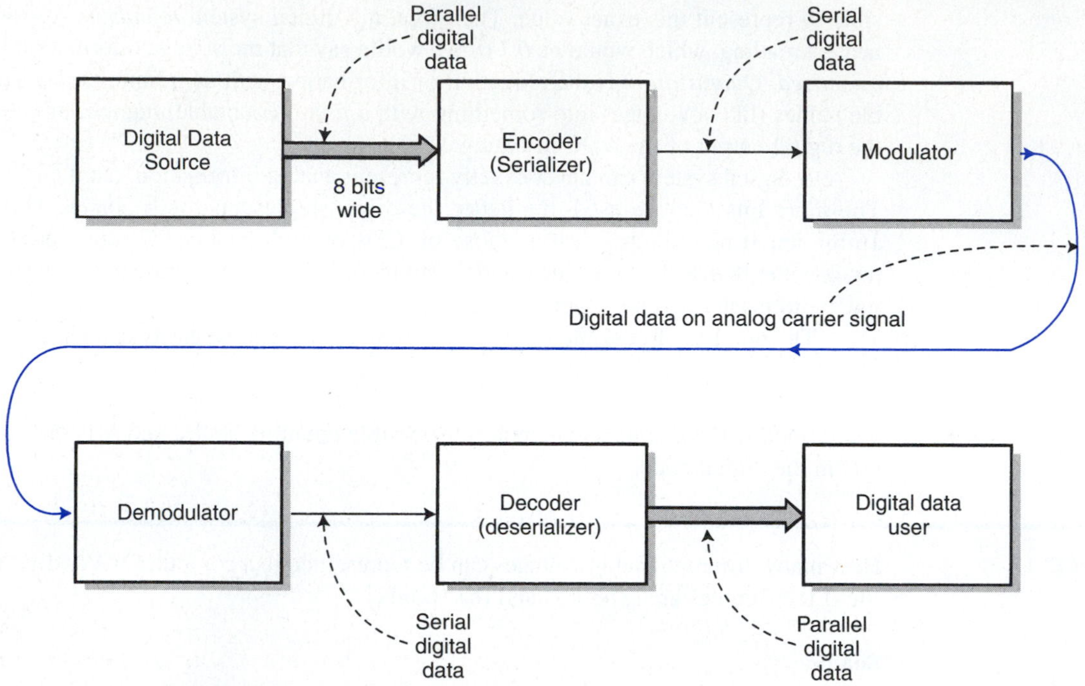

Figure 15–2 A complete digital communication system

is much cheaper to build a one-lane road, especially over a long distance. Notice that the cars line up in a particular order. This is important, since they will need to be assigned the proper lane again when they reach their destination. This is *serial* transmission. In *serial transmission,* one bit is sent at a time. Given equal channel conditions, serial transmission is always slower than parallel transmissions—but is much less expensive!

In Figure 15–2, the *encoder* or *serializer* is responsible for transmitting the digital bits one at a time. We would say that the output of the serializer is *time-division-multiplexed digital data,* or *serial data*. Remember multiplexing? Multiplexing is the sending of more than one piece of information over a communications channel. FM and TV used

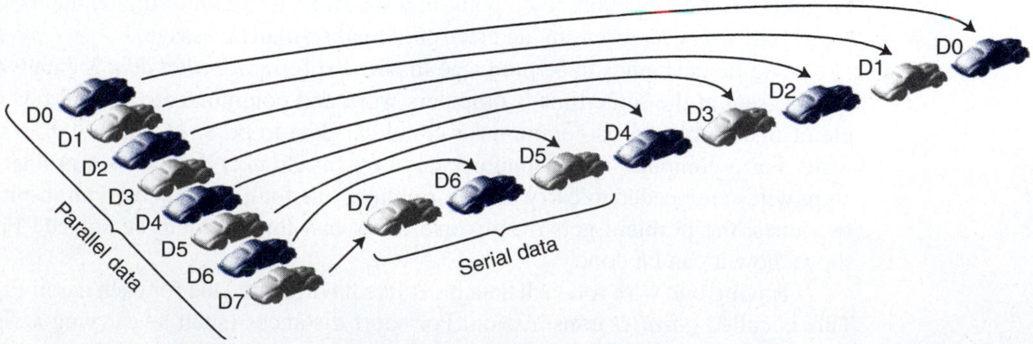

Figure 15–3 Parallel and serial traffic flow

frequency-division multiplexing to carry different parts of their signals. Because each digital bit is assigned a different point in *time,* the serial output signal is considered to be time-division-multiplexed.

Modulation and Demodulation

The serial data signal of Figure 15–2 is almost ready for transmission. In order for the signal to be carried a long distance, it must be placed upon a *carrier* of some type. In other words, the digital signal must be converted into an analog form!

The *analog* carrier can be many things. In a high-speed communication system, the carrier might be a beam of laser light in a fiber-optic "pipe" (discussed in Chapter 18). For communication over a phone line, the carrier may be a sine wave in the range of 300 Hz to 3000 Hz (the frequency range available over a standard dial-up telephone line). The carrier could also be a radio wave, using any of the standard modulation techniques (AM, FM, or PM). Radio wave carriers are often used where it is inconvenient or impossible to run wires, such as between two buildings on a college campus or between an earth station and an orbiting space vehicle.

At the receiving end, the process is simply reversed. The analog carrier is received, and the *demodulator* recovers the serial digital data waveform. A *decoder* or *deserializer* reassembles the parallel data, which is then consumed by the data user (usually another computer).

Measuring Digital Information Flow

Since the units of digital information are binary units or bits, the rate of digital information flow are *bits per second,* or bps. The higher the bps flow, the more information that can be transferred per unit time. Figure 15–4 shows how this works. This is a serial data waveform, with one time slot or *bit cell* assigned to each digital bit. For illustration purposes, the pattern "1010 . . ." is being sent. This pattern is often used to test digital communications devices, since it can be easily generated by the square-wave output of a signal generator.

A *bit cell* is simply the time interval assigned for the transmission of a bit. By knowing the time of each bit, we can calculate the rate of information flow:

$$\text{bps} = \frac{1}{T_b} \qquad \textbf{(15–2)}$$

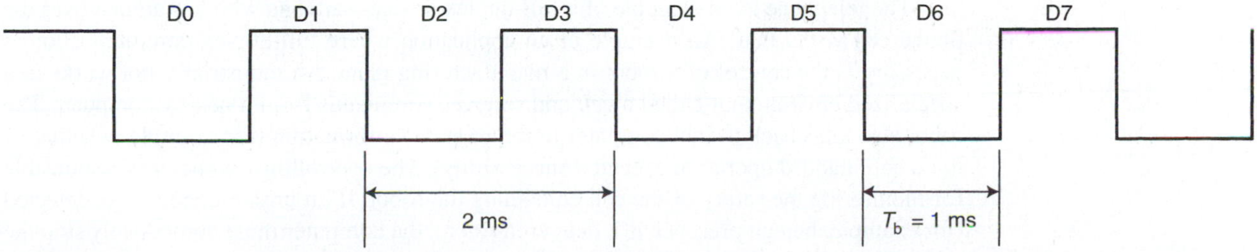

Figure 15–4 A serial data waveform

where T_b is the time interval for one bit, and *bps* is the rate of information flow in bits per second.

EXAMPLE 15–2

In Figure 15–4, compute the data rate in bps and the signal generator frequency that will simulate the given data pattern (alternating 1s and 0s).

Solution

Equation 15–2 calculates data rate:

$$\text{bps} = \frac{1}{T_b} = \frac{1}{1\text{ ms}} = \underline{\underline{1000\text{ bps}}}$$

By examining the waveform, we see that it is merely a square wave with a period of $\underline{2\text{ ms}}$. Therefore, the frequency of the waveform is

$$f = \frac{1}{T} = \frac{1}{2\text{ ms}} = \underline{\underline{500\text{ Hz}}}$$

To simulate a 1000 bps serial data stream, a $\underline{\underline{500\text{ Hz}}}$ square wave should be used. Are you surprised that the frequency is only 500 Hz (and not 1 kHz)? If so, look carefully at the waveform. For each cycle of the square wave, there is both a logic 1 and a logic 0. Each cycle therefore represents *two* bits.

Modes of Communication

The data communication system of Figure 15–2 has the capability to talk in only one direction. For many applications, this is sufficient, and we refer to this as *simplex* communication. A system that communicates in simplex mode can only work in one direction. Broadcasting is a good example of simplex communication: You can hear the DJ on your portable radio, but the DJ can't hear you (at least not through your radio!).

In order to attain two-way communications (which is needed for most applications), we need *two* sets of the blocks of Figure 15–2. The system will now be capable of sending information in both directions. Some two-way systems can only communicate in one direction at a time. For example, a hand held walkie-talkie is designed to transmit only when the push-to-talk button is pressed, and during that time, it can't receive. A system that can communicate in both directions but only one at a time is called a *half-duplex system.*

In some applications, it is necessary to have unrestricted two-way communications. A system that can communicate in both directions at the same time is called a *full-duplex system.* The telephone is an example of a full-duplex device—anyone who has argued over the phone can verify that! An example of an application where full-duplex communication is necessary is the control of a robot in a manufacturing plant. An industrial robot works in a specialized environment called a *cell* and receives commands from a nearby computer. The robot also talks back to that computer to return status information (for example, whether or not a commanded operation executed successfully). The controlling computer is responsible for monitoring the safety of the cell containing the robot. If an unsafe condition is detected (for example, human presence in a dangerous area), the computer must immediately stop the robot. The robot must therefore always be able to "listen" for such a command, even if it is in the process of giving status information back to the computer ("talking").

Section Checkpoint

15–1 What is meant by the term *digital data?*
15–2 What is a binary number?
15–3 Explain the difference between a *bit* and a *byte*.
15–4 How is the representation of information different for digital and analog circuits? Give an example.
15–5 What is a *sample?*
15–6 What is meant by the term *quantization?*
15–7 What is the difference between serial and parallel data?
15–8 Which is faster, serial or parallel data? Why?
15–9 What type of multiplexing is used to create serial data?
15–10 What are the units of digital information flow?
15–11 Explain the difference between simplex, half-duplex, and full-duplex communications.

15–2 NETWORK TOPOLOGIES

A *network* is defined as an organized system for computer communication. You may have already worked on a "networked computer," which usually means a PC connected to a local area network (LAN). There are many possible ways of connecting computers together. The physical layout of a computer network is called its *topology*. Networks can be classified in several ways; the topology (how the network is built) is one of the most fundamental.

Point-to-Point Network

Figure 15–5 shows the simplest possible network topology, a *point-to-point network*. A good example of a point-to-point connection is a dedicated line connecting two large mainframe computers, such as between two branches of a bank. A point-to-point network is sometimes used as the foundation for a *peer-to-peer network*, because the two units (marked A and B in the figure) may be considered to be *equals*. A peer-to-peer network can be formed by connecting two or more computers with equivalent roles. For example, you may have connected two computers together in order to share files through a LAN.

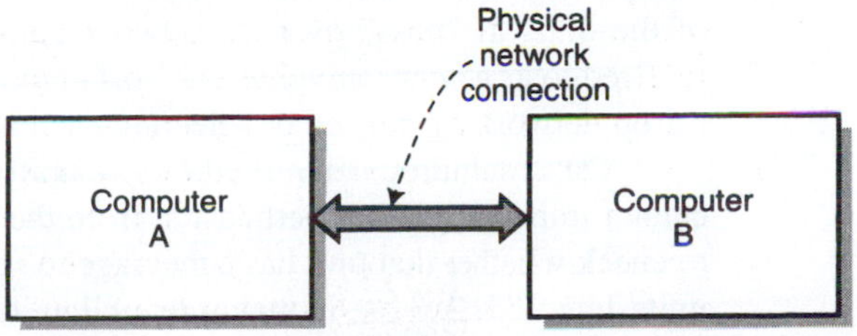

Figure 15–5 A point-to-point network

However, just because a network is physically wired point-to-point doesn't mean that the computers are equal in function (peer-to-peer). For example, when you use a dial-up telephone line to connect to an Internet service provider (ISP), you are creating a virtual point-to-point circuit between your computer and the ISP computer. This is *not* a peer-to-peer relationship; the ISP is providing a service, and your computer is acting as a client (user of the service).

A point-to-point network is often used whenever two computers need to share resources, such as printers, disks, and other storage devices. The primary advantage of this setup is its simplicity; we would say that the *protocol* requirements are very minimal. A *protocol* is a set of rules for communication. For example, a common part of the protocol for speaking in most human languages is to not speak while the other person is talking, to avoid interrupting them. The rules in a computer protocol control communications; they specify things such as when a unit may send a message onto a network, when a unit must "listen," and what a computer should do in case it detects an error in a message. The protocol's rules also determine how to form a message before transmitting it onto the network.

Although the point-to-point network is very simple and easy to implement, it suffers from two shortcomings. First, only two computers are allowed on the network. While this might be useful from a security standpoint, it severely limits the scope of communications. Second, if there are only two computers using the communications line, the cost of providing communications to each computer might be relatively high when compared to other network solutions. For example, suppose that the communications line in Figure 15–5 is a leased T_1 line from the telephone company. This line provides 1.544 Mbps (megabit per second) capacity. A T_1 line can cost over $1000 per month to lease. If the volume of communications between the computers is sufficient, then the capacity of the line will be utilized fully, and the company will be getting its money's worth. It's more likely, though, that the line will be *underutilized* with only two computers on it. *A point-to-point configuration usually does not fully utilize the communications line.*

Note that a peer-to-peer connection of two (or more) personal computers is not exactly a *true* point-to-point connection. Peer-to-peer connection of PCs uses networking protocols that allow multiple units to share the same network wiring as a distributed *bus,* which will be discussed shortly.

Multidrop Network

An improvement over the point-to-point network is the multidrop of Figure 15–6. There are three differences between a point-to-point and multidrop network. First, more computers are allowed onto the same communications line. It now resembles a party-line telephone circuit, where several homes share the same physical telephone line. Second, because there are more units on the line, the rules of communication (protocol) must become more complex. The protocol must now account for more than two computers, and all of the computers must be able to communicate without receiving interference from other stations. The third change allows this. The third difference is the designation of one of the units as "boss" over all network communications. This "boss" computer is often referred to as a *host computer.* The host computer controls all communications on the multidrop network by polling or selecting each individual unit one at a time, in turn.

On a multidrop network, the *slave* units (having addresses 1, 2, and 3 in Figure 15–6) cannot transmit without permission from the host. The host *polls* each unit, one at a time, to check whether that unit has a message to send. Thus, the host will continuously seek out units 1, 2, 3, 1, 2, 3 . . . in that order. All units except the one given permission by the host

Figure 15–6 A multidrop network

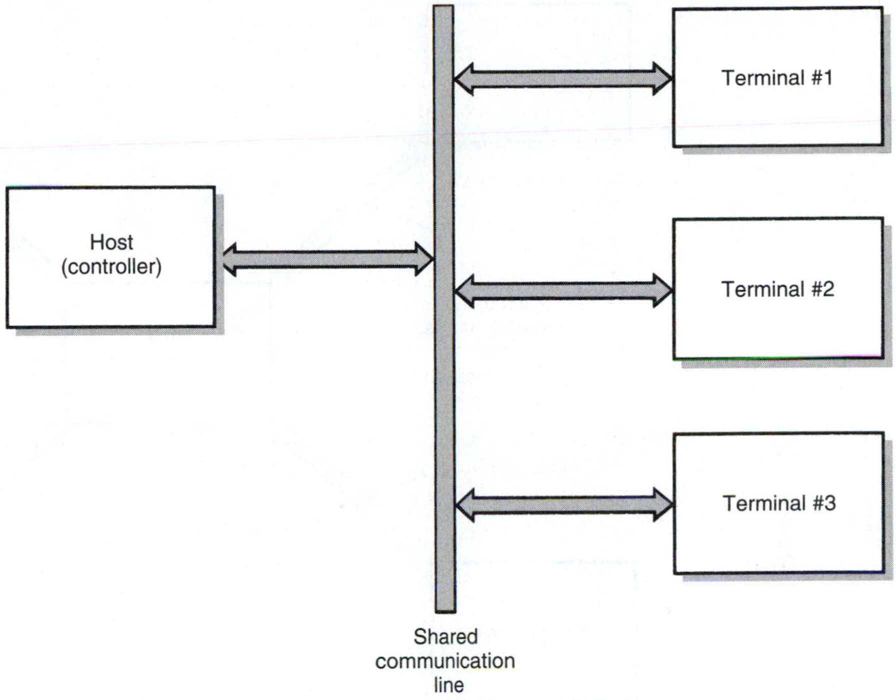

remain silent, so this arrangement prevents the collision of data from two different units. Note that data is exchanged only between the host and a slave; two slave units cannot directly communicate at all, for the host gives permission to talk to only one unit at a time.

The host may have a pending message for one or more of the slave units. In that situation, the host simply sends a *select* message to that unit (notice again that each unit has a unique numerical address). Only the unit that has been selected pays attention to the message; the others ignore it.

The multidrop has as its primary advantage better utilization of the communications line. Many computers can now share the resource, which reduces the per-unit cost. However, from this brief discussion, you can see that the rules of communication (protocol) have become much more complicated than before. Also, there is a hidden problem: What happens if the host computer breaks? Since this one computer has the responsibility of directing all the network activity, the network essentially stops working if the host fails! There is also the problem of a slave unit getting "stuck" in transmit mode. This occasionally happens. If any one unit stays in transmit, the entire network is down! To get a unit "unstuck," a technician normally must manually reset it.

Star Network

The star network of Figure 15–7 is a modification that is intended to improve reliability. Can you see how that is so? If you noticed that there is now one separate communications line for each unit on the network, you're right! On a multidrop network, any one unit can foul things up by hanging in transmit mode. In some applications, that is an unacceptable condition (such as in the control of military apparatus), so the addition of one line per unit

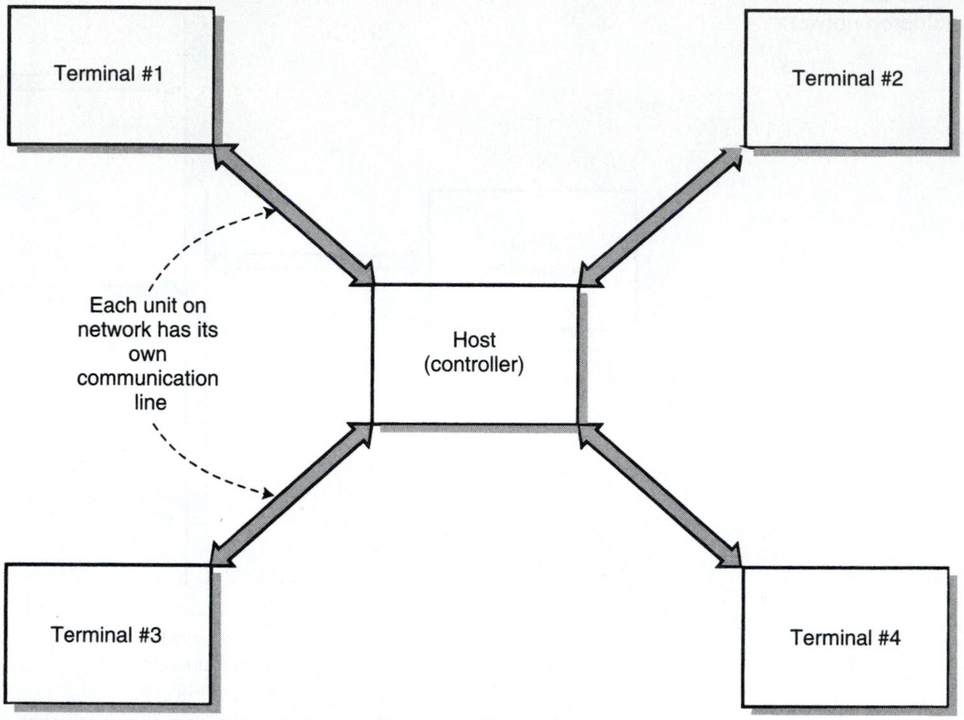

Figure 15–7 A star network

solves that portion of the reliability problem. There is still a centralized host computer that serves all the slave units, so that if this computer fails, the network will stop functioning. However, it is possible to distribute the computing burden of the host among several centralized computers, so that if one of the central computers fails, most of the network will continue to function.

The primary advantage of the star network is its reliability. One renegade slave unit cannot bring this network down. However, the cost of providing communications to each unit may still be high, as it was in the point-to-point network.

Ring Network

The ring network of Figure 15–8 represents a *distributed* approach to the control of communications over a network. By *distributed* we mean that the responsibility for controlling the operation of the network is no longer assigned to a centralized "boss" (host) computer like it was before. All the computers connected on the network equally share the control responsibility. The units on a ring network can be individual personal computers.

The ring network gets its name from the overall shape of the communications path. It looks very much like a giant series circuit. Information always flows in one direction, from one unit to the next, around and around the ring. There is no "boss" on the ring; all of the units are self-responsible about when they can transmit. This is achieved through the use of an electronic "permission slip" called the *token*.

The *token* is a special data pattern (object) that circulates the ring when no one is sending any messages. *The presence of the token on the ring indicates that the network is idle.* Any unit can send a message to any other unit as long as it can capture the token. For

Figure 15–8 A ring network

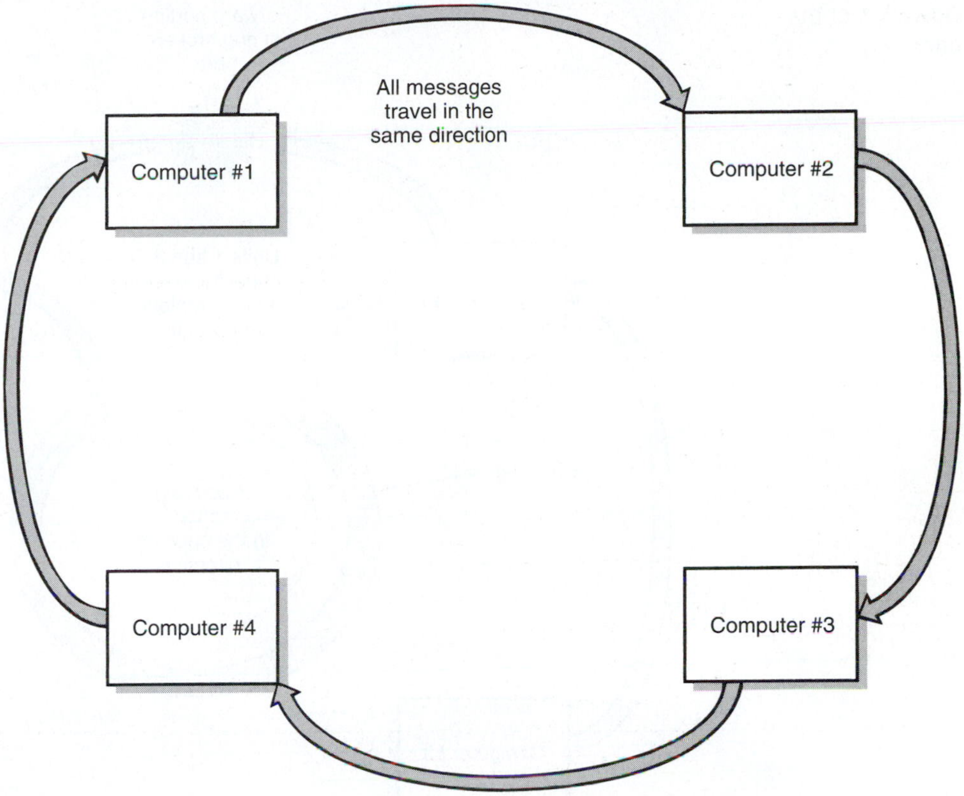

example, suppose that unit 1 needs to send a message to unit 4. The sequence of events would look like this:

1. Unit 1 waits until it receives the circulating token. Upon receiving the token, unit 1 removes the token from circulation and transmits its message (which has address 4 marked as the recipient within the message body).
2. Unit 2 sees the message from unit 1, and since the message is not a token, unit 2 knows it cannot originate any new messages at this time. Therefore, unit 2 repeats the message, which then passes on to unit 3.
3. Unit 3 treats the message in the same way, repeating it for unit 4.
4. Unit 4 recognizes the frame message as having "its" address (4) and proceeds to decipher the message content. Unit 4 also checks the message for errors. Unit 4 then *retransmits* the message back onto the ring, with the "received box" checked and the "error status box" checked. (These two items are actually digital data bits that are reserved as part of the network message format.)
5. Unit 1 receives the message it just sent and (you guessed it) examines the "received" and "error status" bits. If the message was received OK, unit 1 *releases the token* so that other units may communicate. If the message was not received correctly, unit 1 can retransmit it again.

The protocol used to operate a ring network is a little more complex than either of the previous networks, but it is very effective in operation. The primary disadvantage of

Figure 15–9
Operation of the back ring

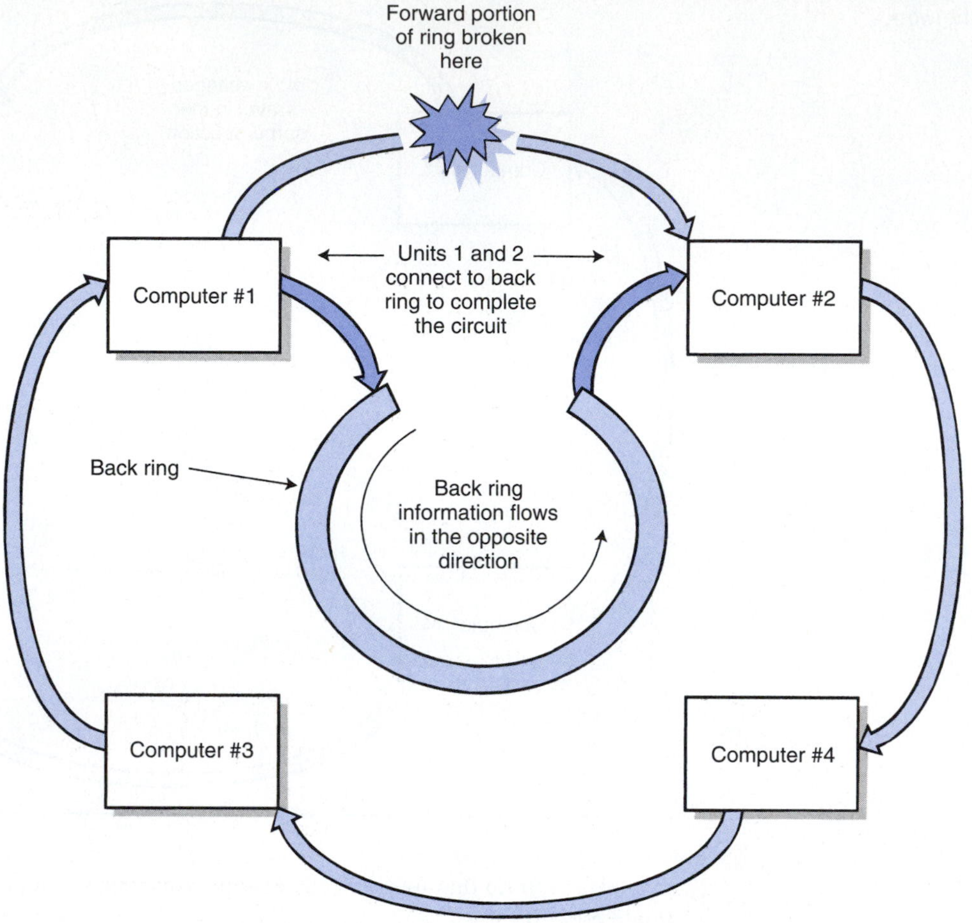

this topology is reliability; if the network is broken at any point, all communications will cease, since a complete circle is needed for the units to talk back and forth.

The IBM token ring network (a local area network used by PCs) is actually a modified ring configuration. The network consists of a *main* and a *back* ring. If a unit detects a communications fault, it will attempt to connect through the redundant back ring. The failure of a single wire does not bother an IBM token ring network. Figure 15–9 shows how this is so.

In the figure, a wire has broken on the *main ring* between units 1 and 2. Momentarily, network communications cease due to the broken connection. At this point, the units on the network begin a programmed strategy to get back on line. Unit 2 senses a loss of dc current on its input pin and connects to the back ring. Unit 1 detects a similar problem, and connects its output to the back ring. Notice that the back ring contains short circuits for all of the units, except 1, so that the messages from 1 can now travel *counterclockwise* on the back ring to unit 2. *The network has "healed" itself, at least temporarily!*

Bus Network

The *bus* network of Figure 15–10 is used in LANs. The *Ethernet* is a bus network. The bus network is well suited to a distributed computing environment, like the ring network. In

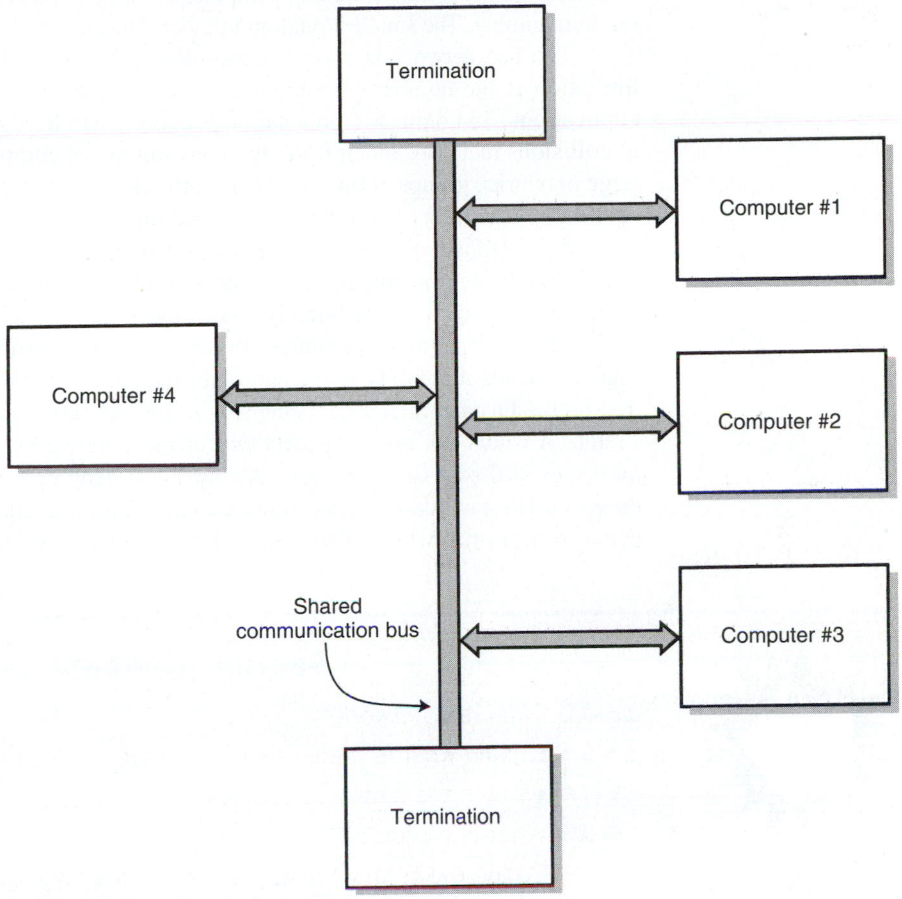

Figure 15–10 A bus network

fact, the network in Figure 15–10 is much like the multidrop setup we discussed earlier, with one major difference: There is no "boss" or host computer for controlling the network communications. As on the ring network, the individual units take full responsibility for their own activity and use a specialized protocol to govern their communications.

The nature of communications on a bus network is very similar to that at a dinner table. At the table, only one person normally speaks at a time and usually addresses one other person by name (though a person can certainly "broadcast" a message for all at the table to hear). The method used to achieve this on a bus network is called CSMA/CD, which stands for *carrier sense multiple access with collision detection*. The operation of CSMA/CD can best be summarized as "listen before you talk." Each unit on a bus network actively listens to communications in progress and waits until no carrier is sensed (no one is transmitting) before transmitting a message onto the network. Since all the computers are on the same bus, they can all hear the outgoing message, but only the computer with the proper network address will respond; the others will remain silent. The receiving computer responds to acknowledge the message, and the network operation then resumes.

It is possible that *two* or more units may attempt to send at the same time. When this happens, we say that a *collision* has occurred on the network. The transmitting units can detect a collision because they can sort of listen while they are transmitting. The signals from the competing computer partially disrupt the outgoing signal, which is immediately detected by both of the competing computers. Both computers stop sending, and each picks a random number. The smallest random number wins, and its holder gets to communicate first.

The bus network is a very economical solution for LAN applications. Its primary limitation is the number of computers that a single "backbone" can support. For many applications, 25 computers on a single backbone can become very slow, because the rate of collisions increases much faster than the number of computers on the network. For very large networks, groups of bus backbone networks are connected through *bridges, switches,* or *routers,* depending upon the communication need.

A *bridge* is a device that connects two networks together and allows them to act as one. It can bridge incompatible networks (such as a token ring and Ethernet) as well. A *switch* is a device that intelligently routes messages between computers on a network. It examines the data link layer address (medium access control or *MAC address*) of the message and sends it out only to the computer with a matching address. This reduces congestion on the bus and increases security. A *router* operates at a higher level than a bridge or switch. A router not only connects two or more networks, but actively controls how the messages will pass between them. A router can deny certain messages passage based on their content and addresses. A router can also translate addressing and message formats between networks, which is something that a bridge or switch can't do.

Section Checkpoint

15–12 What is a *network?*

15–13 Explain what is meant by the *topology* of a computer network.

15–14 What is the simplest network?

15–15 What is a *protocol?*

15–16 Define the terms *poll* and *select* as they are used in a multidrop network.

15–17 Why is a star network more reliable than a multidrop?

15–18 What is the difference between *distributed* and *centralized* control over communications?

15–19 What electronic data pattern (object) gives permission to transmit on a ring network?

15–20 What is the purpose of the back ring in an IBM token ring network?

15–21 What is meant by "CSMA/CD?" What type of network is this used on?

15–22 Why are bridges, switches, and routers needed?

15–3 MODEMS AND DIGITAL MODULATION TECHNIQUES

As we saw in Figure 15–2, a digital message is usually converted to serial digital data and is placed upon an analog carrier prior to transmission over a long distance. This is not *always* true; for example, telephone carrier companies have been rapidly adopting fiber-optic technology as a replacement for the copper trunk lines connecting telephone

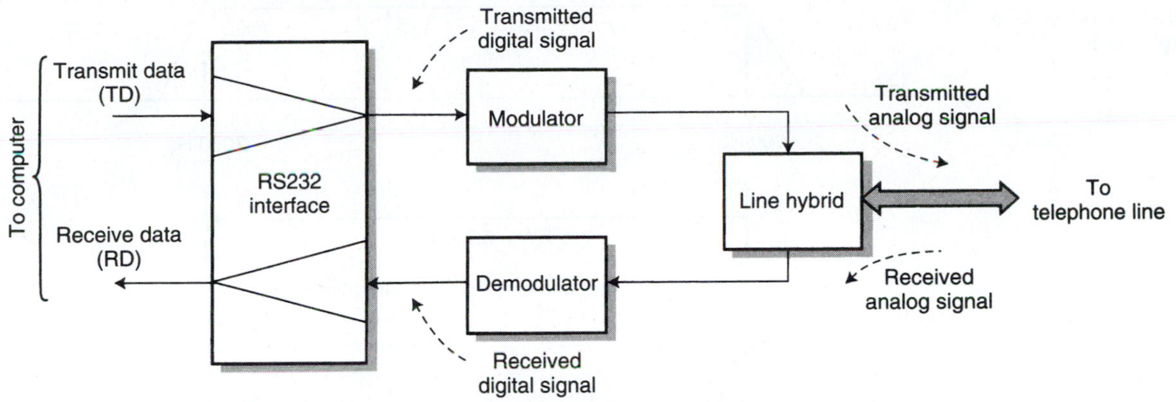

Figure 15–11 Telephone modem block diagram

company central offices and long-distance centers. Fiber optics can carry either analog or digital signals with ease.

There are many cases where digital data must be carried over an analog-only path. Common analog signal paths include conventional voice telephone lines and radio frequency carriers (wireless systems). Whenever a digital signal is to be sent over an analog path, a *modem* is required. The word *modem* is a contraction of the words *modulator* and *demodulator*. A modem capable of both receiving and transmitting will have both sections. A typical telephone line modem block diagram is shown in Figure 15–11.

A telephone modem contains several important sections. The transmit data from the computer passes into the *modulator,* which converts the digital serial data (a stream of 1s and 0s) into an analog signal compatible with the telephone line. We will see shortly how this is done. The *demodulator* section of a modem receives signals from the telephone line (which are on a modulated carrier from another modem). The demodulator extracts the digital information from the incoming carrier signal and passes it on to the computer.

Note that the connection between the computer and modem is a *serial interface*. Recall that serial data is sent one bit at a time. The EIA-232D (also known as RS-232D) is the most popular standard for serial interfacing at low data rates. We will study this interface in a later section.

The *line hybrid* is responsible for interfacing both the modulator and demodulator sections of the modem to the telephone line. It performs two important functions. First, the line hybrid provides *dc isolation* between the modem circuitry and the telephone line, which is required for safety. Second, the line hybrid *isolates* the modulator and demodulator from each other, so that full duplex operation can be achieved if necessary. The modulator and demodulator need to be isolated so that one does not interfere with the other. The line hybrid allows the output signal from the modulator to pass to the telephone line but not to the demodulator. The demodulator is allowed to see only incoming signals from the telephone line.

Telephone Line Characteristics

The characteristics of dial-up telephone lines are standardized. Understanding them will give us insight into how the modulator and demodulator sections of a modem work. Figure 15–12 shows the frequency response of a telephone line.

Figure 15–12
Frequency range of a telephone line

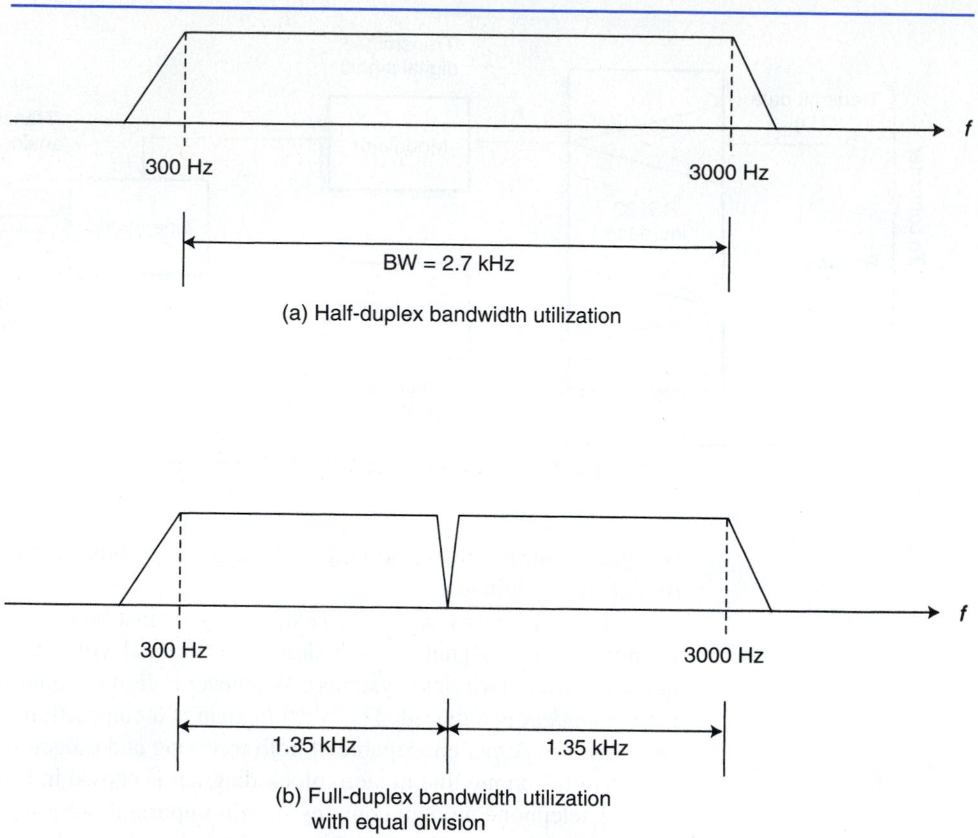

A dial-up telephone connection has a limited frequency response. These values are the result of experimental research by Western Electric. They represent the minimum range needed to reproduce readily intelligible human speech. The lower limit of 300 Hz is the reason why bass tones are lacking in telephone reproduction—they simply aren't needed for understanding spoken words. Notice that dc *cannot* be passed either (dc is 0 Hz, which is *way* below 300 Hz!), which prevents the direct transmission of simple digital signals down the line. For example, a steady TTL logic 1 or 0 would require sending a dc voltage of either 5 V or 0 V, which will not pass.

The difference between the upper limit of 3 kHz and the lower limit of 300 Hz represents the available *bandwidth* of the line, which is 2.7 kHz. In Figure 15–12(a), the entire bandwidth is being used for communication. This is true for simplex and half-duplex communication, where only one modem transmits at a time. For full-duplex communication, the bandwidth must be suitably divided. Most full-duplex modems divide it in two, resulting in the arrangement of Figure 15–12(b). Bandwidth is important; it is one of the factors that controls the amount of information that can be sent through a communications channel.

EXAMPLE 15–3

What will happen when the following signals are passed through a dial-up telephone line?

a. a 100 Hz sine wave

b. a 500 Hz sine wave

c. a 500 Hz square wave (50% duty cycle)

d. a 2 kHz square wave

Solution

By examining each of the above signals in the frequency domain, we can get an idea of what will happen.

a. The 100 Hz sine wave *cannot* pass through the line. It is not in the range of 300 Hz to 3000 Hz.

b. The 500 Hz sine wave passes through with no difficulty; it appears as a 500 Hz sine wave on the other end.

c. The 500 Hz square wave is composed of a 500 Hz fundamental plus an infinite number of odd harmonics at 1500 Hz (3rd harmonic), 2500 Hz (5th harmonic), 3500 Hz (7th harmonic), and so on. Only the 500 Hz, 1500 Hz, and 2500 Hz components make it through the phone line, and the *phase* of these sine wave components is also altered on the way through. The result is a very distorted square wave at the other end of the connection. The wave will have rounded edges and other undesirable changes.

d. The 2 kHz square wave contains a fundamental (2 kHz), 3rd harmonic (6 kHz), and so on. Only the 2 kHz fundamental makes it through the line. *The resulting output at the other end is a 2 kHz sine wave!*

Example 15–3 illustrates two points. First, trying to pass square waves (a simple type of digital signal) through a telephone line directly is a futile exercise. The bandwidth limits and hidden phase shifts destroy the wave's shape. Second, the bandwidth limits partially determine what information will pass through the line. In part (d), the 2 kHz square wave was "magically" transformed into a 2 kHz sine. There was no real magic here; the bandwidth simply would not allow any of the harmonics to pass, leaving only the fundamental sine wave.

This is why the process of *modulation* is necessary. It is possible to communicate digitally over a phone line. The trick is to convert the digital data into something that the telephone line *can* pass. That "something" is, of course, analog sine wave signals. The telephone line can easily pass any sine wave between 300 Hz and 3000 Hz. One of the earliest methods invented for sending digital data through an analog medium was *frequency shift keying,* or *FSK,* as implemented in the *Bell-103 standard.*

Frequency Shift Keying

Figure 15–13 shows the principle behind FSK. FSK is really just a type of frequency modulation. We know that a digital signal cannot pass directly through a phone line. We can still get the *information* represented by the signal to pass if it is represented appropriately. The FSK signal in Figure 15–13 represents a logic 1 or *mark* as a 2225 Hz sine wave and a logic 0 or *space* as a 2025 Hz sine wave. Therefore, a digital bit sequence of 1-0-1 can be sent by sending a 2225 Hz sine wave for the duration of the first bit, then 2025 Hz to represent the second bit (a logic zero), and finally, another 2225 Hz tone to represent the last bit, another logic 1.

The telephone line can pass the 2225 Hz and 2025 Hz sine waves easily, so the *information* is passed. The information is temporarily represented in analog form by the two sine waves. Figure 15–14 shows how this can be done with circuitry. The circuit of

Figure 15–13
Frequency shift keying

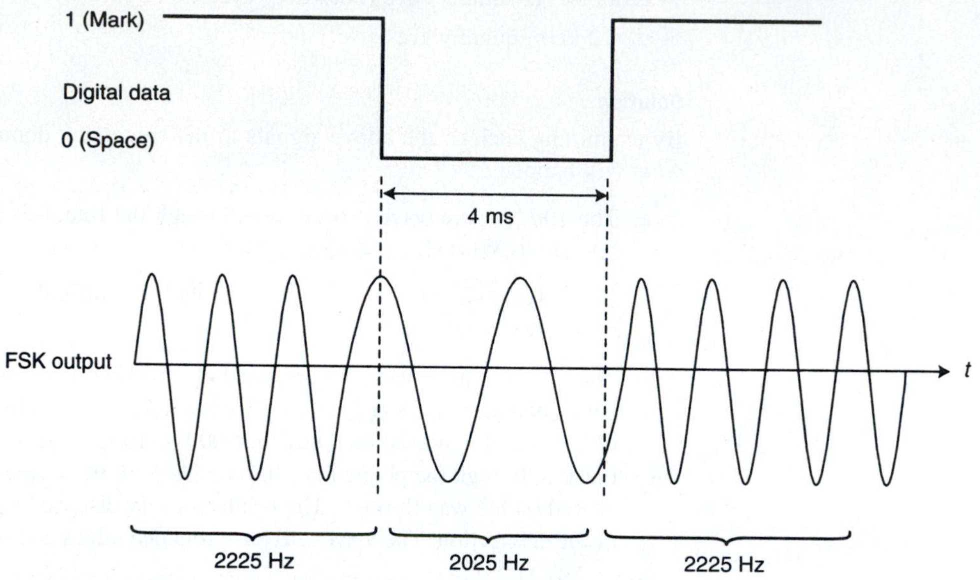

Figure 15–14(a) is really an FM transmitter. The only difference between this FM transmitter and the ones of Chapter 8 is that this unit uses an *audio frequency* carrier in the 2 kHz range. This is because the signal is intended to pass through a telephone line. At audio frequencies, an oscillator can be accurately controlled by an RC time constant, and this is the traditional approach used in FSK modems. (The *typical* approach in modern modems, however, is to replace the modulator and demodulator circuitry with a custom digital signal processor (DSP) chip.)

The original digital signal enters the modulator through an *electronic switch*. The job of this switch is to connect one of two adjustable resistors into the oscillator circuit, depending on the logic state (1 or 0) of the digital signal. One resistor, R_{mark}, is adjusted to

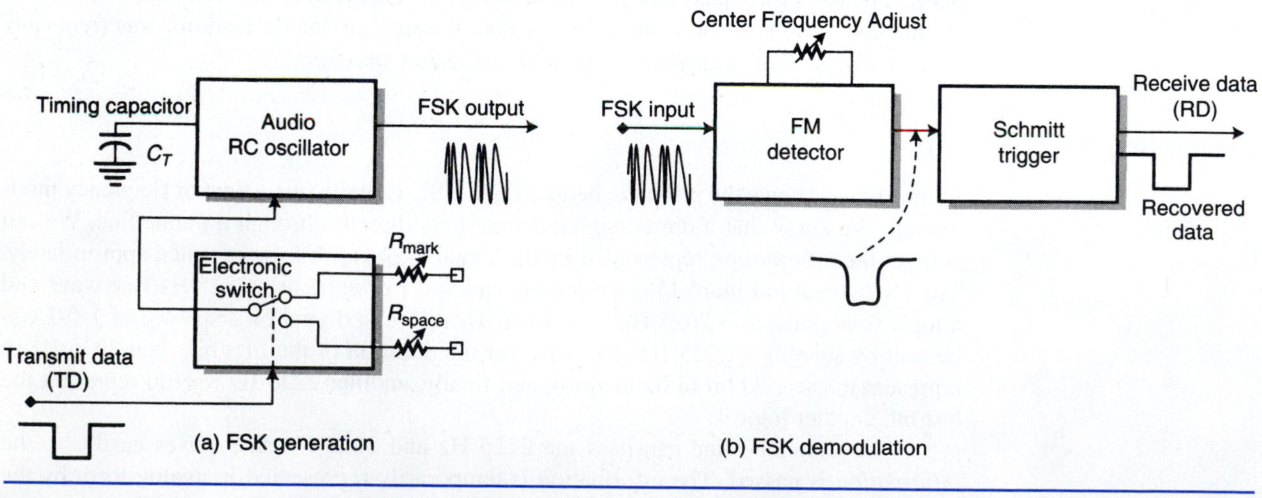

Figure 15–14 FSK modulation and demodulation

provide the mark frequency (2225 Hz) in combination with the timing capacitor C_T in the oscillator circuit. Therefore, when a logic 1 is placed into input pin of the electronic switch, R_{mark} is switched into the circuit, and the oscillator produces the *mark* frequency of 2225 Hz. The opposite happens when a 0 is to be transmitted; the resistor R_{space} is connected in this case. The resulting signal from the oscillator is an FSK signal that can be sent directly down the telephone line.

Reception of the FSK signal involves reversal of the steps. Because FSK is really just frequency modulation, an *FM detector* circuit is used to recover the digital information. An FM detector converts frequency changes back into voltage. In order to ensure that the final recovered data signal is at voltage levels compatible with digital logic (and to clean up any residual noise), it is fed through a Schmitt trigger. The resulting output is a clean digital square wave that is a close replica of the original data waveform.

Bell 103 standard The frequencies of 2225 Hz and 2025 Hz are part of the *Bell 103 standard,* which is an international standard that describes how low-speed FSK modems are to operate. Bell-103 modems operate full-duplex, as shown in Figure 15–15.

Bell-103 modems are of two types, *originate* and *answer*. An originate modem is one that places an outgoing call, and an answer modem is the receiver of a call. A modem can assume either role, depending on whether it has received or placed a call. This distinction is important because different frequency ranges are assigned for each type of modem. The *originate* modem always transmits 1270 Hz (mark, 1) and 1070 Hz (space, 0), while an answer unit always sends 2225 Hz and 2025 Hz as detailed in Figure 15–15. By assigning frequencies in this way, two Bell-103 modems will never disagree about which set of frequencies to use. This works well because a phone line is a full-duplex communications medium.

When a Bell-103 modem is set to originate mode, its modulator is automatically set to produce sine waves at 1270 Hz and 1070 Hz, and its demodulator (receiver) is "tuned" to listen on 2225 Hz and 2025 Hz. Does this make sense? Look at the frequencies produced by the two modems in Figure 15–15. The *originate* modem *transmits* 1270 Hz and 1070 Hz. The *answer* modem *receives* these same two frequencies. The exact opposite is happening in the other direction. The separation of frequencies allows Bell-103 modems to operate full-duplex.

Signaling and BAUD In Figure 15–13, a digital sequence of 1s and 0s is being represented by frequency changes in an analog sine wave carrier. *Signaling* is the process of representing information by producing changes in a carrier signal, and the changes in the carrier signal are called *transitions*. There are only three types of transitions possible—*frequency*

Figure 15–15
Frequencies of operation for Bell-103 modems

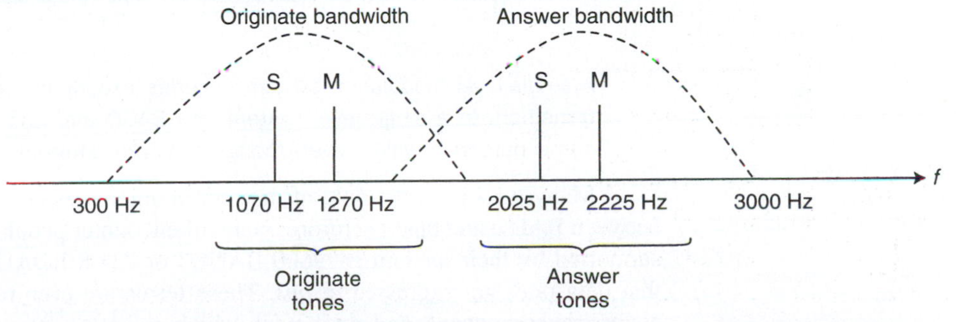

transitions, *phase* transitions, and *amplitude* transitions. In other words, when we create transitions on a carrier signal, we are really just *modulating* the carrier.

The rate at which information is sent is called the *data rate,* and of course, it is measured in bits per second, or bps. We can also measure the rate at which transitions are created on the carrier wave. The *signaling rate* is the number of transitions created per second on the *analog* carrier wave and is measured in units of *BAUD*. A value of 1 BAUD is equal to 1 transition per second. In fact, you can substitute the words "transitions per second" any place you see the word *BAUD* and the meaning will remain unchanged. Equation 15–3 calculates *BAUD*:

$$\text{BAUD} = \frac{1}{T_S} \tag{15–3}$$

where T_S is the time for each transition, sometimes called the *signaling interval.*

There's nothing magical about equation 15–3; in fact, you can probably recognize it as just another variation on the theme "time is one over frequency." Since the units of BAUD are transitions (changes) per second, BAUD itself describes the frequency of changes on the carrier wave.

EXAMPLE 15–4 Calculate the data rate and signaling rate (in BAUD) for Figure 15–13.

Solution

a. The time for each bit is 4 ms by inspection, so equation 15–3 can be used to find the data rate:

$$\text{bps} = \frac{1}{T_b} = \frac{1}{4\text{ ms}} = \underline{\underline{250\text{ bps}}}$$

b. The signaling interval T_S is also 4 ms by inspection, so equation 15–3 can be used:

$$\text{BAUD} = \frac{1}{T_s} = \frac{1}{4\text{ ms}} = \underline{\underline{250\text{ BAUD}}}$$

This is not a particularly fast data rate. In fact, surfing the Web could be a particularly painful exercise at this speed. The numbers we calculated seem unremarkable as well; they're the same. Why is this so?

The reason why the *bps* and *BAUD* values are the same in this case is that *each transition (change) in the carrier wave represents exactly one data bit*. The system of FSK just developed is called a *single-level* modulation system for this reason.

A single-level modulation system encodes exactly one digital bit per analog transition. In a single-level system, the BAUD and bps will always be the same numerical value, even though they are different things.

Many persons in the field of data communications do not understand the difference between BAUD and bps. Therefore, you will encounter people who insist that the data rate supported by their modem is "9600 BAUD" or "28.8 KBAUD," when in fact you know that data rates are expressed in *bps*. These terms are even misused by manufacturers of electronic components and equipment, which makes the matter even worse.

You might wonder why it's important to know the difference between bps and BAUD, especially when they end up with the same value. Here's why: They only have the same value for *single-level* modulation systems, like the Bell-103 FSK modems. In modern high-speed modems, they are very different.

Bandwidth requirements of FSK FSK is not a particularly efficient form of modulation when we examine how it utilizes its frequency space, or bandwidth. We can estimate the bandwidth of a narrow-shift FSK signal (like the one in Example 15–4):

$$BW_{FSK} \approx 5 \text{ BAUD} \qquad (15\text{–}4)$$

This formula is an approximation very much like Carson's rule for FM. Because FSK is really a form of FM, we know that it will really require an *infinite* bandwidth, but because all the sidebands generated are not significant, infinite bandwidth is not required to get a reasonable reproduction of an FSK signal.

Equation 15–4 indirectly tells us something very important about data signals on analog carrier waves. *The bandwidth is directly controlled only by the signaling rate*. The data rate *can* affect the signaling rate, but this is an indirect relationship. If the second part doesn't make perfect sense at this point, don't worry. We will need to develop this idea further to get a good idea of what is really happening.

The bandwidth of a data transmission is directly proportional to the signaling rate in BAUD.

EXAMPLE 15–5

Calculate the bandwidth needed for the FSK signal of Figure 15–13, assuming that the signaling rate is 250 BAUD.

Solution

Equation 15–4 can be directly applied here:

$$BW \approx 5 \text{ BAUD} \approx 5\,(250 \text{ BAUD}) \approx \underline{\underline{1250 \text{ Hz}}}$$

Note that this is very close to *one-half* of the available bandwidth on the telephone line. Why is this important? Figure 15–12 shows why. Bell-103 modems are full-duplex devices, which means that communications must occur in both directions at the same time. The total bandwidth is 2700 Hz, which gives two 1350 Hz "slots" when divided by half. This is slightly larger than the bandwidth needed for 250 BAUD operation.

EXAMPLE 15–6

Calculate the maximum signaling rate possible for FSK, given an available bandwidth of a) 1350 Hz and b) 2700 Hz. Calculate the data rate in each case, given that single-level FSK is in use.

Solution

a. Equation 15–4 can be solved for BAUD; doing this, we get

$$\text{BAUD} \approx \frac{BW}{5} \approx \frac{1350 \text{ Hz}}{5} \approx \underline{\underline{270 \text{ BAUD}}}$$

Since this is a single-level modulation system, the data rate and BAUD will be equal, so by inspection, the maximum data rate will be 270 bps.

b. We can apply the exact same procedure again, and with the increased available bandwidth, we'll get

$$\text{BAUD} \approx \frac{\text{BW}}{5} \approx \frac{2700\text{ Hz}}{5} \approx \underline{\underline{540\text{ BAUD}}}$$

Again, since this is a single-level system, the bps and BAUD will be identical values, and the maximum data rate will be 540 bps.

What is the practical difference between these two cases? In the first case, only one-half of the telephone line's bandwidth has been made available for use. This has limited the signaling rate to 270 BAUD, and likewise, the data rate is limited to 270 bps. (Note that real-world Bell-103 modems will actually operate at 300 bps.) When the bandwidth is "opened up," the maximum signaling rate increases, and the rate of data flow also increases in a like manner. However, even using the maximum bandwidth of the line produces a paltry data rate of 540 bps, which is far below what is actually possible.

The primary limitation of a single-level modulation system is its slow speed.

Multilevel Modulation Systems

Modern high-speed modems use multilevel modulation. This sounds complicated, but it really isn't. Suppose that your friend Jim has an over-the-road shipping company located near the entrance to a major highway. Jim's company ships 55-gallon drums of oil to a distant refining plant, but the road has only two lanes. The city council has ordered Jim to allow only one truck to enter the road per minute, in order to prevent traffic jams.

If Jim loads one barrel of oil onto each truck, how many barrels of oil can he ship in one hour? Since he can send one truck per minute, and there are 60 minutes in an hour, he can ship 60 barrels an hour. How can Jim improve the efficiency of his business?

What if Jim could load *two, three, four,* or more barrels onto each truck before it departs? That would be a reasonable way of improving business. If Jim loaded 2 barrels onto each truck, then he could ship 120 barrels an hour (60 trucks/h times 2 barrels/truck). It would be even better if he could manage 3 or 4 barrels per truck. At 4 barrels per truck, he could ship 240 barrels an hour.

The trucks in this example are *carriers* of cargo and are just like the *transitions* in an analog carrier wave. The *rate* at which trucks leave Jim's business determines how much highway space is needed, just like the *BAUD* (rate of transitioning) determines the *bandwidth* of the modulated analog carrier wave. The drums of oil are analogous to the *data* or *information* being conveyed on a carrier wave.

Note that the number of barrels loaded per truck does not affect the amount of space needed for the trucks in traffic (unless the trucks are made larger in some way). Only the *number of trucks per unit time* controls the amount of highway space ("bandwidth") needed.

We can likewise improve the efficiency of a digital modulation system by loading "more than one barrel per truck." We can make each analog transition represent more than one digital bit. This is the secret of *multilevel modulation,* which is the principle behind all modern high-speed modems.

A multilevel modulation scheme encodes more than one bit per transition.

The data rate and BAUD are no longer the same number in a multilevel modem. They are related by the following equation:

$$\text{bps} = m \times \text{BAUD} \qquad (15\text{–}5)$$

where m is the number of bits per transition or the "level number" of the modulation method.

Multilevel AM A good way of understanding multilevel modulation is to look at a system using AM. Figure 15–16(a) shows an amplitude-shift-keyed (ASK) carrier signal. The only difference between this example and Figure 15–13 is that the *amplitude,* rather than the *frequency,* of the carrier wave is being changed to represent the information. Normally, FM would be used in preference to AM because FM is more immune to noise. Figure 15–16(a) shows a single-level AM system, because there is exactly one bit per transition.

Figure 15–16(b) shows a level-2 ASK signal. This signal is different from the single-level signal in that *two* bits are encoded for each transition. The digital data is fed into the modulator *two* bits at a time. This is equivalent to loading two drums of oil on each of Jim's trucks from the previous example. Let's compare single and multilevel AM numerically.

Figure 15–16 Single and multilevel AM

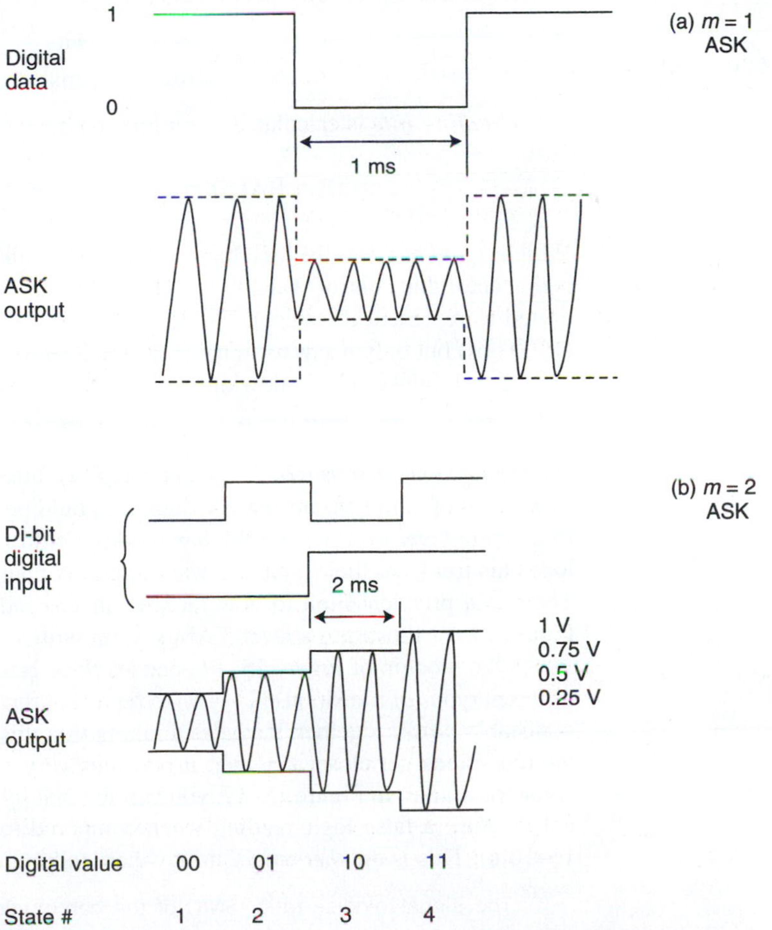

EXAMPLE 15–7 Calculate the data and signaling rates for Figure 15–16(a) and (b). Which signal requires the least bandwidth to transmit?

Solution

a. In Figure 15–16(a), the data rate can be calculated as

$$\text{bps} = \frac{1}{T_b} = \frac{1}{1\text{ ms}} = \underline{\underline{1000\text{ bps}}}$$

And, by inspection, 15–16(a) is a *single-level* system (one bit per transition), so the signaling rate is exactly equal to the data rate, or 1000 BAUD.

b. In 15–16(b), the data rate can be calculated by finding how many bits are transmitted per unit of time. This can be done in terms of equation 15–2, if we closely examine the waveform. There are *two* bits being sent every 2 ms. Therefore, there is a time of 1 ms per bit, and data rate calculates the same as it did before:

$$\text{bps} = \frac{1}{T_b} = \frac{1}{1\text{ ms}} = \underline{\underline{1000\text{ bps}}}$$

Note that we could easily let the arithmetic do the work for us as well. If we divide the number of bits by the time needed to send them, we'll get the same answer:

$$\text{bps} = \frac{\text{bits}}{\text{time}} = \frac{2\text{ bits}}{2\text{ ms}} = \underline{\underline{1000\text{ bps}}}$$

The *signaling rate* is calculated according to equation 15–3:

$$\text{BAUD} = \frac{1}{T_s} = \frac{2}{2\text{ ms}} = \underline{\underline{500\text{ BAUD}}}$$

Which signal requires the least bandwidth to transmit? The rate of transitioning, in BAUD, is the factor that controls bandwidth. The higher the BAUD, the more bandwidth that is required to represent the signal. The level 2 ($m = 2$) system of (b) has the same data rate (1000 bps) but only needs to signal at 500 BAUD to get its message across. Therefore, the level 2 system uses less bandwidth for a given data rate than the single-level system.

The sky is the limit, or is it? Since encoding two bits per transition reduced the bandwidth by a factor of 2, this means that the data rate could be increased by a factor of 2 (compared to a single-level system) and the bandwidth would remain exactly the same. Just as Jim loads his trucks to their limit, we want to put as many bits on each transition as possible. There *is* a practical limit to how far we can go, and there is a *theoretical* limit as well. Figure 15–17 illustrates a level 3 AM system with a "problem."

The modem of Figure 15–17 encodes *three* bits per transition. Now there must be 8 different types of transitions ($2^3 = 8$) to represent these bits, and the transitions are getting noticeably *closer* together. If you're thinking that this could spell trouble, you're right. As the transitions get closer together, it becomes very possible that a *noise* pulse will overcome the signal. In Figure 15–17, you can see that it might not take a very large noise signal to cause a false logic reading when compared to the original ASK system of Figure 15–16(a). This is the *theoretical* limit we are talking about.

The signal-to-noise ratio (*SNR*) of the communications channel, which is the ratio of signal power to noise power, is the factor that limits the number of bits per transition on any modulated carrier wave.

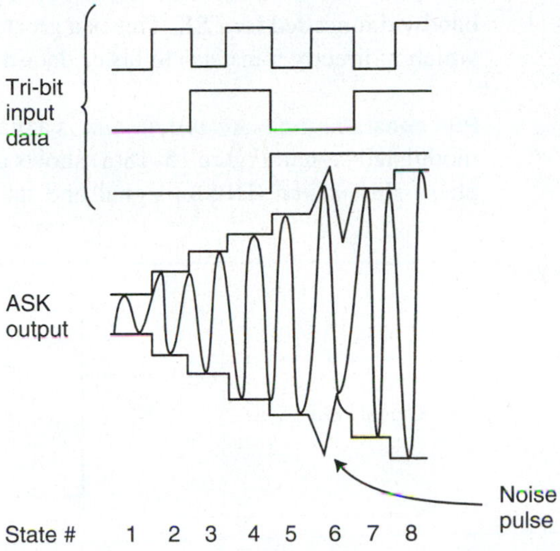

Figure 15–17
Multilevel AM with noise

There is a *practical* limit as well. As we increase the number of bits per transition, the complexity of the modulating and demodulating circuitry increases as well. Most practical telephone modems encode 8 or fewer bits per transition for these reasons.

Total Channel Data-Carrying Capacity

The *capacity* of a communications channel (such as a telephone line) for carrying information can be predicted by the Shannon–Hartley theorem, which is a theoretical rule. The use of this equation is beyond the needs of the technician, and it is stated here only for reference:

$$C = \text{BW} \times \log_2(1 + \text{SNR}) \qquad (15\text{–}6)$$

where C is the channel capacity in bits per second (bps), BW is the channel bandwidth in Hz, and SNR is the ratio of signal-to-noise power for the channel.

This equation gives the *theoretical* information capacity of a communications channel. It tells us that the rate at which data can flow is determined by only two factors. These are the *bandwidth* of the channel (which limits the signaling rate), and the *signal-to-noise ratio,* which limits the number of bits per transition (again, look at the effect of noise on Figure 15–15 to understand how the SNR affects the picture).

The data-carrying capacity of a channel, in bps, is determined by two factors. These are (1) the *bandwidth* and (2) the S/N ratio of the channel.

Modern Modulation Methods

In today's high-speed modems, the most efficient modulation methods are used in order to squeeze out every bit of capacity from a communication channel (usually a telephone line). The most popular modulation methods use *phase-shift keying,* or *PSK,* sometimes combined with AM. PSK is popular, because for a given signaling rate, it uses up much less bandwidth than FSK (which tends to be a bandwidth "hog"). For comparison, the bandwidth of a PSK carrier signal can be approximated by

$$\text{BW}_{\text{PSK}} \approx \text{BAUD} \qquad (15\text{–}7)$$

Compare this with equation 15–4. For a given signaling rate, PSK uses *one-fifth* of the bandwidth needed for FSK. This is a great improvement, as it allows faster signaling rates, which indirectly translates to faster data-transfer rates.

PSK constellation A *constellation* is a diagram showing all the possible states of a phase-modulated signal. Figure 15–18(a) shows a binary (level 1, 1 bit per transition) differential phase-shift-keyed (DPSK) signal and its oscillogram. The representation of the binary

Figure 15–18 Binary DPSK and its constellation

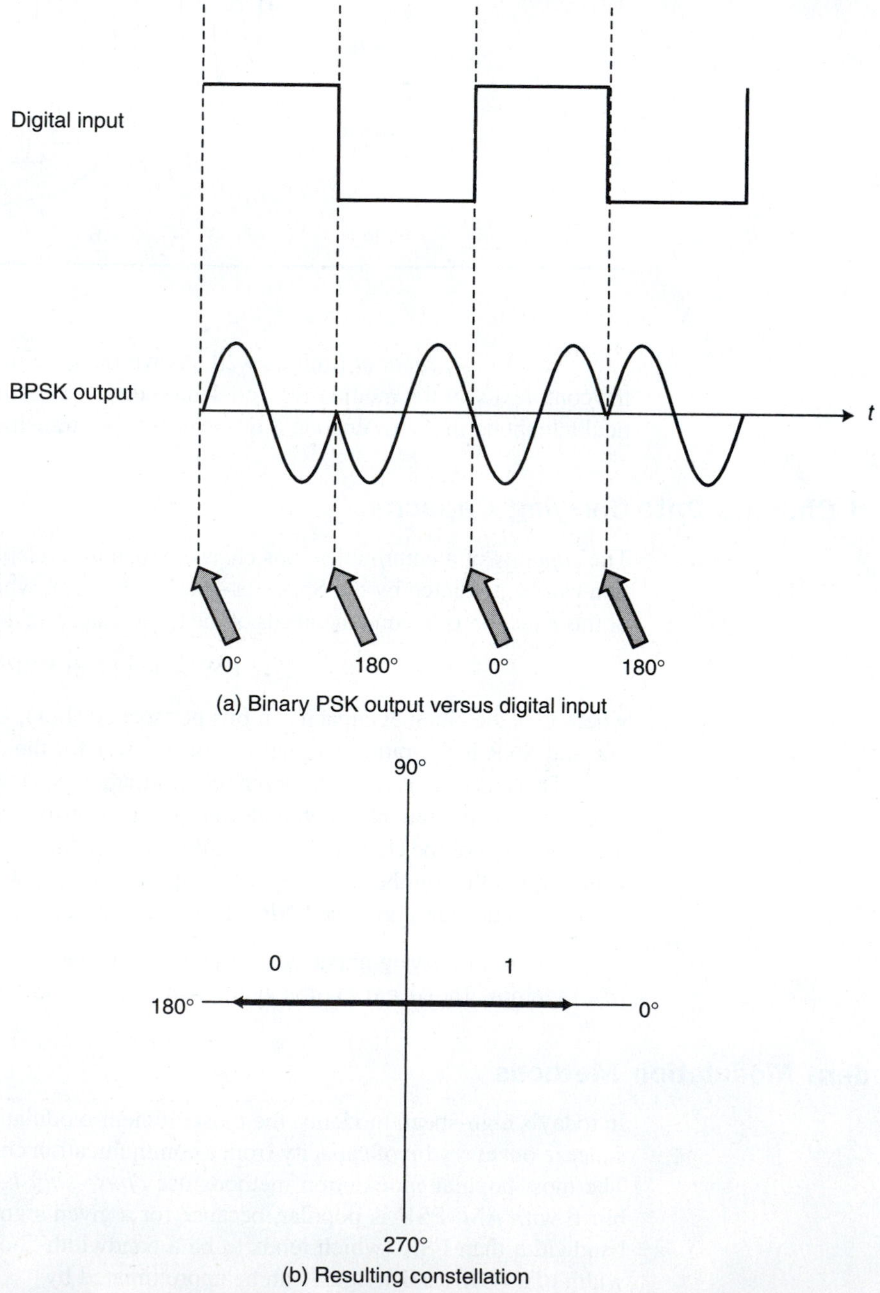

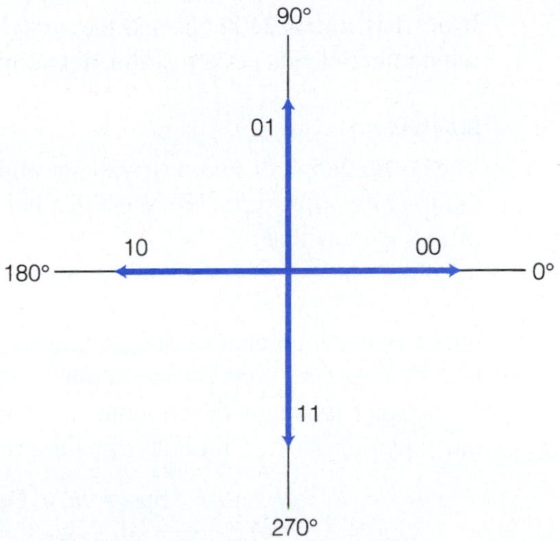

Figure 15–19 A multilevel PSK system

pattern is quite simple for the PSK signal. There is one cycle of carrier for each bit (which helps the receiver to extract clock information), and a logic 1 is represented by a new cycle with *no* phase shift, while a 0 is represented by a new cycle with a 180° shift. The waveforms produced by a *PSK* modulator are unique in appearance because of the sudden phase shifts.

The constellation diagram of Figure 15–18(b) shows the same information but in a much simpler fashion. It looks very much like a phasor diagram. Each of the arrows of Figure 15–18(b) indicates one possible state of the modulator. This modulator has two states because it must encode one digital bit (either a 1 or 0). The choice of 0° and 180° angles is made to simplify the design of modulator and demodulator hardware.

It is possible to encode more than one bit per transition and create a multilevel PSK system, as shown in Figure 15–19. In the figure, we have now increased the number of possible phase angles to four, and because of that, we now encode two bits per transition. This is a level 2 PSK system. It's much easier to think of it in terms of the signal constellation rather than attempt to draw oscillograms; there are too many subtle angles involved in the waveform. Can we push it further? You bet. Figure 15–20 shows yet another multilevel PSK constellation.

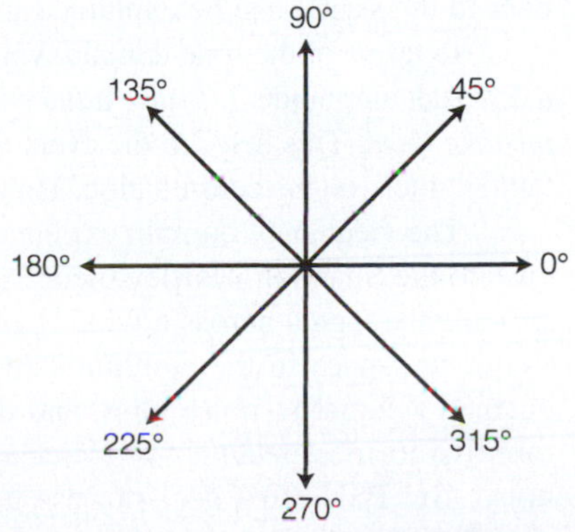

Figure 15–20 Another multilevel PSK system

EXAMPLE 15–8 If the data rate is 2400 bps and the *modulation* method of Figure 15–20 is used, calculate the number of bits per transition, signaling rate, and required bandwidth.

Solution

The constellation of Figure 15–20 has *eight* distinct phase angles. We could say that it has eight *modulation states*. Because this is just like having a digital counter with a modulus of 8, we know that

$$2^m = 8$$

where m is the number of bits per transition. There's no heavy math needed here: $2^3 = 8$, so there must be 3 bits per transition.

Since we know the data rate and number of bits per transition, we can now manipulate equation 15–5 to find the signaling rate.

$$\text{bps} = m \times \text{BAUD}$$

$$\text{BAUD} = \frac{\text{bps}}{m} = \frac{2400\ \text{bps}}{3} = \underline{\underline{800\ \text{BAUD}}}$$

Finally, since this is PSK, we can use equation 15–7 to approximate the bandwidth:

$$\text{BW}_{\text{PSK}} \approx \text{BAUD} \approx \underline{\underline{800\ \text{Hz}}}$$

Compare this to the results we got with FSK and you'll be impressed. We're getting a data rate of 2400 bps while using only 800 Hz of bandwidth. That's not bad at all! And we can do even better than this!

Quadrature amplitude modulation When we have eight or more modulation states in PSK, the effects of noise begin to take over. It doesn't take much noise to cause a received signal constellation to "jitter" back and forth 45°, and in the case of Figure 15–20, that would cause an error in the received data, since the received phase angle would be quite different from that sent. In other words, phase noise will cause the received constellation pattern to rotate back and forth randomly. If the angle of rotation is large enough, the receiver will make an error in demodulating the information. Figure 15–21 shows how PSK can be improved by the addition of amplitude modulation.

Quadrature amplitude modulation, or *QAM,* is used in the latest modems, and employs AM and PM at the same time. The improvement gained by adding an AM component to the signal can be explained in both the time and frequency domains.

In terms of the time domain, you can see that every adjacent phase in Figure 15–21 has a different amplitude. *Because adjacent phases are different in size, the receiver can not easily mistake them.* This helps in the event of phase noise, which shows up as a back-and-forth rotating "jitter" on the constellation. This is probably the most important feature to keep in mind.

The frequency domain explanation makes little sense unless it is examined with regard to the Shannon–Hartley theorem (equation 15–6). The addition of an AM component *spreads* the signal across a wider bandwidth, causing the signal to fill *all* of the available frequency space in the communication channel. Thus, near-maximal use is made of the channel's signal-to-noise ratio, and the available capacity increases. This part is counterintuitive to most technicians; after all, we do know that increasing bandwidth increases noise. But PSK often does not use up very much bandwidth, and the receiver bandwidth may be *wider* than that of the received PSK signal (as in a telephone modem, where the

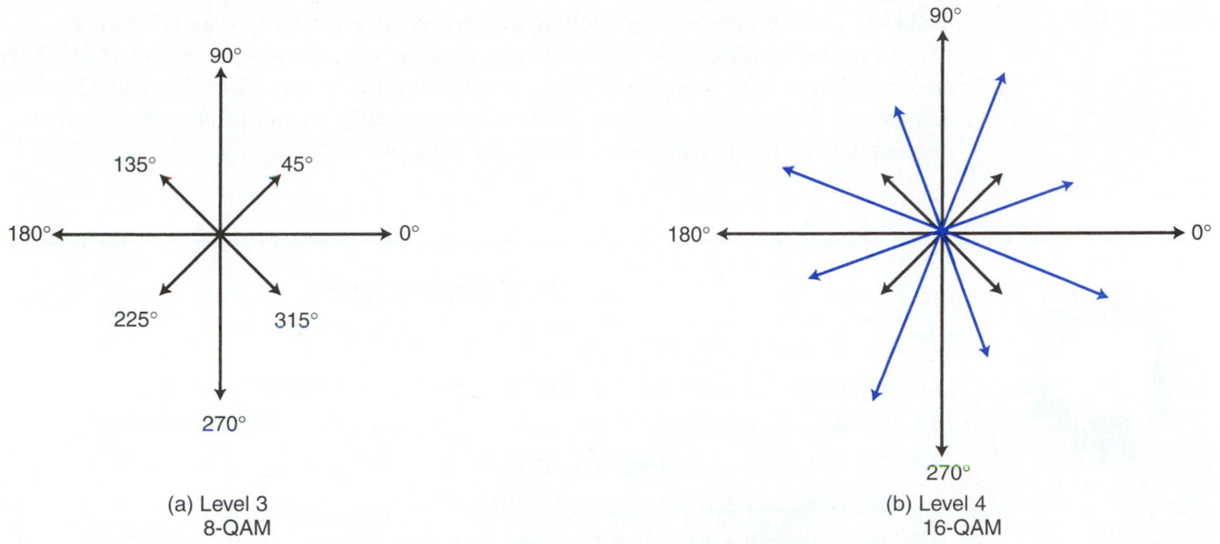

Figure 15–21 Quadrature amplitude modulation

receiver sees *half* of the 2700 Hz bandwidth). We can't control the bandwidth characteristics of a telephone line—we get what the telephone company allows us to have. Reducing bandwidth doesn't make sense in this situation, because we throw away some of the channel's capacity when we do this. Instead, it makes more sense to *fill* the available bandwidth as evenly as possible with our signal. This is the frequency domain effect of QAM.

The frequency domain "spreading" is increased even further by *scrambling* the data prior to modulation. The data scrambler in a modem transmitter mixes (exclusive-ORs) the transmit data with a pseudo-random binary code known to both the transmitter and receiver. The pseudo-random code is synchronized at the receiver and again mixed with the recovered bit stream to unscramble the data.

Modern High-Speed Modem Standards

The quality of telephone connections has steadily increased over the last 20 years, largely thanks to the use of all-digital switching equipment at telephone company facilities. In particular, a signal-to-noise ratio of 35 dB is quite possible on a modem telephone channel. The improved S/N ratio allows modems to encode more bits per transition, which increases the channel capacity and allowable data rates.

The *V32.bis* ("bis" is French for the word *encore,* which means the *second* version of the V32 standard) uses a 128-point constellation. There are 128 modulation states, which means that a V32 modem encodes 7 bits per transition ($2^7 = 128$). Six of these bits are data, the seventh is part of a *Trellis forward error correction code*. V32.bis modems can communicate at 14,400 bps over standard telephone lines while signaling at only 2400 BAUD.

The *V34* standard, introduced in 1995, upgrades the modulation techniques of V32 even further, so that data rates up to 28.8 kbps become possible. In order to achieve these data rates, a V34 modem employs *data compression*. Briefly, data compression methods allow higher data rates by ferreting out redundancies in the transmitted data stream. For example, if a transmitter wants to send a stream of 128 logic 1s, it might send a single 1 followed by a special command telling the receiver that the bit repeats 128 times.

Finally, the *ITU-T V34+* standard increases the maximum data rate limit to 33,600 bps and introduces a continuously variable data rate that varies between 2400 bps and 33,600 bps. Under bad line conditions, V34+ modems automatically accommodate and slow down as needed with no required intervention from the computer systems.

Section Checkpoint

15–23 What are the four major sections of a modem?

15–24 What is the purpose of the line hybrid in a modem?

15–25 What is the frequency range of a standard dial-up telephone line?

15–26 Why can't digital signals be directly passed through a phone line?

15–27 What is FSK? How does it represent digital 1s and 0s?

15–28 Explain the Bell-103 standard. List the four frequencies in this standard and explain each one.

15–29 What is the process of *signaling*?

15–30 Define *BAUD*. How are *bps* and *BAUD* different? (Which one is digital, which one is analog?)

15–31 What is a single-level modulation system?

15–32 What determines the bandwidth of a modulated signal (such as FSK)?

15–33 Why are multilevel modulation systems used? What limitation of single-level systems is overcome by their application?

15–34 What channel parameter limits the number of bits per transition?

15–35 What channel parameter limits the signaling rate?

15–36 What two channel characteristics determine its total information-carrying capacity?

15–37 Why is PSK preferred over FSK for high-speed modems?

15–38 What is a constellation diagram?

15–39 What is QAM? Explain how the addition of amplitude modulation helps improve noise immunity.

15–4 ASYNCHRONOUS DATA AND UARTs

When data is to be sent over a distance, it is first converted from parallel into serial form before being sent to the modulator stage. Once the data has been converted to serial (where only one bit at a time is sent), a problem arises: How can the receiver tell the correct time of arrival for each of the data bits?

In order for the receiver to determine this information, a *clock* signal is required. This is shown in Figure 15–22. Figure 15–22 shows *synchronous* serial data. The data is

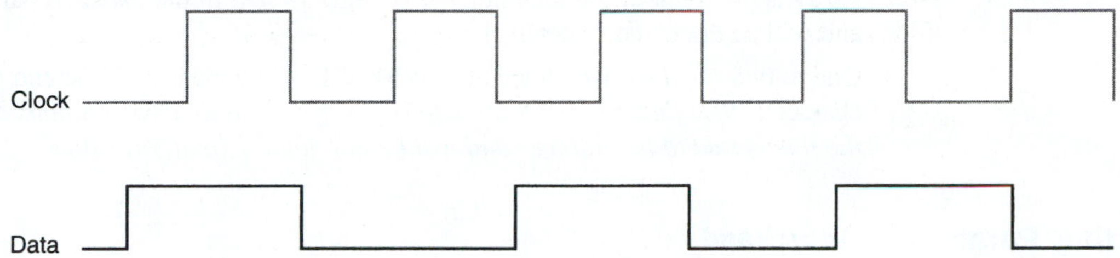

Figure 15–22 Synchronous data transfer

sent one bit at a time, but in addition to the data, a *clock* signal is sent along with it. The receiver uses this clock to synchronize itself with the data. In the figure, the receiver looks for the next data bit on the *rising* edge of the clock signal. Note that there is no apparent way for the receiver to determine when the next *byte* begins (character synchronization). This is usually taken care of automatically by a USART *(Universal Synchronous Asynchronous Receiver Transmitter)* IC.

Synchronous data requires that both a clock signal and a data signal be carried between the transmitter and receiver, which adds slightly to the complexity of modem circuitry. Also, synchronous data can't be "stopped" once a transmission has begun, or the receiver will lose synchronization. Many communication applications use *asynchronous* data. "Asynchronous" means that the data is sent without a clock signal. But wait: A clock signal is required for a receiver to recover the data. Where does the receiver get the clock signal? The answer is that the receiver generates its *own* clock signal and must be programmed with the correct clock frequency beforehand.

In synchronous transmission, a clock must be sent with the data.
An asynchronous receiver must generate its own clock (and the frequency of that clock must be known to the receiver).

Asynchronous Data Formatting

Because no clock is sent with asynchronous data, the receiver must generate its own clock. In addition, there is additional information sent to mark the beginning and end of a character. Because of this added overhead, asynchronous data transfers operate a little more slowly than synchronous transfers, given the same data rate (clock frequencies). Asynchronous transfer is popular because it is much more flexible. For example, asynchronous data need not be sent in a continuous stream. The data flow can start and stop without the receiver losing synchronization. Therefore, it is well suited for sources of intermittent data, such as the results of a person typing on a keyboard.

An asynchronous character consists of the following sequence:

1. A *start bit,* which is always a logic 0 and is one bit-time wide.
2. The string of *data bits* for the character, starting at the *least-significant bit,* or *LSB,* and ending with the most-significant bit. Five to eight data bits can be sent, depending on the protocol (rules of communication). Most applications use eight data bits, because that is one byte.
3. An optional *parity* bit, which is a way to allow the receiver to perform crude error-checking on the received character. Parity can be *none* (none is sent), *odd,* or *even.*

Parity is the count of the total number of logic 1s sent in the data and parity fields and will be discussed in detail shortly.

4. One to two *stop bits,* which are always logic 1s and which mark the end of the character. *Note that whenever no data is being sent on an asynchronous data line, the line is said to be "idling" and will be in a logic 1 (marking) state.*

Transmitting Parameter Shorthand

The settings used by an asynchronous transmitter are called the *parameters* of the transmitter. The parameters include the data rate, number of data bits, parity, and number of stop bits. A shorthand way of expressing these combinations is often used. For example, suppose we are told to set a serial port as follows:

- 9600 bps; 8, N, 1

This information tells us that the data rate is 9600 bps (this will determine the frequency of the receiver and transmitter clock oscillators), that 8 data bits are expected, no parity will be used, and 1 stop bit will be sent. Usually there is a place in software where this information must be entered in order to properly configure the computer for communication. Suppose that you saw the following information:

- 4800 bps; 7, E, 2

The parameter settings are 4800 bps, 7 data bits, even parity, and 2 stop bits.

An Asynchronous Character

Figure 15–23 shows the character A3H (the *H* indicates *hexadecimal,* or base-16 number) sent using the parameters 8,N,1. You can see that each of the parts of the transmitted character are simply sent in order. Note that the data line is *marking* (holding a logic 1) prior

Figure 15–23 A3H sent 8,N,1

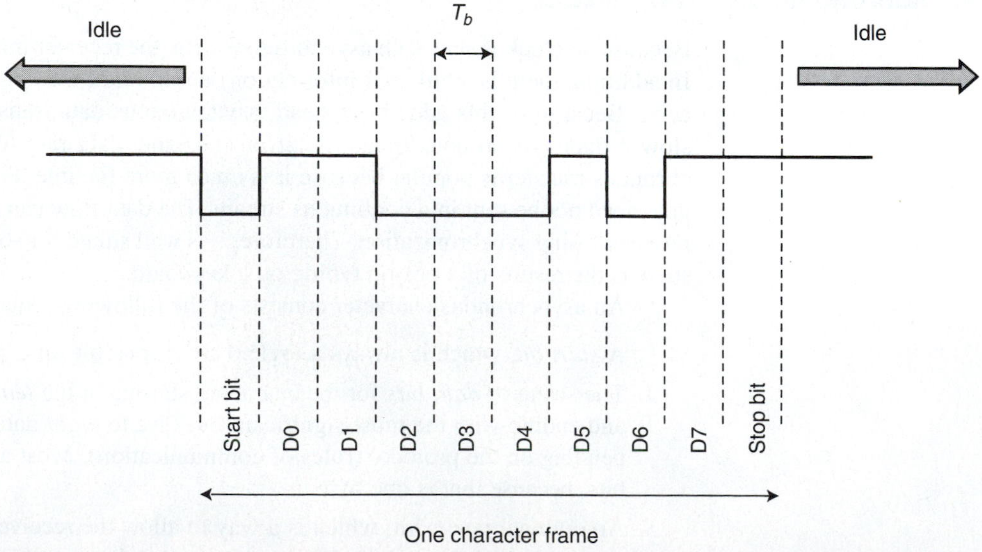

to the sending of the character. This is called the *idle* condition. A technician can immediately recognize an idle line (one which is carrying no data) by connecting an oscilloscope and observing a steady logic 1. The first bit to be sent in the figure is the *start bit,* which is always a logic 0. This is followed by the eight *data bits* (notice that D0 is the least significant bit, and D7 is the most). Finally, the transmission of the *stop bit* completes the character.

Notice that an asynchronous character always begins with start bit (logic 0) and ends with a stop bit (logic 1). Be careful—it is easy to misinterpret the binary data by reversing the bits! When we write the number A3H (as in Figure 15–23), we get the following sequence:

$$A3_{16} =$$

D7	D6	D5	D4	D3	D2	D1	D0
1	0	1	0	0	0	1	1

Data and Character Rates

By definition, the *data rate* of a serial communication is the number of bits transferred per second, as given by equation 15–2. By knowing the data rate of an asynchronous character, we can also calculate the *character rate* of the transmission. The *character rate* is the number of complete characters transmitted in one second. It is calculated as follows:

$$CPS = \frac{1}{T_C} \qquad \textbf{(15–8)}$$

where CPS is the character rate in characters per second, and T_C is the time needed to send one character.

EXAMPLE 15–9

If the data rate is 1000 bps in Figure 15–23, calculate (a) the time for one bit T_b, (b) the time for one character, T_C, and (c) the number of characters that can be sent in one second in CPS.

Solution

a. Equation 15–2 can be used to find the time for each bit:

$$\text{bps} = \frac{1}{T_b}$$

$$T_b = \frac{1}{\text{bps}} = \frac{1}{1000 \text{ bps}} = \underline{\underline{1 \text{ ms}}}$$

b. The character time must be calculated to find the CPS rate. There are *10* bits in each transmitted character. We can learn this by taking "inventory" of the transmitted data as follows:

1	Start bit
8	Data bits
+ 1	Stop bit
10	Total bits in the character

Each bit is 1 ms wide, so the time for a character is

$$T_C = (10 \text{ bits/character})(1 \text{ ms/bit}) = \underline{\underline{10 \text{ ms}}}$$

c. Now that we know the character time, we can use equation 15–8:

$$\text{CPS} = \frac{1}{T_C} = \frac{1}{10 \text{ ms}} = \underline{\underline{100 \text{ CPS}}}$$

Therefore, 100 characters can be sent every second.

EXAMPLE 15–10 Determine the hexadecimal value of the character being sent in Figure 15–24 if the parameters are known to be 7,N,1.

Solution

Simply write out the binary number (remembering that the least significant bit goes first and is written towards the *right*):

D6	D5	D4	D3	D2	D1	D0
1	0	0	0	0	1	1

$$= \underline{\underline{43 \text{ H}}}$$

Error Detection with Parity

Parity is an error-detection method that is based on counting the number of logic 1s sent in a character frame. The count of logic 1s includes the data bits *and* the parity bit. In *even* parity, the total number of logic 1s sent will be an *even* number; in *odd* parity, the number of 1s is an *odd* number.

This sounds confusing, but it is actually quite straightforward. Suppose that we wanted to generate *even* parity for the 8-bit data byte A3H. The result would be worked

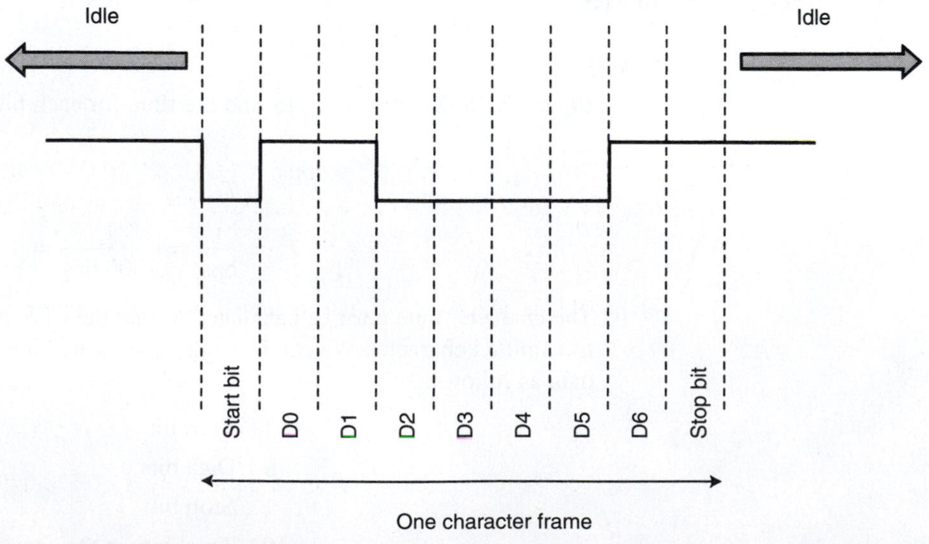

Figure 15–24 An unknown character sent 7,N,1

Figure 15–25
Generating even parity for A3H

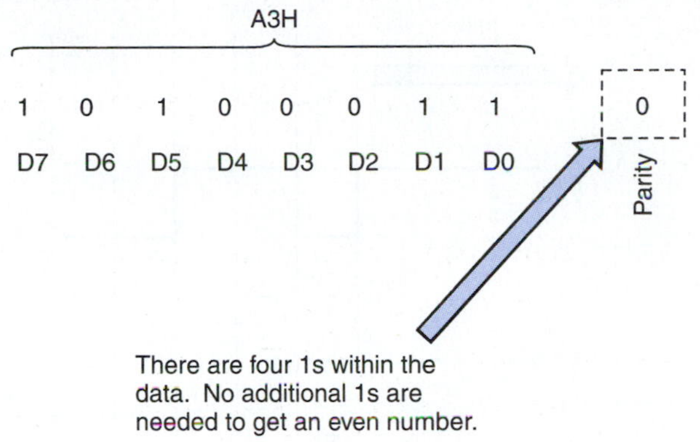
A3H
1 0 1 0 0 0 1 1
D7 D6 D5 D4 D3 D2 D1 D0
0
Parity
There are four 1s within the data. No additional 1s are needed to get an even number.

Figure 15–26
Generating even parity for 23H

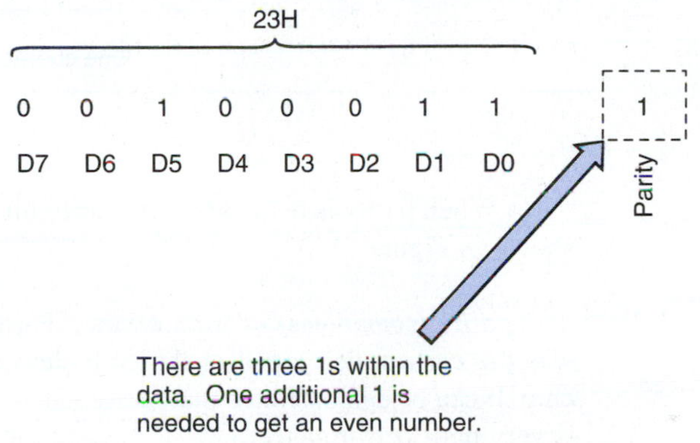
23H
0 0 1 0 0 0 1 1
D7 D6 D5 D4 D3 D2 D1 D0
1
Parity
There are three 1s within the data. One additional 1 is needed to get an even number.

out according to Figure 15–25. In the figure, *four* logic 1s have been sent, which is an even number. Note that this count includes the parity bit. Because there are already an even number of 1s in the data (A3H), the parity bit is set to zero. Look at the formation of even parity for the data 23H in Figure 15–26. As you can see, the data 23H only contains three logic 1s, which is an odd number. To get a total number of 1s that is even, one more is needed; thus the parity bit is set to a logic 1 in this case.

EXAMPLE 15–11 Calculate the *odd* parity bit value for the 8-bit data value of E6H.

Solution

Write E6H in binary and count the number of logic 1s:

$$E6_{16} =$$

D7	D6	D5	D4	D3	D2	D1	D0
1	1	1	0	0	1	1	0

There are *five* 1s in the byte, which is already an odd number. Therefore, the parity bit should be a logic 0.

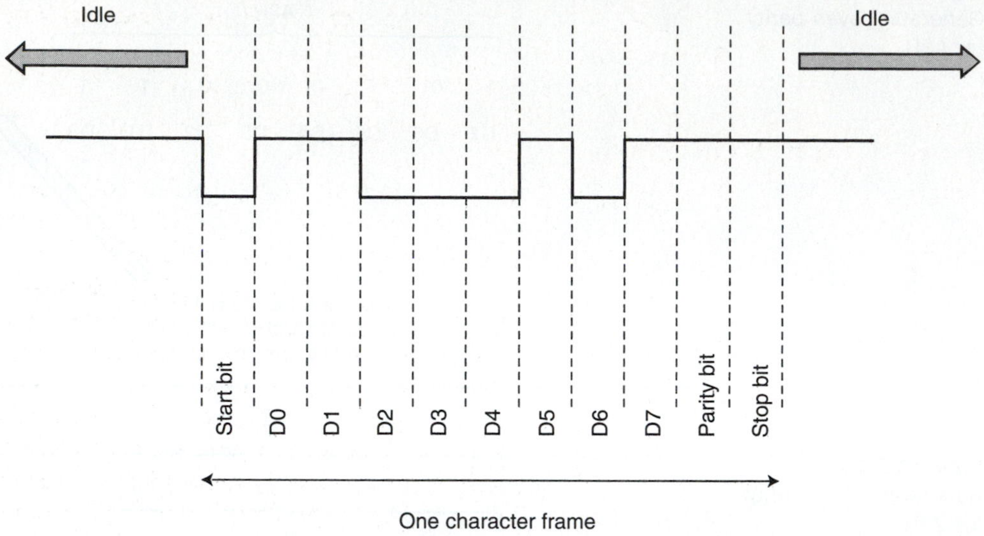

Figure 15–27 A3H sent 8,0,1

When parity is being sent, the parity bit is sent immediately after the last data bit, as shown in Figure 15–27.

How parity (sometimes) detects errors Parity is a very weak error-detection system. It was the earliest developed of the error-detection methods and is built into every UART chip. It can be demonstrated that parity will trap less than 50% of all possible errors, which is very poor performance.

Suppose that a receiving computer is set to receive data with the parameters 8,O,1 and the data of Figure 15–28 comes in. How will the computer detect the error? The

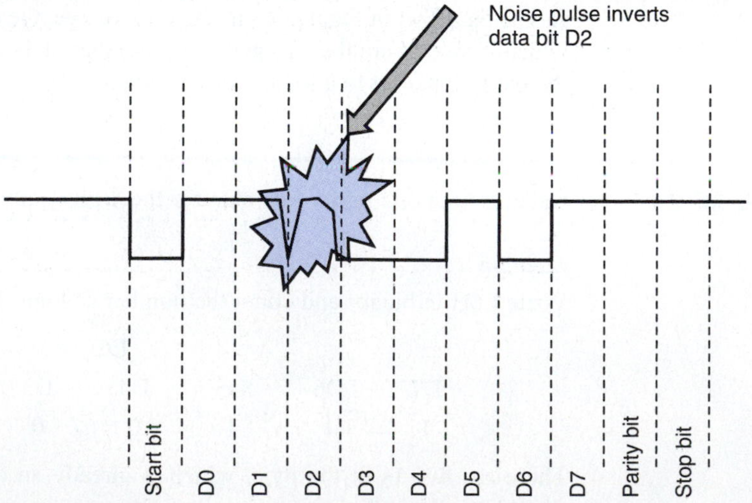

Figure 15–28 An error detected by parity

receiving computer sees data that differs from the original because data bit 2 has been changed by a noise pulse. D2 now reads as a 1 instead of a 0. The total count of received logic 1s is now *six* (don't forget about the parity bit). This is an even number, which doesn't agree with the parameter settings, which specify that all received characters should have *odd* parity. The computer therefore detects the discrepancy.

But wait, you say. What if *two* of the data bits get destroyed by the noise impulse? In that case, the computer sees no error whatsoever. *All it cares about is getting an odd number of logic 1s*. There are *many* combinations of 8 bits that will give an odd number of 1s, but only one of them is the correct data value! This is why parity is considered a weak error-detection method. We will discuss better error-detection methods in the next section.

UART Integrated Circuits

A UART is an integrated circuit designed to convert parallel data to serial (*serialization*) and serial data back to parallel (*deserialization*). There are literally dozens of different types of UART chips, but all of them perform the same basic functions. The latest generations of UARTs are very sophisticated devices and require considerable programming skills to utilize them in practical circuits. Examples of these devices include the Intel 8251A USART, the Motorola 6850 ACIA (*Asynchronous Communications Interface Adapter*), and the Intel 16550 USART (which is used as the basis of the serial port in most personal computers).

There exists a family of simpler UARTs based on a common design originated by General Instruments. The Harris HD6402 is a modern member of this family. It is implemented entirely in CMOS and requires no programming, and thus is an excellent first UART to study. (See Figure 15–29.)

The HD6402 has four major sections. These are the *control, transmitter, receiver,* and *status* groups.

Control section The control section of the UART accepts parameter settings on the *SBS, EPE, PI, CLS1,* and *CLS2* pins. The *CRL* (control register load) pin must be active (high) in order for inputs on the control pins to be active. In many experimental setups, *CRL* is permanently connected to 5 volts and the parameter setting pins are tied high or low, as appropriate. For example, to set the UART for transmitting and receiving with the parameters 8,N,1, the control pins must be set as follows:

- SBS = 0 *(Selects 1 stop bit)*
- PI = 1 *(Inhibits parity when high)*
- EPE = X *(Doesn't care, can be high or low; no parity is selected)*
- CLS1 = 1
- CLS2 = 2 *(Together, CLS1 and CLS2 select the number of data bits)*

Transmitter The UART transmitter converts parallel data (presented on the *TBR1–TBR8* input pins) into serial data on the *TRO* (Transmit Serial Output) pin. The parallel data can come from just about any source; in an experimental setup, it can come from an 8-pole DIP switch. In order to clock the parallel data into the UART, the */TBRL* (active-low) pin is

intersil

HD-6402

CMOS Universal Asynchronous Receiver Transmitter (UART)

March 1997

Features

- 8.0MHz Operating Frequency (HD-6402B)
- 2.0MHz Operating Frequency (HD-6402R)
- Low Power CMOS Design
- Programmable Word Length, Stop Bits and Parity
- Automatic Data Formatting and Status Generation
- Compatible with Industry Standard UARTs
- Single +5V Power Supply
- CMOS/TTL Compatible Inputs

Description

The HD-6402 is a CMOS UART for interfacing computers or microprocessors to an asynchronous serial data channel. The receiver converts serial start, data, parity and stop bits. The transmitter converts parallel data into serial form and automatically adds start, parity and stop bits. The data word length can be 5, 6, 7 or 8 bits. Parity may be odd or even. Parity checking and generation can be inhibited. The stop bits may be one or two or one and one-half when transmitting 5-bit code.

The HD-6402 can be used in a wide range of applications including modems, printers, peripherals and remote data acquisition systems. Utilizing the Intersil advanced scaled SAJI IV CMOS process permits operation clock frequencies up to 8.0MHz (500K Baud). Power requirements, by comparison, are reduced from 300mW to 10mW. Status logic increases flexibility and simplifies the user interface.

Ordering Information

PACKAGE	TEMPERATURE RANGE	2MHz = 125K BAUD	8MHz = 500K BAUD	PKG. NO.
Plastic DIP	-40°C to +85°C	HD3-6402R-9	HD3-6402B-9	E40.6
CERDIP	-40°C to +85°C	HD1-6402R-9	HD1-6402B-9	F40.6
SMD#	-55°C to +125°C	5962-9052501MQA	5962-9052502MQA	F40.6

Pinout

HD-6402 (PDIP, CERDIP)
TOP VIEW

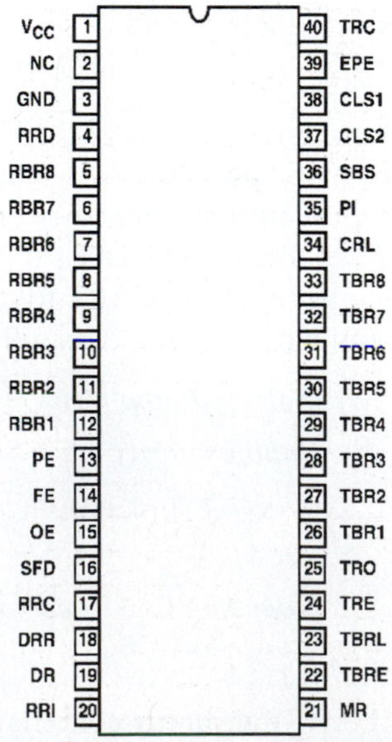

File Number **2956.1**

Figure 15–29 Harris HD6402 UART data sheet (Source: Copyrighted by Intersil Corporation. Used with permission)

Functional Diagram

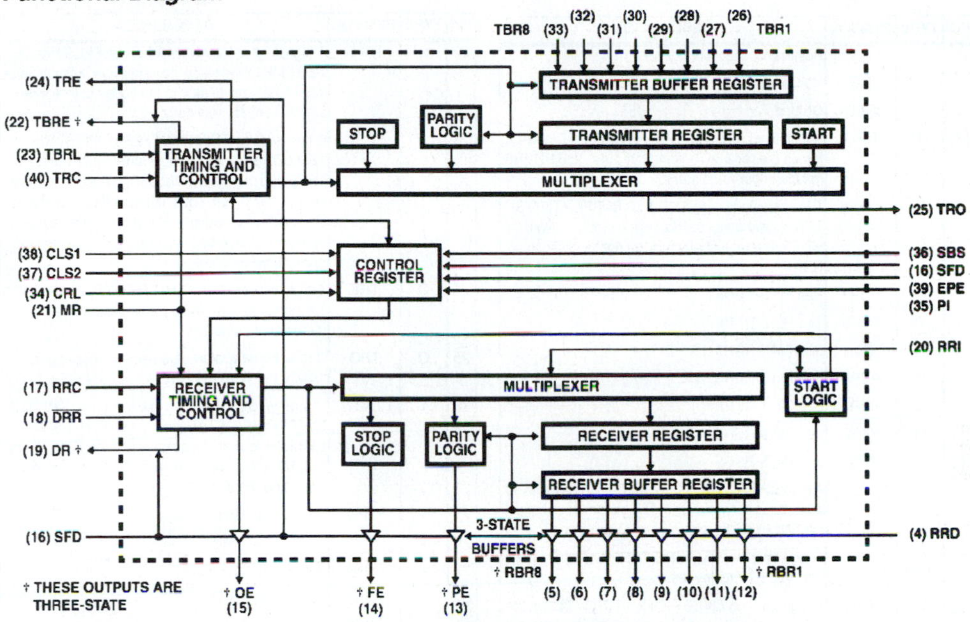

Control Definition

CONTROL WORD					CHARACTER FORMAT			
CLS 2	CLS 1	PI	EPE	SBS	START BIT	DATA BITS	PARITY BIT	STOP BITS
0	0	0	0	0	1	5	ODD	1
0	0	0	0	1	1	5	ODD	1.5
0	0	0	1	0	1	5	EVEN	1
0	0	0	1	1	1	5	EVEN	1.5
0	0	1	X	0	1	5	NONE	1
0	0	1	X	1	1	5	NONE	1.5
0	1	0	0	0	1	6	ODD	1
0	1	0	0	1	1	6	ODD	2
0	1	0	1	0	1	6	EVEN	1
0	1	0	1	1	1	6	EVEN	2
0	1	1	X	0	1	6	NONE	1
0	1	1	x	1	1	6	NONE	2
1	0	0	0	0	1	7	ODD	1
1	0	0	0	1	1	7	ODD	2
1	0	0	1	0	1	7	EVEN	1
1	0	0	1	1	1	7	EVEN	2
1	0	1	X	0	1	7	NONE	1
1	0	1	x	1	1	7	NONE	2
1	1	0	0	0	1	8	ODD	1
1	1	0	0	1	1	8	ODD	2
1	1	0	1	0	1	8	EVEN	1
1	1	0	1	1	1	8	EVEN	2
1	1	1	X	0	1	8	NONE	1
1	1	1	x	1	1	8	NONE	2

Figure 15–29 *(continued)*

Pin Description

PIN	TYPE	SYMBOL	DESCRIPTION
1		V_{CC} †	Positive Voltage Supply
2		NC	No Connection
3		GND	Ground
4	I	RRD	A high level on RECEIVER REGISTER DISABLE forces the receiver holding out-puts RBR1-RBR8 to high impedance state.
5	O	RBR8	The contents of the RECEIVER BUFFER REGISTER appear on these three-state outputs. Word formats less than 8 characters are right justified to RBR1.
6	O	RBR7	See Pin 5-RBR8
7	O	RBR6	See Pin 5-RBR8
8	O	RBR5	See Pin 5-RBR8
9	O	RBR4	See Pin 5-RBR8
10	O	RBR3	See Pin 5-RBR8
11	O	RBR2	See Pin 5-RBR8
12	O	RBR1	See Pin 5-RBR8
13	O	PE	A high level on PARITY ERROR indicates received parity does not match parity programmed by control bits. When parity is inhibited this output is low.
14	O	FE	A high level on FRAMING ERROR indicates the first stop bit was invalid.
15	O	OE	A high level on OVERRUN ERROR indicates the data received flag was not cleared before the last character was transferred to the receiver buffer register.
16	I	SFD	A high level on STATUS FLAGS DISABLE forces the outputs PE, FE, OE, DR, TBRE to a high impedance state.
17	I	RRC	The Receiver register clock is 16X the receiver data rate.
18	I	$\overline{DRR}$	A low level on DATA RECEIVED RESET clears the data received output DR to a low level.
19	O	DR	A high level on DATA RECEIVED indicates a character has been received and transferred to the receiver buffer register.
20	I	RRI	Serial data on RECEIVER REGISTER INPUT is clocked into the receiver register.
21	I	MR	A high level on MASTER RESET clears PE, FE, OE and DR to a low level and sets the transmitter register empty (TRE) to a high level 18 clock cycles after MR falling edge. MR does not clear the receiver buffer register. This input must be pulsed at least once after power up. The HD-6402 must be master reset after power up. The reset pulse should meet V_{IH} and t_{MR}. Wait 18 clock cycles after the falling edge of MR before beginning operation.
22	O	TBRE	A high level on TRANSMITTER BUFFER REGISTER EMPTY indicates the transmitter buffer register has transferred its data to the transmitter register and is ready for new data.
23	I	$\overline{TBRL}$	A low level on TRANSMITTER BUFFER REGISTER LOAD transfers data from inputs TBR1-TBR8 into the transmitter buffer register. A low to high transition on $\overline{TBRL}$ initiates data transfer to the transmitter register. If busy, transfer is automatically delayed so that the two characters are transmitted end to end.
24	O	TRE	A high level on TRANSMITTER REGISTER EMPTY indicates completed transmission of a character including stop bits.
25	O	TRO	Character data, start data and stop bits appear serially at the TRANSMITTER REGISTER OUTPUT.
26	I	TRB1	Character data is loaded into the TRANSMITTER BUFFER REGISTER via inputs TBR1-TBR8. For character formats less than 8 bits the TBR8, 7 and 6 inputs are ignored corresponding to their programmed word length.
27	I	TBR2	See Pin 26-TBR1.
28	I	TBR3	See Pin 26-TBR1.
29	I	TBR4	See Pin 26-TBR1.
30	I	TBR5	See Pin 26-TBR1.
31	I	TBR6	See Pin 26-TBR1.
32	I	TBR7	See Pin 26-TBR1.
33	I	TBR8	See Pin 26-TBR1.
34	I	CRL	A high level on CONTROL REGISTER LOAD loads the control register with the control word. The control word is latched on the falling edge of CRL. CRL may be tied high.
35	I	PI	A high level on PARITY INHIBIT inhibits parity generation, parity checking and forces PE output low.
36	I	SBS	A high level on STOP BIT SELECT selects 1.5 stop bits for 5 character format and 2 stop bits for other lengths.
37	I	CLS2	These inputs program the CHARACTER LENGTH SELECTED (CLS1 low CLS2 low 5 bits) (CLS1 high CLS2 low 6 bits) (CLS1 low CLS2 high 7 bits) (CLS1 high CLS2 high 8 bits.)
38	I	CLS1	See Pin 37-CLS2.
39	I	EPE	When PI is low, a high level on EVEN PARITY ENABLE generates and checks even parity. A low level selects odd parity.
40	I	TRC	The TRANSMITTER REGISTER CLOCK is 16X the transmit data rate.

† A 0.1μF decoupling capacitor from the V_{CC} pin to the GND is recommended.

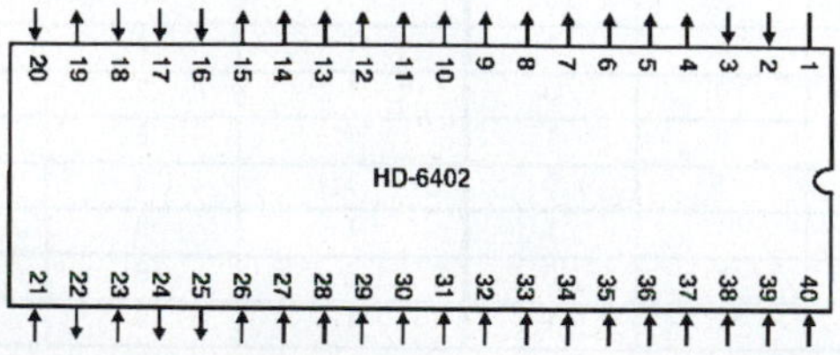

Figure 15–29 *(continued)*

Transmitter Operation

The transmitter section accepts parallel data, formats the data and transmits the data in serial form on the Transmitter Register Output (TRO) terminal (See serial data format). Data is loaded from the inputs TBR1-TBR8 into the Transmitter Buffer Register by applying a logic low on the Transmitter Buffer Register Load (TBRL) input (A). Valid data must be present at least t_{set} prior to and t_{hold} following the rising edge of $\overline{\text{TBRL}}$. If words less than 8 bits are used, only the least significant bits are transmitted. The character is right justified, so the least significant bit corresponds to TBR1 (B).

The rising edge of $\overline{\text{TBRL}}$ clears Transmitter Buffer Register Empty (TBRE). 0 to 1 Clock cycles later, data is transferred to the transmitter register, the Transmitter Register Empty (TRE) pin goes to a low state, TBRE is set high and serial data information is transmitted. The output data is clocked by Transmitter Register Clock (TRC) at a clock rate 16 times the data rate. A second low level pulse on $\overline{\text{TBRL}}$ loads data into the Transmitter Buffer Register (C). Data transfer to the transmitter register is delayed until transmission of the current data is complete (D). Data is automatically transferred to the transmitter register and transmission of that character begins one clock cycle later.

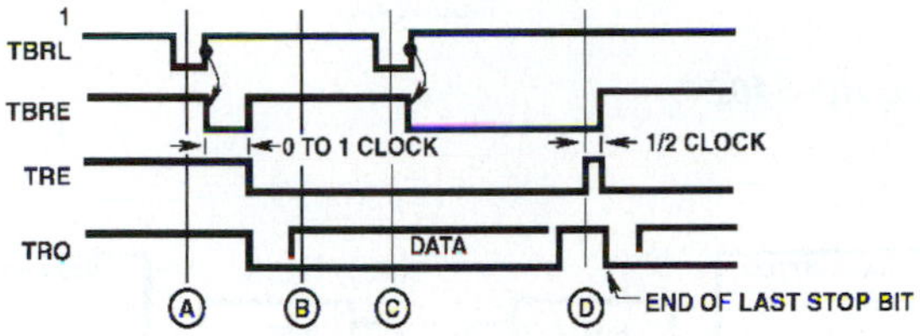

FIGURE 1. TRANSMITTER TIMING (NOT TO SCALE)

Receiver Operation

Data is received in serial form at the Receiver Register Input (RRI). When no data is being received, RRI must remain high. The data is clocked through the Receiver Register Clock (RRC). The clock rate is 16 times the data rate. A low level on Data Received Reset ($\overline{\text{DRR}}$) clears the Data Receiver (DR) line (A). During the first stop bit data is transferred from the Receiver Register to the Receiver Buffer Register (RBR) (B). If the word is less than 8 bits, the unused most significant bits will be a logic low. The output character is right justified to the least significant bit RBR1. A logic high on Overrun Error (OE) indicates overruns. An overrun occurs when DR has not been cleared before the present character was transferred to the RBR. One clock cycle later DR is reset to a logic high, and Framing Error (FE) is evaluated (C). A logic high on FE indicates an invalid stop bit was received, a framing error. A logic high on Parity Error (PE) indicates a parity error.

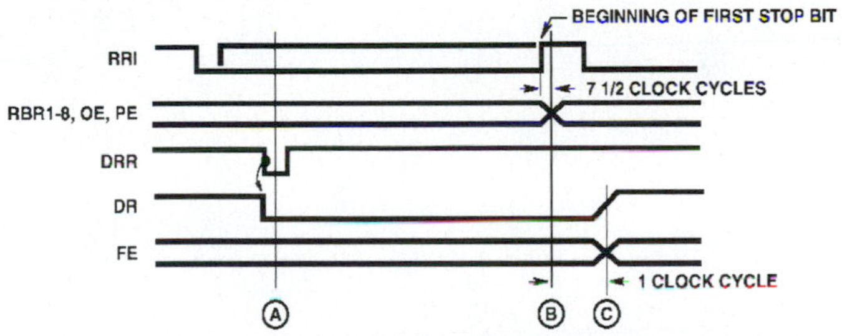

FIGURE 2. RECEIVER TIMING (NOT TO SCALE)

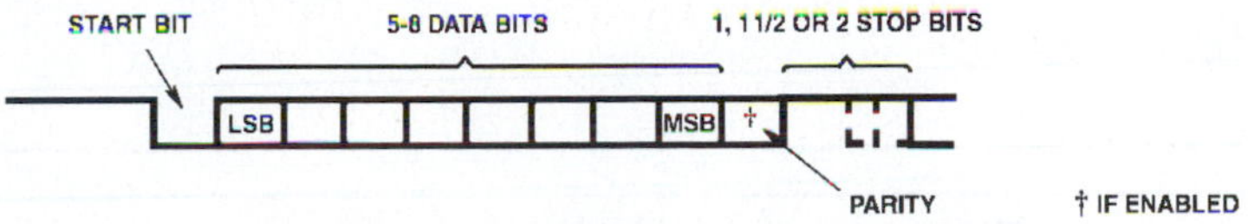

FIGURE 3. SERIAL DATA FORMAT

Figure 15–29 *(continued)*

Start Bit Detection

The receiver uses a 16X clock timing. The start bit could have occurred as much as one clock cycle before it was detected, as indicated by the shaded portion (A). The center of the start bit is defined as clock count 7 1/2. If the receiver clock is a symmetrical square wave, the center of the start bit will be located within ±1/2 clock cycle, ±1/32 bit or 3.125% giving a receiver margin of 46.875%. The receiver begins searching for the next start bit at the center of the first stop bit.

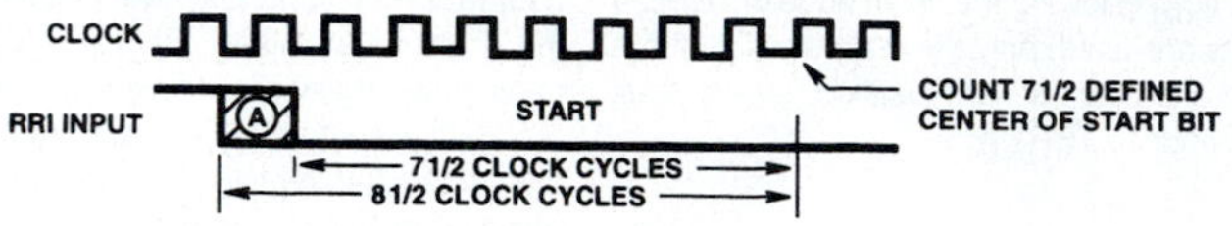

FIGURE 4.

Interfacing with the HD-6402

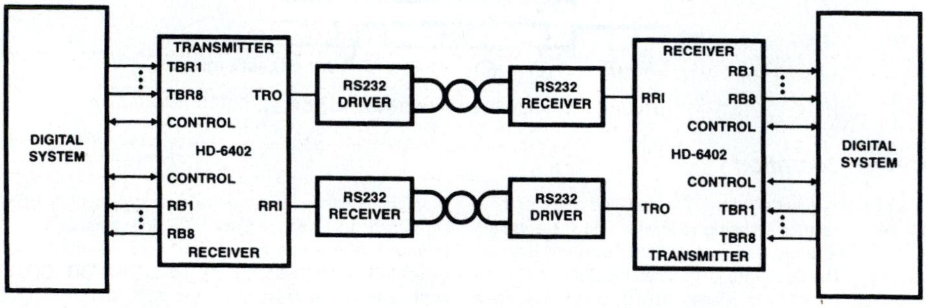

FIGURE 5. TYPICAL SERIAL DATA LINK

Figure 15–29 *(continued)*

Absolute Maximum Ratings

Supply Voltage . . . . +8.0V
Input, Output or I/O Voltage Applied. . . . GND -0.5V to V_{CC} +0.5V
Storage Temperature Range . . . . -65°C to +150°C
Junction Temperature . . . . +175°C
Lead Temperature (Soldering 10s) . . . . +300°C
ESD Classification . . . . Class 1
Typical Derating Factor. . . . . 1mA/MHz Increase in ICCOP

Thermal Information

Thermal Resistance (Typical)	θ_{JA}	θ_{JC}
CERDIP Package	50°C/W	12°C/W
PDIP Package	50°C/W	N/A

Gate Count . . . . 1643 Gates

CAUTION: Stresses above those listed in "Absolute Maximum Ratings" may cause permanent damage to the device. This is a stress only rating and operation of the device at these or any other conditions above those indicated in the operational sections of this specification is not implied.

Operating Conditions

Operating Voltage Range . . . . +4.5V to +5.5V
Operating Temperature Range
HD-6402R-9, HD6402B-9 . . . . -40°C to +85°C

DC Electrical Specifications V_{CC} = 5.0V ± 10%, T_A = -40°C to +85°C (HD-6402R-9, HD-6402B-9)

SYMBOL	PARAMETER	LIMITS MIN	LIMITS MAX	UNITS	CONDITIONS
V_{IH}	Logical "1" Input Voltage	2.0	-	V	V_{CC} = 5.5V
V_{IL}	Logical "0" Input Voltage	-	0.8	V	V_{CC} = 4.5V
II	Input Leakage Current	-1.0	1.0	μA	V_{IN} = GND or V_{CC}, V_{CC} = 5.5V
V_{OH}	Logical "1" Output Voltage	3.0 V_{CC} -0.4	- -	V	I_{OH} = -2.5mA, V_{CC} = 4.5V I_{OH} = -100μA
V_{OL}	Logical "0" Output Voltage	-	0.4	V	I_{OL} = +2.5mA, V_{CC} = 4.5V
I_O	Output Leakage Current	-1.0	1.0	μA	V_O = GND or V_{CC}, V_{CC} = 5.5V
ICCSB	Standby Supply Current	-	100	μA	V_{IN} = GND or V_{CC}; V_{CC} = 5.5V, Output Open
ICCOP	Operating Supply Current (See Note)	-	2.0	mA	V_{CC} = 5.5V, Clock Freq. = 2MHz, V_{IN} = V_{CC} or GND, Outputs Open

NOTE: Guaranteed, but not 100% tested

Capacitance T_A = +25°C

PARAMETER	SYMBOL	CONDITIONS	LIMIT TYPICAL	UNITS
Input Capacitance	CIN	Freq. = 1MHz, all measurements are referenced to device GND	25	pF
Output Capacitance	COUT		25	pF

AC Electrical Specifications V_{CC} = 5.0V ± 10%, T_A = -40°C to +85°C (HD-6402R-9, HD6402B-9)

SYMBOL	PARAMETER	LIMITS HD-6402R MIN	LIMITS HD-6402R MAX	LIMITS HD-6402B MIN	LIMITS HD-6402B MAX	UNITS	CONDITIONS
(1) fCLOCK	Clock Frequency	D.C.	2.0	D.C.	8.0	MHz	C_L = 50pF See Switching Waveform
(2) t_{PW}	Pulse Widths, CRL, DRR, TBRL	150	-	75	-	ns	
(3) t_{MR}	Pulse Width MR	150	-	150	-	ns	
(4) t_{SET}	Input Data Setup Time	50	-	20	-	ns	
(5) t_{HOLD}	Input Data Hold Time	60	-	20	-	ns	
(6) t_{EN}	Output Enable Time	-	160	-	35	ns	

Figure 15–29 *(continued)*

Switching Waveforms

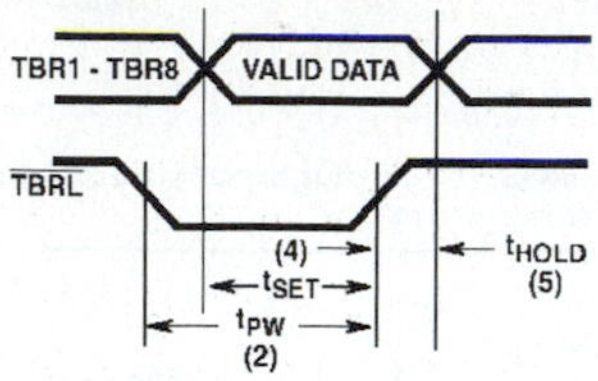

FIGURE 6. DATA INPUT CYCLE

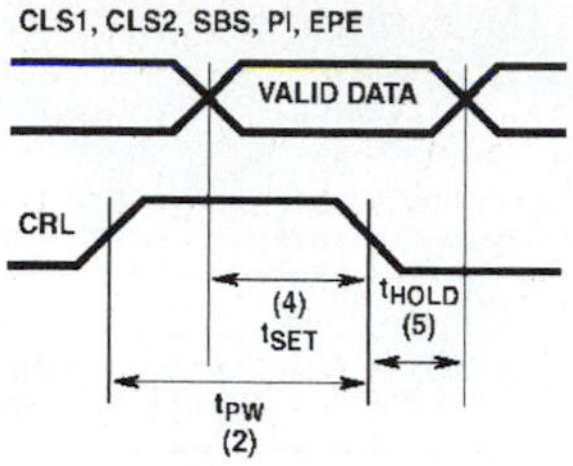

FIGURE 7. CONTROL REGISTER LOAD CYCLE

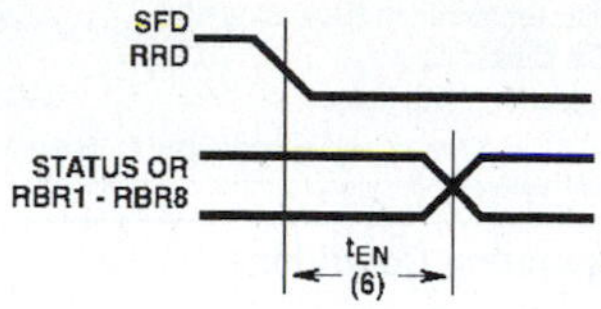

FIGURE 8. STATUS FLAG OUTPUT ENABLE TIME OR DATA OUTPUT ENABLE TIME

A.C. Testing Input, Output Waveform

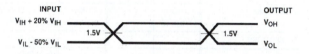

FIGURE 9.

NOTE: A.C. Testing: All input signals must switch between V_{IL} - 50% V_{IL} and V_{IH} + 20% V_{IH}. Input rise and fall times are driven at 1ns/V.

Test Circuit

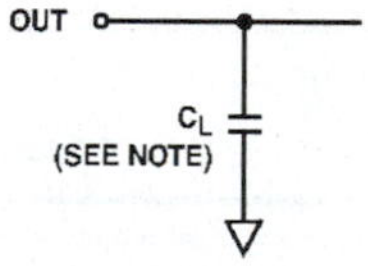

FIGURE 10.

NOTE: Includes stray and jig capacitance, C_L = 50pF.

Figure 15–29 *(continued)*

pulled low while the data is present on the parallel data input pins. Then */TBRL* is driven back high, which begins the transmission of the serial character.

Receiver The purpose of the receiver section is to convert incoming serial data into parallel output. The UART constantly monitors the *RRI* (Receiver Register Input) for activity. The *RRI* pin is where serial data enters the UART. When serial input activity takes place, the UART automatically reassembles the incoming data into parallel form and presents it on the receiver parallel data output pins, *RBR1–RBR8*. Note that these eight output pins are tri-state outputs; to enable them, the *RRD* pin must be driven low. The parallel-data output pins can be connected through buffers (to increase the current capability) to LEDs for experimental monitoring of the receiver. In this application, *RRD* is just tied to ground so that the parallel data outputs (*RBR1–RBR8*) are always enabled.

Status section The *status* section is used to check on the general conditions inside the UART. The *DR* (Data Received) pin goes high to indicate that serial data has been received (and converted to parallel); the *TRE* (Transmitter Register Empty) pin is high when the transmitter is ready for more parallel data. (Remember that the UART is meant to be interfaced with a computer, and the computer can probably present parallel data much faster to the transmitter than the transmitter can convert to serial.)

The status section also contains error indicators for the receiver section. For example, *PE* will go high if data is received with a parity error, and *OE* will go high to indicate an *overrun error* on the receiver (data was received but not removed from the parallel outputs of the UART; the UART has been "overrun" with incoming data).

Clocking requirements Most UARTs use a 16× clock, and the HD6402 is no exception. By this we mean that the clock is set to 16 times the desired data rate. For example, to operate the UART at 300 bps, the clock must be set at (16 × 300) or *4800 Hz*. Two independent clock inputs are provided on the HD6402, *TRC* and *RRC*. These are the transmitter and receiver clock inputs, respectively. They are usually connected to the same clock frequency source. The 16× clock scheme is necessary for the UART to accurately determine the center of each bit cell during reception, which minimizes its sensitivity to jitter in the received data. To locate the center of a bit, the UART internally counts to *8*, because that is one-half of 16. This process is called *center-cell sampling*, and is used in all UARTs.

EXAMPLE 15–12 What control inputs and clock frequency are needed to set the HD6402 UART for the following set of parameters: 9600 bps; 7,E,2

Solution

The frequency of the clock must be 16 times 9600 bps, or 153.6 kHz.

To get 7 data bits, *CLS1* = 0 and *CLS2* = 1.

To get even parity, *PI* = 0 and *EPE* = 1.

Two stop bits are obtained by setting *SBS* = 1.

All of these values are obtained by reviewing the data sheet for the device.

Section Checkpoint

15–40 Explain the difference between *synchronous* and *asynchronous* data.

15–41 How does the clock help in Figure 15–20?

15–42 List the parts of an asynchronous character. What is always at the end of an asynchronous character?

15–43 What is meant by the term *parameters* when referring to serial communication?

15–44 Which data bit is sent first in serial communication?

15–45 What is the difference between data rate and character rate?

15–46 What is *parity*? How does it check for errors?

15–47 Why is parity considered a weak error-detection method?

15–48 What is a UART? List the four sections of the HD6402 UART.

15–5 ERROR DETECTION AND CORRECTION

When a computer receives a data message, the process of communications is nearly complete. Well-designed systems always utilize at least one type of *error detection* to check the validity of the received data. *Error detection* is the set of procedures that allows a receiver to check the "goodness" of data. It is absolutely necessary to have some form of error detection in data communication systems!

Detecting Errors in Data Blocks

Data bytes are almost never sent by themselves. Instead, data is grouped together into *data blocks*. A data block is a group of data that are sent together. It is much more efficient to communicate blocks of data than to send isolated characters. Data blocks are also sometimes called *frames*. Both of these terms mean the same thing.

When a computer receives a data block, it applies an error-detection method to check it. Most error-detection methods are based on simple mathematical ideas, as we will soon see. If the data block is good, the computer sends a *positive acknowledge* or *ACK* signal back to the sender. This tells the sender that the transmitted message was received properly. This is very much like a person saying "OK" during a conversation in order to acknowledge both receipt and understanding of the spoken content.

What if the receiver detects an error? The sending computer needs to retransmit the original message. The listening computer sends a *negative acknowledge,* or *NAK,* signal in this case. The transmitting computer then resends the data block in question. The process will repeat until the receiver says "OK." Systems that ask the sender to repeat messages with errors are called *automatic repeat request* or *ARQ* configurations.

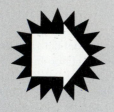

Automatic repeat request ideally requires that a receiver acknowledge *every* transmitted data block. Although this requirement helps to ensure reliability of communications, it also can slow down communications. In a later chapter we will study *windowing,*

which is a technique used by modern protocols to reduce the number of acknowledgment responses needed to transfer a group of data blocks.

Error-Correction Systems

The process of asking a sender to resend a defective message takes extra time, and in some applications that is unacceptable. A *forward error-correcting (FEC)* system is one that has the capability of recovering correct data from bad data blocks. For example, suppose that you receive a handwritten note from a friend at your office who also happens to have bad penmanship. The words are scrawled so badly that you can hardly make them out, but at least you can see:

"L?T'S GO OUT FO? ?IZ?A AT L??CH."

If you concluded that your pal wants to get pizza for lunch, then you have just witnessed a form of forward error correction in action. The reason that you were able to recover meaning from the incomplete phrase is that a lot of the information was distributed and *repeated* within the message. In other words, we would say that the information was *redundantly represented*. Part of this is due to the nature of the English language. For example, you probably guessed that "?IZ?A" was really *pizza* because, in part, that combination of missing letters made sense, and since L??CH was likely to be *lunch,* that made even more sense. A great deal of knowledge must be contained and applied within a receiver to decode messages like this. Fortunately, we can program computers to do much the same thing in a much simpler manner. There is one very important principle here:

> In order for a receiver (or computer) to detect errors within a received message, the information within that message must be represented redundantly (repeatedly).

Both error-detecting (ARQ) and error-correcting (FEC) systems require redundant information in message frames. The difference between them is that an FEC system must have *more* redundant information passed to it within a message, so that it can attempt to reconstruct a "broken" message. For that reason, forward error-correcting systems are generally more complex and may transfer data more slowly than error-detecting systems. So why use forward error correction at all?

Forward error correction is used in situations where there isn't time to resend a message. The most common example is the compact disc (CD) player. Because of the large amount of data contained on a CD, and the allowance for manufacturing tolerances, errors occur in the audio data stream of a CD on a regular basis. There isn't time to reread a defective data block from the disc; that would result in the interruption of the music. A FEC system built into the player corrects the minor errors. A CD therefore does not "skip" unless the data becomes *extremely* corrupt.

Systems for Error Detection

There are two primary error detection systems in use today, the *checksum* and *cyclic redundancy check (CRC)*. Notice that *parity* is not mentioned. Because it lets many errors slip by, parity is seldom used for this purpose.

Most protocol (rules for communication) use a unique *data block structure* to transfer their data. Many protocol use the general form shown in Figure 15–30. The figure is only a *general* layout, but most protocols follow it. The data bytes are sent all

Figure 15–30 General data block format

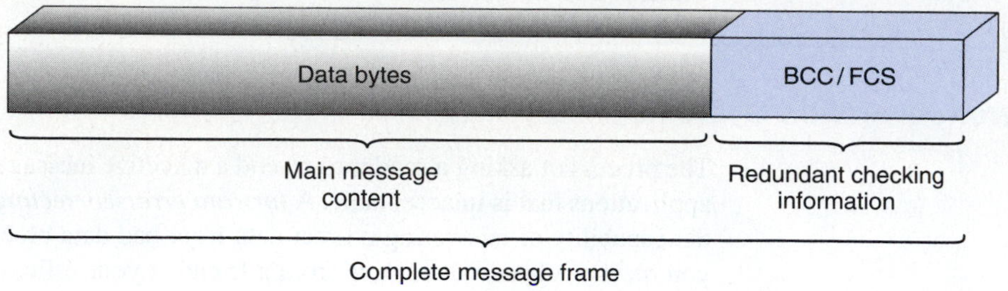
Data bytes
BCC/FCS
Main message content
Redundant checking information
Complete message frame

together first (the number of data bytes depends on the protocol in use), and the *block check character,* or *BCC,* is sent last. (Another name for the BCC is the *frame check sequence,* or *FCS.*)

How Error Detection Works

When a transmitter is getting ready to send a message, it collects the message bytes in a temporary holding place called a *buffer*. The transmitter calculates the BCC according to the rules of the protocol. The BCC is just a number calculated from the message content. Once the BCC has been calculated, the transmitter sends the entire *frame,* which includes the data and BCC, and waits for a reply from the receiver.

Meanwhile, the receiver is decoding the message. Figure 15–31 illustrates what the receiver does. The receiver calculates a *new* BCC (we'll call it BCC′) based on the message data. The receiver also reads the BCC sent from the transmitter. If the message is *good,* the new BCC (BCC′) and the one sent with the message (BCC) will agree, and the receiver will send a positive acknowledge (ACK) back to the sender. If the calculated and received versions of the BCC don't agree, something was wrong with the message and a request will be made for the transmitter to send it again.

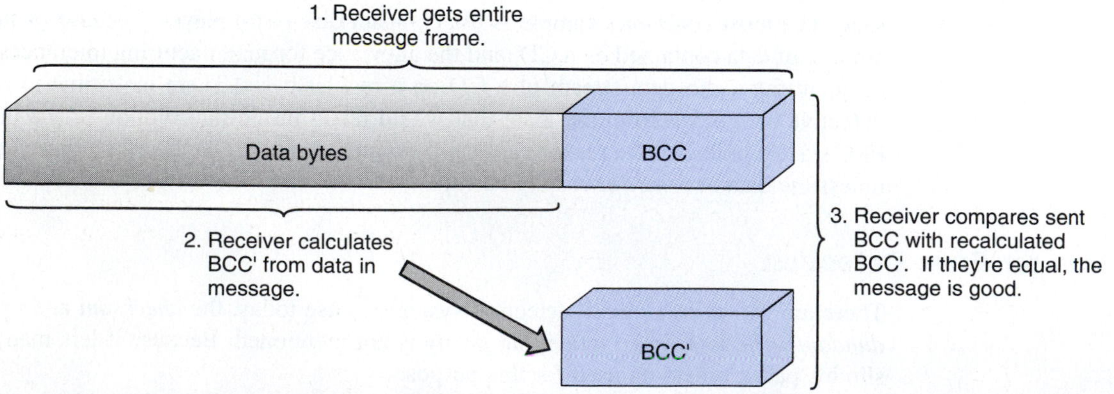
1. Receiver gets entire message frame.
Data bytes
BCC
2. Receiver calculates BCC' from data in message.
BCC'
3. Receiver compares sent BCC with recalculated BCC'. If they're equal, the message is good.

Figure 15–31 Receiver activity

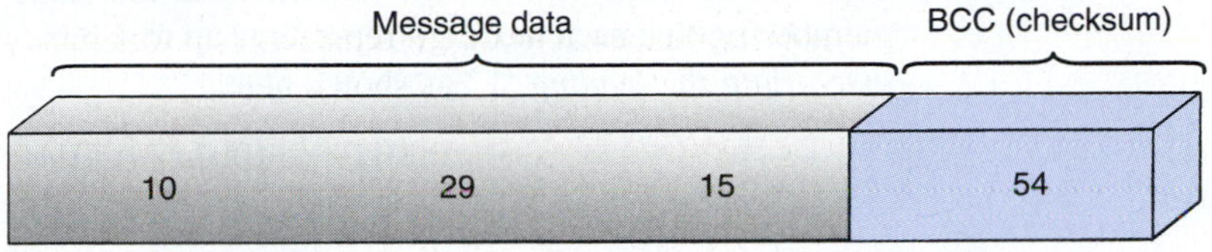

Figure 15–32 A simple message protocol

Checksum Error Detection

The *checksum* error-detection method simply uses the arithmetic sum of all data in the block as the BCC. For example, suppose that we wanted to send messages that were composed of three 2-digit numbers. We might decide to set up the protocol as follows:

- The data block will consist of the three numbers, which are the client data.
- The data block will be followed by a two-digit number that is calculated by finding the sum of the three numbers in the data block. If this number is greater than 99, only the last two digits will be used.

A data block with the contents {10, 29, 15} would therefore look like Figure 15–32.

Suppose that the following data block is received: {10, 29, 15}, with a BCC of 60. How will the receiver respond to this? It will first try to calculate what it thinks the BCC *should* be (BCC′ in Figure 15–29). It will calculate 54 because that is the sum of 10, 29, and 15. However, the transmitted block contains a BCC of 60, and that doesn't agree! The receiver will therefore send *NAK* (negative acknowledge) to request that the data block be sent again.

The only difference between the data block of Figure 15–32 and an actual block that might be sent by a computer is the *number base* in which the arithmetic is being performed. Computers use binary data (base 2) for arithmetic, so the numbers in the data block are actually binary numbers. You will likely see them expressed in hexadecimal (base 16) since that is a convenient number base for humans to use.

EXAMPLE 15–13 Calculate the 8-bit checksum for the hexadecimal numbers {41H, 72H, E3H}. Express the result in hexadecimal.

Solution

The 8-bit checksum is requested, which means that the binary result must have no more than 8 bits. Any higher bits will simply be discarded. To find the checksum, just add the three base-16 numbers:

$$\begin{array}{r} 41\text{H} \\ 72\text{H} \\ +\quad \text{E3H} \\ \hline 196\text{H} \end{array}$$

The sum of these values is 196H, which is too large to fit into 8-bits. However, by remembering that each hex digit represents up to 4 bits, we can easily get an 8-bit answer by *discarding the leading* '1,' as shown here:

0001	1001	0110	
1	9	6	H

Therefore, the 8-bit result is 96H. This reflects exactly how a computer will calculate a checksum.

Limitations of the checksum method There are many types of errors that the checksum method cannot detect. Since it relies on getting a proper sum and nothing else, there are many types of errors that can slip by. Examine the data blocks of Figure 15–33. The data blocks both look perfectly good to the computer—and, in fact, they *might* be good data—or they might not! The top data block is identical to Figure 15–32, except that two of the values have been swapped. The *checksum is not affected by the positions of the items*. This leaves data blocks wide open for all kinds of tampering! In the bottom block, perhaps the right two numbers are in error. One *got* larger, the other got smaller. The "errors" in the two numbers are *complementary* or *opposite*. Again, the checksum method won't detect this type of error.

Cyclic Redundancy Check

A system that *does* trap most errors of the type mentioned is the *cyclic redundancy check*, or CRC. The CRC is probably the most commonly used error-detection method in use today because it is *very* effective at trapping a wide variety of errors. The name of this error-detecting method is very intimidating; however, as a technician, all you really need to know is that a CRC is just the *remainder* of a division process involving all the data in the block. An example best illustrates how CRCs work.

Suppose that we invent a new protocol. This one sends 5-digit numerical messages that are intended as commands to a military outpost. As an error check, we will include a 2-digit BCC at the end of each message to be calculated using a CRC with a fixed divisor of 99. A message with the content "12345" will be calculated as shown in Figure 15–34. The resulting data block looks very much like the previous examples; the only difference

Figure 15–33 Two data blocks for examination

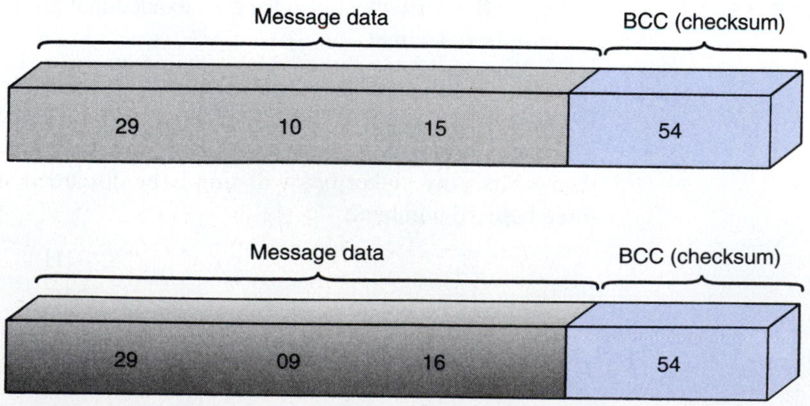

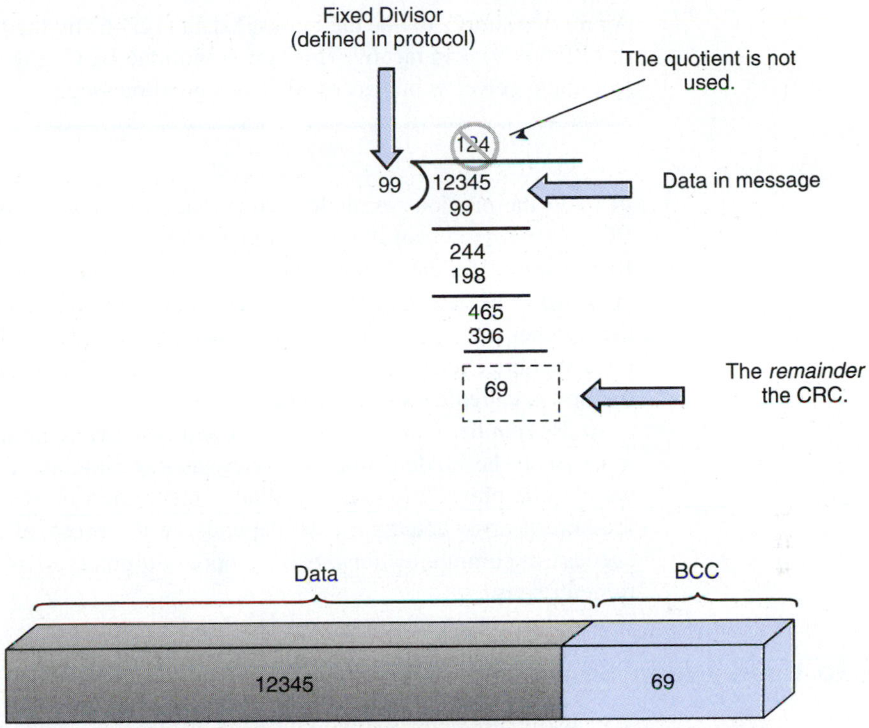

Figure 15–34
Calculating the CRC for message "12345"

is that we're now calculating the BCC in a different way. The CRC is superior to the checksum because it is sensitive to the position of items in a data block. It will catch most transposed items, so it becomes harder to tamper with the block's contents unless the CRC is recalculated. Notice that although the CRC involves division, the quotient of the division isn't used. That is rather strange, but it simplifies the design of CRC calculating logic circuitry considerably.

EXAMPLE 15–14

A data message of {12354} was received, with a BCC of 69. It is known that the message protocol uses a *generator polynomial* of 99. Determine if this is a valid message.

Solution

The *generator polynomial* is just another term meaning "divisor" when it comes to CRCs, and in actual application, it will be a binary number (and you won't ever be calculating it!). What we need to do, then, is just divide the message by 99 and find the remainder. If the remainder agrees with the BCC originally sent (69), then we will accept the message.

The remainder of (12354/99) is in fact 78. This does *not* agree with the BCC that is attached to the block (69), and therefore we will reject this data block.

EXAMPLE 15–15

A data message of {12246} was received, with a BCC of 69. It is known that the message protocol uses a *generator polynomial* of 99. Determine if this is a valid message.

Solution

Again, we simply divide the message data (12246) by the fixed divisor 99. The remainder of (12246/99) is in fact 69. This agrees with the BCC in the original data block, and therefore the receiver would accept this as a good message.

In the previous example, notice that a *different* message than {12345} generated a BCC of 69. How can this be? *Both* messages {12345} and {12246} generate identical BCC values. If a data error occurs that changes one of these messages into the other, the error won't be detected! However, these messages are *99 units* distant from each other on the number line, and that helps to reduce this possibility. In fact, if you experiment with the arithmetic, you will find that the BCC values *repeat* every 99 numbers. This is why this method is called a *cyclic* redundancy check.

Remember that this calculation will be carried out in binary on a computer. It will most likely be hidden from your view as a technician. There is hardly a need for software to display CRC results! What a technician *does* need to understand is that the method of error detection used depends on the protocol used for communications. For successful communications, two or more computers *must* use the same protocol *and* parameter settings.

How to Know When Something is Wrong

When software detects communication errors, it normally and automatically retries communicating with the sender. During this time, the software may appear to be "busy" or "locked," depending on how it is written. There may be a delay from just a fraction of a second to several minutes before an error message is displayed (if you're lucky enough to get even that). *When communications software seems to "stop," it usually means a hard failure of a communication line (broken wire and so forth) or an incorrect parameter setting.*

The Cliff Effect

When the signal quality degrades in an analog system, the output gradually reflects the loss in signal quality. The system continues to work even with marginal signal quality (but it may not be very easy to operate). In contrast, digital communications systems operate differently. A reduction in signal quality causes a digital system to either forward error correct the data (FEC), or request retransmission (ARQ). This is usually transparent to the user. With slight signal problems, the only symptom may be a sluggish system. However, as the communications channel continues to degrade, eventually a point is reached where either the FEC system is overwhelmed (too many errors to correct) or no frames can be received successfully at all. At this point, the system *stops!* (An everyday example of this is the dropping of a cellular telephone connection when the subscriber passes through a marginal reception area.) This is called the *cliff effect* and is unique to digital communication systems. Often as little as a 1 dB difference in signal to noise on the channel can make the difference between a working and nonworking system. *An intermittent system may be operating near the cliff.* Therefore, channel quality checks are an appropriate measure for any digital communications system that intermittently fails.

Section Checkpoint

15–49 What are the two major types of error-handling systems?
15–50 Explain the difference between ARQ and FEC systems.
15–51 What is a data block?
15–52 Explain the purpose of the FCS/BCC at the end of a data block.
15–53 Describe the process occurring in Figure 15–29.
15–54 What is the purpose of the ACK and NAK signals?
15–55 What is the primary limitation of the checksum method?
15–56 Explain how a CRC is calculated; which byproduct becomes the BCC?
15–57 Can a CRC trap all possible errors? Why or why not?
15–58 What is the *cliff effect*?

15–6 THE RS232/EIA232 INTERFACE STANDARDS

When computers began to become compact and widespread (in the 1960s), the need to connect different makes and types of machines together became increasingly important. In the early days of computing, there were no accepted standards for interfacing computers to each other or to peripheral devices like printers and modems. Each manufacturer did that its own way. IBM computers could only be connected to IBM printers and modems. DEC computers could only connect with DEC peripherals. Attempting to connect different brands usually resulted in a nonworking system, accompanied by lots of magic smoke.

The Electronic Industries Association, a consortium of U.S. electronics manufacturers, saw the need for better equipment interfaces. This led to the *RS232* set of standards. The "RS" stands for *recommended standard*. The RS232 standards allow different equipment to be interfaced with a minimum likehood of damage. Over the years, there have been several revisions of the standards. Thus, you will likely see the terms "RS232," "RS232C," "RS232D," and finally "EIA232." These are all the same basic standard, with refinements and renaming of certain pins. All technicians need to know the RS232 standard, for it is a fundamental serial communications interface.

What the Standard Specifies, and What It Guarantees

The RS232 standard specifies three primary things. First, it specifies the *type of electrical connectors* that will be used on serial ports. Second, it specifies the *voltage ranges* for the interface. RS232 signal levels are not TTL compatible. The TTL-logic levels from a UART must be translated to RS232 levels, and the RS232 levels must be translated back to TTL. There are dedicated IC chips such as the Motorola MC1488, MC1489, and Maxim MAX-232 that take care of these details. Last, the RS232 standard specifies the *function* of each signal on the interface. This is important, so that everyone agrees how the signals are to work.

The RS232 standard makes only one guarantee: It promises that if two pieces of equipment are truly compatible with its rules, then they will not harm each other if connected together. That's it. *There is no guarantee that the two devices will communicate at all!* The RS232 standards do not specify the *protocol* for communication. Both devices must use a compatible protocol in order to communicate.

DCEs and DTEs

RS232 devices are divided into two categories, *data terminal equipment,* or *DTEs,* and *data communication equipment,* or *DCEs*. A DTE is usually a computer, and a DCE is usually a modem, printer, or some other peripheral device. There are two important points to remember about this: First, a DTE can be directly connected only to its "opposite," which is a DCE. Two DTEs or two DCEs will not hook up without some custom cable work. Second, the *connectors* for DTEs and DCEs are specified by the standard, and sometimes a technician can identify the type of device by simply looking at the connector.

A DTE normally uses a D-subminiature 25-pin *male* (DB25M) connector. A DCE uses the equivalent *female* connector, a *DB25F.* These connectors are shown in Figure 15–35. In a normal application, a DTE is connected to a DCE with a 25-conductor cable. Figure 15–36 shows two computers connected together with modems. There are actually *two* RS232 interfaces in this figure, one for each computer and its modem. As you can see, there is an RS232 interface circuit in each of the four units. This example shows a long-distance connection between two computer systems, with the telephone system in between as the communications medium. There would be only one DTE and one DCE at a given physical location.

RS232 Voltage Levels

In order to improve the noise immunity of its signals, an RS232 interface uses voltage levels that are quite a bit higher than those of TTL. You might also be surprised to learn

Figure 15–35
D-subminiature connectors for RS232

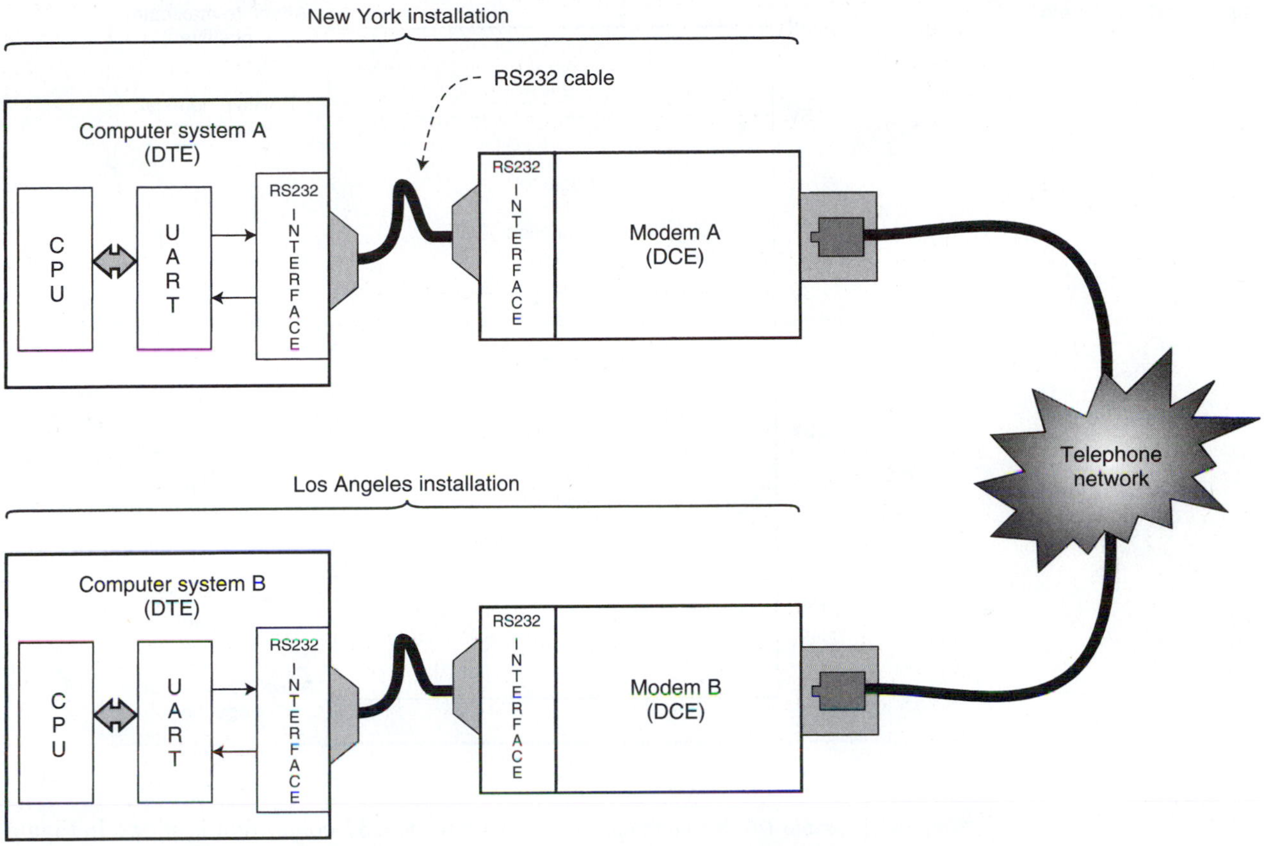

Figure 15–36 Two computers communicating through modems

that they are the *opposite* polarity of TTL signals. There is no good reason for the polarity switch, other than the fact that it was probably a little cheaper to build inverting interface circuits when the standard was first created. Figure 15–37 shows the RS232 signal voltage levels.

Nominal and maximum voltage levels There are two voltage ratings on the interface, the *nominal* (named) and *absolute maximum.* The nominal values are the maximum voltages that should be produced by RS232 output pins; these values are ±15 volts. The *absolute maximum* values are those, which if exceeded may cause the connected RS232 device to smoke! Notice that there is 10 volt safety margin between the nominal and absolute maximum voltages.

At an RS232 receiver (input pin), a logic 1 is represented by any voltage between −3 and −15 volts, and a logic "0" is any voltage between +3 and +15 volts. In between −3 and +3 volts is the *transition region. No signal is allowed to stay within this voltage range. An RS232 pin that measures steadily in this range is not working correctly.*

RS232 pin assignments Not all of the 25 pins on the RS232 interface are used. In fact, an RS232 interface could consist of as few as *two* connected pins, as we'll soon see. Figure 15–38 shows the most important RS232 signals and pins.

Figure 15–37 RS232 signal voltage levels

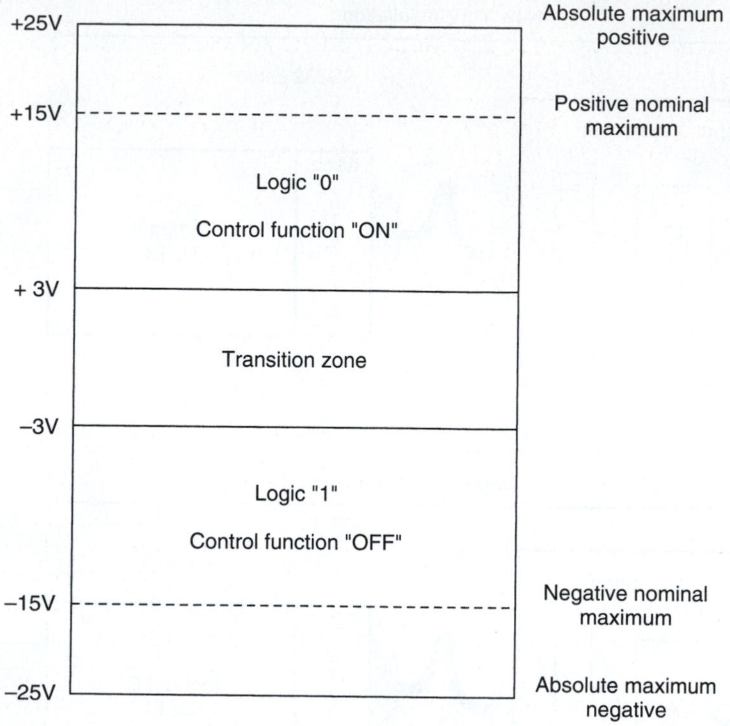

Simplest possible RS232 connection A two-wire RS232 connection is shown in Figure 15–39. This interface can transmit data in only one direction, from the computer (DTE) to the peripheral device (DCE). In other words, it can communicate only in *simplex*. Note how pin 2 on the DTE, *transmit data,* is an *output,* while pin 2 on the DCE, *transmit data,* is an *input*. The naming of functions such as "transmit" and "receive" are all done with respect to the viewpoint of the DTE. Therefore, the DTE transmits data on pin 2, and receives it on pin 3; the DCE actually receives data (from the DTE) on pin 2, and transmits data back to the DTE on pin 3. Figure 15–40 shows a three-wire interface capable of full-duplex communication. The interface has one more signal added, *receive data,* on pin 3. Note again the direction of flow for pins 2 and 3; they're opposites.

EXAMPLE 15–16

An RS232 interface measures –10 volts (steady) on pin 2, and +5 volts (pulsed) on pin 3. Assess the state of communications.

Solution

To understand this problem, we must combine our knowledge of asynchronous data with the RS232 signal levels. The condition on pin 2, which is the *transmit data (TD),* is given as −10 volts *steady*. According to Figure 15–37, that is a logic *1* or *mark*. A serial data line that is steadily marking is *idling*. There is no data being carried on pin 2. The first part of our conclusion is that no data is being sent from the DTE to the DCE.

Pin 3 is the *receive data* pin, and it appears active. The voltage pulse of +5 volts corresponds to a *space,* or logic 0. An asynchronous signal normally does not become a logic

Pin number	RS232C signal name	EIA232D signal name	Function
1	PG—Protective Ground	FG—Frame Ground	Protective ground; connects both chassis for safety
7	SG—Signal Ground	SG—Signal Ground	Ground for signal path; always connected
2	TD—Transmit Data	TD—Transmit Data	Transmitted data from DTE to DCE
3	RD—Receive Data	RD—Receive Data	Received data from DCE to DTE
4	RTS—Request to Send	RTS—Request to Send	Request from DTE for DCE to begin transmitting (in half-duplex modes)
5	CTS—Clear to Send	CTS—Clear to Send	Acknowledgement from DCE to DTE that DCE is ready to accept transmit data (in half-duplex modes)
6	DSR—Data Set Ready	DCE Ready	Signal from DCE to DTE indicating the readiness of DCE to communicate
8	DCD—Data Carrier Detect	RLSD—Receive Line Signal Detect	Signal from DCE to DTE indicating that the DCE's receiver has detected a suitable carrier signal on the line
20	DTR—Data Terminal Ready	DTE Ready	Signal from DTE to DCE that directs DCE to stay online. When DTE turns this signal off, DCE may be forced on-hook (offline).

Figure 15–38 Important RS232 signals

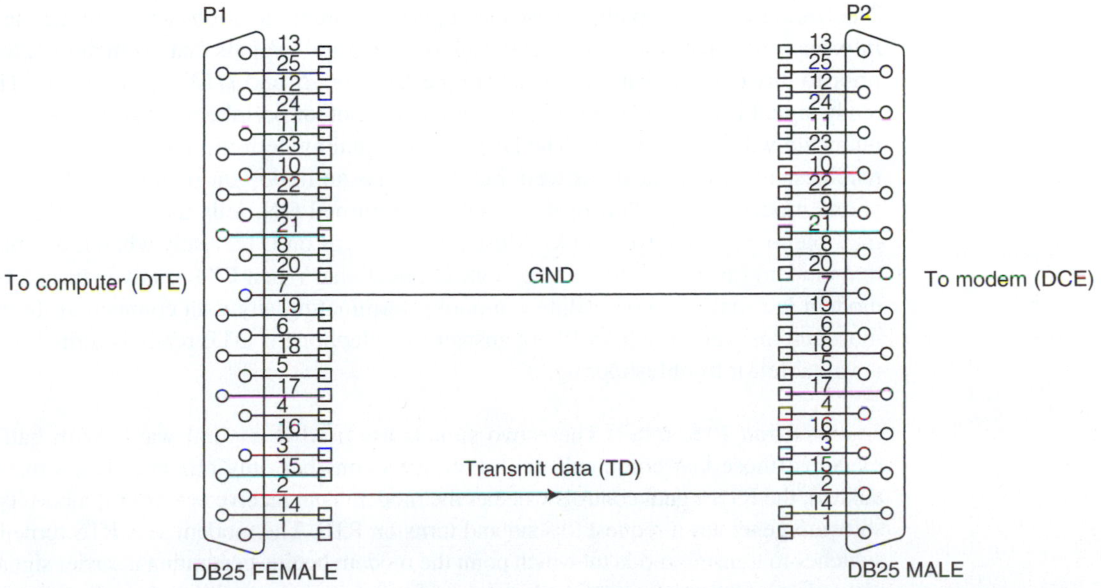

Figure 15–39 A two-wire RS232 interface

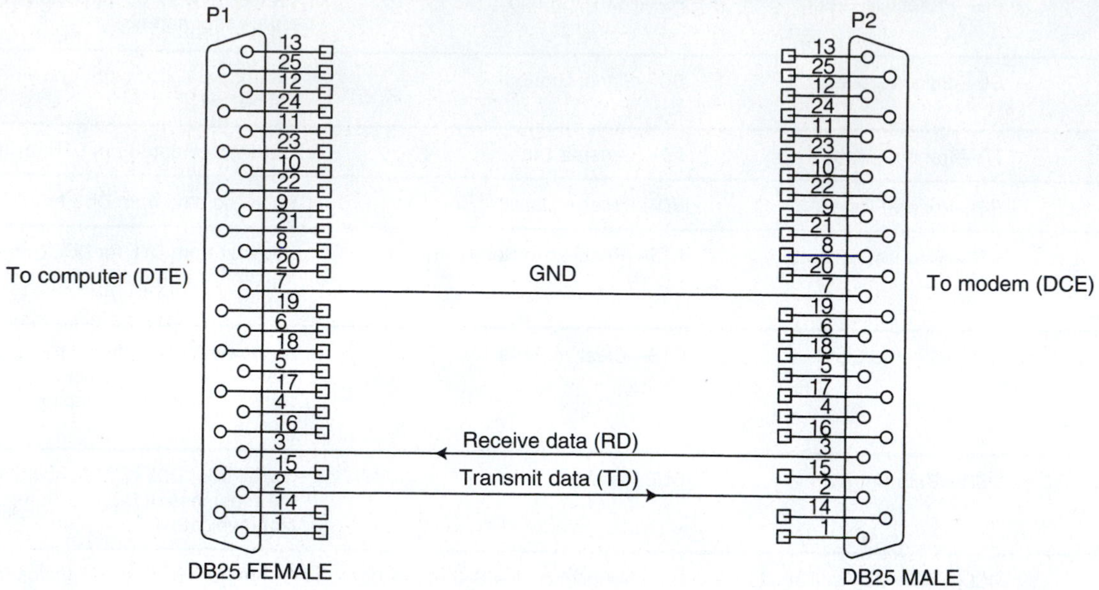

Figure 15–40 A three-wire RS232 interface

0 unless a character is being transmitted. Therefore, there is data flowing from the DCE back to the DTE.

In summary: *The computer is not sending data to the DCE, but the DCE is sending data back to the computer.*

RS232 Control Signals

The RS232 control signals are used to supervise communications and regulate the flow of information on the connection. There are two "master" signals that normally get turned on prior to *any* communication. These are the *DCE ready* and *DTE ready* signals. The DCE ready signal is an indication from the modem (or other peripheral) that it is ready to communicate with the computer. The DTE ready signal by definition indicates the computer's readiness to communicate as well, but it also has an important function with dial-up telephone modems. The DTE ready signal, when turned ON, tells the modem that it is permissible to go online (off-hook). Most software turns on DTE ready when it begins talking to a modem (as in initiating a telephone connection). When DTE ready is turned OFF, the modem is usually forced offline (on-hook), terminating any data connection. In addition, most autoanswer modems will not answer the telephone if DTE ready is turned off, which is valuable for troubleshooting.

Flow control: RTS, CTS These two signals are used in several ways. With half-duplex modems (those that communicate in both directions but only one at a time—like walkie-talkies), the RTS signal controls whether the modem is in receive or transmit mode. Typically, software generates a request to send and turns on RTS. The modem sees RTS turned on and switches to transmit mode (at which point the modem begins generating a carrier signal on the phone line). Once sufficient time has passed for the carrier to stabilize (usually 80 to 150 ms), the modem turns on CTS, which is the green light for the computer to begin sending data.

In full-duplex communications, RTS and CTS are used as *handshaking* signals. A *handshake* is a procedure for the safe transfer of data. A peripheral device may not be able to accept data as fast as the computer can generate it. If the peripheral is not ready for more data, it can turn off CTS. Software in the computer will detect the fall of CTS and will stop sending data until CTS turns back on again. Likewise, RTS is often used by the computer to tell the peripheral that the computer is ready for more data by being ON, or not ready for more data by being OFF. This procedure is called a *RTS/CTS handshake*.

RLSD or DCD The last of the basic I/O pins is the *data carrier detect,* or *receive line signal detect* signal. This is an indication from the modem that the modem has detected a data carrier on the telephone line (or other communications medium). Be careful; when DCD is ON, it doesn't mean that data is flowing. It only shows that the modem "thinks" it detects another modem talking!

Section Checkpoint

15–59 Why is an interfacing standard like RS232 necessary?

15–60 What is guaranteed by the RS232 standard?

15–61 An unknown device has a 25-pin D-sub male connector. What class of RS232 device is it most likely to be?

15–62 What is likely to happen if a voltage greater than ±25 volts is applied to an RS232 signal pin?

15–63 What are the two "master" control signals on the RS232 interface?

15–64 When the DCD/RLSD signal is on, is data flowing? Explain.

15–7 TROUBLESHOOTING DATA COMMUNICATIONS SYSTEMS

Since data communications equipment uses a wide variety of technologies, you might suppose that all sorts of troubles can develop in it. That is certainly an understatement. There is an overwhelming amount of software hidden inside computers, controlling the outcome of each operation. One manufacturer of utility software has estimated that more than 75% of U.S. companies are having moderate to severe problems with the reliability of their computer workstations and networks.

Since data communications uses an intimate combination of software and hardware, it is often hard to tell one type of problem from another. Let's develop some troubleshooting methods for each kind of trouble, then talk about how problems can be prevented.

Hardware Problems

Dale's computer has been acting weird lately. Usually, he arrives at 8 A.M., turns his workstation on, logs into the department's network, and then reviews his latest e-mail over a cup of coffee. About a week ago, Dale arrived as usual, but his machine would not connect to the network!

He called the help desk, and 15 minutes later, an administrator arrived. The machine worked perfectly when the administrator tried logging in. This was very strange. After this happened three days in a row, Dale discovered that his workstation would work fine if it "warmed up" for about 10 minutes.

This is a good example of an *intermittent* communications device. Users will often put up with this sort of behavior for long periods of time—until systems become so unreliable that they are totally useless.

- To troubleshoot an intermittent problem, you should try to reproduce the original conditions as closely as possible. To diagnose PCs and PC hardware, special diagnostic software such as *Check-It*™ is extremely valuable. Listen carefully when a user describes what he or she is doing prior to the problem. Often you will get valuable clues from their comments.

Pam threw her hands up in frustration. "This is the third time today. I'm really sick of this piece of junk!" Pam, the assistant director of the accounting department, was trying to send the weekly report to the company headquarters in Rochester. Her terminal screen was filling with unreadable garbage, and the lights on the nearby modem kept time with the outpouring of gibberish.

A problem like this usually indicates a fault with a modem or telephone line (or other link). Some software will sit and do nothing when it receives garbage on the line, while others (like Pam's) will echo the trash to the screen. Modems can be tested by doing a *loopback test,* which is included in most diagnostic software. A known-good modem can also be quickly substituted. It is also possible that the *telephone line* has some sort of problem. The phone company can verify whether or not that is so—but be careful, their opinion may be biased in their favor!

Don't waste your time trying to repair a modem. They're very inexpensive compared to your time.

- Don't forget to consider *all* devices in the communication path when trouble arises. In the preceeding example, this includes Pam's workstation, her modem, her phone line, *and* the associated hardware at the other end of the connection (in Rochester).

Software Problems

The complexity of the computer software is growing at a boundless rate; the sophistication of the average user is not! Many end users have a difficult time keeping up with the changes in computer software, and many problems with software are "cockpit trouble." Occasionally a user changes an important setting without being aware of it, which can really be hard to find, given that there can be hundreds of possibilities for configuring communications.

- To minimize software problems, end users should be properly trained in the use of the tools. Many companies fail to provide for this. Computer workstation profiles should be centrally developed and tested prior to deployment. Don't use employees as a test bed for a new workstation setup!
- Even a trivial change in the settings for a workstation, such as a data rate or parity mode, can totally disrupt communications. Don't forget to check parameter settings.
- When a problem seems to afflict several computers, seek out what is common to those machines. A critical piece of software may need updating.
- Be patient with users and do not rush when helping them. You will learn from them too!

Prevention is Best

It's best not to have problems in the first place. It has been said that the best system administrators are seldom seen by the users, because their systems operate transparently with a minimum of troubles. The following are very good preventive practices:

- Have a solid data backup plan and stick to it. Your backup plan should include daily, weekly, and monthly media rotations as well as off-site storage (in case of a catastrophe, such as a fire).
- Keep a map of your network and communication system topology and keep this map up-to-date. Do *not* store it on the network!
- Keep hardware up-to-date. In today's world, a state-of-the-art workstation becomes average within six months and outdated (slow) within three years. As hardware ages, its reliability falls.
- Where possible, the security features of workstations should be utilized to limit accidental changes to important settings.
- Adopting standard workstation profile settings for an installation can be very helpful in minimizing software problems. It is much easier to maintain 100 similar workstations than 10 different ones. Good records should be made of profile settings, and archived disk images should be made of each unique workstation profile. Do not store disk images on the network unless they are backed up elsewhere.

SUMMARY

- Information technology is the branch of science dedicated to computers and telecommunications.
- Data communication systems use a wide array of both analog and digital technologies.
- Digital systems never exactly represent analog information. The digital representation is *quantized,* which means it is represented with a finite number of states.
- The two main types of data transmission are serial and parallel. Serial transmission is far more common for long distances.
- Digital data is usually carried over an analog carrier when it is sent a long distance. Any of the standard modulation methods (AM, FM, or PM) can be used to modulate the carrier.
- The units of digital information flow are bits per second, or bps.
- The *topology* of a network is its physical layout. The topology of a network often determines the rules needed for communicating through it.
- A standard dial-up telephone line can only pass 300 Hz to 3000 Hz. To get digital data through a phone line, an analog carrier must be used.
- The Bell 103 standard specifies how FSK (frequency shift keying) will be used over phone lines for full-duplex communication.
- FSK is a form of FM and represents 1s and 0s by different frequency tones.
- BAUD is the rate of signaling and has units of transitions per second.
- The bandwidth required for a transmission is determined by the BAUD and the method of modulation.
- A single-level modulation system encodes one bit per transition. Single-level systems tend to operate at low data rates.
- Multilevel modulation systems represent more than one bit per transition. The number of bits per transition is limited by the SNR of the communications channel.

- According to the Shannon–Hartley theorem, the capacity of a communications channel is controlled by two factors, its *bandwidth* and *signal-to-noise ratio*.
- A *constellation* is a graph showing all the possible signal states for a PSK signal.
- Synchronous data is sent with a clock; asynchronous data is sent without a clock. An asynchronous receiver must create its own clock signal.
- A UART is an IC that transmits and receives asynchronous serial data. Most UARTs are designed to be interfaced with the bus of a microprocessor.
- Parity is a system for detection of errors. It can only detect an odd number of single-bit errors.
- The two types of error-handling systems are ARQ (automatic repeat request) and FEC (forward error correction).
- Data is usually sent in *blocks* or *frames,* with a BCC at the end of the frame for error detection.
- All error-detection systems rely on redundancy in the transmitted data block. Usually the BCC provides this redundancy.
- The *checksum* and *cyclic redundancy check* are two popular methods for detecting errors.
- A checksum is the sum of all data in a block. A CRC is the remainder of a division process where the block data is divided by a fixed number called the *generator polynomial*.
- The RS232 standard is used for serial communications. It does not guarantee that two devices will talk. It only guarantees that they won't damage each other.
- A *protocol* is a set of rules for communication.
- Troubles in data communication systems can be classified as either hardware or software.
- Problems with data communications systems can be minimized by good administration practices.

PROBLEMS

1. For the system of Figure 15–1, give the binary code output that will result for each of the following voltages: 3 V, 3.7 V, 6 V, 6.2 V.
2. What is meant by the term *quantization?*
3. A certain digital system uses 5 bits per sample. How many voltages can it represent?
4. Draw a block diagram of a complete data communications system and using outline form, explain the function of each part.
5. Explain the difference between *serial* and *parallel* data. Give a nonelectronic example to illustrate.
6. What must be done in order to pass digital information through an analog medium (such as a telephone line)?
7. A data waveform has a bit time T_b of 1 μs. Calculate the data rate of this waveform.
8. In order to simulate a "1010 . . ." serial data pattern at 4800 bps, what frequency square wave must be used?
9. Define the terms *simplex, half-duplex,* and *full duplex*. Which mode usually requires the most bandwidth?
10. Draw a diagram of a point-to-point, multidrop, ring, and star network. Which of these has the highest reliability?
11. Define the term *protocol.*
12. Using outline form, explain how a message is originated and delivered on a ring network.
13. What does CSMA/CD stand for? Summarize the operation of units on a CSMA/CD network in one sentence.
14. Draw a block diagram of a telephone modem. For each signal in your diagram, show whether it is analog or digital.
15. What is the frequency response of a standard dial-up telephone line?
16. What will happen when the following signals are passed into a telephone line?

a. a 1000 Hz sine wave
b. a 2.5 kHz square wave
c. a 4000 Hz sine wave

17. What is meant by the terms *mark* and *space?*

18. Draw a spectrogram showing the frequencies transmitted by originate and answer modems according to the Bell 103 standard.

19. Define the term *BAUD*. How are BAUD and bps different?

20. What is a single-level modulation system?

21. A 4800 bps data waveform is being sent into a FSK modulator. What bandwidth will the resulting signal need?

22. A certain communication channel has a 2 kHz bandwidth. What is the maximum signaling rate and data rate possible if single-level FSK is used?

23. What is a multilevel modulation system? What is the advantage of a multilevel system over a single-level system?

24. A certain multilevel modulation system encodes 3 bits per transition and operates at 600 BAUD. What is the actual data rate?

25. What factor limits the number of bits per transition in multilevel systems?

26. What two factors limit the data carrying capacity of a communications channel?

27. Explain why PSK is much more popular than FSK for modern high-speed modems.

28. Draw a signal constellation diagram for a single-level (binary) PSK system. How many modulation states does it have?

29. In order for a PSK modulator to encode 5 bits per transition, how many phase angles must it have?

30. Explain the difference between *synchronous* and *asynchronous* data. Which type is used with UARTs?

31. What is the meaning of each of the following shorthand designations?

a. 8,N,1
b. 5,E,1
c. 7,O,2
d. 6,N,2

32. Draw the asynchronous character 74H sent using parameters 8,N,1. Label each part of the character.

33. Draw the asynchronous character 57H sent using parameters 7,E,2. Label each part of the character.

34. Assuming that the parameters are 8,N,1 in Figure 15–41, determine the hexadecimal data value that is being transmitted.

35. What is the data rate in bps and the character rate in CPS for the character in Figure 15–41?

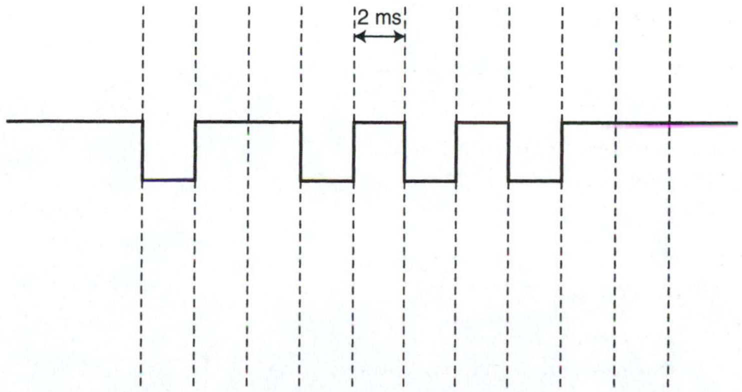

Figure 15–41 An asynchronous character for analysis

36. What is the function of a UART?

37. List the four major sections of the HD6402 UART, explaining the purpose of each one.

38. What control inputs and clock frequency are needed to set the HD6402 UART as follows: 19,200 bps; 8,O,2.

39. What is a data block? Why is data sent in this way?

40. Explain the difference between an ARQ and FEC system. When is forward error correction typically used?

41. Calculate the 8-bit checksum for the hexadecimal data A5H, 7FH, 05H.

42. What is a CRC? How is a CRC calculated?

43. Why is a CRC superior to a checksum?

44. Draw the circuit diagram of a two-wire RS232 interface that will allow a computer to receive data from a peripheral device.

45. Draw a diagram showing the voltage levels for the RS232 interface.

46. What is the absolute maximum voltage that can be placed on any RS232 pin without causing damage?

16

Networking Fundamentals

OBJECTIVES

At the conclusion of this chapter, the reader will be able to:

- describe the function of network hardware devices such as hubs, switches, and routers
- describe the differences between local, metropolitan, and wide area networks
- describe the various media used in local area networks
- list the layers in the OSI model and give the purpose of each
- work with IP addresses, unroutable blocks, subnet masks, and other IP features
- divide IP address blocks using subnetting
- describe the operation of DHCP and ARP in a local area network
- explain the operation of UDP and TCP transport mechanisms
- apply the principles of the OSI model to troubleshooting network systems
- use the *ping* and *traceroute* commands to troubleshoot a network

Computer networks are crucial to the functioning of our modern society. They're found everywhere—in businesses, homes, factories, and even in vehicles. The Internet is the world's largest network, providing global communication to anyone who can access it. By itself, a computer is an island; connected to a network, it can share in all the resources an organization has to offer. Networks both large and small operate on a small number of basic principles. A technician who has mastered these principles can take care of network problems with relative ease.

16–1 NETWORK HARDWARE AND MEDIA

Networks are built for many purposes. *The primary purpose of a network is the sharing of information and resources.* You may have already used a network to share files, a printer, or some other item that several people needed to access. You've probably used e-mail and instant messaging. These applications are just the beginning of what is possible. For example, you can use a network to control the automation in a factory, check the security of your home while you're away, or remotely operate a robot in a hazardous environment. Eventually you'll be able to do these tasks wirelessly from nearly any place in the world, thanks to the advancement of wireless networks. Networks come in many shapes and sizes and use a variety of equipment to accomplish their jobs.

Types of Networks

There are three basic classifications of networks. These are *local area networks* or *LANs, metropolitan area networks* or *MANs,* and *wide area networks* or *WANs*. They are well named. A LAN is a network contained primarily within a single building (or part of a building). Most organizations have one or more LAN systems, usually interconnected. A LAN features simple, inexpensive hardware and can operate at very high speeds (usually 100 Mbps or better; gigabit Ethernet itself can operate at 1000 Mbps or 1 Gbps). In a medium-sized company there may be a separate LAN for each department or office, which improves security (no one can "sniff" the adjacent department's network for sensitive information) and reliability (if one department's network breaks, the rest of the organization remains operative). Most LANs use Ethernet and category 5 or 5e twisted-pair cable to carry the data signals. In general, LANs tend to have a small group of similar computers. Figure 16–1 shows a small LAN.

A MAN is a network contained in more than one building, or is spread across a community. It is often used on college campuses to connect all the departments. Medium and large companies use MANs to connect their offices within the same city. Because of the larger distances in a MAN, twisted-pair copper cable is usually inadequate to carry the signals. The three most popular choices include *FDDI, fiber distributed data interface,* a fiber-optic ring network; *leased lines,* which are provided by a telephone company; and *wireless radio frequency links,* which are used for point-to-point communications between two locations. Because of the specialized hardware (or the use of leased telephone lines),

Figure 16–1 A small Ethernet LAN with workstations, a server, and a printer

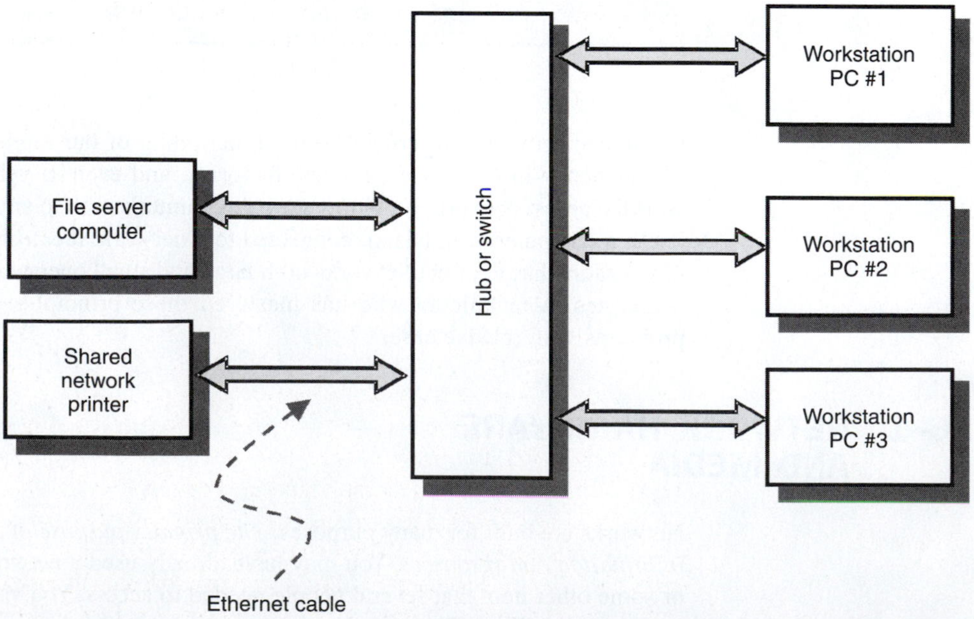

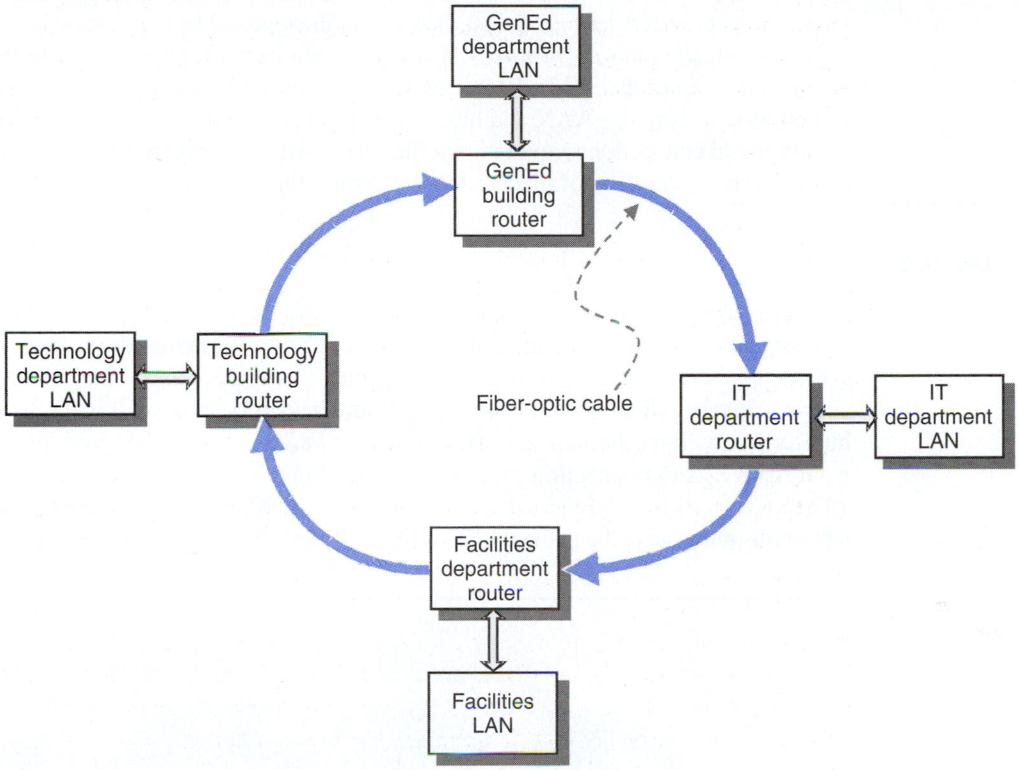

Figure 16–2 A MAN employing an FDDI backbone

MANs are more expensive to operate than LANs. They may still operate at a high data rate as a LAN, but are more likely to be slower if leased lines are involved, or if the volume of traffic is high. Figure 16–2 shows a MAN utilizing an FDDI backbone. The *backbone* of a network is the part that carries the majority of traffic between network destinations.

The MAN in the figure effectively allows each department to communicate with other departments when needed, but also provides important isolation between them. A *router* with an FDDI interface is used at each department to decide what messages should be passed onto the FDDI ring, and which should not. Generally, *broadcast* messages are not repeated by the routers, which helps to prevent a condition called a *broadcast storm* from appearing on the MAN. A broadcast is a message intended to be received by all computers on a network. In a LAN, this is useful, but in a larger network, repetition of broadcasts by routers (or other equipment) can cause the same broadcast message (or a group of messages) to be endlessly repeated on the network; hence a "storm" of broadcast messages appears, which clogs the network with useless traffic. The FDDI ring operates just like the token ring network discussed in Chapter 15. It uses a back ring to help ensure reliability.

As you might have guessed, a *WAN* is a network covering most of a city, several cities, or any larger area. The Internet is the ultimate WAN! A WAN is the most expensive of all networks to operate. Individuals can afford to access the Internet because the cost is divided by the millions of users. It contains many different types of computers and a variety of connections between them, ranging from humble 56 kbps dial-up lines in the backwaters to gigabit-speed OC-768 (optical carrier) fiber optic links capable of operating at 40 Gbps. Even with high-speed backbones, WANs are almost always much slower than LANs and the delays in traffic are quite variable. Most of the structure of the Internet is provided by major telecommunications providers such as Qwest, Sprint, and the regional Bell operating companies.

A *firewall, gateway,* or *router* is needed to connect a LAN or MAN to the Internet. A firewall is a specialized hardware or software device that examines that traffic being passed to and from the WAN and halts messages that could cause harm (for example, an unauthorized connection requested to a file server within an organization). A gateway performs the basic function of placing messages onto the WAN.

Hardware Devices

Networks use a variety of devices to handle the traffic between computers. The three main devices used are *hubs, switches,* and *routers*. A hub is a device used to connect the workstations together on one segment of a LAN. A hub creates a common "party line" (bus) connection between all the devices sharing it. If any device transmits, all other devices on the hub's ports will get the message. Hubs come in various sizes—they may have anywhere from 4 to 32 I/O connections for computers. Hubs also come in two primary speeds, 10 Mbps (for 10-BaseT Ethernet application), and 100 Mbps (for 100-baseT Ethernet). The latter hubs will usually be marked "10/100-BaseT" and can support both speeds. Figure 16–3

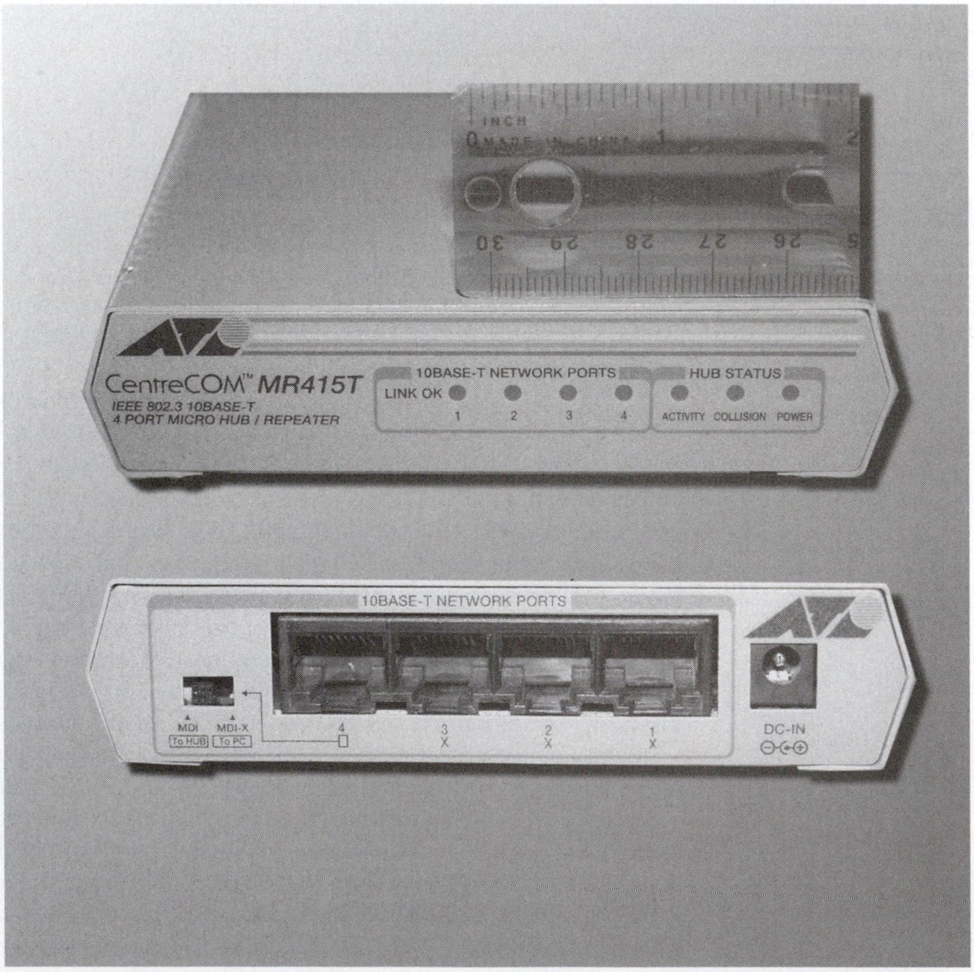

Figure 16–3
A 10-BaseT hub

shows a small 10-BaseT hub that might be a handy addition to a technician's toolkit. Note that ports 1, 2, and 3 are for direct connection to personal computers. Port 4 is the expansion port. To expand the LAN to include more workstations, a second hub can be connected to port 4, making sure to move the switch to the "MDI" position.

Hubs create a common bus connection for computers on a LAN. Although this is certainly a simple way to connect them, it creates two problems. First, since any workstation can intercept messages meant for any other workstation, the security of such a network is not exactly optimal. Second, if two devices transmit at the same time, a *collision* will occur on the wire. You may recall from Chapter 15 that a bus network usually employs CSMA/CD (carrier sense multiple access with collision detection) protocol. Each device must listen to make sure the bus is quiet before transmitting. However, it is possible and very likely that two or more devices will need to talk at the same time when the network is quiet. Since no unit is transmitting, both devices see that it is OK to send and they both begin transmitting. This results in a *collision* on the wire and lost data. Both devices must back off and try again (using a pseudo-random "roll of the dice" to decide which one will go first). Collisions waste valuable time on a network. Excessive collisions can cause a network to grind almost to a halt. For this reason, it is not feasible to connect more than 20–25 workstations on a hub.

For larger LANs (or when security is more critical), *switches* are used. A switch connects just like a hub, but operates in a very different way. When a computer sends a message to a switch, the switch examines the destination MAC address contained within the message. (We will study MAC addresses later. Each computer has a unique MAC address on a LAN.) The switch then forwards the message *only* to the computer with the correct MAC address. The ports for the other computers on the LAN remain silent. This improves security, since a message now flows only to the intended receiver. It also improves speed greatly; since the uninvolved switch ports are silent, the other computers can communicate on them (through the switch) with little or no risk of collisions. A switch is a much better choice for a LAN, even if there are only a few workstations on it. Simple switches require no programming and are simply connected to a network just like a hub. More advanced switches may be programmable and require specialized training to operate.

A *router* is a device that examines the network (usually IP) address of a message and decides which network it should be forwarded to. Routers always have two or more I/O ports and require programming to make them active on a network. Routers have slots where "blades" (I/O cards) can be installed to increase the number of interfaces (and hence, number of networks they can bridge). Routers use *routing tables* to determine where messages should be moved, and *access lists* to decide which messages are allowed through. A routing table is a list of network address destinations with instructions on how to get the message to the destination (which port on the router to send the message out). Routing tables can be configured *statically* (manually, by a programmer) or *dynamically* (using a routing protocol), or a mixture of both. Routers form the backbone of the Internet; without them, the Internet couldn't exist!

Networking Media

A variety of cables are used to connect devices in a network. By far the most popular is *UTP* (unshielded twisted pair) Ethernet because of its low cost and ease of installation. Table 16–1 summarizes the characteristics of the most common media types. The naming system gives the maximum data capacity and signaling method. For example, "10Base2" media operates at a maximum of 10 Mbps and uses baseband signaling (the data signal is sent directly onto the wire without the use of any modulation method).

Table 16–1 Common networking media types and characteristics

Media type	Maximum data rate	Nodes per segment	Maximum length
Coaxial 10Base2, 10Base5	10 Mbps	30 (10Base2) 100 (10Base5)	500 m (10Base5) 200 m (10Base2)
10BaseT (UTP) Category 3 or 5	10 Mbps	2	100 m
100BaseT (UTP) Category 5	100 Mbps	2	100 m
1000BaseTX (UTP) Category 5e	1 Gbps	2	100 m
Fiber optic	> 2 Gbps	2	> 10 km typical
Wireless 802.11(b) (2.4 GHz) 802.11(a) (5.4 GHz)	11 Mbps (802.11b) 54 Mbps (802.11a)	Limited by application; typically 50 or less	Range determined by terrain, antennas; typically 1 km or less

Coaxial cable is largely falling from favor in LAN installations. It is much more expensive and difficult to install when compared to twisted-pair Ethernet. Coaxial cable is installed as shown in Figure 16–4. You may find it in older installations. With coax, a special device called a *vampire tap* is placed on the cable wherever a PC is to be connected. The cable is marked with the acceptable locations for tapping. Cables must also be terminated at the ends with a resistor plug in order to prevent reflections, which will cause false collisions and other problems on the line. It's easy to "break" a coaxial cable network; a moment's carelessness with a connector is all it takes.

UTP is by far the most commonly used media for LANs. It's inexpensive and relatively easy to work with. It comes in various grades, as shown in Table 16–1. Category 3 cable is conventional telephony-grade wire. When LANs were first being installed for businesses in the 1990s, quite a few shortsighted people decided that the existing telephone wiring in their buildings would be great for hooking up the network. However, this type of

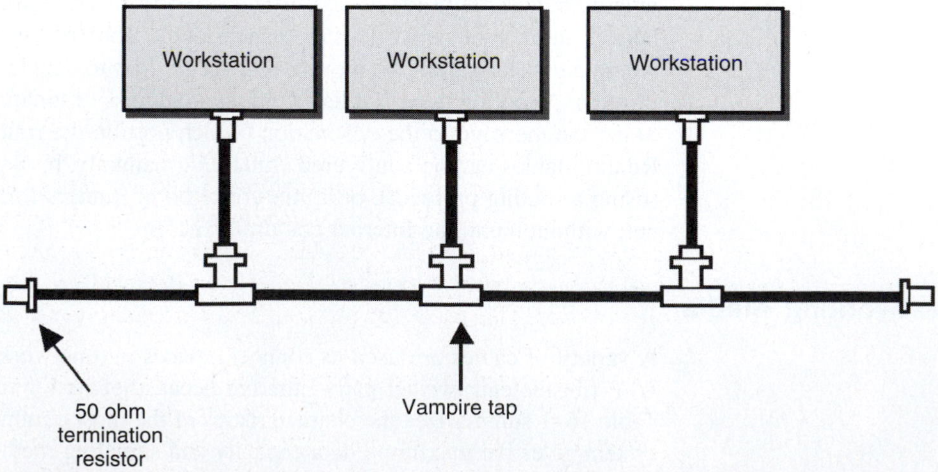

Figure 16–4 Coaxial cable connection in a LAN

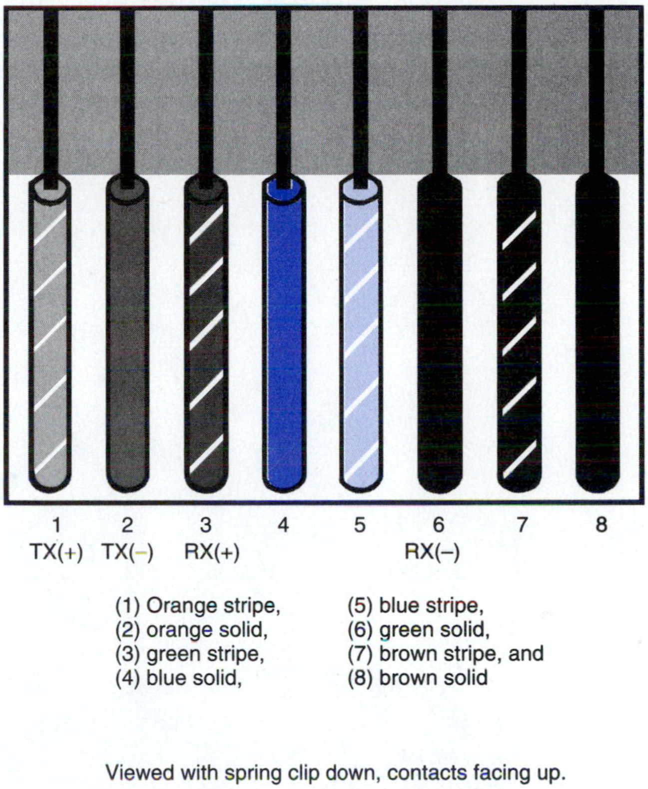

Figure 16–5 Ethernet RJ-45 standard wiring practice

cable is actually poorly suited for data application, and those same people discovered this the hard way (their networks were unreliable!). For current installations, category 5 or even better, category 5e, is preferred. Category 5e can support data rates up to 1 Gbps with modern interface cards.

UTP cable uses a type RJ-45 telephone connector with eight conductors. Four of the conductors are not used except for 1000BaseTX applications. You'll likely be installing these connectors, so make sure to learn the standard color code shown in Figure 16–5. Figure 16–6 shows how to install RJ-45 connectors. In general, UTP cable segments should not exceed 100 meters in length.

UTP cables can be described as *straight, crossover,* or *rolled*. A straight cable has all wires directly connected to the same pin numbers on both ends of the cable. If you wire a Ethernet patch cable on both ends according to Figure 16–5, you'll have a straight cable. Straight cables connect PCs and other devices to hubs, switches, and routers. However, straight cables can't directly connect two PCs. For this application, a *crossover* cable is needed. To build a crossover cable, wire one end according to Figure 16–5; on the other end, *swap the orange and green pairs*. A *rolled* cable is sometimes required to interface two devices. This cable has all the wires in normal order at one end, and in *reverse* order at the other end.

Most organizations use jacket color codes to help identify cables in racks. Blue, green, or gray jackets normally signify *straight* cables, while yellow is typically used for *crossover* cables. Other colors are also used; for example, red is sometimes used for jumpers between switches, or busses passing to router ports. Using a consistent color code

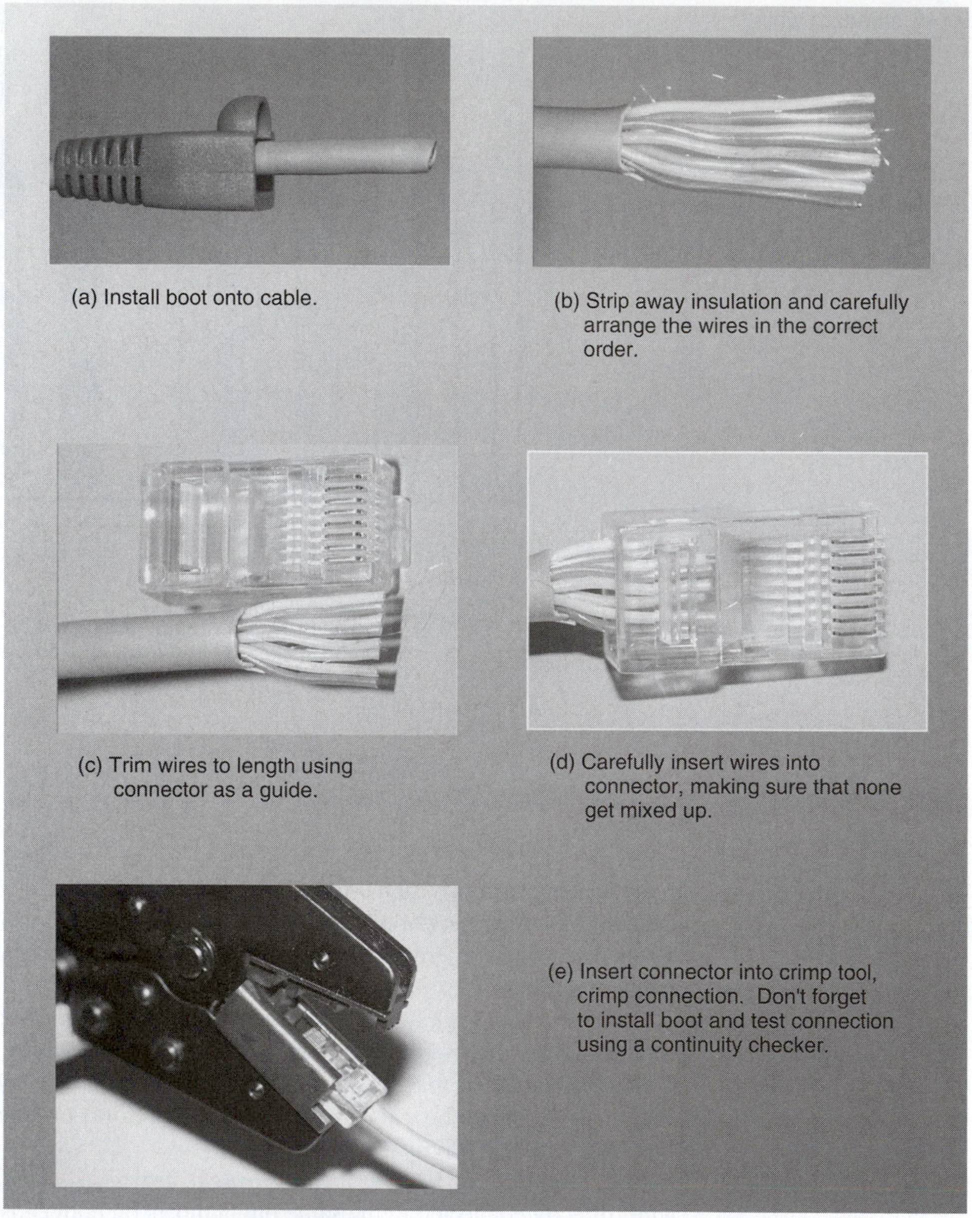

(a) Install boot onto cable.
(b) Strip away insulation and carefully arrange the wires in the correct order.
(c) Trim wires to length using connector as a guide.
(d) Carefully insert wires into connector, making sure that none get mixed up.
(e) Insert connector into crimp tool, crimp connection. Don't forget to install boot and test connection using a continuity checker.

Figure 16–6 How to install RJ-45 connectors

makes it easier to understand where circuits are being routed when you're troubleshooting a network equipment rack.

Fiber-optic cable (discussed in detail in Chapter 18) offers the highest data rates of all the common media and also the largest distances. Special techniques are needed for installation of fiber-optic connectors. A fiber-optic network interface requires *two* fibers, one for receiving and the other for sending. Fiber is practically immune to eavesdropping, so it provides the highest level of security of any network media. You'll often find it used to connect a LAN to a backbone, or for use as a backbone in a FDDI network.

Wireless (sometimes called *WiFi*) network components have advanced rapidly in the last few years. The two predominant types are 802.11(b) and 802.11(g). To connect a computer to an existing 802.11 network, a compatible wireless network card must be installed in it. If the network is using encryption, the encryption key must be programmed into the workstation so that it can access the LAN. Wireless LANs can operate in two modes, *ad hoc* and *access point*. In an ad hoc wireless LAN, each computer directly talks to the other computers. This is useful for sharing files between computers; several computers can form an ad hoc LAN anywhere at any time. The ad hoc LAN has no connection to a wired network unless one or more of the PCs has a separate wired network interface. In access point mode, the wireless cards in the PCs can only talk to a central wireless access point (AP), which grants each device access to the network. The AP is usually connected to a wired LAN so that the computers can access servers, printers, and the Internet.

Section Checkpoint

16–1 List the three types of networks and compare them in terms of speed, cost, and types of hosts likely to be found on them.

16–2 What type of media is used by FDDI?

16–3 What is a good application of FDDI?

16–4 Explain the difference between a switch and a hub.

16–5 What are routers used for?

16–6 Why must CSMA/CD be used on an Ethernet network?

16–7 What is the most popular media for connecting Ethernet networks? Why?

16–8 What media would be used for forming a high-speed connection to a backbone?

16–9 Explain the difference between straight, crossover, and rolled Ethernet cables.

16–10 What are the two most popular WiFi standards? What data rates can they support?

16–2 THE ISO/OSI MODEL

The *OSI* (open systems interconnect) model is a way of representing the function of the various portions of a telecommunication system. It divides systems into seven levels as shown in Figure 16–7.

Many people use the phrase *"Please Do Not Throw Sausage Pizza Away"* in order to help them remember the names of the seven layers. In fact, many systems may not even have or need all seven layers. (The TCP/IP protocol suite is an example of a system employing only five of the layers.) Notice that these layers are stacked. The reason for this is that information passes through these layers in sequence in both directions. This will become more clear as we study the purpose of each layer.

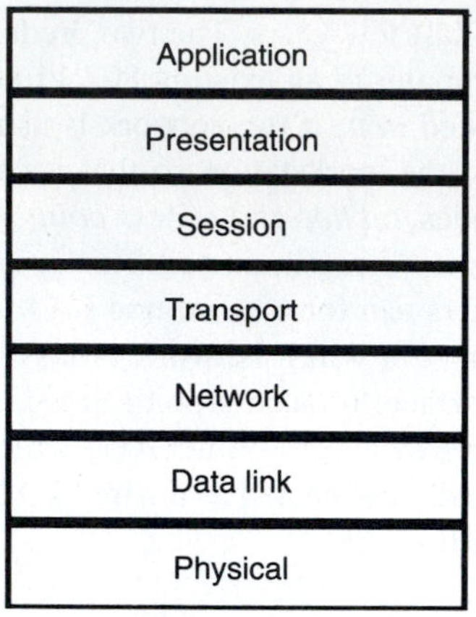

Figure 16–7 Open Systems Interconnect model

Physical Layer

The physical layer includes all data communication *hardware* that is necessary to carry a message. This includes UARTs, RS232 interfaces, modems, transmission lines, fiber optics, and so on. The physical layer also defines the electrical (or optical) representation of binary information and the procedural events that place the encoded signal onto the medium. Everything passes through hardware eventually!

Data Link Layer

This layer is concerned with getting data to flow from one device to another on a network segment. (A *network segment* is a communications channel directly shared by two or more computers.) This involves assembling data into message units called *frames* (think of assembling a collection of pages into an envelope), as well as providing error-checking and error-recovery mechanisms (such as checksums and CRCs). Note that the data link layer merely gets the data moved between two points on a local network. It just oversees the communications, correcting errors as needed.

The data link layer can be divided into two sublayers called *MAC* and *LLC*. The MAC (medium access control) sublayer controls physical access to the communications channel. It provides the mechanisms for controlling who can transmit messages, and when. It also provides local-level addressing in the form of *MAC addresses* (also known as *hardware addresses*). A MAC address is a 48-bit number that represents the local address of the device sending a message. All devices have unique MAC addresses, usually contained in a ROM (read-only memory). The LLC (logical link control) sublayer supervises the transmission of frames and participates in the error recovery process.

TIP MAC addresses are usually expressed as a set of six hexadecimal byte values. The first three bytes represent the manufacturer's ID (assigned to the manufacturer by the Institute of Electrical and Electronic Engineers [IEEE], and the last three are the serial number of the network interface card.

Network Layer

Since a network can be composed of hundreds or thousands of computers, *addresses* are needed on all messages so that data arrives at the proper remote destination when it is sent. This is the function of the *network* layer. This layer forms logical addresses, and takes care of *routing* messages over paths that may require that the message be repeated by many computers along the way.

Addresses at the network layer (normally IPv4 or IPv6) consist of two portions, a *host* address and a *network address*. The network address specifies which physical network segment (anywhere in the world) is to be used as the destination for the message, and the host portion of the address tells which computer on that network should get the message. We will discuss how IP works in greater detail soon.

Transport Layer

This layer is responsible for delivering error-free communication, and it ensures that message packets aren't lost or "dropped." Note that the error-checking provided in this layer is in addition to that provided by the data link layer. The transport layer breaks up large messages from the *session* layer (above it, to be discussed next) for "digestion" by the lower network layers (network, data link, physical). Incoming fragments of messages are reassembled into complete (large) messages by this layer. The three primary Internet protocols that operate at this layer are UDP (user datagram protocol), TCP (transmission control protocol), and ICMP (Internet control message protocol).

Session Layer

The mechanisms within this layer provide a virtual (abstract) connection between two computers called a *session*. It provides needed functions such as name lookup (finding the address of a remote computer by its assigned name using domain name system [DNS] services), security (deciding who can establish a connection, and who can't), and session management. Session management allows several computer processes to share the same communications channel without interference (multiplexing). The session layer uses *source and destination port numbers* to identify and separate communication streams between computers.

Presentation Layer

This layer translates data between the format needed for applications (programs typically operated by an end user) and the session layer (which delivers a virtual connection between two computers). An example of this layer is the *network redirector* in a local area network. The network redirector is a program that is part of the operating system. It makes remote files on a server computer visible to the client computer; it also allows the sharing of printers and other resources over the network. Another example is the translation of graphics (drawing) commands between different computer systems.

Application Layer

This last layer provides services that support end-user applications, such as electronic mail, database access, Internet access, and so on. When a computer program (such as a Web browser or e-mail package) accesses network services, it interacts directly with this layer. The services provided by the application layer are called APIs, or *application programming interfaces*. The most common API implementation for networking is the Berkeley sockets model. In this model, a *socket* is an abstract (virtual) point of connection for a computer program to communicate through, just like a physical socket equipment can be used for passing electrical signals.

How the OSI Layers Work Together

The concepts involved with the seven OSI layers are fairly abstract. It helps to see an example of what they really do. Imagine that Jim wants to send an electronic mail message to his friend Gerry, who lives in Texas. Jim connects his computer to the Internet, using an Internet service provider and launches *Eudora Pro,* an electronic mail *application program*.

Jim types his message to Gerry:

Hi Gerry. I just wanted to let you know that we got the fruitcake you and Lisa sent us for Christmas. By the way, I am still having trouble with my El Camino. It runs really rough when the engine is cold, and it stalls every time I stop at an intersection. After the engine warms up it smooths right out and runs perfectly. Any ideas?
Jim

Jim presses the *SEND* button, and in just a few seconds, the message is on its way to Gerry. What has actually happened here? The *overall* action is that the letter Jim typed ended up on the electronic mail (e-mail) server computer across town at Jim's ISP. That letter then got passed across the country to Gerry's ISP, so that Jerry would be able to retrieve it later. Figure 16–8 shows the activity in terms of the OSI model.

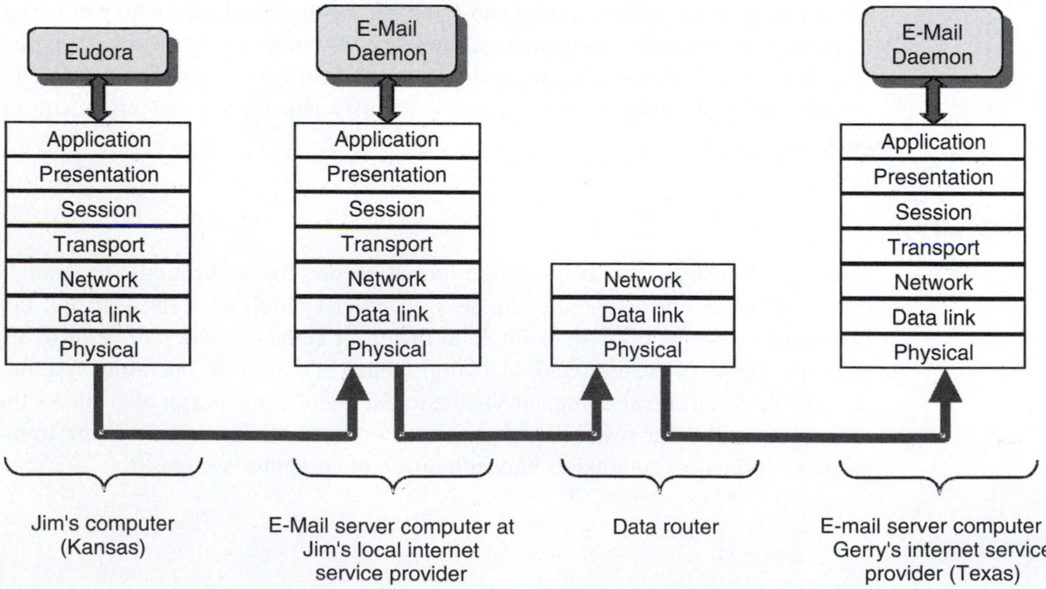

Figure 16–8 Sending electronic mail

The basic electronic sequence of events is as follows (and starts at the top layer, *application* in Jim's computer at left):

1. *Eudora* opened a "connection" to the Internet, through the *application* layer within Jim's computer, using subroutine calls to the operating system entry points. The Internet connection was requested with the *e-mail server computer*.
2. The connection being established, *Eudora* began passing the information and necessary commands to the e-mail server computer. These commands were actually passed to the operating system on Jim's computer. They were translated by the *presentation layer* into the proper format for talking to the ISP computer. The *session* layer simplified matters by allowing the e-mail program to establish a virtual "connection" to the e-mail server computer, and helped find the address of the e-mail server. The session layer also identified which process on the e-mail server would receive the e-mail data by examining the *destination port number* (25, SMTP) supplied by *Eudora* on Jim's computer. The *transport* and *network* layers made sure the message and commands got to the right computer (the e-mail server). The *transport* layer also divided the message into bite-sized chunks for the lower layers. After the message passed through the *transport* layer, it looked a little like this:

   ```
   01 Hi Gerry. I just wanted to . . .
   02 let you know that we got the . . .
   [ . . . More pieces of the message . . . ]
   37 ? Jim
   ```

3. The *data link* layer checked for errors as the pieces of the message passed through it. The *physical* layer was the serial port and modem that connected Jim's computer to the telephone line (he used a dial-up connection at the ISP).
4. At the server computer, the e-mail message and commands moved back up through the layers, being error-checked, reassembled, and transformed as needed until they reached the e-mail *daemon* program. A *daemon* is a program that runs at all times, without the need for human interaction.
5. The e-mail daemon program processed Jim's message and saw that the e-mail needed to be forwarded to Gerry's ISP in Texas, so the message was again disassembled into its parts and sent down through the layers on the e-mail server computer, where it reached the *physical* layer again. This time, the message went out not on a phone line, but onto a T1 Internet line, where it reached a *router*.
6. The *router* examined the logical (IP) address on the e-mail message (it could not tell it was an e-mail, it could only see fragments of text), and found the correct route to the Texas computer. The message went out through *another* branch of the Internet to Gerry's ISP server computer in Texas.
7. The e-mail message arrived in pieces at the Texas computer. It bubbled up through the OSI layers, and was reassembled into one continuous message by the *transport* layer, and it then made its way up to the e-mail daemon program, which recognized the incoming message, and stored it in Gerry's in-box. The message would be retrieved when Gerry connected to the Texas server computer.

There are a *lot* of events taking place during network communications! Notice that every device on a network does not need all the OSI layers. In particular, a router only needs the first three layers, since it merely decides where a message is going to go. The primary activity of a router is at the *network* layer, since routers decide where to send messages based on the IP address within the message packet.

Section Checkpoint

16–11 List the seven ISO/OSI layers, and explain what each one does.

16–12 What three layers are used by routers? What layer does a router work with primarily?

16–13 What is the name of the abstract (virtual) network connection provided by the application layer?

16–14 What is a network redirector?

16–15 Give several examples of programs that interact with the application layer.

16–3 THE INTERNET AND INTERNET PROTOCOL (IP) ADDRESSING

The *Internet* is a global network of computers. Its origins go back to 1957, the year that Russia launched *Sputnik,* the first human-made satellite. The United States government was shocked by this event, and in response, formed *ARPA,* the Advanced Research Projects Agency, as a part of the department of defense. In 1962, the U.S. Air Force commissioned Paul Baran of RAND (a government-controlled corporation) to study how control could be maintained over missiles and bombers in the event of a nuclear attack. This was to be a computer network that could survive a nuclear strike. Baran's proposal involved *decentralizing* the computing power, and developing a *packet-switched network*.

Decentralizing the computers would distribute the burden geographically, so that if a key city was destroyed, the rest of the communication network would be intact to coordinate a counterstrike. A packet-switched network breaks messages up into small *packets* called *datagrams,* which contain origin and destination addresses, and are forwarded through the network of computers until they reach their destination, where they are reassembled into a complete message.

The first actual physical network, called *ARPANET,* wasn't finished until 1969, and it linked four locations: University of California at Los Angeles, SRI (in Stanford), University of California at Santa Barbara, and University of Utah. The data rates were limited to 50 kbps on the links. In the period between 1969 and the present, many technical improvements and innovations came to pass that improved the efficiency of communications. The number of computers connected to the network increased (as of 2003, there were over 600,000,000 host computers online worldwide), and the speed of the communications backbones increased (155 Mbps backbones are now common).

In 1992, CERN released the standards that would form the World Wide Web (WWW), and in 1993, commercial use was allowed. Growth of the Internet (as it was now called) literally exploded, with the number of online hosts doubling every 6 months. Internet access is now considered by many to be a basic utility service, much like the telephone. In fact, most local telephone companies are also ISPs; someday the telephone as we know it will likely become a part of history, having been replaced by digital communications services.

Internet Addressing System

Computers connected to the Internet are assigned a 32-bit *Internet Address,* or *IPv4 address,* or a 128-bit *IPv6 address*. IPv6 is a response to the shortage of IPv4 addresses. An IPv4 address is traditionally written as four decimal numbers separated by periods, even though each of the numbers represents an *octet* or *byte* of information. This format is sometimes called *dotted decimal* notation. For example, one of the hosts at DeVry—Kansas City has the IP address 205.160.208.21. At the time of this writing, most of the Internet is still using IPv4.

An IP address consists of two portions, the *network address* and *host address*. The *network* portion of the address identifies which network segment (worldwide) a message is destined for, and the *host* portion identifies which computer on that network should get the message. Organizations are normally issued *blocks* or groups of IP addresses.

Because some organizations have many computers and others have few, the IPv4 addressing system has some peculiar characteristics. Internet addresses are issued in three groups, class A (for the largest organizations), class B (for large organizations), and class C (for small organizations). This system is called *classful addressing,* and it is the basis of IPv4 addressing. However, it is a wasteful system (about 97% of the available IP addresses are wasted in classful addressing), and is now being supplemented by *CIDR, classless interdomain routing,* a system that allows routing to various network addresses outside of the class-based IPv4 address system.

Class A Internet Addresses

You can identify a class A IP by examining the first number. If it's between 1 and 126, it is a class A address. Very few organizations have class A addresses, because only 126 of them are possible! For a class A Internet address, the last three octets (numbers) give the *host number* within the organization. For example, a computer with the IP 125.6.4.127 is a class A address, and the *host number* of that computer is 6.4.127 (and that is handled within the organization). Because there are three 8-bit digits in the host number with a range of 0–255 for each (0 [all zeroes] and 255 [all ones] are used for two special purposes as we'll soon see), a class A IP address can have up to $2^{24} - 2$ or 16,777,214 host computers associated with it. Only the largest organizations could even approach this number of hosts!

By the way, the first digit in the class A address is called the *network address,* because it uniquely identifies the organization connected to the Internet. There are only 126 class A network addresses available. The "formula" for a class A address looks like this: N.H.H.H. This formula shows us that the first byte is the network address, and the remaining three bytes are the host address.

Class B Internet Addresses

Class B IP addresses start with 128 and end at 191, and use *two* of the octets (bytes) to identify the network (organization) address. Therefore, the range of class B addresses spans 128.1 to 191.254. There 16,384 possible class B network addresses. A class B address has the format N.N.H.H.

The last two bytes in a class B address are the *host number*. For example, the address 185.255.32.1 has the following characteristics:

- A *network address* of 185.255
- A *host number* of 32.1

Because two places (16 bits) are available for the host number in a class B address, an organization of this type can have up to $2^{16} - 2$ or 65,534 different host computers online.

Class C Addresses

Most organizations are small and have class C addresses, which have a first octet in the range of 192 to 223. The first *three* octets give the network address, and the last gives the host number. For example, the address 208.128.98.1 can be interpreted as follows:

- The network address is 208.128.98
- The host number is 1

A class C address holder may have only 254 hosts, since only 8 bits are available for the host number. Its "formula" is N.N.N.H.

You might have noticed that none of the address ranges discussed so far start with 127. The reason is that any address with a first octet of 127 is a *loopback* address for local testing on a single computer. Any IP packets directed by a computer to 127.X.X.X (X = don't care) will be reflected or looped back to the same machine.

Network and Broadcast Addresses

On each network there are two special addresses, the *network* and *broadcast* addresses. The *network address* is calculated by inserting all binary 0s for the host portion of the IP address, and the *broadcast address* is formed by using all binary 1s for the host portion of the address. For example, an organization may possess the class C IPv4 block extending from 208.128.98.0 (the network address) to 208.128.98.255 (the broadcast address). The host addresses on this network are in the range 208.128.98.1 to 208.128.98.254. Be careful about IP address calculations; the number "255" doesn't always mean *broadcast*.

Domain Name System

Internet addresses are difficult for most people to remember and work with, but names are easy to work with. The *domain name system* is a database of Internet server addresses and names. The DNS system is administered by the InterNIC. There are at least eight *primary* or *root domains* that are used as follows:

- .com — Commercial, for-profit organizations
- .edu — Educational institutions
- .gov — Government
- .mil — Military
- .org — Nonprofit organizations
- .net — Internet service providers
- .biz — Business use
- .pro — Professional organizations
- .name — For registration by name

When an Internet user wants to connect with another computer, he or she can do it in two ways. First, he or she can connect with the *physical* Internet address. For example, at a *UNIX* command prompt, one may type either of the following commands to connect with the *Telnet* server at DeVry-KC:

```
telnet kc.devry.edu
telnet 205.160.208.21
```

Likewise, you can instruct a Web browser to go directly to an IP address (instead of an organization name). Try typing the following into the location bar on any Web browser. Why does it work? The reason is that the command bypasses DNS, going directly to the IP address of the Web server. (This won't work for all Web servers, since some sites require the Web browser to transmit a host name in the HTTP GET request.)

```
http://216.239.39.99
```

How DNS Works

When a request is made to connect with another computer by name, the name is passed to a nearby DNS server computer. Some domain names take only one step to "decode" back into a physical IP address. This occurs when the DNS server has direct authority for that domain. For example, the name "devry.edu" can be looked up directly if the lookup is requested directly from the devry.edu DNS server, because this server is responsible for delivering information about the "devry.edu" domain. When the query involves a domain not under the direct authority of a DNS server, a *recursive* process is used. In a recursive process, previous output is used as input for the process (a little like a feedback loop).

For example, let's follow the event chain required to lookup "mail.yahoo.com." First, the root server for the com domain will be asked for the address of "yahoo.com." This server will return the address of the Yahoo DNS server, 66.218.71.198. Next, the Yahoo DNS server is queried for the host "mail.yahoo.com." This step results in the IP address for the "mail" server being returned from the Yahoo DNS server, which can then be directly used by the application program to make a connection to "mail.yahoo.com."

The World Wide Web

The Web has been the subject of much hype, and a lot of people are confused as to what actually constitutes "the Web." You might be surprised to learn that the Web actually doesn't exist at all; it's merely the visible end result of a lot of programs running on computers all over the world, communicating through a physical worldwide network called the Internet (or Net).

The Net is for real. It originally existed as a communications network for the military and scientific communities. Just what is the Net? Basically, the Net is nothing more than a huge network of computers and routers that spans the entire globe. Messages move from point to point in the Net by a systematic electronic relaying system. Since there are so many computers, there are almost always several paths to any destination. The electronic end of the Net chooses a path to get the (hopefully) shortest route to each destination. Computers along this path act as relay stations, receiving the message and repeating it for the next computer or "node." The process continues until the message arrives at its destination.

There are two primary protocols that govern data transmission on the Net. They are TCP (Transmission Control Protocol), and IP (Internet Protocol). Often people lump them together and call them "TCP/IP." The function of TCP is to break a data stream into packets called *datagrams,* and transmit these across the network while checking (and correcting) transmission errors. IP has a much more complex job; its task is to choose the best route to an Internet destination from among the thousands of possible "journeys."

The environment of the Net is a textual one much like the MS-DOS operating system used on PCs. The predominant operating system is called *UNIX* and has a command structure very much like DOS. (*LINUX* is also a dominant operating system on the Net. It is an open-source version of *UNIX* for personal computers utilizing Intel x86 processors.) Each computer can run any operating system it wishes, as long as it knows how to talk to the rest of the Net. The Net is not a graphical environment; it's extremely unfriendly to the novice!

Since the Net is a communications medium unlike any conceived before, people have coined new words to describe it. "Cyberspace" refers to the electronic "space" of the Net, where everything has an abstract computer address (which is really a number) instead of a physical location. "Newbies" are people just starting to learn about the Net. "Spammers" are people who use electronic mail (e-mail) for advertising, a definite breach of "Netiquette" (an unwritten code of conduct subscribed to by the Internet community at large).

So what *is* the Web? Simply put, the Web is a set of protocols (rules for communication) that are implemented in programs on computers across the Net. The primary protocol is the *HyperText Transfer Protocol,* or HTTP for short. HTTP helps specify how data is going to be transferred across the Net, and what needs to be done with the data once it has been received.

To get on the Web, you need a program called a *browser,* and an ISP, who acts as your on-ramp to the Net. The operation of a browser program is really quite simple. The Web is document based. The documents sent during Web activity are very similar to letters written with a word processor; however, there are four major differences.

First, these documents are located in computers all over the world; second, the documents often contain pictures and/or sounds, unlike what you would compose on a typewriter; third, the documents can contain links or references to other documents; and last, these documents (often called Web pages) are written in a special language called HTML, or *HyperText Markup Language*. HTML is a simple language designed to help browsers construct on-screen displays of the Web data they receive. This is no small task, because a Web page author has no way of knowing what type of computer the reader will be using when the page is retrieved, and remember, Web pages can have quite complex mixes of text, graphics, and sound!

Two popular browser programs are Netscape Navigator (Mozilla under Linux) and Microsoft Internet Explorer. There are dozens of others on the market.

Browsers understand HTML. When a Web browser receives an HTML document, it decodes it, resulting in a screen display containing text, graphics, and perhaps even sound and animation. But how does a Web browser know where to get the document from? The answer is the URL, or *Uniform Resource Locator*. You've seen URLs before; it is now common for companies to place a URL at the bottom of an advertisement, such as http://www.toyota.com, or http://www.microsoft.com. URLs must generally be typed in exactly as they're shown, or a browser will either go to the wrong place, or nowhere at all! The one exception to this rule is that the prefix "http://" is often omitted in URLs, since that is a default protocol for Web browsing (your browser will add this to a URL should you leave it out).

Section Checkpoint

16–16 What is the Internet? How many computers are connected to it?

16–17 Why are Internet addresses broken up into three classes?

16–18 How many server computers can each class of Internet presence have online?

16–19 What is CIDR and why is it used?

16–20 Explain the purpose of DNS.

16–21 List the six root domain names, and explain what each is used for.

16–22 Does the World Wide Web physically exist? Explain why or why not.

16–23 What protocol is used in the Internet to route message packets to the proper destination?

16–24 What protocol is used to ensure message delivery in the Internet?

16–4 IP ADDRESSING, SUBNETS AND GATEWAYS

As you learned in a previous section, IP addresses consist of two parts, a *network address* and a *host address*. You might wonder how a computer can tell the difference between them, since there's nothing built into the IP address values that indicates this. The answer is that a separate value called the *subnet mask* contains this information. The subnet mask is a binary number that is the same size as the IPv4 or IPv6 address. It contains logic 1s to mark the network portion of the address, and 0s to indicate the host portion. The subnet mask is assigned by the network administrator. It always begins with a solid or continuous group of 1s and ends with at least two logic 0s.

EXAMPLE 16–1

What is the subnet mask for the following IP addresses: (a) 64.133.132.8; (b) 208.128.98.1. Assume that the networks are corresponding to the classful system discussed in Section 16–2.

Solution

a. The IP address 64.133.132.8 has a first octet of 64, which is in the range of 0 to 126. Therefore this is a class A IP address and it has the following form: NHHH (remember, N = network, H = host). The subnet mask contains all logic 1s for the network portion of the address, and logic 0s for the host portion. In binary, hex, and decimal this subnet mask is:

N	H	H	H
1111 1111	0000 0000	0000 0000	0000 0000
$FF	$00	$00	$00
255	0	0	0

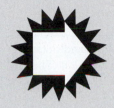

Since subnet masks are traditionally written in decimal, we would report that the subnet mask *SM = 255.0.0.0.*

> ***TIP*** We can write an IP address and subnet mask using CIDR shorthand notation. This IP address would be expressed as 64.133.132.8/8 using this notation. The "/8" tells us that the subnet mask has *eight* consecutive logic 1s. You'll see this notation used extensively in industry.

b. The IP address 208.128.98.1 has a first octet of 208, which is in the range of 192 to 223. Therefore this is a class C IP address and it has the following form: NNNH. This subnet mask is:

N	N	N	H
1111 1111	1111 1111	1111 1111	0000 0000
$FF	$FF	$FF	$00
255	255	255	0

The subnet mask *SM* for the second network is therefore *255.255.255.0*. Again, the CIDR shorthand for this IP would be 208.128.98.1/24, since there are 24 1s in the subnet mask.

The Importance of the Subnet Mask

The subnet mask is a very useful piece of information. Computers on a network use it to determine three very important pieces of information. First, the *network* address can be calculated by performing a bitwise AND of the subnet mask and the IP address of any computer on the network. Second, the *broadcast* address of the network can be found by bitwise OR'ing the ones complement of the subnet mask with any host address on the network (or the network address). Finally, the subnet mask is used to determine whether IP packets are destined for the local network, or a remote (different) network. To send packets outside the network, they must be transmitted to a *gateway*. A gateway is usually a router connected to the Internet. Equations 16–1 and 16–2 show how to find network and broadcast addresses.

$$N = IP \bullet SM \qquad \textbf{(16–1)}$$

where *N* is the network address, *IP* is the IP address of any host or the network IP address, and *SM* is the subnet mask. The symbol "•" means bitwise AND.

$$BC = IP + \sim SM \qquad \textbf{(16–2)}$$

where *BC* is the broadcast address, *IP* is the IP address of any host or the network IP address, and *~SM* is the ones complement of the subnet mask. Let's try an example.

IMPORTANT: The "+" operator in equation 16–2 means bitwise OR, *not* addition!

EXAMPLE 16–2 What are the network and broadcast addresses for host 172.0.0.125 given the subnet mask 255.255.255.0?

Solution

First write the values in hexadecimal and break them into octet (byte) groups. (You can write the values in binary once you've converted to hexadecimal, if desired, but this isn't entirely necessary.) Using equation 16–1, we get:

> *TIP* You can often do binary arithmetic in hexadecimal on a scientific calculator. Check the instructions that come with your calculator.

Host Address:	172.0.0.125	AC.00.00.7D
Subnet mask:	255.255.255.0	FF.FF.FF.00
	(AND) •	________
	= *Network Address*	*AC.00.00.00 = 172.0.0.0*

To find the broadcast address, use equation 16–2:

$$BC = IP + \sim SM \qquad \textbf{(16–2)}$$

This equation says that we must find the *ones complement* of the subnet mask, so first we write the subnet mask in binary:

```
SM = 1111 1111 1111 1111 1111 1111 0000 0000
      F    F    F    F    F    F    0    0
   =   255    .   255   .   255   .    0
```

The broadcast address is obtained by OR'ing the IP address with the one's complement of the subnet mask:

```
~SM  = 0000 0000 0000 0000 0000 0000 1111 1111 (00.00.00.FF)
IP   = 1010 1100 0000 0000 0000 0000 0000 0000 (AC.00.00.00)
(OR) + ________________________________________
BC   = 1010 1100 0000 0000 0000 0000 1111 1111 (AC.00.00.FF)
                                                = 172.0.0.255
```

The broadcast address, *BC = 172.0.0.255*.

Number of Hosts on a Network

The subnet mask indirectly tells us how many host addresses are available on a network. This is important because it tells us how many computers or other devices may be assigned to the network. Recall that the host portion of an IP address corresponds to the number of 0s in the subnet mask. Since this is just a binary number, we can calculate the total number of IP addresses that are available:

$$TOTAL = 2^x \qquad \textbf{(16–3)}$$

where x is the number of "0" subnet mask bits, and *TOTAL* is the number of IP addresses in the network's IP block assignment. Note that this number includes both the *network* and *broadcast* addresses. Can you modify equation 16–3 to account for this? Yes, it must look like:

$$HOSTS = 2^x - 2 \qquad \textbf{(16–4)}$$

Equation 16–4 accounts for the loss of the network and broadcast IP addresses. *All networks have a network and a broadcast address, and it always reduces the number of available host IPs by two.*

EXAMPLE 16–3 How many hosts are possible in the network of Example 16–2?

Solution

Find out how many zeroes are in the subnet mask. The subnet mask is 255.255.255.0, and there are *eight* zeroes. Therefore, equation 16–4 can be used to find the number of hosts:

$$HOSTS = 2^x - 2 = 2^8 - 2 = 256 - 2 = \underline{\underline{254}}$$

This network can therefore support 254 hosts.

Determining the Destination for IP Packets

As you've seen, the subnet mask helps us to identify the network and broadcast addresses for a network system. However, there's another important function. In Figure 16–9, two different networks are connected to the Internet. Both of these networks contain several computers.

Suppose that station A_1 needs to send a packet to station A_2. Since they're on the same LAN (the "A" LAN, top of Figure 16–9), it is not necessary to transmit the message

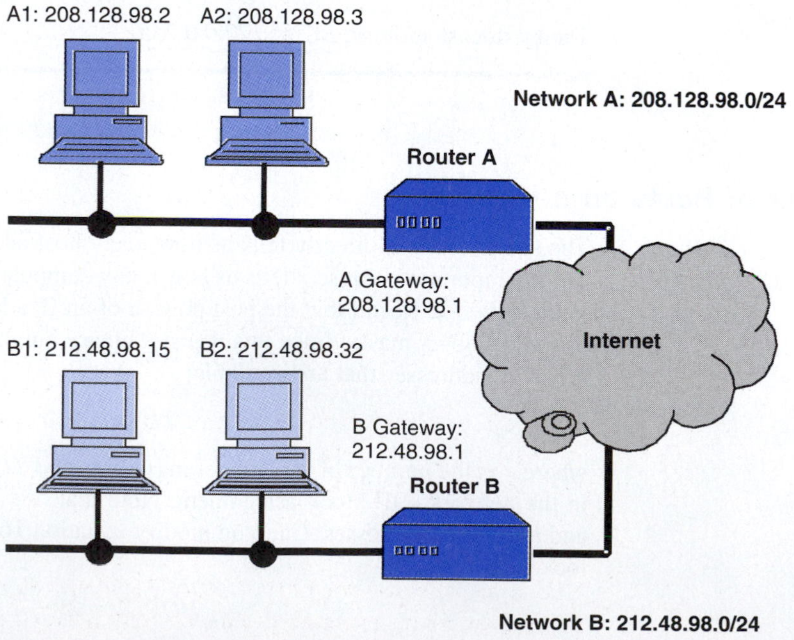

Figure 16–9 Two networks communicating through the Internet

through the Internet. The two workstations are directly connected. But what if station A_1 needs to send a message to station B_1. How will it know that it needs to utilize the gateway? The answer is that it examines the network address of the outgoing packet. To obtain this network address, it uses the subnet mask for its own network. It then compares the resulting network address with its own. If it's the same, then the IP packet can be emitted directly onto the LAN. If it's different, then the IP packet is routed to the gateway and Internet, where it will be correctly routed to network B.

EXAMPLE 16–4

Host A_1 in Figure 16–9 wants to send a message to host B_2. Show how it knows to utilize the gateway to get the message through.

Solution

AND the destination network address with the subnet mask of the destination network (*B* in this case) and compare it to the *A* network address:

$$N = IP \bullet SM = (212.48.98.32) \bullet (255.255.255.0) = \underline{212.48.98.0}$$

This is *different* than the local network address of 208.128.98.0, so the computer automatically transmits the packet to the gateway at 208.128.98.1. The gateway (router) then forwards the packet to the Internet, where other routers will see to it that the packet reaches the *B* network router. The *B* network router then emits the IP packet onto the destination LAN where station B2 can receive it, completing the path.

Dividing Networks

Networks can be divided by a technique called *subnetting*. This is often used to split up large blocks of unused IP addresses. It's equivalent to having a house with a few extra rooms rented out to others. Class A and class B IP address blocks are commonly subnetted by ISPs prior to their allocation to customers. Doing this helps to reduce the waste of precious IP addresses. Subnetting is often done in a small organization to divide a large network into multiple *broadcast domains* for easier management. When a group or *block* of IP addresses is divided, two or more new networks are created as a result. A router must therefore be set up to properly route messages to the new networks. *Dividing a network always results in the loss of some IP addresses.* In the case of an ISP, the loss of a few IP addresses is more than compensated by the "gain" of thousands of IP address blocks that can be leased to customers.

Networks can be divided into power-of-two numbers of slices. In other words, during each subnetting step, a network can be divided into two, four, eight (and so on) equal pieces. Each of the new subnets will have an identical *subnet mask* and number of hosts. To split the network, we add one or more logic 1s to the subnet mask. This step adds one or more network address bits to the IP block. The newly added subnet mask bits tell us the location of the new network address bits. These new bits are then cycled through all their possible binary combinations to produce all the possible subnet addresses.

EXAMPLE 16–5

Subnet the network 207.129.138.0/24. Each network needs to support at least 20 computers. For each new network, give the network address, broadcast address, subnet mask, and the number of hosts.

Solution

First, determine the original subnet mask. According to the notation, it has 24 logic 1s and looks like this:

$$SM = 1111\ 1111\ 1111\ 1111\ 1111\ 1111\ 0000\ 0000$$

Next, determine how many *host* bits will be needed for each subnet. Equation 16–4 tells us that:

$$HOSTS = 2^x - 2$$

where x is the number of bits in the host portion of the IP address. We can substitute various values for x to see how things fit. For example, if we substitute *4* for x, we get $2^4 - 2$ or 14 hosts; this isn't enough computers. We can then try $x = 5$. We get $2^5 - 2$ or *30* hosts per subnet by using five host bits. This will allow for expansion on each subnet.

NOTE: You can mathematically solve for x by: $x = \left\lceil \frac{\log(n + 2)}{\log 2} \right\rceil$ where $\lceil\ \rceil$ is the next largest integer (ceiling) function, and n is the number of hosts needed on the subnet. The $\lceil\ \rceil$ marks tell us to round the result up to the next largest whole number if a fraction is involved.

Now that we know how many host bits are needed, we can determine the new subnet mask for the subnets. Five host bits are needed; therefore, the new subnet mask will have (32 − 5) or 27 logic 1s, as shown below:

These five logic 0s stand for the host portion of the address, which is five bits wide now.

SM′ = 1111 1111 1111 1111 1111 1111 1110 0000 = 255 . 255 . 255 . 224 = "/27"

These three bits were added to split the network into eight parts, each with up to 30 hosts.

Finally, the three new subnet mask bits are substituted back into the IP address and moved through *all possible states* (000 through 111) to give the base address for each network. Since there are three bits, there are 2^3 or 8 new subnets, as follows:

We only need to express the portion of the IP address that we're working with in binary. Everything else can remain in decimal.

SUBNET1 = 207 . 129 . 138 . 0000 0000 = 207 . 129 . 138 . 0
SUBNET2 = 207 . 129 . 138 . 0010 0000 = 207 . 129 . 138 . 32
SUBNET3 = 207 . 129 . 138 . 0100 0000 = 207 . 129 . 138 . 64
SUBNET4 = 207 . 129 . 138 . 0110 0000 = 207 . 129 . 138 . 96
SUBNET5 = 207 . 129 . 138 . 1000 0000 = 207 . 129 . 138 . 128
SUBNET6 = 207 . 129 . 138 . 1010 0000 = 207 . 129 . 138 . 160
SUBNET7 = 207 . 129 . 138 . 1100 0000 = 207 . 129 . 138 . 192
SUBNET8 = 207 . 129 . 138 . 1110 0000 = 207 . 129 . 138 . 224

The subnet mask for all of the subnets is 255.255.255.224 as determined above.

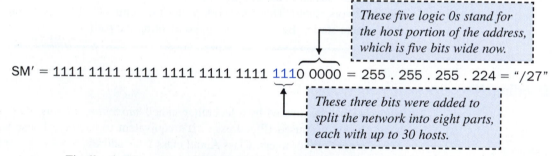

Table 16–2 Subnets for Example 16–5

Subnet number	Network address	Host address range	Broadcast address
1	207 . 129 . 138 . 0	207 . 129 . 138 . 1–207 . 129 . 138 . 30	207 . 129 . 138 . 31
2	207 . 129 . 138 . 32	207 . 129 . 138 . 33–207 . 129 . 138 . 62	207 . 129 . 138 . 63
3	207 . 129 . 138 . 64	207 . 129 . 138 . 65–207 . 129 . 138 . 94	207 . 129 . 138 . 95
4	207 . 129 . 138 . 96	207 . 129 . 138 . 97–207 . 129 . 138 . 126	207 . 129 . 138 . 127
5	207 . 129 . 138 . 128	207 . 129 . 138 . 129–207 . 129 . 138 . 158	207 . 129 . 138 . 159
6	207 . 129 . 138 . 160	207 . 129 . 138 . 161–207 . 129 . 138 . 190	207 . 129 . 138 . 191
7	207 . 129 . 138 . 192	207 . 129 . 138 . 193–207 . 129 . 138 . 222	207 . 129 . 138 . 223
8	207 . 129 . 138 . 224	207 . 129 . 138 . 225–207 . 129 . 138 . 254	207 . 129 . 138 . 255

Finally, we need to find the broadcast address for each subnet. Since we have already expressed the addresses in binary, the easiest way to find the broadcast address is to substitute all 1s for the host address in each subnet. We get:

Host address is all 1s for broadcast address.

SUBNET1_BC = 207.129.138.0001 1111 = 207.129.138.31

We're addressing the first subnet here.

Likewise, the other subnet broadcast addresses can be found:

SUBNET2_BC = 207 . 129 . 138 . 0011 1111 = 207.129.138.63
SUBNET3_BC = 207 . 129 . 138 . 0101 1111 = 207.129.138.95
SUBNET4_BC = 207 . 129 . 138 . 0111 1111 = 207.129.138.127
SUBNET5_BC = 207 . 129 . 138 . 1001 1111 = 207.129.138.159
SUBNET6_BC = 207 . 129 . 138 . 1011 1111 = 207.129.138.191
SUBNET7_BC = 207 . 129 . 138 . 1101 1111 = 207.129.138.223
SUBNET8_BC = 207 . 129 . 138 . 1111 1111 = 207.129.138.255

Note that we didn't need to convert all the octets (bytes) of the IP address into binary in this example. This was done only to make it clear which bits were being manipulated. Table 16–2 summarizes these results.

Subnet Zero and the All Ones Subnet

Two of the subnets in Example 16–5 may not be usable with older, legacy routers and operating systems. These are the first and last subnets (numbers 1 and 8), whose borrowed address bits are 000 (subnet zero) and 111 (all ones subnet). In the modern Internet, this is not a problem. In fact, RFC 1878 (request for comment) specifically states that the practice of excluding the two subnets is obsolete. You can find this RFC on file at http://www.ietf.org/rfc/rfc1878.txt. The potential problem is this: For the all ones subnet, the broadcast address is the same as the broadcast address of the original undivided net.

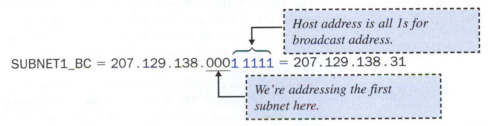

Because of this, misconfigured routers can cause a *broadcast storm,* which is a situation where routers mistakenly repeat broadcast messages between two networks in a continuous loop.

Network Privacy: NAT and Unroutable IP Address Blocks

Because of the shortage of IP addresses, an ISP can't give an individual or organization very many public IPs without strong justification (and some of the customer's money!). For example, suppose that a small electronics service company has 17 computers in-house. All of the computers need Internet access. Will it need to get a parcel of 17 IP addresses from the ISP to allow everyone on the Net? Doing this would be expensive, wasteful (not all the computers may even be in use at the same time), and very unsecure (workstations directly connected to the Internet are likely to be attacked by outsiders. In fact, a certain well-known and popular operating system can be compromised within *minutes* of being connected to the open Internet.).

The solution to this problem is *network address translation* (NAT) with the optional (but typical) use of *unroutable IP address blocks.* Table 16–3 shows the unroutable IP blocks. *Any message packet appearing on the open Internet with an unroutable destination address is automatically dropped by the Internet routers.* This is why these address ranges are called "unroutable."

With NAT, it is possible for dozens of computers inside an organization to share a single public IP address. The NAT process prevents unauthorized persons on the outside from accessing computers inside the NAT firewall. Figure 16–10 shows how this works.

Table 16–3 Unroutable IP blocks

Starting IP address	Ending IP address
10.0.0.0	10.255.255.255
172.16.0.0	172.31.255.255
192.168.0.0	192.168.255.255

Figure 16–10 Network address translation

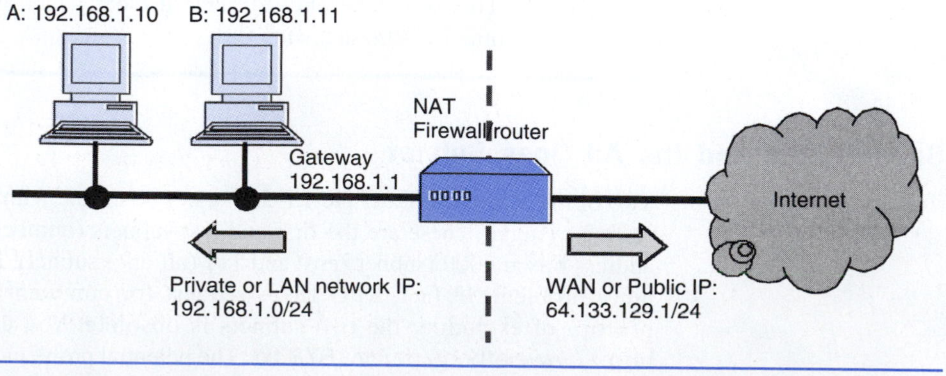

In the network of Figure 16–10, workstations *A* and *B* are assigned unroutable IP addresses. When either device needs to communicate to the Internet, it does so through the NAT firewall by sending its message packets to the *gateway* address of 192.168.1.1. Just like in Figure 16–9, each computer examines the destination IP address and calculates the effective network address when deciding whether the packet is destined for the LAN or a remote network. However, when the NAT firewall receives an outgoing message packet, it removes the source IP address and changes it to one of the WAN or *public* IP addresses before sending the packet out to the Internet. It also saves the original internal private IP address in a lookup table together with the original *source port number,* which identifies which program in the originating computer sent the message. Therefore, the private IP address originally on the packet isn't known to the message recipient. The remote receiver of the message packet sees only the public or WAN IP address of the NAT box, 64.133.129.1, in this case.

How does the reply message from the remote system return to the correct computer? The answer is that the reply will contain a *destination port* value equal to the *translated source port* value provided by the NAT firewall, as well as the WAN IP address of the NAT device. When the NAT device receives an incoming reply, it uses the *destination port* of the reply packet to look up the original internal private IP address in the *NAT table*. It then translates the destination IP address and destination port number back to the correct, initial values: The destination IP will be equal to the original unroutable private IP address of the workstation, and the destination port number will be equal to the original *source port* number provided when the workstation first sent out its message. We will examine port numbers in more detail in the next section.

What if a message comes into the NAT firewall from out of the blue? In other words, what if someone tries to break into the private network? In general, the NAT device will ignore the request because it is not a response to an outgoing message. The NAT box keeps track of all outgoing messages. This would seem to block *all* incoming messages, which isn't very useful! Therefore, a function called *port forwarding* is typically utilized with NAT. When an incoming message with a correct *destination port* number arrives (which specifies a *service* requested from the network), the NAT box can be programmed to forward such requests to a specific internal host IP address. The NAT computer must keep a list of destination (service) ports and internal IP addresses in order for this to work.

EXAMPLE 16–6

Is it possible for more than one computer in the world to have the IP address 10.0.1.10? If this is so, draw a diagram showing how it could happen.

Solution

This is an unroutable IP address. Therefore, it can be possessed by a computer on a LAN, but not on the open Internet. There are probably *thousands* of computers in the world with this IP address because there are *millions* of private LANs using the unroutable IP address blocks. Figure 16–11 shows how this might look.

A note of caution should be introduced here. *Two computers on the same LAN can never have the same IP address*. Most modern operating systems check for this condition before actually setting the IP address on an interface. Having two computers with the same IP address can be a difficult problem to isolate!

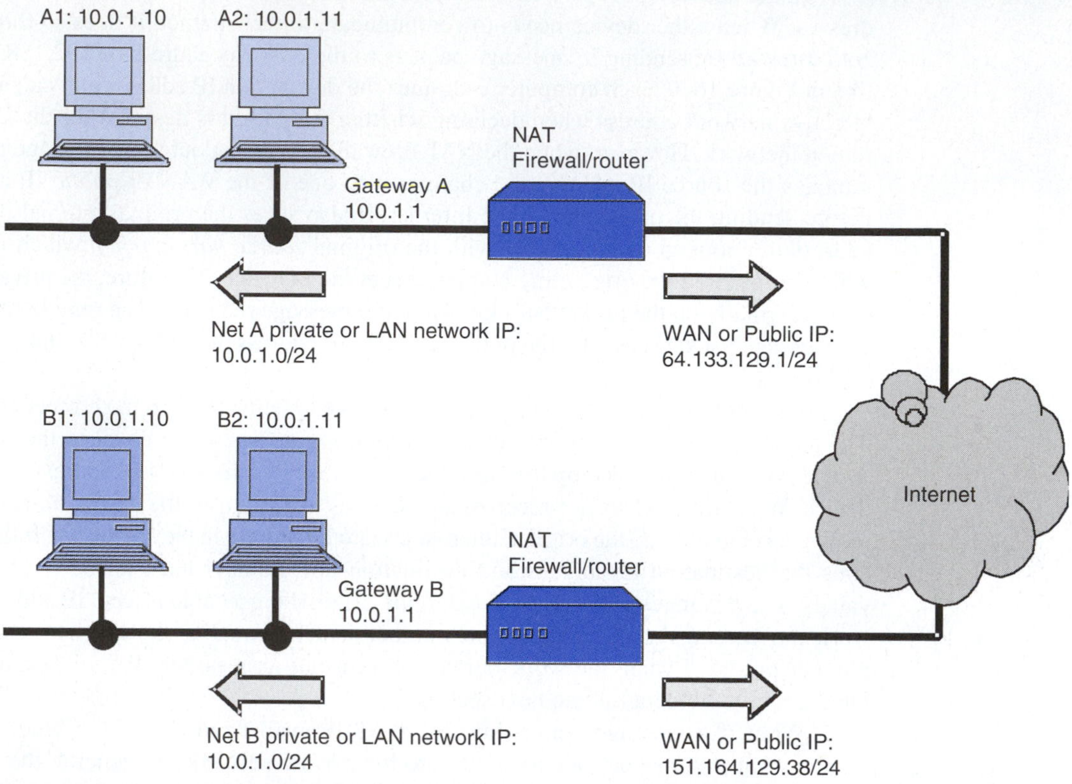

Figure 16–11 Two (or more) computers can have the same private IP address!

Section Checkpoint

16–25 What is the purpose of a subnet mask?

16–26 How big is a subnet mask (in bits) in IPv4 and IPv6?

16–27 What three pieces of information are obtained from the subnet mask?

16–28 What is a gateway?

16–29 Explain how to find the network and broadcast addresses of a network.

16–30 How can you determine the number of hosts on a network from the subnet mask value?

16–31 Why are networks often subnetted?

16–32 How many logic 1s are added to the subnet mask to divide a network into four subnets?

16–33 What are two reasons for the use of NAT? List the unroutable ranges of IP addresses.

16–5 DHCP AND ARP

Automatic IP Address Assignment: DHCP

When an organization has many computers, it soon becomes cumbersome to manually set the IP address of each one. It leads to many problems. For example, John, a system administrator, uses a laptop computer that he takes home every night. He plugs it into both his home network (he's a geek!) and the corporate network. John wants his computer to work on both networks automatically without having to manually switch IP addresses. The *dynamic host configuration protocol,* or *DHCP,* can solve John's problem.

Another problem is common: A new computer is added in a department, or an old one is replaced. In either case, an IP address must be assigned to the machine. If an administrator had to keep track of each machine and its assigned IP address (as well as manually setting all the machines), the time burden would be great. Many companies have *thousands* of computers. This is clearly unacceptable and error-prone!

DHCP provides an automatic method of assigning IP addresses to computers. An administrator sets up a *DHCP server* on the LAN (or a DHCP *relay agent,* which can relay DHCP requests from workstations to a central DHCP server). When a workstation needs to obtain an IP address, it issues a *DHCP discovery* message onto the LAN. If a DHCP server is available, a *DHCP offer* packet is returned. If the workstation can use the available IP address, it then sends a *DHCP request* packet to the server. The server then grants the IP address by sending a *DHCP ACK* message. The acknowledge message contains the same information as in the DHCP offer, except that the *acknowledge* flag is turned on to let the workstation know that it is OK to begin using the IP address.

Many parameters on the workstation can be automatically set by DHCP, not just the IP address. For example, the workstation computer generally needs to know the subnet mask and default gateway for the network, as well as the IP address of preferred DNS (domain name system) servers. All of this information (and more) can be conveyed by DHCP.

Finally, many NAT firewall devices have a built-in DHCP server. This server can be often configured with a web browser pointed at the LAN address of the firewall (often 192.168.1.1).

Address Resolution Protocol: ARP

All messages sent between two computers must pass down through the layers of the protocol stack (OSI model). This means that the data starts at the application layer and is passed down through the transport, network, data link, and physical levels. At the physical level the data consists of bits on the wire (or fiber optic). The receiving computer must reverse the process to get the data back. Recall that the data link layer's responsibility is to get the data onto (and from) the LAN. The data link layer has its own addressing system that is completely separate from the network (usually IP) layer. At the data link layer, MAC (media access control) addresses are used. MAC addresses are 48-bit (6-byte) numbers that uniquely identify each network interface on LAN. They are usually programmed into a ROM (read only memory) on each interface card. Each device must have a different MAC address. The first three bytes of the MAC address are called the organizational unique identifier, or OUI. They are administered by the IEEE and identify the vendor of the NIC card. The last three bytes are the NIC serial number, usually sequentially administered by the vendor.

Therefore, to get a message from one workstation to another, *two* addresses must be known for the destination workstation! These two addresses are the IP address and MAC

Figure 16–12 ARP process on a LAN

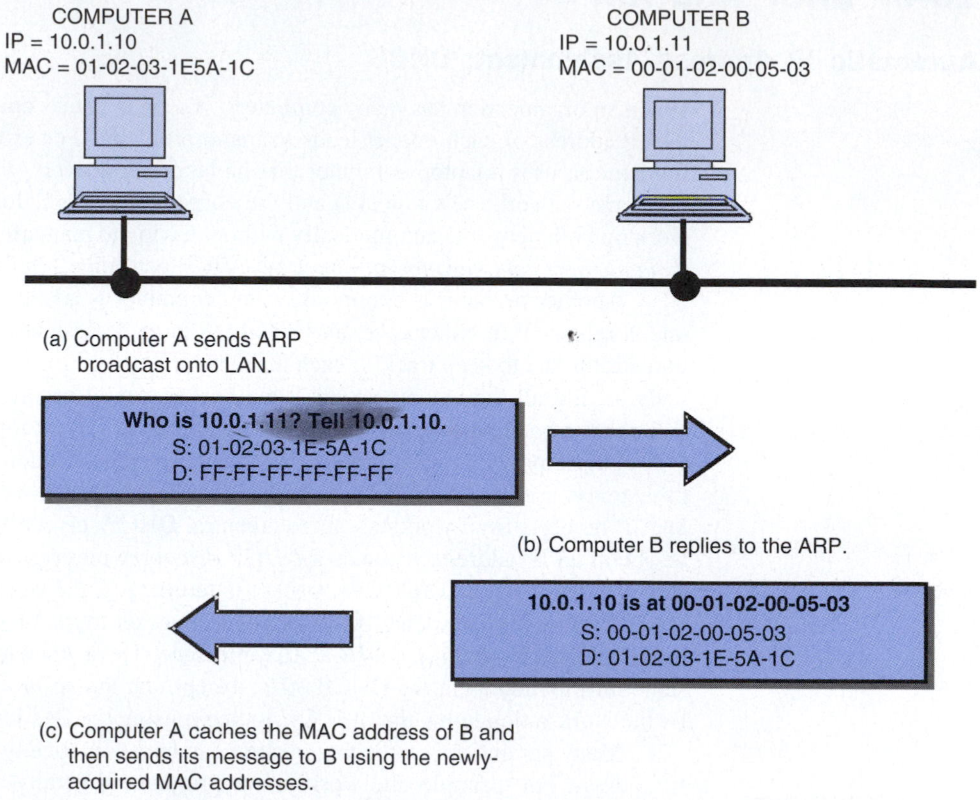
COMPUTER A
IP = 10.0.1.10
MAC = 01-02-03-1E5A-1C
COMPUTER B
IP = 10.0.1.11
MAC = 00-01-02-00-05-03
(a) Computer A sends ARP broadcast onto LAN.
Who is 10.0.1.11? Tell 10.0.1.10.
S: 01-02-03-1E-5A-1C
D: FF-FF-FF-FF-FF-FF
(b) Computer B replies to the ARP.
10.0.1.10 is at 00-01-02-00-05-03
S: 00-01-02-00-05-03
D: 01-02-03-1E-5A-1C
(c) Computer A caches the MAC address of B and then sends its message to B using the newly-acquired MAC addresses.

address of the NIC. *Without knowing the MAC address of the receiver, the data link layer cannot successfully transmit any messages to it.* This leads to a problem: How can a workstation learn the MAC address of a device it wants to talk to? The answer is the *address resolution protocol,* or *ARP*. ARP is used to learn the MAC address of other workstations on the network.

Figure 16–12 shows how the ARP process works. When one device wishes to talk to another but doesn't know the proper MAC address, it first issues an ARP broadcast. The ARP broadcast has the destination MAC address of FF-FF-FF-FF-FF-FF or all 1s. All network interfaces receive this message. Inside the broadcast is a query containing the IP address of computer B. Since computer B has the correct IP address, it formulates an ARP reply and sends it back onto the network. Computer B knows the correct MAC address for computer A at this point. How? Computer A just sent a broadcast, and *all* messages on a LAN have the MAC address of the sender within them. Therefore, computer A receives the reply and stores the MAC address of B in its *ARP cache*. Computer A can now directly send a message to computer B.

The ARP cache is a place where the MAC addresses of other network devices are stored. These are stored for a few minutes in case more communications are going to be needed. This way, the computer doesn't have to ARP for the MAC address of another computer every time it sends it a frame. Figure 16–13 shows how to examine the ARP cache on a Windows 2000 or XP workstation. In the figure you can see that the cache holds the MAC addresses of two devices at IP addresses 192.168.1.1 and 192.168.1.3.

```
Microsoft Windows 2000 [Version 5.00.2195]
(C) Copyright 1985-2000 Microsoft Corp.

C:\ >arp - a

Interface: 192.168.1.100 on Interface 0x1000003
   Internet Address      Physical Address      Type
   192.168.1.1           00-04-5a-ef-7c-d3     dynamic

   192.168.1.3           00-04-5a-64-a0-7e     dynamic
```

Figure 16–13 Examining the ARP cache

Proxy ARP

Figure 16–14 shows how ARP is accomplished through an internetwork. Recall that the data link layer is responsible for getting messages onto a LAN. The network in the figure actually has *two* LAN segments separated by a router.

In Figure 16–14, computer A again wants to send a message to computer B. However, computer B is not on the same network, so computer A must send the IP packet to the *gateway* or router interface 1. To do this, computer A must learn the MAC address of

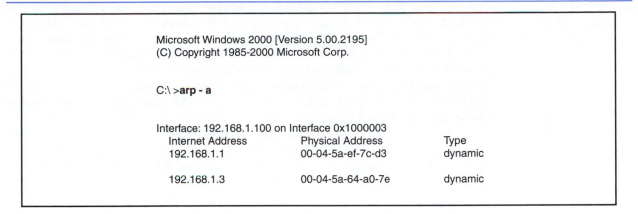

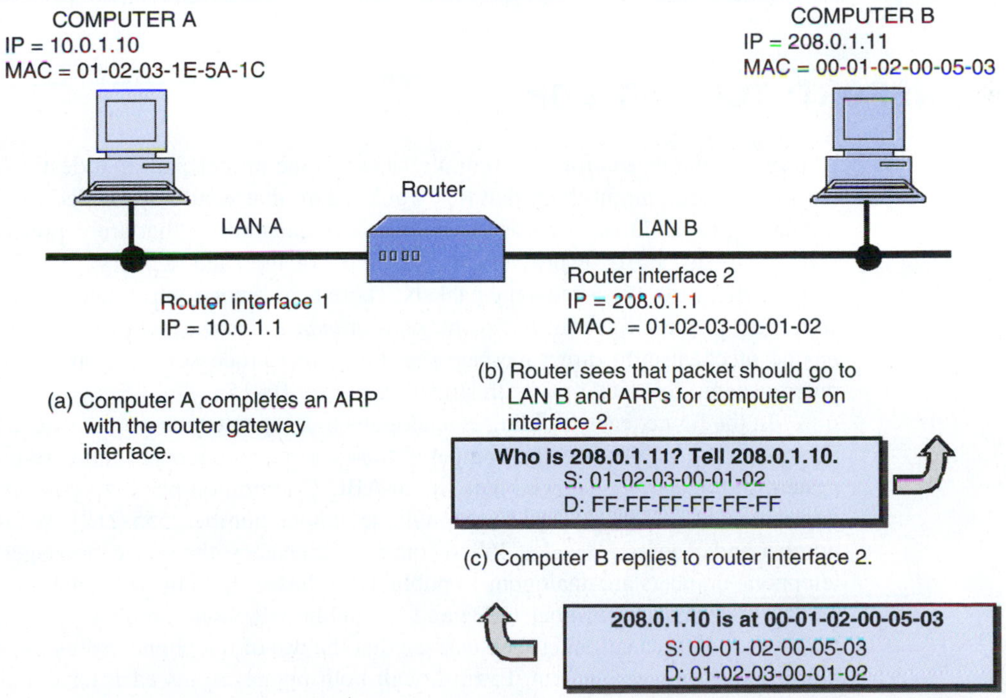

Figure 16–14 Proxy ARP

router interface 1, so it ARPs on LAN A (left-hand side) for the router interface. When the frame leaves computer A, it has the destination IP of computer B but the MAC address of router interface 1! This causes router interface 1 to retrieve the frame (remember, for a frame to be received at the data link layer, it must have the correct MAC address). The router then consults its routing table and learns that computer B's network is on interface 2 (right side of the figure). Router interface 2 must then *proxy ARP* to computer B so it can send the forwarded message to computer B. Thus, *two* ARP sequences had to take place to get the message through the internetwork.

Remember that every time a message is sent onto a LAN, a correct MAC address for the receiver is needed. This MAC address is learned through the ARP process. If a message is handed off between networks (as it certainly will be as it passes through the Internet), multiple proxy ARPs will take place before it reaches its destination.

Section Checkpoint

16–34 What is the purpose of DHCP? What problem does it solve?
16–35 What are MAC addresses? What layer of the OSI model uses them?
16–36 What are the two parts of a MAC address?
16–37 What two addresses are needed to get a message to a workstation?
16–38 Explain the process used during an ARP.
16–39 Demonstrate how to check the ARP cache from the command line.
16–40 When does a *proxy ARP* take place?

16–6 TRANSPORT: TCP AND UDP

IP provides the means for directing messages on the proper path to a destination through the Internet. You might think that this would be all that would be needed to get complete messaging capabilities, but it isn't. IP lacks two capabilities that are supplied by higher-level protocols. First, IP provides *no* guarantee of message delivery. IP only guarantees that it will *try* to deliver message packets. There is no error control mechanism in IP to recover lost packets. Second, IP has no *session management* capability. In other words, IP has no mechanism to direct messages to the correct processes (programs) within a computer system. This situation is illustrated in Figure 16–15.

In the figure, each building is analogous to a separate computer system. Each of the persons in the building can be compared to separate computer programs running on each system. For example, suppose that Jim at ABC Corporation needs to contact the service department at ACME. ACME has a public telephone number, 555-2121, so Jim dials that number and chooses extension 369 to connect with Nancy, the service manager. The public telephone numbers are analogous to public IP addresses held by two computers. Jim gives Nancy his extension number (123) and his public telephone number (555-1212) so that Nancy can return a call later. You can see that the use of telephone extension numbers is a kind of session management. Even though both organizations each have a single public telephone number, multiple simultaneous conversations or "sessions" could be taking place between them, or other companies.

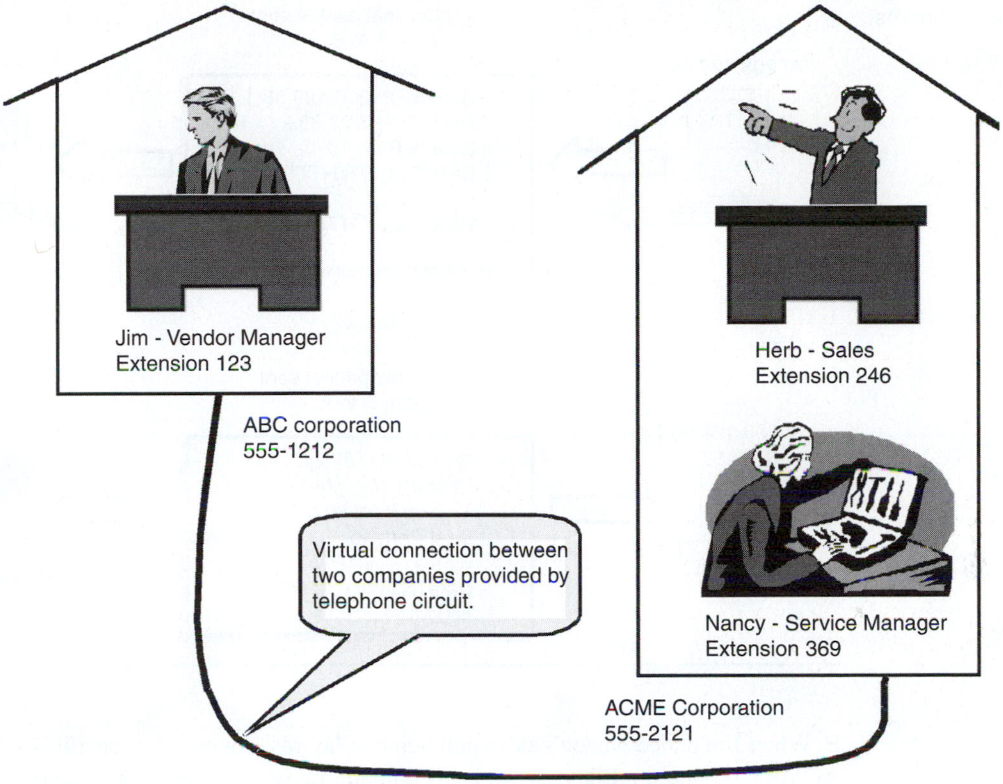

Figure 16–15 Session management analogy

Both UDP and TCP manage communications sessions by the use of *port numbers*. Port numbers are analogous to the extension numbers in the example.[1] There are two types of port numbers, *destination* and *source*. *A destination port number specifies the service being requested from a remote system*. There are standardized assignments for services; a few of these are shown in Table 16–4. A complete listing of well-known ports is given in Appendix

Table 16–4 Well-known TCP port numbers

Port number	Service
13	Daytime—Get date and time from remote system.
21	FTP—File transfer protocol; used to transfer files between systems.
23	Telnet—Opens a terminal (usually command shell) on a remote computer.
25	SMTP—Simple mail transfer protocol. Used to transfer e-mail messages to a server, or from server to server.
53	DNS—Domain name service; used to find the IP address of remote systems.
80	HTTP—Hypertext transfer protocol; used to request and retrieve Web pages.
110	POP3—Post office protocol; used to retrieve mail from a mail server computer.

[1]The analogy becomes weaker for multithreaded applications that have the capability of supporting several simultaneous connections with many clients.

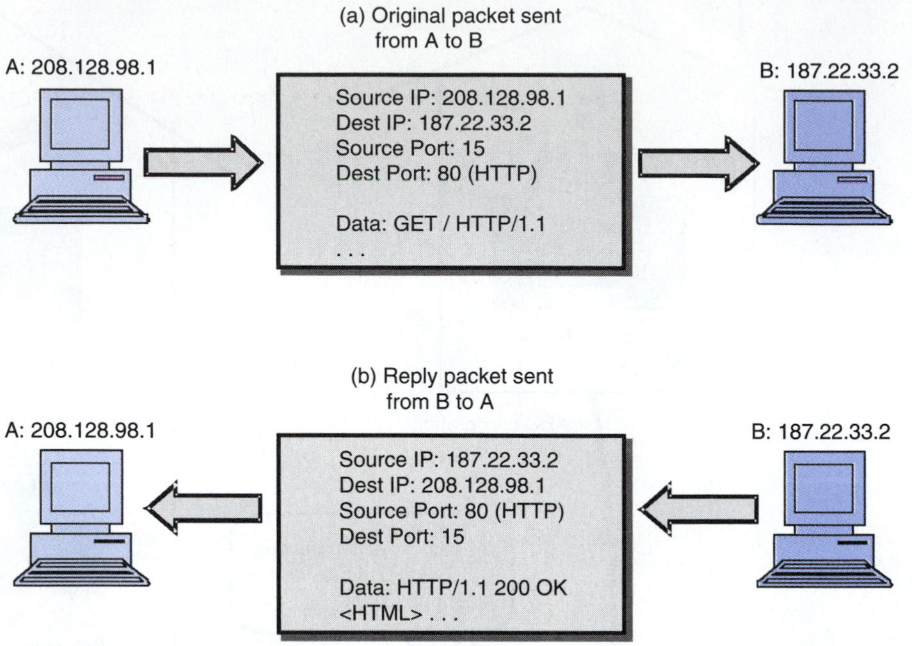

Figure 16–16 Original message packet and its reply

E. When Jim dialed Nancy's extension number, he was requesting a specific service from the ACME Corporation. Likewise, when a computer makes a connection to another computer, it must request a specific service.

When a computer makes a connect request to a remote system, it supplies four pieces of information. These are the source and destination IP addresses, the destination port, and the source port number. The destination port number is the code for the service requested from the remote computer. The source port is a number that represents the *process (program)* that made the connection request on the originating computer.[2] When an IP packet carrying *UDP* or *TCP* session data travels from the originating and remote computers, the port numbers appear as shown in Figure 16–16(a). However, when the remote computer responds back, it transposes both the source and destination IP addresses as well as the port numbers, as shown in Figure 16–16(b). The rule is that the destination port in a TCP or UDP packet always identifies the process that should receive the packet.

Server and Client Model

Both UDP and TCP use a server and client model when they communicate. *In order for two computers to communicate, one must act as a server and the other must behave as a client.* By definition, a server computer awaits incoming connection requests and responds to them, while a client makes outgoing calls to other machines. In Figure 16–16, computer *B* is likely to be a web server. It's set up to respond to information requests on port 80. Port 80 is used to indicate that HTTP is going to be used, which is the standard protocol for retrieving Web pages. There's no magic to this at all. In fact, you can try this exchange out

[2]To be more precise, the source port number is actually derived from the *handle* (a number that refers to an object) of the socket (the UDP or TCP connection) in use by the originating process. The application layer, which is part of the computer's operating system, provides this number.

Figure 16–17 Using telnet to request a Web page from a remote server

```
>telnet google.com 80
GET / HTTP/1.1
HOST: www.google.com
[BLANK LINE]
HTTP/1/1 200 OK
Cache-control: private
Content-type: text/html
Set-Cookie:
PREF=ID=3a5c3ffa13685b6d:TM1073865277:LM1073865
277:S=HxE4g-FQ0W7Gw
jp3; expires=Sun, 17 -Jan-2038 19:14:07 GMT;
path-/; domain =.google.com
Server: GWS/2.1
Content-length: 2690
Date: Sun, 11 Jan 2004 23:54:37 GMT

<html><head><meta http -equiv="content-type"
content="text/html; charset-ISO -8859
-1"><title>Google</title><style><! --

body, td, a, p, .h{font-family: arial, s. . .
```

You type the following, opening a telnet session to the Google Web server. Pay attention to upper and lower case!

for yourself. Figure 16–17 shows how. Simply open up a command (or terminal) window on an Internet-connected machine and type in the commands shown in the figure, followed by a blank line (which completes a request in the HTTP protocol). You'll see the output of Figure 16–17 streaming back to your console, which is the HTTP response data from the web server (and the actual Web page).

> *TIP* The *telnet* utility is very useful for testing remote server programs (at least those that take text commands). Follow the *telnet* command with the port number of the service you wish to connect with.

The User Datagram Protocol (UDP)

UDP is one of two main transport services available for data. It's much simpler than TCP, and is often much quicker. UDP places the user's data onto the network and sends it to the destination with *no* error checking whatsoever! UDP is a *connectionless* service. If a machine is listening on a UDP port, it can receive UDP packets from one or more computers at the same time. If a computer crashes and stops sending data, that's okay; since UDP is connectionless, it doesn't keep track of connections. If a UDP receiver gets a defective packet, it merely discards it. No notification is given to the sender to resend the message! UDP never acknowledges message packets, good or bad. If they get to the destination, fine; if not, that's fine as well! So if it's so unreliable, why is UDP used at all?

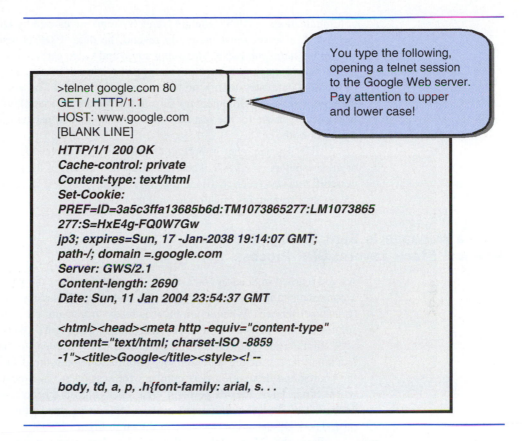

UDP is used for situations that don't require an acknowledgment, or need to be completed at the highest possible speed (even if errors occasionally occur). Most of the communications of DNS (Domain Name System) are carried by UDP. If a communication is missed, a server must manually resend the data. UDP is also extensively used for *streaming audio and video*. Streaming audio and video carry sound and motion pictures in real time and may use a large percentage of a computer and network's capacity. Because of the high data rates, there's no time to resend lost data. These systems either ignore bad datagrams (causing a momentary skip in the picture or sound), use forward error correction to compensate for the lost data (so that a single bad packet doesn't degrade performance), or both.

UDP has one other important characteristic. It is capable of *broadcasting* messages to all computers on a LAN. Therefore, UDP is often used for advertising the existence of network services (such as file and printer shares), as well as for other purposes (such as advertising routing information in an internetwork).

How a Message is Sent through the Protocol Stack Layers: UDP Process

As you learned in Section 16–2, the OSI model describes the various functions needed in a communication system. UDP provides both layer 5 (session management) and layer 4 (transport) services. When a computer is ready to transmit data, the UDP protocol encapsulates the data in its own package, which looks like Figure 16–18(a).

In the figure you can see that UDP adds its own *header* to the data. The UDP header is a small data structure that contains information about the data. It specifically contains the source and destination port numbers for session management and a length indicator word. Since UDP doesn't perform error checking on the data, no other information is needed. The checksum contained in the UDP header checks the *header only,* and not the data. The UDP datagram structure is shown in Figure 16–19.

From the OSI model you'll recall that data moves downward from layer to layer when sending a message. The original data from an application program that needs to be transmitted over the network is presented to layer 7 (application layer) by the software.

Figure 16–18 UDP, IP, and data link headers

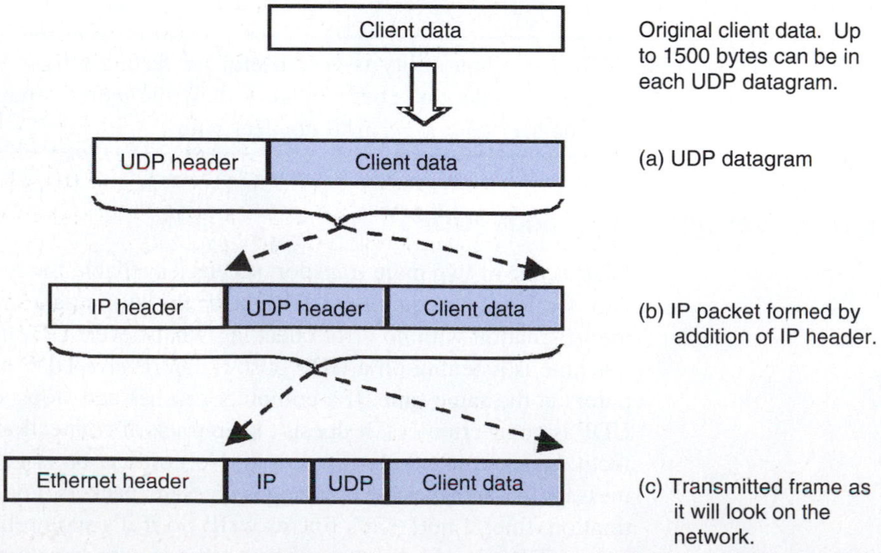

Figure 16–19 UDP datagram structure

Source port (16 bits)	Destination port (16 bits)
UDP length, header = data (16 bits)	UDP checksum
UDP data payload, up to 1500 bytes	

The operating system then presents the data to the UDP implementation, which straddles layers 4 (transport) and 5 (session). What about the presentation layer (layer 6)? The TCP/IP model does not support this layer.

The UDP datagram (Figure 16–19) is then built and passed down to layer 3, the network layer, where it is processed by IP. The packet then looks like Figure 16–18(b). It has both a UDP and an IP header. The IP header holds the source and destination IP addresses plus other vital information for routing the message.

After the IP layer adds its header, the packet is passed down to layer 2, the *data link layer*. Again, layer 2 adds an *Ethernet* header to the data (which now looks like Figure 16–15[c]), and finally, the data is sent out to the LAN through the *physical layer,* which is the network interface card (NIC). Once the data appears on the wire, it can be picked up by another computer. The data link layer adds source and destination *MAC* addresses to the message. For a device to receive the message, its MAC address must match the destination MAC address in the transmitted frame.

The receiving computer reverses the process one layer at a time. First the Ethernet header is removed, and the frame is checked. If it checks okay (and if the MAC address in the Ethernet header matches the receiving computer's NIC), the message is passed to the network layer. The network layer then checks the destination IP address to see if the message is intended for the receiving computer. If it is, the IP header is stripped away and the resulting datagram is sent up to the transport layer, where the UDP header is checked and stripped away, leaving the original data. The destination port number in the UDP header is used to identify which process (program) in the computer should receive the incoming data.

TCP Communications

Transport control protocol, or TCP, is used when reliable two-way communications is needed between two computers. TCP differs from UDP in several important respects. First, all TCP communications are guaranteed. The protocol checks to make sure that each byte is accounted for. No data is lost under TCP. Second, TCP is connection-oriented. In order for TCP communication to take place, the two communicating devices must first go through a three-way connection handshake. TCP does not normally support broadcasting on a LAN. Finally, TCP is a full-duplex protocol. Data can be sent in both directions on the TCP connection at virtually the same time. Because of its high inherent reliability, normal Internet communications normally use TCP.

Within the protocol stack (OSI model), communication is very similar between TCP and UDP. When data is going to be sent using TCP, it is sent to the transport layer so that it can have a TCP header attached, just as in UDP. Therefore, a TCP datagram is handled identically to the UDP version in Figure 16–18. However, much more activity takes place at the transport layer in TCP. To understand this activity we need to study the TCP header structure shown in Figure 16–20.

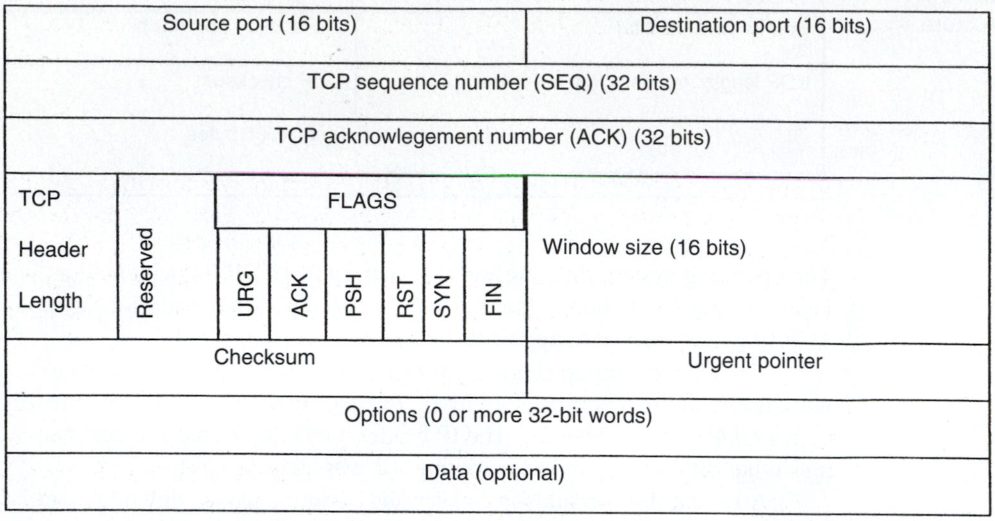

Figure 16–20 TCP header structure

Within the TCP header, the source and destination port numbers serve the same purpose as they did in UDP. They keep track of which programs on both computers are doing the communicating. The remaining information in the TCP header is used to manage the connection and acknowledge the flow of information.

TCP Sequence and Acknowledge Numbers

Since TCP supports two-way communication, two numbers are used to track and acknowledge the flow of data. The *sequence number* is a counter that increases each time more bytes are transmitted. It starts at a random value. For example, if two successive TCP datagrams have the sequence numbers 100000 and 100010, then 10 bytes of data have been sent in the second datagram. The receiver uses the sequence numbers to verify that it hasn't lost any packets (all bytes must be accounted for), and that the received packets are in the correct order. (Recall that the IP layer may deliver multiple copies of the same packet, and occasionally packets may arrive out of order.) The *acknowledge number* is used to acknowledge data from the other end of the connection. *By definition, the acknowledge number is the byte number expected next from the remote system.* This is called *expectational acknowledgment.* Remember that for any TCP conversation, there are two sets of sequence and acknowledge numbers in use at any time, one for each of the two hosts.

EXAMPLE 16–7 Two computers, A and B, have an open TCP connection. Computer A has sent a TCP datagram with the following header information: SEQUENCE=1000, ACKNOWLEDGE=5736. Computer B then responded with a TCP datagram containing SEQUENCE=5736, ACKNOWLEDGE=1100. How many bytes did computer A send to computer B, assuming no errors occurred?

Solution

The beginning sequence number (byte number) from computer A is 1000, and after getting this packet, computer B asked for byte number 1100 (recall the definition of the

ACKNOWLEDGE value is the byte number expected *next*). Therefore, computer B received byte numbers 1000 to 1099 successfully. This is (1100 − 1000) or *100 bytes*.

TCP Flow Control

It's possible that one computer may be able to send more data than a listener is ready to receive. The receiving computer uses the *window size* value to "throttle" the transmitter. By definition, the window size is the maximum number of bytes a TCP sender can transmit without getting an acknowledgment. A TCP receiver can set this value to zero (0) to force a sender to stop sending data. This prevents one computer from overflowing the buffer of another.

TCP Flags

The *flags* are bit values used to manage the TCP connection. The RST (reset) flag is used to reset both sides of a connection. FIN (finish) releases one side of a connection. A datagram with PSH (push) enabled is not buffered by routers on the path; it's sent along immediately. When ACK is 1, a receiver is acknowledging a TCP event (we'll see it in action momentarily). The SYN (synchronize) bit is used during the initiation of a connection. Finally, its possible to send out of sequence packets (for error recovery). The URG (Urgent) flag is set to indicate that out-of-sequence (urgent) data is contained in the packet. When the URG flag is set, the *urgent pointer* gives the offset (byte number) of the out-of-sequence data.

Initiating a TCP Connection

Figure 16–21 shows how a TCP connection is opened. Recall that one machine must be acting as a server,[3] and the other must be a client. The client always originates the connections, and it does so in Figure 16–21 by sending a datagram with the SYN bit turned

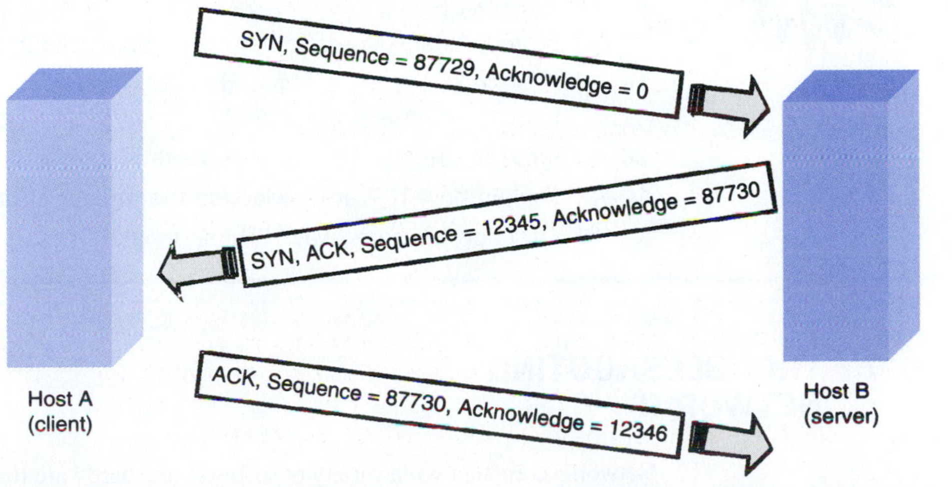

Figure 16–21 Three-way handshake for TCP

[3]To be a server, a computer must *bind* a socket to a port number. Then all incoming TCP connection requests on that port number can be accepted.

on, an acknowledge value of zero (0), and a random value for sequence. The SYN bit tells the server to "synchronize" or start a new connection. The client in Figure 16–21 has chosen an initial sequence number of 87729. This number will be different each time a new TCP connection is started.

If the server wishes to accept the connection, it responds with a second datagram with the SYN and ACK bits turned on. It answers with an acknowledge value equal to one plus the sequence number sent in the first step. This helps the client confirm that the server is synchronized with its transmitted sequence value. In this second step, the server also supplies its *own* sequence value, 12345. Remember that *two* sets of sequence and acknowledge values are in use!

Finally, the client closes the loop by sending a packet with only the ACK bit turned on, and an acknowledge value equal to one plus the server's sequence number. Once this has been received by the server, the connection is completely open and normal communications can take place. This process is called the three-way TCP handshake. It happens each time a TCP connection is opened.

Closing a TCP Connection

Closing a TCP connection can take place in two ways. First, either computer can send a datagram with the FIN (finish) flag turned on. This performs a "half close" on the connection. Once a machine has sent a FIN packet, it will send no more data (but can still receive data from the other end). This is sometimes called a *close and linger* operation, since the computer that sends the FIN can "linger around" and wait for any remaining data from the other side. A computer can also completely close the connection by turning on the RST (reset) bit. This immediately closes the connection (as soon as the other computer gets the datagram).

Section Checkpoint

16–41 What is the purpose of the transport layer of the OSI model?

16–42 What is the difference between UDP and TCP communications?

16–43 Describe how a byte of data from an application travels through the protocol stack layers to get on the wire, assuming UDP transport.

16–44 How do UDP and TCP keep track of sessions? What parts of the header structures are used?

16–45 What is the significance of a destination port number?

16–46 Explain how TCP acknowledges the receipt of data from a sender.

16–47 Describe the three-way TCP handshake.

16–7 TROUBLESHOOTING NETWORKS

Networks contain a wide variety of software and hardware that makes them tick. However, most of the problems encountered in networks are very simple in nature. The use of a consistent troubleshooting approach is very important and leads to much quicker problem resolution than the so-called shotgun methods. Most industry experts recommend using the

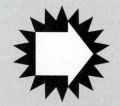

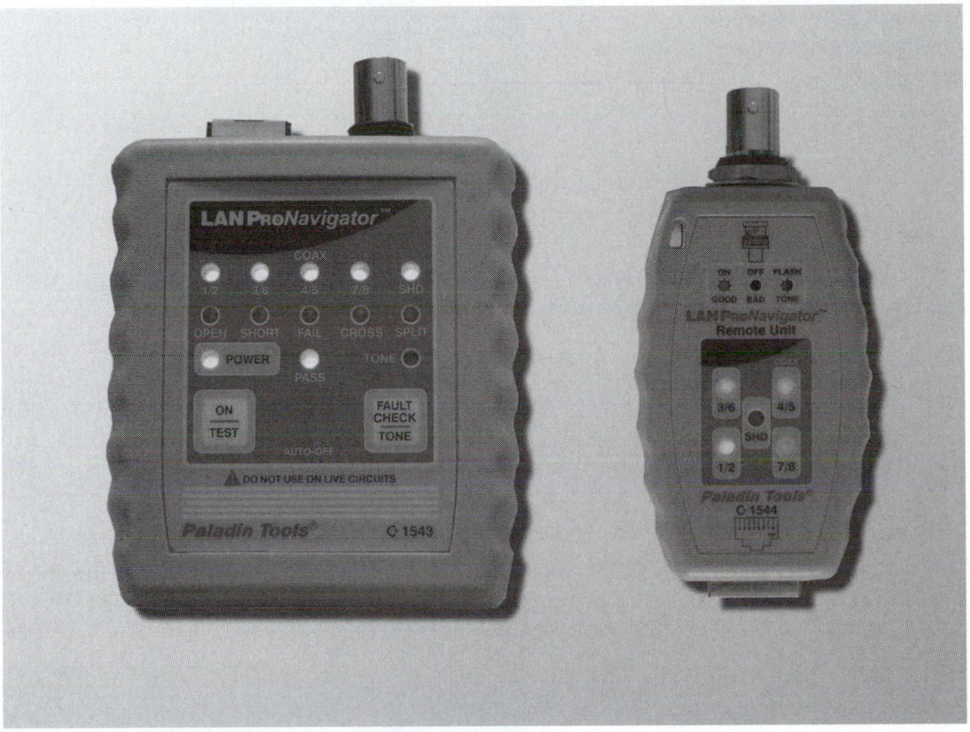

Figure 16–22
Ethernet cable test set

OSI model as a guide to troubleshooting. ***In using the OSI model, always start checking at the physical layer and work your way up from there.*** Test each layer as you proceed. If there's a problem, this approach will not fail to locate it!

Physical Layer Checks

Believe it or not, many would-be troubleshooters forget the simplest of all possible tests at the physical layer—*link lights*. Check the link light located near the Ethernet connector on the network interface card. No link light, no network connection! Broken cables are the most common cause of network connection failure. You can use a cable tester like the one of Figure 16–22 to test cable runs.

The test set in the figure can be used for testing jumpers as well as long cable runs (the remote unit at right is simply connected at the far end of the cable). Also, don't forget to inspect cables for kinks. UTP cable is very susceptible to breakage when it's sharply bent.

Are both devices on the Ethernet cable turned on? Hubs and switches are sometimes placed in odd locations in offices. In one installation a switch was placed behind an employee's desk. Every so often its power plug would get kicked out of the outlet, taking down the network for the entire department!

Data Link Layer Checks

The device drivers for a network interface constitute the data link layer. These are hard to directly test, but at least you can get a confirmation of whether or not the operating system actually thinks the device is OK. In Figure 16–23 the configuration box for a working

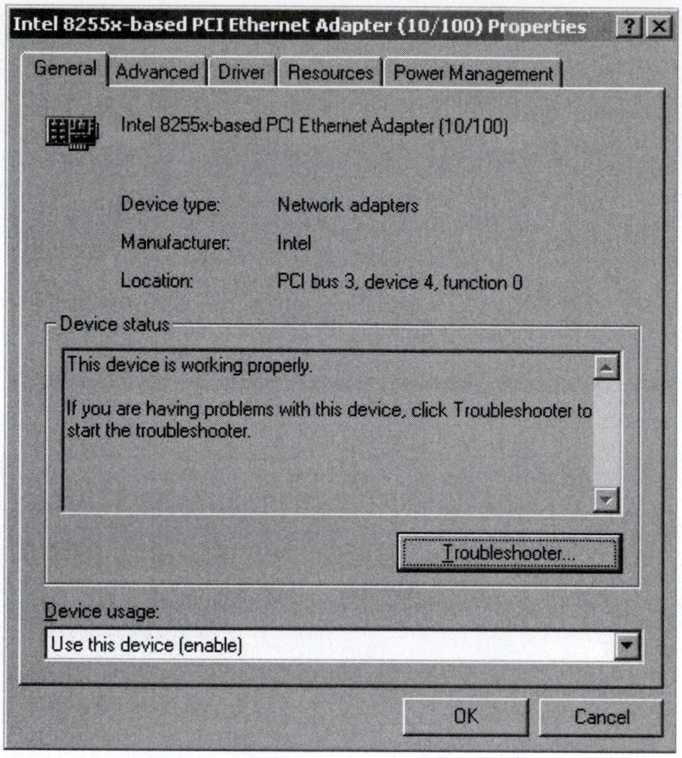

Figure 16–23
Windows 2000 network interface configuration dialog box

network interface has been opened under Windows 2000. You should learn how to check device driver settings for all operating systems found on your organization's network.

TIP Dialogs like Figure 16–23 are not foolproof. It's possible to have a defective NIC and still have a report of a good card from the operating system. Some NIC problems are caused by bad contact between the NIC card and the motherboard socket. Try reseating the card (with the power off) before replacing it.

TIP To check interfaces on a Cisco router, use the command: `show ip interface brief`. This will force the router to summarize the status of all of its IP interfaces. You must be in enable mode to issue this command.

Network Layer Checks

Once you're certain that the physical and data link layers are working, next check the *network* layer. To check the network layer, use the *ping* and *traceroute* commands. *Ping* is a command that sends a special ICMP (Internet Control Message Protocol) packet out to a remote host and waits for an "echo." The remote host normally returns the packet as a ping "echo reply." A successful ping means that your computer is able to send and receive IP packets on the network. Figure 16–24 shows two ways that ping can be used.

Figure 16–24 Two ways of applying the ping utility

```
Microsoft Windows 2000 [Version 5.00.1295]
(c) Copyright 1985-2000 Microsoft Corp.
C:\ >ping 127.0.0.1
Pinging gw9100 [127.0.0.1] with 32 bytes of data:
Reply from 127.0.0.1 bytes=32 time<10ms TTL=128
Reply from 127.0.0.1 bytes=32 time<10ms TTL=128
Reply from 127.0.0.1 bytes=32 time<10ms TTL=128
Reply from 127.0.0.1 bytes=32 time<10ms TTL=128

Ping statistics for 127.0.0.1:
          Packets: sent = 4, received = 4, Lost = 0 (0% loss),
Approximate round trip times in milli-seconds:
          Minimum = 0ms, Maximum = 0ms, Average = 0ms

C:\ >ping faculty.kc.devry.edu

Pinging facult.kc.devery.edu [205.160.208.18] with 32 bytes of data:

Reply from 205.160.208.18: bytes=32 time=50ms TTL=242
Reply from 205.160.208.18: bytes=32 time=40ms TTL=242
Reply from 205.160.208.18: bytes=32 time=40ms TTL=242
Reply from 205.160.208.18: bytes=32 time=40ms TTL=242

Ping statistics for 205.160.208.18:
          Packets: sent = 4, received = 4, Lost = 0 (0% loss),
Approximate round trip times in milli-seconds:
          Minimum = 40ms, Maximum = 50ms, Average = 42ms
```

TIP The address 127.0.0.1 is the IP loopback address, and normally resolves to *localhost* on most machines.

The first *ping* issued in Figure 16–23 actually tests the TCP/IP protocol stack on the local computer. No packet is issued onto the LAN! If you suspect that something is wrong with the *software* setup in the computer, this is a good test.

The second *ping* issued in the figure is attempting to bounce a packet off a remote host. Two OSI layers have been tested by the second ping command. First, the DNS lookup function has been tested—the computer is resolving the domain name "faculty.kc.devry.edu" into the IP address 205.160.208.18. Second, the utility is getting echo replies, so the network layer is successfully communicating with the network.

WARNING A "positive" result from *ping* does not guarantee that you are reaching the remote host you requested! It only indicates that *something* replied to the echo request. Its entirely possible that a different machine (misconfigured with the same IP address, or otherwise malfunctioning) is responding to the *ping*.

The *traceroute* command (`tracert` on Windows systems) is used to follow the path of an IP packet as it travels through an internetwork. It is used when connections fail between two hosts (usually on different networks). It can help you learn where the packets are being dropped. Figure 16–25 shows a *traceroute* being performed. Note that the "-d" option is quite useful: It disables the reverse DNS lookup that *traceroute* normally applies to each node in the trace, which speeds up the process immensely.

In the figure, the route requires 13 hops through Internet routers before reaching the final destination (step 14). If the trace had failed, timeouts would have been shown on the output, with the IP address of the last successful router being evident. At that point you would likely contact the administrator of that particular system for assistance.

Figure 16–25 Using traceroute on a Windows system

```
Microsoft Windows 2000 [Version 5.00.2195]
(C) Copyright 1985-2000 Microsoft Corp.

C:\ >tracert faculty.kc.devry.edu -d

Tracing route to faculty.kc.devery.edu [205.160.208.18]
over a maximum of 30 hops:

  1   10 ms   20 ms   20 ms   66.140.73.254
  2   11ms    20 ms   20 ms   151.164.29.66
  3   10 ms   20 ms   20 ms   151.164.29.230
  4   10 ms   20 ms   21 ms   151.164.188.70
  5   20 ms   30 ms   20 ms   151.164.240.109
  6   20 ms   31 ms   20 ms   151.164.240.89
  7   20 ms   30 ms   30 ms   151.164.189.249
  8   31 ms   20 ms   30 ms   144.228.39.225
  9   20 ms   30 ms   30 ms   144.232.8.245
 10   30 ms   40 ms   40 ms   144.232.18.118
 11   30 ms   40 ms   40 ms   144.232.23.66
 12   40 ms   40 ms   60 ms   160.81.94.144
 13   70 ms   60 ms   70 ms   205.160.208.29
 14   40 ms   40 ms   40 ms   205.160.208.18

Trace complete.
```

TIP The Internet does not guarantee 100% connectivity between all hosts. Occasionally, routers and other network devices fail, and time must be allowed for organizations to make repairs. Sometimes a route must be blocked intentionally to prevent the spread of harmful traffic (such as a denial-of-service attack on a customer site).

Higher Layer Checks

Most network troubles will be in the lower three layers (physical, data link, and network layers). You can test the operation of other layers, often directly. For example, under *UNIX* and *LINUX* operating systems, you can use the *NTTCP* utility to measure the performance of the transport layer. To check the session layer, use *nslookup* to manually perform a DNS lookup on a remote host. For example, issuing:

```
nslookup google.com
```

exercises the DNS service, which is actually a session layer function.

If you find no problems up to this point, then the trouble is very likely to be in the application or application(s) on the machine. What should you do next?

- Check the application settings. Perhaps someone has accidentally changed a setting.
- Try restarting the machine. Sometimes a reboot is needed if the TCP/IP stack code becomes corrupted.
- If some applications work, but others won't, perhaps certain *ports* or *services* are being blocked on the network. Use Appendix E to learn about well-known port numbers, and check the appropriate persons overseeing the portion of the network failing to pass the traffic.

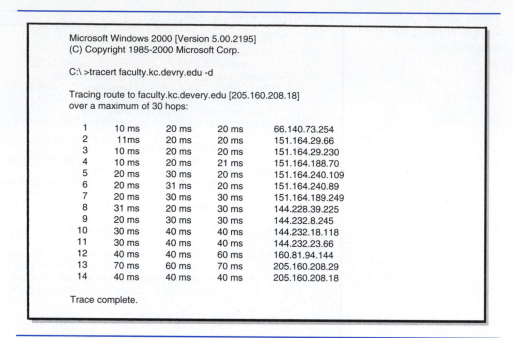

- If your computer is behind a firewall, it may not be able to accept connections from outside the LAN. It's common for incoming access to be blocked for security. You may need to move the computer to the other side of the firewall. A system administrator can also set up port forwarding to the machine, or place it into a "DMZ" (demilitarized zone) where Internet traffic can flow to it freely.

Section Checkpoint

16–48 What layer should always be checked first when troubleshooting a network device?

16–49 Explain how to troubleshoot the physical, data link, and network layers.

16–50 What is the purpose of the *ping* command? What does it *not* guarantee?

16–51 What information can be obtained from *traceroute?*

16–52 Give several tips for troubleshooting applications software.

SUMMARY

- The open systems interconnect (OSI) model contains seven layers, each of which represents a basic function in a communications system. Not all layers are needed by all systems.
- The Internet is a global network of computers. It uses two protocols, TCP and IP, to control communications.
- Internet addresses fall into three main classes, A, B, and C, depending on the number of host computers that an organization possesses.
- The DNS system resolves domain names into physical IP addresses to make it easier for humans to access the Net.
- An IP address consists of a network and host portion. The subnet mask identifies the network portion with 1s and the host portion with 0s.
- The *broadcast* address of a network has all 1s in the host portion.
- The subnet mask can be used to calculate the network and broadcast addresses for any LAN segment.
- When the destination address of a packet is outside the local network, a computer automatically sends the packet to a gateway for forwarding to the Internet.
- DHCP is used to automatically issue IP addresses for workstations on a network. It eliminates the need to manually configure them.
- ARP is used to learn the MAC address of a device if its IP address is known. The MAC address is required for the device to receive any messages from the network.
- TCP and UDP are the two most common transport protocols used in the Internet. Both use a client-server model.
- TCP and UDP use port numbers to manage sessions on the wire. Destination port numbers identify services being requested from servers, and source port numbers identify processes on originating computers. They're roughly equivalent to telephone extension numbers.
- UDP is used where quick, unreliable one-way communication is desired. It performs no error checking.

- TCP fully guarantees the delivery of data from one end to the other. It provides full-duplex communications between computers.
- TCP connections employ a three-way handshake to get them started.
- Networks should be troubleshot using the OSI model, working upward from the physical layer.
- Each OSI layer can be tested using command-line utilities on most computers.

PROBLEMS

1. List the seven layers of the OSI model, and describe the function of each.
2. What does a typical Internet address look like? Give an example. How many bits are in an IP address?
3. For each of the following IP addresses, state the organization class (A, B, or C), the *network address,* and the *host number:*
   a. 105.124.64.25
   b. 135.69.73.2
   c. 64.139.174.156
   d. 222.221.220.219
4. What is the function of DNS? Explain how DNS works.
5. What is the relationship between the World Wide Web (WWW) and the Internet? Which of them physically exists?
6. For the IP address 205.160.23.38/24, calculate the network address, subnet mask, and the broadcast address.
7. How many hosts can be on the network of question 6?
8. A certain host has the IP address 64.133.132.18 and a subnet mask of 255.255.248.0. Write the IP address and subnet mask in CIDR shorthand.
9. Find the subnet masks for the following IP addresses. Express in decimal form.
   a. 151.164.129.2/21
   b. 130.59.23.18/23
10. Calculate the network and broadcast addresses for both networks of question 9.
11. Explain how a computer knows whether or not an IP destination is on its LAN.
12. What device receives IP packets that are to be sent outside of a LAN?
13. Divide the network 208.128.98.0/23 into four equal-size subnets. Give the network address, subnet mask, and broadcast address for each new subnet. Allow both subnet zero and the all ones subnet. *Hint: You need to add two "1" bits to the subnet mask.*
14. How many hosts can be on each of the new subnets of question 13?
15. Divide the network 64.133.16.0/24 into separate networks for use in various departmental LANs within a company. Each LAN needs to support at least 40 hosts. Give the network address, broadcast address, and host address range for each subnet, as well as the new subnet mask for the subnets. Allow both subnet zero and the all ones subnet.
16. List the unroutable IP address blocks. What are they used for?
17. List and explain each step that occurs in a DHCP transaction.
18. What two destination addresses are needed to get a message successfully received by a computer on a network? State the OSI layer that each address belongs to.
19. What protocol is used to discover the MAC address of another workstation on a LAN? Explain how it works.
20. What are two primary protocols used in the transport layer? List them, and compare and contrast them.
21. What are *source* and *destination* port numbers used for? Give a non-network everyday example.

22. An IP packet is received at a server with a destination port of 25. What service is being requested, and what type of traffic will be following? (Use Appendix E if needed.)

23. Give the command line sequence that would be used to test an SMTP server using the *telnet* utility. (Just give the opening command, don't worry about the actual SMTP service commands.)

24. Draw the structure of an Ethernet frame carrying a UDP packet. Explain how each part of the structure is built (which part of the TCP/IP protocol stack builds each part)?

25. A TCP packet is sent from computer A to computer B with the following attributes:

Sequence = 8675309, Acknowledge = 1024, PSH

Computer B responds with a packet that looks like this:

Sequence = 1024, Acknowledge = 8675319, ACK, PSH

How many bytes of data were sent from computer A to computer B?

26. List and explain the steps that are used to establish a TCP connection.

27. In what order should the OSI layers be checked when troubleshooting a network?

28. Explain the procedures for checking the following OSI layers:
   a. Physical
   b. Data link
   c. Network
   d. Transport and higher layers

17

The Global Positioning System

OBJECTIVES

At the conclusion of this chapter, the reader will be able to:

- describe briefly the history of navigation
- explain the orbital parameters of satellites using Kepler's laws
- list and explain the three segments of the current GPS system
- identify navigation signals
- define essential GPS terminology
- describe the operation of various types of GPS receivers
- explain how position is determined
- interpret correction methods as they relate to accuracy of position
- use simple handheld GPS navigational devices
- describe civilian applications for GPS
- troubleshoot civilian GPS installations

Since the beginning of recorded history, humans have been intensely interested in knowing where they are and where they are going. The earliest people probably used stones, logs, or other objects as markers to guide people from place to place. Of course, these markers could be covered by snow, washed away by floods, or tampered with by enemies—and at sea, you couldn't use markers like these at all!

One thing that was relatively constant is the position of the stars, and for a long period of time, the only way of navigating by sea was to use them. This required special skills and was only accurate within a mile or two—and only worked at night under a clear sky. This was hardly good enough when a ship was looking for a safe harbor on a stormy night! The use of lighted beacons and buoys certainly helped, but these devices required maintenance. Buoys could be washed out of position in storms.

During World War II, the first electronic means of navigation were developed. Among these is the *LORAN* system. LORAN transmitters are located along coastal areas, and each LORAN transmitter sends a special timing signal on a LF frequency. A LORAN receiver tunes to three LORAN transmitters and by doing a little arithmetic, it can find its position. LORAN works only in areas where transmitters have been set up and is not very accurate in a moving vehicle. However, LORAN represented a great advance over previous navigation methods.

The United States has been studying space-based navigation since the 1960s. One of the early successful satellite positioning systems was *TRANSIT,* the Navy Navigation Satellite System, which was released for commercial use in 1967. Unfortunately, TRANSIT satellites are in a low earth orbit, and their transmission frequencies of 150 MHz and 400 MHz are fairly susceptible to ionospheric interference. TRANSIT satellites are visible for only about 20 minutes, and a position fix can be obtained once every 1 to 3 hours.

The *Global Positioning System,* or GPS, is the first system designed to provide worldwide, continuous navigation coverage in a highly accurate manner. It was developed by the U.S. Department of Defense and consists of a nominal array of 24 orbiting satellites. (There are actually more than 24 active satellites at present, but only 24 are required for a fully functioning system.) Because of the large number of available satellites, a user can get a position reading at any time, in any location on the globe.

A GPS receiver gives the end user three important pieces of information. First, the receiver derives an accurate value for *time,* which is based upon the atomic clocks in the satellites. Second, the receiver gives the *position,* which is the information most users desire. Finally, by doing a little arithmetic with subsequent position readings, a GPS receiver can report speed or *velocity*.

The possible applications of GPS are endless. Civilians have access to the *C/A (Coarse/Acquisition) code* channel and use GPS during hiking, boating, and other outdoor activities. General Motors uses GPS as part of their OnStar system, which includes a GPS receiver in the automobile—which is linked to a cellular phone. In the event of a serious accident (detected by in-car sensors), the OnStar system automatically dispatches help to the scene.

The military uses GPS for precise navigation and uses the *P (Precision) Code* channel. The P code channel operates at a much higher data rate than the civilian C/A channel and in theory offers better accuracy than the C/A code. Advances in receiver design (such as differential GPS techniques) have diminished the accuracy difference between the C/A and P code channels. The P channel information can be encrypted (it is then referred to as the Y channel), and under this condition it cannot be demodulated by the general public. This would likely occur in a time of war (but could occur at any time) so that enemies would be denied access to precision navigation services.

17–1 SATELLITES AND ORBITS

All satellite communication systems rely upon one or more *space vehicles,* or *satellites,* that are made to orbit the earth in a predetermined way. With GPS satellites this is especially important, for it is important to be able to accurately calculate the position of each satellite. The motion of satellites, planets, and other bodies in space can be described by Kepler's laws of planetary motion.

Kepler's Laws of Planetary Motion

Johannes Kepler published the first two of these laws in 1609, and the third nearly a decade later, in 1618. His laws are as follows:

1. A planet describes an ellipse in its orbit around the sun, with the sun at one focus.
2. A ray directed from the sun to a planet sweeps out equal areas in equal times.
3. The square of the period of a planet's orbit, p, is proportional to the cube of its semimajor axis, a ($p^2 \propto a^3$). (Note that the symbol "$\propto$" means "is proportional to.")

First Law

Kepler's first law is illustrated in Figure 17–1. What it tells us is that planetary orbits are generally *not* perfect circles; they are *ellipses*. An ellipse is a flattened circle.

This first law applies to planets circling the sun, so what about satellites orbiting the Earth? It still largely applies, because the mass of the earth is much larger than that of the orbiting satellite. *Our satellite orbits are elliptical, not circular.*

For bodies that are not so different in mass that appear to "orbit" each other, the motion is much more complicated, and the best analogy is to imagine two dance partners of slightly different weight spinning in a circle, holding onto each other with their arms outstretched on a dance floor. This dance floor has a hidden switch that can turn off the gravity—so once the couple begins spinning, we switch off the gravity and observe their motion. To our surprise, their motion appears "lumpy"—the center of mass of the two dancers changes as they turn in free space, which constantly redefines the "orbit" of each person's orbit around the other. There is a big difference between this example and reality: The gravitational attracting force between two orbiting bodies is under constant change as they move closer and farther from each other, unlike the hand grasps of our surprised dancers!

Figure 17–1 Kepler's first law

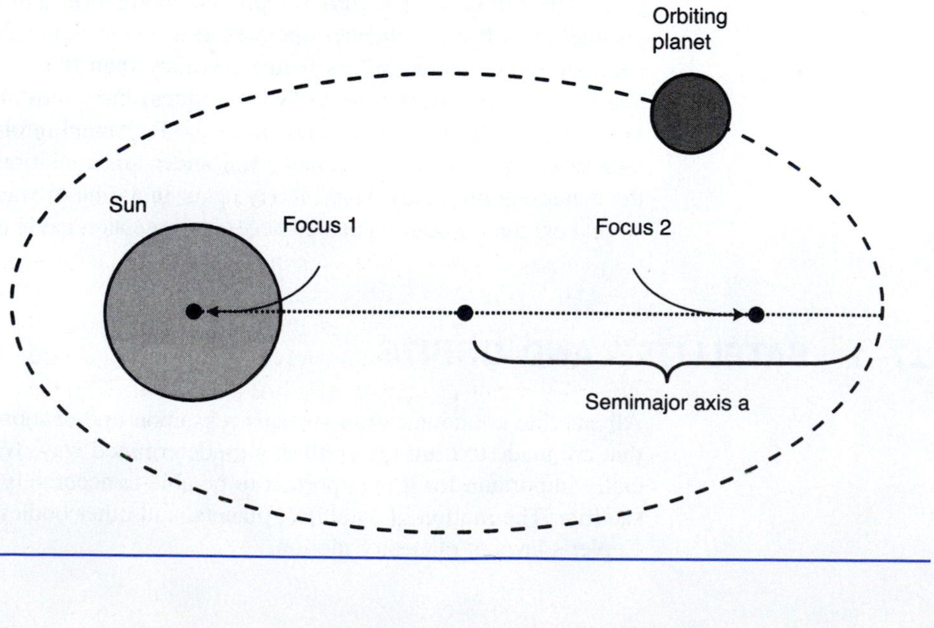

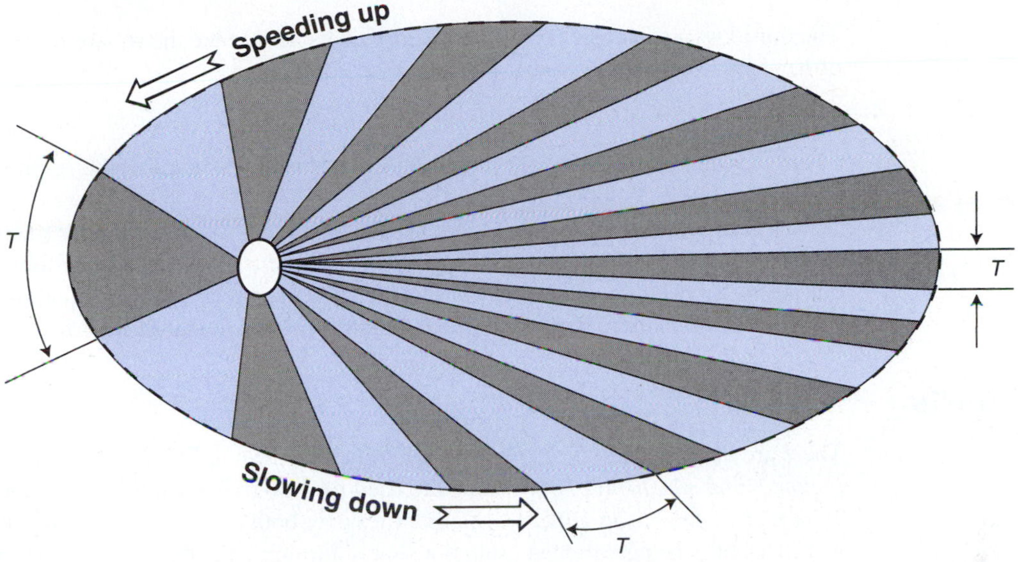

All times *T* are equal, and the area of each slice is equal.

Figure 17–2 Kepler's second law

Second Law

Figure 17–2 demonstrates Kepler's second law. Notice that the areas of each of the pie slices in this figure are equal. Kepler's second law can be condensed by stating that as an orbiting planet approaches the sun, it *speeds up*. An orbiting planet slows down as it moves away from the sun. You might wonder why this is so. The reason is that gravitational attraction builds up rapidly as two bodies approach each other. In order for the orbiting planet to not collide with the sun, it must speed along quickly enough to balance out the centripetal acceleration caused by the gravitational attraction of the sun.

Centripetal acceleration is the acceleration that tends to keep an object moving along a curved (usually circular) path. If you attach a weight to a rope and swing it in a circle, you will feel a force pulling outward on the rope. This is the reaction force to the centripetal accelerating force your muscles are providing in order to hold onto the rope as the weight attempts to fly off in a straight-line path.

As a planet approaches the solar portion of its orbit, it experiences a greater gravitational pull from the sun and moves closer to it. Conservation of momentum predicts that the planet will speed up, since its radius of motion has been reduced. This is why the pie slices in Figure 17–2 have constant area.

The second law tells us that satellites in low orbits must travel much faster than those in high orbits. There is a trade-off here: Satellites in low orbits are much less expensive to launch, can be seen from many places on the earth as they move in their path, and provide stronger signals on earth (because of their closer distances), but their orbits are more likely to be unstable. Low orbits plunge a satellite through part of the atmosphere, which creates drag. High orbits are more expensive but generally more stable; because of the increased distance to earth, however, the signals from satellites in high orbits are much weaker.

Third Law

The third law predicts the orbital period of planets. If we take the square root of both sides of Kepler's relationship, we get the following equation:

$$p \propto a^{3/2} \tag{17–1}$$

where p is the orbital period (in time units), and a is the distance of the semimajor axis of the planet's orbit, as shown in Figure 17–1.

What equation 17–1 tells us is that the orbital *period* of a satellite (the time that it takes to complete one orbital round trip) increases as the height of the satellite is increased, but the relationship is not a linear proportion (the exponent is 3/2 in equation 17–1). We control the orbital period of a satellite by controlling the height of its orbit.

Types of Orbits

There are three primary types of satellite orbits, as shown in Figure 17–3. These are *polar, inclined,* and *equatorial. Polar orbits* are used for low-flying satellite missions such as reconaissance. In such an orbit, the satellite flies over both poles. The satellite path looks like a string that is being wrapped around a slowly turning ball. The exact track (and thus the positions "seen" on earth by the satellite) depend on the orbital altitude.

Inclined orbits are used for communications work where coverage is needed mainly around the earth's middle regions, and continuous communication is not needed. GPS satellites use this type of orbit.

Equatorial orbits are used for *geostationary satellites*. You'll recall that Kepler's third law predicts the orbital period of a satellite in terms of the length of its semimajor axis. For earth satellites, an altitude of 22,300 miles leads to a geostationary orbit, where the satellite takes 24 hours to make one round trip. When placed over the equator in such an orbit, the satellite appears fixed at a point in space, which makes it immensely useful for long-distance communications. About 40% of the earth can be covered by one geostationary satellite; three such satellites can cover about 95% of the earth's surface (the polar regions are hard to reach).

The equatorial orbit locations are prime real estate. They were originally spaced every 4° for 360/4 or 90 "parking spaces." Improvements in technology have resulted in a decrease to 2° spaces, and 1° spaces are not far off.

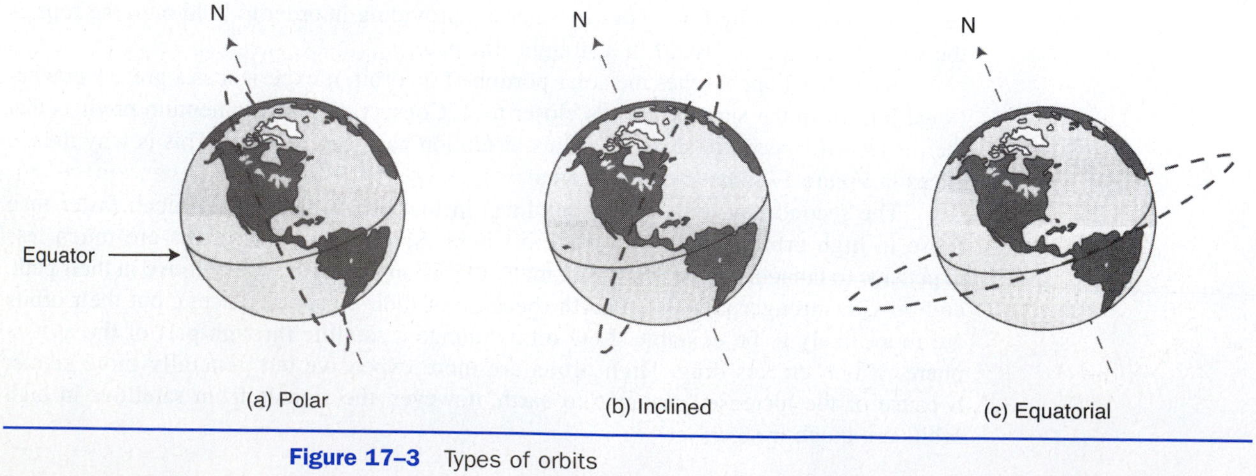

Figure 17–3 Types of orbits

Figure 17–4 The orbits of GPS satellites

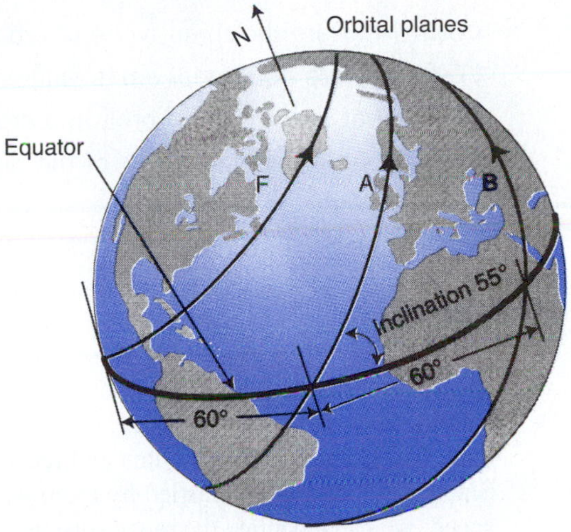

Orbits of GPS Satellites

GPS satellites are placed in orbits that are at 55° inclination with respect to the equator. These satellites orbit within six equally spaced planes (labeled A through F), which appear as overlapping rings circling the globe, as shown in Figure 17–4. There are nominally four satellites in each of these six planes, but they are not equally spaced in order to optimize worldwide visibility.

The semimajor axis of the GPS satellite orbits is approximately 16,156 miles (26,000 km), which is approximately 12,200 miles above the earth's surface. According to Kepler's third law, this leads to an orbital period of slightly less than 12 hours. This means that the satellites complete two round trips every earth day, and because the orbital period is an exact multiple of the period of the earth's rotation, the satellite trajectories (the path of the satellite traced along the earth's surface) repeat daily, which simplifies the calculations needed to find their exact positions.

The height of the GPS satellite orbits gives excellent signal coverage as well but leads to a path loss (signal loss between satellite and GPS receiver on the earth's surface) of around 186 dB. This causes the received GPS signals to be far below the terrestrial noise level on earth (they are approximately −160 dBm at the antenna input of a GPS receiver). Special signal processing techniques are used to compensate for this loss, as we'll soon see.

Section Checkpoint

17–1 What is the advantage of GPS over earlier systems such as LORAN?

17–2 What are the two code channels of GPS? What is each used for?

17–3 State Kepler's three laws of planetary motion.

17–4 Explain how each of Kepler's laws practically applies to satellites.

(continued on p. 680)

17–5 What are the three types of orbits? What is each used for?
17–6 Why are equatorial orbits valuable?
17–7 What is the type of orbit and orbital period of GPS satellites?
17–8 What is one bad effect of the high orbital height of GPS satellites?

17–2 HOW GPS WORKS

The principles behind the operation of GPS are fairly straightforward. In short, GPS operates by measuring the time that elapses between the transmission of special timing signals from the array of GPS satellites and reception of those timing signals at a GPS receiver. Since these signals are carried by radio waves, which travel at an approximately constant speed, we can calculate the range (distance) to each satellite we have locked onto and thus find our position in space. This involves nothing more complicated than that old standby formula from high school, "*distance is rate times time.*" Stated algebraically,

$$D = R \times T \tag{17–2}$$

where D is the distance traveled, R is the rate at which radio wave energy travels (the speed of light, 3×10^8 m/sec), and T is the elapsed time between transmission and reception.

In order to carry on a discussion about how this works, we'll need to assume that we know a few things beforehand. Later on we will see how these details are calculated.

Needed Information

In order for a GPS receiver to calculate its position in space, it needs to know the following information:

- An accurate value for time. The receiver needs to know this so that it can calculate the time of travel for each signal it receives from a satellite. We will see that the receiver can get this from the satellites.
- The position in space of each satellite it is measuring from. This is possible to know because the orbital characteristics of the satellites are known.

How position can be calculated Suppose that we turn on a GPS receiver, and it reads our distance from satellite A as 11,000 miles, as shown in Figure 17–5. With this measurement information, we can be located at an infinite number of points on a sphere surrounding the satellite. Be careful; the sphere of Figure 17–5 doesn't represent the earth. It represents all points that are 11,000 miles away from the center, which is satellite A.

Suppose that we were able to get a fix on another satellite (call it satellite B) and find the distance between ourselves and this second satellite. We would now know that we are 11,000 miles from satellite A and 12,000 miles from satellite B. The picture now looks like Figure 17–6.

There are still an infinite number of points where we could be located, but at least all of these points are now located on the edge of the circle common to satellites A and B in Figure 17–6. This certainly narrows things down a bit. Perhaps we can look at a *third* satellite. If we do so, the result will look like Figure 17–7.

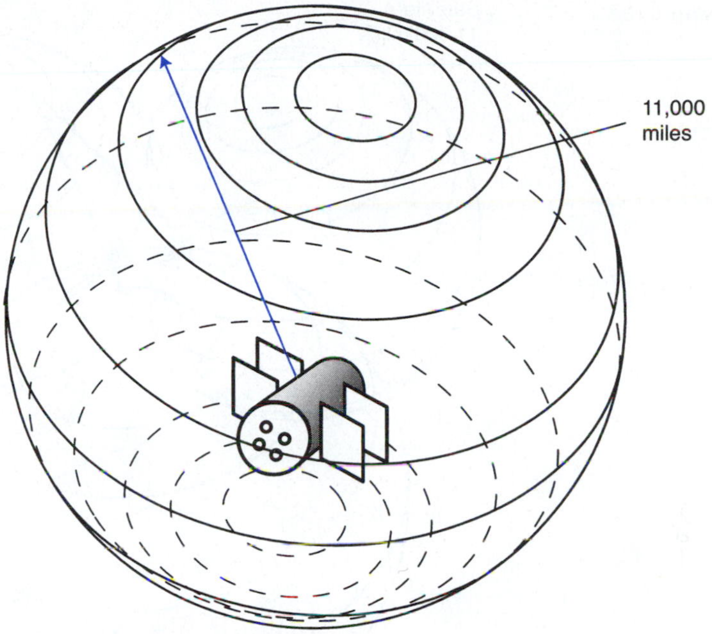

Figure 17–5 11,000 miles from satellite *A*

By reading the distance to satellite *C,* we have now narrowed our position down to one of two possible points. That's much better. In fact, in some applications, this might actually be enough information to plot our position. One of the points might be a totally absurd answer—in which case, we can reject it and use the other solution. Or we might know our altitude, and therefore choose the point that correctly corresponds with that bit of information. In early GPS receivers, these approaches were used. In the early days of GPS (late 1980s), all the planned satellites had not yet been launched, and receivers were

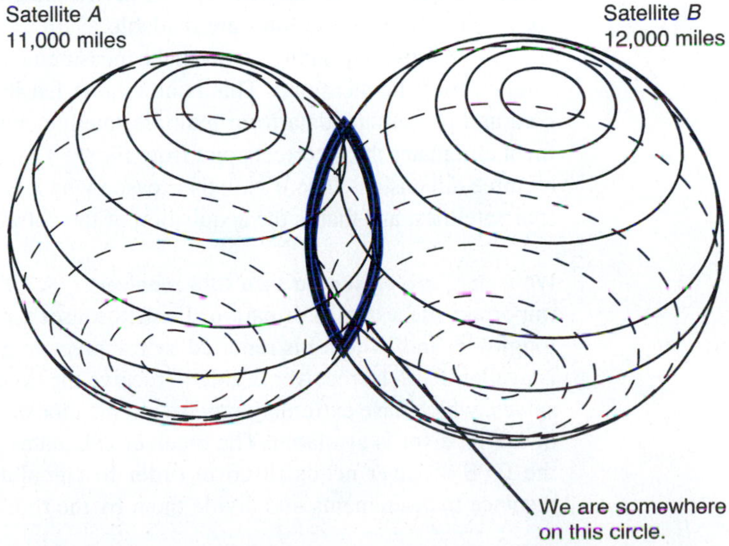

Figure 17–6 The distances to two satellites are known

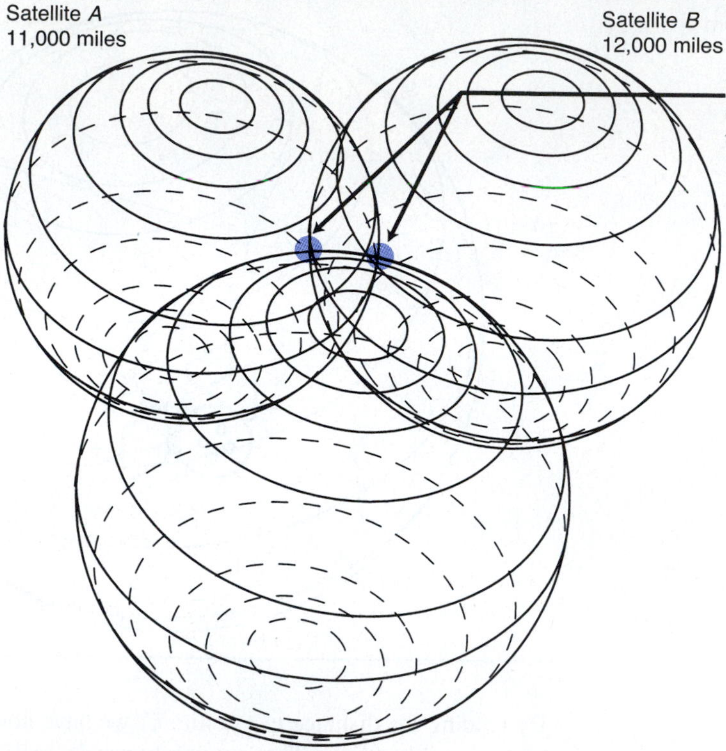

Figure 17–7 The position fix with three satellites

slow to read data from multiple satellites. In those days, receivers had to work under less than optimal conditions, so allowing a receiver to use this last simplification when only three satellites were in view could mean the difference between the user getting a GPS position fix, or not. Current receivers can get a 2-dimensional position (no altitude information) fix when only three satellites are readable.

To remove any ambiguity from the measurement of Figure 17–7, the distance to a fourth satellite must be measured. This is now quite feasible; modern GPS receivers can quickly read and process the data from multiple satellite channels at the same time. This measurement eliminates the incorrect point from Figure 17–7, providing an accurate measurement in all three dimensions. There's another reason why it is desirable to obtain measurements from four satellites, and that is the acquisition of the correct *time,* as we'll soon see.

What data are presented by a GPS receiver? A GPS receiver can give the end user three important pieces of information. First, the user's position in three dimensions (altitude, longitude, and latitude) is reported by the preceding calculation method. Second, the *time* is available, as the receiver is able to receive the time information from the satellites themselves, which have extremely precise atomic clocks on board. Finally, the speed or *velocity* of the receiver is available. The receiver calculates its speed by using equation 17–3. All the GPS receiver needs to do in order to calculate velocity is to take two successive distance measurements and divide them by the time interval between them:

$$v = \frac{\Delta x}{\Delta t} \qquad \textbf{(17–3)}$$

where Δx is the difference between two position readings, and Δt is the difference between the two points in time where these positions were measured.

Getting accurate time Accurate timing is essential to the operation of GPS, because radio waves travel at a very high speed (3×10^8 m/sec, or 186,000 miles/sec). A timing error of just 10 μs represents a distance measurement error of 3000 meters (1.86 miles, or 9840 feet)!

Each of the satellites contains several *atomic clocks*. These are very accurate time-keeping devices that use the natural frequency of either cesium or rhubidium atoms to control their oscillator circuit (much like using a quartz crystal in a conventional oscillator). Because each satellite has several of these clocks on board, their timing errors can be mathematically cancelled to provide superb accuracy (and extra reliability, in case one of them happens to fail). In addition, the satellite clocks are periodically corrected by the master control station located at Falcon Air Force Base, near Colorado Springs, Colorado.

Atomic clocks are very expensive; the satellite clocks cost over $50,000 apiece. If a receiver were required to have such a clock on board, only the very rich could afford to use GPS! However, by using a little algebra, and comparing the measurements from several satellites, a GPS receiver can correct its own internal clock to match that of the satellites. Figure 17–8 shows the effect of timing errors on position.

In the figure, we have restricted the discussion to two dimensions, which makes it easier to draw the diagrams and understand the results. These results will also be valid in

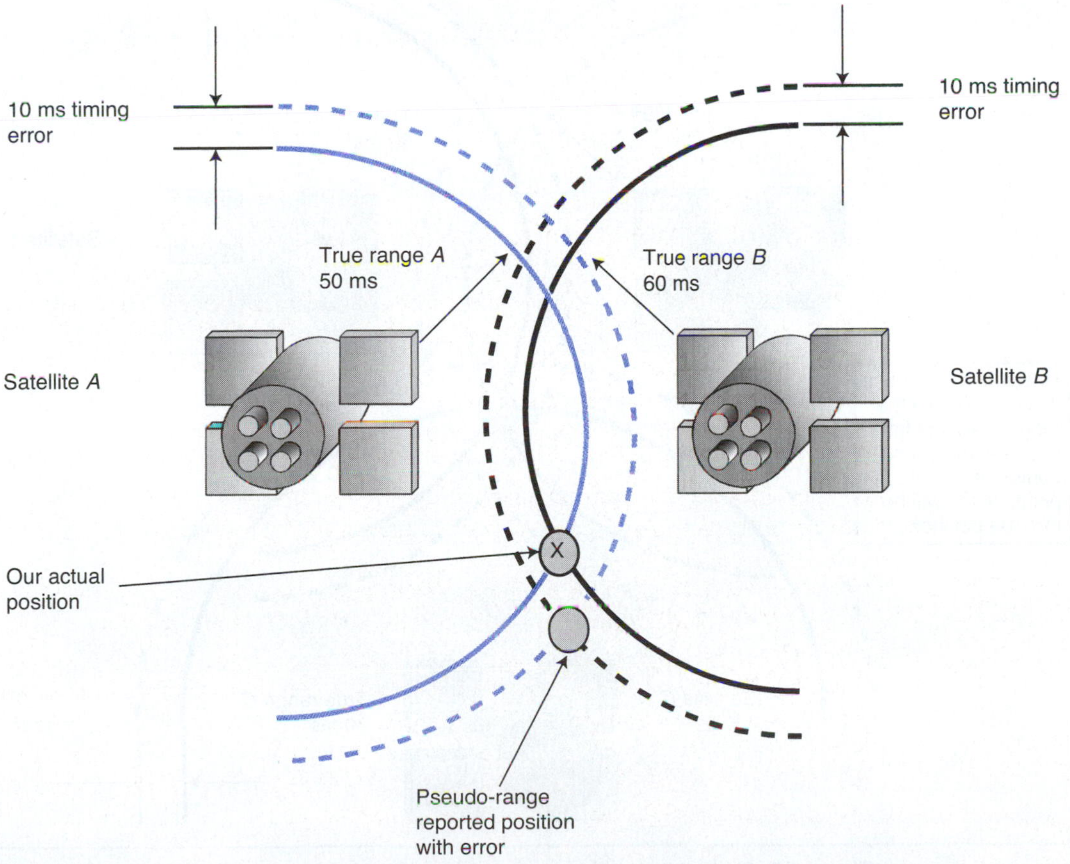

Figure 17–8 Effect of timing errors

three dimensions. We have also shown distances in units of *time,* rather than distance. This makes it easier to see how the corrections can be applied (since we know that distance is obtained by multiplying time and velocity).

Our *actual* position is point *X,* which is one of the two points of intersection from the ranges derived from satellites *A* and *B*. Our receiver has reported that it took 50 ms for the signal to get to us from satellite *A* (a distance of 9300 miles) and 60 ms from satellite *B* (a distance of 11,160 miles). If all of our timing sources (both satellites and the receiver) agree, then this is the point that will be reported.

However, what happens if our receiver clock is 10 ms fast? It will report a *larger* distance to each of the satellites, which is represented by the dashed circles. The receiver now thinks that it is 60 ms (11,160 miles) to satellite *A* and 70 ms (13,020 miles) to satellite *B*. In other words, the 10 ms of clock error caused the range to each satellite to be off by 1860 miles! These ranges containing timing errors are sometimes called *pseudo-ranges,* meaning that they are *false range measurements.*

This problem can be resolved by taking a range measurement from a *third* satellite, as shown in Figure 17–9. If our receiver measures from a third satellite, it will find a

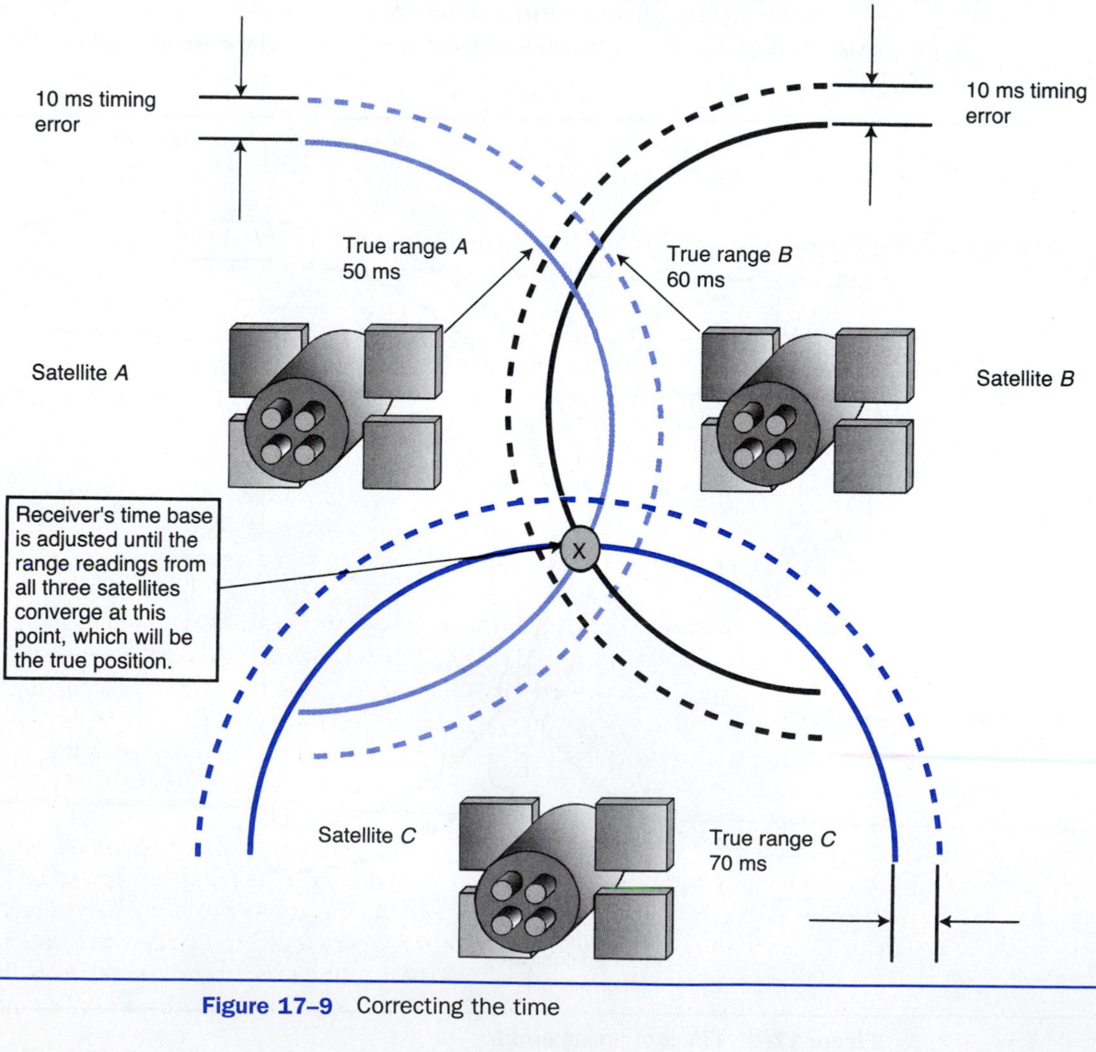

Figure 17–9 Correcting the time

similar timing error (10 ms) in the range measurement. A little geometry will also reveal that the receiver will "discover" that the three pseudo-range circles do not coincide at our actual position. The receiver then "knows" that its internal clock is off and can then make an adjustment to correct itself.

This adjustment is not made by trial and error. Because four satellites are actually used in three dimensions, a system of four equations in four unknowns must be solved to determine the nature of the timing error. These calculations are carried out by the microprocessor in the GPS receiver and are of little consequence to the technician. Once the receiver has corrected its time clock, it can calculate its position with high accuracy.

GPS receivers use a precision crystal oscillator as their internal clock source and constantly perform this correction as they receive data from the satellites. This is one reason that you can really trust the time given on the display of a GPS receiver, as long as the GPS unit has recently locked onto at least four satellites and has been set for the correct time zone. The GPS system uses coordinated universal time (UTC) as its standard, not the regional time of any particular location.

Finding the Satellite Positions

In order to calculate position from the satellite timing signals, the positions of the satellites must be known. Each satellite has a specific and precise *orbit* and *orbital position* that is predictable, and this information is programmed into a lookup table called an *almanac* that is contained within the software in the GPS receiver. The almanac of a GPS receiver contains the planned orbital patterns of all GPS satellites in the receiver's memory.

GPS satellites travel in orbits that are 12,200 miles above the earth's surface and are designed to orbit the earth once every 12 hours. This is part of the GPS master plan, and it allows all the satellites to periodically pass over the Colorado Springs master control station for maintenance and updating. At an altitude of 12,200 miles, a satellite must travel at an average speed of 8,500 mph in order to have a stable orbit. If the satellite moves too fast or slow, it will leave the intended orbit, with disastrous results. Fortunately, at this height, there is no significant atmosphere to interfere with the motion of the satellite, and the primary forces acting on the satellite are the gravitation pulls of the earth, sun, and moon.

Each satellite has a specific identification code, and these codes are programmed into the receiver's almanac. Therefore, when a GPS receiver receives a message from a satellite, it can determine the satellite's position in space by looking up its ID code in the almanac and then applying the almanac information to the satellite's orbital equations.

Ephemeris information In spite of the extreme precision involved in planning the orbits of GPS satellites, orbital changes do occur. Errors build up in the orbital position of a satellite and are caused by the gravitation pulls of the sun and moon and the changing effect of *solar radiation.* Solar radiation tends to exert a push on satellites, but because the earth, sun, and the satellite are in an ever-changing arrangement, this extra push can cause very complicated changes in the orbit, especially when the earth passes between the satellite and the sun (where the solar push temporarily disappears).

Since a GPS satellite passes by one of the Department of Defense monitoring stations twice a day, its altitude, position, and speed can be accurately measured, and this information can be relayed back up to the satellite. The satellite then retransmits this data, called the *ephemeris data,* along with its timing message. The almanac in a receiver gives the *planned* orbital data for each GPS satellite, while the *ephemeris* provides the same information that has been "fine-tuned" by the master control station to reflect the actual orbital conditions of the satellite. By using the ephemeris data, GPS receivers always know the most up-to-date and exact orbital information for each satellite. This is important, for to calculate position accurately, the receiver must be able to calculate the position of each satellite with precision.

Selective availability GPS has the potential to be useful to enemies of the United States, and because of that, the U.S. Department of Defense (DOD) has implemented *selective availability,* or *SA*. When selective availability is enabled, errors are deliberately introduced into the system. For example, random errors can be introduced into the ephemeris and timing messages on the C/A (civilian) channel, which directly degrades the accuracy of GPS measurements. Even with selective ability enabled, innovative GPS receiver manufacturers have found ways of coping with these errors, as we'll see in a later section.

Antispoofing Antispoofing, or AS, is an additional mode that can be selectively turned on or off by the government. The purpose of antispoofing is to prevent an enemy from transmitting false navigation codes on top of the authentic satellite data. When AS is enabled, the P code is further encrypted to form the Y code.

Section Checkpoint

17–9 What two pieces of information must a GPS receiver possess in order to find its position in space?

17–10 Explain how a position "fix" can be obtained from three and four satellites.

17–11 List the three types of information a GPS receiver can provide for the user.

17–12 What type of clocks are used in GPS satellites? Why are these practical for receivers?

17–13 Explain how a GPS receiver finds the correct time.

17–14 How does a GPS receiver know the exact position of each satellite?

17–15 What forces can move a satellite from its planned orbit?

17–16 Explain what the purpose of ephemeris data is. What is the difference between almanac and ephemeris data?

17–17 What is selective availability? How is it implemented?

17–3 GPS SEGMENTS AND NAVIGATION SIGNAL PROCESSING

The GPS system contains an impressive array of technologies. Because of this, understanding how all the pieces of the system fit together can be a little intimidating at first. GPS can be best understood if it is analyzed in pieces or *segments*. The GPS system is divided into three parts. These are the *space segment,* the *control segment,* and the *user segment*.

The GPS satellites make up the *space segment*. The satellites transmit the navigation messages on two different phase-modulated frequencies (we will see why two frequencies are used later).

The *control segment* runs the entire GPS system. It consists of a *master control station* or *MCS,* which is located at Falcon Air Force Base near Colorado Springs, Colorado, and a number of monitoring stations distributed around the globe. The primary mission of the control segment is to track the position and health of all GPS satellites and update the navigation messages transmitted by each satellite. Part of the navigation message includes the *ephemeris data*. The ephemeris is the set of recently computed (and short-lived) information about the GPS satellite's orbit. The ephemeris data is transmitted periodically to each GPS satellite by the master control station, which has received data about each "bird" from the various tracking stations around the world. The satellite then retransmits this information along with the navigation message. In this way, all GPS receivers have exact and up-to-date information about the orbital position of each satellite, which compensates for errors that accumulate in the satellite orbits.

The *user segment* consists of all receivers in use on and near the earth. A GPS receiver has a tremendous computational task; it must lock onto at least four different satellites and compute its position from the received timing signals. Most GPS end users are intimately familiar with their own equipment (the receivers) but have no real idea of the complex network of resources that supports the receiver's operation.

Space Segment

The GPS satellites have been released in *blocks* (groups) since the system's inception in 1978. There are currently 29 active satellites in this segment, divided into *blocks*. Ten Block I satellites were launched between February 1978 and October 1985. These were used as a test bed to demonstrate the capabilities of the system, and all of them have long since been turned off or have failed. The Block I satellites lacked SA capability and could only provide 3–4 days of positioning service without ground station contact.

Nine Block II satellites are in existence, and eight of them are still operational. These are an improvement over the Block I units and have enough memory for 14 days of unsupervised operation. Their design life span is about 7.5 years, and their weight is 1860 pounds when inserted into orbit. These satellites were launched between 1989 and 1990.

The Block IIA satellites were launched between 1990 and 1994 and have expanded memory capabilities. They can provide 180 days of service without contact with the control segment, with some degradation of positional accuracy over time. There are 18 functional Block IIA satellites in orbit.

The Block IIR satellites are designed as replacements for the aging fleet of Block II and Block IIA units; the first of these was unsuccessfully launched in 1997. These satellites are further refined in that they can operate at least 14 days without contact from the control segment and up to 180 days when operating in the autonomous navigation (AUTONAV) mode. Full accuracy is maintained without control segment contact by using intersatellite ranging and communication. The data obtained automatically updates the navigation message from each satellite. Each Block IIR satellite is designed to live for 7.8 years.

Figure 17–10 shows the status of the Block II GPS satellites as of the time of this writing.

Figure 17–10 GPS block II/IIA/IIR satellite status

Launch order	Service vehicle number (SVN)	Launch date	Orbital plane	USSC object number
II-2	13	10 Jun 1989	B3	20061
II-5	17	11 Dec 1989	D3	20361
II-9	15	01 Oct 1990	D2	20830
IIA-10	23	26 Nov 1990	E4	20959
IIA-11	24	04 Jul 1991	D1	21552
IIA-12	25	23 Feb 1992	A2	21890
IIA-14	26	07 Jul 1992	F2	22014
IIA-15	27	09 Sep 1992	A3	22108
IIA-16	32	22 Nov 1992	F1	22231
IIA-17	29	18 Dec 1992	F4	22275
IIA-19	31	30 Mar 1993	C3	22581
IIA-20	37	13 May 1993	C4	22657
IIA-21	39	26 Jun 1993	A1	22700
IIA-22	35	30 Aug 1993	B4	22779
IIA-23	34	26 Oct 1993	D4	22877
IIA-24	36	10 Mar 1994	C1	23027
IIA-25	33	28 Mar 1996	C2	23833
IIA-26	40	16 Jul 1996	E3	23953
IIA-27	30	12 Sep 1996	B2	24320
IIA-28	38	06 Nov 1997	A5	25030
IIR-2	43	23 Jul 1997	F5	24876
IIR-3	46	07 Oct 1999	D2	25933
IIR-4	51	11 May 2000	E1	26360
IIR-5	44	16 Jul 2000	B3	26407
IIR-6	41	10 Nov 2000	F1	26605
IIR-7	54	30 Jan 2001	E4	26690
IIR-8	56	29 Jan 2003	B1	27663
IIR-9	45	31 Mar 2003	D3	27704
IIR-10	47	21 Dec 2003	E2	-

Navigation Message Characteristics

The satellites repeatedly transmit a *navigation message* that is sent at a data rate of 50 bps. The complete navigation message consists of 25 *frames* (data blocks), each containing 1500 bits. Each frame is divided into five 300-bit subframes, which are 10 *words* of 30 bits each. At the 50 bps data rate, it takes 6 seconds to send a subframe, 30 seconds for a frame, and 12.5 minutes for the entire navigation message. Figure 17–11 shows the basic structure of the navigation message frames.

Each subframe starts with the telemetry word and handover word (HOW). The handover word gives critical timing information for extracting the P code. When the HOW is multiplied by 4, it gives the current position, called *x1,* for the next subframe within the P code. The P code doesn't repeat for 266.4 days, so receivers need extra help in locking

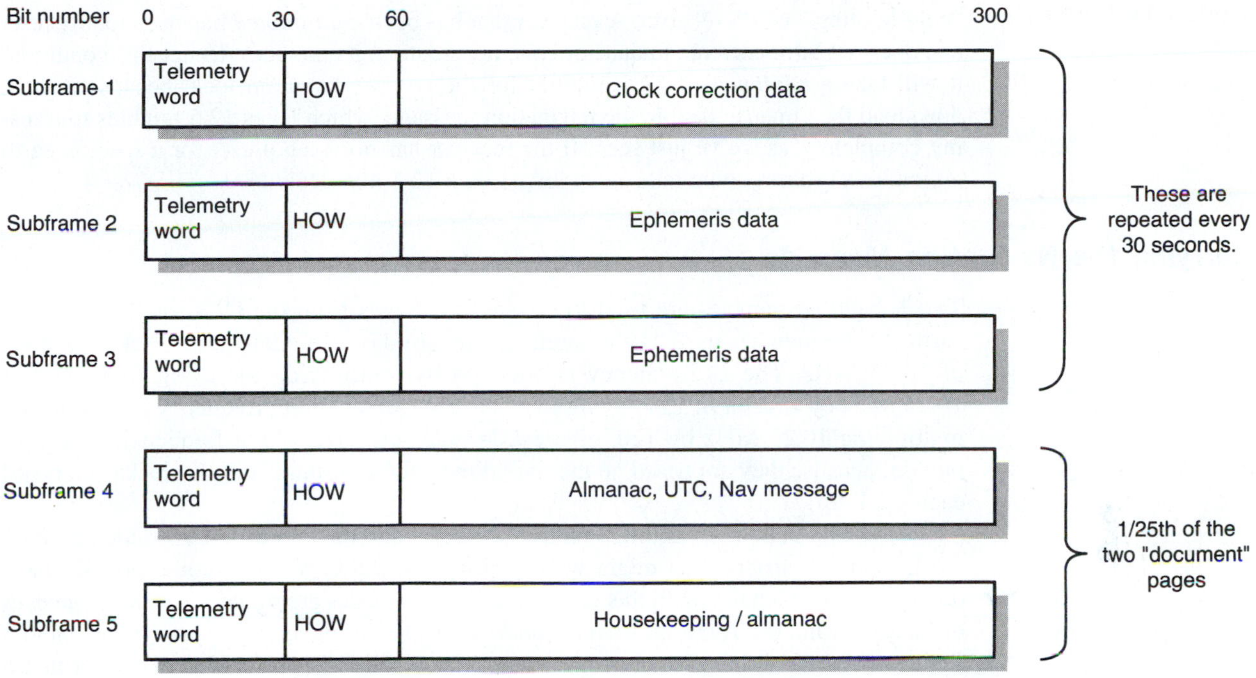

Figure 17–11 Navigation message frame structure

onto it (otherwise it might take a receiver years to lock onto the P code!). Of course, the receiver must be programmed with the P code in order to use the handover word. In this respect, you can see that the C/A code must first be acquired before the P code can be locked onto by a receiver.

Notice that subframes 1, 2, and 3 contain *timely* information. These contain the clock correction factor and satellite ephemeris (recently updated orbital data) that are needed for a receiver to accurately calculate position. These are transmitted with every complete frame (30-second transmission), so their information is available every 30 seconds.

Subframes 4 and 5 are less time critical. They are used to build a long message string containing the general telemetry data and timebase information, as well as the satellite almanac and health status. To construct these messages completely, 25 complete sets of subframes 4 and 5 are needed. Therefore, it takes (25)(30) seconds (12.5 minutes) to send the entire "page" of information transmitted in subframes 4 and 5.

You can think of the contents of subframes 4 and 5 as individual *paragraphs* used to construct two different document "pages," one of which will contain a complete message about the UTC time and almanac (orbital data) for satellites with a service vehicle number (SVN) 25 or greater. The other "page" will contain the almanac data for satellites with SVNs 1 through 24 and internal housekeeping information (such as the operational "health" status of the various satellite subsystems). It always takes 25 complete paragraphs to build these two document pages. Note that with the almanac information from any *one* satellite, the receiver can get a rough indication of the position for any other satellite in the array. This is very useful, as it allows the receiver's internal almanac data to be updated constantly and eliminates the need to reprogram receivers when satellites are taken in or out of service.

The navigation message structure has important implications when it comes to "cold-booting" a GPS receiver. A receiver that has been *cold-booted* has been totally reset and does not have current almanac or ephemeris data in its memory. Under this condition, it will take a *minimum* of 12.5 minutes for the receiver to attain lock, because it must download the almanac data in the navigation message, which takes 12.5 minutes to transmit completely, as we've just seen. If the receiver has not been preset for a specific earth region, this process could take *much* longer than 12.5 minutes!

Carrying the Navigation Message

Two L-band frequencies are used to transmit messages from the GPS satellites back to earth. All frequencies used by the satellites are based on a fundamental clock frequency of 10.23 MHz. The L_1 frequency is obtained by multiplying the clock frequency of 10.23 MHz by a factor of 154, giving $L_1 = 1575.42\ MHz$. The L_2 frequency is obtained by multiplying 10.23 MHz by 120, giving $L_2 = 1227.60\ MHz$. These frequencies are *very* precise, because they are based on the rhubidium and/or cesium "atomic clocks" on board each satellite.

The C/A code is transmitted on the L_1 carrier, and the P code is transmitted on both the L_1 and L_2 carriers. You might wonder just what the C/A and P codes are; we have deliberately avoided that until this point. Both of these codes are *pseudo-random sequences* of binary 1s and 0s. The term *pseudo-random* means that the codes *appear* to be random sequences of 1s and 0s, but they are in fact repeatable and predictable. The fact that a receiver can predict these codes (once it has synchronized to them) is extremely important.

The pseudo-random codes provide three important capabilities: First, they allow *all* the active satellites to transmit on the *same* carrier frequencies (L_1 and L_2) without causing any interference to each other. This is called *CDMA, or code division multiple access,* or *code division multiplexing*. You'll recall that *multiplexing* is a way of sending more than one piece of information on a communications channel.

The second capability that the pseudo-random codes provide is the ability of a receiver to demodulate (or *correlate*) the extremely weak signals from the GPS satellites, thus allowing the use of small receiver antennas. The C/A code energy is about 16 dB *below* the average earth background noise, and the P code is even weaker, about 29 dB below the noise level. Conventional communications is impossible under these conditions; the noise is much stronger than the signal! These codes provide *effective signal gain* at the receiver to pull the navigation message frames out of the noise. This is better than requiring the receivers to have large antennas, which would severely reduce the usefulness of GPS.

The third ability provided by the use of the codes is *security*. The C/A code is public knowledge and repeats every 1 ms (1000 times per second). The P code (which repeats every 266.4 days) can be encrypted at will, forming the Y code. The Y code cannot be demodulated by the general public, and is resistant to jamming.

Generating the Pseudo-Random Code

A digital circuit called a *feedback shift register* is used to generate the C/A and P codes. In reality, a hardware shift register is not used; these codes are generated by a software program running in a DSP (digital signal processor) IC. The receiver runs the same program as the transmitter, so it generates exactly the same pseudo-random bit sequence. Figure 17–12 shows a simple feedback shift register. The exclusive-OR gates in the feedback loop of Figure 17–12 determine what value (1 or 0) will be sent back to the first stage of the

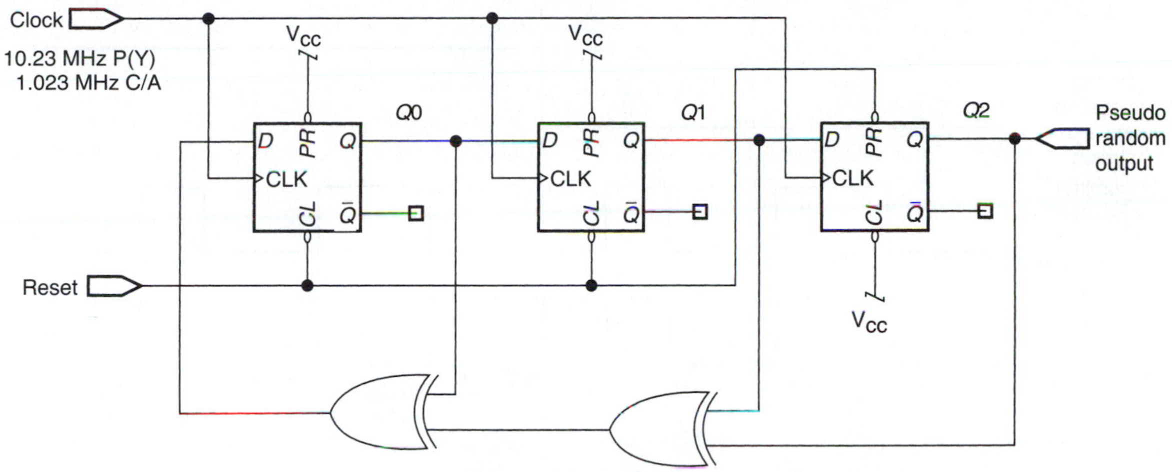

Figure 17–12 A feedback shift register

register. If the initial contents of this shift register are 001_2, then the register state will look like that of Figure 17–13 as it is clocked.

The shift registers used to generate the C/A and P signals are much more complex than the one of Figure 17–12. The C/A register is clocked at 1.023 MHz. The clock rate of the register is referred to as the *chipping rate.* A *chip* is one bit of the pseudo-random code sequence. The code in the C/A register repeats every 1 ms, or 1000 times per second. The P generating register is clocked ten times faster, at 10.23 million chips per second. The P code is much longer, taking 266.4 days to complete an entire cycle.

Modulating the Carrier Signals

As you know, there are three ways that information can be impressed onto a radio frequency carrier. We can alter the *amplitude (AM), frequency (FM),* or *phase (PM)* of the carrier wave to convey the information. Because of its superior noise suppression (and simplicity), *coherent binary phase-shift keying* (BPSK) is used for transmitting the navigation messages from the satellites. In coherent PSK, the phase angle of the carrier is shifted with reference to a *constant* standard of phase. This standard of phase is the unmodulated carrier and is

Figure 17–13 State sequence of feedback shift register

Clock number	Q_0	Q_1	Q_2 (output)
0 (initial state)	0	0	1
1	1	0	0
2	1	1	0
3	0	1	1
4	0	0	1 (begin repeat)
5	1	0	0

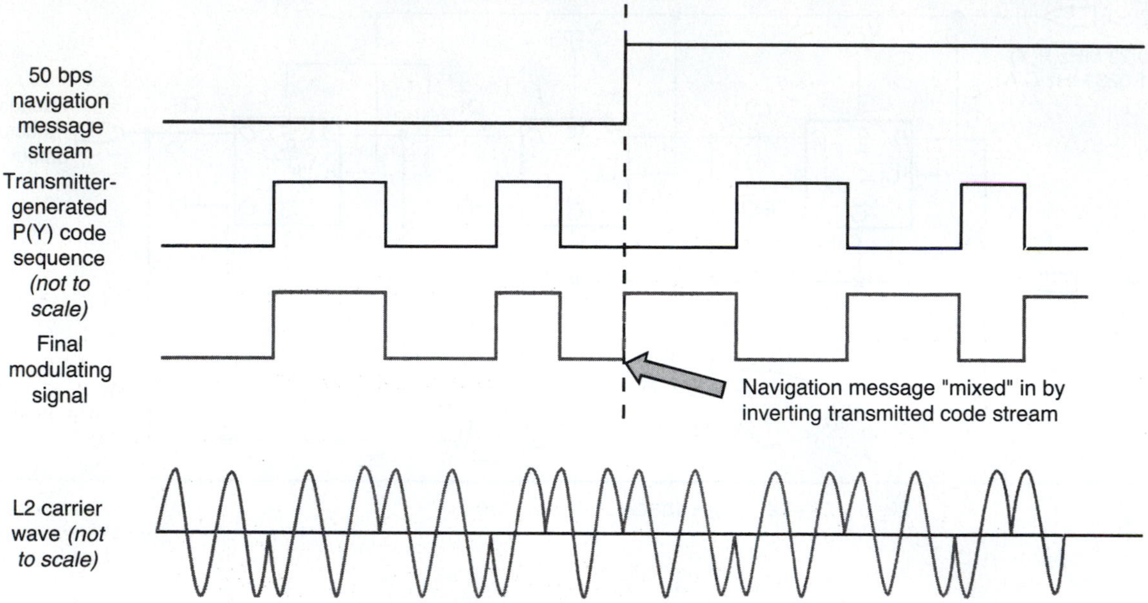

Figure 17–14 Modulation of the L2 carrier

recovered by the receiver during the process of demodulation. Coherent PSK is the opposite of the *differential* phase-shift keying techniques used in Chapter 15.

GPS satellites transmit two radio frequency carriers, L_1 and L_2. The L_1 carrier is impressed with both the C/A and P codes, and the L_2 carrier carries the P code alone. This creates a problem; The L_1 carrier must convey *two* separate pieces of information. This problem is solved by using two 90° shifted (quadrature) carrier signals for L_1. These quadrature-phased signals can be easily separated by a GPS receiver. Figure 17–14 shows a simplified representation of the modulation method used on the L_2 carrier.

There are two steps in the modulation process. First, the *navigation message* is mixed with the pseudo-random code. An exclusive-OR circuit (in software) performs this job. Thus, when the navigation message and pseudo-random code have the same value, the resulting output is a logic 0; when they are different, the output to the phase modulator is a logic 1. The advantage of this process is simplicity; once the pseudo-random code is locked onto by a receiver, the *exact same procedure* will demodulate the navigation message.

Notice that every time there is a switch between a logic 1 and 0 on the final modulating signal, the carrier jumps by 180° of phase. This can be best understood by remembering that this is a *coherent* PSK system. On a logic 0, the modulator is jumping 180° out of phase. On a logic 1, the system is returning to the original (0°) phase value. Successive 1s or 0s cause no added phase shift.

Figure 17–14 is hardly to scale. The P code has a chipping rate of 10.23 MHz, while the navigation data stream has a data rate of only 50 bps. This means that there are 204,600 P code chips for every data bit in the navigation stream. Furthermore, the L_2 carrier signal is at 1227.60 MHz, so there are exactly 120 cycles of the L_2 carrier for each chip in the P code stream and (120)(204,600) or 24,552,000 L_2 carrier cycles for each data bit in the navigation stream.

Figure 17–14 adequately represents the modulation of *one* code stream onto the L_2 carrier, but what about the L_1 carrier, which must carry both the C/A and P code streams? The L_1

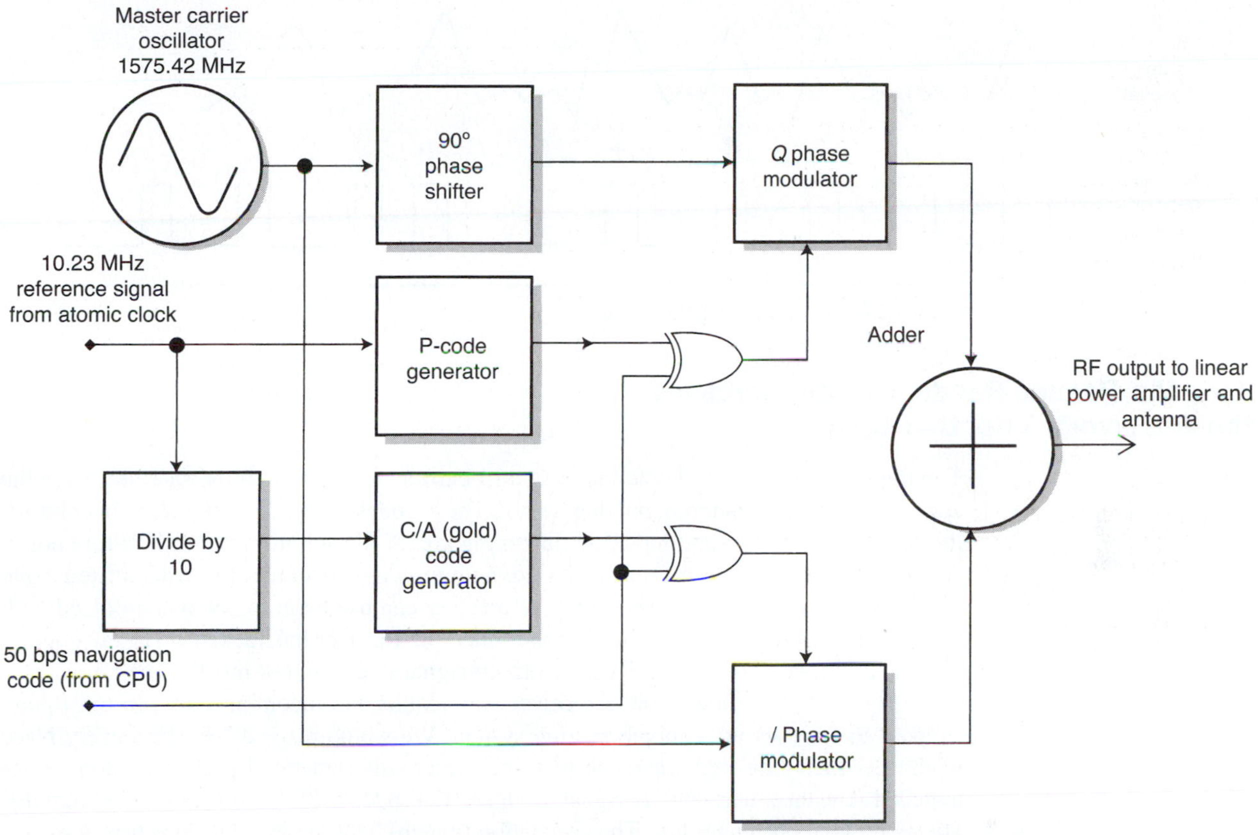

Figure 17–15 Modulating the L_1 carrier

carrier consists of *two* independent phase-modulated sine waves at the *same frequency*. This is accomplished as shown in Figure 17–15. In the figure, the master carrier oscillator operates at 1575.42 MHz. This oscillation frequency is derived by multiplying the 10.23 MHz atomic clock reference signal by 154, which is not shown. The 10.23 MHz reference signal clocks the P-code generator. The P-code output is mixed with the navigation message by an exclusive-OR gate and sent on to the *Q* (quadrature) phase modulator, which receives a 90° advanced version of the L_1 carrier frequency. This generates the P portion of the L_1 carrier.

The C/A information is generated in a similar manner. The 10.23 MHz reference signal is divided by 10 for clocking the C/A code generator, which has a chipping rate of 1.023 MHz. The resulting C/A code is mixed with the navigation message by an exclusive-OR gate, in an identical manner to the P code. The resulting mixture of C/A code and navigation data modulates the in-phase version of the 1575.42 MHz carrier at the *I* (in)-phase modulator, resulting in the complete C/A modulated RF carrier.

An adder combines the two RF signals (which have a 90° phase difference) into a composite signal. Not shown is an RF attenuator that follows the *Q* phase modulator, which lowers the power level of the P code about 13 dB below that of the C/A code output from the *I* modulator.

In practice, the L_1 carrier signal is demodulated at the receiver by extracting the two carrier signals (*I* and *Q*) and separately demodulating the two resulting BPSK data signals, one of which will contain the C/A code, and the other the P code.

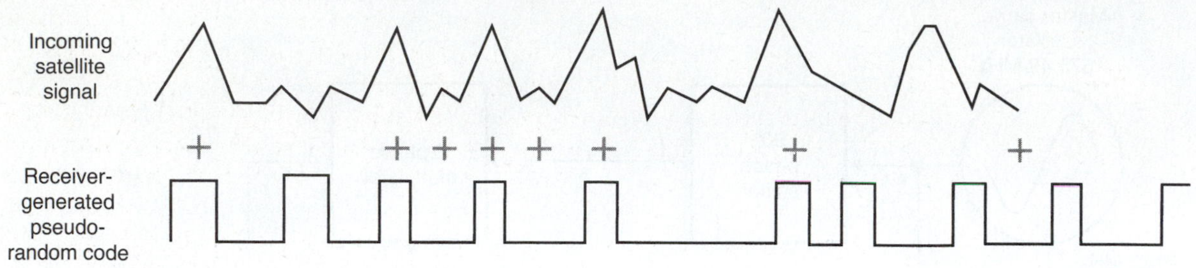

Figure 17–16 Acquiring the code

How the Pseudo-Random Code Increases the Receiver's Effective Gain

The digital codes that modulate the L_1 and L_2 carrier frequencies of the satellites have the appearance of being random, but they're not. These codes are quite predictable. In contrast, the background noise that masks the incoming signal *is* random, because it is true noise. A receiver can use these facts to its advantage. First, because the transmitted code sequences are repeated over and over, the receiver can use them to get synchronized with the data stream from the satellite. Second, once the receiver has locked on to the code, it can use arithmetic averaging on the incoming signal to cancel out most of the noise.

Figure 17–16 shows how a receiver can acquire the incoming code. In the figure, suppose that we set up a simple scoring system. We would place a "+" sign every place where the incoming signal and the receiver's internally generated pseudo-random codes matched. The incoming satellite signal in Figure 17–16 looks like *mostly* noise, and initially, our score would be pretty low. The correlation (match) between the signals is very low.

In a way, this is very much like hearing music being played from a great distance, where you can barely hear it. You know *what* song is playing, but you can just barely hear the words. You know the words to the song by heart (the receiver is programmed with the same pseudo-random code as the satellite), and you know the exact tempo it will be performed at (the receiver clock is equal to the chipping rate of the transmitted signal). If you can just figure out what part of the song is being played at some instant, you can follow it the rest of the way through. In fact, if the song were played over and over, you would probably become *exactly* synchronized to it eventually. This is pretty much what the receiver does, as shown in Figure 17–17.

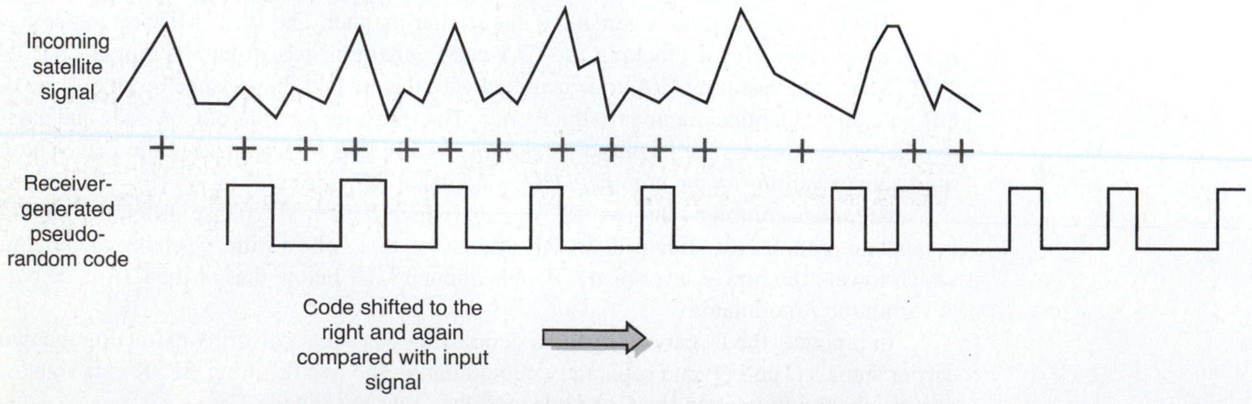

Figure 17–17 The receiver improves the score

When the receiver is attempting to acquire the C/A signal, it takes the score of its internally generated C/A pseudo-random code by comparing it with the signal it is getting from the satellite. If the score is very low, it shifts its internal code over by 1 bit and tries again. Because the C/A code repeats 1000 times a second, the receiver gets many opportunities to try this, and in short order it slides its code into the proper position. The correlation or "score" rises rapidly when this occurs, and the receiver has very likely become synchronized to the code.

Note that every one of the satellites transmits different codes. Because these are pseudo-random codes, they look like normal channel noise. The receiver locks onto the desired satellite by comparing the received signal from the antenna with the C/A code sequence for *that* satellite. The codes from the other satellites just won't correlate, and in fact, they won't bother the receiver at all. *All the satellites therefore can share the same two frequencies (L_1 and L_2) without interference.*

Recovering the Data Stream

The receiver can now work on recovering the data stream. This is possible because the data is sent at a very low rate (50 bps), and it is actually mixed together with the pseudo-random C/A code that is phase-modulating the RF carrier. Figure 17–18 shows how the data is encoded, assuming the absence of noise.

In Figure 17–18, you should be aware that again the drawings are not to scale. The data rate of the C/A channel is 50 bps—there are 50 digital data bits per second. In

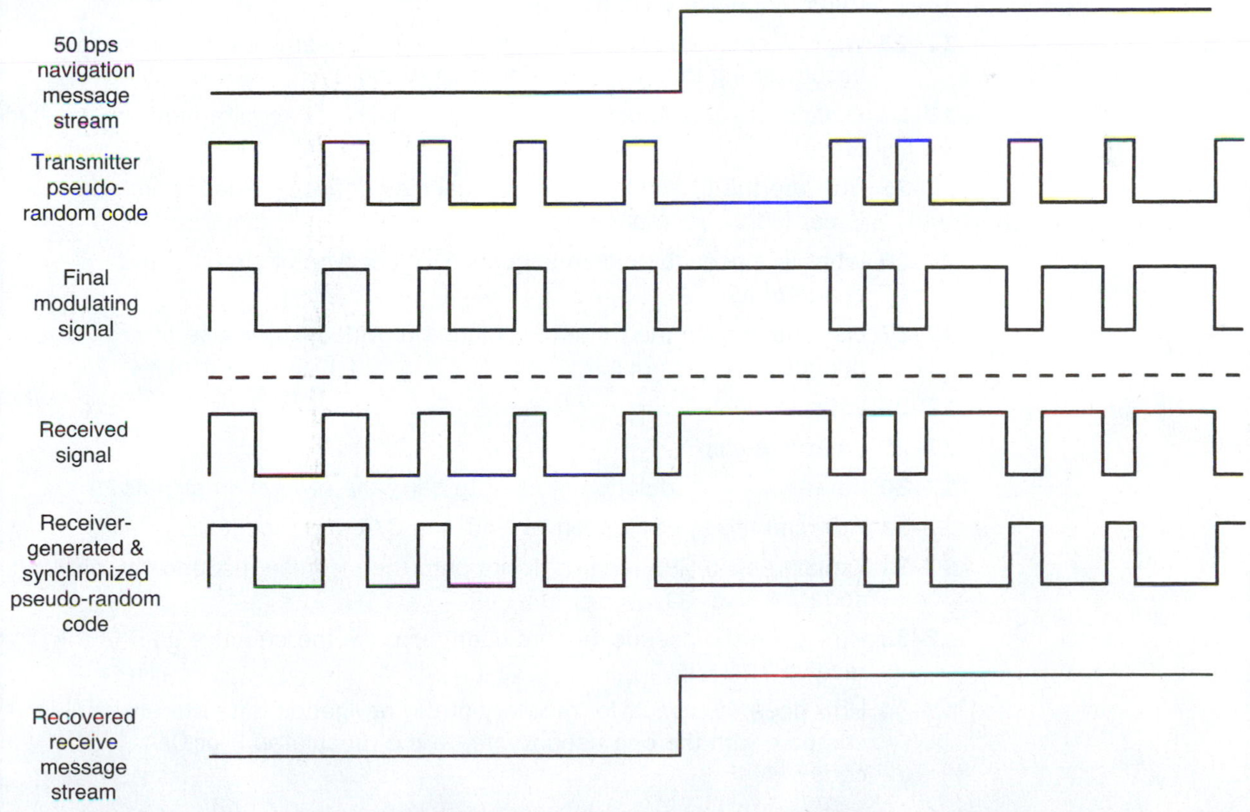

Figure 17–18 Encoding and decoding the satellite data stream

contrast, the chipping rate of the C/A channel is 1.023 MHz. This means that there are actually (1,023,000/50) or 20,460 chips per data bit. In other words, each transmitted data bit occupies 20,460 clocks of the C/A sequence, which is 20 C/A code cycles or *epochs*. An *epoch* is GPS-speak for one repetition of an entire code sequence. The C/A code is easy to lock onto because it repeats its sequence 1000 times per second. Its epoch is 1 ms. Because there are 20 C/A epochs for every data bit, the receiver can *average* the resulting data stream over 20 successive tries. The random component of the noise will tend to approach *zero,* because the average value of random noise is zero. Thus, the averaging process that is allowed by the acquisition of the pseudo-random code acts as an effective *amplifier,* literally lifting the data signal out of the noise.

Section Checkpoint

17–18 List the three segments of the GPS system and explain the function of each one.

17–19 What is the location and function of the master control station?

17–20 Why do GPS satellites require periodic updates from the master control station?

17–21 What is contained in each of the five subframes in a GPS message frame?

17–22 Why is the information in subframes 1 through 3 of a GPS message frame sent with each frame?

17–23 What is contained in the two "documents" assembled by successively accumulating the contents of frames 4 and 5?

17–24 Explain why it is important for a GPS receiver to acquire almanac data.

17–25 How are the L_1 and L_2 carrier frequencies derived? What information is carried on each one?

17–26 What is a pseudo-random sequence? What type of circuit can generate it?

17–27 List and explain the three capabilities provided by the use of pseudo-random codes.

17–28 What is CDMA?

17–29 What is a *chip?*

17–30 What type of modulation is used to carry the navigation signals?

17–31 How can the L_1 carrier convey both the C/A and P codes?

17–32 Explain how a GPS receiver locks onto the incoming pseudo-random code.

17–33 How does the pseudo-random code increase the effective gain of the receiver?

17–34 How does the circuit for recovery of the navigation data stream compare with the one used to create the modulated P or C/A code?

17–4 REDUCING POSITION ERROR: ENHANCED GPS

GPS is certainly a superior navigation technology when compared to its predecessors, but its results are not quite perfect. Because it relies on the reception of radio signals, a number of problems inherent to radio wave propagation can degrade its accuracy. In addition, the intentional errors introduced by selective availability (SA) complicate matters even further. The manufacturers of GPS receivers have been working on many new techniques that improve accuracy. These techniques include *differential GPS (DGPS)* and *carrier-phase GPS*. As of March 2000, selective availability has been at level 0 (disabled), eliminating this source of error for GPS users. The U.S. government has announced that it has no plans to reactivate SA capability in the future.

Sources of Positioning Error

Timing is everything when it comes to processing the received signals from GPS satellites. Timing errors can be caused by the clocks in the receiver and satellites, atmospheric delay of the radio wave transmissions, multipath distortion, and the deliberate manipulation of the GPS timing and data caused by SA (when it's in use).

Each GPS satellite uses several atomic frequency standards as a timing source (10.23 MHz), but even these sources are not perfect. These clocks are monitored and corrected by the master control station on a regular basis, but even so, they can contribute up to about 1.5 meters of error to the distance measurements (a very small value). The clocks in GPS receivers are relatively inexpensive crystal oscillators, and with the techniques discussed earlier in this chapter, they can be closely synchronized with those on the satellites. Because of the wide range of possible receiver designs, it is difficult to estimate the range of errors due to this source.

Atmospheric delay of radio signals is a more significant contributor to timing error, with an estimated error range of 0.5 meters. Radio signals do not always travel at the speed of light. They slow down slightly as they pass through the earth's atmospheric layers. Ionospheric coefficients are included in subframe 4 of the navigation message to assist receivers in compensating for the delay, but the delay is actually hard to predict, since atmospheric conditions change constantly. Some advanced receivers make use of the fact that the ionospheric delay varies with frequency, and they can measure and compare the incoming signals on both the L_1 and L_2 carriers and by doing so, more accurately estimate the ionospheric conditions and delay.

Multipath distortion can further distort the timing signals at a receiver. A signal may not take a direct path to a receiver; it may bounce off buildings or terrain on its way. Such a signal will have obviously traveled farther than normal and will destructively interfere with the signal arriving directly from the satellite. There's little that can be done about this type of distortion, except to locate GPS receivers away from objects that may give false signals. Errors of about 0.5 meter are typical with multipath.

Selective availability is the largest of all the budgeted error sources; originally it was thought that SA would severely limit the civilian applications of GPS. The error range for SA is around 30 meters. Military receivers continue to operate with high accuracy when SA is enabled because they have the key to the cryptographic codes that are used to alter the ephemeris and timing data from the satellites.

With all of these error sources, it might seem impossible to get sub-meter accuracy from civilian GPS receivers—but it is now being done regularly, and centimeter accuracy may not be far off. Let's take a look at two techniques that are currently being employed.

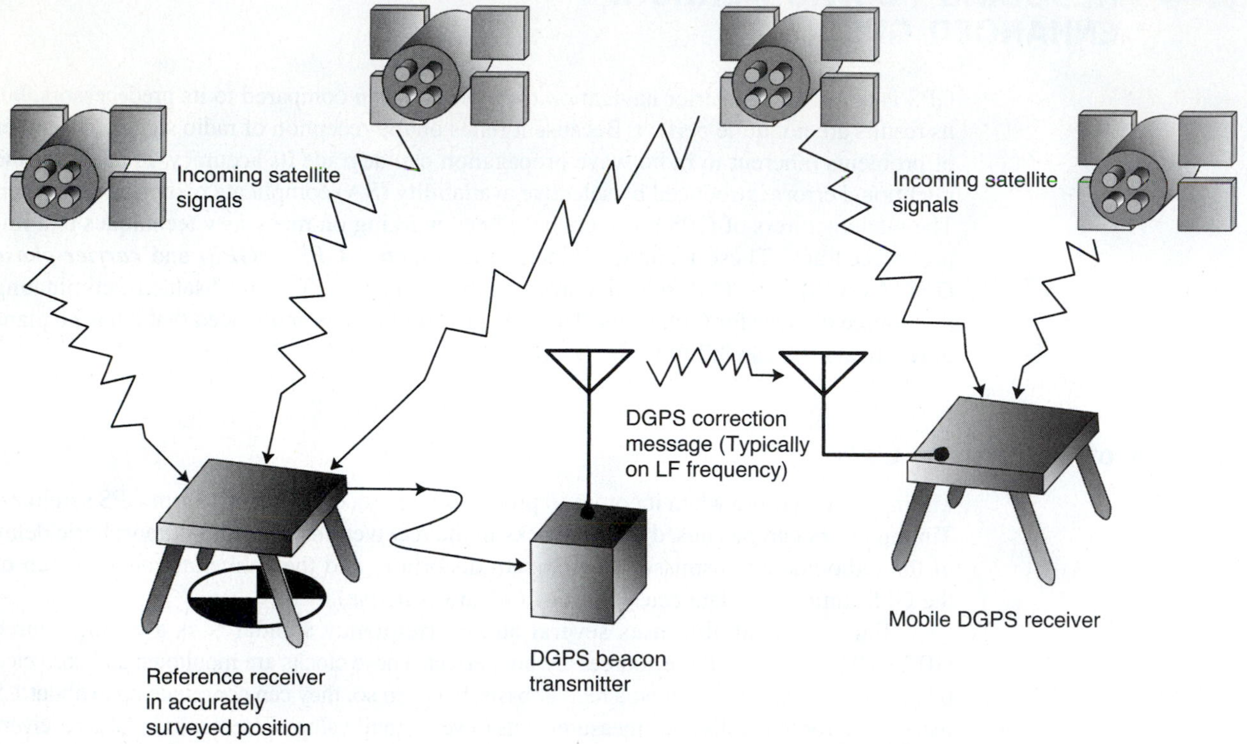

Figure 17–19 Differential GPS

Differential GPS (DGPS)

Horizontal accuracy of better than 1 meter is attainable with differential GPS receivers. The idea behind DGPS is fairly straightforward and is shown in Figure 17–19. DGPS relies on a specially designed reference GPS receiver placed at a precisely known location. This receiver monitors the satellite signals, and because it has been programmed with its precise location, it can calculate the error in the timing signals from each satellite. Because the mobile GPS receivers are geographically close to the reference receiver, they also experience much the same errors. The error information for all the available satellites is collected by the reference receiver and transmitted by the *beacon transmitter*. The mobile GPS receivers receive the beacon transmission on a separate radio frequency, decode the error message, and apply it to their position readings, thus canceling most of the errors, especially that due to SA, which is essentially cancelled.

NAVCEN, the U.S. Coast Guard Navigation Center, operates the Coast Guard Maritime Differential GPS Service, which provides DGPS coverage over most of the coastal United States as well as many inland locations. The purpose of this service is to assist in costal and harbor navigation, which can be tricky for large vessels. Most of the transmissions from NAVCEN facilities are in the LF frequency range. For example, the NAVCEN—Kansas City DGPS transmitting facility has the following characteristics:

- Transmitting beacon antenna location: 39,7.04N;95, 24.53W
- Reference receiver (A) exact location: 39,7.55229N;95, 24.12556W
- Reference transmitter (A) RTCM SC-104 ID: 164
- Reference receiver (B) exact location: 39, 7.55268N;95, 24.08916W

- Reference transmitter (B) RTCM SC-104 ID: 165
- Transmission frequency: 305 kHz
- Data rate: 200 bps

This particular NAVCEN site uses two reference receivers to monitor the GPS satellite transmissions. In order to utilize its services, an end user must possess a DGPS receiver capable of tuning to the transmissions on 305 kHz. This station transmits the corrections in the RTCM SC-104 format, which is accepted by most DGPS receivers. NAVCEN services provide better than 10 meter accuracy.

Commercial DGPS correction services are also widely available. For example, *OMNIStar* provides positioning service (with claimed accuracy of better than 1 meter) to its subscribers over most of the world. *OMNIStar* collects the navigational correction information from reference receivers scattered throughout the world and transmits it down to the end-user via an L-band (1–2 GHz) satellite downlink.

Carrier-Phase GPS

The standard operation of a GPS receiver can be described as *code phase,* because a GPS receiver continually "slides" its internally generated pseudo-random code until it matches up with that of the transmitter. The amount of time that the receiver has to slide its code through is the travel time for the GPS signal and gives the distance to the satellite.

Code-phase GPS works quite well, especially in differential mode, but there is a practical limit to how closely the receiver can lock its pseudo-random code onto that of the transmitter; this is shown in Figure 17–20. The figure shows an exaggerated form of *timing skew,* which is a difference between the timing of the transmitter and receiver. Good GPS receivers can keep the timing difference down in the 1 to 2% range. With a 1% timing skew, the timing error is 1% of the chip time (977.52 μs for C/A channel), or about 9.7752 ns. Since the speed of radio waves is close to 3×10^8 m/sec, a radio wave travels a distance of 2.93 meters in this time. In other words, each 1% of timing skew causes an error of 2.93 meters in the distance reading from *one* satellite. This error is compounded in actual operation, because the receiver must synchronize to the timing signals from several satellites.

A much more precise way of measuring timing is available for each satellite, and it's relatively easy to recover (but hard to measure). That timing reference is the *carrier frequency* modulated by the C/A and P codes. The L_1 carrier frequency of 1.57542 GHz is much faster than the C/A chipping rate of 1.023 MHz—there are 1540 carrier cycles for every C/A chip. If a receiver can lock onto a particular carrier frequency cycle, the error

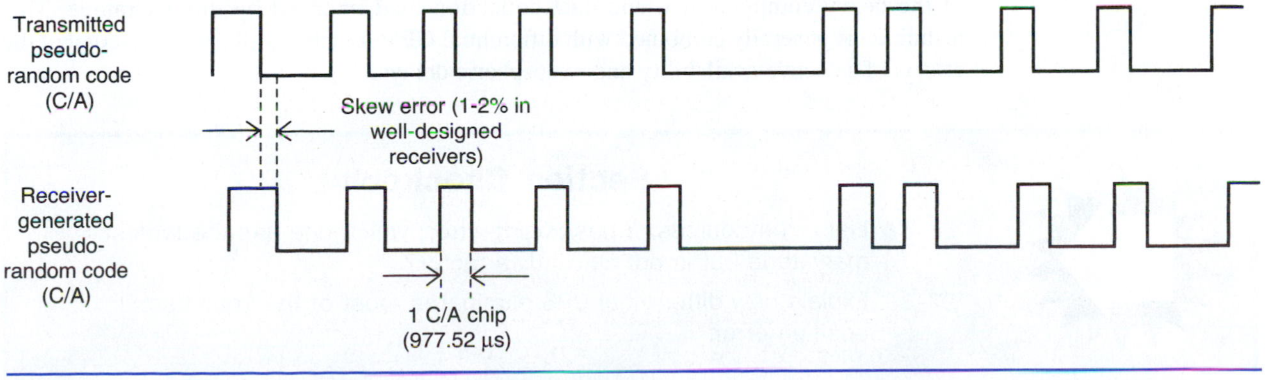

Figure 17–20 Timing skew error

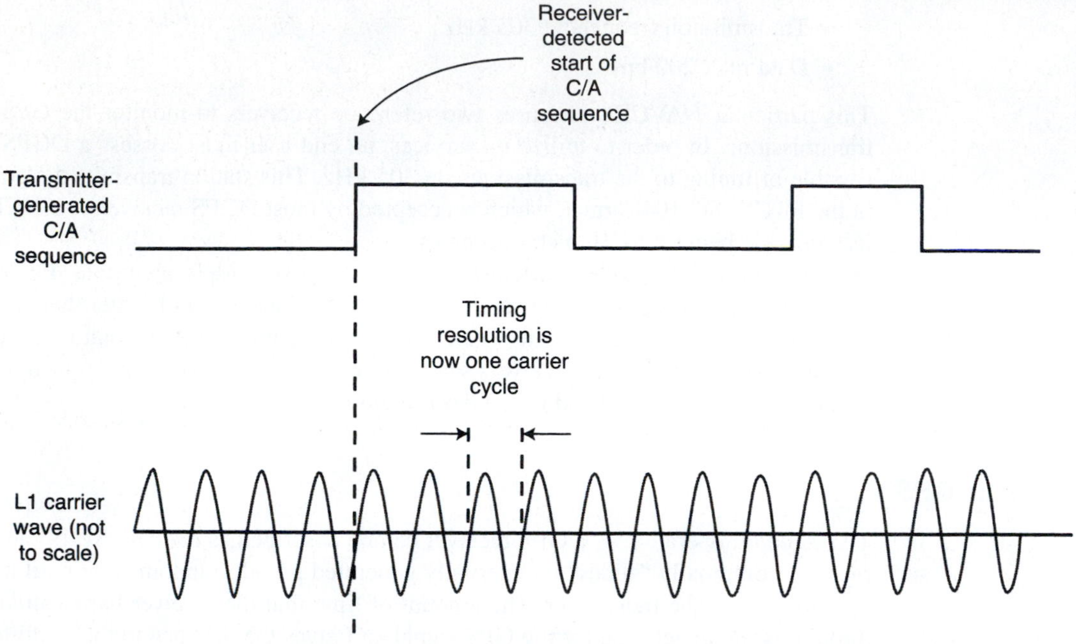

Figure 17–21 The C/A code can help establish carrier phase

can be reduced greatly. The receiver can then determine the range to the satellite by simply counting the number of carrier cycles that have elapsed during the radio signal's trip. The problem is that every carrier cycle looks just like any other; there's nothing to distinguish them. Figure 17–21 shows how the C/A code comes to the rescue.

The trick above is to use the C/A code sequence as a reference or starting point for searching for the right carrier cycle. Searching through 1540 carrier cycles for the "right" one (the time of one C/A chip) is much easier than looking at the entire signal. Thus, the C/A code can be thought of as a starting or reference point for a carrier phase-differential GPS receiver. The actual process used to establish carrier phase is quite a bit more involved, but this is the basic idea behind it. Some advanced receivers can even examine the phase of both the L_1 and L_2 carrier signals and make differential comparisons to reduce errors due to atmospheric factors.

Carrier-phase GPS receivers are now routinely used in applications that require high accuracy, such as surveying. These receivers are not fast in determining position because of the heavy computational and data collection load imposed by this technique. This technique is generally combined with differential GPS, which is still needed to combat the effects of selective availability and atmospheric delays.

Section Checkpoint

17–35 List three sources of positioning error; which one has the largest magnitude but is not currently a factor?

17–36 Explain how differential GPS eliminates most of the sources of position error.

17–37 What is special about the location of a reference receiver for DGPS?

17–38 What signal (in addition to the GPS satellite signals) must a DGPS receiver acquire in order to provide accurate position readings?

17–39 What is the purpose of the Coast Guard Maritime Differential GPS Service? What measurement accuracy is possible with this form of DGPS?

17–40 Why is conventional GPS sometimes called "code-phase GPS?"

17–41 What is timing skew? How does it affect the accuracy of conventional GPS?

17–42 Explain why the L_1 or L_2 carrier phase can provide higher measurement precision than the C/A code.

17–5 TROUBLESHOOTING GPS INSTALLATIONS

GPS receivers are found in a wide variety of settings. As the precision of GPS is improved, more and more applications will be found for this technology. In general, problems can be divided into three basic categories, *reception, interfacing,* and *software*.

Reception Problems

In general, the more satellites that a GPS receiver is able to acquire, the more accurate a position fix it will be able to provide. It stands to reason that anything that degrades the satellite signals before they can reach the receiver can lead to reception problems. Most GPS receivers give a visual indication of how strong the received signals are, and many give a warning message when they detect poor GPS coverage. However, not all GPS receivers are built this way. Some receivers have no display whatsoever—they can only be "read" by a computer using a serial communications interface (such as RS232). In that case, the attached computer and software should be able to track the status of the received satellite signals.

Most problems with reception of the satellite signals can be traced to either an improperly located antenna or a damaged transmission line. At the high frequencies used in GPS, a seemingly insignificant kink or bend in the coax can be enough to seriously degrade coverage, since the damaged point in the coax presents an impedance mismatch and reflects RF energy back to the GPS antenna (instead of letting it pass on to the receiver). GPS antennas should be located in a clear location where they have a clear view of the horizon whenever possible. There must be no objects of any kind placed near the antenna, for they will either severely weaken the incoming signals or cause them to be reflected, resulting in multipath distortion.

Reception difficulties can also be caused by a defective GPS receiver. A technician can test a GPS receiver by substituting a known-good unit in its place. Remember that a brand-new receiver may need 12 minutes or more to become fully functional at first power-on, since it will need to acquire fresh almanac and ephemeris data for all satellites in view and set its internal time clock for the first time. Attempting to troubleshoot a "broken" receiver at the component level is a waste of time for the technician; leave such matters for the receiver manufacturer, who has all the specialized test equipment to ensure that the unit is *really* working correctly after performing repairs.

A differential GPS (DGPS) receiver requires *two* properly functioning antenna systems, one for the L-band GPS satellite signals and another (usually in the 300 kHz band) for reception of the correction signal. When a DGPS receiver fails to decode the correction signal, its accuracy degrades to that of normal GPS. DGPS correction signals sent in the LF band have propagation characteristics similar to that of AM broadcast: They are vertically polarized and cover an area of about a 100 mile radius. The LF DGPS (ground) signals are much more susceptible to noise than the UHF satellite signals. If they cannot be received correctly, the problem may likely be a local noise source, such as a vehicle ignition system. A monitor receiver tuned to the same frequency as the LF DGPS receiver can often be helpful in determining whether poor signal strength or locally generated noise is causing the problem with reception.

Interfacing Problems

The *interface* is the set of connections that allow two (or more) pieces of equipment to communicate. Many GPS receivers use an RS232 interface to communicate with other equipment (such as attached personal computers). The attached equipment must be adjusted for the correction communication parameters and must be compatible with the data output format of the receiver. For example, if the GPS receiver is communicating at 4800 bps, then the attached device must also be set at 4800 bps.

Some DPGS setups use a separate receiver for acquiring the correction signal, which then communicates that information on a regular basis to the main GPS receiver. Failure of the DGPS feature to function at all can be caused by a problem in the interface between these two units or an incorrect setting in one or both units.

Software

Software has a center-stage role in GPS systems. Most of the "works" within a GPS receiver are actually contained within a set of firmware (software in ROM, read-only-memory, chips) that directly controls a digital signal processor (DSP) chip, which performs all the mathematical signal manipulations and calculations in real time (as events happen). In addition, GPS receivers are often connected with computer systems of various types for data logging, map tracking, and other purposes.

The software used must be set up in a manner consistent with the GPS receiver (or receivers) being employed. In embedded GPS applications (such as vehicle locater systems), the end user has no access to the settings and can't "break" anything by fiddling with controls. Figure 17–22 shows a typical GPS receiver built for embedded applications. At the right of the board, the antenna connector is visible. Power and data signals are exchanged through the left-hand connector.

Where personal computers are being used, the user often has an inclination to explore, experiment, and "try things to see what happens." In addition, some PC operating systems are notorious for suddenly "forgetting" important settings for applications, which can cause a previously working system to "break" the next time it's turned on. *The only good defense for these sort of problems is to keep backups of all data* (disk images, for example) and document the correct settings on initial installation. In the event of a suspected software problem, the technician can quickly verify the software settings against those initially recorded.

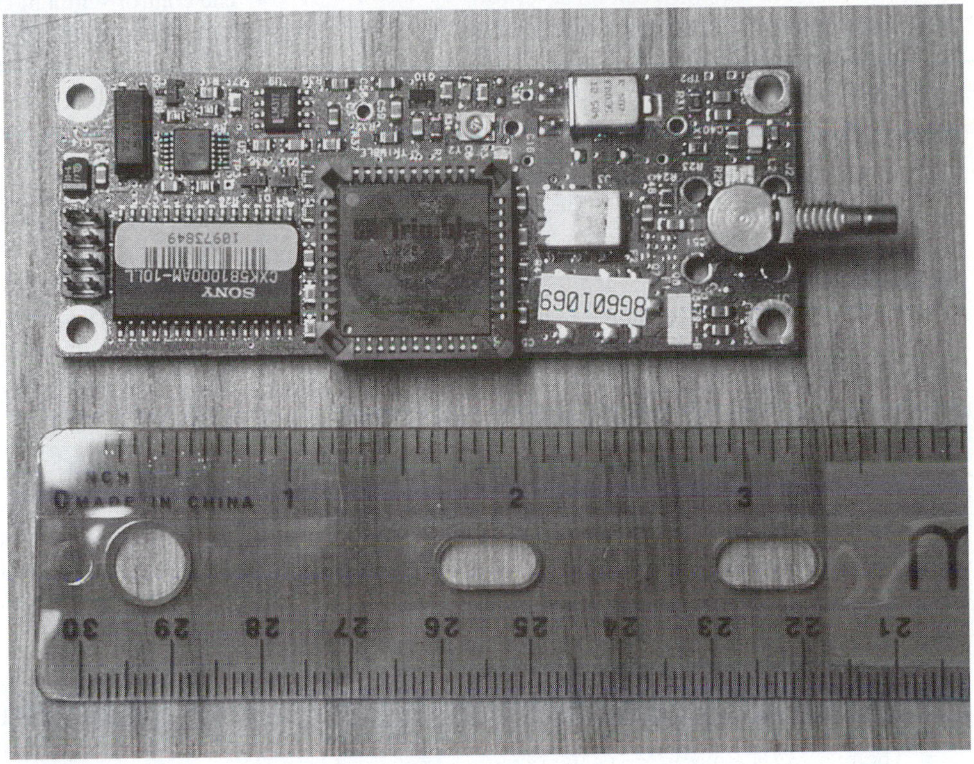

Figure 17–22 A GPS "board" receiver

Section Checkpoint

17–43 What are the three most likely problems to arise in a GPS installation?

17–44 List at least two possible causes for poor GPS reception.

17–45 A customer is using a GPS receiver in a truck, and the coax is being fed through a side window, which has been closed with the coax in place. What problem might arise?

17–46 What is the best way to troubleshoot a suspected GPS receiver?

17–47 Explain how to determine if local noise is interfering with an LF DGPS signal.

17–48 What is an interface? Give an example of an interface problem.

17–49 What is the best defense against software problems where personal computers are involved?

SUMMARY

- The problem of navigation goes back to the beginning of recorded history. GPS is the first worldwide system to provide a universal solution.
- GPS uses two code channels, C/A for civilian application (and for acquiring the P code), and the P code for military navigation.

- The orbits of satellites can be described using Kepler's laws of planetary motion. The mathematical solution of these laws allows GPS receivers to precisely locate GPS satellites in the sky.
- GPS satellites are placed in six different inclined orbits, at 55° with respect to the equator. They are spaced to provide optimum worldwide visibility.
- GPS works by measuring the time-of-arrival of timing signals from the GPS satellites. In order for this to work, the receiver must have very accurate time and an accurate value for the location of at least three satellites.
- By algebraically analyzing the signals from four satellites, the receiver can accurately set its internal time clock.
- A GPS receiver presents the user with three pieces of information: the UTC time, the position (longitude and latitude), and the speed or velocity of the receiver.
- The receiver's almanac gives the planned orbital configuration of all the satellites. The receiver gets precise, up-to-the-minute orbital information for a satellite by downloading its ephemeris data, which is part of the navigation message.
- The master control station (MCS) reviews the orbital tracking of all the GPS satellites and periodically updates their ephemeris data, which preserves high positioning accuracy in the event of orbital drift.
- Selective availability, or SA, is a system whereby deliberate errors are introduced into the timing and ephemeris messages of the satellites in order to degrade the accuracy of GPS receivers. This is a measure to help prevent GPS from being used against the United States or its allies.
- The GPS system consists of the space, control, and user segments.
- The space segment consists of the array of satellites. Each satellite has a unique identifying code (the Service Vehicle Number, or SVN). The satellites are responsible for transmitting the navigation messages.
- The control segment consists of the master control station (MCS) at Falcon Air Force Base and a number of monitoring stations around the globe. The MCS periodically updates the timing and ephemeris data for all the satellites, based on information from the worldwide ground tracking network.
- The user segment consists of all GPS receivers in use on and near the earth.
- The navigation message transmitted by each satellite is sent at 50 bps and consists of 25 frames, which take a total of 12.5 minutes to send.
- Every 30 seconds, one of 25 navigation frames is sent. Each navigation frame contains the clock correction factor and satellite ephemeris data, so that receivers quickly get the most up-to-the-minute information about each satellite.
- Two L-band frequencies are used to transmit navigation messages. These are L_1 at 1575.42 MHz and L_2 at 1227.60 MHz.
- All GPS frequencies are based on an accurate 10.23 MHz timebase, which is multiplied to get the L_1 and L_2 carriers and divided by 10 to produce the C/A clock.
- The navigation message is carried atop two pseudo-random codes, the C/A and P codes. These codes allow all satellites to share the same frequencies as well as provide effective signal gain at the receiver.
- Coherent BPSK is used to modulate both L_1 and L_2 carriers. The L_1 carrier consists of two quadrature-phased sine waves, which carry the C/A and P codes separately.
- Accurate timing is absolutely essential for accurate GPS operation. Atmospheric effects, multipath distortion, and the deliberate timing degradations introduced by SA all reduce the basic accuracy of GPS.
- Differential GPS addresses timing errors by using a reference receiver, which is a GPS receiver whose precise geographic location is known. The reference receiver transmits its correction message to all nearby GPS users.

- Carrier-phase GPS receivers count individual L_1 carrier cycles instead of C/A code chips. This technique greatly increases the precision of measurement and is generally combined with DGPS.
- The three most common problems with GPS installations involve signal reception, interfacing, and software configurations.
- GPS antennas must be located in an open area, away from nearby objects. Transmission lines must be treated with care to avoid signal degradation.
- The best defense against software problems (especially in PC environments) is to make complete backups of working configurations.

PROBLEMS

1. Briefly describe the history of navigation. What is the goal of GPS?
2. List the two code channels used in GPS and describe the use of each one.
3. List Kepler's three laws of planetary motion and explain how each one relates to GPS satellites.
4. What are the three types of satellite orbits? What type is used for GPS satellites, and why?
5. The locations in the equatorial orbit are often referred to as "prime real estate." Why is this so?
6. What information must a GPS receiver have in order to calculate its position in space?
7. Using a simple drawing, show why four satellite measurements are needed for an accurate position fix in three dimensions.
8. List the three pieces of information available from a GPS receiver.
9. If the timing of a GPS receiver is "off" by 20 μs, what approximate error in distance measurement will result (in meters)?
10. What are *pseudo-ranges?*
11. Explain how a GPS receiver acquires accurate time.
12. What information is contained in the *almanac* of a GPS receiver?
13. What is the *ephemeris data?* Explain the difference between the ephemeris and almanac data.
14. What is selective availability (SA)? List two ways that SA can introduce errors.
15. List the three segments of the GPS system and describe the purpose of each one.
16. Where is the *master control station* (MCS) for the GPS system located? What is its purpose?
17. Draw a diagram showing the contents of one frame of the navigation message. Why are the first three subframes retransmitted with each frame?
18. What is the purpose of the handover word (HOW)? Explain why it is needed.
19. What is contained in the two "document pages" assembled by successive reception of the contents of subframes 4 and 5 of the navigation message?
20. Give the L_1 and L_2 frequencies. Explain how they are derived from the 10.23 MHz atomic clock.
21. What is a *pseudo-random sequence?* List three capabilities that these codes provide.
22. What type of circuit is used to generate the C/A and P codes? Draw a simple example of such a circuit.
23. What is a *chip?* What are the chipping rates of the C/A and P codes?
24. What is the data rate of the navigation message stream? How does it compare with the chipping rates of the C/A and P codes?
25. What codes are transmitted on (a) the L_1 carrier and (b) the L_2 carrier? Explain the modulation procedure used for the L_1 carrier.
26. Draw a simple diagram showing how a GPS receiver recovers the navigation message stream.

27. What is the largest contributor to GPS positioning error?

28. Explain the operation of differential GPS. How many receivers are required?

29. How does carrier-phase GPS improve the resolution of measurement?

30. A certain GPS receiver is powered up and passes its self-diagnostic tests but cannot lock onto any satellites. What is the most likely cause of the trouble?

31. A DGPS receiver is being used by a survey team. The unit employs a separate DGPS correction receiver, which is tuned to 305 KHz. The correction receiver indicates that it is receiving valid DGPS corrections. The on-screen display of the receiver indicates that it has locked onto seven strong satellite signals, yet the accuracy indication is only 100 meters. What are *two* possible problems?

32. John is making a cross-country trip by car, taking along his laptop computer (with a mapping program) and a GPS receiver interfaced with the PC. The setup worked perfectly when John left home, but several days into the trip, the map program decided the GPS receiver no longer existed! The GPS receiver works fine by itself and when last checked, locked onto eight satellites. What are two possible problems John should check for?

18

Fiber-Optic and Laser Technology

OBJECTIVES

At the conclusion of this chapter, the reader will be able to:

- describe the construction of fiber-optic cable
- describe the propagation modes in fiber
- utilize fiber specifications (such as numerical aperture) to evaluate fiber performance characteristics
- predict the dispersion and effective bandwidth of a multimode fiber
- describe the operation of LED and laser light sources and their associated drive circuitry
- describe the operation of several light wave demodulation devices
- use electrical and optical test equipment to troubleshoot fiber-optic systems

Transmission lines are used to carry signal energy or information between two points. The traditional forms of these lines (such as twisted wire pairs or coaxial cables) are made of copper wire, and for carrying high-power radio frequency signals over a short distance, they are excellent performers. However, when we want to carry just information (such as a group of telephone conversations) over a long distance without using radio waves (sending it over the air), conventional transmission lines are not the best choice.

Fiber optics are transmission lines designed to convey *light* signals instead of radio signals. In a fiber-optic system, a light wave acts as a carrier signal, and is modulated with information before it's sent down the cable. On the other end, the light wave is "demodulated" and the original information is recovered.

There are several advantages to doing it this way. First, modern fiber-optic cables have much lower signal loss than copper transmission lines, which means that a signal can be carried much farther on an optical cable before it needs amplification. Second, because the frequency of a light wave is very high, much more bandwidth is available on a fiber-optic cable than with a copper line. This means that fiber optics can carry much more information than copper. Also, fiber-optic signals are very secure (it's hard to tamper with a fiber-optic line), and since the signals in the cable are light instead of electricity, they are not affected by external sources of noise. Finally, fiber-optic cables are much lighter than conventional transmission lines and are completely safe for hazardous environments where an explosive atmosphere or nuclear radiation might be present.

There are disadvantages to fiber optics. First, they can be more difficult to work with than copper transmission lines. Special tools are needed to put connectors on fiber-optic cable, and splicing can be a challenge. Second, it is nearly impossible to measure the performance of a fiber-optic connection without specialized optical test equipment. For large companies, these problems are minor when compared with the benefits fiber provides.

18–1 FIBER-OPTIC CONSTRUCTION AND OPERATION

Parts of a Fiber-Optic Cable

The construction of a fiber-optic cable is quite straightforward, as shown in Figure 18–1. A fiber-optic line consists of an *outer jacket,* an inner *cladding,* and a central *core.*

The *outer jacket* provides most of the mechanical strength of the cable. Its purpose is to protect the inner works from contamination by moisture and dirt. The outer jacket is usually made of a tough but flexible plastic. When greater strength is needed, strands of reinforcing fiber (such as kevlar) may also be included inside the outer jacket.

The *cladding* surrounds the inner *core,* and these two components function together to carry light waves down the fiber. The purpose of the core is to carry the light; it can be made of either transparent plastic or glass. The highest-performance fibers use a glass core, which provides much lower loss and wider bandwidth than plastic. The function of the cladding is to help the core contain the light waves. Without the cladding, some of the light would escape the core on the way down the cable, resulting in signal loss. The combined optical characteristics of the core and cladding work together to keep light inside the fiber, as we'll soon see.

Figure 18–2 shows the construction details of a modern single-mode fiber-optic cable.

Characteristics of Light Waves

Light waves are electromagnetic energy, just like radio waves. The primary difference between light and radio waves is *frequency.* Light has a much shorter wavelength than the highest radio frequency and therefore a much higher frequency. The light frequencies used in fiber optics are generally in the *infrared* region, which is invisible to the eye. Infrared light is used because the fiber-optic core materials generally have the lowest signal loss at these frequencies. Figure 18–3 shows the relationship between radio waves and light waves.

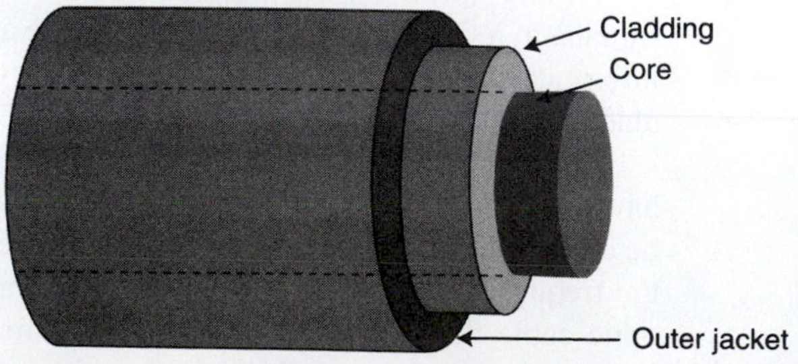

Figure 18–1 Fiber-optic cable construction

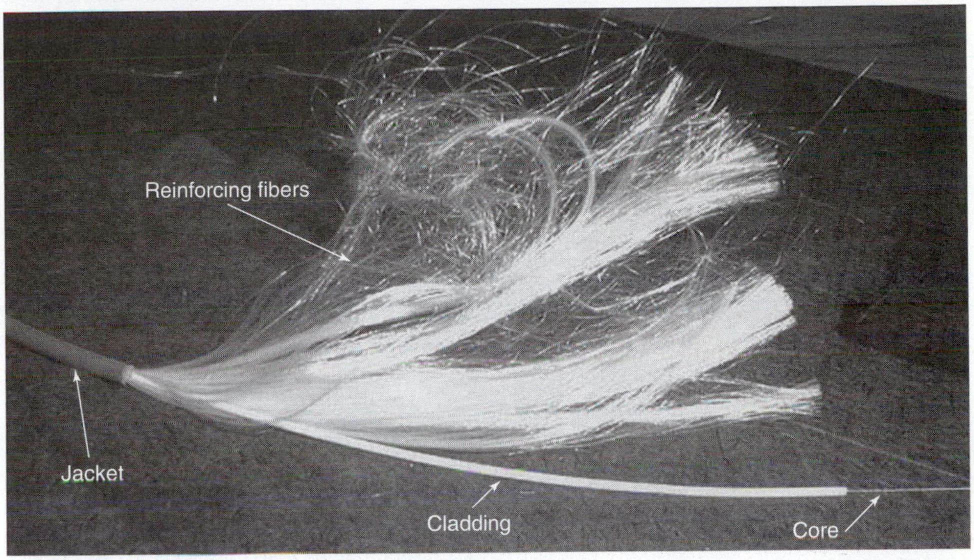

Figure 18–2 Single-mode fiber-optic cable construction

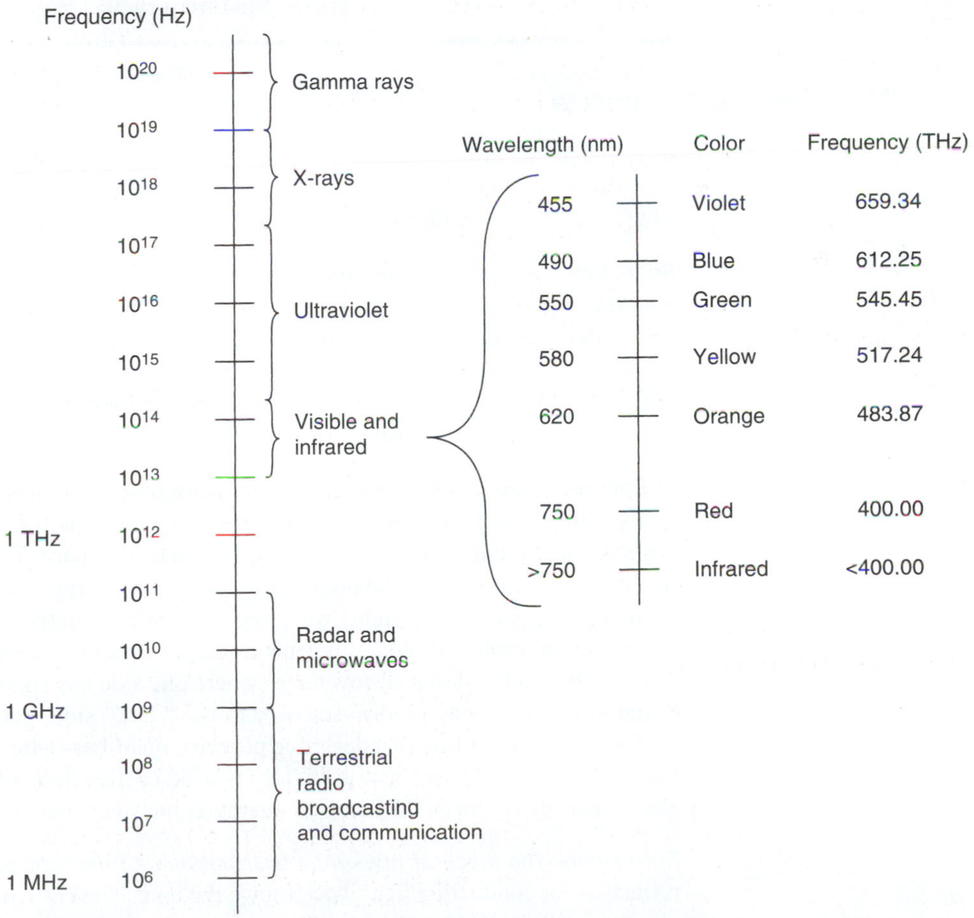

Figure 18–3 Radio and light waves

Figure 18–3 shows that light energy is *much* higher in frequency than radio waves. The lowest light frequency (in the infrared region, at 10 THz) is more than 100 times the highest useful microwave radio frequency (100 GHz). The wavelength of a light wave is calculated in exactly the same way as for a radio wave:

$$\lambda = \frac{v}{f} \tag{18–1}$$

where λ is the wavelength in meters, v is the speed of light (3×10^8 m/sec), and f is the frequency of the wave.

EXAMPLE 18–1 What is the frequency of infrared light with a wavelength of 800 nm?

Solution

Equation 18–1 (the same equation we've used throughout this book) is simply solved for frequency, yielding:

$$f = \frac{v}{\lambda} = \frac{3 \times 10^8 \text{ m/sec}}{800 \times 10^{-9} \text{ m}} = \underline{\underline{3.75 \times 10^{14} \text{ Hz}}} = \underline{\underline{375 \text{ THz}}}$$

This frequency agrees with the data of Figure 18–3. Notice that the visible light spectrum is but a narrow sliver of the available light frequencies!

How Light Waves Propagate

Light tends to move in a straight line unless something interferes with it, changing the direction of motion. There are three ways that a light wave can be made to alter its course. These are *reflection, diffraction,* and *refraction,* as shown in Figure 18–4.

Reflection Reflection is the most intuitive of the propagation modes. When an electromagnetic wave (light or a radio signal) strikes a conductive surface, it bounces off the surface. Good conductors of electricity make good reflectors; poor conductors make poor reflectors. *To make a good optical reflector requires a polished conductive surface*. You'll notice that there is no metal in the optical fiber of Figure 18–1. The reason for this is that fiber optics do not work using ordinary reflection.

Diffraction You've observed *diffraction* many times. Suppose there is one room in your house with the light on, and the door to that room is partially open. The door causes a shadow, but the edges of the shadow are not perfectly sharp. This is caused by the diffraction of light at the edge of the door. It is explained by Huygen's principle, which states that each point on a (radio or light) wavefront can be considered to be a tiny point source of energy, or isotropic radiator. Huygen's principle says that electromagnetic energy doesn't "like" to be confined to a narrow edge (where one side has energy, and the other does not). Some of the wave energy diffuses over to the "dark" side, partially blurring the shadow. Diffraction is not normally a desirable property in a fiber-optic system. Light wave signals that undergo diffraction become slightly blurred or distorted, which limits bandwidth (and also drastically reduces the received energy at the fiber-optic receiver).

Refraction—The mode of fiber-optic transmission Fiber optics rely on the mechanism of refraction for conducting light waves down the core. The basic need in all fiber-optic cables is to keep all the light within the fiber core so that maximum signal will be transferred to the other end. We could accomplish this by surrounding the fiber-optic core with a tube of

Figure 18–4
Reflection, refraction, and diffraction

(a) Reflection

$\theta_r = \theta_i$
(Angle of reflection = Angle of incidence)

θ_i θ_r

Ray entering at angle $\geq \theta_{critical}$ is reflected

Conductive surface

(b) Refraction

$\theta_{critical}$

θ_1

Medium 1
Index of refraction = n_1

Medium 2
Index of refraction = n_2

θ_2

Snell's law gives refraction angles:

$n_1 \sin \theta_1 = n_2 \sin \theta_2$

$\theta_{critical} = \sin^{-1}(n_2/n_1)$

(c) Diffraction

Incident ray

Opaque material

Diffracted rays

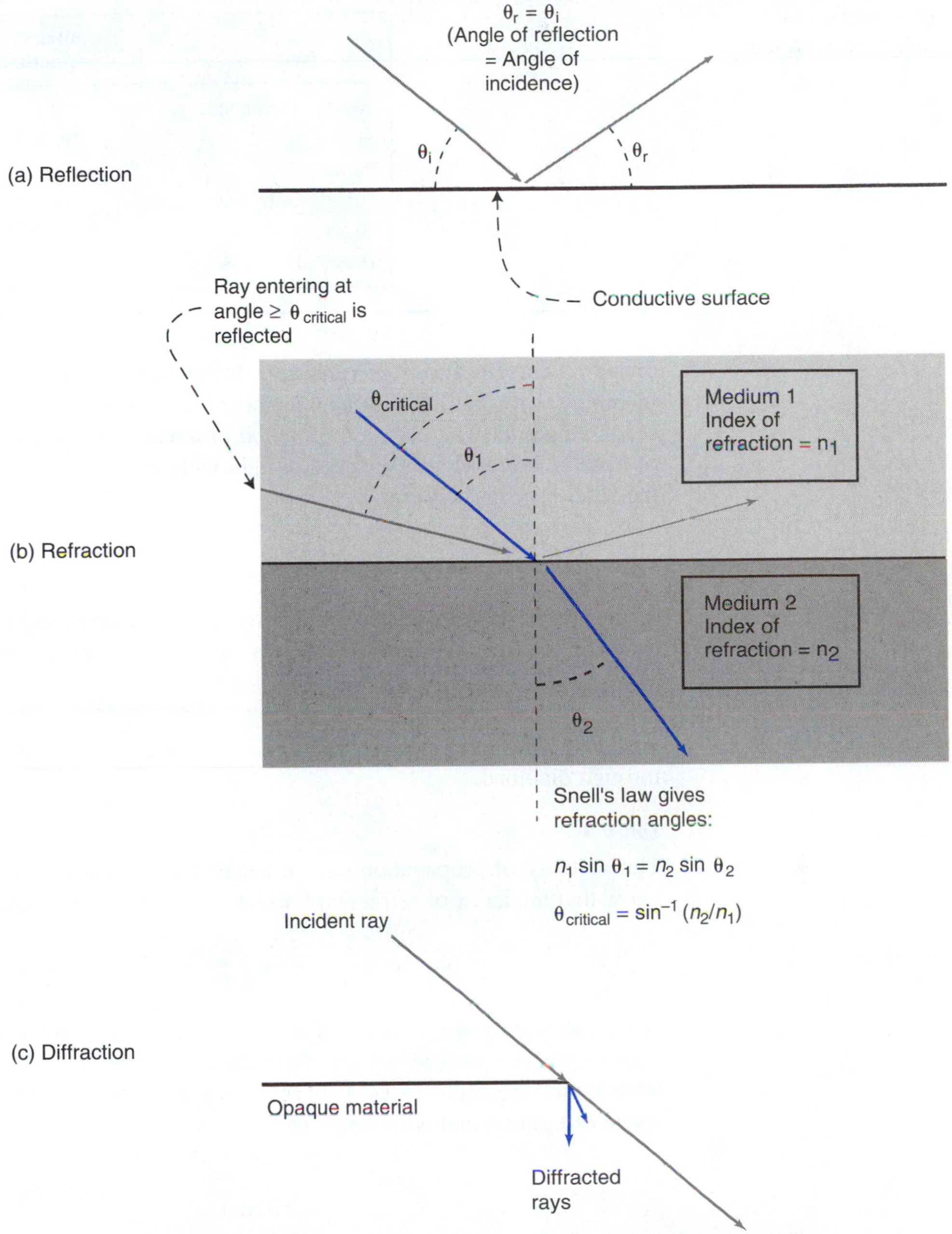

shiny metal (such as polished foil). However, in doing this, we are now relying upon *reflection* of the light to keep it within the core. That is not a very efficient propagation mode, because good reflection requires a metal conductor "mirror" with very low electrical resistance. Our fiber would now be experiencing losses similar to the I^2R losses in a copper transmission line!

However, by using refraction, we can eliminate the need for a metal reflector and still keep most of the signal within the fiber. *Refraction is the bending of a light wave.* When a wave travels from one material into another (such as from air into water), it is *bent*

Figure 18–5 Indices of refraction for selected materials

Material	Index of refraction
Vacuum (free space)	1.0
Air	1.0003
Water	1.33
Fused quartz, S_iO_2	1.46
Glass	1.5
Diamond	2.5

or *refracted*. The index of refraction for each material and the original angle of the incoming wave (ray) determine the amount of bending and the new angle of travel in the second material. The *index of refraction* of a material is defined as the ratio of the speed of light in free space to the speed of light within the material. It is usually symbolized by the letter n in an equation:

$$n = \frac{c}{v} \tag{18–2}$$

where v is the velocity (speed) of light within the material, and c is the speed of light in free space (3×10^8 m/s). Figure 18–5 gives the indices of refraction for various materials.

EXAMPLE 18–2

Calculate the velocity of propagation and wavelength of a ray of green light as it passes through diamond.

Solution

The velocity of propagation can be calculated by rearranging equation 18–2, since we know that the index of refraction for diamond is 2.5 (from Figure 18–5):

$$v = \frac{c}{n} = \frac{3 \times 10^8 \text{ m/s}}{2.5} = \underline{\underline{1.2 \times 10^8 \text{ m/s}}}$$

After calculating the velocity of the waves, the wavelength can be calculated. To find wavelength, we must also know the frequency of the light. The frequency can be calculated by solving equation 18–1 for *frequency* and substituting the free-space values for the speed of light (c) and wavelength (λ):

$$f = \frac{c}{\lambda} = \frac{3 \times 10^8 \text{ m/s}}{550 \text{ nm}} = 5.4545 \times 10^{14} \text{ Hz}$$

The wavelength of the light within the material can now be calculated:

$$\lambda = \frac{v}{f} = \frac{1.2 \times 10^8 \text{ m/s}}{5.4545 \times 10^{14} \text{ Hz}} = \underline{\underline{220 \text{ nm}}}$$

Note that the wavelength got *smaller* as the wave passed through the diamond, but the frequency remained constant. In fact, the two preceding equations can be combined into one simpler relationship:

$$\lambda = \frac{\lambda_{fs}}{n} \tag{18–3}$$

This equation tells us that we can find the wavelength of a signal within a material with a refractive index of n by dividing its free-space wavelength (λ_{fs}) by the refractive index of the material. Applying equation 18–3 gives:

$$\lambda = \frac{\lambda_{fs}}{n} = \frac{550\text{ nm}}{2.5} = \underline{\underline{220\text{ nm}}}$$

That is certainly much simpler, but what does this calculation mean? It simply reminds us that the wavelength of a signal depends on both its frequency and the material it happens to be passing through. Electromagnetic waves (both radio and light waves) always slow down when they enter any material except a perfect vacuum.

Light Wave Propagation in Fiber

Figure 18–6 shows that *two* optical materials are used in a fiber-optic cable, the *core* and the *cladding*. These two materials have slightly different indices of refraction. The cladding always has a slightly *lower* index of refraction than the core. Because of this, when light enters the fiber core at an appropriate angle, it is refracted back towards the center of the core when it tries to escape through the cladding. The light wave experiences

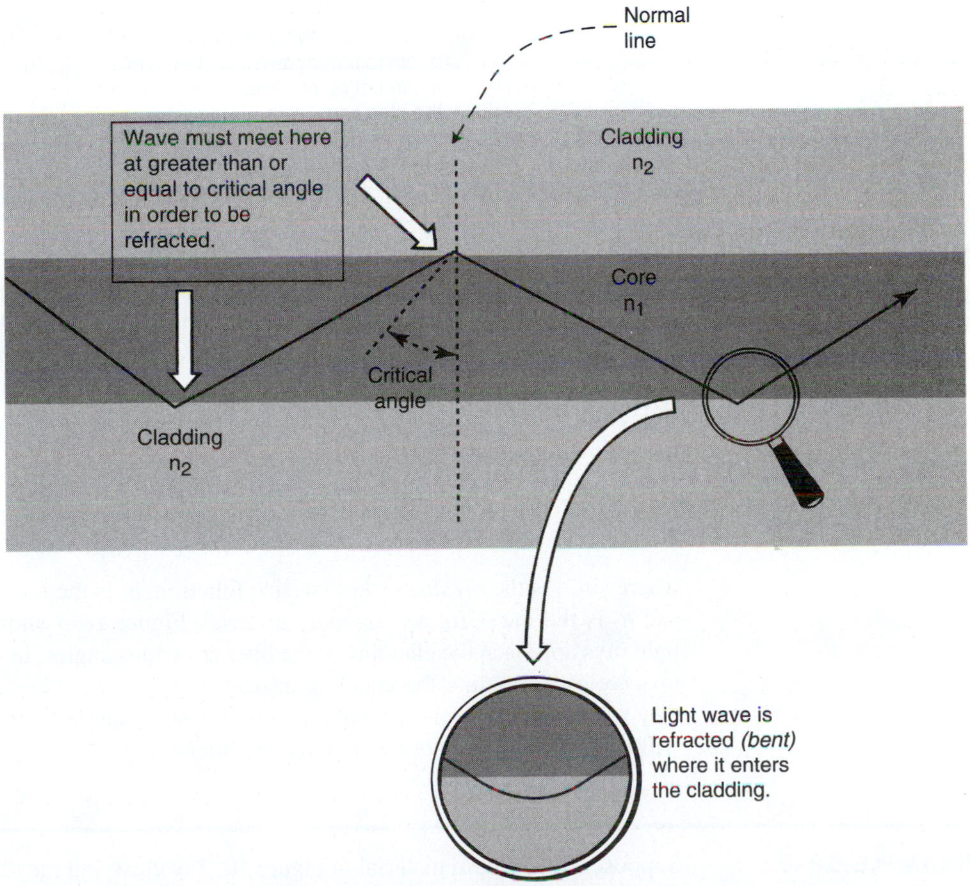

Figure 18–6 Fiber-optic core and cladding

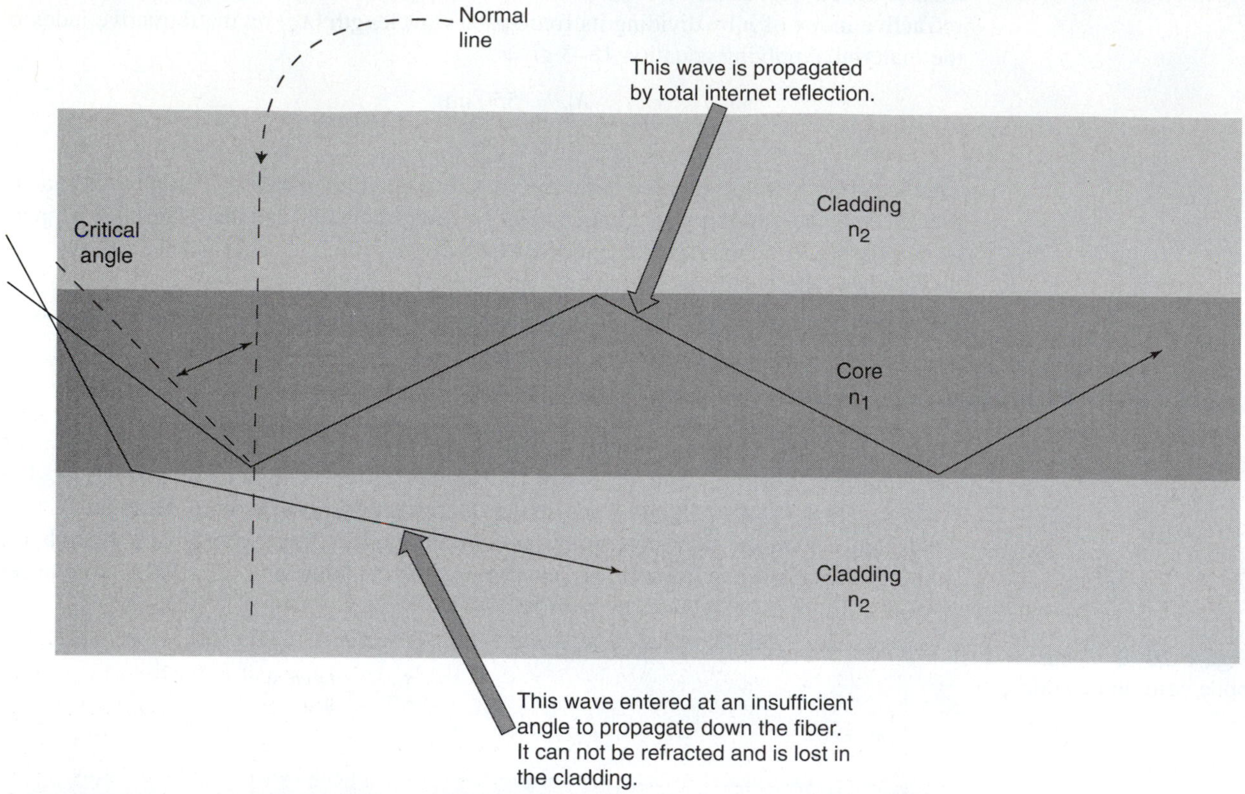
Normal line
This wave is propagated by total internet reflection.
Cladding
n2
Critical angle
Core
n1
Cladding
n2
This wave entered at an insufficient angle to propagate down the fiber. It can not be refracted and is lost in the cladding.

Figure 18–7 Not all waves can be refracted

apparent total internal reflection. We say *apparent,* because the wave is *bent* or *refracted,* not *reflected,* at the junction of the core and cladding.

For the light rays to stay within the fiber of Figure 18–6, they must approach the wall of the cladding at an angle *at least* equal to the *critical angle* for the core-cladding junction. This angle can be calculated using Snell's law. By substituting a value of 90° for Θ_2, we get the following relationship:

$$\Theta_{crit} = \sin^{-1}\frac{n_2}{n_1} \qquad \textbf{(18–4)}$$

where $\sin^{-1}$ is the *arcsine* or inverse sine function, n_1 is the index of refraction for the core, and n_2 is the index for the cladding material. Figure 18–7 shows what happens when the light rays approach the cladding of the fiber at various angles. In the figure, some of the light rays are escaping into the cladding instead of propagating down the core. This is because they are approaching the cladding at too narrow an angle to be refracted. These light rays will be lost; they will not move down the fiber.

EXAMPLE 18–3 Suppose that the core material in Figure 18–7 is glass and the cladding material is a plastic with a refractive index of 1.35. What is the critical angle within the fiber?

Solution

The refractive index of glass is 1.5, as given in Figure 18–5. Equation 18–4 calculates the critical angle:

$$\Theta_{crit} = \sin^{-1}\frac{n_2}{n_1} = \sin^{-1}\frac{1.35}{1.5} = \underline{\underline{64.2^\circ}}$$

Be careful when you perform this calculation. Your calculator must be set to deliver answers in *degrees.* The light rays in our fiber cable must be angled at least 64.2° from the normal line, or they simply won't propagate.

Numerical Aperture and Acceptance Cone

By knowing the indices of refraction of the core and cladding materials, we can calculate two important properties of a fiber-optic cable. These are the *numerical aperture* and *acceptance cone angle.*

The numerical aperture is a dimensionless number that tells us how well a fiber can collect light at its end (where a light source would be coupled). It is calculated as follows:

$$NA = \sqrt{n_1^2 - n_2^2} \qquad \textbf{(18–5)}$$

where n_1 is the index of refraction for the fiber core, and n_2 is the index for the cladding material. The greater the value for numerical aperture, the easier it is for light to get into the fiber. The maximum possible value for NA approaches a value of 1. It might sound like the highest possible NA would be desirable, but this is not always the case. A high value of numerical aperture allows many *propagation modes* to coexist within the fiber. This leads to pulse spreading or *dispersion,* which reduces the available bandwidth. We will study dispersion shortly.

Figure 18–8 shows the *acceptance cone* of a fiber. The acceptance cone is the total angular range of light rays that the fiber can accept and propagate. The calculated *acceptance cone angle* usually gives half of this range. The acceptance cone angle can be directly calculated from the numerical aperture:

$$\Theta_{accept} = \sin^{-1} NA \qquad \textbf{(18–6)}$$

Figure 18–8
Acceptance cone

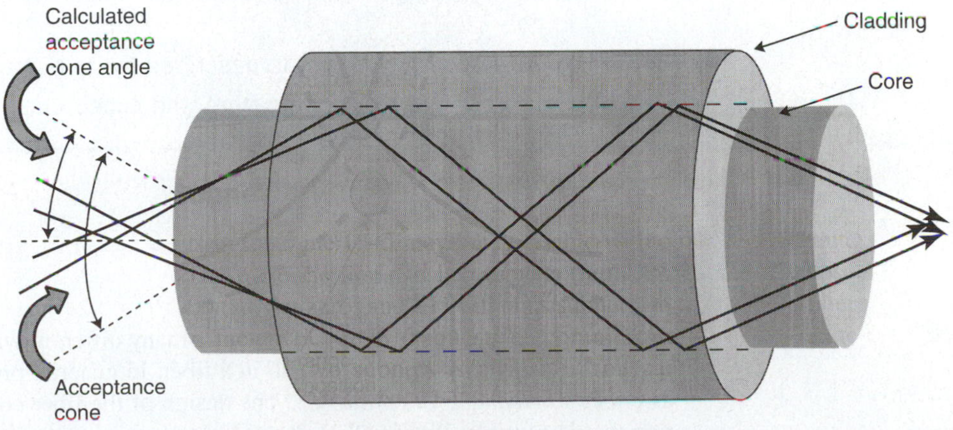

where Θ_{accept} is the acceptance cone angle (half of the range of the actual acceptance cone), and NA is the numerical aperture of the fiber.

EXAMPLE 18–4

Calculate the numerical aperture (NA) and acceptance cone angle for a fiber with a glass core ($n_1 = 1.5$) and a plastic cladding ($n_2 = 1.35$).

Solution

The numerical aperture is calculated according to equation 18–5:

$$NA = \sqrt{n_1^2 - n_2^2} = \sqrt{1.5^2 - 1.35^2} = \underline{\underline{0.654}}$$

The acceptance cone can be directly calculated from this result using equation 18–6:

$$\Theta_{accept} = \sin^{-1} NA = \sin^{-1} 0.654 = \underline{\underline{40.83°}}$$

The total angular range over which the fiber can accept incoming light rays (at least when they're coupled from air to the fiber core) is *twice* the acceptance cone angle, or about 81.66°.

Section Checkpoint

18–1 List several advantages of fiber optics when compared to copper.

18–2 List the three parts of a fiber-optic cable, explaining the purpose of each one.

18–3 What is the difference between light and radio wave energy?

18–4 Why do most fiber-optic systems use infrared light?

18–5 What are the three ways that a light wave can change direction?

18–6 What propagation mechanism keeps light waves inside a fiber-optic core?

18–7 Explain the meaning of the *index of refraction* for an optical material.

18–8 Describe what happens to the wavelength of light as it passes from free space into a material such as glass.

18–9 What is the significance of the *critical angle* for light wave propagation in fiber?

18–10 What property of a fiber is described by its numerical aperture (NA)?

18–11 How are the numerical aperture and acceptance cone angle related?

18–2 PROPAGATION MODES AND DISPERSION

Light waves can move down optical fibers in many different ways. A *propagation mode* is one particular type of light wave path in a fiber. In general, propagation can be classified as either *single mode* or *multimode*. The design of the fiber controls the type of propagation that is possible. Figure 18–9 shows light wave movement in a single mode fiber.

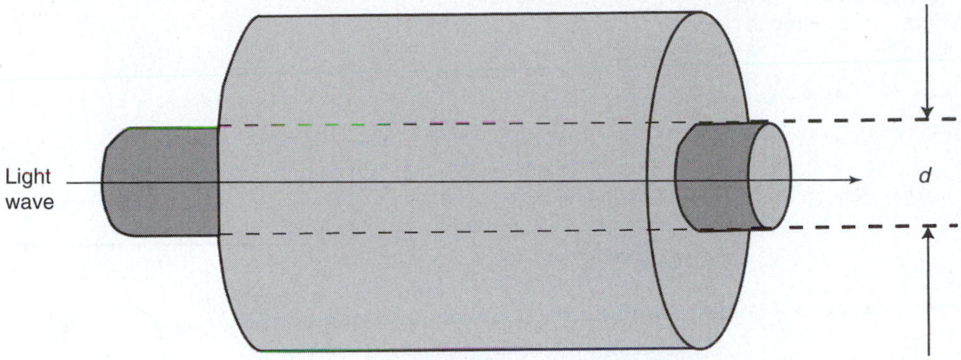

Figure 18–9 Light waves in a single mode fiber

To support single-mode propagation, the fiber diameter *d* must be less than approximately 2 wavelengths of the light frequency to be propagated: $d < 2\lambda$

Single-mode propagation is the simplest possible: The light wave travels straight through (or very nearly so) the fiber from end to end. Because all the light rays arrive in step with each other in a single-mode fiber, very little distortion of the signal occurs. However, single-mode fiber must have a *very* small core diameter, usually less than 10 μm. This small diameter is what restricts operation to a single mode; higher modes just won't fit! Because of the tiny core dimensions, it is more difficult to couple light efficiently into a single-mode fiber. The fiber of Figure 18–2 is a single-mode unit.

Multimode Propagation

When the width of the fiber core is more than approximately two wavelengths of light, multiple propagation modes become possible. These multiple propagation modes complicate the propagation picture, as shown in Figure 18–10. The problem with multimode propagation is that each of the light rays travels a slightly different distance as it moves

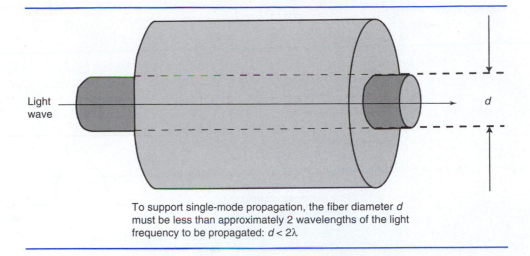

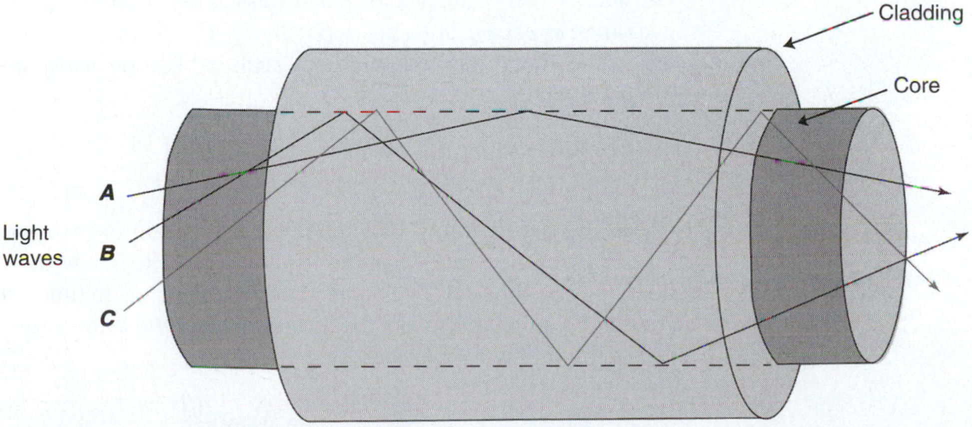

Figure 18–10 Multimode propagation

Figure 18–11 Pulse effects in multimode fibers

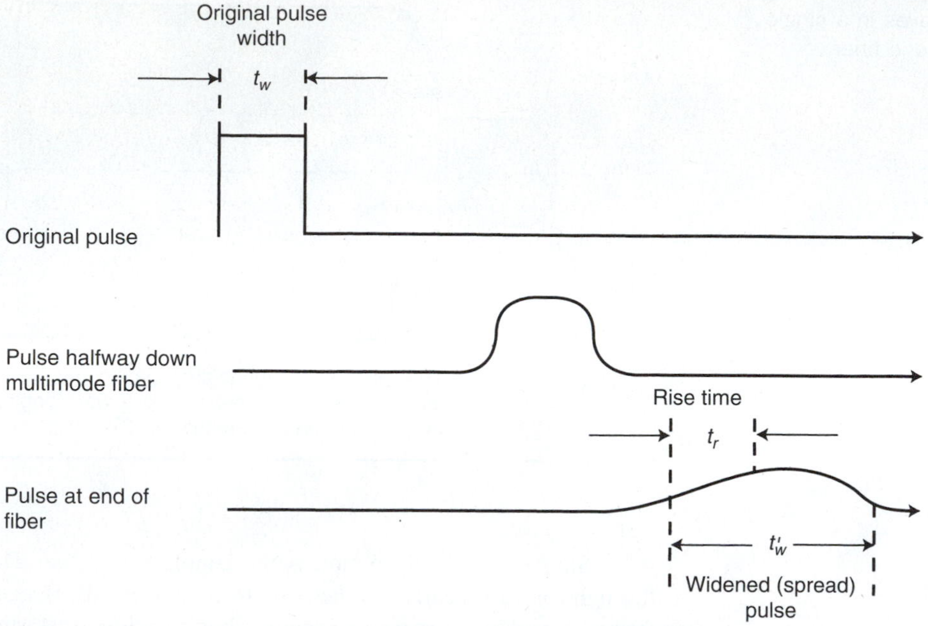

through the fiber. In Figure 18–10, light ray A has the shortest path and arrives first. Rays *B* and *C* are *higher-order modes;* they enter the fiber at steeper angles and take longer to arrive, since they travel a much longer effective distance. All of the light waves do not arrive at the end at the same time! The distortion caused by this effect is called *modal dispersion.* Figure 18–11 shows what happens to the shape of a square pulse as it travels through a multimode fiber.

Modal dispersion causes the edges of digital pulses to be rounded off. The longer the fiber, the worse the distortion becomes. In the figure, the nice square input pulse has already developed rounded edges halfway down the fiber. The dispersion only gets worse as the pulse makes its way to the receiving end. The timing differences between the various lightwave paths are magnified in proportion to the cable length. In this example, the pulse is hardly recognizable at the end; in fact, it appears to have a rise and fall shaped very much like it had passed through an RC low-pass filter circuit! *The widening effect on pulses is what limits the bandwidth of a multimode fiber.* In fact, we can estimate the bandwidth of such a fiber by using an equation from electronic fundamentals:

$$\mathrm{BW} = \frac{0.35}{t_r} \qquad \textbf{(18–7)}$$

where t_r is the rise time (10% to 90% electric field intensity) of the optical signal.

However, to use equation 18–7, we need to know the rise time. For a multimode fiber, the rise time can be estimated as the difference in time between the arrival of the *first* photons of a light pulse and the *last* photons of the same pulse. With trigonometry, we can derive the following relationship:

$$t_r \approx \Delta t \approx \frac{n_1 L}{c}\left(\frac{n_1}{n_2} - 1\right) \qquad \textbf{(18–8)}$$

where t_r is the effective rise time due to modal dispersion, n_1 is the refractive index for the core, n_2 is the cladding index, c is the speed of light, and L is the length of the fiber.

Equation 18–8 tells us two important things. First, it demonstrates that modal dispersion gets progressively worse as the length of a fiber, L, is increased. Second, it shows that the *ratio* of refractive indices for the core and cladding controls the amount of dispersion. The closer together the refractive indices, the less dispersion and the better the pulse shape will appear at the end of the fiber. However, equations 18–5 and 18–6 also tell us that as the refractive indices of the core and cladding get closer in value, the numerical aperture (NA) and acceptance cone of the fiber shrink, making it difficult to get light into the fiber.

EXAMPLE 18–5

Compare the acceptance cone angle and dispersion-limited bandwidth for two different optical fibers. Both fibers are 1 km long. Fiber A has a core with a refractive index of 1.6 and cladding with an index of 1.3. Fiber B has core and cladding indices of 1.6 and 1.5, respectively.

Solution

Fiber A: The numerical aperture must first be calculated according to equation 18–5:

$$\text{NA} = \sqrt{n_1^2 - n_2^2} = \sqrt{1.6^2 - 1.3^2} = 0.933$$

The acceptance cone can be directly calculated from this result using equation 18–6:

$$\Theta_{\text{accept}} = \sin^{-1}\text{NA} = \sin^{-1} 0.933 = \underline{\underline{68.87^\circ}}$$

The amount of modal dispersion can be calculated with equation 18–8:

$$t_r \approx \Delta t \approx \frac{n_1 L}{c}\left(\frac{n_1}{n_2} - 1\right) = \frac{(1.6)(1000\text{ m})}{3 \times 10^8\text{ m/s}}\left(\frac{1.6}{1.3} - 1\right) = 1.23\ \mu\text{s}$$

Knowing this, we can now use equation 18–7 to estimate the bandwidth:

$$\text{BW} = \frac{0.35}{t_r} = \frac{0.35}{1.23\ \mu\text{s}} = \underline{\underline{284.4\text{ kHz}}}$$

Fiber B:

$$\text{NA} = \sqrt{n_1^2 - n_2^2} = \sqrt{1.6^2 - 1.5^2} = 0.557$$

The acceptance cone is

$$\Theta_{\text{accept}} = \sin^{-1}\text{NA} = \sin^{-1} 0.557 = \underline{\underline{33.83^\circ}}$$

The amount of modal dispersion is

$$t_r \approx \Delta t \approx \frac{n_1 L}{c}\left(\frac{n_1}{n_2} - 1\right) = \frac{(1.6)(1000\text{ m})}{3 \times 10^8\text{ m/s}}\left(\frac{1.6}{1.5} - 1\right) = 0.35\ \mu\text{s}$$

The resulting bandwidth is

$$\text{BW} = \frac{0.35}{t_r} = \frac{0.35}{0.35\ \mu\text{s}} = \underline{\underline{984.4\text{ kHz}}}$$

What conclusion can be reached from this example? Fiber B certainly has a much better bandwidth. Fiber B produces less modal dispersion. However, its acceptance cone is not as wide as that of fiber A. Improving one characteristic worsens the other!

Improving the Performance of Multimode Fibers

The dispersion problem with multimode fibers arises because of the difference in time between the various possible light wave paths. *Graded-index fibers* are one solution to the dispersion problem. A graded-index fiber has a special core with a *graded* or *changing* index of refraction, as shown in Figure 18–12.

The basic idea behind the graded-index fiber is as follows. In the higher order propagation modes, light waves "bounce" back and forth between the walls of the core more

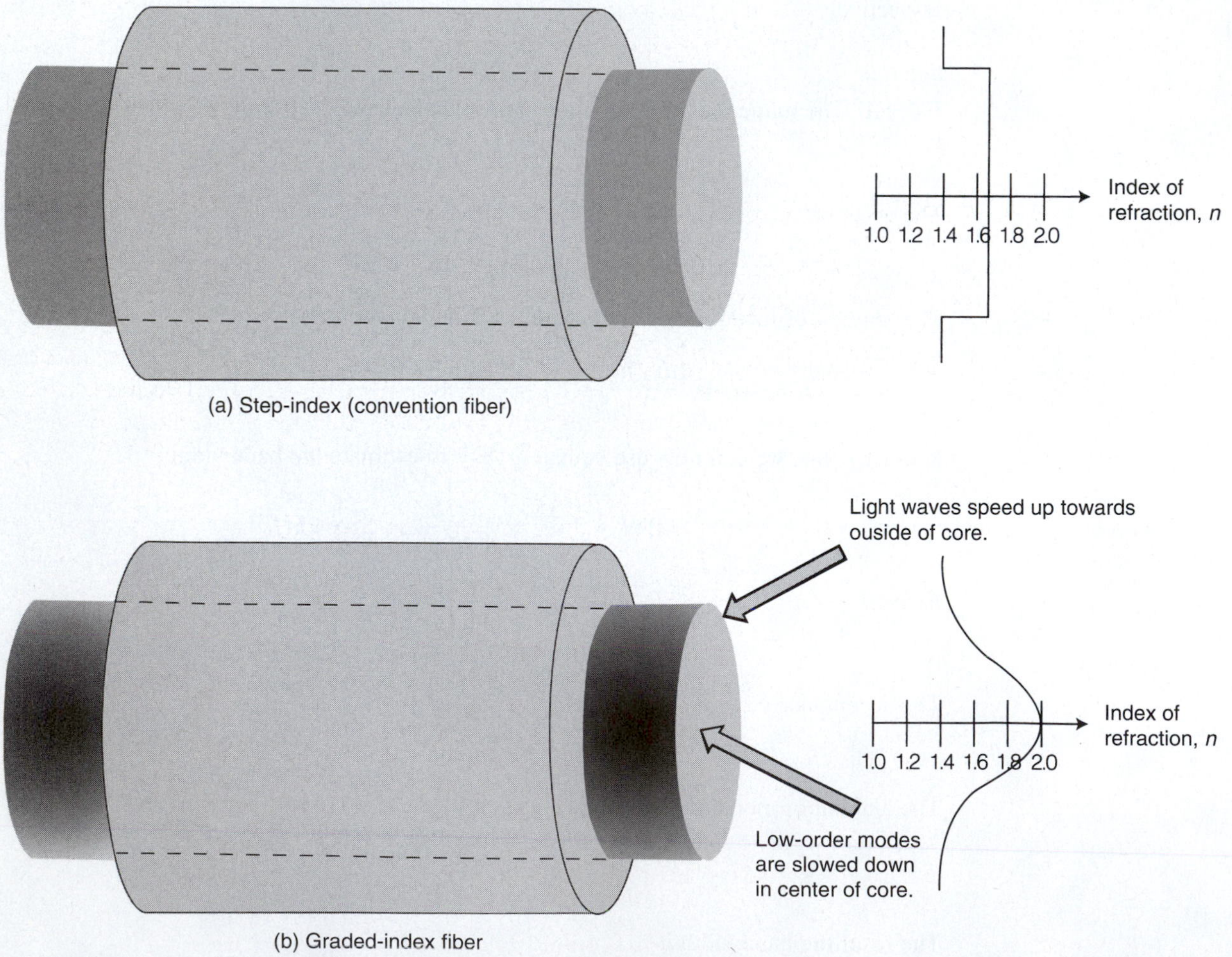

Figure 18–12 Step-index and graded-index fibers

than in low-order modes (those that travel approximately straight through the fiber). *Therefore, to equalize the time taken by each different mode, the lower-order modes must be slowed down in order for the waves from the high-order modes to "catch up."* We know that the index of refraction of a material controls the velocity of wave propagation within it. In a step-index fiber, the velocity of propagation is constant throughout the core. In a graded-index fiber, the center of the core is designed with a higher index of refraction than the outer edges. Therefore, light waves slow down if they pass through the center and speed up as they approach the edge.

The graphs to the right of the fibers in Figure 18–12 show the profile of refractive index for the fiber cores. The graded-index fiber generally uses a *parabolic* profile; you can see that the shape does look somewhat like a parabola. The calculation of this profile involves advanced mathematics far beyond the needs of the technician.

Other Dispersion Modes

There are two other forms of optical distortion that can take place as a light wave passes down a fiber. These are *chromatic dispersion* (sometimes called *material dispersion*) and *waveguide dispersion*. While modal dispersion affects only multimode fibers, these two dispersion sources affect both single-mode and multimode fibers.

Chromatic dispersion The index of refraction for most materials is not a constant. Instead, it depends slightly on the applied frequency (or wavelength). A prism separates white light into colors according to this principle, as shown in Figure 18–13. A prism separates the colors of light because the refractive index of the prism material (glass or plastic) varies with frequency. Therefore, each color is refracted (bent) a different amount and exits at a different angle.

Chromatic dispersion will be present whenever the light source illuminating the fiber possesses more than one frequency of light. No light source is perfect. For example, even though an LED may appear to emit only one color of light, it is actually radiating thousands of different wavelengths (all of which are close together in frequency). Because the velocity of propagation will be different for each of the light frequencies emitted by the LED

Figure 18–13 A prism separating light frequencies

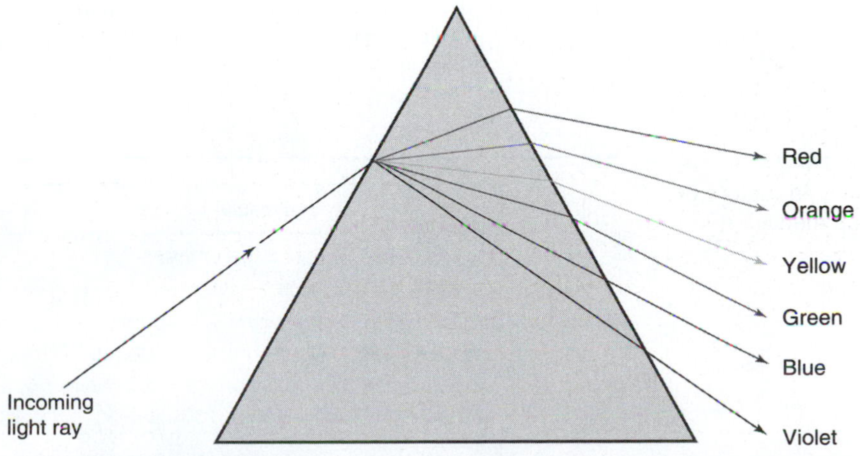

(because the index of refraction depends on frequency), all of the light wave components will not reach the receiver at the same time, resulting in the same type of pulse spreading demonstrated in Figure 18–11. A laser emits a much more narrow range of frequencies than an LED. Therefore, laser light sources produce much smaller amounts of chromatic dispersion in a fiber.

Waveguide dispersion The magnified view of the refraction zone in Figure 18–6 shows an important detail: Light must enter the cladding in order to be refracted back into the core. But wait—the cladding has a lower refractive index than the core. The light waves that pass through the cladding are therefore *sped up* slightly when compared to those passing through the core. This can again cause some of the light waves to arrive at a different time than the main signal, resulting in pulse spreading and bandwidth loss. This form of dispersion is also dependent upon the wavelength of the light.

Careful choice of core and cladding materials minimizes this kind of dispersion. In fact, since this mode is dependent upon wavelength, it is possible to choose a light wavelength that largely cancels the effect.

Fiber Signal Attenuation

The word *attenuate* means to weaken. As signals travel through any transmission medium, they are progressively attenuated. Fiber optics far outperform copper transmission lines in this area, although there are various grades of fiber optics with varying attenuation.

The amount of signal that will be lost in a fiber-optic cable depends on three factors. First is the type of cable and core material used (glass is better than plastic, in general); second, the *wavelength* of the light passing down the cable (most glass fiber has a "sweet spot" around the 1300 nm wavelength where attenuation becomes incredibly small), and finally the *length* of the cable. The decibel loss is directly proportional to the length.

Attenuation in fiber optics is usually rated in decibels per kilometer, or dB/km. Figure 18–14 compares the attenuation of various types of transmission media. Compare the performance of 7/8″ copper coaxial hardline with the single-mode fiber. The copper hardline (a relatively expensive media) costs around $3.00 per foot and gives 48 dB of loss for a 1 km run (the total cost of the transmission line would be around $9800!). In contrast, single-mode glass core fiber-optic cable can be had for under $1.00 per foot and loses only 0.4 dB of the signal! The fiber-optic cable weighs about 2 ounces per foot, for a total

Figure 18–14
Attenuation factors of various media

Type of media	Attenuation, dB/km
RG-213 50 ohm coax, f = 500 MHz	205
7/8″ 50 ohm hardline, f = 500 MHz	48
Graded index, 100 μm core, λ = 850 nm	3.5
Graded index, 100 μm core, λ = 1300 nm	1.5
Single mode, 9 μm glass core, λ = 1300 nm	0.4
Step index (Eska SH-4001), 1000 μm plastic core, λ = 850 nm	220

weight of 410 pounds. In contrast, the hardline weighs about 2 *pounds* per foot and would weigh approximately 6500 pounds! For long-distance signal transmission, fiber-optic cable wins hands down over copper media.

Section Checkpoint

18–12 What are *propagation modes*?

18–13 What property of a single-mode fiber restricts propagation to one mode?

18–14 What is the effect of the different light wave paths of Figure 18–10?

18–15 What is *modal dispersion*? What is its effect on the bandwidth of a fiber?

18–16 Explain the construction of a graded-index fiber. What construction feature reduces modal dispersion?

18–17 Explain the cause of *chromatic dispersion*. Which light source is better at reducing it, laser or LED?

18–18 What causes *waveguide dispersion*?

18–19 How is the energy loss of fiber-optic cables rated? Does it depend on wavelength? (Refer to Figure 18–14.)

18–3 LIGHT WAVE SOURCES AND MODULATION CIRCUITS

In order to produce a light wave signal for use in a fiber-optic system, both LEDs and lasers are popular. LEDs are used in systems where "state-of-the-art" data rates are not necessary. Regardless of whether an LED or a laser is the light source, amplitude modulation is usually employed. Creating FM or PM modulation of a light source is certainly possible but is not an economical option.

LED Characteristics

LEDs are the simplest and most reliable light-emitting device. LEDs for fiber-optic application usually operate in the infrared range, around 800 nm, and provide optical output powers under 1 mW. Figure 18–15 is the data sheet for the Siemens SFH450V, a popular fiber-optic emitter.

The SFH450V is typical of LEDs designed for fiber-optic applications. It is contained in a plastic package designed to act as a receptacle for 1000 μm fiber-optic cable. The device emits infrared light with a wavelength of approximately 950 nm. Figure 2 from the SFH450V data sheet shows that the power output of the LED is controlled by the *current* flowing through it. The higher the current, the more optical output power produced. Notice that the current values are quite a bit larger than might be used for a conventional indicator LED; the device can withstand more than 1000 mA of current in pulse applications and can handle 100 mA continuously. The optical rise time of the device is given as 1000 ns typical.

Figure 18–15
Siemens SFH450V data sheet (Source: Infineon Technologies, A Siemens Company. Reprinted with Permission)

Plastic Fiber Optic Transmitter Diode
Plastic Connector Housing

SFH 450
SFH 450V

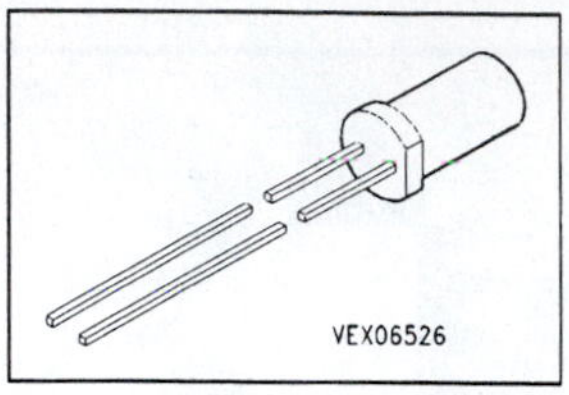

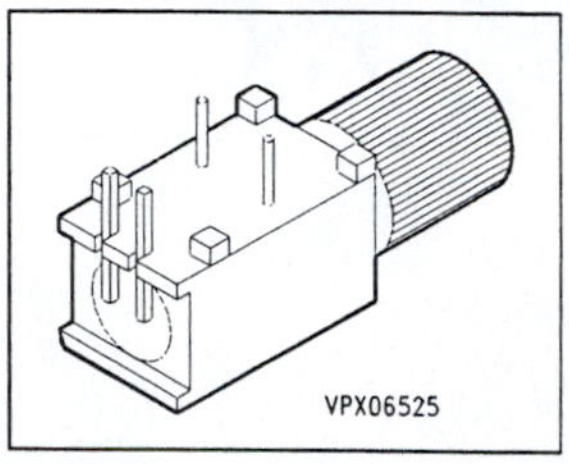

Features

- 2.2 mm aperture holds standard 1000 micron plastic fiber
- No fiber stripping required
- Good linearity
- Molded microlens for efficient coupling

Plastic Connector Housing

- Mounting screw attached to the connector
- Interference-free transmission from light-tight housing
- Transmitter and receiver can be flexibly positioned
- No cross talk
- Auto insertable and wave solderable
- Supplied in tubes

Applications

- Household electronics
- Power electronics
- Optical networks
- Medical instruments
- Automotive electronics
- Light barriers

Type	Ordering Code
SFH 450	Q62702-P1034
SFH 450V	Q62702-P265

Maximum Ratings

Parameter	Symbol	Values	Unit
Operating temperature range	T_{OP}	– 55 ... +100	°C
Storage temperature range	T_{STG}	– 55 ... +100	°C
Junction temperature	T_J	100	°C
Soldering temperature (2 mm from case bottom, $t \leq 5$ s)	T_S	260	°C
Reverse voltage	V_R	5	V

Figure 18–15
(continued)

Maximum Ratings (cont'd)

Parameter	Symbol	Values	Unit
Forward current	I_F	130	mA
Surge current $t \leq 10$ µs, $D = 0$	I_{FSM}	3.5	A
Power dissipation	P_{TOT}	200	mW
Thermal resistance, junction/air	R_{thJA}	375	K/W

Characteristics (T_A = 25 °C)

Parameter	Symbol	Values	Unit
Peak wavelength	λ_{Peak}	950	nm
Spectral bandwidth	$\Delta\lambda$	55	nm
Switching times (R_L = 50 Ω, I_F = 10 mA) 10 % … 90 % 90 % … 10 %	 t_R t_F	 1 1	 µs µs
Capacitance (f = 1 MHz, V_R = 0 V)	C_O	40	pF
Forward voltage (I_F = 10 mA)	V_F	1.3 (≤ 1.5)	V
Output power coupled into plastic fiber (I_F = 10 mA) see **Note 1**	Φ_{IN}	40 … 200	µW
Temperature coefficient Φ_{IN}	TC_Φ	– 0.5	%/K
Temperature coefficient V_F	TC_V	– 1.5	mV/K
Temperature coefficient λ_{Peak}	TC_λ	0.3	nm/K

Note 1: The output power coupled into plastic fiber is measured using a large area detector at the end of a short length of fiber (about 30 cm). This value must not be used for calculating the power budget for a fiber optic system with a long fiber because the numerical aperture of plastic fibers decreases on the first few meters. Therefore the fiber seems to have a higher attenuation over the first few meters compared with the specified value.

Figure 18–15
(continued)

Relative spectral emission $I_{rel} = f(\lambda)$

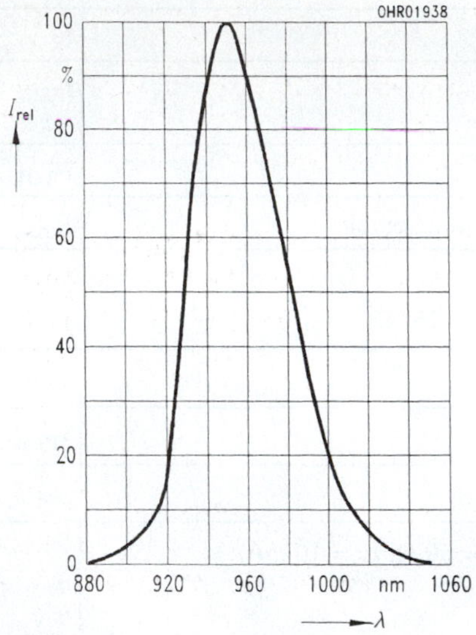

Forward current $I_F = f(V_F)$
single pulse, duration = 20 μs

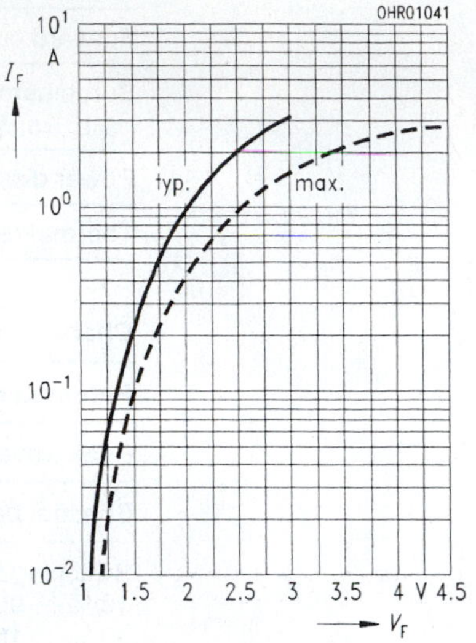

Relative output power $\Phi_{IN}/\Phi_{IN(10\ mA)} = f(I_F)$

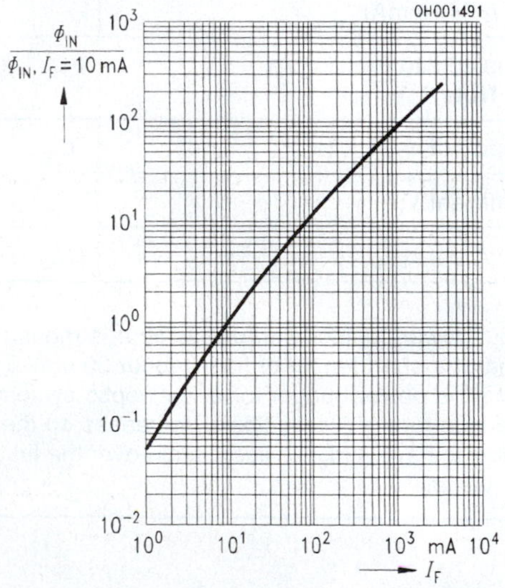

Figure 18–15
(continued)

Maximum permissible forward current
$I_F = f(T_A)$

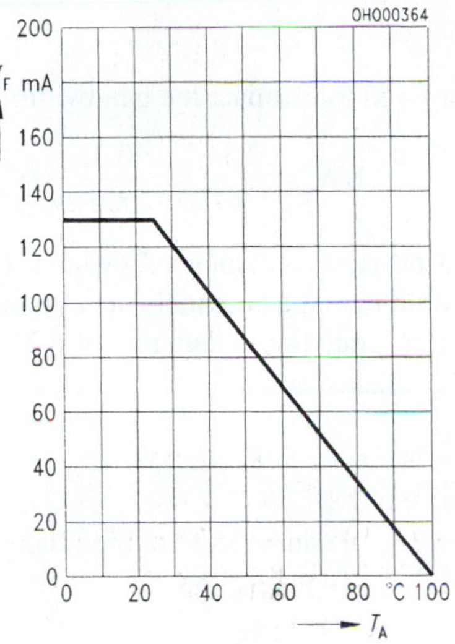

Permissible pulse load $I_F = f(t_p)$,
duty cycle D = parameter, T_A = 25 °C

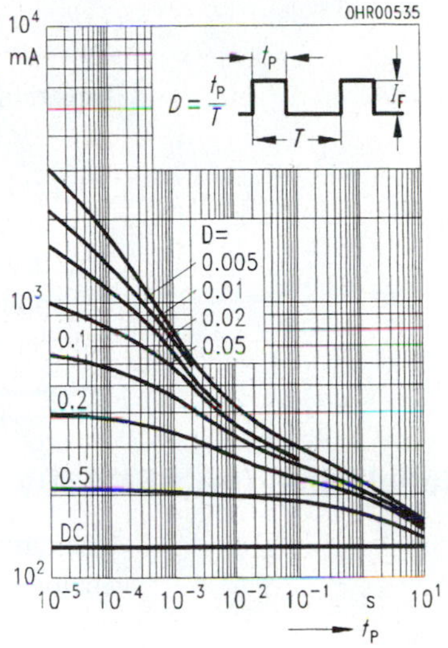

Package Outlines (dimensions in mm, unless otherwise specified)

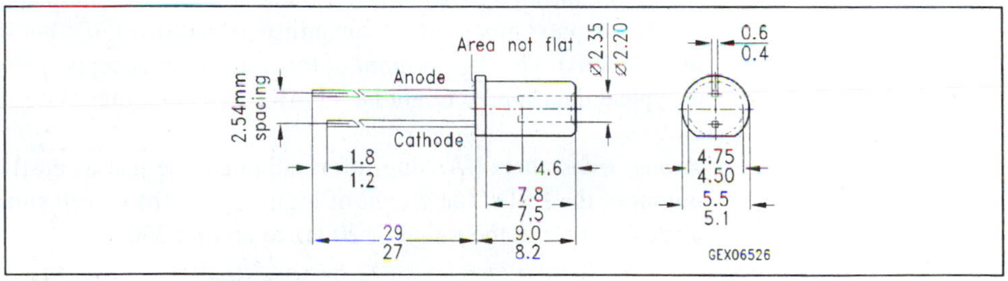

SFH 450

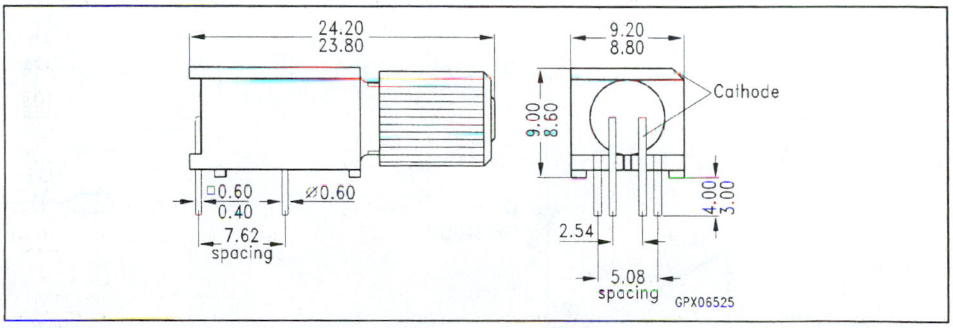

SFH 450V

EXAMPLE 18–6 What bandwidth is available from an SFH450V given the typical rise time of 1000 ns from the datasheet?

Solution

Equation 18–7 can be used to estimate the bandwidth:

$$BW = \frac{0.35}{t_r} = \frac{0.35}{1000\text{ ns}} = \underline{\underline{350\text{ kHz}}}$$

This is a fairly wide bandwidth, but not nearly what is possible with laser emitters. For digital applications, it would hint at a potential data rate of *twice* this figure (there are two digital bits per ac cycle), or a maximum data rate of a 750 kbps.

Modulating the SFH450V

The circuits of Figure 18–16 can be used to modulate the SFH450V with either digital or analog information.

Digital transmitter The digital transmitter of Figure 18–16(a) uses simple on-off keying (OOK) to modulate the lightwave carrier. A logic 1 is represented by turning the light beam *on;* a logic 0 shuts the light off. Inverter U1A, a 74LS04, provides the drive current for the SFH450V infrared LED. Resistor R1 sets the *on* current to approximately 8 mA (this is far below the maximum current capability of the SFH450V but still quite adequate for many applications). The *digital input* of the transmitter accepts *serial* TTL data from any source. A typical data source might be a UART in a computer system.

Analog transmitter Analog information can be just as easily impressed onto the optical output of the LED. The circuit of Figure 18–16(b) is well suited for direct transmission of analog signals in the range of 20 Hz to around 350 kHz.

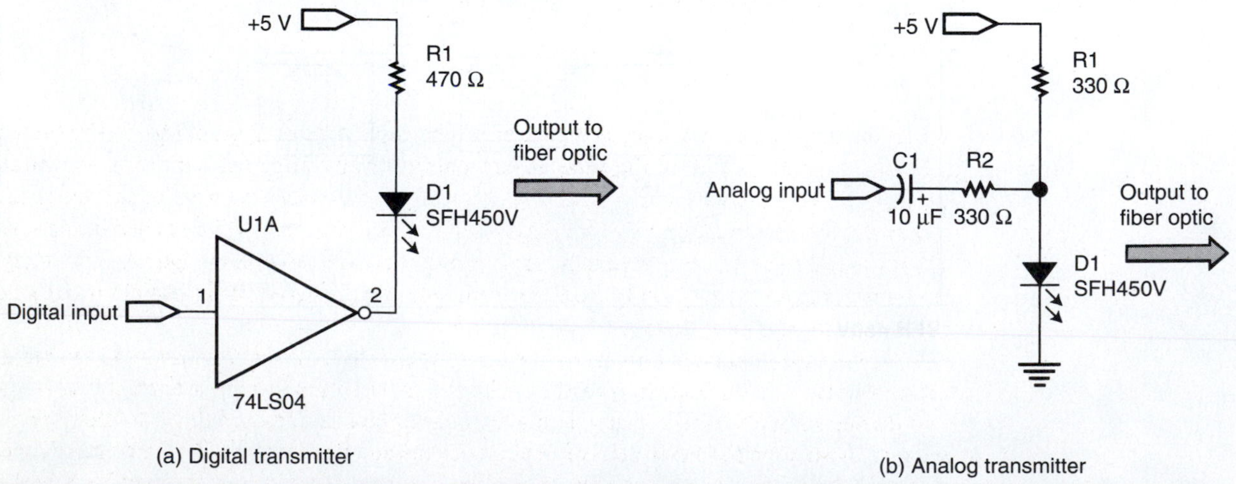

Figure 18–16 Modulation circuitry for the SFH450V

A quiescent current (Q point) of about 12 mA is set up by the current-limiting resistor R_1. This causes LED D_1 to output a steady level of light when there is no information present. By coupling the ac information into D_1 through linearizing resistor R_2, the current in D_1 is made to increase and decrease in step with the information signal, thereby creating *amplitude modulation* of the optical output signal.

This circuit could easily transmit the signal from a microphone if additional amplification were provided (such as an operational amplifier).

Laser Characteristics

The word *laser* is an acronym, standing for *Light Amplification by Stimulated Emission of Radiation.* Laser light sources are used when optical power outputs of more than a few mW are needed. Lasers are of two major types, *gas* and *solid state*. For communication applications, solid-state lasers are probably the most popular. Lasers emit nearly monochromatic light. The word *monochromatic* means "one color." This type of light minimizes the problem of chromatic dispersion.

Laser Operation

Figure 18–17 shows the internal construction of a helium–neon gas laser tube. It consists of a hollow glass tube that has been filled with a mixture of helium and neon gases. One end of the tube has a completely mirrored surface so that no light can escape, and the other has a partially silvered mirror, where a beam of laser light can escape once the device is in operation. A pair of electrodes are provided so that energy can be supplied to the gas inside the tube.

A typical HeNe laser tube requires a voltage of 1000 to 3000 volts dc while it is operating, at a current level of 3 to 8 mA. However, to get the tube conducting at first power-on, a start-up voltage of 5 to 12 kV is needed in order to ionize the gases in the tube. A gas laser tube therefore requires a special power supply able to supply the extra voltage for startup yet regulate the current to a safe value (3–8 mA) once the tube has begun operating.

Because of the high voltages involved, extreme care must be used when working with a gas laser. Typical digital multimeters cannot directly measure the operating voltages of this device; a high-voltage probe must be used.

Laser Theory

When the appropriate voltages are applied to the laser tube of Figure 18–17, the helium and neon gas atoms within the tube absorb energy from the power supply. Some of the atoms move to a higher energy state. This higher energy state is not stable, and the atoms soon want to give up the energy that was put into them. The energy is released in the form of visible light. So far, this action is no different than that of a neon sign. A nice orange glow results from the high-to-low energy transitions of the neon atoms, but not much else. The helium atoms contribute no visible light.

If the tube contained only neon gas, most of the atoms would remain in a low-energy state, because it is difficult to directly excite the neon atoms to higher energy states with electric current. However, the tube contains a mixture of helium and neon. In fact, the gas mixture in the tube is *mostly* helium. Helium atoms are readily excited by the electric current, and something interesting happens: When an atom of helium is "pushed" to a higher energy state by the power supply energy, it eventually must fall back down to its original

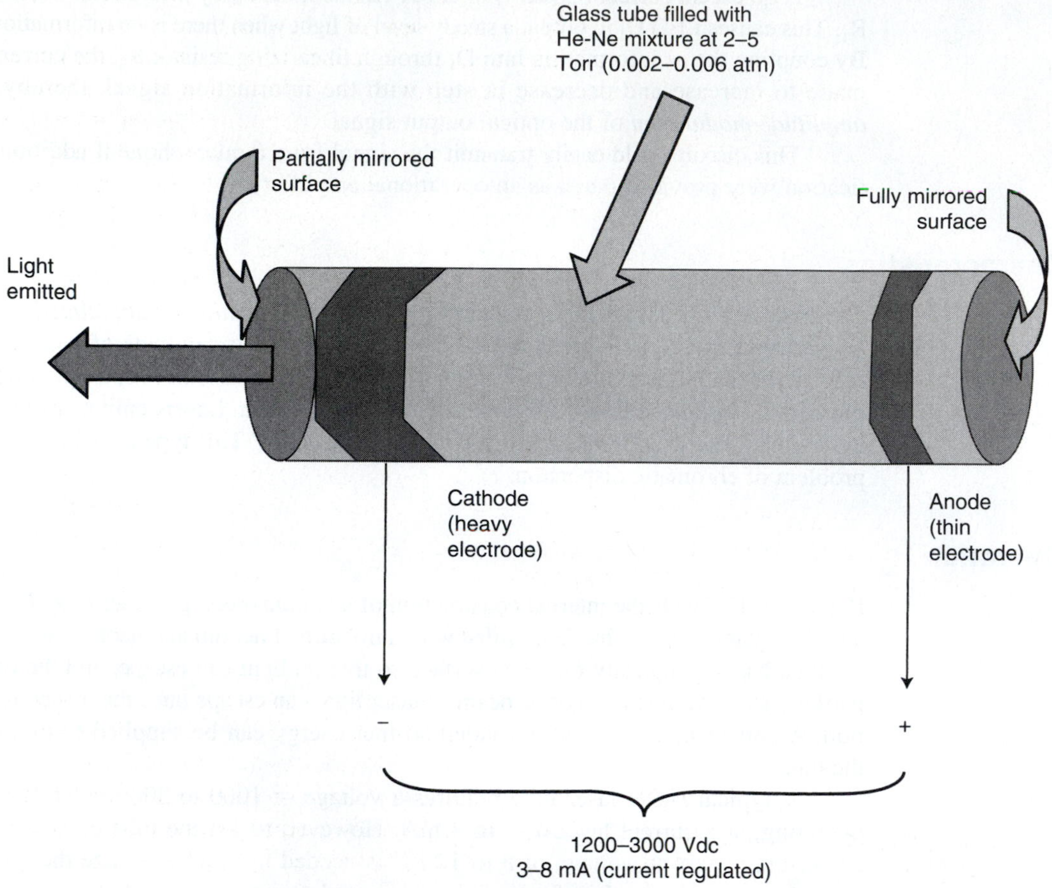

Figure 18–17 Gas laser construction

state. The atom of helium gives up its energy by colliding with a neon atom, which pushes the neon atom into a *very* high energy state. The coupling between the helium (energy givers) and neon (energy users) atoms is very efficient, and if the excitation current is high enough, soon there are more neon atoms in a high-energy state than in a ground state. This condition is called a *population inversion*. Under this condition, the neon atoms give off strong light at *many* different wavelengths (frequencies). The different wavelengths are produced because there are several possible energy levels for the neon atom. The wavelength of light emitted depends on which energy level each atom is excited to.

At this point, we now have a brightly glowing tube of helium–neon gas but still no laser beam. The tube contains light at many different wavelengths (frequencies). Laser light contains mainly *one* frequency. How can we get one frequency from many? A *bandpass filter* will do the trick!

The tube containing the helium–neon gas is made to act as an optical bandpass filter by being constructed to be an *exact* multiple of a number of half-wavelengths of light at the frequency we choose. This frequency must be one of the original frequencies (wavelengths) in the emitted light from the neon atoms. A 4″ HeNe laser tube is *thousands* of half-wavelengths long and is "tuned" to the proper frequency by precision adjustment of one of the reflecting mirrors (this is done by the manufacturer).

The fact that the tube is an integer multiple of a half-wavelength encourages the formation of a very long-standing wave of optical energy within. This standing wave is the result of optical energy reflecting back and forth between the two mirrors inside the laser tube. The energy in the standing wave further encourages more photons to be emitted at the *same wavelength* from the helium–neon mixture. These new photons join the standing wave, which in turn excite even more of the same wavelength to be emitted. *Positive feedback or light amplification is occurring.* The intensity of the light jumps dramatically because of this feedback. This is the process of *lasing.* We essentially have a lightwave amplifier and oscillator! As long as the positive feedback (energy input from the power supply) is greater than the combined losses in the tube (thermal, mirror loss, and so on), the oscillation will continue, and a very intense beam of coherent light will exit the device through the partially silvered mirror.

Disadvantages of Gas Lasers

Gas lasers have largely been replaced by the solid-state laser diode in communications applications. Compared with a laser diode (which is the same size as a typical transistor), a gas laser is bulky. A gas laser also requires a high-voltage power supply and can easily be damaged by vibration and shock. (The mirror alignment in a gas laser is quite delicate. If it is off just a little, the device will not lase.)

Solid-State Laser Diode Operation

Figure 18–18 shows the simplified structure of a laser diode. It is very similar in construction to a conventional light-emitting diode. A laser diode is operated in the forward-bias mode, just like an LED. Current flowing through the junction creates electron-hole pairs, which recombine at the junction, resulting in a photon of light. The choice of material

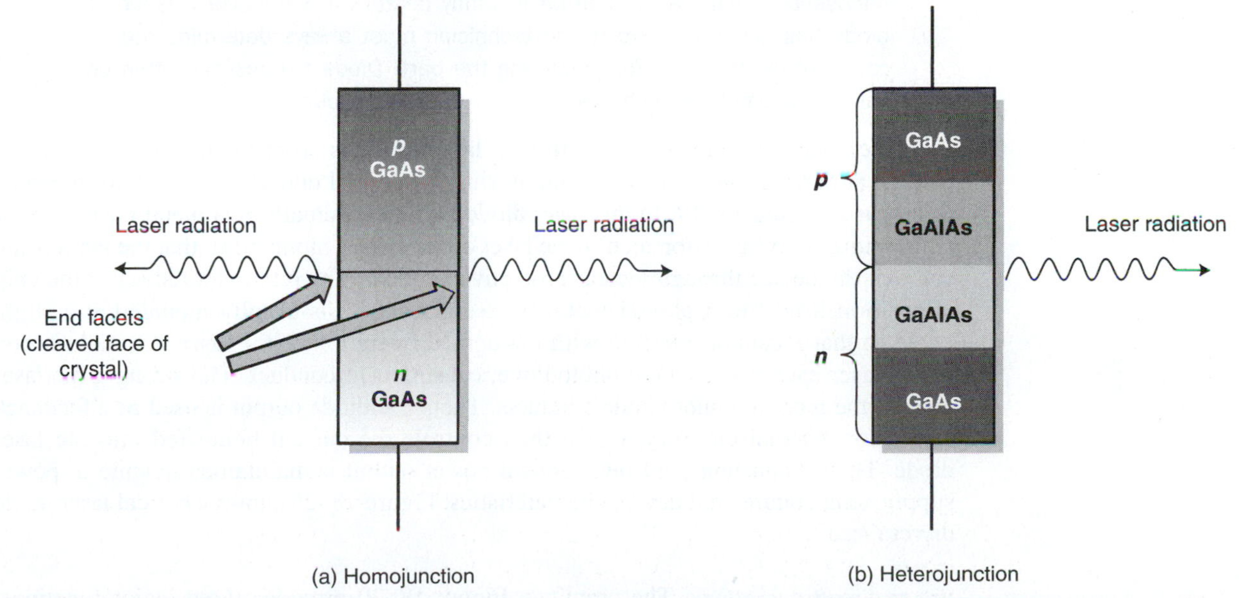

Figure 18–18 Laser diode structures

(such as gallium arsenide) controls the wavelength of the light. As with a gas laser, many discrete wavelengths are actually produced, depending on the semiconductor materials in use and the number of possible atomic excitation states.

The photons are free to travel along the line of the junction, where they meet the cleaved (cut) edge of the crystal. This edge may be simply cut, or it may be micropolished. It forms a partial mirror, and the photons return back into the crystal for another round trip. As in the gas laser, these returning photons set up a standing wave. The length of the crystal must be an integral number of half-wavelengths to reinforce this wave. The standing wave of light within the crystal encourages even more photons to be released at the junction, which again contributes even more optical output. Positive feedback occurs, and the device begins to lase.

The homojunction (*homo* means "same") diode's optical efficiency is not very high; the answer to that is the *heterojunction* laser diode. The additional layers (in this case, gallium arsenide with aluminum) serve to better channel the optical energy back and forth within the crystal, resulting in more power output.

Unlike a gas laser, the output of a solid-state *laser diode* is not a well-defined and narrow beam. Usually, the solid-state laser beam is wedge-shaped and must be shaped by lenses in order to be effective. Solid-state lasers are also electronically very delicate. They are very static sensitive and should be kept in antistatic packaging until installed in a circuit. These devices must also be driven with a precise level of current. Connecting a laser diode directly to a power supply, even through a "current limiter resistor," usually destroys the diode very quickly.

At low current levels, a diode laser acts like an inefficient LED. As the current level is raised, the optical loop gain product exceeds unity and the device begins to lase. At this threshold value of current, the power output rises rapidly. *This value of threshold current varies from diode to diode and is also temperature-dependent.* If too much current is sent through the diode, the optical power output rapidly grows out of control, literally destroying the diode structure (typically the end facets).

Exceeding the rated power output of a laser diode for even a few nanoseconds can cause it to be instantly destroyed. Whenever a laser diode has failed in a circuit, the technician must always determine the cause of the failure before replacing the part! Diode failures can often be caused by problems with power supply or driver circuitry.

Because the control of current in a laser diode is so critical, most laser diodes include an integral *photodiode* for monitoring the optical output. A typical arrangement is shown in Figure 18–19. In this laser diode (which is actually in a metal package very similar to a TO-5 transistor "can"), the laser diode chip is mounted so that the main beam can exit the device through a clear protective window. The rear facet (edge) of the chip is positioned next to a photodetector diode. This diode is usually mounted at a slight angle so that it cannot interfere with the optical operation of the laser. A small portion of the laser energy strikes the photodiode, causing it to conduct. The stronger the laser energy, the more the photodiode conducts. The photodiode output is used as a feedback signal for external circuitry, which then controls the current being fed into the laser diode. Thus, a constant (and safe) optical power output is maintained in spite of power supply, temperature, and device characteristics. Figure 18–20 shows a typical laser diode driver circuit.

Driver circuit operation The circuit in Figure 18–20 provides three major functions: *soft-start, automatic power control,* and *modulation.* The *soft-start* function is provided by R_2, C_1, and Q_1. When power is initially applied, C_1 is discharged, keeping Q_1 the

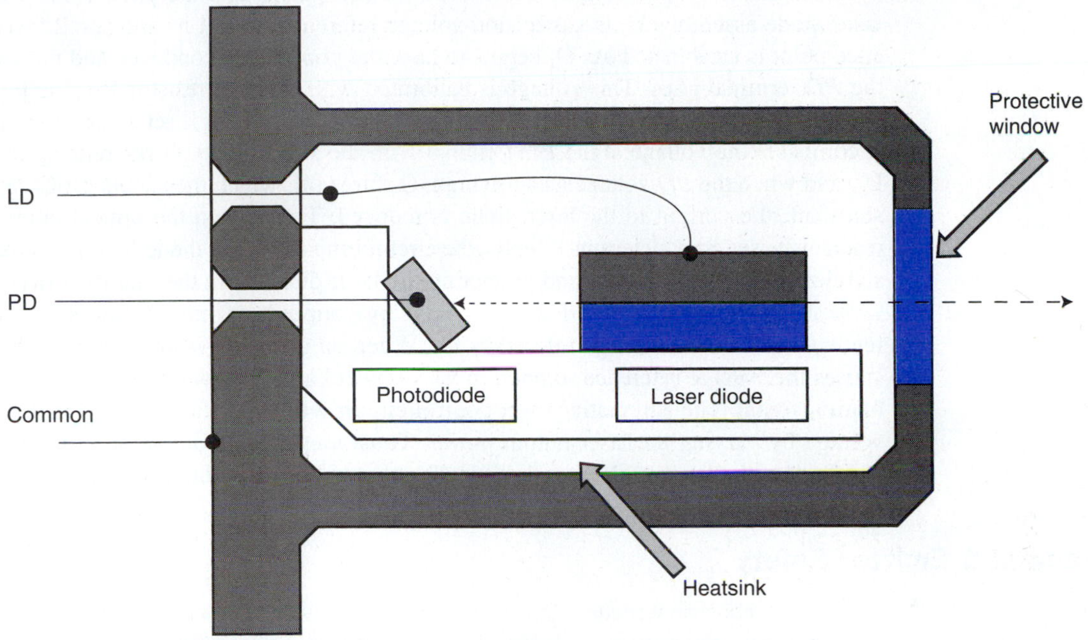

Figure 18–19 Laser diode with integral photodiode

emitter-follower, turned off. Over a period of about 15 ms, the base voltage at Q_1 is allowed to slowly rise as C_1 charges through R_2. Since Q_1 is an emitter-follower, its emitter voltage also slowly rises. Thus, the circuit "soft-starts." The purpose of this is to protect the expensive laser diode from power-on transients (impulses) that might arise in the main power supply.

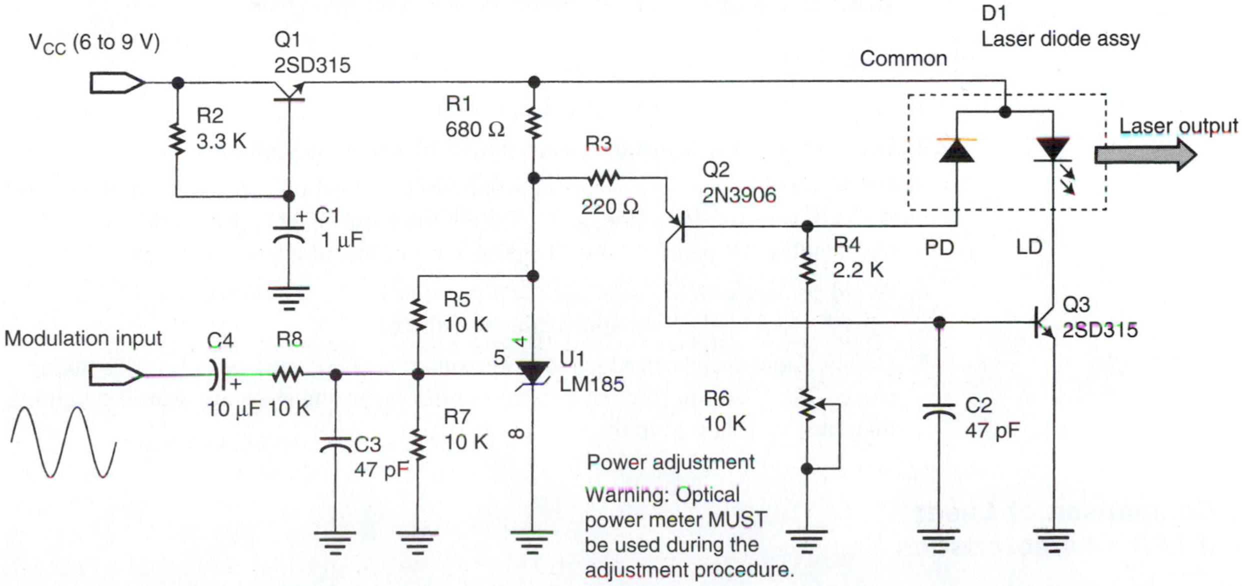

Figure 18–20 A laser diode driver circuit

Automatic power control is achieved by U_1, Q_2, Q_3, and the photodiode within the laser diode assembly. U_1 is a precision voltage reference, and at its anode a 2.5 volt reference point is established. As D_1 begins to lase, the photodiode conducts, and the voltage at the *PD* terminal rises. This voltage is calibrated with variable resistor R_6. The higher the optical output power, the higher the voltage at the *PD* terminal. Q_2 is set up as a "comparator." It compares the voltage at the *PD* terminal with the steady 2.5 volt reference provided by U_1, and when the *PD* voltage gets too high, Q_2 turns off, which then shuts off Q_3. When Q_3 shuts off, the current to the laser diode is reduced. Thus, when the optical output power reaches the set point determined by R_6, the circuit limits the laser diode forward current, thus stabilizing the output power and protecting the laser diode from thermal runaway.

Modulation is achieved in this circuit by coupling an information signal into the feedback pin of the voltage reference, U_1. When an ac signal is coupled into this pin, it causes the voltage reference setpoint to vary upward and downward in step with the information signal. The automatic power control circuit will "track" the change in the setpoint voltage by varying the laser output power. Thus, amplitude modulation will occur. Note that this particular modulation circuit is effective only up to around 100 kHz.

Laser and LED Emitter Safety

The technician working with fiber optics *must* constantly be aware of the safety hazards associated with intense light sources. The light emitted by fiber-optic light sources is usually in the infrared region, which is invisible to the eye. Even a light source that appears dimly lit can be dangerous. A dimly glowing laser or LED may in fact be emitting dangerous levels of invisible infrared radiation, which can cause severe damage to the retina of the eye in a fraction of a second.

There are no pain receptors in the retina, so a person will generally not know that their eyes are being damaged. The damage is cumulative, and vision can be slowly degraded over repeated exposures, or rapidly lost if exposed to high-power sources. The following are *minimum* recommended safety rules. A technician planning on working with these systems is strongly advised to study the available literature for additional safety information.

Basic Safety Rules:

1. Never look into any fiber optic, LED emitter, or laser when the power is on.
2. Do not attempt to adjust the power output of a laser by looking at the light.
3. Wear protective eyewear where appropriate. Note that off-the-shelf sunglasses are not sufficient for protection, and although the lenses may appear dark, they may have "holes" in their frequency response where lots of energy freely passes.
4. Avoid contact with high-voltage supplies; if adjusting the mirrors on gas lasers, use a plastic tool (and do not apply excessive force).
5. Do not allow unauthorized persons into an area where work with lasers is taking place. (There are specific federal safety rules regarding signs and warning lighting that must be followed in the workplace.)

A Comparison of Laser and LED Characteristics

Figure 18–21 summarizes the major differences between laser and LED light sources.

Figure 18–21 A comparison of laser and LED characteristics

Characteristic	LED attribute	Laser diode attribute
Power output	< 1 mW	> 1 mW, power outputs in excess of 1 W possible
Spectral purity	Poor—many different wavelengths emitted	Fair—more narrow spectrum than LED but hardly monochromatic
Data rate	30 Mb/s typical	2.5 Gb/s achievable
Cost	Low—drive circuitry is very simple	Moderate, driver circuits must run closed-loop
Reliability	Typical lifetime in hundreds of thousands of hours. Fails gradually; output falls very slowly over thousands of hours.	Most laser diodes are rated at better than 10,000 hours MTBF (mean time before failure). However, failure is usually sudden and catastrophic.

Section Checkpoint

18–20 What type of modulation is typically used in fiber-optic transmitters?

18–21 What controls the optical power output of an LED?

18–22 Explain the operation of the circuits of Figure 18–16.

18–23 What is the principle of operation behind a gas laser?

18–24 Why is one end of the tube of Figure 18–17 partially silvered?

18–25 How does the HeNe gas laser tube act as a bandpass filter?

18–26 Give two reasons why solid state lasers have largely replaced gas lasers in communications.

18–27 Why should a laser diode be kept in its packaging until installed in a circuit?

18–28 What must a technician do when he or she discovers a failed laser diode in a circuit?

18–29 Explain the purpose of the photodiode in Figure 18–19.

18–30 List five minimum safety rules for working with fiber-optic emitter devices.

18–4 LIGHT WAVE DETECTION

Once a light wave has been modulated and launched down a fiber, the only remaining job is to recover the light beam and extract the information from it. *Photodetectors* are required for this. There are three primary types of photodetectors used with fiber optics.

For low-speed systems (operating under 50 kbps), the *phototransistor* is often used. The primary advantage of a phototransistor is that it provides active amplification of the signal, something that the other photosensing devices don't provide.

Medium-speed systems use *photodiode* detectors. These detectors can operate at speeds up to around 80 MHz (160 Mbps) but require fairly sophisticated signal-conditioning circuitry in order to recover useful signals. They provide no signal gain.

The highest speed solid-state detectors are *PIN photodiodes*. These devices can easily operate at frequencies over 1 GHz (2 Gbps). The recovered signals are weak and cover an extremely wide bandwidth, so high-quality signal conditioning is needed.

Photodiode Operation

Figure 18–22 shows the construction of a typical photodiode unit. The photodiode is constructed in a manner slightly different from a conventional diode. It still consists of P- and N-type silicon, but the P layer of silicon, which forms the anode of the device, has been purposely made very thin. The P layer is so thin, in fact, that light can pass right through it and strike the P–N junction.

You might recall that at the junction of *P* and *N* material in a diode, the charge carriers (holes for *P* material, electrons for *N* material) are very scarce, having been attracted into the *P* and *N* materials. This area is called the *depletion region,* and because of the relative lack of charge carriers, it acts like an *insulator*. In order to get a diode to conduct, we normally apply

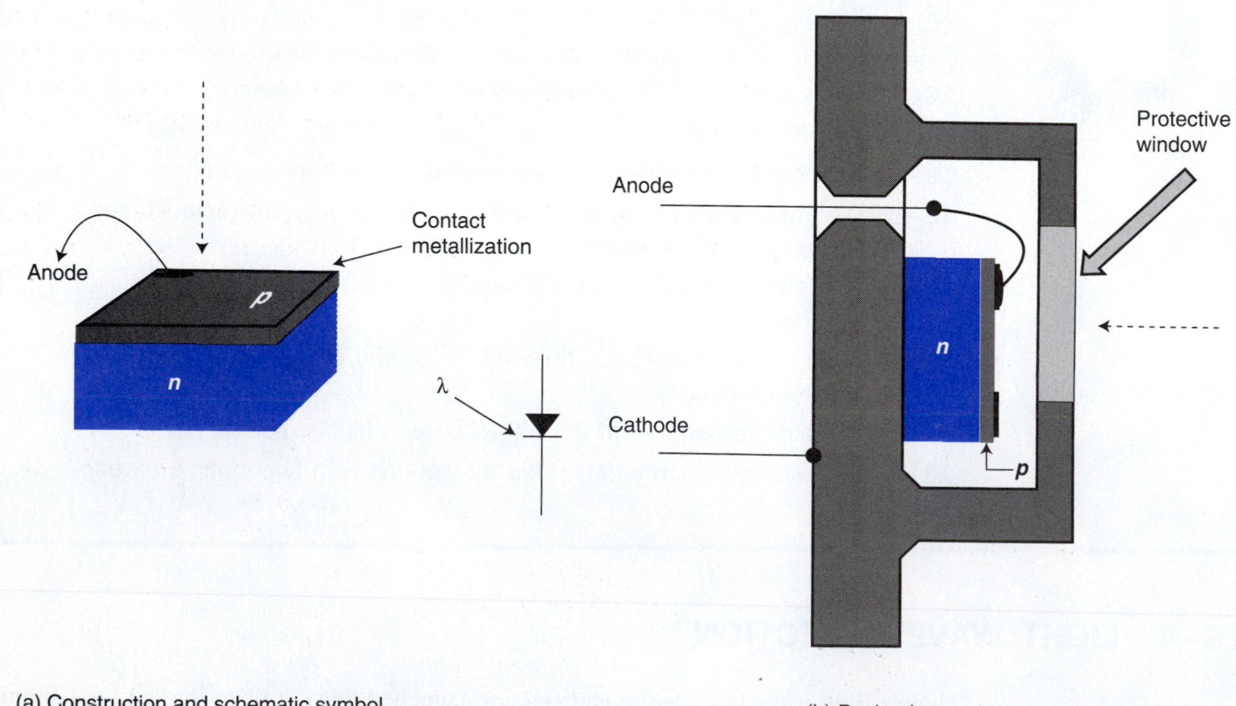

(a) Construction and schematic symbol

(b) Packaging

Figure 18–22 Typical photodiode

forward bias to the device. When we overcome the energy hill of about 0.7 volts (the *barrier potential*), the diode conducts. But because the P–N junction is exposed to light in a photodiode, the energy from incoming light can overcome the energy barrier.

When a photon strikes the region around the P–N junction, it gives up its energy by "shaking loose" an electron-hole pair. This pair recombines rapidly, giving rise to an electrical current. The more light that strikes the junction, the more electron-hole pairs that are created, and the more current that flows.

A photodiode is normally reverse-biased. There are two currents that flow in the device. First, a very small leakage current called the *dark current* flows whether light is present or not. The dark current is the leakage current of the diode junction, and is temperature dependent. The second current that flows is the one that we are interested in, the *light current*. The light current is proportional to the amount of optical power that reaches the chip. The *responsivity* is the rate of change of light current with respect to changing optical input power, typically in units of μA/μW. A typical photodiode for data communications has a responsivity of 0.2 μA/μW, which means that for every microwatt increase of optical input power, the diode current will increase by 0.2 μA. Because the responsivity of photodiodes is quite low, an external amplifier is always required when they are used. Some photodiodes actually include the amplifier on-chip, which reduces noise pickup from wiring.

PIN Photodiodes

A silicon photodiode has a rather large *junction capacitance* associated with the flat P–N structure that is otherwise optimized for gathering light. This capacitance greatly slows the turn-on and turn-off of the device. The *PIN* photodiode structure greatly reduces the capacitance by adding a layer of *I* (intrinsic) silicon, as shown in Figure 18–23. The PIN diode gets its name from the *I* or *intrinsic* layer of silicon that is added between the P and N layers of the junction. An intrinsic semiconductor is one that has not been doped, so it carries a neutral electrical charge and essentially acts as an insulator. In practice, the *I* layer is actually very lightly doped *N* material, but for all practical purposes, it acts as an insulator.

The addition of the *I* layer improves the operation of the diode in two ways. First, the junction capacitance is greatly reduced, which speeds up the device (PIN photodiodes

Figure 18–23 PIN photodiode structure

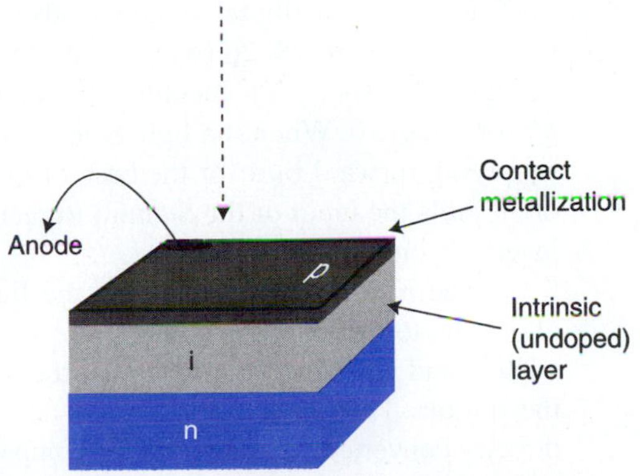

can operate at frequencies in excess of 1 GHz). Second, the *I* layer improves the sensitivity to light and reduces the leakage current. Typical responsivities of PIN photodiodes are in the 0.6 μA/μW range, which is approximately three times more sensitive than conventional photodiodes.

Phototransistor Operation

A phototransistor is little more than a conventional transistor that has been optimized for sensitivity to light. The junctions of most semiconductors are already light-sensitive, so a phototransistor is simply designed for maximum junction exposure. Phototransistors for fiber-optic work are generally built in a plastic case with clear lens covering the semiconductor chip. The SFH350V of Figure 18–24 is a typical phototransistor detector.

The SFH350V is designed in a plastic package similar to the SFH450V emitter of Figure 18–15, so it provides a direct physical connection for the fiber optic. The responsivity of the SFH350V is very high; it is typically 80 μA/μW (it is rated at a minimum of 16 μA/μW), which is more than 100 times more sensitive than a PIN photodiode.

This sensitivity comes at a price, however. Examine the specifications given for *turn-on time* and *turn-off time*. This device takes typically 20 μs to turn off, which translates into an effective bandwidth of only (0.35/20μs) or *17.5 kHz*. When the turn-off and turn-on times of a device are different, use the *longer* of the two times to calculate bandwidth.

The primary advantage of the phototransistor detector is the simplicity of the external circuitry for recovering information from the fiber. Figure 18–25 shows a circuit for recovering analog signals using an SFH350V. The circuit is designed to receive signals from the analog transmitter of Figure 18–16(b). The incoming light wave signal causes a current I_E to flow in the phototransistor. This current is directly proportional to the optical power level of the incoming light wave. Since the light wave signal is getting stronger and weaker in step with the information signal, I_E of the phototransistor also rises and falls in a similar manner. Resistor R_3 is present for simple adjustment of the Q point of the transistor (it should be adjusted for a dc voltage of approximately 2.5 volts at Q_1's emitter.) Resistors R_3 and R_5 convert the varying emitter current back into a varying voltage, which is coupled to the output through dc blocking capacitor C_2. The resulting output signal can be directly amplified by an audio amplifier and sent to a loudspeaker.

Recovery of digital signals is also quite simple, as shown in Figure 18–26. In the receiver of Figure 18–26 [which is designed to be compatible with the digital transmitter of Figure 18–16(a)], the incoming light beam from the fiber is turned *on* for a logic 1, and *off* for a logic 0. When the light beam is *on,* Q_1 (the SFH350V phototransistor) conducts, supplying forward bias for the base of Q_2, which in turn switches on. When Q_2 switches on, it pulls the input of the Schmitt trigger inverter (U2*A*) *low,* resulting in a logic 1 at the inverter's output pin.

The opposite happens when the light beam is turned off. Q_1 switches off, which allows R_3 to turn off Q_2. When Q_2 turns off, its collector-emitter now looks like an open switch, and resistor R_4 can now pull the input of the inverter *high,* resulting in a logic 0 at the output. Thus, a sequence of on-and-off variations in the incoming light beam are directly converted back into a TTL-compatible digital signal, which could be coupled into a UART or other device for recovery of the data.

Figure 18–24 Data sheet of the SFH350V (Source: Infineon Technologies, A Siemens Company. Reprinted with Permission)

Plastic Fiber Optic Phototransistor Detector
Plastic Connector Housing

SFH 350
SFH 350V

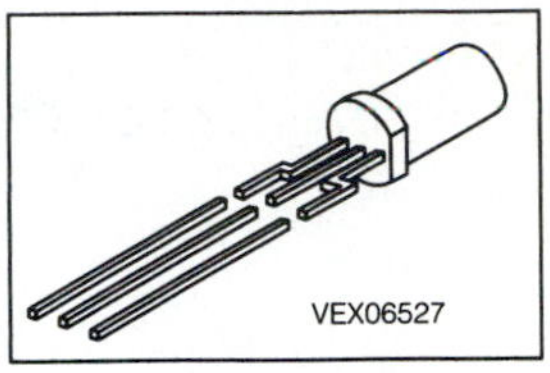

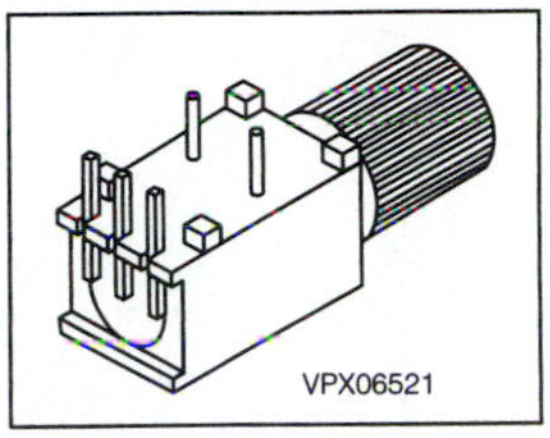

Features

- 2.2 mm aperture holds standard 1000 micron plastic fiber
- No fiber stripping required
- Good linearity
- Sensitive in visible and near IR range
- Molded microlens for efficient coupling

Plastic Connector Housing

- Mounting screw attached to the connector
- Interference-free transmission from light-tight housing
- Transmitter and receiver can be flexibly positioned
- No cross talk
- Auto insertable and wave solderable
- Supplied in tubes

Applications

- Household electronics
- Power electronics
- Optical networks
- Medical instruments
- Automotive electronics
- Light barriers

Type	Ordering Code
SFH 350	Q62702-P1033
SFH 350V	Q62702-P264

Maximum Ratings

Parameter	Symbol	Values	Unit
Operating temperature range	T_{OP}	–55 ... +100	°C
Storage temperature range	T_{STG}	–55 ... +100	°C
Soldering temperature (2 mm from case bottom, $t \leq 5$ s)	T_S	260	°C
Collector-emitter voltage	V_{CE}	50	V

Figure 18–24
(continued)

Maximum Ratings (cont'd)

Parameter	Symbol	Values	Unit
Collector current	I_C	50	mA
Collector peak current ($t \leq 10$ s)	I_{CP}	100	mA
Emitter-bas voltage	V_{EB}	7	V
Reverse voltage	V_R	30	V
Power dissipation $T_A = 25$ °C	P_{tot}	200	mW
Thermal resistance, junction/air	R_{thJA}	375	K/W

Characteristics ($T_A = 25$ °C)

Parameter	Symbol	Values	Unit
Maximum photosensitivity wavelength	λ_{Smax}	850	nm
Photosensitivity spectral range ($S = 10\%\ S_{max}$)	λ	400 … 1100	nm
Dark current ($V_R = 20$ V)	I_R	1 (≤ 10)	nA
Capacitance ($f = 1$ MHz, without light) ($V_{CE} = 0$ V) ($V_{CB} = 0$ V) ($V_{EB} = 0$ V)	 C_{CE} C_{CB} C_{EB}	 10.5 21.5 20.5	 pF pF pF
Rise and fall times of photocurrent ($R_L = 1$ kΩ, $V_{CE} = 5$ V, $I_C = 10$ mA, $\lambda = 959$ nm) 10% … 90% 90% … 10%	 t_R t_F	 20 20	 ms ms
Photocurrent ($V_{CE} = 5$ V, $\Phi_{IN} = 10$ µW coupled from the end of a plastic fiber, $\lambda = 660$ nm)	I_{CE}	0.8 (≥ 0.16)	mA
Forward voltage ($I_F = 50$ mA)	V_F	2.1 (≤ 2.8)	V
Temperature coefficient HFE	TC_{HFE}	0.55	%/K
Temperature coefficient I_{CE} $\lambda = 560 … 660$ nm	TC_I	0.34	%/K
Temperature coefficient I_{CE} $\lambda = 830$ nm	TC_I	0.49	%/K
Temperature coefficient I_{CE} $\lambda = 950$ nm	TC_I	0.66	%/K

Figure 18–24
(continued)

Relative spectral sensitivity $S_{rel} = f(\lambda)$

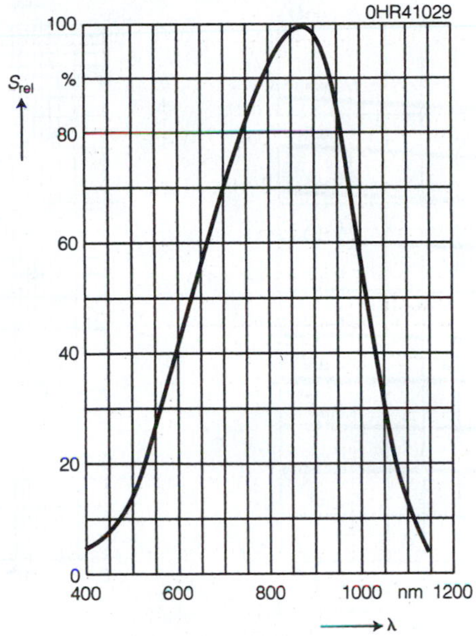

Capacitance $C = f(V_R)$, f = 1 MHz, $E_V = 0$

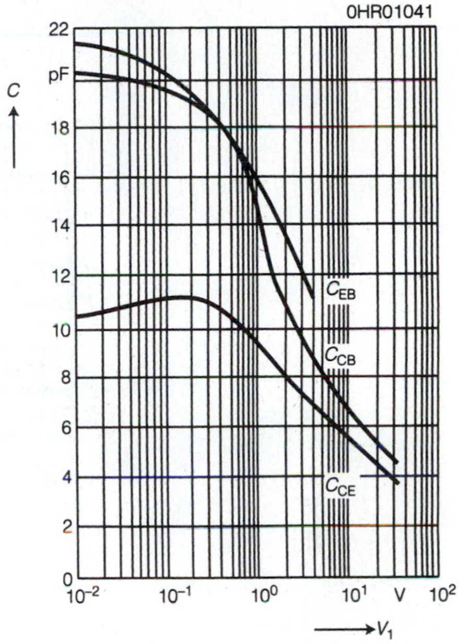

Photocurrent $I_C = f(\Phi_{OUT})$, V_{CE} = 5 V, λ = 560 ... 950 nm

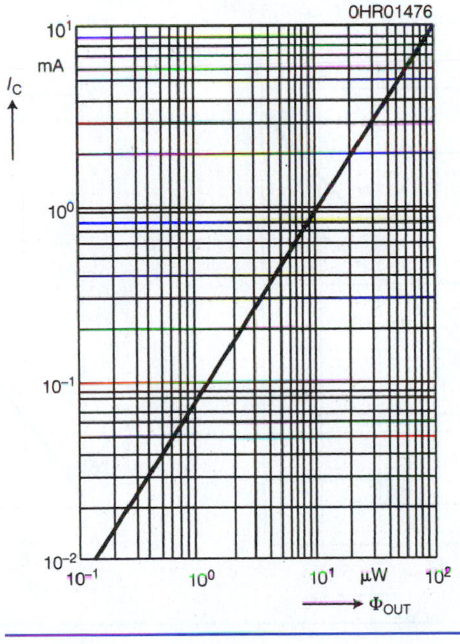

Current gain $HFE = f(I_C)$, V_{CE} = 5 V, T_A = 25 °C

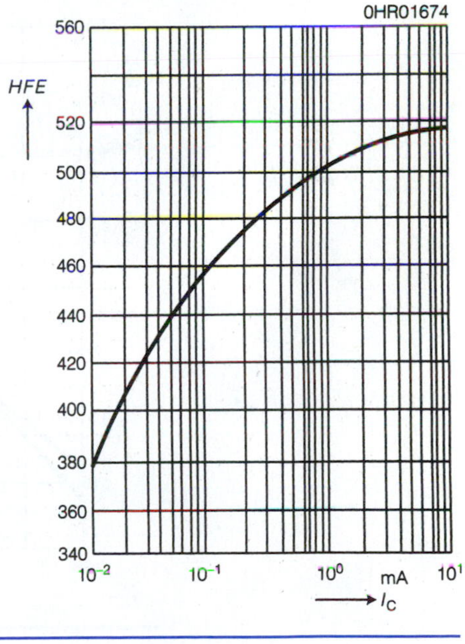

Figure 18–24
(continued)

Output characteristics $I_C = f(V_{CE})$,
I_B = parameter

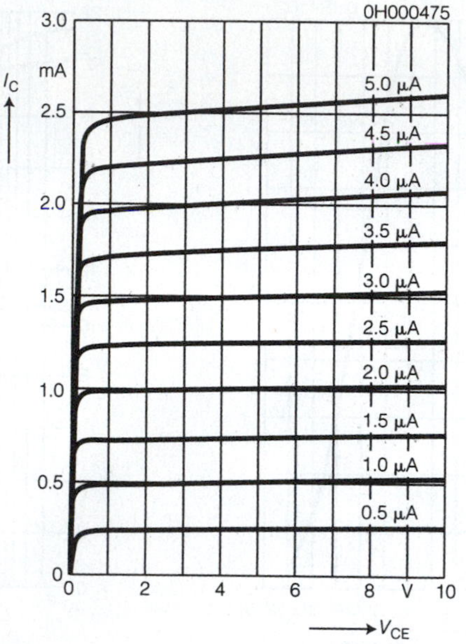

Response time $t = f(I_C)$, V_{CC} = 5 V,
λ = 950 nm

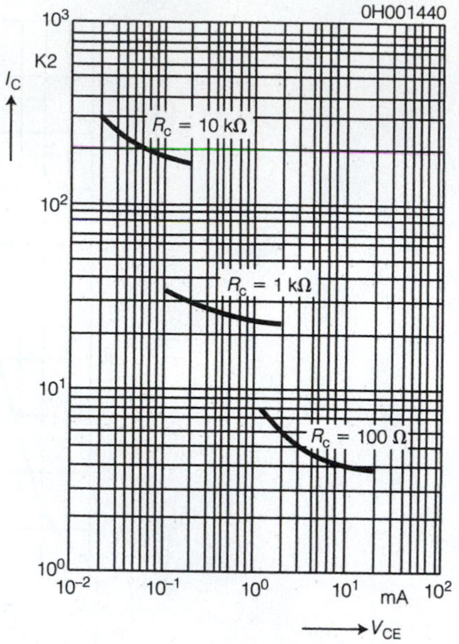

Photocurrent $I_C/I_{C25} = f(T_A)$, V_{CE} = 5 V,
λ = parameter

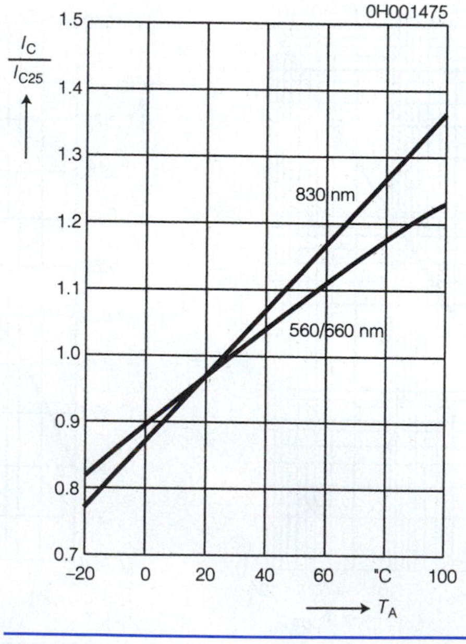

Current gain $HFE/HFE_{25} = f(T_A)$, V_{CE} = 5 V,
I_C = 1 mA

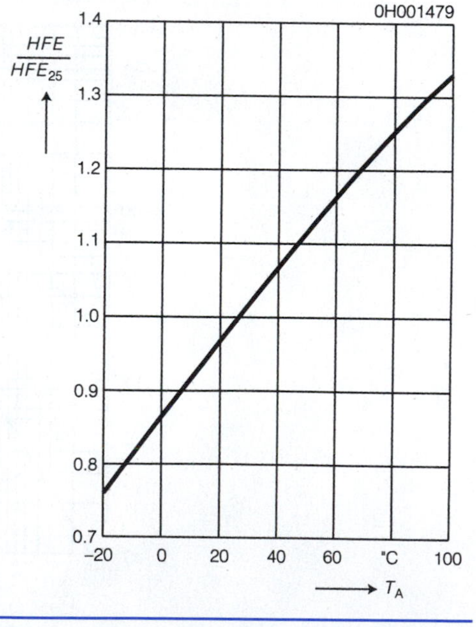

Figure 18–24
(continued)

Package Outlines (dimensions in mm, unless otherwise specified)

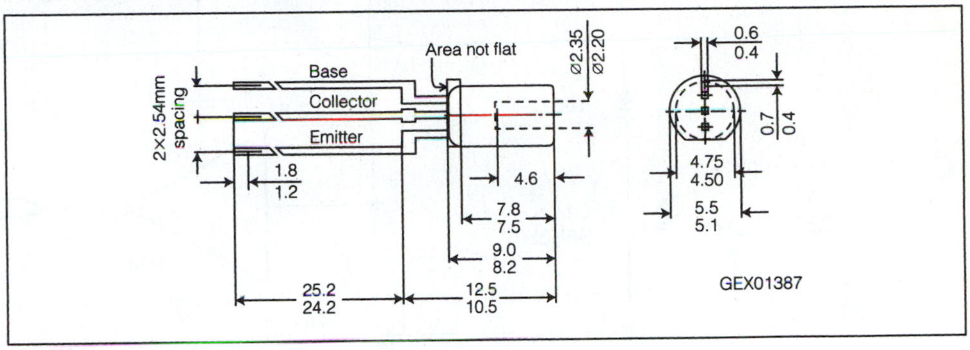

SFH 350

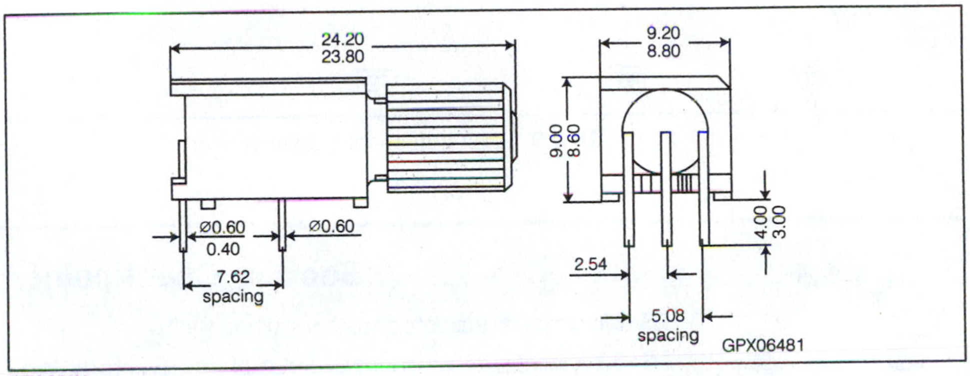

SFH 350V

Figure 18–25
Recovering analog signals with the SFH350V

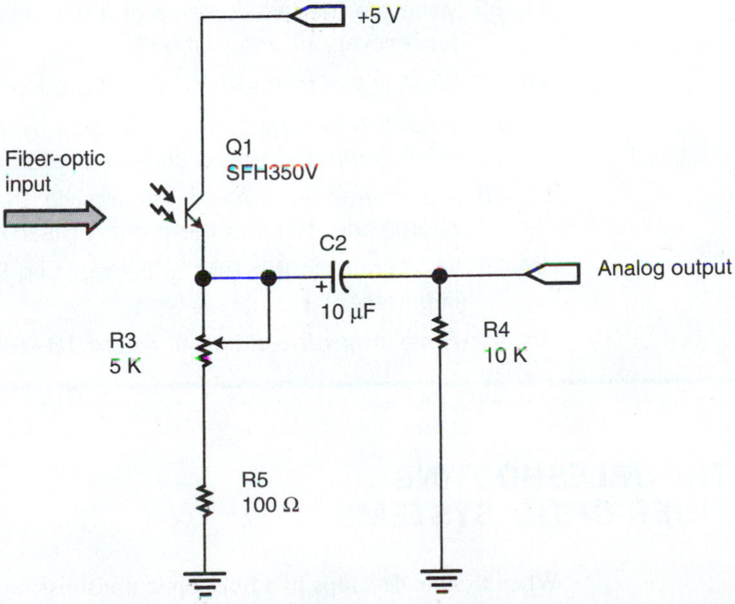

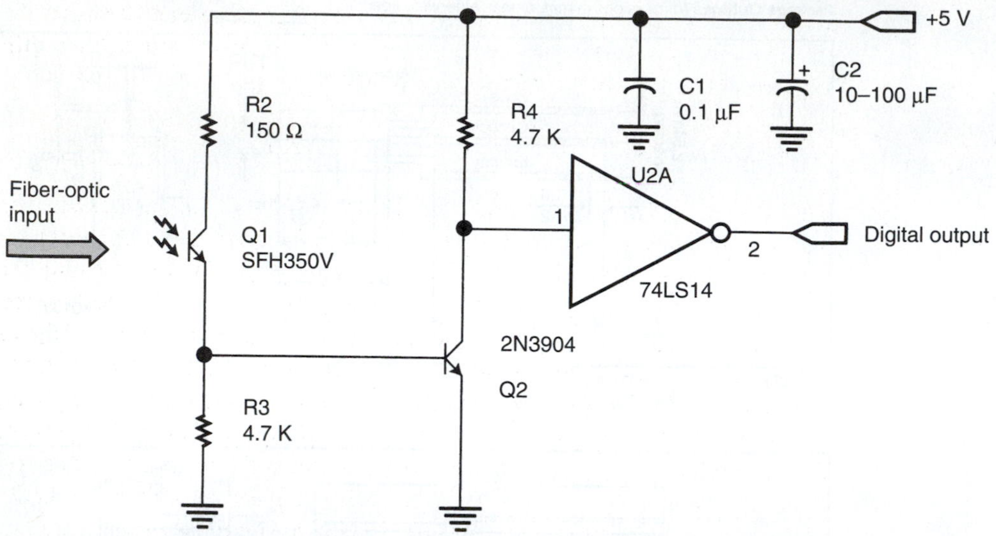

Figure 18–26 Digital fiber-optic demodulator

Section Checkpoint

18–31 What are photodetectors used for?

18–32 List the three most popular photodetectors. Which one is the fastest? Slowest?

18–33 Why is the *P* layer of the photodiode in Figure 18–22 made very thin?

18–34 What happens when a photon strikes the P–N junction of a photodiode?

18–35 What are the two currents of a photodiode? Which one is important for recovery of information?

18–36 What is the *responsivity* of a photodetector? Why is this important?

18–37 What two advantages are provided by the *I* layer of a PIN photodiode when compared with a conventional photodiode?

18–38 A phototransistor has a responsivity that is much higher than a photodiode. What is the primary disadvantage of this device?

18–39 What controls the emitter current I_E in Figure 18–25? Why is this important?

18–40 Explain how the circuit of Figure 18–26 operates.

18–5 TROUBLESHOOTING FIBER-OPTIC SYSTEMS

When trouble develops in a fiber-optic transmission system, it tends to show up in one of two modes. The first mode, which is the easiest to find, is total failure of the communications link. The second type of failure is *decreased quality of transmission,* which often shows up as signal distortion or increased bit-error rates in digital systems (which slows down the flow of traffic as computers are required to resend damaged data packets in response to the errors). The second type of failure can be much more difficult to troubleshoot, especially in complicated installations.

Total Communications Link Failure

The two most common causes of a total failure are physical damage to fiber-optic cables and the loss of power supply voltages. A technician must determine if the problem is in a fiber-optic cable or the fiber-optic interface components (transmitters and receivers).

Buried fiber-optic cables are occasionally damaged by construction crews and property owners, and often it is up to the technician to find the location of the cable damage. This can be done with the use of an *optical time-domain reflectometer (OTDR),* as shown in Figure 18–27. The OTDR operates by emitting a narrow pulse of light and measuring the time it takes for the pulse to echo back. A properly operating fiber-optic cable (with no damage) reflects very little light energy. A defect such as a cut will tend to cause energy to be sent back to the meter, which then calculates the distance to the fault based on the amount of time elapsed between outgoing and incoming pulses and the velocity factor of the fiber optic.

Even though the technician is most likely to be working with prepackaged fiber optic functional blocks (such as complete transmitters and receivers), a check of the power supply voltages to the equipment is still a very good idea in the initial stages of troubleshooting.

There are several ways to determine whether fiber-optic transmitters and receivers are working properly. First, an *optical loopback* connection can be completed between a suspected transmitter (receiver) and a known-good receiver (transmitter). If the loopback works, then the problem is either in the fiber-optic cable or at the other end of the link.

A second method for troubleshooting that is very effective is *component substitution.* A known-good transmitter or receiver module can usually be quickly substituted; if operation is restored, the problem was found. Be careful with this method because it is possible for a system to have *multiple problems.* For example, a transmitter may fail, and coincidentally, someone may damage a buried cable. No amount of component swapping will fix this problem!

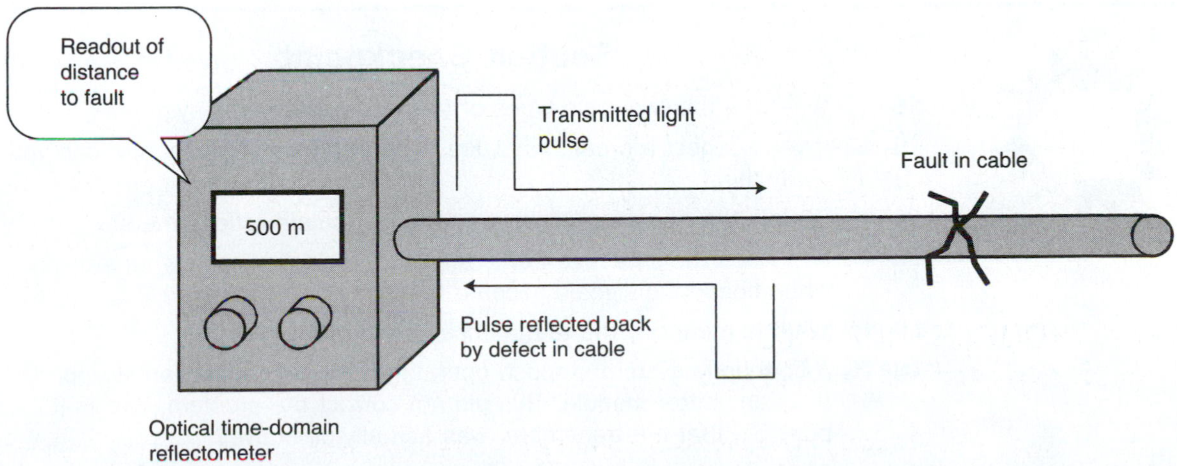

Figure 18–27 Optical TDR operation

The last method of troubleshooting requires an *optical power meter,* which is connected in place of a receiver to determine whether the signal levels are correct (and whether a signal is actually present or not!). An optical power meter can also be connected at the output of a suspected "dead" transmitter. Be careful—a power meter only indicates the presence of optical energy (light); it doesn't tell whether or not the signal is being properly modulated with information.

Decreased Quality of Transmission

In general, anything that abnormally weakens the optical signal between the transmitter and receiver will degrade the reliability of the fiber-optic link. With digital systems, this is often hard to track; because of their on and off nature, digital systems tend to work perfectly until the signal voltages fall below the switching thresholds of the logic family in use. At that point, the system operates quite erratically. The situation is complicated further with the use of digital error-detection and correction codes, which can mask many marginal failures. The system may still seem to operate perfectly (though a bit slower than normal, due to the retransmission of message packets) until the communication link degrades to the point of marginal operation (at which point, things greatly slow down because many messages are being retransmitted!).

A digital communications link that suddenly operates much more slowly than normal may be experiencing a gradual failure. The link slows down due to the retransmission of defective messages.

When a marginal condition is encountered, optical power readings at the receiver and transmitter often are valuable in locating the source of the trouble. If a system has been well documented during its construction, the builders will have recorded or calculated a range of acceptable optical power levels at each point within it. This documentation can be quite valuable when problems later arise.

Don't overlook the "easy" problems when troubleshooting a marginal system. These include crimped cables (an excessively tight bend in a fiber-optic cable effectively lowers its numerical aperture, causing a loss of optical input power) and contaminated or broken optical fittings. These last two problems are especially probable if someone else has attempted to troubleshoot the system before your arrival.

Section Checkpoint

18–41 What are the two main types of failures in fiber-optic systems?

18–42 If you suspect a break in a 1 km buried fiber-optic cable, how can you verify this?

18–43 Explain the operation of an optical time-domain reflectometer.

18–44 Why does the message traffic tend to slow down in a digital system when fiber-optic signals become weak?

18–45 What is meant by the term *optical loopback*?

18–46 A fiber-optic system stopped operating, and the technician swapped the transmitter module. This did not correct the problem. Why is it possible that the transmitter was actually defective?

SUMMARY

- Fiber optics have many advantages over copper media, including lower loss, greater available bandwidth, and safety in hazardous environments.
- A fiber-optic cable consists of a lightwave-conducting core, a lightwave-containing cladding, and in most cases an outer jacket, which affords mechanical protection.
- Light is electromagnetic energy, just like radio waves. It has a much higher frequency and shorter wavelength.
- Infrared light frequencies are used in fiber-optic communication because the core materials transmit these frequencies most efficiently.
- Light waves can be reflected, diffracted, and refracted as they move between two points in space. Refraction is the mode of propagation in fiber optics.
- The index of refraction for a material is the ratio of the velocity of light in free space to the speed of light in the material. The index depends slightly on frequency.
- The core and cladding work together to propagate light in a fiber by the mechanism of refraction.
- The *critical angle* is the minimum angle at which a ray of light must approach the interface of two different materials and be refracted or bent back into the first material. It depends on the indices of refraction for the two materials.
- The numerical aperture (NA) of a fiber is a measure of how easy it is to couple light into it. The maximum value is 1, which is the easiest possible coupling.
- The *acceptance cone angle* gives one-half of the total range of entry angles for light waves into a fiber.
- A *propagation mode* describes a particular type of available light wave path within an optical fiber.
- A *single-mode* fiber allows only one propagation mode, in which light waves move almost straight through.
- In *multimode* fibers, many propagation modes are possible, and the different parts of the optical signal arrive at the receiver at different times.
- *Modal dispersion* is the name given to the distortion of transmitted signals caused by the differences in time-of-arrival caused by the various available propagation modes.
- Modal dispersion limits the available bandwidth of a fiber by stretching out the rise and fall time of digital pulses. It gets worse as a fiber is lengthened.
- *Graded-index* fiber optics reduce modal dispersion by selectively slowing down part of the light wave energy, which causes most of the wave to arrive at the same time.
- Chromatic dispersion is caused by the variance in refractive index with light wavelength (frequency). It is what causes light to be split into colors by a prism.
- Waveguide dispersion is caused by the difference in velocity experienced by portions of the optical signal that pass through the cladding, which is a necessary part of the refraction mechanism.
- Fibers are rated for attenuation in decibels per kilometer or, for inexpensive types, dB per 100 meters.
- LEDs are the simplest fiber-optic emitters but don't produce much optical power (usually less than 1 mW) and are not monochromatic (many wavelengths of light are emitted from them).
- Lasers can emit many watts of power but require sophisticated circuitry to drive them. They are also less reliable than LEDs. However, much more bandwidth is available with lasers.
- A semiconductor laser diode must be treated with care. It is very sensitive to electrostatic discharge (ESD) and can be blown instantly by improper drive currents.
- The drive circuitry for laser diodes usually includes automatic power control, which is derived from a sampling photodiode. The

photodiode is often an integral part of the laser diode.

- Lasers present many safety hazards, especially the invisible types used in fiber optics. Appropriate safety practices must be used at all times.
- Photodiodes, PIN photodiodes, and phototransistors are popular fiber-optic detectors. The PIN photodiode provides the highest data rate among these devices, while the phototransistor allows low data-rate operation with very simple circuitry.
- There are two main types of trouble that appear in fiber-optic systems, *total failure* (easiest to track down) and *decreased quality of transmission*. Total failure is often caused by loss of power supply voltages or damaged cables.
- An optical time-domain reflectometer (OTDR) is handy for localizing faults in fiber-optic cables.
- Substitution of known-good modules is an effective troubleshooting tool. Optical measurements are even better when the appropriate instrumentation is available.

PROBLEMS

1. What are fiber optics? What carries the information in a fiber-optic cable?
2. List several advantages of fiber-optic cable over conventional copper transmission line. Are there any disadvantages of fiber?
3. Draw a diagram showing the parts of a fiber-optic cable. Which parts of the cable are responsible for guiding the light?
4. Calculate the frequency of the following light sources:
   a. reddish-orange, $\lambda = 680$ nm
   b. blue-green, $\lambda = 500$ nm
5. Define each of the following:
   a. reflection
   b. diffraction
   c. refraction
6. In a certain material, light waves travel at 1.5×10^8 m/sec. What is its index of refraction?
7. Calculate the velocity of propagation of light through the following substances:
   a. water
   b. glass
   c. fused quartz
8. Calculate the wavelength of a ray of yellow light if it is passing through a sheet of glass.
9. Calculate the critical angle for the following core-cladding index combinations:
   a. $n_{\text{core}} = 1.57, n_{\text{cladding}} = 1.42$
   b. $n_{\text{core}} = 2.00, n_{\text{cladding}} = 1.80$
   c. $n_{\text{core}} = 1.33, n_{\text{cladding}} = 1.00$
   d. $n_{\text{core}} = 1.75, n_{\text{cladding}} = 1.55$
10. Calculate the numerical aperture (NA) for each of the four combinations of refractive index given in problem 9.
11. Examine the four combinations of refractive index given in problem 9. Find the combination with the *minimum* and *maximum* acceptance cone angles and report these two angles.
12. Define the term *propagation mode.* What are the two major classifications?
13. Explain how an optical fiber can be made to operate in a single mode.
14. What is the primary disadvantage of single-mode fibers?
15. What is *modal dispersion?* What causes this effect in multimode fibers?
16. Calculate the dispersion-limited bandwidth for a step-index optical fiber with a core refractive index of 1.7, a cladding refractive index of 1.2, and a length of 100 m.
17. Explain the difference between graded-index and step-index fibers. How is modal dispersion reduced in a graded-index fiber?
18. Explain the mechanism behind chromatic (material) dispersion. Which light source is likely to cause more chromatic dispersion, LEDs or lasers? Why?

19. What factors control the amount of signal attenuation in a fiber-optic cable?

20. What type of modulation is almost always used with LED and laser light sources?

21. A certain LED is rated with a 20 ns rise time. What bandwidth can it provide?

22. Draw a simple circuit that can place a digital signal onto a fiber, using an LED as the emitter. Explain each part of the circuit.

23. Draw a circuit that can place a low-frequency analog signal onto a fiber, using an LED as the photosource. Explain its operation.

24. Draw a diagram of a helium–neon gas laser and explain its operation using outline form.

25. What are the advantages of solid-state lasers over gas types?

26. Solid-state laser diodes are delicate. What are two ways they can easily be destroyed?

27. When a laser diode has failed in a circuit, what should a technician do?

28. Draw a block diagram of a laser drive circuit. Why is automatic power control necessary?

29. List five safety rules for working with lasers and fiber optics.

30. Draw a diagram of the structure of a typical photodiode. Why is the P region of silicon made to be very thin?

31. Define the term *responsivity* as it applies to photodetector devices.

32. Draw the structure of a PIN photodiode. Explain why it operates much faster than a regular photodiode.

33. A certain phototransistor is rated with a turn-on time of 10 μs and a turn-off time of 25 μs. What bandwidth can this device support?

34. Explain the operation of the digital demodulator circuit of Figure 18–26.

35. What are the two failure modes in fiber-optic communication systems? Which one is easier to troubleshoot?

36. List the two most common causes of total communication failure in a fiber-optic system.

37. What instrument can be used to find the location of a fault in a long fiber-optic line?

38. What is the purpose of an optical power meter? Give an example of where it could be connected in a system to evaluate the system's performance.

39. As a communications link degrades, digital systems tend to operate perfectly at first, then slow down. Why is this usually so?

Appendix A: Decibels

Decibels are units designed for the comparison of voltage or power levels. Decibels allow a wide range of values to be expressed with small, easy-to-handle numbers and also allow addition and subtraction to be used in place of multiplication and division, which speeds calculations.

VOLTAGE GAIN UNITS AND VOLTAGE DECIBELS

The formula for the voltage gain of an amplifier circuit like that of Figure A–1 is as follows:

$$A_v = \frac{V_{\text{out}}}{V_{\text{in}}} \qquad \textbf{(A–1)}$$

where A_v is the *voltage gain* in *volts per volt* (V/V).

Note that although the voltage gain appears to be unitless (the units of voltage in the numerator and denominator cancel), it is not necessarily dimensionless. We encourage the writing of the units "V/V" after the number to emphasize that a *voltage* gain was measured. The voltage gain of the amplifier in Figure A–1 is

$$A_v = \frac{V_{\text{out}}}{V_{\text{in}}} = \frac{20\ \text{V}}{2\ \text{V}} = \underline{\underline{10\ \text{V/V}}}$$

A voltage gain can be expressed in decibels as follows:

$$A_v(\text{dB}) = 20 \log\left(\frac{V_{\text{out}}}{V_{\text{in}}}\right) \qquad \textbf{(A–2)}$$

Figure A–1 A circuit with voltage and power gain

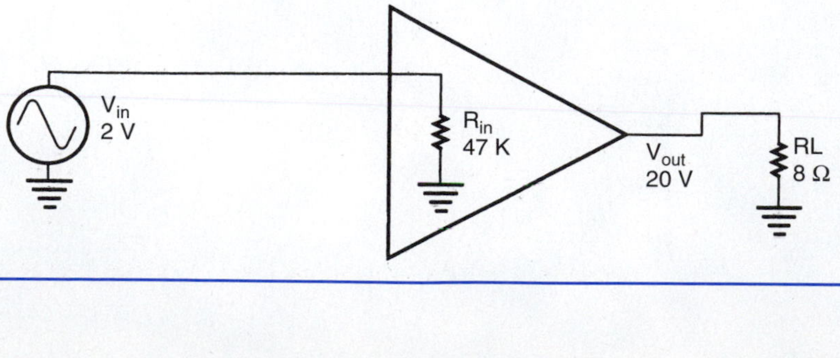

Figure A–2 A circuit with signal loss

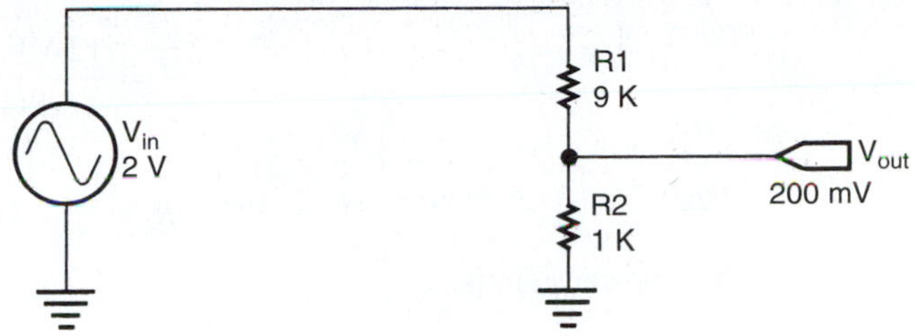

The voltage gain of the amplifier in Figure A–1 is

$$A_v(\text{dB}) = 20 \log\left(\frac{V_{\text{out}}}{V_{\text{in}}}\right) = 20 \log\left(\frac{20\text{ V}}{2\text{ V}}\right) = \underline{\underline{20\text{ dB}}}$$

The circuit of Figure A–1 generates a voltage gain. Its decibel gain is a therefore a positive number. The voltage divider of Figure A–2 generates a voltage *loss*. The output voltage is 10% of the input as determined by the voltage divider rule. The decibel voltage gain of Figure A–2 is

$$A_v(\text{dB}) = 20 \log\left(\frac{V_{\text{out}}}{V_{\text{in}}}\right) = 20 \log\left(\frac{200\text{ mV}}{2\text{ V}}\right) = \underline{\underline{-20\text{ dB}}}$$

The decibel voltage gain is *negative,* indicating a loss of signal.

POWER GAIN UNITS AND POWER DECIBELS

The *power gain* of a circuit is defined by

$$G_p = \frac{P_{\text{out}}}{P_{\text{in}}} \quad \text{(in units of watts/watt, or W/W)} \qquad \textbf{(A–3)}$$

where P_{out} is the power output of the circuit, and P_{in} is the power input.

The *decibel power gain* is calculated by

$$G_p(\text{dB}) = 10 \log\frac{P_{\text{out}}}{P_{\text{in}}} \qquad \textbf{(A–4)}$$

We can calculate the power gain of the circuit of Figure A–1 by first using Ohm's law at both the input and output ports:

$$P_{\text{in}} = \frac{V_{\text{in}}^2}{R_{\text{in}}} \qquad \textbf{(A–5)}$$

where V_{in} is the input voltage, and R_{in} is the input resistance of the amplifier.

$$P_{\text{out}} = \frac{V_{\text{out}}^2}{R_L} \qquad \textbf{(A–6)}$$

where V_{out} is the output voltage, and R_L is the load resistance.

The input and output powers can be calculated:

$$P_{in} = \frac{V_{in}^2}{R_{in}} = \frac{(2\text{ V})^2}{47\text{ k}\Omega} = 85.1\ \mu\text{W}$$

$$P_{out} = \frac{V_{out}^2}{R_L} = \frac{(20\text{ V})^2}{8\ \Omega} = 50\text{ W}$$

The power gain is

$$G_p = \frac{P_{out}}{P_{in}} = \frac{50\text{ W}}{85.1\ \mu\text{W}} = \underline{\underline{5.875 \times 10^5\text{ W/W}}}$$

The decibel power gain can now be calculated according to equation A–4:

$$G_p(\text{dB}) = 10\log\frac{P_{out}}{P_{in}} = 10\log\frac{50\text{ W}}{85.1\ \mu\text{W}} = \underline{\underline{57.7\text{ dB}}}$$

Go back and compare the value that we obtained for *decibel voltage gain* with the *decibel power gain* we just calculated. The numbers are very different—did we make an error in calculation? The answer is surprisingly *no!* The decibel readings are vastly different because the input and output resistances in the circuit are not equal. Because of this, each volt of signal creates a different power level at the input and output. *This is why it is important to mark gain calculations and measurements with power or voltage units, either in W/W or V/V.*

In general, voltage and power decibel calculations should be considered to be incompatible unless the resistance (impedance) levels at all points being measured or calculated is equal. In RF systems, this is often true, as 50 Ω is a standard value for interstage coupling in well-designed systems; however, in RF systems, instrumentation is usually designed to measure power, not voltage, so the technician is likely to be seeing *power* decibels on the instrument readouts.

COMMON DECIBEL POWER RATIOS

Certain power ratios show up over and over again in real-world systems, and the technician would do well to learn the following relationships:

Power ratio, out : in	Decibel value
2:1	3
4:1	6
10:1	10
1:2	–3
1:4	–6
1:10	–10

Note again that a power gain of less than 1 means a power *loss,* and a negative decibel reading will result.

Decibels Add and Subtract

Suppose that we want to estimate the output of a power amplifier that is rated as having a power gain of +23 dB. The input to the amplifier is 500 mW. What should the output power be? Our calculator is back at the office, and we need to know immediately whether the unit is functioning correctly. The panel meter on the amplifier shows a 55 watt output.

A power gain of 23 dB can be thought of as a combination of the following gains from the table:

- +10 dB (10:1 or ×10 power gain)
- +10 dB (10:1 or ×10 power gain)
- +3 dB (2:1 or ×2 power gain)

The 500 mW input signal is therefore subjected to two ×10 multiplications and a ×2 multiplication. We get:

$$P_{out} = 500 \text{ mW} \times 10 \times 10 \times 2 = \underline{\underline{100 \text{ W}}}$$

Our amplifier is sick indeed, because it is only producing 55 watts of output, which is almost 3 dB below what it has been designed to do. Further troubleshooting would be in order!

Appendix B: Bipolar Transistor Fundamentals

BIPOLAR TRANSISTOR BIASING

Bipolar junction transistors (BJTs) can be employed as either switches or amplifiers. When a BJT is employed as a switch, it is desired for the device to either be fully on (saturated) or fully off (cutoff, or open circuited). BJTs are usually biased into midpoint bias when used as class A linear amplifiers, or less than midpoint bias when used in class AB, B, or C. The bias point is located on the dc load line, as shown in Figure B–1.

> *TIP* The base of a transistor is the primary control input. If it appears that the base is being driven by a logic gate output or microcontroller output port pin, then it is likely that the device is being used as a switch. If the base is connected to a voltage divider or similar resistor network, then the transistor is probably being operated as a linear amplifier under midpoint bias.

The saturation current $I_{c(sat)}$ can be calculated as follows:

$$I_{c(sat)} = \frac{V_{CC}}{R_C + R_E} \qquad \textbf{(B–1)}$$

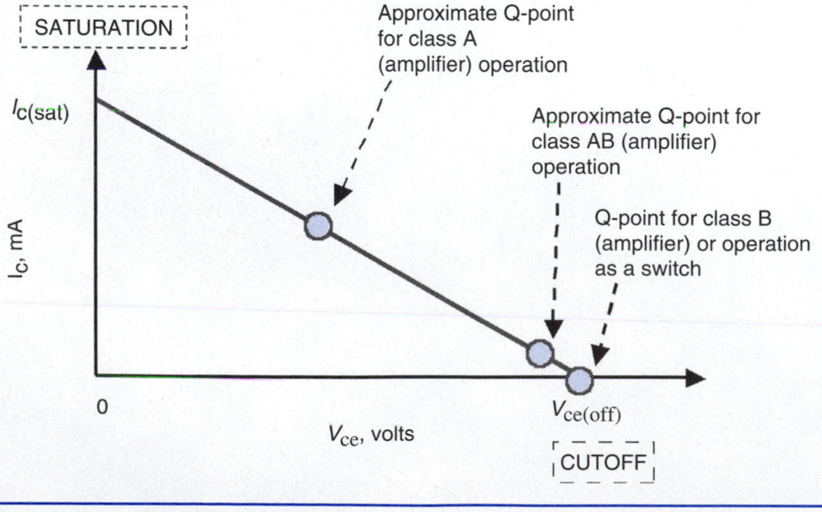

Figure B–1 The dc load line for a bipolar transistor circuit

Figure B–2 NPN transistor switching a motor

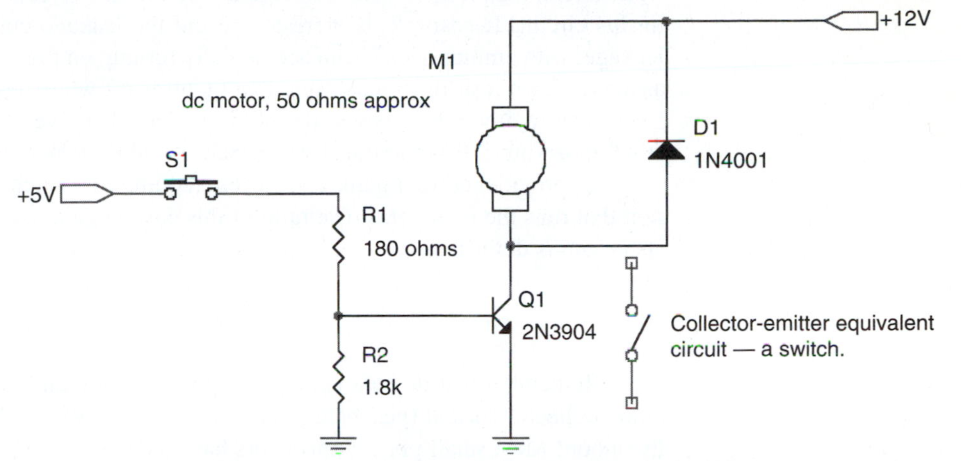

where V_{CC} is the power supply voltage, and R_C and R_E are the collector and emitter dc resistances. This formula is valid no matter how the BJT is being used.

The dc cutoff voltage $V_{ce(off)}$ is normally equal to V_{CC}.

Transistors as Switches

The devices in Figures B–2 and B–3 are being used as switches. One is an NPN device, the other is a PNP device. An NPN silicon transistor requires about 0.6 to 0.7 volts between its base-emitter terminals to turn on. The polarities for a PNP device are simply the opposite.

In Figure B–2, the saturation current for the transistor is:

$$I_{c(sat)} = \frac{V_{CC}}{R_C + R_E} = \frac{12\text{ V}}{50\ \Omega} = \underline{\underline{240\text{ mA}}} \text{ (There is no emitter resistor in this circuit.)}$$

When switch S_1 is depressed, positive voltage is applied to the base-emitter junction through R_1, turning on Q_1. Q_1's emitter-collector terminals then ideally appear as a closed

Figure B–3 A PNP transistor switching on a lamp

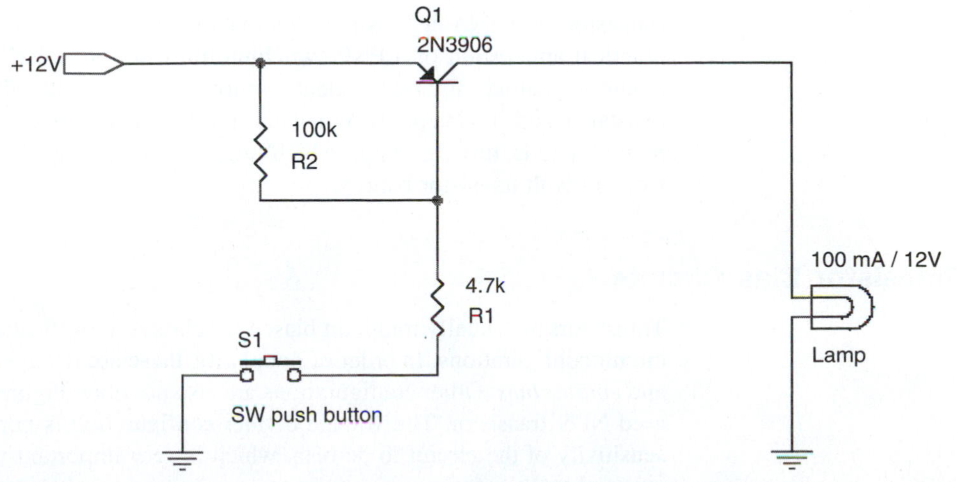

switch. Although it looks like a voltage divider, R_1 and R_2 are *not* performing this function in this circuit. Resistor R_2 is there to prevent the leakage current I_{CBO} (collector to base leakage, with emitter open) from accidentally turning on the transistor. You can think of R_2 as a pull-down resistor to make sure that Q_1 turns off when desired. Diode D_1 is a *catch* or *freewheeling* diode. It prevents the high-voltage inductive transient from motor turn-off from appearing across the transistor, which would destroy the transistor!

Q_1 provides current gain. The current required in switch S_1 is much less than the current that runs the motor (I_C). The ratio of this current gain is called the *dc beta* of the transistor and is defined as:

$$\beta_{dc} = \frac{I_C}{I_B} \quad \text{(B–2)}$$

Remember that dc beta is only an approximation and depends highly upon temperature, collector current (yes, beta is a nonlinear quantity), production run, and the phase of the moon! Most small plastic transistors have a dc beta of at least 100.

Assuming that Q_1 has a beta of 100, will Q_1 fully saturate in this circuit? We can answer that by calculating the base current. By Ohm's law, we can write:

$$I_B = \frac{V_{TH} - V_{BE}}{R_{TH}} \quad \text{(B–3)}$$

where V_{TH} is the Thévenin voltage driving the base, V_{BE} is the base-emitter drop (about 0.7 volt), and R_{TH} is the Thévenin resistance of the base drive circuit. For the circuit of Figure B–1, we get (approximately):

$$I_B = \frac{V_{TH} - V_{BE}}{R_{TH}} = \frac{5\text{ V} - 0.7\text{ V}}{180\ \Omega} = \underline{\underline{23.8\text{ mA}}}$$

The potential collector current is $\beta_{dc} I_B$ or (100)(23.8 mA) or 2.38 A. Because the collector current is limited by the load resistance of 50 Ω to a value much smaller than this, the transistor will become fully saturated and the collector-emitter voltage will be approximately zero (in reality, 100 mV or so for small signal transistors). Again, the transistor has given us current gain. We switched a 240 mA load current with a base current of less than 10% of this value. With efficient design, we can do even better than this.

In Figure B–3, we are switching on a 100 mA lamp with about 2.4 mA of base current. That's not bad! Notice that it takes a *negative* base emitter voltage to turn on a PNP transistor. In this case, R_2 is a *pullup* that ensures turn-off of Q_1. Figure B–4 is a fun and practical application of a PNP switching transistor, a "solar" alarm clock. It uses a CdS (cadmium sulfide) light-dependent resistor. When the CdS cell is in the dark, it has a high resistance and Q_1 is kept off. When the sun illuminates the circuit, the resistance of R_1 dramatically falls, turning on Q_1, which turns on the buzzer. The circuit can operate directly from a 9 volt transistor battery.

Transistor Bias Circuits

Transistors are usually midpoint biased for class A amplification. There are three common circuit configurations. In order of popularity, these are *voltage divider, collector feedback,* and *emitter bias*. Other configurations are also possible. Figure B–5 is a voltage divider biased NPN transistor. The voltage divider configuration is popular because it reduces the sensitivity of the circuit to dc beta, which is very important when operating over a wide range of temperatures.

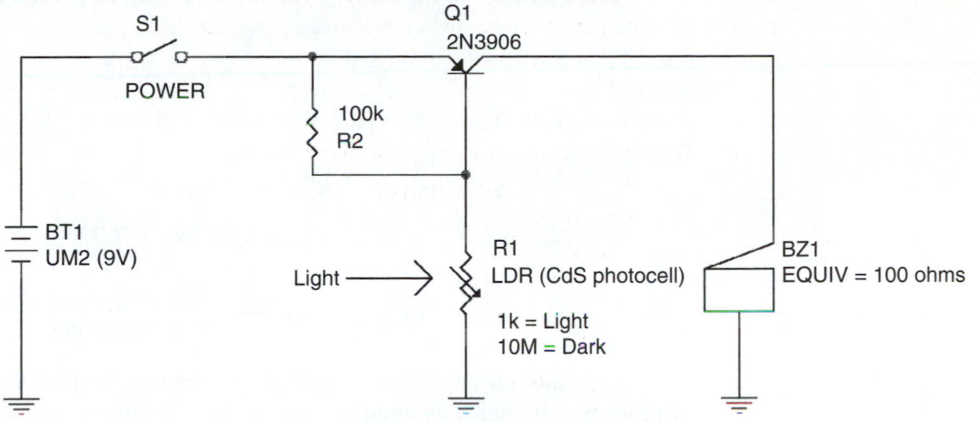

Figure B–4 A "solar" alarm clock

For voltage divider bias, we can estimate the transistor voltages and currents as follows. These estimates assume correct initial design, and a dc beta of at least 100, which is typically true for small signal circuits. The three electrode voltages can be calculated by:

$$V_B \approx V_{CC}\frac{R_1}{R_1 + R_2} \quad \textbf{(B–4)}$$

$$V_E \approx V_B - V_{BE} \approx V_B - 0.7\text{ V (for silicon; use } V_{BE} = 0.3\text{ V for germanium)} \quad \textbf{(B–5)}$$

$$I_C \approx I_E \approx \frac{V_E}{R_E},\text{ where } R_E \text{ is the dc resistance in the emitter circuit} \quad \textbf{(B–6)}$$

$$V_C = V_{CC} - I_C R_C,\text{ where } R_C \text{ is the dc resistance in the collector circuit} \quad \textbf{(B–7)}$$

The position on the load line is sometimes expressed as a percentage called "gamma." A value of 0 for gamma means saturation, and 1 means cutoff. A value of 0.5 means midpoint dc bias. This percentage can be calculated by:

$$\gamma = \frac{V_{CE}}{V_{CC}},\text{ where } V_{CE} \text{ is the collector-emitter voltage.} \quad \textbf{(B–8)}$$

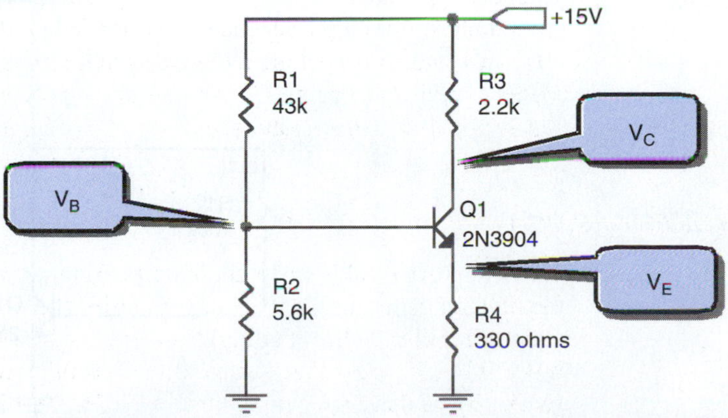

Figure B–5 Voltage divider biasing

The transistor in Figure B–5 has the following dc parameters:

$$V_B \approx V_{CC}\frac{R_1}{R_1 + R_2} \approx 15\text{ V}\frac{5.6\text{ k}}{5.6\text{ k} + 43\text{ k}} = \underline{\underline{1.7\text{ V}}}$$

$$V_E \approx V_B - V_{BE} \approx V_B - 0.7\text{ V} \approx 1.7\text{ V} - 0.7\text{ V} \approx \underline{\underline{1.0\text{ V}}}$$

$$I_C \approx I_E \approx \frac{V_E}{R_E} \approx \frac{1.0\text{ V}}{330\ \Omega} \approx \underline{\underline{3\text{ mA}}}$$

$$V_C = V_{CC} - I_C R_C = 15\text{ V} - (3\text{ mA})(2.2\text{ k}) = \underline{\underline{7.7\text{ V}}}$$

$$\gamma = \frac{V_{CE}}{V_{CC}} = \frac{(7.7\text{ V} - 1.0\text{ V})}{15\text{ V}} = \underline{\underline{0.45}}$$ (The dc Q-point is approximately centered on the load line.)

Collector feedback bias (Figure B–6) is less stable than voltage divider bias because it more directly depends upon beta dc of the transistor. Because of the feedback configuration, the dc equations are a little different. The circuit is used in applications where Q-point stability is not very critical (such as in a capacitor-coupled amplifier). It reduces the number of required resistors.

The dc values for a collector feedback circuit are calculated as follows:

$$V_B \approx V_{BE} \approx 0.7\text{ V}$$ because the emitter is usually tied directly to ground; **(B–9)**

$$I_C = \frac{V_{CC} - V_{BE}}{R_C + R_B/\beta_{dc}}$$ (notice that beta dc is now in the equation) **(B–10)**

The collector voltage is calculated according to equation B–7. For the circuit of Figure B–6 (assuming that beta dc is 100), the dc parameters are:

$$V_B \approx V_{BE} \approx \underline{\underline{0.7\text{ V}}}$$

$$I_C = \frac{V_{CC} - V_{BE}}{R_C + R_B/\beta_{dc}} = \frac{10\text{ V} - 0.7\text{ V}}{1\text{ k} + (200\text{ k}/100)} = \underline{\underline{3.1\text{ mA}}}$$

$$V_C = V_{CC} - I_C R_C = 10\text{ V} - (3.1\text{ mA})(1\text{ k}) = \underline{\underline{6.9\text{ V}}}$$

$$\gamma = \frac{V_{CE}}{V_{CC}} = \frac{(6.9\text{ V} - 0.0\text{ V})}{10\text{ V}} = \underline{\underline{0.69}}$$ (The dc Q-point isn't very well centered at $\beta = 100$.)

Figure B–6 Collector feedback bias

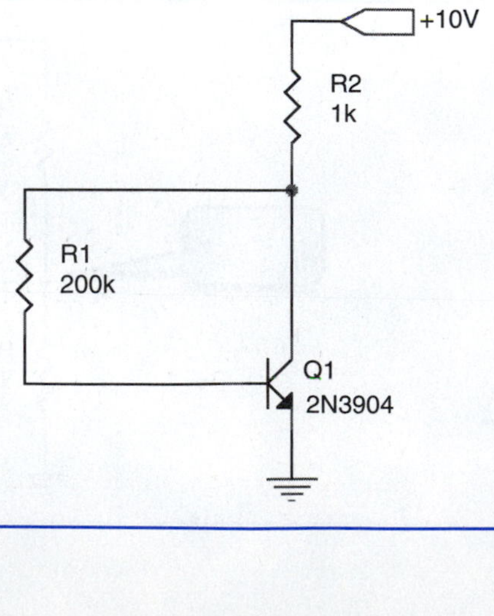

Figure B–7 Emitter bias

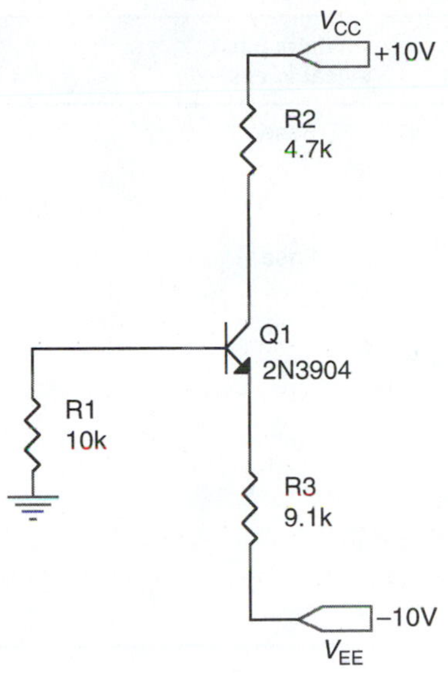

Emitter bias is used when a negative power supply is available. It's handy because the base voltage can be zero volts, which is useful because the input coupling capacitor can be eliminated, which extends the lower limit of frequency response to dc. Figure B–7 shows an NPN transistor with emitter bias.

For an emitter-bias circuit, the dc parameters can be approximated as follows:

$$V_E \approx -V_{BE} \approx -0.7 \text{ V} \quad \textbf{(B–11)}$$

$$V_B \approx 0 \text{ V, because the base is usually coupled directly to ground} \quad \textbf{(B–12)}$$

$$I_C \approx I_E \approx \frac{|V_{EE}| - V_{BE}}{R_E} \quad \textbf{(B–13)}$$

The collector voltage is calculated using equation B–7. For the circuit of Figure B–7, the dc values would be:

$$V_E \approx -V_{BE} \approx \underline{\underline{-0.7 \text{ V}}}$$

$$V_B \approx \underline{\underline{0 \text{ V}}}$$

$$I_C \approx I_E \approx \frac{|V_{EE}| - V_{BE}}{R_E} \approx \frac{10 \text{ V} - 0.7 \text{ V}}{9.1 \text{ k}} \approx \underline{\underline{1 \text{ mA}}}$$

$$V_C = V_{CC} - I_C R_C = 10 \text{ V} - (1 \text{ mA})(4.7 \text{ k}) = \underline{\underline{5.3 \text{ V}}}$$

Transistor Small-Signal Amplifiers

There are three basic circuit configurations used for small-signal amplifiers. They are summarized in Table B–1.

Common Emitter Analysis

A common emitter amplifier is used where maximum power gain is needed, but not at the highest frequencies, since its high-frequency response isn't very good. The equations in

Table B–1 A comparison of BJT amplifier configurations

Amplifier configuration	Where input signal is applied	Where output is obtained from	Major characteristics
Common emitter	Base	Collector 180° phase shift	Best power gain of all configurations—but worst high-frequency response due to Miller effect.
Common collector	Base	Emitter In phase	Voltage gain approximately 1 V/V in all cases. Provides current gain. Better high-frequency response than common-emitter configuration.
Common base	Emitter	Collector In phase	Best high-frequency response of all configurations, but very low input impedance. Usually seen in RF (radio-frequency) applications as a "front end" amplifier.

> The program "cesim.exe" which is located in the \software folder of the CD that comes with this text can be used to evaluate common emitter amplifiers.

Table B–2 can be used to analyze common emitter amplifiers. You probably don't need *all* of the relationships; they've been included for the sake of completeness. The equations refer to the schematic of Figure B–8.

What is the voltage gain, input impedance, and output impedance of the amplifier in Figure B–9? Show all calculations.

First, do the dc analysis:

$$V_B \approx V_{CC}\left(\frac{R_2}{R_1 + R_2}\right) \approx 15\text{ V}\left(\frac{5.6\text{ k}}{5.6\text{ k} + 43\text{ k}}\right) \approx 1.72\text{ V}$$

$$V_E = V_B - V_{BE} = 1.72\text{ V} - 0.7\text{ V} = 1.02\text{ V}$$

$$I_E = V_E/R_E = V_E/R_4 = 1.02\text{ V}/330\ \Omega = 3.1\text{ mA}$$

$$V_C = V_{CC} - I_C R_C = 15\text{ V} - (3.1\text{ mA})(2.2\text{ k}) = 8.15\text{ V}$$

With the dc values in hand, the ac analysis can now be done using the appropriate equations from Table B–2.

$$r'e = \frac{25\text{ mV}}{I_E} = \frac{25\text{ mV}}{3.1\text{ mA}} = 8\ \Omega$$

$$A_V = \frac{r_C}{r_E + r'e} = \frac{R_3 \parallel R_L}{R_4 + r'e} = \frac{1100\ \Omega}{330\ \Omega + 8\ \Omega} = \underline{\underline{3.25\text{ V/V}}}$$

$$Z_{IN} = R_1 \parallel R_2 \parallel Z_{IN(BASE)} = R_1 \parallel R_2 \parallel [\beta(r_E + r'e)] =$$

$$43\text{ k} \parallel 5.6\text{ k} \parallel [100(330\ \Omega + 8\ \Omega)] = \underline{\underline{4321\ \Omega}}$$

$$Z_{OUT} = R_C = R_3 = \underline{\underline{2.2\text{ k}}}$$

Table B–2 Common emitter analysis formulas

Parameter name	Formula	Comment
Voltage gain	$A_v = \frac{r_C}{r'e + r_E}$	r_c is R_L in parallel with R_c. r_E is R_{e1} in the circuit above.
Z_{in}(base)	$Z_{IN(BASE)} = (\beta + 1)(r'e + r_E)$	Input impedance looking into the base electrode. r_E is R_{e1} in the circuit above.
Z_{in}	$Z_{in} = R_{B1} \| R_{B2} \| Z_{IN(BASE)}$	Amplifier total Z_{in} (assumes that frequency is above f_{lco}).
Compliance	$V_{pp} = 2I_{c(q)}r_C$ $V_{pp} = 2V_{ce(q)}\left(\frac{r_C}{r_C + r_E}\right)$	Use the smaller of the two results to determine compliance. The ac load line is optimally centered when the two formulas give the same result. $I_{c(q)}$ is the quiescent (dc) collector current.
AC saturation	$i_{c(sat)} = I_{C(q)} + \frac{V_{ce(q)}}{r_C + r_E}$	Top of ac load line—maximum instantaneous collector current that can flow during each signal ac cycle.
AC cutoff	$v_{CE(off)} = V_{ce(q)} + I_{c(q)}(r_C + r_E)$	Bottom of ac load line—maximum instantaneous collector-emitter voltage (zero current).
Lower cutoff frequency	$f_{lco} = \frac{(\sqrt{2})^{n-1}}{2\pi RC}$	Lower cutoff (−3 dB) frequency from a single or multi-pole high-pass filter of n sections, each with equal RC time constant. When $n = 1$, formula simplifies to the familiar form: $f_{lco} = \frac{1}{2\pi RC}$
Power output	$P_O = \frac{V_{pp^2}}{8 R_L}$	Available power output, sinusoidal excitation. V_{pp} is the compliance of the amplifier.
Efficiency	$\eta = \frac{P_O}{P_i}$	Amplifier efficiency. P_O is the amplifier's power output, and P_i is the total dc power input to the unit. The maximum theoretical efficiency of an RC-coupled amplifier is 6/72 or 8.33%.

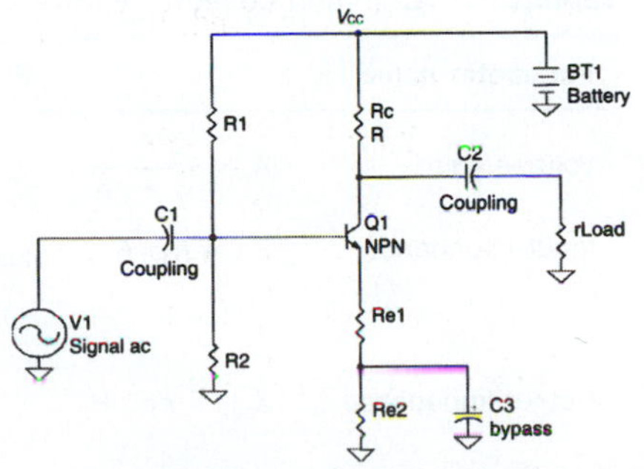

Figure B–8 Generic common emitter amplifier

Figure B–9 A common emitter amplifier for analysis

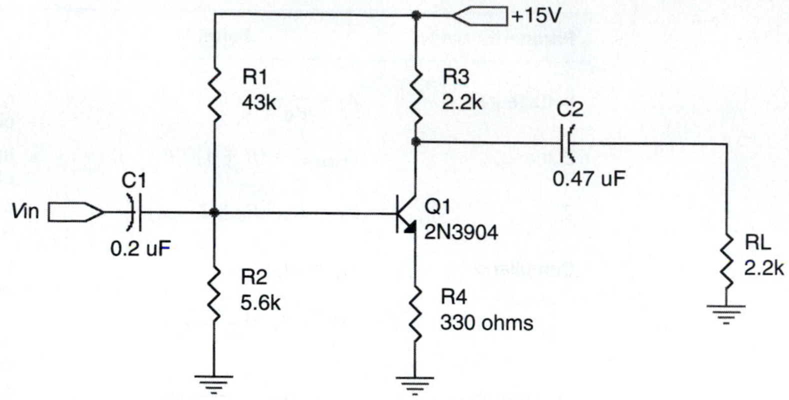

Common Collector Amplifiers

A common collector amplifier is used to provide current gain. It can electronically transform a low-impedance load into a higher impedance. The last stage in an audio amplifier is usually a class AB common collector amplifier, which provides the high level of current needed to drive the loudspeaker. Table B–3 gives the unique equations for common collector analysis.

Analyze the common-collector amplifier of Figure B–10 and determine the following: Voltage gain, input impedance, and output impedance.

First, do the dc analysis:

$$V_B \approx V_{CC}\left(\frac{R_2}{R_1 + R_2}\right) \approx 12\text{ V}\left(\frac{5.6\text{ k}}{5.6\text{ k} + 3.9\text{ k}}\right) \approx 7.07\text{ V}$$

$$V_E = V_B - V_{BE} = 7.07\text{ V} - 0.7\text{ V} = 6.37\text{ V}$$

$$I_E = V_E/R_E = V_E/R_4 = 6.37\text{ V}/330\ \Omega = 19.3\text{ mA}$$

$$V_C = V_{CC} = 12\text{ V}$$

Table B–3 Common collector analysis formulas

Parameter name	Formula	Comment
Voltage gain	$A_V = \dfrac{r_E}{r_E + r'e}$	Voltage gain, in V/V. Usually slightly less than 1.
Input impedance	$Z_{IN} = R_1 \parallel R_2 \parallel Z_{IN(BASE)} = R_1 \parallel R_2 \parallel [\beta(r_E + r'e)]$	Input impedance looking into the amplifier. $Z_{IN(BASE)}$ is the impedance looking into the base alone.
Output impedance	$Z_{OUT} = r'e \parallel R_E$	Output impedance of the amplifier; usually just a few ohms, dominated by transistor $r'e$.

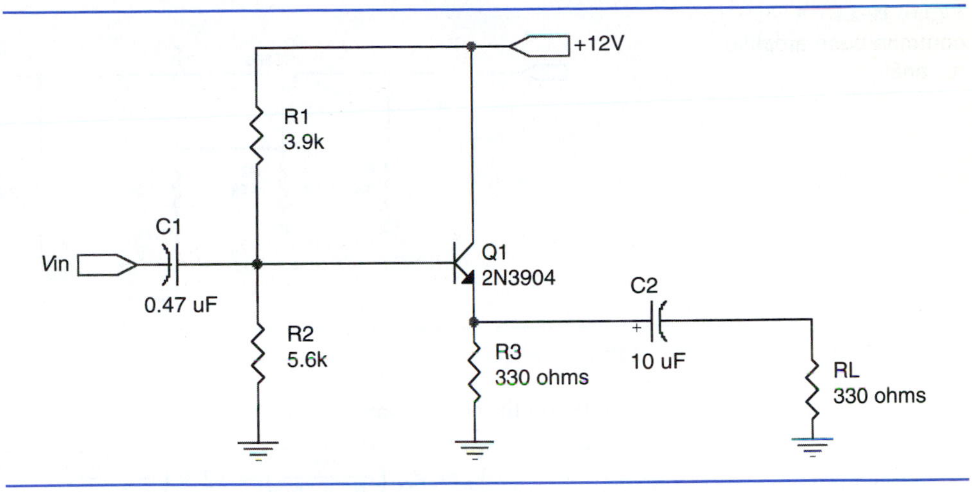

Figure B–10 Common-collector amplifier for analysis

Then, do the ac analysis:

$$r'e = \frac{25\text{ mV}}{I_E} = \frac{25\text{ mV}}{19.3\text{ mA}} = 1.3\ \Omega$$

$$A_V = \frac{r_E}{r_E + r'e} = \frac{R_3 \parallel R_L}{R_3 \parallel R_L + r'e} = \frac{165\ \Omega}{165\ \Omega + 1.3\ \Omega} = \underline{\underline{0.99\text{ V/V}}}$$

$$Z_{IN} = R_1 \parallel R_2 \parallel Z_{IN(BASE)} = R_1 \parallel R_2 \parallel [\beta(r_E + r'e)] = 3.9\text{ k} \parallel 5.6\text{ k} \parallel [100(165\ \Omega + 1.3\ \Omega)] = \underline{\underline{2019\ \Omega}}$$

$$Z_{OUT} = r'e \parallel R_E = 1.3\ \Omega \parallel 330\ \Omega \approx \underline{\underline{1.3\ \Omega}}$$

Common Base Amplifier

The common base amplifier has the best high-frequency response of all three basic configurations. It's often used as a VHF or UHF radio frequency amplifier. It has a very low input impedance (especially at radio frequencies) and doesn't provide as much power gain as a common-emitter design. The low input impedance (sometimes just a few ohms) means that extra impedance matching is needed prior to the amplifier in order to get maximum signal transfer (and power gain) through the stage. Table B–4 shows the unique formulas for common base analysis.

Analyze the amplifier of Figure B–11 and determine the dc voltages at all electrodes of the transistor, the input impedance, and the voltage gain.

Table B–4 Common base analysis formulas

Parameter name	Formula	Comment
Voltage gain	$A_v = \frac{r_C}{r'e}$	r_c is R_L in parallel with R_C.
Z_{IN}	$Z_{IN(BASE)} = r'e \parallel r_E$	Input impedance of amp is dominated by $r'e$ of transistor at low frequencies; usually very capacitive at high frequencies.

Figure B–11 A common base amplifier for analysis

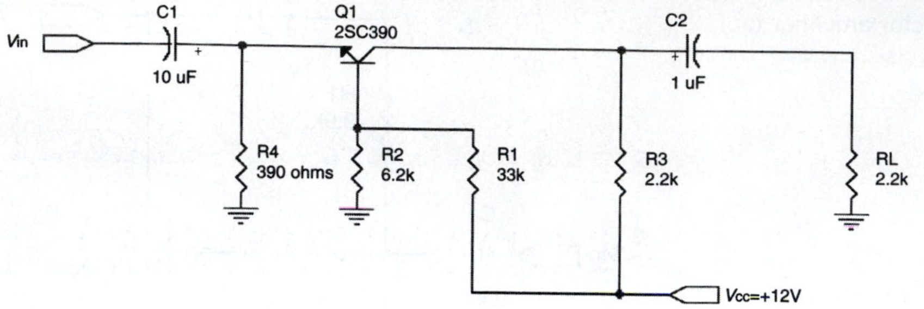

First do the dc analysis, as always:

$$V_B \approx V_{CC}\left(\frac{R_2}{R_1 + R_2}\right) \approx 12\text{ V}\left(\frac{6.2\text{ k}}{6.2\text{ k} + 33\text{ k}}\right) \approx \underline{\underline{1.9\text{ V}}}$$

$$V_E = V_B - V_{BE} = 1.9\text{ V} - 0.7\text{ V} = \underline{\underline{1.2\text{ V}}}$$

$$I_E = V_E/R_E = V_E/R_4 = 1.2\text{ V}/390\ \Omega = \underline{\underline{3\text{ mA}}}$$

$$V_C = V_{CC} - I_C R_C = 12\text{ V} - (3\text{ mA})(2.2\text{ k}) = \underline{\underline{5.4\text{ V}}}$$

Finally, do the ac analysis:

$$r'e = \frac{25\text{ mV}}{I_E} = \frac{25\text{ mV}}{3\text{ mA}} = 8.33\ \Omega$$

$$A_v = \frac{r_C}{r'e} = \frac{R_3 \| R_L}{r'e} = \frac{1100\ \Omega}{8.33\ \Omega} = \underline{\underline{132\text{ V/V}}}\ \text{(A great deal of gain is obtained.)}$$

$$Z_{IN(BASE)} = r'e \| r_E = r'e \| R_4 = 8.33\ \Omega \| 390\ \Omega = \underline{\underline{8.15\ \Omega}}\ \text{(A very low value.)}$$

Appendix C: Amateur Radio Communications

WHAT IS HAM RADIO?

Amateur or "ham" radio is a fascinating hobby that combines the elements of many disciplines, both technical and nontechnical. Amateur radio operators come from all walks of life. Ham radio isn't just about radio; it's about ***communication.*** Hams use many different modes, such as voice, CW (telegraphy or Morse code), television, and packet (computer communication), just to list a few.

The FCC regulates amateur radio and other communications services within the United States. According to 47CFR97.1, the fundamental purposes behind the amateur service are as follows:

- Recognition and enhancement of the value of the amateur service to the public as a voluntary noncommercial communication service, particularly with respect to providing emergency communications.
- Continuation and extension of the amateur's proven ability to contribute to the advancement of the radio art.
- Encouragement and improvement of the amateur service through rules that provide for advancing skills in both the communication and technical phases of the art.
- Expansion of the existing reservoir within the amateur radio service of trained operators, technicians, and electronics experts.
- Continuation and extension of the amateur's unique ability to enhance international goodwill.

WHAT DO HAM OPERATORS DO?

Amateur radio operators engage in a wide variety of activities ranging from highly technical experimental research to public service. Some typical activities include:

- *UHF, Satellite and Microwave Exploration.* Hams communicate well into the microwave region (above 1 GHz). In addition, amateur radio has several OSCARs—orbiting satellites carrying amateur radio. OSCARs relay both analog and digital signals worldwide for hams. The Radio Amateur Satellite Corporation (http://www.amsat.org/) is a worldwide group of hams that share an interest in building, launching, and using amateur satellites.

- *Homebrew Construction.* The most common homemade elements in a modern amateur radio station are antennas; however, many hams have built impressive stations from scratch (even though equipment is readily available, and historically less expensive now than ever). This is a great way to expand your working knowledge of RF and computer technology.
- *Public Service.* Hams provide safety communications for many public events such as races, runs, and other functions. They also work closely with government agencies such as the National Weather Service (where they act as trained weather spotters), FEMA, and the Red Cross (where they aid in disaster recovery). Because of the strong set of skills they possess (radio and computers are just a part of this), hams are well suited for these roles.
- *Computer Communications.* Hams enjoy many computer communications modes including packet radio (a text-based mode that provides error recovery), PSK31 (an efficient, narrowband mode used on HF that uses a PC's soundcard for modulation and demodulation of the digital signal), and even digital voice communications. Many hams write their own communications software.
- *Contesting.* There are dozens of contests held year round for making distant or rare contacts. Sometimes a group of hams will journey to an island or other isolated location with the purpose of establishing communications with as many hams as possible—this is called a DXpedition (DX means "long distance").
- *Miscellaneous Activities.* Hams get together in local clubs for fellowship and training. *Foxhunts* are contests to find hidden transmitters (a useful skill for locating downed aircraft and lost hikers). *Hamfests* are swap meets where equipment is bought, sold, and traded. Some hams act as *volunteer examiners* and administer FCC licensing tests under the auspices of a volunteer examination coordinator (VEC). And of course, many hams still like *talking on the air* with other hams using a microphone or *Morse code* (CW).

WHAT DOES IT TAKE TO BECOME A HAM?

In the United States you can obtain a Technician license by passing a written examination covering basic radio theory and FCC rules. The Morse code is not needed in order to enter the hobby (although it currently is a requirement for the General class and higher licenses). The Technician license will let you talk on the 146 MHz and 222 MHz amateur bands using FM, AM, or SSB. You can find license exams in your area by visiting www.w5yi.org or www.arrl.org. Both of these organizations are VECs.

WHERE CAN I LEARN MORE ABOUT HAM RADIO?

The American Radio Relay League, www.arrl.org, is the premier organization of radio amateurs in the United States. This site contains up to date material about the hobby and information on clubs in your area. Your school may also have an active ham radio group; this is another excellent place to explore. The site http://www.qsl.net/ contains pages for hundreds of ham radio clubs across the country. Finally, if your community doesn't have an amateur radio club, start one!

Appendix D: Well-Known TCP and UDP Port Numbers

The well-known ports are those from 0 through 1023. The latest IANA port assignments can be had from: http://www.iana.org/assignments/port-numbers. Note that although IANA policy is to reserve each port number for both TCP and UDP, both protocols may not necessarily be supported.

Port number	service
1	tcpmux: TCP port service multiplexer
5	rje: Remote Job Entry
7	echo: Echo Service
9	discard: Discard Packet Service (null sink)
11	systat: Users
13	daytime: Daytime Service
17	qotd: Quote Service
18	msp: Message Send Protocol
19	chargen: Ttytst Source
20	ftp-data; Data channel for FTP utility
21	ftp: FTP Service
22	ssh: Secure Shell
23	telnet: Telnet Shell
25	smtp: Simple Mail Transport Protocol
37	time: Time Service
39	rlp: Resource Location Service
42	name: Nameserver Protocol
43	nicname: Whois Protocol
49	tacacs: Login Host Protocol (TACACS)
50	re-mail-ck: Remote Mail Checking Protocol
53	domain: DNS
63	whois++: Whois Protocol
67	bootps: BOOTP Server
68	bootpc: BOOTP Client
69	tftp: Trivial FTP
70	gopher: Internet Gopher
71–74	netrjs-1 to netrjs-4: Remote Job Service

(continued on p. 768)

Port number	service
79	finger: Finger Service
80	http: World Wide Web Protocol
88	kerberos: Kerberos v5
95	supdup
101	hostname: Hostnames (legacy, from sri-nic)
102	iso-tsap: TSAP
105	csnet-ns
107	rtelnet: Remote Telnet
109	pop2: Post Office Protocol v2
110	pop3: Post Office Protocol v3
111	sunrpc: RPC 4.0 portmapper
113	auth: Authentication
115	sftp: Secure FTP
117	uucp-path
119	nntp: USENET News Transfer Protocol
123	ntp: Network Time Protocol
137	netbios-ns: NETBIOS Name Service
138	netbios-dgm: NETBIOS Datagram Service
139	netbios-ssn: NETBIOS Session Service
143	imap: Interim Mail Access Protocol v2
161	snmp: Simple Network Management Protocol
162	snmp-trap: Traps for SNMP
163	cmip-man: ISO management over IP (CMOT)
164	cmip-agent
174	mailq: MAILQ Service
177	xdmcp: X-Window Display Manager Control Protocol
178	nextstep: NeXTStep Window
179	bgp: Border Gateway Routing Protocol
191	prospero: Cliff Neuman's Prospero
194	irc: Internet Relay Chat
199	smux: SNMP Unix Multiplexer
201	at-rtmp: AppleTalk Routing Protocol
202	at-nbp: AppleTalk Name Binding Protocol
204	at-echo: Apple Echo
206	at-zis: AppleTalk Zone Information
209	qmtp: Quick Mail Transfer Protocol
210	z39.50: NISO Z39.50 Database
213	ipx
220	imap3: Interactive Mail Access Protocol
245	link: TtyLink
347	fatserv: Fatmen Server
363	rsvp_tunnel
369	rpc2portmap: Coda Portmapper
370	codaauth2: Coda Authentication Server
372	ulistproc: UNIX Listserv
389	ldap: Lightweight Directory Access Protocol
427	svrloc: Server Location Protocol
434	mobileip-agent
435	mobileip-mn

Port number	service
443	http: Secure HTTP
444	snpp: Simple Network Paging Protocol
445	microsoft-ds
464	kpasswd: Kerberos “passwd”
468	photuris
487	saft: Simple Asynchronous File Transfer
488	gss-http
496	pim-rp-disc
500	isakmp
512	exec: UNIX ‘exec’ Service (TCP)
513	login: UNIX ‘login’ Service (TCP)
513	whod: UNIX ‘whod’ Daemon (UDP)
514	shell: UNIX ‘shell’ (TCP)
514	syslog: UNIX ‘syslog’ (UDP)
515	printer: UNIX Print Spooler
517	talk: UNIX ‘talk’
518	ntalk: UNIX ‘ntalk’
519	utime: UNIX Time Service
520	efs: UNIX ‘efs’
521	ripng: UNIX ‘ripng’
525	timed: UNIX Time Server
538	gdomap: GNUstep distributed objects
535	iiop
540	uucp: UNIX ‘uucp’ Daemon
543	klogin: UNIX “Kerberized” Login
544	ksh: UNIX “Kerberized” rsh
546	dhcpv6-client
547	dhcpv6-server
554	rtsp: Real Time Stream Control Protocol
563	nntps: NNTP over SSL
565	whoami
587	submission: Mail Message Submission
610	npmp-local
611	npmp-gui611
612	hmmp-ind: HMMP Indication / DQS
631	ipp: Internet Printing Protocol
636	ldaps: LDAP over SSL
674	acap
694	ha-cluster: Heartbeat HA-cluster
749	kerberos-adm: Kerberos ‘kadmin’ v5
750	kerberos-iv
765	webster: Network Dictionary Service
767	phonebook: Network Phonebook Service
873	rsync
992	telnets
993	imaps: IMAP over SSL
994	ircs: IRC over SSL
995	pop3s: POP3 over SSL

Appendix E: ASCII, EBCDIC, and Unicode Character Sets

ASCII CODE MATRIX

Least Significant Nybble →

Most Significant Nybble ↓

	0	1	2	3	4	5	6	7	8	9	A	B	C	D	E	F
0	NUL	SOH	STX	ETX	EOT	ENQ	ACK	BEL	BS	HT	LF	VT	FF	CR	SO	SI
1	DEL	DC1	DC2	DC3	DC4	NAK	SYN	ETB	CAN	EM	SUB	ESC	FS	GS	RS	US
2	SP	!	“	#	$	%	&	‘	(	)	*	+	,	-	.	/
3	0	1	2	3	4	5	6	7	8	9	:	;	<	=	>	?
4	@	A	B	C	D	E	F	G	H	I	J	K	L	M	N	O
5	P	Q	R	S	T	U	V	W	X	Y	Z	[		]	^	_
6	‘	a	b	c	d	e	f	g	h	i	j	k	l	m	n	o
7	p	q	r	s	t	u	v	w	x	y	z	{	|	}	~	DEL

INTERPRETING UNICODE

Unicode represents character codes as 16-bit values instead of the usual eight for ASCII or EBCDIC. This allows representation of international character sets. If the MSB (most significant byte) of a Unicode word is zero (00H), then the code in the LSB (least signficant byte) can be interpreted as an ASCII character as shown in the table above. For example, Unicode 0041H is the upper-case letter 'A.' If the MSB is nonzero, then the character set maps to a unique character set; this can be found by visiting http://www.unicode.org. Some computer languages such as Java represent their characters in Unicode.

EBCDIC CODE MATRIX

EBCDIC, the extended binary coded decimal interchange code, is used in older applications designed for mainframe computers. You may see it in certain PC programs that have mainframe communications capabilities. Not all codes are valid EBCDIC and are shown as blanks in the table.

Least Significant Nybble →

Most Significant Nybble ↓

	0	1	2	3	4	5	6	7	8	9	A	B	C	D	E	F
0	NUL	SOH	STX	ETX	SEL	HT	RNL	DEL	GE	SPS	RPT	VT	FF	CR	SO	SI
1	DLE	DC1	DC2	DC3	RES	NL	BS	POC	CAN	EM	UBS	CU1	IFS	IGS	IRS	IUS
2	DS	SOS	FS	WUS	BYP	LF	ETB	ESC	SA	SPE	SM	CSP	MFA	ENQ	ACK	BEL
3			SYN	IR	PP	TRN	NBS	BOT	SBS	IT	RFF	CU3	DC4	NAK		SUB
4	SP	RSP										.	<	(	+	|
5	&										!	$	*	)	;	¬
6	-	/									¦	,	%	-	>	?
7										△	:	#	@	△	=	"
8		a	b	c	d	e	f	g	h	i						
9		j	k	l	m	n	o	p	q	r						
A		-	s	t	u	v	x	x	y	z						
B																
C	(	A	B	C	D	E	F	G	H	I	SHY					
D	)	J	K	L	M	N	O	P	Q	R						
E	\	NSP	S	T	U	V	W	X	Y	Z						
F	0	1	2	3	4	5	6	7	8	9						BO

Glossary

AFC (Automatic Frequency Control): A system for correcting the frequency drift of an oscillator, usually employed in FM receivers.

ALC (Automatic Level Control): A system for controlling and limiting the peak envelope power (PEP) of a SSB exciter to a power level within the dynamic capabilities of the RF power amplifier, implemented by controlling the gain of a low-level stage.

Amplitude Modulation: The process of impressing information onto a carrier signal by varying the carrier signal strength in step with the information.

AMPS (Advanced Mobile Telephone Service): An analog mobile telephone service operating in the 800–900 MHz band.

Angle Modulation: Any form of modulation that causes phase modulation of a carrier signal during the process of modulation. This includes both FM and PM.

Antenna: The portion of a radio communication system that is responsible for converting electrical energy into electromagnetic energy, and vice versa.

Antenna Coupler: A circuit that provides impedance matching and filtering between the final power amplifier and antenna in a radio transmitter.

Antenna Gain: The ratio of radiated power (density) of an antenna compared to that of a reference antenna, which can be either a halfwave dipole (dBd units) or an isotropic radiator (dBi units).

APC (Automatic Power Control): A system for maintaining constant transmitted RF or optical power output under varying conditions (such as a changing supply voltage).

Array: A group of antennas that are arranged to provide a specific pattern of radiation.

ARP (Address Resolution Protocol): A LAN protocol that allows a device to learn the MAC address of a target computer by passing the target IP in a network broadcast message.

ARQ (Automatic Repeat reQuest): An error-handling system that responds to a bad message by requesting another copy of the same message.

ASCII (American Standard Code for Information Interchange): The set of binary codes used to represent printed and control characters on a typewriter keyboard.

Aspect Ratio: The ratio of width to height in an image. It is 4:3 in U.S. television signals.

Asynchronous: Data sent without a clock signal.

Atmospheric Noise: Noise generated within the earth's atmosphere; includes discharges such as lightning, wind and snow static, and so on.

Back Porch: The 3 μs (approximate) flat portion of the video waveform that immediately follows the horizontal synchronization pulse.

Balanced Modulator: An AM modulator that is designed to generate both upper and lower sidebands but no carrier frequency component.

Balun: A balanced-to-unbalanced transformer.

Bandwidth: The difference between the highest and lowest frequencies in a signal. In filter circuit analysis, this difference is measured between the upper and lower cutoff frequencies of the filter.

Barkhausen Criteria: The criteria necessary for an oscillator to start and, if running, continue running. These include a total loop phase shift of 0°, and for the oscillator to start, a loop gain greater than 1.

Baud: The rate at which changes (modulation) are introduced on an analog carrier signal.

Beamwidth: The angle between the two half-power points of an antenna's radiation pattern.

Bessel Identity: A mathematical relationship that describes the relative strength of each frequency component of a phase-modulated sine (cosine) wave in terms of the phase modulation index. The strength is measured relative to the unmodulated wave amplitude.

Binary Phase-Shift Keying (BPSK): A digital modulation scheme where two distinct phase angles are used to represent logic 0 and 1 states. Usually 0° and 180° are chosen.

Bit: A binary unit; theoretically, the smallest possible unit of information. Can hold the values 0 or 1.

Bit Error Rate (BER): The rate at which errors are likely to be produced in a digital system. Expressed as the total number of bits in error divided by the total number of bits transmitted, a ratio.

Block Check Character (BCC): A group of redundant information sent at the end of a data block that can be used by a receiver to test the validity of the data in the block.

Buffer Amplifier: An amplifier that immediately follows the oscillator in a transmitter, whose purpose is to isolate the oscillator from successive transmitter stages.

Capture Effect: The tendency of an FM receiver to be "captured" by the strongest signal at its antenna terminals when two (or more) signals are presented at the same frequency. Only the strongest (captured) signal is demodulated.

Capture Range: The frequency range over which a phase-locked loop can acquire lock if the PLL initial state is free running.

Carrier: An AC energy signal that is changed (modulated) in some way in order to convey information.

Carson's Rule: A quick method for approximating the bandwidth of an FM signal. BW $\approx 2(f_m + \delta)$.

Cavity Resonator: A microwave replacement for a conventional LC resonant circuit. Consists of an empty metal cavity, input and output coupling, and usually some means for tuning by modifying the cavity's dimensions.

CDMA (Code Division Multiple Access): A method of multiplexing whereby all transmitting stations can share the same frequency yet be individually demodulated by locking onto their individual codes.

Ceramic Filter: A relatively inexpensive bandpass filter that is often used in IF amplifiers, and SSB generators (for removal of the undesired sideband).

Characteristic Impedance: In a transmission line, the ratio of voltage to current at any point in the line. For free space (or other electromagnetic wave-propagating medium), the ratio of electric field strength to magnetic field strength at any point. The characteristic impedance of free space is approximately 377 Ω.

Chip: One bit of a pseudo-random code.

Chromatic Dispersion: Same as material dispersion. A form of dispersion (pulse spreading) in optical fiber that is caused by the variance of the refractive index of the various fiber materials with wavelength, resulting in differing times of arrival for different spectral components of the light wave.

CIDR (Classless Interdomain Routing): A system of expressing IP network addresses by giving the network address and subnet mask.

Clapp Oscillator: An oscillator that uses a pair of capacitors to control feedback like a Colpitts oscillator but uses an extra capacitor in the resonant LC circuit to counter the effect of varying capacitance of the active amplifying device.

Classful Addressing: The original Internet addressing scheme that divides IP addresses into class A, B, or C groups depending on the number of hosts needed by an organization. Has been supplanted by CIDR.

CND (Calling Number Delivery): Also known as the caller ID service; provides calling party information as a digital data stream at the beginning of each call.

Coaxial Cable: A transmission line that consists of an inner conductor that is separated from a tubular outer conductor (shield) by an insulating dielectric.

Code Division Multiplexing: The transmission of multiple channels on a common group of frequencies by the use of orthogonal (mutually exclusive) codes (such as Walsh codes). A receiver can lock onto the desired code by synchronizing with it; the other codes appear uncorrelated (as background noise) to the receiver.

CODEC (Coder–Decoder): A hardware device that converts analog to digital (the coder) and back (the decoder).

Collinear Array: An array of antennas that are placed in a straight line and phased at half-wavelength intervals in order to produce an effective gain and narrow beam angle.

Color Burst: The 3.579545 MHz signal that is transmitted briefly (about 9 cycles) on the back porch of a NTSC color video signal. The purpose of the color burst is to resynchronize the color oscillator in the receiver, which is necessary for demodulating the color information in the TV signal.

Colpitts Oscillator: An oscillator that uses a capacitive voltage divider to determine the amount of feedback (feedback ratio).

Constellation: A display of all the possible phases and angles in a digital modulator circuit.

Continuous Wave (CW): An uninterrupted carrier signal. CW power output ratings therefore express continuous possible power output into a load. Also, a designation for a transmission sent using Morse code.

Control Channel: A frequency (or group of frequencies) that is used to transmit supervisory information in a communication system.

Converter: A functional block that converts incoming frequencies up or down, as in the frequency converter section of a superheterodyne receiver.

CPE (Customer Premises Equipment): A CPE provides physical service to a customer, and may be owned by the customer or the service provider.

CPU (Central Processor Unit): The portion of a computer system that is responsible for carrying out the instructions of a program.

Critical Angle: The minimal angle of incidence needed in order to gain propagation of a wave through a medium by refraction.

Crosby Transmitter: An FM transmitter that uses frequency feedback in order to control its center frequency.

Crystal Filter: A filter built using quartz crystals as resonant elements, resulting in a high-quality response with very sharp skirts.

Crystal Oscillator: An oscillator that uses a quartz crystal as the frequency control element.

CSMA/CA (Carrier Sense Multiple Access with Collision Avoidance): The media access method for 802.11 wireless LANs. Because collisions are not detectable on a simplex radio frequency, a collision avoidance algorithm is used by each unit on the LAN.

CSMA/CD (Carrier Sense Multiple Access with Collision Detection): The media access method of Ethernet; devices wait until the channel is quiet before transmitting, and cease transmitting if a collision is detected.

Cyclic Redundancy Check (CRC): An error-checking method that divides the data in a block or frame by a fixed number, returning the remainder as the check word (BCC).

Damper Diode: A diode that conducts during the horizontal retrace in a television receiver, reclaiming some of the energy from the horizontal output transformer and discouraging ringing.

Data Block: A group of data sent together.

Data Communications: The practices involved with the processing and transmission of digital information.

Datagram: Loosely, the equivalent of a data block. More precisely, a data block that has an IP header as it travels through a switched-packet network such as the Internet.

dBm (Decibel Milliwatts): Decibel power with respect to a reference power of 1 mW.

DCE (Data Communication Equipment): Used to designate peripheral devices (such as modems) that are designed to be connected to the RS232 port of a computer system.

DDS (Direct Digital Synthesis): A system of frequency synthesis in which a digital oscillator circuit is implemented as a counter. The output is obtained from a digital-to-analog converter.

Decade: A 10 to 1 ratio of frequencies.

Deemphasis: The attenuation of higher information frequencies, especially in an FM receiver, which reduces the receiver's susceptibility to noise.

Delay Line: A short section of transmission line that slows a signal down, usually so that other portions of the signal can be processed and time-aligned with the delayed signal.

Demodulation: The process of recovering information from a carrier signal.

DHCP (Dynamic Host Configuration Protocol): A protocol that allows workstations on a LAN to automatically acquire IP address assignments and other important network configuration settings.

Diagonal Clipping: Distortion of the information signal in an AM detector that is caused by the inability of the RC time constant in the circuit to keep up with falling slopes in the information signal.

Diffraction: The spreading of light rays at the edge of a surface.

Diplexer: A circuit that allows the signals from two ac sources to be combined without interaction between the sources. Used commonly to add the picture and sound carriers in television broadcast transmitters.

Dipole Antenna: An antenna consisting of two quarter-wavelength sections of conductor.

Directional Coupler: A device for measuring the relative values for forward and reflected power in a transmission line.

Director: A parasitic element in a Yagi-Uda antenna that is placed in front of the driven element with the purpose of focusing the signal in a given direction.

Discriminator: An FM detector. A circuit that converts frequency changes into voltage changes.

Doppler Effect: The frequency shift that occurs when a wave source and receiver are moving relative to each other.

Driven Element: The portion of an antenna that is directly connected to the feedline.

Driver Amplifier: The amplifier stage immediately before the final amplifier stage in a transmitter.

DSB-SC (Double Sideband Suppressed Carrier): An AM signal that has been created in a balanced modulator circuit, which has both upper and lower sidebands but no carrier energy.

DSL (Digital Subscriber Line): A service that provides Internet connectivity to telephone customers over the twisted-pair copper subscriber loop.

DSP (Digital Signal Processing): The science of performing manipulations on analog signals by first converting them to digital, performing arithmetic operations on the signals, then converting them back to analog.

DTE (Data Terminal Equipment): An RS232 designation for computers and other devices that are meant to be connected to peripherals (such as display terminals.)

DTMF (Dual Tone Multiple Frequency): A system of signaling that represents each dialed digit with two unique tones.

Dummy Load: An RF power resistor that is made to be connected to the output of a transmitter for the purpose of off-the-air testing.

Duplexer: A device that allows a transmitter and receiver to share a common antenna system.

EIA-232: The official designation for the RS232 series of standards.

EIRP (Effective Isotropic Radiated Power): EIRP is the product of antenna power gain, G_p (presumably measured at the maxima of the antenna's major lobe), and transmitter power input into the antenna. EIRP is used for calculating the field strengths that will be produced by the transmitter-antenna system.

Envelope Detector: An AM detector consisting of a half-wave diode rectifier and an RC smoothing circuit, which uses rectification to extract the envelope of the AM signal.

Epoch: One complete cycle of a pseudo-random code.

Excess Noise: A noise that is peculiar to field-effect transistors and some types of resistors that is inversely proportional to frequency. This noise grows larger at low frequencies and is most problematic at low audio frequencies.

Exciter: A low-powered transmitter (usually under 100 W) that is meant to be modulated and connected to a larger power amplifier, as in broadcast.

External Noise: Noise that is generated outside of a circuit or system.

Fading: Variations in amplitude at a receiver caused by changes in propagation conditions.

FDDI (Fiber Distributed Data Interface): A standard for transmitting data on optical fiber cables at a rate of around 100 Mbps (about twice as fast as T-3).

FDMA (Frequency Division Multiple Access): Sending multiple pieces of information over a channel by assigning a different carrier frequency to each one. This is the fundamental method of radio.

FEC (Forward Error Correction): A system that attempts to recover damaged messages as they are received, instead of requesting that the damaged messages be resent.

Ferrite: Iron oxide compounds that are used for many purposes in RF circuits, including cores for coils and transformers, antenna cores at LF and MF, as well as signal-diverting components for microwave devices such as circulators.

Fiber Optics: Plastic or glass fibers designed to carry information signals on beams of light.

Finley's Law: In operational amplifiers, the assumption that the output of the amplifier will be in either positive saturation or negative saturation unless the two input voltages are exactly equal. Phase detectors also obey Finley's law. The frequencies at both inputs of a phase detector must be exactly equal or the output of the device goes into saturation.

First Detector: Another name for the mixer in a superheterodyne receiver.

Flicker Noise: Same as excess noise. A noise that is peculiar to field-effect transistors and some types of resistors that is inversely proportional to frequency. This noise grows larger at low frequencies and is most problematic at low audio frequencies.

Flyback Transformer: A high-frequency power transformer used to provide the horizontal deflection signals and high voltage for the second anode in a television receiver.

Flywheel Effect: The effect whereby a tuned LC circuit tends to continue oscillating for a period of time after it is excited.The higher the Q of the circuit, the longer its oscillation (ringing) will continue.

Foster–Seeley Discriminator: An FM detector that converts frequency into phase shifts, then converts the phase shifts into voltages.

Fourier Analysis: The analysis of signals in order to find their frequency (spectral) composition.

Free-Running State: The condition a phase-locked loop (PLL) assumes when no input signal is provided.

Frequency Deviation: The peak frequency change in an FM transmitter, calculated as the maximum frequency minus the carrier frequency, or the carrier frequency minus the minimum frequency.

Frequency Division Multiplexing: Same as FDMA.

Frequency Modulation (FM): A method of impressing information onto a carrier signal whereby the frequency of the carrier is changed in step with the information signal.

Frequency Multiplier: A circuit that is designed to increase the frequency of a signal (usually with the purpose of increasing its deviation) by driving a LC tank circuit tuned to one of the signal's harmonics.

Frequency Synthesizer: A circuit that derives many frequencies from one stable reference frequency.

Front End: The first portions of a radio receiver that process incoming signals from an antenna. Usually includes the preselector and RF amplifier (if used).

Front Porch: The 3 μs (approximate) portion of the video signal that immediately preceeds the horizontal synchronization pulse.

Front to Back Ratio: The ratio of radiated power (density) in the direction of maximum antenna gain (front) divided by the radiated power (density) in the opposite direction (back).

FSK (Frequency Shift Keying): A method of impressing digital data onto a carrier wave by frequency modulation. The carrier assumes two different frequencies, one for a high and the other for a low.

Gateway: A device on a LAN (usually a router) that receives packets that need to be transmitted to a computer outside of the LAN.

Ghost: A portion of a television picture that appears as an afterimage or shadow, caused by reflections in a transmission line or multipath distortion at the receiver.

Ground Wave: The portion of a radio signal that tends to follow the curvature of the earth.

Gunn Diode: A semiconductor diode that produces pulses of microwave current when appropriately biased, making it useful in microwave oscillators.

Gunn Oscillator: An oscillator employing a Gunn diode as the active element.

GSM (Global System for Mobile Communication): A PCS implementation that uses TDMA as its channel multiplexing method. Incompatible with Qualcomm CDMA.

Hamming Code: An error-correcting code that corrects single-bit errors by forming a pointer (address) to the bit in error.

Harmonic: An integer multiple of a fundamental frequency.

Hartley Oscillator: An oscillator that uses a tapped inductor for determination of the feedback ratio.

Hertz Antenna: A half-wave dipole antenna.

High-Level Modulation: Modulation that occurs in the very last stage of a radio transmitter.

Horizontal Resolution: The ability of a television system to resolve closely spaced horizontal details in the image; rated in lines.

Horn Antenna: A low-gain antenna that consists of a gradual widening or opening of a waveguide into free space.

Hub: A device that allows several computers in a LAN to share the same network segment, forming a single collision domain.

ILEC (Incumbent Local Exchange Carrier): The local telephone company, usually controlling the copper subscriber loops for the region.

Image Frequency: An undesired frequency that can be converted to the IF frequency in a superhet receiver.

Information Technology: The science of information management; includes computers, networking, and data communication.

Intelligence: Information, which can be defined as the predictability of an outcome at a receiver. Signals with high predictability (such as pure sine waves) carry little information. Complex signals such as voice (hard to predict) are rich in information content.

Interlacing: The practice of painting a television image in two successive passes, writing odd lines in the first pass and even lines in the next pass. The technique reduces flicker while keeping refresh rates (and bandwidth) low.

Internal Noise: Noise generated within a circuit or system, such as shot noise or thermal noise.

IP (Internet Protocol): The protocol that is used to control the addressing and routing of message packets in the Internet.

IP Address: A 32 bit (IPv4) or 128 bit (IPv6) binary address for a computer; consists of a network and host portion. Also known as a logical address.

Isotropic Source: A theoretical electromagnetic energy source that radiates equally well in all directions in three dimensions.

Johnson Noise: Same as thermal and Brownian noise. A noise generated by the random motion of molecules in a conductor and is uniformly distributed over all frequencies.

Klystron: A vacuum tube amplifier that is used as a narrowband amplifier and oscillator.

LAN (Local Area Network): A group of computers that are connected together within the same building.

Laser (Light Amplification by Stimulated Emission of Radiation): An optical "oscillator" that uses a resonant optical cavity to reinforce a specific frequency (wavelength), resulting in positive feedback and emission of intense light wave energy.

LATA (Local Access and Transport Area): A geographic area that a regional Bell operating company is allowed to provide service in.

Limiter: The circuit in an FM receiver that removes amplitude information from the carrier, thereby reducing noise.

Local Oscillator: The oscillator in a superheterodyne receiver that drives one port of the mixer for the purpose of frequency conversion.

Line Hybrid: A transformer that is used in telephony for cancellation of sidetone, so that the caller's own voice isn't excessively loud in the earpiece.

Lock Range: The total frequency range over which a phase-locked loop (PLL) can remain in lock if its initial state is locked.

Log-Periodic Antenna: A antenna that uses a large number of electrically driven dipole elements for coverage over a wide frequency range.

Loop Antenna: A directional antenna that consists of one or more loops of conductor.

Loopback: A method of testing a system whereby its output is directed back to an input. If the system works under this condition, the problem is likely outside it.

Low-Level Modulation: Modulation that occurs in the early stages of a radio transmitter, before the final amplifier.

Lower Sideband (LSB): The range of frequencies below the carrier that are generated as a result of the process of modulation.

MAC (Media Access Control): The "lower" portion of the data link layer in the ISO open system interconnect model.

MAC Address: The 48-bit address that is contained in a ROM on a network interface card (NIC). Also known as the physical address.

MAC Layer: The portion of the network operating software that controls access to the physical transmission media.

Magnetron: A microwave vacuum tube oscillator that uses resonant cavities to reinforce and collect the microwave signal energy obtained by bending an electron beam into a nearly circular pattern with a magnetic field.

MAN (Metropolitan Area Network): A network contained within a limited geographic area, such as a city.

Marconi Antenna: A vertical antenna that is one-quarter of a wavelength long, usually utilizing the earth or other conductive surface as the missing antenna half.

Mark: A logic 1 or logic high.

Material Dispersion: Same as chromatic dispersion. A form of dispersion (pulse spreading) in optical fiber that is caused by variance of the refractive index of the various fiber materials with wavelength, resulting in differing times of arrival for different spectral components of the light wave.

MCU (Microcontroller Unit): A small microprocessor designed to control a product or system and is built into the device being controlled.

Mechanical Filter: A filter that uses resonant metal discs instead of LC resonant circuits.

Microcontroller: Same as MCU.

Microstrip: A transmission line, formed on a printed circuit board, with precisely controlled dimensions that control its characteristic impedance.

Microwave: Loosely, the frequencies above 1 GHz.

MMIC (Monolithic Microwave Integrated Circuit): An IC chip designed to efficiently amplify a wide range of radio frequencies, including microwaves.

Modal Dispersion: The dispersion or spreading of light waves in a fiber optic that occurs as a result of transit-time differences between various possible light wave paths.

Modem (Modulator/Demodulator): A device that converts digital data signals to analog for transmission and performs the reverse for received signals.

Modulation: The process of impressing information onto a carrier. AM, FM, and PM are the three possible types of modulation.

Modulation Index: A relative measure of the strength or depth of modulation in a transmitter. In AM, the modulation index can be between 0 and 1; in FM, the limits are 0 and infinity (though most broadcast FM seldom exceeds and index of 25).

MUF (Maximum Usable Frequency): The highest frequency that will permit sky-wave propagation between two geographic locations.

Multiplexing: The transmission of more than one piece of information on the same channel or carrier frequency.

NAT (Network Address Translation): A protocol that translates the internal, private IP addresses of computers on a LAN into a public, routable IP address for use on the open Internet.

NBFM (Narrow Band Frequency Modulation): An FM signal with only one pair of significant sidebands, with a resulting modulation index less than or equal to 0.25.

Network: An organized interconnection of computer systems.

Neutralization: The prevention of self-oscillation in an amplifier stage by the application of a carefully controlled amount of negative feedback.

Noise Figure: A measure of the amount of internal noise generated by an amplifier stage. Calculated as NF = dB S/N(in) − dB S/N (out). A noiseless amplifier has a 0 dB noise figure.

Nonlinear: Having a graph or equation that is not a straight line. Many electronic devices act linear for small signal operation but nonlinear for large signals.

Nyquist Theorem: A theorem that states that in order to preserve all the information in a continuous signal, it must be sampled at a frequency at least twice the highest frequency component in the waveform.

Octave: A 2:1 ratio of frequencies.

Orbit: The path of a satellite as it circles the earth. Satellites can be in either polar, inclined, or geostationary orbits.

Orthogonal: At a right angle, or in the case of a code, having zero correlation with another code (zero cross-product). Orthogonal signals can share the same channel and be separated at a receiver.

Oscillator: A circuit that converts dc energy from a power source into ac energy at a controlled frequency.

OSI (Open Systems Interconnect): The OSI is a reference model that categorizes the functions needed in a data communications system.

Overmodulation: Exceeding the maximum allowed modulation index (AM) or deviation (FM) in a transmitter. Overmodulation leads to excessive bandwidth and distortion of the information signal.

Packet: A group of data sent together, usually complete with a header containing the addresses of the sender and receiver plus other housekeeping information.

Parametric Amplifier: An amplifier that provides a power gain by modulating a variable parameter in the circuit. Usually the capacitive reactance is modulated in a paramp.

Parasitic Element: An antenna element that is not electrically connected to the feedline, but contributes to the field pattern of the antenna by reflecting or reradiating a portion of the signal energy.

Parity: A system of error checking that counts the total number of logic 1s in a message.

PCM (Pulse Code Modulation): A method of representing analog signals by conversion of each sample of the original signal to a binary number. The sequence of binary numbers then represents the original waveform.

PCS (Personal Communication Service): A group of digital wireless telephony services residing in the 1800–1900 MHz band.

Peer Layers: Identical open system interconnect (OSI) model layers on two different computer systems that are in virtual contact through the action of the OSI stack.

PEP (Peak Envelope Power): The RMS power of the RF carrier at the crest of the modulation envelope in AM and SSB.

Percent Modulation: In AM, the modulation index expressed as a percentage. In FM, the ratio of deviation to maximum allowed deviation, expressed as a percentage.

Phase Modulation: The impressing of information onto a carrier by varying its phase angle in step with the information signal.

Phased Array: A collection of antennas that are interconnected and arranged to produce a specific radiation pattern.

Phasing Method: A method of generating SSB whereby the undesired sideband is eliminated by the addition of signal components in precise phase relationships.

PLL (Phase-Locked Loop): A subsystem that serves to follow a provided input frequency. Useful for modulation, demodulation, frequency synthesis, and other applications.

PSK (Phase Shift Keying): A digital modulation system that represents the state of a digital data point by phase-modulating the carrier.

Piezoelectric Effect: An effect present in certain crystalline materials (such as quartz) whereby an applied stress induces a voltage across the crystal, and likewise, a voltage applied across the crystal induces mechanical stress (and crystal movement).

Pilot Signal: A signal that is sent along with the information on a carrier for the purpose of helping a receiver either lock onto or demodulate part of the signal. A common example is the 19 kHz pilot used in FM stereo.

Ping (Packet Internetwork Groper): A utility that is used to check for IP connectivity between two computers on a network.

Pixel: The smallest element of a video image; one dot.

Polarization: The orientation of the electric field of a radio wave.

Port: A number used in session management by TCP/IP. The source port number identifies the originating process, and the destination port identifies the target or destination process.

POTS (Plain Old Telephone Service): The analog voice service provided on a copper subscriber loop.

Preemphasis: The boosting of high information frequencies prior to transmission, usually with the intent of improving the high-frequency noise performance of receivers.

Preselector: The portion of a superheterodyne receiver immediately after the antenna, whose main purpose is elimination of the image frequency.

Product Detector: A detector for SSB signals that has two inputs, one for the SSB signal (at the IF frequency) and the other for a steady frequency oscillator, which reinserts the carrier frequency.

Protocol: A set of rules for communication.

Pseudo-random: Having an apparently random but predictable composition.

Pseudo-range: A range reading obtained by a geodetic (GPS) receiver prior to clock correction.

QAM (Quadrature Amplitude Modulation): A system of digital modulation whereby the information is phase-modulated onto the carrier, with a 90° jump between adjacent phase angles.

Quadrature Detector: An FM detector that converts frequency changes into phase changes then measures these phase changes with a digital phase detector circuit (such as an exclusive-OR logic gate).

Quality Factor (*Q*): The ratio of power stored to power lost during one ac cycle in a circuit.

Quantization: The representation of analog information (having an infinite number of possible states) in a form having only a finite number of states (digital number system).

Quieting: A property of FM reception whereby the noise component rapidly shrinks once the incoming signal rises above the receiver's limiting threshold (where all amplitude information is removed from the signal).

RADAR (Radio Detection and Ranging): A set of techniques for determining the range and velocity of a variety of targets by measuring the reflection of radio waves from them.

Radiation: Electromagnetic energy in free space.

Radiation Pattern: The three-dimensional pattern of relative field strength surrounding an antenna system.

Radiation Resistance: The portion of the input impedance of an antenna system that actually results in radiated RF energy when excited by a current.

Raster: The visible portion of a television picture formed by the scan lines.

Ratio Detector: An FM detector that produces a varying ratio of voltages across two resistors that is a copy of the original information signal.

RBOC (Regional Bell Operating Companies): The companies formed after the breakup of AT&T in 1984.

Reactance Modulator: A circuit that converts an input voltage or current into a varying inductive or capacitive reactance.

Reflection: A change in direction in an electromagnetic wave caused by contact with a conductive surface.

Reflector: A parasitic element of an antenna that serves to increase its directionality by reflecting signals back in a forward direction.

Refraction: The bending of an electromagnetic wave as it passes between two mediums with differing indices of refraction.

Refractive Index: The ratio of the speed of light in free space to the actual speed of propagation in a material.

Resolution: The ability of an image-processing system to process small details. In TV systems, resolution is measured in both vertical and horizontal lines.

Router: A network device that directs or routes packets toward correct destinations based on the address information in each packet.

RS232: The Electronic Industries Association Recommended Standard 232. An interfacing standard that specifies the mechanical, electrical, and functional descriptions for a low-speed serial data interface.

SCA (Subsidiary Communication Authorization): A subcarrier authorization that is permitted in the FM broadcast band for the purpose of transmitting telemetry, background music, and other commercial purposes.

Selectivity: The ability of a radio receiver to select a desired signal in the presence of unwanted signals; controlled largely by bandwidth.

Sensitivity: The ability of a radio receiver to process weak signals; typically specified as an antenna input power or voltage that will produce a minimum S/N ratio at the receiver output terminals.

Shape Factor: The ratio of bandwidth at −60 dB response to the bandwith at −6 dB response for a filter circuit.

Shot Noise: Noise that is generated by charges moving through the forward-biased PN junctions of active devices such as diodes and transistors.

Sidetone: Audio in the earpiece of a telephone set (or other device) that is fed back from the local microphone. A low level of sidetone is helpful to telephone users; they can determine that the set is working when they hear a small amount of transmit audio in the earpiece. Excessive sidetone level is annoying to users.

Signaling: The process of modulating an analog carrier, usually for the purpose of conveying digital information. The rate of signaling is expressed in units of BAUD.

Skin Effect: The tendency of currents to flow on or near the skin of conductors at high radio frequencies.

Skip Zone: The geographic area between transmitter and receiver where no signal can be received.

Sky Wave: The portion of a radio wave the propagates by ionospheric refraction.

SLIC (Subscriber Loop Interface Card): A printed circuit card housed at the telephone company end office that contains the circuitry for operating a customer's local loop.

Smith Chart: A graphical design and evaluation tool for transmission lines designed by P. H. Smith in 1938.

SNR (Signal-to-Noise Ratio): The ratio of signal power to noise power on a communications channel.

Socket: A virtual (software) connection between a computer program and a network. Sockets are a layer 7 (application) feature that allow application programs to access the network.

Space: A digital logic 0, or digital "low."

Space Wave: The portion of a radio wave the travels directly from a transmitter to a receiver.

Spectral Mask: A definition of the shape of a transmitted signal in the frequency domain, which includes the overall bandwidth and skirt slopes.

Spectrogram: A frequency-domain display of a signal that shows what frequencies are present, and the amount of energy at each frequency.

Spectrum Analyzer: An instrument that displays a frequency domain picture of signals.

Split: The difference between the transmit and receive frequencies of a communications device. Sometimes called *offset*.

Spur: An undesired frequency output from a transmitter that is not at a harmonic multiple of the carrier frequency.

SSB (Single Sideband): A modified form of AM where the carrier frequency and unwanted sideband are removed, reducing bandwidth and increasing talk-power.

Stagger Tuning: A method of tuning a bandpass filter in which each tuned circuit in the filter is set to a slightly different frequency, thereby providing a wide frequency response.

Standing Wave: An stationary pattern of voltage or current minimum and maximum patterns that develops on a transmission line when reflected energy is present.

Stereo: A system of reproduction employing two distinct audio channels, resulting in more lifelike reproduction.

Stripline: A transmission line fabricated on a circuit board, usually as a conductor over a copper backplane, or even a conductor sandwiched between two copper groundplanes (three layers).

Subnet Mask: A binary number that helps a computer identify the network and host portions of an IP address. Ones in the mask define the network portion; the remaining portion of the address will be the host portion.

Superheterodyne Receiver: A radio receiver that converts the desired incoming RF carrier frequency to a constant IF frequency, resulting in constant selectivity across its operating range.

Surge Impedance: The same thing as the characteristic impedance of a transmission line.

Switch: A device that allows multiple computers to be connected together as a LAN segment to form one broadcast domain. A switch isolates traffic between devices to prevent unwanted collisions, yet forwards broadcasts to all devices.

SWR (Standing Wave Ratio): The ratio of maximum to minimum voltage along a transmission line. When properly matched to a load, a transmission line has a 1:1 SWR (there are no standing waves).

SYN (Synchronize Flag): In TCP, the control signal that is used to establish a new connection.

Sync Separator: A circuit that extracts the synchronization pulses from the video signal in a television receiver.

Synchronous: A form of data transfer where the clock signal is sent along with the data. Usually the clock and data are intermixed according to a fixed encoding rule.

TCP (Transmission Control Protocol): A full-duplex protocol that delivers data through a network as stream of byte data, fully guaranteeing its delivery.

TDR (Time Domain Reflectometry): The practice of measuring the returned echoes of electrical or optical signals on transmission lines, usually done for the purpose of line quality evaluation or troubleshooting.

TDMA (Time Division Multiple Access): Sending multiple pieces of information on a communication channel by sending them one at a time. This is also known as serial communication.

Time Division Multiplexing: Sending multiple pieces of information on the same channel by assigning a different time slot to each information unit.

Thermal Noise: The noise generated by the random motion of molecules in a conductor; same as Johnson and Brownian noise.

Transducer: A device that converts one form of energy into another. An antenna is a common transducer in radio systems.

Transmission Line: A device for transferring energy from one point to another. Includes both electrical (coaxial, for example) and optical (fiber optics).

Traveling Wave: A wave of alternating-current electromagnetic energy that is moving in space. In free space, traveling waves move at the speed of light.

TRF (Tuned Radio Frequency): A receiver that has its tuned circuits all tuned to the frequency of the carrier.

Trimmer Capacitor: A small capacitor, adjustable by screwdriver, that is placed in a tuned circuit for the purpose of fine adjustment by a technician.

Twist: The difference in amplitude between the two tones transmitted by a DTMF encoding device.

TWT (Traveling Wave Tube): A microwave vacuum tube amplifier that provides a high gain over a wide range of frequencies.

UART (Universal Asynchronous Receiver Transmitter): An IC chip that is designed to convert parallel data to serial, and vice versa.

UDP (User Datagram Protocol): An unreliable simplex networking protocol that doesn't guarantee delivery of packets.

Upper Sideband: The range of frequencies above the carrier frequency that is created during the process of modulation.

USART (Universal Synchronous Asynchronous Receiver Transmitter): An IC chip that converts parallel and serial data in both asynchronous and synchronous formats.

UTP (Unshielded Twisted Pair): The media type used in telephone local loops and Ethernet networks.

Varactor Diode: A diode that is optimized for a variable junction capacitance. Varying the reverse bias across the device varies its effective capacitance.

VCO (Voltage Controlled Oscillator): An oscillator whose frequency depends on the application of a dc control voltage.

Velocity Factor: The ratio of the velocity of wave propagation in a transmission line to that of free space.

Vertical Antenna: An antenna that is oriented vertically with respect to the surface of the earth.

Vestigial Sideband: A form of AM transmission in which one of the two sidebands is partially suppressed while the other is sent unaltered. The technique reduces bandwidth requirements but, unlike SSB, preserves the carrier frequency component (CFC) for convenient demodulation at the receiver.

VLAN (Virtual LAN): A logical arrangement that causes several ports of a switch to behave as if they are part of the same LAN segment. VLANs are a programmable feature that allow an administrator to reassign switch ports to various VLAN groups, either by MAC address or switch port number.

VoIP (Voice over IP): A system of telephony that transmits analog voice signals in IP packets through a network (or internetwork).

Walsh Codes: A set of orthogonal (mutually exclusive) codes that are used to code-division multiplex multiple pieces of information on a channel. In general, a Walsh code set with *N* codes requires *N* bits to represent.

WAN (Wide Area Network): A network covering most of a city, several cities, or any larger area.

Waveguide: A hollow metal tube that is used as a transmission line at microwave frequencies.

Waveguide Dispersion: A form of dispersion that results from variation in velocity of propagaton between the core and cladding in a fiber optic.

Wavelength: The distance an electromagnetic wave travels during the time of one ac cycle.

WBFM (Wideband FM): An FM signal that possesses more than one pair of significant sidebands or has a modulation index of greater than 0.25.

White Noise: Noise that contains equal energy at all frequencies except dc. Johnson noise has the same property.

Yagi-Uda Antenna: A directional antenna that consists of a driven element (usually a half-wavelength dipole) and parasitic director and reflector elements.

Index

A

B

C

D

E

F

G

H

I

J

K

L

M

N

O

P

Q

R

S

T

U

V

W

Y

Z